Permutations If $P(n, r)$ (where $r \leq n$) is the number of permutations of n elements taken r at a time, then

$$P(n, r) = \frac{n!}{(n - r)!}.$$

Combinations If $\binom{n}{r}$ denotes the number of combinations of n elements taken r at a time, where $r \leq n$, then

$$\binom{n}{r} = \frac{n!}{(n - r)!r!}.$$

Probability in a Binomial Experiment If p is the probability of success in a single trial of a binomial experiment, the probability of x successes and $n - x$ failures in n independent repeated trials of the experiment is

$$\binom{n}{x} p^x (1 - p)^{n - x}.$$

Mean The mean of the n numbers, $x_1, x_2, x_3, \ldots, x_n$, is

$$\bar{x} = \frac{\sum (x)}{n}.$$

Standard Deviation The standard deviation of the n numbers, $x_1, x_2, x_3, \ldots, x_n$, with mean $\bar{x}$, is

$$s = \sqrt{\frac{\sum (x - \bar{x})^2}{n - 1}}.$$

Binomial Distribution Suppose an experiment is a series of n independent repeated trials, where the probability of a success in a single trial is always p. Let x be the number of successes in the n trials. Then the probability that exactly x successes will occur in n trials is given by

$$\binom{n}{x} p^x (1 - p)^{n - x}.$$

The mean is

$$\mu = np,$$

and the standard deviation is

$$\sigma = \sqrt{np(1 - p)}.$$

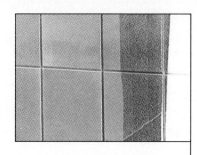

FIFTH EDITION

Mathematics with Applications

IN THE MANAGEMENT, NATURAL, AND SOCIAL SCIENCES

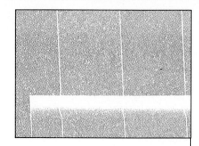

FIFTH EDITION

Mathematics with Applications

IN THE MANAGEMENT, NATURAL, AND SOCIAL SCIENCES

Margaret L. Lial
American River College

Charles D. Miller

Thomas W. Hungerford
Cleveland State University

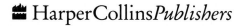

HarperCollins*Publishers*

Student:

To help you make the most of your study time and improve your grades, we have developed the following supplement designed to accompany Lial/Miller/Hungerford *Mathematics with Applications in the Management, Natural, and Social Sciences,* Fifth Edition:

Study Guide and Student's Solutions Manual   0-673-46278-1
by Lial/Miller

You can order a copy at your local bookstore or call HarperCollins Publishers directly at 1-800-782-2665.

Sponsoring Editor: George Duda
Development Editors: Kristen Westman, Adam Bryer
Project Editor: Cathy Wacaser
Art Direction: Julie Anderson
Text and Cover Design: Lucy Lesiak Design, Inc.: Lucy Lesiak
Cover Photo: Peter Bosey
Director of Production: Jeanie A. Berke
Production Assistant: Linda Murray
Compositor: York Graphic Services
Printer and Binder: R. R. Donnelley & Sons Company
Cover Printer: Lehigh Press Lithographers

Mathematics with Applications in the Management, Natural, and Social Sciences, Fifth Edition

Library of Congress Cataloging-in-Publication Data

Lial, Margaret L.
    Mathematics with applications in the management, natural, and
social sciences / Margaret L. Lial, Charles D. Miller, Thomas W.
Hungerford. — 5th ed.
        p. cm.
    Includes index.
    ISBN 0-673-46277-3
    1. Mathematics.   I. Miller, Charles David, 1942–   .
II. Hungerford, Thomas W.   III. Title.
QA37.2.L5   1991
510—dc20   90-20485
                                                                              CIP

ISBN: 0-673-46277-3

        91   92   93   9 8 7 5 4 3

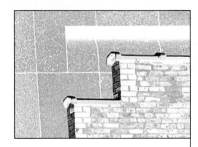

Preface

Mathematics with Applications, Fifth Edition, continues to cover the mathematical topics needed by students in the fields of business management, social science, and natural science. The book is carefully written at a level that makes it accessible to such students. Topics are presented by proceeding from what is already known to new material, from concrete examples to general rules and formulas. There is an emphasis on genuine applications throughout the book. Almost every section includes pertinent applications.

The text is appropriate for a combined finite mathematics and calculus course. It also can be used for a separate course in either finite mathematics or applied calculus as well as for a course in applied college algebra. Previous editions have been used in courses such as Mathematics with Applications, Mathematics for Management and Social Science, Mathematics for Business and Economics, Finite Mathematics, Introduction to Analysis, Algebra with Applications, and College Mathematics.

PREREQUISITES

The only prerequisite assumed is a course in algebra. A thorough review of algebra is given in Chapter 1, and a diagnostic pretest is included in the *Instructor's Guide with Test Bank and Solutions* to help determine for each student how much review is needed. The exercises in this chapter cover a broad range of difficulty and in many sections include challenging exercises.

CHANGES IN THE FIFTH EDITION

The fifth edition preserves the basic format and topic order of the previous editions and builds on their style, clarity, readability, and pedagogy. In general, some rewriting has further improved the exposition. Rearrangement of material provides more clarity for the student and greater flexibility for the instructor. New topics are now introduced exactly at the point where they are needed, and sections not used by all instructors are placed where they can easily be omitted, as noted below. In this edition many examples and exercises have been updated or replaced and several new examples and exercises have been added to freshen the book and make it more interesting and timely. The following paragraphs describe in detail other significant changes in pedagogy and content.

PEDAGOGICAL FEATURES

End-of-chapter summaries have been added which include key terms and symbols, and key concepts.

Calculator usage has been incorporated in a natural fashion throughout the book, as appropriate. Calculator exercises are identified by ⬛ in the margin. The calculator exercises were designed with a scientific calculator in mind, but there is ever-growing interest in graphing calculators. Consequently, we have added **Appendix A: Graphing Calculators.**

Several **challenging problems** that go beyond the examples in the text have been included in most sections. These problems will be identified in the *Instructor's Guide with Test Bank and Solutions*.

Prominent **cautionary notes** have been integrated into the text wherever appropriate to warn students of common errors and misconceptions.

The **Index of Applications,** newly organized by interest area, is provided as a reference tool for instructors and students who may want to investigate specific types of applications.

Additional **business applications** have been provided throughout, and especially in Chapters 1 and 2.

Summary boxes within chapters have been distinguished from the boxed theorems, definitions, and rules to help students review material.

CONTENT FEATURES

A full section is devoted to **factoring.**

The necessary information about **geometric sequences** has been incorporated into the discussion of annuities, and arithmetic sequences have been omitted.

A **summary section for mathematics of finance** has been added. It includes randomly mixed exercises in which the student must determine which formula is needed.

The **graphical approach to linear programming** now presents the technique in one section and applications in the next, rather than combining the two.

The **probability and statistics** chapters have been reorganized. Bayes' Formula immediately follows the discussion of conditional probability at the end of Chapter 8. Combinations and permutations now appear in Chapter 9, where they are first used. Probability distributions and expected value are also in Chapter 9 so that they may be covered easily even if the statistics chapter is omitted.

The **key concepts of calculus** (limits, rate of change, derivatives, maxima and minima, continuity, and the Fundamental Theorem) are now introduced in a way that builds on students' intuition and makes these ideas easier to grasp.

Continuity is now covered at the end of Chapter 11 since some schools prefer to omit this topic.

Integration by substitution appears early in Chapter 13, rather than at the end, which permits a wider variety of examples and exercises throughout the chapter.

Marginal analysis is discussed in more detail.

Multivariate calculus is now in a separate chapter since many schools do not cover this material. Those who want to include it can still follow the order in the previous edition by covering Chapter 14 immediately after Chapter 12.

FEATURES RETAINED FROM PREVIOUS EDITIONS

The following popular features from previous editions have been retained.

- Extensive **examples.**
- A wide variety of **exercises,** keyed to the text.
- Realistic and timely **applications,** many from current journals.
- End-of-chapter **case studies** that demonstrate the use of the topics of the chapter in real-life situations.
- **Problems at the side** help students test their understanding. By working the more than 400 marginal problems as topics are encountered, students can quickly locate their source of difficulty.
- Important rules, definitions, theorems, and summaries are enclosed in boxes and many are highlighted with a title for ease of study and review.
- A second color is used to annotate equations, illuminate troublesome areas, and clarify concepts and processes.

COURSE FLEXIBILITY

The book can be used for a variety of courses, including the following.

Finite Mathematics and Calculus (one year or less) Use the entire book; cover topics from Chapters 1–4 as needed before proceeding to further topics.

Finite Mathematics (one semester or one or two quarters) Use as much of Chapters 1–4 as needed, and then go into the topics of Chapters 5–10 as time permits and local needs require.

Calculus (one semester or quarter) Use Chapters 1–4 as necessary and then use Chapters 11–14.

College Algebra with Applications (one semester or quarter) Use Chapters 1–8 with the topics of Chapters 7 and 8 being optional.

Chapter interdependence is as follows.

Chapter	Prerequisite
1 Fundamentals of Algebra	None
2 Functions and Graphs	None
3 Polynomial and Rational Functions	Chapter 2
4 Exponential and Logarithmic Functions	Chapter 2
5 Mathematics of Finance	Chapter 4
6 Systems of Linear Equations and Matrices	None
7 Linear Programming	Chapters 2, 6
8 Sets and Probability	None
9 Further Topics in Probability	Chapter 8
10 Introduction to Statistics	Chapter 8
11 Differential Calculus	Chapters 2–4
12 Applications of the Derivative	Chapter 11
13 Integral Calculus	Chapters 11–12
14 Multivariate Calculus	Chapters 11–13

SUPPLEMENTS FOR THE INSTRUCTOR

The *Instructor's Guide with Test Bank and Solutions* gives a lengthy set of test questions for each chapter, organized by section, plus answers to all the test questions. It also provides complete solutions to all of the even-numbered text exercises. A list of all challenging problems from exercise sets is included.

The *Instructor's Answer Manual* gives the answers to every text exercise, collected in one convenient location and presented in an easy-to-use format.

The *Scott, Foresman Test Generator for Mathematics* enables instructors to select questions for any section in the text, or to use a ready-made test for each chapter. Instructors may generate tests in multiple-choice or open-response formats, scramble the order of questions while printing, and produce multiple versions of each test (up to 9 with Apple, up to 25 with IBM and Macintosh). The system features printed graphics and accurate mathematics symbols. It also features a preview option that allows instructors to view questions before printing, to regenerate variables, and to replace or skip questions if desired. The IBM version includes an editor that allows instructors to add their own problems to existing data disks.

A set of two-color *Overhead Transparencies* showing charts, figures, and portions of examples is available to accompany lectures.

SUPPLEMENTS FOR THE STUDENT

The *Student's Solutions Manual,* available for purchase by students, provides detailed, worked-out solutions to all of the odd-numbered text exercises.

Computer Applications for Finite Math by Donald R. Coscia is a softbound textbook packaged with two disks (in Apple II and IBM–PC versions) with programs and exercises keyed to the text. The programs allow students to solve meaningful problems without the difficulties of extensive computation. This book bridges the gap between the text and the computer by providing additional explanations and exercises for solution using a microcomputer.

For the calculus portions of the book, some students may find helpful the *Short Calculus Workbook,* prepared by Walter Turner of Western Michigan University. Worked-out examples are presented so that students may follow the solutions step by step. These examples are followed by problems progressing in level of difficulty; only part of the solution is presented, and the student is asked to complete the rest. The *Workbook* supplements the problems presented in the original textbook and provides many additional worked examples.

RELATED BOOKS

Mathematics with Applications, Fifth Edition, is one text within our complete line of Lial/Miller math for management offerings: *Finite Mathematics,* Fourth Edition, *Calculus with Applications,* Fourth Edition, *Calculus with Applications, Brief Version,* and *Finite Mathematics and Calculus with Applications,* Third Edition.

ACKNOWLEDGMENTS

The following instructors reviewed the manuscript and made many helpful suggestions for improvement in this edition.

Bob Beul	*St. Louis University—Metropolitan College*
James E. Carpenter	*Iona College*
Duane E. Deal	*Ball State University*

Carol E. DeVille   *Louisiana Tech University*
George A. Emerson   *National University*
Dauhrice K. Gibson   *Gulf Coast Community College*
Robert E. Goad   *Sam Houston State University*
Katherine J. Huppler   *St. Cloud State University*
Alec Ingraham   *New Hampshire College*
Robert H. Johnston   *Virginia Commonwealth University*
Steve Laroe   *University of Alaska—Fairbanks*
Norman R. Martin   *Northern Arizona University*
Thomas J. Ordoyne   *University of South Carolina at Spartanburg*
Jon M. Plachy   *Metropolitan State College*
Elizabeth Polenzani   *Pasadena City College*
Wayne B. Powell   *Oklahoma State University*
Gordon Shilling   *University of Texas—Arlington*
Pradeep Shukla   *Suffolk University*

We extend thanks to respondents on questionnaires on the fourth edition for their help in suggesting improvements: Charles Ainley, Spokane Falls Community College; Philip G. Buckhiester, North Georgia College; Len Herzstein, Skyline College; Steven Kahn, Anne Arundel Community College; Dennis M. Kern, University of Texas—San Antonio; Arthur Lieberman, Cleveland State University; Charles E. Little, Northern Arizona University; Mary McCarty, Sullivan County Community College; Daniel Marks, Auburn University at Montgomery; Thomas Mickewich, Lake Superior State University; Vernon Nyhoff, Calvin College; Janice Phillipp, Texas Southmost College; Samuel C. Reep, Central Piedmont Community College; Neville Robbins, San Francisco State University; Franklin Sheehan, San Francisco State University; Jimmy Smith, Appalachian State University; Joan M. Spetich, Baldwin-Wallace College; William Stenger, Ambassador College; Brenda Tomulty, Edmonds Community College; Barbara Victor, Illinois Benedictine College; V. S. Vijeyakumer, John Abbott College; Marlene Wand, Kennesaw College; Carlton Woods, Auburn University at Montgomery; and Raymond B. Young, Embry-Riddle Aeronautical University.

Our thanks also go to Jim Eckerman of American River College, who wrote the Appendix on Graphing Calculators; to Jim Walker of American River College, who compiled the Index of Applications; to Phil Smith of American River College, Arthur Lieberman of Cleveland State University, and Dennis Kern of the University of Texas at San Antonio, for their efforts in checking the accuracy of all answers to the exercises; to Paul Eldersveld of the College of DuPage, for coordinating the preparation of the printed supplements; and to our editors, George Duda, Kristen Westman, Adam Bryer, and Cathy Wacaser, who contributed a great deal to the final book.

Margaret L. Lial
Thomas W. Hungerford

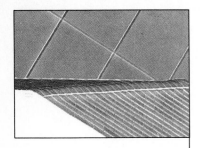

Contents

Chapter 1	**Fundamentals of Algebra** 1
1.1	The Real Numbers 1
1.2	Linear Equations and Applications 12
1.3	Linear Inequalities 23
1.4	Absolute Value Equations and Inequalities 28
1.5	Polynomials 32
1.6	Factoring 36
1.7	Rational Expressions 42
1.8	Integer Exponents 48
1.9	Rational Exponents and Radicals 53
1.10	Quadratic Equations 61
1.11	Polynomial and Rational Inequalities 72
	Chapter 1 Summary 78
	Chapter 1 Review Exercises 79
Case 1	Estimating Seed Demands—The Upjohn Company 84

Chapter 2	**Functions and Graphs** 85
2.1	Functions 86
2.2	Graphs of Functions 93
2.3	Linear Functions 105
2.4	Slope and the Equations of a Line 115
2.5	Applications of Linear Functions 125

Chapter 2 Summary   135

Chapter 2 Review Exercises   136

Case 2   Marginal Cost—Booz, Allen and Hamilton   138

Chapter 3   **Polynomial and Rational Functions**   140

3.1   Quadratic Functions   140

3.2   Applications of Quadratic Functions   148

3.3   Polynomial Functions   153

3.4   Rational Functions   160

Chapter 3 Summary   167

Chapter 3 Review Exercises   167

Chapter 4   **Exponential and Logarithmic Functions**   169

4.1   Exponential Functions   169

4.2   Applications of Exponential Functions   177

4.3   Logarithmic Functions   184

4.4   Applications of Logarithmic Functions   194

Chapter 4 Summary   202

Chapter 4 Review Exercises   203

Case 3   The Van Meegeren Art Forgeries   205

Chapter 5   **Mathematics of Finance**   207

5.1   Simple Interest and Discount   207

5.2   Compound Interest   214

5.3   Annuities   223

5.4   Present Value of an Annuity; Amortization   232

5.5   Applying Financial Formulas   240

Chapter 5 Summary   243

Chapter 5 Review Exercises   244

Case 4   Present Value   247

Chapter 6 | **Systems of Linear Equations and Matrices** 249

6.1 Systems of Linear Equations; The Echelon Method 249

6.2 Solution of Linear Systems by the Gauss-Jordan Method 263

6.3 Basic Matrix Operations 274

6.4 Multiplication of Matrices 283

6.5 Matrix Inverses 293

6.6 Applications of Matrices 301

Chapter 6 Summary 312

Chapter 6 Review Exercises 313

Case 5 | Contagion 317

Case 6 | Leontief's Model of the American Economy 318

Chapter 7 | **Linear Programming** 321

7.1 Graphing Linear Inequalities in Two Variables 321

7.2 Solving Linear Programming Problems Graphically 331

7.3 Applications of Linear Programming 339

7.4 The Simplex Method: Slack Variables and the Pivot 348

7.5 Solving Maximization Problems 356

7.6 Nonstandard Problems; Minimization 367

7.7 Duality 378

Chapter 7 Summary 389

Chapter 7 Review Exercises 390

Case 7 | Merit Pay—The Upjohn Company 394

Case 8 | Making Ice Cream 396

Chapter 8 | **Sets and Probability** 398

8.1 Sets 398

8.2 Applications of Venn Diagrams 407

8.3 Probability 416

8.4 Basic Concepts of Probability 423

8.5 Conditional Probability 438

8.6 Bayes' Formula 452

Chapter 8 Summary   459

Chapter 8 Review Exercises   460

Case 9   Drug Search Probabilities—Smith Kline and French Laboratories   463

Case 10   Making a First Down   463

Case 11   Medical Diagnosis   464

Chapter 9   Further Topics in Probability   465

9.1   Permutations and Combinations   465

9.2   Applications of Counting   478

9.3   Binomial Trials   486

9.4   Markov Chains   491

9.5   Probability Distributions; Expected Value   501

9.6   Decision Making   517

Chapter 9 Summary   521

Chapter 9 Review Exercises   522

Case 12   Optimal Inventory for a Service Truck   526

Chapter 10   Introduction to Statistics   528

10.1   Frequency Distributions   528

10.2   Measures of Central Tendency   536

10.3   Measures of Variation   543

10.4   Normal Distributions   550

10.5   The Binomial Distribution   560

Chapter 10 Summary   566

Chapter 10 Review Exercises   568

Case 13   Inventory Control   570

Chapter 11   Differential Calculus   572

11.1   Limits   572

11.2   Rates of Change   589

11.3   Definition of the Derivative   600

11.4   Techniques for Finding Derivatives   614

11.5   Derivatives of Products and Quotients   628
11.6   The Chain Rule   635
11.7   Derivatives of Exponential and Logarithmic Functions   645
11.8   Continuity and Differentiability   652

Chapter 11 Summary   663
Chapter 11 Review Exercises   665

Chapter 12   **Applications of the Derivative**   668
12.1   Maxima and Minima   668
12.2   The Second Derivative Test   682
12.3   Applications of Maxima and Minima   689
12.4   Curve Sketching (Optional)   700

Chapter 12 Summary   719
Chapter 12 Review Exercises   720
Case 14   The Effect of Oil Price on the Optimal Speed of Ships   722
Case 15   A Total Cost Model for a Training Program   723

Chapter 13   **Integral Calculus**   724
13.1   Antiderivatives; Indefinite Integrals   724
13.2   Integration by Substitution   734
13.3   Area and the Definite Integral   743
13.4   The Fundamental Theorem of Calculus   755
13.5   Applications of Integrals   766
13.6   Tables of Integrals (Optional)   777
13.7   Differential Equations and Applications   779

Chapter 13 Summary   787
Chapter 13 Review Exercises   789
Case 16   Estimating Depletion Dates for Minerals   792
Case 17   How Much Does a Warranty Cost?   794

Chapter 14   **Multivariate Calculus**   796

14.1   Functions of Several Variables   796

14.2   Partial Derivatives   807

14.3   Extrema of Functions of Several Variables   816

Chapter 14 Summary   825

Chapter 14 Review Exercises   826

Appendix A: Graphing Calculators   A-1

Appendix B: Tables   A-9

Table 1 Powers of *e*   A-9

Table 2 Natural Logarithms   A-10

Table 3 Common Logarithms   A-11

Table 4 Compound Interest   A-13

Table 5 Amount of an Annuity   A-15

Table 6 Present Value of an Annuity   A-17

Table 7 Combinations   A-19

Table 8 Areas Under the Normal Curve   A-20

Table 9 Integrals   A-22

Answers to Selected Exercises   A-23

Index   A-68

Index of Applications   A-72

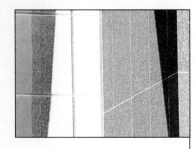

To the Student

FEATURES TO AID YOUR UNDERSTANDING

Side problems, which will help you to learn new concepts and reinforce your understanding, are referred to in the text by numbers within colored squares: **1**. When you see that symbol, you should work the indicated problem at the side before going on.

Special **cautionary notes** will warn you of common errors and misconceptions.

Grey-tinted **summary boxes** within chapters, and extended **end-of-chapter summaries,** which include key terms, symbols, and concepts, and review exercises, will assist your review of the material.

CALCULATOR EXERCISES

Exercises that require the use of a calculator are indicated by ⊞. Also, a few exercises have been included that are suitable to be worked by computer with appropriate software. These are indicated by the symbol ▦.

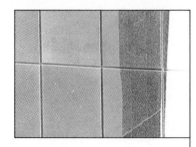

CHAPTER 1

Fundamentals of Algebra

This book deals with the application of mathematics to business, social science, and biology. Almost all of these applications begin with some real-world problem that is solved by first writing appropriate equations or other mathematical relationships that describe the problem. The properties of algebra are then used to simplify the relationships and solve the equations, producing a solution to the real-world problem.

Since algebra is so vital to the study of applications of mathematics, we begin with a review of some of the fundamental ideas of algebra. If you have not used your algebraic skills for some time, it is important for you to study the review material in this chapter carefully; your success in covering the material that follows will depend upon your algebraic skills.

1.1 THE REAL NUMBERS

The various types of numbers used in this book can be explained with a diagram called a **number line.** Draw a number line by choosing any point on a horizontal line and labeling it 0. Then choose any point to the right of 0 and label it 1. The distance between 0 and 1 gives a unit of measure that can be used repeatedly to locate points to the right of 1, labeled 2, 3, 4, and so on, and points to the left of 0, labeled -1, -2, -3, -4, and so on. A number line with several sample numbers located (or **graphed**) on it is shown in Figure 1.1.* ▪**1**

1 Draw a number line and graph the numbers -4, -1, 0, 1, 2.5, and 13/4 on it.

Answer:

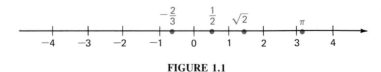

FIGURE 1.1

---

*The use of margin problems is explained in the section ''To the Student'' preceding the table of contents.

Any number that can be associated with a point on the number line is called a **real number.*** Only real numbers will be used in this book. The names of the most common types of real numbers are as follows.

The Real Numbers

Natural (counting) numbers	1, 2, 3, 4, . . .
Whole numbers	0, 1, 2, 3, 4, . . .
Integers	. . . , -3, -2, -1, 0, 1, 2, 3, . . .
Rational numbers	All numbers of the form p/q, where p and q are integers, with $q \neq 0$
Irrational numbers	Real numbers that are not rational

The three dots used in this box show that the numbers continue indefinitely in the same way. The relationships among these types of numbers are shown in Figure 1.2. Notice, for example, that the integers are also rational numbers and real numbers, but the integers are not irrational numbers.

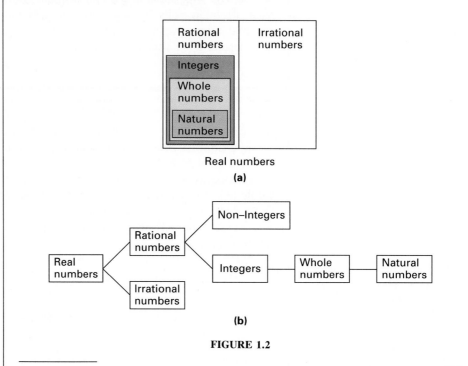

FIGURE 1.2

---

*Not all numbers are real numbers. An example of a number that is not a real number is $\sqrt{-1}$.

2 Name all the types of numbers that apply to the following.

(a) -2

(b) $-5/8$

(c) π

Answer:

(a) Integer, rational, real

(b) Rational, real

(c) Irrational, real

One example of an irrational number is π, the ratio of the circumference of a circle to its diameter. The number π can be approximated by writing $\pi \approx 3.14159$ or $\pi \approx 22/7$ ($\approx$ means "is approximately equal to"), but there is no rational number that is exactly equal to π. Another irrational number can be found by constructing a triangle having a 90° angle, with the two shortest sides each 1 unit long, as shown in Figure 1.3. The third side can be shown to have a length that is irrational (the length is $\sqrt{2}$ units). Many numbers have square roots that are irrational numbers; in fact, if a number is not the square of a rational number, then its square root is irrational.

FIGURE 1.3

All real numbers can be written in decimal form. Rational numbers in decimal form either terminate or repeat. For example,

$$.5, \quad .125, \quad 1.333 \ldots , \quad 2.412412 \ldots$$

are rational numbers. Irrational numbers in decimal form neither terminate nor repeat. There is no way to write an irrational number exactly as a decimal.

▶**EXAMPLE 1**   What kind of number is each of the following?
(a) 6

Consult the list of types of real numbers or Figure 1.2 to see that 6 is a natural number, whole number, integer, rational number, and real number.
(b) 3/4

This number is rational and real.
(c) $\sqrt{8}$

Since 8 is not the square of a rational number, $\sqrt{8}$ is irrational and real. ◀ **2**

Some basic properties of the real numbers follow.

3 Name the property illustrated in each of the following examples.

(a) $(2 + 3) + 9 = (3 + 2) + 9$

(b) $(2 + 3) + 9 = 2 + (3 + 9)$

(c) $(2 + 3) + 9 = 9 + (2 + 3)$

(d) $(4 \cdot 6)p = (6 \cdot 4)p$

(e) $4(6p) = (4 \cdot 6)p$

Answer:

(a) Commutative property

(b) Associative property

(c) Commutative property

(d) Commutative property

(e) Associative property

Properties of the Real Numbers

For all real numbers a, b, and c, the following properties hold true.

Commutative properties $a + b = b + a$ $ab = ba$

Associative properties $(a + b) + c = a + (b + c)$ $(ab)c = a(bc)$

Identity properties   There exists a unique real number 0, called the **additive identity,** such that
$$a + 0 = a \quad \text{and} \quad 0 + a = a.$$
There exists a unique real number 1, called the **multiplicative identity,** such that
$$a \cdot 1 = a \quad \text{and} \quad 1 \cdot a = a.$$

Inverse properties   There exists a unique real number $-a$, called the **additive inverse,** such that
$$a + (-a) = 0 \quad \text{and} \quad (-a) + a = 0.$$
If $a \neq 0$, there exists a unique real number $1/a$, called the **multiplicative inverse,** such that
$$a \cdot \frac{1}{a} = 1 \quad \text{and} \quad \frac{1}{a} \cdot a = 1.$$

Distributive property $a(b + c) = ab + ac$

▶**EXAMPLE 2**   The following statements are examples of the commutative properties. Notice that the order of the numbers changes from one side of the equals sign to the other, so that the order in which two numbers are added or multiplied is unimportant.

(a) $6 + x = x + 6$ **(b)** $(6 + x) + 9 = (x + 6) + 9$

(c) $(6 + x) + 9 = 9 + (6 + x)$ **(d)** $5 \cdot (9 \cdot 8) = (9 \cdot 8) \cdot 5$

(e) $5 \cdot (9 \cdot 8) = 5 \cdot (8 \cdot 9)$ ◀

▶**EXAMPLE 3**   The following statements are examples of the associative properties. Here the order of the numbers does not change, but the placement of the parentheses does change. This means that when three numbers are to be added, the sum of any two can be found first, then that result is added to the remaining number.

(a) $4 + (9 + 8) = (4 + 9) + 8$

(b) $3(9x) = (3 \cdot 9)x$

(c) $(\sqrt{3} + \sqrt{7}) + 2\sqrt{6} = \sqrt{3} + (\sqrt{7} + 2\sqrt{6})$ ◀

4 Name the property illustrated in each of the following examples.

(a) $2 + 0 = 2$

(b) $-\dfrac{1}{4} \cdot (-4) = 1$

(c) $-\dfrac{1}{4} + \dfrac{1}{4} = 0$

(d) $1 \cdot \dfrac{2}{3} = \dfrac{2}{3}$

Answer:

(a) Identity property

(b) Inverse property

(c) Inverse property

(d) Identity property

5 Use the distributive property to complete each of the following.

(a) $4(-2 + 5)$

(b) $2(a + b)$

(c) $-3(p + 1)$

(d) $(8 - k)m$

Answer:

(a) $4(-2) + 4(5)$

(b) $2a + 2b$

(c) $-3p - 3$

(d) $8m - km$

The identity properties give some special properties of the numbers 0 and 1. Since 0 preserves the identity of a real number under addition, 0 is the identity element for addition. In the same way, 1 preserves the identity of a real number under multiplication and is the identity element for multiplication.

▶ **EXAMPLE 4** By the identity properties,
(a) $-8 + 0 = -8$,
(b) $(-9)1 = -9$. ◀

▶ **EXAMPLE 5** By the inverse properties, the statements in parts (a) through (e) are true.
(a) $9 + (-9) = 0$ **(b)** $-15 + 15 = 0$

(c) $6 \cdot \dfrac{1}{6} = 1$ **(d)** $-8 \cdot \left(\dfrac{1}{-8}\right) = 1$

(e) $\dfrac{1}{\sqrt{5}} \cdot \sqrt{5} = 1$ **(f)** There is no real number x such that $0 \cdot x = 1$, so 0 has no inverse for multiplication. ◀ **4**

One of the most important properties of the real numbers, and the only one that involves both addition and multiplication, is the distributive property. The next example shows how this property is applied.

▶ **EXAMPLE 6** By the distributive property,
(a) $9(6 + 4) = 9 \cdot 6 + 9 \cdot 4$
(b) $3(x + y) = 3x + 3y$
(c) $-8(m + 2) = (-8)(m) + (-8)(2) = -8m - 16$
(d) $(5 + x)y = 5y + xy$. ◀ **5**

Note As shown in Example 6(d), the distributive property can also be written as $(a + b)c = ac + bc$.

Comparing two real numbers requires symbols that indicate their order on the number line. The following symbols are used to indicate that one number is greater than or less than another number.

$<$ means *is less than*	$\leq$ means *is less than or equal to*
$>$ means *is greater than*	$\geq$ means *is greater than or equal to*

6 Write *true* or *false* for the following.

(a) $-9 \leq -2$

(b) $8 > -3$

(c) $-14 \leq -20$

Answer:

(a) True

(b) True

(c) False

7 Graph all integers x such that

(a) $-3 < x < 5$

(b) $1 \leq x \leq 5$.

Answer:

(a)
-3-2-1 0 1 2 3 4 5

(b)
0 1 2 3 4 5 6

8 Graph all real numbers x such that

(a) $-5 < x < 1$

(b) $4 < x < 7$.

Answer:

(a)
-5　　　　1

(b)
4　　　　7

The following definitions show how the number line is used to decide which of two given numbers is the greater.

For real numbers a and b,
if a is to the left of b on a number line, then $a < b$;
if a is to the right of b on a number line, then $a > b$.

▶**EXAMPLE 7** Write *true* or *false* for each of the following.
(a) $8 < 12$
This statement says that 8 is less than 12, which is true.
(b) $-6 > -3$
The graph of Figure 1.4 shows that -6 is to the *left* of -3. Thus, $-6 < -3$, and the given statement is false.
(c) $-2 \leq -2$
Since $-2 = -2$, this statement is true. ◀ **6**

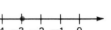

FIGURE 1.4

A number line can be used to draw the graph of a set of numbers, as shown in the next few examples.

▶**EXAMPLE 8** Graph all integers x such that $1 < x < 5$.
The only integers between 1 and 5 are 2, 3, and 4. These integers are graphed on the number line in Figure 1.5 ◀ **7**

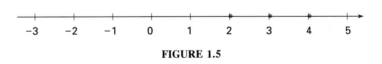

FIGURE 1.5

▶**EXAMPLE 9** Graph all real numbers x such that $1 < x < 5$.
The graph includes all the real numbers between 1 and 5 and not just the integers. Graph these numbers by drawing a heavy line from 1 to 5 on the number line, as in Figure 1.6. Open circles at 1 and 5 show that neither of these points belongs to the graph. ◀ **8**

9 Graph all real numbers x such that

(a) $[4, \infty)$;

(b) $[-2, 1]$.

Answer:

(a)

(b)

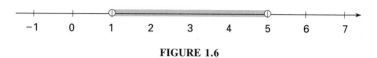

FIGURE 1.6

A set consisting of all the real numbers between two points, such as $1 < x < 5$ in Example 9, is called an **interval.** A special notation called **interval notation** is used to indicate an interval on the number line. For example, the interval including the numbers x, where $-2 < x < 3$, is written as $(-2, 3)$. The parentheses indicate that the numbers -2 and 3 are *not* included. If -2 and 3 are to be included in the interval, square brackets are used, as in $[-2, 3]$. The chart below shows several typical intervals, where $a < b$.

Inequality	Interval Notation	Explanation
$a \le x \le b$	$[a, b]$	Both a and b are included.
$a \le x < b$	$[a, b)$	a is included, b is not.
$a < x \le b$	$(a, b]$	b is included, a is not.
$a < x < b$	(a, b)	Neither a nor b is included.

Interval notation is also used to describe sets such as the set of all numbers x, with $x \ge -2$. This interval is written $[-2, \infty)$.

▶**EXAMPLE 10**   Graph the interval $[-2, \infty)$.

Start at -2 and draw a heavy line to the right, as in Figure 1.7. Use a solid circle at -2 to show that -2 itself is part of the graph. The symbol, ∞, read "infinity," *does not* represent a number. It simply indicates that *all* numbers greater than -2 are in the interval. Similarly, the symbol $(-\infty, 2)$ indicates the set of all numbers x with $x < 2$.   ◀   9

FIGURE 1.7

ORDER OF OPERATIONS   We avoid possible ambiguity when working problems with real numbers by using the following *order of operations,* which has been agreed on as the most useful. This order of operations is used by computers and many calculators.*

_____

*Some calculators may not follow this convention. To see if yours does, use it to work Examples 11–12 and compare the results. If your calculator works differently, consult the instruction manual (or your instructor) for directions.

10 Simplify the following.

(a) $4^2 \div 8 + 3^2 \div 3$

(b) $[-7 + (-9)](-4) - 8(3)$

(c) $\dfrac{-11 - (-12) - 4 \cdot 5}{-2^3 - (-3)(-5)}$

(d) $\dfrac{36 \div 4 \cdot 3 \div 9 + 1}{9 \div (-6) \cdot 8 - 4}$

Answer:

(a) 5

(b) 40

(c) 19/23

(d) $-1/4$

11 Evaluate the following if $m = -5$ and $n = 8$.

(a) $-2mn - 2m^2$

(b) $\dfrac{4(n - 5)^2 - m}{m + n}$

Answer:

(a) 30

(b) $\dfrac{41}{3}$

Order of Operations

If parentheses or square brackets are present:

1. Work separately above and below any fraction bar.
2. Use the rules below within each set of parentheses or square brackets. Starts with the innermost and work outward.

If no parentheses are present:

1. Find all powers and roots, working from left to right.
2. Do any multiplications or divisions in the order in which they occur, working from left to right.
3. Do any additions or subtractions in the order in which they occur, working from left to right.

▶**EXAMPLE 11** Use the order of operations to simplify the following.

(a) $\dfrac{-9(-3) + (-5)}{2(-2)^3 - 5(3)}$

Work separately above and below the fraction bar. Recall that $(-2)^3$ means $(-2)(-2)(-2) = -8$. Follow the order of operations.

$$\frac{-9(-3) + (-5)}{2(-2)^3 - 5(3)} = \frac{-9(-3) + (-5)}{2(-8) - 5(3)} = \frac{27 + (-5)}{-16 - 15} = \frac{22}{-31} = -\frac{22}{31}$$

(b) $-(3 - 5) - [2 - (3^2 - 13)]$

Start with the innermost parentheses. Evaluate the power first: $3^2 = 3 \cdot 3 = 9$. Now follow the order of operations.

$$-(3 - 5) - [2 - (3^2 - 13)] = -(3 - 5) - [2 - (9 - 13)]$$
$$= -(-2) - [2 - (-4)] = 2 - [6] = -4 \quad \blacktriangleleft \boxed{10}$$

▶**EXAMPLE 12** Use the order of operations to evaluate each expression if $x = -2$, $y = 5$, and $z = -3$.

(a) $-4x^2 - 7y + 4z$

Use parentheses when replacing letters with numbers.

$$-4x^2 - 7y + 4z = -4(-2)^2 - 7(5) + 4(-3)$$
$$= -4(4) - 7(5) + 4(-3) = -16 - 35 - 12 = -63$$

(b) $\dfrac{2(x - 5)^2 + 4y}{z + 4} = \dfrac{2(-2 - 5)^2 + 4(5)}{-3 + 4}$

$$= \frac{2(-7)^2 + 20}{1} = 2(49) + 20 = 118 \quad \blacktriangleleft \boxed{11}$$

12 Find the following.

(a) $|-6|$

(b) $-|7|$

(c) $-|-2|$

(d) $|-3 - 4|$

(e) $|2 - 7|$

Answer:

(a) 6

(b) -7

(c) -2

(d) 7

(e) 5

ABSOLUTE VALUE Distance is always given as a nonnegative number. For example, the distance from 0 to -2 on a number line is 2, the same as the distance from 0 to 2. The *absolute value* of a number a gives the distance on the number line from a to 0. Thus, the *absolute value* of both 2 and -2 is 2. Write the absolute value of the real number a as $|a|$. For example, the distance on the number line from 9 to 0 is 9, as is the distance from -9 to 0. (See Figure 1.8.) By definition, $|9| = 9$ and $|-9| = 9$.

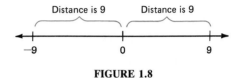

FIGURE 1.8

The facts that $|9| = 9$ and that $|-9| = 9 = -(-9)$ suggest the following algebraic definition of absolute value.

Absolute Value

For any real number a,

$$|a| = a \qquad \text{if } a \geq 0$$

$$|a| = -a \qquad \text{if } a < 0.$$

Note the second part of the definition: for a negative number, say -5, the negative of -5 is the positive number $-(-5) = 5$. Similarly, if a is any negative number, then $-a$ is a *positive* number. Thus, for every real number a, $|a| \geq 0$.

▶**EXAMPLE 13** Find each of the following.

(a) $|5|$
   Since $5 > 0$, then $|5| = 5$.

(b) $|-5| = 5$

(c) $-|-5| = -(5) = -5$
   Evaluate within the absolute value bars first, as we do with parentheses.

(d) $|0| = 0$

(e) $|8 - 9|$
   First simplify the expression inside the absolute value bars.

$$|8 - 9| = |-1| = 1$$

(f) $|-4 - 7| = |-11| = 11$ ◀ **12**

1.1 EXERCISES

Name all the types of numbers that apply to the following. (See Example 1.)

1. 6   **2.** −9   **3.** −7   **4.** 0   **5.** $\dfrac{1}{2}$ **6.** $-\dfrac{5}{11}$ **7.** $\sqrt{7}$ **8.** $-\sqrt{11}$ **9.** π **10.** $\dfrac{1}{\pi}$

Label each of the following as true *or* false.

11. Every integer is a rational number.

12. Every integer is a whole number.

13. Every whole number is an integer.

14. Some whole numbers are not natural numbers.

15. There is a natural number that is not a whole number.

16. Every rational number is a natural number.

17. Every natural number is a rational number.

18. No whole numbers are rational.

Identify the properties that are illustrated in each of the following. Some will require more than one property. Assume all variables represent real numbers. (See Examples 2–6.)

19. $8 \cdot 9 = 9 \cdot 8$

20. $3 + (−3) = 0$

21. $3 + (−3) = (−3) + 3$

22. $0 + (−7) = (−7) + 0$

23. $−7 + 0 = −7$

24. $8 + (12 + 6) = (8 + 12) + 6$

25. $[9(−3)] \cdot 2 = 9[(−3) \cdot 2]$

26. $8(m + 4) = 8m + 8 \cdot 4$

27. $x(y + 2) = xy + 2x$

28. $(7 − y) + 0 = 7 − y$

29. $8(4 + 2) = (2 + 4)8$

30. $x \cdot \dfrac{1}{x} + x \cdot \dfrac{1}{x} = x\left(\dfrac{1}{x} + \dfrac{1}{x}\right)$ (if $x \neq 0$)

Graph each of the following on a number line. (See Example 8.)

31. All integers x such that $−5 < x < 5$

32. All integers x such that $−4 < x < 2$

33. All whole numbers x such that $x \leq 3$

34. All whole numbers x such that $1 \leq x \leq 8$

35. All natural numbers x such that $−1 < x < 5$

36. All natural numbers x such that $x \leq 2$

Graph the following intervals. (See Examples 9 and 10.)

37. $(−\infty, 4)$ **38.** $(5, \infty)$ **39.** $[6, \infty)$ **40.** $[8, \infty)$ **41.** $(−2, \infty)$ **42.** $(−\infty, −4)$

43. $(−5, −3)$ **44.** $(3, 5)$ **45.** $[−3, 6]$ **46.** $[8, 14]$ **47.** $(1, 6]$ **48.** $[−4, 3)$

Perform the indicated operations, using the order of operations given in the text. (See Example 11.)

49. $9 − 4^2 − (−12)$

50. $8^2 − (−4) + 11$

51. $−2(9 − 8) + (−7)(2)^3$

52. $6(−5) − (−3)(2)^4$

53. $(4 − 2^3)(−2 + \sqrt{25})$

54. $[3^2 − (−2)][\sqrt{16} − 2^3]$

55. $\dfrac{−9.23(5.87) + 6.993}{1.225(−8.601) − 148(.0723)}$

56. $\dfrac{189.4(3.221) − 9.447(−8.772)}{4.889[3.177 − 8.291(3.427)]}$

Evaluate each of the following if $p = −2$, $q = 4$, and $r = −5$. (See Example 12.)

57. $−3(p + 5q)$ **58.** $2(q − r)$ **59.** $\dfrac{q + r}{q + p}$ **60.** $\dfrac{3q}{3p − 2r}$ **61.** $\dfrac{\dfrac{q}{4} − \dfrac{r}{5}}{\dfrac{p}{2} + \dfrac{q}{2}}$ **62.** $\dfrac{\dfrac{3r}{10} − \dfrac{5p}{2}}{q + \dfrac{2r}{5}}$

Evaluate each of the following. (See Example 13.)

63. $-|-4|$

64. $-|-2|$

65. $|6 - 4|$

66. $|3 - 17|$

67. $-|12 + (-8)|$

68. $|-6 + (-15)|$

69. $|8 - (-9)|$

70. $|-3 - (-2)|$

71. $|8| - |-4|$

72. $|-9| - |-12|$

73. $-|-4| - |-1 - 14|$

74. $-|6| - |-12 - 4|$

In each of the following problems, fill in the blank with either =, <, or >, so that the resulting statement is true.

75. $|5|$ ____ $|-5|$

76. $|3|$ ____ $|-3|$

77. $-|7|$ ____ $|7|$

78. $-|-4|$ ____ $|4|$

79. $|10 - 3|$ ____ $|3 - 10|$

80. $|6 - (-4)|$ ____ $|-4 - 6|$

81. $|1 - 4|$ ____ $|4 - 1|$

82. $|10 - 8|$ ____ $|8 - 10|$

83. $|-2 + 8|$ ____ $|2 - 8|$

84. $|3 + 1|$ ____ $|-3 - 1|$

85. $|3| \cdot |-5|$ ____ $|3(-5)|$

86. $|3| \cdot |2|$ ____ $|3(2)|$

87. $|3 - 2|$ ____ $|3| - |2|$

88. $|5 - 1|$ ____ $|5| - |1|$

89. In general, if a and b are any real numbers having the same sign (both negative or both positive), is it always true that $|a + b| = |a| + |b|$?

90. If a and b are any real numbers, is it always true that $|a + b| = |a| + |b|$?

91. If a and b are any two real numbers, is it always true that $|a - b| = |b - a|$?

92. For which real numbers b does $|2 - b| = |2 + b|$?

Management   *Use inequality symbols to rewrite each of the following statements, which are based on a recent article in* Business Week *magazine.* Let x represent the unknown in each exercise.*

Example   Rewrite ''Gillette has at least 2/3 of the shaving system's market in North America and Europe'' by letting x represent the fraction of the market. Then $x \geq 2/3$.

93. The new Sensor razor must add at least 4% to Gillette's market share just to recoup its budget.

94. In 1989, at least 65% of Gillette's operating profits came from razors and blades.

95. In 1989, at least 32% of Gillette's revenues came from razors and blades.

96. By the time the Sensor razor reaches the market, Gillette will have spent up to $200 million in research, engineering, and tooling.

97. Gillette will spend up to $110 million on advertising the new blade.

98. Gillette's net profit increase in 1989 was at least $285 million.

Social Science   *Sociologists measure the status of an individual within a society by evaluating for that individual the number x, which gives the percentage of the population with less income than the given person, and the number y, the percentage of the population with less education. The* average status *is defined as $(x + y)/2$, while the individual's* status incongruity *is defined by $|(x - y)/2|$. People with high status incongruities would include unemployed Ph.D.'s (low x, high y) and millionaires who didn't make it past the second grade (high x, low y).*

99. What is the highest possible average status for an individual? The lowest?

100. What is the highest possible status incongruity for an individual? The lowest?

101. Jolene Rizzo makes more money than 56% of the population and has more education than 78%. Find her average status and status incongruity.

102. A popular movie star makes more money than 97% of the population and is better educated than 12%. Find the average status and status incongruity for this individual.

*From ''How a $4 Razor Ends Up Costing $300 Million,'' Business Week, *January 29, 1990, pp. 62–63.*

1 Is -4 a solution of the following equations?

(a) $3x + 5 = -7$

(b) $2x - 3 = 5$

Answer:

(a) Yes

(b) No

1.2 LINEAR EQUATIONS AND APPLICATIONS

One of the main uses of algebra is to solve equations. An **equation** states that two mathematical expressions are equal. Examples of equations include $x + 6 = 9$, $4y + 8 = 12$, and $9z = -36$. The letter in each equation, the unknown, is called the **variable.** Sometimes an equation must be solved that has more than one letter. As a general rule, the first few letters of the alphabet, a, b, c, and so on, are used to represent constants, while letters such as x, y, and z are used for variables.

A **solution** of an equation is a number than can be substituted for the variable in the equation to produce a true statement. For example, substituting the number 9 for x in the equation $2x + 1 = 19$ gives

$$2x + 1 = 19$$
$$2(9) + 1 = 19 \qquad \textbf{Let } x = 9$$
$$18 + 1 = 19. \qquad \textbf{True}$$

This true statement indicates that 9 is a solution of $2x + 1 = 19$. **1**

Equations that can be written in the form $ax + b = c$, where a, b, and c are real numbers, with $a \neq 0$, are called **linear equations.** Examples of linear equations include $5y + 9 = 16$, $8x = 4$, and $-3p + 5 = -8$. Examples of equations that are *not* linear include $|x| = 4$, $2x^2 = 5x + 6$, and $\sqrt{x + 2} = 4$.

The following properties are used to solve equations.

Properties of Equality

For any real numbers a, b, and c,

(a) if $a = b$, then $a + c = b + c$. (The same number may be added to both sides of an equation.)   **Addition property of equality**

(b) if $a = b$ and $c \neq 0$, then $ac = bc$. (The same nonzero number may be multiplied on both sides of an equation.)   **Multiplication property of equality**

Recall that subtraction is defined in terms of addition: for all real numbers a and b,

$$a - b = a + (-b);$$

and division is defined in terms of multiplication: for all real numbers a and all nonzero real numbers b,

$$\frac{a}{b} = a \cdot \frac{1}{b}.$$

2 Solve the following.

(a) $3p - 5 = 19$

(b) $4y + 3 = -5$

(c) $-2k + 6 = 2$

Answer:

(a) 8

(b) -2

(c) 2

3 Solve the following.

(a) $3(m - 6) + 2(m + 4)$
$= 4m - 2$

(b) $-2(y + 3) + 4y$
$= 3(y + 1) - 6$

Answer:

(a) 8

(b) -3

For this reason, special properties for subtraction and division are not needed, as the following examples show.

▶**EXAMPLE 1**   Solve the linear equation $5x - 3 = 12$.

Using the addition property of equality, add 3 to both sides. This isolates the term containing the variable on one side of the equals sign.

$$5x - 3 = 12$$
$$5x - 3 + 3 = 12 + 3 \qquad \textbf{Add 3 to both sides}$$
$$5x = 15$$

To get $1x$ instead of $5x$ on the left, use the fact that $(1/5) \cdot 5 = 1$, and multiply both sides of the equation by $1/5$.

$$5x = 15.$$
$$\frac{1}{5}(5x) = \frac{1}{5}(15) \qquad \textbf{Multiply both sides by } \frac{1}{5}$$
$$1x = 3$$
$$x = 3$$

The solution of the original equation, $5x - 3 = 12$, is 3. Check the solution by substituting 3 for x in the original equation.   ◀ **2**

▶**EXAMPLE 2**   Solve $2k + 3(k - 4) = 2(k - 3)$.

First simplify this equation using the distributive property. By this property, $3(k - 4)$ is $3k - 3 \cdot 4$, or $3k - 12$. Also, $2(k - 3)$ is $2k - 2 \cdot 3$, or $2k - 6$. The equation can now be written as

$$2k + 3(k - 4) = 2(k - 3)$$
$$2k + 3k - 12 = 2k - 6.$$

On the left, $2k + 3k = (2 + 3)k = 5k$, again by the distributive property, which gives

$$5k - 12 = 2k - 6.$$

One way to proceed is to add $-2k$ to both sides.

$$5k - 12 + (-2k) = 2k - 6 + (-2k) \qquad \textbf{Add } -2k \textbf{ to both sides}$$
$$3k - 12 = -6$$
$$3k - 12 + 12 = -6 + 12 \qquad \textbf{Add 12 to both sides}$$
$$3k = 6$$
$$\frac{1}{3}(3k) = \frac{1}{3}(6) \qquad \textbf{Multiply both sides by } \frac{1}{3}$$
$$k = 2$$

The solution is 2. Check this result by substituting 2 for k in the original equation.   ◀ **3**

4 Solve the following.

(a) $\dfrac{x}{2} - \dfrac{x}{4} = 6$

(b) $\dfrac{2x}{3} + \dfrac{1}{2} = \dfrac{x}{4} - \dfrac{9}{2}$

Answer:

(a) 24

(b) −12

The next three examples show how to simplify the solution of linear equations involving fractions. We solve these equations by multiplying both sides of the equation by a **common denominator,** a number that can be divided (with remainder 0) by each denominator in the equation. This step will eliminate the fractions.

Caution   Since this is an application of the multiplication property of *equality,* it can be done *only in an equation.*

▶**EXAMPLE 3**   Solve $\dfrac{r}{10} - \dfrac{2}{15} = \dfrac{3r}{20} - \dfrac{1}{5}$.

Here the denominators are 10, 15, 20, and 5. Each of these numbers can be divided into 60; therefore, 60 is a common denominator. Multiply both sides of the equation by 60.

$$60\left(\frac{r}{10} - \frac{2}{15}\right) = 60\left(\frac{3r}{20} - \frac{1}{5}\right)$$

Use the distributive property to eliminate the denominators.

$$60\left(\frac{r}{10}\right) - 60\left(\frac{2}{15}\right) = 60\left(\frac{3r}{20}\right) - 60\left(\frac{1}{5}\right)$$

$$6r - 8 = 9r - 12$$

Add $-6r$ and 12 to both sides.

$$6r - 8 + (-6r) + 12 = 9r - 12 + (-6r) + 12$$

$$4 = 3r$$

Multiply both sides by 1/3 to get the solution.

$$r = \frac{4}{3}$$

Check this solution in the original equation.   ◀ **4**

▶**EXAMPLE 4**   Solve $\dfrac{4}{3(k + 2)} - \dfrac{k}{3(k + 2)} = \dfrac{5}{3}$.

Multiply both sides of the equation by the common denominator $3(k + 2)$. Here $k \neq -2$, since $k = -2$ would give a 0 denominator, making the fraction meaningless.

$$3(k + 2) \cdot \frac{4}{3(k + 2)} - 3(k + 2) \cdot \frac{k}{3(k + 2)} = 3(k + 2) \cdot \frac{5}{3}$$

5 Solve the equation
$$\frac{5p + 1}{3(p + 1)} = \frac{3p - 3}{3(p + 1)}$$
$$+ \frac{9p - 3}{3(p + 1)}.$$

Answer:
1

6 Solve each equation.

(a) $\dfrac{3p}{p + 1} = 1 - \dfrac{3}{p + 1}$

(b) $\dfrac{8y}{y - 4} = \dfrac{32}{y - 4} - 3$

Answer:
Neither equation has a solution.

Simplify each side and solve for k.

$$4 - k = 5(k + 2)$$
$$4 - k = 5k + 10 \qquad \qquad \text{\textbf{Distributive property}}$$
$$4 - k + k = 5k + 10 + k \qquad \text{\textbf{Add } } k \text{ \textbf{to both sides}}$$
$$4 = 6k + 10$$
$$4 + (-10) = 6k + 10 + (-10) \qquad \text{\textbf{Add } } -10 \text{ \textbf{to both sides}}$$
$$-6 = 6k$$
$$-1 = k \qquad \qquad \text{\textbf{Multiply by } } \dfrac{1}{6}$$

The solution is -1. Substitute -1 for k as a check.

$$\frac{4}{3(-1 + 2)} - \frac{-1}{3(-1 + 2)} \overset{?}{=} \frac{5}{3}$$
$$\frac{4}{3} - \frac{-1}{3} \overset{?}{=} \frac{5}{3}$$
$$\frac{5}{3} = \frac{5}{3}$$

The check shows that -1 is the solution. ◀ **5**

Caution Because the equation in Example 4 has a restriction on k, it is *essential* to check the solution.

▶**EXAMPLE 5** Solve $\dfrac{x}{x - 2} = \dfrac{2}{x - 2} + 2$.

Multiply both sides of the equation by $x - 2$, assuming that $x - 2 \neq 0$. This gives

$$x = 2 + 2(x - 2)$$
$$x = 2 + 2x - 4$$
$$x = 2.$$

Recall the assumption that $x - 2 \neq 0$. Since $x = 2$, we have $x - 2 = 0$, and the multiplication property of equality does not apply. To see this, substitute 2 for x in the original equation—this substitution produces a 0 denominator. Since division by zero is not defined, there is no solution for the given equation. ◀ **6**

7 Solve for x.

(a) $2x - 7y = 3xk$

(b) $8(4 - x) + 6p$
$\quad = -5k - 11yx$

Answer:

(a) $x = \dfrac{7y}{2 - 3k}$

(b) $x = \dfrac{5k + 32 + 6p}{8 - 11y}$

Sometimes an equation with several variables must be solved for one of the variables. This process is called **solving for a specified variable.**

▶**EXAMPLE 6** Solve for x: $3(2x - 5a) + 4b = 4x - 2$.

Use the distributive property to get

$$6x - 15a + 4b = 4x - 2.$$

Treat x as the variable, the other letters as constants. Get all terms with x on one side of the equals sign, and all terms without x on the other side.

$6x - 4x = 15a - 4b - 2$	**Isolate terms with x on the left**
$2x = 15a - 4b - 2$	**Combine terms**
$x = \dfrac{15a - 4b - 2}{2}$	**Multiply by $\dfrac{1}{2}$**

The final equation is solved for x, as required. ◀ **7**

▶**EXAMPLE 7** The formula

$$A = \frac{24f}{b(p + 1)}$$

gives the approximate annual interest rate for a consumer loan paid off with monthly payments.* Here f is the finance charge on the loan, p is the total number of payments, and b is the original balance of the loan. Solve the formula for p.

Treat p as the variable and the other letters as constants. The goal is to isolate p on one side of the equals sign. Begin by multiplying both sides of the formula by $p + 1$. (Since p is the number of payments, p is greater than 0 here, so $p \neq -1$.)

$$A = \frac{24f}{b(p + 1)}$$

$$(p + 1)A = (p + 1)\frac{24f}{b(p + 1)}$$

$$(p + 1)A = \frac{24f}{b}$$

Now think of undoing the operations that have been performed on p. Multiply both sides of the equation by $1/A$ to undo the multiplication by A.

$$\frac{1}{A}(p + 1)A = \frac{1}{A} \cdot \frac{24f}{b}$$

$$p + 1 = \frac{24f}{Ab}$$

_____

*This formula is not accurate enough for the requirements of federal law.

8 Solve $J\left(\dfrac{m}{k} + a\right) = m$ for k.

Answer:

$k = \dfrac{Jm}{m - Ja}$

(We must assume $A \neq 0$. Why is this a very safe assumption here?) Finally, undo the addition of 1 to p by adding -1 on each side to get

$$p = \frac{24f}{Ab} - 1,$$

a result solved for p. ◄ **8**

APPLIED PROBLEMS One of the main reasons for learning mathematics is to be able to use it to solve practical problems. However, for many students, learning how to use mathematical skills in real-world applications is the most difficult task they face. Hints that may help with applications are given in the rest of this section.

A common difficulty with applied problems is trying to do everything at once. It is usually best to attack the problem in stages as follows.

Solving Applied Problems

1. Decide on the unknown. Name it with some variable that you *write down*. Many students try to skip this step. They are eager to get on with the writing of the equation. But this is an important step. If you don't know what the variable represents, how can you write a meaningful equation or interpret a result?
2. Draw a sketch or make a chart, if appropriate, showing the information given in the problem.
3. Decide on a variable expression to represent any other unknowns in the problem. For example, if x represents the width of a rectangle, and you know that the length is one more than twice the width, then *write down* that the length is $1 + 2x$.
4. Using the results of Steps 1–3, write an equation that expresses a condition that must be satisfied.
5. Solve the equation.
6. Check the solution in the words of the *original problem,* not just in the equation you have written.

The following examples illustrate this approach.

►**EXAMPLE 8** If the length of a side of a square is increased by 3 centimeters, the new perimeter is 40 centimeters more than twice the length of the side of the original square. Find the length of a side of the original square.

Step 1 What should the variable represent? To find the length of a side of the original square, let

x = length of a side of the original square.

9 **(a)** A triangle has a perimeter of 45 centimeters. Two of the sides of the triangle are equal in length, with the third side 9 centimeters longer than either of the two equal sides. Find the lengths of the sides of the triangle.

(b) A rectangle has a perimeter which is five times its width. The length is 4 more than the width. Find the length and width of the rectangle.

Answer:

(a) 12 centimeters, 12 centimeters, 21 centimeters

(b) Length is 12, width is 8

Step 2   Draw a sketch, as in Figure 1.9.

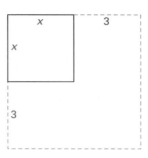

FIGURE 1.9

Step 3   The length of a side of the new square is 3 centimeters more than the length of a side of the old square, so

$x + 3$ = length of a side of the new square.

Now write a variable expression for the new perimeter. Since the perimeter of a square is four times the length of a side,

$4(x + 3)$ = the perimeter of the new square.

Step 4   Now write an equation by looking again at the information in the problem. The new perimeter is 40 more than twice the length of a side of the original square, so the equation is

$$\left(\begin{array}{c}\text{the new}\\\text{perimeter}\end{array}\right) \quad \text{is} \quad 40 \quad \left(\begin{array}{c}\text{more}\\\text{than}\end{array}\right) \quad \left(\begin{array}{c}\text{twice the side of}\\\text{the original square}\end{array}\right)$$

$$4(x + 3) \quad = \quad 40 \quad + \quad 2x.$$

Step 5   Now solve the equation.

$$4(x + 3) = 40 + 2x$$
$$4x + 12 = 40 + 2x$$
$$2x = 28$$
$$x = 14$$

Step 6   Check the solution using the wording of the original problem. The length of a side of the new square would be $14 + 3 = 17$ centimeters; its perimeter would be $4(17) = 68$ centimeters. Twice the length of the side of the original square is $2(14) = 28$ centimeters. Since $40 + 28 = 68$ centimeters, the solution checks with the words of the original problem. ◄ **9**

11. $\dfrac{3x - 2}{7} = \dfrac{x + 2}{5}$

12. $\dfrac{2p + 5}{5} = \dfrac{p + 2}{3}$

13. $\dfrac{x}{3} - 7 = 6 - \dfrac{3x}{4}$

14. $\dfrac{y}{3} + 1 = \dfrac{2y}{5} - 4$

15. $\dfrac{1}{4p} + \dfrac{2}{p} = 3$

16. $\dfrac{2}{t} + 6 = \dfrac{5}{2t}$

17. $\dfrac{m}{2} - \dfrac{1}{m} = \dfrac{6m + 5}{12}$

18. $-\dfrac{3k}{2} + \dfrac{9k - 5}{6} = \dfrac{11k + 8}{k}$

19. $\dfrac{2r}{r - 1} = 5 + \dfrac{2}{r - 1}$

20. $\dfrac{3x}{x + 2} = \dfrac{1}{x + 2} - 4$

21. $\dfrac{4}{x - 3} - \dfrac{8}{2x + 5} + \dfrac{3}{x - 3} = 0$

22. $\dfrac{5}{2p + 3} - \dfrac{3}{p - 2} = \dfrac{4}{2p + 3}$

23. $\dfrac{3}{2m + 4} = \dfrac{1}{m + 2} - 2$

24. $\dfrac{8}{3k - 9} - \dfrac{5}{k - 3} = 4$

Solve each of the following equations for x. (See Example 6.) (In Exercises 29–32, recall that $a^2 = a \cdot a$.)

25. $2(x - a) + b = 3x + a$

26. $5x - (2a + c) = a(x + 1)$

27. $ax + b = 3(x - a)$

28. $4a - ax = 3b + bx$

29. $x = a^2x - ax + 3a - 3$

30. $2a = ax - a - 6x + 6$

31. $a^2x + 3x = 2a^2$

32. $ax + b^2 = bx - a^2$

Solve each equation for the specified variable. Assume all denominators are nonzero. (See Example 7.)

33. $PV = k$ *for V*

34. $i = prt$ *for p*

35. $V = V_0 + gt$ *for g*

36. $S = S_0 + gt^2 + k$ *for g*

37. $A = \dfrac{1}{2}(B + b)h$ *for B*

38. $C = \dfrac{5}{9}(F - 32)$ *for F*

39. $\dfrac{1}{R} = \dfrac{1}{r_1} + \dfrac{1}{r_2}$ *for R*

40. $m = \dfrac{Ft}{v_1 - v_2}$ *for v_2*

Use a calculator to solve each of the following equations. Round to the nearest hundredth.

41. $9.06x + 3.59(8x - 5) = 12.07x + .5612$

42. $-5.74(3.1 - 2.7p) = 1.09p + 5.2588$

43. $\dfrac{2.5x - 7.8}{3.2} + \dfrac{1.2x + 11.5}{5.8} = 6$

44. $\dfrac{4.19x + 2.42}{.05} - \dfrac{5.03x - 9.74}{.02} = 1$

45. $\dfrac{2.63r - 8.99}{1.25} - \dfrac{3.90r - 1.77}{2.45} = r$

46. $\dfrac{8.19m + 2.55}{4.34} - \dfrac{8.17m - 9.94}{1.04} = 4m$

Natural Science *In the metric system of weights and measures, temperature is measured in degrees Celsius (°C) instead of degrees Fahrenheit (°F). To convert back and forth between the two systems, use*

$$C = \dfrac{5(F - 32)}{9} \quad and \quad F = \dfrac{9}{5}C + 32.$$

In each of the following exercises, convert to the other system. Round answers to the nearest tenth of a degree if necessary.

47. 20°C

48. 100°C

49. 59°F

50. 86°F

51. 100°F

52. 350°F

53. 40°C

54. 85°C

Management *Example 7 introduced the formula for the approximate annual interest rate of a loan paid off with monthly payments:*

$$A = \frac{24f}{b(p + 1)}.$$

Use this formula to find the value of the variables not given in each of the following. Round A to the nearest percent and round other variables to the nearest whole numbers. (This formula is not accurate enough for the requirements of federal law.)

55. $f = \$800$, $b = \$4000$, $p = 36$; find A

56. $f = \$60$, $b = \$740$, $p = 12$; find A

57. $A = 8\%$, $b = \$2000$, $p = 36$; find f

58. $A = 5\%$, $b = \$1500$, $p = 24$; find f

59. $A = 6\%$, $f = \$370$, $p = 36$; find b

60. $A = 10\%$, $f = \$490$, $p = 48$; find b

Management *When a loan is paid off early, a portion of the finance charge must be returned to the borrower. By one method of calculating finance charge (called the* rule of 78*), the amount of unearned interest (finance charge to be returned) is given by*

$$u = f \cdot \frac{n(n + 1)}{q(q + 1)}$$

where u represents unearned interest, f is the original finance charge, n is the number of payments remaining when the loan is paid off, and q is the original number of payments. Find the amount of the unearned interest in each of the following.

61. Original finance charge = $800, loan scheduled to run 36 months, paid off with 18 payments remaining

62. Original finance charge = $1400, loan scheduled to run 48 months, paid off with 12 payments remaining

63. Original finance charge = $950, loan scheduled to run 24 months, paid off with 6 payments remaining

64. Original finance charge = $175, loan scheduled to run 12 months, paid off with 3 payments remaining

Solve each applied problem. (See Examples 8–10.)

65. A triangle has a perimeter of 27 centimeters. One side is twice as long as the shortest side. The third side is 7 centimeters longer than the shortest side. Find the length of the shortest side.

66. The length of a rectangle is 3 inches less than twice the width. The perimeter is 54 inches. Find the width.

67. Weijen Luan invests $20,000 received from an insurance settlement in two ways: some at 6%, and some at 8%. Altogether, she makes $1360 per year interest. How much is invested at 8%?

68. Joe Gonzalvez received $52,000 profit from the sale of some land. He invested part at 10% interest, and the rest at 8% interest. He earned a total of $4580 interest per year. How much did he invest at 10%?

69. Matt Whitney won $100,000 in a state lottery. He paid income tax of 30% on the winnings. Of the rest, he invested some at 8 1/2% and some at 6%, making $5450 interest per year. How much is invested at 6%?

70. Mary Collins earned $48,000 from royalties on her cookbook. She paid a 30% income tax on these royalties. The balance was invested in two ways: at 7 1/2% and at 5%. The investments produce $2180 interest income per year. Find the amount invested at 5%.

71. Maria Martinelli bought two plots of land for a total of $120,000. On the first plot, she made a profit of 15%. On the second, she lost 10%. Her total profit was $5500. How much did she pay for each piece of land?

72. Suppose $20,000 is invested at 9%. How much additional money must be invested at 6% to produce a yield of 7.2% on the entire amount invested?

73. The Old Time Goodies Store sells mixed nuts. Cashews sell for $4 per pound, hazelnuts for $3 per pound, and peanuts for $1 per pound. How many pounds of peanuts should be added to 10 pounds of cashews and 8 pounds of hazelnuts to make a mixture which will sell for $2.50 per pound?

74. The Old Time Goodies Store also sells candy. They want to prepare 200 kilograms of a mixture for a special Halloween promotion to sell at $4.84 per kilogram. How much $4 per kilogram candy should be mixed with $5.20 per kilogram candy for the required mix?

Natural Science *Exercises 75 and 76 depend on the idea of the octane rating of gasoline, a measure of its antiknock qualities. Actual gasoline blends are compared to standard fuels. In one measure of octane, a standard fuel is made with only two ingredients: heptane and isooctane. For this fuel, the octane rating is the percent of isooctane; i.e., a gasoline with an octane rating of 98 has the same antiknock properties as a standard fuel that is 98% isooctane.*

75. How many liters of 94 octane gasoline should be mixed with 200 liters of 99 octane gasoline to get a mixture that is 97 octane?

76. A service station has 92 octane and 98 octane gasoline. How many liters of each gasoline should be mixed to provide 12 liters of 96 octane gasoline for a chemistry experiment?

1.3 LINEAR INEQUALITIES

An **inequality** is a statement that two mathematical expressions are *not* equal. Usually an inequality indicates that one expression is less than (or greater than) another. Inequalities are very important in applications. For example, a company wants revenue to be *greater than* costs and must use *less than* the total amount of capital or labor available.

A **linear inequality,** such as $2m + 1 < 7$, is solved by simplifying it to the form $m < k$, for some number k. (Throughout this section, definitions are given only for $<$; but they are equally valid for $>$, $\leq$, or $\geq$.) The following properties are used to simplify an inequality.

Properties of Inequality

For real numbers a, b, and c,

(a) if $a < b$, then $a + c < b + c$
(b) if $a < b$, and if $c > 0$, then $ac < bc$
(c) if $a < b$, and if $c < 0$, then $ac > bc$.

1 (a) First multiply both sides of $-6 < -1$ by 4, and then multiply both sides of $-6 < -1$ by -7.

(b) Multiply both sides of $9 \geq -4$ first by 2, and then by -5.

(c) First add 4, and then add -6 to both sides of $-3 < -1$.

Answer:

(a) $-24 < -4$; $42 > 7$

(b) $18 \geq -8$; $-45 \leq 20$

(c) $1 < 3$; $-9 < -7$

Pay careful attention to part (c): if both sides of an inequality are multiplied by a negative number, the direction of the inequality symbol must be reversed. For example, starting with the true statement $-3 < 5$ and multiplying both sides by the positive number 2 gives

$$-3 \cdot 2 < 5 \cdot 2,$$

or

$$-6 < 10,$$

still a true statement. On the other hand, starting with $-3 < 5$ and multiplying both sides by the negative number -2 gives a true result only if the direction of the inequality symbol is reversed:

$$-3(-2) > 5(-2)$$

$$6 > -10. \quad \blacksquare \, \mathbf{1}$$

2 Solve the following. Graph each solution.

(a) $5z - 11 < 14$

(b) $-3k \leq -12$

(c) $-8y \geq 32$

Answer:

(a) $z < 5$

(b) $k \geq 4$

(c) $y \leq -4$

▶**EXAMPLE 1** Solve $3x + 5 > 11$. Graph the solution.

First, add -5 to both sides.

$$3x + 5 + (-5) > 11 + (-5)$$
$$3x > 6$$

Now multiply both sides by 1/3.

$$\frac{1}{3}(3x) > \frac{1}{3}(6)$$
$$x > 2$$

(Why was the direction of the inequality symbol not changed?) As a check, note that 0, which is not part of the solution, makes the inequality false, while 3, which is part of the solution, makes it true.

$3(0) + 5 > 11$	$3(3) + 5 > 11$
$5 > 11$ **False**	$14 > 11$ **True**

The solution is the interval $(2, \infty)$, which is graphed in Figure 1.10. An open circle at 2 shows that 2 is not included. ◀ **2**

FIGURE 1.10

▶**EXAMPLE 2** Solve $4 - 3y \leq 7 + 2y$.

$$4 - 3y \leq 7 + 2y$$
$$4 - 3y + (-4) \leq 7 + 2y + (-4)$$
$$-3y \leq 3 + 2y$$

Add $-2y$ to both sides. (Remember that *adding* to both sides never changes the direction of the inequality symbol.)

$$-3y + (-2y) \leq 3 + 2y + (-2y)$$
$$-5y \leq 3$$

Multiply both sides by $-1/5$. Since $-1/5$ is negative, change the direction of the inequality symbol.

$$-\frac{1}{5}(-5y) \geq -\frac{1}{5}(3)$$

$$y \geq -\frac{3}{5}$$

3 Solve the following. Graph each solution.

(a) $8 - 6t \geq 2t + 24$

(b) $-4r + 3(r + 1) < 2r$

Answer:

(a) $t \leq -2$

(b) $r > 1$

4 Solve each of the following. Graph each solution.

(a) $9 < k + 5 < 13$

(b) $-6 \leq 2z + 4 \leq 12$

Answer:

(a) $4 < k < 8$

(b) $-5 \leq z \leq 4$

Figure 1.11 shows a graph of the solution, $[-3/5, \infty)$. The solid circle in Figure 1.11 shows that $-3/5$ is included in the solution. ◀ **3**

FIGURE 1.11

▶**EXAMPLE 3** Solve $-2 < 5 + 3m < 20$. Graph the solution.

The inequality $-2 < 5 + 3m < 20$ says that $5 + 3m$ is *between* -2 and 20. Solve this inequality with an extension of the properties given above. Work as follows, first adding -5 to each part.

$$-2 + (-5) < 5 + 3m + (-5) < 20 + (-5)$$
$$-7 < 3m < 15$$

Now multiply each part by $1/3$.

$$-\frac{7}{3} < m < 5$$

A graph of the solution, $(-7/3, 5)$, is given in Figure 1.12. ◀ **4**

FIGURE 1.12

The solutions of some inequalities result in graphs with separate parts, as shown in the next example.

▶**EXAMPLE 4** Solve $3x - 5 < 7$ or $2x - 1 > 13$. Graph the solution.

Begin by solving each inequality separately, keeping the "or" between the two solutions.

$$3x - 5 < 7 \qquad \text{or} \qquad 2x - 1 > 13$$
$$3x < 12 \qquad \text{or} \qquad 2x > 14$$
$$x < 4 \qquad \text{or} \qquad x > 7$$

The final inequality says that x may be any number less than 4 *or* any number greater than 7. The solution includes the numbers graphed in Figure 1.13. There is no shortcut way to write the solution $x < 4$ or $x > 7$. In interval notation the solu-

5 Solve each inequality. Graph each solution.

(a) $5p - 10 < -20$ or $2p + 6 > 8$

(b) $1 - 4y \geq 5$ or $2 - 3y \leq -7$

Answer:

(a) All numbers in $(-\infty, -2)$ or $(1, \infty)$

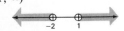

(b) All numbers in $(-\infty, -1]$ or $[3, \infty)$

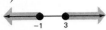

6 In Example 5, what Celsius temperatures correspond to 5°F to 95°F?

Answer:
-15°C to 35°C

tion is written ''all numbers in $(-\infty, 4)$ or $(7, \infty)$,'' which shows that numbers that satisfy the given inequality may be in *either* interval. ◀ **5**

FIGURE 1.13

▶ **EXAMPLE 5**   The formula for converting from Celsius to Fahrenheit temperature is

$$F = \frac{9}{5}C + 32.$$

What Celsius temperature range corresponds to 32°F to 77°F?
  The Fahrenheit temperature range is $32 < F < 77$. Since $F = (9/5)C + 32$,

$$32 < \frac{9}{5}C + 32 < 77.$$

Solve the inequality for C.

$$32 < \frac{9}{5}C + 32 < 77$$

$$0 < \frac{9}{5}C < 45$$

$$0 < C < \frac{5}{9} \cdot 45$$

$$0 < C < 25$$

The corresponding Celsius temperature range is 0°C to 25°C. ◀ **6**

  A product will break even, or produce a profit, only if the revenue R from selling the product at least equals the cost C of producing it, that is if $R \geq C$.

▶ **EXAMPLE 6**   A company analyst has determined that the cost to produce and sell x units of a certain product is $C = 20x + 1000$. The revenue for that product is $R = 70x$. Find the values of x for which the company will break even or make a profit on the product.
  Solve the inequality $R \geq C$.

$$R \geq C$$
$$70x \geq 20x + 1000 \qquad \textbf{Let } R = 70x, \ C = 20x + 1000$$
$$50x \geq 1000$$
$$x \geq 20$$

The company must produce and sell 20 items to break even. ◀

1.3 EXERCISES

Solve each inequality. Graph each solution in Exercises 1–32. (See Examples 1–4.)

1. $6x \leq -18$　　　　**2.** $4m > -32$　　　　**3.** $-3p < 18$　　　　**4.** $-5z \leq 40$

5. $-9a < 0$　　　　**6.** $-4k \geq 0$　　　　**7.** $2x + 1 \leq 9$　　　　**8.** $3y - 2 < 10$

9. $-3p - 2 \geq 1$　　　　**10.** $-5t + 3 \geq -2$　　　　**11.** $6k - 4 < 3k - 1$　　　　**12.** $2a - 2 > 4a + 2$

13. $m - (4 + 2m) + 3 < 2m + 2$　　　　**14.** $2p - (3 - p) \leq -7p - 2$　　　　**15.** $-2(3y - 8) \geq 5(4y - 2)$

16. $5r - (r + 2) \geq 3(r - 1) + 5$　　　　**17.** $3p - 1 < 6p + 2(p - 1)$　　　　**18.** $x + 5(x + 1) > 4(2 - x) + x$

19. $-7 < y - 2 < 4$　　　　**20.** $-3 < m + 6 < 2$　　　　**21.** $8 \leq 3r + 1 \leq 13$

22. $-6 < 2p - 3 \leq 5$　　　　**23.** $-4 \leq \dfrac{2k - 1}{3} \leq 2$　　　　**24.** $-1 \leq \dfrac{5y + 2}{3} \leq 4$

25. $z + 1 \leq 2$ or $z - 5 \geq 1$　　　　**26.** $2y + 2 \leq 5$ or $3y + 4 \geq 22$

27. $6m + 4 \geq 4 + m$ or $2m + 6 < -2 - 2m$　　　　**28.** $5 - 2t > 3$ or $8 - 4t < 2 - 2t$

29. $\dfrac{3}{2}b - 2 < 4$ or $\dfrac{3}{4}b + \dfrac{1}{3} > \dfrac{19}{3}$　　　　**30.** $\dfrac{x}{4} + 3 < \dfrac{9}{4}$ or $\dfrac{x}{2} - 4 > -\dfrac{7}{2}$

31. $\dfrac{3}{5}(2p + 3) \geq \dfrac{1}{10}(5p + 1)$　　　　**32.** $\dfrac{8}{3}(z - 4) \leq \dfrac{2}{9}(3z + 2)$

33. $7.6092k \geq 2.28276$　　　　**34.** $1.3075m < -1.569$

35. $8.0413z - 9.7268 < 1.7251z - .25250$　　　　**36.** $3.2579 + 5.0824k > .76423k + 6.280619$

37. $-(1.42m + 7.63) + 3(3.7m - 1.12) \leq 4.81m - 8.555$　　　　**38.** $3(8.14a - 6.32) - (4.31a - 4.84) > .342a + 9.499$

Find the unknown numbers in each of the following.

39. Five times a number is between -8 and 6.

40. Half a number is between -4 and -1.

41. When 5 is added to three times a number, the result is greater than or equal to 11.

42. If 4 is subtracted from a number, the result is at least 9.

43. One third of a number is added to 2, giving a result at least 8.

44. Seven times a number, minus 5, is no more than 3.

Solve each of the following applied problems. (See Example 5.)

45. A student has a total of 970 points so far in her algebra class. At the end of the course she must have 81% of the 1300 points possible in order to get a B. What is the lowest score she can earn on the 100-point final to get a B in the class?

46. Bill has twice as many dimes as nickels and he has at least 15 coins. At least how many nickels does he have?

47. A nearby business college charges a tuition of $6440 annually. Tom makes no more than $1610 per year in his summer job. What is the least number of summers that he must work in order to make enough for one year's tuition?

48. A nurse must make sure that Ms. Carlson receives at least 30 units of a certain drug each day. This drug comes from red pills or green pills, each of which provides three units of the drug. The patient must have twice as many red pills as green pills. Find the smallest number of green pills that will satisfy the requirement.

Management *In Exercises 49–53, find all values of x where the following products will at least break even. (See Example 6.)*

49. The cost to produce x units of wire is $C = 50x + 5000$, while the revenue is $R = 60x$.

50. The cost to produce x units of squash is $C = 100x + 6000$, while the revenue is $R = 500x$.

51. $C = 85x + 900$; $R = 105x$

52. $C = 70x + 500$; $R = 60x$

53. $C = 1000x + 5000$; $R = 900x$

54. Bill and Cheryl Bradkin went to Portland, Maine, for a week. They needed to rent a car, so they checked out two rental firms. Avis wanted $28 per day, with no mileage fee. Downtown Toyota wanted $108 per week and 14¢ per mile. Let x represent the number of miles that the Bradkins would drive in 1 week. Set up an inequality expressing the rates of the two firms. Then decide how many miles they would have to drive before the Avis car was the better deal.

1.4 ABSOLUTE VALUE EQUATIONS AND INEQUALITIES

Recall from Section 1.1 that the absolute value of the number a, written $|a|$, gives the distance on a number line from a to 0. For example, $|4| = 4$, and $|-7| = 7$. In this section equations and inequalities involving absolute value are discussed.

▶**EXAMPLE 1**   Solve the equation $|x| = 3$.

There are two numbers whose absolute value is 3, namely 3 and -3. The solutions of the given equation are 3 and -3. ◀

▶**EXAMPLE 2**   Solve $|p - 4| = 2$.

This equation will be satisfied if the expression inside the absolute value bars, $p - 4$, equals either 2 or -2:

$$p - 4 = 2 \quad \text{or} \quad p - 4 = -2.$$

Solving these two equations produces

$$p = 6 \quad \text{or} \quad p = 2,$$

so that 6 and 2 are solutions for the original equation. As before, check by substituting in the original equation. ◀ **1**

1 Solve each equation.

(a) $|y| = 9$

(b) $|r + 3| = 1$

(c) $|2k - 3| = 7$

Answer:

(a) 9, -9

(b) -2, -4

(c) 5, -2

▶**EXAMPLE 3**   Solve $|4m - 3| = |m + 6|$.

The quantities in absolute value bars must either be equal or be negatives of one another to satisfy the equation. That is,

$$4m - 3 = m + 6 \quad \text{or} \quad 4m - 3 = -(m + 6)$$
$$3m = 9 \qquad\qquad\qquad 4m - 3 = -m - 6$$
$$m = 3 \qquad\qquad\qquad\qquad 5m = -3$$
$$m = -\frac{3}{5}.$$

2 Solve each equation.

(a) $|r + 6| = |2r + 1|$

(b) $|5k - 7| = |10k - 2|$

Answer:

(a) 5, $-7/3$

(b) -1, 3/5

Check that the solutions for the original equation are 3 and $-3/5$. ◀ **2**

3 Solve each inequality. Graph each solution.

(a) $|x| \le 1$

(b) $|y| \ge 3$

Answer:

(a) $[-1, 1]$

(b) All numbers in $(-\infty, -3]$ or $[3, \infty)$

4 Solve each inequality. Graph each solution.

(a) $|p + 3| < 4$

(b) $|2k - 1| \le 7$

Answer:

(a) $(-7, 1)$

(b) $[-3, 4]$

The next examples show how to solve inequalities with absolute value.

▶ **EXAMPLE 4**   Solve each inequality.

(a) $|x| < 5$

Since absolute value gives the distance from a number to 0, the inequality $|x| < 5$ is true for all real numbers whose distance from 0 is less than 5. This includes all numbers from -5 to 5, or numbers in the interval $(-5, 5)$. A graph of the solution is shown in Figure 1.14.

FIGURE 1.14

(b) $|x| > 5$

In a similar way, the solution of $|x| > 5$ is given by all those numbers whose distance from 0 is *greater* than 5. This includes the numbers satisfying $x < -5$ or $x > 5$. A graph of the solution, all numbers in

$$(-\infty, -5) \quad \text{or} \quad (5, \infty),$$

is shown in Figure 1.15. ◀ **3**

FIGURE 1.15

Example 4 suggests the following generalizations.

Assume a and b are real numbers with $b > 0$.

1. Solve $|a| < b$ by solving $-b < a < b$.
2. Solve $|a| > b$ by solving $a < -b$ or $a > b$.

▶ **EXAMPLE 5**   Solve $|x - 2| < 5$.

Replace a with $x - 2$ and b with 5 in property (1) above. Now solve $|x - 2| < 5$ by solving the inequality

$$-5 < x - 2 < 5.$$

Add 2 to each part, getting the solution

$$-3 < x < 7,$$

which is graphed in Figure 1.16 on the next page. ◀ **4**

5 Solve each inequality. Graph each solution.

(a) $|y - 2| > 5$

(b) $|3k - 1| \geq 2$

(c) $|2 + 5r| - 4 \geq 1$

Answer:

(a) All numbers in $(-\infty, -3)$ or $(7, \infty)$

(b) All numbers in $\left(-\infty, -\dfrac{1}{3}\right]$ or $[1, \infty)$

(c) All numbers in $\left(-\infty, -\dfrac{7}{5}\right]$ or $\left[\dfrac{3}{5}, \infty\right)$

6 Solve each inequality.

(a) $|5m - 2| > -1$

(b) $|2 + 3a| < -3$

(c) $|6 + r| > 0$

Answer:

(a) All real numbers

(b) No solution

(c) All real numbers except -6

FIGURE 1.16

▶**EXAMPLE 6**   Solve $|2 - 7m| - 1 > 4$.

First add 1 on both sides.

$$|2 - 7m| > 5$$

Now use property (2) from above to solve $|2 - 7m| > 5$ by solving the inequality

$$2 - 7m < -5 \quad \text{or} \quad 2 - 7m > 5.$$

Solve each part separately.

$$-7m < -7 \quad \text{or} \quad -7m > 3$$

$$m > 1 \quad \text{or} \quad m < -\frac{3}{7}$$

The solution, all numbers in $\left(-\infty, -\dfrac{3}{7}\right)$ or $(1, \infty)$, is graphed in Figure 1.17. ◀ **5**

FIGURE 1.17

▶**EXAMPLE 7**   Solve $|2 - 5x| \geq -4$.

The absolute value of a number is always nonnegative. Therefore, $|2 - 5x| \geq -4$ is always true, so that the solution is the set of all real numbers. Note that the inequality $|2 - 5x| < -4$ has no solution, since the absolute value of a quantity can never be less than a negative number. ◀ **6**

Absolute value inequalities can be used to indicate how far a number may be from a given number. The next example illustrates this use of absolute value.

▶**EXAMPLE 8**   Write each statement using absolute value.

(a) k is at least 4 units from 1.

If k is at least 4 units from 1, then the distance from k to 1 is greater than or equal to 4. See Figure 1.18(a). Since k may be on either side of 1 on the number line, k may be less than -3 or greater than 5. Write this statement using absolute value as follows:

$$|k - 1| \geq 4.$$

7 Write each statement using absolute value.

(a) m is at least 3 units from 5.

(b) t is within .01 of 4.

Answer:

(a) $|m - 5| \geq 3$

(b) $|t - 4| \leq .01$

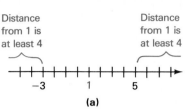

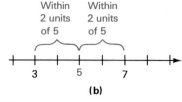

FIGURE 1.18

(b) p is within 2 units of 5.

   This statement means that the difference between p and 5 must be less than or equal to 2. See Figure 1.18(b). Using absolute value notation, the statement is written as

$$|p - 5| \leq 2. \quad \blacktriangleleft \quad \boxed{7}$$

1.4 EXERCISES

Solve each of the following equations. (See Examples 1–3.)

1. $|a - 2| = 1$

2. $|x - 3| = 2$

3. $|3m - 1| = 2$

4. $|4p + 2| = 5$

5. $|5 - 3x| = 3$

6. $|-3a + 7| = 3$

7. $\left|\dfrac{z - 4}{2}\right| = 5$

8. $\left|\dfrac{m + 2}{2}\right| = 7$

9. $\left|\dfrac{5}{r - 3}\right| = 10$

10. $\left|\dfrac{3}{2h - 1}\right| = 4$

11. $\left|\dfrac{6y + 1}{y - 1}\right| = 3$

12. $\left|\dfrac{3a - 4}{2a + 3}\right| = 1$

13. $|2k - 3| = |5k + 4|$

14. $|p + 1| = |3p - 1|$

15. $|4 - 3y| = |7 + 2y|$

16. $|2 + 5a| = |4 - 6a|$

Solve each of the following inequalities. Graph each solution. (See Examples 4–7.)

17. $|x| \leq 3$

18. $|y| \leq 10$

19. $|m| > 1$

20. $|z| > 5$

21. $|a| < -2$

22. $|b| > -5$

23. $|x| - 3 \leq 7$

24. $|r| + 3 \leq 10$

25. $|2x + 5| < 3$

26. $\left|x - \dfrac{1}{2}\right| < 2$

27. $|3m - 2| > 4$

28. $|4x - 6| > 10$

29. $|3z + 1| \geq 7$

30. $|8b + 5| \geq 7$

31. $\left|5x + \dfrac{1}{2}\right| - 2 < 5$

32. $\left|x + \dfrac{2}{3}\right| + 1 < 4$

Find all unknown numbers in each of the following.

33. The absolute value of a number is no more than 6.

34. If the absolute value of a number is found, the result is at least 5.

35. If 6 is added to four times a number, the absolute value of the result is no more than 1.

36. Add -2 to twice a number. The absolute value of the result is less than or equal to 5.

37. The absolute value of the sum of a number and 2 is found. Eight is subtracted from the absolute value. The final result is at least 4.

38. Three times a number is added to 8. The absolute value of this sum is found. Two is added to the absolute value. The final result is no more than 7.

Write each of the following statements using absolute value. (See Example 8.)

39. *x* is within 4 units of 2.

40. *m* is no more than 8 units from 9.

41. *z* is no less than 2 units from 12.

42. *p* is at least 5 units from 9.

43. *k* is 6 units from 9.

44. *r* is 5 units from 3.

45. If *x* is within .0004 units of 2, then *y* is within .00001 units of 7.

46. *y* is within .001 units of 10 whenever *x* is within .02 units of 5.

1.5 POLYNOMIALS

As mentioned in Section 1.1,

$$3^2 = 3 \cdot 3 \quad \text{and} \quad 2^3 = 2 \cdot 2 \cdot 2.$$

This section gives a more general meaning to the symbol a^n. First, recall that *n* is the **exponent** (or **power**) in a^n, and *a* is the **base.**

If *n* is a natural number and *a* is any real number, then

$$a^n = a \cdot a \cdot a \cdots a,$$

where *a* appears as a factor *n* times.

▶**EXAMPLE 1** Evaluate the following.

(a) $8^3 = 8 \cdot 8 \cdot 8 = 512$
Read 8^3 as "8 cubed."

(b) $5^2 = 5 \cdot 5 = 25$
Read 5^2 as "5 squared."

(c) $4^5 = 4 \cdot 4 \cdot 4 \cdot 4 \cdot 4 = 1024$
Read 4^5 as "4 to the fifth power."

(d) $\left(\dfrac{3}{4}\right)^2 = \dfrac{3}{4} \cdot \dfrac{3}{4} = \dfrac{9}{16}$ ◀ **1**

1 Evaluate the following.

(a) 6^3

(b) 5^4

(c) 1^7

(d) $\left(\dfrac{2}{5}\right)^3$

Answer:

(a) 216

(b) 625

(c) 1

(d) 8/125

Caution A common error in using exponents occurs with expressions such as $4 \cdot 3^2$. The exponent of 2 applies only to the base 3, so that

$$4 \cdot 3^2 = 4 \cdot 3 \cdot 3 = 36.$$

On the other hand,

$$(4 \cdot 3)^2 = (4 \cdot 3)(4 \cdot 3) = 12 \cdot 12 = 144,$$

and so

$$4 \cdot 3^2 \neq (4 \cdot 3)^2.$$

2 Evaluate the following.

(a) $3 \cdot 6^2$

(b) $5 \cdot 4^3$

(c) -3^6

(d) $(-3)^6$

(e) $-2 \cdot 3^5$

Answer:

(a) 108

(b) 320

(c) -729

(d) 729

(e) -486

3 Simplify the following.

(a) $5^3 \cdot 5^6$

(b) $(-3)^4 \cdot (-3)^{10}$

(c) $(5p)^2 \cdot (5p)^8$

Answer:

(a) 5^9

(b) $(-3)^{14}$

(c) $(5p)^{10}$

Be careful, too, to distinguish between expressions like -2^4 and $(-2)^4$.

$$-2^4 = -(2^4) = -(2 \cdot 2 \cdot 2 \cdot 2) = -16$$
$$(-2)^4 = (-2)(-2)(-2)(-2) = 16$$

and so

$$-2^4 \neq (-2)^4. \quad \boxed{2}$$

By the definition of an exponent,

$$3^4 \cdot 3^2 = (3 \cdot 3 \cdot 3 \cdot 3)(3 \cdot 3) = 3^6.$$

This suggests the following property for the product of two powers of a number.

If m and n are natural numbers and a is a real number, then

$$a^m \cdot a^n = a^{m+n}.$$

▶**EXAMPLE 2**   Simplify the following.

(a) $7^4 \cdot 7^6 = 7^{4+6} = 7^{10}$

(b) $(-2)^3 \cdot (-2)^5 = (-2)^{3+5} = (-2)^8$

(c) $(3k)^2 \cdot (3k)^3 = (3k)^5$

(d) $(m + n)^2 \cdot (m + n)^5 = (m + n)^7$ ◀ **3**

A polynomial is an algebraic expression like

$$5x^4 + 2x^3 + 6x, \quad 8m^3 + 9m^2 - 6m + 3, \quad 10p, \quad \text{or} \quad -9.$$

More formally, a **polynomial** in one variable is an expression of the form

$$a_n x^n + a_{n-1} x^{n-1} + \ldots + a_1 x + a_0,$$

where $a_0, a_1, a_2, \ldots,$ and a_n are real numbers, n is a natural number, and $a_n \neq 0$. For example, the polynomial

$$8x^3 + 9x^2 - 6x + 3$$

is of the form

$$a_n x^n + a_{n-1} x^{n-1} + \ldots + a_1 x + a_0,$$

with $n = 3$, $a_n = a_3 = 8$, $a_{n-1} = a_2 = 9$, $a_1 = -6$, and $a_0 = 3$. Expressions that are *not* polynomials include

$$8x^3 + \frac{6}{x}, \quad \frac{9 + x}{2 - x}, \quad \text{and} \quad \frac{-p^2 + 5p + 3}{2p - 1}.$$

4 Add or subtract.

(a) $(-2x^2 + 7x + 9)$
$+ (3x^2 + 2x - 7)$

(b) $(4x + 6) - (13x - 9)$

(c) $(9x^3 - 8x^2 + 2x)$
$- (9x^3 - 2x^2 - 10)$

Answer:

(a) $x^2 + 9x + 2$

(b) $-9x + 15$

(c) $-6x^2 + 2x + 10$

An expression involving only multiplication (or division), such as $9p^4$, is called a **term.** The number 9 is called the **coefficient,** p is the **variable,** and 4 is the **exponent.** (Exponents are discussed in more detail later in this chapter.) The **degree of a term** with only one variable is the exponent on the variable. For example, the term $9p^4$ has degree 4. The **degree of a polynomial** is the highest degree of any of its terms. Thus, the degree of $-p^2 + 5p + 3$ is 2.

Two terms having the same variable and the same exponent are called **like terms;** other terms are called **unlike terms.** Polynomials can be added or subtracted by using the distributive property to combine like terms. Only like terms can be combined. For example,

$$12y^4 + 6y^4 = (12 + 6)y^4 = 18y^4$$

and

$$-2m^2 + 8m^2 = (-2 + 8)m^2 = 6m^2.$$

The polynomial $8y^4 + 2y^5$ has unlike terms, so it cannot be further simplified. Polynomials are subtracted using the fact that $a - b = a + (-b)$. The next example shows how to add and subtract polynomials by combining terms.

▶**EXAMPLE 3**   Add or subtract as indicated.

(a) $(8x^3 - 4x^2 + 6x) + (3x^3 + 5x^2 - 9x + 8)$

Combine like terms.

$(8x^3 - 4x^2 + 6x) + (3x^3 + 5x^2 - 9x + 8)$

$= (8x^3 + 3x^3) + (-4x^2 + 5x^2) + (6x - 9x) + 8$ **Commutative property and associative property**

$= 11x^3 + x^2 - 3x + 8$ **Distributive property**

(b) $(-4x^4 + 6x^3 - 9x^2 - 12) + (-3x^3 + 8x^2 - 11x + 7)$

$= -4x^4 + 3x^3 - x^2 - 11x - 5$

(c) $(2x^2 - 11x + 8) - (7x^2 - 6x + 2)$

Use the definition of subtraction: $a - b = a + (-b)$. Here a and b are polynomials, and $-b$ is

$$-(7x^2 - 6x + 2) = -7x^2 + 6x - 2.$$

Now perform the subtraction.

$(2x^2 - 11x + 8) - (7x^2 - 6x + 2)$

$= (2x^2 - 11x + 8) + (-7x^2 + 6x - 2)$

$= -5x^2 - 5x + 6$ ◀ **4**

The distributive property is also used to multiply polynomials. For example, the product of $8x$ and $6x - 4$ is found as follows.

$8x(6x - 4) = 8x(6x) - 8x(4)$ **Distributive property**

$= 48x^2 - 32x$ $x \cdot x = x^2$

5 Find the following products.

(a) $-6r(2r - 5)$

(b) $(8m + 3)(m^4 - 2m^2 + 6m)$

(c) $(5k - 1)(2k + 3)$

(d) $(7z - 3)(2z + 5)$

Answer:

(a) $-12r^2 + 30r$

(b) $8m^5 + 3m^4 - 16m^3 + 42m^2 + 18m$

(c) $10k^2 + 13k - 3$

(d) $14z^2 + 29z - 15$

6 Write an expression for profit if revenue is $7x^2 - 3x + 8$ and cost is $3x^2 + 5x - 2$.

Answer:
$4x^2 - 8x + 10$

The product of $3p - 2$ and $5p + 1$ can be found by using the distributive property twice.

$$(3p - 2)(5p + 1) = (3p - 2)(5p) + (3p - 2)(1)$$
$$= 3p(5p) - 2(5p) + 3p(1) - 2(1)$$
$$= 15p^2 - 10p + 3p - 2$$
$$= 15p^2 - 7p - 2$$

A shortcut for finding the product of two binomials like $(3p - 2)(5p + 1)$ is called **FOIL** for **F**irst, **O**uter, **I**nner, **L**ast. For example,

$$\begin{array}{cccc} \textbf{F} & \textbf{O} & \textbf{I} & \textbf{L} \end{array}$$
$$(5m + 6)(m - 4) = 5m(m) + 5m(-4) + 6(m) + 6(-4)$$
$$= 5m^2 - 20m + 6m - 24$$
$$= 5m^2 - 14m - 24.$$

With practice you should be able to add the outer and inner products mentally and write the three terms of the product without the intermediate steps.

▶**EXAMPLE 4**   Find each product.

(a) $2p^3(3p^2 - 2p + 5) = 2p^3(3p^2) + 2p^3(-2p) + 2p^3(5)$
$$= 6p^5 - 4p^4 + 10p^3$$

(b) $(3k - 2)(k^2 + 5k - 4) = 3k(k^2 + 5k - 4) - 2(k^2 + 5k - 4)$
$$= 3k^3 + 15k^2 - 12k - 2k^2 - 10k + 8$$
$$= 3k^3 + 13k^2 - 22k + 8$$

(c) $(6t + 5)(5t - 1) = 30t^2 + 19t - 5$

Use FOIL. Mentally combine the outer and inner products to get $-6t + 25t = 19t$. ◀ **5**

The profit from the sales of an item is the difference between the revenue R received from the sales and the cost C to sell the item. Thus, profit P is found from the equation $P = R - C$.

▶**EXAMPLE 5**   Suppose the cost to sell x compact discs is $2x^2 - 2x + 10$ and the revenue from the sale of x discs is $5x^2 + 12x - 1$. Write an expression for the profit.

Since $P = R - C$, the profit is

$$P = (5x^2 + 12x - 1) - (2x^2 - 2x + 10)$$
$$= 3x^2 + 14x - 11. ◀ \textbf{6}$$

1.5   EXERCISES

Evaluate each of the following. (See Example 1.)

1. 5^4

2. 4^3

3. $\left(\dfrac{4}{5}\right)^3$

4. $\left(\dfrac{2}{3}\right)^4$

5. -5^2

6. -3^4

7. $(-2)^6$

8. $(-2)^4$

9. $5 \cdot 2^3$

10. $3 \cdot 4^2$

11. $-2 \cdot 3^4$

12. $-3 \cdot 2^5$

Simplify each of the following. Leave answers with exponents. (See Example 2.)

13. $2^4 \cdot 2^3$

14. $3^8 \cdot 3^3$

15. $(-5)^2 \cdot (-5)^5$

16. $(-4)^4 \cdot (-4)^6$

17. $(-3)^5 \cdot 3^4$

18. $8^2 \cdot (-8)^3$

19. $(2z)^5 \cdot (2z)^6$

20. $(6y)^3 \cdot (6y)^5$

Add or subtract as indicated. (See Example 3.)

21. $(8m + 9) + (6m - 3)$

22. $(-7p - 11) + (8p + 5)$

23. $(2x^2 - 6x + 11) + (-3x^2 + 7x - 2)$

24. $(-3a^2 + 2a - 5) + (7a^2 + 2a + 9)$

25. $(-4y^2 - 3y + 8) - (2y^2 - 6y - 2)$

26. $(7b^2 + 2b - 5) - (3b^2 + 2b - 6)$

27. $(2x^3 - 2x^2 + 4x - 3) - (2x^3 + 8x^2 - 1)$

28. $(3y^3 + 9y^2 - 11y + 8) - (-4y^2 + 10y - 6)$

29. $(.613x^2 - 4.215x + .892) - .47(2x^2 - 3x + 5)$

30. $.83(5r^2 - 2r + 7) - (7.12r^2 + 6.423r - 2)$

Find each of the following products. (See Example 4.)

31. $3p(2p - 5)$

32. $4y(8y + 1)$

33. $-9m(2m^2 + 3m - 1)$

34. $2a(4a^2 - 6a + 3)$

35. $(3z + 5)(4z^2 - 2z + 1)$

36. $(2k + 3)(4k^3 - 3k^2 + k)$

37. $(6k - 1)(2k - 3)$

38. $(8r + 3)(r - 1)$

39. $(3y + 5)(2y - 1)$

40. $(2a - 5)(4a + 3)$

41. $(5r - 3s)(5r + 4s)$

42. $(9k + q)(2k - q)$

43. $(.012x - .17)(.3x + .54)$

44. $(6.2m - 3.4)(.7m + 1.3)$

45. $2p - 3[4p - (3p + 1)]$

46. $5k - [k + (-3 + 5k)]$

47. $(3x - 1)(x + 2) - (2x + 5)^2$

48. $(4x + 3)(2x - 1) - (x + 3)^2$

Write an expression for profit given the following expressions for revenue and cost. (See Example 5.)

49. Revenue: $5x^3 - 3x + 1$; cost: $4x^2 + 5x$

50. Revenue: $3x^3 + 2x^2$; cost: $x^3 - x^2 + x + 10$

51. Revenue: $2x^2 - 4x + 50$; cost: $x^2 + 3x + 10$

52. Revenue: $10x^2 + 8x + 12$; cost: $2x^2 - 3x + 20$

1.6 FACTORING

The reverse of polynomial multiplication, where a polynomial is written as a product of other polynomials, is called **factoring.** For example, one way to factor the number 18 is to write it as the product $9 \cdot 2$. Since $18 = 9 \cdot 2$, both 9 and 2 are called **factors** of 18. The integer factors of 18 are 2, 9, -2, -9, 6, 3, -6, -3, 18, 1, -18, -1.

1 Factor out the greatest common factor.

(a) $12r + 9k$

(b) $75m^2 + 100n^2$

(c) $6m^4 - 9m^3 + 12m^2$

(d) $3(2k + 1)^3 + 4(2k + 1)^4$

Answer:

(a) $3(4r + 3k)$

(b) $25(3m^2 + 4n^2)$

(c) $3m^2(2m^2 - 3m + 4)$

(d) $(2k + 1)^3(7 + 8k)$

The distributive property is used to factor polynomials. It is customary to restrict factoring to *integer* coefficients in order to avoid an infinite number of possible factors.

The algebraic expression $15m + 45$ is made up of two terms, $15m$ and 45. Each of these terms can be divided by 15. In fact, $15m = 15 \cdot m$ and $45 = 15 \cdot 3$. By the distributive property,

$$15m + 45 = 15 \cdot m + 15 \cdot 3 = 15(m + 3).$$

Both 15 and $m + 3$ are factors of $15m + 45$. Since 15 divides into all terms of $15m + 45$ and is the largest number that will do so, it is called the **greatest common factor** for the polynomial $15m + 45$. The process of writing $15m + 45$ as $15(m + 3)$ is called **factoring out** the greatest common factor.

▶**EXAMPLE 1**   Factor out the greatest common factor.

(a) $12p - 18q$

Both $12p$ and $18q$ are divisible by 6, and

$$12p - 18q = 6 \cdot 2p - 6 \cdot 3q = 6(2p - 3q).$$

(b) $8x^3 - 9x^2 + 15x$

Each of these terms is divisible by x.

$$8x^3 - 9x^2 + 15x = (8x^2) \cdot x - (9x) \cdot x + 15 \cdot x$$
$$= x(8x^2 - 9x + 15)$$

(c) $5(4x - 3)^3 + 2(4x - 3)^2$

The quantity $(4x - 3)^2$ is a common factor. Factoring it out gives

$$5(4x - 3)^3 + 2(4x - 3)^2 = (4x - 3)^2[5(4x - 3) + 2]$$
$$= (4x - 3)^2(20x - 15 + 2)$$
$$= (4x - 3)^2(20x - 13). \quad ◀ \quad \boxed{1}$$

A polynomial may not have a greatest common factor (other than 1), and yet may still be factorable. For example, the polynomial $x^2 + 5x + 6$ can be factored as $(x + 2)(x + 3)$. Check this: find the product $(x + 2)(x + 3)$; you should get $x^2 + 5x + 6$.

Starting with a polynomial such as $x^2 + 5x + 6$, how can it be written as the product $(x + 2)(x + 3)$? Note that $2 \cdot 3 = 6$ and $2 + 3 = 5$. To factor a polynomial of the form $x^2 + bx + c$, we should look for two numbers whose sum is b and whose product is c.

▶**EXAMPLE 2**   Factor $y^2 + 8y + 15$.

Factor this polynomial by finding two numbers whose *product* is 15, and whose *sum* is 8. Use trial and error to find these numbers. Begin by listing all pairs of integers having a product of 15. As you do this, also form the sum of the numbers.

2 Factor the following.

(a) $m^2 + 11m + 30$

(b) $h^2 + 10h + 9$

(c) $a^2 - 7a + 10$

(d) $r^2 - 5r - 14$

Answer:

(a) $(m + 5)(m + 6)$

(b) $(h + 9)(h + 1)$

(c) $(a - 5)(a - 2)$

(d) $(r - 7)(r + 2)$

Products	Sums
$15 \cdot 1 = 15$	$15 + 1 = 16$
$5 \cdot 3 = 15$	$5 + 3 = 8$
$(-1) \cdot (-15) = 15$	$-1 + (-15) = -16$
$(-5) \cdot (-3) = 15$	$-5 + (-3) = -8$

Only the numbers 3 and 5 have a product of 15 and a sum of 8, so $y^2 + 8y + 15$ factors as

$$y^2 + 8y + 15 = (y + 3)(y + 5).$$

The result can also be written as $(y + 5)(y + 3)$. ◀

▶ **EXAMPLE 3** Factor $p^2 - 8p - 20$.

Find two numbers whose product is -20 and whose sum is -8. Make a list of all pairs of integers whose product is -20. Choose from this list that pair whose sum is -8. You should find the pair -10 and 2; the product of these numbers is -20 and their sum is -8. Therefore,

$$p^2 - 8p - 20 = (p - 10)(p + 2). \quad ◀ \boxed{2}$$

An extension of the method used in Examples 2 and 3 can be used to factor a polynomial of the form $ax^2 + bx + c$, such as $6x^2 + 17x + 5$. In this case, since $ac = 6 \cdot 5 = 30$, look for factors of 30 with a sum of 17. Since the sum is to be positive, only positive factors need be considered.

Factors of 30	Sums
$6 \cdot 5$	$6 + 5 = 11$
$3 \cdot 10$	$3 + 10 = 13$
$2 \cdot 15$	$2 + 15 = 17$
$1 \cdot 30$	$1 + 30 = 31$

Now replace $17x$ with $2x + 15x$, writing the polynomial as

$$6x^2 + 17x + 5$$
$$6x^2 + 2x + 15x + 5.$$

Group the first two terms and the last two terms, and factor out a common factor from each group.

$$(6x^2 + 2x) + (15x + 5)$$
$$2x(3x + 1) + 5(3x + 1).$$

Finally, factor out the common binomial factor $3x + 1$.

$$6x^2 + 17x + 5 = (3x + 1)(2x + 5).$$

3 Factor the following.

(a) $3k^2 + k - 2$

(b) $3m^2 + 5mn - 2n^2$

(c) $6p^2 + 13pq - 5q^2$

(d) $4p^2 - 8p + 12$

Answer:

(a) $(3k - 2)(k + 1)$

(b) $(3m - n)(m + 2n)$

(c) $(2p + 5q)(3p - q)$

(d) $4(p^2 - 2p + 3)$

▶**EXAMPLE 4**   Factor each of the following polynomials.

(a) $6p^2 - 7p - 5$

Look for factors of $6(-5) = -30$ with a sum of -7.

Factors of −30	Sums
$-30 \cdot 1$	$-30 + 1 = -29$
$-15 \cdot 2$	$-15 + 2 = -13$
$-6 \cdot 5$	$-6 + 5 = -1$
$-3 \cdot 10$	$-3 + 10 = 7$
$-10 \cdot 3$	$-10 + 3 = -7$

Write the polynomial as

$$6p^2 - 7p - 5 = (6p^2 - 10p) + (3p - 5)$$
$$= 2p(3p - 5) + 1(3p - 5) \qquad \text{Factor each group}$$
$$= (3p - 5)(2p + 1). \qquad \text{Factor out the common binomial factor}$$

(b) $2x^2 + 9xy - 5y^2$

Find factors of $(2x^2)(-5y^2) = -10x^2y^2$ with a sum of $9xy$.

Factors	Sums
$2xy \cdot -5xy$	$2xy - 5xy = -3xy$
$10xy \cdot -1xy$	$10xy - 1xy = 9xy$

Write the polynomial as $2x^2 + 10xy - 1xy - 5y^2$, group the first two terms and the last two terms, and factor each group. Be careful with the minus signs when factoring the second group.

$$2x^2 + 9xy - 5y^2 = 2x^2 + 10xy - 1xy - 5y^2$$
$$= (2x^2 + 10xy) + (-1xy - 5y^2) \qquad \text{Group in pairs}$$
$$= 2x(x + 5y) - y(x + 5y) \qquad \text{Factor each group}$$
$$= (x + 5y)(2x - y) \qquad \text{Factor out the common binomial}$$

(c) $3r^2 + 18r + 21$

Notice that this polynomial has a common factor of 3, so $3r^2 + 18r + 21 = 3(r^2 + 6r + 7)$. Now try to factor the trinomial $r^2 + 6r + 7$. Since the middle term is positive, only positive factors can be used. The only positive factors of 7 are 7 and 1. But $7 + 1 = 8$, not 6, so $r^2 + 6r + 7$ cannot be factored. The polynomial $3r^2 + 18r + 21$ is factored completely as $3(r^2 + 6r + 7)$. ◀ **3**

Caution Remember to always look first for a common factor. Also, in Example 4(a), notice that it is important to include the common factor of 1 from the second group. Without the 1 in $2p + 1$, the result would be $(3p - 5)(2p)$ which does *not* equal $6p^2 - 7p - 5$.

The method shown above can be used to find the following special factorizations which occur so often that they should be memorized.

$$x^2 - y^2 = (x + y)(x - y) \qquad \text{\textbf{Difference of two squares}}$$
$$\left. \begin{array}{l} x^2 + 2xy + y^2 = (x + y)^2 \\ x^2 - 2xy + y^2 = (x - y)^2 \end{array} \right\} \qquad \text{\textbf{Perfect squares}}$$

▶**EXAMPLE 5**   Factor each of the following.

(a) $4m^2 - 9$

Notice that $4m^2 - 9$ is the difference of two squares, since $4m^2 = (2m)^2$ and $9 = 3^2$. Use the pattern for the difference of two squares, letting $2m$ replace x and 3 replace y. Then the pattern $x^2 - y^2 = (x + y)(x + y)$ becomes

$$4m^2 - 9 = (2m)^2 - 3^2$$
$$= (2m + 3)(2m - 3).$$

(b) $128p^2 - 98q^2$

First factor out the common factor of 2.

$$128p^2 - 98q^2 = 2(64p^2 - 49q^2)$$
$$= 2[(8p)^2 - (7q)^2]$$
$$= 2(8p + 7q)(8p - 7q)$$

(c) $x^2 + 36$

The *sum* of two squares usually cannot be factored. To see this, check some possibilities.

$$(x + 6)(x + 6) = (x + 6)^2 = x^2 + 12x + 36$$
$$(x + 4)(x + 9) = x^2 + 13x + 36$$

Any product of two binomials will always have a middle term unless it is the *difference* of two squares.

(d) $a^2 + 12a + 36$

Since the first and last terms are perfect squares, this trinomial may be a perfect square. Check the middle term to see. By the pattern given earlier, it should be $2xy$. Here a replaces x and 6 replaces y in the pattern, so the middle term should be $2(a)(6) = 12a$. The middle term *is* $12a$, so the trinomial can be factored as

$$a^2 + 12a + 36 = (a + 6)^2.$$

4 Factor the following.

(a) $9p^2 - 49$

(b) $y^2 + 100$

(c) $m^2 - 8mn + 16n^2$

(d) $100k^2 - 60kp + 9p^2$

(e) $9r^2 + 12r + 4 - t^2$

Answer:

(a) $(3p + 7)(3p - 7)$

(b) Cannot be factored

(c) $(m - 4n)^2$

(d) $(10k - 3p)^2$

(e) $(3r + 2 + t)(3r + 2 - t)$

5 Factor the following.

(a) $a^3 + 1000$

(b) $z^3 - 64$

(c) $1000m^3 - 27z^3$

Answer:

(a) $(a + 10)(a^2 - 10a + 100)$

(b) $(z - 4)(z^2 + 4z + 16)$

(c) $(10m - 3z)(100m^2 + 30mz + 9z^2)$

(e) $9p^2 - 24pq + 16q^2$

Since $9p^2 = (3p)^2$ and $16q^2 = (4q)^2$, let $3p$ replace x and $4q$ replace y in the pattern for a perfect square. The middle term is

$$24pq = 2(3p)(4q),$$

so the trinomial is factored as

$$9p^2 - 24pq + 16q^2 = (3p - 4q)^2.$$

(f) $4z^2 + 12z + 9 - w^2$

Notice that the first three terms can be factored as a perfect square.

$$4z^2 + 12z + 9 - w^2 = (2z + 3)^2 - w^2$$

Written in this form, the expression is the difference of squares, which can be factored as

$$(2z + 3)^2 - w^2 = [(2z + 3) + w][(2z + 3) - w]$$
$$= (2z + 3 + w)(2z + 3 - w). \blacktriangleleft \boxed{4}$$

Finally, two more special factorizations occur from time to time.

$$x^3 - y^3 = (x - y)(x^2 + xy + y^2) \qquad \textbf{Difference of two cubes}$$
$$x^3 + y^3 = (x + y)(x^2 - xy + y^2) \qquad \textbf{Sum of two cubes}$$

▶**EXAMPLE 6** Factor each of the following.

(a) $k^3 - 8$

Use the pattern for the difference of two cubes, since $k^3 = (k)^3$ and $8 = (2)^3$, to get

$$k^3 - 8 = k^3 - 2^3 = (k - 2)(k^2 + 2k + 4).$$

(b) $m^3 + 125 = m^3 + 5^3 = (m + 5)(m^2 - 5m + 25)$

(c) $8k^3 - 27z^3 = (2k)^3 - (3z)^3 = (2k - 3z)(4k^2 + 6kz + 9z^2)$ ◀ **5**

1.6 EXERCISES

Factor out the greatest common factor in each of the following. (See Example 1.)

1. $4z + 4$ **2.** $9y - 9$ **3.** $8x + 6y + 4z$ **4.** $15p + 9q + 12s$

5. $m^3 - 9m^2 + 6m$ **6.** $y^3 + 6y^2 + 8y$ **7.** $8a^3 - 16a^2 + 24a$ **8.** $3y^3 + 24y^2 + 9y$

9. $25p^4 - 20p^3q + 100p^2q^2$ **10.** $60m^4 - 120m^3n + 50m^2n^2$ **11.** $2(5x - 1)^2 + 8(5x - 1)^3$

12. $5(2x + 7)^4 - 3(2x + 7)^3$ **13.** $9(x - 4)^5 - (x - 4)^3$ **14.** $3(x + 6)^2 + 2(x + 6)^4$

Factor each of the following. Factor out the greatest common factor as necessary. (See Examples 1–5.)

15. $x^2 + 4x - 5$

16. $y^2 + y - 72$

17. $6a^2 - 48a - 120$

18. $r^2 + rs - 42s^2$

19. $x^2 - 64$

20. $9p^2 - 24p + 16$

21. $3m^3 + 12m^2 + 9m$

22. $3r^2 - r - 2$

23. $b^2 - 8b + 7$

24. $y^2 - 6yx + 8x^2$

25. $m^2 - 6mn + 9n^2$

26. $4p^2 - 9$

27. $3p^2 - 7p + 10$

28. $8r^2 + r + 6$

29. $9m^2 - 25$

30. $a^2 - 10ab + 25b^2$

31. $a^2 + 4ab + 5b^2$

32. $r^2 + r - 20$

33. $2x^2 - 5x - 3$

34. $4y^2 + y - 3$

35. $3k^2 + 2k - 8$

36. $2a^2 - 17a + 30$

37. $21m^2 + 13mn + 2n^2$

38. $20y^2 + 39yx - 11x^2$

39. $121a^2 - 100$

40. $12s^2 + 11st - 5t^2$

41. $5a^2 - 7ab - 6b^2$

42. $100a^2 + 9$

43. $y^2 - 4yz - 21z^2$

44. $15y^2 + y - 2$

45. $9x^2 + 64$

46. $y^2 + 20yx + 100x^2$

47. $z^2 + 14zy + 49y^2$

48. $144m^2 - 81$

49. $24a^4 + 10a^3b - 4a^2b^2$

50. $18x^5 + 15x^4z - 75x^3z^2$

51. $6x^2 + x - 1$

52. $16m^2 + 40m + 25$

53. $y^2 + 10y + 25 - z^2$

54. $9k^2 + 6k + 1 - p^2$

55. $m^2 - n^2 + 2n - 1$

56. $a^2 - 4b^2 - 4b - 1$

57. $3x^4(x^2 + 9)^3 + 2x^2(x^2 + 9)^4$

58. $5m^3(m^3 - 1)^2 - 3m^5(m^3 - 1)^3$

Factor each of the following. (See Example 6.)

59. $a^3 - 216$

60. $b^3 + 125$

61. $8r^3 - 27s^3$

62. $1000p^3 + 27q^3$

63. $64m^3 + 125$

64. $216y^3 - 343$

65. $1000y^3 - z^3$

66. $125p^3 + 8q^3$

---

1 What values of the variable make each denominator equal 0?

(a) $\dfrac{5}{x - 3}$

(b) $\dfrac{2x - 3}{4x - 1}$

(c) $\dfrac{x + 2}{x}$

Answer:

(a) 3

(b) 1/4

(c) 0

1.7 RATIONAL EXPRESSIONS

Later chapters of the book contain work with algebraic fractions. Examples of these fractions, called **rational expressions,** include

$$\frac{8}{x - 1}, \quad \frac{3x^2 + 4x}{5x - 6}, \quad \text{and} \quad \frac{2 + \dfrac{1}{y}}{y}.$$

Since rational expressions involve quotients, it is important to keep in mind values of the variables that make denominators 0. For example, 1 cannot be used as a replacement for x in the first rational expression above, and 6/5 cannot be used in the second one, since these values make the respective denominators equal 0. **1**

The rules for operations with rational expressions are the usual rules for fractions.

2 Write each of the following in lowest terms.

(a) $\dfrac{12k + 36}{18}$

(b) $\dfrac{15m + 30m^2}{5m}$

(c) $\dfrac{2p^2 + 3p + 1}{p^2 + 3p + 2}$

Answer:

(a) $\dfrac{2(k + 3)}{3}$ or $\dfrac{2k + 6}{3}$

(b) $3(1 + 2m)$ or $3 + 6m$

(c) $\dfrac{2p + 1}{p + 2}$

Operations with Rational Expressions

For all mathematical expressions P, $Q \neq 0$, R, and $S \neq 0$,

(a) $\dfrac{P}{Q} = \dfrac{PS}{QS}$ **Fundamental property**

(b) $\dfrac{P}{Q} \cdot \dfrac{R}{S} = \dfrac{PR}{QS}$ **Multiplication**

(c) $\dfrac{P}{Q} + \dfrac{R}{Q} = \dfrac{P + R}{Q}$ **Addition**

(d) $\dfrac{P}{Q} - \dfrac{R}{Q} = \dfrac{P - R}{Q}$ **Subtraction**

(e) $\dfrac{P}{Q} \div \dfrac{R}{S} = \dfrac{P}{Q} \cdot \dfrac{S}{R}$, $R \neq 0$. **Division**

The following examples illustrate these operations.

▶**EXAMPLE 1** Write each of the following rational expressions in lowest terms (so that the numerator and denominator have no common factor with integer coefficients except 1 or -1).

(a) $\dfrac{12m}{-18}$

Both $12m$ and -18 are divisible by 6. By operation (a) above,

$$\frac{12m}{-18} = \frac{2m \cdot 6}{-3 \cdot 6}$$

$$= \frac{2m}{-3}$$

$$= -\frac{2m}{3}.$$

(b) $\dfrac{8x + 16}{4} = \dfrac{8(x + 2)}{4} = \dfrac{4 \cdot 2(x + 2)}{4} = \dfrac{2(x + 2)}{1} = 2(x + 2).$

The numerator, $8x + 16$, was factored so that the common factor could be identified. The answer could also be written as $2x + 4$, if desired.

(c) $\dfrac{k^2 + 7k + 12}{k^2 + 2k - 3} = \dfrac{(k + 4)(k + 3)}{(k - 1)(k + 3)} = \dfrac{k + 4}{k - 1}$ ◀ **2**

The values of k in Example 1(c) are restricted to $k \neq 1$ and $k \neq -3$. From now on, such restrictions will be assumed when working with rational expressions.

3 Multiply.

(a) $\dfrac{3r^2}{5} \cdot \dfrac{20}{9r}$

(b) $\dfrac{y-4}{y^2-2y-8} \cdot \dfrac{y^2-4}{3y}$

Answer:

(a) $\dfrac{4r}{3}$

(b) $\dfrac{y-2}{3y}$

▶**EXAMPLE 2**   **(a)** Multiply $\dfrac{2}{3} \cdot \dfrac{y}{5}$.

Use operation (b) above; multiply the numerators and then the denominators.

$$\frac{2}{3} \cdot \frac{y}{5} = \frac{2 \cdot y}{3 \cdot 5} = \frac{2y}{15}$$

The result, $2y/15$, is in lowest terms.

(b) $\dfrac{3y+9}{6} \cdot \dfrac{18}{5y+15}$

Factor where possible.

$$\frac{3y+9}{6} \cdot \frac{18}{5y+15} = \frac{3(y+3)}{6} \cdot \frac{18}{5(y+3)}$$

$$= \frac{3 \cdot 18(y+3)}{6 \cdot 5(y+3)} \qquad \begin{array}{l}\textbf{Multiply numerators}\\ \textbf{and denominators}\end{array}$$

$$= \frac{3 \cdot 6 \cdot 3(y+3)}{6 \cdot 5(y+3)} \qquad \mathbf{18 = 6 \cdot 3}$$

$$= \frac{3 \cdot 3}{5} \qquad\qquad \textbf{Write in lowest terms}$$

$$= \frac{9}{5}$$

(c) $\dfrac{m^2+5m+6}{m+3} \cdot \dfrac{m^2+m-6}{m^2+3m+2}$

$$= \frac{(m+2)(m+3)}{m+3} \cdot \frac{(m-2)(m+3)}{(m+2)(m+1)} \qquad \textbf{Factor}$$

$$= \frac{(m+2)(m+3)(m-2)(m+3)}{(m+3)(m+2)(m+1)} \qquad \textbf{Multiply}$$

$$= \frac{(m-2)(m+3)}{m+1} \qquad \textbf{Lowest terms}$$

$$= \frac{m^2+m-6}{m+1} \qquad ◀ \; \boxed{3}$$

▶**EXAMPLE 3**   **(a)** Divide $\dfrac{8x}{5} \div \dfrac{11x^2}{20}$.

As shown in operation (e) above, invert the second expression and multiply.

$$\frac{8x}{5} \div \frac{11x^2}{20} = \frac{8x}{5} \cdot \frac{20}{11x^2} \qquad \textbf{Invert and multiply}$$

$$= \frac{8x \cdot 20}{5 \cdot 11x^2} \qquad \textbf{Multiply}$$

$$= \frac{32}{11x} \qquad \textbf{Lowest terms}$$

4 Divide.

(a) $\dfrac{5m}{16} \div \dfrac{m^2}{10}$

(b) $\dfrac{2y - 8}{6} \div \dfrac{5y - 20}{3}$

(c) $\dfrac{m^2 - 2m - 3}{m(m + 1)} \div \dfrac{m + 4}{5m}$

Answer:

(a) $\dfrac{25}{8m}$

(b) $\dfrac{1}{5}$

(c) $\dfrac{5(m - 3)}{m + 4}$

(b) $\dfrac{9p - 36}{12} \div \dfrac{5(p - 4)}{18}$

$$= \dfrac{9p - 36}{12} \cdot \dfrac{18}{5(p - 4)} \qquad \textbf{Invert and multiply}$$

$$= \dfrac{9(p - 4)}{12} \cdot \dfrac{18}{5(p - 4)} \qquad \textbf{Factor}$$

$$= \dfrac{27}{10} \qquad \qquad \begin{array}{l} \textbf{Multiply and write} \\ \textbf{in lowest terms} \end{array} \blacktriangleleft \ \boxed{4}$$

▶ **EXAMPLE 4**   Add or subtract as indicated.

(a) $\dfrac{4}{5k} - \dfrac{11}{5k}$

As operation (d) above shows, when two rational expressions have the same denominators, subtract by subtracting the numerators and keeping the common denominator.

$$\dfrac{4}{5k} - \dfrac{11}{5k} = \dfrac{4 - 11}{5k} = -\dfrac{7}{5k}$$

(b) $\dfrac{7}{p} + \dfrac{9}{2p} + \dfrac{1}{3p}$

These three denominators are different; operation (c) for addition requires the same denominators. Find a common denominator, one which can be divided by p, $2p$, and $3p$. A common denominator here is $6p$. Rewrite each rational expression, using operation (a), with a denominator of $6p$. Then, using operation (c), add the numerators and keep the common denominator.

$$\dfrac{7}{p} + \dfrac{9}{2p} + \dfrac{1}{3p} = \dfrac{6 \cdot 7}{6 \cdot p} + \dfrac{3 \cdot 9}{3 \cdot 2p} + \dfrac{2 \cdot 1}{2 \cdot 3p} \qquad \textbf{Operation (a)}$$

$$= \dfrac{42}{6p} + \dfrac{27}{6p} + \dfrac{2}{6p}$$

$$= \dfrac{42 + 27 + 2}{6p} \qquad \textbf{Operation (c)}$$

$$= \dfrac{71}{6p}$$

(c) $\dfrac{k^2}{k^2 - 1} - \dfrac{2k^2 - k - 3}{k^2 + 3k + 2}$

Factor the denominators to find a common denominator.

$$\dfrac{k^2}{k^2 - 1} - \dfrac{2k^2 - k - 3}{k^2 + 3k + 2} = \dfrac{k^2}{(k + 1)(k - 1)} - \dfrac{2k^2 - k - 3}{(k + 1)(k + 2)}$$

5 Add or subtract.

(a) $\dfrac{3}{4r} + \dfrac{8}{3r}$

(b) $\dfrac{1}{m-2} - \dfrac{3}{2(m-2)}$

(c) $\dfrac{p+1}{p^2-p} + \dfrac{p^2-1}{p^2+p-2}$

Answer:

(a) $\dfrac{41}{12r}$

(b) $\dfrac{-1}{2(m-2)}$

(c) $\dfrac{p^3+p^2+2p+2}{p(p-1)(p+2)}$

The common denominator is $(k+1)(k-1)(k+2)$. Write each fraction with the common denominator.

$$\frac{k^2}{(k+1)(k-1)} - \frac{2k^2-k-3}{(k+1)(k+2)}$$

$$= \frac{k^2(k+2)}{(k+1)(k-1)(k+2)} - \frac{(2k^2-k-3)(k-1)}{(k+1)(k-1)(k+2)}$$

$$= \frac{k^3+2k^2-(2k^2-k-3)(k-1)}{(k+1)(k-1)(k+2)} \qquad \text{Subtract fractions}$$

$$= \frac{k^3+2k^2-(2k^3-3k^2-2k+3)}{(k+1)(k-1)(k+2)} \qquad \begin{array}{l}\text{Multiply}\\ (2k^2-k-3)(k-1)\end{array}$$

$$= \frac{k^3+2k^2-2k^3+3k^2+2k-3}{(k+1)(k-1)(k+2)} \qquad \text{Polynomial subtraction}$$

$$= \frac{-k^3+5k^2+2k-3}{(k+1)(k-1)(k+2)} \qquad \text{Combine terms} \blacktriangleleft \boxed{5}$$

COMPLEX FRACTIONS   Any quotient of two rational expressions is called a **complex fraction.** Complex fractions can often be simplified by the methods shown in the following examples.

▶ **EXAMPLE 5**   Simplify each complex fraction.

(a) $\dfrac{6 - \dfrac{5}{k}}{1 + \dfrac{5}{k}}$

Multiply both numerator and denominator by the common denominator k.

$$\frac{6-\dfrac{5}{k}}{1+\dfrac{5}{k}} = \frac{k\left(6-\dfrac{5}{k}\right)}{k\left(1+\dfrac{5}{k}\right)} \qquad \text{Multiply by } \frac{k}{k}$$

$$\frac{k\left(6-\dfrac{5}{k}\right)}{k\left(1+\dfrac{5}{k}\right)} = \frac{6k-k\left(\dfrac{5}{k}\right)}{k+k\left(\dfrac{5}{k}\right)} \qquad \text{Distributive property}$$

$$= \frac{6k-5}{k+5} \qquad \text{Simplify}$$

6 Simplify each complex fraction.

(a) $\dfrac{t - \dfrac{1}{t}}{2t + \dfrac{3}{t}}$

(b) $\dfrac{\dfrac{m}{m+2} + \dfrac{1}{m}}{\dfrac{1}{m} - \dfrac{1}{m+2}}$

Answer:

(a) $\dfrac{t^2 - 1}{2t^2 + 3}$

(b) $\dfrac{m^2 + m + 2}{2}$

(b) $\dfrac{\dfrac{a}{a+1} + \dfrac{1}{a}}{\dfrac{1}{a} + \dfrac{1}{a+1}}$

Multiply both numerator and denominator by the common denominator of all the fractions, in this case $a(a + 1)$. Doing so gives

$$\frac{\dfrac{a}{a+1} + \dfrac{1}{a}}{\dfrac{1}{a} + \dfrac{1}{a+1}} = \frac{\left(\dfrac{a}{a+1} + \dfrac{1}{a}\right)a(a+1)}{\left(\dfrac{1}{a} + \dfrac{1}{a+1}\right)a(a+1)} = \frac{a^2 + (a+1)}{(a+1) + a} = \frac{a^2 + a + 1}{2a + 1}.$$

As an alternative method of solution, first perform the indicated additions in the numerator and denominator, and then divide.

$$\frac{\dfrac{a}{a+1} + \dfrac{1}{a}}{\dfrac{1}{a} + \dfrac{1}{a+1}} = \frac{\dfrac{a^2 + 1(a+1)}{a(a+1)}}{\dfrac{1(a+1) + 1(a)}{a(a+1)}} = \frac{\dfrac{a^2 + a + 1}{a(a+1)}}{\dfrac{2a+1}{a(a+1)}}$$

$$= \frac{a^2 + a + 1}{a(a+1)} \cdot \frac{a(a+1)}{2a+1} = \frac{a^2 + a + 1}{2a+1} \quad \blacktriangleleft \boxed{6}$$

1.7 EXERCISES

Write each of the following in lowest terms. Factor as necessary. (See Example 1.)

1. $\dfrac{6m}{24}$

2. $\dfrac{8k}{56}$

3. $\dfrac{7z^2}{14z}$

4. $\dfrac{32y^2}{16y}$

5. $\dfrac{25p^3}{10p^2}$

6. $\dfrac{14z^3}{6z^2}$

7. $\dfrac{8k + 16}{9k + 18}$

8. $\dfrac{20r + 10}{30r + 15}$

9. $\dfrac{3(t + 5)}{(t + 5)(t - 3)}$

10. $\dfrac{-8(y - 4)}{(y + 2)(y - 4)}$

11. $\dfrac{8x^2 + 16x}{4x^2}$

12. $\dfrac{36y^2 + 72y}{9y}$

13. $\dfrac{m^2 - 4m + 4}{m^2 + m - 6}$

14. $\dfrac{r^2 - r - 6}{r^2 + r - 12}$

15. $\dfrac{x^2 + 3x - 4}{x^2 - 1}$

16. $\dfrac{z^2 - 5z + 6}{z^2 - 4}$

17. $\dfrac{8m^2 + 6m - 9}{16m^2 - 9}$

18. $\dfrac{6y^2 + 11y + 4}{3y^2 + 7y + 4}$

Multiply or divide as indicated in each of the following. Write all answers in lowest terms. (See Examples 2 and 3.)

19. $\dfrac{9k^2}{25} \cdot \dfrac{5}{3k}$

20. $\dfrac{21m^3}{9m} \cdot \dfrac{12m^2}{7m}$

21. $\dfrac{15p^3}{9p^2} \div \dfrac{6p}{10p^2}$

22. $\dfrac{3r^2}{9r^3} \div \dfrac{8r^3}{6r}$

23. $\dfrac{a + b}{2p} \cdot \dfrac{12}{5(a + b)}$

24. $\dfrac{3(x - 1)}{y} \cdot \dfrac{2y}{7(x - 1)}$

25. $\dfrac{2k + 8}{6} \div \dfrac{3k + 12}{2}$

26. $\dfrac{5m + 25}{10} \cdot \dfrac{12}{6m + 30}$

27. $\dfrac{9y - 18}{6y + 12} \cdot \dfrac{3y + 6}{15y - 30}$

28. $\dfrac{12r + 24}{36r - 36} \div \dfrac{6r + 12}{8r - 8}$

29. $\dfrac{4a + 12}{2a - 10} \div \dfrac{a^2 - 9}{a^2 - a - 20}$

30. $\dfrac{6r - 18}{9r^2 + 6r - 24} \cdot \dfrac{12r - 16}{4r - 12}$

31. $\dfrac{k^2 - k - 6}{k^2 + k - 12} \cdot \dfrac{k^2 + 3k - 4}{k^2 + 2k - 3}$

32. $\dfrac{n^2 - n - 6}{n^2 - 2n - 8} \div \dfrac{n^2 - 9}{n^2 + 7n + 12}$

33. $\dfrac{m^2 + 3m + 2}{m^2 + 5m + 4} \div \dfrac{m^2 + 5m + 6}{m^2 + 10m + 24}$

34. $\dfrac{y^2 + y - 2}{y^2 + 3y - 4} \div \dfrac{y^2 + 3y + 2}{y^2 + 4y + 3}$

35. $\dfrac{2m^2 - 5m - 12}{m^2 - 10m + 24} \div \dfrac{4m^2 - 9}{m^2 - 9m + 18}$

36. $\dfrac{6n^2 - 5n - 6}{6n^2 + 5n - 6} \cdot \dfrac{12n^2 - 17n + 6}{12n^2 - n - 6}$

Add or subtract as indicated in each of the following. Write all answers in lowest terms. (See Example 4.)

37. $\dfrac{2}{3y} - \dfrac{1}{4y}$

38. $\dfrac{6}{11z} - \dfrac{5}{2z}$

39. $\dfrac{a + 1}{2} - \dfrac{a - 1}{2}$

40. $\dfrac{y + 6}{5} - \dfrac{y - 6}{5}$

41. $\dfrac{3}{p} + \dfrac{1}{2}$

42. $\dfrac{9}{r} - \dfrac{2}{3}$

43. $\dfrac{2}{y} - \dfrac{1}{4}$

44. $\dfrac{6}{11} + \dfrac{3}{a}$

45. $\dfrac{1}{6m} + \dfrac{2}{5m} + \dfrac{4}{m}$

46. $\dfrac{8}{3p} + \dfrac{5}{4p} + \dfrac{9}{2p}$

47. $\dfrac{1}{m - 1} + \dfrac{2}{m}$

48. $\dfrac{8}{y + 2} - \dfrac{3}{y}$

49. $\dfrac{8}{3(a - 1)} + \dfrac{2}{a - 1}$

50. $\dfrac{5}{2(k + 3)} + \dfrac{2}{k + 3}$

51. $\dfrac{2}{5(k - 2)} + \dfrac{3}{4(k - 2)}$

52. $\dfrac{11}{3(p + 4)} - \dfrac{5}{6(p + 4)}$

53. $\dfrac{2}{x^2 - 2x - 3} + \dfrac{5}{x^2 - x - 6}$

54. $\dfrac{3}{m^2 - 3m - 10} + \dfrac{5}{m^2 - m - 20}$

55. $\dfrac{2y}{y^2 + 7y + 12} - \dfrac{y}{y^2 + 5y + 6}$

56. $\dfrac{-r}{r^2 - 10r + 16} - \dfrac{3r}{r^2 + 2r - 8}$

57. $\dfrac{3k}{2k^2 + 3k - 2} - \dfrac{2k}{2k^2 - 7k + 3}$

58. $\dfrac{4m}{3m^2 + 7m - 6} - \dfrac{m}{3m^2 - 14m + 8}$

In each of the following exercises, simplify the complex fraction. (See Example 5.)

59. $\dfrac{1 + \dfrac{1}{x}}{1 - \dfrac{1}{x}}$

60. $\dfrac{2 - \dfrac{2}{y}}{2 + \dfrac{2}{y}}$

61. $\dfrac{\dfrac{1}{x + h} - \dfrac{1}{x}}{h}$

62. $\dfrac{\dfrac{1}{(x + h)^2} - \dfrac{1}{x^2}}{h}$

63. $\dfrac{1 + \dfrac{1}{1 - b}}{1 - \dfrac{1}{1 + b}}$

64. $\dfrac{m - \dfrac{1}{m^2 - 4}}{\dfrac{1}{m + 2}}$

1.8 INTEGER EXPONENTS

In Section 1.5 we showed that

$$a^m \cdot a^n = a^{m+n}.$$

In this section, we will develop further definitions and properties of exponents. We begin with a rule for quotients. By definition,

$$\frac{6^5}{6^2} = \frac{6 \cdot 6 \cdot 6 \cdot 6 \cdot 6}{6 \cdot 6} = 6 \cdot 6 \cdot 6 = 6^3.$$

1 Evaluate the following.

(a) 17^0

(b) 30^0

(c) $(-10)^0$

(d) $-(12)^0$

Answer:

(a) 1

(b) 1

(c) 1

(d) -1

Since there are 5 factors of 6 in the numerator and 2 factors of 6 in the denominator, the quotient has $5 - 2 = 3$ factors of 6. In general, if $m > n$, then

$$\frac{a^m}{a^n} = a^{m-n}.$$

Using this rule,

$$\frac{3^5}{3^5} = 3^{5-5} = 3^0.$$

On the other hand, $3^5/3^5 = 1$. Therefore, it is reasonable to define $3^0 = 1$.

If a is any nonzero real number, then

$$a^0 = 1.$$

The symbol 0^0 is not defined.

▶**EXAMPLE 1**   Evaluate the following.

(a) $6^0 = 1$

(b) $(-9)^0 = 1$

(c) $-(4^0) = -(1) = -1$ ◀ **1**

If m is less than n in the quotient rule given above, the result has a negative exponent. For example,

$$\frac{3^2}{3^4} = 3^{2-4} = 3^{-2}.$$

However, by definition

$$\frac{3^2}{3^4} = \frac{3 \cdot 3}{3 \cdot 3 \cdot 3 \cdot 3} = \frac{1}{3^2},$$

which suggests that 3^{-2} should be defined to be $1/3^2$. Thus, we have the following definition.

If n is a natural number, and if $a \neq 0$, then

$$a^{-n} = \frac{1}{a^n}.$$

2 Evaluate the following.

(a) 6^{-2}

(b) -6^{-3}

(c) -3^{-4}

(d) $\left(\dfrac{5}{8}\right)^{-1}$

(e) $\left(\dfrac{1}{2}\right)^{-4}$

Answer:

(a) $1/36$

(b) $-1/216$

(c) $-1/81$

(d) $8/5$

(e) 16

3 Evaluate the following.

(a) $\left(\dfrac{7}{3}\right)^{-2}$

(b) $\left(\dfrac{5}{9}\right)^{-3}$

Answer:

(a) $9/49$

(b) $729/125$

▶**EXAMPLE 2** Evaluate the following.

(a) $3^{-2} = \dfrac{1}{3^2} = \dfrac{1}{9}$

(b) $5^{-4} = \dfrac{1}{5^4} = \dfrac{1}{625}$

(c) $9^{-1} = \dfrac{1}{9^1} = \dfrac{1}{9}$

(d) $-4^{-2} = -\dfrac{1}{4^2} = -\dfrac{1}{16}$

(e) $\left(\dfrac{3}{4}\right)^{-1} = \dfrac{1}{\left(\dfrac{3}{4}\right)^1} = \dfrac{1}{\dfrac{3}{4}} = \dfrac{4}{3}$

(f) $\left(\dfrac{2}{3}\right)^{-3} = \dfrac{1}{\left(\dfrac{2}{3}\right)^3} = \dfrac{1}{\dfrac{2^3}{3^3}} = 1 \cdot \dfrac{3^3}{2^3} = \dfrac{3^3}{2^3} = \dfrac{27}{8}$ ◀ **2**

Parts (e) and (f) of Example 2 involve work with fractions that can lead to error. For a useful shortcut with such fractions, use the properties of division of rational numbers and the definition of a negative exponent to get

$$\left(\dfrac{a}{b}\right)^{-n} = \dfrac{1}{\left(\dfrac{a}{b}\right)^n} = \dfrac{1}{\dfrac{a^n}{b^n}} = 1 \cdot \dfrac{b^n}{a^n} = \dfrac{b^n}{a^n} = \left(\dfrac{b}{a}\right)^n.$$

For all positive integers n and nonzero real numbers a and b,

$$\left(\dfrac{a}{b}\right)^{-n} = \left(\dfrac{b}{a}\right)^n.$$

▶**EXAMPLE 3** Evaluate each of the following.

(a) $\left(\dfrac{1}{3}\right)^{-2} = \left(\dfrac{3}{1}\right)^2 = 9$

(b) $\left(\dfrac{3}{4}\right)^{-3} = \left(\dfrac{4}{3}\right)^3 = \dfrac{64}{27}$ ◀ **3**

The definitions and properties discussed above are summarized below, together with three power rules that follow from the definition of an exponent. These rules and definitions should be memorized.

4 Simplify the following.

(a) $9^6 \cdot 9^{-4}$

(b) $\dfrac{8^7}{8^{-3}}$

(c) $\dfrac{14^9}{14^{12}}$

(d) $(13^4)^{-3}$

(e) $(6y)^4$

(f) $\left(\dfrac{3}{4}\right)^{-2}$

Answer:

(a) 9^2

(b) 8^{10}

(c) $1/14^3$

(d) $1/13^{12}$

(e) $6^4 y^4$

(f) $4^2/3^2$

5 Simplify the following. Give answers with only positive exponents. Assume all variables represent nonzero real numbers.

(a) $\dfrac{(t^{-1})^2}{t^{-5}}$

(b) $\dfrac{(3z)^{-1} z^4}{z^2}$

(c) $(p^{-5}q)^{-1}$

(d) $\dfrac{(3a)^{-2}(b^2 c^3)^4}{(ab)^{-3} c^4}$

Answer:

(a) t^3

(b) $z/3$

(c) p^5/q

(d) $\dfrac{ab^{11}c^8}{3^2}$

Definitions and Properties of Exponents

For any integers m and n, and any real numbers a and b for which the following exist,

(a) $a^m \cdot a^n = a^{m+n}$ **Product property**

(b) $\dfrac{a^m}{a^n} = a^{m-n}$ **Quotient property**

(c) $(a^m)^n = a^{mn}$

(d) $(ab)^m = a^m \cdot b^m$

(e) $\left(\dfrac{a}{b}\right)^m = \dfrac{a^m}{b^m}$ **Power properties**

(f) $a^0 = 1$

(g) $a^{-n} = \dfrac{1}{a^n}$

(h) $\left(\dfrac{a}{b}\right)^{-n} = \left(\dfrac{b}{a}\right)^n .$

▶ **EXAMPLE 4**   Use the properties of exponents to simplify each of the following. Write answers with positive exponents.

(a) $7^{-4} \cdot 7^6 = 7^2$ Property (a)

(b) $\dfrac{9^{14}}{9^{-6}} = 9^{20}$ Property (b)

(c) $(2m^3)^4 = 2^4 \cdot (m^3)^4 = 2^4 m^{12}$ Properties (c) and (d)

(d) $(3x)^4 = 3^4 x^4$ Property (d)

(e) $\left(\dfrac{9}{7}\right)^{-6} = \left(\dfrac{7}{9}\right)^6 = \dfrac{7^6}{9^6}$ Properties (h) and (e)   ◀ **4**

▶ **EXAMPLE 5**   Use the definition of negative exponents and the properties of exponents to simplify each of the following expressions. Give answers with only positive exponents. Assume all variables represent nonzero real numbers.

(a) $\dfrac{(m^3)^{-2}}{m^4} = \dfrac{m^{-6}}{m^4} = m^{-6-4} = m^{-10} = \dfrac{1}{m^{10}}$

(b) $\dfrac{(2y)^3 y^{-1}}{y^2} = \dfrac{2^3 y^3 y^{-1}}{y^2} = \dfrac{2^3 y^2}{y^2} = 2^3$

(c) $(x^2 y^{-3})^4 = (x^2)^4 (y^{-3})^4 = x^8 y^{-12} = \dfrac{x^8}{y^{12}}$

(d) $\dfrac{(2y)^{-3}(y^2 z)^3}{(yz)^{-2} z^3} = \dfrac{2^{-3} y^{-3} y^6 z^3}{y^{-2} z^{-2} z^3} = \dfrac{2^{-3} y^3 z^3}{y^{-2} z} = 2^{-3} y^5 z^2 = \dfrac{y^5 z^2}{2^3}$ ◀ **5**

6 Simplify. Give answers with positive exponents.

(a) $\dfrac{3^{-5} \cdot 3^{-2}}{3^{-6} \cdot 3^{3}}$

(b) $4^{-1} + 2^{-2}$

(c) $\dfrac{3^{-1} + 2^{-2}}{2^{-2}}$

Answer:

(a) $1/3^4$ (b) $1/2$ (c) $7/3$

▶**EXAMPLE 6** Simplify. Write the result with positive exponents.

(a) $\dfrac{2^{-3} \cdot 2^{-1}}{2^{4} \cdot 2^{-7}} = \dfrac{2^{-4}}{2^{-3}} = 2^{-4-(-3)} = 2^{-1} = \dfrac{1}{2}$

(b) $2^{-1} + 3^{-1} = \dfrac{1}{2} + \dfrac{1}{3} = \dfrac{5}{6}$

(c) $\dfrac{2^{-1} - 3^{-2}}{2^{-1}} = \dfrac{\dfrac{1}{2} - \dfrac{1}{3^2}}{\dfrac{1}{2}} = \dfrac{\dfrac{1}{2} - \dfrac{1}{9}}{\dfrac{1}{2}} = \dfrac{\dfrac{9}{18} - \dfrac{2}{18}}{\dfrac{1}{2}} = \dfrac{7}{18} \cdot \dfrac{2}{1} = \dfrac{7}{9}$ ◀ **6**

1.8 EXERCISES

Evaluate each of the following. Write all answers without exponents. (See Examples 1-3.)

1. 7^3 **2.** 4^2 **3.** 8^{-1} **4.** 9^{-2} **5.** 2^{-3} **6.** 3^{-4} **7.** 5^{-1}

8. 6^{-3} **9.** 8^{-3} **10.** 12^{-2} **11.** $\left(\dfrac{1}{2}\right)^{-3}$ **12.** $\left(\dfrac{1}{5}\right)^{-3}$ **13.** $\left(\dfrac{2}{7}\right)^{-2}$ **14.** $\left(\dfrac{4}{5}\right)^{-3}$

Simplify each of the following. Write all answers using only positive exponents. (See Examples 4 and 5.)

15. $\dfrac{3^{-4}}{3^2}$ **16.** $\dfrac{9^{-2}}{9^5}$ **17.** $\dfrac{2^{-5}}{2^{-2}}$ **18.** $\dfrac{14^{-3}}{14^{-8}}$ **19.** $\dfrac{6^{-1}}{6}$

20. $\dfrac{15}{15^{-1}}$ **21.** $4^{-3} \cdot 4^6$ **22.** $5^{-9} \cdot 5^{10}$ **23.** $7^{-5} \cdot 7^{-2}$ **24.** $9^{-1} \cdot 9^{-3}$

25. $\dfrac{8^9 \cdot 8^{-7}}{8^{-3}}$ **26.** $\dfrac{5^{-4} \cdot 5^6}{5^{-1}}$ **27.** $\dfrac{10^8 \cdot 10^{-10}}{10^4 \cdot 10^2}$ **28.** $\dfrac{2^{-4} \cdot 2^{-3}}{2^6 \cdot 2^{-5}}$ **29.** $\left(\dfrac{5^{-6} \cdot 5^3}{5^{-2}}\right)^{-1}$

30. $\left(\dfrac{8^{-3} \cdot 8^4}{8^{-2}}\right)^{-2}$

Simplify each of the following. Assume all variables represent positive real numbers. Write answers with only positive exponents. (See Examples 4 and 5.)

31. $\dfrac{x^4 \cdot x^3}{x^5}$ **32.** $\dfrac{y^9 \cdot y^7}{y^{13}}$ **33.** $\dfrac{(4k^{-1})^2}{2k^{-5}}$ **34.** $\dfrac{(3z^2)^{-1}}{z^5}$

35. $(a^4b^{-2})^{-5}$ **36.** $(c^{-2}d^3)^{-4}$ **37.** $(2p^{-1})^3 \cdot (5p^2)^{-2}$ **38.** $(5^{-1}m^2)^{-3} \cdot (3m^{-2})^4$

39. $\dfrac{7^{-1} \cdot 7r^{-3}}{7^2 \cdot (r^{-2})^2}$ **40.** $\dfrac{12^3 \cdot 12^5 \cdot y^{-2}}{12^{-1} \cdot (y^{-3})^{-2}}$ **41.** $\dfrac{6k^{-4} \cdot (3k^{-1})^{-2}}{2^3 \cdot k^2}$ **42.** $\dfrac{8p^{-3} \cdot (4p^2)^{-2}}{p^{-5}}$

43. $\dfrac{(2x)^{-2}(x^{-1}y)^{-3}}{(xy)^{-2}y^2}$ **44.** $\dfrac{(3z^{-1})^2(z^2w)^{-2}}{(2z^{-1}w^{-2})^3}$ **45.** $(m^4p^{-2})^2 \cdot (m^{-1}p^2)$ **46.** $(z^3x^5)^{-1} \cdot (z^{-4}x^{-2})$

47. $(5a^2b^{-3})^{-1} \cdot (3^{-1}a^{-2}b^2)^{-2}$ **48.** $(2p^2q^{-3})^{-2} \cdot (7p^{-1}q^2)^{-1}$

Evaluate each expression, letting $a = 2$, $b = -3$, and $c = 0$.

49. $a^3 + b$ **50.** $b^3 + a^2$ **51.** $-b^2 + 3(c + 5)$ **52.** $2a^2 - b^4$ **53.** $a^7 b^9 c^5$ **54.** $12a^{15}b^6c^8$

55. $a^b + b^a$ **56.** $b^c + a^c$ **57.** $a^{-1} + b^{-1}$ **58.** $b^{-2} - a^{-1}$ **59.** $a^{-b} + b^{-c}$ **60.** $a^{2b} + b^b$

Perform all indicated operations and write answers with positive exponents. Assume all variables represent positive real numbers. (See Example 6.)

61. $2^{-1} - 3^{-1}$ **62.** $4^{-1} + 5^{-1}$ **63.** $(6^{-1} + 2^{-1})^{-1}$ **64.** $(5^{-1} - 10^{-1})^{-1}$

65. $\dfrac{3^{-2} - 4^{-1}}{4^{-1}}$ **66.** $\dfrac{6^{-1} + 5^{-2}}{6^{-1}}$ **67.** $\dfrac{a^{-1} + b^{-1}}{(ab)^{-1}}$ **68.** $\dfrac{p^{-1} - q^{-1}}{(pq)^{-2}}$

69. $(a + b)^{-1}(a^{-1} + b^{-1})$ **70.** $(m^{-1} + n^{-1})^{-1}$

1.9 RATIONAL EXPONENTS AND RADICALS

In this section the definition of a^n will be extended to include rational values of n, such as 1/2 and 5/4, and not just integer values. In order to do this, we need some terminology.

There are two numbers whose square is 16, 4 and -4. The positive one, 4, is called the **square root** (or second root) of 16.* Similarly, there are two numbers whose fourth power is 16, 2 and -2. We call 2 the **fourth root** of 16. This suggests the following generalization.

If n is even, the ***n*th root of *a*** is the positive real number whose nth power is a.

All nonnegative numbers have nth roots for every natural number n, but *no negative number has a real, even nth root*. For example there is no real number whose square is -16, so -16 has no square root.

We say that the **cube root** (or third root) of 8 is 2 because $2^3 = 8$. Similarly, since $(-2)^3 = -8$, we say -2 is the cube root of -8. Again, we can generalize.

If n is odd, the ***n*th root of *a*** is the real number whose nth power is a.

Every real number has an nth root for every *odd* natural number n.

---

*Sometimes the positive square root is called the *principal* square root.

1 Evaluate the following.

(a) $16^{1/2}$

(b) $16^{1/4}$

(c) $-256^{1/2}$

(d) $(-256)^{1/2}$

(e) $-8^{1/3}$

(f) $243^{1/5}$

Answer:

(a) 4

(b) 2

(c) -16

(d) Not a real number

(e) -2

(f) 3

2 Evaluate the following.

(a) $16^{3/4}$

(b) $25^{5/2}$

(c) $32^{7/5}$

(d) $100^{3/2}$

Answer:

(a) 8

(b) 3125

(c) 128

(d) 1000

We can now define rational exponents. If they are to have the same properties as integer exponents, we want $a^{1/2}$ to be a number such that

$$(a^{1/2})^2 = a^{1/2} \cdot a^{1/2} = a^{1/2+1/2} = a^1 = a.$$

Thus, $a^{1/2}$ should be a number whose square is a and it is reasonable to *define* $a^{1/2}$ to be the square root of a (if it exists). Similarly, $a^{1/3}$ is defined to be the cube root of a, and $a^{1/n}$ is defined to be the nth root of a (if it exists).

▶**EXAMPLE 1**   Evaluate the following roots.

(a) $36^{1/2} = 6$ since $6^2 = 36$

(b) $-100^{1/2} = -10$

(c) $-(225^{1/2}) = -15$

(d) $625^{1/4} = 5$ since $5^4 = 625$

(e) $(-1296)^{1/4}$ is not a real number, but $-1296^{1/4} = -6$ since $6^4 = 1296$

(f) $(-27)^{1/3} = -3$

(g) $-32^{1/5} = -2$ ◀ **1**

For more general rational exponents, the symbol $a^{m/n}$ should be defined so that the properties for exponents still hold. For the first power property to hold,

$$(a^{1/n})^m \text{ must equal } a^{m/n}.$$

This suggests the following definition.

For all integers m and all positive integers n, and for all real numbers a for which $a^{1/n}$ exists,

$$a^{m/n} = (a^{1/n})^m.$$

▶**EXAMPLE 2**   Evaluate the following.

(a) $27^{2/3} = (27^{1/3})^2$
$= 3^2 = 9$

(b) $32^{2/5} = (32^{1/5})^2$
$= 2^2 = 4$

(c) $64^{4/3} = (64^{1/3})^4$
$= 4^4 = 256$

(d) $25^{3/2} = (25^{1/2})^3$
$= 5^3 = 125$ ◀ **2**

3 Simplify. Assume all variables represent positive real numbers.

(a) $6^{2/5} \cdot 6^{3/5}$

(b) $\dfrac{8^{2/3} \cdot 8^{-4/3}}{8^2}$

(c) $3x^{1/4} \cdot 5x^{5/4}$

(d) $\left(\dfrac{2k^{1/3}}{p^{5/4}}\right)^2 \cdot \left(\dfrac{4k^{-2}}{p^5}\right)^{3/2}$

(e) $a^{5/8}(2a^{3/8} + a^{-1/8})$

Answer:

(a) 6

(b) $1/8^{8/3}$ or $1/2^8$

(c) $15x^{3/2}$

(d) $32/(p^{10}k^{7/3})$

(e) $2a + a^{1/2}$

It can be shown that all the properties given earlier for integer exponents also apply to rational exponents.

▶**EXAMPLE 3** Simplify the following. Assume all variables represent positive real numbers.

(a) $\dfrac{27^{1/3} \cdot 27^{5/3}}{27^3} = \dfrac{27^{1/3+5/3}}{27^3}$ **Product property**

$\qquad\qquad = \dfrac{27^2}{27^3} = 27^{2-3}$ **Quotient property**

$\qquad\qquad = 27^{-1} = \dfrac{1}{27}$ **Definition of negative exponent**

(b) $81^{5/4} \cdot 4^{-3/2} = (81^{1/4})^5 \cdot (4^{1/2})^{-3}$ **Definition of $a^{m/n}$**
$\qquad\qquad = 3^5 \cdot 2^{-3}$ **Definition of $a^{1/n}$**

$\qquad\qquad = \dfrac{3^5}{2^3}$ or $\dfrac{243}{8}$ **Definition of negative exponent**

(c) $6y^{2/3} \cdot 2y^{1/2} = 12y^{2/3+1/2} = 12y^{7/6}$

(d) $\left(\dfrac{3m^{5/6}}{y^{3/4}}\right)^2 \cdot \left(\dfrac{8y^{-3}}{m^6}\right)^{2/3} = \dfrac{3^2m^{5/3}}{y^{3/2}} \cdot \dfrac{8^{2/3}y^{-2}}{m^4}$ **Power properties**

$\qquad\qquad = \dfrac{9m^{5/3}}{y^{3/2}} \cdot \dfrac{4}{m^4y^2}$ **Definition of a^{-n}**

$\qquad\qquad = \dfrac{36m^{(5/3)-4}}{y^{(3/2)+2}}$ **Product and quotient properties**

$\qquad\qquad = \dfrac{36m^{-7/3}}{y^{7/2}} = \dfrac{36}{m^{7/3}y^{7/2}}$ **Definition of a^{-n}**

(e) $m^{2/3}(m^{7/3} + 2m^{1/3}) = (m^{2/3+7/3} + 2m^{2/3+1/3}) = m^3 + 2m$ ◀ **3**

Most roots of integers are irrational numbers that can be approximated with a calculator. A calculator with a square root key, $\boxed{\sqrt{x}}$, will give not only square roots, but by pressing it twice, fourth roots; by pressing it a third time, eighth roots; and so on (since $(x^{1/2})^{1/2} = x^{1/4}$, the fourth root of x, and $[(x^{1/2})^{1/2}]^{1/2} = x^{1/8}$, the eighth root of x).

Some calculators have a key labeled $\boxed{y^x}$ (or $\boxed{x^y}$). This key allows you to find any (appropriate) number to any power. For example, to evaluate $81^{5/4}$, enter 81, press y^x, and enter 1.25 (for 5/4). You should get 243. However, the calculator may not always give the value of an expression in the most useful form. In Example 3(b) we found that $81^{5/4} \cdot 4^{-3/2} = 3^5/2^3$ or 243/8. The calculator gives it as 30.375, which may not be as useful. Some calculators give an error message if you try to calculate a negative number to an odd power (such as $(-8)^{1/3}$). In that case, use a positive base and change the sign of the answer.

4 Simplify.

(a) $\sqrt[3]{27}$

(b) $\sqrt[4]{625}$

(c) $\sqrt[6]{64}$

(d) $\sqrt[3]{\dfrac{64}{125}}$

Answer:

(a) 3

(b) 5

(c) 2

(d) 4/5

RADICALS The nth root of a was denoted above as $a^{1/n}$. An alternative notation for nth roots uses **radicals.**

If n is an even natural number and $a \geq 0$, or n is an odd natural number,

$$a^{1/n} = \sqrt[n]{a}.$$

The symbol $\sqrt{}$ is a **radical sign,** the number a is the **radicand,** and n is the **index** of the radical. The familiar symbol $\sqrt{a}$ is used instead of $\sqrt[2]{a}$.

▶**EXAMPLE 4** Simplify the following.

(a) $\sqrt[4]{16} = 16^{1/4} = 2$

(b) $\sqrt[5]{-32} = -2$

(c) $\sqrt[3]{1000} = 10$

(d) $\sqrt[6]{\dfrac{64}{729}} = \dfrac{2}{3}$ ◀ **4**

The symbol $a^{m/n}$ also can be written in an alternative notation using radicals.

For all rational numbers m/n and all real numbers a for which $\sqrt[n]{a}$ exists,

$$a^{m/n} = (\sqrt[n]{a})^m \quad \text{or} \quad a^{m/n} = \sqrt[n]{a^m}.$$

Notice that $\sqrt[n]{x^n}$ cannot be written simply as x when n is even. For example, if $x = -5$,

$$\sqrt{x^2} = \sqrt{(-5)^2} = \sqrt{25} = 5 \neq x.$$

However, $|-5| = 5$, so that $\sqrt{x^2} = |x|$ when x is -5. This is true in general.

For any real number a and any natural number n,

$$\sqrt[n]{a^n} = |a| \text{ if } n \text{ is even}$$

and

$$\sqrt[n]{a^n} = a \text{ if } n \text{ is odd.}$$

To avoid this difficulty that $\sqrt[n]{a^n}$ is not necessarily equal to a, we shall assume that all variables in radicands represent only nonnegative numbers, as they usually do in applications.

5 Simplify.

(a) $\sqrt{3} \cdot \sqrt{27}$

(b) $\sqrt{\dfrac{3}{49}}$

(c) $\sqrt[5]{\sqrt[4]{3}}$

Answer:

(a) 9

(b) $\dfrac{\sqrt{3}}{7}$

(c) $\sqrt[20]{3}$

The properties of exponents can be written with radicals as shown below.

For all real numbers a and b and positive integers m and n for which all indicated roots exist,

(a) $\sqrt[n]{a} \cdot \sqrt[n]{b} = \sqrt[n]{ab}$

(b) $\dfrac{\sqrt[n]{a}}{\sqrt[n]{b}} = \sqrt[n]{\dfrac{a}{b}}$ $(b \neq 0)$

(c) $\sqrt[m]{\sqrt[n]{a}} = \sqrt[mn]{a}.$

To see why property (c) is true, write the expression using exponents.

$$\sqrt[m]{\sqrt[n]{a}} = (a^{1/n})^{1/m} = a^{(1/n)(1/m)} = a^{1/(mn)} = \sqrt[mn]{a}$$

▶ **EXAMPLE 5** Simplify the following.

(a) $\sqrt{6} \cdot \sqrt{54} = \sqrt{6 \cdot 54} = \sqrt{324} = 18$
Alternatively, simplify $\sqrt{54}$ first.
$$\sqrt{6} \cdot \sqrt{54} = \sqrt{6} \cdot \sqrt{9 \cdot 6}$$
$$= \sqrt{6} \cdot 3\sqrt{6} = 3 \cdot 6 = 18$$

(b) $\sqrt{\dfrac{7}{64}} = \dfrac{\sqrt{7}}{\sqrt{64}} = \dfrac{\sqrt{7}}{8}$

(c) $\sqrt[7]{\sqrt[3]{2}} = \sqrt[21]{2}$ ◀ **5**

SIMPLIFYING RADICALS When working with numbers, it is customary to write them in their simplest form. For example, 10/2 is written as 5, −9/6 is written as −3/2, and 4/20 is written as 1/5. Expressions with radicals also should be written in their simplest form. The *simplified form* of a radical is defined as follows.

Simplified Radicals

An expression with radicals is said to be simplified when all the following conditions are satisfied.

1. The radicand has no factors with powers greater than or equal to the index of the radical.
2. The index of the radical and the power of the radicand have no common factor except 1.
3. All denominators are rational numbers.
4. All indicated operations have been performed (if possible).

6 Simplify.

(a) $\sqrt{80}$

(b) $\sqrt[3]{32m^4p^6q^2}$

Answer:

(a) $4\sqrt{5}$

(b) $2mp^2\sqrt[3]{4mq^2}$

7 Simplify. Assume all variables represent positive real numbers.

(a) $5\sqrt[3]{4x^2z} - 3\sqrt[3]{4x^2z}$

(b) $\sqrt{27m^3n} + 2m\sqrt{12mn}$

Answer:

(a) $2\sqrt[3]{4x^2z}$

(b) $7m\sqrt{3mn}$

8 Multiply.

(a) $(\sqrt{5} - \sqrt{2})(3 + \sqrt{2})$

(b) $(\sqrt{3} + \sqrt{7})(\sqrt{3} - \sqrt{7})$

Answer:

(a) $3\sqrt{5} + \sqrt{10} - 3\sqrt{2} - 2$

(b) -4

▶**EXAMPLE 6** Simplify each of the following. Assume all variables represent positive real numbers.

(a) $\sqrt{175} = \sqrt{25 \cdot 7} = \sqrt{25} \cdot \sqrt{7} = 5\sqrt{7}$

(b) $\sqrt[3]{81x^5y^7z^6} = \sqrt[3]{27 \cdot 3 \cdot x^3 \cdot x^2 \cdot y^3 \cdot y^3 \cdot y \cdot z^3 \cdot z^3}$

$= \sqrt[3]{(3^3x^3y^3y^3z^3z^3)(3x^2y)}$

$= 3xyyzz\sqrt[3]{3x^2y} = 3xy^2z^2\sqrt[3]{3x^2y}$ ◀ **6**

Radicals with the same radicand and the same index, such as $3\sqrt[4]{11pq}$ and $-7\sqrt[4]{11pq}$, are called **like radicals.** Only like radicals can be added or subtracted. As shown in part (b) of the next example, it is sometimes necessary to simplify radicals before adding or subtracting. As before, assume that all variables represent only positive numbers.

▶**EXAMPLE 7** Simplify the following.

(a) $3\sqrt[4]{11pq} - 7\sqrt[4]{11pq} = (3 - 7)\sqrt[4]{11pq} = -4\sqrt[4]{11pq}$

(b) $\sqrt{98x^3y} + 3x\sqrt{32xy}$

First remove all perfect square factors from under the radical. Then use the distributive property, as follows.

$\sqrt{98x^3y} + 3x\sqrt{32xy} = \sqrt{49 \cdot 2 \cdot x^2 \cdot x \cdot y} + 3x\sqrt{16 \cdot 2 \cdot x \cdot y}$

$= 7x\sqrt{2xy} + (3x)(4)\sqrt{2xy}$

$= 7x\sqrt{2xy} + 12x\sqrt{2xy} = 19x\sqrt{2xy}$ ◀ **7**

Multiplying radical expressions is much like multiplying polynomials.

▶**EXAMPLE 8** Multiply the following.

(a) $(\sqrt{2} + 3)(\sqrt{8} - 5) = \sqrt{2}(\sqrt{8}) - \sqrt{2}(5) + 3\sqrt{8} - 3(5)$ **FOIL**

$= \sqrt{16} - 5\sqrt{2} + 3(2\sqrt{2}) - 15$

$= 4 - 5\sqrt{2} + 6\sqrt{2} - 15$

$= -11 + \sqrt{2}$

(b) $(\sqrt{7} - \sqrt{10})(\sqrt{7} + \sqrt{10}) = (\sqrt{7})^2 - (\sqrt{10})^2$

$= 7 - 10 = -3$ ◀ **8**

RATIONALIZING THE DENOMINATOR Before calculators were easily available, it was useful to **rationalize denominators** (write denominators with no radicals) because it was easy to calculate results like $\sqrt{2}/2 = 1.414/2 = .707$ mentally, but more difficult to compute $1/\sqrt{2} = 1/1.414$ without pencil and paper. However, there are other good reasons to rationalize denominators (and sometimes numerators). Some examples will occur in Chapter 11 of this text. The process of rationalizing the denominator is explained in the next examples.

9 Rationalize the denominator.

(a) $\dfrac{2}{\sqrt{5}}$

(b) $\sqrt[3]{\dfrac{3}{2}}$

Answer:

(a) $\dfrac{2\sqrt{5}}{5}$

(b) $\dfrac{\sqrt[3]{12}}{2}$

10 Rationalize the denominator.

(a) $\dfrac{4}{1+\sqrt{3}}$

(b) $\dfrac{1-\sqrt{2}}{1+\sqrt{2}}$

Answer:

(a) $-2 + 2\sqrt{3}$

(b) $-3 + 2\sqrt{2}$

▶**EXAMPLE 9** Rationalize each denominator.

(a) $\dfrac{4}{\sqrt{3}}$

To rationalize the denominator, multiply by 1 in the form $\sqrt{3}/\sqrt{3}$ so that the denominator is $\sqrt{3} \cdot \sqrt{3} = 3$, a rational number.

$$\frac{4}{\sqrt{3}} \cdot \frac{\sqrt{3}}{\sqrt{3}} = \frac{4\sqrt{3}}{3}$$

(b) $\sqrt[4]{\dfrac{3}{5}}$

Start by using the fact that the radical of a quotient can be written as the quotient of radicals.

$$\sqrt[4]{\frac{3}{5}} = \frac{\sqrt[4]{3}}{\sqrt[4]{5}}$$

Multiply numerator and denominator by $\sqrt[4]{5^3}$ to rationalize the denominator. Use this number so that the denominator will be $\sqrt[4]{5} \cdot \sqrt[4]{5^3} = \sqrt[4]{5^4} = 5$, a rational number. This multiplication gives

$$\frac{\sqrt[4]{3}}{\sqrt[4]{5}} = \frac{\sqrt[4]{3} \cdot \sqrt[4]{5^3}}{\sqrt[4]{5} \cdot \sqrt[4]{5^3}} = \frac{\sqrt[4]{3 \cdot 5^3}}{\sqrt[4]{5^4}} = \frac{\sqrt[4]{375}}{5}. \quad ◀ \; \boxed{9}$$

▶**EXAMPLE 10** Rationalize the denominator of $\dfrac{1}{1 - \sqrt{2}}$.

A useful approach here is to multiply both the numerator and denominator by the **conjugate*** of the denominator, in this case $1 + \sqrt{2}$. As suggested by Example 8(b), the product $(1 - \sqrt{2})(1 + \sqrt{2})$ is rational.

$$\frac{1}{1 - \sqrt{2}} = \frac{1(1 + \sqrt{2})}{(1 - \sqrt{2})(1 + \sqrt{2})} = \frac{1 + \sqrt{2}}{1 - 2}$$

$$= \frac{1 + \sqrt{2}}{-1} = -1 - \sqrt{2} \quad ◀ \; \boxed{10}$$

1.9 EXERCISES

Evaluate each of the following. Write all answers without exponents. (See Examples 1–2.)

1. $81^{1/2}$　　**2.** $16^{1/4}$　　**3.** $27^{1/3}$　　**4.** $81^{1/4}$　　**5.** $8^{2/3}$　　**6.** $9^{3/2}$

7. $1000^{2/3}$　　**8.** $64^{3/2}$　　**9.** $-125^{2/3}$　　**10.** $-125^{4/3}$　　**11.** $\left(\dfrac{4}{9}\right)^{1/2}$　　**12.** $\left(\dfrac{16}{25}\right)^{1/2}$

---

*The conjugate of $a\sqrt{m} + b\sqrt{n}$ is $a\sqrt{m} - b\sqrt{n}$.

13. $\left(\dfrac{64}{27}\right)^{1/3}$ **14.** $\left(\dfrac{8}{125}\right)^{1/3}$ **15.** $16^{-5/4}$ **16.** $625^{-1/4}$ **17.** $\left(\dfrac{27}{64}\right)^{-1/3}$ **18.** $\left(\dfrac{121}{100}\right)^{-3/2}$

19. $2^{-1} + 4^{-1}$ **20.** $2^{-2} + 3^{-2}$ **21.** $(3^2 + 4^2)^{1/2}$ **22.** $(16^{1/2} + 25^{1/2})^2$

Simplify each of the following. Write all answers using only positive exponents. Assume all variables represent positive real numbers. (See Example 3.)

23. $2^{1/2} \cdot 2^{3/2}$ **24.** $5^{3/8} \cdot 5^{5/8}$ **25.** $27^{2/3} \cdot 27^{-1/3}$ **26.** $9^{-3/4} \cdot 9^{1/4}$ **27.** $\dfrac{4^{2/3} \cdot 4^{5/3}}{4^{1/3}}$

28. $\dfrac{3^{-5/2} \cdot 2^{3/2}}{3^{7/2} \cdot 3^{-9/2}}$ **29.** $\dfrac{6^{2/5} \cdot 6^{-4/5}}{6^2 \cdot 6^{-1/5}}$ **30.** $\dfrac{3^{-5/3} \cdot 3^{2/3}}{3^{-1} \cdot 3^{4/3}}$ **31.** $\dfrac{4^{-2/3} \cdot 4^{1/5}}{4^{5/3}}$ **32.** $\dfrac{8^{1/4} \cdot 8^{-3/4}}{8^{-5/3}}$

33. $(2p)^{1/2} \cdot (2p^3)^{1/3}$ **34.** $(5k^2)^{3/2} \cdot (5k^{1/3})^{3/4}$ **35.** $\dfrac{(mn)^{1/5}(m^2n)^{-2/5}}{m^{1/3}n}$ **36.** $\dfrac{(r^2s)^{3/2}(rs^2)^{1/4}}{r^3s}$

37. $x^{3/4}(x^2 - 3x^3)$ **38.** $p^{2/3}(2p^{1/3} + 5p)$ **39.** $2z^{1/2}(3z^{-1/2} + z^{1/2})$ **40.** $4a^{2/3}(2a^{1/3} - 3a^{3/4})$

41. $(m + 2)^{1/2}[4(m + 2)^{-1/2} - 3(m + 2)^{3/2}]$ **42.** $(k - 1)^{-1/2}[2(k - 1)^{3/2} + (k - 1)^{1/2}]$

Simplify each of the following. Assume that all variables represent positive real numbers. (See Examples 4–6 and 9.)

43. $\sqrt[3]{125}$ **44.** $\sqrt[4]{81}$ **45.** $\sqrt[4]{1296}$ **46.** $\sqrt[3]{343}$ **47.** $\sqrt[5]{-3125}$ **48.** $\sqrt[7]{-128}$

49. $\sqrt{50}$ **50.** $\sqrt{45}$ **51.** $-\sqrt[4]{32}$ **52.** $-\sqrt[4]{243}$ **53.** $-\sqrt{\dfrac{9}{5}}$ **54.** $\sqrt{\dfrac{3}{8}}$

55. $-\sqrt[3]{\dfrac{3}{2}}$ **56.** $-\sqrt[3]{\dfrac{4}{5}}$ **57.** $\sqrt[4]{\dfrac{3}{2}}$ **58.** $\sqrt[4]{\dfrac{32}{81}}$ **59.** $\sqrt{24 \cdot 3^2 \cdot 2^4}$ **60.** $\sqrt{18 \cdot 2^3 \cdot 4^2}$

61. $\sqrt{8z^5x^8}$ **62.** $\sqrt{24m^6n^5}$ **63.** $\sqrt{m^2n^7p^8}$ **64.** $\sqrt{81p^5x^2y^6}$ **65.** $\sqrt{\dfrac{2}{3x}}$ **66.** $\sqrt{\dfrac{5}{3p}}$

67. $\sqrt{\dfrac{x^5y^3}{3^2}}$ **68.** $\sqrt{\dfrac{g^3h^5}{r^3}}$ **69.** $\sqrt[3]{\sqrt{4}}$ **70.** $\sqrt[4]{\sqrt[3]{2}}$ **71.** $\sqrt[6]{\sqrt[3]{x}}$ **72.** $\sqrt[8]{\sqrt[4]{x}}$

Simplify each of the following. Assume that all variables represent positive real numbers. (See Examples 7 and 8.)

73. $4\sqrt{3} - 5\sqrt{12} + 3\sqrt{75}$ **74.** $2\sqrt{5} - 3\sqrt{20} + 2\sqrt{45}$ **75.** $\sqrt{50} - 8\sqrt{8} + 4\sqrt{18}$

76. $6\sqrt{27} - 3\sqrt{12} + 5\sqrt{48}$ **77.** $3\sqrt{28p} - 4\sqrt{63p} + \sqrt{112p}$ **78.** $9\sqrt{8k} + 3\sqrt{18k} - \sqrt{32k}$

79. $3\sqrt[3]{16} - 4\sqrt[3]{2}$ **80.** $\sqrt[3]{2} - \sqrt[3]{16} + 2\sqrt[3]{54}$ **81.** $2\sqrt[3]{3} + 4\sqrt[3]{24} - \sqrt[3]{81}$

82. $\sqrt[3]{32} - 5\sqrt[3]{4} + \sqrt[3]{108}$ **83.** $(\sqrt{2} + 3)(\sqrt{2} - 3)$ **84.** $(\sqrt{5} + \sqrt{2})(\sqrt{5} - \sqrt{2})$

85. $(\sqrt[3]{11} - 1)(\sqrt[3]{11^2} + \sqrt[3]{11} + 1)$ **86.** $(\sqrt[3]{7} + 3)(\sqrt[3]{7^2} - 3\sqrt[3]{7} + 9)$ **87.** $(3\sqrt{2} + \sqrt{3})(2\sqrt{3} - \sqrt{2})$

88. $(4\sqrt{5} - 1)(3\sqrt{5} + 2)$

Rationalize the denominator of each of the following. Assume that all variables represent non-negative numbers and that no denominators are 0. (See Example 10.)

89. $\dfrac{3}{1 - \sqrt{2}}$ **90.** $\dfrac{2}{1 + \sqrt{5}}$ **91.** $\dfrac{4 - \sqrt{2}}{2 - \sqrt{2}}$ **92.** $\dfrac{\sqrt{3} - 1}{\sqrt{3} - 2}$ **93.** $\dfrac{p}{\sqrt{p} + 2}$ **94.** $\dfrac{\sqrt{r}}{3 - \sqrt{r}}$

95. Simplify: $\dfrac{a}{\sqrt{a^2 - 4}} + \dfrac{3\sqrt{a^2 - 4}}{a}$

96. Simplify: $\dfrac{\sqrt{p^2 - 9} - \dfrac{1}{\sqrt{p^2 - 9}}}{p + 3}$

97. Management The price of a certain type of solar heater is approximated by p, where

$$p = 2x^{1/2} + 3x^{2/3}$$

and x is the number of units supplied. Find the price when the supply is 64 units.

98. Management The demand for a certain commodity and the price are related by the equation

$$p = 1000 - 200x^{-2/3} \quad (x > 0)$$

where x is the number of units of the product demanded. Find the price when the demand is 27.

Social Science *In our system of government, the president is elected by the* electoral college, *and not by individual voters. Because of this, smaller states have a greater voice in the selection of a president than they would otherwise.*

Two political scientists have studied the problems of campaigning for president under the current system and have concluded that candidates should allot their money according to the formula

$$\frac{\text{amount for}}{\text{large state}} = \left(\frac{E_{\text{large}}}{E_{\text{small}}}\right)^{3/2} \times \text{amount for a small state.}$$

Here E_{large} represents the electoral vote of the large state, and E_{small} represents the electoral vote of the small state. Find the amount that should be spent in each of the following larger states if \$1,000,000 is spent in the small state and the following statements are true.

99. The large state has 48 electoral votes and the small state has 3.

100. The large state has 36 electoral votes and the small state has 4.

Natural Science *A Delta Airlines map gives a formula for calculating the visible distance from a jet plane to the horizon. On a clear day, this distance is approximated by $D = 1.22x^{1/2}$, where x is altitude in feet, and D is distance to the horizon in miles. Find D for an altitude of*

101. 5000 feet; **102.** 10,000 feet; **103.** 30,000 feet; **104.** 40,000 feet.

Natural Science *(These problems require a calculator with a y^x key.) The Galápagos Islands are a chain of islands ranging in size from .2 to 2249 square miles. A biologist has shown that the number of different land-plant species on an island in this chain is related to the size of the island by $S = 28.6A^{.32}$, where A is the area of an island in square miles and S is the number of different plant species on that island. Estimate S (rounding to the nearest whole number) for islands of area*

105. 1 square mile **106.** 25 square miles; **107.** 300 square miles; **108.** 2000 square miles.

1.10 QUADRATIC EQUATIONS

An equation with 2 as the highest exponent on the variable is called **quadratic.** Some examples of quadratic equations are

$$2x^2 + 5x - 1 = 0, \quad 3m^2 = 6 - m, \quad y^2 = 4, \quad \text{and} \quad 2m^2 - m = 0.$$

If a, b, and c are real numbers, with $a \neq 0$, then

$$ax^2 + bx + c = 0,$$

is a **quadratic equation.**

1 Solve the following equations.

(a) $(y - 6)(y + 2) = 0$

(b) $(5k - 3)(k + 5) = 0$

(c) $(2r - 9)(3r + 5) = 0$

Answer:

(a) $6, -2$

(b) $3/5, -5$

(c) $9/2, -5/3$

2 Solve the following.

(a) $y^2 + 3y = 10$

(b) $2r^2 + 9r = 5$

(c) $3k^2 = 2k + 8$

Answer:

(a) $2, -5$

(b) $1/2, -5$

(c) $-4/3, 2$

(If $a = 0$, the equation becomes $bx + c = 0$, which is linear.) A quadratic equation written in the form $ax^2 + bx + c = 0$ is said to be in **standard form.**

The simplest method of solving a quadratic equation, but one that is not always easily applied, is by factoring. This method depends on the **zero-factor property.**

Zero-Factor Property

If a and b are real numbers, with $ab = 0$, then $a = 0$ or $b = 0$ or both.

▶**EXAMPLE 1** Solve the equation $(x - 4)(3x + 7) = 0$.

By the zero-factor property, the product $(x - 4)(3x + 7)$ can equal 0 only if at least one of the factors equals 0. That is, the product equals zero only if $x - 4 = 0$ or $3x + 7 = 0$. Solving each of these equations separately will give the solutions of the original equation.

$$x - 4 = 0 \quad \text{or} \quad 3x + 7 = 0$$
$$x = 4 \quad \text{or} \qquad 3x = -7$$
$$x = -\frac{7}{3}$$

The solutions of the equation $(x - 4)(3x + 7) = 0$ are 4 and $-7/3$. Check these solutions by substitution in the original equation. ◀ **1**

▶**EXAMPLE 2** Solve $6r^2 + 7r = 3$.

Write the equation in standard form as

$$6r^2 + 7r - 3 = 0.$$

Now factor $6r^2 + 7r - 3$ to get

$$(3r - 1)(2r + 3) = 0.$$

By the zero-factor property, the product $(3r - 1)(2r + 3)$ can equal 0 only if

$$3r - 1 = 0 \quad \text{or} \quad 2r + 3 = 0.$$

Solve each of these equations separately to find that the solutions of the original equation are 1/3 and $-3/2$. Check these solutions by substitution in the original equation. ◀ **2**

An equation such as $x^2 = 5$ has two solutions, $\sqrt{5}$ and $-\sqrt{5}$. The same idea is true in general.

3 Solve each equation.

(a) $p^2 = 21$

(b) $(m + 7)^2 = 15$

(c) $(2k - 3)^2 = 5$

Answer:

(a) $\pm\sqrt{21}$

(b) $-7 \pm \sqrt{15}$

(c) $(3 \pm \sqrt{5})/2$

Square-Root Property

If $b > 0$, then the solutions of $x^2 = b$ are $\sqrt{b}$ and $-\sqrt{b}$.

The two solutions are sometimes abbreviated $\pm\sqrt{b}$.

▶**EXAMPLE 3** Solve each equation.

(a) $m^2 = 17$

By the square-root property, the solutions are $\sqrt{17}$ and $-\sqrt{17}$, abbreviated $\pm\sqrt{17}$.

(b) $(y - 4)^2 = 11$

Use a generalization of the square-root property, working as follows:

$$(y - 4)^2 = 11$$
$$y - 4 = \sqrt{11} \quad \text{or} \quad y - 4 = -\sqrt{11}$$
$$y = 4 + \sqrt{11} \qquad\qquad y = 4 - \sqrt{11}.$$

Abbreviate the solutions as $4 \pm \sqrt{11}$. ◀ **3**

As suggested by Example 3(b), any quadratic equation can be solved using the square-root property if the equation can be written in the form $(x + n)^2 = k$. For instance, to write $2x^2 - 12x + 1 = 0$ in this form, we need to write the left side as a perfect square, $(x + n)^2$. First add -1 to both sides of the equation, then multiply both sides by $1/2$, so that x^2 has a coefficient of 1.

$$2x^2 - 12x + 1 = 0$$
$$2x^2 - 12x \quad\;\; = -1$$
$$x^2 - 6x \quad\;\;\; = -\frac{1}{2} \qquad (*)$$

Adding 9 to both sides will make the left side a perfect square, so the equation can be written in the desired form.

$$x^2 - 6x + 9 = -\frac{1}{2} + 9$$
$$(x - 3)^2 = \frac{17}{4}$$

Use the square-root property to complete the solution.

$$x - 3 = \pm\sqrt{\frac{17}{4}} = \frac{\pm\sqrt{17}}{2}$$

$$x = 3 \pm \frac{\sqrt{17}}{2} = \frac{6 \pm \sqrt{17}}{2}$$

The two solutions are $\dfrac{6 + \sqrt{17}}{2}$ and $\dfrac{6 - \sqrt{17}}{2}$.

Notice that the number 9 added to both sides of equation (*) is found by taking $(1/2)(6) = 3$ and squaring it: $3^2 = 9$. This always works because

$$\left(x + \frac{b}{2}\right)^2 = x^2 + 2\left(\frac{b}{2}\right)x + \left(\frac{b}{2}\right)^2 = x^2 + bx + \left(\frac{b}{2}\right)^2.$$

The process of changing $2x^2 - 12x + 1 = 0$ to $(x - 3)^2 = 17/4$ is called **completing the square.***

▶**EXAMPLE 4**   Solve $9z^2 - 12z - 1 = 0$.

Get a coefficient of 1 in the given equation by multiplying both sides by 1/9.

$$z^2 - \frac{4}{3}z - \frac{1}{9} = 0$$

Add 1/9 on both sides.

$$z^2 - \frac{4}{3}z = \frac{1}{9}$$

Find half of $-4/3$ and square it.

$$\frac{1}{2}\left(-\frac{4}{3}\right) = -\frac{2}{3} \quad \text{and} \quad \left(-\frac{2}{3}\right)^2 = \frac{4}{9}$$

Add 4/9 to both sides of the equation to get

$$z^2 - \frac{4}{3}z + \frac{4}{9} = \frac{1}{9} + \frac{4}{9}.$$

Factoring on the left and adding on the right gives

$$\left(z - \frac{2}{3}\right)^2 = \frac{5}{9}.$$

---

*Completing the square is discussed further in Section 3.1.

4 Solve by completing the square.

(a) $m^2 + 2m = 3$

(b) $6r^2 - r = 2$

Answer:

(a) $1, -3$

(b) $-1/2, 2/3$

Now use the square-root property and simplify the square root to get

$$z - \frac{2}{3} = \sqrt{\frac{5}{9}} \quad \text{or} \quad z - \frac{2}{3} = -\sqrt{\frac{5}{9}}$$

$$z - \frac{2}{3} = \frac{\sqrt{5}}{3} \quad \text{or} \quad z - \frac{2}{3} = -\frac{\sqrt{5}}{3}$$

$$z = \frac{2}{3} + \frac{\sqrt{5}}{3} \quad \text{or} \quad z = \frac{2}{3} - \frac{\sqrt{5}}{3}.$$

These two solutions may be written as

$$z = \frac{2 + \sqrt{5}}{3} \quad \text{or} \quad z = \frac{2 - \sqrt{5}}{3},$$

abbreviated as $(2 \pm \sqrt{5})/3$. ◀ **4**

The method of completing the square can be used on the general quadratic equation,

$$ax^2 + bx + c = 0 \quad (a \neq 0),$$

to convert it to one whose solution can be found by the square-root property. This will give a general formula for solving any quadratic equation. Going through the necessary algebra produces the following important result.

Quadratic Formula

The solutions of the quadratic equation $ax^2 + bx + c = 0$, where $a \neq 0$, are given by

$$x = \frac{-b \pm \sqrt{b^2 - 4ac}}{2a}.$$

▶ **EXAMPLE 5** Solve $x^2 - 4x - 5 = 0$ by the quadratic formula.

The equation is already in standard form (it has 0 alone on one side of the equals sign) so that the letters a, b, and c of the quadratic formula can be identified. The coefficient of the squared term gives the value of a; here $a = 1$. Also, $b = -4$ and $c = -5$. (Be careful to get the correct signs.) Substitute these values into the quadratic formula.

$$x = \frac{-(-4) \pm \sqrt{(-4)^2 - 4(1)(-5)}}{2(1)} \qquad \text{Let } a = 1, b = -4, c = -5$$

$$= \frac{4 \pm \sqrt{16 + 20}}{2} \qquad (-4)^2 = (-4)(-4) = 16$$

$$x = \frac{4 \pm 6}{2} \qquad \sqrt{16 + 20} = \sqrt{36} = 6$$

5 Use the quadratic formula to solve each of the following equations.

(a) $3x^2 + 11x - 4 = 0$

(b) $2z^2 - 7z - 4 = 0$

Answer:

(a) $1/3$, -4

(b) $-1/2$, 4

The $\pm$ sign represents the two solutions of the equation. First use $+$ and then use $-$ to find each of the solutions.

$$x = \frac{4 + 6}{2} = \frac{10}{2} = 5 \quad \text{or} \quad x = \frac{4 - 6}{2} = \frac{-2}{2} = -1$$

The two solutions are 5 and -1. Check by substituting each value in the original equation. ◀ **5**

▶**EXAMPLE 6** Solve $x^2 + 1 = 4x$.

First add $-4x$ to both sides, to get 0 alone on the right side.

$$x^2 - 4x + 1 = 0$$

Now identify the letters, a, b, and c. Here $a = 1$, $b = -4$, and $c = 1$. Substitute these numbers into the quadratic formula.

$$x = \frac{-(-4) \pm \sqrt{(-4)^2 - 4(1)(1)}}{2(1)}$$

$$= \frac{4 \pm \sqrt{16 - 4}}{2}$$

$$= \frac{4 \pm \sqrt{12}}{2}$$

Simplify the solutions by writing $\sqrt{12}$ as $\sqrt{4 \cdot 3} = \sqrt{4} \cdot \sqrt{3} = 2\sqrt{3}$. Substituting $2\sqrt{3}$ for $\sqrt{12}$ gives

$$= \frac{4 \pm 2\sqrt{3}}{2}$$

$$= \frac{2(2 \pm \sqrt{3})}{2} \qquad \textbf{Factor } 4 \pm 2\sqrt{3}$$

$$x = 2 \pm \sqrt{3}.$$

The two solutions are $2 + \sqrt{3}$ and $2 - \sqrt{3}$.

The exact values of the solutions are $2 + \sqrt{3}$ and $2 - \sqrt{3}$. In many cases a decimal approximation of these solutions is needed. Use a calculator to find that $\sqrt{3} \approx 1.732$, so that (to the nearest thousandth) the solutions are

$$2 + \sqrt{3} \approx 2 + 1.732 = 3.732$$

or

$$2 - \sqrt{3} \approx 2 - 1.732 = .268. \quad ◀$$

6 Find exact and approximate solutions for the following.

(a) $y^2 - 2y = 2$

(b) $x^2 - 6x + 4 = 0$

Answer:

(a) Exact: $1 + \sqrt{3}$, $1 - \sqrt{3}$; approximate: 2.732, $-.732$

(b) Exact: $3 + \sqrt{5}$, $3 - \sqrt{5}$; approximate: 5.236, $.764$

7 Solve the following equations.

(a) $9k^2 - 6k + 1 = 0$

(b) $4m^2 + 28m + 49 = 0$

(c) $2x^2 - 5x + 5 = 0$

Answer:

(a) $1/3$

(b) $-7/2$

(c) No real number solutions

Caution In Example 6, be careful to factor $4 + 2\sqrt{3}$ *before* dividing by the denominator 2.

$$\frac{4 + 2\sqrt{3}}{2} \neq \frac{2 + 2\sqrt{3}}{1} \quad \boxed{6}$$

The above quadratic equations all had two different solutions. This is not always the case, as the following example shows.

▶**EXAMPLE 7** Solve the following.

(a) $9x^2 - 30x + 25 = 0$

Here $a = 9$, $b = -30$, and $c = 25$. By the quadratic formula,

$$x = \frac{-(-30) \pm \sqrt{(-30)^2 - 4(9)(25)}}{2(9)}$$

$$= \frac{30 \pm \sqrt{900 - 900}}{18}$$

$$= \frac{30 \pm 0}{18}$$

$$= \frac{30}{18} = \frac{5}{3}.$$

The given equation has only one solution. (The rational solution indicates that this equation could have been solved by factoring.)

(b) $x^2 - 6x + 10 = 0$

Since $a = 1$, $b = -6$, and $c = 10$,

$$x = \frac{-(-6) \pm \sqrt{(-6)^2 - 4(1)(10)}}{2(1)}$$

$$= \frac{6 \pm \sqrt{36 - 40}}{2}$$

$$= \frac{6 \pm \sqrt{-4}}{2}.$$

Recall from Section 1.9 that $\sqrt{-4}$ is not a real number, so there are no *real number* solutions to this equation. ◀ **7**

8 Solve each applied problem.

(a) A plane traveling at a constant rate goes 810 kilometers with the wind, then turns around and travels for 720 kilometers against the wind. If the speed of the wind is 15 kilometers per hour and the total flight took 6 hours, find the speed of the plane.

(b) The length of a picture is 2 inches more than the width. It is mounted on a mat that extends 2 inches beyond the picture on all sides. What are the dimensions of the picture if the area of the mat is 99 square inches?

Answer:

(a) 255 kph

(b) 5 inches by 7 inches

▶**EXAMPLE 8**   A landscape architect wants to make an exposed gravel border of uniform width around a small shed behind a company plant. The shed is 10 feet by 6 feet. He has enough gravel to cover 36 square feet. How wide should the border be?

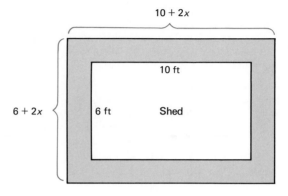

FIGURE 1.19

A sketch of the shed with border is shown in Figure 1.19. Let x represent the width of the border. Then the width of the large rectangle is $6 + 2x$ and its length is $10 + 2x$. The area of the large rectangle is $(6 + 2x)(10 + 2x)$. The area occupied by the shed is $6 \cdot 10 = 60$. The area of the border is found by subtracting the area of the shed from the area of the large rectangle. This difference should be 36 square feet, giving the equation

$$(6 + 2x)(10 + 2x) - 60 = 36.$$

Solve this equation with the following sequence of steps.

$$60 + 32x + 4x^2 - 60 = 36$$
$$4x^2 + 32x - 36 = 0$$
$$x^2 + 8x - 9 = 0$$
$$(x + 9)(x - 1) = 0$$

The solutions are -9 and 1. The number -9 cannot be the width of the border, so the solution is to make the border 1 foot wide. ◀ **8**

Some equations that are not quadratic can be solved as quadratic equations by making a suitable substitution. Such equations are called **quadratic in form.**

▶**EXAMPLE 9**   Solve $4m^4 - 9m^2 + 2 = 0$.
Use the substitutions

$$x = m^2 \quad \text{and} \quad x^2 = m^4$$

9 Solve each equation.

(a) $9x^4 - 23x^2 + 10 = 0$

(b) $4x^4 = 7x^2 - 3$

Answer:

(a) $-\sqrt{5/3}, \sqrt{5/3}, -\sqrt{2}, \sqrt{2}$

(b) $-\sqrt{3}/2, \sqrt{3}/2, -1, 1$

to rewrite the equation as

$$4x^2 - 9x + 2 = 0,$$

a quadratic equation that can be solved by factoring.

$$4x^2 - 9x + 2 = 0$$
$$(x - 2)(4x - 1) = 0$$
$$x - 2 = 0 \quad \text{or} \quad 4x - 1 = 0$$
$$x = 2 \quad \text{or} \quad 4x = 1$$
$$x = \frac{1}{4}$$

Since $x = m^2$,

$$m^2 = 2 \quad \text{or} \quad m^2 = \frac{1}{4}$$

$$m = \pm\sqrt{2} \quad \text{or} \quad m = \pm\frac{1}{2}.$$

There are four solutions: $-\sqrt{2}, \sqrt{2}, -1/2,$ and $1/2$. ◀ **9**

In Example 9, we saw that a fourth degree equation had 4 solutions. Every equation of degree n (for natural numbers n) has up to n real number solutions.

▶**EXAMPLE 10** Solve $4 + \dfrac{1}{z - 2} = \dfrac{3}{2(z - 2)^2}$.

First, substitute u for $z - 2$ to get

$$4 + \frac{1}{z - 2} = \frac{3}{2(z - 2)^2}$$

$$4 + \frac{1}{u} = \frac{3}{2u^2}.$$

Now multiply both sides of the equation by the common denominator $2u^2$; then solve the resulting quadratic equation.

$$2u^2\left(4 + \frac{1}{u}\right) = 2u^2\left(\frac{3}{2u^2}\right)$$

$$8u^2 + 2u = 3$$

$$8u^2 + 2u - 3 = 0$$

$$(4u + 3)(2u - 1) = 0$$

$$4u + 3 = 0 \quad \text{or} \quad 2u - 1 = 0$$

$$u = -\frac{3}{4} \quad \text{or} \quad u = \frac{1}{2}$$

10 Solve each equation.

(a) $\dfrac{13}{3(p+1)} + \dfrac{5}{3(p+1)^2} = 2$

(b) $1 + \dfrac{1}{a} = \dfrac{5}{a^2}$

Answer:

(a) $-4/3,\ 3/2$

(b) $(-1 + \sqrt{21})/2,$
$(-1 - \sqrt{21})/2$

11 Solve each of the following equations for the indicated variable. Assume no variable equals 0.

(a) $k = mp^2 - bp$ for p

(b) $r = \dfrac{APk^2}{3}$ for k

Answer:

(a) $\dfrac{b \pm \sqrt{b^2 + 4mk}}{2m}$

(b) $\pm \sqrt{\dfrac{3r}{AP}}$ or $\dfrac{\pm \sqrt{3rAP}}{AP}$

Since $u = z - 2,$

$$z - 2 = -\frac{3}{4} \qquad \text{or} \qquad z - 2 = \frac{1}{2}$$

$$z = -\frac{3}{4} + 2 \qquad\qquad z = \frac{1}{2} + 2$$

$$z = \frac{5}{4} \qquad \text{or} \qquad z = \frac{5}{2}.$$

The solutions are 5/4 and 5/2. Check both solutions in the original equation. ◀ **10**

The next example shows how to solve an equation for a specified variable when the equation is quadratic in that variable.

▶**EXAMPLE 11** Solve $v = mx^2 + x$ for x. (Assume $m \neq 0$.)
    The equation is quadratic in x because of the x^2 term. Use the quadratic formula, first writing the equation in standard form.

$$v = mx^2 + x$$
$$0 = mx^2 + x - v$$

Let $a = m$, $b = 1$, and $c = -v$. Then the quadratic formula gives

$$x = \frac{-1 \pm \sqrt{1^2 - 4(m)(-v)}}{2m}$$

$$x = \frac{-1 \pm \sqrt{1 + 4mv}}{2m} \qquad ◀ \ \boxed{11}$$

1.10 EXERCISES

Solve each of the following equations. (See Examples 1 and 2.)

1. $(y - 5)(y + 4) = 0$ **2.** $(m + 3)(m - 1) = 0$ **3.** $x^2 + 5x + 6 = 0$ **4.** $y^2 - 3y + 2 = 0$

5. $x^2 = 3 + 2x$ **6.** $3m + 4 = m^2$ **7.** $m^2 + 16 = 8m$ **8.** $y^2 + 49 = 14y$

9. $2k^2 - k = 10$ **10.** $6x^2 = 7x + 5$ **11.** $6x^2 - 5x = 4$ **12.** $9s^2 + 12s = -4$

13. $m(m - 7) = -10$ **14.** $z(2z + 7) = 4$ **15.** $9x^2 - 16 = 0$ **16.** $25y^2 - 64 = 0$

17. $16x^2 - 16x = 0$ **18.** $12y^2 - 48y = 0$

Solve the following equations by the square-root property or by completing the square. (See Examples 3 and 4.)

19. $x^2 = 29$ **20.** $p^2 = 37$ **21.** $(m - 3)^2 = 5$ **22.** $(p + 2)^2 = 7$

23. $(3k - 1)^2 = 19$ **24.** $(4t + 5)^2 = 3$ **25.** $q^2 + 2q = 8$ **26.** $m^2 + 5m - 6 = 0$

27. $4z^2 - 4z = 1$ **28.** $11p^2 - 11p + 1 = 0$

Solve each of the following equations. If the solutions involve square roots, give both the exact and approximate solutions. (See Examples 1–7.)

29. $3x^2 - 5x + 1 = 0$ **30.** $2x^2 - 7x + 1 = 0$ **31.** $2m^2 = m + 4$ **32.** $p^2 + p - 1 = 0$

33. $k^2 - 10k = -20$ **34.** $r^2 = 13 - 12r$ **35.** $2x^2 + 12x + 5 = 0$ **36.** $5x^2 + 4x = 1$

37. $2r^2 - 7r + 5 = 0$ **38.** $8x^2 = 8x - 3$ **39.** $6k^2 - 11k + 4 = 0$ **40.** $8m^2 - 10m + 3 = 0$

41. $x^2 + 3x = 10$ **42.** $2x^2 = 3x + 5$ **43.** $2x^2 - 7x + 30 = 0$ **44.** $3k^2 + k = 6$

45. $5m^2 + 5m = 0$ **46.** $8r^2 + 16r = 0$

Give all real number solutions of the following equations. (See Examples 9 and 10. Hint: in Exercise 53, let $u = p - 3$.)

47. $m^4 - 3m^2 - 10 = 0$ **48.** $2t^4 + 3t^2 - 2 = 0$ **49.** $6y^4 = y^2 + 15$

50. $4a^4 = 2 - 7a^2$ **51.** $b^4 - 3b^2 - 5 = 0$ **52.** $2x^4 - 2x^2 - 1 = 0$

53. $6(p - 3)^2 + 5(p - 3) - 6 = 0$ **54.** $12(q + 4)^2 - 13(q + 4) - 4 = 0$ **55.** $2(p + 5)^2 - 3(p + 5) = 20$

56. $4(2y - 3)^2 - (2y - 3) - 3 = 0$ **57.** $1 + \dfrac{7}{2a} = \dfrac{15}{2a^2}$ **58.** $5 - \dfrac{4}{k} - \dfrac{1}{k^2} = 0$

59. $-\dfrac{2}{3z^2} + \dfrac{1}{3} + \dfrac{8}{3z} = 0$ **60.** $2 + \dfrac{5}{x} + \dfrac{1}{x^2} = 0$

In **(a)** *add or subtract as indicated. In* **(b)** *solve the equation.*

61. (a) $\dfrac{6}{r} - \dfrac{5}{r - 2} - 1$ **(b)** $\dfrac{6}{r} = \dfrac{5}{r - 2} - 1$ **62. (a)** $\dfrac{8}{z - 1} - \dfrac{5}{z} - \dfrac{2z}{z - 1}$ **(b)** $\dfrac{8}{z - 1} = \dfrac{5}{z} + \dfrac{2z}{z - 1}$

63. (a) $\dfrac{1}{2k} + \dfrac{3}{k - 1} - 5$ **(b)** $\dfrac{1}{2k} + \dfrac{3}{k - 1} = -5$ **64. (a)** $\dfrac{4}{p^2 + 3p} - \dfrac{2}{p + 3} + \dfrac{1}{2}$ **(b)** $\dfrac{4}{p^2 + 3p} - \dfrac{2}{p + 3} = \dfrac{1}{2}$

Solve the following problems. (See Example 8.)

65. A shopping center has a rectangular area of 40,000 square yards enclosed on three sides for a parking lot. The length is 200 yards more than twice the width. Find the length and width of the lot.

66. An ecology center wants to set up an experimental garden. It has 300 meters of fencing to enclose a rectangular area of 5000 square meters. Find the length and width of the rectangle.

67. Fred went into a frame-it-yourself shop. He wanted a frame 3 inches longer than wide. The frame he chose extends 1.5 inches beyond the picture on each side. Find the outside dimensions of the frame if the area of the unframed picture is 70 square inches.

68. Joan wants to buy a rug for a room that is 12 feet by 15 feet. She wants to leave a uniform strip of floor around the rug. She can afford 108 square feet of carpeting. What dimensions should the rug have?

69. Management The manager of a bicycle shop knows that the cost of selling x bicycles is $C = 20x + 60$ and the revenue from selling x bicycles is $R = x^2 - 8x$. Find the break-even value of x.

70. Management A company that produces breakfast cereal has found that its operating cost in dollars is $C = 40x + 150$ and its revenue in dollars is $R = 65x - x^2$. For what value(s) of x will the company break even?

71. Physical Science If a ball is thrown upward with an initial velocity of 64 feet per second, then its height after t seconds is $h = 64t - 16t^2$. In how many seconds will the ball reach

(a) 64 ft?   **(b)** 28 ft?
(c) Why are two answers possible?

72. Physical Science A particle moves horizontally with its distance from a starting point in centimeters at t seconds given by $d = 11t^2 - 10t$.

(a) How long will it take before the particle returns to the starting point?
(b) When will the particle be 100 centimeters from the starting point?

Solve each of the following equations for the indicated variable. Assume all denominators are nonzero, and that all variables represent positive real numbers. (See Example 11.)

73. $S = \dfrac{1}{2}gt^2$ for t

74. $a = \pi r^2$ for r

75. $L = \dfrac{d^4 k}{h^2}$ for h

76. $F = \dfrac{kMv^2}{r}$ for v

77. $S = S_0 + gt^2 + k$ for t

78. $P = \dfrac{E^2 R}{(r + R)^2}$ for R

79. $S = 2\pi rh + 2\pi r^2$ for r

80. $pm^2 - 8qm + \dfrac{1}{r} = 0$ for m

1.11 POLYNOMIAL AND RATIONAL INEQUALITIES

We begin our discussion of polynomial inequalities with **quadratic inequalities;** inequalities of the form $ax^2 + bx + c > 0$ (or $<$, or $\leq$, or $\geq$). The highest exponent on the variable is always 2. Examples of quadratic inequalities include

$$x^2 - x - 12 < 0, \quad 3y^2 + 2y \geq 0, \quad \text{and} \quad m^2 \leq 4.$$

A method of solving quadratic inequalities is shown in the next few examples.

▶**EXAMPLE 1** Solve the quadratic inequality $x^2 - x - 12 < 0$.

The solution of this inequality will include all numbers that make $x^2 - x - 12$ *negative* (< 0). Thus, we are concerned only with the *sign* of the expression. The sign of the expression will change from negative to positive or positive to negative only on either side of those x values that make the expression 0. To find these x values; we solve the *equation* $x^2 - x - 12 = 0$.

$$x^2 - x - 12 = 0$$
$$(x + 3)(x - 4) = 0$$
$$x + 3 = 0 \quad \text{or} \quad x - 4 = 0$$
$$x = -3 \quad \text{or} \quad x = 4$$

1 Solve the following. Graph the solution.

(a) $y^2 + 2y - 3 < 0$

(b) $2p^2 + 3p - 2 < 0$

Answer:

(a) $(-3, 1)$

(b) $(-2, 1/2)$

These numbers, -3 and 4, separate the number line into three regions, as shown in Figure 1.20.

FIGURE 1.20

Since the polynomial will change sign only at a point where its value is 0, the sign of the polynomial will be the same (either positive or negative) throughout a given region. To find the sign of the polynomial in each region, substitute any value from a region for x in the polynomial. For example,

in region A, let $x = -5$: $(-5)^2 - (-5) - 12 = 18 > 0$;

in region B, let $x = 0$: $(0)^2 - (0) - 12 = -12 < 0$;

in region C, let $x = 5$: $(5)^2 - (5) - 12 = 8 > 0$.

The only region where $x^2 - x - 12$ is negative (< 0) is region B, so the solution is $(-3, 4)$, as shown in Figure 1.21. ◀ **1**

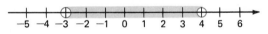

FIGURE 1.21

▶**EXAMPLE 2** Solve the quadratic inequality $r^2 + 3r \geq 4$.

First rewrite the inequality so that one side is 0.

$$r^2 + 3r \geq 4$$
$$r^2 + 3r - 4 \geq 0 \qquad \text{Add } -4 \text{ to both sides}$$

Now solve the corresponding equation.

$$r^2 + 3r - 4 = 0$$
$$(r - 1)(r + 4) = 0$$
$$r = 1 \quad \text{or} \quad r = -4$$

Test a number from each region shown in Figure 1.22.

Let $x = -5$ from region A: $(-5)^2 + 3(-5) - 4 = 6 > 0$.

Let $x = 0$ from region B: $(0)^2 + 3(0) - 4 = -4 < 0$.

Let $x = 2$ from region C: $(2)^2 + 3(2) - 4 = 6 > 0$.

2 Solve each inequality. Graph each solution.

(a) $k^2 + 2k - 15 \geq 0$

(b) $3m^2 + 7m \geq 6$

Answer:

(a) All numbers in $(-\infty, -5]$ or $[3, \infty)$

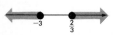

(b) All numbers in $(-\infty, -3]$ or $[2/3, \infty)$

3 Solve each inequality. Graph each solution.

(a) $m^3 - 9m > 0$

(b) $2k^3 - 50k \leq 0$

Answer:

(a) All numbers in $(-3, 0)$ or $(3, \infty)$

(b) All numbers in $(-\infty, -5]$ or $[0, 5]$

We want the inequality to be ≥ 0, that is, positive or 0. The solution includes numbers in region A and in region C, as well as -4 and 1, the endpoints. The solution, which includes all numbers in the intervals $(-\infty, -4]$ or $[1, \infty)$, is graphed in Figure 1.22. ◀ **2**

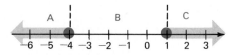

FIGURE 1.22

Although the inequality in the next example is not a quadratic inequality, it can be solved in the same way. Any **polynomial inequality** in the form $P(x) > 0$ or $P(x) < 0$ that can be factored can be solved in this way.

▶**EXAMPLE 3**   Solve $q^3 - 4q > 0$.

Solve the corresponding equation by factoring.

$$q^3 - 4q = 0$$
$$q(q^2 - 4) = 0$$
$$q(q + 2)(q - 2) = 0$$
$$q = 0 \quad \text{or} \quad q + 2 = 0 \quad \text{or} \quad q - 2 = 0$$
$$q = 0 \qquad\qquad q = -2 \qquad\qquad q = 2$$

These three numbers separate the number line into the four regions shown in Figure 1.23. Test a number from each region.

$$\text{If } q = -3, (-3)^3 - 4(-3) = -15 < 0.$$
$$\text{If } q = -1, (-1)^3 - 4(-1) = 3 > 0.$$
$$\text{If } q = 1, (1)^3 - 4(1) = -3 < 0.$$
$$\text{If } q = 3, (3)^3 - 4(3) = 15 > 0.$$

The numbers that make the polynomial > 0, or positive, are in the intervals

$$(-2, 0) \quad \text{or} \quad (2, \infty),$$

as graphed in Figure 1.23. ◀ **3**

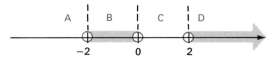

FIGURE 1.23

RATIONAL INEQUALITIES Inequalities with quotients of algebraic expressions are called **rational inequalities.** These inequalities can be solved in much the same way as polynomial inequalities.

▶**EXAMPLE 4**   Solve the rational inequality $\dfrac{5}{x+4} \geq 1$.

Write the corresponding equation and get one side equal to 0.

$$\frac{5}{x+4} = 1$$

$$\frac{5}{x+4} - 1 = 0$$

Write the left side as a single fraction.

$$\frac{5}{x+4} - \frac{x+4}{x+4} = 0 \qquad \textbf{Get a common denominator}$$

$$\frac{5 - (x+4)}{x+4} = 0 \qquad \textbf{Subtract fractions}$$

$$\frac{5 - x - 4}{x+4} = 0 \qquad \textbf{Distributive property}$$

$$\frac{1 - x}{x+4} = 0$$

The quotient can change sign only when the denominator is 0 or when the numerator is 0. This happens when

$$1 - x = 0 \quad \text{or} \quad x + 4 = 0$$
$$x = 1 \quad \text{or} \quad x = -4.$$

As in the earlier examples, test a number from each of the regions determined by 1 and -4 in the original inequality.

$$\text{Let } x = -5: \ \frac{5}{-5+4} = \frac{5}{-1} = -5 \ngeq 1.$$

$$\text{Let } x = 0: \ \frac{5}{0+4} = \frac{5}{4} \geq 1.$$

$$\text{Let } x = 2: \ \frac{5}{2+4} = \frac{5}{6} \ngeq 1.$$

The test shows that numbers in $(-4, 1)$ satisfy the inequality. With a quotient, the endpoints must be considered individually to make sure that no denominator is 0. In this inequality, -4 makes the denominator 0, while 1 satisfies the given inequality. Write the solution as $(-4, 1]$. ◀

4 Solve each inequality.

(a) $\dfrac{3}{x - 2} \geq 4$

(b) $\dfrac{p}{1 - p} < 3$

Answer:

(a) $(2, 11/4]$

(b) All numbers in $(-\infty, 3/4)$ or $(1, \infty)$

5 Solve each rational inequality.

(a) $\dfrac{3y - 2}{2y + 5} < 1$

(b) $\dfrac{3c - 4}{2 - c} \geq -5$

Answer:

(a) $(-5/2, 7)$

(b) All numbers in $(-\infty, 3]$ or $(2, \infty)$

Caution As suggested by Example 4, be very careful with the endpoints of the intervals in the solution of rational inequalities. **4**

▶**EXAMPLE 5** Solve $\dfrac{2x - 1}{3x + 4} < 5$.

Write the corresponding equation, then get 0 on one side. Begin by subtracting 5 on both sides and combining the terms on the left into a single fraction.

$$\frac{2x - 1}{3x + 4} = 5$$

$$\frac{2x - 1}{3x + 4} - 5 = 0 \qquad \text{Get 0 on one side}$$

$$\frac{2x - 1 - 5(3x + 4)}{3x + 4} = 0 \qquad \text{Subtract}$$

$$\frac{-13x - 21}{3x + 4} = 0 \qquad \text{Combine terms}$$

Set the numerator and denominator each equal to 0 and solve the two equations.

$$-13x - 21 = 0 \qquad \text{or} \quad 3x + 4 = 0$$

$$x = -\frac{21}{13} \quad \text{or} \qquad x = -\frac{4}{3}$$

Use the values $-21/13$ and $-4/3$ to divide the number line into three intervals. Test a number from each interval in the original inequality. The quotient is negative for numbers in $(-\infty, -21/13)$ or $(-4/3, \infty)$. Neither endpoint satisfies the given inequality. ◀

Caution Never treat rational inequalities as if they were equations, as in this **incorrect** "solution."

$$\frac{x + 3}{x - 1} > -2$$

$$x + 3 > -2(x - 1) \qquad \text{Multiply both sides by } x - 1$$

$$x + 3 > -2x + 2$$

$$3x > -1$$

$$x > -\frac{1}{3}$$

According to this, 0 should be a solution. But putting $x = 0$ in the original inequality produces the false statement $-3 > -2$. The source of the error is multiplying by $x - 1$. This quantity is negative for some values of x and positive for others. To solve this problem correctly by this method, you must consider two separate cases and reverse the direction of the inequality when $x - 1$ is negative. **5**

1.11 EXERCISES

Solve each of the following quadratic inequalities. Graph each solution. (See Examples 1 and 2.)

1. $(m + 2)(m - 4) < 0$ **2.** $(k - 1)(k + 2) > 0$ **3.** $(t + 6)(t - 1) \geq 0$ **4.** $(y - 2)(y + 3) \leq 0$

5. $y^2 - 3y + 2 < 0$ **6.** $z^2 - 4z - 5 \leq 0$ **7.** $2k^2 + 7k - 4 > 0$ **8.** $6r^2 - 5r - 4 > 0$

9. $q^2 - 7q + 6 \leq 0$ **10.** $2k^2 - 7k - 15 \leq 0$ **11.** $6m^2 + m > 1$ **12.** $10r^2 + r \leq 2$

13. $2y^2 + 5y \leq 3$ **14.** $3a^2 + a > 10$ **15.** $x^2 \leq 25$ **16.** $y^2 \geq 4$

17. $p^2 - 16p > 0$ **18.** $r^2 - 9r < 0$

Solve the following inequalities. (See Example 3.)

19. $t^3 - 16t \leq 0$ **20.** $z^3 - 64z \leq 0$ **21.** $2r^3 - 98r > 0$

22. $3m^3 - 27m > 0$ **23.** $4m^3 + 7m^2 - 2m > 0$ **24.** $6p^3 - 11p^2 + 3p > 0$

Solve the following rational inequalities. (See Examples 4 and 5.)

25. $\dfrac{m - 3}{m + 5} \leq 0$ **26.** $\dfrac{r + 1}{r - 4} > 0$ **27.** $\dfrac{k - 1}{k + 2} > 1$ **28.** $\dfrac{a - 5}{a + 2} < -1$ **29.** $\dfrac{3}{x - 6} \leq 2$

30. $\dfrac{1}{k - 2} < \dfrac{1}{3}$ **31.** $\dfrac{2y + 3}{y - 5} \leq 1$ **32.** $\dfrac{a + 2}{3 + 2a} \leq 5$ **33.** $\dfrac{7}{k + 2} \geq \dfrac{1}{k + 2}$ **34.** $\dfrac{5}{p + 1} > \dfrac{12}{p + 1}$

Management *The commodity market is very unstable; money can be made or lost quickly when invested in soybeans, wheat, pork bellies, and so on. Suppose that an investor kept track of her total profit, P, at time t, measured in months, after she began investing and found that*

$$P = 4t^2 - 29t + 30.$$

For example, 4 months after she began investing, her total profits were

$$P = 4 \cdot 4^2 - 29 \cdot 4 + 30 \qquad \text{Let } t = 4$$
$$= 4 \cdot 16 - 116 + 30$$
$$= 64 - 116 + 30$$
$$P = -22,$$

so that she was $22 "in the hole" after 4 months.

35. Find her total profit after 8 months.

36. Find her total profit after 10 months.

37. Find when she has just broken even. (Let $P = 0$, and solve the resulting equation.)

38. Find the time intervals where she has been ahead. (Solve the inequality $4t^2 - 29t + 30 > 0$.)

39. An analyst has found that his company's profits, in hundreds of thousands of dollars, are given by

$$P = 3x^2 - 35x + 50,$$

where x is the amount, in hundreds of dollars, spent on advertising. For what values of x does the company make a profit?

40. The manager of a large apartment complex has found that the profit is given by

$$P = -x^2 + 250x - 15,000,$$

where x is the number of units rented. For what values of x does the complex produce a profit?

41. **Physical Science** A projectile is fired from ground level. After t seconds its height above the ground is $220t - 16t^2$ feet. For what time period is the projectile at least 624 feet above the ground?

42. **Physical Science** Suppose the velocity of an object is given by

$$v = 2t^2 - 5t - 12,$$

where t is time in seconds. (Here t can be positive or negative.) Find the intervals where the velocity is negative.

KEY TERMS AND SYMBOLS

1.1 $\approx$ is approximately equal to
π pi
$|a|$ absolute value of a
number line
graph
real number
natural (counting) number
whole number
integer
rational number
irrational number
properties of the real numbers
additive inverse
multiplicative inverse
interval
interval notation
order of operations
absolute value

1.2 variable
linear equation
properties of equality
solving for a specified variable

1.3 $<$ is less than
$\leq$ is less than or equal to
$>$ is greater than
$\geq$ is greater than or equal to
linear inequality
properties of inequality

1.5 a^n a to the power n
exponent or power
base
polynomial
term
coefficient
degree of a term
degree of a polynomial

like terms
FOIL

1.6 factoring
factor
greatest common factor
difference of two squares
perfect square
difference of two cubes
sum of two cubes

1.7 rational expression
operations with rational
   expressions
complex fraction

1.8 $a^{-n} = \dfrac{1}{a^n}$ negative exponent

properties of exponents

1.9 $a^{1/n}$ nth root of a
$\sqrt{a}$ square root of a
$\sqrt[n]{a}$ nth root of a
radical
radical sign
radicand
index
simplified radical
like radicals
rationalizing the denominator
conjugate

1.10 quadratic equation
zero-factor property
square-root property
completing the square
quadratic formula

1.11 quadratic inequality
polynomial inequality
rational inequality

KEY CONCEPTS

Absolute Value	Assume a and b are real numbers with $b > 0$. The solutions of $|a| = b$ are $a = b$ or $a = -b$. The solutions of $|a| < b$ are $-b < a < b$. The solutions of $|a| > b$ are $a < -b$ or $a > b$.
Factoring	$x^2 - y^2 = (x + y)(x - y)$ $\qquad x^3 - y^3 = (x - y)(x^2 + xy + y^2)$ $x^2 + 2xy + y^2 = (x + y)^2$ $\qquad x^3 + y^3 = (x + y)(x^2 - xy + y^2)$ $x^2 - 2xy + y^2 = (x - y)^2$
Rules for Radicals	Let a and b be real numbers, n and k be positive integers, and m be any integer for which the following exist. $a^{m/n} = \sqrt[n]{a^m} = (\sqrt[n]{a})^m \qquad \sqrt[n]{a^n} = |a|$ if n is even $\qquad \sqrt[n]{a^n} = a$ if n is odd $\sqrt[n]{a} \cdot \sqrt[n]{b} = \sqrt[n]{ab} \qquad\qquad \dfrac{\sqrt[n]{a}}{\sqrt[n]{b}} = \sqrt[n]{\dfrac{a}{b}} \quad (b \neq 0) \qquad \sqrt[k]{\sqrt[n]{a}} = \sqrt[kn]{a}$
Rules for Exponents	Let a, b, r, and s be any real numbers for which the following exist. $a^{-r} = \dfrac{1}{a^r} \qquad a^0 = 1 \qquad \left(\dfrac{a}{b}\right)^{-r} = \left(\dfrac{b}{a}\right)^r \qquad a^{1/r} = \sqrt[r]{a}$ $a^r \cdot a^s = a^{r+s} \qquad\qquad (a^r)^s = a^{rs} \qquad\qquad \left(\dfrac{a}{b}\right)^r = \dfrac{a^r}{b^r}$ $\dfrac{a^r}{a^s} = a^{r-s} \qquad\qquad (ab)^r = a^r b^r$
Solving Quadratic Equations	Let a, b, and c be real numbers. Factoring: If $ab = 0$, then $a = 0$ or $b = 0$ or both. Square-Root Property: If $b > 0$, then the solutions of $x^2 = b$ are $\sqrt{b}$ and $-\sqrt{b}$. Quadratic Formula: The solutions of $ax^2 + bx + c$ $(a \neq 0)$ are $x = \dfrac{-b \pm \sqrt{b^2 - 4ac}}{2a}$.

CHAPTER 1 REVIEW EXERCISES

Name the numbers from the list -12, -6, $-9/10$, $-\sqrt{7}$, $-\sqrt{4}$, 0, $1/8$, $\pi/4$, 6, $\sqrt{11}$ *that are*

1. natural numbers; **2.** whole numbers; **3.** integers;

4. rational numbers; **5.** irrational numbers; **6.** real numbers.

For Exercises 7–16, choose all words from the following list which apply: natural numbers, whole numbers, integers, rational numbers, irrational numbers, real numbers, undefined.

7. 22   **8.** 0   **9.** $\sqrt{36}$ **10.** $-\sqrt{25}$ **11.** -1 **12.** $\dfrac{5}{8}$ **13.** $\sqrt{15}$

14. $-\dfrac{6}{0}$ **15.** $\dfrac{3\pi}{4}$ **16.** $\pi - \pi$

Write the following numbers in numerical order from smallest to largest.

17. $-7, -3, 8, \pi, -2, 0$ **18.** $\dfrac{5}{6}, \dfrac{1}{2}, -\dfrac{2}{3}, -\dfrac{5}{4}, -\dfrac{3}{8}$

19. $|6 - 4|, -|-2|, |8 + 1|, -|3 - (-2)|$ **20.** $\sqrt{7}, -\sqrt{8}, -|\sqrt{16}|, |-\sqrt{12}|$

Write without absolute value bars.

21. $-|-6| + |3|$ **22.** $|-5| + |-9|$ **23.** $7 - |-8|$ **24.** $|-2| - |-7 + 3|$

Graph each of the following on a number line.

25. $x \geq -3$ **26.** $-4 < x \leq 6$ **27.** $x < -2$ **28.** $x \leq 1$

Use the order of operations to simplify.

29. $(-6 + 2 \cdot 5)(-2)$ **30.** $-4(-7 - 9 \div 3)$ **31.** $\dfrac{-8 + (-6)(-3) \div 9}{6 - (-2)}$ **32.** $\dfrac{20 \div 4 \cdot 2 \div 5 - 1}{-9 - (-3) - 12 \div 3}$

Solve each equation.

33. $4k - 11 + 3k = 2k - 1$ **34.** $2(k + 3) - 4(1 - k) = 6$ **35.** $2m + 7 = 3m + 1$

36. $4k - 2(k - 1) = 12$ **37.** $5y - 2(y + 4) = 3(2y + 1)$ **38.** $\dfrac{x - 3}{2} = \dfrac{2x + 1}{3}$

39. $\dfrac{p}{2} - \dfrac{3p}{4} = 8 + \dfrac{p}{3}$ **40.** $\dfrac{2r}{5} - \dfrac{r - 3}{10} = \dfrac{3r}{5}$ **41.** $\dfrac{2z}{5} - \dfrac{4z - 3}{10} = \dfrac{-z + 1}{10}$

42. $\dfrac{p}{p + 2} - \dfrac{3}{4} = \dfrac{2}{p + 2}$ **43.** $\dfrac{2m}{m - 3} = \dfrac{6}{m - 3} + 4$ **44.** $\dfrac{15}{k + 5} = 4 - \dfrac{3k}{k + 5}$

Solve for x.

45. $9px - 2 = x$ **46.** $11x - 5y = 2yx$ **47.** $3(x + 2b) + a = 2x - 6$

48. $9x - 11(k + p) = x(a - 1)$ **49.** $\dfrac{x}{m - 2} = kx - 3$ **50.** $r^2x - 5x = 3r^2$

Management *Solve each of the following problems.*

51. A stereo is on sale for 15% off. The sale price is $425. What was the original price?

52. To make a special mix for Valentine's Day, the owner of a candy store wants to combine chocolate hearts which sell for $5 per pound with candy kisses which sell for $3.50 per pound. How many pounds of each kind should be used to get 30 pounds of a mix that can be sold for $4.50 per pound?

53. A real estate firm invests the $100,000 proceeds from a sale in two ways. The first portion is invested in a shopping center that provides an annual return of 8%. The rest is invested in a small apartment building with an annual return of 5%. The firm wants an annual income of $6800 from these investments. How much should be put into each investment?

54. For a particular line of tools, a hardware store has average monthly costs in dollars of $C = 55x + 180$ per unit and corresponding revenue of $R = 100x$ per unit, where x is the number of units sold. How many units must be sold for the store to break even on those tools for the month?

Solve each of the following inequalities.

55. $-9x < 4x + 7$

56. $11y \geq 2y - 8$

57. $-5z - 4 \geq 3(2z - 5)$

58. $-(4a + 6) < 3a - 2$

59. $3r - 4 + r > 2(r - 1)$

60. $7p - 2(p - 3) \leq 5(2 - p)$

61. $5 \leq 2x - 3 \leq 7$

62. $-8 < 3a - 5 < -2$

Solve each equation.

63. $|a + 4| = 7$

64. $|-y + 2| = 3$

65. $\left|\dfrac{r - 5}{3}\right| = 6$

66. $\left|\dfrac{2 - 3a}{7}\right| = 9$

67. $\left|\dfrac{8r - 1}{2}\right| = 7$

68. $\left|\dfrac{3p - 5}{3}\right| = 2$

69. $|5r - 1| = |2r + 3|$

70. $|k + 7| = |3k - 8|$

Solve each inequality.

71. $|m| \leq 7$

72. $|r| < 2$

73. $|p| > 3$

74. $|z| > -1$

75. $|b| \leq -1$

76. $|5m - 8| \leq 2$

77. $|7k - 3| < 5$

78. $|2p - 1| > 2$

79. $|3r + 7| > 5$

80. $|2 - 5y| < -1$

Perform each of the following operations.

81. $(-9m^2 + 11m - 7) + (2m^2 - 12m + 4)$

82. $(3q^3 - 9q^2 + 6) + (4q^3 - 8q + 3)$

83. $(5r^4 - 6r^2 + 2r) - (-3r^4 + 2r^2 - 9r)$

84. $(-7z^3 + 8z^2 - z) - (z^3 + 4z^2 + 10z)$

85. $-(r^5 + 2r^4 + 8r^2) + 3(r^4 + 9r^2)$

86. $2(3y^6 - 9y^2 + 2y) - (5y^6 - 10y^2 - 4y)$

87. $(8y - 7)(2y + 7)$

88. $(9k + 2)(4k - 3)$

89. $(7z + 10y)(3z - 5y)$

90. $(2r + 11s)(4r - 9s)$

91. $(3k - 5m)^2$

92. $(4a - 3b)^2$

93. $(5x - 2)^3$

94. $(3m + 2)^4$

95. $(3w - 2)(5w^2 - 4w + 1)$

96. $(2k + 5)(3k^3 - 4k^2 + 8k - 2)$

Factor as completely as possible.

97. $7z^2 - 9z^3 + z$

98. $6m^3 + 4m^2 - 3m$

99. $12p^5 - 8p^4 + 20p^3$

100. $15y^7 - 25y^4 + 50y^3$

101. $r^2 + rp - 42p^2$

102. $z^2 - 6zk - 16k^2$

103. $6m^2 - 13m - 5$

104. $4k^2 + 11k - 3$

105. $3m^2 - 8m - 35$

106. $16p^2 + 24p + 9$

107. $144p^2 - 169q^2$

108. $81z^2 - 25x^2$

109. $8y^3 - 1$

110. $125a^3 + 216$

Perform each operation.

111. $\dfrac{6p}{7} \cdot \dfrac{28p}{15}$

112. $\dfrac{9r^2}{16} \div \dfrac{27r}{32}$

113. $\dfrac{m^2 - 2m - 3}{m(m - 3)} \div \dfrac{m + 1}{4m}$

114. $\dfrac{3m-9}{8m} \cdot \dfrac{16m+24}{15}$

115. $\dfrac{k^2+k}{8k^3} \cdot \dfrac{4}{k^2-1}$

116. $\dfrac{x^2+x-2}{x^2+5x+6} \div \dfrac{x^2+3x-4}{x^2+4x+3}$

117. $\dfrac{5}{2r} + \dfrac{6}{r}$

118. $\dfrac{3}{8y} + \dfrac{1}{2y} + \dfrac{7}{9y}$

119. $\dfrac{8}{r-1} - \dfrac{3}{r}$

120. $\dfrac{1}{4y} + \dfrac{8}{5y}$

121. $\dfrac{\frac{1}{p}+\frac{1}{q}}{1-\frac{1}{pq}}$

122. $\dfrac{\frac{2}{r}-\frac{3}{5}}{\frac{1}{r}+\frac{4}{5}}$

Simplify each of the following. Write all answers without negative exponents. Assume all variables represent positive real numbers.

123. 4^{-2} **124.** 6^{-1} **125.** 5^{-3} **126.** 8^{-2} **127.** $\left(\dfrac{3}{4}\right)^{-3}$

128. $\left(\dfrac{5}{8}\right)^{-2}$ **129.** $6^4 \cdot 6^{-3}$ **130.** $7^4 \cdot 7^{-6}$ **131.** $\dfrac{8^{-5}}{8^{-3}}$ **132.** $\dfrac{6^{-2}}{6^3}$

133. $\dfrac{9^4 \cdot 9^{-5}}{(9^{-2})^2}$ **134.** $\dfrac{k^4 \cdot k^{-3}}{(k^{-2})^{-3}}$ **135.** $4^{-1} + 2^{-1}$ **136.** $3^{-2} + 3^{-1}$ **137.** $125^{2/3}$

138. $128^{3/7}$ **139.** $9^{-5/2}$ **140.** $\left(\dfrac{144}{49}\right)^{-1/2}$ **141.** $\dfrac{5^{1/3} \cdot 5^{1/2}}{5^{3/2}}$ **142.** $\dfrac{2^{3/4} \cdot 2^{-1/2}}{2^{1/4}}$

143. $(3a^2)^{1/2} \cdot (3^2 a)^{3/2}$ **144.** $(4p)^{2/3} \cdot (2p^3)^{3/2}$ **145.** $\sqrt[3]{27}$ **146.** $\sqrt[4]{625}$

147. $\sqrt[5]{-32}$ **148.** $\sqrt[6]{-64}$ **149.** $\sqrt{24}$ **150.** $\sqrt{63}$

151. $\sqrt[3]{54p^3q^5}$ **152.** $\sqrt[4]{64a^5b^3}$ **153.** $\sqrt{\dfrac{5n^2}{6m}}$ **154.** $\sqrt{\dfrac{3x^3}{2z}}$

155. $\sqrt[3]{\sqrt{5}}$ **156.** $\sqrt[4]{\sqrt[3]{10}}$ **157.** $2\sqrt{3} - 5\sqrt{12}$ **158.** $8\sqrt{7} + 2\sqrt{28}$

159. $5\sqrt[3]{16r^2s^3} - s\sqrt[3]{54r^2}$ **160.** $2\sqrt[4]{16mn^2} - \sqrt[4]{81mn^2}$ **161.** $(\sqrt{5}-1)(\sqrt{5}+1)$ **162.** $(\sqrt{7}-\sqrt{3})(\sqrt{7}+\sqrt{3})$

163. $(2\sqrt{5}-\sqrt{3})(\sqrt{5}+2\sqrt{3})$ **164.** $(4\sqrt{7}+\sqrt{2})(3\sqrt{7}-\sqrt{2})$ **165.** $\dfrac{\sqrt{2}}{1+\sqrt{3}}$ **166.** $\dfrac{4+\sqrt{2}}{4-\sqrt{5}}$

Solve each equation. Give only real number solutions.

167. $x^2 = 7$ **168.** $p^2 = 23$ **169.** $(b+7)^2 = 5$ **170.** $(2p+1)^2 = 7$

171. $x^2 - 4x + 3 = 0$ **172.** $y^2 + 9y + 8 = 0$ **173.** $2p^2 + 3p = 2$ **174.** $2y^2 = 15 + y$

175. $x^2 - 2x = 2$ **176.** $r^2 + 4r = 1$ **177.** $2m^2 - 12m = 11$ **178.** $9k^2 + 6k = 2$

179. $2a^2 + a - 15 = 0$ **180.** $12x^2 = 8x - 1$ **181.** $2q^2 - 11q = 21$ **182.** $3x^2 + 2x = 16$

183. $6k^4 + k^2 = 1$ **184.** $21p^4 = 2 + p^2$ **185.** $2x^4 = 7x^2 + 15$ **186.** $3m^4 + 20m^2 = 7$

187. $3 = \dfrac{13}{z} + \dfrac{10}{z^2}$ **188.** $1 + \dfrac{13}{p} + \dfrac{40}{p^2} = 0$

189. $2 + \dfrac{15}{x-1} + \dfrac{18}{(x-1)^2} = 0$ **190.** $\dfrac{5}{(2t+1)^2} = 2 - \dfrac{9}{2t+1}$

Solve each equation for the specified variable.

191. $p = \dfrac{E^2R}{(r + R)^2}$ for r

192. $p = \dfrac{E^2R}{(r + R)^2}$ for E

193. $K = s(s - a)$ for s

194. $kz^2 - hz - t = 0$ for z

Solve each inequality.

195. $r^2 + r - 6 < 0$

196. $y^2 + 4y - 5 \geq 0$

197. $2z^2 + 7z \geq 15$

198. $3k^2 \leq k + 14$

199. $8a^2 + 10a > 3$

200. $9m^2 + 13m > 10$

201. $3r^2 - 5r \leq 0$

202. $2q^2 + q > 0$

203. $\dfrac{m + 2}{m} \leq 0$

204. $\dfrac{q - 4}{q + 3} > 0$

205. $\dfrac{5}{p + 1} > 2$

206. $\dfrac{6}{a - 2} \leq -3$

207. $\dfrac{2}{r + 5} \leq \dfrac{3}{r - 2}$

208. $\dfrac{1}{z - 1} > \dfrac{2}{z + 1}$

Work each applied problem.

209. A recreation director wants to fence off a rectangular playground beside an apartment building. The building forms one boundary, so she needs to fence only the other three sides. The area of the playground is to be 11,250 square meters. She has enough material to build 325 meters of fence. Find the length and width of the playground.

210. Two cars leave an intersection at the same time. One travels north and the other heads west traveling 10 mph faster. After 1 hour they are 50 miles apart. What were their speeds?

The Upjohn Company has a subsidiary that buys seeds from farmers and then resells them. Each spring the firm contracts with farmers to grow the seeds. The firm must decide on the number of acres that it will contract for. The problem faced by the company is that the demand for seeds is not constant, but fluctuates from year to year. Also, the number of tons of seeds produced per acre varies, depending on weather and other factors. In an attempt to decide the number of acres that should be planted in order to maximize profits, a company mathematician created a model of the variables involved in determining the number of acres to plant.

The analysis of this model required advanced methods that we will not go into. We can however, give the conclusion. If $500 \leq AX + Q \leq 1500$ tons, the number of acres that will maximize profit in the long run is found by solving the equation

$$\frac{AX + Q}{1000} - \frac{1}{2} = \frac{(S - C_p)X - C_A}{(S - C_p + C_c)X}$$

for A. The variables in the equation are:

A = number of acres of land contracted by the company

X = quantity of seed produced per acre of land (in tons)

Q = quantity of seed in inventory from previous years

S = selling price per ton of seed

C_p = variable cost (production, marketing, etc.) per ton of seed

C_c = cost to carry over 1 ton of seed from previous year

C_A = variable cost per acre of land.

To advise management of the number of acres of seed to contract for, the mathematician studied past records to find the values of the various variables. From these records and from predictions of future trends, it was concluded that $S = \$10,000$ per ton, $X = .1$ ton per acre (on the average), $Q = 200$ tons, $C_p = \$5000$ per ton, $C_A = \$100$ per acre, $C_c = \$3000$ per ton.

EXERCISES

1. Solve the equation for A.

2. How many acres should be planted?

3. How many tons of seed will be produced?

4. Find the total revenue that will be received from the sale of the seeds.

---

*Based on work by David P. Rutten, Senior Mathematician, The Upjohn Company, Kalamazoo, Michigan.

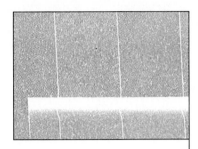

CHAPTER 2

Functions and Graphs

Functions are an extremely useful way of describing many real-world situations in which the value of one quantity varies with, depends on, or determines the value of another. For example, the graph in Figure 2.1 shows the temperature (in degrees Fahrenheit) at each time during a single day (with time measured in hours after midnight).

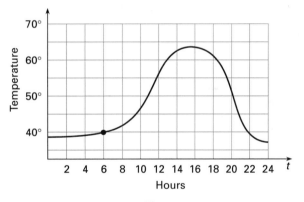

FIGURE 2.1

The graph relates the two variables, time and temperature. With each time t for $0 \le t \le 24$, the graph associates a temperature. For instance, the time $t = 6$ is associated with the temperature $40°$ since the point $(6, 40)$ is on the graph. This process of associating with each value of one variable (time) a specific value of a second variable (temperature) is an example of a *function*.

The first part of this chapter presents the basic ideas of functions, functional notation, and graphs of functions. In the last three sections we study the special case of linear functions (those whose graphs are straight lines). Nonlinear functions will be considered in Chapters 3 and 4.

2.1 FUNCTIONS

The cost of leasing an office in an office building depends on the number of square feet of space in the office. The relationship between area and rent in a particular building is shown in this table.

Area (square feet)	Rent (dollars)
350	630
690	1242
850	1530
1000	1800
1400	2520

This is an example of a *function* which relates the two variables area and rent. The area is the **independent variable** and the rent (which depends on the area) is the **dependent variable.**

Another example of a function is a mutual fund that provides a 6% annual return, which means that an investment of x dollars produces a return of 6% of x (that is, $.06x$) dollars. For instance, an investment of $7200 dollars produces a return of

$$6\% \text{ of } \$7200 = .06(\$7200) = \$432.$$

The general situation can be described by the formula

$$y = .06x,$$

in which x is the investment (in dollars) and y is the annual return (in dollars). Here x is the independent variable and y is the dependent variable (because the value of x determines the value of y).

Each of these examples involves two sets of numbers.

First Set	Second Set
areas	rents
investments	returns

In each case, there is a *rule* which assigns to each number in the first set a specific number in the second set (the table is the rule for areas and rents and the formula is the rule for investment and return). In the general case, we have the following definition.

A **function** is a rule that assigns to each element of a set X exactly one element from a set Y.

►**EXAMPLE 1** (a) The correspondence in Figure 2.2 is a function from the set $X = \{12, 0, -3, 7, 11\}$ to the set $Y = \{30, 27, 9, 3, 11\}$. This function assigns to the number 12 in X the number 27 in Y. In general, it assigns to each number x in X the number y in Y that is at the end of the arrow that begins at x.

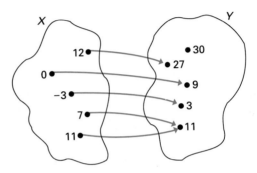

FIGURE 2.2

Note that *every* number in X is assigned to exactly one number in Y (although some numbers in X may be assigned to the same number in Y). Some numbers in Y may not have any number in X assigned to them. The numbers in the set X are the values of the independent variable and the numbers in Y the values of the dependent variable.

(b) The correspondence in Figure 2.3 is *not* a function because the number 28 in X is assigned to *two* numbers in Y (19 and 27). ◄

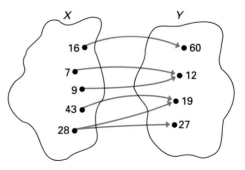

FIGURE 2.3

1 Do the following define functions?

(a) The correspondence defined by the rule $y = x^2 + 5$.

(b) The $\boxed{1/x}$ key on a calculator

(c) A computer, x and the users of the computer, y

Answer:

(a) Yes

(b) Yes

(c) No

▶**EXAMPLE 2**   Which of the following describe functions?

(a) The optical reader at the checkout counter of the supermarket that converts codes to prices.

    For each code, the reader produces exactly one price, so this is a function.

(b) The correspondence given by entering a number in a calculator and pressing the $\boxed{x^2}$ key.

    This is a function because the calculator produces just one x^2 value for each x value entered. The possible input numbers are the values of the independent variable and the possible output numbers are the values of the dependent variable.

(c) Assign to each number x the number y given by this table.

x	1	1	2	2	3	3
y	3	-3	5	-5	8	-8

    Since at least one x-value corresponds to more than one y-value, this table does not define a function.

(d) Assign to each number x the number y given by this equation: $y = 3x - 5$.

    Since the equation determines a unique value of y for each value of x, it defines a function. ◀ **1**

    The preceding examples show that there are many ways to define a function. Almost all of the functions used in this book will be defined by equations, as in part (d) of Example 2, and x *will be assumed to represent the independent variable*.

    As we have seen, each function involves a correspondence between a set of input numbers (the values of the independent variable) and a set of output numbers (the values of the dependent variable). These sets are given special names.

    The set of all values of the independent variable of a function is called the **domain** of the function.

    The set of all corresponding values of the dependent variable is called the **range** of the function.

    The domain and range may or may not be the same set. For instance, in the function given by the x^2 key on a calculator, the domain consists of all decimal numbers (positive, negative, or 0) that can be entered in the calculator, but the range consists only of *nonnegative* decimals (since $x^2 \geq 0$ for every x). In the function given by the equation $y = 3x - 5$, both the domain and range are the set of all real numbers because of the following *agreement on domains*.

    Unless otherwise stated, assume that the domain of all functions defined by an equation is the largest set of real numbers that are meaningful replacements for the independent variable.

2 Do the following define functions?

(a) $y = -6x + 1$

(b) $y = x^2$

(c) $x = y^2 - 1$

(d) $y < x + 2$

Answer:

(a) Yes

(b) Yes

(c) No

(d) No

3 Give the domain and range.

(a) $y = 3x + 1$

(b) $y = x^2$

Answer:

(a) Both are the set of all real numbers

(b) Domain: set of all real numbers; range $[0, \infty)$

For example, suppose

$$y = \frac{-4x}{2x - 3}.$$

Any real number can be used for x except $x = 3/2$, which makes the denominator equal 0. By the agreement on domains, the domain of this function is the set of all real numbers except 3/2.

▶**EXAMPLE 3**   Decide whether each of the following equations defines a function. Give the domain and range of any functions.
(a) $y = -4x + 11$
  For a given value of x, calculating $-4x + 11$ produces exactly one value of y. (For example, if $x = -7$, then $y = -4(-7) + 11 = 39$.) Since one value of the independent variable leads to exactly one value of the dependent variable, $y = -4x + 11$ is a function. Both x and y may take on any real-number values, so both the domain and range are the set of all real numbers, written $(-\infty, \infty)$.
(b) $y^2 = x$
  Suppose $x = 36$. Then $y^2 = x$ becomes $y^2 = 36$, from which $y = 6$ or $y = -6$. Since one value of the independent variable x can lead to two values of the dependent variable y, $y^2 = x$ does not define a function.   ◀ **2**

▶**EXAMPLE 4**   Find the domain and range for each of the functions defined as follows.
(a) $y = x^4$
  Any number may be raised to the fourth power so the domain is $(-\infty, \infty)$. Since $x^4 \geq 0$ for every value of x, the range is the interval $[0, \infty)$.
(b) $y = \sqrt{6 - x}$.
  For y to be a real number, $6 - x$ must be nonnegative. This happens only when $6 - x \geq 0$, or $6 \geq x$, making the domain the interval $(-\infty, 6]$. The range is $[0, \infty)$ because $\sqrt{6 - x}$ is always nonnegative.
(c) $y = \dfrac{1}{x + 3}$
  Since the denominator cannot be 0, $x \neq -3$ and the domain consists of all numbers in the intervals.

$$(-\infty, -3) \quad \text{or} \quad (-3, \infty).$$

Because the numerator can never be 0, $y \neq 0$. There are no other restrictions on y, so the range consists of the numbers in $(-\infty, 0)$ or $(0, \infty)$.   ◀ **3**

4 Let $f(x) = -3x - 8$. Find the following.

(a) $f(0)$

(b) $f(-1)$

(c) $f(5)$

Answer:

(a) -8

(b) -5

(c) -23

FUNCTIONAL NOTATION The letters f, g, and h frequently are used to represent functions. For example, f might be used to name the function defined by $y = 5 - 3x$. For a given value of x in the domain of a function f, there is exactly one corresponding value of y in the range. To emphasize that y is obtained by applying function f to the element x, replace y with the symbol $f(x)$, read "f of x" or "f at x." Here x is the independent variable; either y or $f(x)$ represents the dependent variable.

Using $f(x)$ to replace y in the equation $y = 5 - 3x$ gives

$$f(x) = 5 - 3x.$$

If 2 is chosen as a value of x, $f(x)$ becomes

$$f(2) = 5 - 3(2)$$
$$f(2) = -1.$$

In a similar manner,

$$f(-4) = 5 - 3(-4) = 17, \quad f(0) = 5, \quad f(-6) = 23,$$

and so on. **4**

It may be helpful to think of a function as a machine—such as a computer or calculator—that takes an input x from the domain and uses it to produce an output $f(x)$, as shown in Figure 2.4.

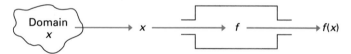

FIGURE 2.4

In this context, think of a function rule, such as

$$f(x) = 3x + 4 \quad \text{or} \quad g(x) = \sqrt{x^2 + 1},$$

as a set of directions telling the machine what should be done to each input number x. For instance, the rule of g says "square the number, add 1, and take the square root of the result."

▶**EXAMPLE 5** Let $g(x) = \sqrt{x^2 + 1}$. Find each of the following.

(a) $g(3)$

Replace x with 3.

$$g(3) = \sqrt{3^2 + 1} = \sqrt{10}$$

5 Let $f(x) = 5x^2 - 2x + 1$.
Find the following.

(a) $f(1)$

(b) $f(3)$

(c) $f(m)$

Answer:

(a) 4

(b) 40

(c) $5m^2 - 2m + 1$

(b) $g(a)$

Follow the directions given by the rule of g by replacing x by a.

$$g(a) = \sqrt{a^2 + 1}$$

(c) $g(a + 2)$

To see what g does to the number $a + 2$, replace x by $a + 2$ in the rule of g.

$$g(a + 2) = \sqrt{(a + 2)^2 + 1}$$
$$= \sqrt{(a^2 + 4a + 4) + 1}$$
$$= \sqrt{a^2 + 4a + 5} \quad \blacktriangleleft$$

The following use of functional notation is quite common. It won't cause confusion if you remember that the rule of a function is a set of directions.

To determine what the function does to a quantity, replace x by that quantity in the rule of the function.

▶**EXAMPLE 6** Let $g(x) = -x^2 + 4x - 5$. Find each of the following.

(a) $g(x + h)$

Replace x by the quantity $x + h$ in the rule of g.

$$g(x + h) = -(x + h)^2 + 4(x + h) - 5$$
$$= -(x^2 + 2xh + h^2) + (4x + 4h) - 5$$
$$= -x^2 - 2xh - h^2 + 4x + 4h - 5$$

(b) $g\left(\dfrac{2}{x}\right)$

$$g\left(\frac{2}{x}\right) = -\left(\frac{2}{x}\right)^2 + 4\left(\frac{2}{x}\right) - 5 = -\frac{4}{x^2} + \frac{8}{x} - 5 \quad \blacktriangleleft$$

Caution Functional notation is *not* the same as ordinary algebraic notation. You cannot simplify an expression such as $f(x + h)$ by writing $f(x) + f(h)$. To see why, consider the function $f(x) = x^2$.

$$f(3 + 2) = f(5) = 5^2 = 25$$

But $f(3) = 3^2 = 9$ and $f(2) = 2^2 = 4$, so that

$$f(3) + f(2) = 9 + 4 = 13.$$

Hence, $f(3 + 2) \neq f(3) + f(2)$. **5**

6 A developer estimates that the total cost of building x large apartment complexes in a year is approximated by

$$A(x) = x^2 + 80x + 60,$$

where $A(x)$ represents the cost in hundred thousands of dollars. Find the cost of building

(a) 4 complexes;

(b) 10 complexes.

Answer:

(a) \$39,600,000

(b) \$96,000,000

▶**EXAMPLE 7** Suppose the sales of a small company have been estimated to be

$$S(x) = 125 + 80x,$$

where $S(x)$ represents the total sales in thousands of dollars in year x, with $x = 0$ representing 1990. Estimate the sales in each of the following years.

(a) 1990

Since $x = 0$ corresponds to 1990, the sales for 1990 are given by $S(0)$. Substituting 0 for x gives

$$S(0) = 125 + 80(0) \qquad \text{Let } x = 0$$
$$= 125.$$

Since $S(x)$ represents sales in thousands of dollars, the sales would be estimated as 125×1000, or \$125,000, in 1990.

(b) 1994

To estimate sales in 1994, let $x = 4$.

$$S(4) = 125 + 80(4) = 125 + 320 = 445,$$

so that sales should be about \$445,000 in 1994. ◀ **6**

2.1 EXERCISES

Which of the following rules define y as a function of x? (See Examples 1–3.)

1. $y = 8x - 3$ **2.** $y = -4 + x$ **3.** $y = x^2$ **4.** $y = \sqrt{x}$ **5.** $x = |y|$ **6.** $x = y^4 - 1$

7. $y = \dfrac{1}{x + 3}$ **8.** $x = \dfrac{-2}{y}$ **9.** $x = \dfrac{4}{y - 1}$ **10.** $y = \dfrac{1}{2 - x}$ **11.** $x = \sqrt{y + 2}$ **12.** $y = |x + 4|$

Give the domain and range of each function defined as follows. (See Examples 3 and 4.)

13. $f(x) = 2x$ **14.** $f(x) = x + 2$ **15.** $f(x) = x^4$ **16.** $f(x) = (x - 1)^2$

17. $f(x) = |2 - x|$ **18.** $f(x) = |x - 3|$ **19.** $f(x) = \dfrac{1}{1 - x}$ **20.** $f(x) = \dfrac{-2}{x - 4}$

For each of the following functions, find (a) $f(4)$, (b) $f(-3)$, (c) $f(0)$, (d) $f(a)$. (See Example 5.)

21. $f(x) = 3x + 2$ **22.** $f(x) = 5x - 6$ **23.** $f(x) = -2x - 4$ **24.** $f(x) = -3x + 7$

25. $f(x) = 6$ **26.** $f(x) = 0$ **27.** $f(x) = 2x^2 + 4x$ **28.** $f(x) = x^2 - 2x$

29. $f(x) = -x^2 + 5x + 1$ **30.** $f(x) = -x^2 - x + 5$ **31.** $f(x) = \sqrt{x + 3}$ **32.** $f(x) = \sqrt{5 - x}$

For each of the following functions, find (a) $f(a)$, (b) $f(-r)$, (c) $f(m + 3)$, (d) $f(a + b)$. (See Examples 5 and 6.)

33. $f(x) = 2x - 3$ **34.** $f(x) = 3x + 4$ **35.** $f(x) = 2x^2 - x$ **36.** $f(x) = 6 - x^2$

37. $f(x) = x^3 + 1$ **38.** $f(x) = 2 - x^3$ **39.** $f(x) = \dfrac{3}{x - 1}$ **40.** $f(x) = \dfrac{-1}{2 + x}$

41. $f(x) = \sqrt{2x}$ **42.** $f(x) = \sqrt{4 - x}$

Management   *Work the following exercises. (See Example 7.)*

43. To rent a midsized car from Avis costs $40 per day or fraction of a day. If you pick up the car in Boston and drop it off in Utica, there is a fixed $40 charge. Let $C(x)$ represent the cost of renting the car for x days, taking it from Boston to Utica. Find each of the following.

(a) $C\left(\dfrac{3}{4}\right)$ **(b)** $C\left(\dfrac{9}{10}\right)$ **(c)** $C(1)$

(d) $C\left(1\dfrac{5}{8}\right)$ **(e)** $C\left(2\dfrac{1}{9}\right)$

44. A chain-saw rental firm charges $7 per day or fraction of a day to rent a saw, plus a fixed fee of $4 for resharpening the blade. Let $S(x)$ represent the cost of renting a saw for x days. Find each of the following.

(a) $S\left(\dfrac{1}{2}\right)$ **(b)** $S(1)$ **(c)** $S\left(1\dfrac{1}{4}\right)$ **(d)** $S\left(3\dfrac{1}{2}\right)$

(e) $S(4)$ **(f)** $S\left(4\dfrac{1}{10}\right)$ **(g)** $S\left(4\dfrac{9}{10}\right)$

45. Suppose the revenue of a small company that sells by mail is approximated by

$$R(t) = 1000 + 50(t + 1),$$

where $R(t)$ represents revenue in thousands of dollars. Here t is time in years, with $t = 0$ representing the year 1991. Find the estimated revenue in each of the following years. (See Example 5.)

(a) 1991   **(b)** 1992   **(c)** 1993   **(d)** 1994

The following functions have multipart rules. To find $f(-3)$, for example, you must first determine which part of the rule applies to $x = -3$. In each exercise, find $f(-3)$, $f(0)$, $f(2)$, $f(4)$, and $f(10)$.

46. $f(x) = \begin{cases} 3x & \text{if } x \le 0 \\ 15 & \text{if } 0 < x < 4 \\ x^2 & \text{if } x \ge 4 \end{cases}$

47. $f(x) = \begin{cases} x^2 & \text{if } x < -1 \\ x + 2 & \text{if } -1 \le x \le 4 \\ 10 & \text{if } x > 4 \end{cases}$

2.2 GRAPHS OF FUNCTIONS

It is sometimes convenient to think of a function as a set of ordered pairs, in which each number in the domain is paired with its corresponding functional value in the range. For instance, some of the ordered pairs that make up the function $f(x) = x^2$ are

$$(2, 4), \quad (-3, 9), \quad (5, 25), \quad (½, ¼), \quad \text{and} \quad (0, 0).$$

In each pair, the first coordinate is a number in the domain and the second coordinate is the value of the function f at that number.

We can draw a picture of a function, called its **graph,** by plotting each of the ordered pairs of the function in the plane. To do this, we use a **Cartesian coordinate system,** as shown in Figure 2.5. The horizontal number line, or **x-axis,** represents the elements from the domain of the function, and the vertical, or **y-axis,** represents the elements from the range. The point where the number lines cross is the zero point on both of these number lines; this point is called the **origin.**

1 Locate $(-1, 6)$, $(-3, -5)$, $(4, -3)$, $(0, 2)$, and $(-5, 0)$ on a coordinate system.

Answer:

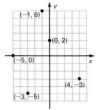

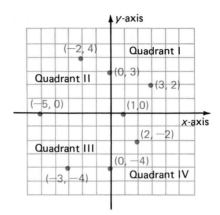

FIGURE 2.5

Each point in a Cartesian coordinate system corresponds to an ordered pair of real numbers. Several points and their corresponding ordered pairs are shown in Figure 2.5. In the point $(-2, 4)$, for example, -2 is the **x-coordinate** and 4 is the **y-coordinate.** From now on, instead of referring to "the point corresponding to the ordered pair $(-2, 4)$," we will say "the point $(-2, 4)$."

The x-axis and y-axis divide the plane into four parts, or **quadrants.** For example, Quadrant I includes points whose x- and y-coordinates are both positive. The quadrants are numbered as shown in Figure 2.5. The points of the axes themselves belong to no quadrant.   **1**

Most functions we shall study have infinitely many numbers in their domains, and hence, infinitely many points on their graphs. Since you can only plot a finite number of points, you should follow this procedure.

1. Select a few numbers in the domain of f (include both negative and positive ones when possible) and compute the corresponding values of $f(x)$.
2. Plot the points $(x, f(x))$ computed in step one. Use these points and any other information you may have about the function to make an "educated guess" about the shape of the entire graph.
3. Unless you have information to the contrary, assume that the graph is continuous (unbroken) wherever it is defined.

▶**EXAMPLE 1**   Graph the function $f(x) = 3 - 2x$.

If $x = -2$, then $f(-2) = 3 - 2(-2) = 7$, giving the corresponding ordered pair $(-2, 7)$. Using additional values of x in the equation gives the ordered pairs listed below.

x	-2	-1	0	1	2	3
y	7	5	3	1	-1	-3
Ordered Pair	$(-2, 7)$	$(-1, 5)$	$(0, 3)$	$(1, 1)$	$(2, -1)$	$(3, -3)$

2 Let $f(x) = 4x - 7$. Graph this function.

Answer:

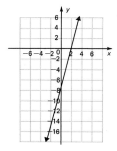

These ordered pairs are graphed in Figure 2.6(a). Their pattern suggests that the entire graph is a straight line, part of which is shown in Figure 2.6(b). Later in this chapter we shall provide a proof of this fact. The arrows in the graph indicate that the line continues on forever in the indicated directions. ◀ **2**

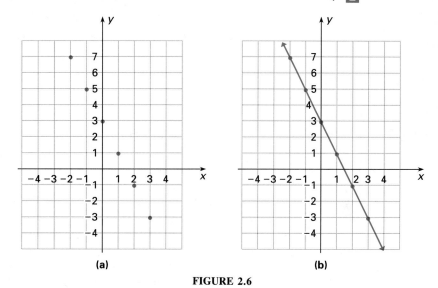

(a)   (b)

FIGURE 2.6

When applying the graphing technique used in Example 1, it is often a good idea to find the points where the graph crosses the axes. An ***x*-intercept** of a graph is the *x*-coordinate of the point where the graph crosses the *x*-axis (the *y*-coordinate of this point is 0 since it is on the axis). Similarly, a ***y*-intercept** is the *y*-coordinate of the point where the graph crosses the *y*-axis (the *x*-coordinate of this point is necessarily 0).* For example, the intercepts are marked on the graphs in Figure 2.7.

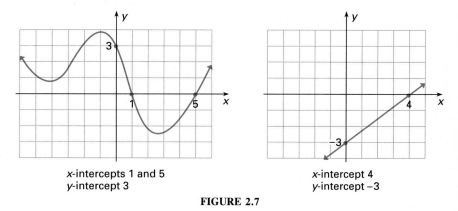

x-intercepts 1 and 5   *x*-intercept 4
y-intercept 3   *y*-intercept −3

FIGURE 2.7

*Some instructors prefer to define the intercepts to be the *points* where the graph crosses the axes. Either definition provides essentially the same information.

▶ **EXAMPLE 2**   Graph each function.
(a) $f(x) = |x|$

First find the intercepts. To find the y-intercepts, set $x = 0$ in the rule of the function (because the y-axis consists of points with x-coordinate 0).

$$y = |x|$$
$$y = |0| = 0$$

To find the x-intercepts, set $y = 0$ (because points on the x-axis have 0 y-coordinates).

$$y = |x|$$
$$0 = |x|, \text{ which implies } x = 0$$

Both intercepts give the ordered pair (0, 0). Now choose other values for x, the independent variable, some positive and some negative. For example, if $x = 2$, $y = 2$. Also if $x = -2$, $y = 2$. Continuing in this way, we produce the following table of values.

x	-3	-2	-1	0	1	2	3
y	3	2	1	0	1	2	3

Plotting the points from the table and connecting them gives the graph in Figure 2.8. Note that the domain is $(-\infty, \infty)$ and the range is $[0, \infty)$.

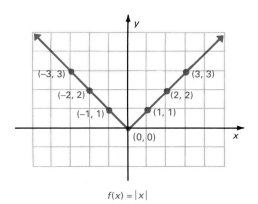

$f(x) = |x|$

FIGURE 2.8

(b) $f(x) = |3x + 4|$

Find the intercepts as in part (a). If $x = 0$, $y = 4$, giving the point (0, 4). If $y = 0$,

3 Graph each function.

(a) $f(x) = 2 - |x|$

(b) $f(x) = |5x - 7|$

Answer:

(a)

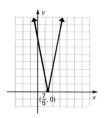

(b)

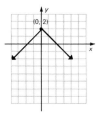

$$f(x) = y = |3x + 4|$$
$$0 = |3x + 4|$$
$$3x + 4 = 0$$
$$x = -\frac{4}{3}.$$

The x-intercept is $-4/3$, which gives the point $(-4/3, 0)$. Find a few other ordered pairs with x-values on either side of the x-intercept $-4/3$, as in the table in Figure 2.9. Plot these points and complete the graph as in Figure 2.9. This function also has domain $(-\infty, \infty)$ and range $[0, \infty)$. ◀ **3**

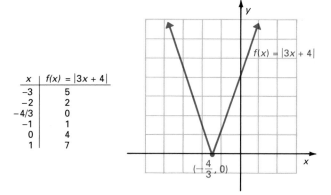

x	$f(x) = |3x + 4|$
-3	5
-2	2
$-4/3$	0
-1	1
0	4
1	7

FIGURE 2.9

The functions graphed in Example 2 are called **absolute value functions.** The graphs of the absolute value functions in Example 2 are made up of portions of two different lines. Such functions are called **piecewise functions.** Other piecewise functions are defined with different equations for different parts of the domain.

▶**EXAMPLE 3** Graph each function.

(a) $f(x) = \begin{cases} x + 1 & \text{if } x \le 2 \\ -2x + 7 & \text{if } x > 2 \end{cases}$

For $x \le 2$, use the equation $y = x + 1$ to find the ordered pairs in the first table below, including the endpoint value $x = 2$ for one of the pairs. For $x > 2$, use the equation $y = -2x + 7$ to find the ordered pairs in the second table.

$x \le 2$

x	-2	0	2
$y = x + 1$	-1	1	3

$x > 2$

x	2	3	4
$y = -2x + 7$	3	1	-1

4 Graph $f(x)$, where

$$f(x) = \begin{cases} -2x + 5 & \text{if } x < 2 \\ x - 4 & \text{if } x \ge 2. \end{cases}$$

Answer:

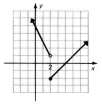

Note that even though 2 is not in the interval $x > 2$, we found the ordered pair for that endpoint because the graph will extend right up to that point. Since this endpoint (2, 3) agrees with the endpoint for the interval $x \le 2$, the two parts of the graph are joined at that point as shown in Figure 2.10.

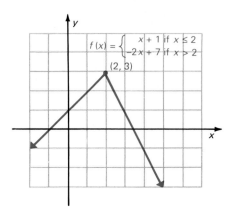

FIGURE 2.10

(b) $f(x) = \begin{cases} x & \text{if } x \le 0 \\ x + 1 & \text{if } x > 0 \end{cases}$

The ordered pairs $(-2, -2)$, $(-1, -1)$, and $(0, 0)$ satisfy $y = x$. The endpoint is $(0, 0)$. For $y = x + 1$, some ordered pairs are $(1, 2)$ and $(2, 3)$. The ordered pair $(0, 1)$ is an endpoint, but is *not* part of the graph, as indicated by the open circle in Figure 2.11. ◀ **4**

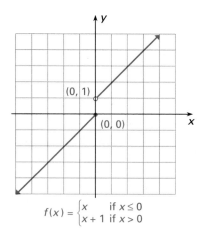

FIGURE 2.11

5 Graph $y = [\frac{1}{2}x + 1]$.

Answer:

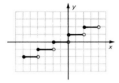

6 Assume that the post office charges 30¢ per ounce, or fraction of an ounce, to mail a letter. Graph the ordered pairs (ounces, cost).

Answer:

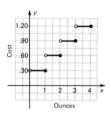

The **greatest integer function,** written $y = [x]$, is defined by saying that $[x]$ is the greatest integer less than or equal to x. For example, $[8] = 8$, $[-5] = -5$, $[\pi] = 3$, $[12\frac{1}{9}] = 12$, $[-2.001] = -3$, and so on.

▶**EXAMPLE 4**   Graph $y = [x]$.

For x in the interval $0 \le x < 1$, the value of $[x] = 0$. For values of x where $1 \le x < 2$, $[x] = 1$, and so on. Thus, the graph, as shown in Figure 2.12, consists of a series of line segments. In each case, the left endpoint of the segment is included, and the right endpoint is excluded. The domain of the function is the set of all real numbers, and the range is the set of integers.   ◀ **5**

The greatest integer function, graphed in Figure 2.12, is an example of a **step function.**

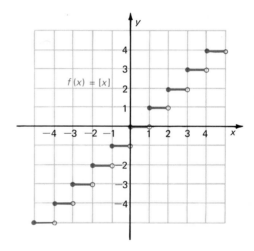

FIGURE 2.12

▶**EXAMPLE 5**   An overnight delivery service charges $25 for a package weighing up to 2 pounds. For each additional pound there is an additional charge of $3. Let $D(x)$ represent the cost to send a package weighing x pounds. Graph $D(x)$ for x in the interval (0, 6].

For x in the interval (0, 2], $y = 25$. For x in (2, 3], $y = 25 + 3 = 28$. For x in (3, 4], $y = 28 + 3 = 31$, and so on. The graph is shown in Figure 2.13.   ◀ **6**

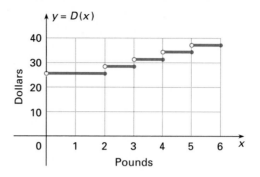

FIGURE 2.13

► **EXAMPLE 6** Business graphs are often made up of portions of straight lines. For example, the graph of Figure 2.14 shows one prediction of interest rates for two kinds of real estate loans for the period of the 1980s. (Interest rates from 1966 are included for comparison.) To get this graph, points are plotted for different years and connected with straight line segments. The graph suggests that mortgages for single family homes (the lower graph) averaged about 14% in 1980, a little over 11% in 1985, and 10% in 1989. ◄

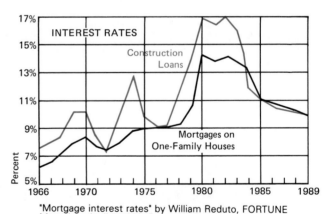

"Mortgage interest rates" by William Reduto, FORTUNE Magazine, April 7, 1980. Reprinted by permission.

FIGURE 2.14

The graphs of many functions do not consist of straight line segments. In such cases it may be necessary to plot additional points to determine a pattern.

▶**EXAMPLE 7**   Graph $g(x) = \sqrt{x + 1}$.

Since the rule of the function is defined only when $x + 1 \geq 0$ (that is, when $x \geq -1$), the domain of g is the interval $[-1, \infty)$. By using a calculator, we obtain the table and graph in Figure 2.15. ◀

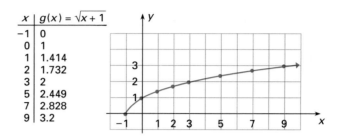

x	$g(x) = \sqrt{x + 1}$
−1	0
0	1
1	1.414
2	1.732
3	2
5	2.449
7	2.828
9	3.2

FIGURE 2.15

If a graph is to represent a function, each value of x from the domain must lead to exactly one value of y. In the graph in Figure 2.16, the domain value x_1 leads to *two* y-values, y_1 and y_2. Since the given x-value corresponds to two different y-values, this is not the graph of a function. This example suggests the **vertical line test** for the graph of a function.

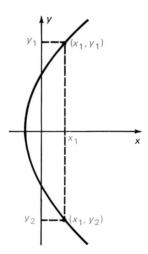

FIGURE 2.16

Vertical Line Test

If a vertical line intersects a graph at more than one point, the graph is not the graph of a function.

7 Does this graph represent a function?

Answer:
No

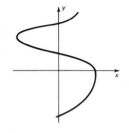

▶**EXAMPLE 8**   Use the vertical line test to decide which of the graphs in Figure 2.17 are graphs of functions.

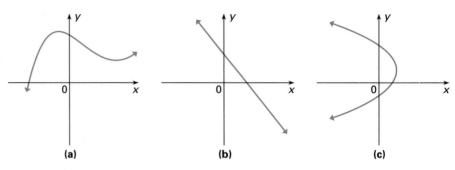

(a) (b) (c)

FIGURE 2.17

(a) Every vertical line intersects this graph in at most one point, so this is the graph of a function.
(b) Again, each vertical line intersects the graph in at most one point, showing that this is the graph of a function.
(c) It is possible for a vertical line to intersect the graph in part (c) twice. This is not the graph of a function. ◀ **7**

2.2   EXERCISES

Which of the following are graphs of functions? (See Example 8.)

1.

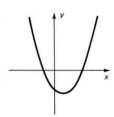

2.

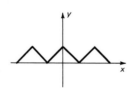

3.

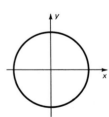

4.

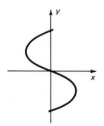

5.

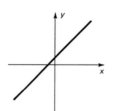

6.

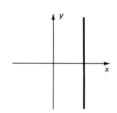

Graph each of the following. (See Examples 1 and 2.)

7. $y = |x + 1|$ **8.** $y = |x - 1|$ **9.** $y = |2 - x|$ **10.** $y = |-3 - x|$ **11.** $y = |5x + 4|$

12. $3y = |x + 2|$ **13.** $y = -|x|$ **14.** $y = |x| - 1$ **15.** $y = |x| + 4$ **16.** $y = 2|x| - 1$

Graph each of the following functions. (See Example 3.)

17. $y = \begin{cases} x - 1 & \text{if } x \le 3 \\ 2 & \text{if } x > 3 \end{cases}$

18. $y = \begin{cases} 6 - x & \text{if } x \le 3 \\ 3x - 6 & \text{if } x > 3 \end{cases}$

19. $y = \begin{cases} 4 - x & \text{if } x < 2 \\ 1 + 2x & \text{if } x \ge 2 \end{cases}$

20. $y = \begin{cases} -2 & \text{if } x \ge 1 \\ 2 & \text{if } x < 1 \end{cases}$

21. $y = \begin{cases} 2x + 1 & \text{if } x \ge 0 \\ x & \text{if } x < 0 \end{cases}$

22. $y = \begin{cases} 5x - 4 & \text{if } x \ge 1 \\ x & \text{if } x < 1 \end{cases}$

23. $y = \begin{cases} 2 + x & \text{if } x < -4 \\ -x & \text{if } -4 \le x \le 5 \\ 3x & \text{if } x > 5 \end{cases}$

24. $y = \begin{cases} -2x & \text{if } x < -3 \\ 3x + 1 & \text{if } -3 \le x < 2 \\ -4x & \text{if } x \ge 2 \end{cases}$

25. $y = \begin{cases} |x| & \text{if } x > -2 \\ x & \text{if } x \le -2 \end{cases}$

26. $y = \begin{cases} |x| - 1 & \text{if } x > -1 \\ x - 1 & \text{if } x \le -1 \end{cases}$

Graph each of the following functions. (See Example 4.)

27. $y = [-x]$ **28.** $y = [2x]$ **29.** $y = [2x - 1]$ **30.** $y = [3x + 1]$ **31.** $y = [3x]$ **32.** $y = [3x] + 1$

33. $y = [3x] - 1$ **34.** $y = x - [x]$

Graph each of the following functions. (See Example 7.)

35. $y = \sqrt{x}$ **36.** $y = \sqrt{x - 1}$ **37.** $y = \sqrt{x} + 1$ **38.** $y = \sqrt{x^2 + 1}$ **39.** $y = x^2$ **40.** $y = x^2 - 2$

Solve the following exercises. (See Examples 5 and 6.)

41. The charge to rent a Haul-It-Yourself Trailer is $25 plus $2 per hour or portion of an hour. Find the cost to rent a trailer for
(a) 2 hours;   (b) 1.5 hours;   (c) 4 hours;
(d) 3.7 hours.
(e) Graph the ordered pairs (hours, cost).

42. A delivery company charges $3 plus 50¢ per mile or part of a mile. Find the cost for a trip of
(a) 3 miles;   (b) 3.4 miles;   (c) 3.9 miles;
(d) 5 miles.
(e) Graph the ordered pairs (miles, cost).
(f) Is this a function?

43. A college typing service charges $3 plus $7 per hour or fraction of an hour. Graph the ordered pairs (hours, cost).

44. A parking garage charges $1 plus 50¢ per hour or fraction of an hour. Graph the ordered pairs (hours, cost).

45. A car rental costs $37 for 1 day, which includes 50 free miles. Each additional 25 miles, or portion, costs $10. Graph the ordered pairs (miles, cost).

46. For a lift truck rental of no more than 3 days, the charge is $300. An additional charge of $75 is made for each day or portion of a day after 3. Graph the ordered pairs (days, cost).

Solve the following problems.

47. Natural Science When a diabetic takes long-acting insulin, the insulin reaches its peak effect on the blood sugar level in about 3 hours. This effect remains fairly constant for 5 hours, then declines, and is very low until the next injection. In a typical patient, the level of blood sugar might be given by the following function.

$$i(t) = \begin{cases} 40t + 100 & \text{if } 0 \le t \le 3 \\ 220 & \text{if } 3 < t \le 8 \\ -80t + 860 & \text{if } 8 < t \le 10 \\ 60 & \text{if } 10 < t \le 24 \end{cases}$$

Here $i(t)$ is the blood sugar level, in appropriate units, at time t measured in hours from the time of the injection. Chuck takes his insulin at 6 A.M. Find the blood sugar level at each of the following times.

(a) 7 A.M.   **(b)** 9 A.M.   **(c)** 10 A.M.
(d) Noon   **(e)** 2 P.M.   **(f)** 5 P.M.
(g) Midnight   **(h)** Graph $y = i(t)$.
(i) From the graph in part (h), at what time(s) is Chuck's blood sugar level highest? lowest?

48. **Natural Science**   A factory begins emitting particulate matter into the atmosphere at 8 A.M. each workday, with the emissions continuing until 4 P.M. The level of pollutants, $P(t)$, measured by a monitoring station 1/2 mile away is approximated as follows, where t represents the number of hours since 8 A.M.

$$P(t) = \begin{cases} 75t + 100 & \text{if} \quad 0 \le t \le 4 \\ 400 & \text{if} \quad 4 < t < 8 \\ -100t + 1200 & \text{if} \quad 8 \le t \le 10 \\ -\dfrac{50}{7}t + \dfrac{1900}{7} & \text{if} \quad 10 < t < 24 \end{cases}$$

Find the level of pollution at
(a) 9 A.M.   **(b)** 11 A.M.   **(c)** 5 P.M.
(d) 7 P.M.   **(e)** Midnight.   **(f)** Graph $y = P(t)$.
(g) From the graph in part (f), at what time(s) is the pollution level highest? lowest?

49. **Management**   The **elasticity of demand** is the percent by which the demand for a product changes as price changes. For example, an elasticity of -1 means that demand changes at the same rate as price, so that a 10% price increase will cause a 10% drop in demand. An elasticity of $-.4$ means that a 10% price increase will cause a .4 of $10\% = 4\%$ drop in demand. Recently, there has been much controversy over the elasticity of demand for gasoline. It is now agreed that the short-term elasticity is small (you still have to get to work tomorrow), but the longer term elasticity is much higher (your next car will be much more fuel efficient). One recent projection of gasoline elasticity is as follows, where $e(t)$ represents elasticity at time t measured in years from some base year.

$$e(t) = \begin{cases} -.10 & \text{for} \quad 0 \le t \le 2 \\ -.25 & \text{for} \quad 2 < t \le 5 \\ -.50 & \text{for} \quad 5 < t \le 8 \\ -.75 & \text{for} \quad 8 < t \le 12 \\ -1.10 & \text{for} \quad 12 < t \le 16 \end{cases}$$

For example, a 10% price increase now will cause a .75 of $10\% = 7.5\%$ drop in demand in years 9 through 12.
(a) Graph $y = e(t)$.
(b) Give the domain and range for e.

50. **Management**   Normally, an increase in price will produce an increase in the supply of an item. However, there are a few items where an increase in price produces a *decrease* in supply. An example is labor—in developing countries with few consumer goods available, an increase in wage rates can actually lead to workers putting in fewer hours. Such situations produce *backward-bending* supply curves. Some economists now feel that oil may be on a backward-bending supply curve—as the price increases, some exporting nations can get sufficient revenue for their needs with the sale of fewer barrels of oil. To see how these curves work, let x be the number of millions of barrels of oil produced in some fixed time period, and let p be the price per barrel. Based on certain published data, the supply curve must be graphed as follows.
(a) For $10 \le x \le 12$, graph $p = (5/2)x$.
(b) For $12 < x \le 14$, graph $p = x + 18$.
(c) For $32 \le p \le 38$, graph $x = 14$.
(d) For $11 \le x \le 14$, graph $p = -2x + 66$.

51. The graph below, from *Business Week* magazine, February 22, 1990 (page 36), shows the number of aircraft with airfones since 1984.*
(a) Is this graph that of a function?
(b) What does the domain represent?
(c) Estimate the range.

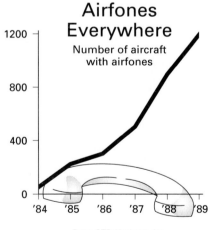

Airfones Everywhere
Number of aircraft with airfones

Data: GTE Airphone, Inc.

---

*"Ground to Airfone: Prepare for Hostile Fire" reprinted from the February 22, 1990 issue of *Business Week* by special permission.

1 Graph the function given by $2x - 3y = 12$.

Answer:

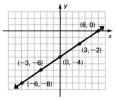

2.3 LINEAR FUNCTIONS

In this section we begin a study of *linear functions,* a type of function that has many important applications.

A **linear function** is a function whose rule can be written in the form

$$f(x) = ax + b$$

for some constants a and b.

Here are some examples of linear functions.

$$f(x) = 2x + 3, \quad g(x) = -3x + 4, \quad \text{and} \quad h(x) = x - 2$$

Both $f(x) = 3x$ and $g(x) = -5$ are linear functions (in which one of the constants a or b is 0).

It can be shown that the graph of every linear function is a straight line. Since a straight line is completely determined by any two points on the line, only two points are needed to obtain the graph of a linear function.

▶ **EXAMPLE 1**   The equation $x + 2y = 6$ determines a linear function since it can be rewritten as

$$2y = -x + 6$$

$$y = -\frac{1}{2}x + 3, \quad \text{or equivalently,} \quad f(x) = -\frac{1}{2}x + 3.$$

By letting $x = 0$, we see that $(0, 3)$ is on the graph, and by letting $x = 2$, we see that $(2, 2)$ is on the graph. Plotting the points $(0, 3)$ and $(2, 2)$ and drawing the line they determine produces the graph in Figure 2.18 ◀ **1**

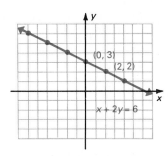

FIGURE 2.18

When graphing linear functions, it is often useful to use the x- and y-intercepts to determine the two points to graph.

2 Find the intercepts of each of the following. Graph the lines.

(a) $3x + 4y = 12$

(b) $5x - 2y = 8$

Answer:

(a) x-intercept 4
y-intercept 3

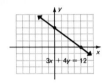

(b) x-intercept 8/5
y-intercept -4

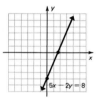

▶**EXAMPLE 2** Use the intercepts to draw the graph of $f(x) = -2x + 5$.

Find the y-intercept by letting $x = 0$ in the equation $y = -2x + 5$ and solving for y.

$$y = -2x + 5$$
$$y = -2(0) + 5 \qquad \text{Let } x = 0$$
$$y = 5$$

The y-intercept is 5, leading to the ordered pair $(0, 5)$. In the same way, the x-intercept is found by letting $y = 0$ and solving for x.

$$0 = -2x + 5 \qquad \text{Let } y = 0$$
$$2x = 5 \qquad \text{Add } 2x \text{ to both sides}$$
$$x = \frac{5}{2} = 2\frac{1}{2} \qquad \text{Multiply both sides by } \frac{1}{2}$$

The x-intercept is 5/2, or $2\frac{1}{2}$, so the graph goes through $(5/2, 0)$. These two intercepts lead to the graph of Figure 2.19.

As a check, a third point can be found by choosing another value of x (or y) and finding the corresponding value of the other variable. Check that $(1, 3)$, $(2, 1)$, $(3, -1)$, and $(4, -3)$, among other points, satisfy the equation $y = -2x + 5$ and lie on the line of Figure 2.19 ◀ **2**

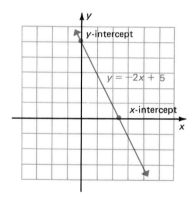

FIGURE 2.19

The next example shows a graph that does not cross the x-axis, and thus has no x-intercept.

3 Graph $y = -5$. Does this graph represent a function?

Answer:
Yes

4 Graph $x = 4$. Is this a function?

Answer:
No

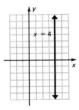

▶**EXAMPLE 3** Graph $f(x) = -3$.

The equation $f(x) = y = -3$, or equivalently, $y = 0x - 3$, always gives the same y value, -3, for any value of x. Therefore, no value of x will make $y = 0$, so the graph has no x-intercept. Since $y = -3$ is a linear function with a straight line graph, and since the graph cannot cross the x-axis, the line must be parallel to the x-axis. For any value of x, the value of y is -3, and so the graph is the horizontal line parallel to the x-axis, with y-intercept -3, as shown in Figure 2.20. As the vertical line test shows, the graph is the graph of a function. ◀ **3**

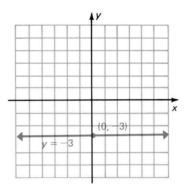

FIGURE 2.20

Example 3 suggests the following.

> The graph of $y = k$, where k is a real number, is the horizontal line having y-intercept k.

The graph of a function of the form $y = k$ is a horizontal line. Although an equation of the form $x = k$ is *not* the equation of a linear function, it does have a line for a graph, as the next example shows.

▶**EXAMPLE 4** Graph $x = -1$.

Obtain the graph of $x = -1$ by completing some ordered pairs using the equivalent form, $x = 0y - 1$. For example, $(-1, 0)$, $(-1, 1)$, $(-1, 2)$, and $(-1, 4)$ are some ordered pairs satisfying the equation $x = -1$. (The first component of these ordered pairs is always -1, which is what $x = -1$ means.) Here, more than one second component corresponds to the same first component, -1. As the graph of Figure 2.21 shows, a vertical line can cut this graph in more than one point. (In fact, a vertical line cuts the graph in an infinite numbers of points.) This confirms that $x = -1$ is not a function. ◀ **4**

5 Graph

(a) $x = 5y$;

(b) $5x = y$.

Answer:

(a)

(b)

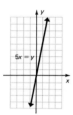

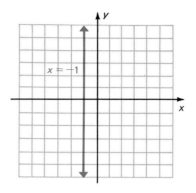

FIGURE 2.21

Example 4 suggests the following.

> The graph of $x = k$, where k is a real number, is the vertical line having x-intercept k.

As shown in Example 4, $x = -1$ is a linear equation that does not define a linear function. Linear equations of this form, $x = k$, where k is a real number, are the only linear equations that do not define linear functions.

▶ **EXAMPLE 5** Graph $f(x) = -3x$.

Begin by looking for the x-intercept. If $y = 0$, then since $y = f(x)$,

$$0 = -3x \qquad \text{Let } y = 0$$

$$0 = x, \qquad \text{Multiply both sides by } -\frac{1}{3}$$

giving the ordered pair $(0, 0)$. Letting $x = 0$ leads to exactly the same ordered pair, $(0, 0)$. Two different points are needed to determine a straight line, and the intercepts have led to only one point. Get a second point by choosing some other value of x (or y). For example, if $x = 2$,

$$y = -3x = -3(2) = -6,$$

giving the ordered pair $(2, 6)$. These two ordered pairs, $(0, 0)$ and $(2, -6)$, were used to get the graph in Figure 2.22. ◀ **5**

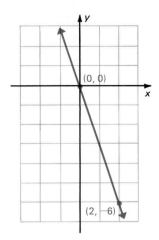

FIGURE 2.22

Linear functions can be very useful in setting up a mathematical model for a real-life situation. In almost every case, linear (or any other reasonably simple) functions provide only *approximations* to the real-world situations. However, these approximations are often remarkably useful.

SUPPLY AND DEMAND The supply of and demand for an item are usually related to its price. Producers will supply large numbers of the item at a high price, but consumer demand will be low. As the price of the item decreases, consumer demand increases, but producers are less willing to supply large numbers of the item. The curves that show the quantity that will be supplied at a given price and the quantity that will be demanded at a given price are called **supply and demand curves.** Linear functions often appear as supply and demand curves, as in the next example.

▶**EXAMPLE 6** Greg Odjakjian, an economist, has studied the supply and demand for aluminum siding and has determined that price per unit,* p, and demand, x, are related by the linear function

$$p = 60 - \frac{3}{4}x.$$

_____

*An appropriate unit here might be, for example, one thousand square feet of siding.

6 Suppose price and demand are related by $p = 100 - 4x$.

(a) Find the price if the demand is 10 units.

(b) Find the demand if the price is $80.

Answer:

(a) $60

(b) 5 units

(a) Find the demand at a price of $40 per unit.

Let $p = 40$.

$$p = 60 - \frac{3}{4}x$$

$$40 = 60 - \frac{3}{4}x \qquad \text{Let } p = 40$$

$$-20 = -\frac{3}{4}x \qquad \text{Add } -60 \text{ on both sides}$$

$$\frac{80}{3} = x \qquad \text{Multiply both sides by } -\frac{4}{3}$$

At a price of $40 per unit, 80/3 (or $26\frac{2}{3}$) units will be demanded; this gives the ordered pair (80/3, 40). (It is customary to write the ordered pairs so that price comes second.)

(b) Find the price if the demand is 32 units.

Let $x = 32$.

$$p = 60 - \frac{3}{4}x$$

$$p = 60 - \frac{3}{4}(32) \qquad \text{Let } x = 32$$

$$p = 60 - 24$$

$$p = 36$$

With a demand of 32 units, the price is $36. This gives the ordered pair (32, 36).

(c) Graph $p = 60 - \frac{3}{4}x$.

Use the ordered pairs (80/3, 40) and (32, 36) to get the demand graph shown in Figure 2.23. Only the portion of the graph in Quadrant I is shown, since the supply and demand functions are meaningful only for positive values of p and x. **6**

(d) From Figure 2.23, at a price of $30, what quantity is demanded?

Price is located on the y-axis. Look for 30 on the y-axis and read across to where the line $y = 30$ crosses the demand graph. As the graph shows, this occurs where the demand is 40.

(e) At what price will 60 units be demanded?

Quantity is located on the x-axis. Find 60 on the x-axis and read up to where the vertical line $x = 60$ crosses the demand graph. This occurs where the price is $15 per unit.

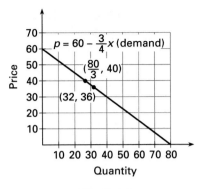

FIGURE 2.23

(f) What quantity is demanded at a price of $60 per unit?

The point $(0, 60)$ on the demand graph shows that the demand is 0 at a price of $60 (that is, there is no demand at such a high price). ◀

▶ **EXAMPLE 7**   Suppose that the economist of Example 6 concludes that the price and supply of siding are related by

$$p = \frac{3}{4}x.$$

(a) Find the supply if the price is $60 per unit.

$$60 = \frac{3}{4}x \qquad \text{Let } p = 60$$

$$80 = x$$

If the price is $60 per unit, then 80 units will be supplied to the marketplace. This gives the ordered pair $(80, 60)$.

(b) Find the price per unit if the supply is 16 units.

$$p = \frac{3}{4}(16) = 12 \qquad \text{Let } x = 16$$

If the supply is 16 units, then the price is $12 per unit. This gives the ordered pair $(16, 12)$.

(c) Graph $p = \frac{3}{4}x$.

Use the ordered pairs $(80, 60)$ and $(16, 12)$ to get the supply graph shown in Figure 2.24.

(d) Use the graph in Figure 2.24 to find the price at which 60 units will be supplied.

Quantity is located on the *x*-axis. The vertical line *x* = 60 crosses the supply graph when the price is $45 per unit. ◄

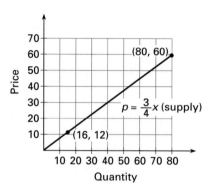

FIGURE 2.24

Graphing both the supply and demand curves on the same axes make it easy to answer certain questions.

▶**EXAMPLE 8** The supply and demand curves of Examples 6 and 7 are shown in Figure 2.25, with the supply curve in color.

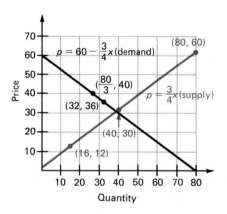

FIGURE 2.25

(a) Use Figure 2.25 to determine whether there is a surplus or shortage of supply at a price of $40 per unit.

7 The demand for a certain commodity is related to the price by $p = 80 - (2/3)x$. The supply is related to the price by $p = (4/3)x$. Find

(a) the equilibrium demand;

(b) the equilibrium price.

Answer:

(a) 40

(b) 160/3

Find 40 on the y-axis. Read across to where the line $y = 40$ crosses the supply graph. Since the supply graph lies to the right of the demand graph at that point, supply is greater than demand and there is a surplus of supply.
(b) For what quantity does demand exceed supply?

The demand graph lies above the supply graph so demand is greater than supply when the quantity supplied or demanded is in the interval [0, 40). ◀

As shown in the graphs of Figure 2.25, both the supply and the demand functions pass through the point (40, 30). If the price of the siding is more than $30, the supply will exceed the demand. At a price less than $30, the demand will exceed the supply. Only at a price of $30 will demand and supply be equal. For this reason, $30 is called the *equilibrium price*. When the price is $30, demand and supply both equal 40 units, the *equilibrium supply* or *equilibrium demand*. By definition, the **equilibrium price** of a commodity is the price at the point where the supply and demand graphs for that commodity cross. The **equilibrium demand** is the demand at that same point; the **equilibrium supply** is the supply at that point. By definition, the equilibrium supply and the equilibrium demand are equal.

▶**EXAMPLE 9**   Use algebra to find the equilibrium supply, demand, and price for the aluminum siding. (See Examples 6 and 7.)

The equilibrium supply is found when the prices from both supply and demand are equal. From Example 6, $p = 60 - (3/4)x$; in Example 7, $p = (3/4)x$. Set these two expressions for p equal to get the linear equation

$$60 - \frac{3}{4}x = \frac{3}{4}x$$

$$240 - 3x = 3x \qquad \text{Multiply both sides by 4}$$

$$240 = 6x \qquad \text{Add } 3x \text{ to both sides}$$

$$40 = x.$$

The equilibrium supply is 40 units, the same answer found above. This is also the equilibrium demand. To find the equilibrium price, substitute 40 for x in either the supply or demand function and solve for p. As we saw above, the equilibrium price is $30. ◀ **7**

2.3 EXERCISES

Graph each of the following linear equations. Identify any that are not linear functions. (See Examples 1–5.)

1. $y = 2x + 1$

2. $y = 3x - 1$

3. $y = 3x + 2$

4. $y = x + 5$

5. $3y + 4x = 12$

6. $4y + 5x = 10$

7. $y = -2$

8. $x = 4$

9. $6x + y = 12$

10. $x + 3y = 9$

11. $x - 5y = 4$

12. $2y + 5x = 20$

13. $x + 5 = 0$

14. $y - 4 = 0$

15. $5y - 3x = 12$

16. $2x + 7y = 14$

17. $8x + 3y = 10$

18. $9y - 4x = 12$

19. $y = 2x$

20. $y = -5x$

21. $y = -4x$

22. $y = 2x$

23. $x - 3y = 0$

24. $x + 4y = 0$

Management *Use the supply and demand curves graphed below to answer Exercises 25–28. (See Examples 6–8.)*

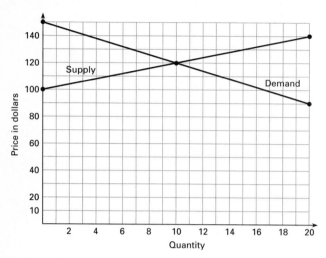

25. At what price are 20 items supplied?

26. At what price are 20 items demanded?

27. Find the equilibrium supply and the equilibrium demand.

28. Find the equilibrium price.

Management *Work the following exercises. (See Examples 6–9.)*

29. Suppose that the demand and price for a certain model of electric can opener are related by

$$p = 16 - \frac{5}{4}x,$$

where p is price, in dollars, and x is demand. Find the price for a demand of
(a) 0 units;   **(b)** 4 units;   **(c)** 8 units.
Find the demand for the electric can opener at a price of
(d) \$6;   **(e)** \$11;   **(f)** \$16.
(g) Graph $p = 16 - (5/4)x$.
Suppose the price and supply of the can openers are related by

$$p = \frac{3}{4}x,$$

where x represents the supply, and p the price. Find the supply when the price is
(h) \$0;   **(i)** \$10;   **(j)** \$20.

(k) Graph $p = (3/4)x$ on the same axes used for part (g).
(l) Find the equilibrium supply.
(m) Find the equilibrium price.

30. Let the supply and demand function for strawberry-flavored licorice be

$$\text{supply: } p = \frac{3}{2}x \quad \text{and} \quad \text{demand: } p = 81 - \frac{3}{4}x.$$

(a) Graph these on the same axes.
(b) Find the equilibrium demand.
(c) Find the equilibrium price.

31. Let the supply and demand functions for butter pecan ice cream be given by

$$\text{supply: } p = \frac{2}{5}x \quad \text{and} \quad \text{demand: } p = 100 - \frac{2}{5}x.$$

(a) Graph these on the same axes.
(b) Find the equilibrium demand.
(c) Find the equilibrium price.

32. Let the supply and demand functions for sugar be given by

$$\text{supply: } p = 1.4x - .6$$

and $\text{demand: } p = -2x + 3.2.$

(a) Graph these on the same axes.
(b) Find the equilibrium demand.
(c) Find the equilibrium price.

33. In a recent issue of *Business Week,* the president of Insta-Tune, a chain of franchised automobile tune-up shops, says that people who buy a franchise and open a shop pay a weekly fee of

$$y = .07x + \$135$$

to company headquarters. Here y is the fee and x is the total amount of money taken in during the week by the tune-up center. Find the weekly fee if x is
(a) \$0;   **(b)** \$1000;   **(c)** \$2000;   **(d)** \$3000.
(e) Graph y.

34. In a recent issue of *The Wall Street Journal,* we are told that the relationship between the amount of money that an average family spends on food eaten at home, x, and the amount of money it spends on eating out, y, is approximated by the model $y = .36x$. Find y if x is
(a) \$40;   **(b)** \$80;   **(c)** \$120.   **(d)** Graph y.

Management *Recall that when revenue and cost functions are given, the break-even point is the x-value where revenue equals cost.*

35. For x thousand policies, an insurance company claims that their monthly revenue in dollars is given by $R(x) = 125x$ and their monthly cost in dollars is given by $C(x) = 100x + 5000$.
  (a) Find the break-even point.
  (b) Graph the revenue and cost functions on the same axis.
  (c) From the graph, estimate the revenue and cost when $x = 100$ (100 thousand policies).

36. The owners of a parking lot have determined that their weekly revenue and cost in dollars are given by $R(x) = 80x$ and $C(x) = 50x + 2400$, where x is the number of long-term parkers.
  (a) Find the break-even point.
  (b) Graph $R(x)$ and $C(x)$ on the same axes.
  (c) From the graph, estimate the revenue and cost when there are 60 long-term parkers.

2.4 SLOPE AND THE EQUATIONS OF A LINE

As mentioned in the previous section, the graph of a straight line is completely determined by two different points on the line. The graph of a straight line also can be drawn knowing only *one* point on the line *if* the "steepness" of the line is known, too. The number that represents the "steepness" of a line is called the *slope* of that line.

To see how slope is defined, start with Figure 2.26, which shows a line passing through the two different points $(x_1, y_1) = (-3, 5)$ and $(x_2, y_2) = (2, -4)$. The difference in the two x values,

$$x_2 - x_1 = 2 - (-3) = 5$$

in this example, is called the **change in x.** The Greek letter Δ (delta) is used to denote change. The symbol Δx (read "delta x") represents the change in x. In the same way, Δy represents the **change in y.** In this example,

$$\Delta y = y_2 - y_1 = -4 - 5 = -9.$$

The **slope** of the line through the two points (x_1, y_1) and (x_2, y_2), where $x_1 \neq x_2$, is defined as the quotient of the change in y and the change in x, or

$$\text{slope} = \frac{\text{change in } y}{\text{change in } x} = \frac{\Delta y}{\Delta x} = \frac{y_2 - y_1}{x_2 - x_1}.$$

1 Find the slope of the line through

(a) $(6, 11)$, $(-4, -3)$;

(b) $(-3, 5)$, $(-2, 8)$.

Answer:

(a) $7/5$

(b) 3

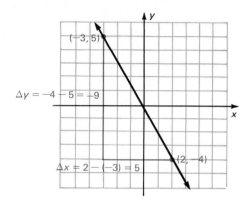

FIGURE 2.26

The slope of the line in Figure 2.26 is

$$\text{slope} = \frac{\Delta y}{\Delta x}$$

$$= \frac{-4 - 5}{2 - (-3)}$$

$$= -\frac{9}{5}.$$

Using similar triangles from geometry, it can be shown that the slope is independent of the choice of points on the line. That is, the same value of the slope will be obtained for *any* choice of two different points on the line.

▶**EXAMPLE 1** Find the slope of the line through the points $(-7, 6)$ and $(4, 5)$.
Let $(x_1, y_1) = (-7, 6)$. Then $(x_2, y_2) = (4, 5)$. Use the definition of slope.

$$\text{slope} = \frac{\Delta y}{\Delta x} = \frac{5 - 6}{4 - (-7)} = -\frac{1}{11}$$

The slope also could have been found by letting $(x_1, y_1) = (4, 5)$ and $(x_2, y_2) = (-7, 6)$. In that case,

$$\text{slope} = \frac{6 - 5}{-7 - 4} = \frac{1}{-11} = -\frac{1}{11},$$

the same answer. ◀ **1**

Caution When finding the slope of a line, be careful to subtract the x-values and y-values in the same order. For instance, consider the slope of the line through the points (2, 3) and (4, 9).

Correct	Incorrect
2nd point − 1st point	2nd point − 1st point
$\dfrac{9-3}{4-2} = 3$	$\dfrac{9-3}{2-4} = -3$
2nd point − 1st point	1st point − 2nd point
1st point − 2nd point	1st point − 2nd point
$\dfrac{3-9}{2-4} = 3$	$\dfrac{3-9}{4-2} = -3.$
1st point − 2nd point	2nd point − 1st point

The slope of a line is a measure of the steepness of the line. Figure 2.27 shows examples of lines with different slopes. Lines with positive slopes go up as x goes from left to right along the x-axis, while lines with negative slopes go down as x goes from left to right.

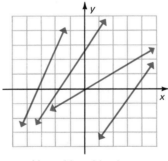

Lines with positive slope

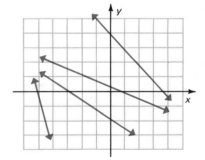

Lines with negative slope

FIGURE 2.27

▶**EXAMPLE 2** Find the slope of the line $3x - 4y = 12$.

The slope is found using two different points on the line. The intercepts can be used for the two points. First let $x = 0$ and then let $y = 0$ to find the intercepts.

If $x = 0$,

$3x - 4y = 12$

$3(0) - 4y = 12$ **Let $x = 0$**

$-4y = 12$

$y = -3$

If $y = 0$

$3x - 4y = 12$

$3x - 4(0) = 12$ **Let $y = 0$**

$3x = 12$

$x = 4$

2 Find the slope of

(a) $8x + 5y = 9$;

(b) $2x - 7y = 6$;

(c) $8y = x$.

Answer:

(a) $-8/5$

(b) $2/7$

(c) $1/8$

3 Use $x = 0$ and $y = 0$ to get two ordered pairs and then find the slope of each line.

(a) $2x + 3y = 5$

(b) $4x + 5 = -6y$

Answer:

(a) $(0, 5/3)$ and $(5/2, 0)$; $-2/3$

(b) $(0, -5/6)$ and $(-5/4, 0)$; $-2/3$

4 Tell if the lines in each of the following pairs are *parallel, perpendicular,* or *neither.*

(a) $x - 2y = 6$ and $2x + y = 5$

(b) $3x + 4y = 8$ and $x + 3y = 2$

(c) $2x - y = 7$ and $2y = 4x - 5$

Answer:

(a) Perpendicular

(b) Neither

(c) Parallel

These results give the ordered pairs $(0, -3)$ and $(4, 0)$. Now the slope can be found from the definition.

$$\text{slope} = \frac{0 - (-3)}{4 - 0} = \frac{3}{4} \quad \blacktriangleleft \quad \boxed{2}$$

We shall assume the following facts without proof.

> Two nonvertical lines are parallel whenever they have the same slope.
>
> Two lines, where neither is vertical, are perpendicular whenever the product of their slopes is -1.

▶**EXAMPLE 3** Tell whether each of the following pairs of lines are *parallel, perpendicular,* or *neither*.

(a) $2x + 3y = 5$ and $4x + 5 = -6y$

Find the slope of each line by first finding two points on each line. The intercepts are a good choice. $\boxed{3}$

From Problem 3 at the side, both slopes are $-2/3$, so the lines are parallel.

(b) $3x = y + 7$ and $x + 3y = 4$

Verify that the slope of $3x = y + 7$ is 3 and the slope of $x + 3y = 4$ is $-1/3$. Since $3(-1/3) = -1$, these lines are perpendicular.

(c) $x + y = 4$ and $x - 2y = 3$

The slope of the first line is -1 and of the second line is $1/2$. Since the slopes are not equal and their product is $-1/2$, not -1, the lines are neither parallel nor perpendicular. $\blacktriangleleft \quad \boxed{4}$

SLOPE-INTERCEPT FORM A generalization of the method of Example 2 can be used to find the equation of a line, given its y-intercept and slope. Assume that a line has y-intercept b, so that it goes through $(0, b)$. Let the slope of the line be m. If (x, y) is any point on the line *other* than $(0, b)$, using the definition of slope with the points $(0, b)$ and (x, y) gives

$$m = \frac{y - b}{x - 0}$$

$$m = \frac{y - b}{x}$$

$$mx = y - b$$

from which

$$y = mx + b.$$

This result is summarized as follows.

5 Find an equation for the line with

(a) y-intercept -3 and slope 2/3;

(b) y-intercept 1/4 and slope $-3/2$.

Answer:

(a) $y = \dfrac{2}{3}x - 3$

(b) $y = -\dfrac{3}{2}x + \dfrac{1}{4}$

6 Find the slope and y-intercept for

(a) $x + 4y = 6$;

(b) $3x - 2y = 1$.

Answer:

(a) Slope $-1/4$; y-intercept 3/2

(b) Slope 3/2; y-intercept $-1/2$

Slope-Intercept Form

If a line has slope m and y-intercept b, then

$$y = mx + b$$

is the **slope-intercept form** of the equation of the line.

▶**EXAMPLE 4** Find an equation for the line with y-intercept 7/2 and slope $-5/2$.
Use the slope-intercept form with $b = 7/2$ and $m = -5/2$.

$$y = mx + b$$
$$y = -\frac{5}{2}x + \frac{7}{2} \quad ◀ \boxed{5}$$

The slope-intercept form of the equation of a line shows that the slope can be found from the equation by solving for y. In this form, the coefficient of x is the slope and the constant term is the y-intercept.

For example, we found in Example 2 that the slope of the line $3x - 4y = 12$ is 3/4. This slope can also be found by solving for y.

$$3x - 4y = 12$$
$$-4y = -3x + 12 \qquad \text{Add } -3x$$
$$y = \frac{3}{4}x - 3 \qquad \text{Divide by } -4$$

The coefficient of x gives the slope, 3/4.

▶**EXAMPLE 5** Find the slope and y-intercept for each of the following lines.
(a) $5x - 3y = 1$
Solve for y.
$$5x - 3y = 1$$
$$-3y = -5x + 1$$
$$y = \frac{5}{3}x - \frac{1}{3}$$

The slope is 5/3 and the y-intercept is $-1/3$.
(b) $-9x + 6y = 2$
Solve for y.
$$-9x + 6y = 2$$
$$6y = 9x + 2$$
$$y = \frac{3}{2}x + \frac{1}{3}$$

The slope is 3/2 and the y-intercept is 1/3. ◀ $\boxed{6}$

7 Use the slope and
y-intercept to graph

(a) $2x + 5y = 10$;

(b) $-x + y = 4$.

Answer:

(a)

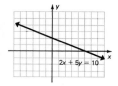

(b)

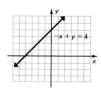

The slope and y-intercept of a line can be used to draw the graph of the line, as shown in the next example.

▶ **EXAMPLE 6**   Use the slope and y-intercept to graph $3x - 2y = 2$.
Solve for y.

$$3x - 2y = 2$$
$$-2y = -3x + 2$$
$$y = \frac{3}{2}x - 1$$

The slope is $3/2$ and the y-intercept is -1.

To draw the graph, first locate the y-intercept -1, as shown in Figure 2.28. Use the slope to find a second point on the graph. If m represents the slope, then

$$m = \frac{\Delta y}{\Delta x} = \frac{3}{2}.$$

If x changes by positive 2 units ($\Delta x = 2$), then y will change by positive 3 units ($\Delta y = 3$). Find the second point by starting at the y-intercept graphed in Figure 2.28 and moving 2 units to the right and 3 units up. Once this second point is located, a line can be drawn through it and the y-intercept. Since $3/2 = -3/-2$, the second point also can be located by moving 2 units to the left (-2) and 3 units down (-3). This is a good way to check the line found by using $m = 3/2$.   ◀   **7**

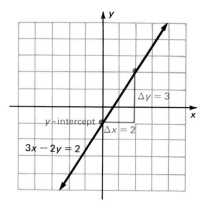

FIGURE 2.28

POINT-SLOPE FORM   The slope-intercept form of the equation of a line involves the slope and the y-intercept. Sometimes, however, the slope of a line is known, together with one point (perhaps *not* the y-intercept) that the line goes through. The *point-slope form* of the equation of a line is used to find an equation in this case. Let

8 Find an equation of the line having the given slope and going through the given point.

(a) $m = -3/5$, $(5, -2)$

(b) $m = 1/3$, $(6, 8)$

Answer:

(a) $5y = -3x + 5$

(b) $3y = x + 18$

(x_1, y_1) be any fixed point on the line and let (x, y) represent any other point on the line. If m is the slope of the line, then, by the definition of slope,

$$\frac{y - y_1}{x - x_1} = m,$$

or

$$y - y_1 = m(x - x_1).$$

Point-Slope Form

If a line has slope m and passes through the point (x_1, y_1), then

$$y - y_1 = m(x - x_1)$$

is the **point-slope form** of the equation of the line.

▶**EXAMPLE 7**   Find an equation of the line with the given slope and going through the given point.

(a) $(-4, 1)$, $m = -3$

Use the point-slope form since a point the line goes through, together with the slope of the line, is known. Substitute the values $x_1 = -4$, $y_1 = 1$, and $m = -3$ into the point-slope form.

$$y - y_1 = m(x - x_1)$$
$$y - 1 = -3[x - (-4)] \qquad \text{Let } y_1 = 1, \, m = -3, \, x_1 = -4$$
$$y - 1 = -3(x + 4)$$
$$y - 1 = -3x - 12 \qquad\qquad \text{Distributive property}$$
$$y = -3x - 11$$

(b) $(3, -7)$, $m = 5/4$

$$y - y_1 = m(x - x_1)$$
$$y - (-7) = \frac{5}{4}(x - 3) \qquad \text{Let } y_1 = -7, \, m = \frac{5}{4}, \, x_1 = 3$$
$$y + 7 = \frac{5}{4}(x - 3)$$
$$4(y + 7) = 5(x - 3) \qquad \text{Multiply both sides by 4}$$
$$4y + 28 = 5x - 15 \qquad \text{Distributive property}$$
$$4y = 5x - 43 \qquad ◀ \; \boxed{8}$$

The point-slope form can also be used to find an equation of a line given two different points that the line goes through. The procedure for doing this is shown in the next example.

9 Find an equation of the line through

(a) (2, 3) and (−4, 6);

(b) (−8, 2) and (3, −6).

Answer:

(a) $2y = -x + 8$

(b) $11y = -8x - 42$

10 Find an equation of the line through (−2, 5) and (7, 5).

Answer:

$y = 5$

11 Find an equation of the line through (−5, 1) and (−5, 7).

Answer:

$x = -5$

▶**EXAMPLE 8** Find an equation of the line through (5, 4) and (−10, −2).

Begin by using the definition of slope to find the slope of the line that passes through the two points.

$$\text{slope} = m = \frac{-2 - 4}{-10 - 5} = \frac{-6}{-15} = \frac{2}{5}$$

Use $m = 2/5$ and either of the given points in the point-slope form. If $(x_1, y_1) = (5, 4)$, then

$$y - y_1 = m(x - x_1)$$

$$y - 4 = \frac{2}{5}(x - 5) \qquad \text{Let } y_1 = 4, m = \frac{2}{5}, x_1 = 5$$

$$5(y - 4) = 2(x - 5) \qquad \textbf{Multiply both sides by 5}$$

$$5y - 20 = 2x - 10 \qquad \textbf{Distributive property}$$

$$5y = 2x + 10.$$

Check that the result is the same when $(x_1, y_1) = (-10, -2)$. ◀ **9**

▶**EXAMPLE 9** Find an equation of the line through (8, −4) and (−2, −4).

First find the slope.

$$m = \frac{-4 - (-4)}{-2 - 8} = \frac{0}{-10} = 0$$

Choose, say, (8, −4) as (x_1, y_1), and use the point-slope form.

$$y - y_1 = m(x - x_1)$$

$$y - (-4) = 0(x - 8) \qquad \textbf{Let } y_1 = -4, m = 0, x_1 = 8$$

$$y + 4 = 0 \qquad\qquad 0(x - 8) = 0$$

$$y = -4 \blacktriangleleft$$

As shown in the previous section, $y = -4$ represents a horizontal line with y-intercept −4. **10**

▶**EXAMPLE 10** Find an equation of the line through (4, 3) and (4, −6).

Begin by finding the slope.

$$m = \frac{-6 - 3}{4 - 4} = \frac{-9}{0}$$

Division by 0 is not defined, so the slope is undefined. Graphing the given ordered pairs (4, 3) and (4, −6) and drawing a line through them gives a vertical line. From the last section, vertical lines have equations of the form $x = k$, where k can be any real number. Since the x-coordinate of the two ordered pairs given above is 4, the desired equation is $x = 4$. ◀ **11**

Examples 9 and 10 suggest the following.

The slope of every horizontal line is 0.

The slope of every vertical line is undefined.

A summary of the equations of lines discussed in this section follows.

Equation	Description
$ax + by = c$	If $a \neq 0$ and $b \neq 0$, line has x-intercept c/a and y-intercept c/b.
$x = k$	**Vertical line,** x-intercept k, no y-intercept, undefined slope
$y = k$	**Horizontal line,** y-intercept k, no x-intercept, slope 0
$y = mx + b$	**Slope-intercept form,** slope m, y-intercept b
$y - y_1 = m(x - x_1)$	**Point-slope form,** slope m, line passes through (x_1, y_1)

2.4 EXERCISES

In each of the following exercises, find the slope, if it exists, of the line through the given pair of points. (See Example 1.)

1. $(-8, 6)$, $(2, 4)$ **2.** $(-3, 2)$, $(5, 9)$ **3.** $(-1, 4)$, $(2, 6)$ **4.** $(3, -8)$, $(4, 1)$

5. The origin and $(-4, 6)$ **6.** The origin and $(8, -2)$ **7.** $(-2, 9)$, $(-2, 11)$ **8.** $(7, 4)$, $(7, 12)$

9. $(3, -6)$, $(-5, -6)$ **10.** $(5, -11)$, $(-9, -11)$

Tell whether each pair of lines is parallel, perpendicular, *or* neither. *(See Example 3.)*

11. $4x - 3y = 6$ and $3x + 4y = 8$ **12.** $2x - 5y = 7$ and $15y = 5 + 6x$ **13.** $3x + 2y = 8$ and $6y = 5 - 9x$

14. $x - 3y = 4$ and $y = 1 - 3x$ **15.** $4x = 2y + 3$ and $2y = 2x + 3$ **16.** $2x - y = 6$ and $x - 2y = 4$

17. $2x - y = 9$ and $x = 2y$ **18.** $4x = 2y + 7$ and $y = 2x$

Find the slope and y-intercept of each of the following lines. (See Examples 2 and 5.)

19. $y = 3x + 4$ **20.** $y = -3x + 2$ **21.** $y + 4x = 8$ **22.** $y - x = 3$

23. $3x + 4y = 5$ **24.** $2x - 5y = 8$ **25.** $3x + y = 0$ **26.** $y - 4x = 0$

27. $2x + 5y = 0$ **28.** $3x - 4y = 0$ **29.** $y = 8$ **30.** $y = -4$

31. $y + 2 = 0$ **32.** $y - 3 = 0$ **33.** $x = -8$ **34.** $x = 3$

Graph the line through the given point with the given slope. (See Example 6.)

35. $(-4, 2)$, $m = \dfrac{2}{3}$ **36.** $(3, -2)$, $m = \dfrac{3}{4}$ **37.** $(-5, -3)$, $m = -2$ **38.** $(-1, 4)$, $m = 2$

39. $(8, 2)$, $m = 0$ **40.** $(2, -4)$, $m = 0$ **41.** $(6, -5)$, undefined slope   **42.** $(-8, 9)$, undefined slope

43. $(0, -2)$, $m = \dfrac{3}{4}$ **44.** $(0, -3)$, $m = \dfrac{2}{5}$ **45.** $(5, 0)$, $m = \dfrac{1}{4}$ **46.** $(-9, 0)$, $m = \dfrac{5}{2}$

Find an equation for the line with the given y-intercept and slope. (See Example 4.)

47. 4, $m = -\dfrac{3}{4}$ **48.** -3, $m = \dfrac{2}{3}$ **49.** -2, $m = -\dfrac{1}{2}$ **50.** $\dfrac{3}{2}$, $m = \dfrac{1}{4}$

51. $\dfrac{5}{4}$, $m = \dfrac{3}{2}$ **52.** $-\dfrac{3}{8}$, $m = \dfrac{3}{4}$

Find an equation for each of the following lines. (See Examples 4 and 7–10.)

53. Through $(-4, 1)$, $m = 2$ **54.** Through $(5, 1)$, $m = -1$ **55.** Through $(0, 3)$, $m = -3$

56. Through $(-2, 3)$, $m = 3/2$ **57.** Through $(3, 2)$, $m = 1/4$ **58.** Through $(0, 1)$, $m = -2/3$

59. Through $(-1, 1)$ and $(2, 5)$ **60.** Through $(4, -2)$ and $(6, 8)$ **61.** Through $(9, -6)$ and $(12, -8)$

62. Through $(-5, 2)$ and $(7, 5)$ **63.** Through $(-8, 4)$ and $(-8, 6)$ **64.** Through $(2, -5)$ and $(4, -5)$

65. Through $(-1, 3)$ and $(0, 3)$ **66.** Through $(2, 9)$ and $(2, -9)$

Many real-world situations can be approximately described by a straight line graph. One way to find the equation of such a straight line is to use two typical data points from the graph and the point-slope form of the equation of a line. In Exercises 67–71, assume that the data in each problem can be approximated fairly closely by a straight line. Use the given information to find the equation of the line. Find the slope of each of the lines.

67. Management A company finds that it can make a total of 20 solar heaters for $13,900, while 10 heaters cost $7500. Let y be the total cost to produce x solar heaters.

68. Management The sales of a small company were $27,000 in its second year of operation and $63,000 in its fifth year. Let y represent sales in year x.

69. Social Science According to research done by the political scientist James March, if the Democrats win 45% of the two-party vote for the House of Representatives, they win 42.5% of the seats. If the Democrats win 55% of the vote, they win 67.5% of the seats. Let y be the percent of seats won and x the percent of the two-party vote.

70. Social Science If the Republicans win 45% of the two-party vote, they win 32.5% of the seats (see Exercise 69.) If they win 60% of the vote, they get 70% of the seats. Let y represent the percent of the seats and x the percent of the vote.

71. Natural Science When a certain industrial pollutant is introduced into a river, the reproduction of catfish declines. In a given period of time, 3 tons of the pollutant results in a fish population of 37,000. Also, 12 tons of pollutant produce a fish population of 28,000. Let y be the fish population when x tons of pollutant are introduced into the river.

72. Natural Science A person's tibia bone goes from ankle to knee. A male with a tibia 40 centimeters in length will have a height of 177 centimeters, while a male's tibia 43 centimeters in length corresponds to a height of 185 centimeters.

(a) Write a linear equation showing how the height of a male, h, relates to the length of his tibia, t.

(b) Estimate the height of a male having a tibia of length 38 centimeters; 45 centimeters.

(c) Estimate the length of the tibia of a male for a height of 190 centimeters.

73. Natural Science The radius bone goes from the wrist to the elbow. A female whose radius bone is 24 centimeters long would be 167 centimeters tall, while a radius of 26 centimeters in a female corresponds to a height of 174 centimeters.

(a) Write a linear equation showing how the height of a female, h, corresponds to the length of her radius bone, r.

(b) Estimate the height of a female having a radius of length 23 centimeters; 27 centimeters.

(c) Estimate the length of a radius bone of a female for a height of 170 centimeters.

74. Show that b gives the y-intercept in the slope-intercept form $y = mx + b$.

2.5 APPLICATIONS OF LINEAR FUNCTIONS

When mathematics is used to solve a real-world problem, the first step is to write one or more mathematical relationships (equations or inequalities) that describe the situation. In this phase of problem solving, if the result is to be useful, it is important to include all the pertinent information. Setting up these relationships requires a solid understanding of the situation to be described, as well as a knowledge of the most useful mathematics available. In this section we show how linear functions can be applied to a variety of real-world situations.

TEMPERATURE One of the most common linear relationships found in everyday situations deals with temperature. Recall that water freezes at 32° Fahrenheit and 0° Celsius, while it boils at 212° Fahrenheit and 100° Celsius. The ordered pairs (0, 32) and (100, 212) are graphed in Figure 2.29 on axes showing Fahrenheit (F) as a function of Celcius (C). The line joining them is the graph of the function.

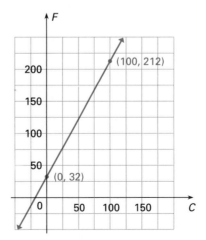

FIGURE 2.29

1 Find a linear equation describing the sales for Company B from Example 2.

Answer:
$y = 3000x + 5000$

▶**EXAMPLE 1** Derive an equation relating F and C.

To derive the required linear equation, first find the slope using the given ordered pairs, (0, 32) and (100, 212).

$$m = \frac{212 - 32}{100 - 0} = \frac{9}{5}$$

The F-intercept of the graph is 32, so by the slope-intercept form the equation of the line is

$$F = \frac{9}{5}C + 32.$$

This equation can be rewritten to give C in terms of F.

$$C = \frac{5}{9}(F - 32) \quad ◀$$

SALES ANALYSIS It is common to compare the change in sales of two companies by comparing the rates at which these sales change. If the sales of the two companies can be approximated by linear functions, we can use the work of the last section to find the rate of change of the dependent variable compared to the change in the independent variable.

▶**EXAMPLE 2** The chart below shows sales in two different years for two different companies.

Company	Sales in 1988	Sales in 1991
A	$10,000	$16,000
B	5000	14,000

A study of company records suggests that the sales of both companies have increased linearly (that is, the sales can be closely approximated by a linear function). Find a linear equation describing the sales for Company A.

To find a linear equation describing the sales, let $x = 0$ represent 1988 so that 1991 corresponds to $x = 3$. Then, by the chart above, the line representing the sales for Company A passes through the points (0, 10,000) and (3, 16,000). The slope of the line through these points is

$$\frac{16,000 - 10,000}{3 - 0} = 2000.$$

Using the point-slope form of the equation of a line gives

$$y - 10,000 = 2000(x - 0)$$
$$y = 2000x + 10,000$$

as an equation describing the sales of Company A. ◀ **1**

2 A certain new anticholesterol drug is related to the blood cholesterol level by the linear model

$$y = 280 - 3x,$$

when x is the dosage of the drug (in grams) and y is the blood cholesterol level.

(a) Find the blood cholesterol level if 12 grams of the drug are administered.

(b) In general, an increase of 1 gram in the dose causes what change in the blood cholesterol level?

Answer:

(a) 244

(b) A decrease of 3 units

AVERAGE RATE OF CHANGE   Notice that the sales for Company A in Example 2 increased from $10,000 to $16,000 over the period from 1988 to 1991, representing a total increase of $6000 in 3 years.

$$\text{Average rate of change in sales} = \frac{\$6000}{3} = \$2000 \text{ per year}$$

This is the same as the slope found in Example 2. Verify that the average annual rate of change in sales for Company B over the 3-year period also agrees with the slope of the equation found in Problem 1 at the side. Management needs to watch the rate of change in sales closely in order to be aware of any unfavorable trends. If the rate of change is decreasing, then sales growth is slowing down, and this trend may require some response.

Example 2 illustrates a useful fact about linear functions. If $f(x) = mx + b$ is a *linear* function, then the **average rate of change** in y with respect to x (the change in y divided by the corresponding change in x) is the slope of the line $y = mx + b$. In particular, the average rate of change of a linear function is constant. (See Exercises 33 and 34 in this section for examples of functions where the average rate of change is not constant.)

▶ **EXAMPLE 3**   Suppose that a researcher has concluded that a dosage of x grams of a certain stimulant causes a rat to gain

$$y = 2x + 50$$

grams of weight, for appropriate values of x. If the researcher administers 30 grams of the stimulant, how much weight will the rat gain? What is the average rate of change of weight gain?

Let $x = 30$. The rat will gain

$$y = 2(30) + 50 = 110$$

grams of weight.

The average rate of change of weight gain with respect to the amount of stimulant is given by the slope of the line. The slope of $y = 2x + 50$ is 2, so that when the dose is increased by 1 gram the change in weight gain is an increase of 2 grams. ◀ **2**

COST ANALYSIS   The cost of manufacturing an item commonly consists of two parts. The first is a **fixed cost** for designing the product, setting up a factory, training workers, and so on. Within broad limits, the fixed cost is constant for a particular product and does not change as more items are made. The second part is a *cost per item* for labor, materials, packing, shipping, and so on. The total value of this second cost *does* depend on the number of items made.

3 The cost in dollars to produce x kilograms of chocolate candy is given by $C(x)$, where in dollars

$$C(x) = 3.5x + 800.$$

Find each of the following.

(a) The fixed cost

(b) The total cost for 12 kilograms

(c) The marginal cost per kilogram

(d) The marginal cost of the 40th kilogram

Answer:

(a) $800

(b) $842

(c) $3.50

(d) $3.50

▶ **EXAMPLE 4** Suppose that the cost of producing clock-radios can be approximated by the linear model

$$C(x) = 12x + 100.$$

where $C(x)$ is the cost in dollars to produce x radios. The cost to produce 0 radios is

$$C(0) = 12(0) + 100 = 100,$$

or $100. This amount, $100, is the fixed cost.

Once the company has invested the fixed cost into the clock-radio project, what is the additional cost per radio? To find out, first find the cost of a total of 5 radios:

$$C(5) = 12(5) + 100 = 160,$$

or $160. The cost of 6 radios is

$$C(6) = 12(6) + 100 = 172,$$

or $172. The sixth radio costs $172 - $160 = $12 to produce. In the same way, the 81st radio costs $C(81) - C(80) = \$1072 - \$1060 = \$12$ to produce. In fact, the $(n + 1)$st radio costs

$$C(n + 1) - C(n) = [12(n + 1) + 100] - [12n + 100] = 12,$$

or $12 to produce. Since each additional radio costs $12 to produce, $12 is the variable cost per radio. The number 12 is also the slope of the cost function, $C(x) = 12x + 100$. ◀

Example 4 can easily be generalized. Suppose the total cost to make x items is given by the linear cost function $C(x) = mx + b$. Then the fixed cost (the cost which occurs even if no items are produced) is found by letting $x = 0$.

$$C(0) = m \cdot 0 + b = b$$

Thus, the fixed cost is the y-intercept of the cost function.

In economics, **marginal cost** is the rate of change of cost. Marginal cost is important to management in making decisions in areas such as cost control, pricing, and production planning. If the cost function is $C(x) = mx + b$, then its graph is a straight line with slope m. Since the slope represents the average rate of change, the marginal cost is the number m. **3**

In Example 4, the marginal cost (slope) 12 was also the cost of producing one more radio. We now show that the same thing is true for any linear cost function $C(x) = mx + b$. If n items have been produced, the cost of the $(n + 1)$st item is the difference between the cost of producing $n + 1$ items and the cost of producing n items, namely, $C(n + 1) - C(n)$. Since $C(x) = mx + b$,

$$\text{Cost of one more item} = C(n + 1) - C(n)$$
$$= [m(n + 1) + b] - [mn + b]$$
$$= mn + m + b - mn - b = m$$
$$= \text{marginal cost.}$$

4 The total cost of producing 10 units of a business calculator is $220. The marginal cost per calculator is $14. Find the cost function, $C(x)$, if it is linear.

Answer:
$C(x) = 14x + 80$

This discussion is summarized as follows.

In a **linear cost function** $C(x) = mx + b$, m represents the marginal cost and b the fixed cost. The marginal cost is the cost of producing one more item.

Conversely, if the fixed cost is b and the marginal cost is always the same constant m, then the cost function for producing x items is $C(x) = mx + b$.

If $C(x)$ is the total cost to manufacture x items, then the **average cost** per item is given by

$$\overline{C}(x) = \frac{C(x)}{x}.$$

In Example 4, the average cost per clock-radio is

$$\overline{C}(x) = \frac{C(x)}{x} = \frac{12x + 100}{x} = 12 + \frac{100}{x}.$$

Note what happens to the term $100/x$ as x gets larger.

$$\overline{C}(100) = 12 + \frac{100}{100} = 12 + 1 = \$13$$

$$\overline{C}(500) = 12 + \frac{100}{500} = 12 + \frac{1}{5} = \$12.20$$

$$\overline{C}(1000) = 12 + \frac{100}{1000} = 12 + \frac{1}{10} = \$12.10$$

As more and more items are produced, $100/x$ gets smaller and the average cost per item decreases.

▶**EXAMPLE 5**   The marginal cost for raising a certain type of fruit fly for laboratory study is $12 per unit of fruit flies, while the cost to produce 100 units is $1500.
(a) Find the cost function $C(x)$, given that it is linear.

Since the cost function is linear, it can be written in the form $C(x) = mx + b$. The marginal cost is $12 per unit, which gives the value for m, leading to $C(x) = 12x + b$. To find b, use the fact that the cost of producing 100 units of fruit flies is $1500, or $C(100) = 1500$. Substituting $x = 100$ and $C(x) = 1500$ into $C(x) = 12x + b$ gives

$$C(x) = 12x + b$$
$$1500 = 12 \cdot 100 + b$$
$$1500 = 1200 + b$$
$$300 = b.$$

The cost function is $C(x) = 12x + 300$, where the fixed cost is $300. **4**

5 In Problem 4, at the side, find the average cost per calculator to produce 100 calculators.

Answer:
$14.80

(b) Find the average cost per unit to produce 50 units and 500 units.

The average cost per unit is

$$\overline{C}(x) = \frac{C(x)}{x} = \frac{12x + 300}{x} = 12 + \frac{300}{x}.$$

If 50 units of fruit flies are produced, the average cost is

$$\overline{C}(50) = 12 + \frac{300}{50} = 18,$$

or $18 per unit. Producing 500 units of fruit flies will lead to an average cost of

$$\overline{C}(500) = 12 + \frac{300}{500} = 12\frac{3}{5},$$

or $12.60 per unit. ◀ **5**

BREAK-EVEN ANALYSIS   The **revenue** $R(x)$ from selling x units of a product is the product of the price per unit p and the number of units sold (demand) x, so that

$$R(x) = px.$$

The corresponding **profit** $P(x)$ is the difference between revenue $R(x)$ and cost $C(x)$. That is,

$$P(x) = R(x) - C(x).$$

A company can make a profit only if the revenue received from its customers exceeds the cost of producing its goods and services. The number of units at which revenue just equals cost is the **break-even point.**

▶ **EXAMPLE 6**   A firm producing poultry feed finds that the total cost $C(x)$ of producing x units is given by

$$C(x) = 20x + 100.$$

Management plans to charge $24 per unit for the feed.
(a) How many units must be sold for the firm to break even?

The firm will break even (no profit and no loss) as long as revenue just equals cost, or $R(x) = C(x)$. From the given information, since $R(x) = px$ and $p = \$24$,

$$R(x) = 24x.$$

Substituting for $R(x)$ and $C(x)$ in the equation $R(x) = C(x)$ gives

$$24x = 20x + 100,$$

from which $x = 25$. The firm breaks even by selling 25 units. The graphs of $C(x) = 20x + 100$ and $R(x) = 24x$ are shown in Figure 2.30. The break-even point (where $x = 25$) is shown on the graph. If the company produces more than 25 units (if $x > 25$), it makes a profit. If it produces less than 25 units, it loses money.

6 For a certain magazine,
$C(x) = .70x + 1200$, where x is
the number of magazines sold.
The magazine sells for $1 per
copy. Find the break-even point.

Answer:
4000 magazines

(b) What is the profit if 100 units of feed are sold?
Use the formula for profit $P(x)$.

$$P(x) = R(x) - C(x)$$
$$= 24x - (20x + 100)$$
$$= 4x - 100$$

Then $P(100) = 4(100) - 100 = 300$. The firm will make a profit of $300 from the sale of 100 units of feed. ◀ 6

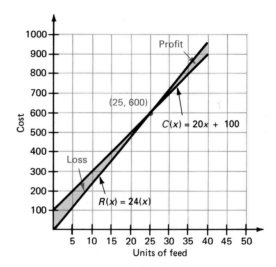

FIGURE 2.30

2.5 EXERCISES

1. **Management** Suppose the sales of a particular brand of electric guitar are

$$S(x) = 300x + 2000,$$

where $S(x)$ represents the number of guitars sold in year x, with $x = 0$ corresponding to 1988. Find the sales in each of the following years.
(a) 1990 **(b)** 1991 **(c)** 1992 **(d)** 1988
(e) Find the annual rate of change of the sales.

2. **Natural Science** If the population of ants in an anthill is

$$A(x) = 1000x + 6000,$$

where $A(x)$ represents the number of ants present at the end of month x, find the number of ants present at the end of each of the following months. Let $x = 0$ represent June.

(a) June **(b)** July **(c)** August **(d)** December
(e) What is the monthly rate of change of the number of ants?

3. **Natural Science** Let $N(x) = -5x + 100$ represent the number of bacteria (in thousands) present in a certain tissue culture at time x, measured in hours, after an antibacterial spray is introduced into the environment. Find the number of bacteria present at each of the following times.
(a) $x = 0$ **(b)** $x = 6$ **(c)** $x = 20$
(d) What is the hourly rate of change in the number of bacteria? Interpret the negative sign in the answer.

4. Let $R(x) = -8x + 240$ represent the number of students in a large business mathematics class, where x represents the number of hours of study required weekly. Find the num-

ber of students present at each of the following levels of required study.

(a) $x = 0$ (b) $x = 5$ (c) $x = 10$

(d) What is the rate of change of the number of students in the class with respect to the number of hours of study. Interpret the negative sign in the answer.

(e) The professor in charge of the class likes to have exactly 16 students. How many hours of study must be required in order to have exactly 16 students? (Hint: find a value of x such that $R(x) = 16$.)

5. **Management** Assume that the sales of a certain automobile parts company are approximated by a linear function. Suppose that sales were $200,000 in 1984 and $1,000,000 in 1991. Let $x = 0$ represent 1984 and $x = 7$ represent 1991.

(a) Find the equation giving the company's yearly sales.

(b) Find the sales in 1986.

(c) Estimate the sales in 1993.

6. **Management** Assume that the sales of a certain appliance dealer are approximated by a linear function. Suppose that sales were $850,000 in 1986 and $1,262,500 in 1991. Let $x = 0$ represent 1986.

(a) Find the equation giving the dealer's yearly sales.

(b) What were the dealer's sales in 1989?

(c) Estimate sales in 1994.

7. Over a recent 3-year period, medical expenses in the U.S. rose in a linear pattern from 10.3% of the gross national product in year 0 to 11.2% in year 3.

(a) Assuming that this change in medical expenses continues to be linear, write a linear function describing the percent of the gross national product devoted to medical expenses, y, in terms of the year, x.

(b) Give the average rate of change of medical expenses from year 0 to 3. Compare it with the slope of the line in part (a). What do you find?

8. The number of children in the U.S. from 5 to 13 years old decreased from 31.2 million in 1980 to 30.3 million in 1986.

(a) Write a linear function describing this population y in terms of year x for the given period.

(b) What was the average rate of change in this population over the period from 1980 to 1986?

9. **Management** Suppose the number of bottles of a vitamin, $V(x)$, on hand at the beginning of the day in a health food store is given by

$$V(x) = 600 - 20x,$$

where $x = 1$ corresponds to June 1 and x is measured in days. If the store is open every day of the month, find the number of bottles on hand at the beginning of each of the following days.

(a) June 6 (b) June 12 (c) June 24

(d) When will the last bottle from this stock be sold?

(e) What is the average daily rate of change of this stock?

10. **Social Science** In psychology, the just-noticeable-difference (JND) for some stimulus is defined as the amount by which the stimulus must be increased so that a person will perceive it as having just barely increased. For example, suppose a research study indicates that a line 40 centimeters in length must be increased to 42 centimeters before a subject thinks that it is longer. In this case, the JND would be $42 - 40 = 2$ centimeters. In a particular experiment, the JND (y) is given by

$$y = .03x,$$

where x represents the original length of the line. Find the JND for lines having the following lengths.

(a) 10 centimeters (b) 20 centimeters

(c) 50 centimeters (d) 100 centimeters

(e) Find the rate of change in the JND with respect to the original length of the line.

Management *Write a cost function for each of the following. Identify all variables used. (See Example 5.)*

11. A chain-saw rental firm charges $12 plus $1 per hour.

12. A trailer-hauling service charges $45 plus $2 per mile.

13. A parking garage charges 35¢ plus 30¢ per half hour.

14. For a 1-day rental, a car rental firm charges $14 plus 6¢ per mile.

Management *Assume that each of the following can be expressed as a linear cost function. Find the appropriate cost function in each case. (See Example 5.)*

15. Fixed cost, $100; 50 items cost $1600 to produce.

16. Fixed cost, $400; 10 items cost $650 to produce.

17. Fixed cost, $1000; 40 items cost $2000 to produce.

18. Fixed cost, $8500; 75 items cost $11,875 to produce.

19. Marginal cost, $50; 80 items cost $4500 to produce.

20. Marginal cost, $120; 100 items cost $15,800 to produce.

21. Marginal cost, $90; 150 items cost $16,000 to produce.

22. Marginal cost, $120; 700 items cost $96,500 to produce.

23. Management The manager of a local restaurant told us that his marginal cost per cup of coffee is $.125. If his fixed cost is $20, find the total cost of producing the following numbers of cups of coffee.
(a) 1000 cups   (b) 1001 cups

24. Management In deciding whether or not to set up a new manufacturing plant, company analysts have decided that a reasonable function for the total cost to produce x items is

$$C(x) = 500,000 + 4.75x.$$

(a) Find the total cost to produce 100,000 items.
(b) Find the marginal cost per item of the items to be produced in this plant.

Management *Assume that each of the following can be expressed as a linear cost function. Find* (a) *the cost function;* (b) *the revenue function;* (c) *the break-even point. (See Example 6.)*

	Fixed Cost	Marginal Cost per Item	Item Sells for
25.	$500	$10	$35
26.	$180	$11	$20
27.	$250	$18	$28
28.	$1500	$30	$80
29.	$2700	$100	$125
30.	$165	$20	$25

Find the average cost per item for x items in Exercises 31 and 32. (See Example 5.)

31. $C(x) = 800 + 20x$ (a) 10   (b) 50   (c) 200

32. $C(x) = 500,000 + 4.75x$ (a) 1000   (b) 5000
(c) 10,000

The average rate of change *of a function f(x) as x changes from c to d is* $\dfrac{f(d) - f(c)}{d - c}$, *which is the slope of the straight line joining the points (c, f(c)) and (d, f(d)) on the graph of f. In Exercises 33 and 34 you are asked to find average rates of change for some nonlinear functions.*

33. Management The graph below shows the total sales, y, in thousands of dollars from the distribution of x thousand catalogs. Find the average rate of change of sales with respect to the number of catalogs distributed for the following changes in x.
(a) 10 to 20   (b) 10 to 40
(c) 20 to 30   (d) 30 to 40

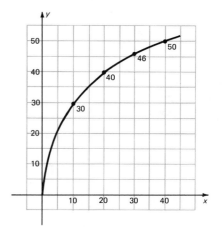

34. Management The graph below shows annual sales (in units) of a typical product. Sales increase slowly at first to some peak, hold steady for a while, and then decline as the product goes out of style. Find the average annual rate of change in sales for the following changes in years.
(a) 1 to 3   (b) 2 to 4   (c) 3 to 6
(d) 5 to 7   (e) 7 to 9   (f) 8 to 11
(g) 9 to 10   (h) 10 to 12

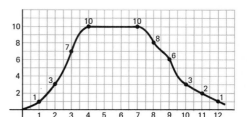

35. The cost to produce x units of wire is $C(x) = 50x + 5000$, while the revenue is $R(x) = 60x$. Find the break-even point and the revenue at the break-even point.

36. The cost to produce x units of squash is given by $C(x) = 100x + 6000$, while the revenue is $R(x) = 500x$. Find the break-even point.

Management *Suppose that you are the manager of a firm. You are considering the manufacture of a new product, so you ask the accounting department to produce cost estimates and the sales department to produce sales estimates. After you receive the data, you must decide whether to go ahead with production of the new product. Analyze the following data (find a break-even point) and then decide what you would do. (See Example 6.)*

37. $C(x) = 105x + 6000$; $R(x) = 250x$; no more than 400 units can be sold.

38. $C(x) = 275x + 5625$; $R(x) = 320x$; no more than 200 units can be sold.

39. $C(x) = 80x + 7000$; $R(x) = 95x$; no more than 400 units can be sold.

40. $C(x) = 65x + 9500$; $R(x) = 80x$; no more than 600 units can be sold.

41. $C(x) = 140x + 3000$; $R(x) = 125x$ (Hint: what does a negative value of x mean?)

42. $C(x) = 1750x + 95,000$; $R(x) = 1750x$

43. The sales of a certain furniture company in thousands of dollars are shown in the chart below.

x (year)	y (sales)
0	48
1	59
2	66
3	75
4	80
5	90

(a) Graph this data, plotting years on the x-axis and sales on the y-axis. (Note that the data points can be closely approximated by a straight line.)

(b) Draw a line through the points (2, 66) and (5, 90). The other four points should be close to this line. (These two points were selected as "best" representing the line that could be drawn through the data points.)

(c) Use the two points of (b) to find an equation for the line that approximates the data.

(d) Complete the following chart.

Year	Sales (actual)	Sales (predicted from equation of (c))	Difference, Actual Minus Predicted
0			
1			
2			
3			
4			
5			

(e) Use the result of (c) to predict sales in year 7.

(f) Do the same for year 9.

44. **Social Science** Most people are not very good at estimating the passage of time. Some people's estimations are too fast, and others' are too slow. One psychologist has constructed a mathematical model for actual time as a function of estimated time: if y represents actual time and x estimated time, then

$$y = mx + b,$$

where m and b are constants that must be determined experimentally for each person.

Suppose that for a particular person, $m = 1.25$ and $b = -5$. Find y if x is

(a) 30 minutes; (b) 60 minutes;

(c) 120 minutes; (d) 180 minutes.

Suppose that for another person, $m = .85$ and $b = 1.2$. Find y if x is

(e) 15 minutes; (f) 30 minutes;

(g) 60 minutes; (h) 120 minutes.

For this same person, find x if y is

(i) 60 minutes; (j) 90 minutes.

KEY TERMS AND SYMBOLS

2.1 $f(x)$ "f of x"
independent variable
dependent variable
function
domain
range

2.2 graph
Cartesian coordinate system
x-coordinate
y-coordinate
quadrant
x-intercept
y-intercept
absolute value function
piecewise function
greatest integer function
step function
vertical line test

2.3 linear function
supply and demand curves
equilibrium price
equilibrium supply/demand

2.4 Δx change in x
Δy change in y
m slope
slope-intercept form
point-slope form

2.5 average rate of change
fixed cost
marginal cost
linear cost function
average cost
revenue
profit
break-even point

KEY CONCEPTS

A **function** is a rule that assigns to each element of a set X exactly one element from a set Y.

The **domain** of a function is the set of all values of the independent variable.

The **range** of a function is the set of all values of the dependent variable.

If a vertical line intersects the graph of a function in more than one point, the graph is not that of a function.

If $p = f(x)$ gives the price per unit when x units can be supplied, and $p = g(x)$ gives the price per unit when x units are demanded, then the **equilibrium price, supply,** and **demand** occur at the x-value such that $f(x) = g(x)$.

The **slope** of the line through the points (x_1, y_1) and (x_2, y_2), where $x_1 \neq x_2$ is $m = \dfrac{y_2 - y_1}{x_2 - x_1}$.

Nonvertical **parallel lines** have the same slope and **perpendicular lines** have slopes with a product of -1.

The line with equation $ax + by = c$ has x-intercept c/a and y-intercept c/b.

The line with equation $x = k$ is vertical with x-intercept k, no y-intercept, and undefined slope.

The line with equation $y = k$ is horizontal, with y-intercept k, no x-intercept, and slope 0.

The line with equation $y = mx + b$ has slope m and y-intercept b.

The line with equation $y - y_1 = m(x - x_1)$ has slope m and goes through (x_1, y_1).

A **linear cost function** has equation $C(x) = mx + b$ where m is the **marginal cost** (the cost of producing one more item) and b is the **fixed cost.** If $R(x)$ is the revenue function, the **break-even point** is the x-value such that $C(x) = R(x)$.

CHAPTER 2   REVIEW EXERCISES

Graph each of the following, assuming y is a function of x. Give the domain and range.

1. $2x - 5y = 10$ **2.** $3x + 7y = 21$ **3.** $f(x) = (2x + 1)(x - 1)$ **4.** $f(x) = (x + 4)(x + 3)$

5. $f(x) = -2 + x^2$ **6.** $f(x) = 3x^2 - 7$ **7.** $f(x) = \dfrac{2}{x^2 + 1}$ **8.** $f(x) = \dfrac{-3 + x}{x + 10}$

9. $y + 1 = 0$ **10.** $f(x) = 3$

For each function defined as follows, find **(a)** $f(6)$, **(b)** $f(-2)$, **(c)** $f(p)$, **(d)** $f(r + 1)$.

11. $f(x) = 4x - 1$ **12.** $f(x) = 3 - 4x$ **13.** $f(x) = -x^2 + 2x - 4$ **14.** $f(x) = 8 - x - x^2$

15. Let $f(x) = 5x - 3$ and $g(x) = -x^2 + 4x$. Find each of the following.
    (a) $f(-2)$ **(b)** $g(3)$ **(c)** $g(-4)$ **(d)** $f(5)$ **(e)** $g(-k)$ **(f)** $g(3m)$ **(g)** $g(k - 5)$ **(h)** $f(3 - p)$

Graph each of the following functions.

16. $f(x) = -|x|$ **17.** $f(x) = |x| - 3$ **18.** $f(x) = -|x| - 2$ **19.** $f(x) = -|x + 1| + 3$

20. $f(x) = 2|x - 3| - 4$ **21.** $f(x) = [x - 3]$ **22.** $f(x) = [\frac{1}{2}x - 2]$

23. $f(x) = \begin{cases} -4x + 2 & \text{if } x \le 1 \\ 3x - 5 & \text{if } x > 1 \end{cases}$ **24.** $f(x) = \begin{cases} 3x + 1 & \text{if } x < 2 \\ -x + 4 & \text{if } x \ge 2 \end{cases}$ **25.** $f(x) = \begin{cases} |x| & \text{if } x < 3 \\ 6 - x & \text{if } x \ge 3 \end{cases}$

---

26. A trailer hauling service charges $45, plus $2 per mile or part of a mile. Graph the ordered pairs (miles, cost).

27. Let f be a function that gives the cost to rent a floor polisher for x hours. The cost is a flat $3 for cleaning the polisher plus $4 per day or fraction of a day for using the polisher.
    (a) Graph f.
    (b) Give the domain and range of f.

28. Assume that it costs 30¢ to mail a letter weighing 1 ounce or less, with each additional ounce, or portion of an ounce, costing 27¢. Let $C(x)$ represent the cost to mail a letter weighing x ounces. Find the cost of mailing letters of the following weights.
    (a) 3.4 ounces   **(b)** 1.02 ounces
    (c) 5.9 ounces   **(d)** 10 ounces
    (e) Graph C.
    (f) Give the domain and range for C.

Graph each of the following lines. Find the slope in Exercises 29–36, and find an equation of the line in Exercises 37–40.

29. $y = 4x + 3$ **30.** $y = 6 - 2x$ **31.** $3x - 5y = 15$

32. $2x + 7y = 14$ **33.** $x + 2 = 0$ **34.** $y = 1$

35. $y = 2x$ **36.** $x + 3y = 0$ **37.** Through $(2, -4)$, $m = 3/4$

38. Through $(0, 5)$, $m = -2/3$ **39.** Through $(-4, 1)$, $m = 3$ **40.** Through $(-3, -2)$, $m = -1$

41. The supply and demand for a certain commodity are related by

  supply: $p = 6x + 3$;   demand: $p = 19 - 2x$,

where p represents the price at a supply or demand, respectively, of x units. Find the supply and the demand when the price is
(a) 10;   (b) 15;   (c) 18.

(d) Graph both the supply and the demand functions on the same axes.
(e) Find the equilibrium price.
(f) Find the equilibrium supply; the equilibrium demand.

42. For a particular product, 72 units will be supplied at a price of 6, while 104 units will be supplied at a price of 10. Write a linear supply function for this product.

Find the slope of the line through each of the following pairs of points.

43. $(-2, 5)$ and $(4, 7)$ **44.** $(4, -1)$ and $(3, -3)$ **45.** $(0, 0)$ and $(11, -2)$ **46.** $(0, 0)$ and $(0, 7)$

Find the slope of each of the following lines.

47. $2x + 3y = 15$ **48.** $4x - y = 7$ **49.** $x + 4 = 9$ **50.** $3y - 1 = 14$

51. $y = x$ **52.** $y = -2$ **53.** $x + 5y = 0$ **54.** $y - 3x = 0$

Find an equation for each of the following lines.

55. Through $(5, -1)$, slope 2/3   **56.** Through $(8, 0)$, slope $-1/4$ **57.** Through $(5, -2)$ and $(1, 3)$

58. Through $(2, -3)$ and $(-3, 4)$ **59.** Undefined slope, through $(-1, 4)$ **60.** Slope 0, through $(-2, 5)$

61. x-intercept -3, y-intercept 5   **62.** x-intercept $-2/3$, y-intercept 1/2

Find each of the following linear cost functions.

63. Eight units of paper cost $300; fixed cost is $60.

64. Fixed cost is $2000; 36 units cost $8480.

65. Twelve units cost $445; 50 units cost $1585.

66. Thirty units cost $1500; 120 units cost $5640.

67. Eighty units cost $2735; 175 units cost $5775.

68. The graph at the right shows revenues and operating income in millions of dollars for Phoenix Technologies, Ltd. for several years.* Estimate the average rate of change of revenues or income over the following intervals.
(a) Revenue: 1986–90   (b) Revenue: 1989–90
(c) Income: 1986–90   (d) Income: 1988–90

69. The cost of producing x units of a product is $C(x)$, where $C(x) = 20x + 100$. The product sells for $40 per unit.
(a) Find the break-even point.
(b) What revenue will the company receive if it sells just the number of units from part (a)?

70. A product can be sold for $25 per unit. The cost of producing x units in a certain plant is $C(x) = 24x + 5000$. Should the item be produced?

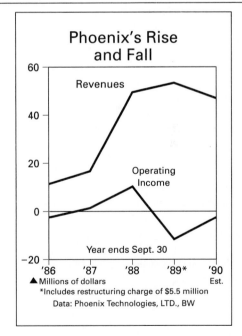

Phoenix's Rise and Fall

Revenues

Operating Income

Year ends Sept. 30

'86 '87 '88 '89* '90
Est.

▲ Millions of dollars
*Includes restructuring charge of $5.5 million
Data: Phoenix Technologies, LTD., BW

---

*"Phoenix Keeps Getting Its Feathers Ruffled," *Business Week*, January 22, 1990.

Booz, Allen and Hamilton is a large management consulting firm. One of the services it provides to client companies is profitability studies, which show ways in which the client can increase profit levels. The client company requesting the analysis presented in this case is a large producer of a staple food. The company buys from farmers, and then processes the food in its mills, resulting in a finished product. The company sells both at retail and under its own brands, and in bulk to other companies who use the product in the manufacture of convenience foods.

The client company has been reasonably profitable in recent years, but the management retained Booz, Allen and Hamilton to see whether its consultants could suggest ways of increasing company profits. The management of the company had long operated with the philosophy of trying to process and sell as much of its product as possible, since they felt this would lower the average processing cost per unit sold. However, the consultants found that the client's fixed mill costs were quite low, and that, in fact, processing extra units made the cost per unit start to increase. (There are several reasons for this: the company must run three shifts, machines break down more often, and so on.)

In this case, we shall discuss the marginal cost of two of the company's products. The marginal cost (cost of producing an extra unit) of production for product A was found by the consultants to be approximated by the linear function

$$y = .133x + 10.09,$$

where x is the number of units produced (in millions) and y is the marginal cost.

For example, at a level of production of 3.1 million units, an additional unit of product A would cost about

$$y = .133(3.1) + 10.09$$
$$\approx \$10.50.\dagger$$

At a level of production of 5.7 million units, an extra unit costs $10.85. Figure 1 shows a graph of the marginal cost function from $x = 3.1$ to $x = 5.7$, the domain to which the function above was found to apply.

The selling price for product A is $10.73 per unit, so that, as shown on the graph of Figure 1, the company was losing money on many units of the product that it sold. Since the selling price could not be raised if the company was to remain competitive, the consultants recommended that production of product A be cut.

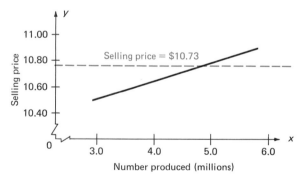

FIGURE 1

*Case study, "Marginal Cost—Booz, Allen and Hamilton" supplied by John R. Dowdle of Booz, Allen & Hamilton, Inc. Reprinted by permission.

†The symbol "$\approx$" means *is approximately equal to*.

For product B, the Booz, Allen and Hamilton consultants found a marginal cost function given by

$$y = .0667x + 10.29,$$

with x and y as defined above. Verify that at a production level of 3.1 million units, the marginal cost is $10.50, while at a production level of 5.7 million units, the marginal cost is $10.67. Since the selling price of this product is $9.65, the consultants again recommended a cutback in production.

The consultants ran similar cost analyses of other products made by the company, and then issued their recommendations: the company should reduce total production by 2.1 million units. The analysts predicted that this would raise profits for the products under discussion from $8.3 million annually to $9.6 million, which is very close to what actually happened when the client took this advice.

EXERCISES

1. At what level of production, x, was the marginal cost of a unit of product A equal to the selling price?

2. Graph the marginal cost function for product B from $x = 3.1$ million units to $x = 5.7$ million units.

3. Find the number of units for which marginal cost equals the selling price for product B.

4. For product C, the marginal cost of production is

$$y = .133x + 9.46.$$

   (a) Find the marginal cost at a level of production of 3.1 million units; of 5.7 million units.
   (b) Graph the marginal cost function.
   (c) For a selling price of $9.57, find the level of production for which the cost equals the selling price.

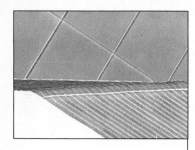

CHAPTER 3

Polynomial and Rational Functions

While the linear and piecewise functions of Chapter 2 are suitable for many different types of applications, not all real-world situations can be adequately approximated by these functions. In many cases, a curve is needed, and not a straight line. One important set of curves comes from the **polynomial functions,** those defined by polynomial expressions. The first two sections of this chapter discuss polynomial functions defined by quadratic expressions; more general polynomial functions are discussed in the third section. Rational functions, defined by quotients of polynomials, are discussed in Section 3.4.

3.1 QUADRATIC FUNCTIONS

By definition, a linear function is given by $f(x) = ax + b$, for real numbers a and b. Another type of function that is very useful in applications is a *quadratic function,* in which the independent variable is squared. A quadratic function is an especially good model for many situations with a maximum or a minimum function value. Quadratic functions also may be used to describe supply and demand curves, cost, revenue, and profit, as well as other quantities.

A function f is a **quadratic function** if
$$f(x) = ax^2 + bx + c,$$
where a, b, and c are real numbers with $a \neq 0$.

The restriction $a \neq 0$ is made to guarantee that the function is not linear.

1 Graph each of the following parabolas.

(a) $f(x) = x^2 - 4$

(b) $f(x) = x^2 + 5$

Answer:

(a)

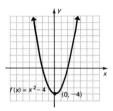

(b)

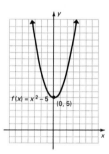

▶**EXAMPLE 1**   Graph the quadratic function defined by

$$f(x) = x^2.$$

The function defined by $f(x) = x^2$ is the simplest of all quadratic functions. (To see that $f(x) = x^2$ is quadratic, let $a = 1$, $b = 0$, and $c = 0$ in the general quadratic function defined by $f(x) = ax^2 + bx + c$.) Graph the equation $y = x^2$ by choosing several negative, zero, and positive values for x, and then finding the corresponding values for y, as in the table with Figure 3.1. The resulting points can then be plotted, with a smooth curve drawn through them, as in Figure 3.1. The domain of the function defined by $f(x) = x^2$ is the set of all real numbers, while the range is the set of all nonnegative real numbers. ◀

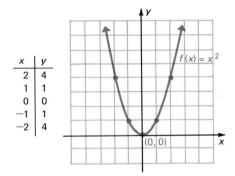

FIGURE 3.1

The curve in Figure 3.1 is called a **parabola.** The lowest point on this parabola, $(0, 0)$, is called the **vertex.** It can be shown that every quadratic fuction has a graph that is a parabola. If the graph of Figure 3.1 were folded in half along the y-axis, the two halves of the parabola would match exactly. This means that the graph of a quadratic function is *symmetric* about a vertical line through the vertex: this line is the **axis of the parabola.** The axis of the parabola in Figure 3.1 is $x = 0$, the y-axis.

Parabolas have many useful properties. Cross sections of radar dishes and spotlights form parabolas. Discs often visible on the sidelines of televised football games are microphones having reflectors with parabolic cross sections. These microphones are used by the television networks to pick up the shouted signals of the quarterbacks.

▶**EXAMPLE 2**   Graph the quadratic function defined by $f(x) = x^2 + 2$.

For any value of x that might be chosen, the corresponding value of y will be 2 more than for the parabola graphed in Figure 3.1. This means that the graph of the parabola $f(x) = x^2 + 2$, as shown in Figure 3.2, is shifted 2 units upward compared to the graph of $f(x) = x^2$. For example, $(2, 4)$ and $(-1, 1)$ are on the graph of $f(x) = x^2$, while the corresponding points on the graph of $f(x) = x^2 + 2$ are $(2, 6)$ and $(-1, 3)$. Also, the point $(0, 2)$ is the vertex of the parabola $f(x) = x^2 + 2$, while $(0, 0)$ is the vertex of the parabola $f(x) = x^2$. The axis is still the line $x = 0$. ◀ **1**

2 Graph $f(x) = -(x - 3)^2$.

Answer:

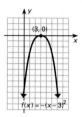

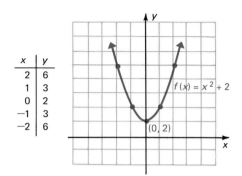

x	y
2	6
1	3
0	2
−1	3
−2	6

FIGURE 3.2

▶**EXAMPLE 3**   Graph the function defined by $f(x) = -(x + 2)^2$.

The table in Figure 3.3 shows several ordered pairs that belong to the function. These ordered pairs show that the graph of $f(x) = -(x + 2)^2$ is "upside down" in comparison to $f(x) = x^2$ (because of the negative sign) and also shifted 2 units to the left. Here the vertex, $(-2, 0)$, is the *highest* point on the graph. The axis of this parabola is the line $x = -2$.   ◀ **2**

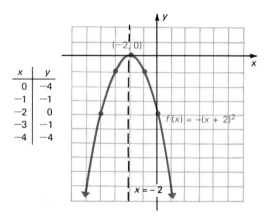

x	y
0	−4
−1	−1
−2	0
−3	−1
−4	−4

FIGURE 3.3

As shown in Example 2, the graph of $f(x) = x^2 + 2$ was shifted upward 2 units in comparison with the graph of $f(x) = x^2$. Similarly, in Example 3, the graph of $f(x) = -(x + 2)^2$ is shifted 2 units to the left when compared to $f(x) = x^2$, and the parabola opens downward. Both upward (or downward) and side-to-side shifts can occur in the same parabola.

3 Graph the following.

(a) $f(x) = (x + 4)^2 - 3$

(b) $f(x) = -2(x - 3)^2 + 1$

Answer:

(a)

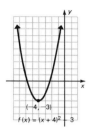

(b)

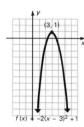

The facts about parabolas developed in the examples are summarized here.

> If f is a quadratic function defined by $y = a(x - h)^2 + k$, then the graph of the function f is a parabola having its vertex at (h, k) and axis of symmetry $x = h$.
>
> If $a > 0$, the parabola opens upward; if $a < 0$, it opens downward. If $0 < |a| < 1$, the parabola is "broader" than $y = x^2$, while if $|a| > 1$, the parabola is "narrower" than $y = x^2$.

▶ **EXAMPLE 4**   Graph $f(x) = 2(x - 3)^2 - 5$.

This parabola has vertex $(3, -5)$ and opens upward. The axis is the line $x = 3$. The coefficient 2 causes the values of y to increase more rapidly than in the parabola $f(x) = x^2$, so that the graph in this example is "narrower" than the graph of $f(x) = x^2$. Figure 3.4 shows a table of ordered pairs satisfying $f(x) = 2(x - 3)^2 - 5$, as well as the graph of the function. Since every parabola is symmetric about its axis, for each ordered pair on one side of the axis, symmetry can be used to find corresponding ordered pairs on the other side. Thus, the ordered pair $(1, 3)$ suggests the ordered pair $(5, 3)$ on the other side of the axis of symmetry. Similarly, $(2, -3)$ and $(4, -3)$ are symmetric with respect to the axis.   ◀ **3**

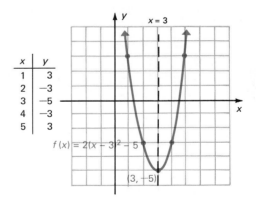

x	y
1	3
2	-3
3	-5
4	-3
5	3

$f(x) = 2(x - 3)^2 - 5$

FIGURE 3.4

Since the vertex of a parabola is the highest or lowest point on the parabola, it is the most important point for graphing or for applications of quadratic functions. The vertex and axis of a parabola can be found quickly if the equation of the parabola is in the form

$$f(x) = a(x - h)^2 + k$$

for real numbers $a \neq 0$, h and k. An equation not given in this form can be converted to it by a process called *completing the square,* first discussed in Section 1.10. This process is reviewed in the following example.

4 Complete the square for each of the following. Then graph the parabola.

(a) $f(x) = x^2 - 6x + 11$

(b) $f(x) = x^2 + 8x + 18$

Answer:

(a) $f(x) = (x - 3)^2 + 2$

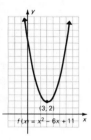

(b) $f(x) = (x + 4)^2 + 2$

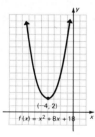

▶**EXAMPLE 5** Graph $f(x) = x^2 - 2x + 3$ by completing the square.

As mentioned, this parabola can be graphed by first completing the square to rewrite the equation in the form $f(x) = a(x - h)^2 + k$. Begin by writing,

$$f(x) = x^2 - 2x + 3$$

as

$$f(x) = (x^2 - 2x \quad) + 3.$$

The expression inside the parentheses must be written as the square of some quantity. Do this by taking half the coefficient of x, $(1/2)(-2) = -1$, and then squaring this result: $(-1)^2 = 1$.* Now add and subtract 1 inside the parentheses. This gives

$$f(x) = (x^2 - 2x + 1 - 1) + 3$$

or, by the associative and commutative properties,

$$f(x) = (x^2 - 2x + 1) + (-1 + 3).$$

Factor $x^2 - 2x + 1$ as $(x - 1)^2$, to get

$$f(x) = (x - 1)^2 + 2.$$

By this result, the graph of the given function is a parabola with the vertex at $(1, 2)$. Since $a = 1$ is positive, the parabola opens upward. Find a few additional points by choosing, say, $x = 2$ and $x = 3$ to get the ordered pairs $(2, 3)$ and $(3, 6)$. By symmetry, the ordered pairs $(0, 3)$ and $(-1, 6)$ are also on the graph. Use these points to get the graph shown in Figure 3.5. ◀ **4**

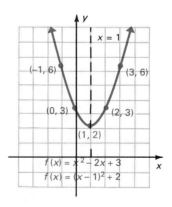

FIGURE 3.5

*Take *half* the coefficient of x since $(x + b)^2 = x^2 + 2bx + b^2$. The constant b can be found by setting $2b$ equal to the coefficient of x and solving. Find b^2 by squaring b.

5 Complete the square and graph the following.

(a) $f(x) = 3x^2 - 12x + 14$

(b) $f(x) = -x^2 + 6x - 12$

Answer:

(a) $f(x) = 3(x - 2)^2 + 2$

$f(x) = 3x^2 - 12x + 14$

(2, 2)

(b) $f(x) = -(x - 3)^2 - 3$

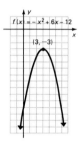

$f(x) = -x^2 + 6x - 12$

(3, −3)

▶**EXAMPLE 6**   Graph $f(x) = -2x^2 + 12x - 19$.

Get $f(x)$ in the form $f(x) = a(x - h)^2 + k$ by first factoring -2 from $-2x^2 + 12x$.

$$f(x) = -2x^2 + 12x - 19 = -2(x^2 - 6x) - 19$$

Take half of -6: $(1/2)(-6) = -3$. Square this result: $(-3)^2 = 9$. Add and subtract 9 inside the parentheses.

$$y = -2(x^2 - 6x + 9 - 9) - 19$$
$$y = -2(x^2 - 6x + 9) + (-2)(-9) - 19 \qquad \text{**Distributive property**}$$

Simplify and factor to get

$$y = -2(x - 3)^2 - 1.$$

Since $a = -2$ is negative, the parabola opens downward. The vertex is at $(3, -1)$, and the parabola is narrower than $y = x^2$. Use these results and plot additional ordered pairs as needed to get the graph of Figure 3.6.   ◀ **5**

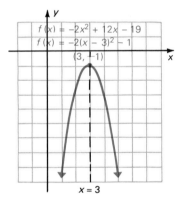

FIGURE 3.6

Rewriting the general equation $f(x) = ax^2 + bx + c$ in the form $f(x) = a(x - h)^2 + k$ by completing the square, we can show that

$$h = \frac{-b}{2a} \quad \text{and} \quad k = f(h).$$

Using these results with the function from Example 6, with $a = -2$ and $b = 12$, gives

$$h = \frac{-12}{2(-2)} = 3 \quad \text{and} \quad k = f(3) = -2(3)^2 + 12(3) - 19$$
$$= -18 + 36 - 19$$
$$= -1.$$

Thus, the vertex is $(3, -1)$, the same result as in Example 6.

6 Given $f(x) = x^2 - 7x + 12$, find the following.

(a) The intercepts

(b) The vertex and axis

Answer:

(a) $x = 3$, $x = 4$, $y = 12$

(b) $(7/2, -1/4)$; $x = 7/2$

In Chapter 2 we used the intercepts to graph linear functions. The intercepts also are useful in graphing quadratic functions.

▶**EXAMPLE 7**   Graph $f(x) = x^2 - x - 6$.

Find the intercepts by setting each variable equal to 0.

Set $f(x) = y = 0$:

$$0 = x^2 - x - 6$$
$$0 = (x + 2)(x - 3)$$
$$x + 2 = 0 \quad \text{or} \quad x - 3 = 0$$
$$x = -2 \qquad x = 3$$

The x-intercepts are -2 and 3.

Set $x = 0$:

$$f(x) = y = 0^2 - 0 - 6 = -6$$

The y-intercept is -6.

Since $a = 1$ and $b = -1$, the x-value of the vertex is

$$\frac{-b}{2a} = \frac{-(-1)}{2 \cdot 1} = \frac{1}{2}.$$

The y-value of the vertex is

$$f\left(\frac{1}{2}\right) = \left(\frac{1}{2}\right)^2 - \frac{1}{2} - 6 = -\frac{25}{4}.$$

The vertex is $(1/2, -25/4)$ and the axis of the parabola is $x = 1/2$. Plot the vertex, the three intercepts, and use symmetry to complete the graph, shown in Figure 3.7. ◀ **6**

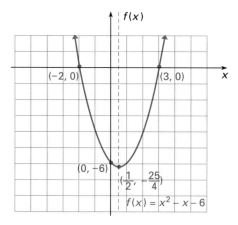

FIGURE 3.7

3.1 EXERCISES

1. Graph the functions defined in parts (a)–(d) on the same coordinate system.
   (a) $f(x) = 2x^2$ **(b)** $f(x) = 3x^2$
   (c) $f(x) = \frac{1}{2}x^2$ **(d)** $f(x) = \frac{1}{3}x^2$
   (e) How does the coefficient affect the shape of the graph?

2. Graph the functions defined in parts (a)–(d) on the same coordinate system.
   (a) $f(x) = \frac{1}{2}x^2$ **(b)** $f(x) = -\frac{1}{2}x^2$
   (c) $f(x) = 4x^2$ **(d)** $f(x) = -4x^2$
   (e) What effect does the minus sign have on the graph?

3. Graph the functions defined in parts (a)–(d) on the same coordinate system.
   (a) $f(x) = x^2 + 2$ **(b)** $f(x) = x^2 - 1$
   (c) $f(x) = x^2 + 1$ **(d)** $f(x) = x^2 - 2$
   (e) How do these graphs differ from the graph of $f(x) = x^2$?

4. Graph the functions defined in parts (a)–(d) on the same coordinate system.
   (a) $f(x) = (x - 2)^2$ **(b)** $f(x) = (x + 1)^2$
   (c) $f(x) = (x + 3)^2$ **(d)** $f(x) = (x - 4)^2$
   (e) How do these graphs differ from the graph of $f(x) = x^2$?

Graph the following parabolas. Find the vertex and axis of symmetry of each. (See Examples 1–7.)

5. $f(x) = -x^2$

6. $f(x) = -\frac{1}{2}x^2$

7. $f(x) = -x^2 + 1$

8. $f(x) = x^2 - 3$

9. $f(x) = 3x^2 - 2$

10. $f(x) = -2x^2 + 4$

11. $f(x) = (x + 2)^2$

12. $f(x) = (x - 3)^2$

13. $f(x) = -(x - 4)^2$

14. $f(x) = -(x + 5)^2$

15. $f(x) = -2(x - 3)^2$

16. $f(x) = -3(x + 4)^2$

17. $f(x) = (x - 1)^2 - 3$

18. $f(x) = (x - 2)^2 + 1$

19. $f(x) = -(x + 4)^2 + 2$

20. $f(x) = -(x + 1)^2 - 3$

21. $f(x) = x^2 - 4x + 6$

22. $f(x) = x^2 + 6x + 3$

23. $f(x) = x^2 + 12x + 1$

24. $f(x) = x^2 - 10x + 3$

25. $f(x) = 2x^2 + 4x + 1$

26. $f(x) = 3x^2 + 6x + 5$

27. $f(x) = 2x^2 - 4x + 5$

28. $f(x) = -3x^2 + 24x - 46$

29. $f(x) = -x^2 + 6x - 6$

30. $f(x) = -x^2 + 2x + 5$

Determine the following for each of the following functions.   **(a)** *Find the x-intercepts.*
(b) *For what values of f(x) is x = 0?*   **(c)** *For what values of x is f(x) a minimum?*
(d) *What is the minimum value of f(x)?*

31. $f(x) = x^2 - 2x - 15$

32. $f(x) = x^2 - 8x + 16$

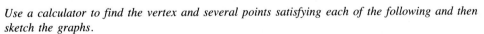

Use a calculator to find the vertex and several points satisfying each of the following and then sketch the graphs.

33. $f(x) = .14x^2 + .56x - .3$

34. $f(x) = .82x^2 + 3.28x - .4$

35. $f(x) = -.09x^2 - 1.8x + .5$

36. $f(x) = -.35x^2 + 2.8x - 3$

1 When a company sells x units of a product, its profit is $P(x) = -2x^2 + 40x + 280$. Find

(a) the number of units that should be sold so that maximum profit is received;

(b) the maximum profit.

Answer:

(a) 10 units

(b) $480

3.2 APPLICATIONS OF QUADRATIC FUNCTIONS

The fact that the vertex of a parabola of the form $y = ax^2 + bx + c$ is the highest or lowest point on the graph can be used in applications to find a maximum or a minimum value. When $a > 0$, the graph opens upward, so the function has a minimum. When $a < 0$, the graph opens downward, producing a maximum.

▶ **EXAMPLE 1** Ms. Tilden owns and operates Aunt Emma's Blueberry Pies. She has hired a consultant to analyze her business operations. The consultant tells her that her profits, $P(x)$, from the sale of x units of pies, are given by

$$P(x) = 120x - x^2.$$

How many units of pies should be sold in order to maximize profit? What is the maximum profit?

The profit function can be rewritten as $P(x) = -x^2 + 120x + 0$. The graph of P is a parabola with vertex (60, 3600) opening downward. Since the parabola opens downward, the vertex leads to *maximum* profit. Figure 3.8 shows the portion of the profit function in Quadrant I, the only quadrant of interest here. (She cannot sell a negative number of pies, and she is not interested in a negative profit.) The maximum profit of $3600 is obtained when 60 units of pies are sold. Here the profit increases as more and more pies are sold up to the point $x = 60$, and then decreases as more and more pies are sold past this point. ◀ **1**

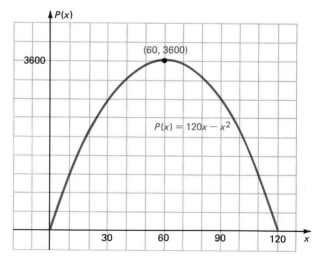

FIGURE 3.8

▶**EXAMPLE 2**   Suppose that the price and demand for an item are related by

$$p = 150 - 6x^2, \qquad \textbf{Demand function}$$

where p is the price and x is the number of items demanded (in hundreds). The price and supply are related by

$$p = 10x^2 + 2x, \qquad \textbf{Supply function}$$

where x is the supply of the item (in hundreds). Find the equilibrium demand (and supply) and the equilibrium price.

The graphs of both of these equations are parabolas; these parabolas can be graphed as shown in Section 3.1, giving the results shown in Figure 3.9. Only those portions of the graphs that lie in the first quadrant are included, since neither supply, demand, nor price can be negative.

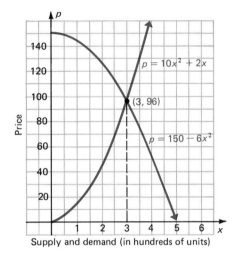

FIGURE 3.9

The equilibrium demand and supply, by definition, occur at the same price p. Setting $150 - 6x^2$, where x is the demand, equal to $10x^2 + 2x$, where x is the supply, gives an equation that leads to the equilibrium supply and demand.

$$150 - 6x^2 = 10x^2 + 2x$$

Write this quadratic equation in standard form as follows.

$$0 = 16x^2 + 2x - 150 \qquad \textbf{Add } -150 \textbf{ and } 6x^2 \textbf{ to both sides}$$

$$0 = 8x^2 + x - 75 \qquad \textbf{Multiply both sides by } \frac{1}{2}$$

2 The price and demand for an item are related by $p = 32 - x^2$, while price and supply are related by $p = x^2$. Find

(a) the equilibrium supply;

(b) the equilibrium price.

Answer:

(a) 4

(b) 16

This equation can be solved by the quadratic formula given in Section 1.10. Here $a = 8$, $b = 1$, and $c = -75$.

$$x = \frac{-1 \pm \sqrt{1 - 4(8)(-75)}}{2(8)}$$

$$= \frac{-1 \pm \sqrt{1 + 2400}}{16} \qquad -4(8)(-75) = 2400$$

$$= \frac{-1 \pm 49}{16} \qquad \sqrt{1 + 2400} = \sqrt{2401} = 49$$

$$x = \frac{-1 + 49}{16} = \frac{48}{16} = 3 \quad \text{or} \quad x = \frac{-1 - 49}{16} = -\frac{50}{16} = -\frac{25}{8}$$

It is not possible to make $-25/8$ units, so discard that answer, and use only $x = 3$. Since x represents supply and demand (in hundreds), equilibrium demand (and supply) is $3 \cdot 100$, or 300 units. Find the equilibrium price by substituting 3 for x in either the supply or the demand function. Using the supply function gives

$$p = 10x^2 + 2x$$
$$p = 10 \cdot 3^2 + 2 \cdot 3 \qquad \text{Let } x = 3$$
$$= 10 \cdot 9 + 6$$
$$p = 96. \quad \blacktriangleleft \quad \boxed{2}$$

▶**EXAMPLE 3** The rental manager of a small apartment complex with 16 units has found from experience that each \$40 increase in the monthly rent results in an empty apartment. All 16 apartments will be rented at a monthly rent of \$500. How many \$40 increases will produce maximum monthly income for the complex?

Let x represent the number of \$40 increases. Then the number of apartments rented will be $16 - x$. Also, the monthly rent per apartment will be $500 + 40x$. (There are x \$40 increases for a total increase of $40x$.) The monthly income, $I(x)$, is given by the number of apartments rented times the rent per apartment, so

$$I(x) = (16 - x)(500 + 40x)$$
$$= 8000 + 640x - 500x - 40x^2$$
$$= 8000 + 140x - 40x^2.$$

Since the value of x that will produce maximum income occurs at the vertex, use the methods of Section 3.1 to see that the vertex is $(7/4, 16245/2)$, or $(1.75, 8122.50)$. Since x represents the number of \$40 increases and each \$40 increase causes 1 empty apartment, x must be a whole number, either 1 or 2. Try $x = 1$ and $x = 2$ in the quadratic equation to see which one gives the best result.

$$\text{If } x = 1, \ I(x) = -40(1)^2 + 140(1) + 8000 = 8100.$$

$$\text{If } x = \mathbf{2}, \ I(x) = -40(2)^2 + 140(2) + 8000 = \mathbf{8120}.$$

These results show that the best answer is to charge rents of $500 + 2(40) = 580$, or $580, and leave 2 apartments vacant. As a check, $14(580) = 8120$, as found above. ◀

3.2 EXERCISES

Work each of the following problems. (See Example 1.)

1. **Management** George Duda runs a sandwich shop. By studying data for his past costs, he has found that a mathematical model describing the cost of operating his shop is given by

   $$C(x) = x^2 - 10x + 40,$$

   where $C(x)$ is the daily cost to make x units of sandwiches.
   (a) Complete the square for $C(x)$.
   (b) Graph the parabola resulting from part (a).
   (c) From the vertex of the parabola, find the number of units of sandwiches George must sell to produce minimum cost.
   (d) What is the minimum cost?

2. **Management** Josette Skelnik runs a taco stand. She has found that her profits are approximated by

   $$P(x) = -x^2 + 60x - 400,$$

   where $P(x)$ is the profit from selling x units of tacos.
   (a) Complete the square for $P(x)$.
   (b) Graph the parabola.
   (c) Find the number of units of tacos that Josette should make to produce maximum profit.
   (d) What is the maximum profit?

3. **Management** French fries produce a tremendous profit (150% to 300%) for many fast food restaurants. Management, therefore, desires to maximize the number of bags sold. Suppose that a mathematical model connecting p, the profit per day from french fries (in hundreds of dollars), and x, the price per bag (in dimes), is $p = -x^2 + 6x - 1$.
   (a) Find the price per bag that leads to maximum profit.
   (b) What is the maximum profit?

4. **Natural Science** A researcher in physiology has decided that a good mathematical model for the number of impulses fired after a nerve has been stimulated is given by $y = -x^2 + 20x - 60$, where y is the number of responses per millisecond and x is the number of milliseconds since the nerve was stimulated.
   (a) When will the maximum firing rate be reached?
   (b) What is the maximum firing rate?

5. **Natural Science** If an object is thrown upward with an initial velocity of 32 feet per second, then its height, in feet, above the ground after t seconds is given by

   $$h = 32t - 16t^2.$$

   Find the maximum height attained by the object. Find the number of seconds it takes for the object to hit the ground.

6. The number of mosquitoes in millions in a certain area, $M(x)$, depends on the June rainfall in inches, x, approximately as follows.

   $$M(x) = 10x - x^2$$

   Find the rainfall that will produce the maximum number of mosquitoes. What is the maximum number of mosquitoes?

7. Glenview Community College wants to construct a rectangular playing field on land bordered on one side by a building. The college maintenance department has 320 ft of fencing with which to enclose the other three sides. What should the dimensions of the field be if all the fencing is to be used and the field is to have maximum area?

8. In Exercise 7, what dimensions would produce the maximum area if all four sides were to be fenced?

Work each of the following problems. (See Example 2.)

9. Management   Suppose the price, p, is related to the demand, x, where x is measured in hundreds of items, by

$$p = 640 - 5x^2.$$

Find p when the demand is
(a) 0;   **(b)** 5;   **(c)** 10.
(d) Graph $p = 640 - 5x^2$.
Suppose that the price and supply of the item are related by $p = 5x^2$.
(e) Graph $p = 5x^2$ on the axes used in part (d).
(f) Find the equilibrium supply.
(g) Find the equilibrium price.

Management   *Suppose that the supply and demand of a certain textbook are related to price by*

$$\text{supply: } p = \frac{1}{5}x^2; \quad \text{demand: } p = -\frac{1}{5}x^2 + 40.$$

10. Estimate the demand at a price of
(a) 10;   **(b)** 20;   **(c)** 30;   **(d)** 40.
(e) Graph $p = -(1/5)x^2 + 40$.
Approximate the supply at a price of
(f) 5;   **(g)** 10;   **(h)** 20;   **(i)** 30.
(j) Graph $p = (1/5)x^2$ on the axes used in part (e).

11. Find the equilibrium demand in Exercise 9.

12. Find the equilibrium price in Exercise 10.

13. The manager of a bicycle shop has found that, at a price (in dollars) of $p(x) = 150 - \dfrac{x}{4}$ per bicycle, x bicycles will be sold.
(a) Find an expression for the total revenue from the sale of x bicycles. (Hint: revenue = demand × price.)
(b) Find the number of bicycle sales that leads to maximum revenue.
(c) Find the maximum revenue.

14. If the price (in dollars) of a videotape is $p(x) = 40 - \dfrac{x}{10}$, then x tapes will be sold.
(a) Find an expression for the total revenue from the sale of x tapes.
(b) Find the number of tapes that will produce maximum revenue.
(c) Find the maximum revenue.

15. Management   The demand for a certain type of cosmetic is given by

$$p = 500 - x,$$

where p is the price when x units are demanded.
(a) Find the revenue, $R(x)$, that would be obtained at a demand of x.
(b) Graph the revenue function, $R(x)$.
(c) From the graph of the revenue function, estimate the price that will produce the maximum revenue.
(d) What is the maximum revenue?

Work each problem. (See Example 3.)

16. Management   A charter flight charges a fare of $200 per person, plus $4 per person for each unsold seat on the plane. If the plane holds 100 passengers and if x represents the number of unsold seats, find the following.
(a) An expression for the total revenue received for the flight. (Hint: multiply the number of people flying, $100 - x$, by the price per ticket.)
(b) The graph for the expression of part (a).
(c) The number of unsold seats that will produce the maximum revenue.
(d) The maximum revenue.

17. The revenue of a charter bus company depends on the number of unsold seats. If 100 seats are sold, the price is $50. Each unsold seat increases the cost per passenger by $1. Let x represent the number of unsold seats.
(a) Write an expression for the number of seats that are sold.
(b) Write an expression for the price per seat.
(c) Write an expression for the revenue.
(d) Find the number of unsold seats that will produce the maximum revenue.
(e) Find the maximum revenue.

18. A group has rented a banquet hall for a class reunion. If 100 people attend, the cost of the hall per person will be $2. The cost per person will increase by $.10 for each person less than 100. What will be the maximum total cost for renting the hall if no more than 100 people attend? In this case, how many will attend and what will the cost per person be?

19. Natural Science   Between the months of June and October, the percent of maximum possible chlorophyll production in a leaf is approximated by $C(x)$, where

$$C(x) = 10x + 50.$$

Here x is time in months, with $x = 1$ representing June. From October through December, $C(x)$ is approximated by

$$C(x) = -20(x - 5)^2 + 100,$$

where x is as above. Find the percent of maximum possible chlorophyll production in each of the following months.

(a) June   **(b)** July   **(c)** September
(d) October   **(e)** November   **(f)** December

20. Natural Science   Use your results in Exercise 19 to sketch a graph of $y = C(x)$ from June through December. In which month is chlorophyll production a maximum?

21. A culvert is shaped like a parabola, 18 centimeters across the top and 12 centimeters deep. How wide is the culvert 8 centimeters from the top?

22. Let x be in the interval $0 \leq x \leq 1$. Use a graph to suggest that the product $x(1 - x)$ is always less than or equal to 1/4. For what values of x does the product equal 1/4?

Recall that profit equals revenue minus cost. In Exercises 23 and 24 find the following.

(a) The break-even point (to the nearest tenth)
(b) The x-value that makes revenue a maximum
(c) The x-value that makes profit a maximum
(d) The maximum profit
(e) For what x-values will a loss occur?
(f) For what x-values will a profit occur?

23. $R(x) = 60x - 2x^2$; $C(x) = 20x + 80$

24. $R(x) = 75x - 3x^2$; $C(x) = 21x + 65$

3.3 POLYNOMIAL FUNCTIONS

We have now discussed linear and quadratic functions and found their graphs. Both these functions are special types of *polynomial functions,* functions defined by a polynomial expression.

A function f is a **polynomial function of degree n,** where n is a nonnegative integer, if

$$f(x) = a_n x^n + a_{n-1} x^{n-1} + \ldots + a_1 x + a_0,$$

where $a_n, a_{n-1}, \ldots, a_1$ and a_0 are real numbers with $a_n \neq 0$.

For $n = 1$, a polynomial function takes the form

$$f(x) = a_1 x + a_0,$$

a linear function. A linear function, therefore, is a polynomial function of degree 1. (An exception: a linear function of the form $f(x) = a_0$ for a real number a_0 is a polynomial function of degree 0.) A polynomial function of degree 2 is a quadratic function that can be written in the form

$$f(x) = a_2 x^2 + a_1 x + a_0.$$

Perhaps the simplest polynomial functions are those defined by an expression of the form $y = x^n$, so some of these functions are graphed in the next examples.

1 Graph $f(x) = x^3 - 5$.

Answer:

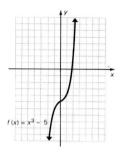

2 Graph the following.

(a) $f(x) = x^5 - 2$

(b) $f(x) = -x^3 + 3$

Answer:

(a)

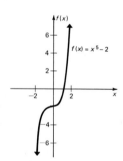

(b)

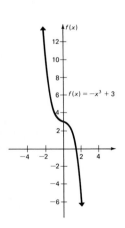

▶**EXAMPLE 1** Graph $f(x) = x^3$.

First, find several ordered pairs belonging to the graph. Be sure to choose some negative x-values, $x = 0$, and some positive x-values to get representative ordered pairs. Find as many ordered pairs as you need in order to see the shape of the graph. For example, you may need to locate additional points between $(1, 1)$ and $(2, 8)$. Use $x = 5/4$, $x = 3/2$, and $x = 7/4$ to find such points. Then plot the ordered pairs and draw a smooth curve through them, getting the graph in Figure 3.10. ◀ **1**

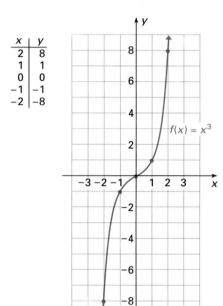

FIGURE 3.10

▶**EXAMPLE 2** Graph each of the following.

(a) $f(x) = (1/2)x^3$

The graph will be "broader" than that of $f(x) = x^3$, but will have the same general shape. The graph goes through the points $(-2, -4)$, $(-1, -1/2)$, $(0, 0)$, $(1, 1/2)$, and $(2, 4)$. See Figure 3.11.

(b) $f(x) = (3/2)x^4$

The following table gives some typical ordered pairs.

x	-2	-1	0	1	2
y	24	3/2	0	3/2	24

The graph is shown in Figure 3.12. ◀ **2**

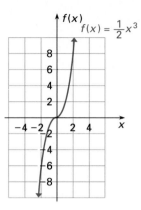

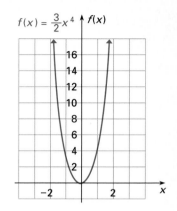

FIGURE 3.11   **FIGURE 3.12**

The graphs above suggest that the domain of every polynomial function is the set of all real numbers. The range of a polynomial function of odd degree (1, 3, 5, 7, and so on) is also the set of all real numbers. Graphs typical of polynomial functions of odd degree are shown in Figure 3.13.

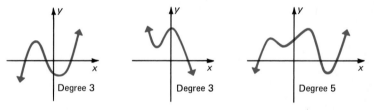

FIGURE 3.13

A polynomial function of even degree has a range that takes the form $y \leq k$ or $y \geq k$, for some real number k. Figure 3.14 shows two graphs of typical polynomial functions of even degree.

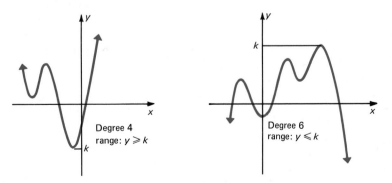

FIGURE 3.14

3 Identify the *x*-intercepts of each graph.

(a)

(b)

Answer:

(a) −2, 1, 2

(b) −3

It is difficult to graph most polynomial functions without the use of calculus. A large number of points must be plotted to get a reasonably accurate graph. However, if the expression defining a polynomial function can be factored, its graph can be approximated without plotting too many points, as shown in the next examples. Also, these examples show the *x*-intercepts used in drawing the graph. Recall that *x*-intercepts are found by letting $f(x) = 0$. **3**

▶**EXAMPLE 3** Graph $f(x) = (2x + 3)(x - 1)(x + 2)$.

Multiplying out the expression on the right would show that f is a third degree polynomial, called a *cubic* polynomial. Sketch the graph of f by first finding any *x*-intercepts; do this by setting $f(x) = 0$.

$$f(x) = 0$$
$$(2x + 3)(x - 1)(x + 2) = 0$$

Solve this equation by placing each of the three factors equal to 0.

$$2x + 3 = 0 \quad \text{or} \quad x - 1 = 0 \quad \text{or} \quad x + 2 = 0$$
$$x = -\frac{3}{2} \qquad\qquad x = 1 \qquad\qquad x = -2$$

The three numbers, $-3/2$, 1, and -2, divide the *x*-axis into four regions:

$$x < -2, \quad -2 < x < -\frac{3}{2}, \quad -\frac{3}{2} < x < 1, \quad \text{and} \quad 1 < x.$$

These regions are shown in Figure 3.15.

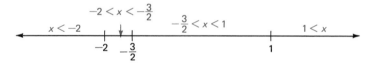

FIGURE 3.15

The sign of $f(x)$ can change from positive to negative or negative to positive only at an *x*-intercept, since $f(x) = 0$ at that point. Thus, in each of the regions in Figure 3.15 the values of $f(x)$ are either always positive or always negative. Find the sign of $f(x)$ in each region by selecting an *x*-value in each region and substituting to determine if the values of the function are positive or negative in that region. A typical selection of test points and the results of the tests are shown below.

Region	Test Point	Value of $f(x)$	Sign of $f(x)$
$x < -2$	-3	-12	negative
$-2 < x < -\dfrac{3}{2}$	$-\dfrac{7}{4}$	$\dfrac{11}{32}$	positive
$-\dfrac{3}{2} < x < 1$	0	-6	negative
$1 < x$	2	28	positive

4 Graph
$f(x) = (3x - 4)(x + 2)(x - 3)$.

Answer:

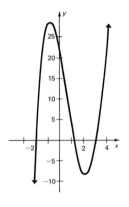

When the values of $f(x)$ are negative, the graph is below the x-axis, and when the values of $f(x)$ are positive, the graph is above the x-axis, so that the graph looks something like the sketch in Figure 3.16. If necessary, the sketch could be improved by plotting additional points in each region. ◀ **4**

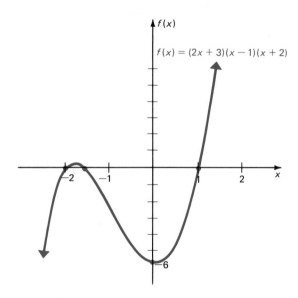

FIGURE 3.16

▶ **EXAMPLE 4**   Sketch the graph of $f(x) = 3x^4 + x^3 - 2x^2$.
The polynomial can be factored as follows.

$$3x^4 + x^3 - 2x^2 = x^2(3x^2 + x - 2)$$
$$= x^2(3x - 2)(x + 1)$$

Find the x-intercepts by letting $f(x) = 0$. The polynomial is 0 at $x = 0$, $x = 2/3$, and $x = -1$. These three values divide the x-axis into four regions:

$$x < -1, \quad -1 < x < 0, \quad 0 < x < \frac{2}{3}, \quad \text{and} \quad \frac{2}{3} < x.$$

Determine the sign of $f(x)$ in each region by substituting an x-value from each region into the function to get the following information.

5 Graph
$f(x) = 3x^3 + 5x^2 - 2x$.

Answer:

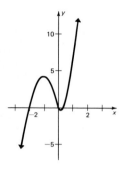

Region	Test Point	Value of f(x)	Sign of f(x)	Location Relative to Axis
$x < -1$	-2	32	positive	above
$-1 < x < 0$	$-\dfrac{1}{2}$	$-\dfrac{7}{16}$	negative	below
$0 < x < \dfrac{2}{3}$	$\dfrac{1}{2}$	$-\dfrac{3}{16}$	negative	below
$\dfrac{2}{3} < x$	1	2	positive	above

With the values of x used for the test points and the corresponding values of $f(x)$, sketch the graph as shown in Figure 3.17. ◀ **5**

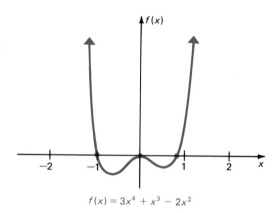

$$f(x) = 3x^4 + x^3 - 2x^2$$

FIGURE 3.17

It is difficult to graph polynomial functions that cannot be factored without calculus. Section 12.4 discusses the use of calculus techniques for graphing polynomials.

3.3 EXERCISES

Graph each of the following polynomial functions. (See Examples 1–4.)

1. $f(x) = x^3 + 2$

2. $f(x) = x^3 - 1$

3. $f(x) = x^4$

4. $f(x) = x^4 - 2$

5. $f(x) = 2x(x - 3)(x + 2)$

6. $f(x) = -x(2x + 1)(x - 3)$

7. $f(x) = (x + 2)(x - 3)(x + 4)$

8. $f(x) = (x - 3)(x - 1)(x + 1)$

9. $f(x) = x^2(x - 2)(x + 3)$

10. $f(x) = x^2(x + 1)(x - 1)$

11. $f(x) = (x + 2)(x - 2)(x + 1)(x - 1)$

12. $f(x) = (x - 3)(x + 3)(x - 2)(x + 2)$

13. $f(x) = x^3 - 7x^2 + 6x$

14. $f(x) = x^3 + x^2 - 6x$

15. $f(x) = x^3 - 2x^2 - 8x$

16. $f(x) = x^3 + 3x^2 - 4x$

17. $f(x) = x^4 - 5x^2$

18. $f(x) = x^4 - 2x^2$

19. $f(x) = x^4 - 7x^2 + 12$

20. $f(x) = x^4 - 7x^2 + 6$

21. $f(x) = 8x^4 - 2x^3 - x^2$

22. $f(x) = 6x^4 - x^3 - x^2$

In Exercises 23–26, use a calculator to find the indicated points. Graph each function by plotting the points and connecting them to get the graph.

23 Natural Science The polynomial function defined by

$$A(x) = -.015x^3 + 1.058x$$

gives the approximate alcohol concentration (in tenths of a percent) in an average person's bloodstream x hours after drinking about eight ounces of 100 proof whiskey. The function is approximately valid for $0 \leq x < 8$. Find the following values.

(a) $A(1)$ **(b)** $A(2)$ **(c)** $A(4)$
(d) $A(6)$ **(e)** $A(8)$ **(f)** Graph $A(x)$.
(g) Using the graph you drew for part (f), estimate the time of maximum alcohol concentration.
(h) In one state, a person is legally drunk if the blood alcohol concentration exceeds .15%. Use the graph of part (f) to estimate the period in which this average person is legally drunk.

24. Natural Science A technique for measuring cardiac output depends on the concentration of a dye after a known amount is injected into a vein near the heart. In a normal heart, the concentration of the dye at time x (in seconds) is given by the function defined by

$$g(x) = -.006x^4 + .140x^3 - .053x^2 + 1.79x.$$

(a) Find the following: $g(0)$; $g(1)$; $g(2)$; $g(3)$.
(b) Graph $g(x)$ for $x \geq 0$. (Hint: the only intercept is $(0, 0)$.)

25 Natural Science The pressure of the oil in a reservoir tends to drop with time. By taking sample pressure readings for a particular oil reservoir, petroleum engineers have found that the change in pressure is given by

$$P(t) = t^3 - 18t^2 + 81t,$$

where t is time in years from the date of the first reading.
(a) Find the following: $P(0)$; $P(3)$; $P(7)$; $P(10)$.
(b) Graph $P(t)$.
(c) For what time period is the change in pressure (drop) increasing? decreasing?

26 Natural Science During the early part of the 20th century, the deer population of the Kaibab Plateau in Arizona experienced a rapid increase because hunters had reduced the number of natural predators. The increase in population depleted the food resources and eventually caused the population to decline. For the period from 1905 to 1930, the deer population was approximated by

$$D(x) = -.125x^5 + 3.125x^4 + 4000,$$

where x is time in years from 1905.
(a) Find the following: $D(0)$; $D(5)$; $D(10)$; $D(15)$; $D(20)$; $D(25)$.
(b) Graph $D(x)$.
(c) From the graph, over what period of time (from 1905 to 1930) was the population increasing? relatively stable? decreasing?

In many applications, it is necessary to find the maximum or minimum values of a function. (Finding such values is one of the main uses of calculus.) We can find approximate maximum or minimum values of polynomial functions for given intervals by first evaluating the function at the left endpoint of the given interval. Then add .1 to the value of x and reevaluate the polynomial. Keep doing this until the right endpoint of the interval is reached. Then identify the approximate maximum and minimum value for the polynomial on the interval.

27. $f(x) = x^3 + 4x^2 - 8x - 8;\ .3 \leq x \leq 1$

28. $f(x) = 2x^3 - 5x^2 - x + 1;\ -1 \leq x \leq 0$

29. $f(x) = x^4 - 7x^3 + 13x^2 + 6x - 28;\ -2 \leq x \leq -1$

30. $f(x) = x^4 - 7x^3 + 13x^2 + 6x - 28;\ 2 \leq x \leq 3$

Get a table of ordered pairs over the given interval at intervals of .5 for the functions defined as follows. Then graph the function.

31. $f(x) = x^3 + 3x^2 - 2x + 1$ for $-3 \leq x \leq 1.5$

32. $f(x) = -3x^4 - 2x^3 + x^2 + x$ for $-3 \leq x \leq 1.5$

33. $f(x) = -x^4 + 2x^3 - 3x^2 + 4x - 7$ for $-1 \leq x \leq 2$

34. $f(x) = 2x^3 - 3x^2 + 4x - 7$ for $-1 \leq x \leq 2$

3.4 RATIONAL FUNCTIONS

A *rational function* is defined by an expression that is a quotient of two polynomials with denominator not zero.

A function f is a **rational function** if

$$f(x) = \frac{P(x)}{Q(x)}, \quad Q(x) \neq 0,$$

where $P(x)$ and $Q(x)$ are polynomials.

Since any values of x which make $Q(x) = 0$ are excluded from the domain of $f(x)$, a rational function usually has a graph with one or more breaks in it. The next few examples show how to graph rational functions.

▶**EXAMPLE 1** Graph the rational function defined by $y = \dfrac{2}{1+x}$.

This function is undefined for $x = -1$, since -1 leads to a 0 denominator. For this reason, the graph of this function will not intersect the vertical line $x = -1$. Since x can take on any value except -1, the values of x can approach -1 as closely as desired from either side of -1, as shown in the following table of values.

x approaches -1
↓

x	-1.5	-1.2	-1.1	-1.01	$-.99$	$-.9$	$-.8$	$-.5$
$1+x$	$-.5$	$-.2$	$-.1$	$-.01$	$.01$	$.1$	$.2$	$.5$
$\dfrac{2}{1+x}$	-4	-10	-20	-200	200	20	10	4

↑
$|y|$ **gets larger and larger**

The table above suggests that as x gets closer and closer to -1 from either side, the sum $1 + x$ gets closer and closer to 0, and $|2/(1 + x)|$ gets larger and larger. The part of the graph in Figure 3.18 near $x = -1$ shows this behavior. The vertical line $x = -1$ that is approached by the curve is called a *vertical asymptote*.

As $|x|$ gets larger and larger, $y = 2/(1 + x)$ gets closer and closer to 0, as shown in the table below.

x	-101	-11	-2	0	9	99
$1+x$	-100	-10	-1	1	10	100
$y = \dfrac{2}{1+x}$	$-.02$	$-.2$	-2	2	$.2$	$.02$

1 Graph the following.

(a) $f(x) = \dfrac{3}{5 - x}$

(b) $f(x) = \dfrac{-4}{x + 4}$

Answer:

(a)

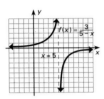

(b)

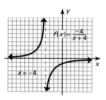

The horizontal line $y = 0$ is called a *horizontal asymptote* for this graph. Using the asymptotes and plotting the intercept and several points (shown with the figure) gives the graph of Figure 3.18. ◄ **1**

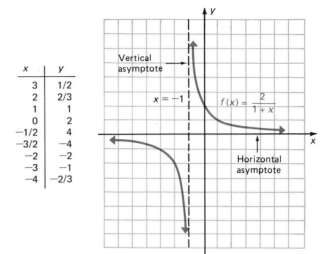

x	y
3	1/2
2	2/3
1	1
0	2
−1/2	4
−3/2	−4
−2	−2
−3	−1
−4	−2/3

FIGURE 3.18

Example 1 suggests the following conclusion.

If a number k makes the denominator 0 but the numerator nonzero in the expression defining a rational function, then the line $x = k$ is a **vertical asymptote** for the graph of the function.

Also, wherever the values of y approach but do not equal some number k as $|x|$ gets larger and larger, the line $y = k$ is a **horizontal asymptote** for the graph.

► **EXAMPLE 2** Graph $f(x) = \dfrac{3x + 2}{2x + 4}$.

Find the vertical asymptote by setting the denominator equal to 0, then solving for x.

$$2x + 4 = 0$$
$$x = -2$$

The line $x = -2$ is a vertical asymptote. Next, find the intercepts.

Let **x = 0.**   Let **y = 0.**

$$y = \frac{3 \cdot 0 + 2}{2 \cdot 0 + 4} = \frac{1}{2} \qquad 0 = \frac{3x + 2}{2x + 4}$$

$$3x + 2 = 0$$

$$x = -\frac{2}{3}$$

2 Graph the following.

(a) $f(x) = \dfrac{3x + 5}{x - 3}$

(b) $f(x) = \dfrac{2 - x}{x + 3}$

Answer:

(a)

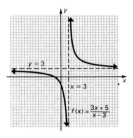

(b)

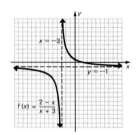

The *y*-intercept is 1/2. The *x*-intercept is −2/3. Now find additional ordered pairs on either side of the vertical asymptote. A table of values, given with the graph in Figure 3.19, gives several ordered pairs. Choosing larger and larger *x*-values gives (12, 1.36), (20, 1.41), (50, 1.46), and so on. Choosing smaller and smaller *x*-values gives (−12, 1.7), (−20, 1.61), (−50, 1.54), and so on. The corresponding points are approaching the horizontal asymptote *y* = 1.5 or *y* = 3/2. Using the asymptotes and plotting several points including the intercepts, leads to the graph of Figure 3.19. ◀

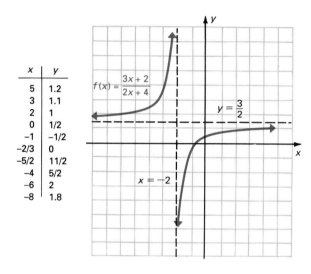

x	y
5	1.2
3	1.1
2	1
0	1/2
−1	−1/2
−2/3	0
−5/2	11/2
−4	5/2
−6	2
−8	1.8

FIGURE 3.19

In Example 2, the horizontal asymptote of

$$y = \frac{3x + 2}{2x + 4} \quad \text{was} \quad y = \frac{3}{2}.$$

In general, the horizontal asymptote of

$$y = \frac{ax + b}{cx + d} \quad \text{is} \quad y = \frac{a}{c}.* \quad \boxed{2}$$

The final two examples show applications of rational functions.

---

*The reason for this will be shown in Section 11.1.

3 Using the function of Example 3, find the cost to remove the following percents of pollutants.

(a) 70%

(b) 85%

(c) 90%

(d) 98%

Answer:

(a) $35,000

(b) About $73,000

(c) About $101,000

(d) About $221,000

▶**EXAMPLE 3** In many situations involving environmental pollution, much of the pollutant can be removed from the air or water at a fairly reasonable cost, but the last, small part of the pollutant can be very expensive to remove.

Cost as a function of the percentage of pollutant removed from the environment can be calculated for various percentages of removal, with a curve fitted through the resulting data points. This curve then leads to a function that approximates the situation. Rational functions often are a good choice for these **cost-benefit functions.**

For example, suppose a cost-benefit function is given by

$$f(x) = \frac{18x}{106 - x},$$

where $f(x)$ or y is the cost (in thousands of dollars) of removing x percent of a certain pollutant. The domain of x is the set of all numbers from 0 to 100, inclusive; any amount of pollutant from 0% to 100% can be removed. To remove 100% of the pollutant here would cost

$$y = \frac{18(100)}{106 - 100} = 300,$$

or $300,000. Check that 95% of the pollutant can be removed for $155,000, 90% for $101,000, and 80% for $55,000. Using these points, as well as others that could be obtained from the function above, leads to the graph shown in Figure 3.20. ◀ **3**

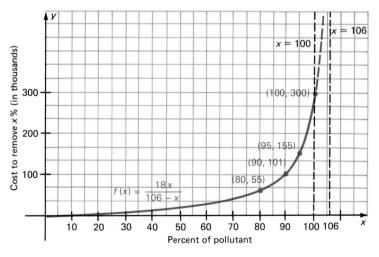

FIGURE 3.20

4 Rework Example 4 with the product-exchange function

$$y = \frac{60000 - 10x}{60 + x}$$

to find the maximum amount of each wine that can be produced.

Answer:
6000 tons of red, 1000 tons of white

In management, **product-exchange functions** give the relationship between quantities of two items that can be produced by the same machine or factory. For example, an oil refinery can produce gasoline, heating oil, or a combination of the two; a winery can produce red wine, white wine, or a combination of the two. The next example discusses a product-exchange function.

▶**EXAMPLE 4** The product-exchange function for the Golden Grape Winery for red wine x and white wine y, in tons, is

$$y = \frac{100{,}000 - 50x}{1000 + x}.$$

Graph the function in Quadrant I and find the maximum quantity of each kind of wine that can be produced.

We first find the intercepts. For $x = 0$, the y-intercept is 100. Let $y = 0$ to get $100{,}000 - 50x = 0$, from which the x-intercept is 2000. Since we are interested only in the Quadrant I portion of the graph, we can find a few more points in Quadrant I and complete the graph as shown in Figure 3.21.

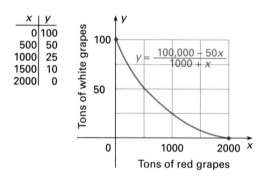

x	y
0	100
500	50
1000	25
1500	10
2000	0

FIGURE 3.21

The maximum value of y occurs when $x = 0$, so the maximum amount of white wine that can be produced is 100 tons, given by the y-intercept. The x-intercept gives the maximum amount of red wine that can be produced, 2000 tons. ◀ **4**

3.4 EXERCISES

Graph each of the following. Give the equations of the asymptotes. (See Examples 1–2.)

1. $f(x) = \dfrac{1}{x + 2}$

2. $f(x) = \dfrac{1}{x - 1}$

3. $f(x) = \dfrac{-4}{x - 3}$

4. $f(x) = \dfrac{-1}{x + 3}$

5. $f(x) = \dfrac{2}{x}$

6. $f(x) = \dfrac{-3}{x}$

7. $f(x) = \dfrac{2}{3 + 2x}$

8. $f(x) = \dfrac{4}{5 + 3x}$

9. $f(x) = \dfrac{3x}{x - 1}$

10. $f(x) = \dfrac{4x}{3 - 2x}$

11. $f(x) = \dfrac{x}{x - 9}$ **12.** $f(x) = \dfrac{3x}{x - 4}$ **13.** $f(x) = \dfrac{x + 1}{x - 4}$ **14.** $f(x) = \dfrac{x - 3}{x + 5}$ **15.** $f(x) = \dfrac{2x - 1}{4x + 2}$

16. $f(x) = \dfrac{3x - 6}{6x - 1}$ **17.** $f(x) = \dfrac{1 - 2x}{5x + 20}$ **18.** $f(x) = \dfrac{6 - 3x}{4x + 12}$ **19.** $f(x) = \dfrac{-x - 4}{3x + 6}$ **20.** $f(x) = \dfrac{-x + 8}{2x + 5}$

---

21. Management Suppose that the average cost per unit, $C(x)$, of producing x units of margarine is given by

$$C(x) = \frac{500}{x + 30}.$$

Find the average cost per unit of producing each of the following quantities.

(a) 10 units **(b)** 20 units **(c)** 50 units
(d) 70 units **(e)** 100 units
(f) Graph $y = C(x)$.

22. Natural Science Suppose a cost-benefit model (see Example 4) is given by

$$f(x) = \frac{6.5x}{102 - x},$$

where $f(x)$ is the cost in thousands of dollars of removing x percent of a certain pollutant. Find the cost of removing the following percents of pollutants.

(a) 0% **(b)** 50% **(c)** 80% **(d)** 90%
(e) 95% **(f)** 99% **(g)** 100%
(h) Graph the function.

23. Natural Science Suppose a cost-benefit model is given by

$$f(x) = \frac{6.7x}{100 - x},$$

where $f(x)$ is the cost in thousands of dollars of removing x percent of a given pollutant. Find the cost of removing each of the following percents of pollutants.

(a) 50% **(b)** 70% **(c)** 80% **(d)** 90%
(e) 95% **(f)** 98% **(g)** 99%
(h) Is it possible, according to this model to remove *all* the pollutant?
(i) Graph the function.

24. Natural Science In Exercise 23, what percent of pollutant can be removed for
(a) $10,000? **(b)** $38,000?

25. Social Science The average waiting time in a line (or queue) before getting served is given by

$$W = \frac{S(S - A)}{A},$$

where A is the average rate that people arrive at the line and S is the average service time. At a certain fast food restaurant, the average service time is 3 minutes. Find W for each of the following average arrival times.

(a) 1 minute **(b)** 2 minutes **(c)** 2.5 minutes
(d) What is the vertical asymptote?
(e) Graph the equation on the interval $(0, 3]$.
(f) What happens to W when $A > 3$? What does this mean?

26. Natural Science To calculate the drug dosage for a child, pharmacists may use the formula

$$d(x) = \frac{Dx}{x + 12},$$

where x is the child's age in years and D is the adult dosage. The adult dosage of Naldecon is 70 milligrams.
(a) What is the vertical asymptote for the graph of this function?
(b) What is the horizontal asymptote?
(c) Graph $d(x)$.

27. Antique car fans often enter their cars in a *concours d'elegance* in which a maximum of 100 points can be awarded to a particular car. Points are awarded for the general attractiveness of the car. Based on a recent article in *Business Week*, we constructed the following mathematical model for the cost, in thousands of dollars, of restoring a car so that it will win x points.

$$C(x) = \frac{10x}{49(101 - x)}$$

Find the cost of restoring a car so that it will win
(a) 99 points; **(b)** 100 points.

28. Physical Science In electronics, the circuit gain is given by

$$G(R) = \frac{R}{r + R},$$

where R is the resistance of a temperature sensor in the circuit and r is a constant. Let $r = 1000$ ohms.
(a) Find any vertical asymptotes of the graph of the function.
(b) Find any horizontal asymptotes of the graph of the function.
(c) Graph $G(R)$.

Management *Sketch the Quadrant I portion of the graph of each of the functions defined as follows, and then estimate the maximum quantities of each product that can be produced. (Hint: look at the intercepts.) (See Example 4.)*

29. The product-exchange function for gasoline, x, and heating oil, y, in hundreds of gallons per day, is

$$y = \frac{125,000 - 25x}{125 + 2x}.$$

30. A drug factory found the product-exchange function for a red tranquilizer, x, and a blue tranquilizer, y, is

$$y = \frac{900,000,000 - 30,000x}{x + 90,000}.$$

Management *In recent years the economist Arthur Laffer has been a center of controversy because of his **Laffer curve**, an idealized version of which is shown to the right. According to this curve,* increasing *a tax rate, say from x_1 percent to x_2 percent on the graph, can actually lead to a* decrease *in government revenue. All economists agree on the endpoints—0 revenue at tax rates of both 0% and 100%, but there is much disagreement on the location of the rate x_1 that produces maximum revenue.*

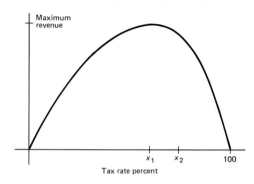

31. Suppose an economist studying the Laffer curve produced the rational function

$$f(x) = \frac{60x - 6000}{x - 120},$$

where $y = f(x)$ is government revenue in millions from a tax rate of x percent, and $50 \le x \le 100$. Find the revenue from a tax rate of
(a) 50%; **(b)** 60%; **(c)** 80%;
(d) 100% **(e)** Graph the function.

32. Suppose the economist studies a different tax, this time producing

$$f(x) = \frac{80x - 8000}{x - 110},$$

with $y = f(x)$ giving government revenue in tens of millions of dollars for a tax rate of x percent, with the function valid for $55 \le x \le 100$. Find the revenue from a tax rate of
(a) 55%; **(b)** 60%; **(c)** 70%; **(d)** 90%;
(e) 100%. **(f)** Graph the function.

Use a calculator to find several ordered pairs and then sketch the graphs of the functions defined as follows. Identify any horizontal or vertical asymptotes.

33. $f(x) = \dfrac{-2x^2 + x - 1}{2x + 3}$ **34.** $f(x) = \dfrac{3x + 2}{x^2 - 4}$ **35.** $f(x) = \dfrac{2x^2 - 5}{x^2 - 1}$ **36.** $f(x) = \dfrac{4x^2 - 1}{x^2 + 1}$

KEY TERMS AND SYMBOLS

3.1 polynomial function
quadratic function
parabola
vertex
axis of the parabola

3.4 rational function
vertical asymptote
horizontal asymptote
cost-benefit function
product-exchange function

KEY CONCEPTS

The **quadratic function** defined by $y = a(x - h)^2 + k$ has a graph that is a **parabola** with vertex (h, k) and axis of symmetry $x = h$. The parabola opens upward if $a > 0$, downward if $a < 0$. If the equation is in the form $f(x) = ax^2 + bx + c$, the vertex is $\left(-\dfrac{b}{2a}, f\left(-\dfrac{b}{2a}\right)\right)$.

A **polynomial function** of degree higher than two can be roughly sketched by finding the x-intercepts, dividing the number line into regions determined by the x-intercepts, then finding the sign of the function ($+$ or $-$) in each region.

If a number k makes the denominator of a rational function 0, but the numerator nonzero, then the line $x = k$ is a **vertical asymptote** for the graph.

Whenever the values of y approach, but never equal, some number k as $|x|$ gets larger and larger, the line $y = k$ is a **horizontal asymptote** for the graph.

Many **rational functions** can be roughly sketched by finding the asymptotes, then plotting a few points in each region determined by the vertical asymptotes and the x-intercepts.

CHAPTER 3   REVIEW EXERCISES

Graph each of the following. Give the vertex and axis of each.

1. $f(x) = x^2 - 4$

2. $f(x) = 6 - x^2$

3. $f(x) = -(x - 1)^2$

4. $f(x) = (x + 2)^2$

5. $f(x) = 3(x + 1)^2 - 5$

6. $f(x) = -(1/4)(x - 2)^2 + 3$

7. $f(x) = -(x + 3)^2 - 2$

8. $f(x) = (x - 4)^2 + 3$

9. $f(x) = x^2 - 4x + 2$

10. $f(x) = x^2 + 2x - 3$

11. $f(x) = -x^2 + 6x - 3$

12. $f(x) = -x^2 - 4x + 1$

13. $f(x) = 4x^2 - 8x + 3$

14. $f(x) = 2x^2 + 4x - 3$

15. $f(x) = -3x^2 - 12x - 8$

16. $f(x) = -3x^2 - 6x + 2$

17. $f(x) = -x^2 + 5x - 2$

18. $f(x) = -x^2 - 3x + 4$

Find the maximum or minimum value for the following.

19. $f(x) = x^2 - 4x + 1$

20. $f(x) = x^2 + 6x + 3$

21. $f(x) = -3x^2 - 12x - 1$

22. $f(x) = -2x^2 + 4x - 3$

23. $f(x) = 4x^2 - 8x + 3$

24. $f(x) = -3x^2 - 6x + 2$

Solve each of the following.

25. The commodity market is very unstable; money can be made or lost quickly when investing in soybeans, wheat, pork bellies, and the like. Suppose that an investor kept track of her total profit, P, at time t, measured in months, after she began investing and found that $P = 4t^2 - 29t + 30$. Find the time intervals where she has been ahead. (Hint: $t > 0$ in this case.)

26. Suppose the velocity of an object is given by $v = 2t^2 - 5t - 12$, where t is time in seconds. (Here t can be positive or negative.) Find the intervals where the velocity is negative.

27. An analyst has found that his company's profits, in hundreds of thousands of dollars, are given by $P = 3x^2 - 35x + 50$, where x is the amount, in hundreds of dollars, spent on advertising. Decide for what values of x the company makes a profit.

28. The manager of a large apartment complex has found that the profit is given by $P = -x^2 + 250x - 15{,}000$, where x is the number of units rented. Decide for what values of x the complex produces a profit.

29. Find the rectangular region of maximum area that can be enclosed with 200 meters of fencing.

30. Find the rectangular region of maximum area that can be enclosed with 200 meters of fencing if no fencing is needed along one side of the region.

Graph each of the following polynomial functions.

31. $f(x) = x^3 - 2$ **32.** $f(x) = 4 - x^3$ **33.** $f(x) = -x^4 + 1$ **34.** $f(x) = 2 + x^4$

35. $f(x) = -(x - 3)^3$ **36.** $f(x) = (x + 1)^3$ **37.** $f(x) = (x + 2)^4$ **38.** $f(x) = -(x - 4)^4$

39. $f(x) = x(2x - 1)(x + 2)$ **40.** $f(x) = 3x(3x + 2)(x - 1)$ **41.** $f(x) = 2x^3 - 3x^2 - 2x$

42. $f(x) = x^3 - 3x^2 - 4x$ **43.** $f(x) = x^4 - 5x^2 - 6$ **44.** $f(x) = x^4 - 7x^2 - 8$

Graph each of the following rational functions.

45. $f(x) = \dfrac{1}{x - 3}$ **46.** $f(x) = \dfrac{-2}{x + 4}$ **47.** $f(x) = \dfrac{-3}{2x - 4}$ **48.** $f(x) = \dfrac{5}{3x + 7}$ **49.** $f(x) = \dfrac{5x + 5}{3x - 5}$ **50.** $f(x) = \dfrac{3 - x}{2x - 9}$

---

51. A cost-benefit curve for pollution control is given by

$$y = \frac{9.2x}{106 - x},$$

where y is the cost in thousands of dollars of removing x percent of a specific industrial pollutant. Find y for each of the following values of x.

(a) $x = 50$ (b) $x = 98$
(c) What percent of the pollutant can be removed for $22,000$?

52. The average cost to make x units of a product is given by $A(x)$, where

$$A(x) = \frac{6x}{2x + 5},$$

with x representing the number of units made. Find the average cost to make
(a) 10 units;   (b) 30 units;
(c) 50 units;   (d) 100 units.

53. The cost and revenue functions for a frozen yogurt shop are given by

$$C(x) = \frac{400x + 400}{x + 4} \quad \text{and} \quad R(x) = 100x.$$

Graph $C(x)$ and $R(x)$ on the same axes and answer the following questions.
(a) What is the break-even point for this shop?
(b) If the profit function is given by $P(x)$, does $P(1)$ represent a profit or a loss?
(c) Does $P(4)$ represent a profit or a loss?

54. The supply and demand functions for the yogurt shop in Exercise 53 are as follows:

$$\text{supply: } p = \frac{x^2}{4} + 25; \quad \text{demand: } p = \frac{500}{x},$$

where p is price in dollars for x units of yogurt.
(a) Graph both functions on the same axes, and from the graph, estimate the equilibrium point.
(b) Give the x-intervals where supply exceeds demand.
(c) Give the x-intervals where demand exceeds supply.

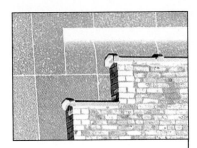

CHAPTER 4

Exponential and Logarithmic Functions

Exponential functions may be the single most important category of functions used in practical applications. These functions, and the closely related logarithmic functions, are used to describe growth and decay, which are important ideas in management, social science, and biology.

The following problem illustrates the behavior of exponential functions. A person contracts to work for a period of thirty days and is given the option of receiving a lump sum of $1000 or receiving $.01 on the first day of work, $.02 on the second, $.04 on the third, $.08 on the fourth, and so on, with each day's pay double that of the previous day. While the first choice may appear more lucrative (after all, the *total* amount earned by the second method after seven days is only $1.27), the second method actually will earn the worker over five million dollars on the thirtieth day alone, and over ten million dollars for the month!

A function that defines the amount of payment on the *n*th day in the example above is

$$f(n) = .01(2^{n-1}).$$

The base 2 appears as a result of the doubling that occurs. A function in which the variable appears in the exponent is an *exponential function*. Exponential functions tend to increase or decrease dramatically. The function in the example above is an increasing function. An example of a decreasing function is one that gives the amount of a radioactive substance present as time passes, since the amount gradually diminishes.

For many years *logarithms* were used primarily to assist in involved calculations. Current technology has made this use of logarithms obsolete, but *logarithmic functions* play an important role in many applications of mathematics. We shall see in this chapter that exponential and logarithmic functions are intimately related.

4.1 EXPONENTIAL FUNCTIONS

In earlier chapters we have discussed functions involving expressions such as x^2, $(2x + 1)^3$, or $x^{3/4}$, where the variable or variable expression is the base of an exponential expression, and the exponent is a constant. In an exponential function, the variable is in the exponent and the base is a constant.

1 Graph $f(x) = \left(\dfrac{1}{3}\right)^x$.

Answer:

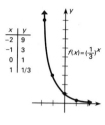

x	y
-2	9
-1	3
0	1
1	1/3

$f(x) = (\frac{1}{3})^x$

An **exponential function** with base a is defined as

$$f(x) = a^x,$$

where $a > 0$ and $a \neq 1$.

(If $a = 1$, the function is the constant function $f(x) = 1$.)

▶ **EXAMPLE 1** Graph the following exponential functions.

(a) $f(x) = 2^x$

Begin by making a table of values of x and y, as shown in Figure 4.1(a). Then plot these points and draw a smooth curve through them, getting the graph shown in Figure 4.1(a). The graph approaches the negative x-axis but will never touch it, since *every* power of 2 is positive. This makes the x-axis a horizontal asymptote. The graph suggests that as x gets larger and larger, values of 2^x also get larger and larger. However, there is no vertical asymptote.

(b) $g(x) = 2^{-x}$

By the properties of exponents,

$$2^{-x} = \frac{1}{2^x} = \left(\frac{1}{2}\right)^x.$$

Construct a table of values and draw a smooth curve through the resulting points. (See Figure 4.1(b)). ◀ **1**

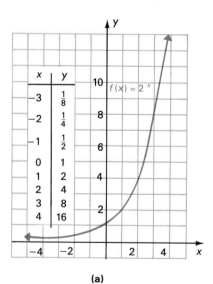

x	y
-3	$\frac{1}{8}$
-2	$\frac{1}{4}$
-1	$\frac{1}{2}$
0	1
1	2
2	4
3	8
4	16

$f(x) = 2^x$

(a)

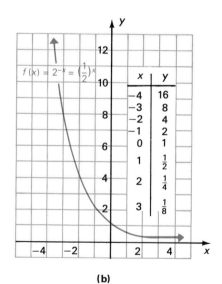

$f(x) = 2^{-x} = \left(\frac{1}{2}\right)^x$

x	y
-4	16
-3	8
-2	4
-1	2
0	1
1	$\frac{1}{2}$
2	$\frac{1}{4}$
3	$\frac{1}{8}$

(b)

FIGURE 4.1

2 Graph $f(x) = 2^{x+1}$.

Answer:

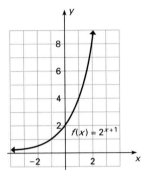

As suggested by the graphs in Figure 4.1, the domain of an exponential function defined by an expression of the form $y = a^x$ includes all real numbers, and the range includes all positive real numbers. The base a is restricted to positive values, with negative or zero bases not allowed. For example, the domain of $f(x) = (-4)^x$ could not include such numbers as $x = 1/2$ because $(-4)^{1/2} = \sqrt{-4}$, which is not a real number. The resulting graph would be at best a series of separate points having little practical use.

The graph of $f(x) = 2^x$ in Figure 4.1(a) is typical of the graphs of $y = a^x$, where $a > 1$. For larger values of a, the graph rises to the right more steeply, but the general shape is similar to that of $f(x) = 2^x$. When $0 < a < 1$, the graph decreases to the right like the graph of $g(x) = (1/2)^x$ in Figure 4.1(b). The graphs of $f(x) = 2^x$ and $g(x) = 2^{-x}$ are mirror images of each other with respect to the y-axis. The graphs of several typical exponential functions in Figure 4.2 illustrate these statements.

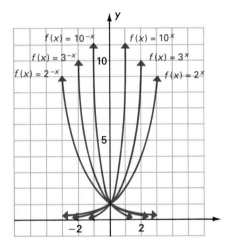

FIGURE 4.2

▶ **EXAMPLE 2**   Graph $f(x) = 3^{1-x}$.

To find points to plot, choose values of x that make the exponent positive, zero, and negative. Since $1 - x = 0$ if $x = 1$, choose some other x-values less than 1 and some greater than 1. See the table of values with the graph in Figure 4.3. ◀ **2**

3 Graph $f(x) = \left(\frac{1}{2}\right)^{-x^2}$.

Answer:

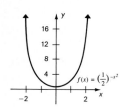

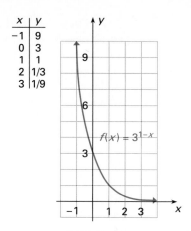

FIGURE 4.3

▶**EXAMPLE 3** Graph $f(x) = 2^{-x^2}$.

Using a calculator to find several ordered pairs and then plotting them gives the graph in Figure 4.4. Both the negative sign and the fact that x is squared affect the shape of the graph, with the final result looking quite different from the typical graphs of exponential functions in Figures 4.1 and 4.2. Graphs such as this are important in probability, where the normal curve has an equation similar to the one in this example. ◀ **3**

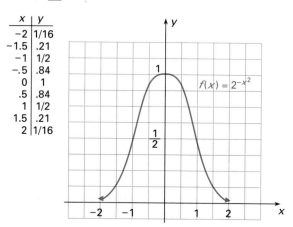

FIGURE 4.4

The graphs of typical exponential functions in Figure 4.2 suggest that a given value of x leads to exactly one value of a^x and each value of a^x corresponds to exactly one value of x. Because of this, an equation with a variable in the exponent, called an **exponential equation,** can often be solved using the following property.

4 Solve each equation.

(a) $6^x = 6^4$

(b) $3^{2x} = 3^9$

(c) $5^{-4x} = 5^3$

Answer:

(a) 4

(b) 9/2

(c) −3/4

5 Solve each equation.

(a) $8^{2x} = 4$

(b) $5^{3x} = 25^4$

(c) $36^{-2x} = 6$

Answer:

(a) 1/3

(b) 8/3

(c) −1/4

6 The number of organisms present at time t is given by

$$f(t) = 75(2)^{.5t}.$$

Find the number of organisms present at

(a) $t = 0$;

(b) $t = 2$;

(c) $t = 4$.

Answer:

(a) 75

(b) 150

(c) 300

If $a > 0$, $a \neq 1$, and $a^x = a^y$, then $x = y$.

Caution Both bases must be the same. The value $a = 1$ is excluded since $1^2 = 1^3$, for example, even though $2 \neq 3$.

As an example, solve $2^{3x} = 2^7$ using this property, as follows:

$$2^{3x} = 2^7$$
$$3x = 7$$
$$x = \frac{7}{3}. \quad \boxed{4}$$

▶**EXAMPLE 4** Solve $9^x = 27$.

First rewrite both sides of the equation so the bases are the same. Since $9 = 3^2$ and $27 = 3^3$,

$$(3^2)^x = 3^3$$
$$3^{2x} = 3^3$$
$$2x = 3$$
$$x = \frac{3}{2}. \quad ◀ \boxed{5}$$

You may wish to review Section 1.9 on rational exponents at this time.

As noted earlier, exponential functions have many practical applications. For example, in situations which involve growth or decay of a population, the size of the population at a given time t often is determined by an exponential function of t.

▶**EXAMPLE 5** The oxygen consumption of yearling salmon (in appropriate units) increases exponentially with speed of swimming according to

$$f(x) = 100(3)^{.6x},$$

where x is the speed in feet per second. Find each of the following.

(a) $f(0)$

Substitute 0 for x.

$$f(0) = 100(3)^{.6(0)} = 100(3)^0 = 100 \cdot 1 = 100$$

When the fish are still (their speed is 0) their oxygen consumption is 100 units.

(b) $f(5)$

Replace x with 5.

$$f(5) = 100(3)^{.6(5)} = 100(3)^3 = 2700$$

A speed of 5 feet per second increases the oxygen consumption to 2700 units. ◀ **6**

4.1 EXERCISES

Many of the exercises in this section are designed to be worked with the aid of a calculator.

Which of the following are exponential functions?

1. $f(x) = 5^x - 1$ **2.** $f(x) = 2x^5$ **3.** $f(x) = 4x^3 - 1$ **4.** $f(x) = 2 \cdot 3^{5x-1}$

Let $f(x) = (2/3)^x$. Find each of the following values.

5. $f(2)$ **6.** $f(-1)$ **7.** $f\left(\dfrac{1}{2}\right)$ **8.** $f(0)$

Graph each of the following exponential functions. (See Examples 1–3.)

9. $f(x) = 3^x$ **10.** $f(x) = 4^x$ **11.** $f(x) = 3^{-x}$

12. $f(x) = 4^{-x}$ **13.** $f(x) = \left(\dfrac{1}{4}\right)^x$ **14.** $f(x) = \left(\dfrac{1}{3}\right)^x$

15. $f(x) = 3^{2x}$ **16.** $f(x) = 4^{x/2}$ **17.** $f(x) = 2^{-x/2}$

18. $f(x) = 2^{-2x}$ **19.** $f(x) = 3^{x+1}$ **20.** $f(x) = 3^{2x-1}$

21. $f(x) = 10 - 5 \cdot 2^{-x}$ **22.** $f(x) = 100 - 80 \cdot 2^{-x}$ **23.** $f(x) = x \cdot 2^x$

24. $f(x) = x^2 \cdot 2^x$

Solve each of the following equations. (See Example 4.)

25. $5^x = 25$ **26.** $3^x = \dfrac{1}{9}$ **27.** $2^x = \dfrac{1}{8}$

28. $4^x = 64$ **29.** $a^x = a^2 \ (a > 0)$ **30.** $a^x = \dfrac{1}{a^2} \ (a > 0)$

31. $16^x = 64$ **32.** $\left(\dfrac{1}{8}\right)^x = 8$ **33.** $\left(\dfrac{3}{4}\right)^x = \dfrac{16}{9}$

34. $5^{-2x} = \dfrac{1}{25}$ **35.** $3^{x-1} = 9$ **36.** $16^{-x+1} = 8$

37. $25^{-2x} = 3125$ **38.** $16^{x+2} = 64^{2x-1}$ **39.** $81^{-2x} = 3^{x-1}$

40. $7^{-x} = 49^{x+3}$ **41.** $2^{|x|} = 16$ **42.** $5^{-|x|} = \dfrac{1}{25}$

43. $2^{x^2-4x} = \dfrac{1}{16}$ **44.** $5^{x^2+x} = 1$ **45.** $8^{x^2} = 2^{5x}$

46. $9^x = 3^{x^2}$

Work the following exercises.

47. If $1 is deposited into an account paying 5% per year compounded annually, then after t years the account will contain

$$y = (1 + .05)^t = (1.05)^t$$

dollars.

(a) Use a calculator to complete the following table.

t	0	1	2	3	4	5	6	7	8	9	10
y	1					1.28					1.63

(b) Graph $y = (1.05)^t$.

48. If money loses value at the rate of 4% per year, the value of $1 in t years is given by

$$y = (1 - .04)^t = (.96)^t.$$

(a) Use a calculator to complete the following table.

t	0	1	2	3	4	5	6	7	8	9	10
y	1					.82					.66

(b) Graph $y = (.96)^t$.

49. If money loses value, then it takes more dollars to buy the same item. Use the results of Exercise 48(a) to answer the following questions.

(a) Suppose a house costs $65,000 today. Estimate the cost of a similar house in 10 years. (Hint: solve the equation $.66t = \$65,000$.)

(b) Estimate the cost of a $20 textbook in 8 years.

50. **Social Science** Under certain conditions, the number of individuals of a species that is newly introduced into an area can double every year. That is, if t represents the number of years since the species was introduced into the area and y represents the number of individuals, then

$$y = 6 \cdot 2^t$$

if 6 animals were introduced into the area originally.

(a) Complete the following table.

t	0	1	2	3	4	5	6	7	8	9	10
y	6					192					6144

(b) Graph $y = 6 \cdot 2^t$.

51. **Social Science** City planners have determined that the population of a city is given by

$$P(t) = 1,000,000(2^{.2t}),$$

where t represents time measured in years. Find each of the following values.

(a) $P(0)$ **(b)** $P(5/2)$ **(c)** $P(5)$ **(d)** $P(10)$

(e) Graph $P(t)$.

52. **Natural Science** Suppose the quantity in grams of a radioactive substance present at time t is

$$Q(t) = 500(3^{-.5t}),$$

where t is measured in months. Find the quantity present at each of the following times.

(a) $t = 0$ **(b)** $t = 2$ **(c)** $t = 4$ **(d)** $t = 10$

(e) Graph $Q(t)$.

Management *The scrap value of a machine is the value of the machine at the end of its useful life. By one method of calculating scrap value, where it is assumed a constant percentage of value is lost annually, the scrap value S is given by*

$$S = C(1 - r)^n,$$

where C is the original cost, n is the useful life of the machine in years, and r is the constant annual percentage of value lost. Find the scrap value for each of the following machines.

53. Original cost, $54,000; life, 8 years; annual rate of value loss, 12%

54. Original cost, $178,000; life, 11 years; annual rate of value loss, 14%

55. **Management** Midwest Creations finds that its total sales $T(x)$, in thousands, from the distribution of x catalogs, where x is measured in thousands, is approximated by

$$T(x) = \frac{2500}{1 + 24 \cdot 2^{-x/4}}.$$

Find the total sales resulting from the distribution of

(a) 0 catalogs; **(b)** 5000 catalogs;

(c) 24,000 catalogs; **(d)** 49,000 catalogs.

56. **Natural Science**   *Escherichia coli* is a strain of bacteria that occurs naturally in many different situations. Under certain conditions, the number of these bacteria present in a colony is given by

$$E(t) = E_0 \cdot 2^{t/30},$$

where $E(t)$ is the number of bacteria present t minutes after the beginning of an experiment, and E_0 is the number present when $t = 0$. Let $E_0 = 1,000,000$ and use a calculator with a y^x key to find the number of bacteria at the following times.

(a) $t = 5$ (b) $t = 10$ (c) $t = 15$
(d) $t = 20$ (e) $t = 30$ (f) $t = 60$
(g) $t = 90$ (h) $t = 120$

57. **Social Science**   A person learning certain skills involving repetition tends to learn quickly at first. Then learning tapers off and approaches some upper limit. Suppose the number of characters per minute a typesetter can produce is given by

$$p(t) = 250 - 120(2.8)^{-.5t},$$

where t is the number of months the typesetter has been in training. Find each of the following.

(a) $p(2)$ (b) $p(4)$ (c) $p(10)$ (d) Graph $p(t)$.

58. Suppose the domain of $y = 2^x$ is restricted to include only rational values of x (which are the only values discussed prior to this section). Describe in words the resulting graph.

59. In our definition of exponential function we ruled out negative values of a. However, in a textbook on mathematical economics, published by a well-known publisher, the author obtained a ''graph'' of $y = (-2)^x$ by plotting the following points and drawing a smooth curve through them.

x	-4	-3	-2	-1	0	1	2	3
y	$\dfrac{1}{16}$	$-\dfrac{1}{8}$	$\dfrac{1}{4}$	$-\dfrac{1}{2}$	1	-2	4	-8

The graph oscillates very neatly from positive to negative values of y. Comment on this approach. (This example shows the dangers of point plotting when drawing graphs.)

Natural Science   *Generally speaking, the larger an organism, the greater its complexity (as measured by counting the number of different types of cells present). The figure shows an estimate of the maximum number of cells (or the largest volume) in various organisms plotted against the number of types of cells found in those organisms.**

60. Use the graph to estimate the maximum number of cells and the corresponding volume for each of the following organisms.

(a) whale   (b) sponge   (c) green alga

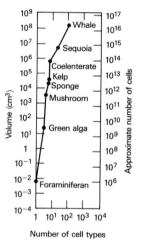

61. From the graph, estimate the number of cell types in each of the following organisms.

(a) mushroom   (b) kelp   (c) sequoia

*From *On Size and Life* by Thomas A. McMahon and John Tyler Bonner. Copyright © 1983 by Thomas A. McMahon and John Tyler Bonner. Reprinted by permission of W. H. Freeman and Company.

1 Use a calculator to find each of the following.

(a) $e^{.06}$

(b) $e^{-.06}$

(c) $e^{2.30}$

(d) $e^{-2.30}$

Answer:

(a) 1.06184

(b) .94176

(c) 9.97418

(d) .10026

4.2 APPLICATIONS OF EXPONENTIAL FUNCTIONS

A certain irrational number, denoted e, arises naturally in a variety of mathematical situations (much the same way that the number π appears when you consider the area of a circle).* To nine decimal places,

$$e = 2.718281828.$$

Perhaps the single most useful exponential function is the function defined by $f(x) = e^x$.

A calculator should be used to evaluate powers of e. For instance, to find $e^{.14}$, enter .14 and press the $\boxed{e^x}$ key (or press $\boxed{\text{Inv}}$ $\boxed{\text{ln}}$, depending on the calculator).† The display will show (to five places) 1.15027. Most calculators allow you to use negative exponents with the e^x key, but if yours doesn't, just remember that for any positive k, $e^{-k} = 1/e^k$. **1**

In Figure 4.5, the functions defined by

$$g(x) = 2^x, \quad f(x) = e^x, \quad \text{and} \quad h(x) = 3^x$$

are graphed for comparison.

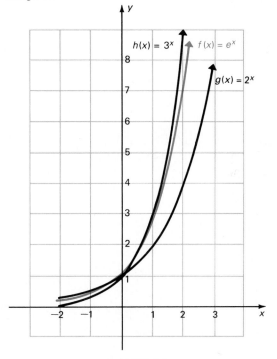

FIGURE 4.5

---

*See page 178 for an example.

†If these directions don't work with your calculator, consult the instruction booklet.

2 In Example 1, find the population of the city after

(a) 1 year;

(b) 3 years;

(c) 7 years;

(d) 10 years.

Answer:

(a) About 10,400

(b) About 11,300

(c) About 13,200

(d) About 14,900

In many situations in biology, economics, and the social sciences, a quantity changes at a rate proportional to the quantity present. In such cases, the amount present at time t is a function of t called the *exponential growth function*.

Exponential Growth Function

Let y_0 be the amount or number of a quantity present at time $t = 0$. Then, under certain conditions, the amount present at time t is given by

$$f(t) = y_0 e^{kt}$$

for some constant k.

The next example illustrates exponential growth.

▶**EXAMPLE 1**   Suppose the population of a midwestern city is

$$P(t) = 10,0000e^{.04t},$$

where t represents time measured in years. The population at time $t = 0$ is

$$P(0) = 10,000e^{(.04)0}$$
$$= 10,000e^0$$
$$= 10,000(1)$$
$$= 10,000.$$

The population of the city is 10,000 at time $t = 0$, written $P_0 = 10,000$. The population of the city at year $t = 5$ is

$$P(5) = 10,000e^{(.04)5}$$
$$= 10,000e^{.2}.$$

A calculator shows that $e^{.2} = 1.22140$ (to five decimal places), so that

$$P(5) = 10,000(1.22140) = 12,214.$$

In 5 years the population of the city will be about 12,000. ◀ **2**

▶**EXAMPLE 2**   Suppose the amount, y, of a certain radioactive substance present at time t is given by

$$y = 1000e^{-.1t},$$

where t is measured in days and y is measured in grams.
(a) How much of the substance will be present at time $t = 0$?

3 Suppose the number of bacteria in a culture at time t is

$$y = 500e^{.4t},$$

where t is measured in hours.

(a) How many bacteria are present at $t = 0$?

(b) How many bacteria are present at $t = 10$?

Answer:

(a) 500

(b) About 27,300

4 Suppose that sales of a product are given by $S(x) = 1200 - 800e^{-.5x}$. Find

(a) $S(0)$;

(b) $S(2)$;

(c) $S(4)$;

(d) $S(10)$.

(e) Find an equation of the horizontal asymptote for the graph.

(f) Graph $y = S(x)$.

Answer:

(a) 400

(b) 906

(c) 1090

(d) 1190

(e) $y = 1200$

(f)

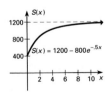

When $t = 0$,

$$y = 1000e^{-.1(0)}$$
$$= 1000 \cdot 1 = 1000,$$

with $y_0 = 1000$ grams.

(b) How much will be present at time $t = 5$?

$$y = 1000e^{-.1(5)}$$
$$= 1000e^{-.5}$$
$$\approx 1000(.60653)$$
$$= 606.53 \text{ grams} \quad \blacktriangleleft \quad \boxed{3}$$

▶**EXAMPLE 3** Sales of a new product often grow rapidly at first and then begin to level off with time. For example, suppose the sales, $S(x)$, in some appropriate unit, of a new model typewriter are approximated by

$$S(x) = 1000 - 800e^{-x},$$

where x represents the number of years the typewriter has been on the market. Calculate S_0, $S(1)$, $S(2)$, and $S(4)$. Graph $S(x)$.

Find S_0 by letting $x = 0$.

$$S_0 = S(0) = 1000 - 800 \cdot 1 = 200 \qquad e^{-x} = e^{-0} = 1$$

Using a calculator,

$$S(1) = 1000 - 800e^{-1}$$
$$\approx 1000 - 294.3$$
$$= 705.7,$$

which rounds to 706.

In the same way, verify that $S(2) = 892$ and $S(4) = 985$. Plotting several such points leads to the graph shown in Figure 4.6. Notice that as x increases, e^{-x} decreases and approaches 0. Thus, $S(x) = 1000 - 800e^{-x}$ approaches $S(x) = 1000 - 0 = 1000$. This means that the graph approaches the horizontal asymptote $y = 1000$. As the graph suggests, sales will tend to level off with time and gradually approach a level of 1000 units. ◀ **4**

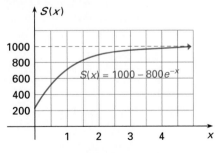

FIGURE 4.6

5 Suppose the value of the assets of a certain company at time t are given by

$$f(t) = 100{,}000 - 75{,}000e^{-.2t},$$

where t is measured in years. Find $f(t)$ for the following values of t.

(a) $t = 0$

(b) $t = 5$

(c) $t = 10$

(d) $t = 25$

(e) Find the horizontal asymptote.

(f) Graph $f(t)$.

Answer:

(a) 25,000

(b) 72,410

(c) 89,850

(d) 99,500

(e) $y = 100{,}000$

(f)

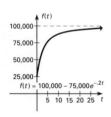

▶**EXAMPLE 4**   Assembly line operations tend to have a high turnover of employees, forcing the companies involved to spend much time and effort in training new workers. It has been found that a worker new to the operation of a certain task on the assembly line will produce items according to

$$P(x) = 25 - 25e^{-.3x}$$

where $P(x)$ is the number of items produced by the worker on day x. How many items will be produced by a new worker on day 8?

Evaluate $P(8)$.

$$P(8) = 25 - 25e^{-.3(8)}$$
$$= 25 - 25e^{-2.4}$$
$$\approx 25 - 2.3 = 22.7$$

On the eighth day, the worker can be expected to produce about 23 items. Plotting several such points leads to the graph of $P(x)$ shown in Figure 4.7.   ◀   **5**

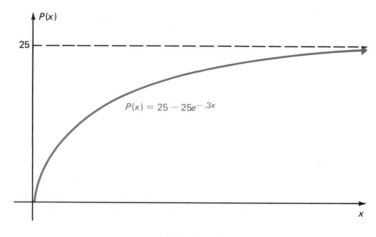

FIGURE 4.7

Graphs such as the one in Figure 4.7 are called **learning curves.** According to such a graph, a new worker tends to learn quickly at first; then learning tapers off and approaches some upper limit. This is characteristic of the learning of certain types of skills involving the repetitive performance of the same task. Learning curves are examples of *limited growth functions.*

▶**EXAMPLE 5**   Under certain conditions the total number of facts of a certain kind that are remembered is approximated by

$$N(t) = y_0\left(\frac{1 + e}{1 + e^{t+1}}\right)$$

where $N(t)$ is the number of facts remembered at time t, measured in days, and y_0 is the number of facts remembered initially. Graph $y = N(t)$.

6 In Example 5, find

(a) $N(2)$;

(b) $N(3)$;

(c) $N(5)$.

Answer:

(a) $.18y_0$

(b) $.07y_0$

(c) $.009y_0$

Plot some points to help draw the graph. For example, if $t = 0$,

$$N(0) = y_0\left(\frac{1 + e}{1 + e^1}\right) = y_0(1) = y_0.$$

If $t = 1$,

$$N(1) = y_0\left(\frac{1 + e}{1 + e^2}\right) \approx .44y_0.$$

Plotting several such points (see the problem at the side) leads to the graph of Figure 4.8. Graphs such as this are called **forgetting curves.** ◄ **6**

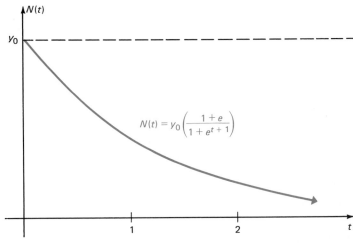

FIGURE 4.8

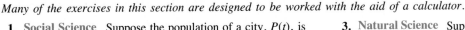

4.2 EXERCISES

Many of the exercises in this section are designed to be worked with the aid of a calculator.

1. **Social Science** Suppose the population of a city, $P(t)$, is given by

$$P(t) = 1,000,000e^{.02t},$$

where t represents time measured in years. Find each of the following values.
(a) P_0 **(b)** $P(2)$ **(c)** $P(4)$ **(d)** $P(10)$

2. **Natural Science** A population of lice, $L(t)$, is given by

$$L(t) = 100e^{.1t},$$

where t is measured in months. Find each of the following values.
(a) L_0 **(b)** $L(1)$ **(c)** $L(6)$ **(d)** $L(12)$

3. **Natural Science** Suppose the quantity, $Q(t)$, measured in grams, of a radioactive substance present at time t is given by

$$Q(t) = 500e^{-.05t},$$

where t is measured in days. Find the quantity present at each of the following times.
(a) $t = 0$ **(b)** $t = 4$ **(c)** $t = 8$ **(d)** $t = 20$

4. **Natural Science** The amount of a chemical in grams that will dissolve in a solution is given by

$$C(t) = 10e^{.02t},$$

where t is the temperature of the solution. Find each of the following.
(a) $C(0°)$ **(b)** $C(10°)$ **(c)** $C(30°)$ **(d)** $C(100°)$

5. Natural Science Let the number of bacteria, $B(t)$, present in a certain culture be given by

$$B(t) = 25,000e^{.2t},$$

where t is time measured in hours, and $t = 0$ corresponds to noon. Find the number of bacteria present at each of the following times.
(a) Noon   **(b)** 1 P.M.   **(c)** 2 P.M.   **(d)** 5 P.M.

6. Natural Science When a bactericide is introduced into a certain culture, the number of bacteria present, $D(t)$, is given by

$$D(t) = 50,000e^{-.01t},$$

where t is time measured in hours. Find the number of bacteria present at each of the following times.
(a) $t = 0$ **(b)** $t = 5$ **(c)** $t = 20$ **(d)** $t = 50$

7. In Chapter 5, it is shown that P dollars compounded continuously (every instant) at an annual rate of interest i would amount to

$$A = Pe^{ni}$$

at the end of n years. How much would $20,000 amount to at 8% compounded continuously for the following number of years?
(a) 1 year   **(b)** 5 years   **(c)** 10 years

8. The number (in hundreds) of fish in a small commercial pond is given by

$$F(t) = 27 - 15e^{-.8t},$$

where t is in years. Find each of the following.
(a) F_0 **(b)** $F(1)$ **(c)** $F(5)$ **(d)** $F(10)$

9. The number of words per minute that an average typist can type is given by

$$W(t) = 60 - 30e^{-.5t},$$

where t is time in months after the beginning of a typing class. Find each of the following.
(a) W_0 **(b)** $W(1)$ **(c)** $W(4)$ **(d)** $W(6)$

10. Management Sales of a new model can opener are approximated by

$$S(x) = 5000 - 4000e^{-x},$$

where x represents the number of years that the can opener has been on the market, and $S(x)$ represents sales in thousands. Find each of the following.
(a) $S(0)$ **(b)** $S(1)$ **(c)** $S(2)$
(d) $S(5)$ **(e)** $S(10)$
(f) Find the horizontal asymptote for the graph.
(g) Graph $y = S(x)$.

11. Management Assume that a person new to an assembly line will produce

$$P(x) = 500 - 500e^{-x}$$

items per day, where x is time measured in days. Find each of the following.
(a) $P(0)$ **(b)** $P(1)$ **(c)** $P(2)$
(d) $P(5)$ **(e)** $P(10)$
(f) Find the horizontal asymptote for the graph.
(g) Graph $y = P(x)$.

12. Management Experiments have shown that sales of a product, under relatively stable market conditions but in the absence of promotional activities such as advertising, tend to decline at a constant yearly rate. This rate of sales decline varies considerably from product to product, but seems to remain the same for any particular product. The sales decline can often be expressed by

$$S(t) = S_0e^{-at},$$

where $S(t)$ is the rate of sales at time t measured in years, S_0 is the rate of sales at time $t = 0$, and a is the sales decay constant. Suppose the sales decay constant for a particular product is $a = .10$. Let $S_0 = 50,000$. Find
(a) $S(1)$;   **(b)** $S(3)$.

13. Management In Exercise 12, suppose $S_0 = 80,000$ and $a = .05$. Find
(a) $S(2)$;   **(b)** $S(10)$.

14. Social Science A sociologist has shown that the fraction $y(t)$ of people in a group who have heard a rumor after t days is approximated by

$$y(t) = \frac{y_0e^{kt}}{1 - y_0(1 - e^{kt})},$$

where y_0 is the fraction of people who have heard the rumor at time $t = 0$, and k is a constant. A graph of $y(t)$ for a particular value of k is shown in the figure.
(a) If $k = .1$ and $y_0 = .05$, find $y(10)$.
(b) If $k = .2$ and $y_0 = .10$, find $y(5)$.

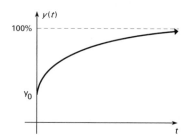

15. The higher a student's grade-point average, the fewer applications the student must send to medical schools (other things being equal). Using information given in a guidebook for prospective medical school students, we constructed the following mathematical model of the number of applications that a student should send out:

$$y = 540e^{-1.3x},$$

where y is the number of applications that should be sent out by a person whose grade-point average is x. Here $2.0 \le x \le 4.0$. Use a calculator with a y^x key to find the number of applications that should be sent out by students having a grade-point average of
(a) 2.0; **(b)** 2.5; **(c)** 3.0; **(d)** 3.5; **(e)** 3.9; **(f)** 4.0.

16. **Natural Science** Many environmental situations place effective limits on the growth of the numbers of an organism in an area. Many such limited growth situations are described by the *logistic function,* defined by

$$G(t) = \frac{m \cdot G_0}{G_0 + (m - G_0)e^{-kmt}},$$

where G_0 is the initial number present, m is the maximum possible size of the population, and k is a positive constant. Assume $G_0 = 1000$, $m = 25{,}000$, and $k = .04$.
(a) Find $G(5)$.   **(b)** Find $G(10)$.

Management *The National Basketball Association revenues increased dramatically in the decade of the 1980s. Using data from the February 5, 1990 issue of* Business Week *magazine (page 65), we constructed the function*

$$R(t) = 100e^{.18t}$$

to approximate annual revenues of the association, where $R(t)$ is revenue in millions of dollars in year t, with $t = 0$ representing 1981.*

17. Use this function to estimate revenue in each of the following years. Compare the estimates with the numbers on the graph at the top of the next column.
(a) 1981   **(b)** 1986   **(c)** 1990

18. For players' salaries, we constructed the function

$$P(t) = 170e^{.16t},$$

where $P(t)$ represents salaries paid in thousands of dollars in year t, with $t = 0$ representing 1981. In the second graph, use this function to estimate salaries in each of the following years.
(a) 1981   **(b)** 1986   **(c)** 1990

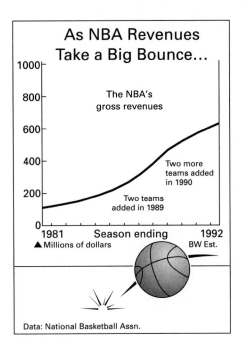

As NBA Revenues Take a Big Bounce...

The NBA's gross revenues

Two more teams added in 1990

Two teams added in 1989

1981   Season ending   1992
▲ Millions of dollars   BW Est.

Data: National Basketball Assn.

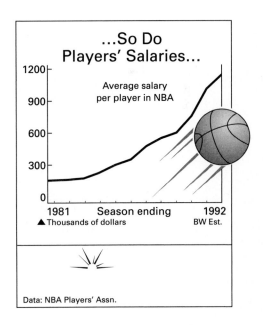

...So Do Players' Salaries...

Average salary per player in NBA

1981   Season ending   1992
▲ Thousands of dollars   BW Est.

Data: NBA Players' Assn.

---

*From ''The NBA is Paying Out Like There's No Tomorrow,'' *Business Week,* February 5, 1990, page 64–66.

Natural Science   *Newton's Law of Cooling says that the rate at which a body cools is proportional to the difference in temperature between the body and an environment into which it is introduced. Using calculus, the temperature $F(t)$ of the body at time t after being introduced into an environment having constant temperature T_0 is*

$$F(t) = T_0 + Ce^{-kt},$$

where C and k are constants. Use this result in Exercises 19–24.

19. Find the temperature of an object when $t = 4$ if $T_0 = 125$, $C = .8$, and $k = .2$.

20. Find the temperature of an object when $t = 9$ if $T_0 = 180$, $C = .5$, and $k = .6$.

21. A piece of metal is heated to 300°C and then placed in a cooling liquid at 50°C. After 4 minutes the metal has cooled to 175°C. Find its temperature after 12 minutes.

22. Boiling water, at 100°C, is placed in a freezer at 0°C. The temperature of the water is 50°C after 24 minutes. Find the temperature of the water after 96 minutes.

23. A volcano discharges lava at 800°C. The surrounding air has a temperature of 20°C. The lava cools to 410°C in 5 hours. Find its temperature after 15 hours.

24. Paisley refuses to drink coffee cooler than 95°F. She makes coffee with a temperature of 170°F in a room with a temperature of 70°F. The coffee cools to 120°F in 10 minutes. What is the longest time she can let the coffee sit before she drinks the coffee?

Natural Science   *The pressure of the atmosphere, $p(h)$, in pounds per square inch, is given by*

$$p(h) = p_0 e^{-kh},$$

where h is the height above sea level and p_0 and k are constants. The pressure at sea level is 15 pounds per square inch and the pressure is 9 pounds per square inch at a height of 12,000 feet.

25. Find the pressure at an altitude of 6000 feet.

26. What would be the pressure encountered by a spaceship at an altitude of 150,000 feet?

4.3 LOGARITHMIC FUNCTIONS

Until the development of computers and calculators, logarithms were the only effective tool for large-scale numerical computations. They are no longer needed for this, but logarithmic functions still play a crucial role in many applications.

Logarithms are simply a *new language for old ideas*—essentially a special case of exponents.

Definition of Logarithm

For $a > 0$ and $a \neq 1$,

$$y = \log_a x \quad \text{means} \quad a^y = x.$$

Read $y = \log_a x$ as "y is the logarithm of x to the base a." For example, the exponential statement $2^4 = 16$ can be translated into the equivalent logarithmic statement $4 = \log_2 16$. Thus, $\log_a x$ is an *exponent;* it is the answer to the question

"to what power must a be raised to produce x?"

1 Write the logarithmic form of

(a) $5^3 = 125$;

(b) $3^{-4} = 1/81$;

(c) $8^{2/3} = 4$.

Answer:

(a) $\log_5 125 = 3$

(b) $\log_3 (1/81) = -4$

(c) $\log_8 4 = 2/3$

2 Write the exponential form of

(a) $\log_{16} 4 = 1/2$;

(b) $\log_3 (1/9) = -2$;

(c) $\log_{16} 8 = 3/4$.

Answer:

(a) $16^{1/2} = 4$

(b) $3^{-2} = 1/9$

(c) $16^{3/4} = 8$

This key definition should be memorized. It is important to remember the location of the base and exponent in each part of the definition.

$$\text{logarithmic form:} \quad y = \log_a x$$

exponent ↓, base ↑

$$\text{exponential form:} \quad a^y = x$$

exponent ↓, base ↑

▶ **EXAMPLE 1** This example shows several statements written in both exponential and logarithmic forms.

Exponential Form	Logarithmic Form
(a) $3^2 = 9$	$\log_3 9 = 2$
(b) $(1/5)^{-2} = 25$	$\log_{1/5} 25 = -2$
(c) $10^5 = 100,000$	$\log_{10} 100,000 = 5$
(d) $4^{-3} = 1/64$	$\log_4 (1/64) = -3$
(e) $2^{-4} = 1/16$	$\log_2 (1/16) = -4$
(f) $e^0 = 1$	$\log_e 1 = 0$

◀ **1** **2**

For a given *positive* value of x, the definition of logarithm leads to exactly one value of y, so that $y = \log_a x$ defines a logarithmic function of base a. (The base a must be positive, with $a \neq 1$.)

If $a > 0$ and $a \neq 1$, the **logarithmic function** with base a is defined as

$$f(x) = \log_a x.$$

The graphs of the exponential function $f(x) = 2^x$ and the logarithmic function $g(x) = \log_2 x$ are shown in Figure 4.9. The graphs show that $f(3) = 2^3 = 8$, while $g(8) = \log_2 8 = 3$. Thus, $f(3) = 8$ and $g(8) = 3$. Also, $f(2) = 4$ and $g(4) = 2$. In fact, for any number m, if $f(m) = p$, then $g(p) = m$. Functions related in this way are called **inverses** of each other. The graphs also show that the domain of the exponential function (the set of real numbers) is the range of the logarithmic function. Also, the range of the exponential function (the set of positive real numbers) is the domain of the logarithmic function. Every logarithmic function is the inverse of some exponential function. This means that we can graph logarithmic functions by rewriting them as exponential functions using the definition of logarithm. The

3 (a) Graph $f(x) = 3^x$ and $g(x) = \log_3 x$ on the same axes.

(b) Graph $f(x) = (1/3)^x$ and $g(x) = \log_{1/3} x$ on the same axes.

Answer:

(a)

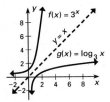

(b)

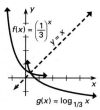

graphs in Figure 4.9 show a characteristic of inverse functions: their graphs are mirror images about the line $y = x$. A more complete discussion of inverse functions is given in most standard intermediate algebra and college algebra books.

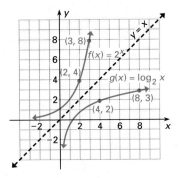

FIGURE 4.9

To graph a logarithmic function, rewrite it in exponential form, as in the next example.

▶ **EXAMPLE 2**   Graph $g(x) = \log_{1/2} x$.

This graph is found by rewriting $y = \log_{1/2} x$ in exponential form as $(1/2)^y = x$. Then choose several values of y, since y is now the independent variable, and find the corresponding values of x. See the table of values shown below. Use the results to get the graph of $g(x) = \log_{1/2} x$ in Figure 4.10. Again, for comparison, a graph of $f(x) = (1/2)^x$ is included. These two graphs are mirror images with respect to $y = x$. ◀ **3**

	$x = \left(\dfrac{1}{2}\right)^y$	
y	x	(x, y)
-2	$\left(\dfrac{1}{2}\right)^{-2} = 4$	$(4, -2)$
-1	$\left(\dfrac{1}{2}\right)^{-1} = 2$	$(2, -1)$
0	$\left(\dfrac{1}{2}\right)^{0} = 1$	$(1, 0)$
1	$\left(\dfrac{1}{2}\right)^{1} = \dfrac{1}{2}$	$\left(\dfrac{1}{2}, 1\right)$
2	$\left(\dfrac{1}{2}\right)^{2} = \dfrac{1}{4}$	$\left(\dfrac{1}{4}, 2\right)$

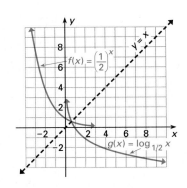

FIGURE 4.10

Typical graphs of logarithmic and exponential functions are shown below.

Exponential Functions

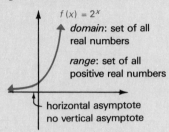

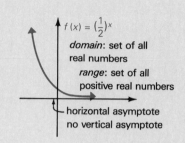

Logarithmic Functions

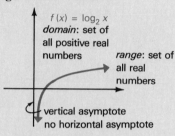

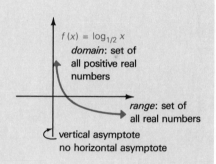

The usefulness of logarithmic functions depends in large part on the following *properties of logarithms*.

Properties of Logarithms

Let x and y be any positive real numbers and r be any real number. Let a be a positive real number, $a \neq 1$. Then

(a) $\log_a xy = \log_a x + \log_a y$;   (b) $\log_a \dfrac{x}{y} = \log_a x - \log_a y$;

(c) $\log_a x^r = r \log_a x$;   (d) $\log_a a = 1$;

(e) $\log_a 1 = 0$;   (f) $\log_a a^y = y$;

(g) $a^{\log_a x} = x$.

Note Because these properties are so useful, they should be memorized.

4 Use the properties of logarithms to rewrite and evaluate each of the following, given $\log_3 7 \approx 1.77$ and $\log_3 5 \approx 1.46$.

(a) $\log_3 35$

(b) $\log_3 7/5$

(c) $\log_3 25$

(d) $\log_3 3$

(e) $\log_3 1$

Answer:

(a) 3.23

(b) .31

(c) 2.92

(d) 1

(e) 0

5 Simplify, using the properties of logarithms.

(a) $\log_a 5x + \log_a 3x^4$

(b) $\log_a 3p - \log_a 5q$

(c) $4 \log_a k - 3 \log_a m$

Answer:

(a) $\log_a 15x^5$

(b) $\log_a 3p/(5q)$

(c) $\log_a k^4/m^3$

To prove property (a), let

$$m = \log_a x \quad \text{and} \quad n = \log_a y.$$

Then, by the definition of logarithm,

$$a^m = x \quad \text{and} \quad a^n = y.$$

Multiply to get

$$a^m \cdot a^n = x \cdot y,$$

or, by a property of exponents,

$$a^{m+n} = xy.$$

Use the definition of logarithm to rewrite this last statement as

$$\log_a xy = m + n.$$

Replace m with $\log_a x$ and n with $\log_a y$ to get

$$\log_a xy = \log_a x + \log_a y.$$

Properties (b) and (c) can be proven in a similar way. Since $a^1 = a$ and $a^0 = 1$, properties (d)–(g) come from the definition of logarithm.

▶**EXAMPLE 3** Using the properties of logarithms, if $\log_6 7 \approx 1.09$ and $\log_6 5 \approx .9$,
(a) $\log_6 35 = \log_6 7 \cdot 5 = \log_6 7 + \log_6 5 \approx 1.09 + .9 = 1.99;$
(b) $\log_6 5/7 = \log_6 5 - \log_6 7 \approx -.19;$
(c) $\log_6 5^3 = 3 \log_6 5 \approx 3(.9) = 2.7;$
(d) $\log_6 6 = 1;$
(e) $\log_6 1 = 0.$ ◀ **4**

▶**EXAMPLE 4** If all the following variable expressions represent positive numbers, then for $a > 0$, $a \neq 1$,
(a) $\log_a x + \log_a (x - 1) = \log_a x(x - 1);$
(b) $\log_a \dfrac{x^2 + 4}{x + 6} = \log_a (x^2 + 4) - \log_a (x + 6);$
(c) $\log_a 9x^5 = \log_a 9 + \log_a x^5 = \log_a 9 + 5 \log_a x.$ ◀ **5**

Caution There is no logarithm property that allows you to simplify the logarithm of a sum, such as $\log_a (x^2 + 4)$. In particular, $\log_a (x^2 + 4)$ is *not* equal to $\log_a x^2 + \log_a 4$. Property (a) of logarithms in the box above shows that $\log_a x^2 + \log_a 4 = \log_a 4x^2.$

6 Find each common logarithm.

(a) log 27

(b) log 1089

(c) log .00426

Answer:

(a) 1.4314

(b) 3.0370

(c) −2.3706

Historically, one of the main applications of logarithms has been as an aid to numerical calculation. The properties of logarithms and a table of logarithms were used to simplify many numerical problems. With today's widespread use of calculators, this application of logarithms has declined. Since our number system has base 10, logarithms to base 10 were most convenient for numerical calculations, so base 10 logarithms were called **common logarithms.** Common logarithms are still useful in other applications. For simplicity,

$$\log_{10} x \quad \text{is abbreviated} \quad \log x.$$

With this notation,
$$\log 1000 = \log 10^3 = 3,$$
$$\log 100 = \log 10^2 = 2,$$
$$\log 1 = \log 10^0 = 0,$$
$$\log .01 = \log 10^{-2} = -2, \quad \text{and so on.}$$

Many calculators have a $\boxed{\log x}$ key for evaluating common logarithms.

▶**EXAMPLE 5** Use a calculator to find the following common logarithms.
(a) log 127

Enter 127 and press the log x key.* The display will read (to five decimal places) 2.10037. So,
$$\log 127 = 2.10037.$$

(b) log 40,120

Entering 40,120 and pressing the log x key shows that log 40,120 = 4.60336.

(c) log .058 = −1.23657

Notice that the logarithm of a number between 0 and 1 is negative. Refer to Figures 4.9 and 4.10 to see why. ◀ **6**

One application of common logarithms is their use in measuring the intensity of earthquakes. The magnitude of an earthquake on the Richter Scale is determined by the maximum amplitude recorded by a seismometer and the distance of that seismometer from the epicenter of the earthquake. A unit change in magnitude corresponds to a change in amplitude by a factor of 10. Thus, common logarithms are used to express the magnitude. This means that an earthquake with a Richter Scale rating of 4 is ten times more powerful than one with a rating of 3. (See Section 4.4 Exercises 63 and 64.)

Most practical applications of logarithms use the number e as base. (Recall that to seven decimal places, $e = 2.7182818$.) Logarithms to base e are called **natural logarithms,** and

$$\log_e x \quad \text{is abbreviated} \quad \ln x.$$

Read ln x as "el-en x."

---

*Some calculators may require a different sequence of keystrokes; check your instruction booklet.

7 Find the following.

(a) ln 6.1

(b) ln 20

(c) ln .8

(d) ln .1

Answer:

(a) 1.8083

(b) 2.9957

(c) −.2231

(d) −2.3026

8 Solve each equation.

(a) $\log_x 6 = 1$

(b) $\log_5 25 = m$

(c) $\log_{27} x = 2/3$

Answer:

(a) 6

(b) 2

(c) 9

Although common logarithms may seem more "natural" than logarithms to base e, there are several good reasons for using natural logarithms instead. The most important reason is discussed in Section 4.4.

▶**EXAMPLE 6** Use a calculator to find the following logarithms.
(a) ln 85
Enter 85, press the $\boxed{\ln x}$ key, and read the result: 4.4427.*
(b) ln 36 = 3.5835
(c) ln .01 = −4.6052 ◀ **7**

Equations involving logarithms are often solved by using the fact that a logarithmic equation can be rewritten (with the definition of logarithm) as an exponential equation. In other cases, the properties of logarithms may be useful in simplifying an equation involving logarithms.

▶**EXAMPLE 7** Solve each of the following equations.
(a) $\log_x 8/27 = 3$
First, use the definition of logarithm and write the expression in exponential form.

$$x^3 = \frac{8}{27} \qquad \text{Definition of logarithm}$$

$$x^3 = \left(\frac{2}{3}\right)^3 \qquad \text{Write } \frac{27}{3} \text{ as a cube}$$

$$x = \frac{2}{3} \qquad \text{Set bases equal}$$

The solution is 2/3.
(b) $\log_4 x = 5/2$
In exponential form, the given statement becomes

$$4^{5/2} = x \qquad \text{Definition of logarithm}$$
$$(4^{1/2})^5 = x \qquad \text{Definition of rational exponent}$$
$$2^5 = x$$
$$32 = x.$$

The solution is 32. ◀ **8**

In the next example, properties of logarithms are needed to solve equations.

---

*Your calculator may require a different sequence of keystrokes. Check your instruction booklet if you do not have a key labeled ln x.

9 Solve each equation.

(a) $\log_5 x + 2 \log_5 x = 3$

(b) $\log_6 (a + 2)$
$\quad - \log_6 \dfrac{a - 7}{5} = 1$

Answer:

(a) 5

(b) 52

▶**EXAMPLE 8**   Solve each equation.

(a) $\log_2 x - \log_2 (x - 1) = 1$

By the properties of logarithms, the left-hand side can be simplified as

$$\log_2 x - \log_2 (x - 1) = \log_2 \frac{x}{x - 1}.$$

The original equation then becomes

$$\log_2 \frac{x}{x - 1} = 1.$$

Use the definition of logarithm to write this last result in exponential form.

$$\frac{x}{x - 1} = 2^1 = 2$$

Solve this equation.

$$\frac{x}{x - 1} \cdot (x - 1) = 2(x - 1)$$
$$x = 2(x - 1)$$
$$x = 2x - 2$$
$$-x = -2$$
$$x = 2$$

The domain of logarithmic functions includes only positive real numbers, so it is *necessary* to check this proposed solution in the original equation.

$$\log_2 x - \log_2 (x - 1) \overset{?}{=} 1$$
$$\log_2 2 - \log_2 (2 - 1) \overset{?}{=} 1 \qquad \text{Let } x = 2$$
$$1 - 0 = 1 \qquad \text{Definition of logarithm}$$

The solution, 2, checks.   **9**

(b) $\log x - \log 8 = 1$

Use a property of logarithms to combine the terms on the left side.

$$\log x - \log 8 = \log \frac{x}{8}$$

The equation becomes

$$\log \frac{x}{8} = 1.$$

Recall $\log x$ means $\log_{10} x$, so

$$\log \frac{x}{8} = \log_{10} \frac{x}{8} = 1$$

$$\frac{x}{8} = 10^1 = 10 \qquad \textbf{Definition of logarithm}$$

$$x = 80.$$

Since $x = 80$ is in the domain of $\log x$, the solution is 80. ◀

4.3 EXERCISES

Write each of the following in logarithmic form. (See Example 1.)

1. $2^3 = 8$ **2.** $5^2 = 25$ **3.** $3^4 = 81$ **4.** $6^3 = 216$ **5.** $\left(\frac{1}{3}\right)^{-2} = 9$ **6.** $\left(\frac{3}{4}\right)^{-2} = \frac{16}{9}$

Write each of the following in exponential form. (See Example 1.)

7. $\log_2 128 = 7$ **8.** $\log_3 81 = 4$ **9.** $\log_5 \frac{1}{25} = -2$ **10.** $\log_2 \frac{1}{8} = -3$

11. $\log 10{,}000 = 4$ **12.** $\log .00001 = -5$

Evaluate each of the following.

13. $\log 1000$ **14.** $\log 100$ **15.** $\log .01$ **16.** $\log .0001$ **17.** $\log_5 25$ **18.** $\log_9 81$

19. $\log_4 64$ **20.** $\log_6 216$ **21.** $\log_2 \frac{1}{4}$ **22.** $\log_3 \frac{1}{27}$ **23.** $\log_e \sqrt{e}$ **24.** $\log_e \frac{1}{e}$

25. Complete the following table of values for $y = \log_3 x$.

x	1/27						9
y	−3	−2	−1	0	1	2	3

Graph $y = \log_3 x$ using the same scale on both axes.

26. Complete the following table of values for $y = 3^x$.

x	−3	−2	−1	0	1	2	3
y		1/9					27

Graph $y = 3^x$ on the same axes you used in Exercise 25. Compare the two graphs. How are they related?

Graph each of the following. (See Example 2.)

27. $f(x) = \log_4 x$ **28.** $f(x) = \log x$ **29.** $f(x) = \log_3 (x - 1)$ **30.** $f(x) = \log_2 (1 + x)$

Use the properties of logarithms and write each of the following as a sum, difference, or product. Assume all variable expressions represent positive real numbers.

31. $\log_3 \frac{2}{5}$ **32.** $\log_4 \frac{6}{7}$ **33.** $\log_9 7m$ **34.** $\log_5 8p$

35. $\log_3 \frac{3x}{5k}$ **36.** $\log_7 \frac{11p}{13y}$ **37.** $\log_k \frac{pq^2}{m}$ **38.** $\log_z \frac{x^5 y^3}{3}$

39. $\log_r (5m + 7n)$ **40.** $\log_y (8k + 7z)$ **41.** $\log_3 \frac{5\sqrt{2}}{\sqrt[4]{7}}$ **42.** $\log_2 \frac{9\sqrt[3]{5}}{\sqrt[4]{3}}$

Use a calculator to find each of the following logarithms. (See Examples 5 and 6.)

43. log 60.4   **44.** log 801   **45.** log .00156   **46.** ln 35

47. ln 58,500   **48.** ln .397   **49.** log 56 − log 8   **50.** ln 320 + ln 5

Express each of the following as the logarithm of a single number or expression. Assume all variables represent positive numbers. (See Example 4.)

51. log 15 − log 3   **52.** log 4 + log 8 − log 2   **53.** 3 ln 2 + 2 ln 3   **54.** $2 \ln 5 - \frac{1}{2} \ln 25$

55. 3 log x − 2 log y **56.** 2 log u + 3 log w − 6 log v

57. ln $(3x + 2)$ + ln $(x + 4)$ **58.** 2 ln $(x + 1)$ − ln $(x + 2)$

59. 3 log x − 2 log $(x + 1)^2$ + $\frac{1}{2}$ log $(x + 2)$ **60.** 2 log y + 3 log z − log $(y + z)$

Suppose $\log_b 2 = a$ and $\log_b 3 = c$. Use the properties of logarithms to find the following logarithms.

61. $\log_b 8$ **62.** $\log_b 24$ **63.** $\log_b 54$ **64.** $\log_b 144$ **65.** $\log_b (72b)$ **66.** $\log_b (4b^2)$

Solve each of the following equations. (See Examples 7 and 8.)

67. $\log_x 25 = -2$ **68.** $\log_x \dfrac{1}{16} = -2$ **69.** $\log_9 27 = m$ **70.** $\log_8 4 = z$

71. $\log_y 8 = \dfrac{3}{4}$ **72.** $\log_r 7 = \dfrac{1}{2}$ **73.** $\log_3 (5x + 1) = 2$ **74.** $\log_5 (9x - 4) = 1$

75. log x − log $(x + 3)$ = −1   **76.** log m − log $(m - 4)$ = −2

77. $\log_3 (y + 2) = \log_3 (y - 7) + \log_3 4$ **78.** $\log_8 (z - 6) = 2 - \log_8 (z + 15)$

79. ln k − ln $(k + 1)$ = ln 5   **80.** ln $(5 + 4y)$ − ln $(3 + y)$ = ln 3

81. $7 + 2 \log_5 (2x - 1) = 9$ **82.** $3 \log_2 (5x - 2) - 1 = 5$

Work the following exercises.

 83. Management Suppose the sales of a certain product are approximated by

$$S(t) = 125 + 83 \log(5t + 1),$$

where $S(t)$ is sales in thousands of dollars t years after the product was introduced on the market. Find
(a) $S(0)$;   **(b)** $S(2)$;   **(c)** $S(4)$;   **(d)** $S(31)$.
(e) Graph $y = S(t)$.

 84. Natural Science The population of an animal species that is introduced into a certain area may grow rapidly at first but then grow more slowly as time goes on. A logarithmic function can provide an excellent description of such growth. Suppose that the population of foxes, $F(t)$, in an area t months after the foxes were introduced there is

$$F(t) = 500 \log (2t + 3).$$

Find the population of foxes at the following times.
(a) When they are first released into the area (that is, when $t = 0$)
(b) After 3 months   **(c)** After 15 months
(d) Graph $y = F(t)$.

Management *In many applications, data is graphed on a* logarithmic scale, *where differences between successive measurements are not always the same. For example, the graph from the July 30, 1979, issue of* Fortune *magazine (page 58),* shows the price/performance ratio for various*

---

*Graph adapted from ''Price vs. Performance'' from How the 4300 Fits I.B.M.'s New Strategy by Bro Uttal from *Fortune*, July 30, 1979. Copyright © 1979 Time, Inc. All rights reserved. Reprinted by permission.

models of IBM computers. Notice on the vertical scale that the distance from 100 *to* 500 *is the same as the distance from* 1000 *to* 5000. *This is characteristic of a graph drawn with logarithmic scales.*

85. Estimate the price for a 4300 model computer producing a performance index of 10. Do the same for a System/370 (use the upper graph).

86. To locate a performance index of 50 on the horizontal scale, first find that log 50 ≈ 1.7, so that 50 would be located about .7, or 70% of the way from 10 to 100. Locate 50 on the horizontal axis, and then estimate the price for a model 4300 computer with a performance index of 50.

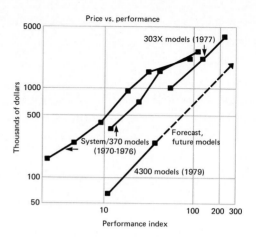

4.4 APPLICATIONS OF LOGARITHMIC FUNCTIONS

This section first shows an additional method of solving exponential and logarithmic equations and then shows several applications using this method. Some of these applications depend on the following result.

Let x and y be positive numbers. Let b be a positive number, $b \neq 1$.

$$\text{If } x = y, \text{ then } \log_b x = \log_b y.$$

$$\text{If } \log_b x = \log_b y, \text{ then } x = y.$$

For convenience, base e is generally used in these applications.

▶**EXAMPLE 1** Solve $3^x = 5$.

Since 3 and 5 cannot easily be written with the same base, the methods of Section 4.1 cannot be used to solve this equation. Instead use the result given above and take natural logarithms of both sides.

$$3^x = 5$$

$$\ln 3^x = \ln 5$$

$$x \ln 3 = \ln 5 \qquad \text{Property (c) of logarithms}$$

$$x = \frac{\ln 5}{\ln 3}$$

1 Solve each equation. Round to the nearest thousandth.

(a) $2^x = 7$

(b) $5^m = 50$

(c) $3^y = 17$

Answer:

(a) 2.807

(b) 2.431

(c) 2.579

2 Solve each equation. Round to the nearest thousandth.

(a) $e^{.1x} = 11$

(b) $e^{-x/2} = 5$

(c) $e^{3+x} = .893$

Answer:

(a) 23.979

(b) -3.219

(c) -3.113

From a calculator, ln 5 ≈ 1.6094 and ln 3 ≈ 1.0986, so

$$x \approx \frac{1.6094}{1.0986} \approx 1.465,$$

rounded to the nearest thousandth.

A calculator with a y^x key can be used to check this answer. Evaluate $3^{1.465}$; the result should be approximately 5. This step verifies that, to the nearest thousandth, the solution of the given equation is 1.465. ◀ **1**

By the definition of logarithm, ln $e = 1$. This fact, along with properties of logarithms, can be used to solve equations involving powers of e, as the next example shows.

▶**EXAMPLE 2** Solve each equation, and round solutions to the nearest thousandth.

(a) $e^{.04m} = 13$

Take natural logarithms of both sides.

$$e^{.04m} = 13$$
$$\ln e^{.04m} = \ln 13$$

Use property (c) of logarithms to rewrite this last equation as

$$.04m \ln e = \ln 13.$$

Since ln $e = 1$, the equation becomes

$$.04m(1) = \ln 13$$
$$.04m = \ln 13$$
$$m = \frac{\ln 13}{.04}$$
$$m \approx \frac{2.5649}{.04}$$
$$m = 64.123 \quad \text{(rounded)}.$$

(b) $e^{2-x} = .798$

Work as above.

$$\ln e^{2-x} = \ln .798$$
$$(2 - x) \ln e = \ln .798 \qquad \textbf{Property (c) of logarithms}$$
$$2 - x = \ln .798 \qquad \textbf{ln } e = 1$$
$$-x = -2 + \ln .798$$
$$x = 2 - \ln .798$$
$$x \approx 2 - (-.226)$$
$$x = 2.226 \quad ◀ \quad \textbf{2}$$

3 Solve each equation. Round to the nearest thousandth.

(a) $6^m = 3^{2m-1}$

(b) $5^{6a-3} = 2^{4a+1}$

Answer:

(a) 2.710

(b) .802

▶**EXAMPLE 3** Solve $3^{2x-1} = 4^{x+2}$.

Taking natural logarithms on both sides gives

$$\ln 3^{2x-1} = \ln 4^{x+2}.$$

Now use property (c) of logarithms.

$$(2x - 1)\ln 3 = (x + 2)\ln 4$$
$$2x \ln 3 - \ln 3 = x \ln 4 + 2 \ln 4 \quad \text{Distributive property}$$
$$2x \ln 3 - x \ln 4 = 2 \ln 4 + \ln 3 \quad \text{Addition property of equality}$$

Factor out x on the left to get

$$x(2 \ln 3 - \ln 4) = 2 \ln 4 + \ln 3$$

or
$$x = \frac{2 \ln 4 + \ln 3}{2 \ln 3 - \ln 4}.$$

Using properties of logarithms,

$$x = \frac{\ln 16 + \ln 3}{\ln 9 - \ln 4}$$

or, finally,
$$x = \frac{\ln 48}{\ln \dfrac{9}{4}} = \frac{\ln 48}{\ln 2.25}.$$

This quotient could be approximated by a decimal if desired.

$$x = \frac{\ln 48}{\ln 2.25} \approx 4.774$$

To the nearest thousandth, the solution is 4.774. ◀ **3**

▶**EXAMPLE 4** Solve $3e^{x^2} = 600$.

First divide each side by 3 to get

$$e^{x^2} = 200.$$

Now take natural logarithms on both sides; then use properties of logarithms.

$$e^{x^2} = 200$$
$$\ln e^{x^2} = \ln 200$$
$$x^2 \ln e = \ln 200$$
$$x^2 = \ln 200 \qquad \ln e = 1$$
$$x = \pm\sqrt{\ln 200}$$
$$x \approx \pm 2.302.$$

4 Solve each equation. Round to the nearest thousandth.

(a) $8e^{k^2} = 16$

(b) $e^{x^2+1} = 35$

(c) $e^{2x^2-3} = 9$

Answer:

(a) $\pm.833$

(b) ±1.599

(c) ±1.612

5 Solve each equation.

(a) $\log_2 (p + 9) - \log_2 p$
$= \log_2 (p + 1)$

(b) $\log_3 (m + 1)$
$- \log_3 (m - 1) = \log_3 m$

Answer:

(a) 3

(b) $1 + \sqrt{2} \approx 2.414$

The solutions are ±2.302, rounding to the nearest thousandth. (The symbol $\pm$ is used as a shortcut for writing the two solutions, 2.302 and -2.302.) ◀ **4**

The result given at the beginning of this section, with the properties of logarithms from Section 4.3, is useful in solving logarithmic equations, as shown in the next examples.

▶**EXAMPLE 5**   Solve $\log (x + 4) - \log (x + 2) = \log x$.

Using property (b) of logarithms, rewrite the equation as

$$\log \frac{x + 4}{x + 2} = \log x.$$

Then

$$\frac{x + 4}{x + 2} = x$$
$$x + 4 = x(x + 2)$$
$$x + 4 = x^2 + 2x$$
$$x^2 + x - 4 = 0.$$

By the quadratic formula,

$$x = \frac{-1 \pm \sqrt{1 + 16}}{2},$$

so that

$$x = \frac{-1 + \sqrt{17}}{2} \quad \text{or} \quad x = \frac{-1 - \sqrt{17}}{2}.$$

Log x cannot be evaluated for $x = (-1 - \sqrt{17})/2$, since this number is negative and not in the domain of log x. By substitution, verify that $(-1 + \sqrt{17})/2$ is a solution. ◀ **5**

▶**EXAMPLE 6**   Suppose that $A(t)$, the amount of a certain radioactive substance present at time t, is given by

$$A(t) = 1000e^{-.1t},$$

where t is measured in days and $A(t)$ in grams. At time $t = 0$,

$$A(0) = 1000e^{-.1(0)}$$
$$= 1000$$

grams of the substance present. Also,

6 The amount of a substance present at time t (in hours) is given by

$$A(t) = 530e^{-.2t},$$

and $A(t)$ is measured in grams. How much of the substance will remain after 5 hours? That is, find $A(5)$.

Answer:
About 195 grams

7 Find the half-life of the substance in Problem 6 above.

Answer:
About 3.5 hours

$$A(5) = 1000e^{-.1(5)}$$
$$= 1000e^{-.5}$$
$$\approx 1000(.60653)$$
$$= 606.53,$$

so that about 607 grams are still present after 5 days. Now let us find the half-life of the substance. (The **half-life** of a radioactive substance is the time it takes for exactly half the sample to decay.)

Find the half-life by finding a value of t such that $A(t) = (1/2)(1000) = 500$ grams. That is, find the half-life by solving the equation

$$500 = 1000e^{-.1t}.$$

First, divide both sides by 1000, obtaining

$$\frac{1}{2} = e^{-.1t}.$$

Now take natural logarithms of both sides. This gives

$$\ln \frac{1}{2} = \ln e^{-.1t}.$$

Using property (c) of logarithms.

$$\ln \frac{1}{2} = (-.1t)(\ln e),$$

and since $\ln e = 1$,

$$\ln \frac{1}{2} = -.1t,$$

or

$$t = \frac{\ln \frac{1}{2}}{-.1} = \frac{\ln .5}{-.1} \approx 6.9.$$

It will take about 6.9 days for half the sample to decay. ◀ **6** **7**

▶ **EXAMPLE 7** Carbon 14, also known as radiocarbon, is a radioactive form of carbon that is found in all living plants and animals. After a plant or animal dies, the radiocarbon disintegrates with a half-life of approximately 5600 years. Scientists can determine the age of the remains by comparing the amount of radiocarbon with the amounts present in living plants and animals. This technique is called *carbon dating*. The amount of radiocarbon present after t years is given by

$$y = y_0 e^{-(\ln 2)(1/5600)t},$$

where y_0 is the amount present in living plants and animals.

8 What is the age of a specimen in which $y = (1/3)y_0$?

Answer:
About 8880 years

(a) Verify the formula for $t = 5600$.
Substitute 5600 for t. Then

$$y = y_0e^{-(\ln 2)(1/5600)5600} = y_0e^{-\ln 2} = y_0(e^{\ln 2})^{-1} = y_0(2)^{-1} = \frac{1}{2}y_0.$$

This result is correct. Since 5600 years is the half-life of carbon 14, half of the initial amount should remain after 5600 years.

(b) A round table hanging in Winchester Castle (England) was alleged to belong to King Arthur, who lived in the 5th century. A chemical analysis recently showed that the table had 91% of the amount of radiocarbon present in living wood. How old is the table?

The amount of radiocarbon present in the round table after y years is $.91y_0$. Therefore, in the equation

$$y = y_0e^{-(\ln 2)(1/5600)t}$$

replace y with $.91y_0$ and solve for t.

$$.91y_0 = y_0e^{-(\ln 2)(1/5600)t}$$

$$.91 = e^{-(\ln 2)(1/5600)t} \qquad \text{Divide both sides by } y_0$$

$$\ln .91 = \ln e^{-(\ln 2)(1/5600)t} \qquad \text{Take logarithms on both sides}$$

$$\ln .91 = -(\ln 2)(1/5600)t \qquad \text{Property (c) of logarithms}$$

$$t = \frac{(5600)\ln .91}{-\ln 2} \approx 760$$

The table is about 760 years old and therefore could not have belonged to King Arthur. ◄ **8**

The most important applications of exponential and logarithmic functions for the fields of management and economics are given in Chapter 5 on the mathematics used in finance. These include compound interest and annuities.

4.4 EXERCISES

Solve each of the following equations. Round to the nearest thousandth. (See Examples 1–5.)

1. $3^x = 6$

2. $4^x = 12$

3. $7^x = 8$

4. $13^p = 55$

5. $3^{a+2} = 5$

6. $5^{2-x} = 12$

7. $6^{1-2k} = 8$

8. $2^{k-3} = 11$

9. $5 \cdot 4^{3m-1} = 5 \cdot 12^{m+2}$

10. $8 \cdot 3^{2m-5} = 8 \cdot 13^{m-1}$

11. $e^{k-1} = 4$

12. $e^{2-y} = 12$

13. $2e^{5a+2} = 8$

14. $10e^{3z-7} = 5$

15. $2^x = -3$

16. $(1/4)^p = -4$

17. $\left(1 + \dfrac{r}{2}\right)^5 = 9$

18. $\left(1 + \dfrac{n}{4}\right)^3 = 12$

19. $100(1 + .02)^{3+n} = 150$

20. $500(1 + .05)^{p/4} = 200$

21. $2^{x^2-1} = 12$

22. $3^{2-x^2} = 4$

23. $2(e^x + 1) = 10$

24. $5(e^{2x} - 2) = 15$

25. $\log (t - 1) = 1$

26. $\log q^2 = 1$

27. $\log (x - 3) = 1 - \log x$

28. $\log (z - 6) = 2 - \log (z + 15)$

29. $\ln (y + 2) = \ln (y - 7) + \ln 4$

30. $\ln p - \ln (p + 1) = \ln 5$

31. $\ln (3x - 1) - \ln (2 + x) = \ln 2$

32. $\ln (8k - 7) - \ln (3 + 4k) = \ln (9/11)$

33. $\ln (5 + 4y) - \ln (3 + y) = \ln 3$

34. $\ln m + \ln (2m + 5) = \ln 7$

35. $\ln x + 1 = \ln (x - 4)$

36. $\ln (4x - 2) = \ln 4 - \ln (x - 2)$

37. $2 \ln (x - 3) = \ln (x + 5) + \ln 4$

38. $\ln (k + 5) + \ln (k + 2) = \ln 14k$

39. $\log_5 (r + 2) + \log_5 (r - 2) = 1$

40. $\log_4 (z + 3) + \log_4 (z - 3) = 1$

41. $\log_3 (a - 3) = 1 + \log_3 (a + 1)$

42. $\log w + \log (3w - 13) = 1$

43. $\log_2 \sqrt{2y^2} - 1 = 1/2$

44. $\log_2 (\log_2 x) = 1$

45. $\log z = \sqrt{\log z}$

46. $\log x^2 = (\log x)^2$

47. $\log_x 5.87 = 2$

48. $\log_x 11.9 = 3$

49. $1.8^{p+4} = 9.31$

50. $3.7^{5z-1} = 5.88$

Work the following exercises. (See Example 6.)

51. Natural Science The amount of a certain radioactive specimen present at time t (measured in seconds) is given by

$$A(t) = 5000e^{-.02t},$$

where $A(t)$ is the amount measured in grams. Find each of the following.
(a) $A(0)$ (b) $A(5)$ (c) $A(20)$
(d) The half-life of the specimen

52. Social Science The population of a small mining town has been decreasing, with the population at time t in years given by

$$P(t) = 8000e^{-.04t}.$$

Find each of the following.
(a) $P(0)$ (b) $P(5)$ (c) $P(10)$
(d) How many years will it take for the town to lose half its population?
(e) When will the population be down to 2000 people?

53. Natural Science Find the half-life for the following radioactive substances if the amount of the substance present after t days is as follows.
(a) $A(t) = 5000e^{-.03t}$
(b) $A(t) = 2350e^{-.08t}$
(c) $A(t) = 18{,}000e^{-.0002t}$

54. Natural Science The number of bacteria in a certain culture, $B(t)$, is approximated by

$$B(t) = 250{,}000e^{-.04t},$$

where t is time measured in hours. Find the following.
(a) $B(0)$ (b) $B(5)$ (c) $B(20)$
(d) The time it will take until only 125,000 bacteria are present
(e) The time it will take until only 25,000 bacteria are present

Using Example 7, find the age of a specimen for each of the following.

55. $y = .8y_0$

56. $y = .4y_0$

57. $y = .1y_0$

58. $y = .01y_0$

59. Natural Science A large cloud of radioactive debris from a nuclear explosion has floated over the Pacific Northwest, contaminating much of the hay supply. Consequently, farmers in the area are concerned that the cows who eat this hay will give contaminated milk. (The tolerance level for radioactive iodine in milk is 0.) The percent of the initial amount of radioactive iodine still present in the hay after t days is approximated by $P(t)$, which is given by the function

$$P(t) = 100e^{-.1t}.$$

(a) Find the percent remaining after 4 days.
(b) Find the percent remaining after 10 days.
(c) Some scientists feel that the hay is safe after the percent of radioactive iodine has declined to 10% of the original amount. Solve the equation $10 = 100e^{-.1t}$ to find the number of days before the hay may be used.
(d) Other scientists believe that the hay is not safe until the level of radioactive iodine has declined to only 1% of the original level. Find the number of days that this would take.

60. Social Science The number of years, $N(r)$, since two independently evolving languages split off from a common ancestral language is approximated by

$$N(r) = -5000 \ln r,$$

where r is the proportion of the words from the ancestral language that is common to both languages now. Find each of the following.
(a) $N(.9)$ (b) $N(.5)$ (c) $N(.3)$

(d) How many years have elapsed since the split if 70% of the words of the ancestral language are common to both languages today?

(e) If two languages split off from a common ancestral language about 1000 years ago, find r.

Natural Science *For Exercises 61–64, recall that* $\log x$ *represents the common (or base 10) logarithm of x.*

61. The loudness of sounds is measured in a unit called a decibel. To do this, a very faint sound, called the *threshold sound*, is assigned an intensity I_0. If a particular sound has intensity I, then the decibel rating of this louder sound is

$$10 \cdot \log \frac{I}{I_0}.$$

Find the decibel ratings of sounds having the following intensities.

(a) $100I_0$ **(b)** $1000I_0$

(c) $100{,}000I_0$ **(d)** $1{,}000{,}000I_0$

62. Find the decibel ratings of the following sounds having intensities as given. Round answers to the nearest whole number.

(a) Whisper, $115I_0$

(b) Busy street, $9{,}500{,}000I_0$

(c) Heavy truck, 20 meters away, $1{,}200{,}000{,}000I_0$

(d) Rock music, $895{,}000{,}000{,}000I_0$

(e) Jetliner at takeoff, $109{,}000{,}000{,}000{,}000I_0$

63. The intensity of an earthquake, measured on the *Richter Scale*, is given by

$$\log \frac{I}{I_0},$$

where I_0 is the intensity of an earthquake of a certain (small) size. Find the Richter Scale ratings of earthquakes having intensity

(a) $1000I_0$;   **(b)** $1{,}000{,}000I_0$;   **(c)** $100{,}000{,}000I_0$.

 64. (a) In 1989, the San Francisco region experienced an earthquake with a Richter Scale rating of 7.1. Use a calculator with a y^x key to express the intensity of this earthquake as a multiple of I_0. (See Exercise 63.)

(b) The San Francisco earthquake of 1906 had a Richter Scale rating of 8.3. Express the intensity of this earthquake as a multiple of I_0.

(c) Compare the intensity of the two San Francisco earthquakes discussed above.

 To find the maximum permitted levels of certain pollutants in fresh water, the EPA has established the functions defined in Exercises 65–66, where M(h) is the maximum permitted level of pollutant for a water hardness of h milligrams per liter. Find M(h) in each case. (These results give the maximum permitted average concentration in micrograms per liter for a 24-hour period.)

65. Copper: $M(h) = e^r$, where $r = .65 \ln h - 1.94$ and $h = 9.7$.

66. *Lead*: $M(h) = e^r$, where $r = 1.51 \ln h - 3.37$ and $h = 8.4$.

The following graphs are plotted on a logarithmic *scale, where differences between successive measurements are not always the same. Data that do not plot in a linear pattern on the usual Cartesian axes often form a linear pattern when plotted on a logarithmic scale. Notice that on the vertical scale, the distance from 1 to 2 is not the same as the distance from 2 to 3, and so on. This is characteristic of a graph drawn with logarithmic scales.*

67. The graph below gives the rate of oxygen consumption for resting guinea pigs of various sizes. This rate is proportional to body mass raised to the power .67.* Estimate the oxygen consumption for a guinea pig with body mass of .3 kilograms. Do the same for one with body mass of .7.

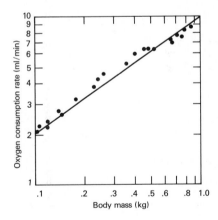

---

*Figures for Exercises 67 and 68 are from *On Size and Life* by Thomas A. McMahon and John Tyler Bonner. Copyright © 1983 by Thomas A. McMahon and John Tylor Bonner. Reprinted by permission of W. H. Freeman and Company.

68. The graph below gives the world weight-lifting records as $\log W_T$, plotted against the logarithm of body weight. Here W_T is the total weight lifted in three lifts: the press, the snatch, and the clean-and-jerk. The numbers beside each point indicate the body weight class, in pounds.

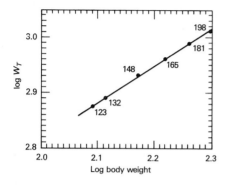

(a) Find the record for a weight of 150 pounds and for a weight of 165 pounds. (Use base 10 logarithms.)

(b) Find the body weight that corresponds to a record of 750 pounds.

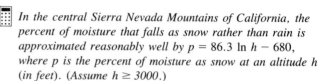 *In the central Sierra Nevada Mountains of California, the percent of moisture that falls as snow rather than rain is approximated reasonably well by $p = 86.3 \ln h - 680$, where p is the percent of moisture as snow at an altitude h (in feet). (Assume $h \geq 3000$.)*

69. Find the percent of moisture that falls as snow at the following altitudes.

(a) 3000 ft   (b) 4000 ft   (c) 7000 ft

70. Graph p.

CHAPTER 4   SUMMARY

KEY TERMS AND SYMBOLS

4.1   exponential function
exponential equation

4.2 e irrational number $\approx$ 2.718281828
exponential growth function
learning curve
forgetting curve

4.3 $\log_a x$ base a logarithm of x
$\log x$ base 10 logarithm of x

$\ln x$ base e logarithm of x
logarithm
logarithmic function
inverses
properties of logarithms
common logarithm
natural logarithm

4.4   half-life

KEY CONCEPTS

An important application of exponents is the **exponential growth function,** defined as $f(t) = y_0 e^{kt}$, where y_0 is the amount of a quantity present at time $t = 0$, $e \approx 2.71828$, and k is a constant.

The **logarithm** of x to the base a is defined as follows. For $a > 0$ and $a \neq 1$, $y = \log_a x$ means $a^y = x$. Thus, $\log_a x$ is an *exponent,* the power to which a must be raised to produce x.

Properties of Logarithms

Let x, y, and a be positive real numbers, $a \neq 1$, and let r be any real number.

$$\log_a xy = \log_a x + \log_a y \qquad \log_a \frac{x}{y} = \log_a x - \log_a y$$

$$\log_a x^r = r \log_a x \qquad \log_a 1 = 0$$

$$\log_a a = 1 \qquad a^{\log_a x} = x$$

$$\log_a a^y = y$$

Solving Exponential and Logarithmic Equations

Let $a > 0$, $a \neq 1$.

If $a^x = a^y$, then $x = y$ and   if $x = y$, then $a^x = a^y$.

If $x = y$, then $\log_a x = \log_a y$, $x > 0$, $y > 0$.

If $\log_a x = \log_a y$, then $x = y$, $x > 0$, $y > 0$.

CHAPTER 4   REVIEW EXERCISES

Solve each of the following equations.

1. $2^{3x} = \dfrac{1}{8}$

2. $\left(\dfrac{9}{16}\right)^x = \dfrac{3}{4}$

3. $9^{2y-1} = 27^y$

4. $\dfrac{1}{2} = \left(\dfrac{b}{4}\right)^{1/4}$

Graph each of the following.

5. $f(x) = 5^x$

6. $f(x) = 5^{-x}$

7. $f(x) = \log_5 x$

8. $f(x) = \log_{1/5} x$

---

9. A company finds that its new workers produce
$$P(x) = 100 - 100e^{-.8x}$$
items per day, after x days on the job. Find each of the following.
(a) $P(0)$ **(b)** $P(1)$ **(c)** $P(5)$
(d) How many items per day would you expect an experienced worker to produce?

10. The amount of a certain radioactive material, in grams, present after t days, is given by
$$A(t) = 800e^{-.04t}.$$
Find $A(t)$ if
(a) $t = 0$;   **(b)** $t = 5$.

Write each of the following using logarithms.

11. $2^6 = 64$

12. $3^{1/2} = \sqrt{3}$

13. $e^{.09} = 1.09417$

14. $10^{1.07918} = 12$

Write each of the following using exponents.

15. $\log_2 32 = 5$

16. $\log_{10} 100 = 2$

17. $\ln 82.9 = 4.41763$

18. $\log 15.46 = 1.18921$

Evaluate each of the following.

19. $\log_3 81$ **20.** $\log_8 16$ **21.** $\log_{32} 16$ **22.** $\log_{25} 5$ **23.** $\log_{100} 1000$ **24.** $\log_{1/2} 4$

Find each of the following common logarithms.

25. $\log 18$ **26.** $\log 3/4$ **27.** $\log .83$ **28.** $\log 6050$

Find each of the following natural logarithms.

29. $\ln 6.2$ **30.** $\ln 700$ **31.** $\ln 483$ **32.** $\ln .504$

Simplify using the properties of logarithms.

33. $\log_5 3k + \log_5 7k^3$ **34.** $\log_3 2y^3 - \log_3 8y^2$ **35.** $2 \cdot \log_2 x - 3 \cdot \log_2 m$ **36.** $5 \cdot \log_4 r - 3 \cdot \log_4 r^2$

Solve each equation. Round to the nearest thousandth.

37. $8^p = 19$ **38.** $3^z = 11$ **39.** $5 \cdot 2^{-m} = 35$ **40.** $2 \cdot 15^{-k} = 18$

41. $e^{-5-2x} = 5$ **42.** $e^{3x-1} = 12$ **43.** $10^{2r-3} = 17$ **44.** $8^{9y-4} = 15$

45. $6^{2-m} = 2^{3m+1}$ **46.** $5^{3r-1} = 6^{2r+5}$ **47.** $(1 + .003)^k = 1.089$ **48.** $(1 + .094)^z = 2.387$

49. $4 \cdot 3^{x^2} = 15$ **50.** $3 \cdot 5^{p^2} = 28$ **51.** $\log (m + 2) = 1$ **52.** $\log x^2 = 2$

53. $\log_2 (3k - 2) = 4$ **54.** $\log_5 \left(\dfrac{5z}{z - 2}\right) = 2$ **55.** $\log x + \log (x + 3) = 1$

56. $\log_2 r + \log_2 (r - 2) = 3$ **57.** $\log (p - 1) = 1 + \log p$ **58.** $\ln (m + 3) - \ln m = \ln 2$

59. $2 \ln (y + 1) = \ln (y^2 - 1) + \ln 5$ **60.** $\log_3 k + \log_3 (k + 2) = \log_3 8$ **61.** $3 + 2 \log_4 (2x - 1) = 1$

62. $\log_2 (1 - 3x) - 4 = 2$

63. The height, in meters, of the members of a certain tribe is approximated by

$$h = .5 + \log t,$$

where t is the tribe member's age in years, $1 \le t \le 20$. Find the height of a tribe member of age
(a) 2 years;   (b) 5 years;
(c) 10 years;   (d) 20 years.

64. The turnover of legislators is a problem of interest to political scientists. One model of legislative turnover in the U.S. House of Representatives is given by

$$M = 434e^{-.08t},$$

where M is the number of continuously serving members at time t.* This model is based on the 1965 membership of the House. Thus, 1965 corresponds to $t = 0$, 1966 to $t = 1$, and so on. Find the number of continuously serving members in each of the following years.
(a) 1969   (b) 1973   (c) 1979

---

*Excerpt from "Exponential Models of Legislative Turnover" by Thomas W. Casstevens. Reprinted by permission of COMAP, Arlington, MA.

After the liberation of Belgium at the end of World War II, officials began a search for Nazi collaborators. One person arrested as a collaborator was a minor painter, H. A. Van Meegeren; he was charged with selling a valuable painting by the Dutch artist Vermeer (1632–75) to the Nazi Hermann Goering. He defended himself from the very serious charge of collaboration by claiming that the painting was a fake—he had forged it himself.

He also claimed that the beautiful and famous painting "Disciples at Emmaus," as well as several other supposed Vermeers, was his own work. To prove this, he did another "Vermeer" in his prison cell. An international panel of experts was assembled, which pronounced as forgeries all the "Vermeers" in question.

Many people would not accept the verdict of this panel for the painting "Disciples at Emmaus"; it was felt to be too beautiful to be the work of a minor talent such as Van Meegeren. In fact, the painting was declared genuine by a noted art scholar and sold for $170,000. The question of the authenticity of this painting continued to trouble art historians, who began to insist on conclusive proof one way or the other. This proof was given by a group of scientists at Carnegie-Mellon University, using the idea of radioactive decay.

The dating of objects is based on radioactivity; the atoms of certain radioactive elements are unstable, and within a given time period a fixed fraction of such atoms will spontaneously disintegrate, forming atoms of a new element. If t_0 represents some initial time, N_0 represents the number of atoms present at time t_0, and N represents the number present at some later time t, then it can be shown (using physics and calculus) that

$$t - t_0 = \frac{1}{\lambda} \cdot \ln \frac{N_0}{N},$$

where λ is a "decay constant" that depends on the radioactive substance under consideration.

If t_0 is the time that the substance was formed or made, then $t - t_0$ is the age of the item. Thus, the age of an item is given by

$$\frac{1}{\lambda} \cdot \ln \frac{N_0}{N}.$$

The decay constant λ can be readily found, as can N, the number of atoms present now. The problem is N_0—we can't find a value for this variable. However, it is possible to get reasonable ranges for the values of N_0. This is done by studying the white lead in the painting. This pigment has been used by artists for over 2000 years. It contains a small amount of the radioactive substance lead-210 and an even smaller amount of radium-226.

Radium-226 disintegrates through a series of intermediate steps to produce lead-210. The lead-210, in turn, decays to form polonium-210. This last process, lead-210 to polonium-210, has a half-life of 22 years. That is, in 22 years half the initial quantity of lead-210 will decay to polonium-210.

When lead ore is processed to form white lead, most of the radium is removed with other waste products. Thus, most of the supply of lead-210 is cut off, with the remainder beginning to decay very rapidly. This process continues until the lead-210 in the white lead is once more in equilibrium with the small amount of radium then present. Let $y(t)$ be the amount of lead-210 per gram of white lead present at time t since manufacture of the pigment. Let r represent the number of disintegrations of radium-226 per minute per gram of white lead. (Actually, r is a function of time, but the half life of radium-226 is so long in comparison to the time interval in question that we assume it to be a constant.) If λ is the decay constant for lead-210, then it can be shown that

$$y(t) = \frac{r}{\lambda}[1 - e^{-\lambda(t-t_0)}] + y_0 e^{-\lambda(t-t_0)}. \tag{1}$$

*From "The Van Meegeren Art Forgeries" by Martin Braun from *Applied Mathematical Sciences,* Vol. 15. Copyright © 1975. Published by Springer-Verlag New York, Inc. Reprinted by permission.

All variables in this result can be evaluated except y_0, the amount of lead-210 present originally. To get around this problem, we use the fact that the original amount of lead-210 was in radioactive equilibrium with the larger amount of radium-226 in the ore from which the metal was extracted. We therefore take samples of different ores and compute the rate of disintegration of radium-226. The results are as shown in the table.

Location of Ore	Disintegrations of Radium-226 per Minute per gram of White Lead
Oklahoma	4.5
S.E. Missouri	.7
Idaho	.18
Idaho	2.2
Washington	140.0
British Columbia	.4

The numbers in the table vary from .18 to 140—quite a range. Since the number of disintegrations is proportional to the amount of lead-210 present originally, we must conclude that y_0 also varies over a tremendous range. Thus, equation (1) cannot be used to obtain even a crude estimate of the age of a painting. However, we want to distinguish only between a modern forgery and a genuine painting that would be 300 years old.

To do this, we observe that if the painting is very old compared to the 22 year half-life of lead-210, then the amount of radioactivity from the lead-210 will almost equal the amount of radioactivity from the radium-226. On the other hand, if the painting is modern, then the amount of radioactivity from the lead-210 will be much greater than the amount from the radium-226.

We want to know if the painting is modern or 300 years old. To find out, let $t - t_0 = 300$ in equation (1), getting

$$\lambda y_0 = \lambda \cdot y(t) \cdot e^{300\lambda} - r(e^{300\lambda} - 1) \qquad \textbf{(2)}$$

after some rearrangement of terms. If the painting is modern, then λy_0 should be a very large number; λy_0 represents the number of disintegrations of the lead-210 per minute per gram of white lead at the time of manufacture. By studying samples of white lead, we can conclude that λy_0 should never be anywhere near as large as 30,000. Thus, we use equation (2) to calculate λy_0; if our result is greater than 30,000 we conclude that the painting is a modern forgery. The details of this calculation are left for the exercises.

EXERCISES

1. To calculate λ, use the formula

$$\lambda = \frac{\ln 2}{\text{half-life}}.$$

   Find λ for lead-210, whose half-life is 22 years.

2. For the painting "Disciples at Emmaus," the current rate of disintegration of the lead-210 was measured and found to be $\lambda \cdot y(t) = 8.5$. Also, r was found to be .8. Use this information and equation (2) to calculate λy_0. Based on your results, what do you conclude about the age of the painting?

The table below lists several other possible forgeries. Decide which of them must be modern forgeries.

Title	$\lambda \cdot y(t)$	r
3. "Washing of Feet"	12.6	.26
4. "Lace Maker"	1.5	.4
5. "Laughing Girl"	5.2	.6
6. "Woman Reading Music"	10.3	.3

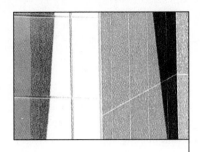

CHAPTER 5

Mathematics of Finance

Not too many years ago, some large corporations could borrow money at 3%, and home buyers could obtain loans at $4\frac{1}{2}$%. Today, however, even the largest corporations must pay at least 10% for their money, and credit customers pay 18% or more to banks and retail outlets. Thus, it is important that both corporate managers and consumers understand *interest,* the cost of borrowing money.

The formulas for interest are developed in this chapter. A calculator (equipped with a y^x key) is needed to use many of them. (Alternatively, Tables 4–6 in the back of the book may be used, but they are less convenient and may not be as accurate).

5.1 SIMPLE INTEREST AND DISCOUNT

Interest is the fee paid for the use of someone else's money. For example, you might pay interest to a bank for money you borrow or the bank might pay you interest on the money in your savings account. The amount of money that is borrowed or deposited is called the **principal.** The fee paid as interest depends on the interest **rate** and the length of **time** for which you have the use of the money. Unless stated otherwise, the time, $t,$ is measured in years and the interest rate, $r,$ is in percent per year, expressed as a decimal; for instance, 8% = .08 or 9.5% = .095.

There are two common ways of computing interest, the first of which we study in this section. **Simple interest** is interest paid only on the amount deposited and not on past interest.

The **simple interest,** $I,$ for t years on an amount of P dollars at a rate of interest r per year for t years is

$$I = Prt.$$

1 Find the simple interest for the following.

(a) $1000 at 8% for 2 years

(b) $5500 at 10.5% for $1\frac{1}{2}$ years

(c) $12,200 at 11.75% for 7 months

Answer:

(a) $160

(b) $866.25

(c) $836.21

▶**EXAMPLE 1** The simple interest on $15,000 at $12\frac{1}{4}$% for 11 months is found from the formula $I = Prt$, with $P = $15,000$, $r = 12\frac{1}{4}$% $= .1225$, and $t = 11/12$ of a year.

$$I = Prt$$

$$I = $15,000(.1225)\left(\frac{11}{12}\right)$$

$$I = $1684.38 \text{ rounded to the nearest cent} \quad ◀ \quad \boxed{1}$$

If you deposit P dollars at interest rate r for t years, your money earns interest of $I = Prt$. When this interest is added to your deposit, the total amount in your account is

$$\text{principal} + \text{interest} = P + Prt = P(1 + rt).$$

This amount is called the **future value** of P dollars at interest rate r for t years. When loans are involved, the future value is often called the **maturity value** and represents the total amount of principal and interest that must be repaid.

> The **future value** or **maturity value**, A, of P dollars for t years at a rate of interest r per year is
>
> $$A = P(1 + rt).$$

▶**EXAMPLE 2** Find the maturity value for each of the following loans at simple interest.

(a) A loan of $2500 to be repaid in 8 months with interest of 12.1%

The loan is for 8 months, or $8/12 = 2/3$ of a year. The maturity value is

$$A = P(1 + rt)$$

$$A = 2500\left[1 + .121\left(\frac{2}{3}\right)\right]$$

$$\approx 2500[1 + .08067] \approx 2701.67,$$

or $2701.67. (Interest is rounded to the nearest cent, as is customary in financial problems). Since the maturity value is the sum of principal and interest, the amount of interest paid on this loan is given by:

$$\text{maturity value} - \text{principal} = $2701.67 - $2500 = $201.67.$$

2 Find the maturity value of each loan.

(a) $10,000 at 10% for 6 months

(b) $8970 at 11% for 9 months

(c) $95,106 at 9.8% for 76 days

Answer:

(a) $10,500

(b) $9710.03

(c) $97,073.64

3 Find the present value of the following future amounts. Assume 12% interest.

(a) $7500 in 1 year

(b) $89,000 in 5 months

(c) $164,200 in 125 days

Answer:

(a) $6696.43

(b) $84,761.90

(c) $157,632.00

(b) A loan of $11,280 for 85 days at 11% interest

It is common to assume 360 days in a year when working with simple interest. We shall usually make such an assumption in this book. The maturity value in this example is

$$A = 11,280\left[1 + .11\left(\frac{85}{360}\right)\right] \approx 11,280[1.0259722] = 11,572.97,$$

or $11,572.97. ◀ **2**

Part (b) of Example 2 assumed 360 days in a year. Interest found using 360 days is called **ordinary interest,** while interest found using 365 days is **exact interest.**

PRESENT VALUE   A sum of money that can be deposited today to yield some larger amount in the future is called the **present value** of that future amount. Let P be the present value of some amount A at some time t (in years) in the future. Assume a simple interest rate r. As above, the future value of this sum is

$$A = P(1 + rt).$$

Dividing each side by $1 + rt$ gives the following formula for present value.

$$P = \frac{A}{1 + rt}$$

▶**EXAMPLE 3**   Find the present value of the following future amounts at simple interest.

(a) $10,000 in 1 year, if interest is 8%

Here $A = 10,000$, $t = 1$, and $r = .08$. Use the formula for present value.

$$P = \frac{10,000}{1 + (.08)(1)} = \frac{10,000}{1.08} \approx 9259.26$$

If $9259.26 were deposited today at 8% interest, a total of $10,000 would be in the account in 1 year. These two sums, $9259.26 today, and $10,000 in a year, are equivalent (at 8%) because the first amount becomes the other in a year.

(b) $32,000 in 4 months at 9% interest

$$P = \frac{32,000}{1 + (.09)\left(\frac{4}{12}\right)} = \frac{32,000}{1.03} = 31,067.96$$

A deposit of $31,067.96 today, at 9% interest, would produce $32,000 in 4 months. ◀ **3**

4 Mark McFarland is owed $19,500 by Glenn Wilson. The money will be paid in 11 months, with no interest. If the current interest rate is 13%, how much should McFarland be willing to accept today in settlement of the debt?

Answer:

$17,423.68

▶**EXAMPLE 4** Because of a court settlement, Charlie Dawkins owes $5000 to Arnold Parker. The money must be paid in 10 months, with no interest. Suppose Dawkins wishes to pay the money today. What amount should Parker be willing to accept? Assume an interest rate of 11%.

The amount that Parker should be willing to accept is given by the present value:

$$P = \frac{5000}{1 + (.11)\left(\dfrac{10}{12}\right)} \approx \frac{5000}{1.09167} = 4580.14.$$

Parker should be willing to accept $4580.14 today in settlement of the obligation. ◀ **4**

▶**EXAMPLE 5** If you borrow $40,000 today and are required to pay $41,400 in 4 months to pay off the loan and interest, what is the simple interest rate?

Here $P = 40,000$, $A = 41,400$, and $t = 4/12 = 1/3$. We can find the interest rate r by using the equation

$$A = P(1 + rt).$$

$$41,400 = 40,000\left(1 + r \cdot \frac{1}{3}\right)$$

$$41,400 = 40,000 + \frac{40,000r}{3}$$

$$1400 = \frac{40,000r}{3}$$

$$40,000r = 3 \cdot 1400 = 4200$$

$$r = \frac{4200}{40,000} = .105$$

Therefore, the interest rate is 10.5%. ◀

SIMPLE DISCOUNT NOTES The loans discussed up to this point are called **simple interest notes,** where interest on the face value of the loan is added to the loan and paid at maturity. Another common type of note, called a **simple discount note,** has the interest deducted in advance from the amount of a loan before giving the *balance* to the borrower. The *full* value of the note must be paid at maturity. The money that is deducted is called the **bank discount** or just the **discount,** and the money actually received by the borrower is called the **proceeds.**

5 Craig Johns signs an agreement at his bank to pay the bank $25,000 in 5 months. The bank charges a 13% discount rate. Find the amount of the discount and the amount Johns actually receives.

Answer:
$1354.17; $23,645.83

6 Refer to Problem 5 at the side above, and find the effective rate of interest paid by Johns.

Answer:
$13.7% (to the nearest tenth).

▶**EXAMPLE 6**   Elizabeth Macias needs a loan from her bank and agrees to pay $8500 to her banker in 9 months. The banker subtracts a discount of 12% and gives the balance to Macias. Find the amount of the discount and the proceeds.

The discount is found in the same way that simple interest is found, except that it is based on the amount to be repaid.

$$\text{Discount} = 8500(.12)\left(\frac{9}{12}\right) = 765.00$$

The proceeds are found by subtracting the discount from the original amount.

$$\text{Proceeds} = \$8500 - \$765.00 = \$7735.00 \quad ◀ \quad \boxed{5}$$

In Example 6, the borrower was charged a discount of 12%. However, 12% is *not* the interest rate paid, since 12% applies to the $8500, while the borrower actually received only $7735. To find the rate of interest actually paid by the borrower, called the **effective rate,** work as in the next example.

▶**EXAMPLE 7**   Find the actual rate of interest paid by Macias in Example 6.

To find the actual rate of interest paid by Macias in Example 6, use the formula for simple interest, $I = Prt$, with r the unknown. Since the borrower received only $7735 and must repay $8500, $I = 8500 - 7735 = 765$. Here, $P = 7735$ and $t = 9/12 = .75$. Substitute these values into $I = Prt$.

$$I = Prt$$
$$765 = 7735(r)(.75)$$
$$\frac{765}{7735(.75)} = r$$
$$.132 \approx r$$

The effective interest rate paid by the borrower is about 13.2%. ◀ $\boxed{6}$

Let D represent the amount of discount on a loan. Then $D = Art$, where A is the maturity value of the loan (the amount borrowed plus interest), and r is the stated rate of interest. The amount actually received, the proceeds, can be written as $P = A - D$, or $P = A - Art = A(1 - rt)$.

The formulas for discount are summarized below.

If D is the discount on a loan having a maturity value A at a rate of interest r for t years, and if P represents the proceeds, then

$$P = A - D \quad \text{or} \quad P = A(1 - rt).$$

7 A firm accepts a $21,000 note due in 7 months with interest of 12.5%. Suppose the firm discounts the note at a bank 75 days before it is due. Find the amount the firm would receive if the bank charges a 13.4% discount rate.

Answer:
$21,902.25

One common use of discount interest is in *discounting a note,* a process by which a promissory note due at some time in the future can be converted to cash now.

▶ **EXAMPLE 8**   Kent Merrill owes $4250 to Don Chinn. The loan is payable in 1 year at 12% interest. Chinn needs cash to buy a new car, so 3 months before the loan is payable he goes to his bank to have the loan discounted. The bank charges a 14% discount fee. Find the amount of cash he will receive from the bank.

First find the maturity value of the loan, the amount Merrill must pay to Chinn. By the formula for maturity value,

$$A = P(1 + rt)$$
$$A = 4250[1 + (.12)(1)]$$
$$= 4250 (1.12) = 4760,$$

or $4760.

The bank applies its discount rate to this total:

$$\text{Amount of discount} = 4760(.14)\left(\frac{3}{12}\right) = 166.60.$$

(Remember that the loan was discounted 3 months before it was due.) Chinn actually receives

$$\$4760 - \$166.60 = \$4593.40$$

in cash from the bank. Three months later, the bank would get $4760 from Merrill.   ◀ **7**

5.1 EXERCISES

 Most of the exercises in this section are designed to be worked with a calculator.

Find the simple interest. (See Example 1.)

1. $1000 at 12% for 1 year

2. $4500 at 16% for 1 year

3. $25,000 at 11% for 9 months

4. $3850 at 7% for 8 months

5. $1974 at 16.2% for 7 months

6. $3724 at 14.1% for 11 months

Find the simple interest. Assume a 360-day year and 30 days in each month.

7. $12,000 at 14% for 72 days

8. $38,000 at 9% for 216 days

9. $5147.18 at 17.3% for 58 days

10. $2930.42 at 13.9% for 123 days

11. $7980 at 15%; the loan was made May 7 and is due on September 19

12. $5408 at 10%; the loan was made August 16 and is due on December 30

Find the simple interest. Assume 365 days in a year, and use the exact number of days in a month. (Assume 28 days in February.)

13. $7800 at 16%; made on July 7 and due October 25

14. $11,000 at 15%; made on February 19 and due May 31

15. $2579 at 17.6%; made on October 4 and due March 15

16. $37,098 at 9.2%; made on September 12 and due July 30

Find the present value of the future amounts. Assume 360 days in a year and 30 days in each month. (See Example 3.)

17. $15,000 in 8 months, money earns 16%

18. $48,000 in 9 months, money earns 14%

19. $5276 in 3 months, money earns 7.4%

20. $6892 in 7 months, money earns 8.2%

21. $15,402 in 125 days, money earns 9.3%

22. $29,764 in 310 days, money earns 11.4%

Find the proceeds for the following simple discount notes. Assume 360 days in a year and 30 days in each month. (See Example 6.)

23. $7150, discount rate 16%, length of loan 11 months

24. $9450, discount rate 18%, length of loan 7 months

25. $358, discount rate 11.6%, length of loan 183 days

26. $509, discount rate 13.2%, length of loan 238 days

Find the interest rate on the following simple discount notes. (See Example 7.)

27. $6000, discount rate 12%, length of loan 6 months

28. $9500, discount rate 15%, length of loan 8 months

29. $17,400, discount rate 9.5%, length of loan 1 year

30. $11,250, discount rate 13.5%, length of loan 3 months

Work the following exercises.

31. Donna Sharp borrowed $25,900 from her father to start a flower shop. She repaid him after 11 months, with interest of 12.4%. Find the total amount she repaid.

32. A corporation accountant forgot to pay the firm's income tax of $725,896.15 on time. The government charged a penalty of 17.7% interest for the 34 days the money was late. Find the total amount, tax and penalty, that was paid. (Use a 365-day year.)

33. A bank offers a $20,000 certificate of deposit that pays $21,850 after 1 year. What interest rate is the bank paying?

34. A bond sells for $800 and will be redeemed for $830 in 6 months. What is the simple interest rate paid by this bond?

35. Tuition of $1769 will be due when the spring term begins, in 4 months. What amount should a student deposit today, at 6.25%, to have enough to pay the tuition?

36. A firm of attorneys has ordered seven new IBM typewriters, at a cost of $2104 each. The machines will not be delivered for 7 months. What amount could the firm deposit in an account paying 15.42% to have enough to pay for the machines?

37. A manufacturer's representative makes a large sale and will be paid a commission in 3 months, when the goods are delivered. She plans to use the money to buy a new computer for her office. The computer currently sells for $14,500, but the price is expected to increase to $16,000 soon. So she borrows $14,500 at 13% simple interest for 3 months and buys the computer now. How much interest will she pay and how much money will she save overall by buying the computer now?

38. An investor wants her $6000 to earn as much interest as possible during the next 6 months. Should she put it all in a savings account that pays 10.5% simple interest or should she put $2500 in a bond paying 16% simple interest and the remainder in a savings account paying 7% simple interest?

39. Roy Gerard needs $5196 to pay for remodeling work on his house. He plans to repay the loan in 10 months. His bank loans money at a discount rate of 14%. How much should he borrow to pay for the work?

40. Mun Kang decides to go back to college. To get to school she buys a small car for $6100. She decides to borrow the money at the bank, where they charge a 12.8% discount rate. If she will repay the loan in 7 months, find the amount of the loan.

41. Marge Prullage signs a $4200 note at the bank. The bank charges a 13.2% discount rate. Find the net proceeds if the note is for 10 months. Find the effective interest rate charged by the bank.

42. A bank charges a 13.1% discount rate on a $1000 note for 90 days. Find the effective rate.

43. Patrick Kelley owes $7000 to the Eastside Music Shop. He has agreed to pay the amount in 7 months, at an interest rate of 12%. Two months before the loan is due to be paid, the store discounts it at the bank. The bank charges a 13.7% discount rate. How much money does the store receive?

44. A building contractor gives a $13,500 note to a plumber. The note is due in 9 months, with interest of 15%. Three months after the note is signed, the plumber discounts it at the bank. The bank charges a 17.1% discount rate. How much money does the plumber actually receive?

5.2 COMPOUND INTEREST

Simple interest is normally used for loans or investments of a year or less. For longer periods *compound interest* is used. With **compound interest,** interest is charged (or paid) on interest as well as on principal. For example, if $1000 is deposited at 8% compound interest, then the interest for the first year is $1000(.08) = \$80$, just as with simple interest, so that the account balance is $1080 at the end of the year. During the second year, interest is paid on the entire $1080 (not just on the original $1000 as would be the case with simple interest), so the amount in the account at the end of the second year is $1080 + 1080(.08) = 1166.40$. This is more than simple interest would produce. **1**

To find a formula for compound interest, suppose that P dollars are deposited at interest rate r per year. The amount A on deposit after 1 year is given by the formula for simple interest maturity value

$$A = P(1 + rt), \tag{1}$$

with $t = 1$: $\qquad\qquad A = P(1 + r \cdot 1) = P(1 + r).$

If the deposit earns compound interest, the interest during the second year is paid on the total amount on deposit at the end of the first year, $P(1 + r)$. Applying formula (1) with $P(1 + r)$ in place of P and $t = 1$ shows that at the end of 2 years the amount on deposit is

$$A = [P(1 + r)](1 + r) = P(1 + r)^2.$$

Using formula (1) again, with $P(1 + r)^2$ in place of P and $t = 1$, shows that the amount on deposit at the end of the third year is

$$P(1 + r)^3.$$

Continuing in this way, we see that the total amount on deposit after t years is $P(1 + r)^t$, which is called the **compound amount.**

Caution Notice the similarities and differences between this formula for compound interest and the formula for simple interest.

$$\text{Compound interest:} \quad A = P(1 + r)^t$$
$$\text{Simple interest:} \quad A = P(1 + rt)$$

1 Use the formula

$$A = P(1 + rt)$$

to find the amount in the account after 2 years at 8% simple interest.

Answer:
$1160

2 Suppose $17,000 is deposited at 8% compounded annually for 11 years.

(a) Find the compound amount.

(b) Find the amount of interest earned.

Answer:

(a) $39,637.87

(b) $22,637.87

Both formulas involve the annual interest rate r (in decimal form) and the time t in years. The important distinction between the two formulas is that in the compound interest formula, the number of years t is an *exponent,* so that money grows much more rapidly when interest is compounded.

Interest can be compounded more than once a year. Common **compounding periods** include *semiannually* (two periods per year), *quarterly* (four periods per year), *monthly* (twelve periods) and *daily* (usually 365 periods per year). To find the interest rate per period, i, *divide* the interest rate per year by the number of compounding periods per year. The total number of compounding periods, n, is found by *multiplying* the number of years t by the number of compounding periods per year. Then the general formula in the next box can be derived in much the same way as the formula above.

If P dollars are deposited for n compounding periods at a rate of interest i per period, the **compound amount** A is

$$A = P(1 + i)^n.$$

▶**EXAMPLE 1** Suppose $1000 is deposited for 6 years in an account paying 8% per year compounded annually.
(a) Find the compound amount.
Since there is one compounding period per year, in the formula above $P = 1000$, $i = 8\%/1 = .08$, and $n = 6 \cdot 1 = 6$. The compound amount is

$$A = P(1 + i)^n$$
$$A = 1000(1 + .08)^6 = 1000(1.08)^6.$$

The number $(1.08)^6$ can be found with a calculator, as shown in Section 4.1: $(1.08)^6 \approx 1.58687$, and so

$$A = \$1000(1.58687) = 1586.87.$$

(b) Find the actual amount of interest earned.
Subtract the initial deposit from the compound amount.

$$\text{Amount of interest} = \$1586.87 - \$1000 = \$586.87 \quad \blacktriangleleft \quad \boxed{2}$$

▶**EXAMPLE 2** Find the amount of interest earned by a deposit of $1000 for 6 years at 6% compounded quarterly.
Interest compounded quarterly is compounded four times a year. In 6 years, there are $4 \cdot 6 = 24$ quarters, or 24 periods. Interest of 6% per year is 6%/4, or 1.5%, per quarter. The compound amount is

$$1000(1 + .015)^{24} = 1000(1.015)^{24}.$$

3 Find the compound amount.

(a) $10,000 at 12% compounded semiannually for 7 years

(b) $36,000 at 12% compounded monthly for 2 years

Answer:

(a) $22,609.04

(b) $45,710.45

4 **(a)** Find the final amount on deposit and the interest earned if $2500 is deposited into an account paying 6% compounded quarterly. The money is deposited for 5 years.

(b) Suppose $4000 is deposited at 6% compounded semiannually for 10 years. Find the final amount on deposit.

Answer:

(a) $3367.14; $867.14

(b) $7224.44

Using a calculator,

$$A = 1000(1.015)^{24} = 1000(1.42950) = 1429.50.$$

The compound amount is $1429.50 and the interest earned is $1429.50 − $1000 = $429.50. ◄ **3**

▶**EXAMPLE 3** Find the compound amount if $900 is deposited at 8% compounded semiannually for 8 years.

In 8 years there are $8 \cdot 2 = 16$ semiannual periods. If interest is 8% per year, then $8\%/2 = 4\%$ is earned per semiannual period. The compound amount is

$$A = 900(1.04)^{16} = 900(1.87298) = 1685.68,$$

or $1685.68. ◄ **4**

The more often interest is compounded within a given time period, the more interest will be earned. Surprisingly, however, there is a limit on the amount of interest, no matter how often it is compounded. To see this, suppose that $1 is invested at 100% interest per year, compounded n times per year. Then the interest rate (in decimal form) is 1.00 and the interest rate per period is $1/n$. According to the formula (with $P = 1$), the compound amount at the end of 1 year will be $A = \left(1 + \dfrac{1}{n}\right)^n$. Using a computer with double precision accuracy gives the following results for various values of n.

Interest Is Compounded	n	$\left(1 + \dfrac{1}{n}\right)^n$
Annually	1	$\left(1 + \dfrac{1}{1}\right)^1 = 2$
Semiannually	2	$\left(1 + \dfrac{1}{2}\right)^2 = 2.25$
Quarterly	4	$\left(1 + \dfrac{1}{4}\right)^4 \approx 2.4414$
Monthly	12	$\left(1 + \dfrac{1}{12}\right)^{12} \approx 2.6130$
Daily	365	$\left(1 + \dfrac{1}{365}\right)^{365} \approx 2.71457$
Hourly	8760	$\left(1 + \dfrac{1}{8760}\right)^{8760} \approx 2.718127$
Every minute	525,600	$\left(1 + \dfrac{1}{525,600}\right)^{525,600} \approx 2.7182792$
Every second	31,536,000	$\left(1 + \dfrac{1}{31,536,000}\right)^{31,536,000} \approx 2.7182818$

5 Find the compound amount, assuming continuous compounding.

(a) $12,000 at 10% for 5 years

(b) $22,876 at 7.2% for 9 years

Answer:

(a) $19,784.66

(b) $43,732.36

Since interest is rounded to the nearest penny, the compound amount never exceeds $2.72, no matter how big n is.

The table above suggests that as n takes larger and larger values, then the corresponding values of $\left(1 + \dfrac{1}{n}\right)^n$ get closer and closer to a specific real number, whose decimal expansion begins 2.71828 . . . . This is indeed the case, as is shown in calculus, and the number 2.71828 . . . is denoted e.*

The preceding example is typical of what happens when interest is compounded n times per year, with larger and larger values of n. It can be shown that no matter what interest rate or principal is used, there is always an upper limit on the compound amount, which is called the compound amount from **continuous compounding.** Not surprisingly, it involves the number e.

Continuous Compounding

The compound amount A for a deposit of P dollars at interest rate r per year compounded continuously for t years is given by

$$A = Pe^{rt}.$$

Many calculators have an e^x key for computing powers of e.

▶**EXAMPLE 4** Suppose $5000 is invested at an annual rate of 8% compounded continuously for 5 years. Find the compound amount.

In the formula for continuous compounding, let $P = 5000$, $r = .08$ and $t = 5$. Then a calculator with an e^x key shows that

$$A = 5000e^{(.08)5} = 5000e^{.4} = \$7459.12.$$

You can readily verify that daily compounding would have produced a compound amount about 32¢ less. ◀ **5**

EFFECTIVE RATE The preceding examples show that the same **stated** (or **nominal**) **rate** of interest can produce different amounts of interest, depending on the number of compoundings per year. If interest is compounded m times per year at rate r, then the **effective rate** is defined to be the rate which, when compounded annually, would produce the same amount of interest.

*Applications of the exponential function $f(x) = e^x$ are discussed in Section 4.2.

6 Find the effective rate corresponding to a nominal rate of

(a) 12% compounded monthly;

(b) 8% compounded quarterly.

Answer:

(a) 12.68%

(b) 8.24%

7 Find the effective rate corresponding to a nominal rate of

(a) 15% compounded monthly;

(b) 10% compounded quarterly.

Answer:

(a) 16.08%

(b) 10.38%

▶**EXAMPLE 5** Find the effective rate corresponding to a nominal rate of 6% compounded semiannually.

A calculator shows that $100 at 6% compounded semiannually will grow to

$$A = 100\left(1 + \frac{.06}{2}\right)^2 = 100(1.03)^2 = \$106.09.$$

Thus, the actual amount of compound interest is $106.09 − $100 = $6.09. Now if you earn $6.09 interest on $100 in 1 year with annual compounding, your rate is $6.09/100 = .0609 = 6.09\%$. Thus, the effective rate here is 6.09%. ◀ **6**

In the preceding example we found the effective rate by dividing compound interest for 1 year by the original principal. The same thing can be done with any principal P and rate r compounded m times per year.

$$\text{Effective rate} = \frac{\text{compound interest}}{\text{principal}}$$

$$= \frac{\text{compound amount} - \text{principal}}{\text{principal}}$$

$$= \frac{P\left(1 + \dfrac{r}{m}\right)^m - P}{P} = \frac{P\left[\left(1 + \dfrac{r}{m}\right)^m - 1\right]}{P}$$

$$= \left(1 + \frac{r}{m}\right)^m - 1.$$

The **effective rate** corresponding to a stated rate of interest r per year compounded m times per year is

$$\left(1 + \frac{r}{m}\right)^m - 1.$$

▶**EXAMPLE 6** Find the effective rate of a loan with interest of 9% compounded monthly.

Use the formula in the box, with $r = .09$ and $m = 12$. The effective rate is

$$\left(1 + \frac{.09}{12}\right)^{12} - 1 = (1.0075)^{12} - 1 = .0938,$$

or 9.38%. ◀ **7**

8 Find P in Example 8 if the interest rate is

(a) 6%;

(b) 10%.

Answer:

(a) $4483.55

(b) $3725.53

▶**EXAMPLE 7**   Bank A is now lending money at 13.2% interest compounded annually. The rate at Bank B is 12.6% compounded monthly and the rate at Bank C is 12.7% compounded quarterly. If you need to borrow money, at which bank will you pay the least interest?

Compare the effective rates.

$$\text{Bank A:} \quad \left(1 + \frac{.132}{1}\right)^1 - 1 = .132 = 13.2\%$$

$$\text{Bank B:} \quad \left(1 + \frac{.126}{12}\right)^{12} - 1 \approx .13354 = 13.354\%$$

$$\text{Bank C:} \quad \left(1 + \frac{.127}{4}\right)^4 - 1 \approx .13318 = 13.318\%$$

The lowest effective interest rate is at Bank A, which has the highest nominal rate. ◀

PRESENT VALUE WITH COMPOUND INTEREST   The formula for compound interest, $A = P(1 + i)^n$, has four variables, A, P, i, and n. Given the values of any three of these variables, the value of the fourth can be found. In particular, if A (the future amount), i, and n are known, then P can be found. Here P is the amount that should be deposited today to produce A dollars in n periods.

▶**EXAMPLE 8**   Joan Nakamura must pay a lump sum of $6000 in 5 years. What amount deposited today at 8% compounded annually will grow to $6000 in 5 years?

Here $A = 6000$, $i = .08$, $n = 5$, and P is unknown. Substituting these values into the formula for the compound amount gives

$$6000 = P(1.08)^5 \approx P(1.469328)$$

and

$$P \approx \frac{6000}{1.469328} = 4083.50.$$

If Nakamura deposits $4083.50 for 5 years in an account paying 8% compounded annually, she will have $6000 when she needs it. ◀ **8**

As the last example shows, $6000 in 5 years is the same as $4083.50 today (if money can be deposited at 8% compounded annually). Recall from the first section that an amount that can be deposited today to yield a given sum in the future is called the *present value* of this future sum.

Generalizing from Example 8, the present value of A dollars compounded at an interest rate i per period for n periods is

$$P = \frac{A}{(1 + i)^n} \quad \text{or} \quad P = A(1 + i)^{-n}.$$

9 Find the present value in Example 9 if money is deposited at 10% compounded semiannually.

Answer:
$6648.33

10 Find the number of years it will take for $500 to increase to $750 in an account paying 6% interest compounded semiannually.

Answer:
About 7 years

Compare this with the present value of an amount at simple interest r for t years, given in the previous section:

$$P = \frac{A}{1 + rt}.$$

▶**EXAMPLE 9** Find the present value of $16,000 in 9 years if money can be deposited at 6% compounded semiannually.

In 9 years there are $2 \cdot 9 = 18$ semiannual periods. A rate of 6% per year is 3% in each semiannual period. Apply the formula with $A = 16,000$, $i = .03$, and $n = 18$.

$$P = \frac{A}{(1 + i)^n} = \frac{16,000}{(1 + .03)^{18}} \approx \frac{16,000}{1.702433} \approx 9398.31$$

A deposit of $9398.31 today, at 6% compounded semiannually, will produce a total of $16,000 in 9 years. ◀ **9**

The formula for compound amount can also be solved for n.

▶**EXAMPLE 10** If you deposit $700 in an account that pays 8% interest compounded quarterly, how many years will it take for your balance to reach $1000?

Since 8% is compounded quarterly, the interest rate i is .02 per period. We must find n in the equation

$$A = P(1 + .02)^n,$$

where $A = 1000$ and $P = 700$. That is,

$$1000 = 700(1.02)^n$$

or equivalently,

$$(1.02)^n = 1000/700 \approx 1.42857.$$

One way to find n is to use logarithms and a calculator, as in Chapter 4. A reasonable approximation can be found by reading down the 2% column in Table 4 until you find the closest entry to 1.42857, which is 1.42825. This number corresponds to 18 periods. Therefore, the approximate amount of time is $18/4 = 4.5$ years for the balance to reach $1000. ◀ **10**

When interest is compounded continuously, the present value can be found by solving the continuous compounding formula $A = Pe^{rt}$ for P.

▶**EXAMPLE 11**   How much must be deposited today in an account paying 7.5% interest compounded continuously in order to have $10,000 in 4 years?

Here the future value is $A = 10,000$, the interest rate is $r = .075$, and the number of years is $t = 4$. The present value P is found as follows.

$$A = Pe^{rt}$$
$$10,000 = Pe^{(.075)4}$$
$$10,000 = Pe^{.3}$$
$$P = \frac{10,000}{e^{.3}} \approx \$7408.18 \quad ◀$$

5.2 EXERCISES

▦ *Most of the exercises in this section are designed to be worked with a calculator.*

Find the compound amount when the following deposits are made. (See Examples 1–3.)

1. $4500 at 8% compounded annually for 20 years

2. $810 at 8% compounded annually for 12 years

3. $470 at 12% compounded semiannually for 12 years

4. $15,000 at 8% compounded semiannually for 11 years

5. $7500 at 10% compounded quarterly for 9 years

6. $8000 at 8% compounded quarterly for 4 years

Find the amount of interest earned by the following deposits. (See Examples 1–2.)

7. $6000 at 8% compounded annually for 8 years

8. $21,000 at 6% compounded annually for 5 years

9. $43,000 at 10% compounded semiannually for 9 years

10. $7500 at 8% compounded semiannually for 5 years

11. $2196.58 at 10.8% compounded quarterly for 4 years

12. $4915.73 at 11.6% compounded quarterly for 3 years

Find the present value of the following sums. (See Examples 8 and 9.)

13. $4500 at 8% compounded annually for 9 years

14. $11,500 at 8% compounded annually for 12 years

15. $15,902.74 at 9.8% compounded annually for 7 years

16. $27,159.68 at 12.3% compounded annually for 11 years

17. $2000 at 8% compounded semiannually for 8 years

18. $2000 at 12% compounded semiannually for 8 years

19. $8800 at 12% compounded quarterly for 5 years

20. $7500 at 8% compounded quarterly for 9 years

21. If money can be invested at 8% compounded quarterly, which is larger: $1000 now or $1210 in 5 years?

22. If money can be invested at 6% compounded annually, which is larger: $10,000 now or $15,000 in 10 years?

Find the compound amount if $20,000 is invested at 8% compounded continuously for the following number of years. (See Example 4.)

23. 1   24. 5   25. 10   26. 15   27. 3   28. 7

Find the effective rate corresponding to the following nominal rates. (See Examples 5–7.)

29. 4% compounded semiannually

32. 10% compounded semiannually

30. 8% compounded quarterly

33. 12% compounded semiannually

31. 8% compounded semiannually

34. 12% compounded quarterly

Find the present value of $17,200 for the following number of years, if money can be deposited at 11.4% compounded continuously.

35. 2 **36.** 4 **37.** 7 **38.** 10

Under certain conditions, Swiss banks pay negative interest—they charge you. (You didn't think all that secrecy was free?) Suppose a bank "pays" −2.4% interest compounded annually. Use a calculator and find the compound amount for a deposit of $150,000 after the following.

39. 2 years **40.** 4 years **41.** 8 years **42.** 12 years

Work the following exercises.

43. Vu Le deposited $18,000 in an account paying 12% compounded quarterly. How much interest will he earn in 7 years?

44. Ann Fridl-Marshall placed $25,000 in an account paying 10% compounded semiannually. How much interest will she earn in 5 years?

45. Electric rates are rising by 4% per year. If the average monthly bill is now $45, what will be the average monthly bill in 10 years?

46. The population of a certain state is now 12 million and is increasing at an annual rate of 6%. What will the population of the state be in 5 years? In 10 years?

47. The average price of a house nationally is now $78,000. Housing prices are increasing at a rate of 7% per year. What will the average price of a house be in 12 years?

48. Annual tuition at a public university is now $2200 per year and is expected to increase at an annual rate of 8%. Room and board now costs $3600 per year and is expected to increase at an annual rate of 4%. What will be the total cost per year (tuition, room and board) in 15 years?

49. One bank pays 8% compounded semiannually; another pays 8% compounded quarterly. Find the difference in the interest earned in 5 years on a deposit of $24,000.

50. Which is a more profitable rate for your investment: 12.5% compounded daily, 12.8% compounded monthly, or 13.5% compounded annually?

51. Luisa Gonzales needs $18,000 in 3 years.
  (a) What amount can Gonzales deposit today, at 8% compounded annually, so that she will have the needed money?
  (b) How much interest will be earned?
  (c) Suppose Gonzales can deposit only $12,000 today. How much would she be short of the $18,000 in 3 years?

52. Brent Alderman must pay $27,000 in settlement of an obligation in 3 years.
  (a) What amount can he deposit today, at 12% compounded monthly, to have enough?
  (b) How much interest will he earn?
  (c) Suppose Alderman can deposit only $12,000 today. How much would he be short of the needed $27,000 in 3 years?

53. A movie theater ticket now costs $6. If prices continue to rise at 8% per year, approximately how many years will it take for the price to reach $10?

54. A pair of designer jeans now costs $45 and the price is expected to increase 12% each year. Approximately how many years from now will the price be $100?

55. Use the techniques of Section 4.4 or experiment with a calculator to determine approximately how long it will take the general level of prices to double if the average annual inflation rate is
  (a) 4%; (b) 6%; (c) 8%; (d) 12%

56. Round off your answers in Exercise 55 to the nearest year and compare them with the numbers 72/4, 72/6, 72/8, 72/12. Use this evidence to state a "rule of thumb" for determining approximate doubling time. This rule, which has long been used by bankers, is called the *rule of 72*.

57. (a) The consumption of electricity has increased historically at 6% per year. If it continues to increase at this rate indefinitely, find the number of years before the electric utilities would need to double their generating capacity.
  (b) Suppose a conservation campaign coupled with higher rates caused the demand for electricity to increase at only 2% per year, as it has recently. Find the number of years before the utilities would need to double generating capacity.

1 Write the first four terms of the geometric sequence with $a = 5$ and $r = -2$. Then find the seventh term.

Answer:
$5, -10, 20, -40; 320$

5.3 ANNUITIES

So far in this chapter, only lump sum deposits and payments have been discussed. But many financial situations involve a sequence of equal payments at regular intervals, such as weekly deposits in a savings account or monthly payments on a mortgage or car loan. In order to develop formulas to deal with periodic payments like these, we must first discuss sequences.

GEOMETRIC SEQUENCES If a and r are fixed nonzero numbers, then the infinite list of numbers, $a, ar, ar^2, ar^3, ar^4, \ldots$ is called a **geometric sequence.** For instance, if $a = 5$ and $r = 2$, we have the sequence

$$5, 5 \cdot 2, 5 \cdot 2^2, 5 \cdot 2^3, 5 \cdot 2^4, \ldots,$$

or

$$5, 10, 20, 40, 80, \ldots.$$

In the sequence $a, ar, ar^2, ar^3, ar^4, \ldots$, the number a is the first term of the sequence, ar the second term, ar^2 the third term, ar^3 the fourth term, and so on. Thus, for any $n \geq 1$,

$$ar^{n-1} \text{ is the } n\text{th term of the sequence.}$$

Each term in the sequence is r times the preceding term. The number r is called the **common ratio** of the sequence.

▶**EXAMPLE 1** Find the seventh term of the geometric sequence $6, 24, 96, \ldots$.
Here $a = 6$ and $r = 24/6 = 4$. Using ar^{n-1}, with $n = 7$, the seventh term is

$$ar^6 = 6(4)^6 = 6(4096) = 24{,}576. \quad ◀$$

▶**EXAMPLE 2** Write the first five terms of the geometric sequence with $a = 100$ and $r = .1$.
The first five terms are

$$100, 100(.1), 100(.1)^2, 100(.1)^3, 100(.1)^4,$$

or

$$100, 10, 1, .1, .01. \quad ◀ \quad \boxed{1}$$

Next we find the sum S_n of the first n terms of a geometric sequence. That is, we find S_n, where

$$S_n = a + ar + ar^2 + ar^3 + \ldots + ar^{n-1}. \tag{1}$$

If $r = 1$, this is easy since

$$S_n = \underbrace{a + a + a + \ldots + a}_{n \text{ terms}} = na.$$

2 (a) Find S_4 and S_7 for the geometric sequence 5, 15, 45, 135, ...

(b) Find S_5 for the geometric sequence having $a = -3$ and $r = -5$.

Answer:

(a) $S_4 = 200$, $S_7 = 5465$

(b) $S_5 = -1563$

If $r \neq 1$, multiply both sides of equation (1) by r, obtaining

$$rS_n = ar + ar^2 + ar^3 + ar^4 + \ldots + ar^n. \qquad (2)$$

Now subtract corresponding sides of equation (1) from equation (2):

$$
\begin{array}{rl}
rS_n = & ar + ar^2 + ar^3 + \ldots + ar^{n-1} + ar^n \\
- \quad S_n = & -(a + ar + ar^2 + ar^3 + \ldots + ar^{n-1}) \\
\hline
rS_n - S_n = & ar^n - a \\
\end{array}
$$

$$S_n(r - 1) = a(r^n - 1)$$

$$S_n = \frac{a(r^n - 1)}{r - 1}.$$

Hence, we have this useful formula.

If a geometric sequence has first term a and common ratio r, then the **sum** of the first n terms is given by

$$S_n = \frac{a(r^n - 1)}{r - 1}, \quad r \neq 1.$$

▶**EXAMPLE 3** Find the sum of the first six terms of the geometric sequence 3, 12, 48, ...

Here $a = 3$ and $r = 4$. Find S_6 by the result above.

$$S_6 = \frac{3(4^6 - 1)}{4 - 1} \qquad \text{Let } n = 6, a = 3, r = 4$$

$$= \frac{3(4096 - 1)}{3}$$

$$= 4095 \quad \blacktriangleleft \quad \boxed{2}$$

ORDINARY ANNUITIES A sequence of equal payments made at equal periods of time is called an **annuity.** If the payments are made at the end of the time period, and if the frequency of payments is the same as the frequency of compounding, the annuity is called an **ordinary annuity.** The time between payments is the **payment period,** and the time from the beginning of the first payment period to the end of the last period is called the **term** of the annuity. The **future value** of the annuity, the final sum on deposit, is defined as the sum of the compound amounts of all the payments, compounded to the end of the term.

For example, suppose $1500 is deposited at the end of the year for the next 6 years in an account paying 8% per year compounded annually. Figure 5.1 shows this annuity schematically.

3 Repeat these steps for an annuity of $2000 at the end of each year for 3 years. Assume interest of 8% compounded annually.

(a) The first deposit of $2000 produces a total of _____.

(b) The second deposit becomes _____ .

(c) No interest is earned on the third deposit, so the total in the account is _____ .

Answer:

(a) $2332.80

(b) $2160.00

(c) $6492.80

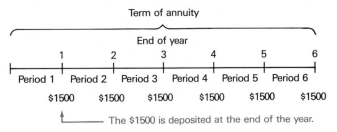

FIGURE 5.1

To find the future value of this annuity, look separately at each of the $1500 payments. The first of these payments will produce a compound amount of

$$1500(1 + .08)^5 = 1500(1.08)^5.$$

Use 5 as the exponent instead of 6 since the money is deposited at the *end* of the first year and earns interest for only 5 years. The second payment of $1500 will produce a compound amount of $1500(1.08)^4$. As shown in Figure 5.2, the future value of the annuity is

$$1500(1.08)^5 + 1500(1.08)^4 + 1500(1.08)^3 + 1500(1.08)^2$$
$$+ 1500(1.08)^1 + 1500.$$

(The last payment earns no interest at all.) If you read this in reverse order, you see that it is just the sum of the first six terms of a geometric sequence with $a = 1500$, $r = 1.08$, and $n = 6$. Therefore, the sum is

$$\frac{a(r^n - 1)}{r - 1} = \frac{1500[(1.08)^6 - 1]}{1.08 - 1} = \$11{,}003.89. \quad \boxed{3}$$

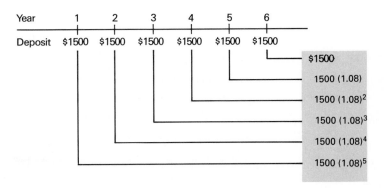

FIGURE 5.2

To generalize this result, suppose that payments of R dollars each are deposited into an account at the *end of each period* for n *periods,* at a rate of interest i *per period.* The first payment of R dollars will produce a compound amount of $R(1 + i)^{n-1}$ dollars, the second payment will produce $R(1 + i)^{n-2}$ dollars, and so on; the final payment earns no interest and contributes just R dollars to the total. If S represents the future value of the annuity, then (as shown in Figure 5.3),

$$S = R(1 + i)^{n-1} + R(1 + i)^{n-2} + R(1 + i)^{n-3} + \ldots + R(1 + i) + R$$

or, written in reverse order,

$$S = R + R(1 + i)^1 + R(1 + i)^2 + \ldots + R(1 + i)^{n-1}.$$

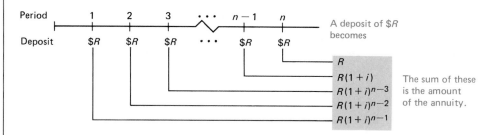

FIGURE 5.3

This result is the sum of the first n terms of the geometric sequence having first term R and common ratio $1 + i$. Using the formula for the sum of the first n terms of a geometric sequence,

$$S = \frac{R[(1 + i)^n - 1]}{(1 + i) - 1} = \frac{R[(1 + i)^n - 1]}{i} = R\left[\frac{(1 + i)^n - 1}{i}\right].$$

The quantity in brackets is commonly written $s_{\overline{n}|i}$ (read "s-angle-n at i"), so that

$$S = R \cdot s_{\overline{n}|i}.$$

Values of $s_{\overline{n}|i}$ can be found using a calculator or from Table 5 at the back of this book.

The next box summarizes this work.

The **future value S of an ordinary annuity** of n payments of R dollars each at the end of consecutive interest periods with interest compounded at a rate of interest i per period is

$$S = R\left[\frac{(1 + i)^n - 1}{i}\right] \quad \text{or} \quad S = R \cdot s_{\overline{n}|i}.$$

4 Johnson Building Materials deposits $2500 at the end of each year into an account paying 8% per year compounded annually. Find the total amount on deposit after

(a) 6 years;

(b) 10 years.

Answer:

(a) $18,339.82

(b) $36,216.41

5 Helen's Dry Goods deposits $5800 at the end of each quarter for 4 years. Find the final amount on deposit if the deposits earn

(a) 8% compounded quarterly;

(b) 6% compounded quarterly.

Answer:

(a) $108,107.85

(b) $104,007.75

6 Ms. Black deposits $800 at the beginning of each 6-month period for 5 years. Find the final amount if the account pays

(a) 6% compounded semiannually;

(b) 8% compounded semiannually.

Answer:

(a) $9446.24

(b) $9989.08

▶**EXAMPLE 4** Tom Bleser is an athlete who feels that his playing career will last 7 years. To prepare for his future, he deposits $22,000 at the end of each year for 7 years in an account paying 6% compounded annually. How much will he have on deposit after 7 years?

His payments form an ordinary annuity with $R = 22,000$, $n = 7$, and $i = .06$. The future value of this annuity (by the formula above) is

$$S = 22,000\left[\frac{(1.06)^7 - 1}{.06}\right] . = \$184,664.43. \quad ◀ \quad \boxed{4}$$

▶**EXAMPLE 5** Suppose $1000 is deposited at the end of each 6-month period for 5 years in an account paying 16% compounded semiannually. Find the future value of the annuity.

Interest of $16\%/2 = 8\%$ is earned semiannually. In 5 years there are $5 \cdot 2 = 10$ semiannual periods. Find $s_{\overline{10}|.08} \approx 14.48656$ from Table 5 or a calculator. The future value is

$$S = 1000\left(\frac{(1.08)^{10} - 1}{.08}\right) \approx 1000(14.48656) = \$14,486.56. \quad ◀ \quad \boxed{5}$$

ANNUITIES DUE The formula developed above is for *ordinary annuities*—those with payments at the *end* of each time period. These results can be modified slightly to apply to **annuities due**—annuities where payments are made at the *beginning* of each time period.

To find the future value of an annuity due, treat each payment as if it were made at the *end* of the *preceding* period. That is, find $s_{\overline{n}|i}$ for *one additional period;* to compensate for this, subtract the amount of one payment. Thus, the **future value of an annuity due** of n payments of R dollars each at the beginning of consecutive interest periods with interest compounded at the rate of i per period is

$$S = R\left[\frac{(1 + i)^{n+1} - 1}{i}\right] - R \quad \text{or} \quad S = R \cdot s_{\overline{n+1}|i} - R.$$

▶**EXAMPLE 6** Find the future value of an annuity due if payments of $500 are made at the beginning of each quarter for 7 years in an account paying 12% compounded quarterly.

In 7 years there are 28 quarterly periods. Add one period to get $n = 29$, and use the formula with $i = 12\%/4 = 3\%$.

$$S = 500\left[\frac{(1.03)^{29} - 1}{.03}\right] - 500$$

$$S \approx 500(45.21885) - 500 = \$22,109.43$$

The account will contain a total of $22,109.43 after 7 years. ◀ \boxed{6}

7 Francisco Arce needs $8000 in 6 years so that he can go on an archeological dig. He wants to deposit equal payments at the end of each quarter so that he will have enough to go on the dig. Find the amount of each payment if the bank pays

(a) 12% compounded quarterly;

(b) 8% compounded quarterly.

Answer:

(a) $232.38

(b) $262.97

SINKING FUNDS Just as the formula $A = P(1 + i)^n$ can be solved for any of its variables, the formula for the future value of an annuity can be used to find the values of variables other than S. In Example 7 below, the amount of money wanted at the end, S, is given, and we need to find R, the amount of each payment.

▶**EXAMPLE 7** Betsy Martini wants to buy an expensive video camera 3 years from now. She plans to deposit an equal amount at the end of each quarter for 3 years in order to accumulate enough money to pay for the camera. The camera will cost $2400, and the bank pays 6% interest compounded quarterly. Find the amount of each of the 12 deposits she will make.

This example describes an ordinary annuity with $S = 2400$, $i = .015$ (6%/4 = 1.5%), and $n = 3 \cdot 4 = 12$ periods. The unknown here is the amount of each payment, R. By the formula for the amount of an annuity,

$$2400 = R \cdot s_{\overline{12}|.015} = R\left(\frac{(1.015)^{12} - 1}{.015}\right).$$

A calculator shows that

$$2400 \approx R(13.04121)$$

$$R \approx \frac{2400}{13.04121} = \$184.03. \quad ◀ \quad \boxed{7}$$

The annuity in Example 7 is a **sinking fund:** a fund set up to receive periodic payments. These periodic payments ($184.03 in the example), together with the interest earned by the payments, are designed to produce a given sum at some time in the future. As another example, a sinking fund might be set up to receive money that will be needed to pay off the principal on a loan at some time in the future.

▶**EXAMPLE 8** The Stockdales are close to retirement. They agree to sell an antique urn to the local museum for $17,000. Their tax adviser suggests that they defer receipt of this money until they retire, 5 years in the future. (At that time, they might well be in a lower tax bracket.) Find the amount of each payment the museum must make into a sinking fund so that it will have the necessary $17,000 in 5 years. Assume that the museum can earn 8% compounded annually on its money. Also, assume that the payments are made annually.

These payments are the periodic payments into an ordinary annuity. The annuity will amount to $17,000 in 5 years at 8% compounded annually. Using the formula and a calculator,

$$17,000 = R \cdot s_{\overline{5}|.08} = R\left(\frac{(1.08)^5 - 1}{.08}\right)$$

or

8 A firm must pay off $40,000 worth of bonds in 7 years. Find the amount of each annual payment to be made into a sinking fund if money earns 6% compounded annually.

Answer:
$4765.40

$$17,000 \approx R(5.86660)$$

$$R \approx \frac{17,000}{5.86660} = 2897.76,$$

or $2897.76. If the museum deposits $2897.76 at the end of each year for 5 years in an account paying 8% compounded annually, it will have the total amount that it needs. This result is shown in the following *sinking fund table*. An explanation of its construction is given below.

Payment Number	Amount of Deposit	Interest Earned	Total in Account
1	$2897.76	$ 0	$ 2897.76
2	2897.76	231.82	6027.34
3	2897.76	482.19	9407.29
4	2897.76	752.58	13,057.63
5	2897.76	1044.61	17,000.00

Payment 1 is made at the end of the first year and earns interest of .08(2897.76) = $231.82 during the second year. When payment 2 is made, the total in the account is the sum of the first year total (last column of line 1), the second payment (second column of line 2) and the interest earned by the first payment (third column of line 2), as shown in the last column of line 2. This total earns interest of .08(6027.34) = $482.19 during the next year. This interest (third column of line 3) plus the total at the end of payment 2 (last column of line 2) plus the third payment (second column of line 3) produces the total $9407.29 in the last column of line 3.

The remaining lines of the table are obtained similarly. The last payment in a sinking fund table may differ slightly from the others because of earlier rounding. ◄ **8**

5.3 EXERCISES

 Most of the exercises in this section are designed to be worked with a calculator.

Find the fifth term of each of the following geometric sequences. (See Example 1.)

1. $a = 3, r = 2$ **2.** $a = 5, r = 3$ **3.** $a = -8, r = 3$ **4.** $a = -6, r = 2$

5. $a = 1, r = -3$ **6.** $a = 12, r = -2$ **7.** $a = 1024, r = \frac{1}{2}$ **8.** $a = 729, r = \frac{1}{3}$

Find the sum of the first four terms for each of the following geometric sequences. (See Example 3.)

9. $a = 1, r = 2$ **10.** $a = 3, r = 3$ **11.** $a = 5, r = \frac{1}{5}$

12. $a = 6$, $r = \dfrac{1}{2}$

13. $a = 128$, $r = -\dfrac{3}{2}$

14. $a = 81$, $r = -\dfrac{2}{3}$

Find each of the following values.

15. $s_{\overline{12}|.05}$

16. $s_{\overline{20}|.06}$

17. $s_{\overline{16}|.04}$

18. $s_{\overline{40}|.02}$

19. $s_{\overline{20}|.01}$

20. $s_{\overline{18}|.015}$

21. $s_{\overline{15}|.04}$

22. $s_{\overline{30}|.015}$

Find the future value of the following ordinary annuities. Interest is compounded annually. (See Example 4.)

23. $R = 100$, $i = .06$, $n = 4$

24. $R = 1000$, $i = .06$, $n = 5$

25. $R = 10{,}000$, $i = .05$, $n = 19$

26. $R = 100{,}000$, $i = .08$, $n = 23$

27. $R = 8500$, $i = .06$, $n = 30$

28. $R = 11{,}200$, $i = .08$, $n = 25$

29. $R = 46{,}000$, $i = .06$, $n = 32$

30. $R = 29{,}500$, $i = .05$, $n = 15$

Find the future value of each of the following ordinary annuities. Payments are made and interest is compounded as given. (See Example 5.)

31. $R = 9200$, 10% interest compounded semiannually for 7 years

32. $R = 3700$, 8% interest compounded semiannually for 11 years

33. $R = 800$, 6% interest compounded semiannually for 12 years

34. $R = 4600$, 8% interest compounded quarterly for 9 years

35. $R = 15{,}000$, 12% interest compounded quarterly for 6 years

36. $R = 42{,}000$, 10% interest compounded semiannually for 12 years

Find the future value of each annuity due. Assume that interest is compounded annually. (See Example 6.)

37. $R = 600$, $i = .06$, $n = 8$

38. $R = 1400$, $i = .08$, $n = 10$

39. $R = 20{,}000$, $i = .08$, $n = 6$

40. $R = 4000$, $i = .06$, $n = 11$

Find the future value of each annuity due. (See Example 6.)

41. Payments of $1000 made at the beginning of each year for 9 years at 8% compounded annually

42. $750 deposited at the beginning of each year for 15 years at 6% compounded annually

43. $100 deposited at the beginning of each quarter for 9 years at 12% compounded quarterly

44. $1500 deposited at the beginning of each semiannual period for 11 years at 6% compounded semiannually

Find the periodic payment that will amount to the given sums under the given conditions. (See Examples 7 and 8.)

45. $S = \$10{,}000$, interest is 5% compounded annually, payments made at the end of each year for 12 years

46. $S = \$100{,}000$, interest is 8% compounded semiannually, payments made at the end of each semiannual period for 9 years

47. $S = \$50{,}000$, interest is 12% compounded quarterly, payments made at the end of each quarter for 8 years

48. $S = \$147{,}200$, interest is 12% compounded monthly, payments made at the end of each month for 2 years

Work the following exercises.

49. Pat Dillon deposits $12,000 at the end of each year for 9 years in an account paying 8% interest compounded annually.
   (a) Find the final amount she will have on deposit.
   (b) Pat's brother-in-law works in a bank that pays 6% compounded annually. If she deposits her money in this bank instead of the one above, how much will she have in her account?
   (c) How much would Dillon lose over 9 years by using her brother-in-law's bank?

50. Charlie is saving for a computer. At the end of each month he puts $60 in a savings account that pays 8% interest compounded monthly. How much is in the account after 3 years?

51. Hassi is paid on the first day of the month and $80 is automatically deducted from his pay and deposited in a savings account. If the account pays 7.5% interest compounded monthly, how much will be in the account after 3 years and 9 months?

52. A father opened a savings account for his daughter on the day she was born, depositing $1000. Each year on her birthday he deposits another $1000, making the last deposit on her twenty-first birthday. If the account pays 9.5% interest compounded annually, how much is in the account at the end of the day on the daughter's twenty-first birthday?

53. A 45-year-old man puts $1000 in a retirement account at the end of each quarter until he reaches the age of 60 and makes no further deposits. If the account pays 11% interest compounded quarterly, how much will be in the account when the man retires at age 65?

54. At the end of each quarter a 50-year-old woman puts $1200 in a retirement account that pays 7% interest compounded quarterly. When she reaches age 60, she withdraws the entire amount and places it in a mutual fund that pays 9% interest compounded monthly. From then on she deposits $300 in the mutual fund at the end of each month. How much is in the account when she reaches age 65?

55. Jasspreet Kaur deposits $2435 at the beginning of each semiannual period for 8 years in an account paying 6% compounded semiannually. She then leaves that money alone, with no further deposits, for an additional 5 years. Find the final amount on deposit after the entire 13-year period.

56. Chuck deposits $10,000 at the beginning of each year for 12 years in an account paying 5% compounded annually. He then puts the total amount on deposit in another account paying 6% compounded semiannually for another 9 years. Find the final amount on deposit after the entire 21-year period.

57. Ray Berkowitz needs $10,000 in 8 years.
  (a) What amount should he deposit at the end of each quarter at 8% compounded quarterly so that he will have his $10,000?
  (b) Find Berkowitz's quarterly deposit if the money is deposited at 6% compounded quarterly.

58. Harv's Meats knows that it must buy a new deboner machine in 4 years. The machine costs $12,000. In order to accumulate enough money to pay for the machine, Harv decides to deposit a sum of money at the end of each 6 months in an account paying 6% compounded semiannually. How much should each payment be?

59. Barb Silverman wants to buy an $18,000 car in 6 years. How much money must she deposit at the end of each

quarter in an account paying 12% compounded quarterly so that she will have enough to pay for her car?

Suppose a 40-year-old person deposits $2000 per year in an Individual Retirement Account until age 65. Find the total in the account with the following assumptions of interest rates. (Assume semiannual compounding, with payment of $1000 made at the end of each semiannual period.)

60. 4%   **61.** 6%   **62.** 8%   **63.** 10%

Find the amount of each payment to be made into a sinking fund so that enough will be present to accumulate the following amounts. (See Example 7.)

64. $8500, money earns 8% compounded annually, 7 annual payments

65. $11,000, money earns 6% compounded semiannually, for 6 years

66. $75,000, money earns 6% compounded semiannually, for $4\frac{1}{2}$ years

67. $50,000, money earns 10% compounded quarterly, for $2\frac{1}{2}$ years

68. $25,000, money earns 12% compounded quarterly, for $3\frac{1}{2}$ years

69. $6000, money earns 8% compounded monthly, for 3 years

70. $9000, money earns 12% compounded monthly, for $2\frac{1}{2}$ years

Find the final amount of the ordinary annuities in Exercises 71 and 72.

71. $892.17 each month for 2 years in an account paying 7.5% interest compounded monthly

72. $2476.32 each quarter for 5.5 years at 7.81% interest compounded quarterly.

Refer to Example 8 for Exercises 73 and 74.

73. Jill Streitsel sells some land in Nevada. She will be paid a lump sum of $60,000 in 7 years. Until then, the buyer pays 8% interest, paid quarterly.
  (a) Find the amount of each quarterly interest payment.
  (b) The buyer sets up a sinking fund so that enough money will be present to pay off the $60,000. The buyer wants to make semiannual payments into the sinking fund; the account pays 6% compounded semiannually. Find the amount of each payment into the fund.
  (c) Prepare a table showing the amount in the sinking fund after each deposit.

74. Jeff Reschke bought a rare stamp for his collection. He agreed to pay a lump sum of $4000 after 5 years. Until then, he pays 6% interest.

(a) Find the amount of each semiannual interest payment.

(b) Reschke sets up a sinking fund so that enough money will be present to pay off the $4000. He wants to make annual payments into the fund. The account pays 8% compounded annually. Find the amount of each payment into the fund.

(c) Prepare a table showing the amount in the sinking fund after each deposit.

5.4 PRESENT VALUE OF AN ANNUITY; AMORTIZATION

Suppose that at the end of each year, for the next 10 years, $500 is deposited in a savings account paying 7% interest compounded annually. This is an example of an ordinary annuity. The **present value** of this annuity is the amount that would have to be deposited in one lump sum today (at the same compound interest rate) in order to produce exactly the same balance at the end of 10 years. We can find a formula for the present value of an annuity as follows.

Suppose deposits of R dollars are made at the end of each period for n periods at interest rate i per period. Then the amount in the account after n periods is the future value of this annuity:

$$S = R \cdot s_{\overline{n}|i} = R\left[\frac{(1 + i)^n - 1}{i}\right].$$

On the other hand, if P dollars are deposited today at the same compound interest rate i, then at the end of n periods, the amount in the account is $P(1 + i)^n$. If P is the present value of the annuity, this amount must be the same as the amount S in the formula above, that is,

$$P(1 + i)^n = R\left[\frac{(1 + i)^n - 1}{i}\right].$$

To solve this equation for P, multiply both sides by $(1 + i)^{-n}$.

$$P = R(1 + i)^{-n}\left[\frac{(1 + i)^n - 1}{i}\right]$$

Use the distributive property; also recall that $(1 + i)^{-n}(1 + i)^n = 1$.

$$P = R\left[\frac{(1 + i)^{-n}(1 + i)^n - (1 + i)^{-n}}{i}\right]$$

$$P = R\left[\frac{1 - (1 + i)^{-n}}{i}\right]$$

1 What lump sum deposited today at 8% compounded annually will yield the same total amount as payments of

(a) $650 at the end of each year for 9 years?

(b) $17,580 at the end of each year for 14 years?

Answer:

(a) $4060.48

(b) $144,933.69

2 What lump sum deposited today would be equivalent to equal payments of $1000 at the end of each quarter for 4 years if the money is deposited at 8% compounded quarterly?

Answer:
$13,577.71

The quantity in brackets is abbreviated as $a_{\overline{n}|i}$, (read "*a*-angle-*n* at *i*"), so

$$a_{\overline{n}|i} = \frac{1 - (1 + i)^{-n}}{i}.$$

Here is a summary of what we have done.

The present value P of an annuity of n payments of R dollars each at end of consecutive interest periods with interest compounded at a rate interest i per period is

$$P = R\left[\frac{1 - (1 + i)^{-n}}{i}\right] \quad \text{or} \quad P = R \cdot a_{\overline{n}|i}.$$

▶**EXAMPLE 1** What lump sum deposited today at 8% interest compounded annually will yield the same total amount as payments of $1500 at the end of each year for 12 years, also at 8% interest compounded annually?

We want to find the present value of an annuity of $1500 per year for 12 years at 8% compounded annually. A calculator shows that

$$P = R \cdot a_{\overline{12}|.08} = 1500\left[\frac{1 - (1 + .08)^{-12}}{.08}\right]$$

$$\approx 1500(7.53608) = 11,304.12.$$

Thus, a lump sum deposit of $11,304.12 today at 8% compounded annually will yield the same total after 12 years as deposits of $1500 at the end of each year for 12 years at 8% compounded annually. ◀ **1**

▶**EXAMPLE 2** Mr. Jones and Ms. Gonsalez are both graduates of the Forestvire Institute of Technology. They both agree to contribute to the endowment fund of FIT. Mr. Jones says that he will give $500 at the end of each year for 9 years. Ms. Gonsalez prefers to give a lump sum today. What lump sum can she give that will equal the present value of Mr. Jones' annual gifts, if the endowment fund earns 7.5% compounded annually?

Here $R = 500$, $n = 9$, and $i = .075$ and we have

$$P = R \cdot a_{\overline{9}|.075} = 500\left[\frac{1 - (1.075)^{-9}}{.075}\right]$$

$$\approx 500(6.37889) = 3189.44.$$

Therefore, Ms. Gonsalez must donate a lump sum of $3189.44 today. ◀ **2**

3 Find the monthly payment in Example 3 if the loan is for 48 months.

Answer:
$131.67

4 Kristen Westman buys a small business for $174,000. She agrees to pay off the cost in payments at the end of each semiannual period for 7 years, with interest of 10% compounded semiannually on the unpaid balance. Find the amount of each payment.

Answer:
$17,578.17

One of the most important uses of annuities is in determining the equal monthly payments needed to pay off a loan, as illustrated in the next example.

▶**EXAMPLE 3** A car costs $6000. After a down payment of $1000, the balance will be paid off in 36 monthly payments with interest of 12% per year on the unpaid balance. Find the amount of each payment.

A single lump sum payment of $5000 today would pay off the loan. So, $5000 is the present value of an annuity of 36 monthly payments with interest of 12%/12 = 1% per month. Thus, $P = 5000$, $n = 36$, $i = .01$, and we must find the monthly payment R in the formula

$$P = R\left[\frac{1 - (1 + i)^{-n}}{i}\right]$$

$$5000 = R\left[\frac{1 - (1.01)^{-36}}{.01}\right] \approx R(30.10751)$$

$$R = \frac{5000}{30.10751} = 166.07$$

A monthly payment of $166.07 will be needed. ◀ **3**

AMORTIZATION A loan is **amortized** if both the principal and interest are paid by a sequence of equal periodic payments. In Example 3 above, a loan of $5000 at 12% interest compounded monthly could be amortized by paying $166.07 per month for 36 months.

The periodic payment needed to amortize a loan may be found, as in Example 3, by solving the present value equation for R. A loan of P dollars at interest rate i per period may be amortized in n equal periodic payments of R dollars made at the end of each period, where

$$R = \frac{P}{a_{\overline{n}|i}} = \frac{P}{\left[\dfrac{1 - (1 + i)^{-n}}{i}\right]} = \frac{Pi}{1 - (1 + i)^{-n}}.$$

▶**EXAMPLE 4** A speculator agrees to pay $15,000 for a parcel of land; this amount, with interest, will be paid over 4 years, with semiannual payments, at an interest rate of 12% compounded semiannually. Find the amount of each payment.

Here $P = 15,000$ and the semiannual interest rate is $i = 12\%/2 = 6\% = .06$. The number of semiannual payments is $n = 2 \cdot 4 = 8$. Therefore,

$$R = \frac{15,000}{a_{\overline{8}|.06}} = \frac{15,000}{\left[\dfrac{1 - (1.06)^{-8}}{.06}\right]} \approx \frac{15,000}{6.20979} = 2415.54.$$

Each payment is $2415.54. ◀ **4**

5 If the mortgage in Example 5 runs for 15 years, find **(a)** the monthly payment; **(b)** the total amount of interest paid.

Answer:

(a) $819.21

(b) $69,457.80

▶**EXAMPLE 5** The Perez family buys a house for $94,000 with a down payment of $16,000. They take out a 30-year mortgage for $78,000 at an annual interest rate of 9.6%.

(a) Find the amount of the monthly payment needed to amortize this loan.

Here $P = 78,000$ and the monthly interest rate is $9.6\%/12 = .096/12 = .008.$* The number of monthly payments is $12 \cdot 30 = 360$. Therefore,

$$R = \frac{78,000}{a_{\overline{360}|.008}} = \frac{78,000}{\left[\dfrac{1 - (1.008)^{-360}}{.008}\right]} \approx \frac{78,000}{117.90229} = 661.56.$$

Monthly payments of $661.56 are required to amortize the loan.

(b) Find the total amount of interest paid when the loan is amortized over 30 years.

The Perez family makes 360 payments of $661.56 each, for a total of $238,161.60. Since the amount of the loan was $78,000, the total interest paid is

$$\$238,161.60 - \$78,000 = \$160,161.60.$$

This large amount of interest is typical of what happens with a long mortgage. A 15-year mortgage would have higher payments but would involve significantly less interest. **5**

(c) Find the part of the first payment that is interest and the part that is applied to reducing the debt.

As we saw in part (a), the monthly interest rate is .008. During the first month, the entire $78,000 is owed. Interest on this amount for 1 month is found by the formula for simple interest.

$$I = Prt = 78,000(.008)(1) = 624$$

At the end of the month, a payment of $661.56 is made; since $624 of this is interest, a total of

$$\$661.56 - \$624 = \$37.56$$

is applied to the reduction of the original debt. ◀

AMORTIZATION SCHEDULES In the preceding example, 360 payments are made to amortize a $78,000 loan. The loan balance after the first payment is reduced by only $37.56, which is much less than $(1/360)(78,000) \approx \$216.67$. Therefore, even though equal *payments* are made to amortize a loan, the loan *balance* does not decrease in equal steps. This fact is very important if a loan is paid off early.

---

*Mortgage rates are quoted in terms of annual interest, but it is always understood that the monthly rate is 1/12 of the annual rate and that interest is compounded monthly.

▶**EXAMPLE 6**   Ellen Ryan borrows $1000 for 1 year at 12% annual interest compounded monthly.

(a) What is her monthly loan payment?

The monthly interest rate is 12%/12 = 1% = .01, so that her payment is

$$R = \frac{1000}{a_{\overline{12}|.01}} = \frac{1000}{\left[\dfrac{1-(1.01)^{-12}}{.01}\right]} \approx \frac{1000}{11.25508} = \$88.85.$$

(b) After making three payments, she decides to pay off the remaining balance all at once. How much must she pay?

Since nine payments remain to be paid, they can be thought of as an annuity consisting of nine payments of $88.85 at 1% interest per period. The present value of this annuity is

$$88.85\left[\frac{1-(1.01)^{-9}}{.01}\right] = 761.09.$$

So Ellen's remaining balance, computed by this method, is $761.09.

An alternative method of figuring the balance is to consider the payments already made as an annuity of three payments. At the beginning, the present value of this annuity was

$$88.85\left[\frac{1-(1.01)^{-3}}{.01}\right] = 261.31.$$

So she still owes the difference $1000 − $261.31 = $738.69. Furthermore, she owes the interest on this amount for 3 months, for a total of

$$(738.69)(1.01)^3 = \$761.07.$$

This balance due differs from the one obtained by the first method by 2¢ because the monthly payment and the other calculations were rounded to the nearest penny.   ◀

Although most people wouldn't quibble about a 2¢ difference in the balance due in Example 6, the difference in other cases (larger amounts or longer terms) might be more than that. A bank or business must keep its books accurately to the nearest penny, so it must determine the balance due in cases like this unambiguously and exactly. This is done by means of an **amortization schedule,** which lists how much of each payment is interest and how much goes to reduce the balance, and how much is owed after *each* payment.

▶**EXAMPLE 7**   Determine the exact amount Ellen Ryan in Example 6 owes after three monthly payments.

An amortization table for the loan is shown below. It is obtained as follows. The annual interest rate is 12% compounded monthly, so the interest rate per month is 12%/12 = 1% = .01. When the first payment is made, 1 month's interest, namely .01(1000) = $10, is owed. Subtracting this from the $88.85 payment leaves $78.85 to be applied to repayment. Hence, the principal at the end of the first payment period is 1000 − 78.85 = $921.15, as shown in the ''payment 1'' line of the chart.

When payment 2 is made, 1 month's interest on $921.15 is owed, namely .01(921.15) = $9.21. Subtracting this from the $88.85 payment leaves $79.64 to reduce the principal. Hence, the principal at the end of payment 2 is 921.15 − 79.64 = $841.51. The interest portion of payment 3 is based on this amount and the remaining lines of the table are found in a similar fashion.

Payment Number	Amount of Payment	Interest for Period	Portion to Principal	Principal at End of Period
0	—	—	—	1000.00
1	88.85	10.00	78.85	921.15
2	88.85	9.21	79.64	841.51
3	88.85	8.42	80.43	761.08
4	88.85	7.61	81.24	679.84
5	88.85	6.80	82.05	597.79
6	88.85	5.98	82.87	514.92
7	88.85	5.15	83.70	431.22
8	88.85	4.31	84.54	346.68
9	88.85	3.47	85.38	261.30
10	88.85	2.61	86.24	175.06
11	88.85	1.75	87.10	87.96
12	88.84	.88	87.96	0

The schedule shows that after three payments, she still owes $761.08, an amount that differs slightly from that obtained by either method in Example 6. ◀

The amortization schedule in Example 7 is typical. In particular, note that all payments are the same except the last one. It is often necessary to adjust the amount of the final payment to account for rounding off earlier and to insure that the final balance is exactly 0.

An amortization schedule also shows how the periodic payments are applied to interest and principal. The amount going to interest decreases with each payment, while the amount going to reduce the principal owed increases with each payment.

5.4 EXERCISES

Most of the exercises in this section are designed to be worked with a calculator.

Find each of the following values.

1. $a_{\overline{15}|.06}$ **2.** $a_{\overline{10}|.03}$ **3.** $a_{\overline{18}|.04}$ **4.** $a_{\overline{30}|.01}$ **5.** $a_{\overline{16}|.01}$ **6.** $a_{\overline{32}|.02}$

7. $a_{\overline{6}|.015}$ **8.** $a_{\overline{18}|.015}$

Find the present value of each of the following ordinary annuities. (See Examples 1 and 2.)

9. Payments of $1000 are made annually for 9 years at 8% compounded annually.

10. Payments of $5000 are made annually for 11 years at 6% compounded annually.

11. Payments of $890 are made annually for 16 years at 8% compounded annually.

12. Payments of $1400 are made annually for 8 years at 8% compounded annually.

13. Payments of $10,000 are made semiannually for 15 years at 10% compounded semiannually.

14. Payments of $50,000 are made quarterly for 10 years at 8% compounded quarterly.

15. Payments of $15,806 are made quarterly for 3 years at 10.8% compounded quarterly.

16. Payments of $18,579 are made every 6 months for 8 years at 9.4% compounded semiannually.

Find the lump sum deposited today that will yield the same total amount as payments of $10,000 at the end of each year for 15 years, at each of the given interest rates. (See Examples 1 and 2.)

17. 4%, compounded annually

18. 5%, compounded annually

19. 6%, compounded annually

20. 8%, compounded annually

21. 12%, compounded annually

22. 10%, compounded annually

23. In his will the late Mr. Hudspeth said that each child in his family could have an annuity of $2000 at the end of each year for 9 years, or the equivalent present value. If money can be deposited at 8% compounded annually, what is the present value?

24. A 50-year-old man plans to retire at age 62. How much money must he deposit today at 9% interest compounded annually to guarantee that when he retires there will be $250,000 in his account?

25. Lynn Chiang buys a new car costing $6000. She agrees to make payments at the end of each monthly period for 4 years.
   (a) If she pays 12% interest on the unpaid balance, what is the amount of each payment?
   (b) Find the total amount of interest Chiang will pay.

26. Bargain Barn sells a stereo system for $600 down and monthly payments of $30 for the next 3 years. If the interest rate is 1.25% per month on the unpaid balance, find
   (a) the cost of the stereo system;
   (b) the total amount of interest paid.

27. What sum deposited today at 5% compounded annually for 8 years will provide the same amount as $1000 deposited at the end of each year for 8 years at 6% compounded annually?

28. What lump sum deposited today at 8% compounded quarterly for 10 years will yield the same final amount as deposits of $4000 at the end of each 6-month period for 10 years at 6% compounded semiannually?

Find the payment necessary to amortize each of the following loans. Interest is calculated on the unpaid balance. (See Examples 3–4.)

29. $1000, 9 annual payments at 8%

30. $2500, 6 quarterly payments at 8%

31. $41,000, 10 semiannual payments at 10%

32. $90,000, 12 annual payments at 8%

33. $140,000, 15 quarterly payments at 12%

35. $5500, 24 monthly payments at 12%

34. $7400, 18 semiannual payments at 6%

36. $45,000, 36 monthly payments at 12%

Find the monthly house payment necessary to amortize the following loans. Interest is calcu-lated on the unpaid balance. (See Example 5.)

37. $49,560 at 10.75% for 25 years

39. $53,762 at 12.45% for 30 years

38. $70,892 at 11.11% for 30 years

40. $96,511 at 10.57% for 25 years

41. Sheridan Whiteside buys a house for $285,000. He pays $60,000 down and takes out a mortgage at 9.5% on the balance. Find his monthly payment and the total amount of interest he will pay if the length of the mortgage is
(a) 15 years;   (b) 20 years;   (c) 25 years.

42. A new car at Dealer A is priced at $11,000; you must pay $3000 down and pay the balance in 48 monthly payments at an annual interest rate of 12% compounded monthly. At Dealer B, the car costs $10,600, with a $3000 down payment and 48 monthly payments at an annual interest rate of 15% compounded monthly. Which is a better deal?

43. An insurance firm pays $4000 for a new printer for its computer. It amortizes the loan for the printer in 4 annual payments at 8% interest on the unpaid balance. Prepare an amortization schedule for this machine.

44. Large semitrailer trucks cost $72,000 each. Ace Trucking buys such a truck and agrees to pay for it by a loan which will be amortized with 9 semiannual payments at 16% interest on the unpaid balance. Prepare an amortization schedule for this truck.

45. One retailer charges $1048 for a correcting electric typewriter. A firm of tax accountants buys 8 of these machines. They make a down payment of $1200 and agree to amortize the balance with monthly payments at 18% interest on the unpaid balance for 4 years. Prepare an amortization schedule showing the first 6 payments.

46. When Teresa Flores opened her law office, she bought $14,000 worth of law books and $7200 worth of office furniture. She paid $1200 down and agreed to amortize the balance with semiannual payments for 5 years at 16% interest on the unpaid balance. Prepare an amortization schedule for this purchase.

47. The Adams family bought a house for $81,000. They paid $20,000 down and took out a 30-year mortgage for the balance at 11%. Use one of the methods in Example 6 to estimate the mortgage balance after 100 payments.

48. After making 180 payments on their mortgage, the Adams family of Exercise 47 sells their house for $136,000. They must pay closing costs of $3700 plus 2.5% of the sale price of the house. Approximately how much money will they receive at the close of the sale after their current mortgage is paid off and closing costs are deducted?

Use the amortization table in Example 7 to answer the following questions.

49. How much of the fourth payment is interest?

50. How much of the tenth payment is interest?

51. How much of the second payment is used to reduce the debt?

52. How much of the eleventh payment is used to reduce the debt?

53. How much interest is paid in the first 4 months of the loan?

54. How much interest is paid in the last 4 months of the loan?

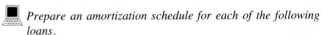 *Prepare an amortization schedule for each of the following loans.*

55. $37,947.50 with 8.5% interest on the unpaid balance, to be paid with equal annual payments over 10 years

56. $4835.80 at 9.25% interest on the unpaid balance, to be repaid in 5 years in equal semiannual payments

5.5 APPLYING FINANCIAL FORMULAS

We have presented a lot of new formulas in this chapter. By answering the following questions, you can decide which formula to use for a particular problem.

1. Is simple or compound interest involved?

   Simple interest is normally used for investments or loans of a year or less; compound interest is normally used in all other cases.

2. If simple interest is being used, what is being sought: interest amount, future value, present value, or discount?

3. If compound interest is being used, does it involve a lump sum (single payment) or an annuity (sequence of payments)?

   (a) For a lump sum,

      (i) Is ordinary compound interest or continuous interest involved?

      (ii) What is being sought: present value, future value, number of periods at interest, or effective rate?

   (b) For an annuity,

      (i) Is it an ordinary annuity (payment at the end of each period) or an annuity due (payment at the beginning of each period)?

      (ii) What is being sought: present value, future value, or payment amount?

Once you have answered these questions, choose the appropriate formula from the Chapter Summary and work the problem. As a final step, consider whether the answer you get makes sense. For instance, present value should always be less than future value. The amount of interest or the payments in an annuity should be fairly small compared to the total future value.

The formulas presented in this chapter are summarized in the Key Concepts at the end of this chapter.

5.5 EXERCISES

 For the exercises in this section, round money amounts to the nearest cent, time to the nearest day, and rates to the nearest percent. Most of the exercises require a calculator with a y^x key.

Find the present value of $82,000 *for the following number of years, if the money can be deposited at 12% compounded quarterly.*

1. 3 years   **2.** 5 years   **3.** 7 years   **4.** 10 years

Find the amount of the payment necessary to amortize the following amounts.

5. $72,000, 11 annual payments at 12%

6. $4250, 13 quarterly payments at 12%

7. $58,000, 23 quarterly payments at 16%

8. $112,000, 32 monthly payments at 12%

Find the compound amount for each of the following.

9. $1000 at 10% compounded annually for 17 years

10. $4792.35 at 12% compounded semiannually for $5\frac{1}{2}$ years

11. $2500 at 16% compounded quarterly for $3\frac{3}{4}$ years

12. $23,500 at 18% compounded monthly for 30 months

Find the simple interest for each of the following. Assume 360 days in a year.

13. $2500 at 8.5% for 2 years

14. $42,500 at 11.75% for 10 months

15. $32,662 at 8.882% for 225 days

16. $56,009 at 13.221% for 179 days

Find the amount of interest earned by each of the following deposits.

17. $22,500 at 12% compounded quarterly for $5\frac{1}{4}$ years

18. $53,142 at 18% compounded monthly for 32 months

Find the compound amount and the amount of interest earned by a deposit of $32,750 at 10% compounded continuously for the following number of years.

19. 2 years

20. 5 years

21. $7\frac{1}{2}$ years

22. 9.2 years

Find the present value of the following future amounts. Assume 360 days in a year and use simple interest.

23. $50,000 for 11 months, money earns 14%

24. $17,320 for 9 months, money earns 13.5%

25. $122,300 for 138 days, money earns 11.75%

26. $51,800 for 312 days, money earns 12.801%

Find the proceeds for the following. Assume 360 days in a year and use simple interest.

27. $72,113 for 11 months, discount rate 15%

28. $23,561 for 112 days, discount rate 14.33%

29. $267,100 for 271 days, discount rate 15.72%

30. $52,430 for 8 months, discount rate 12.5%

Find the amount of each of the following annuities.

31. $1000 is deposited at the end of each year for 12 years, money earns 10% compounded annually

32. $2500 is deposited at the end of each semiannual period for $5\frac{1}{2}$ years, money earns 12% compounded semiannually

33. $250 is deposited at the end of each quarter for $7\frac{1}{4}$ years, money earns 12% compounded quarterly

34. $800 is deposited at the end of each month for $1\frac{1}{4}$ years, money earns 18% compounded monthly

35. $100 is deposited at the beginning of each year for 5 years, money earns 8% compounded yearly

36. $5200 is deposited at the beginning of each quarter for $3\frac{3}{4}$ years, money earns 16% compounded quarterly

Find the amount of each payment to be made to a sinking fund to accumulate the indicated amounts.

37. $10,000, money earns 10% compounded annually, 7 annual payments

38. $42,000, money earns 12% compounded quarterly, 13 quarterly payments

39. $100,000, money earns 12% compounded semiannually, 9 semiannual payments

40. $53,000, money earns 18% compounded monthly, 35 monthly payments

Find the present value of each ordinary annuity.

41. Payments of $1200 are made annually for 7 years at 10% compounded annually

42. Payments of $500 are made semiannually at 10% compounded semiannually for $3\frac{1}{2}$ years

43. Payments of $1500 are made quarterly for $5\frac{1}{4}$ years at 12% compounded quarterly

44. Payments of $905.43 are made monthly for $2\frac{11}{12}$ years at 18% compounded monthly

Prepare an amortization schedule for each of the following loans. Interest is calculated on the unpaid balance.

45. $8500 loan repaid in semiannual payments for $3\frac{1}{2}$ years at 12%

46. $40,000 loan repaid in quarterly payments for $1\frac{3}{4}$ years at 16%

Solve the following applied problems.

47. A 10-month loan of $42,000 at simple interest produces $5320 interest. Find the interest rate.

48. Willa Burke deposits $803.47 at the end of each quarter for $3\frac{3}{4}$ years in an account paying 12% compounded quarterly. Find the final amount in the account and the amount of interest earned.

49. Makarim Wibison owes $7850 to a relative. He has agreed to pay the money in 5 months at simple interest of 10%. One month before the loan is due, the relative discounts the loan at the bank. The bank charges a 13.2% discount rate. How much money does the relative receive?

50. A small resort must add a swimming pool to compete with a new resort built nearby. The pool will cost $28,000. The resort borrows the money and agrees to repay it with equal payments at the end of each quarter for $6\frac{1}{2}$ years at an interest rate of 12% compounded quarterly. Find the amount of each payment.

51. According to the terms of a divorce settlement, one spouse must pay the other a lump sum of $2800 in 17 months. What lump sum can be invested today, at 12% compounded monthly, so that enough will be available for the payment?

52. An accountant loans $28,000 at simple interest to her business. The loan is at 11.5% and earns $3255 interest. Find the time of the loan in months.

53. A firm of attorneys deposits $5000 of profit sharing money at the end of each semiannual period for $7\frac{1}{2}$ years. Find the final amount in the account if the deposit earns 10% compounded semiannually. Find the amount of interest earned.

54. Find the principal that must be invested at 11.25% simple interest to earn $937.50 interest in 8 months.

55. In 3 years Ms. Thompson must pay a pledge of $7500 to her college's building fund. What lump sum can she deposit today, at 10% compounded semiannually, so that she will have enough to pay the pledge?

56. The owner of Eastside Hallmark borrows $48,000 to expand the business. The money will be repaid in equal payments at the end of each year for 7 years. Interest is 10%. Find the amount of each payment.

57. To buy a new computer, Mark Nguyen borrows $3250 from a friend at 9% interest compounded annually for 4 years. Find the compound amount he must pay back at the end of the 4 years.

58. A small business invests some spare cash for 3 months at 12.2% simple interest and earns $244 in interest. Find the amount of the investment.

59. When the Lee family bought their home, they borrowed $115,700 at 10.5% compounded monthly for 25 years. If they make all 300 payments, repaying the loan on schedule, how much interest will they pay? Assume the last payment is the same as the previous ones.

60. Suppose $84,720 is deposited for 7 months and earns $8055.46 in interest. Find the rate of interest.

CHAPTER 5 SUMMARY

KEY TERMS AND SYMBOLS

5.1 interest
principal
rate
time
simple interest
future value (maturity value)
ordinary interest
exact interest
present value
discount (bank discount)
proceeds
effective rate

5.2 compound interest
compound amount
compounding period

continuous compounding
nominal rate (stated rate)
effective rate

5.3 geometric sequence
common ratio
annuity
ordinary annuity
payment period
term of an annuity
future value of an annuity
annuity due
sinking fund

5.4 present value of an annuity
amortize a loan
amortization schedule

KEY CONCEPTS

Simple Interest

The **simple interest** I on an amount of P dollars for t years at interest rate r per year is $I = Prt$.

The **future value** A of P dollars at simple interest rate r for t years is $A = P(1 + rt)$.

The **present value** P of a future amount of A dollars at simple interest rate r for t years is

$$P = \frac{A}{1 + rt}.$$

If D is the **discount** on a loan having maturity value A at simple interest rate r for t years, then $D = Art$. If D is the discount and P the **proceeds** of a loan having maturity value A at simple interest rate r for t years, then $P = A - D$ or $P = A(1 - rt)$.

Compound Interest

If P dollars is deposited for n time periods at compound interest rate i per period, the **compound amount (future value)** A is $A = P(1 + i)^n$.

The **present value** P of A dollars at compound interest rate i per period for n periods is

$$P = \frac{A}{(1 + i)^n} = A(1 + i)^{-n}.$$

The **effective rate** corresponding to a stated interest rate r per year, compounded m times per year, is

$$\left[1 + \frac{r}{m}\right]^m - 1.$$

Continuous Compound Interest	If P dollars is deposited for t years at interest rate r per year, compounded continuously, the **compound amount (future value)** A is $A = Pe^{rt}$.

The **present value** P of A dollars at interest rate r per year compounded continuously for t years is

$$P = \frac{A}{e^{rt}}.$$

Annuities	The **future value S of an ordinary annuity** of n payments of R dollars each at the end of consecutive interest periods with interest compounded at rate i per period is

$$S = R\left[\frac{(1 + i)^n - 1}{i}\right] \quad \text{or} \quad S = R \cdot s_{\overline{n}|i}.$$

The **present value P of an ordinary annuity** of n payments of R dollars each at the end of consecutive interest periods with interest compounded at rate i per period is

$$P = R\left[\frac{1 - (1 + i)^{-n}}{i}\right] \quad \text{or} \quad P = R \cdot a_{\overline{n}|i}.$$

The **future value S of an annuity due** of n payments of R dollars each at the beginning of consecutive interest periods with interest compounded at rate i per period is

$$S = R\left[\frac{(1 + i)^{n+1} - 1}{i}\right] - R \quad \text{or} \quad S = R \cdot s_{\overline{n+1}|i} - R.$$

CHAPTER 5   REVIEW EXERCISES

These exercises will require a calculator with a y^x key.

Find the simple interest.

1. $15,903 at 8% for 8 months

2. $4902 at 9.5% for 11 months

3. $42,368 at 15.22% for 5 months

4. $3478 at 7.4% for 88 days (assume a 360-day year)

5. $2390 at 18.7% from May 3 to July 28 (assume 365 days in a year)

6. $69,056.12 at 15.5% from September 13 to March 25 of the following year (assume a 365-day year)

Find the present value of the future amounts. Assume 360 days in a year; use simple interest.

7. $25,000 in 10 months, money earns 7%

8. $459.57 in 7 months, money earns 8.5%

9. $80,612 in 128 days, money earns 6.77%

Find the proceeds. Assume 360 days in a year.

10. $56,882, discount rate 9%, length of loan 5 months

11. $802.34, discount rate 18.6%, length of loan 11 months

12. $12,000, discount rate 12.09%, length of loan 145 days

Work the following exercises.

13. Tom Wilson owes $5800 to his mother. He has agreed to pay the money in 10 months at an interest rate of 10%. Three months before the loan is due, his mother discounts the loan at the bank. The bank charges a 14.45% discount rate. How much money does his mother receive?

14. Larry DiCenso needs $9812 to buy new equipment for his business. The bank charges a discount of 14%. Find the amount of DiCenso's loan if he borrows the money for 7 months.

Find the effective annual rate for the following bank discount rates. (Assume a 1-year loan of $1000).

15. 14%   **16.** 17.5%

Find the following compound amounts.

17. $1000 at 8% compounded annually for 9 years

18. $2800 at 6% compounded annually for 10 years

19. $19,456.11 at 12% compounded semiannually for 7 years

20. $312.45 at 6% compounded semiannually for 16 years

21. $1900 at 12% compounded quarterly for 9 years

22. $57,809.34 at 12% compounded quarterly for 5 years

23. $2500 at 12% compounded monthly for 3 years

24. $11,702.55 at 10% compounded monthly for 4 years

Find the amount of interest earned by each of the following deposits.

25. $3954 at 8% compounded annually for 12 years

26. $12,699.36 at 6% compounded semiannually for 7 years

27. $7801.72 at 12% compounded quarterly for 5 years

28. $48,121.91 at 9% compounded monthly for 2 years

29. $12,903.45 at 12.37% compounded quarterly for 29 quarters

30. $34,677.23 at 9.72% compounded monthly for 32 months

Find the present value of the following amounts if money can be invested at the given rate.

31. $5000 in 9 years, 8% compounded annually

32. $12,250 in 5 years, 12% compounded semiannually

33. $42,000 in 7 years, 18% compounded monthly

34. $17,650 in 4 years, 16% compounded quarterly

35. $1347.89 in 3.5 years, 13.77% compounded semiannually

36. $2388.90 in 44 months, 12.93% compounded monthly

Work the following exercises.

37. In 4 years Mr. Heeren must pay a pledge of $5000 to his church's building fund. What lump sum can he invest today, at 12% compounded semiannually, so that he will have enough to pay his pledge?

38. Joann Pujals must make an alimony payment of $1500 in 15 months. What lump sum can she invest today, at 13% compounded monthly, so that she will have enough to make the payment?

Write out the first five terms of the following geometric sequences.

39. $a = 2$, $r = 3$ **40.** $a = 4$, $r = \frac{1}{2}$

Find the sixth term of the following geometric sequences.

41. $a = -3$, $r = 2$ **42.** $a = -2$, $r = -2$

Find the sum of the first four terms of the geometric sequences in Exercises 43 and 44.

43. $a = -3$, $r = 3$ **44.** $a = 8000$, $r = -\frac{1}{2}$ **45.** Find $s_{\overline{30}|.01}$.   **46.** Find $s_{\overline{20}|.05}$.

In Exercises 47–54, find the future value of each annuity.

47. $500 is deposited at the end of each 6-month period for 8 years; money earns 6% compounded semiannually

48. $1288 is deposited at the end of each year for 14 years; money earns 8% compounded annually

49. $4000 is deposited at the end of each quarter for 7 years; money earns 8% compounded quarterly

50. $233 is deposited at the end of each month for 4 years; money earns 12% compounded monthly

51. $672 is deposited at the beginning of each quarter for 7 years; money earns 12% compounded quarterly

52. $11,900 is deposited at the beginning of each month for 13 months; money earns 12% compounded monthly

53. An Hoang deposits $491 at the end of each quarter for 9 years. The account pays 9.4% compounded quarterly.

54. J. Euclid deposits $1526.38 at the beginning of each 6-month period in an account paying 8.6% compounded semiannually. How much will be in the account after 5 years?

Find the amount of each payment to be made into a sinking fund to accumulate the indicated amounts.

55. $6500, money earns 8% compounded annually, 6 annual payments

56. $57,000, money earns 6% compounded semiannually, for $8\frac{1}{2}$ years

57. $233,188, money earns 9.7% compounded quarterly, for $7\frac{3}{4}$ years

58. $1,056,788, money earns 8.12% compounded monthly, for $4\frac{1}{2}$ years

Find the present value of each of the following ordinary annuities.

59. Payments of $850 are made annually for 4 years at 8% compounded annually.

60. Payments of $1500 are made quarterly for 7 years at 6% compounded quarterly

61. Payments of $4210 are made semiannually for 8 years at 8.6% compounded semiannually

62. Payments of $877.34 are made monthly for 17 months at 7.4% compounded monthly.

Work the following exercises.

63. Vicki Manchester borrows $20,000 from the bank to help her expand her business. She agrees to repay the money in equal payments at the end of each year for 9 years. Interest is at 8.9% per year on the unpaid balance. Find the amount of each payment.

64. Ken Murrill wants to expand his pharmacy. He does this by taking out a loan of $49,275 from the bank at an interest rate of 12.2% on the unpaid balance. He repays the loan in 48 monthly payments. Find the amount of each payment needed to amortize the loan.

Find the amount of the payment needed to amortize the following loans. Assume that interest is calculated on the unpaid balance.

65. $80,000 loan, 8% interest, 9 annual payments

66. $3200 loan, 13% interest, 10 quarterly payments

67. $32,000 loan, 9.4% interest, 17 quarterly payments

68. $51,607 loan, 13.6% interest, 32 monthly payments

Find the monthly house payments for the following mortgages. Interest is calculated on the unpaid balance.

69. $56,890 at 14.74% for 25 years

70. $77,110 at 10.45% for 30 years

For many years, the most popular mortgages were those for 25 or 30 years. Recently, however, accelerated mortgages, with 15-year payoffs, have become popular. With these mortgages, a slightly higher monthly payment produces a paid-off loan after 15 years. In Exercises 71 and 72, assume the last payment is the same as the previous ones.

71. Assume an $80,000 mortgage at 12% compounded monthly. Find the monthly payment and the total amount of interest paid over the following number of years.
(a) 30   **(b)** 25   **(c)** 15

72. Assume a $125,000 mortgage at 10% compounded monthly. Find the monthly payment and the total amount of interest paid over the following number of years.
(a) 30   **(b)** 25   **(c)** 15

Prepare amortization schedules for the following loans. Interest is calculated on the unpaid balance.

73. $5000 at 10% semiannually for 3 years, semiannual payments at the end of each period

74. $12,500 at 12% quarterly for 2 years, quarterly payments at the end of each period

The Southern Pacific Railroad, with lines running from Oregon to Louisiana, is one of the country's most profitable railroads. The railroad has vast landholdings (granted by the government in the last half of the nineteenth century) and is diversified into trucking, pipelines, and data transmission.

The railroad was recently faced with a decision on the fate of an old bridge which crosses the Kings River in central California. The bridge is on a minor line which carries very little traffic. Just north of the Southern Pacific bridge is another bridge, owned by the Santa Fe Railroad; it too carries little traffic. The Southern Pacific had two alternatives: it could replace its own bridge or it could negotiate with the Sante Fe for the rights to run trains over its bridge. In the second alternative a yearly fee would be paid to the Santa Fe, and new connecting tracks would be built. The situation is shown in the figure.

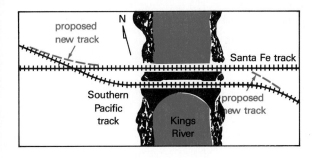

To find the better of these two alternatives, the railroad used the following approach.

1. Calculate estimated expenses for each alternative.
2. Calculate annual cash flows in after-tax dollars. At a 48% corporate tax rate, $1 of expenses actually costs the railroad $.52, and $1 of revenue can bring a maximum of $.52 in profit. Cash flow for a given year is found by the following formula.

$$\text{Cash flow} = -.52 \text{ (operating and maintenance expenses)}$$
$$+.52 \text{ (savings and revenue)}$$
$$+.48 \text{ (depreciation)}$$

_____
*Based on information supplied by The Southern Pacific Transportation Company, San Francisco.

3. Calculate the net present values of all cash flows for future years. The formula used is

$$\text{Net present value} = \Sigma(\text{cash flow in year } i)(1 + k)^{1-i}$$

where i is a particular year in the life of the project and k is the assumed annual rate of interest. (Recall: Σ indicates a sum.) The net present value is found for interest rates from 0% to 20%.
4. The interest rate that leads to a net present value of $0 is called the _rate of return_ on the project.

Let us now see how these steps worked out in practice.

Alternative 1: Operate over the Sante Fe bridge First, estimated expenses were calculated.

1976	Work done by Southern Pacific on Santa Fe track	$27,000
1976	Work by Southern Pacific on its own track	11,600
1976	Undepreciated portion of cost of current bridge	97,410
1976	Salvage value of bridge	12,690
1977	Annual maintenance of Santa Fe track	16,717
1977	Annual rental fee to Santa Fe	7,382

From these figures and others not given here, annual cash flows and net present values were calculated. The following table was then prepared.

Interest Rate, %	Net Present Value
0	$85,731
4	67,566
8	53,332
12	42,008
16	32,875
20	25,414

Although the table does not show a net present value of $0, the interest rate that leads to that value is 44%. This is the rate of return for this alternative.

Alternative 2: Build a new bridge Again, estimated expenses were calculated.

1976	Annual maintenance	$ 2,870
1976	Annual inspection	120
1976	Repair trestle	17,920
1977	Install bridge	189,943
1977	Install walks and handrails	15,060
1978	Repaint steel (every 10 years)	10,000
1978	Repair trusses (every 10 years, increases by $200)	2,000
1981	Replace ties	31,000
1988	Repair concrete (every 10 years)	400
2021	Replace ties	31,000

After cash flows and net present values were calculated, the following table was prepared.

Interest Rate, %	Net Present Value
0	$399,577
4	96,784
7.8	0
8	−3,151
12	−43,688
16	−62,615
20	−72,126

In this alternative the net present value is $0 at 7.8%, the rate of return.

Based on this analysis, the first alternative, renting from the Santa Fe, is clearly preferable.

EXERCISES

Find the cash flow in each of the following years. Use the formula given above.

1. Alternative 1, year 1977, operating expenses $6228, maintenance expenses $2976, savings $26,251, depreciation $10,778

2. Alternative 1, year 1984, same as Exercise 1, but depreciation is only $1347

3. Alternative 2, year 1976, maintenance $2870, operating expenses $6386, savings $26,251, no depreciation

4. Alternative 2, year 1980, operating expenses $6228, maintenance expenses $2976, savings $10,618, depreciation $6736

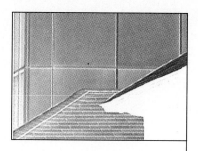

CHAPTER 6

Systems of Linear Equations and Matrices

Many applications of mathematics require finding the solutions of two or more equations. The solutions must satisfy *all* of the equations. A set of equations related in this way is called a **system of equations.** In this chapter we will discuss systems of equations, introduce the idea of a *matrix,* and then show how matrices are used to solve systems of equations.

6.1 SYSTEMS OF LINEAR EQUATIONS; THE ECHELON METHOD

An animal feed is made from three ingredients: corn, soybeans, and cottonseed. One hundred units of each ingredient provides the number of units of protein, fat, and fiber shown in the table. For example, the entries in the first column—25, 40, and 30—mean that 100 units of corn provide 25 units of protein, 40 units of fat, and 30 units of fiber.

	Corn	*Soybeans*	*Cottonseed*
Protein	25	40	20
Fat	40	20	30
Fiber	30	20	10

Now suppose we need to know the number of units of each ingredient that should be used to make a feed that contains 2200 units of protein, 2800 units of fat, and 1800 units of fiber. To find out, we let x represent the required number of units

1 (a) Check that $x = 40$, $y = 15$, and $z = 30$ is a solution of the first equation.

$$25x + 40y + 20z = 2200$$
$$25(\_) + 40(\_) + 20(\_) = 2200$$
$$\_ + \_ + \_ = 2200$$
$$\_ = 2200$$

Do these three numbers satisfy the equation?

(b) Do these numbers satisfy all of the equations?

Answer:

(a) 40; 15; 30; 1000; 600; 600; 2200; yes

(b) Yes

2 Decide whether $(-1, 4, 3)$ is a solution for each of the following equations.

(a) $x + y + z = 6$

(b) $3x - y + 4z = 5$

(c) $4x + 3y + 9z = 36$

Answer:

(a) Yes

(b) Yes

(c) No

of corn, y the number of units of soybeans, and z the number of units of cottonseed. Since the total amount of protein is to be 2200 units,

$$25x + 40y + 20z = 2200.$$

The feed must supply 2800 units of fat, so

$$40x + 20y + 30z = 2800,$$

and 1800 units of fiber, so

$$30x + 20y + 10z = 1800.$$

To solve this problem, we must find values of x, y, and z that satisfy this system of three equations. Verify that $x = 40$, $y = 15$, and $z = 30$ is a solution of the system, since these numbers satisfy all three equations. In fact, this is the only solution of this system. Many practical problems lead to such systems of *first-degree equations.* **1**

A **linear equation in n unknowns** is an equation that can be written in the form

$$a_1x_1 + a_2x_2 + \cdots + a_nx_n = k,$$

where $a_1, a_2, \ldots,$ and k are all real numbers, with at least one of the a_is not zero. For example,

$2x + 3y = 6$	is a linear equation in two unknowns,
$4x - 3y + 2z = 8$	is a linear equation in three unknowns,
$6x - y + 7z + 2w = 1$	is a linear equation in four unknowns,

and so on.

A solution of the linear equation

$$a_1x_1 + a_2x_2 + \cdots + a_nx_n = k$$

is a sequence of numbers $s_1, s_2, \ldots, s_n,$ such that

$$a_1s_1 + a_2s_2 + \cdots + a_ns_n = k.$$

A solution of an equation is usually written between parentheses as $(s_1, s_2, \ldots, s_n)$. For example, $(1, 6, 2)$ is a solution of the equation $3x_1 + 2x_2 - 4x_3 = 7$, since $3(1) + 2(6) - 4(2) = 7$. This is an extension of the idea of an ordered pair, which was introduced in Chapter 2. A solution of a first-degree equation in two unknowns is an ordered pair, and the graph of the equation is a straight line. For this reason, all first-degree equations are also called linear equations. In this section we develop a method for solving a system of first-degree equations. Although the discussion will be confined to equations with only a few variables, the method of solution can be extended to systems with many variables. **2**

Because the graph of a linear equation in two variables is a straight line, there are three possibilities for the solution of a system of two linear equations in two variables.

1. The two graphs are lines intersecting at a single point. The coordinates of this point give the only solution of the system. (See Figure 6.1(a).)
2. The graphs are distinct parallel lines. When this is the case, the system is **inconsistent;** that is, there is no solution common to both equations. (See Figure 6.1(b).)
3. The graphs are the same line. Here the equations are said to be **dependent,** since any solution of one equation is also a solution of the other. There are an infinite number of solutions. (See Figure 6.1(c).)

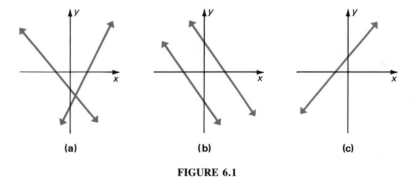

(a)   (b)   (c)

FIGURE 6.1

A linear equation in two variables is usually solved by the **elimination** or **addition method,** as illustrated in Example 1.

▶**EXAMPLE 1** Solve the system

$$3x - 4y = 1 \tag{1}$$
$$2x + 3y = 12. \tag{2}$$

To eliminate one variable by addition of the two equations, the coefficients of either x or y in the two equations must be additive inverses. For example, let us choose to eliminate x. We multiply both sides of equation (1) by 2 and both sides of equation (2) by -3 to get

$$6x - 8y = 2 \tag{3}$$
$$-6x - 9y = -36. \tag{4}$$

3 Solve the system of equations

$$3x + 2y = -1$$
$$5x - 3y = 11.$$

Draw the graph of each equation on the same axes.

Answer:
$(1, -2)$

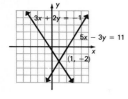

Adding these equations gives a new equation with just one variable, y. This equation can be solved for y.

$$6x - 8y = 2$$
$$\underline{-6x - 9y = -36}$$
$$-17y = -34$$
$$y = 2$$

To find the corresponding value of x, substitute 2 for y in either of equations (1) or (2). We choose equation (1).

$$3x - 4(2) = 1$$
$$3x - 8 = 1$$
$$x = 3$$

The solution of the given system is $(3, 2)$. The graphs of both equations of the system are shown in Figure 6.2. They intersect at $(3, 2)$, the solution of the system. ◀ **3**

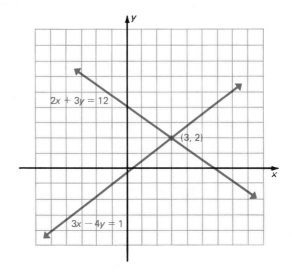

FIGURE 6.2

▶**EXAMPLE 2** Solve the system

$$3x - 2y = 4$$
$$-6x + 4y = 7.$$

Eliminate x by multiplying both sides of the first equation by 2 and adding the results to the second equation.

4 Solve the system

$$x - y = 4$$
$$2x - 2y = 3.$$

Draw the graph of each equation on the same axes.

Answer:

No solution, inconsistent system

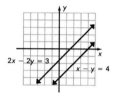

$$6x - 4y = 8$$
$$\underline{-6x + 4y = 7}$$
$$0 = 15$$

In the sum, both variables are eliminated. The result is a false statement, a signal that these two equations have no common solution. The system is inconsistent and has no solution. As Figure 6.3 shows, the graphs of the equations of the system produce two distinct parallel lines. ◀ **4**

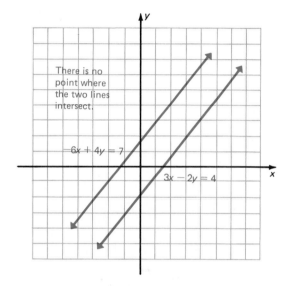

FIGURE 6.3

▶ **EXAMPLE 3** Solve the system

$$-4x + y = 2$$
$$8x - 2y = -4.$$

Eliminate x by multiplying both sides of the first equation by 2 and adding the results to the second equation.

$$-8x + 2y = 4$$
$$\underline{8x - 2y = -4}$$
$$0 = 0$$

Again, both variables are eliminated from the second equation. The result is a statement that is true for any values of the variable, indicating that the two equations have the same graph. When this happens, there is an infinite number of solutions for

5 Solve each of the following systems.

(a) $3x - 4y = 13$
       $12x - 16y = 52$

(b) $2x + 7y = 8$
       $-4x - 14y = 12$

Answer:

(a) All ordered pairs that satisfy the equation $3x - 4y = 13$ (or $12x - 16y = 52$), dependent system

(b) No solution, inconsistent system

the system. All the ordered pairs that satisfy the equation $-4x + y = 2$ (or $8x - 2y = -4$) are solutions. See Figure 6.4 ◀ **5**

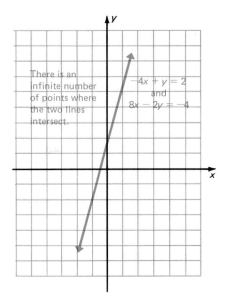

There is an infinite number of points where the two lines intersect.

$-4x + y = 2$ and $8x - 2y = -4$

FIGURE 6.4

In larger systems, with more equations and more variables, there also may be exactly one solution, no solutions, or infinitely many solutions. If no solution satisfies every equation in the system, the system is inconsistent, and if there are infinitely many solutions that satisfy all the equations in the system, the equations are dependent.

THE ECHELON METHOD Although the elimination method can be used to solve systems with more than two variables, it becomes a complicated process, even for systems of three variables. To solve a large system, we use properties of algebra to change, or transform, the system into a simpler *equivalent* system. An **equivalent system** is one that has the same solutions as the given system.

Applying any of the following **transformations** to a system of equations produces an equivalent system.

1. Interchange any two equations.
2. Multiply both sides of an equation by a nonzero real number.
3. Replace any equation by the sum of itself and a multiple of any other equation in the system.

6 Verify that $x = 2$, $y = 1$ is the solution of the system

$$x - 3y = -1$$
$$3x + 2y = 8.$$

(a) Replace the second equation by the sum of itself and -3 times the first equation.

(b) What is the solution of the system in part (a)?

Answer:

(a) The system becomes

$$x - 3y = -1$$
$$11y = 11$$

(b) $x = 2$, $y = 1$

A formal proof will not be given here, but the example in Problem 6 at the side will illustrate the fact that transformations of these types do produce equivalent systems. **6**

The **echelon method** is a systematic way of using the three transformations to transform a system into an equivalent system having the form

$$x + by + cz = d$$
$$y + ez = f$$
$$z = g,$$

where b, c, d, e, f, and g are real numbers. Since the last equation gives the value of z, the complete solution can be found by substitution. The advantage of the echelon method is that it can be carried out with matrix notation (discussed later), making it possible to solve systems of equations by computer.

▶ **EXAMPLE 4** Use the echelon method to solve the system

$$2x + y - z = 2$$
$$x + 3y + 2z = 1$$
$$x + y + z = 2.$$

Use transformations to get an equivalent system that has the form shown above. In this new system, the coefficient of x in the first equation should be 1. Use the first transformation to interchange the first two equations. (As an alternative approach, a coefficient of 1 could be found by multiplying both sides of the first equation by 1/2.) The new system follows. (A shorthand notation in color shows the steps involved. The letter R represents the rows of the system.)

$$x + 3y + 2z = 1 \qquad \text{Interchange } R_1, R_2$$
$$2x + y - z = 2$$
$$x + y + z = 2$$

Replace the second equation above with an equation without x by multiplying both sides of the first equation by -2 and adding the results to the second equation. This process gives the new system

$$x + 3y + 2z = 1$$
$$-5y - 5z = 0 \qquad -2R_1 + R_2$$
$$x + y + z = 2.$$

7 Use the echelon method to solve each system.

(a) $2x + y = -1$
$x + 3y = 2$

(b) $2x - y + 3z = 2$
$x + 2y - z = 6$
$-x - y + z = -5$

Answer:

(a) $(-1, 1)$

(b) $(3, 1, -1)$

Make the coefficient of y equal to 1 in the second equation by multiplying both sides by $-1/5$.

$$x + 3y + 2z = 1$$
$$y + z = 0 \qquad -\frac{1}{5}R_2$$
$$x + y + z = 2$$

Replace the third equation with an equation not containing x by multiplying both sides of the first equation by -1 and adding the results to the third equation to get the new system

$$x + 3y + 2z = 1$$
$$y + z = 0$$
$$-2y - z = 1. \qquad -1R_1 + R_3$$

Replace the third equation with an equation not containing y by multiplying both sides of the second equation by 2 and adding the results to the third equation.

$$x + 3y + 2z = 1$$
$$y + z = 0$$
$$z = 1. \qquad 2R_2 + R_3$$

The third equation of this system gives the value of $z : z = 1$. Substitute 1 for z in the second equation to get $y = -1$. Finally, substitute 1 for z and -1 for y in the first equation to get $x = 2$. The solution for the given system is $(2, -1, 1)$. ◀ **7**

In summary, to solve a linear system in n variables by the echelon method, perform the following steps *using the three transformations* given earlier.

The Echelon Method of Solving a Linear System

1. If possible, arrange the equations so that there is an x_1 term in the first equation, an x_2 term in the second equation, and so on.
2. Make the coefficient of the first variable equal to 1 in the first equation and 0 in the other equations.
3. Make the coefficient of the second variable equal to 1 in the second equation and 0 in all remaining equations.
4. Make the coefficient of the third variable equal to 1 in the third equation and 0 in all remaining equations.
5. Continue in this way until the last equation is of the form $x_n = k$, where k is a constant.

PARAMETERS The systems of equations discussed so far have had the same number of equations as variables. Such systems have either one solution, no solution, or infinitely many solutions. Some mathematical models lead to systems of equations with fewer equations than variables, however. Such systems have either infinitely many solutions or no solution.

▶**EXAMPLE 5**   Solve the system

$$2x - 3y + 4z = 6 \qquad (1)$$
$$x - 2y + z = 9 \qquad (2)$$
$$3x - 6y + 3z = 27. \qquad (3)$$

Since the third equation here is a multiple of the second equation, this system really has only the two equations

$$2x - 3y + 4z = 6 \qquad (1)$$
$$x - 2y + z = 9. \qquad (2)$$

Use the steps of the echelon method as far as possible. Exchange the two equations so that the coefficient of x is 1 in the first equation of the system. This gives

$$x - 2y + z = 9 \qquad \text{Interchange } R_1 \text{ and } R_2 \qquad (2)$$
$$2x - 3y + 4z = 6. \qquad (1)$$

To get a new equation without x, multiply both sides of equation (2) by -2 and add to equation (1). The new system is

$$x - 2y + z = 9 \qquad (2)$$
$$y + 2z = -12 \qquad -2R_1 + R_2 \qquad (4)$$

Since there are only two equations, it is not possible to continue with the echelon method. To complete the solution, solve equation (4) for y.

$$y = -2z - 12$$

Now substitute the result for y in equation (2), and solve for x.

$$x - 2y + z = 9 \qquad (2)$$
$$x - 2(-2z - 12) + z = 9$$
$$x + 4z + 24 + z = 9$$
$$x + 5z = -15$$
$$x = -5z - 15$$

Each choice of a value for z leads to values for x and y. For example,

if $z = 1$,
then $x = -5(1) - 15 = -20$ and $y = -2(1) - 12 = -14$;

if $z = -6$,
then $x = 15$ and $y = 0$;

if $z = 0$,
then $x = -15$ and $y = -12$.

8 Use the following values of z to find additional solutions for the system of Example 5.

(a) $z = 7$

(b) $z = -14$

(c) $z = 5$

Answer:

(a) $(-50, -26, 7)$

(b) $(55, 16, -14)$

(c) $(-40, -22, 5)$

There are infinitely many solutions for the original system, since z can take on infinitely many values. The solutions are all triples

$$(-5z - 15, -2z - 12, z),$$

where z is any real number. ◀ **8**

Since both x and y in Example 5 were expressed in terms of z, the variable z is called a **parameter.** If we solved the system in a different way, x or y could be the parameter. The system in Example 5 had one more variable than equations. If there are two more variables than equations, there usually will be two parameters, and so on.

Now let us investigate the situation in which there are fewer variables than equations. For example, consider the system

$$2x + 3y = 8 \qquad (1)$$
$$x - y = 4 \qquad (2)$$
$$5x + y = 7, \qquad (3)$$

which has three equations in two variables. Since each of these equations has a line as its graph, there are various possibilities: three lines that intersect at a common point, three lines that cross at three different points, three lines of which two are the same line so that the intersection would be a point, three lines of which two are parallel, or three lines that are all parallel so that there would be no intersection, and so on. As in the case of n equations with n variables, the possibilities result in a unique solution, no solution, or infinitely many solutions.

To solve the system using the echelon method, begin by exchanging equations (1) and (2) to get a 1 for the coefficient of x in the first equation.

$$x - y = 4 \qquad \textbf{Interchange } R_1 \textbf{ and } R_2 \qquad (2)$$
$$2x + 3y = 8 \qquad (1)$$
$$5x + y = 7 \qquad (3)$$

Now multiply both sides of equation (2) by -2 and add the result to equation (1) to get $5y = 0$. The new system is

$$x - y = 4 \qquad (2)$$
$$5y = 0 \qquad -2R_1 + R_2 \qquad (4)$$
$$5x + y = 7. \qquad (3)$$

Multiply both sides of equation (4) by 1/5 to get $y = 0$. Substituting $y = 0$ leads to $x = 4$ in equation (2) and $x = 7/5$ in equation (3), a contradiction. This contradiction shows that the system has no solution. See Figure 6.5, which shows the graphs of the three equations of the system.

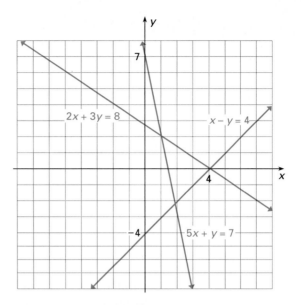

FIGURE 6.5

APPLICATIONS The mathematical techniques in this text will be useful to you only if you are able to apply them to practical problems. To do this, always begin by reading the problem carefully. Next, identify what must be found. Let each unknown quantity be represented by a variable. (It is a good idea to *write down* exactly what each variable represents.) Now reread the problem, looking for all necessary data. Write that down, too. Finally, look for one or more sentences that lead to equations or inequalities. The next example illustrates these steps.

▶**EXAMPLE 6**   A bank teller has a total of 70 bills in five-, ten-, and twenty-dollar denominations. The number of fives is three times the number of tens, while the total value of the money is $960. Find the number of each type of bill.

A chart can be a useful way to organize the information given in a problem. Start by indicating the three types of bills in the first column. Choose variables to represent the number of each type of bill in the second column. Then write expressions for the value per bill of each type and the total value for each type.

	Number of Bills	Value of Each Type	Total Value of Each Type
Fives	x	5	$5x$
Tens	y	10	$10y$
Twenties	z	20	$20z$
Totals	70		960

9 Write a system of equations to solve Example 6 if the number of tens is 1/2 the number of fives, and there are a total of 85 bills worth $900.

Answer:

$$x + y + z = 85$$
$$x - 2y \quad\quad = 0$$
$$5x + 10y + 20z = 900$$

10 Find the equation of the parabola $y = ax^2 + bx + c$ that goes through the points (0, 4), (2, 0), and (1, 1).

Answer:

$$y = x^2 - 4x + 4$$

From the first column of the chart, since the total number of bills is 70,

$$x + y + z = 70.$$

From the third column,

$$5x + 10y + 20z = 960.$$

To get one more equation, go back to the wording of the problem. We have not used the sentence that says that the number of fives, x, is three times the number of tens, y. This gives

$$x = 3y \quad \text{or} \quad x - 3y = 0.$$

Now the system can be written as

$$x + \quad y + \quad z = \quad 70$$
$$x - \quad 3y \quad\quad = \quad 0$$
$$5x + 10y + 20z = 960.$$

Solving by the echelon method gives the solution $x = 24$, $y = 8$, and $z = 38$. Thus, the teller had 24 fives, 8 tens, and 38 twenties. Check this answer in the words of the original problem. ◀ **9**

▶ **EXAMPLE 7** Find the equation of the parabola $y = ax^2 + bx + c$ that goes through the points (2, 15), (−3, 45), and (0, 3).

Since (2, 15) is on the graph, $x = 2$, $y = 15$ satisfy the equation $y = ax^2 + bx + c$, so that

$$15 = a(2^2) + b(2) + c.$$

Similarly, since (−3, 45) and (0, 3) are on the graph,

$$45 = a(-3)^2 + b(-3) + c \quad \text{and} \quad 3 = a(0^2) + b(0) + c.$$

Rewriting these equations produces the system

$$4a + 2b + c = 15$$
$$9a - 3b + c = 45$$
$$c = \quad 3$$

This system can be solved by the methods of this section; its solutions are $a = 4$, $b = -2$, and $c = 3$. Hence, the equation of the parabola through the given points is $y = 4x^2 - 2x + 3$. ◀ **10**

6.1 EXERCISES

Use the elimination method to solve each of the following systems of two equations in two un-knowns. (See Examples 1–3.)

1. $x + y = 9$
$2x - y = 0$

2. $4x + y = 9$
$3x - y = 5$

3. $5x + 3y = 7$
$7x - 3y = -19$

4. $2x + 7y = -8$
$-2x + 3y = -12$

5. $3x + 2y = -6$
$5x - 2y = -10$

6. $-6x + 2y = 8$
$5x - 2y = -8$

7. $2x - 3y = -7$
$5x + 4y = 17$

8. $4m + 3n = -1$
$2m + 5n = 3$

9. $5p + 7q = 6$
$10p - 3q = 46$

10. $12s - 5t = 9$
$3s - 8t = -18$

11. $6x + 7y = -2$
$7x - 6y = 26$

12. $2a + 9b = 3$
$5a + 7b = -8$

13. $3x + 2y = 5$
$6x + 4y = 8$

14. $9x - 5y = 1$
$-18x + 10y = 1$

15. $4x - y = 9$
$-8x + 2y = -18$

16. $3x + 5y + 2 = 0$
$9x + 15y + 6 = 0$

In Exercises 17–20, first multiply both sides of each equation by its common denominator to eliminate the fractions. Then use the elimination method to solve.

17. $\dfrac{x}{2} + \dfrac{y}{3} = 8$

$\dfrac{2x}{3} + \dfrac{3y}{2} = 17$

18. $\dfrac{x}{5} + 3y = 31$

$2x - \dfrac{y}{5} = 8$

19. $\dfrac{x}{2} + y = \dfrac{3}{2}$

$\dfrac{x}{3} + y = \dfrac{1}{3}$

20. $x + \dfrac{y}{3} = -6$

$\dfrac{x}{5} + \dfrac{y}{4} = -\dfrac{7}{4}$

Use the echelon method to solve each of the following systems of three equations in three un-knowns. Check your answers. (See Example 4.)

21. $x + y + z = 2$
$2x + y - z = 5$
$x - y + z = -2$

22. $2x + y + z = 9$
$-x - y + z = 1$
$3x - y + z = 9$

23. $x + 3y + 4z = 14$
$2x - 3y + 2z = 10$
$3x - y + z = 9$

24. $4x - y + 3z = -2$
$3x + 5y - z = 15$
$-2x + y + 4z = 14$

25. $x + 2y + 3z = 8$
$3x - y + 2z = 5$
$-2x - 4y - 6z = 5$

26. $3x - 2y - 8z = 1$
$9x - 6y - 24z = -2$
$x - y + z = 1$

27. $2x - 4y + z = -4$
$x + 2y - z = 0$
$-x + y + z = 6$

28. $4x - 3y + z = 9$
$3x + 2y - 2z = 4$
$x - y + 3z = 5$

29. $x + 4y - z = 6$
$2x - y + z = 3$
$3x + 2y + 3z = 16$

30. $3x + y - z = 7$
$2x - 3y + z = -7$
$x - 4y + 3z = -6$

31. $5m + n - 3p = -6$
$2m + 3n + p = 5$
$-3m - 2n + 4p = 3$

32. $2r - 5s + 4t = -35$
$5r + 3s - t = 1$
$r + s + t = 1$

33. $a - 3b - 2c = -3$
$3a + 2b - c = 12$
$-a - b + 4c = 3$

34. $2x + 2y + 2z = 6$
$3x - 3y - 4z = -1$
$x + y + 3z = 11$

Solve each of the following systems of equations. Let z be the arbitrary variable. (See Example 5.)

35. $5x + 3y + 4z = 19$
$3x - y + z = -4$

36. $3x + y - z = 0$
$2x - y + 3z = -7$

37. $x + 2y + 3z = 11$
$2x - y + z = 2$

38. $-x + y - z = -7$
$2x + 3y + z = 7$

39. $x + y - z = -20$
$2x - y + z = 11$

40. $4x + 3y + z = 1$
$-2x - y + 2z = 0$

For each of the following, write a system of equations and then solve. (See Example 7.)

41. Find a and b so that the line $ax + by = 5$ contains the points $(-2, 1)$ and $(-1, -2)$.

42. Find m and b so that the line $y = mx + b$ contains the points $(4, 6)$ and $(-5, -3)$.

43. Find a, b, and c so that the graph of $y = ax^2 + bx + c$ contains the points $(2, 3)$, $(-1, 0)$, and $(-2, 2)$.

44. Find a, b, and c so that the points $(2, 14)$, $(0, 0)$ and $(-1, -1)$ lie on the graph of $y = ax^2 + bx + c$.

Write a system of equations for each of the following problems; then solve the system. (See Example 6.)

45. **Management**   The price at which grapefruit is sold affects the consumer demand for grapefruit according to the equation

$$2p = -.2q + 5,$$

where p is the price per pound (in dollars) at which consumers will demand q thousand pounds of grapefruit. The amount q of grapefruit that producers will supply at price p is governed by the equation $5p = .3q + 5.3$. Find the equilibrium quantity and the equilibrium price. (That is, find the quantity and price at which supply equals demand, or the values of p and q that satisfy both equations.)

46. **Management**   The consumer demand for grapes is given by the equation $10p = -3q + 48.9$ and the producer's supply by the equation $5p = .3q + 3.75$, where p and q are as in Exercise 45. Find the equilibrium quantity and the equilibrium price.

47. An investor has stock in USAir and BPAmerica worth a total of $16,000. The USAir stock currently sells for $30 a share and the BPAmerica for $70 a share. If BPAmerica stock triples in value and USAir goes up 50%, the investor's stock will be worth $34,500. How many shares of each stock does she have?

48. If 20 pounds of rice and 10 pounds of potatoes cost $16.20 and 30 pounds of rice and 12 pounds of potatoes cost $23.04, how much will 10 pounds of rice and 50 pounds of potatoes cost?

49. An apparel shop sells skirts for $45 and blouses for $35. Its entire stock is worth $51,750. But sales are slow and only half the skirts and two thirds of the blouses are sold, for a total of $30,600. How many skirts and blouses are left in the store?

50. A working couple earned a total of $4352. The wife earned $64 per day; the husband earned $8 per day less. Find the number of days each worked if the total number of days worked by both was 72.

51. **Management**   A company produces two models of bicycles, model 201 and model 301. Model 201 requires 2 hours of assembly time and model 301 requires 3 hours of assembly time. The parts for model 201 cost $25 per bike and the parts for model 301 cost $30 per bike. If the company has a total of 34 hours of assembly time and $365 available per day for these two models, how many of each can be made in a day?

52. **Management**   Midtown Manufacturing Company makes two products, plastic plates and plastic cups. Both require time on two machines: each unit of plates—1 hour on machine A and 2 hours on machine B; each unit of cups—3 hours on machine A and 1 hour on machine B. Both machines operate 15 hours a day. How many units of each product can be produced in a day under these conditions?

53. Juanita invests $10,000, received from her grandmother, in three ways. With one part, she buys mutual funds which offer a return of 8% per year. The second part, which amounts to twice the first, is used to buy government bonds at 9% per year. She puts the rest in the bank at 5% annual interest. The first year her investments bring a return of $830. How much did she invest in each way?

54. **Management**   To get the necessary funds for a planned expansion, a small company took out three loans totaling $25,000. The company was able to borrow some of the money at 16%. They borrowed $2,000 more than one half the amount of the 16% loan at 20%, and the rest at 18%. The total annual interest was $4440. How much did they borrow at each rate?

55. **Natural Science**   A hospital dietician is planning a special diet for a certain patient. The total amount per meal of food groups A, B, and C must equal 400 grams. The diet should include one third as much of group A as of group B, and the sum of the amounts of group A and group C should equal twice the amount of group B. How many grams of each food group should be included?

56. **Natural Science**   Three brands of fertilizer are available that provide nitrogen, phosphoric acid, and soluble potash to the soil. One bag of each brand provides the following units of each nutrient.

		Nutrient	
Brand	Nitrogen	Phosphoric Acid	Potash
A	1	3	2
B	2	1	0
C	3	2	1

For ideal growth, the soil in a certain country needs 18 units of nitrogen, 23 units of phosphoric acid, and 13 units of potash per acre. How many bags of each brand of fertilizer should be used per acre for ideal growth?

57. **Management** At a pottery factory, fuel consumption for heating the kilns varies with the size of the order being fired. In the past, the company recorded these figures.

x = Number of Platters	y = Fuel Cost per Platter
6	$2.80
8	2.48
10	2.24

(a) Find an equation of the form $y = ax^2 + bx + c$ whose graph contains the three points corresponding to the data in the table. (See Example 7.)

(b) How many platters should be fired at one time in order to minimize the fuel cost per platter? What is the minimum fuel cost per platter?

58. **Management** The business analyst for Midtown Manufacturing wants to find an equation that can be used to project sales of a relatively new product. For the years 1988, 1989, and 1990, sales were $15,000, $32,000, and $123,000, respectively.

(a) Graph the sales for the years 1988, 1989, and 1990, letting the year 1988 equal 0 on the x-axis. Let the values on the vertical axis be in thousands. (For example, the point (1989, 32,000) will be graphed as (1, 32).)

(b) Find the equation of the straight line $ax + by = c$ through the points for 1988 and 1990.

(c) Find the equation of the parabola $y = ax^2 + bx + c$ through the three given points.

(d) Find the projected sales for 1993 first by using the equation from part (b) and then by using the equation from part (c). If you were to estimate sales of the product in 1993, which result would you choose? Why?

6.2 SOLUTION OF LINEAR SYSTEMS BY THE GAUSS-JORDAN METHOD

In the last section we used the echelon method to solve linear systems of equations. Since the variables are always the same, we really need to keep track of just the coefficients and the constants. For example, look at the system solved in Example 4 of the previous section:

$$2x + y - z = 2$$
$$x + 3y + 2z = 1$$
$$x + y + z = 2.$$

This system can be written in an abbreviated form as

$$\begin{bmatrix} 2 & 1 & -1 & 2 \\ 1 & 3 & 2 & 1 \\ 1 & 1 & 1 & 2 \end{bmatrix}.$$

Such a rectangular array of numbers enclosed by brackets is called a **matrix** (plural: **matrices**). Each number in the array is an **element** or **entry.** To separate the constants in the last column of the matrix from the coefficients of the variables, use a vertical line, producing the following **augmented matrix.**

$$\left[\begin{array}{ccc|c} 2 & 1 & -1 & 2 \\ 1 & 3 & 2 & 1 \\ 1 & 1 & 1 & 2 \end{array}\right]$$

The rows of the augmented matrix can be transformed in the same way as the equations of the system, since the matrix is just a shortened form of the system. The following **row operations** on the augmented matrix correspond to the transformations of systems of equations given earlier.

For any augmented matrix of a system of equations, the following **row operations** produce an augmented matrix of an equivalent system:

1. interchanging any two rows;
2. multiplying the elements of a row by any nonzero real number;
3. replacing a row by adding a multiple of the elements of another row to the corresponding elements of the row.

Row operations, like the transformations of systems of equations, are reversible. If they are used to change matrix A to matrix B, then it is possible to use row operations to transform B back into A. In addition to their use in solving equations, row operations are very important in the simplex method of Chapter 7. The row operations will be abbreviated with the same notation used earlier for equations.

By the first row operation, the matrix

$$\left[\begin{array}{ccc|c} 1 & 3 & 5 & 6 \\ 0 & 1 & 2 & 3 \\ 2 & 1 & -2 & -5 \end{array}\right] \text{ becomes } \left[\begin{array}{ccc|c} 0 & 1 & 2 & 3 \\ 1 & 3 & 5 & 6 \\ 2 & 1 & -2 & -5 \end{array}\right] \quad \text{Interchange } R_1, R_2$$

by exchanging the first two rows. Row three is left unchanged.

The second row operation can be used to change

$$\left[\begin{array}{ccc|c} 1 & 3 & 5 & 6 \\ 0 & 1 & 2 & 3 \\ 2 & 1 & -2 & -5 \end{array}\right] \text{ to } \left[\begin{array}{ccc|c} -2 & -6 & -10 & -12 \\ 0 & 1 & 2 & 3 \\ 2 & 1 & -2 & -5 \end{array}\right] \quad -2R_1$$

by multiplying the elements of the first row of the original matrix by -2, with rows two and three left unchanged.

1 Perform the following row operations on the matrix

$$\begin{bmatrix} -1 & 5 \\ 3 & -2 \end{bmatrix}.$$

(a) $2R_1$

(b) $-3R_1 + R_2$

(c) $2R_2 + R_1$

Answer:

(a) $\begin{bmatrix} -2 & 10 \\ 3 & -2 \end{bmatrix}$

(b) $\begin{bmatrix} -1 & 5 \\ 6 & -17 \end{bmatrix}$

(c) $\begin{bmatrix} 5 & 1 \\ 3 & -2 \end{bmatrix}$

Using the third row operation,

$$\begin{bmatrix} 1 & 3 & 5 & | & 6 \\ 0 & 1 & 2 & | & 3 \\ 2 & 1 & -2 & | & -5 \end{bmatrix} \text{ becomes } \begin{bmatrix} -1 & 2 & 7 & | & 11 \\ 0 & 1 & 2 & | & 3 \\ 2 & 1 & -2 & | & -5 \end{bmatrix} \quad -1R_3 + R_1$$

by working as follows. First multiply each element in the third row of the original matrix by -1 and then add the results to the corresponding elements in the first row of that matrix.

$$\begin{bmatrix} 1 + 2(-1) & 3 + 1(-1) & 5 + (-2)(-1) & | & 6 + (-5)(-1) \\ 0 & 1 & 2 & | & 3 \\ 2 & 1 & -2 & | & -5 \end{bmatrix}$$

$$= \begin{bmatrix} -1 & 2 & 7 & | & 11 \\ 0 & 1 & 2 & | & 3 \\ 2 & 1 & -2 & | & -5 \end{bmatrix}$$

Caution In the last example, note that again rows two and three are left unchanged, *even though the elements of row three were used to transform row one.* **1**

THE GAUSS-JORDAN METHOD The *Gauss-Jordan method* is an extension of the echelon method of solving systems. Before the Gauss-Jordan method can be used, the system must be in proper form: the terms with variables should be on the left and the constants on the right in each equation, with the variables in the same order in each equation. Although systems with two variables and two equations may be solved by other methods, we use one here as a simple example to illustrate the Gauss-Jordan method.

▶**EXAMPLE 1** Solve the system $3x - 4y = 1$
$$5x + 2y = 19.$$

The system is already in the proper form. The procedure for solving the system is parallel to the echelon method except for the last step. The goal of the Gauss-Jordan method is to end up with a matrix of the form

$$\begin{bmatrix} 1 & 0 & | & m \\ 0 & 1 & | & n \end{bmatrix},$$

which corresponds to the system of equations $x = m$
$$y = n,$$

giving the solution of the original system. We compare the echelon method and the Gauss-Jordan method in the following table.

2 Use the Gauss-Jordan method to solve the system

$$x + 2y = 11$$
$$-4x + y = -8.$$

Give the shorthand notation and the new matrix in (b)–(d).

(a) Set up the augmented matrix.

(b) Get 0 in row two, column one.

(c) Get 1 in row two, column two.

(d) Finally, get 0 in row one, column two.

(e) The solution for the system is _____ .

Answer:

(a) $\begin{bmatrix} 1 & 2 & | & 11 \\ -4 & 1 & | & -8 \end{bmatrix}$

(b) $4R_1 + R_2$; $\begin{bmatrix} 1 & 2 & | & 11 \\ 0 & 9 & | & 36 \end{bmatrix}$

(c) $\frac{1}{9}R_2$; $\begin{bmatrix} 1 & 2 & | & 11 \\ 0 & 1 & | & 4 \end{bmatrix}$

(d) $-2R_2 + R_1$; $\begin{bmatrix} 1 & 0 & | & 3 \\ 0 & 1 & | & 4 \end{bmatrix}$

(e) $(3, 4)$

Echelon Method	*Gauss-Jordan Method*

Echelon Method

$$3x - 4y = 1$$
$$5x + 2y = 19$$

Multiply both sides of the first equation by 1/3 so that x has a coefficient of 1.

$$x - \tfrac{4}{3}y = \tfrac{1}{3} \qquad \tfrac{1}{3}R_1$$
$$5x + 2y = 19$$

Eliminate x from the second equation by adding -5 times the first equation to the second equation.

$$x - \tfrac{4}{3}y = \tfrac{1}{3}$$
$$\tfrac{26}{3}y = \tfrac{52}{3} \qquad -5R_1 + R_2$$

Multiply both sides of the second equation by 3/26 to get $y = 2$.

$$x - \tfrac{4}{3}y = \tfrac{1}{3}$$
$$y = 2 \qquad \tfrac{3}{26}R_2$$

Substitute $y = 2$ into the first equation and solve for x to get $x = 3$.

$$x = 3$$
$$y = 2$$

Gauss-Jordan Method

Rewrite the system as an augmented matrix.

$$\begin{bmatrix} 3 & -4 & | & 1 \\ 5 & 2 & | & 19 \end{bmatrix}$$

Multiply each element of row one by 1/3.

$$\begin{bmatrix} 1 & -\tfrac{4}{3} & | & \tfrac{1}{3} \\ 5 & 2 & | & 19 \end{bmatrix} \qquad \tfrac{1}{3}R_1$$

Add -5 times the elements of row one to the elements of row two.

$$\begin{bmatrix} 1 & -\tfrac{4}{3} & | & \tfrac{1}{3} \\ 0 & \tfrac{26}{3} & | & \tfrac{52}{3} \end{bmatrix} \qquad -5R_1 + R_2$$

Multiply the elements of row two by 3/26.

$$\begin{bmatrix} 1 & -\tfrac{4}{3} & | & \tfrac{1}{3} \\ 0 & 1 & | & 2 \end{bmatrix} \qquad \tfrac{3}{26}R_2$$

Multiply the elements of row two by 4/3 and add to the elements of row one.

$$\begin{bmatrix} 1 & 0 & | & 3 \\ 0 & 1 & | & 2 \end{bmatrix} \qquad \tfrac{4}{3}R_2 + R_1$$

The solution of the system, $(3, 2)$, can be read directly from the last column of the final matrix.

In both cases, check this solution by substitution in both equations of the original system. ◀

When using the Gauss-Jordan method to solve a system of three equations with three variables, the final matrix will have the form

$$\begin{bmatrix} 1 & 0 & 0 & | & m \\ 0 & 1 & 0 & | & n \\ 0 & 0 & 1 & | & p \end{bmatrix},$$

where m, n, and p are real numbers. The solution $x = m$, $y = n$, and $z = p$ can be read from this final form of the matrix.

When using row operations to transform a matrix, it is usually best to work column by column from left to right. For each column, the first change should produce a 1 in the main diagonal. Next, perform the steps that give zeros in the remainder of the column. Then proceed to the next column. **2**

3 Continue the solution of the system in Example 3 as follows. Give the shorthand notation and the matrix for each step.

(a) Get 1 in row two, column two.

(b) Get 0 in row one, column two.

(c) Now get 0 in row three, column two.

Answer:

(a) $\frac{1}{6}R_2$;

$$\begin{bmatrix} 1 & -1 & 5 & | & -6 \\ 0 & 1 & -\frac{8}{3} & | & \frac{14}{3} \\ 0 & 4 & -3 & | & 11 \end{bmatrix}$$

(b) $1R_2 + R_1$;

$$\begin{bmatrix} 1 & 0 & \frac{7}{3} & | & -\frac{4}{3} \\ 0 & 1 & -\frac{8}{3} & | & \frac{14}{3} \\ 0 & 4 & -3 & | & 11 \end{bmatrix}$$

(c) $-4R_2 + R_3$;

$$\begin{bmatrix} 1 & 0 & \frac{7}{3} & | & -\frac{4}{3} \\ 0 & 1 & -\frac{8}{3} & | & \frac{14}{3} \\ 0 & 0 & \frac{23}{3} & | & -\frac{23}{3} \end{bmatrix}$$

(Solution continued in the text.)

▶**EXAMPLE 2**   Use the Gauss-Jordan method to solve the system

$$\begin{aligned} x \quad\quad + 5z &= -6 + y \\ 3x + 3y \quad\quad &= 10 + z \\ x + 3y + 2z &= \;\;5. \end{aligned}$$

The system must first be rewritten in proper form as follows.

$$\begin{aligned} x - \;\;y + 5z &= -6 \\ 3x + 3y - \;\;z &= \;\;10 \\ x + 3y + 2z &= \;\;\;5 \end{aligned}$$

Begin the solution by writing the augmented matrix of the linear system.

$$\begin{bmatrix} 1 & -1 & 5 & | & -6 \\ 3 & 3 & -1 & | & 10 \\ 1 & 3 & 2 & | & 5 \end{bmatrix}$$

The first element in column one is already 1. Get 0 for the second element in column one by multiplying each element in the first row by -3 and adding the results to the corresponding elements in row two.

$$\begin{bmatrix} 1 & -1 & 5 & | & -6 \\ 0 & 6 & -16 & | & 28 \\ 1 & 3 & 2 & | & 5 \end{bmatrix} \quad -3R_1 + R_2$$

Now, change the first element in row three to 0 by multiplying each element of the first row by -1 and adding the results to the corresponding elements of the third row (again using row operation (3)).

$$\begin{bmatrix} 1 & -1 & 5 & | & -6 \\ 0 & 6 & -16 & | & 28 \\ 0 & 4 & -3 & | & 11 \end{bmatrix} \quad -1R_1 + R_3$$

This transforms the first column. Transform the second column in a similar manner.   **3**

4 Use the Gauss-Jordan method to solve

$$x + y - z = 6$$
$$2x - y + z = 3$$
$$-x + y + z = -4.$$

Answer:
$(3, 1, -2)$

Complete the solution by transforming the third column.

$$\begin{bmatrix} 1 & 0 & \frac{7}{3} & | & -\frac{4}{3} \\ 0 & 1 & -\frac{8}{3} & | & \frac{14}{3} \\ 0 & 0 & 1 & | & -1 \end{bmatrix} \quad \frac{3}{23}R_3$$

$$\begin{bmatrix} 1 & 0 & 0 & | & 1 \\ 0 & 1 & -\frac{8}{3} & | & \frac{14}{3} \\ 0 & 0 & 1 & | & -1 \end{bmatrix} \quad -\frac{7}{3}R_3 + R_1$$

$$\begin{bmatrix} 1 & 0 & 0 & | & 1 \\ 0 & 1 & 0 & | & 2 \\ 0 & 0 & 1 & | & -1 \end{bmatrix} \quad \frac{8}{3}R_3 + R_2$$

The linear system associated with this last augmented matrix is

$$x \quad\quad = 1$$
$$y \quad = 2$$
$$z = -1,$$

and the solution is $(1, 2, -1)$. ◀ **4**

Note Notice that the first two row operations are used to get the ones and the third row operation is used to get the zeros.

Use the **Gauss-Jordan method** to solve a linear system by performing the following steps.

1. Write each equation with variable terms in the same order on the left and constants on the right.
2. Write the augmented matrix that corresponds to the system.
3. Use row operations to transform the first column so that the first element is 1 and the remaining elements are 0.
4. Use row operations to transform the second column so that the second element is 1 and the remaining elements are 0.
5. Use row operations to transform the third column so that the third element is 1 and the remaining elements are 0.
6. Continue in this way as far as possible.
7. Read the solution from the final matrix.

5 Solve each system.

(a) $x - y = 4$
    $-2x + 2y = 1$

(b) $3x - 4y = 0$
    $2x + y = 0$

Answer:

(a) No solution

(b) $(0, 0)$

▶**EXAMPLE 3**   Use the Gauss-Jordan method to solve the system

$$x + y = 2$$
$$2x + 2y = 5.$$

Begin by writing the augmented matrix.

$$\begin{bmatrix} 1 & 1 & | & 2 \\ 2 & 2 & | & 5 \end{bmatrix}$$

Get 0 for the first element in row two by multiplying the numbers in row one by -2 and adding the results to the corresponding elements in row two.

$$\begin{bmatrix} 1 & 1 & | & 2 \\ 0 & 0 & | & 1 \end{bmatrix} \quad -2R_1 + R_2$$

The next step is to get a 1 for the second element in row two. Since this is impossible, we cannot go further. This matrix gives the system

$$x + y = 2$$
$$0x + 0y = 1.$$

The second equation simplifies to $0 = 1$, so the system is inconsistent and has no solution. The row $[0 \quad 0 \mid 1]$ is a signal that the system is inconsistent. ◀ **5**

▶**EXAMPLE 4**   Use the Gauss-Jordan method to solve the system

$$x + 2y - z = 0$$
$$3x - y + z = 6$$
$$-2x - 4y + 2z = 0.$$

Start with the augmented matrix.

$$\begin{bmatrix} 1 & 2 & 1 & | & 0 \\ 3 & -1 & 1 & | & 6 \\ -2 & -4 & 2 & | & 0 \end{bmatrix}.$$

The first element in row one is 1. Use row operations to get zeros in the rest of column one.

$$\begin{bmatrix} 1 & 2 & -1 & | & 0 \\ 0 & -7 & 4 & | & 6 \\ -2 & -4 & 2 & | & 0 \end{bmatrix} \quad -3R_1 + R_2$$

$$\begin{bmatrix} 1 & 2 & -1 & | & 0 \\ 0 & -7 & 4 & | & 6 \\ 0 & 0 & 0 & | & 0 \end{bmatrix} \quad 2R_1 + R_3$$

6 Continue the solution of the system of Example 4 as follows.

(a) Get 1 in row two, column two.

(b) Get 0 in row one, column two.

Answer:

(a) $\begin{bmatrix} 1 & 2 & -1 & | & 0 \\ 0 & 1 & -\frac{4}{7} & | & -\frac{6}{7} \\ 0 & 0 & 0 & | & 0 \end{bmatrix}$

(b) $\begin{bmatrix} 1 & 0 & \frac{1}{7} & | & \frac{12}{7} \\ 0 & 1 & -\frac{4}{7} & | & -\frac{6}{7} \\ 0 & 0 & 0 & | & 0 \end{bmatrix}$

7 Use the Gauss-Jordan method to solve the following.

(a) $3x + 9y = -6$
$-x - 3y = 2$

(b) $2x + 9y = 12$
$4x + 18y = 5$

Answer:

(a) y arbitrary,
$x = -3y - 2$
or $(-3y - 2, y)$

(b) No solution

The row of all zeros in the last matrix indicates that two of the equations are dependent. Notice in the original system the third equation is -2 times the first equation, so these equations are dependent. **6**

Use the result of Problem 6(b) at the side to write the system

$$x + \frac{1}{7}z = \frac{12}{7}$$

$$y - \frac{4}{7}z = -\frac{6}{7}.$$

Solving the first equation for x and the second for y gives the solution

$$z \text{ arbitrary}$$

$$y = -\frac{6}{7} + \frac{4}{7}z$$

$$x = \frac{12}{7} - \frac{1}{7}z,$$

or $(12/7 - z/7, -6/7 + 4z/7, z)$. ◀ **7**

▶ **EXAMPLE 5** The U-Drive Rent-a-Truck Company plans to spend 3 million dollars on 200 new trucks. Each van will cost \$10,000, each small truck, \$15,000, and each large truck, \$25,000. Past experience shows that they need twice as many vans as small trucks. How many of each kind of vehicle can they buy?

Let x be the number of vans, y the number of small trucks, and z the number of large trucks. Then $x + y + z = 200$. The cost of x vans at \$10,000 each is $10,000x$, the cost of y small trucks is $15,000y$, and the cost of z large trucks is $25,000z$, so that $10,000x + 15,000y + 25,000z = 3,000,000$. Dividing this equation on each side by 5000 makes it $2x + 3y + 5z = 600$. Finally, the number of vans is twice the number of small trucks: $x = 2y$, or equivalently, $x - 2y = 0$.

To solve the system

$$x + y + z = 200$$

$$2x + 3y + 5z = 600$$

$$x - 2y \quad = 0,$$

we form the augmented matrix and use the indicated row operations.

8 In Example 5, suppose the U-Drive Company can spend only 2 million dollars on 150 new trucks, and that they need three times as many vans as small trucks. Write a system of equations to express these conditions.

Answer:

$$\begin{aligned} x + y + z &= 150 \\ 2x + 3y + 5z &= 400 \\ x - 3y &= 0 \end{aligned}$$

$$\left[\begin{array}{ccc|c} 1 & 1 & 1 & 200 \\ 2 & 3 & 5 & 600 \\ 1 & -2 & 0 & 0 \end{array}\right]$$

$$\left[\begin{array}{ccc|c} 1 & 1 & 1 & 200 \\ 0 & 1 & 3 & 200 \\ 0 & -3 & -1 & -200 \end{array}\right] \quad \begin{array}{l} -2R_1 + R_2 \\ \\ -R_1 + R_3 \end{array}$$

$$\left[\begin{array}{ccc|c} 1 & 0 & -2 & 0 \\ 0 & 1 & 3 & 200 \\ 0 & 0 & 8 & 400 \end{array}\right] \quad \begin{array}{l} -R_2 + R_1 \\ \\ 3R_2 + R_3 \end{array}$$

$$\left[\begin{array}{ccc|c} 1 & 0 & -2 & 0 \\ 0 & 1 & 3 & 200 \\ 0 & 0 & 1 & 50 \end{array}\right] \quad \dfrac{1}{8}R_3$$

$$\left[\begin{array}{ccc|c} 1 & 0 & 0 & 100 \\ 0 & 1 & 0 & 50 \\ 0 & 0 & 1 & 50 \end{array}\right] \quad \begin{array}{l} 2R_3 + R_1 \\ \\ -3R_3 + R_2 \end{array}$$

The final matrix corresponds to the system

$$\begin{aligned} x &= 100 \\ y &= 50 \\ z &= 50. \end{aligned}$$

Therefore, U-Drive should buy 100 vans, 50 small trucks, and 50 large trucks. ◀ **8**

Although the examples have used only systems with two equations and two variables or three equations and three variables, the Gauss-Jordan method can be used for any system with n equations and n variables. In fact, the method can be used with n equations and m variables, $n \leq m$. Notice that the system in Example 4 is actually equivalent to a system of just two equations and three variables. While the method does become tedious, even with three equations and three variables, it is very suitable for use by computers. A computer can produce the solution to a fairly large system very quickly.

6.2 EXERCISES

Use the indicated row operations to change each matrix.

1. Interchange R_1 and R_3.

$$\left[\begin{array}{ccc|c} 2 & 3 & 5 & 4 \\ 6 & 9 & 1 & 10 \\ 1 & 4 & 3 & 8 \end{array}\right]$$

2. Interchange R_2 and R_3.

$$\left[\begin{array}{ccc|c} 1 & 4 & 8 & 12 \\ 1 & 6 & 9 & 3 \\ 0 & 5 & 8 & 15 \end{array}\right]$$

3. Replace R_1 by $\frac{1}{3}R_1$.

$$\begin{bmatrix} 3 & 6 & 12 & \Big| & 18 \\ 0 & 5 & 2 & \Big| & 9 \\ 4 & 7 & 8 & \Big| & 15 \end{bmatrix}$$

4. Replace R_3 by $\frac{1}{4}R_3$.

$$\begin{bmatrix} 1 & 0 & 3 & \Big| & 6 \\ 0 & 1 & 7 & \Big| & 5 \\ 0 & 0 & 4 & \Big| & 8 \end{bmatrix}$$

5. Replace R_1 by $-4R_2 + R_1$.

$$\begin{bmatrix} 1 & 4 & 2 & \Big| & 9 \\ 0 & 1 & 5 & \Big| & 14 \\ 0 & 3 & 8 & \Big| & 16 \end{bmatrix}$$

6. Replace R_3 by $-5R_1 + R_3$.

$$\begin{bmatrix} 1 & 3 & 4 & \Big| & 20 \\ 0 & 6 & 5 & \Big| & 30 \\ 5 & 9 & 12 & \Big| & 15 \end{bmatrix}$$

Use the Gauss-Jordan method to solve each of the following systems of equations. (See Examples 1–4.)

7.
$\begin{aligned} x + 2y &= 5 \\ 2x + y &= -2 \end{aligned}$

8.
$\begin{aligned} 3x - 2y &= 4 \\ 3x + y &= -2 \end{aligned}$

9.
$\begin{aligned} y &= 5 - 4x \\ 2x &= 3 - y \end{aligned}$

10.
$\begin{aligned} 4x - 2y &= 3 \\ -2x + 3y &= 1 \end{aligned}$

11.
$\begin{aligned} x + 2y &= 1 \\ 2x + 4y &= 3 \end{aligned}$

12.
$\begin{aligned} x - y &= 1 \\ -x + y &= -1 \end{aligned}$

13.
$\begin{aligned} x - z &= -3 \\ y + z &= 9 \\ x + z &= 7 \end{aligned}$

14.
$\begin{aligned} x + 3y - 6z &= 7 \\ 2x - y + 2z &= 0 \\ x + y + 2z &= -1 \end{aligned}$

15.
$\begin{aligned} x &= 1 - y \\ 2x &= z \\ 2z &= -2 - y \end{aligned}$

16.
$\begin{aligned} 3x + 5y - z &= 0 \\ 4x - y + 2z &= 1 \\ -6x - 10y + 2z &= 0 \end{aligned}$

17.
$\begin{aligned} x + y &= -1 \\ y + z &= 4 \\ x + z &= 1 \end{aligned}$

18.
$\begin{aligned} x + y - z &= 6 \\ 2x - y + z &= -9 \\ x - 2y + 3z &= 1 \end{aligned}$

19.
$\begin{aligned} y &= x - 1 \\ y &= 6 + z \\ z &= -1 - x \end{aligned}$

20.
$\begin{aligned} x - 2y + z &= 5 \\ 2x + y - z &= 2 \\ -2x + 4y - 2z &= 2 \end{aligned}$

21.
$\begin{aligned} 2x + 3y + z &= 9 \\ 4x + y - 3z &= -7 \\ 6x + 2y - 4z &= -8 \end{aligned}$

22.
$\begin{aligned} 5x - 4y + 2z &= 4 \\ 10x + 3y - z &= 27 \\ 15x - 5y + 3z &= 25 \end{aligned}$

23.
$\begin{aligned} 3x + 2y - z &= -16 \\ 6x - 4y + 3z &= 12 \\ 3x + 3y + z &= -11 \end{aligned}$

24.
$\begin{aligned} 4x - 2y - 3z &= -23 \\ -4x + 3y + z &= 11 \\ 8x - 5y + 4z &= 6 \end{aligned}$

Work the following problems. (See Example 5.)

25. Management An auto manufacturer sends cars from two plants, I and II, to dealerships A and B located in a midwestern city. Plant I has a total of 28 cars to send, and plant II has 8. Dealer A needs 20 cars, and dealer B needs 16. Transportation costs based on the distance of each dealership from each plant are $220 from I to A, $300 from I to B, $400 from II to A, and $180 from II to B. The manufacturer wants to limit transportation costs to $10,640. How many cars should be sent from each plant to each of the two dealerships? Use the Gauss-Jordan method to find the solution.

26. Management A knitting shop ordered yarn from three suppliers, I, II, and III. One month the shop ordered a total of 100 units of yarn from these suppliers. The delivery costs were $80, $50, and $65 per unit for the orders from suppliers I, II, and III, respectively, with total delivery

costs of $5990. The shop ordered the same amount from suppliers I and III. How many units were ordered from each supplier? Use the Gauss-Jordan method to find the solution.

 In Exercises 27–30, write a system of equations and then use a calculator or computer to solve the system.

27. Natural Science Natural Brand plant food is made from three chemicals. The mix must include 10.8% of the first chemical and the other two chemicals must be in the ratio of 4 to 3 as measured by weight. How much of each chemical is required to make 750 kilograms of the plant food?

28. Natural Science Three species of bacteria are fed three foods, I, II, and III. A bacterium of the first species consumes 1.3 units each of foods I and II and 2.3 units of food III each day. A bacterium of the second species consumes 1.1 units of food I, 2.4 units of food II, and 3.7 units of food III each day. A bacterium of the third species consumes 8.1 units of I, 2.9 units of II, and 5.1 units of III each day. If 16,000 units of I, 28,000 units of II, and 44,000 units of III are supplied each day, how many of each species can be maintained in this environment?

29. **Natural Science**   A lake is stocked each spring with three species of fish, A, B, and C. Three foods, I, II, and III, are available in the lake. Each fish of species A requires 1.32 units of food I, 2.9 units of food II, and 1.75 units of food III on the average each day. Species B fish each require 2.1 units of food I, .95 units of food II, and .6 units of food III daily. Species C fish require .86, 1.52, and 2.01 units of I, II, and III per day, respectively. If 490 units of food I, 897 units of food II, and 653 units of food III are available daily, how many of each species should be stocked?

30. **Management**   A company produces three combinations of mixed vegetables which sell in 1 kilogram packages. Italian style combines .3 kilograms of zucchini, .3 of broccoli, and .4 of carrots. French style combines .6 kilograms of broccoli and .4 of carrots. Oriental style combines .2 kilograms of zucchini, .5 of broccoli, and .3 of carrots. The company has a stock of 16,200 kilograms of zucchini, 41,400 kilograms of broccoli, and 29,400 kilograms of carrots. How many packages of each style should they prepare to use up their supplies?

31. **Management**   A manufacturer purchases a part for use at both of its two plants—one in Roseville, California, the other in Akron, Ohio. The part is available in limited quantities from two suppliers. Each supplier has 75 units available. The Roseville plant needs 40 units and the Akron plant requires 75 units. The first supplier charges $70 per unit delivered to Roseville and $90 per unit delivered to Akron. Corresponding costs from the second supplier are $80 and $120. The manufacturer wants to order a total of 75 units from the first, less expensive, supplier, with the remaining 40 units to come from the second supplier. If the company spends $10,750 to purchase the required number of units for the two plants, find the number of units that should be purchased from each supplier for each plant as follows.
(a) Assign variables to the four unknowns.
(b) Write a system of five equations with the four variables. (Not all equations will involve all four variables.)
(c) Use the Gauss-Jordan method to solve the system of equations.

32. **Social Science**   At rush hours, substantial traffic congestion is encountered at the traffic intersections shown in the figure. *(The streets are all one way.)*

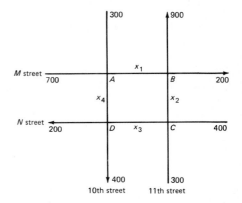

The city wishes to improve the signals at these corners so as to speed the flow of traffic. The traffic engineers first gather data. As the figure shows, 700 cars per hour come down M Street to intersection A; 300 cars per hour come to intersection A on 10th Street. A total of x_1 of these cars leave A on M Street, while x_4 cars leave A on 10th Street. The number of cars entering A must equal the number leaving, so that

$$x_1 + x_4 = 700 + 300$$
or $$x_1 + x_4 = 1000.$$

For intersection B, x_1 cars enter B on M Street, and x_2 cars enter B on 11th Street. The figure shows that 900 cars leave B on 11th while 200 leave on M. We have

$$x_1 + x_2 = 900 + 200$$
$$x_1 + x_2 = 1100.$$

At intersection C, 400 cars enter on N Street, 300 on 11th Street, while x_2 leave on 11th Street and x_3 leave on N Street. This gives

$$x_2 + x_3 = 400 + 300$$
$$x_2 + x_3 = 700.$$

Finally, intersection D has x_3 cars entering on N and x_4 entering on 10th. There are 400 leaving D on 10th and 200 leaving on N, so that

$$x_3 + x_4 = 400 + 200$$
$$x_3 + x_4 = 600.$$

(a) Use the four equations to set up an augmented matrix, and then use the Gauss-Jordan method to solve it. (Hint: keep going until you get a row of all zeros.)

(b) Since you got a row of all zeros, the system of equations does not have a unique solution. Write three equations, corresponding to the three nonzero rows of the matrix.

(c) Solve each of the equations for x_4.

(d) One of your equations should have been $x_4 = 1000 - x_1$. What is the largest possible value of x_1 so that x_4 is not negative? What is the largest value of x_4 so that x_1 is not negative?

(e) Your second equation should have been $x_4 = x_2 - 100$. Find the smallest possible value of x_2 so that x_4 is not negative.

(f) For the third equation, $x_4 = 600 - x_3$, find the largest possible values of x_3 and x_4 so that neither variable is negative.

(g) Look at your answers for parts (d)-(f). What is the maximum value of x_4 so that all the equations are satisfied and all variables are nonnegative? of x_3? of x_2? of x_1?

6.3 BASIC MATRIX OPERATIONS

As shown in Section 6.2, matrices can be used to represent systems of linear equations. The study of matrices has been of interest to mathematicians, chemists, physicists, and engineers for some time. During the last 50 years, the use of matrices has gained increasing importance in the fields of management, natural science, and social science because matrices provide a convenient way to organize data, as Example 1 demonstrates.

▶**EXAMPLE 1** The EZ Life Company manufactures sofas and armchairs in three models, A, B, and C. The company has regional warehouses in New York, Chicago, and San Francisco. In its August shipment, the company sends 10 model A sofas, 12 model B sofas, 5 model C sofas, 15 model A chairs, 20 model B chairs, and 8 model C chairs to each warehouse.

This data might be organized by first listing it as follows.

| Sofas | 10 model A | 12 model B | 5 model C |
| Chairs | 15 model A | 20 model B | 8 model C |

Alternatively, we might tabulate the data in a chart.

		Model		
		A	B	C
Furniture	Sofa	10	12	5
	Chair	15	20	8

1 Rewrite this information in a matrix with three rows and two columns.

Answer:

$$\begin{bmatrix} 10 & 15 \\ 12 & 20 \\ 5 & 8 \end{bmatrix}$$

2 Give the size of each of the following matrices.

(a) $\begin{bmatrix} 2 & 1 & -5 & 6 \\ 3 & 0 & 7 & -4 \end{bmatrix}$

(b) $\begin{bmatrix} 1 & 2 & 3 \\ 4 & 5 & 6 \\ 9 & 8 & 7 \end{bmatrix}$

Answer:

(a) 2×4

(b) 3×3

3 Use the numbers 2, 5, -8, 4 to write

(a) a row matrix;

(b) a column matrix;

(c) a square matrix.

Answer:

(a) [2   5 -8 4]

(b) $\begin{bmatrix} 2 \\ 5 \\ -8 \\ 4 \end{bmatrix}$

(c) $\begin{bmatrix} 2 & 5 \\ -8 & 4 \end{bmatrix}$ or $\begin{bmatrix} 2 & -8 \\ 5 & 4 \end{bmatrix}$

(Other answers are possible.)

With the understanding that the numbers in each row refer to the furniture type (sofa, chair) and the numbers in each column refer to the model (A, B, C), the same information can be given by a matrix, as follows.

$$M = \begin{bmatrix} 10 & 12 & 5 \\ 15 & 20 & 8 \end{bmatrix} \quad \blacktriangleleft \boxed{1}$$

Matrices often are named with capital letters, as in Example 1. A matrix is classified by its size, that is, by the number of rows and columns that it contains. For example, matrix M above has two rows and three columns. This matrix is called a 2×3 (read "2 by 3") matrix. By definition, a matrix with m rows and n columns is size $m \times n$. The number of rows is always given first.

▶**EXAMPLE 2**   **(a)** The matrix $\begin{bmatrix} 6 & 5 \\ 3 & 4 \\ 5 & -1 \end{bmatrix}$ is a 3×2 matrix.

(b) $\begin{bmatrix} 5 & 8 & 9 \\ 0 & 5 & -3 \\ -4 & 0 & 5 \end{bmatrix}$ is a 3×3 matrix.

(c) [1   6   5 -2 5] is a 1×5 matrix.

(d) $\begin{bmatrix} 3 \\ -5 \\ 0 \\ 2 \end{bmatrix}$ is a 4×1 matrix. $\blacktriangleleft \boxed{2}$

A matrix with the same number of rows as columns is called a **square matrix.** The matrix in Example 2(b) above is a square matrix, as are

$$\begin{bmatrix} -5 & 6 \\ 8 & 3 \end{bmatrix} \quad \text{and} \quad \begin{bmatrix} 0 & 0 & 0 & 0 \\ -2 & 4 & 1 & 3 \\ 0 & 0 & 0 & 0 \\ -5 & -4 & 1 & 8 \end{bmatrix}.$$

A matrix containing only one row is called a **row matrix** or **row vector.** The matrix in Example 2(c) is a row matrix, as are

[5   8],   [6 -9 2],   and   [-4 0   0   0].

A matrix of only one column, such as in Example 2(d), is a **column matrix** or **column vector.**   **3**

4 Give the values of the variables that make each of the following statements true.

(a) $\begin{bmatrix} x & 2 \\ 5 & y \end{bmatrix} = \begin{bmatrix} 6 & p \\ q & -1 \end{bmatrix}$

(b) $\begin{bmatrix} 1 & 2 & x \end{bmatrix} = \begin{bmatrix} y \\ 2 \\ 8 \end{bmatrix}$

Answer:

(a) $x = 6$, $y = -1$, $p = 2$, $q = 5$

(b) Can never be true

Equality for matrices is defined as follows.

Two matrices are **equal** if they are the same size and if corresponding elements are equal.

Using this definition, the matrices

$$\begin{bmatrix} 2 & 1 \\ 3 & -5 \end{bmatrix} \text{ and } \begin{bmatrix} 1 & 2 \\ -5 & 3 \end{bmatrix}$$

are not equal (even though they contain the same elements and are the same size) since corresponding elements differ.

▶**EXAMPLE 3**   **(a)** From the definition of matrix equality given above, the only way that the statement

$$\begin{bmatrix} 2 & 1 \\ p & q \end{bmatrix} = \begin{bmatrix} x & y \\ -1 & 0 \end{bmatrix}$$

can be true is if $2 = x$, $1 = y$, $p = -1$, and $q = 0$.
(b) The statement

$$\begin{bmatrix} x \\ y \end{bmatrix} = \begin{bmatrix} 1 \\ 4 \\ 0 \end{bmatrix}$$

can never be true, since the two matrices are different sizes. (One is 2×1 and the other is 3×1.) ◀ **4**

ADDITION The matrix given in Example 1,

$$M = \begin{bmatrix} 10 & 12 & 5 \\ 15 & 20 & 8 \end{bmatrix},$$

shows the August shipment from the EZ Life plant to each of its warehouses. If matrix N below gives the September shipment to the New York warehouse, what is the total shipment for each item of furniture to the New York warehouse for these two months?

5 Find each sum when possible.

(a) $\begin{bmatrix} 2 & 5 & 7 \\ 3 & -1 & 4 \end{bmatrix}$

$\quad + \begin{bmatrix} -1 & 2 & 0 \\ 10 & -4 & 5 \end{bmatrix}$

(b) $\begin{bmatrix} 1 \\ 2 \\ 3 \end{bmatrix} + \begin{bmatrix} 2 & -1 \\ 4 & 5 \\ 6 & 0 \end{bmatrix}$

(c) $[5 \quad 4 \quad -1]$

$\quad + [-5 \quad 2 \quad 3]$

Answer:

(a) $\begin{bmatrix} 1 & 7 & 7 \\ 13 & -5 & 9 \end{bmatrix}$

(b) Not possible

(c) $[0 \quad 6 \quad 2]$

$$N = \begin{bmatrix} 45 & 35 & 20 \\ 65 & 40 & 35 \end{bmatrix}$$

If 10 model A sofas were shipped in August and 45 in September, then altogether 55 model A sofas were shipped in the 2 months. Adding the other corresponding entries gives a new matrix, Q, that represents the total shipment for the 2 months.

$$Q = \begin{bmatrix} 55 & 47 & 25 \\ 80 & 60 & 43 \end{bmatrix}$$

It is convenient to refer to Q as the "sum" of M and N.

The way these two matrices were added illustrates the following definition of addition of matrices.

The **sum** of two $m \times n$ matrices X and Y is the $m \times n$ matrix $X + Y$ in which each element is the sum of the corresponding elements of X and Y.

It is important to remember that only matrices that are the same size can be added.

▶**EXAMPLE 4** Find each sum if possible.

(a) $\begin{bmatrix} 5 & -6 \\ 8 & 9 \end{bmatrix} + \begin{bmatrix} -4 & 6 \\ 8 & -3 \end{bmatrix} = \begin{bmatrix} 5 + (-4) & -6 + 6 \\ 8 + 8 & 9 + (-3) \end{bmatrix} = \begin{bmatrix} 1 & 0 \\ 16 & 6 \end{bmatrix}$

(b) The matrices

$$A = \begin{bmatrix} 5 & 8 \\ 6 & 2 \end{bmatrix} \quad \text{and} \quad B = \begin{bmatrix} 3 & 9 & 1 \\ 4 & 2 & 5 \end{bmatrix}$$

are different sizes, so it is not possible to find the sum $A + B$. ◀ **5**

▶**EXAMPLE 5** The September shipments from the EZ Life Company to the New York, San Francisco, and Chicago warehouses are given in matrices N, S, and C below.

$$N = \begin{bmatrix} 45 & 35 & 20 \\ 65 & 40 & 35 \end{bmatrix}, \quad S = \begin{bmatrix} 30 & 32 & 28 \\ 43 & 47 & 30 \end{bmatrix}, \quad C = \begin{bmatrix} 22 & 25 & 38 \\ 31 & 34 & 35 \end{bmatrix}$$

What was the total amount shipped to the three warehouses in September?

6 From the result of Example 5, find the total number of the following shipped to the three warehouses.

(a) Model A chairs

(b) Model B sofas

(c) Model C chairs

Answer:

(a) 139

(b) 92

(c) 100

The total of the September shipments is represented by the sum of the three matrices N, S, and C.

$$N + S + C = \begin{bmatrix} 45 & 35 & 20 \\ 65 & 40 & 35 \end{bmatrix} + \begin{bmatrix} 30 & 32 & 28 \\ 43 & 47 & 30 \end{bmatrix} + \begin{bmatrix} 22 & 25 & 38 \\ 31 & 34 & 35 \end{bmatrix}$$

$$= \begin{bmatrix} 97 & 92 & 86 \\ 139 & 121 & 100 \end{bmatrix}$$

For example, from this sum the total number of model C sofas shipped to the three warehouses in September was 86. ◀ **6**

As mentioned in Section 1.1, the additive inverse of the real number a is $-a$; a similar definition is given for the additive inverse of a matrix.

The **additive inverse** (or *negative*) of a matrix X is the matrix $-X$ in which each element is the additive inverse of the corresponding element of X.

If

$$A = \begin{bmatrix} 1 & 2 & 3 \\ 0 & -1 & 5 \end{bmatrix} \quad \text{and} \quad B = \begin{bmatrix} -2 & 3 & 0 \\ 1 & -7 & 2 \end{bmatrix},$$

then by the definition of the additive inverse of a matrix,

$$-A = \begin{bmatrix} -1 & -2 & -3 \\ 0 & 1 & -5 \end{bmatrix} \quad \text{and} \quad -B = \begin{bmatrix} 2 & -3 & 0 \\ -1 & 7 & -2 \end{bmatrix}.$$

By the definition of matrix addition, for each matrix X, the sum $X + (-X)$ is a **zero matrix,** O, whose elements are all zeros. There is an $m \times n$ zero matrix for each pair of values of m and n. Zero matrices have the following *identity property:*

If O is the $m \times n$ zero matrix, and A is any $m \times n$ matrix, then

$$A + O = O + A = A.$$

Compare this with the identity property for real numbers: for any real number a, $a + 0 = 0 + a = a$.

SUBTRACTION The **subtraction** of matrices can be defined in a manner comparable to subtraction for real numbers.

7 Find each of the following differences when possible.

(a) $\begin{bmatrix} 2 & 5 \\ -1 & 0 \end{bmatrix} - \begin{bmatrix} 6 & 4 \\ 3 & -2 \end{bmatrix}$

(b) $\begin{bmatrix} 1 & 5 & 6 \\ 2 & 4 & 8 \end{bmatrix} - \begin{bmatrix} 2 & 1 \\ 10 & 3 \end{bmatrix}$

(c) $[5 \quad -4 \quad 1]$
$\quad - [6 \quad 0 \quad -3]$

Answer:

(a) $\begin{bmatrix} -4 & 1 \\ -4 & 2 \end{bmatrix}$

(b) Not possible

(c) $[-1 \quad -4 \quad 4]$

For two $m \times n$ matrices X and Y, the **difference** of X and Y is the $m \times n$ matrix $X - Y$ in which each element is the difference of the corresponding elements of X and Y, or, equivalently,

$$X - Y = X + (-Y).$$

According to this definition, matrix subtraction can be performed by subtracting corresponding elements. For example, using A, B, and $-B$ as defined above,

$$A - B = \begin{bmatrix} 1 & 2 & 3 \\ 0 & -1 & 5 \end{bmatrix} - \begin{bmatrix} -2 & 3 & 0 \\ 1 & -7 & 2 \end{bmatrix}$$

$$= \begin{bmatrix} 1-(-2) & 2-3 & 3-0 \\ 0-1 & -1-(-7) & 5-2 \end{bmatrix}$$

$$= \begin{bmatrix} 3 & -1 & 3 \\ -1 & 6 & 3 \end{bmatrix}.$$

▶**EXAMPLE 6** (a) $[8 \quad 6 \quad -4] - [3 \quad 5 \quad -8] = [5 \quad 1 \quad 4]$

(b) The matrices

$$\begin{bmatrix} -2 & 5 \\ 0 & 1 \end{bmatrix} \text{ and } \begin{bmatrix} 3 \\ 5 \end{bmatrix}$$

are different sizes and cannot be subtracted. ◀ **7**

▶**EXAMPLE 7** During September the Chicago warehouse of the EZ Life Company shipped out the following numbers of each model.

$$K = \begin{bmatrix} 5 & 10 & 8 \\ 11 & 14 & 15 \end{bmatrix}$$

What was the Chicago warehouse inventory on October 1, taking into account only the number of items received and sent out during the month?

The number of each kind of item received during September is given by matrix C from Example 5; the number of each model sent out during September is given by matrix K above. The October 1 inventory will be represented by the matrix $C - K$ as shown below.

$$\begin{bmatrix} 22 & 25 & 38 \\ 31 & 34 & 35 \end{bmatrix} - \begin{bmatrix} 5 & 10 & 8 \\ 11 & 14 & 15 \end{bmatrix} = \begin{bmatrix} 17 & 15 & 30 \\ 20 & 20 & 20 \end{bmatrix} ◀$$

▶**EXAMPLE 8** A drug company is testing 200 patients to see if Painoff (a new headache medicine) is effective. Half the patients receive Painoff and half receive a placebo. The data on the first 50 patients is summarized in this matrix:

<u>**Pain Relief Obtained**</u>

	Yes	No
Patient took **Painoff**	22	3
Patient took **placebo**	8	17

For example, row 2 shows that of the people who took the placebo, 8 got relief, but 17 did not. The test was repeated on three more groups of 50 patients each, with the results summarized by these matrices.

$$\begin{bmatrix} 21 & 4 \\ 6 & 19 \end{bmatrix} \qquad \begin{bmatrix} 19 & 6 \\ 10 & 15 \end{bmatrix} \qquad \begin{bmatrix} 23 & 2 \\ 3 & 22 \end{bmatrix}$$

The total results of the test can be obtained by adding these four matrices.

$$\begin{bmatrix} 22 & 3 \\ 8 & 17 \end{bmatrix} + \begin{bmatrix} 21 & 4 \\ 6 & 19 \end{bmatrix} + \begin{bmatrix} 19 & 6 \\ 10 & 15 \end{bmatrix} + \begin{bmatrix} 23 & 2 \\ 3 & 22 \end{bmatrix} = \begin{bmatrix} 85 & 15 \\ 27 & 73 \end{bmatrix}$$

Since 85 of 100 patients got relief with Painoff and only 27 of 100 with the placebo, it appears that Painoff is effective. ◀

6.3 EXERCISES

Mark each of the following statements as true *or* false. *If false, tell why.*

1. $\begin{bmatrix} 1 & 3 \\ 5 & 7 \end{bmatrix} = \begin{bmatrix} 1 & 5 \\ 3 & 7 \end{bmatrix}$

2. $\begin{bmatrix} 1 \\ 2 \\ 3 \end{bmatrix} = [1 \quad 2 \quad 3]$

3. $\begin{bmatrix} x \\ y \end{bmatrix} = \begin{bmatrix} 3 \\ 5 \end{bmatrix}$ if $x = 3$ and $y = 5$.

4. $\begin{bmatrix} 3 & 5 & 2 & 8 \\ 1 & -1 & 4 & 0 \end{bmatrix}$ is a 4 × 2 matrix.

5. $\begin{bmatrix} 1 & 9 & -4 \\ 3 & 7 & 2 \\ -1 & 1 & 0 \end{bmatrix}$ is a square matrix.

6. $\begin{bmatrix} 2 & 4 & -1 \\ 3 & 7 & 5 \\ 0 & 0 & 0 \end{bmatrix} = \begin{bmatrix} 2 & 4 & -1 \\ 3 & 7 & 5 \end{bmatrix}$

Find the size of each of the following. Identify any square, column, or row matrices. (See Example 2.) Give the additive inverse (or negative) of each matrix.

7. $\begin{bmatrix} -4 & 8 \\ 2 & 3 \end{bmatrix}$

8. $\begin{bmatrix} -9 & 6 & 2 \\ 4 & 1 & 8 \end{bmatrix}$

9. $\begin{bmatrix} -6 & 8 & 0 & 0 \\ 4 & 1 & 9 & 2 \\ 3 & -5 & 7 & 1 \end{bmatrix}$

10. $[8 \quad -2 \quad 4 \quad 6 \quad 3]$

11. $\begin{bmatrix} 2 \\ 4 \end{bmatrix}$

12. $[-9]$

Find the values of the variables in each of the following. (See Example 3.)

13. $\begin{bmatrix} 2 & 1 \\ 4 & 8 \end{bmatrix} = \begin{bmatrix} x & 1 \\ y & z \end{bmatrix}$

14. $\begin{bmatrix} -5 \\ y \end{bmatrix} = \begin{bmatrix} -5 \\ 8 \end{bmatrix}$

15. $\begin{bmatrix} x + 6 & y + 2 \\ 8 & 3 \end{bmatrix} = \begin{bmatrix} -9 & 7 \\ 8 & k \end{bmatrix}$

16. $\begin{bmatrix} 9 & 7 \\ r & 0 \end{bmatrix} = \begin{bmatrix} m - 3 & n + 5 \\ 8 & 0 \end{bmatrix}$

17. $\begin{bmatrix} -7 + z & 4r & 8s \\ 6p & 2 & 5 \end{bmatrix} + \begin{bmatrix} -9 & 8r & 3 \\ 2 & 5 & 4 \end{bmatrix} = \begin{bmatrix} 2 & 36 & 27 \\ 20 & 7 & 12a \end{bmatrix}$

18. $\begin{bmatrix} -a + 2 & 3z + 1 & 5m \\ 4k & 0 & 3 \end{bmatrix} + \begin{bmatrix} 3a & 2z & 5m \\ 2k & 5 & 6 \end{bmatrix} = \begin{bmatrix} 10 & -14 & 80 \\ 10 & 5 & 9 \end{bmatrix}$

Perform the indicated operations where possible. (See Examples 4 and 6.)

19. $\begin{bmatrix} 1 & 2 & 5 & -1 \\ 3 & 0 & 2 & -4 \end{bmatrix} + \begin{bmatrix} 8 & 10 & -5 & 3 \\ -2 & -1 & 0 & 0 \end{bmatrix}$

20. $\begin{bmatrix} 1 & 5 \\ 2 & -3 \\ 3 & 7 \end{bmatrix} + \begin{bmatrix} 2 & 3 \\ 8 & 5 \\ -1 & 9 \end{bmatrix}$

21. $\begin{bmatrix} 1 & 5 & 7 \\ 2 & 2 & 3 \end{bmatrix} + \begin{bmatrix} 4 & 8 & -7 \\ 1 & -1 & 5 \end{bmatrix}$

22. $\begin{bmatrix} 2 & 4 \\ -8 & 1 \end{bmatrix} + \begin{bmatrix} 9 & -3 \\ 8 & 5 \end{bmatrix}$

23. $\begin{bmatrix} 1 & 3 & -2 \\ 4 & 7 & 1 \end{bmatrix} + \begin{bmatrix} 3 & 0 \\ 6 & 4 \\ -5 & 2 \end{bmatrix}$

24. $\begin{bmatrix} 1 & 3 & -2 \\ 4 & 7 & 1 \end{bmatrix} - \begin{bmatrix} 3 & 6 & -5 \\ 0 & 4 & 2 \end{bmatrix}$

25. $\begin{bmatrix} 2 & 8 & 12 & 0 \\ 7 & 4 & -1 & 5 \\ 1 & 2 & 0 & 10 \end{bmatrix} - \begin{bmatrix} 1 & 3 & 6 & 9 \\ 2 & -3 & -3 & 4 \\ 8 & 0 & -2 & 17 \end{bmatrix}$

26. $\begin{bmatrix} 2 & 1 \\ 5 & -3 \\ -7 & 2 \\ 9 & 0 \end{bmatrix} + \begin{bmatrix} 1 & -8 & 0 \\ 5 & 3 & 2 \\ -6 & 7 & -5 \\ 2 & -1 & 0 \end{bmatrix}$

27. $\begin{bmatrix} -4x + 2y & -3x + y \\ 6x - 3y & 2x - 5y \end{bmatrix} + \begin{bmatrix} -8x + 6y & 2x \\ 3y - 5x & 6x + 4y \end{bmatrix}$

28. $\begin{bmatrix} 4k - 8y \\ 6z - 3x \\ 2k + 5a \\ -4m + 2n \end{bmatrix} - \begin{bmatrix} 5k + 6y \\ 2z + 5x \\ 4k + 6a \\ 4m - 2n \end{bmatrix}$

Using matrices

$$O = \begin{bmatrix} 0 & 0 \\ 0 & 0 \end{bmatrix}, P = \begin{bmatrix} m & n \\ p & q \end{bmatrix}, T = \begin{bmatrix} r & s \\ t & u \end{bmatrix}, \text{ and } X = \begin{bmatrix} x & y \\ z & w \end{bmatrix},$$

verify that the statements in Exercises 29–33 are true.

29. $X + T$ is a 2×2 matrix. (See Example 1.)

30. $X + T = T + X$ (Commutative property of addition of matrices)

31. $X + (T + P) = (X + T) + P$ (Associative property of addition of matrices)

32. $X + (-X) = O$ (Inverse property of addition of matrices)

33. $P + O = P$ (Identity property of addition of matrices)

34. Which of the above properties are valid for matrices that are not square?

35. **Management** An investment group planning a shopping center decided to include a market, a barber shop, a variety store, a drug store, and a bakery. They estimated the initial cost and the guaranteed rent (both in dollars per square foot) for each type of store, respectively, as follows. Initial cost: 18, 10, 8, 10, and 10; guaranteed rent: 2.7, 1.5, 1.0, 2.0, and 1.7. Write this information first as a 5×2 matrix and then as a 2×5 matrix. (See Example 1.)

36. **Management** In reviewing its circulation and advertising, a weekly magazine found that in 1970 its circulation was 770,000, it had 2817 pages of advertising, and the gross advertising revenue was $9,718,180. Comparable figures for 1975 were 1,300,000; 3693; and $18,537,057. The figures for 1980 were 1,700,000; 3809; and $29,950,700; and for 1985 they were 2,000,000; 3450; and $42,598,770. Write this information first as a 4×3 matrix and then as a 3×4 matrix.

37. **Natural Science** A dietician prepares a diet specifying the amounts a patient should eat of four basic food groups: group I, meats; group II, fruits and vegetables; group III, breads and starches; group IV, milk products. Amounts are given in "exchanges" which represent 1 ounce (meat), 1/2 cup (fruits and vegetables), 1 slice (bread), 8 ounces (milk), or other suitable measurements.
  (a) The number of "exchanges" for breakfast for each of the four food groups, respectively, are 2, 1, 2, and 1; for lunch, 3, 2, 2, and 1; and for dinner, 4, 3, 2, and 1. Write a 3×4 matrix using this information.

  (b) The amounts of fat, carbohydrates, and protein in each food group respectively are as follows.

    Fat: 5, 0, 0, 10
    Carbohydrates: 0, 10, 15, 12
    Protein: 7, 1, 2, 8

  Use this information to write a 4×3 matrix.

  (c) There are 8 calories per unit of fat, 4 calories per unit of carbohydrates, and 5 calories per unit of protein; summarize this data in a 3×1 matrix.

38. **Natural Science** At the beginning of a laboratory experiment, five baby rats measured 5.6, 6.4, 6.9, 7.6, and 6.1 centimeters in length, and weighed 144, 138, 149, 152, and 146 grams, respectively.
  (a) Write a 2×5 matrix using this information.

  (b) At the end of 2 weeks, their lengths were 10.2, 11.4, 11.4, 12.7, and 10.8 centimeters, and they weighed 196, 196, 225, 250, and 230 grams. Write a 2×5 matrix with this information.

  (c) Use matrix subtraction and the matrices found in (a) and (b) to write a matrix that gives the amount of change in length and weight for each rat. (See Examples 5, 7, and 8.)

  (d) The following week the rats gained as shown in the matrix below.

$$\begin{array}{c} \text{Length} \\ \text{Weight} \end{array} \begin{bmatrix} 1.8 & 1.5 & 2.3 & 1.8 & 2.0 \\ 25 & 22 & 29 & 33 & 20 \end{bmatrix}$$

  What were their lengths and weights at the end of this week?

39. **Management** There are three FastShop convenience stores in East Podunk. This week, Store I sold 88 loaves of bread, 48 quarts of milk, 16 jars of peanut butter, and 112 pounds of cold cuts. Store II sold 105 loaves of bread, 72 quarts of milk, 21 jars of peanut butter, and 147 pounds of cold cuts. Store III sold 60 loaves of bread, 40 quarts of milk, no peanut butter, and 50 pounds of cold cuts.
  (a) Use a 3×4 matrix to express the sales information for the three stores.

  (b) During the following week, sales on these products at Store I increased by 25%; sales at Store II increased by 1/3; and sales at Store III increased by 10%. Write the sales matrix for that week.

  (c) Write a matrix that represents total sales over the 2-week period.

40. Management   A toy company has plants in Boston, Chicago, and Seattle that manufacture toy rockets and robots. The matrix below gives the production costs (in dollars) for each item at the Boston plant:

$$\begin{array}{cc} & \begin{array}{cc} \text{Rockets} & \text{Robots} \end{array} \\ \begin{array}{c} \text{Material} \\ \text{Labor} \end{array} & \begin{bmatrix} 4.27 & 6.94 \\ 3.45 & 3.65 \end{bmatrix} \end{array}$$

(a) In Chicago, a rocket costs $4.05 for materials and $3.27 for labor; a robot costs $7.01 for material and $3.51 for labor. In Seattle, material costs are $4.40 for rockets and $6.90 for robots; labor costs are $3.54 for rockets and $3.76 for robots. Write the production cost matrices for Chicago and Seattle.

(b) Assume each plant makes the same number of each item. Write a matrix that expresses the average production costs for all three plants.

(c) Suppose labor costs increase by .11 per item in Chicago and material costs there increase by .37 for a rocket and .42 for a robot. What is the new production cost matrix for Chicago?

(d) After the Chicago cost increases, the Boston plant is closed and production divided evenly among the other two plants. What is the matrix that now expresses the average production costs for the entire country?

6.4 MULTIPLICATION OF MATRICES

Suppose one of the EZ Life Company warehouses receives the following order, written in matrix form, where the entries have the same meaning as in the previous section.

$$\begin{bmatrix} 5 & 4 & 1 \\ 3 & 2 & 3 \end{bmatrix}$$

Later, the store that sent the order asks the company to send five more of the same order. The five new orders can be written as one matrix by multiplying each element in the matrix by 5, giving the product

$$5\begin{bmatrix} 5 & 4 & 1 \\ 3 & 2 & 3 \end{bmatrix} = \begin{bmatrix} 25 & 20 & 5 \\ 15 & 10 & 15 \end{bmatrix}.$$

In work with matrices, a real number, like the 5 in the product above, is called a **scalar.**

The **product** of a scalar k and a matrix X is the matrix kX, each of whose elements is k times the corresponding element of X.

For example,

$$(-3)\begin{bmatrix} 2 & -5 \\ 1 & 7 \end{bmatrix} = \begin{bmatrix} -6 & 15 \\ -3 & -21 \end{bmatrix}.$$

1 In this example of the EZ Life Company, find the total value of the New York sets if model A sofas sell for $1200, model B for $1600, and model C for $1300.

Answer:

$27,100

Next we shall define the product of two matrices. To understand the reasoning behind the definition of matrix multiplication, look again at the EZ Life Company. Suppose sofas and chairs of the same model are often sold as sets with matrix W showing the number of each model set in each warehouse.

$$\begin{array}{c} \\ \text{New York} \\ \text{Chicago} \\ \text{San Francisco} \end{array} \begin{array}{ccc} A & B & C \\ \begin{bmatrix} 10 & 7 & 3 \\ 5 & 9 & 6 \\ 4 & 8 & 2 \end{bmatrix} \end{array} = W$$

If the selling price of a model A set is $800, of a model B set $1000, and of a model C set $1200, find the total value of the sets in the New York warehouse as follows.

Type	Number of Sets		Price of Set		Total
A	10	×	$ 800	=	$ 8000
B	7	×	1000	=	7000
C	3	×	1200	=	3600
			Total for New York		$18,600

The total value of the three kinds of sets in New York is $18,600. **1**
The work done in the table above is summarized as follows:

$$10(\$800) + 7(\$1000) + 3(\$1200) = \$18,600.$$

In the same way, the Chicago sets have a total value of

$$5(\$800) + 9(\$1000) + 6(\$1200) = \$20,200,$$

and in San Francisco, the total value of the sets is

$$4(\$800) + 8(\$1000) + 2(\$1200) = \$13,600.$$

The selling prices can be written as a column matrix, P, and the total value in each location as a column matrix, V.

$$\begin{bmatrix} 800 \\ 1000 \\ 1200 \end{bmatrix} = P \quad \text{and} \quad \begin{bmatrix} 18,600 \\ 20,200 \\ 13,600 \end{bmatrix} = V$$

Consider how the first row of the matrix W and the single column P lead to the first entry of V.

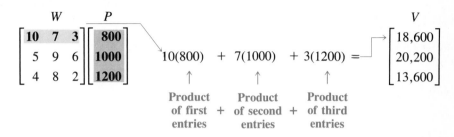

$$10(800) \quad + \quad 7(1000) \quad + \quad 3(1200) \quad =$$

Product of first entries + Product of second entries + Product of third entries

Similarly, adding the products of corresponding entries in the second row of W and the column P produces the second entry in V. The third entry in V is obtained in the same way by using the third row of W and column P. This suggests that it is reasonable to *define* the product WP to be V.

$$WP = \begin{bmatrix} 10 & 7 & 3 \\ 5 & 9 & 6 \\ 4 & 8 & 2 \end{bmatrix} \begin{bmatrix} 800 \\ 1000 \\ 1200 \end{bmatrix} = \begin{bmatrix} 18,600 \\ 20,200 \\ 13,600 \end{bmatrix} = V$$

Note the sizes of the matrices here: the product of a 3×3 matrix and a 3×1 matrix is a 3×1 matrix.

With this model in mind, we define the **product of a row and a column** (with the same number of entries in each) to be the *number* obtained by multiplying the corresponding entries (first by first, second by second, and so on) and adding the results. Then **matrix multiplication** is defined as follows.

Let A be an $m \times n$ matrix and let B be an $n \times k$ matrix. The **product matrix** AB is the $m \times k$ matrix whose entry in the i-th row and j-th column is

the product of the i-th row of A and the j-th column of B.

Caution Be careful when multiplying matrices. Remember that the number of *columns* of A must equal the number of *rows* of B in order to get the product matrix AB. The final product will have as many rows as A and as many columns as B.

2 Matrix A is 4×6 and matrix B is 2×4.

(a) Can AB be found? If so, give its order.

(b) Can BA be found? If so, give its order.

Answer:

(a) No

(b) Yes; 2×6

3 Find the product AB given

$$A = \begin{bmatrix} 2 & 4 \\ 5 & 6 \end{bmatrix} \text{ and}$$

$$B = \begin{bmatrix} -3 \\ 4 \end{bmatrix}.$$

Answer:

$$AB = \begin{bmatrix} 10 \\ 9 \end{bmatrix}$$

▶**EXAMPLE 1** Suppose matrix A is 2×2 and matrix B is 2×4. Can the product AB be calculated? What is the order of the product?

The following diagram helps decide the answers to these questions.

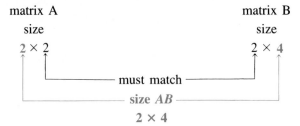

The product of A and B can be calculated because A has two columns and B has two rows. The product will be a 2×4 matrix. ◀ **2**

▶**EXAMPLE 2** Find the product AB given

$$A = \begin{bmatrix} 2 & 3 & -1 \\ 4 & 2 & 2 \end{bmatrix} \quad \text{and} \quad B = \begin{bmatrix} 1 \\ 8 \\ 6 \end{bmatrix}.$$

Since matrix A is 2×3 and matrix B is 3×1, matrix AB can be found and will be a 2×1 matrix.

Step 1 Multiply the elements of the first row of A and the corresponding elements of the column of B.

$$\begin{bmatrix} 2 & 3 & -1 \\ 4 & 2 & 2 \end{bmatrix} \begin{bmatrix} 1 \\ 8 \\ 6 \end{bmatrix} \qquad 2 \cdot 1 + 3 \cdot 8 + (-1) \cdot 6 = 20$$

Therefore, 20 is the first row entry of the product matrix AB.

Step 2 Multiply the elements of the second row of A and the corresponding elements of B.

$$\begin{bmatrix} 2 & 3 & -1 \\ 4 & 2 & 2 \end{bmatrix} \begin{bmatrix} 1 \\ 8 \\ 6 \end{bmatrix} \qquad 4 \cdot 1 + 2 \cdot 8 + 2 \cdot 6 = 32$$

Step 3 Write the product using the two entries we found above.

$$AB = \begin{bmatrix} 2 & 3 & -1 \\ 4 & 2 & 2 \end{bmatrix} \begin{bmatrix} 1 \\ 8 \\ 6 \end{bmatrix} = \begin{bmatrix} 20 \\ 32 \end{bmatrix} \quad ◀ \ \boxed{3}$$

4 Find the product CD given

$$C = \begin{bmatrix} 1 & 3 & 5 \\ 2 & -4 & -1 \end{bmatrix} \text{ and}$$

$$D = \begin{bmatrix} 2 & -1 \\ 4 & 3 \\ 1 & -2 \end{bmatrix}.$$

Answer:

$$CD = \begin{bmatrix} 19 & -2 \\ -13 & -12 \end{bmatrix}$$

▶ **EXAMPLE 3**   Find the product CD given

$$C = \begin{bmatrix} -3 & 4 & 2 \\ 5 & 0 & 4 \end{bmatrix} \quad \text{and} \quad D = \begin{bmatrix} -6 & 4 \\ 2 & 3 \\ 3 & -2 \end{bmatrix}.$$

Here matrix C is 2×3 and matrix D is 3×2, so matrix CD can be found and will be 2×2.

Step 1

$$\begin{bmatrix} -3 & 4 & 2 \\ 5 & 0 & 4 \end{bmatrix} \begin{bmatrix} -6 & 4 \\ 2 & 3 \\ 3 & -2 \end{bmatrix} \qquad (-3) \cdot (-6) + 4 \cdot 2 + 2 \cdot 3 = 32$$

Step 2

$$\begin{bmatrix} -3 & 4 & 2 \\ 5 & 0 & 4 \end{bmatrix} \begin{bmatrix} -6 & 4 \\ 2 & 3 \\ 3 & -2 \end{bmatrix} \qquad (-3) \cdot 4 + 4 \cdot 3 + 2 \cdot (-2) = -4$$

Step 3

$$\begin{bmatrix} -3 & 4 & 2 \\ 5 & 0 & 4 \end{bmatrix} \begin{bmatrix} -6 & 4 \\ 2 & 3 \\ 3 & -2 \end{bmatrix} \qquad 5 \cdot (-6) + 0 \cdot 2 + 4 \cdot 3 = -18$$

Step 4

$$\begin{bmatrix} -3 & 4 & 2 \\ 5 & 0 & 4 \end{bmatrix} \begin{bmatrix} -6 & 4 \\ 2 & 3 \\ 3 & -2 \end{bmatrix} \qquad 5 \cdot 4 + 0 \cdot 3 + 4 \cdot (-2) = 12$$

Step 5   The product is

$$CD = \begin{bmatrix} -3 & 4 & 2 \\ 5 & 0 & 4 \end{bmatrix} \begin{bmatrix} -6 & 4 \\ 2 & 3 \\ 3 & -2 \end{bmatrix} = \begin{bmatrix} 32 & -4 \\ -18 & 12 \end{bmatrix}. \qquad ◀ \ \boxed{4}$$

5 Give the size of each of the following products that can be found.

(a) $\begin{bmatrix} 2 & 4 \\ 6 & 8 \end{bmatrix} \begin{bmatrix} 1 & 2 & 3 \\ 0 & -1 & 2 \end{bmatrix}$

(b) $\begin{bmatrix} 1 & 2 \\ 5 & 10 \\ 12 & 7 \end{bmatrix} \begin{bmatrix} 2 & 4 \\ 3 & 6 \\ 9 & 1 \end{bmatrix}$

(c) $\begin{bmatrix} 5 \\ 2 \\ 4 \end{bmatrix} \begin{bmatrix} 1 & 0 & 6 \end{bmatrix}$

Answer:

(a) 2×3

(b) Not possible

(c) 3×3

▶**EXAMPLE 4** Find BA given

$$A = \begin{bmatrix} 1 & -3 \\ 7 & 2 \end{bmatrix} \quad \text{and} \quad B = \begin{bmatrix} 1 & 0 & -1 \\ 3 & 1 & 4 \end{bmatrix}.$$

Since B is a 2×3 matrix and A is a 2×2 matrix, the product BA cannot be found. ◀ **5**

Note In Example 4, although BA cannot be found, matrix AB can. Since A is 2×2 and B is 2×3, AB will be a 2×3 matrix.

Matrix multiplication has some similarities with the multiplication of numbers. (Exercises 27–30 at the end of this section illustrate this fact.)

For any matrices A, B, C, such that all the indicated sums and products exist, matrix multiplication is associative and distributive.

$$A(BC) = (AB)C \qquad A(B + C) = AB + AC \qquad (B + C)A = BA + CA$$

However, there are important differences between matrix multiplication and multiplication of numbers. (See Exercises 31–34 at the end of this section.) In particular, matrix multiplication is *not* commutative.

If A and B are matrices such that the products AB and BA exist,

$$AB \text{ may not equal } BA.$$

▶**EXAMPLE 5** A contractor builds three kinds of houses, models A, B, and C, with a choice of two styles, Spanish or contemporary. Matrix P shows the number of each kind of house planned for a new 100-home subdivision.

	Spanish	Contemporary
Model A	0	30
Model B	10	20
Model C	20	20

$= P$

The amounts for each of the exterior materials used depend primarily on the style of the house. These amounts are shown in matrix Q. (Concrete is in cubic yards, lumber in units of 1000 board feet, brick in 1000s, and shingles in units of 100 square feet.)

$$\begin{array}{c} & \text{Concrete} \quad \text{Lumber} \quad \text{Brick} \quad \text{Shingles} \\ \begin{array}{c} \text{Spanish} \\ \text{Contemporary} \end{array} & \left[\begin{array}{cccc} 10 & 2 & 0 & 2 \\ 50 & 1 & 20 & 2 \end{array}\right] = Q \end{array}$$

Matrix R gives the cost for each kind of material.

$$\begin{array}{c} & \text{Cost per Unit} \\ \begin{array}{c} \text{Concrete} \\ \text{Lumber} \\ \text{Brick} \\ \text{Shingles} \end{array} & \left[\begin{array}{c} 20 \\ 180 \\ 60 \\ 25 \end{array}\right] = R \end{array}$$

(a) What is the total cost for each model house?

First find PQ. The product PQ shows the amount of each material needed for each model house.

$$PQ = \left[\begin{array}{cc} 0 & 30 \\ 10 & 20 \\ 20 & 20 \end{array}\right]\left[\begin{array}{cccc} 10 & 2 & 0 & 2 \\ 50 & 1 & 20 & 2 \end{array}\right]$$

$$= \begin{array}{c} \text{Concrete} \quad \text{Lumber} \quad \text{Brick} \quad \text{Shingles} \\ \left[\begin{array}{cccc} 1500 & 30 & 600 & 60 \\ 1100 & 40 & 400 & 60 \\ 1200 & 60 & 400 & 80 \end{array}\right] \begin{array}{c} \text{Model A} \\ \text{Model B} \\ \text{Model C} \end{array} \end{array}$$

Now multiply PQ and R, the cost matrix, to get the total cost for each model house.

$$\begin{array}{c} & & \text{Cost} \\ \left[\begin{array}{cccc} 1500 & 30 & 600 & 60 \\ 1100 & 40 & 400 & 60 \\ 1200 & 60 & 400 & 80 \end{array}\right] & \left[\begin{array}{c} 20 \\ 180 \\ 60 \\ 25 \end{array}\right] = & \left[\begin{array}{c} 72,900 \\ 54,700 \\ 60,800 \end{array}\right] \begin{array}{c} \text{Model A} \\ \text{Model B} \\ \text{Model C} \end{array} \end{array}$$

(b) How much of each of the four kinds of material must be ordered?

The totals of the columns of matrix PQ will give a matrix whose elements represent the total amounts of each material needed for the subdivision. Call this matrix T, and write it as a row matrix.

$$T = [3800 \quad 130 \quad 1400 \quad 200]$$

(c) What is the total cost for material?

Find the total cost of all the materials by taking the product of matrix T, the matrix showing the total amounts of each material, and matrix R, the cost matrix.

6 Let matrix A be

$$\begin{array}{c} \text{Vitamin} \\ \begin{array}{cc} & \begin{array}{ccc} \text{C} & \text{E} & \text{K} \end{array} \\ \text{Brand} \begin{array}{c} \text{X} \\ \text{Y} \end{array} & \left[\begin{array}{ccc} 2 & 7 & 5 \\ 4 & 6 & 9 \end{array}\right] \end{array} \end{array}$$

and matrix B be

$$\begin{array}{c} \text{Cost} \\ \begin{array}{cc} & \begin{array}{cc} \text{X} & \text{Y} \end{array} \\ \text{Vitamin} \begin{array}{c} \text{C} \\ \text{E} \\ \text{K} \end{array} & \left[\begin{array}{cc} 12 & 14 \\ 18 & 15 \\ 9 & 10 \end{array}\right] \end{array} \end{array}.$$

(a) What quantities do matrices A and B represent?

(b) What quantities does the product AB represent?

(c) What quantities does the product BA represent?

Answer:

(a) $A = $ brand/vitamin, $B = $ vitamin/cost

(b) $AB = $ brand/cost

(c) Not meaningful, although the product BA can be found

(To multiply these and get a 1×1 matrix, representing total cost, we must multiply a 1×4 matrix by a 4×1 matrix. This is why T was written as a row matrix in (b) above.)

$$TR = [3800 \quad 130 \quad 1400 \quad 200]\left[\begin{array}{c} 20 \\ 180 \\ 60 \\ 25 \end{array}\right] = [188,400]$$

(d) Suppose the contractor builds the same number of homes in five subdivisions. Calculate the total amount of each material for each model for all five subdivisions. Multiply PQ by the scalar 5, as follows.

$$5\left[\begin{array}{cccc} 1500 & 30 & 600 & 60 \\ 1100 & 40 & 400 & 60 \\ 1200 & 60 & 400 & 80 \end{array}\right] = \left[\begin{array}{cccc} 7500 & 150 & 3000 & 300 \\ 5500 & 200 & 2000 & 300 \\ 6000 & 300 & 2000 & 400 \end{array}\right] \blacktriangleleft$$

We can introduce a notation to help keep track of the quantities a matrix represents. For example, we can say that matrix P, from Example 5, represents models/styles, matrix Q represents styles/materials, and matrix R represents materials/cost. In each case, the meaning of the rows is written first and the columns second. When we found the product PQ in Example 5, the rows of the matrix represented models and the columns represented materials. Therefore, we can say the matrix product PQ represents models/materials. The common quantity, styles, in both P and Q was eliminated in the product PQ. Do you see that the product $(PQ)R$ represents models/cost?

In practical problems this notation helps decide in what order to multiply two matrices so that the results are meaningful. In Example 5(c) we could have found either product RT or product TR. However, since T represents subdivisions/materials and R represents materials/cost, the product TR gives subdivisions/cost. **6**

6.4 EXERCISES

In each of the following exercises, the dimensions of two matrices A and B are given. Find the dimensions of the product AB and the product BA, whenever these products exist. (See Examples 1 and 4.)

1. A is 2×2, B is 2×2 **2.** A is 3×3, B is 3×3 **3.** A is 4×2, B is 2×4 **4.** A is 3×1, B is 1×3

5. A is 3×5, B is 5×2 **6.** A is 4×3, B is 3×6 **7.** A is 4×2, B is 3×4 **8.** A is 7×3, B is 2×7

Let $A = \left[\begin{array}{cc} -2 & 4 \\ 0 & 3 \end{array}\right]$ *and* $B = \left[\begin{array}{cc} -6 & 2 \\ 4 & 0 \end{array}\right]$. *Find each of the following.*

9. $2A$ **10.** $-3B$ **11.** $-4B$ **12.** $5A$ **13.** $-4A + 5B$ **14.** $3A - 10B$

Find each of the following matrix products where possible. (See Examples 1–4.)

15. $\begin{bmatrix} 1 & 2 \\ 3 & 4 \end{bmatrix} \begin{bmatrix} -1 \\ 7 \end{bmatrix}$

16. $\begin{bmatrix} -1 & 5 \\ 7 & 0 \end{bmatrix} \begin{bmatrix} 6 \\ 2 \end{bmatrix}$

17. $\begin{bmatrix} 2 & 2 & -1 \\ 3 & 0 & 1 \end{bmatrix} \begin{bmatrix} 0 & 2 \\ -1 & 4 \\ 0 & 2 \end{bmatrix}$

18. $\begin{bmatrix} -9 & 2 & 1 \\ 3 & 0 & 0 \end{bmatrix} \begin{bmatrix} 2 \\ -1 \\ 4 \end{bmatrix}$

19. $\begin{bmatrix} -4 & 1 \\ 2 & -3 \end{bmatrix} \begin{bmatrix} 1 & 0 \\ 0 & 1 \end{bmatrix}$

20. $\begin{bmatrix} 1 & 0 \\ 0 & 1 \end{bmatrix} \begin{bmatrix} 3 & -2 \\ 1 & -5 \end{bmatrix}$

21. $\begin{bmatrix} 1 & 0 & 0 \\ 0 & 1 & 0 \\ 0 & 0 & 1 \end{bmatrix} \begin{bmatrix} 3 & -5 & 7 \\ -2 & 1 & 6 \\ 0 & -3 & 4 \end{bmatrix}$

22. $\begin{bmatrix} -8 & 9 & 2 \\ 3 & -4 & 7 \\ -1 & 6 & 3 \end{bmatrix} \begin{bmatrix} 1 & 0 & 0 \\ 0 & 1 & 0 \\ 0 & 0 & 1 \end{bmatrix}$

23. $\begin{bmatrix} 1 & 2 \\ 3 & 4 \end{bmatrix} \begin{bmatrix} -1 & 5 \\ 7 & 0 \end{bmatrix}$

24. $\begin{bmatrix} -1 & 5 \\ 7 & 0 \end{bmatrix} \begin{bmatrix} 1 & 2 \\ 3 & 4 \end{bmatrix}$

25. $\begin{bmatrix} -2 & -3 & 7 \\ 1 & 5 & 6 \end{bmatrix} \begin{bmatrix} 1 \\ 2 \\ 3 \end{bmatrix}$

26. $\begin{bmatrix} 6 \\ 5 \\ 4 \end{bmatrix} \begin{bmatrix} -1 & 1 & 1 \end{bmatrix}$

27. $\left(\begin{bmatrix} 4 & 3 \\ 1 & 2 \\ 0 & -5 \end{bmatrix} \begin{bmatrix} 2 & -2 \\ 1 & -1 \end{bmatrix} \right) \begin{bmatrix} 10 \\ 0 \end{bmatrix}$

28. $\begin{bmatrix} 4 & 3 \\ 1 & 2 \\ 0 & -5 \end{bmatrix} \left(\begin{bmatrix} 2 & -2 \\ 1 & -1 \end{bmatrix} \begin{bmatrix} 10 \\ 0 \end{bmatrix} \right)$

29. $\begin{bmatrix} 2 & -2 \\ 1 & -1 \end{bmatrix} \left(\begin{bmatrix} 4 & 3 \\ 1 & 2 \end{bmatrix} + \begin{bmatrix} 7 & 0 \\ -1 & 5 \end{bmatrix} \right)$

30. $\begin{bmatrix} 2 & -2 \\ 1 & -1 \end{bmatrix} \begin{bmatrix} 4 & 3 \\ 1 & 2 \end{bmatrix} + \begin{bmatrix} 2 & -2 \\ 1 & -1 \end{bmatrix} \begin{bmatrix} 7 & 0 \\ -1 & 5 \end{bmatrix}$

In Exercises 31–33, use the matrices

$$A = \begin{bmatrix} -3 & -9 \\ 2 & 6 \end{bmatrix} \quad \text{and} \quad B = \begin{bmatrix} 4 & 6 \\ 2 & 3 \end{bmatrix}.$$

31. Show that $AB \neq BA$. Hence, matrix multiplication is not commutative.

32. Show that $(A + B)^2 \neq A^2 + 2AB + B^2$.

33. Show that $(A + B)(A - B) \neq A^2 - B^2$.

34. Show that $D^2 = D$, where

$$D = \begin{bmatrix} 1 & 0 & 0 \\ \frac{1}{2} & 0 & \frac{1}{2} \\ 0 & 0 & 1 \end{bmatrix}.$$

Given matrices

$$P = \begin{bmatrix} m & n \\ p & q \end{bmatrix}, \quad X = \begin{bmatrix} x & y \\ z & w \end{bmatrix}, \quad T = \begin{bmatrix} r & s \\ t & u \end{bmatrix},$$

verify that the statements in Exercises 35–39 are true.

35. $(PX)T = P(XT)$ (Associative property: see Exercises 27 and 28)

36. $P(X + T) = PX + PT$ (Distributive property: see Exercises 29 and 30)

37. PX is a 2×2 matrix.

38. $k(X + T) = kX + kT$ for any real number k

39. $(k + h)P = kP + hP$ for any real numbers k and h

40. Management Burger Barn's three locations sell hamburgers, fries, and soft drinks. Barn I sells 900 burgers, 600 orders of fries, and 750 soft drinks each day. Barn II sells 1500 burgers a day and Barn III sells 1150. Soft drink sales number 900 a day at Barn II and 825 a day at Barn III. Barn II sells 950 and Barn III sells 800 orders of fries per day.
(a) Write a 3×3 matrix S that displays daily sales figures for all locations.
(b) Burgers cost $1.50 each, fries $.90 an order and soft drinks $.60 each. Write a 1×3 matrix P that displays the prices.
(c) What matrix product displays the daily revenue at each of the three locations?
(d) What is the total daily revenue from all locations?

41. Management The Bread Box, a small neighborhood bakery, sells four main items: sweet rolls, bread, cake, and pie. The amount of certain major ingredients required to make these items is given in matrix A.

	Eggs	Flour*	Sugar*	Butter*	Milk*	
	1	4	$\frac{1}{4}$	$\frac{1}{4}$	1	Sweet rolls (dozen)
$A =$	0	3	0	$\frac{1}{4}$	0	Bread (loaves)
	4	3	2	1	1	Cake (1)
	0	1	0	$\frac{1}{3}$	0	Pie (1)

The cost (in cents) for each ingredient when purchased in large lots and in small lots is given by matrix B.

Cost (in cents)

	Large Lot	Small Lot	
	5	5	Eggs
	8	10	Flour†
$B =$	10	12	Sugar†
	12	15	Butter†
	5	6	Milk†

(a) Use matrix multiplication to find a matrix representing the comparative costs per item under the two purchase options.

*Measured in cups.
†Cost per cup.

Suppose a day's orders consist of 20 dozen sweet rolls, 200 loaves of bread, 50 cakes, and 60 pies.
(b) Represent these orders as a 1×4 matrix and use matrix multiplication to write as a matrix the amount of each ingredient required to fill the day's orders.
(c) Use matrix multiplication to find a matrix representing the costs under the two purchase options to fill the day's orders.

42. Let I be the matrix $I = \begin{bmatrix} 1 & 0 \\ 0 & 1 \end{bmatrix}$, and let matrices P, X, and T be defined as in Exercises 35–39.
(a) Find IP, PI, IX.
(b) Without calculating, guess what the matrix IT might be.
(c) Suggest a reason for naming a matrix such as I an identity matrix.

43. Management The EZ Life Company buys three machines costing $90,000, $60,000, and $120,000, respectively. They plan to depreciate the machines over a 3-year period using the sum-of-the-years'-digits method, where the company assumes that the machines will have 20% of their value left at the end of 3 years. Thus they will compute the depreciation of 80% of the cost using rates of 1/2, 1/3, and 1/6, respectively, for the 3-year period.
(a) Write a column matrix C to represent the cost of these three assets.
(b) To get a matrix that represents 80% of the costs, let

$$P = \begin{bmatrix} .8 & 0 & 0 \\ 0 & .8 & 0 \\ 0 & 0 & .8 \end{bmatrix}.$$

Then PC is the desired matrix. Calculate PC.
(c) Write a row matrix R representing the three rates given above.
(d) Find $(PC)R$. This product matrix (which should be 3×3) represents the depreciation charge for each asset for each year of the 3-year period.

44. In Exercise 37, Section 6.3, label the matrices found in parts (a), (b), and (c) respectively X, Y, and Z.
(a) Find the product matrix XY. What do the entries of this matrix represent?
(b) Find the product matrix YZ. What do the entries represent?

In Exercises 45 and 46 show that the equation $AX = B$ represents a linear system of two equations in two unknowns. Solve the system and substitute in the matrix equation to check your results.

45. $A = \begin{bmatrix} 2 & 4 \\ 1 & 3 \end{bmatrix}$, $X = \begin{bmatrix} x_1 \\ x_2 \end{bmatrix}$, $B = \begin{bmatrix} 2 \\ -1 \end{bmatrix}$

46. $A = \begin{bmatrix} 1 & 2 \\ -3 & 5 \end{bmatrix}$, $X = \begin{bmatrix} x_1 \\ x_2 \end{bmatrix}$, $B = \begin{bmatrix} -4 \\ 12 \end{bmatrix}$

Use a computer or calculator and the following matrices to find the matrix products in Exercises 47–53.

$$A = \begin{bmatrix} 2 & 3 & -1 & 5 & 10 \\ 2 & 8 & 7 & 4 & 3 \\ -1 & -4 & -12 & 6 & 8 \\ 2 & 5 & 7 & 1 & 4 \end{bmatrix} \quad B = \begin{bmatrix} 9 & 3 & 7 & -6 \\ -1 & 0 & 4 & 2 \\ -10 & -7 & 6 & 9 \\ 8 & 4 & 2 & -1 \\ 2 & -5 & 3 & 7 \end{bmatrix}$$

$$C = \begin{bmatrix} -6 & 8 & 2 & 4 & -3 \\ 1 & 9 & 7 & -12 & 5 \\ 15 & 2 & -8 & 10 & 11 \\ 4 & 7 & 9 & 6 & -2 \\ 1 & 3 & 8 & 23 & 4 \end{bmatrix} \quad D = \begin{bmatrix} 5 & -3 & 7 & 9 & 2 \\ 6 & 8 & -5 & 2 & 1 \\ 3 & 7 & -4 & 2 & 11 \\ 5 & -3 & 9 & 4 & -1 \\ 0 & 3 & 2 & 5 & 1 \end{bmatrix}$$

47. CD

48. AC

49. CA

50. DC

51. Is $AC = CA$?

52. Is $CD = DC$?

53. Find $C + D$, $(C + D)B$, CB, DB, and $CB + DB$. Does $(C + D)B = CB + DB$?

---

1 Let $A = \begin{bmatrix} 3 & -2 \\ 4 & -1 \end{bmatrix}$

and $I = \begin{bmatrix} 1 & 0 \\ 0 & 1 \end{bmatrix}$.

Find IA and AI.

Answer:

$IA = \begin{bmatrix} 3 & -2 \\ 4 & -1 \end{bmatrix} = A$

and $AI = \begin{bmatrix} 3 & -2 \\ 4 & -1 \end{bmatrix} = A$

6.5 MATRIX INVERSES

In Section 6.3, a zero matrix was defined with properties similar to those of the real number zero, the identity for addition. Recall from Section 1.1 that the real number 1 is the identity element for multiplication of real numbers: for any real number a, $a \cdot 1 = 1 \cdot a = a$. In this section, an **identity matrix I** is defined that has properties similar to those of the number 1. This identity matrix is then used to find the multiplicative inverse of any square matrix that has an inverse.

If I is to be the identity matrix, the products AI and IA must both equal A. The 2×2 identity matrix that satisfies these conditions is

$$I = \begin{bmatrix} 1 & 0 \\ 0 & 1 \end{bmatrix}. \quad \boxed{1}$$

To check that I, as defined above, is really the 2×2 identity matrix, let

$$A = \begin{bmatrix} a & b \\ c & d \end{bmatrix}.$$

Then AI and IA should both equal A.

$$AI = \begin{bmatrix} a & b \\ c & d \end{bmatrix} \begin{bmatrix} 1 & 0 \\ 0 & 1 \end{bmatrix} = \begin{bmatrix} a(1) + b(0) & a(0) + b(1) \\ c(1) + d(0) & c(0) + d(1) \end{bmatrix} = \begin{bmatrix} a & b \\ c & d \end{bmatrix} = A$$

$$IA = \begin{bmatrix} 1 & 0 \\ 0 & 1 \end{bmatrix} \begin{bmatrix} a & b \\ c & d \end{bmatrix} = \begin{bmatrix} 1(a) + 0(c) & 1(b) + 0(d) \\ 0(a) + 1(c) & 0(b) + 1(d) \end{bmatrix} = \begin{bmatrix} a & b \\ c & d \end{bmatrix} = A$$

This verifies that I has been defined correctly. (It can also be shown that I is the only 2×2 identity matrix.)

The identity matrices for 3×3 matrices and 4×4 matrices, respectively, are

$$I = \begin{bmatrix} 1 & 0 & 0 \\ 0 & 1 & 0 \\ 0 & 0 & 1 \end{bmatrix} \quad \text{and} \quad I = \begin{bmatrix} 1 & 0 & 0 & 0 \\ 0 & 1 & 0 & 0 \\ 0 & 0 & 1 & 0 \\ 0 & 0 & 0 & 1 \end{bmatrix}.$$

By generalizing, an identity matrix can be found for any n by n matrix: this identity matrix will have 1s on the main diagonal from upper left to lower right, with all other entries equal to 0.

Recall that the multiplicative inverse of the nonzero real number a is $1/a$. The product of a and its multiplicative inverse $1/a$ is 1. Now we try a similar thing with matrices: given a matrix A, is there a matrix A^{-1} (read "A-inverse") satisfying both

$$AA^{-1} = I \quad \text{and} \quad A^{-1}A = I?$$

It turns out that this matrix inverse A^{-1} can often be found using the row operations of Section 6.2.

Caution Only square matrices have inverses, but not every square matrix has an inverse. If an inverse exists, it is unique. That is, any given square matrix has no more than one inverse. Note that the symbol A^{-1} does not mean $1/A$; the symbol A^{-1} is just the notation for the inverse of matrix A.

▶**EXAMPLE 1** Given matrices A and B below, decide if they are inverses.

$$A = \begin{bmatrix} 2 & 3 \\ 1 & 8 \end{bmatrix} \quad B = \begin{bmatrix} -1 & 3 \\ 1 & -2 \end{bmatrix}$$

2 Given

$$A = \begin{bmatrix} 1 & 2 \\ 4 & 6 \end{bmatrix}$$

and

$$B = \begin{bmatrix} -3 & 1 \\ 2 & -\frac{1}{2} \end{bmatrix},$$

decide if they are inverses.

Answer:
Yes, since $AB = BA = I$.

The matrices are inverses if AB and BA both equal I.

$$AB = \begin{bmatrix} 2 & 3 \\ 1 & 8 \end{bmatrix} \begin{bmatrix} -1 & 3 \\ 1 & -2 \end{bmatrix} = \begin{bmatrix} 1 & 0 \\ 7 & -13 \end{bmatrix} \neq I$$

Since $AB \neq I$, the two matrices are not inverses of each other. ◀ **2**

The augmented matrices written earlier, such as

$$\begin{bmatrix} 1 & 2 & | & 2 \\ 3 & 4 & | & 5 \end{bmatrix},$$

are really a combination of two matrices written as one. The vertical bar separates the two original matrices

$$\begin{bmatrix} 1 & 2 \\ 3 & 4 \end{bmatrix} \quad \text{and} \quad \begin{bmatrix} 2 \\ 5 \end{bmatrix}.$$

As an example of finding the multiplicative inverse of a matrix, let us look for the inverse of

$$A = \begin{bmatrix} 2 & 4 \\ 1 & -1 \end{bmatrix}.$$

Let the unknown inverse matrix be

$$A^{-1} = \begin{bmatrix} x & y \\ z & w \end{bmatrix}.$$

By the definition of matrix inverse, $AA^{-1} = I$, or

$$AA^{-1} = \begin{bmatrix} 2 & 4 \\ 1 & -1 \end{bmatrix} \begin{bmatrix} x & y \\ z & w \end{bmatrix} = \begin{bmatrix} 1 & 0 \\ 0 & 1 \end{bmatrix}.$$

Use matrix multiplication to get

$$\begin{bmatrix} 2x + 4z & 2y + 4w \\ x - z & y - w \end{bmatrix} = \begin{bmatrix} 1 & 0 \\ 0 & 1 \end{bmatrix}.$$

Setting corresponding elements equal gives the system of equations

$$2x + 4z = 1 \tag{1}$$
$$2y + 4w = 0 \tag{2}$$
$$x - z = 0 \tag{3}$$
$$y - w = 1. \tag{4}$$

Since equations (1) and (3) involve only x and z, while equations (2) and (4) involve only y and w, these four equations lead to two systems of equations,

$$2x + 4z = 1 \qquad 2y + 4w = 0$$
$$\text{and}$$
$$x - z = 0 \qquad y - w = 1.$$

Writing the two systems as augmented matrices gives

$$\begin{bmatrix} 2 & 4 & | & 1 \\ 1 & -1 & | & 0 \end{bmatrix} \quad \text{and} \quad \begin{bmatrix} 2 & 4 & | & 0 \\ 1 & -1 & | & 1 \end{bmatrix}.$$

Each of these systems can be solved by the Gauss-Jordan method. However, since the elements to the left of the vertical bar are identical, the two systems can be combined into one matrix

$$\begin{bmatrix} 2 & 4 & | & 1 & 0 \\ 1 & -1 & | & 0 & 1 \end{bmatrix}$$

and solved simultaneously as follows. Interchange the two rows to get a 1 in the upper left corner.

$$\begin{bmatrix} 1 & -1 & | & 0 & 1 \\ 2 & 4 & | & 1 & 0 \end{bmatrix} \qquad \textbf{Interchange } R_1, R_2$$

Multiply row one by -2 and add the results to row two to get

$$\begin{bmatrix} 1 & -1 & | & 0 & 1 \\ 0 & 6 & | & 1 & -2 \end{bmatrix}. \qquad -2R_1 + R_2$$

Now, to get a 1 in the second row, second column position, multiply row two by 1/6.

$$\begin{bmatrix} 1 & -1 & | & 0 & 1 \\ 0 & 1 & | & \frac{1}{6} & -\frac{1}{3} \end{bmatrix} \qquad \frac{1}{6}R_2$$

Finally, add row two to row one to get a 0 in the second column above the 1.

$$\begin{bmatrix} 1 & 0 & | & \frac{1}{6} & \frac{2}{3} \\ 0 & 1 & | & \frac{1}{6} & -\frac{1}{3} \end{bmatrix} \qquad R_2 + R_1$$

The numbers in the first column to the right of the vertical bar give the values of x and z. The second column gives the values of y and w. That is,

$$\begin{bmatrix} 1 & 0 & | & x & y \\ 0 & 1 & | & z & w \end{bmatrix} = \begin{bmatrix} 1 & 0 & | & \frac{1}{6} & \frac{2}{3} \\ 0 & 1 & | & \frac{1}{6} & -\frac{1}{3} \end{bmatrix}$$

so that
$$A^{-1} = \begin{bmatrix} x & y \\ z & w \end{bmatrix} = \begin{bmatrix} \frac{1}{6} & \frac{2}{3} \\ \frac{1}{6} & -\frac{1}{3} \end{bmatrix}.$$

3 (a) Find A^{-1} if

$$A = \begin{bmatrix} 2 & 2 \\ 4 & 1 \end{bmatrix}.$$

(b) Check your answer by finding AA^{-1} and $A^{-1}A$.

Answer:

(a) $\begin{bmatrix} -\frac{1}{6} & \frac{1}{3} \\ \frac{2}{3} & -\frac{1}{3} \end{bmatrix}$

(b) Both equal $\begin{bmatrix} 1 & 0 \\ 0 & 1 \end{bmatrix}$.

4 (a) Complete this step.

(b) Write this row transformation as _____.

Answer:

(a)

$$\begin{bmatrix} 1 & 0 & 1 & | & 1 & 0 & 0 \\ 0 & -2 & -3 & | & -2 & 1 & 0 \\ 0 & 0 & -3 & | & -3 & 0 & 1 \end{bmatrix}$$

(b) $-3R_1 + R_3$

Check by multiplying A and A^{-1}. The result should be I.

$$AA^{-1} = \begin{bmatrix} 2 & 4 \\ 1 & -1 \end{bmatrix}\begin{bmatrix} \frac{1}{6} & \frac{2}{3} \\ \frac{1}{6} & -\frac{1}{3} \end{bmatrix} = \begin{bmatrix} \frac{1}{3} + \frac{2}{3} & \frac{4}{3} - \frac{4}{3} \\ \frac{1}{6} - \frac{1}{6} & \frac{2}{3} + \frac{1}{3} \end{bmatrix} = \begin{bmatrix} 1 & 0 \\ 0 & 1 \end{bmatrix} = I.$$

Verify that $A^{-1}A = I$, also. **3**

This procedure for finding the inverse of a matrix can be generalized as follows.

To obtain an **inverse matrix** A^{-1} for any $n \times n$ matrix A for which A^{-1} exists, follow these steps.

1. From the augmented matrix $[A \,|\, I]$ where I is the $n \times n$ identity matrix.
2. Perform row operations on $[A \,|\, I]$ to get a matrix of the form $[I \,|\, B]$.
3. Matrix B is A^{-1}.

▶**EXAMPLE 2**   Find A^{-1} if $A = \begin{bmatrix} 1 & 0 & 1 \\ 2 & -2 & -1 \\ 3 & 0 & 0 \end{bmatrix}$.

First write the augmented matrix $[A \,|\, I]$.

$$[A \,|\, I] = \begin{bmatrix} 1 & 0 & 1 & | & 1 & 0 & 0 \\ 2 & -2 & -1 & | & 0 & 1 & 0 \\ 3 & 0 & 0 & | & 0 & 0 & 1 \end{bmatrix}$$

The augmented matrix already has 1 in the upper left-hand corner as needed, so begin by selecting the row operation which will result in a 0 for the first element in row two. Multiply row one by -2 and add the result to row two. This gives

$$\begin{bmatrix} 1 & 0 & 1 & | & 1 & 0 & 0 \\ 0 & -2 & -3 & | & -2 & 1 & 0 \\ 3 & 0 & 0 & | & 0 & 0 & 1 \end{bmatrix}. \qquad -2R_1 + R_2$$

Get 0 for the first element in row three by multiplying row one by -3 and adding to row three. **4**

5 (a) Complete this step.

(b) Write this row transformation as _____.

Answer:

(a)

$$\begin{bmatrix} 1 & 0 & 0 & | & 0 & 0 & \frac{1}{3} \\ 0 & 1 & \frac{3}{2} & | & 1 & -\frac{1}{2} & 0 \\ 0 & 0 & 1 & | & 1 & 0 & -\frac{1}{3} \end{bmatrix}$$

(b) $-1R_3 + R_1$

6 (a) Find A^{-1} if

$$A = \begin{bmatrix} 2 & 8 \\ -1 & -5 \end{bmatrix}.$$

(b) Check this answer by finding AA^{-1} and $A^{-1}A$.

Answer:

(a) $\begin{bmatrix} \frac{5}{2} & 4 \\ -\frac{1}{2} & -1 \end{bmatrix}$

(b) Each product equals

$$\begin{bmatrix} 1 & 0 \\ 0 & 1 \end{bmatrix}.$$

Get 1 for the second element in row two by multiplying row two of the matrix found in Problem 4 at the side by $-1/2$, obtaining the new matrix

$$\begin{bmatrix} 1 & 0 & 1 & | & 1 & 0 & 0 \\ 0 & 1 & \frac{3}{2} & | & 1 & -\frac{1}{2} & 0 \\ 0 & 0 & -3 & | & -3 & 0 & 1 \end{bmatrix}. \qquad -\frac{1}{2}R_2$$

Get 1 for the third element in row three by multiplying row three by $-1/3$, with the result

$$\begin{bmatrix} 1 & 0 & 1 & | & 1 & 0 & 0 \\ 0 & 1 & \frac{3}{2} & | & 1 & -\frac{1}{2} & 0 \\ 0 & 0 & 1 & | & 1 & 0 & -\frac{1}{3} \end{bmatrix}. \qquad -\frac{1}{3}R_3$$

Now get 0 for the third element in row one by multiplying row three by -1 and adding to row one. **5**

Get 0 for the third element in row two by multiplying row three of the matrix found in Problem 5 at the side by $-3/2$ and adding to row two.

$$\begin{bmatrix} 1 & 0 & 0 & | & 0 & 0 & \frac{1}{3} \\ 0 & 1 & 0 & | & -\frac{1}{2} & -\frac{1}{2} & \frac{1}{2} \\ 0 & 0 & 1 & | & 1 & 0 & -\frac{1}{3} \end{bmatrix} \qquad -\frac{3}{2}R_3 + R_2$$

This last transformation gives the desired inverse:

$$A^{-1} = \begin{bmatrix} 0 & 0 & \frac{1}{3} \\ -\frac{1}{2} & -\frac{1}{2} & \frac{1}{2} \\ 1 & 0 & -\frac{1}{3} \end{bmatrix}.$$

Confirm this by forming the products $A^{-1}A$ and AA^{-1}. Each should equal I. ◀ **6**

▶ **EXAMPLE 3** Find A^{-1} if $A = \begin{bmatrix} 2 & -4 \\ 1 & -2 \end{bmatrix}$.

Using row operations to transform the first column of the augmented matrix

$$\begin{bmatrix} 2 & -4 & | & 1 & 0 \\ 1 & -2 & | & 0 & 1 \end{bmatrix}$$

results in the following matrices.

7 Find the inverse, if it exists, of each matrix below.

(a) $\begin{bmatrix} 8 & 4 \\ 6 & 3 \end{bmatrix}$ **(b)** $\begin{bmatrix} 0 & 1 \\ 1 & 0 \end{bmatrix}$

(c) $\begin{bmatrix} 6 & -1 & 2 \\ 4 & 1 & 3 \\ -3 & \frac{1}{2} & -1 \end{bmatrix}$

Answer:

(a) and **(c)** have no inverse;
(b) is its own inverse.

$$\begin{bmatrix} 1 & -2 & \Big| & \frac{1}{2} & 0 \\ 1 & -2 & \Big| & 0 & 1 \end{bmatrix} \quad \frac{1}{2}R_1$$

$$\begin{bmatrix} 1 & -2 & \Big| & \frac{1}{2} & 0 \\ 0 & 0 & \Big| & -\frac{1}{2} & 1 \end{bmatrix} \quad -1R_1 + R_2$$

At this point, the matrix should be changed so that the second element of row two will be 1. Since that element is now 0, there is no way to complete the desired transformation. What is wrong? Remember, near the beginning of this section we mentioned that some matrices do not have inverses. Matrix A is an example of a matrix that has no inverse. In this case, there is no matrix A^{-1} such that $AA^{-1} = A^{-1}A = I$. ◀ **7**

6.5 EXERCISES

Use row operation (3) to change each of the following matrices as indicated.

1. $\begin{bmatrix} 2 & 4 \\ 4 & 7 \end{bmatrix}$; $-2R_1 + R_2$

2. $\begin{bmatrix} -1 & 4 \\ 7 & 0 \end{bmatrix}$; $7R_1 + R_2$

3. $\begin{bmatrix} 1 & \frac{1}{5} \\ 5 & 2 \end{bmatrix}$; $-5R_1 + R_2$

4. $\begin{bmatrix} 5 & 7 \\ 2 & -4 \end{bmatrix}$; $\frac{4}{7}R_1 + R_2$

5. $\begin{bmatrix} 1 & 5 & 6 \\ -2 & 3 & -1 \\ 4 & 7 & 0 \end{bmatrix}$; $2R_1 + R_2$

6. $\begin{bmatrix} 2 & 5 & 6 \\ 4 & -1 & 2 \\ 3 & 7 & 1 \end{bmatrix}$; $-6R_3 + R_1$

7. $\begin{bmatrix} -3 & 1 & -4 \\ 2 & 1 & 3 \\ -7 & 5 & 2 \end{bmatrix}$; $-5R_2 + R_3$

8. $\begin{bmatrix} 4 & 10 & -8 \\ 7 & 4 & 3 \\ -1 & 1 & 0 \end{bmatrix}$; $-4R_3 + R_2$

Decide whether the given matrices are inverses of each other. (Check to see if their product is I. See Example 1.)

9. $\begin{bmatrix} 2 & 3 \\ 1 & 1 \end{bmatrix}$ and $\begin{bmatrix} -1 & 3 \\ 1 & -2 \end{bmatrix}$

10. $\begin{bmatrix} 5 & 7 \\ 2 & 3 \end{bmatrix}$ and $\begin{bmatrix} 3 & -7 \\ -2 & 5 \end{bmatrix}$

11. $\begin{bmatrix} 2 & 1 \\ 3 & 2 \end{bmatrix}$ and $\begin{bmatrix} 2 & 1 \\ -3 & 2 \end{bmatrix}$

12. $\begin{bmatrix} -1 & 2 \\ 3 & -5 \end{bmatrix}$ and $\begin{bmatrix} -5 & -2 \\ -3 & -1 \end{bmatrix}$

13. $\begin{bmatrix} 1 & 2 & 0 \\ 0 & 1 & 0 \\ 0 & 1 & 0 \end{bmatrix}$ and $\begin{bmatrix} 1 & -2 & 0 \\ 0 & 1 & 0 \\ 0 & -1 & 1 \end{bmatrix}$

14. $\begin{bmatrix} 0 & 1 & 0 \\ 0 & 0 & -2 \\ 1 & -1 & 0 \end{bmatrix}$ and $\begin{bmatrix} 1 & 0 & 1 \\ 1 & 0 & 0 \\ 0 & -1 & 0 \end{bmatrix}$

15. $\begin{bmatrix} 1 & 3 & 3 \\ 1 & 4 & 3 \\ 1 & 3 & 4 \end{bmatrix}$ and $\begin{bmatrix} 7 & -3 & -3 \\ -1 & 1 & 0 \\ -1 & 0 & 1 \end{bmatrix}$

16. $\begin{bmatrix} -1 & 0 & 2 \\ 3 & 1 & 0 \\ 0 & 2 & -3 \end{bmatrix}$ and $\begin{bmatrix} -\frac{1}{5} & \frac{4}{15} & -\frac{2}{15} \\ \frac{3}{5} & \frac{1}{5} & \frac{2}{5} \\ \frac{2}{5} & \frac{2}{15} & -\frac{1}{15} \end{bmatrix}$

Find the inverse, if it exists, for each of the following matrices. (See Examples 2 and 3.)

17. $\begin{bmatrix} 1 & -1 \\ 2 & 0 \end{bmatrix}$

18. $\begin{bmatrix} -1 & 2 \\ -2 & -1 \end{bmatrix}$

19. $\begin{bmatrix} 3 & -1 \\ -5 & 2 \end{bmatrix}$

20. $\begin{bmatrix} -1 & -2 \\ 3 & 4 \end{bmatrix}$

21. $\begin{bmatrix} -6 & 4 \\ -3 & 2 \end{bmatrix}$

22. $\begin{bmatrix} 5 & 10 \\ -3 & -6 \end{bmatrix}$

23. $\begin{bmatrix} 1 & 0 & 0 \\ 0 & -1 & 0 \\ 1 & 0 & 1 \end{bmatrix}$

24. $\begin{bmatrix} 1 & 0 & 1 \\ 0 & -1 & 0 \\ 2 & 1 & 1 \end{bmatrix}$

25. $\begin{bmatrix} -1 & -1 & -1 \\ 4 & 5 & 0 \\ 0 & 1 & -3 \end{bmatrix}$

26. $\begin{bmatrix} 2 & 0 & 4 \\ 3 & 1 & 5 \\ -1 & 1 & -2 \end{bmatrix}$

27. $\begin{bmatrix} 1 & 2 & 3 \\ -3 & -2 & -1 \\ -1 & 0 & 1 \end{bmatrix}$

28. $\begin{bmatrix} 2 & 0 & 4 \\ 1 & 0 & -1 \\ 3 & 0 & -2 \end{bmatrix}$

29. $\begin{bmatrix} 2 & 4 & 6 \\ -1 & -4 & -3 \\ 0 & 1 & -1 \end{bmatrix}$

30. $\begin{bmatrix} 2 & 2 & -4 \\ 2 & 6 & 0 \\ -3 & -3 & 5 \end{bmatrix}$

31. $\begin{bmatrix} 1 & -2 & 3 & 0 \\ 0 & 1 & -1 & 1 \\ -2 & 2 & -2 & 4 \\ 0 & 2 & -3 & 1 \end{bmatrix}$

32. $\begin{bmatrix} 1 & 1 & 0 & 2 \\ 2 & -1 & 1 & -1 \\ 3 & 3 & 2 & -2 \\ 1 & 2 & 1 & 0 \end{bmatrix}$

In Exercises 33–37, let $A = \begin{bmatrix} a & b \\ c & d \end{bmatrix}$. Show that each of the statements in Exercises 33–35 is true.

33. $IA = A$ **34.** $AI = A$ **35.** $A \cdot O = O$

36. Find A^{-1}. (Assume $ad - bc \neq 0$.) Show that $AA^{-1} = I$.

37. Show that $A^{-1}A = I$.

38. Using the definitions and properties listed in this section, show that for square matrices A and B of the same order, if $AB = O$ and if A^{-1} exists, then $B = O$.

 Use a computer and matrices C and D to find the inverses in Exercises 39–44.

$$C = \begin{bmatrix} -6 & 8 & 2 & 4 & -3 \\ 1 & 9 & 7 & -12 & 5 \\ 15 & 2 & -8 & 10 & 11 \\ 4 & 7 & 9 & 6 & -2 \\ 1 & 3 & 8 & 23 & 4 \end{bmatrix} \quad D = \begin{bmatrix} 5 & -3 & 7 & 9 & 2 \\ 6 & 8 & -5 & 2 & 1 \\ 3 & 7 & -4 & 2 & 11 \\ 5 & -3 & 9 & 4 & -1 \\ 0 & 3 & 2 & 5 & 1 \end{bmatrix}$$

39. C^{-1} **40.** $(CD)^{-1}$ **41.** D^{-1}

42. Is $C^{-1}D^{-1} = (CD)^{-1}$?

43. Find $C^{-1}C$. This product *should* equal the identity matrix I, but doesn't *quite* equal I, because of *round-off error*.

44. Find DD^{-1}. Does this product exactly equal I?

6.6 APPLICATIONS OF MATRICES

This section gives a variety of applications of matrices.

SOLVING SYSTEMS WITH MATRICES Consider this system of linear equations.

$$2x - 3y = 4$$
$$x + 5y = 2$$

Let

$$A = \begin{bmatrix} 2 & -3 \\ 1 & 5 \end{bmatrix}, \quad X = \begin{bmatrix} x \\ y \end{bmatrix}, \quad B = \begin{bmatrix} 4 \\ 2 \end{bmatrix}.$$

Since

$$AX = \begin{bmatrix} 2 & -3 \\ 1 & 5 \end{bmatrix}\begin{bmatrix} x \\ y \end{bmatrix} = \begin{bmatrix} 2x - 3y \\ x + 5y \end{bmatrix} \quad \text{and} \quad B = \begin{bmatrix} 4 \\ 2 \end{bmatrix},$$

the original system is equivalent to the single matrix equation $AX = B$. Similarly, any system of linear equations can be written as a matrix equation $AX = B$. The matrix A is called the **coefficient matrix.**

A matrix equation $AX = B$ can be solved if A^{-1} exists. Assuming A^{-1} exists and using the facts that $A^{-1}A = I$ and $IX = X$ along with the associative property of mulitplication of matrices gives

$$AX = B$$
$$A^{-1}(AX) = A^{-1}B \qquad \text{Multiply both sides by } A^{-1}$$
$$(A^{-1}A)X = A^{-1}B \qquad \text{Associative property}$$
$$IX = A^{-1}B \qquad \text{Inverse property}$$
$$X = A^{-1}B. \qquad \text{Identity property}$$

When multiplying by matrices on both sides of a matrix equation, be careful to multiply in the same order on both sides of the equation, since multiplication of matrices is not commutative (unlike multiplication of real numbers). This discussion is summarized below.

A system of equations $AX = B$, where A is the matrix of coefficients, X is the matrix of variables, and B is the matrix of constants, is solved by first finding A^{-1}. Then, if A^{-1} exists, $X = A^{-1}B$.

This method is most practical in cases where A^{-1} is known or where several systems with the same coefficient matrix, but different constants, are to be solved. Then just one inverse matrix must be found.

▶**EXAMPLE 1**   Use the inverse of the coefficient matrix to solve the following systems.

(a) $-x - 2y + 2z = 9$
    $2x + y - z = -3$
    $3x - 2y + z = -6$

(b) $-x - 2y + 2z = 3$
    $2x + y - z = 3$
    $3x - 2y + z = 7$

(c) $-x - 2y + 2z = 12$
    $2x + y - z = 0$
    $3x - 2y + z = 18$

Notice that the three systems all have the same matrix of coefficients and the same matrix of variables.

$$A = \begin{bmatrix} -1 & -2 & 2 \\ 2 & 1 & -1 \\ 3 & -2 & 1 \end{bmatrix} \quad \text{and} \quad X = \begin{bmatrix} x \\ y \\ z \end{bmatrix}$$

Find A^{-1} first, then use it to solve all four systems.

To find A^{-1}, we start with matrix

$$[A \mid I] = \begin{bmatrix} -1 & -2 & 2 & | & 1 & 0 & 0 \\ 2 & 1 & -1 & | & 0 & 1 & 0 \\ 3 & -2 & 1 & | & 0 & 0 & 1 \end{bmatrix}$$

and use row operations to get $[I \mid A^{-1}]$, from which

$$A^{-1} = \begin{bmatrix} \frac{1}{3} & \frac{2}{3} & 0 \\ \frac{5}{3} & \frac{7}{3} & -1 \\ \frac{7}{3} & \frac{8}{3} & -1 \end{bmatrix}.$$

Now we can solve each of the three systems by using $X = A^{-1}B$.

(a) Here the matrix of coefficients is

$$B = \begin{bmatrix} 9 \\ -3 \\ -6 \end{bmatrix}.$$

Since $X = A^{-1}B$,

$$X = \begin{bmatrix} \frac{1}{3} & \frac{2}{3} & 0 \\ \frac{5}{3} & \frac{7}{3} & -1 \\ \frac{7}{3} & \frac{8}{3} & -1 \end{bmatrix} \begin{bmatrix} 9 \\ -3 \\ -6 \end{bmatrix} = \begin{bmatrix} 1 \\ 14 \\ 19 \end{bmatrix}.$$

From this result, $x = 1$, $y = 14$, and $z = 19$, and the solution is $(1, 14, 19)$.

1 **(a)** Write the matrix of coefficients, the matrix of variables, and the matrix of constants for the system

$$2x + 6y = -14$$
$$-x - 2y = 3.$$

(b) Solve the system in part (a) by using the inverse of the coefficient matrix.

Answer:

(a) $A = \begin{bmatrix} 2 & 6 \\ -1 & -2 \end{bmatrix}$,

$X = \begin{bmatrix} x \\ y \end{bmatrix}$,

$B = \begin{bmatrix} -14 \\ 3 \end{bmatrix}$

(b) $(5, -4)$

(b) The matrix of coefficients is $B = \begin{bmatrix} 3 \\ 3 \\ 7 \end{bmatrix}$.

$$X = A^{-1}B = \begin{bmatrix} \frac{1}{3} & \frac{2}{3} & 0 \\ \frac{5}{3} & \frac{7}{3} & -1 \\ \frac{7}{3} & \frac{8}{3} & -1 \end{bmatrix} \begin{bmatrix} 3 \\ 3 \\ 7 \end{bmatrix} = \begin{bmatrix} 3 \\ 5 \\ 8 \end{bmatrix}$$

The solution is $(3, 5, 8)$.

(c) This time, $B = \begin{bmatrix} 12 \\ 0 \\ 18 \end{bmatrix}$.

$$X = A^{-1}B = \begin{bmatrix} \frac{1}{3} & \frac{2}{3} & 0 \\ \frac{5}{3} & \frac{7}{3} & -1 \\ \frac{7}{3} & \frac{8}{3} & -1 \end{bmatrix} \begin{bmatrix} 12 \\ 0 \\ 18 \end{bmatrix} = \begin{bmatrix} 4 \\ 2 \\ 10 \end{bmatrix}$$

The solution is $(4, 2, 10)$. ◀ **1**

INPUT-OUTPUT ANALYSIS An interesting application of matrix theory to economics was developed by Nobel Prize winner Wassily Leontief. His application of matrices to the interdependencies in an economy is called **input-output** analysis. In practice, input-output analysis is very complicated with many variables. We shall discuss only simple examples with a few variables.

Input-output models are concerned with the production and flow of goods (and perhaps services). In an economy with n basic commodities (or sectors), the production of each commodity uses some (perhaps all) of the commodities in the economy as inputs. The amounts of each commodity used in the production of 1 unit of each commodity can be written as an $n \times n$ matrix A, called the **technological** or **input-output matrix** of the economy.

▶**EXAMPLE 2** Suppose a simplified economy involves just three commodity categories: agriculture, manufacturing, and transportation, all in appropriate units. Production of 1 unit of agriculture requires 1/2 unit of manufacturing and 1/4 unit of transportation. Production of 1 unit of manufacturing requires 1/4 unit of agriculture and 1/4 unit of transportation; while production of 1 unit of transportation requires 1/3 unit of agriculture and 1/4 unit of manufacturing. Write the input-output matrix of this economy.

2 Write a 2×2 technological matrix in which 1 unit of electricity requires 1/2 unit of water and 1/3 unit of electricity, while 1 unit of water requires no water but 1/4 unit of electricity.

Answer:

	Elec.	Water
Elec.	$\frac{1}{3}$	$\frac{1}{4}$
Water	$\frac{1}{2}$	0

The matrix is shown below.

$$
\begin{array}{cc}
 & \text{Output} \\
\text{Input}\quad\begin{array}{l}
\text{Agriculture}\\
\text{Manufacturing}\\
\text{Transportation}
\end{array} &
\begin{array}{ccc}
\text{Agricul-}\atop\text{ture} & \text{Manufac-}\atop\text{turing} & \text{Trans-}\atop\text{portation}
\end{array}\\
\end{array}
$$

$$
\begin{array}{l}
\\
\text{Agriculture}\\
\text{Manufacturing}\\
\text{Transportation}
\end{array}
\left[
\begin{array}{ccc}
0 & \frac{1}{4} & \frac{1}{3}\\
\frac{1}{2} & 0 & \frac{1}{4}\\
\frac{1}{4} & \frac{1}{4} & 0
\end{array}
\right] = A
$$

The first column of the input-output matrix represents the amount of each of the three commodities consumed in the production of 1 unit of agriculture. The second column gives the corresponding amounts required to produce 1 unit of manufacturing, and the last column gives the amounts needed to produce 1 unit of transportation. (Although it is unrealistic, perhaps, that production of 1 unit of a commodity requires none of that commodity, the simpler matrix involved is useful for our purposes.) ◀ **2**

Another matrix used with the input-output matrix is a matrix giving the amount of each commodity produced, called the **production matrix,** or the **vector of gross output.** In an economy producing n commodities, the production matrix can be represented by a column matrix x with entries $x_1, x_2, x_3, \ldots, x_n$.

▶ **EXAMPLE 3**   In Example 2, suppose the production matrix is

$$
X = \begin{bmatrix} 60 \\ 52 \\ 48 \end{bmatrix}.
$$

Then 60 units of agriculture, 52 units of manufacturing, and 48 units of transportation are produced. As 1/4 unit of agriculture is used for each unit of manufacturing produced, $1/4 \times 52 = 13$ units of agriculture must be used up in the "production" of manufacturing. Similarly, $1/3 \times 48 = 16$ units of agriculture will be used up in the "production" of transportation. Thus $13 + 16 = 29$ units of agriculture are used for production in the economy. Look again at the matrices A and X. Since X gives the number of units of each commodity produced and A gives the amount (in units) of each commodity used to produce 1 unit of the various commodities, the matrix product AX gives the amount of each commodity used up in production.

$$
AX = \begin{bmatrix} 0 & \frac{1}{4} & \frac{1}{3} \\ \frac{1}{2} & 0 & \frac{1}{4} \\ \frac{1}{4} & \frac{1}{4} & 0 \end{bmatrix}\begin{bmatrix} 60 \\ 52 \\ 48 \end{bmatrix} = \begin{bmatrix} 29 \\ 42 \\ 28 \end{bmatrix}
$$

This product shows that 29 units of agriculture, 42 units of manufacturing, and 28 units of transporation are used to produce 60 units of agriculture, 52 units of manufacturing, and 48 units of transportation. ◀

3 (a) Write a 2×1 matrix X to represent gross production of 9000 units of electricity and 12,000 units of water.

(b) Find AX using A from the last problem.

(c) Find D using $D = X - AX$.

Answer:

(a) $\begin{bmatrix} 9000 \\ 12,000 \end{bmatrix}$

(b) $\begin{bmatrix} 6000 \\ 4500 \end{bmatrix}$

(c) $\begin{bmatrix} 3000 \\ 7500 \end{bmatrix}$

We have seen that the matrix product AX represents the amount of each commodity used in the production process. The remainder (if any) must be enough to satisfy the demand for the various commodities from outside the production system. In an n-commodity economy, this demand can be represented by a **demand matrix** D with entries $d_1, d_2, \ldots, d_n$. The difference between the production matrix, X, and the amount, AX, used in the production process must equal the demand, D, or

$$D = X - AX.$$

In Example 3,

$$D = \begin{bmatrix} 60 \\ 52 \\ 48 \end{bmatrix} - \begin{bmatrix} 29 \\ 42 \\ 28 \end{bmatrix} = \begin{bmatrix} 31 \\ 10 \\ 20 \end{bmatrix}.$$

This result shows that production of 60 units of agriculture, 52 units of manufacturing, and 48 units of transportation would satisfy a demand of 31, 10, and 20 units of each, respectively. **3**

In practice, A and D usually are known and X must be found. That is, we need to decide what amounts of production are necessary to satisfy the required demands. Matrix algebra can be used to solve the equation $D = X - AX$ for X.

$$D = X - AX$$
$$D = IX - AX \qquad \textbf{Identity property}$$
$$D = (I - A)X \qquad \textbf{Distributive property}$$

If the matrix $I - A$ has an inverse, then

$$X = (I - A)^{-1}D.$$

▶**EXAMPLE 4**   Suppose, in the 3-commodity economy of Examples 2 and 3, there is a demand for 516 units of agriculture, 258 units of manufacturing, and 129 units of transportation. What should production of each commodity be?

The demand matrix is

$$D = \begin{bmatrix} 516 \\ 258 \\ 129 \end{bmatrix}.$$

Find the production matrix by first calculating $I - A$.

$$I - A = \begin{bmatrix} 1 & 0 & 0 \\ 0 & 1 & 0 \\ 0 & 0 & 1 \end{bmatrix} - \begin{bmatrix} 0 & \frac{1}{4} & \frac{1}{3} \\ \frac{1}{2} & 0 & \frac{1}{4} \\ \frac{1}{4} & \frac{1}{4} & 0 \end{bmatrix} = \begin{bmatrix} 1 & -\frac{1}{4} & -\frac{1}{3} \\ -\frac{1}{2} & 1 & -\frac{1}{4} \\ -\frac{1}{4} & -\frac{1}{4} & 1 \end{bmatrix}$$

4 A simple economy depends on just two products, beer and pretzels.

(a) Suppose 1/2 unit of beer and 1/2 unit of pretzels are needed to make 1 unit of beer, and 3/4 unit of beer is needed to make 1 unit of pretzels. Write the technological matrix A for the economy.

(b) Find $I - A$.

(c) Find $(I - A)^{-1}$.

(d) Find the gross production X that will be needed to get a net production of

$$Y = \begin{bmatrix} 100 \\ 1000 \end{bmatrix}.$$

Answer:

(a) $\begin{bmatrix} \frac{1}{2} & \frac{3}{4} \\ \frac{1}{2} & 0 \end{bmatrix}$

(b) $\begin{bmatrix} \frac{1}{2} & -\frac{3}{4} \\ -\frac{1}{2} & 1 \end{bmatrix}$

(c) $\begin{bmatrix} 8 & 6 \\ 4 & 4 \end{bmatrix}$

(d) $\begin{bmatrix} 6800 \\ 4400 \end{bmatrix}$

Using row operations, find the inverse of $I - A$.

$$(I - A)^{-1} = \begin{bmatrix} 1.40 & .50 & .59 \\ .84 & 1.36 & .62 \\ .56 & .47 & 1.30 \end{bmatrix}$$

(The entries are rounded to two decimal places.) Since $X = (I - A)^{-1}D$,

$$X = \begin{bmatrix} 1.40 & .50 & .59 \\ .84 & 1.36 & .62 \\ .56 & .47 & 1.30 \end{bmatrix} \begin{bmatrix} 516 \\ 258 \\ 129 \end{bmatrix} = \begin{bmatrix} 928 \\ 864 \\ 578 \end{bmatrix}$$

(rounded to the nearest whole numbers).

From the last result, we see that production of 928 units of agriculture, 864 units of manufacturing, and 578 units of transportation is required to satisfy demands of 516, 258, and 129 units, respectively. ◀

▶**EXAMPLE 5** An economy depends on two basic products, wheat and oil. To produce 1 metric ton of wheat requires .25 metric tons of wheat and .33 metric tons of oil. Production of 1 metric ton of oil consumes .08 metric tons of wheat and .11 metric tons of oil. Find the production which will satisfy a demand of 500 metric tons of wheat and 1000 metric tons of oil.

The input-output matrix, A, and the matrix $I - A$, are

$$A = \begin{bmatrix} .25 & .08 \\ .33 & .11 \end{bmatrix} \quad \text{and} \quad I - A = \begin{bmatrix} .75 & -.08 \\ -.33 & .89 \end{bmatrix}.$$

Next, calculate $(I - A)^{-1}$.

$$(I - A)^{-1} = \begin{bmatrix} 1.39 & .13 \\ .51 & 1.17 \end{bmatrix} \quad \text{(rounded)}$$

To find the production matrix X, use the equation $X = (I - A)^{-1}D$, with

$$D = \begin{bmatrix} 500 \\ 1000 \end{bmatrix}.$$

The production matrix is

$$X = \begin{bmatrix} 1.39 & .13 \\ .51 & 1.17 \end{bmatrix} \begin{bmatrix} 500 \\ 1000 \end{bmatrix} = \begin{bmatrix} 815 \\ 1425 \end{bmatrix}.$$

Production of 815 metric tons of wheat and 1425 metric tons of oil is required to satisfy the indicated demand. ◀ **4**

5 Write the message "*when*" using 2 × 1 matrices.

Answer:

$$\begin{bmatrix} 23 \\ 8 \end{bmatrix}, \begin{bmatrix} 5 \\ 14 \end{bmatrix}$$

CODE THEORY Governments need sophisticated methods of coding and decoding messages. One example of such an advanced code uses matrix theory. Such a code takes the letters in the words and divides them into groups. (Each space between words is treated as a letter; punctuation is disregarded.) Then, numbers are assigned to the letters of the alphabet. For our purposes, let the letter *a* correspond to 1, *b* to 2, and so on. Let the number 27 correspond to a space between words.

For example, the message

mathematics is for the birds

can be divided into groups of three letters each.

mat   hem   ati   cs–   is–   for   –th   e–b   ird   s– –

(We used – to represent a space between words.) We now write a column matrix for each group of three symbols using the corresponding numbers, as determined above, instead of letters. For example, the letters *mat* can be encoded as

$$\begin{bmatrix} 13 \\ 1 \\ 20 \end{bmatrix}.$$

The coded message then consists of the 3 × 1 column matrices:

$$\begin{bmatrix} 13 \\ 1 \\ 20 \end{bmatrix}, \begin{bmatrix} 8 \\ 5 \\ 13 \end{bmatrix}, \begin{bmatrix} 1 \\ 20 \\ 9 \end{bmatrix}, \begin{bmatrix} 3 \\ 19 \\ 27 \end{bmatrix}, \begin{bmatrix} 9 \\ 19 \\ 27 \end{bmatrix}, \begin{bmatrix} 6 \\ 15 \\ 18 \end{bmatrix}, \begin{bmatrix} 27 \\ 20 \\ 8 \end{bmatrix}, \begin{bmatrix} 5 \\ 27 \\ 2 \end{bmatrix}, \begin{bmatrix} 9 \\ 18 \\ 4 \end{bmatrix}, \begin{bmatrix} 19 \\ 27 \\ 27 \end{bmatrix}.$$ **5**

We can further complicate the code by choosing a matrix that has an inverse (in this case a 3 × 3 matrix, call it *M*) and finding the products of this matrix and each of the above column matrices. The size of each group, the assignment of numbers to letters, and the choice of matrix *M* must all be predetermined.

Suppose we choose

$$M = \begin{bmatrix} 1 & 3 & 3 \\ 1 & 4 & 3 \\ 1 & 3 & 4 \end{bmatrix}.$$

If we find the products of *M* and the column matrices above, we have a new set of column matrices,

$$\begin{bmatrix} 1 & 3 & 3 \\ 1 & 4 & 3 \\ 1 & 3 & 4 \end{bmatrix}\begin{bmatrix} 13 \\ 1 \\ 20 \end{bmatrix} = \begin{bmatrix} 76 \\ 77 \\ 96 \end{bmatrix}, \text{ and so on.}$$

6 Use the matrix given below to find the 2 × 1 matrices to be transmitted for the message you encoded in Problem 5 at the side.

$$\begin{bmatrix} 2 & 1 \\ 5 & 0 \end{bmatrix}$$

Answer:

$$\begin{bmatrix} 54 \\ 115 \end{bmatrix}, \begin{bmatrix} 24 \\ 25 \end{bmatrix}$$

The entries of these matrices can then be transmitted to an agent as the message 76, 77, 96, and so on. **6**

When the agent receives the message, it is divided into groups of numbers with each group formed into a column matrix. After multiplying each column matrix by the matrix M^{-1}, the message can be read.

Although this type of code is relatively simple, it is actually difficult to break. Many complications are possible. For example, a long message might be placed in groups of 20, thus requiring a 20 × 20 matrix for coding and decoding. Finding the inverse of such a matrix would require an impractical amount of time if calculated by hand. For this reason some of the largest computers are used by government agencies involved in coding.

ROUTING The diagram in Figure 6.6 shows the roads connecting four cities. Another way of representing this information is shown in matrix A, where the entries represent the number of roads connecting two cities without passing through another city.* For example, from the diagram we see that there are two roads connecting city 1 to city 4 without passing through either city 2 or 3. This information is entered in row one, column four and again in row four, column one of matrix A.

$$A = \begin{bmatrix} 0 & 1 & 2 & 2 \\ 1 & 0 & 1 & 0 \\ 2 & 1 & 0 & 1 \\ 2 & 0 & 1 & 0 \end{bmatrix}$$

Note that there are zero roads connecting each city to itself. Also, there is one road connecting cities 3 and 2.

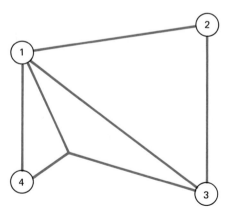

FIGURE 6.6

*From *Matrices with Applications,* section 3.2, example 5, by Hugh G. Campbell. Copyright © 1968, pp. 50–51. Adapted by permission of Prentice-Hall, Englewood Cliffs, New Jersey.

How many ways are there to go from city 1 to city 2, for example, by going through exactly one other city? Since we must go through one other city, we must go through either city 3 or city 4. On the diagram in Figure 6.6, we see that we can go from city 1 to city 2 through city 3 in two ways. We can go from city 1 to city 3 in two ways and then from city 3 to city 2 in one way, so there are $2 \cdot 1 = 2$ ways to get from city 1 to city 2 through city 3. It is not possible to go from city 1 to city 2 through city 4, because there is no direct route between cities 4 and 2.

The matrix A^2 gives the number of ways to travel between any two cities by passing through exactly one other city. Multiply matrix A by itself, to get A^2. Let the first row, second column entry of A^2 be b_{12}. (We use a_{ij} to denote the entry in the i-th row and j-th column of matrix A.) The entry b_{12} is found as follows.

$$b_{12} = a_{11}a_{12} + a_{12}a_{22} + a_{13}a_{32} + a_{14}a_{42}$$
$$= 0 \cdot 1 + 1 \cdot 0 + 2 \cdot 1 + 2 \cdot 0$$
$$= 2$$

The first product $0 \cdot 1$ in the calculations above represents the number of ways to go from city 1 to city 1, (0), and then from city 1 to city 2, (1). The 0 result indicates that such a trip does not involve a third city. The only nonzero product $(2 \cdot 1)$ represents the two routes from city 1 to city 3 and the one route from city 3 to city 2 which result in the $2 \cdot 1$ or 2 routes from city 1 to city 2 by going through city 3.

Similarly, A^3 gives the number of ways to travel between any two cities by passing through exactly two cities. Also, $A + A^2$ represents the total number of ways to travel between two cities with at most one intermediate city.

The diagram can be given many other interpretations. For example, the lines could represent lines of mutual influence between people or nations, or they could represent communication lines such as telephone lines.

6.6 EXERCISES

For each of the following, solve the matrix equation $AX = B$ for X. (See Example 1.)

1. $A = \begin{bmatrix} 1 & 3 \\ -2 & 4 \end{bmatrix}$, $B = \begin{bmatrix} 15 \\ 10 \end{bmatrix}$

2. $A = \begin{bmatrix} 2 & -2 \\ 1 & 0 \end{bmatrix}$, $B = \begin{bmatrix} -4 \\ 3 \end{bmatrix}$

3. $A = \begin{bmatrix} -2 & 4 \\ 3 & -1 \end{bmatrix}$, $B = \begin{bmatrix} 40 & -20 \\ 80 & 20 \end{bmatrix}$

4. $A = \begin{bmatrix} 5 & 10 \\ 8 & 14 \end{bmatrix}$, $B = \begin{bmatrix} -15 & 10 \\ 20 & 5 \end{bmatrix}$

5. $A = \begin{bmatrix} 1 & 0 & 2 \\ -1 & 1 & 0 \\ 3 & 0 & 4 \end{bmatrix}$, $B = \begin{bmatrix} 8 \\ 4 \\ -6 \end{bmatrix}$

6. $A = \begin{bmatrix} 2 & 4 & 0 \\ 1 & -2 & 0 \\ 0 & 0 & 3 \end{bmatrix}$, $B = \begin{bmatrix} 72 \\ -24 \\ 48 \end{bmatrix}$

7. $A = \begin{bmatrix} -3 & 0 & 6 \\ 1 & 1 & 0 \\ 0 & 2 & 5 \end{bmatrix}$, $B = \begin{bmatrix} 12 & 3 \\ -6 & 0 \\ 0 & -3 \end{bmatrix}$

8. $A = \begin{bmatrix} -2 & 2 & 0 \\ 3 & 1 & 0 \\ 0 & 0 & 1 \end{bmatrix}$, $B = \begin{bmatrix} 8 & -16 \\ 0 & 32 \\ -11 & 5 \end{bmatrix}$

Use matrix algebra to solve the following matrix equations for X. Then use the given matrices to find X and check your work.

9. $N = X - MX$, $N = \begin{bmatrix} 8 \\ -12 \end{bmatrix}$, $M = \begin{bmatrix} 0 & 1 \\ -2 & 1 \end{bmatrix}$

10. $A = BX + X$, $A = \begin{bmatrix} 4 & 6 \\ -2 & 2 \end{bmatrix}$, $B = \begin{bmatrix} -2 & -2 \\ 3 & 3 \end{bmatrix}$

Solve each of the systems of equations in Exercises 11–22 using the inverse of the coefficient matrix. The inverses for several of the systems were found in Exercises 23–32 of the previous section. (See Example 1.)

11. $\begin{aligned} -x - y - z &= 1 \\ 4x + 5y &= -2 \\ y - 3z &= 3 \end{aligned}$

12. $\begin{aligned} 2x \quad + 4z &= -8 \\ 3x + y + 5z &= 2 \\ -x + y - 2z &= 4 \end{aligned}$

13. $\begin{aligned} 2x + 4y + 6z &= 4 \\ -x - 4y - 3z &= 8 \\ y - z &= -4 \end{aligned}$

14. $\begin{aligned} 2x + 2y - 4z &= 12 \\ 2x + 6y &= 16 \\ -3x - 3y + 5z &= -20 \end{aligned}$

15. $\begin{aligned} x + 2y + 3z &= 5 \\ 2x + 3y + 2z &= 2 \\ -x - 2y - 4z &= -1 \end{aligned}$

16. $\begin{aligned} x + y - 3z &= 4 \\ 2x + 4y - 4z &= 8 \\ -x + y + 4z &= -3 \end{aligned}$

17. $\begin{aligned} x + 2y &= -10 \\ -x + 4z &= 8 \\ -y + z &= 4 \end{aligned}$

18. $\begin{aligned} x + z &= 3 \\ y + 2z &= 8 \\ -x + y &= 4 \end{aligned}$

19. $\begin{aligned} 2x - 2y &= 5 \\ 4y + 8z &= 7 \\ x + 2z &= 1 \end{aligned}$

20. $\begin{aligned} 6x + 5y &= 10 \\ x - z &= 3 \\ -12x - 10y &= -20 \end{aligned}$

21. $\begin{aligned} x - 2y + 3z \quad &= 4 \\ y - z + w &= -8 \\ -2x + 2y - 2z + 4w &= 12 \\ 2y - 3z + w &= -4 \end{aligned}$

22. $\begin{aligned} x + y \quad + 2w &= 3 \\ 2x - y + z - w &= 3 \\ 3x + 3y + 2z - 2w &= 5 \\ x + 2y + z \quad &= 3 \end{aligned}$

Find the production matrix given the following input-output and demand matrices. (See Examples 4 and 5.)

23. $A = \begin{bmatrix} \frac{1}{2} & \frac{2}{5} \\ \frac{1}{4} & \frac{1}{5} \end{bmatrix}$, $D = \begin{bmatrix} 2 \\ 4 \end{bmatrix}$

24. $A = \begin{bmatrix} \frac{1}{5} & \frac{1}{25} \\ \frac{3}{5} & \frac{1}{20} \end{bmatrix}$, $D = \begin{bmatrix} 3 \\ 10 \end{bmatrix}$

25. $A = \begin{bmatrix} .1 & .03 \\ .07 & .6 \end{bmatrix}$, $D = \begin{bmatrix} 5 \\ 10 \end{bmatrix}$

26. $A = \begin{bmatrix} .01 & .03 \\ .05 & .05 \end{bmatrix}$, $D = \begin{bmatrix} 100 \\ 200 \end{bmatrix}$

27. $A = \begin{bmatrix} .4 & 0 & .3 \\ 0 & .8 & .1 \\ 0 & .2 & .4 \end{bmatrix}$, $D = \begin{bmatrix} 1 \\ 3 \\ 2 \end{bmatrix}$

28. $A = \begin{bmatrix} .1 & .5 & 0 \\ 0 & .3 & .4 \\ .1 & .2 & .1 \end{bmatrix}$, $B = \begin{bmatrix} 10 \\ 4 \\ 2 \end{bmatrix}$

 In Exercises 29 and 30, refer to Example 5.

29. Management If the demand is changed to 690 metric tons of wheat and 920 metric tons of oil, how many units of each commodity should be produced?

30. Management Change the technological matrix so that production of 1 metric ton of wheat requires 1/5 metric ton of oil (and no wheat), and the production of 1 metric ton of oil requires 1/3 metric ton of wheat (and no oil). To satisfy the same demand matrix, how many units of each commodity should be produced?

31. Management A simplified economy has only two industries, the electric company and the gas company. Each dollar's worth of the electric company's output requires .40 of its own output and .50 of the gas company's output.

Each dollar's worth of the gas company's output requires .25 of its own output and .60 of the electric company's output. What should the production of electricity and gas be (in dollars) if there is a $12 million demand for gas and a $15 million demand for electricity?

Use a computer to work Exercises 32 and 33, which refer to Examples 2 and 3.

32. Management Suppose 1/4 unit of manufacturing and 1/2 unit of transportation are required to produce 1 unit of agriculture; 1/2 unit of agriculture and 1/4 unit of transportation to produce 1 unit of manufacturing; and 1/4 unit of agriculture and 1/4 unit of manufacturing to produce 1 unit of transportation. How many units of each commodity should be produced to satisfy a demand of 1000 units for each commodity?

33. Management In Exercise 32, find the number of units of each commodity that should be produced to satisfy a demand for 500 units of each commodity.

34. Management A primitive economy depends on two basic goods, yams and pork. Production of 1 bushel of yams requires 1/4 bushel of yams and 1/2 of a pig. To produce 1 pig requires 1/6 bushel of yams. Find the amount of each commodity that should be produced to get
(a) 1 bushel of yams and 1 pig;
(b) 100 bushels of yams and 70 pigs.

35. Use the methods of the text to encode the message

Arthur is here.

Break the message into groups of two letters and use the matrix

$$M = \begin{bmatrix} -1 & 2 \\ 2 & 5 \end{bmatrix}.$$

36. Encode the following message by breaking it into groups of two letters and using the matrix of Exercise 35.

Attack at dawn unless too cold.

37. Finish encoding the message given in the text.

38. Use matrix A in the discussion on routing in the text to find A^2. Then answer the following questions. How many ways are there to travel from
(a) City 1 to city 3 by passing through exactly one city?
(b) City 2 to city 4 by passing through exactly one city?
(c) City 1 to city 3 by passing through at most one city?
(d) City 2 to city 4 by passing through at most one city?

39. Find A^3. (See Exercise 38.) Then answer the following questions.
(a) How many ways are there to travel between cities 1 and 4 by passing through exactly two cities?
(b) How many ways are there to travel between cities 1 and 4 by passing through at most two cities?

40. Management A small telephone system connects three cities. There are four lines between cities 3 and 2, three lines connecting city 3 with city 1, and two lines between cities 1 and 2.
(a) Write a matrix B to represent this information.
(b) Find B^2.

(c) How many lines which connect cities 1 and 2 go through exactly one other city (city 3)?
(d) How many lines which connect cities 1 and 2 go through at most one other city?

41. Management The figure shows four southern cities served by Delta Airlines.

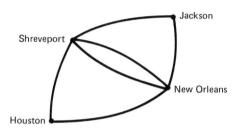

(a) Write a matrix to represent the number of nonstop routes between cities.
(b) Find the number of one-stop flights between Houston and Jackson.
(c) Find the number of flights between Houston and Shreveport which require at most one stop.
(d) Find the number of one-stop flights between New Orleans and Houston.

42. Natural Science The figure shows a food web. The arrows indicate the food sources of each population. For example, cats feed on rats and on mice.

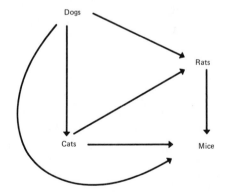

(a) Write a matrix C in which each row and corresponding column represents a population in the food chain. Enter a 1 when the population in a given row feeds on the population in the given column.
(b) Calculate and interpret C^2.

KEY TERMS AND SYMBOLS

system of equations

6.1 linear equation in n unknowns
inconsistent system
dependent equations
elimination method (addition method)
equivalent systems
transformations
echelon method
parameter

6.2 $\begin{bmatrix} a & b & c \\ d & e & f \end{bmatrix}$ matrix (matrices)
element
entry
augmented matrix
row operations
Gauss-Jordan method

6.3 square matrix
row matrix (row vector)

column matrix (column vector)
equal matrices
sum of matrices
additive inverse of a matrix
zero matrix
subtraction (difference) of matrices

6.4 scalar
product of a scalar and a matrix
product of matrices

6.5 identity matrix
inverse matrix

6.6 coefficient matrix
input-output model
technological matrix (input-output matrix)
production matrix (vector of gross output)
demand matrix
code theory
routing theory

KEY CONCEPTS

The following **transformations** are used to change a system of equations to an equivalent system with a solution that is obvious.
1. Interchange any two equations.
2. Multiply both sides of an equation by a nonzero real number.
3. Replace any equation by the sum of itself and a multiple of any other equation in the system.

The **echelon method** is a systematic way of using the three transformations to transform a system into an equivalent system in the form
$$x + by + cz = d$$
$$y + ez = f$$
$$z = g,$$
where b, c, d, e, f, and g are real numbers. See Section 6.1 for details.

The following **matrix row operations** correspond to the transformations of systems of equations.
1. Interchange any two rows.
2. Multiply the elements of any row by any nonzero real number.
3. Replace a row by adding a multiple of the elements of another row to the corresponding elements of the row.

The **Gauss-Jordan method** is an extension of the echelon method of solving systems of linear equations that uses row operations. See Section 6.2 for details.

Operations on Matrices

The **sum** of two $m \times n$ matrices X and Y is the $m \times n$ matrix $X + Y$ in which each element is the sum of the corresponding elements of X and Y. The **difference** of two $m \times n$ matrices X and Y is the $m \times n$ matrix $X - Y$ in which each element is the difference of the corresponding elements of X and Y.

The **product** of a scalar k and a matrix X is the matrix kX, with each element k times the corresponding element of X.

The **product** matrix AB of an $m \times n$ matrix A and an $n \times k$ matrix B is the $m \times k$ matrix whose entry in the i-th row and j-th column is the product of the i-th row of A and the j-th column of B.

The **inverse** matrix A^{-1} for any $n \times n$ matrix A for which A^{-1} exists is found as follows. Form the augmented matrix $[A \mid I]$; perform row operations on $[A \mid I]$ to get the matrix $[I \mid A^{-1}]$.

CHAPTER 6 REVIEW EXERCISES

Use the echelon method to solve each of the following linear systems. Identify any dependent or inconsistent systems.

1. $3x - 5y = -18$
$2x + 7y = 19$

2. $6x + 5y = 53$
$4x - 3y = 29$

3. $\dfrac{2}{3}x - \dfrac{3}{4}y = 13$
$\dfrac{1}{2}x + \dfrac{2}{3}y = -5$

4. $3x + 7y = 10$
$18x + 42y = 50$

5. $4x + 8y = 15$
$\dfrac{4}{3}x + \dfrac{8}{3}y = 5$

6. $.9x - .2y = .8$
$.3x + .7y = 4.1$

7. $2x - 3y + z = -5$
$x + 4y + 2z = 13$
$5x + 5y + 3z = 14$

8. $x - 3y = 12$
$2y + 5z = 1$
$4x + z = 25$

9. A student bought some candy bars, paying 25¢ each for some and 50¢ each for others. The student bought a total of 22 bars, paying a total of $8.50. How many of each kind of bar did he buy?

10. Five hundred seventy-five people attend a ball game and total ticket sales are $2575. If adult tickets cost $5 and children's tickets are $3, how many adults attended the game? How many children?

11. Donna Sharp wins $50,000 in a lottery. She invests part of the money at 6%, twice as much at 7%, and the balance at 9%. Total annual interest is $3800. How much is invested at each rate?

12. A collection of nickels, dimes, and quarters totals $8.20. The number of nickels and dimes together is twice the number of quarters. The value of the nickels is one third the value of the dimes. How many of each kind of coin are there?

13. A social service agency provides counseling, meals, and shelter to clients referred by sources I, II, and III. Clients from source I require an average of $100 for food, $250 for shelter, and no counseling. Source II clients require an average of $100 for counseling, $200 for food, and nothing for shelter. Source III clients require an average of $100 for counseling, $150 for food, and $200 for shelter. The agency has funding of $25,000 for counseling, $50,000 for food, and $32,500 for shelter. How many clients from each source can be served?

Find solutions for the following systems in terms of the arbitrary variable z.

14. $3x - 4y + z = 2$
$2x + y - 4z = 1$

15. $2x + 3y - z = 5$
$-x + 2y + 4z = 8$

Use the Gauss-Jordan method to solve each of the following systems.

16. $2x + 3y = 10$
$-3x + y = 18$

17. $5x + 2y = -10$
$3x - 5y = -6$

18. $3x + y = -7$
$x - y = -5$

19. $x - z = -3$
$y + z = 6$
$2x - 3z = -9$

20. $2x - y + 4z = -1$
$-3x + 5y - z = 5$
$2x + 3y + 2z = 3$

21. $5x - 8y + z = 1$
$3x - 2y + 4z = 3$
$10x - 16y + 2z = 3$

22. Each week at a furniture factory, there are 2000 work hours available in the construction department, 1400 work hours in the painting department, and 1300 work hours in the packing department. Producing a chair requires 2 hours of construction, 1 hour of painting, and 2 hours for packing. Producing a table requires 4 hours of construction, 3 hours of painting, and 3 hours for packing. Producing a chest requires 8 hours of construction, 6 hours of painting, and 4 hours for packing. If all available time is used in every department, how many of each item are produced each week?

For each of the following, find the size of the matrices, find the values of any variables, and identify any square, row, or column matrices.

23. $\begin{bmatrix} 2 & 3 \\ 5 & q \end{bmatrix} = \begin{bmatrix} a & b \\ c & 9 \end{bmatrix}$

24. $\begin{bmatrix} 2 & x \\ y & 6 \\ 5 & z \end{bmatrix} = \begin{bmatrix} a & -1 \\ 4 & 6 \\ p & 7 \end{bmatrix}$

25. $[m \quad 4 \quad z \quad -1] = [12 \quad k \quad -8 \quad r]$

26. $\begin{bmatrix} a+5 & 3b & 6 \\ 4c & 2+d & -3 \\ -1 & 4p & q-1 \end{bmatrix} = \begin{bmatrix} -7 & b+2 & 2k-3 \\ 3 & 2d-1 & 4l \\ m & 12 & 8 \end{bmatrix}$

27. $\begin{bmatrix} 6 & m \\ k & 5-x \end{bmatrix} + \begin{bmatrix} r & m+3 \\ k+1 & 2x \end{bmatrix} = \begin{bmatrix} 8 & 10 \\ 12 & -2 \end{bmatrix}$

28. $\begin{bmatrix} 3p & 1 & y \\ q & 5 & z \end{bmatrix} - \begin{bmatrix} 3+p & r & 2y+1 \\ q+2 & m & 10 \end{bmatrix}$
$= \begin{bmatrix} 4p & 2r & 5y \\ 3q & 2m & -z \end{bmatrix}$

29. The activities of a grazing animal can be classified roughly into three categories: grazing, moving, and resting. Sup-

pose horses spend 8 hours grazing, 8 moving, and 8 resting; cattle spend 10 grazing, 5 moving, and 9 resting; sheep spend 7 grazing, 10 moving, and 7 resting; and goats spend 8 grazing, 9 moving, and 7 resting. Write this information as a 4×3 matrix.

30. The New York Stock Exchange reports in the daily newspapers give the dividend, price-to-earnings ratio, sales (in hundreds of shares), last price, and change in price for each company. Write the following stock reports as a 4×5 matrix. American Telephone & Telegraph: 5, 7, 2532, $52\frac{3}{8}$, $-\frac{1}{4}$, General Electric: 3, 9, 1464, 56, $+\frac{1}{8}$. Gulf Oil: 2.50, 5, 4974, 41, $-1\frac{1}{2}$. Sears: 1.36, 10, 1754, 18, $+\frac{1}{2}$.

Given the matrices.

$$A = \begin{bmatrix} 4 & 10 \\ -2 & -3 \\ 6 & 9 \end{bmatrix}, \quad B = \begin{bmatrix} 2 & 3 & -2 \\ 2 & 4 & 0 \\ 0 & 1 & 2 \end{bmatrix},$$

$$C = \begin{bmatrix} 5 & 0 \\ -1 & 3 \\ 4 & 7 \end{bmatrix}, \quad D = \begin{bmatrix} 6 \\ 1 \\ 0 \end{bmatrix}, \quad E = [1 \quad 3 \quad -4],$$

$$F = \begin{bmatrix} -1 & 4 \\ 3 & 7 \end{bmatrix}, \quad G = \begin{bmatrix} 2 & 5 \\ 1 & 6 \end{bmatrix},$$

find each of the following (if possible).

31. $-A$

32. $-E$

33. $C + 2A$

34. $2G - 4F$

35. $2A - 5C$

36. $3B + C$

37. Refer to Exercise 30. Write a 4×2 matrix using the sales and price changes for the four companies. The next day's sales and price changes for the same four companies were 2310, 1258, 5061, 1812 and $-1/4$, $-1/4$, $+1/2$, $+1/2$, respectively. Write a 4×2 matrix using these new sales and price change figures. Use matrix addition to find the total sales and price changes for the two days.

38. An oil refinery in Tulsa sent 110,000 gallons of oil to a Chicago distributor, 73,000 to a Dallas distributor, and 95,000 to an Atlanta distributor. Another refinery in New Orleans sent the following amounts to the same three distributors: 85,000, 108,000, 69,000. The next month the two refineries sent the same distributors new shipments of oil as follows: from Tulsa, 58,000 to Chicago, 33,000 to Dallas, and 80,000 to Atlanta; from New Orleans, 40,000, 52,000, and 30,000, respectively.
(a) Write the monthly shipments from the two distributors to the three refineries as 3×2 matrices.
(b) Use matrix addition to find the total amounts sent to the refineries from each distributor.

Use the matrices given above Exercise 31 to find each of the following (if possible).

39. AF **40.** AC **41.** ED **42.** BD **43.** AFG **44.** EBC

45. An office supply manufacturer makes two kinds of paper clips, standard and extra large. To make a unit of standard paper clips requires 1/4 hour on a cutting machine and 1/2 hour on a machine that shapes the clips. A unit of extra large paper clips requires 1/3 hour on each machine.

(a) Write this information as a 2×2 matrix (machine/size)
(b) If the cutting machine is operated for 12 hours and the shaping machine for 6 hours, use matrix multiplication to find out how many of each type of clip can be made. (Hint: write the hours as a 1×2 matrix.)

46. Jane Cunningham buys shares of three stocks. Their cost per share and dividend earnings per share are $32, $23, $54, and $1.20, $1.49, and $2.10, respectively. She buys 50 shares of the first stock, 20 shares of the second, and 15 shares of the third.
(a) Write the cost per share and earnings per share of the stocks as a 3×2 matrix.
(b) Write the number of shares of each stock as a 1×3 matrix.
(c) Use matrix multiplication to find the total cost and total dividend earnings of these stocks.

47. Give an example (other than ones in the text or exercises) of 2×2 matrices A and B such that $AB \neq BA$.

48. Give an example of matrices A and B such that AB exists, but BA does not exist.

Find the inverse of each of the following matrices that has an inverse.

49. $\begin{bmatrix} -4 & 2 \\ 0 & 3 \end{bmatrix}$ **50.** $\begin{bmatrix} 2 & 1 \\ 5 & 3 \end{bmatrix}$ **51.** $\begin{bmatrix} 6 & 4 \\ 3 & 2 \end{bmatrix}$ **52.** $\begin{bmatrix} 2 & 0 \\ -1 & 5 \end{bmatrix}$

53. $\begin{bmatrix} 2 & 0 & 4 \\ 1 & -1 & 0 \\ 0 & 1 & -2 \end{bmatrix}$ **54.** $\begin{bmatrix} 2 & -1 & 0 \\ 1 & 0 & 1 \\ 1 & -2 & 0 \end{bmatrix}$ **55.** $\begin{bmatrix} 2 & 3 & 5 \\ -2 & -3 & -5 \\ 1 & 4 & 2 \end{bmatrix}$ **56.** $\begin{bmatrix} 1 & 3 & 6 \\ 4 & 0 & 9 \\ 5 & 15 & 30 \end{bmatrix}$

Refer again to the matrices given above Exercise 31 to find each of the following (if possible).

57. G^{-1} **58.** B^{-1} **59.** $(A + C)^{-1}$ **60.** $(F - G)^{-1}$ **61.** $F^{-1} - G^{-1}$

62. If c and d are nonzero real numbers, find the inverse of the matrix
$$\begin{bmatrix} c & 0 \\ 0 & d \end{bmatrix}.$$

Solve each of the following matrix equations $AX = B$ for X.

63. $A = \begin{bmatrix} 2 & 4 \\ -1 & -3 \end{bmatrix}, \quad B = \begin{bmatrix} 8 \\ 3 \end{bmatrix}$

64. $A = \begin{bmatrix} 1 & 3 \\ -2 & 4 \end{bmatrix}, \quad B = \begin{bmatrix} 15 \\ 10 \end{bmatrix}$

65. $A = \begin{bmatrix} 1 & 0 & 2 \\ -1 & 1 & 0 \\ 3 & 0 & 4 \end{bmatrix}, \quad B = \begin{bmatrix} 8 \\ 4 \\ -6 \end{bmatrix}$

66. $A = \begin{bmatrix} 2 & 4 & 0 \\ 1 & -2 & 0 \\ 0 & 0 & 3 \end{bmatrix}, \quad B = \begin{bmatrix} 72 \\ -24 \\ 48 \end{bmatrix}$

Use the method of matrix inverses to solve each of the following equations.

67. $x + y = 4$
$2x + 3y = 10$

68. $5x - 3y = -2$
$2x + 7y = -9$

69. $2x + y = 5$
$3x - 2y = 4$

70. $x - 2y = 7$
$3x + y = 7$

71. $x + y + z = 1$
$2x - y = -2$
$3y + z = 2$

72. $x = -3$
$y + z = 6$
$2x - 3z = -9$

73. $3x - 2y + 4z = 4$
$4x + y - 5z = 2$
$-6x + 4y - 8z = -2$

74. $x + 2y = -1$
$3y - z = -5$
$x + 2y - z = -3$

Solve each of the following problems by any method.

75. A gold merchant has some 12 carat gold (12/24 pure gold), and some 22 carat gold (22/24 pure). How many grams of each could be mixed to get 25 grams of 15 carat gold?

76. A chemist has some 40% acid solution and some 60% solution. How many liters of each should be used to get 40 liters of a 45% solution?

77. How many pounds of tea worth $4.60 a pound should be mixed with tea worth $6.50 a pound to get 10 pounds of a mixture worth $5.74 a pound?

78. A solution of a drug with a strength of 5% is to be mixed with a 15% solution to get 50 milliliters of an 8% solution. How many milliliters of the 5% solution should be used?

79. A boat travels at a constant speed a distance of 57 km downstream in 3 hours, then turns around and travels 55 km upstream in 5 hours. What is the speed of the boat and of the current?

80. The cashier at an amusement park has a total of $2480, made up of fives, tens, and twenties. The total number of bills is 290, and the value of the tens is $60 more than the value of the twenties. How many of each type of bill does the cashier have?

81. Ms. Tham invests $50,000 three ways—at 8%, $8\frac{1}{2}$%, and 11%. In total, she receives $4436.25 per year in interest. The interest from the 11% investment is $80 more than the interest on the 8% investment. Find the amount she has invested at each rate.

82. Tickets to a band concert cost $2 for children, $3 for teenagers, and $5 for adults. 570 people attended the concert and total ticket receipts were $1950. Three fourths as many teenagers as children attended. How many children, teenagers, and adults were at the concert?

Find the production matrix given the following input-output and demand matrices.

83. $A = \begin{bmatrix} .01 & .05 \\ .04 & .03 \end{bmatrix}$, $D = \begin{bmatrix} 200 \\ 300 \end{bmatrix}$

84. $A = \begin{bmatrix} .2 & .1 & .3 \\ .1 & 0 & .2 \\ 0 & 0 & .4 \end{bmatrix}$, $D = \begin{bmatrix} 500 \\ 200 \\ 100 \end{bmatrix}$

85. Given the input-output matrix $A = \begin{bmatrix} 0 & \frac{1}{4} \\ \frac{1}{2} & 0 \end{bmatrix}$ and the demand matrix $D = \begin{bmatrix} 2100 \\ 1400 \end{bmatrix}$, find each of the following.
(a) $I - A$ **(b)** $(I - A)^{-1}$
(c) the production matrix X

86. An economy depends on two commodities, goats and cheese. It takes 2/3 of a unit of goats to produce 1 unit of cheese and 1/2 unit of cheese to produce 1 unit of goats.
(a) Write the input-output matrix for this economy.
(b) Find the production required to satisfy a demand of 400 units of cheese and 800 units of goats.

87. In a simple economic model, a country has two industries: agriculture and manufacturing. To produce $1 of agricultural output requires .10 of agricultural output and .40 of manufacturing output. To produce $1 of manufacturing output requires .70 of agricultural output and .20 of manufacturing output. If agricultural demand is $60,000 and manufacturing demand is $20,000, what must each industry produce? (Round answers to the nearest whole number.)

88. The matrix below represents the number of direct flights between four cities.

	A	B	C	D
A	0	1	0	1
B	1	0	0	1
C	0	0	0	1
D	1	1	1	0

(a) Find the number of one-stop flights between cities A and C.
(b) Find the total number of flights between cities B and C that are either direct or one-stop.
(c) Find the matrix that gives the number of two-stop flights between these cities.

Let $M = \begin{bmatrix} 2 & 5 \\ -3 & 1 \end{bmatrix}$.

89. (a) Use M to encode the following message: It is too late.
(b) What matrix should be used to decode this message?

Suppose that three people have contracted a contagious disease.* A second group of five people may have been in contact with the three infected persons. A third group of six people may have been in contact with the second group. We can form a 3×5 matrix P with rows representing the first group of three and columns representing the second group of five. We enter a one in the corresponding position if a person in the first group has contact with a person in the second group. These direct contacts are called *first-order contacts*. Similarly, we form a 5×6 matrix Q representing the first-order contacts between the second and third group. For example, suppose

$$P = \begin{bmatrix} 1 & 0 & 0 & 1 & 0 \\ 0 & 0 & 1 & 1 & 0 \\ 1 & 1 & 0 & 0 & 0 \end{bmatrix} \quad \text{and}$$

$$Q = \begin{bmatrix} 1 & 1 & 0 & 1 & 1 & 1 \\ 0 & 0 & 0 & 0 & 1 & 0 \\ 0 & 0 & 0 & 0 & 0 & 0 \\ 0 & 1 & 0 & 1 & 0 & 0 \\ 1 & 0 & 0 & 0 & 1 & 0 \end{bmatrix}.$$

From matrix P we see that the first person in the first group had contact with the first and fourth persons in the second group. Also, none of the first group had contact with the last person in the second group.

A *second-order contact* is an indirect contact between persons in the first and third group through some person in the second group. The product matrix PQ indicates these contacts. Verify that the second row, fourth column entry of PQ is 1. That is, there is one second-order contact between the second person in group one and the fourth person in group three. Let a_{ij} denote the element in the i-th row and j-th column of the matrix PQ. By looking at the products that form a_{24} below, we see that the common contact was with the fourth individual in group two. (The p_{ij} are entries in P, and the q_{ij} are entries in Q.)

$$a_{24} = p_{21}q_{14} + p_{22}q_{24} + p_{23}q_{34} + p_{24}q_{44} + p_{25}q_{54}$$
$$= 0 \cdot 1 + 0 \cdot 0 + 1 \cdot 0 + 1 \cdot 1 + 0 \cdot 1$$
$$= 1$$

The second person in group one and the fourth person in group three both had contact with the fourth person in group two.

This idea could be extended to third, fourth-, and larger-order contacts. It indicates a way to use matrices to trace the spread of a contagious disease. It could also pertain to the dispersal of ideas or anything that might pass from one individual to another.

EXERCISES

1. Find the second-order contact matrix PQ mentioned in the text.

2. How many second-order contacts were there between the second contagious person and the third person in the third group?

3. Is there anyone in the third group who has had no contacts at all with the first group?

4. The totals of the columns in PQ give the total number of second-order contacts per person, while the column totals in P and Q give the total number of first-order contacts per person. Which person(s) in the third group had the most contacts, counting first- and second-order contacts?

---

*"First and Second Order Contact to a Contagious Disease" from *Finite Mathematics with Applications to Business, Life Sciences, and Social Sciences* by Stanley Grossman.

In the April 1965 issue of *Scientific American,* Wassily Leontief explained his input-output system using the 1958 American economy as an example.* He divided the economy into 81 sectors, grouped into six families of related sectors. In order to keep the discussion reasonably simple, we will treat each family of sectors as a single sector and so, in effect, work with a six-sector model. The sectors are listed in Table 1.

Table 1

Sector	Examples
Final Nonmetal (FN)	Furniture, processed food
Final Metal (FM)	Household appliances, motor vehicles
Basic Metal (BM)	Machine-shop products, mining
Basic Nonmetal (BN)	Agriculture, printing
Energy (E)	Petroleum, coal
Services (S)	Amusements, real estate

The workings of the American economy in 1958 are described in the input-output table (Table 2) based on Leontief's figures. We will demonstrate the meaning of Table 2 by considering the first left-hand column of numbers. The numbers in this column mean that 1 unit of final nonmetal production requires the consumption of .170 unit of (other) final nonmetal production, .003 unit of final metal output, .025 unit of basic metal products, and so on down the column. Since the unit of measurement that Leontief used for this table is millions of dollars, we conclude that the production of $1 million worth of final nonmetal production consumes $.170 million, or $170,000, worth of other final nonmetal products, $3000 of final metal products, $25,000 of basic metal products, and so on. Similarly, the entry in the column headed FM and opposite S of .074 means that $74,000 worth of input from the service industries goes into the production of $1 million worth of final metal products, and the number .358 in the column headed E and opposite E means that $358,000 worth of energy must be consumed to produce $1 million worth of energy.

By the underlying assumption of Leontief's model, the production of n units (n = any number) of final nonmetal production consumes .170n units of final nonmetal output, .003n units of final metal output, .025n units of basic metal production, and

---

*Adapted from *Applied Finite Mathematics* by Robert F. Brown and Brenda W. Brown. Copyright © 1977 by Robert F. Brown and Brenda W. Brown. Reprinted by permission.

Table 2

	FN	FM	BM	BN	E	S
FN	.170	.004	0	.029	0	.008
FM	.003	.295	.018	.002	.004	.016
BM	.025	.173	.460	.007	.011	.007
BN	.348	.037	.021	.403	.011	.048
E	.007	.001	.039	.025	.358	.025
S	.120	.074	.104	.123	.173	.234

so on. Thus, production of $50 million worth of products from the final nonmetal section of the 1958 American economy required (.170)(50) = 8.5 units ($8.5 million) worth of final nonmetal input, (.003)(50) = .15 units of final metal input, (.025)(50) = 1.25 units of basic metal production, and so on.

Example 1 According to the simplified input-output table for the 1958 American economy, how many dollars worth of final metal products, basic nonmetal products, and services are required to produce $120 million worth of basic metal products?

Each unit ($1 million worth) of basic metal products requires .018 units of final metal products because the number in the BM column of the table opposite FM is .018. Thus, $120 million, or 120 units, requires (.108)(120) = 2.16 units, or $2.16 million worth of final metal products. Similarly, 120 units of basic metal production uses (.021)(120) = 2.52 units of basic nonmetal production and (.104)(120) = 12.48 units of services, or $2.52 million and $12.48 million worth of basic nonmetal output and services, respectively. ◀

The Leontief model also involves a *bill of demands,* that is, a list of requirements for units of output beyond that required for its inner workings as described in the input-output table. These demands represent exports, surpluses, government and individual consumption, and the like. The bill of demands (in millions) for the simplified version of the 1958 American economy we have been using is shown below.

FN	$99,640
FM	$75,548
BM	$14,444
BN	$33,501
E	$23,527
S	$263,985

We can now use the methods developed in Section 6.6 to answer this question: how many units of output from each sector are needed in order to run the economy and fill the bill of demands? The units of output from each sector required to run the economy and fill the bill of demands is unknown, so we denote them by variables. In our example, there are six quantities which are, at the moment, unknown. The number of units of final nonmetal production required to solve the problem will be our first unknown, because this sector is represented by the first row of the input-output matrix. The unknown quantity of final nonmetal units will be represented by the symbol x_1. Following the same pattern, we represent the unknown quantities in the following manner.

x_1 = units of final nonmetal production required
x_2 = units of final metal production required
x_3 = units of basic metal production required
x_4 = units of basic nonmetal production required
x_5 = units of energy required
x_6 = units of services required

These six numbers are the quantities we are attempting to calculate, and the variables will make up the entries in our production matrix.

To find these numbers, first let A be the 6×6 matrix corresponding the input-output table.

$$A = \begin{bmatrix} .170 & .004 & 0 & .029 & 0 & .008 \\ .003 & .295 & .018 & .002 & .004 & .016 \\ .025 & .173 & .460 & .007 & .011 & .007 \\ .348 & .037 & .021 & .403 & .011 & .048 \\ .007 & .001 & .039 & .025 & .358 & .025 \\ .120 & .074 & .104 & .123 & .173 & .234 \end{bmatrix}$$

A is the input-output matrix. The bill of demands leads to a 6×1 demand matrix D, and X is the matrix of unknowns.

$$D = \begin{bmatrix} 99,640 \\ 75,548 \\ 14,444 \\ 33,501 \\ 23,527 \\ 263,985 \end{bmatrix} \quad \text{and} \quad X = \begin{bmatrix} x_1 \\ x_2 \\ x_3 \\ x_4 \\ x_5 \\ x_6 \end{bmatrix}$$

To use the formula $D = (I - A)X$, or $X = (I - A)^{-1}D$, we need to find $I - A$.

$$I - A = \begin{bmatrix} 1 & 0 & 0 & 0 & 0 & 0 \\ 0 & 1 & 0 & 0 & 0 & 0 \\ 0 & 0 & 1 & 0 & 0 & 0 \\ 0 & 0 & 0 & 1 & 0 & 0 \\ 0 & 0 & 0 & 0 & 1 & 0 \\ 0 & 0 & 0 & 0 & 0 & 1 \end{bmatrix}$$

$$- \begin{bmatrix} .170 & .004 & 0 & .029 & 0 & .008 \\ .003 & .295 & .018 & .002 & .004 & .016 \\ .025 & .173 & .460 & .007 & .011 & .007 \\ .348 & .037 & .021 & .403 & .011 & .048 \\ .007 & .001 & .039 & .025 & .358 & .025 \\ .120 & .074 & .104 & .123 & .173 & .234 \end{bmatrix}$$

$$= \begin{bmatrix} .830 & -.004 & 0 & -.029 & 0 & -.008 \\ -.003 & .705 & -.018 & -.002 & -.004 & -.016 \\ -.025 & -.173 & .540 & -.007 & -.011 & -.007 \\ -.348 & -.037 & -.021 & .597 & -.011 & -.048 \\ -.007 & -.001 & -.039 & -.025 & .642 & -.025 \\ -.120 & -.074 & -.104 & -.123 & -.173 & .766 \end{bmatrix}$$

By the method given in Section 6.5, we find the inverse (actually an approximation).

$$(I - A)^{-1} = \begin{bmatrix} 1.234 & .014 & .006 & .064 & .007 & .018 \\ .017 & 1.436 & .057 & .012 & .020 & .032 \\ .071 & .465 & 1.877 & .019 & .045 & .031 \\ .751 & .134 & .100 & 1.740 & .066 & .124 \\ .060 & .045 & .130 & .082 & 1.578 & .059 \\ .339 & .236 & .307 & .312 & .376 & 1.349 \end{bmatrix}$$

Therefore,

$$X = (I - A)^{-1}D =$$

$$\begin{bmatrix} 1.234 & .014 & .006 & .064 & .007 & .018 \\ .017 & 1.436 & .057 & .012 & .020 & .032 \\ .071 & .465 & 1.877 & .019 & .045 & .031 \\ .751 & .134 & .100 & 1.740 & .066 & .124 \\ .060 & .045 & .130 & .082 & 1.578 & .059 \\ .339 & .236 & .307 & .312 & .376 & 1.349 \end{bmatrix} \begin{bmatrix} 99,640 \\ 75,548 \\ 14,444 \\ 33,501 \\ 23,527 \\ 263,985 \end{bmatrix}$$

$$= \begin{bmatrix} 131,161 \\ 120,324 \\ 79,194 \\ 178,936 \\ 66,703 \\ 426,542 \end{bmatrix}.$$

From this result,

$$\begin{aligned} x_1 &= 131,161 \\ x_2 &= 120,324 \\ x_3 &= 79,194 \\ x_4 &= 178,936 \\ x_5 &= 66,703 \\ x_6 &= 426,542. \end{aligned}$$

In other words, it would require 131,161 units ($131,161 million worth) of final nonmetal production, 120,324 units of final metal output, 79,194 units of basic metal products, and so on to run the 1958 American economy and completely fill the stated bill of demands.

EXERCISES

1. A much simplified version of Leontief's 42-sector analysis of the 1947 American economy divides the economy into just 3 sectors: agriculture, manufacturing, and the household (i.e., the sector of the economy that produces labor). It is summarized in the following input-output table.

	Agriculture	Manufacturing	Household
Agriculture	.245	.102	.051
Manufacturing	.099	.291	.279
Household	.433	.372	.011

The bill of demands (in billions of dollars) is shown below.

Agriculture	2.88
Manufacturing	31.45
Household	30.91

(a) Write the input-output matrix A, the demand matrix D, and the matrix X.
(b) Compute $I - A$.
(c) Check that $(I - A)^{-1} = \begin{bmatrix} 1.454 & .291 & .157 \\ .533 & 1.763 & .525 \\ .837 & .791 & 1.278 \end{bmatrix}$ is an approximation to the inverse of $I - A$ by calculating $(I - A)^{-1}(I - A)$.
(d) Use the matrix of part (c) to compute X.
(e) Explain the meaning of the numbers in X in dollars.

2. An analysis of the 1958 Israeli ecomony* is here simplified by grouping the economy into three sectors: agriculture, manufacturing, and energy. The input-output table is given below.

	Agriculture	Manufacturing	Energy
Agriculture	.293	0	0
Manufacturing	.014	.207	.017
Energy	.044	.010	.216

Exports (in thousands of Israeli pounds) were as follows.

Agriculture	138,213
Manufacturing	17,597
Energy	1,786

(a) Write the input-output matrix A and the demand (export) matrix D.
(b) Compute $I - A$.
(c) Check that $(I - A)^{-1} = \begin{bmatrix} 1.414 & 0 & 0 \\ .027 & 1.261 & .027 \\ .080 & .016 & 1.276 \end{bmatrix}$ is an approximation to the inverse of $I - A$ by calculating $(I - A)^{-1}(I - A)$.
(d) Use the matrix of part (c) to determine the number of Israeli pounds worth of agricultural products, manufactured goods, and energy required to run this model of the Israeli economy and export the stated value of products.

---

*Wassily Leontief, Input-Output Economics (New York: Oxford University Press, 1966), pp. 54–57.

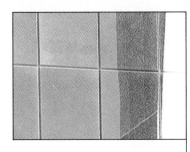

CHAPTER 7

Linear Programming

Many realistic problems involve inequalities. For example, a factory may have no more than 200 workers on a shift and must manufacture at least 3000 units at a cost of no more than $35 each. How many workers should it have per shift in order to produce the required units at minimal cost? *Linear programming* is a method for finding the optimal (best possible) solution for such problems, if there is one.

In this chapter we shall study two methods of solving linear programming problems: the graphical method and the simplex method. The graphical method requires a knowledge of **linear inequalities,** those involving only first-degree polynomials in x and y. So we begin with a study of such inequalities.

7.1 GRAPHING LINEAR INEQUALITIES IN TWO VARIABLES

Examples of linear inequalities in two variables include

$$x + 2y < 4, \quad 3x + 2y > 6, \quad \text{and} \quad 2x - 5y \geq 10.$$

A solution of a linear inequality is an ordered pair that satisfies the inequality. For example (4, 4) is a solution of

$$3x - 2y \leq 6.$$

(Check by substituting 4 for x and 4 for y.) Since a linear inequality has an infinite number of solutions for every choice of a value for x, the best way to show the solution is with a graph. The graph of an inequality in two variables consists of all the points in the plane whose coordinates are solutions of the inequality.

▶**EXAMPLE 1**   Graph the linear inequality $3x - 2y \leq 6$.

Because of the "=" portion of $\leq$, the points of the line $3x - 2y = 6$ satisfy the linear inequality $3x - 2y \leq 6$ and are part of its graph. As in Chapter 2, find the intercepts by first letting $x = 0$ and then letting $y = 0$; use these points to get the graph of $3x - 2y = 6$ shown in Figure 7.1.

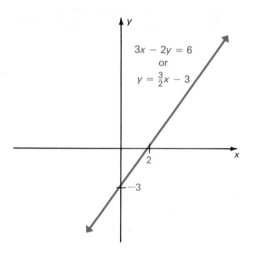

FIGURE 7.1

The points on the line satisfy "$3x - 2y$ *equals* 6." The points satisfying "$3x - 2y$ *is less than* 6" can be found by first solving $3x - 2y \leq 6$ for y.

$$3x - 2y \leq 6$$
$$-2y \leq -3x + 6$$
$$y \geq \frac{3}{2}x - 3 \qquad \text{Multiply by } -1/2; \\ \text{reverse the inequality}$$

As shown in Figure 7.2, the points *above* the line $3x - 2y = 6$ satisfy

$$y > \frac{3}{2}x - 3,$$

while those below the line satisfy

$$y < \frac{3}{2}x - 3.$$

1 Graph.

(a) $2x + 5y \leq 10$

(b) $x - y \geq 4$

Answer:

(a)

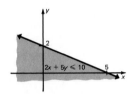

(b)

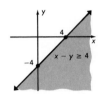

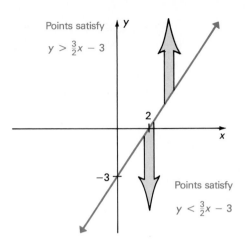

FIGURE 7.2

The line itself is the **boundary.** In summary, the inequality $3x - 2y \leq 6$ is satisfied by all points *on or above* the line $3x - 2y = 6$. Indicate the points above the line by shading, as in Figure 7.3. The line and shaded region of Figure 7.3 make up the graph of the linear inequality $3x - 2y \leq 6$. ◀ **1**

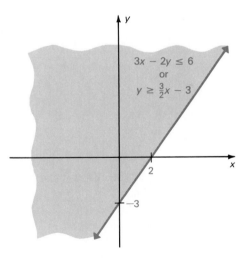

FIGURE 7.3

After an inequality is solved for y, the inequality symbol tells whether the points above, on, or below the boundary satisfy the inequality. An alternative method that does not require first solving the inequality for y involves the use of a test point. This method is used in Example 2.

2 Graph.

(a) $2x + 3y > 12$

(b) $3x - 2y < 6$

Answer:

(a)

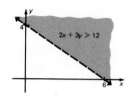

(b)

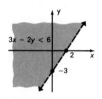

▶**EXAMPLE 2** Graph $x + 4y < 4$.

The boundary here is the line $x + 4y = 4$. Since the points on this line do not satisfy $x + 4y < 4$, the line is drawn dashed, as in Figure 7.4. Decide whether to shade the region above the line or the region below the line by choosing as a test point any point not on the boundary line. For example, we choose the point $(1, 0)$, which is not on the line $x + 4y = 4$. Substitute 1 for x and 0 for y in the given inequality.

$$x + 4y < 4$$
$$1 + 4(0) < 4 \qquad \text{Let } x = 1, y = 0$$
$$1 < 4 \qquad \text{True}$$

Since the result $1 < 4$ is true, the test point $(1, 0)$ is on the side of the line where all the points satisfy $x + 4y < 4$. For this reason, shade the side containing $(1, 0)$, as in Figure 7.4.

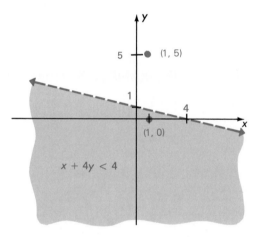

FIGURE 7.4

If a different test point is chosen, say $(1, 5)$, and substituted in $x + 4y < 4$, we obtain $1 + 4 \cdot 5 < 4$, which is *false*. Therefore, the side of the boundary line that does *not* contain $(1, 5)$ is the set of solutions, as shown in Figure 7.4. Thus, either test point leads to the same conclusion. ◀ **2**

Note If it is not on the boundary, $(0, 0)$ is usually the best choice for the test point because it makes the calculations as simple as possible, which avoids mistakes.

As these examples suggest, the graph of a linear inequality is a region in the plane, perhaps including the line that is the boundary of the region. Each of the shaded regions is an example of a **half-plane,** a region on one side of a line. For example, in Figure 7.5 line r divides the plane into half-planes P and Q. The points of line r belong to neither P nor Q. Line r is the boundary of each half-plane.

3 Graph each of the following.

(a) $x \geq 3$

(b) $y - 3 \leq 0$

Answer:

(a)

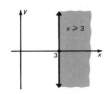

(b)

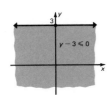

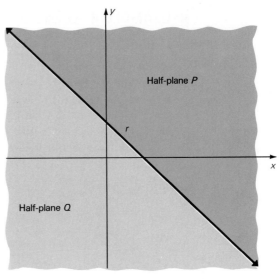

FIGURE 7.5

▶**EXAMPLE 3**   Graph each of the following.

(a) $x \leq -1$

Recall that the graph of $x = -1$ is the vertical line through $(-1, 0)$. The points where $x < -1$ lie in the half-plane to the left of the boundary line, as shown in Figure 7.6(a).

(b) $y \geq 2$

The graph of $y = 2$ is the horizontal line through $(0, 2)$. The points where $y \geq 2$ lie in the half-plane above this boundary line, as shown in Figure 7.6(b). ◀ **3**

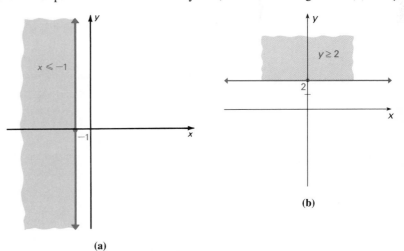

(a)

(b)

FIGURE 7.6

4 Graph $2x < y$.

Answer:

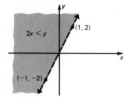

▶**EXAMPLE 4** Graph $x \geq 3y$.

Start by graphing the boundary, $x = 3y$. If $x = 0$, then $y = 0$, giving the point $(0, 0)$. Setting $y = 0$ would produce $(0, 0)$ again. To get a second point on the line, choose another number for x. Choosing $x = 3$ gives $y = 1$, so that $(3, 1)$ is a point on the line $x = 3y$. The points $(0, 0)$ and $(3, 1)$ lead to the line graphed in Figure 7.7. Let us choose $(6, 1)$ as a test point to decide which half-plane to shade. (Any point that is not on the line $x = 3y$ may be used.) Replacing x with 6 and y with 1 in the original inequality gives a true statement, so the half-plane containing $(6, 1)$ is shaded, as shown in Figure 7.7. ◀ **4**

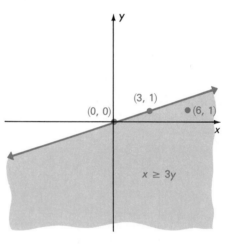

FIGURE 7.7

The steps used to graph a linear inequality are summarized below.

Graphing a Linear Inequality

1. Draw the graph of the boundary line. Make the line solid if the inequality involves $\leq$ or $\geq$; make the line dashed if the inequality involves $<$ or $>$.
2. Decide which half-plane to shade: either
   (a) solve the inequality for y, getting $y \geq mx + b$ or $y \leq mx + b$; shade the region above the line $y = mx + b$ if $y \geq mx + b$; shade the region below the line if $y \leq mx + b$; or
   (b) choose any point not on the line as a test point; shade the half-plane that includes the test point if the test point satisfies the original inequality; otherwise, shade the half-plane on the other side of the boundary line.

5 Graph the system
$x + y \le 6$, $2x + y \ge 4$.

Answer:

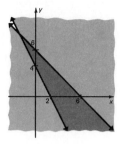

SYSTEMS OF INEQUALITIES   Realistic problems often involve many inequalities. For example, a manufacturing problem might produce inequalities resulting from production requirements, as well as inequalities about cost requirements. A set of at least two inequalities is called a **system of inequalities.** The graph of a system of inequalities is made up of all those points that satisfy all the inequalities of the system at the same time. The next example shows how to graph a system of inequalities.

▶**EXAMPLE 5**   Graph the system

$$y < -3x + 12$$
$$x < 2y.$$

First, graph the solution of $y < -3x + 12$. The boundary, the line with equation $y = -3x + 12$, is dashed. The points below the boundary satisfy $y < -3x + 12$. Now graph the solution of $x < 2y$ on the same axes. Again, the boundary line $x = 2y$ is dashed. Use a test point to see that the region above the boundary satisfies $x < 2y$. The heavily shaded region in Figure 7.8 shows all the points that satisfy both inequalities of the system.   ◀ **5**

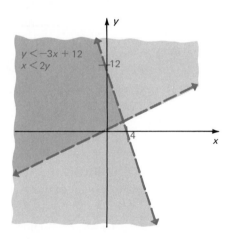

FIGURE 7.8

The heavily shaded region of Figure 7.8 is sometimes called the **region of feasible solutions,** or just the **feasible region,** since it is made up of all the points that satisfy (are feasible for) each inequality of the system.

▶**EXAMPLE 6**   Graph the feasible region for the system

$$2x - 5y \le 10$$
$$x + 2y \le 8$$
$$x \ge 0, \quad y \ge 0.$$

6 Graph the feasible region of the system

$x + 4y \leq 8$, $x - y \geq 3$

$x \geq 0$, $y \geq 0$.

Answer:

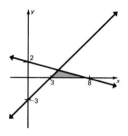

On the same axes, graph each inequality by graphing the boundary and choosing the appropriate half-plane. Graph the solid boundary line $2x - 5y = 10$ by first locating the intercepts that give the points $(5, 0)$ and $(0, -2)$. Use a test point to see that $2x - 5y < 10$ is satisfied by the points above the boundary. In the same way, the solid boundary line $x + 2y = 8$ goes through $(8, 0)$ and $(0, 4)$. A test point will show that the graph of $x + 2y < 8$ includes all points below the boundary. The inequalities $x \geq 0$ and $y \geq 0$ restrict the feasible region to the first quadrant. Find the feasible region by locating the intersection (overlap) of *all* the half-planes. This feasible region is shaded in Figure 7.9. ◄ **6**

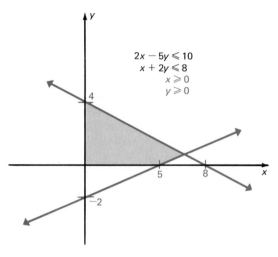

FIGURE 7.9

APPLICATIONS As we shall see in the rest of this chapter, many realistic problems lead to systems of linear inequalities. The next example is typical of such problems.

►**EXAMPLE 7** Midtown Manufacturing Company makes plastic plates and cups, both of which require time on two machines. A unit of plates requires 1 hour on machine A and 2 on machine B, while a unit of cups requires 3 hours on machine A and 1 on machine B. Each machine is operated for at most 15 hours per day. Write a system of inequalities expressing these conditions and graph the feasible region.

Let x represent the number of units of plates to be made, and y represent the number of units of cups. Then make a chart that summarizes the given information.

	Number	Time on Machine	
	Made	A	B
Plates	x	1	2
Cups	y	3	1
Maximum time available		15	15

On machine A, x units of plates require a total of $1 \cdot x = x$ hours while y units of cups require $3 \cdot y = 3y$ hours. Since machine A is available no more than 15 hours a day,

$$x + 3y \le 15. \qquad \text{Machine A}$$

The requirement that machine B be used no more than 15 hours a day gives

$$2x + y \le 15. \qquad \text{Machine B}$$

It is not possible to produce a negative number of cups or plates, so that

$$x \ge 0 \quad \text{and} \quad y \ge 0.$$

The feasible region for this system of inequalities is shown in Figure 7.10. ◀

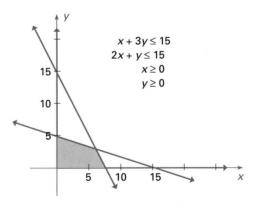

$$x + 3y \le 15$$
$$2x + y \le 15$$
$$x \ge 0$$
$$y \ge 0$$

FIGURE 7.10

7.1 EXERCISES

Graph each of the following linear inequalities. (See Examples 1–4.)

1. $x + y \le 2$ **2.** $y \le x + 1$ **3.** $x \ge 3 + y$ **4.** $y \ge x - 3$ **5.** $4x - y < 6$

6. $3y + x > 4$ **7.** $3x + y < 6$ **8.** $2x - y > 2$ **9.** $x + 3y \ge -2$ **10.** $2x + 3y \le 6$

11. $4x + 3y > -3$ **12.** $5x + 3y > 15$ **13.** $2x - 4y < 3$ **14.** $4x - 3y < 12$ **15.** $x \le 5y$

16. $2x \ge y$ **17.** $-3x < y$ **18.** $-x > 6y$ **19.** $y < x$ **20.** $y > -2x$

21. $y \le -2$ **22.** $x \ge 3$

Graph the feasible region for the following systems of inequalities. (See Examples 5 and 6.)

23. $x - y \ge 0$
      $x \le 4$

24. $3x + 2y \ge 0$
      $y \ge 2x$

25. $3x + 2y \ge 18$
      $x - 2y \ge 8$

26. $2x + y \ge 4$
      $4x - 5y \le 20$

27. $x + y \le 1$
      $x - y \ge 2$

28. $3x - 4y < 6$
      $2x + 5y > 15$

29. $2x - y < 1$
      $3x + y < 6$

30. $x + 3y \le 6$
      $2x + 4y \ge 7$

31. $-x - y < 5$
      $2x - y < 4$

32. $6x - 4y > 8$
      $3x + 2y > 4$

33. $x + y \le 4$
      $x - y \le 5$
      $4x + y \le -4$

34. $3x - 2y \ge 6$
      $x + y \le -5$
      $y \le 4$

35. $-2 < x < 3$
      $-1 \le y \le 5$
      $2x + y < 6$

36. $-2 < x < 2$
      $y > 1$
      $x - y > 0$

37. $2y + x \ge -5$
      $y \le 3 + x$
      $x \ge 0$
      $y \ge 0$

38. $2x + 3y \le 12$
      $2x + 3y > -6$
      $3x + y < 4$
      $x \ge 0$
      $y \ge 0$

39. $3x + 4y > 12$
      $2x - 3y < 6$
      $0 \le y \le 2$
      $x \ge 0$

40. $0 \le x \le 9$
      $x - 2y \ge 4$
      $3x + 5y \le 30$
      $y \ge 0$

Write a system of inequalities and graph the feasible region for each of the following problems. In Exercises 41 and 42, first complete the chart. (See Example 7.)

41. A small pottery shop makes two kinds of planters, glazed and unglazed. The glazed type requires 1/2 hour to throw on the wheel and 1 hour in the kiln. The unglazed type takes 1 hour to throw on the wheel and 6 hours in the kiln. The wheel is available for at most 8 hours a day and the kiln is available for at most 20 hours per day.

	Number	Hours Wheel	Hours Kiln
Glazed	x		
Unglazed	y		
Maximum number of hours available			

42. Carmella and Walt produce handmade shawls and afghans. They spin the yarn, dye it, and then weave it. A shawl requires 1 hour of spinning, 1 hour of dyeing, and 1 hour of weaving. An afghan needs 2 hours of spinning, 1 of dyeing, and 4 of weaving. Together, they spend at most 8 hours spinning, 6 hours dyeing, and 14 hours weaving.

	Number	Hours Spinning	Hours Dyeing	Hours Weaving
Shawls	x			
Afghans	y			
Maximum number of hours available		8	6	14

43. Management Southwestern Oil supplies two distributors located in the Northwest. One distributor needs at least 3000 barrels of oil, and the other needs at least 5000 barrels. Southwestern can send out at most 10,000 barrels. Let x = the number of barrels of oil sent to distributor 1 and y = the number sent to distributor 2.

44. Management The California Almond Growers have 2400 boxes of almonds to be shipped from their plant in Sacramento to Des Moines and San Antonio. The Des Moines market needs at least 1000 boxes, while the San Antonio market must have at least 800 boxes. Let x = the number of boxes to be shipped to Des Moines and y = the number of boxes to be shipped to San Antonio.

45. Management   The loan department in a bank has set aside $25 million for commercial and home loans. The bank's policy is to allocate at least four times as much money to home loans as to commercial loans. The bank's return is 12% on a home loan and 10% on a commercial loan. The manager of the loan department wants to earn a return of at least $2.8 million on these loans. Let $x =$ the amount for home loans and $y =$ the amount for commercial loans.

46. Management   A cement manufacturer produces at least 3.2 million barrels of cement annually. He is told by the Environmental Protection Agency that his operation emits 2.5 pounds of dust for each barrel produced. The EPA has ruled that annual emissions must be reduced to 1.8 million pounds. To do this, the manufacturer plans to replace the present dust collectors with two types of electronic precipitators. One type would reduce emissions to .5 pounds per barrel and would cost 16¢ per barrel. The other would reduce the dust to .3 pounds per barrel and would cost 20¢ per barrel. The manufacturer does not want to spend more than .8 million dollars on the precipitators. He needs to know how many barrels he should produce with each type. Let $x =$ the number of barrels in millions produced with the first type and $y =$ the number of barrels in millions produced with the second type.

7.2 SOLVING LINEAR PROGRAMMING PROBLEMS GRAPHICALLY

Many problems in business, biology, and economics involve finding the optimum value (either the maximum or the minimum) of a function, subject to certain restrictions. The function to be optimized is called the **objective function.** In a **linear programming** problem, the maximum or minimum value of the objective function is found subject to a set of restrictions, or **constraints,** given by linear inequalities. When only two variables are involved, the solution to a linear programming problem can be found by first graphing the set of constraints, then finding the feasible region as discussed in the previous section. This method is explained in the following example.

▶**EXAMPLE 1**   Find the maximum and minimum values of the objective function $z = 2x + 5y$, subject to the following constraints.

$$3x + 2y \leq 6$$
$$-2x + 4y \leq 8$$
$$x + \ y \geq 1$$
$$x \geq 0, \quad y \geq 0$$

The feasible region is graphed in Figure 7.11. Every point in the feasible region satisfies all the constraints. However, we want to find those points that produce the maximum possible value of the objective function. To see how to find this maximum value, let us add to the graph of Figure 7.11 lines that represent the objective

1 Suppose the objective function in Example 1 is changed to $z = 5x + 2y$.

(a) Sketch the graphs of the objective function when $z = 0$, $z = 5$, and $z = 10$ on the region of feasible solutions given in Figure 7.11.

(b) From the graph, decide what values of x and y will maximize the objective function.

Answer:

(a)

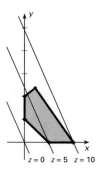

(b) $(2, 0)$

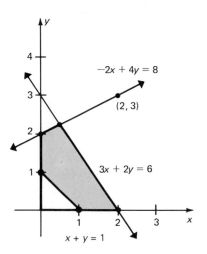

FIGURE 7.11

function $z = 2x + 5y$ for various sample values of z. By choosing the values 0, 5, 10, and 15 for z, the objective function becomes (in turn)

$$0 = 2x + 5y, \quad 5 = 2x + 5y, \quad 10 = 2x + 5y, \quad 15 = 2x + 5y.$$

These four lines are graphed in Figure 7.12. (All the lines are parallel because they have the same slope.) The figure shows that z cannot take on the value 15 because the graph for $z = 15$ is entirely outside the feasible region. The maximum possible value of z will be obtained from a line parallel to the others and between the lines representing the objective function when $z = 10$ and $z = 15$. The value of z will be as large as possible and all constraints will be satisfied if this line just touches the feasible region. This occurs at point A. The coordinates of point A are found by solving the system

$$3x + 2y = 6$$
$$-2x + 4y = 8$$

to get $(1/2, 9/4)$. The value of z at this point is

$$z = 2x + 5y$$
$$= 2\left(\frac{1}{2}\right) + 5\left(\frac{9}{4}\right)$$
$$z = 12\frac{1}{4}.$$

The maximum possible value of z is $12\frac{1}{4}$. Similarly, the minimum value of z occurs at point B, which has coordinates $(1, 0)$. The minimum value of z is $2(1) + 5(0) = 2$. ◀ **1**

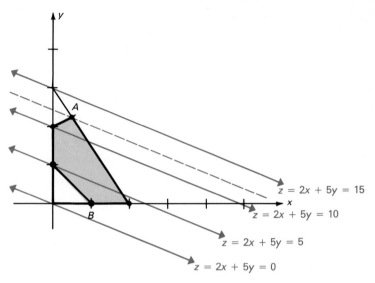

FIGURE 7.12

Points such as A and B in Example 1 are called corner points. A **corner point** is a point in the feasible region where the boundary lines of two constraints cross.

The feasible region in Figure 7.11 is **bounded**, since the region is enclosed by boundary lines on all sides. Linear programming problems with bounded regions always have solutions. However, if Example 1 did not include the constraint $3x + 2y \le 6$, the feasible region would be **unbounded**, and there would be no way to *maximize* the value of the objective function.

Some general conclusions can be drawn from the method of solution used in Example 1. Figure 7.13 shows various feasible regions and the lines that result from

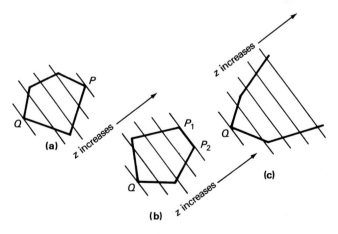

FIGURE 7.13

2 **(a)** Identify the corner points in the graph.

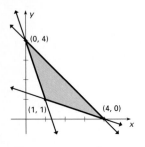

(b) Which corner point would minimize $z = 2x + 3y$?

Answer:

(a) (0, 4), (1, 1), (4, 0)

(b) (1, 1)

various values of z. (We assume the lines are in order from left to right as z increases.) In part (a) of the figure, the objective function takes on its minimum value at corner point Q and its maximum value at P. The minimum is again at Q in part (b), but the maximum occurs at P_1 or P_2, or any point on the line segment connecting them. Finally, in part (c), the minimum value occurs at Q, but the objective function has no maximum value because the feasible region is unbounded.

The preceding discussion suggests the truth of the **corner point theorem**.

Corner Point Theorem

If the feasible region is bounded, then the objective function has both a maximum and a minimum value and each occurs at one or more corner points.

If the feasible region is unbounded, the objective function may not have a maximum or minimum. But if a maximum or minimum value exists, it will occur at one or more corner points.

This theorem simplifies the job of finding an optimum value: First, graph the feasible region and find all corner points. Then test each corner point in the objective function. Finally, identify the corner point producing the optimum solution. For unbounded regions, first decide whether or not the required optimum can be found;

With the theorem, the problem in Example 1 could have been solved by identifying the five corner points of Figure 7.11: (0, 1), (0, 2), (1/2, 9/4), (2, 0), and (1, 0). Then, substituting each of these points into the objective function $z = 2x + 5y$ would identify the corner points that produce the maximum or the minimum value of z.

Corner Point	Value of $z = 2x + 5y$
(0, 1)	$2(0) + 5(1) = 5$
(0, 2)	$2(0) + 5(2) = 10$
$\left(\dfrac{1}{2}, \dfrac{9}{4}\right)$	$2\left(\dfrac{1}{2}\right) + 5\left(\dfrac{9}{4}\right) = 12\left(\dfrac{1}{4}\right)$ (maximum)
(2, 0)	$2(2) + 5(0) = 4$
(1, 0)	$2(1) + 5(0) = 2$ (minimum)

From these results, the corner point (1/2, 9/4) yields the maximum value of $12\frac{1}{4}$ and the corner point (1, 0) gives the minimum value of 2. These are the same values found earlier. **2**

A summary of the steps in solving a linear programming problem by the graphical method is given here.

Solving a Linear Programming Problem Graphically

1. Write the objective function and all necessary constraints.
2. Graph the feasible region.
3. Determine the coordinates of each of the corner points.
4. Find the value of the objective function at each corner point.
5. For a bounded region, the solution is given by the corner point producing the optimum value of the objective function.
6. For an unbounded region, check that a solution actually exists. If it does, it will occur at a corner point.

▶**EXAMPLE 2** Sketch the feasible region for the following set of constraints:

$$3y - 2x \geq 0$$
$$y + 8x \leq 52$$
$$y - 2x \leq 2$$
$$x \geq 3.$$

Then find the maximum and minimum values of the objective function $z = 5x + 2y$.

The graph in Figure 7.14 shows that the feasible region is bounded. The corner points were found by solving systems of two equations using the methods of Chapter 6. For example, the point (5, 12) is the intersection of the lines corresponding to the equations $y + 8x = 52$ and $y - 2x = 2$, and the solution of the system

$$y + 8x = 52$$
$$y - 2x = 2$$

is the ordered pair (5, 12). Also, the corner point (6, 4) was found by solving the system

$$y + 8x = 52$$
$$3y - 2x = 0.$$

Use the corner points from the graph to find the maximum and minimum values of the objective function.

Corner Point	Value of $z = 5x + 2y$	
(3, 2)	$5(3) + 2(2) = 19$	(minimum)
(6, 4)	$5(6) + 2(4) = 38$	
(5, 12)	$5(5) + 2(12) = 49$	(maximum)
(3, 8)	$5(3) + 2(8) = 31$	

3 Use the region of feasible solutions in the sketch to find the following.

(a) The values of x and y that maximize $z = 2x - y$.

(b) The maximum value of $z = 2x - y$.

(c) The values of x and y that minimize $z = 4x + 3y$.

(d) The minimum value of $z = 4x + 3y$.

Answer:

(a) (5, 2)

(b) 8

(c) (1, 1)

(d) 7

4 The sketch shows a feasible region. Let $z = x + 3y$. Use the sketch to find the values of x and y that

(a) minimize z;

(b) maximize z.

Answer:

(a) (3, 1)

(b) (5, 4)

The minimum value of $z = 5x + 2y$ is 19 at the corner point (3, 2). The maximum value is 49 at (5, 12). ◀ **3**

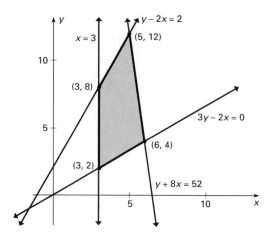

FIGURE 7.14

▶**EXAMPLE 3** Solve the following linear programming problem.

$$\text{Minimize} \qquad z = x + 2y$$
$$\text{subject to:} \qquad x + y \le 10$$
$$3x + 2y \ge 6$$
$$x \ge 0, \quad y \ge 0.$$

The feasible region is shown in Figure 7.15. From the figure, the corner points are (0, 3), (0, 10), (10, 0), and (2, 0). These corner points give the following values of z.

Corner Point	Value of $z = x + 2y$
(0, 3)	$0 + 2(3) = 6$
(0, 10)	$0 + 2(10) = 20$
(10, 0)	$10 + 2(0) = 10$
(2, 0)	$2 + 2(0) = 2$ **(minimum)**

The minimum value of z is 2; it occurs at (0, 2). ◀ **4**

5 The sketch below shows a region of feasible solutions. From the sketch decide what ordered pair would minimize z.

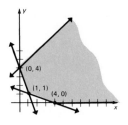

Answer:
(1, 1)

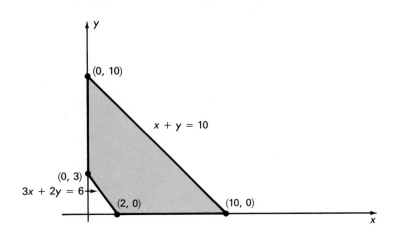

FIGURE 7.15

▶ **EXAMPLE 4** Solve the following linear programming problem.

$$\text{Minimize} \qquad z = 2x + 4y$$
$$\text{subject to:} \qquad x + 2y \geq 10$$
$$3x + \ y \geq 10$$
$$x \geq 0, \quad y \geq 0.$$

Figure 7.16 shows the feasible region with corner points (0, 10), (2, 4), and (10, 0). Find the value of z for each corner point.

Corner Point	Value of $z = 2x + 4y$
(0, 10)	$2(0) + 4(10) = 40$
(2, 4)	$2(2) + 4(4) = 20$ (minimum)
(10, 0)	$2(10) + 4(0) = 20$ (minimum)

In this case, both (2, 4) and (10, 0), as well as all the points on the boundary line between them, give the same optimum value of z. There is an infinite number of equally "good" values of x and y which give the same minimum value of the objective function $z = 2x + 4y$. This minimum value is 20. ◀ **5**

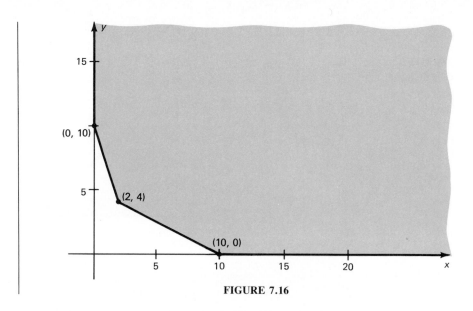

FIGURE 7.16

7.2 EXERCISES

Exercises 1–6 show regions of feasible solutions. Use these regions to find maximum and minimum values of each given objective function. (See Examples 1–2.)

1. $z = 3x + 5y$

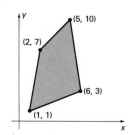

2. $z = 6x + y$

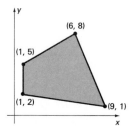

3. $z = .40x + .75y$

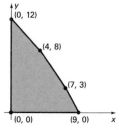

4. $z = .35x + 1.25y$

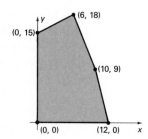

5. $z = 2x + 3y$

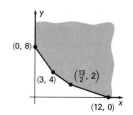

6. $z = 5x + 6y$

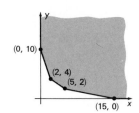

Find the maximum and minimum values of $z = 4x + 5y$ (if possible) for each of the following sets of constraints. (See Examples 2–4.)

7.
$x + y \leq 5$
$2x + y \geq 4$
$x \geq 0, \quad y \geq 0$

8.
$x + 2y \geq 2$
$-2x + y \leq 1$
$x \leq 2$

9.
$-x + 2y \leq 6$
$3x + y \geq 3$
$x \geq 0, \quad y \geq 0$

10.
$-3x + 4y \leq 8$
$5x - 2y \leq 10$
$x \geq 0, \quad y \geq 0$

11.
$y \leq 10$
$10x - 3y \leq 50$
$x \geq 0, \quad y \geq 0$

12.
$3x + 2y \geq 6$
$x + 2y \geq 4$
$x \geq 0, \quad y \geq 0$

Use graphical methods to solve Exercises 13–20. (See Examples 2–4.)

13. Maximize
subject to:

$z = 5x + 2y$
$2x + 3y \leq 6$
$4x + y \leq 6$
$x \geq 0, \quad y \geq 0.$

14. Minimize
subject to:

$z = x + 3y$
$x + y \leq 10$
$5x + 2y \geq 20$
$-x + 2y \geq 0$
$x \geq 0, \quad y \geq 0.$

15. Maximize
subject to:

$z = 2x + y$
$3x - y \geq 12$
$x + y \leq 15$
$x \geq 2, \quad y \geq 5.$

16. Maximize
subject to:

$z = x + 3y$
$2x + 3y \leq 100$
$5x + 4y \leq 200$
$x \geq 10, \quad y \geq 20.$

17. Maximize
subject to:

$z = 4x + 2y$
$x - y \leq 10$
$5x + 3y \leq 75$
$x \geq 0, \quad y \geq 0.$

18. Maximize
subject to:

$z = 4x + 5y$
$10x - 5y \leq 100$
$20x + 10y \geq 150$
$x \geq 0, \quad y \geq 0.$

19. Find values of $x \geq 0$ and $y \geq 0$ which maximize $z = 10x + 12y$ subject to each of the following sets of constraints.

(a) $x + y \leq 20$
$x + 3y \leq 24$

(b) $3x + y \leq 15$
$x + 2y \leq 18$

(c) $2x + 5y \geq 22$
$4x + 3y \leq 28$
$2x + 2y \leq 17$

20. Find values of $x \geq 0$ and $y \geq 0$ that minimize $z = 3x + 2y$ subject to each of the following sets of constraints.

(a) $10x + 7y \leq 42$
$4x + 10y \geq 35$

(b) $6x + 5y \geq 25$
$2x + 6y \geq 15$

(c) $x + 2y \geq 10$
$2x + y \geq 12$
$x - y \leq 8$

7.3 APPLICATIONS OF LINEAR PROGRAMMING

In this section we show several applications of linear programming problems with two variables.

▶**EXAMPLE 1** A 4-H Club member raises only geese and pigs. She wants to raise no more than 16 animals including no more than 10 geese. She spends $15 to raise a goose and $45 to raise a pig, and has $540 available for this project. Find the maximum profit she can make if each goose produces a profit of $6 and each pig a profit of $20.

Let x represent the number of geese to be produced, and let y represent the number of pigs. Start by summarizing the information of the problem in a table.

	Number	Cost to Raise	Profit Each
Geese	x	$ 15	$ 6
Pigs	y	45	20
Maximum available	16	$540	

Use this table to write the necessary constraints. Since the total number of animals cannot exceed 16, the first constraint is

$$x + y \leq 16.$$

"No more than 10 geese" leads to

$$x \leq 10.$$

The cost to raise x geese at $15 per goose is $15x$ dollars, while the cost for y pigs at $45 each is $45y$ dollars. Since only $540 is available,

$$15x + 45y \leq 540.$$

Dividing both sides by 15 gives the equivalent inequality

$$x + 3y \leq 36.$$

The number of geese and pigs cannot be negative, so

$$x \geq 0, \quad y \geq 0.$$

The 4-H club member wants to know the number of geese and the number of pigs that should be raised for maximum profit. Each goose produces a profit of $6, and each pig, $20. If z represents total profit, then

$$z = 6x + 20y$$

is the objective function, which is to be maximized.

We must solve the following linear programming problem.

Maximize $\quad z = 6x + 20y \qquad$ **Objective function**

subject to: $\qquad x + y \leq 16$

$\qquad\qquad\qquad x \leq 10$

$\qquad\qquad x + 3y \leq 36 \qquad$ **Constraints**

$\qquad x \geq 0, \quad y \geq 0.$

Using the methods of the previous section, graph the feasible region for the system of inequalities given by the constraints as in Figure 7.17.

1 Find the corner points P and Q in Figure 7.17.

Answer:
$P = (6, 10)$
$Q = (10, 6)$

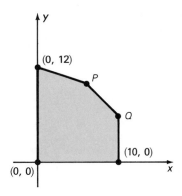

FIGURE 7.17

The corner points (0, 12), (0, 0), and (10, 0) can be read directly from the graph. The coordinates of each of the other corner points can be found by solving a system of linear equations. **1**

Test each corner point in the objective function to find the maximum profit.

Corner Point	$z = 6x + 20y$
(0, 12)	$6(0) + 20(12) = 240$ (**maximum**)
(6, 10)	$6(6) + 20(10) = 236$
(10, 6)	$6(10) + 20(6) = 180$
(10, 0)	$6(10) + 20(0) = \ \ 60$
(0, 0)	$6(0) + 20(0) = \ \ \ \ 0$

The maximum of 240 occurs at (0, 12). Thus, 12 pigs and no geese will produce a maximum profit of $240. ◀

▶**EXAMPLE 2**   An office manager needs to purchase new filing cabinets. He knows that Ace cabinets cost $40 each, require 6 square feet of floor space, and hold 8 cubic feet of files. On the other hand, each Excello cabinet costs $80, requires 8 square feet of floor space, and holds 12 cubic feet. His budget permits him to spend no more than $560 on files, while the office has room for no more than 72 square feet of cabinets. The manager desires the greatest storage capacity within the limitations imposed by funds and space. How many of each type cabinet should he buy?

Let x represent the number of Ace cabinets to be bought and let y represent the number of Excello cabinets. The information given in the problem can be summarized as follows.

2 Find the corner point labeled Q on the region of feasible solutions given below.

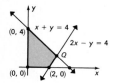

Answer:

$\left(\dfrac{8}{3}, \dfrac{4}{3}\right)$

	Number	Cost of Each	Space Required	Storage Capacity
Ace	x	$ 40	6 sq ft	8 cu ft
Excello	y	$ 80	8 sq ft	12 cu ft
Maximum available		$560	72 sq ft	

Write the constraints imposed by cost and space.

$$40x + 80y \le 560 \qquad \text{Cost}$$
$$6x + 8y \le 72 \qquad \text{Floor space}$$

Since the number of cabinets cannot be negative, $x \ge 0$ and $y \ge 0$. The objective function to be maximized gives the amount of storage capacity provided by some combination of Ace and Excello cabinets. From the information in the chart, the objective function is

$$\text{Storage space} = z = 8x + 12y.$$

In summary, the given problem has produced the following linear programming problem.

$$\begin{aligned} \text{Maximize} \qquad & z = 8x + 12y \\ \text{subject to:} \qquad & 40x + 80y \le 560 \\ & 6x + 8y \le 72 \\ & x \ge 0, \quad y \ge 0. \end{aligned}$$

A graph of the feasible region is shown in Figure 7.18. Three of the corner points can be identified from the graph as (0, 0), (0, 7), and (12, 0). The fourth corner point, labeled Q in the figure, can be found by solving the system of equations

$$40x + 80y = 560$$
$$6x + 8y = 72.$$

By solving this system, we find that Q is the point (8, 3). **2**

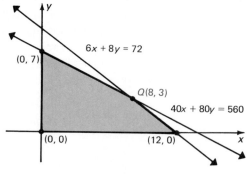

FIGURE 7.18

3 A popular cereal combines oats and corn. At least 27 tons of the cereal is to be made. For the best flavor, the amount of corn should be at least twice the amount of oats. Corn costs $200 per ton and oats cost $300 per ton. How much of each grain should be used to minimize the cost?

(a) Make a chart to organize the information given in the problem.

(b) Write an equation for the objective function.

(c) Write four inequalities for the constraints.

Answer:

(a)

	Number of Tons	Cost/Ton
Oats	x	$300
Corn	y	200
	27	

(b) $z = 300x + 200y$

(c) $x + y \geq 27$
$\quad\quad y \geq 2x$
$\quad\quad x \geq 0$
$\quad\quad y \geq 0$

Use the corner point theorem to find the maximum value of z.

Corner Point	Value of $z = 8x + 12y$
(0, 0)	0
(0, 7)	84
(12, 0)	96
(8, 3)	100 (maximum)

The objective function, which represents storage space, is maximized when $x = 8$ and $y = 3$. The manager should buy 8 Ace cabinets and 3 Excello cabinets. ◄ **3**

▶**EXAMPLE 3**   Certain laboratory animals must have at least 30 grams of protein and at least 20 grams of fat per feeding period. These nutrients come from food A, which costs 18¢ per unit and supplies 2 grams of protein and 4 of fat, and food B, with 6 grams of protein and 2 of fat, costing 12¢ per unit. Food B is bought under a long-term contract requiring that at least 2 units of B be used per serving. How much of each food must be bought to produce minimum cost per serving?

Let x represent the amount of food A needed, and y the amount of food B. Use the given information to produce the following table.

Food	Number of Units	Grams of Protein	Grams of Fat	Cost
A	x	2	4	18¢
B	y	6	2	12¢
Minimum required		30	20	

The linear programming problem can be stated as follows.

$$\text{Minimize} \quad z = .18x + .12y$$
$$\text{subject to:} \quad 2x + 6y \geq 30$$
$$4x + 2y \geq 20$$
$$y \geq 2$$
$$x \geq 0, \quad y \geq 0.$$

(The constraint $y \geq 0$ is redundant because of the constraint $y \geq 2$.) A graph of the feasible region with the corner points identified is shown in Figure 7.19.

4 A popular cereal combines oats and corn. At least 27 tons of the cereal is to be made. For the best flavor, the amount of corn should be at least twice the amount of oats. Corn costs $200 per ton and oats cost $300 per ton. How much of each grain should be used to minimize the cost?

(a) Graph the feasible region and find the corner points.

(b) Determine the minimum value of the objective function and the point where it occurs.

Answer:

(a)

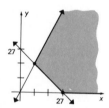

Corner points: (27, 0), (9, 18)

(b) $6300 at (9, 18)

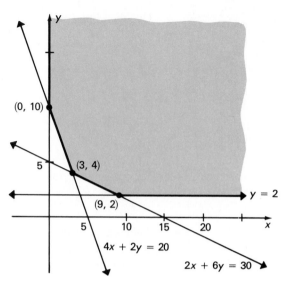

FIGURE 7.19

Find the corner points by solving the systems of equations for each pair of intersecting boundary lines. Use the corner point theorem to find the minimum value of z as shown in the chart below.

Corner Points	$z = .18x + .12y$
(0, 10)	$.18(0) + .12(10) = 1.20$
(3, 4)	$.18(3) + .12(4) = 1.02$ (minimum)
(9, 2)	$.18(9) + .12(2) = 1.86$

The minimum value of 1.02 occurs at (3, 4). Thus, 3 units of A and 4 units of B will produce a minimum cost of $1.02 per serving. ◄ **4**

The feasible region in Figure 7.19 is an unbounded feasible region—the region extends indefinitely to the upper right. With this region it would not be possible to *maximize* the objective function, because the total cost of the food could always be increased by encouraging the animals to eat more.

7.3 EXERCISES

Write the constraints in Exercises 1–6 as linear inequalities. Identify all variables used. (Not all the information is used in Exercises 5 and 6.) (See Examples 1–3.)

1. Product A requires 2 hours on machine I, while product B needs 3 hours on the same machine. The machine is available at most 45 hours per week.

2. A cow requires a third of an acre of pasture and a sheep needs a quarter acre. A rancher wants to use at least 120 acres of pasture.

3. Juan Garcia needs at least 25 units of vitamin A per day. Green pills provide 4 units and red pills provide 1.

4. Sandra Wong spends 3 hours selling a small computer and 5 hours selling a larger model. She works no more than 45 hours per week.

5. Coffee costing $6 a pound is to be mixed with coffee costing $5 a pound to get at least 50 pounds of a mixture.

6. A tank in an oil refinery holds 120 gallons. The tank contains a mixture of light oil worth $1.25 per gallon and heavy oil worth $.80 per gallon.

Solve the following linear programming problems. (See Examples 1–3.)

7. Management The Miers Company produces small engines for several manufacturers. The company receives orders from two assembly plants for their Topflight engine. Plant I needs at least 50 engines, and plant II needs at least 27 engines. The company can send at most 85 engines to these two assembly plants. It costs $20 per engine to ship to plant I and $35 per engine to ship to plant II. How many engines should be shipped to each plant to minimize shipping costs? What is the minimum cost?

8. Management A manufacturer of refrigerators must ship at least 100 refrigerators to its two West Coast warehouses. Each warehouse holds a maximum of 100 refrigerators. Warehouse A holds 25 refrigerators already, and warehouse B has 20 on hand. It costs $12 to ship a refrigerator to warehouse A and $10 to ship one to warehouse B. How many refrigerators should be shipped to each warehouse to minimize costs? What is the minimum cost?

9. Management A company is considering two insurance plans with the types of coverage (in thousands of dollars) and premiums per thousand dollars as shown below.

		Coverage		
		Fire/Theft	Liability	Premium
Policy	A	$10	$ 80	$50
	B	$15	$120	$40

The company wants at least $100,000 fire/theft insurance and at least $1,000,000 liability insurance from these plans. How many units should be purchased from each plan to minimize the cost of the premiums?

10. Management Seall Manufacturing Company makes color television sets. It produces a bargain set that sells for $100 profit and a deluxe set that sells for $150 profit. On the assembly line the bargain set requires 3 hours, and the deluxe set takes 5 hours. The cabinet shop spends 1 hour on the cabinet for the bargain set and 3 hours on the cabinet for the deluxe set. Both sets require 2 hours of time for testing and packing. On a particular production run, the Seall Company has available 3900 work hours on the assembly line, 2100 work hours in the cabinet shop, and 2200 work hours in the testing and packing department. How many sets of each type should it produce to make maximum profit? What is the maximum profit?

11. Management A machine shop manufactures two types of bolts. Each type of bolt requires three machines for its manufacture. The time required on each machine is shown in the table below.

		Machine Group		
		I	II	III
Bolts	Type 1	.1 min	.1 min	.1 min
	Type 2	.1 min	.4 min	.5 min

Production schedules are made up one day at a time. In a day there are 240, 720, and 160 minutes available, respectively, on these machines. Type 1 bolts sell for 10¢ and type 2 bolts for 12¢. How many of each type of bolt should be manufactured per day to maximize revenue? What is the maximum revenue?

12. Management The manufacturing process requires that oil refineries must manufacture at least 2 gallons of gasoline for every gallon of fuel oil. To meet the winter demand for fuel oil, at least 3 million gallons a day must be produced. The demand for gasoline is no more than 6.4 million gallons per day. If the refinery sells gasoline for $1.25 per gallon and fuel oil for $1 per gallon, how much of each should be produced to maximize revenue? Find the maximum revenue.

13. Management A candy company has 100 kilograms of chocolate-covered nuts and 125 kilograms of chocolate-covered raisins to be sold as two different mixes. One mix will contain 1/2 nuts and 1/2 raisins and will sell for $6 per kilogram. The other mix will contain 1/3 nuts and 2/3 raisins and will sell for $4.80 per kilogram. How many kilograms of each mix should the company prepare for maximum revenue? Find the maximum revenue.

14. Management A small country can grow only two crops for export, coffee and cocoa. The country has 500,000 hectares of land available for the crops. Long-term contracts require that at least 100,000 hectares be devoted to coffee and at least 200,000 hectares to cocoa. Cocoa must be processed locally, and production bottlenecks limit cocoa to 270,000 hectares. Coffee requires two workers per hectare, with cocoa requiring five. No more than 1,750,000 people are available for working with these crops. Coffee produces a profit of $220 per hectare and cocoa a profit of $310 per hectare. How many hectares should the country devote to each crop in order to maximize profit? Find the maximum profit.

15. Management The Mostpure Milk Company gets milk from two dairies and then blends the milk to get the desired amount of butterfat for the company's premier product. If Dairy I can supply at most 50 gallons of milk averaging 3.7% butterfat, and Dairy II can supply at most 80 gallons of milk averaging 3.2% butterfat, how much milk from each supplier should Mostpure use to get at most 100 gallons of milk with maximum butterfat? What is the maximum butterfat?

16. Management A greeting card manufacturer has 400 copies of a particular card in warehouse I and 500 copies of the same card in warehouse II. A greeting card shop in San Jose orders 350 copies of the card, and another shop in Memphis orders 300 copies. The shipping costs per copy to these shops from the two warehouses are shown in the following table.

		Destination	
		San Jose	Memphis
Warehouse	I	.25	.15
	II	.20	.30

How many copies should be shipped to each city from each warehouse to minimize shipping costs? What is the minimum cost? (Hint: use x, $350 - x$, y, and $300 - y$ as the variables.)

17. Management A pension fund manager decides to invest at most $40 million in U.S. Treasury Bonds paying 12% annual interest and in mutual funds paying 8% annual interest. He plans to invest at least $20 million in bonds and at least $15 million in mutual funds. How much should be invested in each to maximize annual interest? What is the maximum annual interest?

18. Natural Science Mark, who is ill, takes vitamin pills. Each day he must have at least 16 units of vitamin A, 5 units of vitamin B_1, and 20 units of vitamin C. He can choose between pill #1, which contains 8 units of A, 1 of B_1, and 2 of C, and pill #2, which contains 2 units of A, 1 of B_1, and 7 of C. Pill #1 costs 15¢, and pill #2 costs 30¢. How many of each pill should he buy in order to minimize his cost? What is the minimum cost?

19. Natural Science A certain predator requires at least 10 units of protein and 8 units of fat per day. One prey of species I provides 5 units of protein and 2 units of fat; one prey of species II provides 3 units of protein and 4 units of fat. Capturing and digesting each species II prey requires 3 units of energy, and capturing and digesting each species I prey requires 2 units of energy. How many of each prey would meet the predator's daily food requirements with the least expenditure of energy? Are the answers reasonable? How could they be interpreted?

20. Natural Science Ms. Oliveras was given the following advice. She should supplement her daily diet with at least 6000 USP units of vitamin A, at least 195 milligrams of

Vitamin C, and at least 600 USP units of vitamin D. Ms. Oliveras finds that Mason's Pharmacy carries Brand X vitamin pills at 5¢ each and Brand Y vitamins at 4¢ each. Each Brand X pill contains 3000 USP units of A, 45 milligrams of C, and 75 USP units of D, while Brand Y pills contain 1000 USP units of A, 50 milligrams of C, and 200 USP units of D. What combination of vitamin pills should she buy to obtain the least possible cost? What is the least possible cost per day?

21. **Natural Science** Sing, who is dieting, requires two food supplements, I and II. He can get these supplements from two different products, A and B, as shown in the following table.

		Grams of Supplement per Serving	
		I	II
Product	A	3	2
	B	2	4

Sing's physician has recommended that he include at least 15 grams of each supplement in his daily diet. If product A costs 25¢ per serving and product B costs 40¢ per serving, how can he satisfy his dietetic requirements most economically? Find the minimum cost.

22. **Management** In a small town in South Carolina, zoning rules require that the window space (in square feet) in a house be at least one-sixth of the space used up by solid walls. The monthly cost to heat the house is 20¢ for each square foot of solid walls and 80¢ for each square foot of windows. Find the maximum total area (windows plus walls) if $160 is available to pay for heat.

The importance of linear programming is shown by the inclusion of linear programming problems on most qualification examinations for Certified Public Accountants. Exercises 23–25 are reprinted from one such examination.*

*Material from *Uniform CPA Examinations and Unofficial Answers,* copyright © 1973, 1974, 1975 by the American Institute of Certified Public Accountants, Inc., is reprinted with permission.

The Random Company manufactures two products, Zeta and Beta. Each product must pass through two processing operations. All materials are introduced at the start of Process No. 1. There are no work-in-process inventories. Random may produce either one product exclusively or various combinations of both products subject to the following constraints.

	Process No. 1	Process No. 2	Contribution Margin Per Unit
Hours required to produce 1 unit of:			
Zeta	1 hour	1 hour	$4.00
Beta	2 hours	3 hours	5.25
Total capacity in hours per day	1000 hours	1275 hours	

A shortage of technical labor has limited Beta production to 400 units per day. There are no constraints on the production of Zeta other than the hour constraints in the above schedule. Assume that all relationships between capacity and production are linear.

23. Given the objective to maximize total contribution margin, what is the production constraint for Process No. 1?
    (a) Zeta + Beta ≤ 1000
    (b) Zeta + 2 Beta ≤ 1000
    (c) Zeta + Beta ≥ 1000
    (d) Zeta + 2 Beta ≥ 1000

24. Given the objective to maximize total contribution margin, what is the labor constraint for production of Beta?
    (a) Beta ≤ 400   (b) Beta ≥ 400
    (c) Beta ≤ 425   (d) Beta ≥ 425

25. What is the objective function of the data presented?
    (a) Zeta + 2 Beta = $9.25
    (b) $4.00 Zeta + 3($5.25) Beta = total contribution margin
    (c) $4.00 Zeta + $5.25 Beta = total contribution margin
    (d) 2($4.00) Zeta + 3($5.25) Beta = total contribution margin

7.4 THE SIMPLEX METHOD: SLACK VARIABLES AND THE PIVOT

In the previous sections, linear programming problems were solved by the graphical method. The graphical method illustrates the basic ideas of linear programming, but it is practical only for problems with two variables. For problems with more than two variables, or with two variables and many constraints, the simplex method is used.* This method grew from a practical problem faced by George B. Danzig in 1947. Danzig was concerned with finding the least expensive way to allocate supplies for the United States Air Force. The simplex method is a very powerful application of mathematics and is still an active area of research.

The **simplex method** starts with the selection of one corner point (often the origin) from the feasible region. Then, in a systematic way, another corner point is found which improves the value of the objective function. Finally, an optimum solution is reached or it can be seen that none exists.

The simplex method requires a number of steps. We have divided the presentation of these steps into two parts. First, in this section, the problem is set up and the method begun; and then, in the next section the method is completed.

Because the simplex method is used for problems with many variables, it usually is not convenient to use letters such as x, y, z, or w as variable names. Instead, the symbols x_1 (read "x-sub-one"), x_2, x_3, and so on, are used. In the simplex method, all constraints must be expressed in the linear form

$$a_1x_1 + a_2x_2 + a_3x_3 + \cdots \leq b,$$

where x_1, x_2, x_3, . . . are variables, a_1, a_2, . . . , a_n are coefficients, and b is a constant.

We first discuss the simplex method for linear programming problems in *standard maximum form*.

Standard Maximum Form

A linear programming problem is in **standard maximum form** if

1. the objective function is to be maximized;
2. all variables are nonnegative ($x_i \geq 0$, $i = 1, 2, 3, . . .$);
3. all constraints involve $\leq$;
4. the constants on the right side in the constraints are all nonnegative ($b \geq 0$).

(In Sections 7.6 and 7.7 we discuss problems that do not meet all of these conditions.)

---

*A computer diskette with the simplex method is included in D. R. Coscia, *Computer Applications in Finite Mathematics and Calculus,* Scott, Foresman and Co., 1985.

1 Rewrite the following set of constraints as equations by adding slack variables.

$$x_1 + x_2 + x_3 \le 12$$
$$2x_1 + 4x_2 \qquad \le 15$$
$$x_2 + 3x_3 \le 10$$

Answer:

$$x_1 + x_2 + x_3 + x_4 = 12$$
$$2x_1 + 4x_2 \qquad + x_5 = 15$$
$$x_2 + 3x_3 + x_6 = 10$$

We begin the simplex method by converting the constraints, which are linear inequalities, into linear equations. This is done by adding a nonnegative variable, called a **slack variable,** to each constraint. For example, convert the inequality $x_1 + x_2 \le 10$ into an equation by adding the slack variable x_3, to get

$$x_1 + x_2 + x_3 = 10, \quad \text{where } x_3 \ge 0.$$

The inequality $x_1 + x_2 \le 10$ says that the sum $x_1 + x_2$ is less than or perhaps equal to 10. The variable x_3 "takes up any slack" and represents the amount by which $x_1 + x_2$ fails to equal 10. For example, if $x_1 + x_2$ equals 8, then x_3 is 2. If $x_1 + x_2 = 10$, the value of x_3 is 0.

Caution   A different slack variable must be used for each constraint.

▶**EXAMPLE 1**   Restate the following linear programming problem by introducing slack variables.

$$\begin{aligned}
\text{Maximize} \qquad & z = 3x_1 + 2x_2 + x_3 \\
\text{subject to:} \qquad & 2x_1 + x_2 + x_3 \le 150 \\
& 2x_1 + 2x_2 + 8x_3 \le 200 \\
& 2x_1 + 3x_2 + x_3 \le 320
\end{aligned}$$

with $x_1 \ge 0, \quad x_2 \ge 0, \quad x_3 \ge 0.$

Rewrite the three constraints as equations by adding slack variables x_4, x_5, and x_6, one for each constraint. By this process, the problem is restated as

$$\begin{aligned}
\text{Maximize} \qquad & z = 3x_1 + 2x_2 + x_3 \\
\text{subject to:} \qquad & 2x_1 + x_2 + x_3 + x_4 \qquad\qquad = 150 \\
& 2x_1 + 2x_2 + 8x_3 \qquad + x_5 \qquad = 200 \\
& 2x_1 + 3x_2 + x_3 \qquad\qquad + x_6 = 320
\end{aligned}$$

with $x_1 \ge 0, \quad x_2 \ge 0, \quad x_3 \ge 0, \quad x_4 \ge 0, \quad x_5 \ge 0, \quad x_6 \ge 0.$ ◀ **1**

Adding slack variables to the constraints converts a linear programming problem into a system of linear equations. These equations should have all variables on the left of the equals sign and all constants on the right. All the equations of Example 1 satisfy this condition except for the objective function, $z = 3x_1 + 2x_2 + x_3$, which may be written with all variables on the left as

$$-3x_1 - 2x_2 - x_3 + z = 0.$$

Now the equations of Example 1 can be written as the following augmented matrix.

$$
\begin{array}{ccccccc}
x_1 & x_2 & x_3 & x_4 & x_5 & x_6 & z \\
\end{array}
$$

$$
\left[
\begin{array}{ccccccc|c}
2 & 1 & 1 & 1 & 0 & 0 & 0 & 150 \\
2 & 2 & 8 & 0 & 1 & 0 & 0 & 200 \\
2 & 3 & 1 & 0 & 0 & 1 & 0 & 320 \\
\hline
-3 & -2 & -1 & 0 & 0 & 0 & 1 & 0
\end{array}
\right]
$$

indicators

This matrix is called the **initial simplex tableau.** The numbers in the bottom row, which are from the objective function, are called **indicators** (except for the 0 at the far right). The column headed z will never change in all future work, and will be omitted from now on.

▶**EXAMPLE 2**   Set up the initial simplex tableau for the following problem:

A farmer has 100 acres of available land which he wishes to plant with a mixture of potatoes, corn, and cabbage. It costs him $400 to produce an acre of potatoes, $160 to produce an acre of corn, and $280 to produce an acre of cabbage. He has a maximum of $20,000 to spend. He makes a profit of $120 per acre of potatoes, $40 per acre of corn, and $60 per acre of cabbage. How many acres of each crop should he plant to maximize his profit?

Begin by summarizing the given information as follows.

Crop	Number of Acres	Cost per Acre	Profit per Acre
Potatoes	x_1	$400	$120
Corn	x_2	160	40
Cabbage	x_3	280	60
Maximum available	100	$20,000	

If the number of acres allotted to each of the three crops is represented by x_1, x_2, and x_3, respectively, then the constraints of the example can be expressed as

$$
\begin{aligned}
x_1 + \quad x_2 + \quad x_3 &\le \quad 100 \qquad \text{Number of acres} \\
400x_1 + 160x_2 + 280x_3 &\le 20{,}000 \qquad \text{Production costs}
\end{aligned}
$$

where x_1, x_2, and x_3 are all nonnegative. The first of these constraints says that $x_1 + x_2 + x_3$ is less than or perhaps equal to 100. Use x_4 as the slack variable, giving the equation

$$
x_1 + x_2 + x_3 + x_4 = 100.
$$

Here x_4 represents the amount of the farmer's 100 acres that will not be used. (x_4 may be 0 or any value up to 100.)

2 Set up the initial simplex tableau for the following linear programming problem:

Maximize $z = 2x_1 + 3x_2$
subject to: $x_1 + 2x_2 \le 85$
$2x_1 + x_2 \le 92$
$x_1 + 4x_2 \le 104$
with $x_1 \ge 0$, $x_2 \ge 0$.

Answer:

$$\begin{bmatrix} 1 & 2 & 1 & 0 & 0 & \vline & 85 \\ 2 & 1 & 0 & 1 & 0 & \vline & 92 \\ 1 & 4 & 0 & 0 & 1 & \vline & 104 \\ \hline -2 & -3 & 0 & 0 & 0 & \vline & 0 \end{bmatrix}$$

In the same way, the constraint $400x_1 + 160x_2 + 280x_3 \le 20{,}000$ can be converted into an equation by adding a slack variable, x_5:

$$400x_1 + 160x_2 + 280x_3 + x_5 = 20{,}000.$$

The slack variable x_5 represents any unused portion of the farmer's \$20,000 capital. (Again, x_5 may have any value from 0 to 20,000.)

The objective function represents the profit. The farmer wants to maximize

$$z = 120x_1 + 40x_2 + 60x_3.$$

The linear programming problem can now be stated as follows:

Maximize $\quad z = 120x_1 + 40x_2 + 60x_3$

subject to: $\qquad x_1 + x_2 + x_3 + x_4 \qquad\quad = \quad 100$
$\qquad\qquad 400x_1 + 160x_2 + 280x_3 \qquad + x_5 = 20{,}000$

with $x_1 \ge 0$, $x_2 \ge 0$, $x_3 \ge 0$, $x_4 \ge 0$, $x_5 \ge 0$.

Rewrite the objective function as $-120x_1 - 40x_2 - 60x_3 + z = 0$, to get the following initial simplex tableau.

$$\begin{array}{ccccc} x_1 & x_2 & x_3 & x_4 & x_5 \\ \end{array}$$
$$\begin{bmatrix} 1 & 1 & 1 & 1 & 0 & \vline & 100 \\ 400 & 160 & 280 & 0 & 1 & \vline & 20{,}000 \\ \hline -120 & -40 & -60 & 0 & 0 & \vline & 0 \end{bmatrix} \blacktriangleleft \boxed{2}$$

The maximization problem in Example 2 consists of a system of two equations (describing the constraints) in five variables, together with the objective function. As with the graphical method, it is necessary to solve this system to find corner points of the region of feasible solutions. To produce a single, distinct solution, the number of variables must equal the number of equations in the system.

For example, the system of equations in Example 2 has an infinite number of solutions since there are more variables than equations. To see this, solve the system for x_4 and x_5.

$$x_4 = 100 - x_1 - x_2 - x_3$$
$$x_5 = 20{,}000 - 400\,x_1 - 160x_2 - 280x_3$$

Each choice of values for x_1, x_2, and x_3 gives corresponding values for x_4 and x_5 that produce a solution of the system. But only some of these solutions are feasible. In a feasible solution all variables must be nonnegative. To get a unique feasible solution, we set three of the five variables equal to 0. (Some choices of three variables may not result in a feasible solution.) In general, if there are m equations, then m variables can be nonzero. These are called **basic variables,** and the corresponding solutions are called **basic feasible solutions.** Each basic feasible solution

3 Read a solution from the matrix shown below.

$$
\begin{array}{ccccc}
x_1 & x_2 & x_3 & x_4 & x_5 \\
\end{array}
$$
$$
\left[\begin{array}{ccccc|c}
1 & 2 & 0 & 1 & 10 & 25 \\
0 & 6 & 1 & 3 & 1 & 50 \\
\end{array} \right]
$$

Answer:
$x_1 = 25$, $x_2 = 0$, $x_3 = 50$, $x_4 = 0$, $x_5 = 0$

corresponds to a corner point. In particular, if we choose the solution with $x_1 = 0$, $x_2 = 0$, and $x_3 = 0$, then $x_4 = 100$ and $x_5 = 20{,}000$. This solution, which corresponds to the corner point at the origin, is hardly optimal. It produces a profit of \$0 for the farmer, since the objective function becomes

$$-120(0) - 40(0) - 60(0) + 0x_4 + 0x_5 + z = 0.$$

In the next section we will use the simplex method to start with this solution and improve it to find the maximum possible profit.

Each step of the simplex method produces a solution that corresponds to a corner point of the region of feasible solutions. These solutions can be read directly from the matrix, as shown in the next example.

▶**EXAMPLE 3** Read a solution from the matrix below.

$$
\begin{array}{ccccc}
x_1 & x_2 & x_3 & x_4 & x_5 \\
\end{array}
$$
$$
\left[\begin{array}{ccccc|c}
2 & 1 & 8 & 5 & 0 & 27 \\
9 & 0 & 3 & 12 & 1 & 45 \\
\hline
-2 & 0 & -4 & 0 & 0 & 0 \\
\end{array} \right]
$$

The variables x_2 and x_5 are basic variables. They can quickly be identified because the columns for these variables (above the line) are columns from an identity matrix. The 1 in the x_2 column is in the first row. This means that if x_1, x_3, and x_4 are 0, $x_2 = 27$ (the last number in the first row). Also, $x_5 = 45$ from the last number in the second row. The solution is thus $x_1 = 0$, $x_2 = 27$, $x_3 = 0$, $x_4 = 0$, and $x_5 = 45$, with $z = 0$ from the information in the last row. ◀ **3**

PIVOTS Feasible solutions read directly from the initial simplex tableau are seldom optimal. It is necessary to proceed to other feasible solutions (corresponding to other corner points of the feasible region) until the optimum solution is found. To get these other solutions, use the row operations from Chapter 6 to change the tableau by using one of the nonzero entries of the tableau as a **pivot.** Pivoting, explained in the next example, produces a new tableau leading to another solution of the system of equations obtained from the original problem.

▶**EXAMPLE 4** Pivot about the indicated 2 of the initial simplex tableau

$$
\begin{array}{cccccc}
x_1 & x_2 & x_3 & x_4 & x_5 & x_6 \\
\end{array}
$$
$$
\left[\begin{array}{cccccc|c}
2 & 1 & 1 & 1 & 0 & 0 & 150 \\
2 & 2 & 8 & 0 & 1 & 0 & 200 \\
2 & 3 & 1 & 0 & 0 & 1 & 320 \\
\hline
-3 & -2 & -1 & 0 & 0 & 0 & 0 \\
\end{array} \right].
$$

4 Pivot about the indicated 2 in each simplex tableau.

(a)

$$\begin{array}{ccccc} x_1 & x_2 & x_3 & x_4 & x_5 \\ \left[\begin{array}{ccccc|c} 4 & 2 & 0 & 1 & 0 & 200 \\ 3 & 1 & 1 & 0 & 1 & 300 \end{array}\right] \end{array}$$

(b)

$$\begin{array}{cccc} x_1 & x_2 & x_3 & x_4 \\ \left[\begin{array}{cccc|c} 1 & 3 & 1 & 0 & 20 \\ 4 & 2 & 0 & 1 & 50 \end{array}\right] \end{array}$$

Answer:

(a)

$$\begin{array}{ccccc} x_1 & x_2 & x_3 & x_4 & x_5 \\ \left[\begin{array}{ccccc|c} 2 & 1 & 0 & \frac{1}{2} & 0 & 100 \\ 1 & 0 & 1 & -\frac{1}{2} & 1 & 200 \end{array}\right] \end{array}$$

(b)

$$\begin{array}{cccc} x_1 & x_2 & x_3 & x_4 \\ \left[\begin{array}{cccc|c} -5 & 0 & 1 & -\frac{3}{2} & -55 \\ 2 & 1 & 0 & \frac{1}{2} & 25 \end{array}\right] \end{array}$$

To pivot about the indicated 2, change x_1 into a basic variable by getting a 1 where the 2 is now and changing all other entries in the x_1 column to 0. Start by multiplying each entry of row one by 1/2.

$$\begin{array}{cccccc} x_1 & x_2 & x_3 & x_4 & x_5 & x_6 \\ \left[\begin{array}{cccccc|c} 1 & \frac{1}{2} & \frac{1}{2} & \frac{1}{2} & 0 & 0 & 75 \\ 2 & 2 & 8 & 0 & 1 & 0 & 200 \\ 2 & 3 & 1 & 0 & 0 & 1 & 320 \\ \hline -3 & -2 & -1 & 0 & 0 & 0 & 0 \end{array}\right] \end{array} \quad \frac{1}{2}R_1$$

Now get 0 in row two, column one by multiplying each entry in row one by -2 and adding the result to the corresponding entry in row two.

$$\begin{array}{cccccc} x_1 & x_2 & x_3 & x_4 & x_5 & x_6 \\ \left[\begin{array}{cccccc|c} 1 & \frac{1}{2} & \frac{1}{2} & \frac{1}{2} & 0 & 0 & 75 \\ 0 & 1 & 7 & -1 & 1 & 0 & 50 \\ 2 & 3 & 1 & 0 & 0 & 1 & 320 \\ \hline -3 & -2 & -1 & 0 & 0 & 0 & 0 \end{array}\right] \end{array} \quad -2R_1 + R_2$$

Change the 2 in row three, column one to a 0 by a similar process.

$$\begin{array}{cccccc} x_1 & x_2 & x_3 & x_4 & x_5 & x_6 \\ \left[\begin{array}{cccccc|c} 1 & \frac{1}{2} & \frac{1}{2} & \frac{1}{2} & 0 & 0 & 75 \\ 0 & 1 & 7 & -1 & 1 & 0 & 50 \\ 0 & 2 & 0 & -1 & 0 & 1 & 170 \\ \hline -3 & -2 & -1 & 0 & 0 & 0 & 0 \end{array}\right] \end{array} \quad -2R_1 + R_3$$

Finally, change the indicator -3 to 0 by multiplying each entry in row one by 3 and adding the result to the corresponding entry in row four.

$$\begin{array}{cccccc} x_1 & x_2 & x_3 & x_4 & x_5 & x_6 \\ \left[\begin{array}{cccccc|c} 1 & \frac{1}{2} & \frac{1}{2} & \frac{1}{2} & 0 & 0 & 75 \\ 0 & 1 & 7 & -1 & 1 & 0 & 50 \\ 0 & 2 & 0 & -1 & 0 & 1 & 170 \\ \hline 0 & -\frac{1}{2} & \frac{1}{2} & \frac{3}{2} & 0 & 0 & 225 \end{array}\right] \end{array} \quad 3R_1 + R_4$$

The final result is a new simplex tableau which gives the solution $x_1 = 75$, $x_2 = 0$, $x_3 = 0$, $x_4 = 0$, $x_5 = 50$, and $x_6 = 170$. Substituting these results into the objective function gives

$$0(75) - \frac{1}{2}(0) + \frac{1}{2}(0) + \frac{3}{2}(0) + 0(50) + 0(170) + z = 225,$$

or $z = 225$. (This shows that the value of z is always the number in the lower right-hand corner.) ◀ **4**

In the simplex method, this process is repeated until an optimum solution is found, if one exists. In the next section, we discuss how to decide where to pivot to improve the value of the objective function and how to tell when an optimum solution has been reached or does not exist.

7.4 EXERCISES

Convert each of the inequalities in Exercises 1–4 into equations by adding a slack variable. (See Example 1.)

1. $x_1 + 2x_2 \leq 6$ **2.** $3x_1 + 5x_2 \leq 100$ **3.** $2x_1 + 4x_2 + 3x_3 \leq 100$ **4.** $8x_1 + 6x_2 + 5x_3 \leq 250$

For Exercises 5–8, (a) determine the number of slack variables needed; (b) name them; (c) use slack variables to convert each constraint into a linear equation. (See Example 1.)

5. Maximize $z = 10x_1 + 12x_2$
   subject to: $4x_1 + 2x_2 \leq 20$
                 $5x_1 + x_2 \leq 50$
                 $2x_1 + 3x_2 \leq 25$
                 $x_1 \geq 0, \quad x_2 \geq 0.$

6. Maximize $z = 1.2x_1 + 3.5x_2$
   subject to: $2.4x_1 + 1.5x_2 \leq 10$
                 $1.7x_1 + 1.9x_2 \leq 15$
                 $x_1 \geq 0, \quad x_2 \geq 0.$

7. Maximize $z = 8x_1 + 3x_2 + x_3$
   subject to: $7x_1 + 6x_2 + 8x_3 \leq 118$
                 $4x_1 + 5x_2 + 10x_3 \leq 220$
                 $x_1 \geq 0, \quad x_2 \geq 0, \quad x_3 \geq 0.$

8. Maximize $z = 12x_1 + 15x_2 + 10x_3$
   subject to: $2x_1 + 2x_2 + x_3 \leq 8$
                 $x_1 + 4x_2 + 3x_3 \leq 12$
                 $x_1 \geq 0, \quad x_2 \geq 0, \quad x_3 \geq 0.$

Write the solution that can be read from Exercises 9–12. (See Example 3.)

9.

x_1	x_2	x_3	x_4	x_5	
2	2	0	3	1	15
3	4	1	6	0	20
−2	−1	0	1	0	10

10.

x_1	x_2	x_3	x_4	x_5	
0	2	1	1	3	5
1	5	0	1	2	8
0	−2	0	1	1	10

11.

x_1	x_2	x_3	x_4	x_5	x_6	
6	2	1	3	0	0	8
2	2	0	1	0	1	7
2	1	0	3	1	0	6
−3	−2	0	2	0	0	12

12.

x_1	x_2	x_3	x_4	x_5	x_6	
0	2	0	1	2	2	3
0	3	1	0	1	2	2
1	4	0	0	3	5	5
0	−4	0	0	4	3	20

Pivot as indicated in the simplex tableaus in Exercises 13–18. Read the solution from the final result. (See Example 4.)

13.

x_1	x_2	x_3	x_4	x_5	
1	2	4	1	0	56
2	[2]	1	0	1	40
−1	−3	−2	0	0	0

14.

x_1	x_2	x_3	x_4	x_5	
5	4	1	1	0	50
3	3	[2]	0	1	40
−1	−2	−4	0	0	0

15.

$$\begin{array}{cccccc} x_1 & x_2 & x_3 & x_4 & x_5 & x_6 \end{array}$$
$$\begin{bmatrix} 2 & 2 & 1 & 1 & 0 & 0 & | & 12 \\ 1 & 2 & 3 & 0 & 1 & 0 & | & 45 \\ 3 & 1 & 1 & 0 & 0 & 1 & | & 20 \\ \hline -2 & -1 & -3 & 0 & 0 & 0 & | & 0 \end{bmatrix}$$

16.

$$\begin{array}{cccccc} x_1 & x_2 & x_3 & x_4 & x_5 & x_6 \end{array}$$
$$\begin{bmatrix} 4 & 2 & 3 & 1 & 0 & 0 & | & 22 \\ 2 & 2 & 5 & 0 & 1 & 0 & | & 28 \\ 1 & 3 & 2 & 0 & 0 & 1 & | & 45 \\ \hline -3 & -2 & -4 & 0 & 0 & 0 & | & 0 \end{bmatrix}$$

17.

$$\begin{array}{cccccc} x_1 & x_2 & x_3 & x_4 & x_5 & x_6 \end{array}$$
$$\begin{bmatrix} 1 & 1 & 1 & 1 & 0 & 0 & | & 60 \\ 3 & 1 & 2 & 0 & 1 & 0 & | & 100 \\ 1 & 2 & 3 & 0 & 0 & 1 & | & 200 \\ \hline -1 & -1 & -2 & 0 & 0 & 0 & | & 0 \end{bmatrix}$$

18.

$$\begin{array}{ccccccc} x_1 & x_2 & x_3 & x_4 & x_5 & x_6 & x_7 \end{array}$$
$$\begin{bmatrix} 1 & 2 & 3 & 1 & 1 & 0 & 0 & | & 115 \\ 2 & 1 & 8 & 5 & 0 & 1 & 0 & | & 200 \\ 1 & 0 & 1 & 0 & 0 & 0 & 1 & | & 50 \\ \hline -2 & -1 & -1 & -1 & 0 & 0 & 0 & | & 0 \end{bmatrix}$$

Introduce slack variables as necessary and then write the initial simplex tableau for the linear programming problems in Exercises 19–24. (See Example 2.)

19. Maximize $z = 5x_1 + x_2$
subject to: $2x_1 + 3x_2 \le 6$
$4x_1 + x_2 \le 6$
$5x_1 + 2x_2 \le 15$
$x_1 \ge 0, \quad x_2 \ge 0.$

20. Maximize $z = x_1 + 3x_2$
subject to: $x_1 + x_2 \le 10$
$5x_1 + 2x_2 \le 20$
$x_1 + 2x_2 \le 36$
$x_1 \ge 0, \quad x_2 \ge 0.$

21. Maximize $z = x_1 + 5x_2 + 10x_3$
subject to: $2x_1 + 4x_2 + x_3 \le 18$
$x_1 + 6x_2 + 2x_3 \le 45$
$5x_1 + 7x_2 + 3x_3 \le 60$
$x_1 \ge 0, \quad x_2 \ge 0, \quad x_3 \ge 0.$

22. Maximize $z = 5x_1 + 3x_2 + 4x_3$
subject to: $4x_1 + 3x_2 + 2x_3 \le 60$
$3x_1 + 4x_2 + x_3 \le 24$
$x_1 \ge 0, \quad x_2 \ge 0, \quad x_3 \ge 0.$

23. Maximize $z = 6x_1 + 2x_2 + 3x_3$
subject to: $x_1 + 2x_2 + 3x_3 \le 10$
$2x_1 + x_2 + x_3 \le 8$
$3x_1 + 2x_3 \le 6$
$x_1 \ge 0, \quad x_2 \ge 0, \quad x_3 \ge 0.$

24. Maximize $z = 5x_1 - x_2 + 3x_3$
subject to: $3x_1 + 2x_2 + x_3 \le 36$
$x_1 + 4x_2 + x_3 \le 24$
$x_1 - x_2 - x_3 \le 32$
$x_1 \ge 0, \quad x_2 \ge 0, \quad x_3 \ge 0.$

Management *Set up Exercises 25–30 for solution by the simplex method; that is, express the linear constraints and objective function, add slack variables, and set up the initial simplex tableau. (See Example 2.)*

25. A candy company has 100 kilograms of chocolate-covered nuts and 125 kilograms of chocolate-covered raisins to be sold as two different mixtures. One mix will contain 1/2 nuts and 1/2 raisins and will sell for $6 per kilogram. The other mix will contain 1/3 nuts and 2/3 raisins, and will sell for $4.80 per kilogram. How many kilograms of each mix should the company prepare for maximum revenue? (This is Exercise 13, Section 7.3.)

26. Seall Manufacturing Company makes color television sets. It produces a bargain set that sells for $100 profit and a deluxe set that sells for $150 profit. On the assembly line, the bargain set requires 3 hour's work, while the deluxe set takes 5 hours. The cabinet shop spends 1 hour on the cabinet for the bargain set and 3 hours on the cabinet for the deluxe set. Both sets require 2 hours of time for testing and packing. On a particular production run the Seall Company has available 3900 work hours on the assembly line, 2100 work hours in the cabinet shop, and 2200 work hours in the testing and packing department. How many sets of each type should it produce to make maximum profit? What is the maximum profit? (See Exercise 10, Section 7.3.)

27. A small boat manufacturer builds three types of fiberglass boats: prams, runabouts, and trimarans. The pram sells at a profit of $75, the runabout at a profit of $90, and the trimaran at a profit of $100. The factory is divided into two sections. Section A does the molding and construction work, while section B does the painting, finishing, and equipping. The pram takes 1 hour in section A and 2 hours in section B. The runabout takes 2 hours in A and 5 hours in B. The trimaran takes 3 hours in A and 4 hours in B. Section A has a total of 6240 hours available and section B has 10,800 hours available for the year. The manufacturer has ordered a supply of fiberglass that will build at most 3000 boats, figuring the average amount used per boat. How many of each type of boat should be made to produce maximum profit? What is the maximum profit?

28. Caroline's Quality Candy Confectionery is famous for fudge, chocolate cremes, and pralines. Its candy-making equipment is set up to make 100-pound batches at a time. Currently there is a chocolate shortage and the company can get only 120 pounds of chocolate in the next shipment. On a week's run, the confectionery's cooking and processing equipment is available for a total of 42 machine hours. During the same period the employees have a total of 56 work hours available for packaging. A batch of fudge requires 20 pounds of chocolate while a batch of cremes uses 25 pounds of chocolate. The cooking and processing take 120 minutes for fudge, 150 minutes for chocolate cremes, and 200 minutes for pralines. The packaging times measured in minutes per 1-pound box are 1, 2, and 3, respec-

tively, for fudge, cremes, and pralines. Determine how many batches of each type of candy the confectionery should make, assuming that the profit per pound box is 50¢ on fudge, 40¢ on chocolate cremes, and 45¢ on pralines. What is the maximum profit?

29. A cat breeder has the following amounts of cat food: 90 units of tuna, 80 units of liver, and 50 units of chicken. To raise a Siamese cat, the breeder must use 2 units of tuna, 1 of liver, and 1 of chicken per day, while raising a Persian cat requires 1, 2, and 1 units, respectively, per day. If a Siamese cat sells for $12 while a Persian cat sells for $10, how many of each should be raised in order to obtain maximum gross income? What is the maximum gross income?

30. Banal, Inc., produces art for motel rooms. Its painters can turn out mountain scenes, seascapes, and pictures of clowns. Each painting is worked on by three different artists, T, D, and H. Artist T works only 25 hours per week, while D and H work 45 and 40 hours per week, respectively. Artist T spends 1 hour on a mountain scene, 2 hours on a seascape, and 1 hour on a clown. Corresponding times for D and H are 3, 2, and 2 hours, and 2, 1, and 4 hours respectively. Banal makes $20 on a mountain scene, $18 on a seascape, and $22 from a clown. The head painting packer can't stand clowns, so that no more than 4 clown paintings may be done in a week. Find the number of each type of painting that should be made weekly in order to maximize profit. Find the maximum possible profit.

7.5 SOLVING MAXIMIZATION PROBLEMS

In the previous section we showed how to prepare a linear programming problem for solution. First, we converted the constraints to linear equations with slack variables; then we used the coefficients from the linear equations and the objective function to write an augmented matrix. Finally, we used the pivot to go from one corner point to another corner point in the region of feasible solutions.

Now we are ready to put all this together and produce an optimum value for the objective function. To see how this is done, let us complete Example 2 from Section 7.4, the example about the farmer.

In the previous section, we set up the following simplex tableau.

$$\begin{array}{ccccc} x_1 & x_2 & x_3 & x_4 & x_5 \\ \begin{bmatrix} 1 & 1 & 1 & 1 & 0 & | & 100 \\ 400 & 160 & 280 & 0 & 1 & | & 20{,}000 \\ \hline -120 & -40 & -60 & 0 & 0 & | & 0 \end{bmatrix} \end{array}$$

This tableau leads to the solution $x_1 = 0$, $x_2 = 0$, $x_3 = 0$, $x_4 = 100$, and $x_5 = 20{,}000$, with x_4 and x_5 as the basic variables. These values produce a value of 0 for z. Since a value of 0 for the farmer's profit is not an optimum, we will try to improve this value.

The coefficients of x_1, x_2, and x_3 in the objective function are nonzero, so the profit could be improved by making any one of these variables take on a nonzero value in a solution. To decide which variable to use, look at the indicators in the initial simplex tableau above. The coefficient of x_1, -120, is the "most negative" of the indicators. This means that x_1 has the largest coefficient in the objective function, so that profit is increased the most by increasing x_1.

As we saw earlier, because there are two equations in the system, only two of the five variables can be basic variables (and be nonzero). If x_1 is nonzero in the solution, then x_1 will be a basic variable. This means that either x_4 or x_5 no longer will be a basic variable. To decide which variable will no longer be basic, start with the equations of the system.

$$\begin{aligned} x_1 + \quad x_2 + \quad x_3 + x_4 \quad\quad &= \quad 100 \\ 400x_1 + 160x_2 + 280x_3 \quad\quad + x_5 &= 20{,}000, \end{aligned}$$

and solve for x_4 and x_5, respectively.

$$\begin{aligned} x_4 &= 100 - x_1 - x_2 - x_3 \\ x_5 &= 20{,}000 - 400x_1 - 160x_2 - 280x_3 \end{aligned}$$

Only x_1 is being changed to a nonzero value; both x_2 and x_3 keep the value 0. Replacing x_2 and x_3 with 0 gives

$$\begin{aligned} x_4 &= 100 - x_1 \\ x_5 &= 20{,}000 - 400x_1. \end{aligned}$$

Since both x_4 and x_5 must remain nonnegative, there is a limit to how much the value of x_1 can be increased. The equation $x_4 = 100 - x_1$ (or $x_4 = 100 - 1x_1$) shows that x_1 cannot exceed 100/1, or 100. The second equation, $x_5 = 20{,}000 - 400x_1$, shows that x_1 cannot exceed 20,000/400, or 50. To satisfy both these condi-

1 Find the value of $z = 120x_1 + 40x_2 + 60x_3 + 0x_4 + 0x_5$ if $x_1 = 50$, $x_2 = 0$, $x_3 = 0$, $x_4 = 50$, $x_5 = 0$.

Answer:
6000

tions, x_1 cannot exceed 50, the smaller of 50 and 100. If we let x_1 take the value 50, then $x_1 = 50$, $x_2 = 0$, $x_3 = 0$, and $x_5 = 0$. Since $x_4 = 100 - x_1$, then

$$x_4 = 100 - 50 = 50. \quad \boxed{1}$$

This solution gives a profit of

$$z = 120x_1 + 40x_2 + 60x_3 + 0x_4 + 0x_5$$
$$= 120(50) + 40(0) + 60(0) + 0(50) + 0(0) = 6000,$$

or $6000.

The same result could have been found from the initial simplex tableau given above. To use the tableau, select the most negative indicator. (If no indicator is negative, then the value of the objective function cannot be improved.)

$$
\begin{array}{ccccc}
x_1 & x_2 & x_3 & x_4 & x_5 \\
\end{array}
$$

$$
\left[
\begin{array}{ccccc|c}
1 & 1 & 1 & 1 & 0 & 100 \\
400 & 160 & 280 & 0 & 1 & 20{,}000 \\
\hline
-120 & -40 & -60 & 0 & 0 & 0 \\
\end{array}
\right]
$$

most negative indicator

The most negative indicator identifies the variable whose value is to be made nonzero. To find the variable that is now basic and will become nonbasic, calculate the quotients that were found above. Do this by dividing each number from the far right column of the tableau by the corresponding number from the column with the most negative indicator.

Quotients

$100/1 = 100$

smaller $\rightarrow$ $20{,}000/400 = 50$

$$
\begin{array}{ccccc}
x_1 & x_2 & x_3 & x_4 & x_5 \\
\end{array}
$$

$$
\left[
\begin{array}{ccccc|c}
1 & 1 & 1 & 1 & 0 & 100 \\
400 & 160 & 280 & 0 & 1 & 20{,}000 \\
\hline
-120 & -40 & -60 & 0 & 0 & 0 \\
\end{array}
\right]
$$

The smaller quotient is 50, from the second row. This identifies 400 as the pivot. Using 400 as the pivot, perform the appropriate row operations to get the second simplex tableau. First, get 1 in the pivot position by multiplying each element of the second row by 1/400. To make the other entries in the pivot column zeros, you must add multiples of the *pivot row* to the other rows. Multiply each of the entries in the second row by -1 and add the results to the corresponding entries in the first row, to get a 0 above the pivot. Get a 0 below the pivot, as the first indicator, in a similar way. The new tableau is

2 For the simplex tableau below, find the following.

(a) The pivot

(b) The second tableau

(c) The solution shown in the second tableau

(d) Can the solution be improved?

$$\begin{array}{ccccc} x_1 & x_2 & x_3 & x_4 & x_5 \\ \left[\begin{array}{ccccc|c} 1 & 2 & 6 & 1 & 0 & 16 \\ 1 & 3 & 0 & 0 & 1 & 25 \\ -1 & -4 & -3 & 0 & 0 & 0 \end{array}\right] \end{array}$$

Answer:

(a) 2

(b)

$$\begin{array}{ccccc} x_1 & x_2 & x_3 & x_4 & x_5 \\ \left[\begin{array}{ccccc|c} \frac{1}{2} & 1 & 3 & \frac{1}{2} & 0 & 8 \\ -\frac{1}{2} & 0 & -9 & -\frac{3}{2} & 1 & 1 \\ 1 & 0 & 9 & 2 & 0 & 32 \end{array}\right] \end{array}$$

(c) $(0, 8, 0, 0, 1)$

(d) No

$$\begin{array}{ccccc} x_1 & x_2 & x_3 & x_4 & x_5 \\ \left[\begin{array}{ccccc|c} 0 & .6 & .3 & 1 & -.0025 & 50 \\ 1 & .4 & .7 & 0 & .0025 & 50 \\ 0 & 8 & 24 & 0 & .3 & 6000 \end{array}\right] \end{array} \begin{array}{l} -1R_2 + R_1 \\ \frac{1}{400}R_2 \\ 120R_2 + R_3 \end{array}$$

and the solution read from this tableau is

$$x_1 = 50, \quad x_2 = 0, \quad x_3 = 0, \quad x_4 = 50, \quad x_5 = 0,$$

the same result found above. The entry 6000 (in color) in the lower right corner of the tableau gives the value of the objective function for this solution:

$$z = \$6000.$$

None of the indicators in the final simplex tableau are negative, which means that the value of z cannot be improved beyond $6000. To see why, recall that the last row gives the coefficients of the objective function. Including the coefficient of 1 for the z column which was dropped gives

$$0x_1 + 8x_2 + 24x_3 + 0x_4 + .3x_5 + z = 6000,$$

or

$$z = 6000 - 0x_1 - 8x_2 - 24x_3 - 0x_4 - .3x_5.$$

Since x_2, x_3, and x_5 are 0, $z = 6000$, but if any of these three variables were to increase, z would decrease.

This result suggests that the optimal solution has been found as soon as no indicators are negative. As long as an indicator is negative, the value of the objective function can be improved. If any indicators are negative, we just find a new pivot and use row operations, repeating the process until no negative indicators remain.

We can finally state the solution to the problem about the farmer: the optimum value of z is 6000, where $x_1 = 50$, $x_2 = 0$, $x_3 = 0$, $x_4 = 50$, and $x_5 = 0$. That is, the farmer will make a maximum profit of $6000 by planting 50 acres of potatoes. Another 50 acres should be left unplanted. It may seem strange that leaving assets unused can produce a maximum profit, but such results actually occur often. **2**

In summary, the following steps are involved in solving a standard maximum linear programming problem by the simplex method.

Simplex Method

1. Determine the objective function.
2. Write all necessary constraints.
3. Convert each constraint into an equation by adding slack variables.
4. Set up the initial simplex tableau.
5. Locate the most negative indicator. If there are two such indicators, choose one.
6. Form the necessary quotients to find the pivot. Disregard any negative quotients or quotients with a 0 denominator. The smallest nonnegative quotient gives the location of the pivot. If all quotients must be disregarded, no maximum solution exists.* If two quotients are equally the smallest, let either determine the pivot.†
7. Use row operations to change the pivot to 1 and all other numbers in that column to 0.
8. If the indicators are all positive or 0, this is the final tableau. If not, go back to Step 5 above and repeat the process until a tableau with no negative indicators is obtained.
9. Read the solution from this final tableau. The maximum value of the objective function is the number in the lower right corner of the final tableau.

▶ **EXAMPLE 1**   To compare the simplex method with the graphical method, the simplex method is used to solve the problem of Example 2, Section 7.3. The graph is shown again in Figure 7.20. The objective function to be maximized was

$$z = 8x_1 + 12x_2. \qquad \text{Storage space}$$

(Since we are using the simplex method, x_1 and x_2 are used as variables instead of x and y.) The constraints were as follows.

$$40x_1 + 80x_2 \leq 560 \qquad \text{Cost}$$
$$6x_1 + 8x_2 \leq 72 \qquad \text{Floor space}$$
$$x_1 \geq 0, \quad x_2 \geq 0$$

Add a slack variable to each constraint.

$$40x_1 + 80x_2 + x_3 = 560$$
$$6x_1 + 8x_2 + x_4 = 72$$

---

*Some special circumstances are noted at the end of Section 7.6.

†It may be that the first choice of a pivot does not produce a solution. In that case, try the other choice.

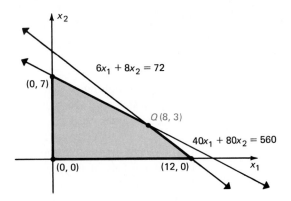

FIGURE 7.20

Write the initial simplex tableau.

$$\begin{array}{cccc} x_1 & x_2 & x_3 & x_4 \end{array}$$

$$\begin{bmatrix} 40 & 80 & 1 & 0 & 560 \\ 6 & 8 & 0 & 1 & 72 \\ \hline -8 & -12 & 0 & 0 & 0 \end{bmatrix}$$

This tableau leads to the solution $x_1 = 0$, $x_2 = 0$, $x_3 = 560$, and $x_4 = 72$, with $z = 0$, which corresponds to the origin in Figure 7.20. The most negative indicator is -12. The necessary quotients are

$$\frac{560}{80} = 7 \quad \text{and} \quad \frac{72}{8} = 9.$$

The smaller quotient is 7, giving 80 as the pivot. Use row operations to get the new tableau.

$$\begin{bmatrix} \frac{1}{2} & 1 & \frac{1}{80} & 0 & 7 \\ 2 & 0 & -\frac{1}{10} & 1 & 16 \\ \hline -2 & 0 & \frac{3}{20} & 0 & 84 \end{bmatrix} \quad \begin{array}{l} \frac{1}{80}R_1 \\ -8R_1 + R_2 \\ 12R_1 + R_3 \end{array}$$

The solution from this tableau is $x_1 = 0$, $x_2 = 7$, $x_3 = 0$, and $x_4 = 16$, with $z = 84$, which corresponds to the corner point $(0, 7)$ in Figure 7.20. Because of the indicator -2, the value of z can be improved. The new quotients are

$$\frac{7}{\frac{1}{2}} = 14 \quad \text{and} \quad \frac{16}{2} = 8.$$

Since 8 is smaller, use the 2 in row two, column one as pivot to get the final tableau.

3 A linear programming problem has the initial tableau given below. Use the simplex method to solve the problem.

$$\begin{array}{cccc}x_1 & x_2 & x_3 & x_4\end{array}$$
$$\left[\begin{array}{cccc|c} 1 & 1 & 1 & 0 & 40 \\ 2 & 1 & 0 & 1 & 24 \\ \hline -300 & -200 & 0 & 0 & 0 \end{array}\right]$$

Answer:
$x_1 = 0$, $x_2 = 24$, $x_3 = 16$, $x_4 = 0$, $z = 4800$

$$\left[\begin{array}{cccc|c} 0 & 1 & \frac{3}{80} & -\frac{1}{4} & 3 \\ 1 & 0 & -\frac{1}{20} & \frac{1}{2} & 8 \\ 0 & 0 & \frac{1}{20} & 1 & 100 \end{array}\right] \quad \begin{array}{l} -\frac{1}{2}R_2 + R_1 \\ \frac{1}{2}R_2 \\ 2R_2 + R_3 \end{array}$$

Here the solution is $x_1 = 8$, $x_2 = 3$, $x_3 = 0$, and $x_4 = 0$, with $z = 100$. This solution, which corresponds to the corner point $(8, 3)$ in Figure 7.20, is the same as the solution found earlier. ◀

Each simplex tableau in Example 1 gave a solution corresponding to one of the corner points of the feasible region. As shown in Figure 7.21, the first solution corresponded to the origin, with $z = 0$. By choosing the appropriate pivot, we moved systematically to a new corner point, $(0, 7)$, which improved the value of z to 84. The next tableau took us to $(8, 3)$, producing the optimum value of $z = 100$. There was no reason to test the last corner point, $(12, 0)$, since the optimum value of z was found before that point was reached. **3**

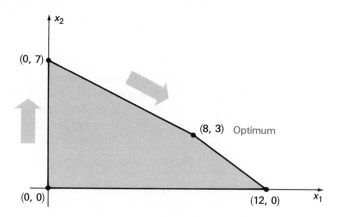

FIGURE 7.21

The next example shows how to handle zero or a negative number in the column containing the pivot.

▶**EXAMPLE 2** Find the pivot for the following initial simplex tableau.

$$\begin{array}{ccccc}x_1 & x_2 & x_3 & x_4 & x_5\end{array}$$
$$\left[\begin{array}{ccccc|c} 1 & -2 & 1 & 0 & 0 & 100 \\ 3 & 4 & 0 & 1 & 0 & 200 \\ 5 & 0 & 0 & 0 & 1 & 150 \\ \hline -10 & -25 & 0 & 0 & 0 & 0 \end{array}\right]$$

4 Find the pivot for the following tableau.

$$\begin{array}{ccccc} x_1 & x_2 & x_3 & x_4 & x_5 \\ \left[\begin{array}{ccccc|c} 0 & 1 & 1 & 0 & 0 & 50 \\ -2 & 3 & 0 & 1 & 0 & 78 \\ 2 & 4 & 0 & 0 & 1 & 65 \\ \hline -5 & -3 & 0 & 0 & 0 & 0 \end{array}\right] \end{array}$$

Answer:
2 (in first column)

The most negative indicator is -25. To find the pivot, find the quotients formed by the entries in the right-most column and in the x_2 column: $100/(-2)$, $200/4$, and $150/0$. We can disregard the quotient $150/0$, which comes from the equation

$$5x_1 + 0x_2 + 0x_3 + 0x_4 + x_5 = 150.$$

If $x_1 = 0$, the equation reduces to

$$0x_2 + x_5 = 150 \quad \text{or} \quad x_5 = 150 - 0x_2.$$

Since the pivot is in the x_2 column, x_5 cannot become 0. In general, disregard quotients with a 0 denominator.

The quotients predict the value of a variable in the solution, and thus cannot be negative. In this example, the quotient $100/(-2) = -50$ comes from the equation

$$x_1 - 2x_2 + x_3 + 0x_4 + 0x_5 = 100.$$

For $x_1 = 0$, this becomes

$$x_3 = 100 + 2x_2.$$

Thus, increases in x_2 produce increases in x_3. So the nonnegative constraint on x_3 does not limit the increase in x_2. Since x_2 must be nonnegative anyway, this equation tells us nothing new. For this reason, disregard negative quotients also.

The only usable quotient is $200/4 = 50$, making 4 the pivot. If all the quotients are either negative or have 0 denominators, no unique optimum solution will be found. Such a situation indicates an unbounded feasible region. The quotients, then, tell whether or not an optimum solution exists. ◄ **4**

Although linear programming problems with more than a few variables would seem to be very complex, in practical applications many entries in the simplex tableau are zeros.

7.5 EXERCISES

In Exercises 1–6, the initial tableau of a linear programming problem is given. Use the simplex method to solve each problem. (See Examples 1 and 2.)

1.
$$\begin{array}{ccccc} x_1 & x_2 & x_3 & x_4 & x_5 \\ \left[\begin{array}{ccccc|c} 1 & 2 & 4 & 1 & 0 & 8 \\ 2 & 2 & 1 & 0 & 1 & 10 \\ \hline -2 & -5 & -1 & 0 & 0 & 0 \end{array}\right] \end{array}$$

2.
$$\begin{array}{ccccc} x_1 & x_2 & x_3 & x_4 & x_5 \\ \left[\begin{array}{ccccc|c} 2 & 2 & 1 & 1 & 0 & 10 \\ 1 & 2 & 3 & 0 & 1 & 15 \\ \hline -3 & -2 & -1 & 0 & 0 & 0 \end{array}\right] \end{array}$$

3.

x_1	x_2	x_3	x_4	x_5	
1	3	1	0	0	12
2	1	0	1	0	10
1	1	0	0	1	4
−2	−1	0	0	0	0

4.

x_1	x_2	x_3	x_4	x_5	x_6	
2	2	1	1	0	0	50
1	1	3	0	1	0	40
4	2	5	0	0	1	80
−2	−3	−5	0	0	0	0

5.

x_1	x_2	x_3	x_4	x_5	x_6	
2	2	8	1	0	0	40
4	−5	6	0	1	0	60
2	−2	6	0	0	1	24
−14	−10	−12	0	0	0	0

6.

x_1	x_2	x_3	x_4	x_5	
3	2	4	1	0	18
2	1	5	0	1	8
−1	−4	−2	0	0	0

Use the simplex method to solve Exercises 7–16.

7. Maximize $z = 2x_1 + 5x_2$
subject to: $x_1 + 2x_2 \le 8$
$4x_1 + x_2 \le 16$
$x_1 \ge 0, \quad x_2 \ge 0.$

8. Maximize $z = 10x_1 + 12x_2$
subject to: $4x_1 + 2x_2 \le 20$
$5x_1 + x_2 \le 50$
$2x_1 + 2x_2 \le 24$
$x_1 \ge 0, \quad x_2 \ge 0.$

9. Maximize $z = 8x_1 + 3x_2 + x_3$
subject to: $x_1 + 6x_2 + 8x_3 \le 118$
$x_1 + 5x_2 + 10x_3 \le 220$
$x_1 \ge 0, \quad x_2 \ge 0, \quad x_3 \ge 0.$

10. Maximize $z = 12x_1 + 15x_2 + 5x_3$
subject to: $2x_1 + 2x_2 + x_3 \le 8$
$x_1 + 4x_2 + 3x_3 \le 12$
$x_1 \ge 0, \quad x_2 \ge 0, \quad x_3 \ge 0.$

11. Maximize $z = -4x_1 + 2x_2 + x_3$
subject to: $-2x_1 + 3x_2 + 3x_3 \le 100$
$3x_1 + 5x_2 + 10x_3 \le 150$
$x_1 \ge 0, \quad x_2 \ge 0, \quad x_3 \ge 0.$

12. Maximize $z = 4x_1 + x_2 + 3x_3$
subject to: $x_1 + 3x_3 \le 6$
$6x_1 + 3x_2 + 12x_3 \le 40$
$x_1 \ge 0, \quad x_2 \ge 0, \quad x_3 \ge 0.$

13. Maximize $z = 300x_1 + 200x_2 + 100x_3$
subject to: $x_1 + x_2 + x_3 \le 100$
$2x_1 + 3x_2 + 4x_3 \le 320$
$2x_1 + x_2 + x_3 \le 160$
$x_1 \ge 0, \quad x_2 \ge 0, \quad x_3 \ge 0.$

14. Maximize $z = 5x_1 + 4x_2 + 6x_3$
subject to: $x_1 + x_2 + 2x_3 \le 180$
$3x_1 + 2x_2 + x_3 \le 300$
$x_1 + 2x_2 + 2x_3 \le 240$
$x_1 \ge 0, \quad x_2 \ge 0, \quad x_3 \ge 0.$

15. Maximize $z = x_1 + 2x_2 + x_3 + 5x_4$
subject to: $x_1 + 2x_2 + x_3 + x_4 \le 50$
$3x_1 + x_2 + 2x_3 + x_4 \le 100$
$x_1 \ge 0, \quad x_2 \ge 0, \quad x_3 \ge 0, \quad x_4 \ge 0.$

16. Maximize $z = x_1 + x_2 + 4x_3 + 5x_4$
subject to: $x_1 + 2x_2 + 3x_3 + x_4 \le 115$
$2x_1 + x_2 + 8x_3 + 5x_4 \le 200$
$x_1 + x_3 \le 50$
$x_1 \ge 0, \quad x_2 \ge 0, \quad x_3 \ge 0, \quad x_4 \ge 0.$

Set up and solve Exercises 17–24 by the simplex method.

17. Social Science Jayanta is working to raise money for the homeless by sending information letters and making follow-up calls to local labor organizations and church groups. She discovered that each church group requires 2 hours of letter writing and 1 hour of follow-up, while for each labor union she needs 2 hours of letter writing and 3 hours of follow-up. Jayanta can raise $100 from each church group and $200 from each union local, and she has a maximum of 16 hours of letter-writing time and a maximum of 12 hours of follow-up time available per month. Determine the most profitable mixture of groups she should contact and the most money she can raise in a month.

18. **Natural Science** A biologist has 500 kilograms of nutrient A, 600 kilograms of nutrient B, and 300 kilograms of nutrient C. These nutrients will be used to make four types of food whose contents (in percent of nutrient per kilogram of food) and whose "growth values" are as shown below.

Food	Nutrient (%)			Growth value
	A	B	C	
P	0	0	100	90
Q	0	75	25	70
R	37.5	50	12.5	60
S	62.5	37.5	0	50

How many kilograms of each food should be produced in order to maximize total growth value? Find the maximum growth value.

19. **Management** The Soul Sounds Recording Company produces three types of compact discs: jazz, blues, and reggae. Each jazz disc requires 6 hours of recording, 12 hours of mixing, and 2 hours of editing. Each blues disc requires 6 hours of recording, 6 hours of mixing, and 4 hours of editing. Each reggae disc requires 3 hours of recording, 6 hours of mixing, and 1 hour of editing. The recording studio is available 69 hours per week; staff to operate the mixing board are available 78 hours per week; the editor has at most 40 hours per week for work. Soul Sound receives $6 for each jazz and reggae disc and $8 for each blues disc. How many of each type of disc should the company process each week to maximize their income? What is the maximum income?

20. **Management** A baker has 150 units of flour, 90 units of sugar, and 150 units of raisins. A loaf of raisin bread requires 1 unit of flour, 1 of sugar, and 2 of raisins, while a raisin cake needs 5, 2, and 1 units, respectively. If raisin bread sells for $1.05 a loaf and raisin cake for $2.40 each, how many of each should be baked so that gross income is maximized? What is the maximum gross income?

21. **Management** Super Souvenir Company makes paperweights, plaques, and ornaments. Each paperweight requires 8 units of plastic, 3 units of metal, and 2 units of paint. Each plaque requires 4 units of plastic, and 1 unit each of metal and paint. Each ornament requires 2 units each of plastic and metal, and 1 unit of paint. They make a profit of $3 on each paperweight and each ornament, and $4 on each plaque. If 36 units of plastic, 24 units of metal,

and 30 units of paint are available today, how many of each kind of souvenir should be made in order to maximize profit?

22. **Management** Caroline's Quality Candy Confectionery is famous for fudge, chocolate cremes, and pralines. Its candy-making equipment is set up to make 100-pound batches at a time. Currently there is a chocolate shortage and the company can get only 120 pounds of chocolate in the next shipment. On a week's run, the confectionery's cooking and processing equipment is available for a total of 42 machine hours. During the same period the employees have a total of 56 work hours available for packaging. A batch of fudge requires 20 pounds of chocolate while a batch of cremes uses 25 pounds of chocolate. The cooking and processing take 120 minutes for fudge, 150 minutes for chocolate cremes, and 200 minutes for pralines. The packaging times measured in minutes per 1-pound box are 1, 2, 3, respectively, for fudge, cremes, and pralines. Determine how many batches of each type of candy the confectionery should make, assuming that the profit per pound box is 50¢ on fudge, 40¢ on chocolate cremes, and 45¢ on pralines. Also, find the maximum profit for the week. (See Exercise 28, Section. 7.4.)

23. **Management** A manufacturer of bicycles builds one-, three-, and ten-speed models. The bicycles need both aluminum and steel. The company has available 91,800 units of steel and 42,000 units of aluminum. The one-, three-, and ten-speed models need, respectively, 20, 30, and 40 units of steel and 12, 21, and 16 units of aluminum. How many of each type of bicycle should be made in order to maximize profit if the company makes $8 per one-speed bike, $12 per three-speed, and $24 per ten-speed? What is the maximum possible profit?

24. **Social Science** A political party is planning a half-hour television show. The show will have 3 minutes of direct requests for money from viewers. Three of the party's politicians will be on the show—a senator, a congresswoman, and a governor. The senator, a party "elder statesman," demands that he be on at least twice as long as the governor. The total time taken by the senator and the governor must be at least twice the time taken by the congresswoman. Based on a preshow survey, it is believed that 40, 60, and 50 (in thousands) viewers will watch the program for each minute the senator, congresswoman, and governor, respectively, are on the air. Find the time that should be allotted to each politician in order to get the maximum number of viewers. Find the maximum number of viewers.

Management *The next two problems come from past CPA examinations.* Select the appropriate answer for each question.*

25. The Ball Company manufactures three types of lamps, labeled A, B, and C. Each lamp is processed in two departments, I and II. Total available man-hours per day for departments I and II are 400 and 600, respectively. No additional labor is available. Time requirements and profit per unit for each lamp type are as follows.

	A	B	C
Man-hours in I	2	3	1
Man-hours in II	4	2	3
Profit per unit	$5	$4	$3

The company has assigned you as the accounting member of its profit planning committee to determine the numbers of types of A, B, and C lamps that it should produce in order to maximize its total profit from the sale of lamps. The following questions relate to a linear programming model that your group has developed.

(a) The coefficients of the objective function would be
  (1) 4, 2, 3;   **(2)** 2, 3, 1;
  (3) 5, 4, 3;   **(4)** 400, 600.
(b) The constraints in the model would be
  (1) 2, 3, 1;   **(2)** 5, 4, 3;
  (3) 4, 2, 3;   **(4)** 400, 600.
(c) The constraint imposed by the available man-hours in department I could be expressed as
  (1) $4X_1 + 2X_2 + 3X_3 \leq 400$;
  (2) $4X_1 + 2X_2 + 3X_3 \geq 400$;
  (3) $2X_1 + 3X_2 + 1X_3 \leq 400$;
  (4) $2X_1 + 3X_2 + 1X_3 \geq 400$.

26. The Golden Hawk Manufacturing Company wants to maximize the profits on products A, B, and C. The contribution margin for each product follows.

Product	Contribution Margin
A	$2
B	5
C	4

The production requirements and departmental capacities, by departments, are as follows.

	Production Requirements by Product (hours)			Departmental Capacity (total hours)
Department	A	B	C	
Assembling	2	3	2	30,000
Painting	1	2	2	38,000
Finishing	2	3	1	28,000

(a) What is the profit-maximization formula for the Golden Hawk Company?
  (1) $2A + $5B + $4C = X$ (where X = profit)
  (2) $5A + 8B + 5C \leq 96,000$
  (3) $2A + $5B + $4C \leq X$
  (4) $2A + $5B + $4C = 96,000$
(b) What is the constraint for the Painting Department of the Golden Hawk Company?
  (1) $1A + 2B + 2C \geq 38,000$
  (2) $2A + $5B + $4C \geq 38,000$
  (3) $1A + 2B + 2C \leq 38,000$
  (4) $2A + 3B + 2C \leq 30,000$

Determine the constraints and the objective function for Exercises 27 and 28, then use a calculator or a computer to solve each problem.

27. Management A manufacturer makes two products, toy trucks and toy fire engines. Both are processed in four different departments, each of which has a limited capacity. The sheet metal department can handle at least $1\frac{1}{2}$ times as many trucks as fire engines. The truck assembly department can handle at most 6700 trucks per week, while the fire engine assembly department assembles at most 5500 fire engines weekly. The painting department, which finishes both toys, has a maximum capacity of 12,000 per week. If the profit is $8.50 for a toy truck and $12.10 for a toy fire engine, how many of each item should the company produce to maximize profit?

28. The average weights of the three species stocked in the lake referred to in Section 6.2, Exercise 29, are 1.62, 2.14, and 3.01 kilograms for species A, B, and C, respectively. If the largest amounts of food that can be supplied each day are given as in Exercise 29, how should the lake be stocked to maximize the weight of the fish supported by the lake?

7.6 NONSTANDARD PROBLEMS; MINIMIZATION

So far we have used the simplex method to solve linear programming problems in standard maximum form only. In this section, this work is extended to include linear programming problems with mixed $\leq$ and $\geq$ constraints. The solution of these problems then gives a method of solving minimization problems. (An alternative approach for solving minimization problems is given in the next section.)

PROBLEMS WITH $\leq$ AND $\geq$ CONSTRAINTS   Suppose a new constraint is added to the farmer problem in Example 2 of Section 7.4: to satisfy orders from regular buyers, the farmer must plant a total of at least 60 acres of the three crops. This constraint introduces the new inequality

$$x_1 + x_2 + x_3 \geq 60.$$

As before, this inequality must be rewritten as an equation in which the variables all represent nonnegative numbers. The inequality $x_1 + x_2 + x_3 \geq 60$ means that

$$x_1 + x_2 + x_3 - x_6 = 60$$

for some nonnegative variable x_6. (Remember that x_4 and x_5 are the slack variables in the problem.)

The new variable, x_6, is called a **surplus variable.** The value of this variable represents the excess number of acres (over 60) that may be planted. Since the total number of acres planted is to be no more than 100 but at least 60, the value of x_6 can vary from 0 to 40.

We must now solve the system of equations

$$
\begin{aligned}
x_1 + \quad x_2 + \quad x_3 + x_4 \qquad\qquad\qquad &= \quad 100 \\
400x_1 + 160x_2 + 280x_3 \qquad + x_5 \qquad\quad &= 20{,}000 \\
x_1 + \quad x_2 + \quad x_3 \qquad\qquad - x_6 \quad &= \quad 60 \\
-120x_1 - \ 40x_2 - \ 60x_3 \qquad\qquad\quad + z &= \quad\ \ 0
\end{aligned}
$$

with x_1, x_2, x_3, x_4, x_5, and x_6 all nonnegative.

Set up the initial simplex tableau. (The z column is omitted as before.)

$$
\begin{array}{cccccc}
x_1 & x_2 & x_3 & x_4 & x_5 & x_6 \\
\end{array}
$$

$$
\left[
\begin{array}{cccccc|c}
1 & 1 & 1 & 1 & 0 & 0 & 100 \\
400 & 160 & 280 & 0 & 1 & 0 & 20{,}000 \\
1 & 1 & 1 & 0 & 0 & -1 & 60 \\
\hline
-120 & -40 & -60 & 0 & 0 & 0 & 0 \\
\end{array}
\right]
$$

This tableau gives the solution

$$x_1 = 0, \quad x_2 = 0, \quad x_3 = 0, \quad x_4 = 100, \quad x_5 = 20{,}000, \quad x_6 = -60.$$

But this is not a feasible solution, since x_6 is negative. All the variables in any feasible solution must be nonnegative if the solution is to correspond to a corner point of the region of feasible solutions.

When a negative value of a variable appears in the solution, we use row operations to transform the matrix until a solution is found in which all variables are nonnegative. Here the difficulty is caused by the -1 in row three of the matrix. We do not have the third column of the usual 3×3 identity matrix. To correct this, use row operations to change any column that has nonzero entries (such as the x_1, x_2, or x_3 columns) to one in which the third-row entry is 1 and the other entries are 0. The choice of a column is arbitrary. We will choose the x_2 column, and if this choice does not lead to a feasible solution, we will try one of the other columns.

The third-row entry in the x_2 column is already 1. Using row operations to get zeros in the rest of the column gives the following tableau.

$$
\begin{array}{cccccc}
x_1 & x_2 & x_3 & x_4 & x_5 & x_6 \\
\end{array}
$$

$$
\left[
\begin{array}{cccccc|c}
0 & 0 & 0 & 1 & 0 & 1 & 40 \\
240 & 0 & 120 & 0 & 1 & 160 & 10,400 \\
1 & 1 & 1 & 0 & 0 & -1 & 60 \\
\hline
-80 & 0 & -20 & 0 & 0 & -40 & 2400 \\
\end{array}
\right]
\begin{array}{l}
-1R_3 + R_1 \\
-160R_3 + R_2 \\
\\
40R_3 + R_4
\end{array}
$$

This tableau gives the solution

$$x_1 = 0, \quad x_2 = 60, \quad x_3 = 0, \quad x_4 = 40, \quad x_5 = 10,400, \quad \text{and} \quad x_6 = 0.$$

which is feasible. The process of applying row operations to get a feasible solution is called **phase I** of the solution. When a feasible solution is reached, phase I is ended and **phase II** can begin. In phase II, the simplex method is applied as usual to find the optimal feasible solution. Here, for the first step of phase II, the pivot is 240.

$$
\begin{array}{cccccc}
x_1 & x_2 & x_3 & x_4 & x_5 & x_6 \\
\end{array}
$$

$$
\left[
\begin{array}{cccccc|c}
0 & 0 & 0 & 1 & 0 & 1 & 40 \\
\boxed{240} & 0 & 120 & 0 & 1 & 160 & 10,400 \\
1 & 1 & 1 & 0 & 0 & -1 & 60 \\
-80 & 0 & -20 & 0 & 0 & -40 & 2400 \\
\end{array}
\right]
$$

$$
\begin{array}{cccccc}
x_1 & x_2 & x_3 & x_4 & x_5 & x_6 \\
\end{array}
$$

$$
\left[
\begin{array}{cccccc|c}
0 & 0 & 0 & 1 & 0 & 1 & 40 \\
1 & 0 & .5 & 0 & .004 & .667 & 43.3 \\
0 & 1 & .5 & 0 & -.004 & -1.667 & 16.7 \\
0 & 0 & 20 & 0 & .32 & 13.4 & 5864 \\
\end{array}
\right]
\begin{array}{l}
\\
\dfrac{1}{240}R_2 \\
-1R_2 + R_3 \\
80R_2 + R_4
\end{array}
$$

The second matrix above has been obtained from the first by standard row operations; some numbers in it have been rounded. This final tableau gives

$$x_1 = 43.3, \quad x_2 = 16.7, \quad x_3 = 0, \quad x_4 = 40, \quad x_5 = 0, \quad \text{and} \quad x_6 = 0.$$

1 Give the solution from the
first tableau.

Answer:

$x_1 = 0$, $x_2 = 0$, $x_3 = -60$,
$x_4 = 120$.

For maximum profit with this new constraint, the farmer should plant 43.3 acres of
potatoes, 16.7 acres of corn, and no cabbage. Forty acres of the 100 available
should not be planted. The profit will be $5864, less than the $6000 profit if he
planted only 50 acres of potatoes. Because of the additional constraint that at least
60 acres must be planted, the profit is reduced

▶**EXAMPLE 1**   Maximize $z = 10x_1 + 8x_2$

subject to: $4x_1 + 4x_2 \geq 60$

$2x_1 + 5x_2 \leq 120$

$x_1 \geq 0, \quad x_2 \geq 0$

Add slack or surplus variables to the constraints as needed to get the following
system.

$$4x_1 + 4x_2 - x_3 \qquad\qquad = 60$$
$$2x_1 + 5x_2 \qquad + x_4 \qquad = 120$$
$$-10x_1 - 8x_2 \qquad\qquad + z = 0$$

Now write the first simplex tableau.

$$
\begin{array}{cccc}
x_1 & x_2 & x_3 & x_4 \\
\end{array}
$$

$$
\left[
\begin{array}{cccc|c}
4 & 4 & -1 & 0 & 60 \\
2 & 5 & 0 & 1 & 120 \\
\hline
-10 & -8 & 0 & 0 & 0 \\
\end{array}
\right] \quad \boxed{1}
$$

The solution, from Problem 1 at the side, is not feasible because of the -60. In
phase I, use row transformations to modify the tableau to produce a feasible solution. We need a column with 1 in the first row and zeros in the rest of the rows. Let
us choose the x_1 column. Multiply the entries in the first row by 1/4 to get 1 in the
top row of the column. Then use row transformations to get zeros in the other rows
of that column.

$$
\begin{array}{cccc}
x_1 & x_2 & x_3 & x_4 \\
\end{array}
$$

$$
\left[
\begin{array}{cccc|c}
1 & 1 & -\frac{1}{4} & 0 & 15 \\
0 & 3 & \frac{1}{2} & 1 & 90 \\
\hline
0 & 2 & -\frac{5}{2} & 0 & 150 \\
\end{array}
\right]
\begin{array}{l}
\frac{1}{4}R_1 \\
-2R_1 + R_2 \\
10R_1 + R_3 \\
\end{array}
$$

This solution,

$$x_1 = 15, \quad x_2 = 0, \quad x_3 = 0, \quad \text{and} \quad x_4 = 90,$$

is feasible, so phase I is complete. Perform phase II by completing the solution in
the usual way. The pivot is 1/2. The next tableau is as follows.

2 The first tableau of a linear programming problem is given below. Use the second column to complete phase I of the solution and give the feasible solution that results.

$$\begin{bmatrix} 3 & 1 & 1 & 0 & | & 50 \\ 4 & 2 & 0 & -1 & | & 70 \\ \hline -8 & -10 & 0 & 0 & | & 0 \end{bmatrix}$$

Answer:

$$\begin{bmatrix} 1 & 0 & 1 & \frac{1}{2} & | & 15 \\ 2 & 1 & 0 & -\frac{1}{2} & | & 35 \\ \hline 12 & 0 & 0 & -5 & | & 350 \end{bmatrix}$$

Solution: $x_1 = 0$, $x_2 = 35$, $x_3 = 15$, $x_4 = 0$

$$\begin{array}{cccc} x_1 & x_2 & x_3 & x_4 \end{array}$$
$$\begin{bmatrix} 1 & \frac{5}{2} & 0 & \frac{1}{2} & | & 60 \\ 0 & 6 & 1 & 2 & | & 180 \\ \hline 0 & 17 & 0 & 5 & | & 600 \end{bmatrix}$$

No indicators are negative, so the optimum value of the objective function is

$$z = 600 \quad \text{when } x_1 = 60 \quad \text{and} \quad x_2 = 0. \quad \blacktriangleleft \quad \boxed{2}$$

The approach discussed above is used to solve problems where the constraints are mixed $\leq$ and $\geq$ inequalities. The method also can be used to solve a problem where the constant term is negative.

MINIMIZATION PROBLEMS The definition of a problem in standard maximum form was given earlier in this chapter. Now we can define a linear programming problem in **standard minimum form** as follows.

Standard Minimum Form

A linear programming problem is in standard minimum form if:

1. the objective function is to be minimized;
2. all variables are nonnegative;
3. all constraints involve $\geq$;
4. all the constants on the right side in the constraints are nonnegative.

The difference between maximization and minimization problems is in conditions 1 and 3: in problems stated in standard minimum form, the objective function is to be *minimized* rather than maximized, and all constraints must have $\geq$ instead of $\leq$.

Problems in standard minimum form can be solved with the method of surplus variables presented above. To solve a problem in standard minimum form, first observe that the minimum of an objective function is the same number as the *maximum* of the *negative* of the function. For example, if $w = 3y_1 + 2y_2$, then $z = -w = -3y_1 - 2y_2$. For corner points $(0, 3)$, $(6, 0)$, and $(2, 1)$, the same ordered pair that minimizes w maximizes z, as shown below.

	$w = 3y_1 + 2y_2$	$z = -3y_1 - 2y_2$
$(0, 3)$	6 (minimum)	-6 (maximum)
$(6, 0)$	18	-18
$(2, 1)$	8	-8

The next example illustrates the procedure described above. We use y_1 and y_2 as variables and w for the objective function as a reminder that this is a minimizing problem.

► EXAMPLE 2   Minimize $w = 3y_1 + 2y_2$

subject to: $y_1 + 3y_2 \geq 6$

$2y_1 + y_2 \geq 3$

$y_1 \geq 0, \quad y_2 \geq 0.$

Change this to a maximization problem by letting z equal the *negative* of the objective function: $z = -w$. Then find the *maximum* value of z.

$$z = -w = -3y_1 - 2y_2$$

The problem can now be stated as follows.

Maximize $z = -3y_1 - 2y_2$

subject to: $y_1 + 3y_2 \geq 6$

$2y_1 + y_2 \geq 3$

$y_1 \geq 0, \quad y_2 \geq 0.$

For phase I of the solution, subtract surplus variables and set up the first tableau.

$$\begin{array}{cccc} y_1 & y_2 & y_3 & y_4 \\ \end{array}$$
$$\left[\begin{array}{cccc|c} 1 & 3 & -1 & 0 & 6 \\ 2 & 1 & 0 & -1 & 3 \\ \hline 3 & 2 & 0 & 0 & 0 \end{array} \right]$$

The solution, $y_1 = 0$, $y_2 = 0$, $y_3 = -6$, and $y_4 = -3$, contains negative numbers. Row operations must be used to get a tableau with a feasible solution. Let us use the y_1 column since it already has a 1 in the first row. Begin by getting a 0 in the second row, and then a 0 in the third row.

$$\begin{array}{cccc} y_1 & y_2 & y_3 & y_4 \\ \end{array}$$
$$\left[\begin{array}{cccc|c} 1 & 3 & -1 & 0 & 6 \\ 0 & -5 & 2 & -1 & -9 \\ \hline 0 & -7 & 3 & 0 & -18 \end{array} \right] \begin{array}{l} \\ -2R_1 + R_2 \\ -3R_1 + R_3 \end{array}$$

The solution is $y_1 = 6$, $y_2 = 0$, $y_3 = 0$, $-y_4 = -9$ (or $y_4 = 9$), which is a feasible solution. Now, start phase II, finding the solution as usual by the simplex method. The pivot is -5.

3 Minimize $w = 2y_1 + 3y_2$
subject to:

$$y_1 + y_2 \geq 10$$
$$2y_1 + y_2 \geq 16$$
$$y_1 \geq 0, \quad y_2 \geq 0.$$

Answer:
$y_1 = 10$, $y_2 = 0$, $y_3 = 0$,
$y_4 = 4$; $w = 20$

$$
\begin{array}{cccc}
y_1 & y_2 & y_3 & y_4 \\
\end{array}
$$
$$
\left[
\begin{array}{cccc|c}
1 & 0 & \frac{1}{5} & -\frac{3}{5} & \frac{3}{5} \\
0 & 1 & -\frac{2}{5} & \frac{1}{5} & \frac{9}{5} \\
0 & 0 & \frac{1}{5} & \frac{7}{5} & -\frac{27}{5}
\end{array}
\right]
\begin{array}{l}
-3R_2 + R_1 \\
-\frac{1}{5}R_2 \\
7R_2 + R_3
\end{array}
$$

The solution is

$$y_1 = \frac{3}{5}, \quad y_2 = \frac{9}{5}, \quad y_3 = 0, \quad \text{and} \quad y_4 = 0.$$

This solution is feasible, and the tableau has no negative indicators. Since $z = -27/5$ and $z = -w$, then $w = 27/5$ is the minimum value, which is obtained when $y_1 = 3/5$ and $y_2 = 9/5$. ◄ **3**

Let us summarize the steps involved in the phase I and phase II method used to solve nonstandard problems in this section.

Solving Nonstandard Problems

1. If necessary, convert the problem to a maximum problem.
2. Add slack variables and subtract surplus variables as needed.
3. Write the initial simplex tableau.
4. If the solution from this tableau is not feasible, use row operations to get a feasible solution (phase I).
5. After a feasible solution is reached, solve by the simplex method (phase II).

▶**EXAMPLE 3**   A college textbook publisher has received orders from two colleges, C_1 and C_2. C_1 needs at least 500 books, and C_2 needs at least 1000. The publisher can supply the books from either of two warehouses. Warehouse W_1 has 900 books available and warehouse W_2 has 700. The costs to ship a book from each warehouse to each college are given below.

		To	
		C_1	C_2
From	W_1	$1.20	1.80
	W_2	$2.10	1.50

How many books should be sent from each warehouse to each college to minimize the shipping costs?

To begin, let

$$y_1 = \text{the number of books shipped from } W_1 \text{ to } C_1;$$
$$y_2 = \text{the number of books shipped from } W_2 \text{ to } C_1;$$
$$y_3 = \text{the number of books shipped from } W_1 \text{ to } C_2;$$
$$y_4 = \text{the number of books shipped from } W_2 \text{ to } C_2.$$

C_1 needs at least 500 books, so

$$y_1 + y_2 \geq 500.$$

Similarly,

$$y_3 + y_4 \geq 1000.$$

Since W_1 has 900 books available and W_2 has 700 available,

$$y_1 + y_3 \leq 900 \quad \text{and} \quad y_2 + y_4 \leq 700.$$

The company wants to minimize shipping costs, so the objective function is

$$w = 1.20y_1 + 2.10y_2 + 1.80y_3 + 1.50y_4.$$

Now write the problem as a system of linear equations, adding slack or surplus variables as needed, and let $z = -w$.

$$
\begin{array}{rcrcrcrcrcrcl}
y_1 + & y_2 & & & - y_5 & & & & & & & = & 500 \\
 & & y_3 + & y_4 & & - y_6 & & & & & & = & 1000 \\
y_1 & & + & y_3 & & & + y_7 & & & & & = & 900 \\
 & y_2 & & + & y_4 & & & + y_8 & & & & = & 700 \\
1.20y_1 + & 2.10y_2 + & 1.80y_3 + & 1.50y_4 & & & & & + z & = & & & 0
\end{array}
$$

Set up the first simplex tableau.

y_1	y_2	y_3	y_4	y_5	y_6	y_7	y_8	
1	1	0	0	−1	0	0	0	500
0	0	1	1	0	−1	0	0	1000
1	0	1	0	0	0	1	0	900
0	1	0	1	0	0	0	1	700
1.20	2.10	1.80	1.50	0	0	0	0	0

The indicated solution is

$$y_5 = -500, \quad y_6 = -1000, \quad y_7 = 900, \quad y_8 = 700,$$

which is not feasible since y_5 and y_6 are negative.

Starting with the second row of the matrix, where the constant is 1000, let us choose y_4 to replace y_6 as a basic variable, because only one constraint row must be changed. The next simplex tableau is as follows.

$$
\begin{array}{cccccccc}
y_1 & y_2 & y_3 & y_4 & y_5 & y_6 & y_7 & y_8 \\
\end{array}
$$

$$
\left[
\begin{array}{cccccccc|c}
1 & 1 & 0 & 0 & -1 & 0 & 0 & 0 & 500 \\
0 & 0 & 1 & 1 & 0 & -1 & 0 & 0 & 1000 \\
1 & 0 & 1 & 0 & 0 & 0 & 1 & 0 & 900 \\
0 & 1 & -1 & 0 & 0 & 1 & 0 & 1 & -300 \\
\hline
1.2 & 2.1 & .3 & 0 & 0 & 1.5 & 0 & 0 & -1500 \\
\end{array}
\right]
$$

The solution at this point is $y_4 = 1000$, $y_5 = -500$, $y_7 = 900$, and $y_8 = -300$, which is not feasible. Continue with phase I, choosing y_1, to replace y_5 as a basic variable. The result is the following tableau.

$$
\begin{array}{cccccccc}
y_1 & y_2 & y_3 & y_4 & y_5 & y_6 & y_7 & y_8 \\
\end{array}
$$

$$
\left[
\begin{array}{cccccccc|c}
1 & 1 & 0 & 0 & -1 & 0 & 0 & 0 & 500 \\
0 & 0 & 1 & 1 & 0 & -1 & 0 & 0 & 1000 \\
0 & -1 & 1 & 0 & 1 & 0 & 1 & 0 & 400 \\
0 & -1 & 1 & 0 & 0 & -1 & 0 & -1 & 300 \\
\hline
0 & .9 & .3 & 0 & 1.2 & 1.5 & 0 & 0 & -2100 \\
\end{array}
\right]
$$

The new solution is $y_1 = 500$, $y_4 = 1000$, $y_7 = 400$, and $y_8 = -300$, which still is not feasible. This time, replace the basic variable y_8 with y_3, getting the following tableau.

$$
\begin{array}{cccccccc}
y_1 & y_2 & y_3 & y_4 & y_5 & y_6 & y_7 & y_8 \\
\end{array}
$$

$$
\left[
\begin{array}{cccccccc|c}
1 & 1 & 0 & 0 & -1 & 0 & 0 & 0 & 500 \\
0 & 1 & 0 & 1 & 0 & 0 & 0 & 1 & 700 \\
0 & 0 & 0 & 0 & 1 & 1 & 1 & 1 & 100 \\
0 & -1 & 1 & 0 & 0 & -1 & 0 & -1 & 300 \\
\hline
0 & 1.2 & 0 & 0 & 1.2 & 1.8 & 0 & .3 & -2190 \\
\end{array}
\right]
$$

The solution now is $y_1 = 500$, $y_3 = 300$, $y_4 = 700$, and $y_7 = 100$, which is feasible. Since there are no negative indicators, the solution is optimal, and phase II is not needed. The publisher should ship 500 books from W_1 to C_1, 300 books from W_1 to C_2, and 700 books from W_2 to C_2 for a minimum shipping cost of \$2190. ◀

We certainly have not covered all the possible complications that can arise in using the simplex method. Some of the difficulties include the following:

1. Some of the constraints may be *equations* instead of inequalities. In this case, *artificial variables* must be used.
2. Occasionally, a transformation will cycle—that is, produce a "new" solution which was an earlier solution in the process. These situations are known as *degeneracies* and special methods are available for handling them.
3. It may not be possible to convert a nonfeasible basic solution to a feasible basic solution. In that case, no solution can satisfy all the constraints. Graphically, this means there is no region of feasible solutions.

(These difficulties are covered in more detail in advanced texts. One example of a text that you might find helpful is *An Introduction to Management Science,* Third Edition, by David R. Anderson, Dennis J. Sweeney, and Thomas A. Williams, 1988, West Publishing Company.)

Two linear programming models in actual use, one on merit pay, the other on making ice cream, are presented at the end of this chapter. These models illustrate the usefulness of linear programming. In most real applications, the number of variables is so large that these problems could not be solved without the use of a method, like the simplex method, that can be adapted to a computer.

7.6 EXERCISES

Rewrite each system of inequalities, adding slack variables or subtracting surplus variables as necessary. (See Example 1.)

1. $2x_1 + 3x_2 \leq 8$
$x_1 + 4x_2 \geq 7$

2. $5x_1 + 8x_2 \leq 10$
$6x_1 + 2x_2 \geq 7$

3. $x_1 + x_2 + x_3 \leq 100$
$x_1 + x_2 + x_3 \geq 75$
$x_1 + x_2 \geq 27$

4. $2x_1 + x_3 \leq 40$
$x_1 + x_2 \geq 18$
$x_1 + x_3 \geq 20$

Convert Exercises 5–8 into maximization problems. (See Example 2.)

5. Minimize $w = 4y_1 + 3y_2 + 2y_3$
    subject to: $y_1 + y_2 + y_3 \geq 5$
                $y_1 + y_2 \geq 4$
           $2y_1 + y_2 + 3y_3 \geq 15$
    $y_1 \geq 0, \quad y_2 \geq 0, \quad y_3 \geq 0.$

6. Minimize $w = 8y_1 + 3y_2 + y_3$
    subject to: $7y_1 + 6y_2 + 8y_3 \geq 18$
                $4y_1 + 5y_2 + 10y_3 \geq 20$
    $y_1 \geq 0, \quad y_2 \geq 0, \quad y_3 \geq 0.$

7. Minimize $w = y_1 + 2y_2 + y_3 + 5y_4$
    subject to: $y_1 + y_2 + y_3 + y_4 \geq 50$
             $3y_1 + y_2 + 2y_3 + y_4 \geq 100$
    $y_1 \geq 0, \quad y_2 \geq 0, \quad y_3 \geq 0, \quad y_4 \geq 0.$

8. Minimize $w = y_1 + y_2 + 4y_3$
    subject to: $y_1 + 2y_2 + 3y_3 \geq 115$
           $2y_1 + y_2 + y_3 \leq 200$
            $y_1 + y_3 \geq 50$
    $y_1 \geq 0, \quad y_2 \geq 0, \quad y_3 \geq 0.$

Use the simplex method to solve Exercises 9–14. (See Examples 1 and 2.)

9. Maximize $z = 12x_1 + 10x_2$
subject to: $x_1 + 2x_2 \geq 24$
$x_1 + x_2 \leq 40$
$x_1 \geq 0, \quad x_2 \geq 0.$

10. Maximize $z = 6x_1 + 8x_2$
subject to: $3x_1 + 4x_2 \geq 48$
$2x_1 + 4x_2 \leq 60$
$x_1 \geq 0, \quad x_2 \geq 0.$

11. Find $x_1 \geq 0$, $x_2 \geq 0$, and $x_3 \geq 0$ such that
$$x_1 + x_2 + x_3 \leq 150$$
$$x_1 + x_2 + x_3 \geq 100$$
and $z = 2x_1 + 5x_2 + 3x_3$ is maximized.

12. Find $x_1 \geq 0$, $x_2 \geq 0$, and $x_3 \geq 0$ such that
$$x_1 + x_2 + 2x_3 \leq 38$$
$$2x_1 + x_2 + x_3 \geq 24$$
and $z = 3x_1 + 2x_2 + 2x_3$ is maximized.

13. Find $x_1 \geq 0$ and $x_2 \geq 0$ such that
$$x_1 + x_2 \leq 100$$
$$x_1 + x_2 \geq 50$$
$$2x_1 + x_2 \leq 110$$
and $z = 2x_1 + 3x_2$ is maximized.

14. Find $x_1 \geq 0$ and $x_2 \geq 0$ such that
$$x_1 + 2x_2 \leq 18$$
$$x_1 + 3x_2 \geq 12$$
$$2x_1 + 2x_2 \leq 24$$
and $z = 5x_1 + 10x_2$ is maximized.

Solve each of the following by the two-phase method.

15. Find $y_1 \geq 0$, $y_2 \geq 0$ such that
$$10y_1 + 5y_2 \geq 100$$
$$20y_1 + 10y_2 \geq 150$$
and $w = 4y_1 + 5y_2$ is minimized.

16. Minimize $w = 3y_1 + 2y_2$
subject to: $2y_1 + 3y_2 \geq 60$
$y_1 + 4y_2 \geq 40$
$y_1 \geq 0, \quad y_2 \geq 0.$

17. Minimize $w = 2y_1 + y_2 + 3y_3$
subject to: $y_1 + y_2 + y_3 \geq 100$
$2y_1 + y_2 \geq 50$
$y_1 \geq 0, \quad y_2 \geq 0, \quad y_3 \geq 0.$

18. Minimize $w = 3y_1 + 2y_2$
subject to: $y_1 + 2y_2 \geq 10$
$y_1 + y_2 \geq 8$
$2y_1 + y_2 \geq 12$
$y_1 \geq 0, \quad y_2 \geq 0.$

Use the simplex method to solve Exercises 19–29. (See Example 3.)

19. Management Southwestern Oil supplies two distributors in the Northwest from two outlets. Distributor D_1 needs at least 3000 barrels of oil, and distributor D_2 needs at least 5000 barrels. The two outlets can each furnish 5000 barrels of oil. The costs per barrel to send the oil are given below.

From	To D_1	D_2
S_1	$30	$20
S_2	$25	$22

How should the oil be supplied to minimize shipping costs?

20. Mark, who is ill, takes vitamin pills. Each day he must have at least 16 units of vitamin A, 5 units of vitamin B_1, and 20 units of vitamin C. He can choose between pill #1 which costs 10¢ and contains 8 units of A, 1 of B_1, and 2 of C, and pill #2 which costs 20¢ and contains 2 units of A, 1 of B_1, and 7 of C. How many of each pill should he buy in order to minimize his cost? (See Exercise 18, Section 7.3.)

21. Management A bank has set aside a maximum of $25 million for commercial and home loans. The bank's policy is to loan at least four times as much for home loans as for commercial loans. Because of prior commitments, at least $10 million will be used for these two types of loans. The bank earns 12% on home loans and 10% on commercial loans. What amount of money should be loaned out for each type of loan to maximize the interest income?

22. Sam, who is dieting, requires two food supplements, I and II. He can get these supplements from two different products, A and B, as shown in the following table.

	Supplement (grams per serving)	
	I	II
Product A	3	2
B	2	4

Sam's physician has recommended that he include at least 15 grams of supplement I but no more than 12 grams of II in his daily diet. If product A costs 25¢ per serving and product B costs 40¢ per serving, how can he satisfy his requirements most economically?

23. **Management** Brand X Canners produce canned whole tomatoes and tomato sauce. This season, they have available 3,000,000 kilograms of tomatoes for these two products. To meet the demands of regular customers, they must produce at least 80,000 kilograms of sauce and 800,000 kilograms of whole tomatoes. The cost per kilogram is $4 to produce canned whole tomatoes and $3.25 to produce tomato sauce. How many kilograms of tomatoes should they use for each product to minimize cost?

24. **Management** A brewery produces regular beer and a lower-carbohydrate "light" beer. Steady customers of the brewery buy 12 units of regular beer and 10 units of light beer. While setting up the brewery to produce the beers, the management decides to produce extra beer, beyond that needed to satisfy the steady customers. The cost per unit of regular beer is $36,000 and the cost per unit of light beer is $48,000. The number of units of light beer should not exceed twice the number of units of regular beer. At least 20 additional units of beer can be sold. How much of each type beer should be made so as to minimize total production costs?

25. The chemistry department at a local college decides to stock at least 800 small test tubes and 500 large test tubes. It wants to buy at least 1500 test tubes to take advantage of a special price. Since the small tubes are broken twice as often as the larger, the department will order at least twice as many small tubes as large. If the small test tubes cost 15¢ each and the large ones, made of a cheaper glass, cost 12¢ each, how many of each size should they order to minimize cost?

26. **Management** Topgrade Turf lawn seed mixture contains three types of seeds: bluegrass, rye, and bermuda. The costs per pound of the three types of seed are 20¢, 15¢, and 5¢. In each mixture there must be at least 20% bluegrass seed and the amount of bermuda must be no more than the amount of rye. To fill current orders, the company must make at least 5000 pounds of the mixture. How much of each kind of seed should be used to minimize cost?

27. **Management** Virginia Keleske has decided to invest a $100,000 inheritance in government securities that earn 7% per year, municipal bonds that earn 6% per year, and mutual funds that earn an average of 10% per year. She will spend at least $40,000 on government securities, and she wants at least half the inheritance to go to bonds and mutual funds. How much should be invested in each way to maximize the interest yet meet the constraints? What is the maximum interest she can earn?

28. **Natural Science** A biologist must make a nutrient for her algae. The nutrient must contain the three basic elements D, E, and F, and must contain at least 10 kilograms of D, 12 kilograms of E, and 20 kilograms of F. The nutrient is made from three ingredients, I, II, and III. The quantity of D, E, and F in 1 unit of each of the ingredients is as given in the following chart.

1 Unit of Ingredient	Contains the Following Elements in Kilograms			Cost of 1 Unit of Ingredient
	D	E	F	
I	4	3	0	4
II	1	2	4	7
III	10	1	5	5

How many units of each ingredient are required to meet her needs at minimum cost?

29. **Management** The manufacturer of a popular personal computer has orders from two dealers. Dealer D_1 wants at least 32 computers, and dealer D_2 wants at least 20 computers. The manufacturer can fill the orders from either of two warehouses, W_1 or W_2. W_1 has 25 of the computers on hand, and W_2 has 30. The costs (in dollars) to ship one computer to each dealer from each warehouse are given below.

		To	
		D_1	D_2
From	W_1	14	22
	W_2	12	10

How should the orders be filled to minimize shipping costs?

Determine the constraints and the objective function for Exercises 30 and 31, then use a calculator or a computer to solve each problem.

30. Management Natural Brand plant food is made from three chemicals. (See Section 6.2, Exercise 27.) In a batch of the plant food there must be at least 81 kilograms of the first chemical and the other two chemicals must be in the ratio of 4 to 3. If the three chemicals cost $1.09, $.87, and $.65 per kilogram, respectively, how much of each should be used to minimize the cost of producing at least 750 kilograms of the plant food?

31. Management A company is developing a new additive for gasoline. The additive is a mixture of three liquid ingredients, I, II, and III. For proper performance, the total amount of additive must be at least 10 ounces per gallon of gasoline. However, for safety reasons, the amount of additive should not exceed 15 ounces per gallon of gasoline. At least 1/4 ounce of ingredient I must be used for every ounce of ingredient II and at least 1 ounce of ingredient III must be used for every ounce of ingredient I. If the cost of I, II, and III is $.30, $.09, and $.27 per ounce, respectively, find the mixture of the three ingredients that pro-

duces the minimum cost of the additive. How much of the additive should be used per gallon of gasoline?

 Solve the following linear programming problem, which has both "greater than" and "less than" constraints.

32. Management A popular soft drink called Sugarlo, which is advertised as having a sugar content of no more than 10%, is blended from five ingredients, each of which has some sugar content. Water may also be added to dilute the mixture. The sugar content of the ingredients and their costs per gallon are given below.

	Ingredient					
	1	2	3	4	5	Water
Sugar content(%)	.28	.19	.43	.57	.22	0
Cost ($/gal.)	.48	.32	.53	.28	.43	.04

At least .01 of the content of Sugarlo must come from ingredients 3 or 4, .01 must come from ingredients 2 or 5, and .01 from ingredients 1 or 4. How much of each ingredient should be used in preparing 15,000 gallons of Sugarlo to minimize the cost?

7.7 DUALITY

In this section we discuss an alternative method of solving minimizing problems in standard form. An interesting connection exists between standard maximizing and standard minimizing problems: any solution of a standard maximizing problem produces the solution of an associated standard minimizing problem, or vice versa. Each of these associated problems is called the **dual** of the other. One advantage of duals is that standard minimizing problems can be solved by the simplex methods already discussed. Let us explain the idea of a dual with an example. (This is similar to Example 2 in the previous section; compare this method of solution with the one given there.)

 EXAMPLE 1 Minimize $w = 8y_1 + 16y_2$

subject to:
$$y_1 + 5y_2 \geq 9$$
$$2y_1 + 2y_2 \geq 10$$
$$y_1 \geq 0, \quad y_2 \geq 0.$$

(As mentioned earlier, y_1 and y_2 are used as variables and w as the objective function as a reminder that this is a minimization problem.) Without considering

1 Use the corner points in Figure 7.22(a) to find the minimum value of $w = 8y_1 + 16y_2$ and where it occurs.

Answer:
48 when $y_1 = 4$, $y_2 = 1$

2 Use Figure 7.22(b) to find the maximum value of $z = 9x_1 + 10x_2$ and where it occurs.

Answer:
48 when $x_1 = 2$, $x_2 = 3$

slack variables just yet, write the augmented matrix of the system of inequalities, and include the coefficients of the objective function (not their negatives) as the last row in the matrix.

$$\begin{bmatrix} 1 & 5 & | & 9 \\ 2 & 2 & | & 10 \\ 8 & 16 & | & 0 \end{bmatrix}$$

Look now at the following new matrix, obtained from the one above by interchanging rows and columns.

$$\begin{bmatrix} 1 & 2 & | & 8 \\ 5 & 2 & | & 16 \\ 9 & 10 & | & 0 \end{bmatrix}$$

The *rows* of the first matrix (for the minimizing problem) are the *columns* of the second matrix.

The entries in this second matrix could be used to write the following maximizing problem in standard form (again ignoring the fact that the numbers in the last row are not negative).

$$\begin{aligned} \text{Maximize} \quad & z = 9x_1 + 10x_2 \\ \text{subject to:} \quad & x_1 + 2x_2 \leq 8 \\ & 5x_1 + 2x_2 \leq 16 \\ & x_1 \geq 0, \quad x_2 \geq 0. \end{aligned}$$

Figure 7.22(a) shows the region of feasible solutions for the minimization problem given above, while Figure 7.22(b) shows the region of feasible solutions for the maximization problem produced by exchanging rows and columns. ◀ **1** **2**

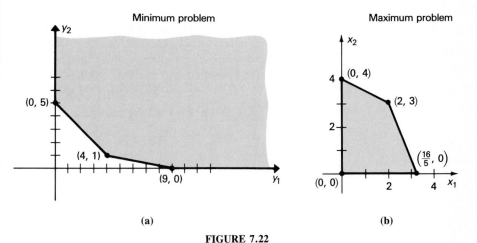

(a)   (b)

FIGURE 7.22

The two feasible regions in Figure 7.22 are different and the corner points are different, but the values of the objective functions found in Problems 1 and 2 at the side are equal—both are 48. An even closer connection between the two problems is shown by using the simplex method to solve the maximization problem given above.

$$\text{Maximization problem}$$

$$
\begin{array}{cccc}
x_1 & x_2 & x_3 & x_4
\end{array}
$$
$$
\left[
\begin{array}{cccc|c}
1 & 2 & 1 & 0 & 8 \\
5 & 2 & 0 & 1 & 16 \\
\hline
-9 & -10 & 0 & 0 & 0
\end{array}
\right]
$$

$$
\begin{array}{cccc}
x_1 & x_2 & x_3 & x_4
\end{array}
$$
$$
\left[
\begin{array}{cccc|c}
\frac{1}{2} & 1 & \frac{1}{2} & 0 & 4 \\
4 & 0 & -1 & 1 & 8 \\
\hline
-4 & 0 & 5 & 0 & 40
\end{array}
\right]
\quad
\begin{array}{l}
\frac{1}{2}R_1 \\
-2R_1 + R_2 \\
10R_1 + R_3
\end{array}
$$

$$
\begin{array}{cccc}
x_1 & x_2 & x_3 & x_4
\end{array}
$$
$$
\left[
\begin{array}{cccc|c}
0 & 1 & \frac{5}{8} & -\frac{1}{8} & 3 \\
1 & 0 & -\frac{1}{4} & \frac{1}{4} & 2 \\
\hline
0 & 0 & 4 & 1 & 48
\end{array}
\right]
\quad
\begin{array}{l}
-\frac{1}{2}R_2 + R_1 \\
\frac{1}{4}R_2 \\
4R_2 + R_3
\end{array}
$$

The maximum is 48 when $x_1 = 2$, $x_2 = 3$.

Notice that the solution to the *minimization problem* (namely, $y_1 = 4$, $y_2 = 1$) is found in the bottom row and slack variable columns of the final simplex tableau for the maximization problem. This result suggests that standard minimization problems can be solved by forming the dual standard maximization problem, solving it by the simplex method, and then reading the solution for the minimization problem from the bottom row of the final simplex tableau.

Before using this method to solve a minimization problem, let us find the duals of some typical linear programming problems. The process of exchanging the rows and columns of a matrix, which is used to find the dual, is called **transposing** the matrix, and each of the two matrices is the **transpose** of the other.

▶**EXAMPLE 2**   Find the transpose of each matrix.

(a) $A = \begin{bmatrix} 2 & -1 & 5 \\ 6 & 8 & 0 \\ -3 & 7 & -1 \end{bmatrix}$

3 Give the transpose of each matrix.

(a) $\begin{bmatrix} 2 & 4 \\ 6 & 3 \\ 1 & 5 \end{bmatrix}$

(b) $\begin{bmatrix} 4 & 7 & 10 \\ 3 & 2 & 6 \\ 5 & 8 & 12 \end{bmatrix}$

Answer:

(a) $\begin{bmatrix} 2 & 6 & 1 \\ 4 & 3 & 5 \end{bmatrix}$

(b) $\begin{bmatrix} 4 & 3 & 5 \\ 7 & 2 & 8 \\ 10 & 6 & 12 \end{bmatrix}$

Write the rows of matrix A as the columns of the transpose.

$$\text{Transpose of } A = \begin{bmatrix} 2 & 6 & -3 \\ -1 & 8 & 7 \\ 5 & 0 & -1 \end{bmatrix}$$

(b) The transpose of $\begin{bmatrix} 1 & 2 & 4 & 0 \\ 2 & 1 & 7 & 6 \end{bmatrix}$ is $\begin{bmatrix} 1 & 2 \\ 2 & 1 \\ 4 & 7 \\ 0 & 6 \end{bmatrix}$. ◀ **3**

▶**EXAMPLE 3** Write the dual of the following standard minimization linear programming problems.

(a) Minimize $w = 10y_1 + 8y_2$

subject to: $y_1 + 2y_2 \geq 2$

$y_1 + y_2 \geq 5$

$y_1 \geq 0, \quad y_2 \geq 0.$

Begin by writing the augmented matrix for the given problem.

$$\begin{bmatrix} 1 & 2 & | & 2 \\ 1 & 1 & | & 5 \\ 10 & 8 & | & 0 \end{bmatrix}$$

Form the transpose of this matrix to get

$$\begin{bmatrix} 1 & 1 & | & 10 \\ 2 & 1 & | & 8 \\ 2 & 5 & | & 0 \end{bmatrix}$$

The dual problem is stated from this second matrix as follows (using x instead of y.)

Maximize $z = 2x_1 + 5x_2$

subject to: $x_1 + x_2 \leq 10$

$2x_1 + x_2 \leq 8$

$x_1 \geq 0, \quad x_2 \geq 0.$

4 Write the dual of the following linear programming problem. Minimize $w = 2y_1 + 5y_2 + 6y_3$ subject to:

$$2y_1 + 3y_2 + y_3 \geq 15$$
$$y_1 + y_2 + 2y_3 \geq 12$$
$$5y_1 + 3y_2 \geq 10$$
$$y_1 \geq 0, \quad y_2 \geq 0, \quad y_3 \geq 0.$$

Answer:
Maximize $z = 15x_1 + 12x_2 + 10x_3$ subject to:

$$2x_1 + x_2 + 5x_3 \leq 2$$
$$3x_1 + x_2 + 3x_3 \leq 5$$
$$x_1 + 2x_2 \leq 6$$
$$x_1 \geq 0, \quad x_2 \geq 0, \quad x_3 \geq 0.$$

(b) Minimize $w = 7y_1 + 5y_2 + 8y_3$

subject to:

$$3y_1 + 2y_2 + y_3 \geq 10$$
$$y_1 + y_2 + y_3 \geq 8$$
$$4y_1 + 5y_2 \geq 25$$
$$y_1 \geq 0, \quad y_2 \geq 0, \quad y_3 \geq 0.$$

The dual problem is stated as follows.

Maximize $z = 10x_1 + 8x_2 + 25x_3$

subject to:

$$3x_1 + x_2 + 4x_3 \leq 7$$
$$2x_1 + x_2 + 5x_3 \leq 5$$
$$x_1 + x_2 \leq 8$$
$$x_1 \geq 0, \quad x_2 \geq 0, \quad x_3 \geq 0. \blacktriangleleft \quad \boxed{4}$$

In Example 3, all the constraints of the standard minimization problems were $\geq$ inequalities, while all those in the dual maximization problems were $\leq$ inequalities. This is generally the case; inequalities are reversed when the dual problem is stated.

The following table shows the close connection between a problem and its dual.

Given Problem	*Dual Problem*
m variables	n variables
n constraints	m constraints
Coefficients from objective function	Constants
Constants	Coefficients from objective function

The next theorem, whose proof requires advanced methods, guarantees that a standard minimization problem can be solved by forming a dual standard maximization problem.

Theorem of Duality

The objective function w of a minimizing linear programming problem takes on a minimum value if and only if the objective function z of the corresponding dual maximizing problem takes on a maximum value. The maximum value of z equals the minimum value of w.

This method is illustrated in the following example. (This was also Example 2 of the previous section; compare this solution with the one given there.)

▶**EXAMPLE 4** Minimize $w = 3y_1 + 2y_2$

subject to: $y_1 + 3y_2 \geq 6$

$2y_1 + y_2 \geq 3$

$y_1 \geq 0, \quad y_2 \geq 0.$

Use the given information to write the matrix

$$\begin{bmatrix} 1 & 3 & | & 6 \\ 2 & 1 & | & 3 \\ 3 & 2 & | & 0 \end{bmatrix}$$

Transpose to get the following matrix for the dual problem.

$$\begin{bmatrix} 1 & 2 & | & 3 \\ 3 & 1 & | & 2 \\ 6 & 3 & | & 0 \end{bmatrix}$$

Write the dual problem from this matrix, as follows.

Maximize $z = 6x_1 + 3x_2$

subject to: $x_1 + 2x_2 \leq 3$

$3x_1 + x_2 \leq 2$

$x_1 \geq 0, \quad x_2 \geq 0.$

Solve this standard maximization problem using the simplex method. Start by introducing slack variables to give the system

$$x_1 + 2x_2 + x_3 \qquad\qquad = 3$$
$$3x_1 + x_2 \qquad + x_4 \qquad = 2$$
$$-6x_1 - 3x_2 - 0x_3 - 0x_4 + z = 0$$

with $x_1 \geq 0, \quad x_2 \geq 0, \quad x_3 \geq 3, \quad x_4 \geq 0.$

The first tableau for this system is given below, with the pivot as indicated.

Quotients		x_1	x_2	x_3	x_4	
3/1 = 3		1	2	1	0	3
2/3		3	1	0	1	2
		−6	−3	0	0	0

The simplex method gives the following final tableau.

	x_1	x_2	x_3	x_4	
0	1	$\frac{3}{5}$	$-\frac{1}{5}$	$\frac{7}{5}$
1	0	$-\frac{1}{5}$	$\frac{2}{5}$	$\frac{1}{5}$
0	0	$\frac{3}{5}$	$\frac{9}{5}$	$\frac{27}{5}$

5 Minimize $w = 10y_1 + 8y_2$
subject to:

$$y_1 + 2y_2 \geq 2$$
$$y_1 + y_2 \geq 5$$
$$y_1 \geq 0, \quad y_2 \geq 0.$$

Answer:
$y_1 = 0$, $y_2 = 5$, for a minimum of 40

The last row of this final tableau shows that the solution of the given standard minimization problem is as follows:

The minimum value of $w = 3y_1 + 2y_2$, subject to the given constraints, is 27/5 and occurs when $y_1 = 3/5$ and $y_2 = 9/5$.

The minimum value of w, 27/5, is the same as the maximum value of z. ◄ **5**

Let us summarize the steps in solving a standard minimum linear programming problem by the method of duals.

Solving Minimum Problems with Duals

1. Find the dual standard maximum problem.
2. Solve the maximum problem using the simplex method.
3. The minimum value of the objective function w is the maximum value of the objective function z.
4. The optimum solution is given by the entries in the bottom row of the columns corresponding to the slack variables.

FURTHER USES OF THE DUAL The dual is useful not only in solving minimization problems, but also in seeing how small changes in one variable will affect the value of the objective function. For example, suppose an animal breeder needs at least 6 units per day of nutrient A and at least 3 units of nutrient B and that the breeder can choose between two different feeds, feed 1 and feed 2. Find the minimum cost for the breeder if each bag of feed 1 costs \$3 and provides 1 unit of nutrient A and 2 units of B, while each bag of feed 2 costs \$2 and provides 3 units of nutrient A and 1 of B.

If y_1 represents the number of bags of feed 1 and y_2 represents the number of bags of feed 2, the given information leads to

$$\text{Minimize} \qquad w = 3y_1 + 2y_2$$
$$\text{subject to:} \qquad y_1 + 3y_2 \geq 6$$
$$2y_1 + y_2 \geq 3$$
$$y_1 \geq 0, \quad y_2 \geq 0.$$

This standard minimization linear programming problem is the one we solved in Example 4 of this section. In that example, we formed the dual and reached the following final tableau:

$$
\begin{array}{cccc}
x_1 & x_2 & x_3 & x_4 \\
\end{array}
$$
$$
\left[
\begin{array}{cccc|c}
0 & 1 & \frac{3}{5} & -\frac{1}{5} & \frac{7}{5} \\
1 & 0 & -\frac{1}{5} & \frac{2}{5} & \frac{1}{5} \\
0 & 0 & \frac{3}{5} & \frac{9}{5} & \frac{27}{5} \\
\end{array}
\right]
$$

6 The final tableau of the dual of the problem about filing cabinets in Example 2, Section 7.3 and Example 1, Section 7.5 is given below.

(a) What are the imputed amounts of storage for each unit of cost and floor space?

(b) What are the shadow values of the cost and the floor space?

$$
\begin{array}{cccc}
x_1 & x_2 & x_3 & x_4 \\
\end{array}
$$

$$
\left[
\begin{array}{cccc|c}
0 & 1 & \frac{1}{2} & -\frac{1}{4} & 1 \\
1 & 0 & -\frac{1}{20} & -\frac{1}{80} & \frac{1}{20} \\
0 & 0 & 8 & 3 & 100 \\
\end{array}
\right]
$$

Answer:

(a) Cost: 28 sq ft
floor space: 72 sq ft

(b) $\dfrac{1}{20}$, 1

This final tableau shows that the breeder will obtain minimum feed costs by using 3/5 bag of feed 1 and 9/5 bag of feed 2 per day, for a daily cost of 27/5 = 5.40 dollars.

Now look at the data from the feed problem shown in the table below.

	Units of Nutrient (per bag)		Cost per Bag
	A	B	
Feed 1	1	2	$3
Feed 2	3	1	$2
Minimum nutrient needed	6	3	

If x_1 and x_2 are the cost per unit of nutrients A and B, the constraints of the dual problem can be stated as follows.

$$\text{Cost of feed 1: } x_1 + 2x_2 \leq 3$$
$$\text{Cost of feed 2: } 3x_1 + x_2 \leq 2$$

The solution of the dual problem, which maximizes nutrients, can be read from the final tableau:

$$x_1 = \frac{1}{5} = .20 \quad \text{and} \quad x_2 = \frac{7}{5} = 1.40,$$

which means that a unit of nutrient A costs 1/5 of a dollar = $.20, while a unit of nutrient B costs 7/5 dollars = $1.40. The minimum daily cost, $5.40, is found by the following procedure.

$$
\begin{aligned}
(\$.20 \text{ per unit of A}) \times (6 \text{ units of A}) &= \$1.20 \\
+ (\$1.40 \text{ per unit of B}) \times (3 \text{ units of B}) &= \$4.20 \\
\hline
\text{Minimum daily cost} &= \$5.40
\end{aligned}
$$

The numbers .20 and 1.40 are called the **shadow costs** of the nutrients. These two numbers from the dual, $.20 and $1.40, also allow the breeder to estimate feed costs for "small" changes in nutrient requirements. For example, an increase of 1 unit in the requirement for each nutrient would produce a total cost of

$5.40	6 units of A, 3 of B
.20	1 extra unit of A
1.40	1 extra unit of B
$7.00	Total cost per day **6**

7.7 EXERCISES

Find the transpose of each matrix. (See Example 2.)

1. $\begin{bmatrix} 1 & 2 & 3 \\ 3 & 2 & 1 \\ 1 & 10 & 0 \end{bmatrix}$

2. $\begin{bmatrix} 2 & 5 & 8 & 6 & 0 \\ 1 & -1 & 0 & 12 & 14 \end{bmatrix}$

3. $\begin{bmatrix} -1 & 4 & 6 & 12 \\ 13 & 25 & 0 & 4 \\ -2 & -1 & 11 & 3 \end{bmatrix}$

4. $\begin{bmatrix} 1 & 11 & 15 \\ 0 & 10 & -6 \\ 4 & 12 & -2 \\ 1 & -1 & 13 \\ 2 & 25 & -1 \end{bmatrix}$

State the dual problem but do not solve it. (See Example 3.)

5. Minimize $w = 2y_1 + 4y_2$
subject to: $3y_1 + y_2 \ge 4$
$y_1 + 2y_2 \ge 6$
$y_1 \ge 0, \quad y_2 \ge 0.$

6. Minimize $w = 8y_1 + 3y_2 + y_3$
subject to: $7y_1 + 6y_2 + 8y_3 \ge 18$
$4y_1 + 5y_2 + 10y_3 \ge 20$
$y_1 \ge 0, \quad y_2 \ge 0.$

7. Minimize $w = y_1 + 3y_2 + 4y_3$
subject to: $3y_1 + 4y_2 + 6y_3 \ge 8$
$y_1 + 5y_2 + 2y_3 \ge 12$
$y_1 \ge 0, \quad y_2 \ge 0.$

8. Minimize $w = 8y_1 + 3y_2$
subject to: $y_1 + y_2 \ge 17$
$3y_1 + 6y_2 \ge 21$
$2y_1 + 4y_2 \ge 19$
$y_1 \ge 0, \quad y_2 \ge 0.$

9. Minimize $w = 3y_1 + 4y_2$
subject to: $y_1 + 7y_2 \ge 18$
$4y_1 + y_2 \ge 15$
$5y_1 + 3y_2 \ge 20$
$y_1 \ge 0, \quad y_2 \ge 0.$

10. Minimize $w = 4y_1 + 3y_2 + 2y_3$
subject to: $y_1 + y_2 + y_3 \ge 5$
$y_1 + y_2 \ge 4$
$2y_1 + y_2 + 3y_3 \ge 15$
$y_1 \ge 0, \quad y_2 \ge 0, \quad y_3 \ge 0.$

11. Minimize $w = y_1 + y_2 + 4y_3$
subject to: $y_1 + 2y_2 + 3y_3 \ge 115$
$2y_1 + y_2 + 8y_3 \ge 200$
$y_1 + y_3 \ge 50$
$y_1 \ge 0, \quad y_2 \ge 0, \quad y_3 \ge 0.$

12. Minimize $w = y_1 + 2y_2 + y_3 + 5y_4$
subject to: $y_1 + y_2 + y_3 + y_4 \ge 50$
$3y_1 + y_2 + 2y_3 + y_4 \ge 100$
$y_1 \ge 0, \quad y_2 \ge 0, \quad y_3 \ge 0, \quad y_4 \ge 0.$

Use the simplex method to solve. (See Example 4.)

13. Minimize $w = 4y_1 + 5y_2$
subject to: $10y_1 + 5y_2 \ge 100$
$20y_1 + 10y_2 \ge 150$
$y_1 \ge 0, \quad y_2 \ge 0.$

14. Minimize $w = 3y_1 + 2y_2$
subject to: $2y_1 + 3y_2 \ge 60$
$y_1 + 4y_2 \ge 40$
$y_1 \ge 0, \quad y_2 \ge 0.$

15. Minimize
$$w = 2y_1 + y_2 + 3y_3$$
subject to:
$$y_1 + y_2 + y_3 \geq 100$$
$$2y_1 + y_2 \qquad \geq 50$$
$$y_1 \geq 0, \quad y_2 \geq 0, \quad y_3 \geq 0.$$

16. Minimize
$$w = 3y_1 + 2y_2$$
subject to:
$$y_1 + 2y_2 \geq 10$$
$$y_1 + y_2 \geq 8$$
$$2y_1 + y_2 \geq 12$$
$$y_1 \geq 0, \quad y_2 \geq 0.$$

17. Minimize
$$w = y_1 + y_2 + 3y_3$$
subject to:
$$2y_1 + y_2 + y_3 \geq 8$$
$$y_1 + 2y_2 + 4y_3 \geq 12$$
$$y_1 \geq 0, \quad y_2 \geq 0, \quad y_3 \geq 0.$$

18. Minimize
$$w = 20y_1 + 12y_2 + 40y_3$$
subject to:
$$y_1 + y_2 + 5y_3 \geq 20$$
$$2y_1 + y_2 + y_3 \geq 30$$
$$y_1 \geq 0, \quad y_2 \geq 0, \quad y_3 \geq 0.$$

19. Minimize
$$w = 12y_1 + 10y_2 + 7y_3$$
subject to:
$$2y_1 + y_2 + y_3 \geq 7$$
$$y_1 + 2y_2 + y_3 \geq 4$$
$$y_1 \geq 0, \quad y_2 \geq 0, \quad y_3 \geq 0.$$

20. Minimize
$$w = 4y_1 + 2y_2 + y_3$$
subject to:
$$y_1 + y_2 + y_3 \geq 4$$
$$3y_1 + y_2 + 3y_3 \geq 6$$
$$y_1 + y_2 + 3y_3 \geq 5$$
$$y_1 \geq 0, \quad y_2 \geq 0, \quad y_3 \geq 0.$$

21. Minimize
$$w = 3y_1 + y_2 + 4y_3$$
subject to:
$$2y_1 + y_2 + y_3 \geq 6$$
$$y_1 + 2y_2 + y_3 \geq 8$$
$$2y_1 + y_2 + 2y_3 \geq 12$$
$$y_1 \geq 0, \quad y_2 \geq 0, \quad y_3 \geq 0.$$

22. Minimize
$$w = 6y_1 + 4y_2 + 2y_3$$
subject to:
$$2y_1 + 2y_2 + y_3 \geq 2$$
$$y_1 + 3y_2 + 2y_3 \geq 3$$
$$y_1 + y_2 + 2y_3 \geq 4$$
$$y_1 \geq 0, \quad y_2 \geq 0, \quad y_3 \geq 0.$$

23. Management A furniture company makes tables and chairs. Union contracts require that the total number of tables and chairs produced must be at least 60 per week. Sales experience has shown that at least 1 table must be made for every 3 chairs that are made. If it costs $152 to make a table and $40 to make a chair, how many of each should be produced each week to minimize the cost. What is the minimum cost?

24. Sing, who is dieting, requires two food supplements, I and II. He can get these supplements from two different products, A and B, as shown in the following table.

		Supplement (grams per serving)	
		I	II
Product	A	3	2
	B	2	4

Sing's physician has recommended that he include at least 15 grams of each supplement in his daily diet. If product A costs 25¢ per serving and product B costs 40¢ per serving, how can he satisfy his requirements most economically?

25. Management An animal food must provide at least 50 units of vitamins and 60 calories per serving. One gram of soymeal provides 2 units of vitamins and 5 calories. One gram of meat byproducts provides 6.5 units of vitamins and 3 calories. One gram of grain provides 5 units of vitamins and 10 calories. If a gram of soymeal costs 4¢, a gram of meat byproducts 5¢, and a gram of grain 5¢, what mixture of these three ingredients will provide the required vitamins and calories at minimum cost?

26. Management Brand X canners produce canned corn, beans, and carrots. Labor contracts require them to produce at least 1000 cases per month. Based on past sales, they should produce at least twice as many cases of corn as of beans. At least 340 cases of carrots must be produced to fulfill commitments to a major distributor. It costs $10 to produce a case of beans, $15 to produce a case of corn, and $25 to produce a case of carrots. How many cases of each vegetable should be produced to minimize costs?

27. Refer to the end of Section 7.7, to the text on minimizing the daily cost of feeds.
 (a) Find a combination of feeds that will cost $7.00 and give 7 units of A and 4 units of B.
 (b) Use the dual variables to predict the daily cost of feed if the requirements change to 5 units of A and 4 units of B. Find a combination of feeds to meet these requirements at the predicted price.

28. Management   A small toy manufacturing firm has 200 squares of felt, 600 ounces of stuffing, and 90 feet of trim available to make two types of toys, a small bear and a monkey. The bear requires 1 square of felt and 4 ounces of stuffing. The monkey requires 2 squares of felt, 3 ounces of stuffing, and 1 foot of trim. The firm makes $1 profit on each bear and $1.50 profit on each monkey. The linear program to maximize profit is

$$\begin{aligned}\text{Maximize} \quad & x_1 + 1.5x_2 = z \\ \text{subject to:} \quad & x_1 + 2x_2 \le 200 \\ & 4x_1 + 3x_2 \le 600 \\ & x_2 \le 90.\end{aligned}$$

The final simplex tableau is

$$\begin{bmatrix} 0 & 1 & .8 & -.2 & 0 & 40 \\ 1 & 0 & -.6 & .4 & 0 & 120 \\ 0 & 0 & -.8 & .2 & 1 & 50 \\ 0 & 0 & .6 & .1 & 0 & 180 \end{bmatrix}$$

 (a) What is the corresponding dual problem?
 (b) What is the optimal solution to the dual problem?
 (c) Use the shadow values to estimate the profit the firm will make if their supply of felt increases to 210 squares.

 (d) How much profit will the firm make if their supply of stuffing is cut to 590 ounces and their supply of trim is cut to 80 feet?

29. Refer to the problem about the farmer, solved at the beginning of Section 7.5.
 (a) Give the dual problem.
 (b) Use the shadow values to estimate the farmer's profit if land is cut to 90 acres but capital increases to $21,000.
 (c) Suppose the farmer has 110 acres but only $19,000. Find the optimum profit and the planting strategy that will produce this profit.

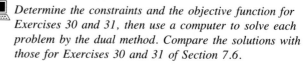 *Determine the constraints and the objective function for Exercises 30 and 31, then use a computer to solve each problem by the dual method. Compare the solutions with those for Exercises 30 and 31 of Section 7.6.*

30. Management   Natural Brand plant food is made from three chemicals. In a batch of the plant food there must be at least 81 kilograms of the first chemical and the other two chemicals must be in the ratio of 4 to 3. If the three chemicals cost $1.09, $.87, and $.65 per kilogram, respectively, how much of each should be used to minimize the cost of producing at least 750 kilograms of the plant food?

31. Management   A company is developing a new additive for gasoline. The additive is a mixture of three liquid ingredients, I, II, III. For proper performance, the total amount of additive must be at least 10 ounces per gallon of gasoline. However, for safety reasons, the amount of additive should not exceed 15 ounces per gallon of gasoline. At least 1/4 ounce of ingredient I must be used for every ounce of ingredient II and at least 1 ounce of ingredient III must be used for every ounce of ingredient I. If the cost of I, II, and III is $.30, $.09, and $.27 per ounce, respectively, find the mixture of the three ingredients that produces the minimum cost of the additive. How much of the additive should be used per gallon of gasoline?

KEY TERMS AND SYMBOLS

	linear inequality	slack variable
7.1	boundary	initial simplex tableau
	half-plane	indicator
	system of inequalities	basic variables
	region of feasible solutions	basic feasible solutions
	(feasible region)	pivot
7.2	objective function	**7.6** surplus variable
	linear programming	phase I
	constraints	phase II
	corner point	standard minimum form
	corner point theorem	
	bounded feasible region	**7.7** dual
	unbounded feasible region	transpose of a matrix
		theorem of duality
7.4	standard maximum form	shadow costs

KEY CONCEPTS

Graphing a Linear Inequality	Graph the boundary line as a solid line if the inequality includes ''or equal'', a broken line otherwise. Shade the half-plane that includes a test point that makes the inequality true. The graph of a system of inequalities, called the **region of feasible solutions,** includes all points that satisfy all the inequalities of the system at the same time.
Solving Linear Programming Problems	**Graphically:** Determine the objective function and all necessary constraints. Graph the region of feasible solutions. The maximum or minimum value will occur at one or more of the corner points of this region.
	Simplex Method: Determine the objective function and all necessary constraints. Convert each constraint into an equation by adding slack variables. Set up the initial simplex tableau. Locate the most negative indicator. Form the quotients to determine the pivot. Use row operations to change the pivot to 1 and all other numbers in that column to 0. If the indicators are all positive or 0, this is the final tableau. If not, choose a new pivot and repeat the process until no indicators are negative. Read the solution from the final tableau. The optimum value of the objective function is the number in the lower right corner of the final tableau. For problems with **mixed constraints,** add surplus variables as well as slack variables. In phase I, use row operations to transform the matrix until the solution is feasible. In phase II, use the simplex method as described above. For **minimum** problems, let the objective function be w and set $-w = z$. Then proceed as with mixed constraints.
Solving Minimum Problems With Duals	Find the dual maximum problem. Solve the dual using the simplex method. The minimum value of the objective function w is the maximum value of the dual objective function z. The optimal solution is found in the entries in the bottom row of the columns corresponding to the slack variables.

Graph each of the following linear inequalities.

1. $y \geq 2x + 3$ **2.** $3x - y \leq 5$ **3.** $3x + 4y \leq 12$ **4.** $2x - 6y \geq 18$ **5.** $y \geq x$ **6.** $y \leq 3$

Graph the solution of each of the following systems of inequalities.

7. $x + y \leq 6$
$2x - y \geq 3$

8. $4x + y \geq 8$
$2x - 3y \leq 6$

9. $-4 \leq x \leq 2$
$-1 \leq y \leq 3$
$x + y \leq 4$

10. $2 \leq x \leq 5$
$1 \leq y \leq 7$
$x - y \leq 3$

11. $x + 3y \geq 6$
$4x - 3y \leq 12$
$x \geq 0$
$y \geq 0$

12. $x + 2y \leq 4$
$2x - 3y \leq 6$
$x \geq 0$
$y \geq 0$

Set up a system of inequalities for each of the following problems; then graph the region of feasible solutions.

13. A bakery makes both cakes and cookies. Each batch of cakes requires 2 hours in the oven and 3 hours in the decorating room. Each batch of cookies needs $1\frac{1}{2}$ hours in the oven and 2/3 of an hour in the decorating room. The oven is available no more than 15 hours a day, while the decorating room can be used no more than 13 hours a day.

14. A company makes two kinds of pizza, basic and plain. Basic contains cheese and beef, while plain contains onions and beef. The company sells at least 3 units a day of basic, and at least 2 units of plain. The beef costs $5 per unit for basic, and $4 per unit for plain. They can spend no more than $50 per day on beef. Dough for basic is $2 per unit, while dough for plain is $1 per unit. The company can spend no more than $16 per day on dough.

Use the given regions to find the maximum and minimum values of the objective function
$z = 2x + 4y.$

15.

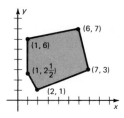

16.

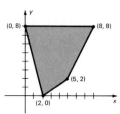

Use the graphical method to solve Exercises 17–22.

17. Maximize $z = 2x + 4y$
     subject to: $3x + 2y \leq 12$
                $5x + y \geq 5$
                $x \geq 0, \quad y \geq 0.$

18. Find $x \geq 0$ and $y \geq 0$ such that
$$8x + 9y \geq 72$$
$$6x + 8y \geq 72$$
and $w = 3x + 2y$ is minimized.

19. Find $x \geq 0$ and $y \geq 0$ such that
$$x + y \leq 50$$
$$2x + y \geq 20$$
$$x + 2y \geq 30$$
and $w = 4x + 2y$ is minimized.

20. Maximize $z = 4x + 3y$
     subject to: $2x + 7y \leq 14$
                $2x + 3y \leq 10$
                $x \geq 0, \quad y \geq 0.$

21. How many batches of cakes and cookies should the bakery of Exercise 13 make in order to maximize profits if cookies produce a profit of $20 per batch and cakes produce a profit of $30 per batch?

22. How many units of each kind of pizza should the company of Exercise 14 make in order to maximize profits if basic sells for $20 per unit and plain for $15 per unit?

For Exercises 23–26, (a) select appropriate variables, (b) write the objective function, (c) write the constraints as inequalities.

23. Roberta Hernandez sells three items, A, B, and C, in her gift shop. Each unit of A costs her $5 to buy, $1 to sell, and $2 to deliver. For each unit of B, the costs are $3, $2, and $1, respectively, and for each unit of C the costs are $6, $2, and $5, respectively. The profit on A is $4, on B it is $3, and on C, $3. How many of each should she get to maximize her profit if she can spend $1200 to buy, $800 on selling costs, and $500 on delivery costs?

24. An investor is considering three types of investment: a high-risk venture into oil leases with a potential return of 15%, a medium-risk investment in bonds with a 9% return, and a relatively safe stock investment with a 5% return. He has $50,000 to invest. Because of the risk, he will limit his investment in oil leases and bonds to 30% and his investment in oil leases and stock to 50%. How much should he invest in each to maximize his return, assuming investment returns are as expected?

25. The Aged Wood Winery makes two white wines, Fruity and Crystal, from two kinds of grapes and sugar. The wines require the following amounts of each ingredient per gallon and produce a profit per gallon as shown below.

	Grape A (bushels)	Grape B (bushels)	Sugar (pounds)	Profit (dollars)
Fruity	2	2	2	12
Crystal	1	3	1	15

The winery has available 110 bushels of grape A, 125 bushels of grape B, and 90 pounds of sugar. How much of each wine should be made to maximize profit?

26. A company makes three sizes of plastic bags: 5 gallon, 10 gallon and 20 gallon. The production time in hours for cutting, sealing, and packaging a unit of each size is shown below.

Size	Cutting	Sealing	Packaging
5 gallon	1	1	2
10 gallon	1.1	1.2	3
20 gallon	1.5	1.3	4

There are at most 8 hours available each day for each of the three operations. If the profit on a unit of 5-gallon bags is $1, 10-gallon bags is $.90, and 20-gallon bags is $.95, how many of each size should be made per day to maximize the profit?

For each of the following problems, (a) add slack variables and (b) set up the first simplex tableau.

27. Maximize $z = 5x_1 + 3x_2$
subject to: $2x_1 + 5x_2 \leq 50$
           $x_1 + 3x_2 \leq 25$
        $4x_1 + x_2 \leq 18$
           $x_1 + x_2 \leq 12$
   $x_1 \geq 0, \quad x_2 \geq 0.$

28. Maximize $z = 25x_1 + 30x_2$
subject to: $3x_1 + 5x_2 \leq 47$
          $x_1 + x_2 \leq 25$
        $5x_1 + 2x_2 \leq 35$
        $2x_1 + x_2 \leq 30$
   $x_1 \geq 0, \quad x_2 \geq 0.$

29. Maximize $z = 5x_1 + 8x_2 + 6x_3$
subject to: $x_1 + x_2 + x_3 \leq 90$
      $2x_1 + 5x_2 + x_3 \leq 120$
      $x_1 + 3x_2 \leq 80$
   $x_1 \geq 0, \quad x_2 \geq 0, \quad x_3 \geq 0.$

30. Maximize $z = 2x_1 + 3x_2 + 4x_3$
subject to: $x_1 + x_2 + x_3 \leq 100$
      $2x_1 + 3x_2 \leq 500$
      $x_1 + 2x_3 \leq 350$
   $x_1 \geq 0, \quad x_2 \geq 0, \quad x_3 \geq 0.$

For each of the following, use the simplex method to solve the maximizing linear programming problems with initial tableaus as given.

31.

x_1	x_2	x_3	x_4	x_5	
1	2	3	1	0	28
2	4	1	0	1	32
−5	−2	−3	0	0	0

32.

x_1	x_2	x_3	x_4	
2	1	1	0	10
1	3	0	1	16
−2	−3	0	0	0

33.

x_1	x_2	x_3	x_4	x_5	x_6	
1	2	2	1	0	0	50
3	1	0	0	1	0	20
1	0	2	0	0	1	15
−5	−3	−2	0	0	0	0

34.

x_1	x_2	x_3	x_4	x_5	
3	6	1	0	0	28
1	1	0	1	0	12
2	1	0	0	1	16
−1	−2	0	0	0	0

Convert the following problems into maximization problems.

35. Minimize $w = 10y_1 + 15y_2$
subject to: $y_1 + y_2 \geq 17$
$5y_1 + 8y_2 \geq 42$
$y_1 \geq 0, \quad y_2 \geq 0.$

36. Minimize $w = 20y_1 + 15y_2 + 18y_3$
subject to: $2y_1 + y_2 + y_3 \geq 112$
$y_1 + y_2 + y_3 \geq 80$
$y_1 + y_2 \geq 45$
$y_1 \geq 0, \quad y_2 \geq 0, \quad y_3 \geq 0.$

37. Minimize $w = 7y_1 + 2y_2 + 3y_3$
subject to: $y_1 + y_2 + 2y_3 \geq 48$
$y_1 + y_2 \geq 12$
$y_3 \geq 10$
$3y_1 + y_3 \geq 30$
$y_1 \geq 0, \quad y_2 \geq 0, \quad y_3 \geq 0.$

Use the simplex method to solve the following mixed constraint problems.

38. Maximize $z = 2x_1 + 4x_2$
subject to: $3x_1 + 2x_2 \leq 12$
$5x_1 + x_2 \geq 5$
$x_1 \geq 0, \quad x_2 \geq 0.$

39. Minimize $w = 4y_1 + 2y_2$
subject to: $y_1 + y_2 \leq 50$
$2y_1 + y_2 \geq 20$
$y_1 + 2y_2 \geq 30$
$y_1 \geq 0, \quad y_2 \geq 0.$

The following tableaus are the final tableaus of minimizing problems solved by letting $w = -z$. Give the solution and the minimum value of the objective function for each problem.

40.

0	1	0	2	5	0	17
0	0	1	3	1	1	25
1	0	0	4	2	$\frac{1}{2}$	8
0	0	0	2	5	0	−427

41.

0	0	2	1	0	6	6	92
1	0	3	0	0	0	2	47
0	1	0	0	0	1	0	68
0	0	4	0	1	0	3	35
0	0	5	0	0	2	9	−1957

The following tableaus are the final tableaus of minimizing problems solved by the method of duals. State the solution and the minimum value of the objective function for each problem.

42.
$$\begin{bmatrix} 1 & 0 & 0 & 3 & 1 & 2 & | & 12 \\ 0 & 0 & 1 & 4 & 5 & 3 & | & 5 \\ 0 & 1 & 0 & -2 & 7 & -6 & | & 8 \\ 0 & 0 & 0 & 5 & 7 & 3 & | & 172 \end{bmatrix}$$

43.
$$\begin{bmatrix} 0 & 0 & 1 & 6 & 3 & 1 & | & 2 \\ 1 & 0 & 0 & 4 & -2 & 2 & | & 8 \\ 0 & 1 & 0 & 10 & 7 & 0 & | & 12 \\ 0 & 0 & 0 & 9 & 5 & 8 & | & 62 \end{bmatrix}$$

44.
$$\begin{bmatrix} 1 & 0 & 7 & -1 & | & 100 \\ 0 & 1 & 1 & 3 & | & 27 \\ 0 & 0 & 7 & 2 & | & 640 \end{bmatrix}$$

45. Solve Exercise 23.

46. Solve Exercise 24.

47. Solve Exercise 25.

48. Solve Exercise 26.

Solve the following minimization problems.

49. A contractor builds boathouses in two basic models, the atlantic and pacific. Each atlantic model requires 1000 feet of framing lumber, 3000 cubic feet of concrete, and $2000 for advertising. Each pacific model requires 2000 feet of framing lumber, 3000 cubic feet of concrete, and $3000 for advertising. Contracts call for using at least 8000 feet of framing lumber, 18,000 cubic feet of concrete, and $15,000 worth of advertising. If the total spent on each atlantic model is $3000 and the total spent on each pacific model is $4000, how many of each model should be built to minimize costs?

50. A steel company produces two types of alloys. A run of type I requires 3000 pounds of molybdenum and 2000 tons of iron ore pellets as well as $2000 in advertising. A run of type II requires 3000 pounds of molybdenum and 1000 tons of iron ore pellets as well as $3000 in advertising. Total costs are $15,000 on a run of type I and $6000 on a run of type II. Because of various contracts, the company must use at least 18,000 pounds of molybdenum and 7000 tons of iron ore pellets and spend at least $14,000 on advertising. How much of each type should be produced to minimize costs?

Individuals doing the same job within the management of a company often receive different salaries. These salaries may differ because of the length of service of an employee, the productivity of an individual worker, and so on. However, for each job there is usually an established minimum and maximum salary.

Many companies make annual reviews of the salary of each of their management employees. At these reviews, an employee may receive a general cost of living increase, an increase based on merit, both, or neither.

In this case, we look at a mathematical model for distributing merit increases in an optimum way. An individual who is due for salary review may be described as shown in the figure below. Here i represents the number of the employee whose salary is being reviewed.

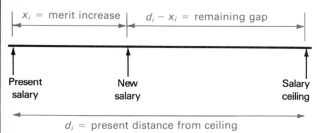

d_i = present distance from ceiling

Here the salary ceiling is the maximum salary for the job classification, x_i is the merit increase to be awarded to the individual ($x_i \geq 0$), d_i is the present distance of the current salary from the salary ceiling, and the difference, $d_i - x_i$, is the remaining gap.

We let w_i be a measure of the relative worth of the individual to the company. This is the most difficult variable of the model to actually calculate. One way to evaluate w_i is to give a rating sheet to a number of co-workers and supervisors of employee i. An average rating can be obtained and then divided by the highest rating received by any employee with that same job. This number then gives the worth of employee i relative to all other employees with that job.

The best way to allocate money available for merit pay increases is to minimize

$$\sum_{i=1}^{n} w_i(d_i - x_i).$$

This sum is found by multiplying the relative worth of employee i and the salary gap of employee i after any merit increase. Here n represents the total number of employees who have this job. One constraint here is that the total of all merit increases cannot exceed P, the total amount available for merit increases. That is,

$$\sum_{i=1}^{n} x_i \leq P.$$

Also, the increases for an employee must not put that employee over the maximum salary. That is, for employee i,

$$x_i \leq d_i.$$

We can simplify the objective function, using rules from algebra.

$$\sum_{i=i}^{n} w_i(d_i - x_i) = \sum_{i=1}^{n} (w_i d_i - w_i x_i)$$

$$= \sum_{i=1}^{n} w_i d_i - \sum_{i-1}^{n} w_i x_i$$

*Based on part of a paper by Jack Northam, Head, Mathematical Services Department, The Upjohn Company, Kalamazoo, Michigan.

For a given individual, w_i and d_i are constant. Therefore, the sum of the $w_i d_i$ is some constant, say Z, and

$$\sum_{i=i}^{n} w_i(d_i - x_i) = Z - \sum_{i=1}^{n} w_i x_i.$$

We want to minimize the sum on the left; we can do so by *maximizing* the sum on the right. (Why?) Thus, the original model simplifies to maximizing.

$$\sum_{i=1}^{n} w_i x_i,$$

subject to

$$\sum_{i=1}^{n} x_i \leq P \quad \text{and} \quad x_i \leq d_i$$

for each i.

EXERCISES

Here are current salary information and job evaluation averages for six employees who have the same job. The salary ceiling is $1700 per month.

Employee Number	Evaluation Average	Current Salary
1	570	$1600
2	500	1550
3	450	1500
4	600	1610
5	520	1530
6	565	1420

1. Find w_i for each employee by dividing that employee's evaluation average by the highest evaluation average.

 2. Use the simplex method to find the merit increase for each employee. Assume that P is 400.

3. What are some of the limitations and advantages of solving this problem by a strictly mathematical approach?

The first step in the commercial manufacture of ice cream is to blend several ingredients (such as dairy products, eggs, and sugar) to obtain a mix that meets the necessary minimum quality restrictions regarding butterfat content, serum solids, and so on.

Usually, many different combinations of ingredients may be blended to obtain a mix of the necessary quality. Within this range of possible substitutes, the firm desires the combination that produces minimum total cost. This problem is quite suitable for solution by linear programming methods.

A "mid-quality" line of ice cream requires the following minimum percentages of constituents by weight.

Fat	16.00%
Serum solids	8.00
Sugar solids	16.00
Egg solids	.35
Stabilizer	.25
Emulsifier	.15
Total	40.75%

The balance of the mix is water. A batch of mix is made by blending a number of ingredients, each of which contains one or more of the necessary constituents, and nothing else. The chart on the facing page shows the possible ingredients and their costs.

In setting up the mathematical model, use $c_1, c_2, \ldots, c_{14}$ as the cost per unit of ingredients, and $x_1, x_2, \ldots, x_{14}$ for the quantities of ingredients. The table shows the composition of each ingredient. For example, 1 pound of ingredient 1 (the 40% cream) contains .400 pounds of fat and .054 pounds of serum solids, with the balance being water.

The ice cream mix is made up in batches of 100 pounds at a time. For a batch to contain 16% fat, at least $100(.16) = 16$

*From *Linear Programming in Industry: Theory and Application, An Introduction* by Sven Danø, Fourth revised and enlarged edition. Copyright © 1974 by Springer-Verlag/Wein. Reprinted by permission of Springer-Verlag New York, Inc.

pounds of fat is necessary. Fat is contained in ingredients 1, 2, 3, 4, 5, 6, 10, and 11. The requirement of at least 16 pounds of fat produces the constraint

$$.400x_1 + .230x_2 + .805x_3 + .800x_4 + .998x_5 + .040x_6 + .500x_{10} + .625x_{11} \geq 16.$$

Similar constraints can be obtained for the other constituents. Because of the minimum requirements, the total of the ingredients will be at least 40.75 pounds, or

$$x_1 + x_2 + \cdots + x_{13} \geq 40.75.$$

The balance of the 100 pounds is water, making $x_{14} \leq 59.25$.

The table shows that ingredients 12 and 13 must be used—there is no alternate way of getting these constituents into the final mix. For a 100-pound batch of ice cream, $x_{12} = .25$ pound and $x_{13} = .15$ pound. Removing x_{12} and x_{13} as variables permits a substantial simplification of the model; the problem is now reduced to the following system of four constraints:

$$.400x_1 + .230x_2 + .805x_3 + .800x_4 + .998x_5 + .040x_6 + .500x_{10} + .625x_{11} \geq 16$$
$$.054x_1 + .069x_2 + .025x_4 + .078x_6 + .280x_7 + .970x_8 \geq 8$$
$$.707x_9 + .100x_{10} \geq 16$$
$$.350x_{10} + .315x_{11} \geq .35.$$

The objective function, which is to be minimized, is

$$c = .298x_1 + .174x_2 + \cdots + 1.090x_{11}.$$

As usual, $x_1 \geq 0, x_2 \geq 0, \ldots, x_{14} \geq 0$.

Solving this linear programming problem with the simplex method gives

$$x_2 = 67.394, \quad x_8 = 3.459, \quad x_9 = 22.483, \quad x_{10} = 1.000.$$

The total cost of these ingredients is $14.094. To this, we must add the cost of ingredients 12 and 13, producing a total cost of

$$14.094 + (.600)(.25) + (.420)(.15) = 14.307,$$

or $14.307.

	Ingredients								
	1	*2*	*3*	*4*	*5*	*6*	*7*	*8*	*9*
	40% cream	*23% cream*	*Butter*	*Plastic cream*	*Butter oil*	*4% milk*	*Skim condensed milk*	*Skim milk powder*	*Liquid sugar*
Cost ($/lb)									
	.298	.174	.580	.576	.718	.045	.052	.165	.061
Constituents 1. Fat	.400	.230	.805	.800	.998	.040			
2. Serum solids	.054	.069		.025		.078	.280	.970	
3. Sugar solids									.707
4. Egg solids									
5. Stabilizer									
6. Emulsifier									

	Ingredients					
	10	*11*	*12*	*13*	*14*	
	Sugared egg yolk	*Powdered egg yolk*	*Stabilizer*	*Emulsifier*	*Water*	
Cost ($/lb)						
	.425	1.090	.600	.420	0	*Requirements*
Constituents 1. Fat	.500	.625				16.00
2. Serum solids						8.00
3. Sugar solids	.100					16.00
4. Egg solids	.350	.315				.35
5. Stabilizer			1			.25
6. Emulsifier				1		.15

EXERCISES

1. Find the mix of ingredients needed for a 100-pound batch of the following grades of ice cream.

  a. "Generic," sold in a white box, with at least 14% fat and at least 6% serum solids (all other ingredients the same)

  b. "Premium," sold with a funny foreign name, with at least 17% fat and at least 16.5% sugar solids (all other ingredients the same)

2. Solve the problem given in this case by using a computer.

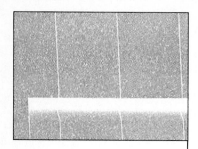

CHAPTER 8

Sets and Probability

This chapter and the next explore the basic concepts of probability theory, an area of mathematics that has become increasingly important in management and in the biological and social sciences. Probability theory allows us to deal with situations that involve uncertainty, as typified by statements like these.

There is a 30% chance of rain tomorrow.

The vaccine is effective with 99% of the patients.

The probability of an earthquake has increased.

The basic concepts of probability are discussed in this chapter and applications of probability are discussed in the next chapter.

Sets and operations on sets are the basic tools for studying probability, so we begin with them.

8.1 SETS

Think of a **set** as a collection of objects. For example, a set of coins might contain one of each type of coin now put out by the United States government. Another set might include all the students in a class. The set containing the numbers 3, 4, and 5 is written

$$\{3, 4, 5\},$$

where set braces, { }, are used to enclose the numbers belonging to the set. The numbers, 3, 4, and 5 are called the **elements** or **members** of this set. To show that 4 is an element of the set {3, 4, 5}, use the symbol $\in$ and write

$$4 \in \{3, 4, 5\},$$

read ''4 is an element of the set containing 3, 4, and 5.''

1 Write *true* or *false*.

(a) $9 \in \{8, 4, -3, -9, 6\}$

(b) $4 \notin \{3, 9, 7\}$

(c) If $M = \{0, 1, 2, 3, 4\}$, then $0 \in M$.

Answer:

(a) False

(b) True

(c) True

2 List the elements in the following sets.

(a) $\{x \mid x$ is a counting number more than 5 and less than 8$\}$

(b) $\{x \mid x$ is an integer, $-3 < x \le 1\}$

Answer:

(a) $\{6, 7\}$

(b) $\{-2, -1, 0, 1\}$

Also, $5 \in \{3, 4, 5\}$. Place a slash through the symbol $\in$ to show that 8 is *not* an element of this set.

$$8 \notin \{3, 4, 5\}$$

This is read "8 is not an element of the set $\{3, 4, 5\}$."

Sets are often named with capital letters, so that if

$$B = \{5, 6, 7\},$$

then, for example, $6 \in B$ and $10 \notin B$. **1**

Two sets are **equal** if they contain exactly the same elements. The sets $\{5, 6, 7\}$, $\{7, 6, 5\}$, and $\{6, 5, 7\}$ all contain exactly the same elements and are equal. In symbols,

$$\{5, 6, 7\} = \{7, 6, 5\} = \{6, 5, 7\}.$$

Sets that do not contain exactly the same elements are *not equal*. For example, the sets $\{5, 6, 7\}$ and $\{5, 6, 7, 8\}$ do not contain exactly the same elements and are not equal. We show this by writing

$$\{5, 6, 7\} \ne \{5, 6, 7, 8\}.$$

Sometimes we describe a set by a common property of its elements rather than by a list of elements. This common property can be expressed with **set-builder notation;** for example,

$$\{x \mid x \text{ has property } P\}$$

(read "the set of all elements x such that x has property P") represents the set of all elements x having some property P.

▶ **EXAMPLE 1** Write the elements belonging to each of the following sets.
(a) $\{x \mid x$ is a natural number less than 5$\}$
   The natural numbers less than 5 make up the set $\{1, 2, 3, 4\}$.
(b) $\{x \mid x$ is a state that touches Florida$\} = \{$Alabama, Georgia$\}$ ◀ **2**

The **universal set** in a particular discussion is a set that contains all of the objects being discussed. In grade-school arithmetic, for example, the set of whole numbers might be the universal set, whereas in a college calculus class the universal set might be the set of all real numbers. When it is necessary to consider the universal set being used, it will be clearly specified or easily understood from the context of the problem.

3 Write *true* or *false*.

(a) $\{3, 4, 5\} \subseteq \{2, 3, 4, 6\}$

(b) $\{1, 2, 5, 8\} \subseteq$
$\{1, 2, 5, 8, 10, 11\}$

(c) $\{3, 6, 9, 10\} \subseteq$
$\{3, 9, 11, 13\}$

Answer:

(a) False

(b) True

(c) False

Sometimes every element of one set also belongs to another set. For example, if

$$A = \{3, 4, 5, 6\}$$

and

$$B = \{2, 3, 4, 5, 6, 7, 8\},$$

then every element of A is also an element of B. This is an example of the following definition.

A set A is a **subset** of a set B (written $A \subseteq B$) provided that every element of A is also an element of B.

▶**EXAMPLE 2**   **(a)** $\{5, 6, 9, 10\} \subseteq \{5, 6, 7, 8, 9, 10, 11\}$ because every element of the first set is also an element of the second.
(b) The set $A = \{2, 3, 4\}$ is *not* a subset of the set $B = \{3, 4, 5, 6, 7\}$ because one element of A (namely, 2) is not an element of B. To indicate that A is not a subset of B we write $A \nsubseteq B$. ◀   **3**

Sometimes a set has no elements. Some examples are the set of woman presidents of the United States, the set of counting numbers less than 1, and the set of men more than 10 feet tall. A set with no elements is called the **empty set.** The symbol $\emptyset$ is used to represent the empty set.

Every set A is a subset of itself because the statement "every element of A is also an element of A" is always true. It is also true that the empty set is a subset of every set.*

For any set A,

$$\emptyset \subseteq A \quad \text{and} \quad A \subseteq A.$$

Caution   Be careful to distinguish between the symbols 0, $\emptyset$, and $\{0\}$. The symbol 0 represents a *number;* $\emptyset$ represents a *set* with no elements; and $\{0\}$ represents a *set* with one element, the number 0. Also, note that although $\emptyset$ is sometimes used for the number zero in computer science, in mathematics it always represents an empty set.

---

*This fact is not intuitively obvious to most people. If you wish, you can think of it as a convention that we agree to adopt in order to simplify the statements of several results later.

4 List all subsets of
{w, x, y, z}.

Answer:
∅, {w}, {x}, {y}, {z}, {w, x},
{w, y}, {w, z}, {x, y}, {x, z},
{y, z}, {w, x, y}, {w, x, z},
{w, y, z}, {x, y, z}, {w, x, y, z}

▶**EXAMPLE 3** List all possible subsets for each of the following sets.
(a) {7, 8}

There are 4 subsets of {7, 8}:

$$∅, \quad \{7\}, \quad \{8\}, \quad \{7, 8\}.$$

(b) {a, b, c}

There are 8 subsets of {a, b, c}:

$$∅, \quad \{a\}, \quad \{b\}, \quad \{c\}, \quad \{a, b\}, \quad \{a, c\}, \quad \{b, c\}, \quad \{a, b, c\}. \quad ◀$$

In Example 3, the subsets of {7, 8} and the subsets of {a, b, c} were founded by trial and error. An alternative method uses a **tree diagram,** a systematic way of listing all the subsets of a given set. Figures 8.1(a) and (b) show tree diagrams for finding the subsets of {7, 8} and {a, b, c}. **4**

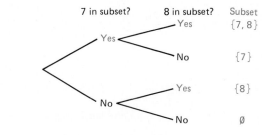

(a) Find the subsets of {7, 8}

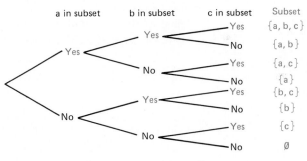

(b) Find the subsets of {a, b, c}

FIGURE 8.1

By using the fact that there are two possibilities for each element (either it's in the subset or it's not) we have found that a set with 2 elements has 4 ($=2^2$) subsets and a set with 3 elements has 8 ($=2^3$) subsets. Similar arguments work for any finite set and lead to this conclusion.

A set of n distinct elements has 2^n subsets.

5 Find the number of subsets for each of the following sets.

(a) {1, 2, 3}

(b)

{−6, −5, −4, −3, −2, −1, 0}

(c) {6}

Answer:

(a) 8   **(b)** 128   **(c)** 2

6 Refer to sets A, B, C, and U in the diagram.

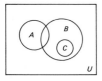

(a) Is $A \subseteq B$?

(b) Is $C \subseteq B$?

(c) Is $C \subseteq U$?

(d) Is $\emptyset \subseteq A$?

Answer:

(a) No

(b) Yes

(c) Yes

(d) Yes

7 Let U = {a, b, c, d, e, f, g}, with K = {c, d, f, g} and R = {a, c, d, e, g}. Find

(a) K';

(b) R'.

Answer:

(a) {a, b, e}

(b) {b, f}

▶**EXAMPLE 4** Find the number of subsets for each of the following sets.

(a) {3, 4, 5, 6, 7}

Since this set has 5 elements, it has 2^5 or 32 subsets.

(b) {−1, 2, 3, 4, 5, 6, 12, 14}

This set has 8 elements and therefore has 2^8 or 256 subsets.

(c) $\emptyset$

Since the empty set has 0 elements, it has $2^0 = 1$ subset, $\emptyset$ itself. ◀ **5**

Figure 8.2 shows a set A which is a subset of a set B, since A is entirely in B. The rectangle represents the universal set, U. Such diagrams, called **Venn diagrams,** are used to illustrate relationships among sets. **6**

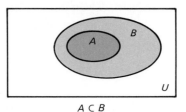

$$A \subset B$$

FIGURE 8.2

Given a set A and a universal set U, the set of all elements of U which do *not* belong to A is called the **complement** of set A. For example, if set A is the set of all the female students in your class and U is the set of all students in the class, then the complement of A would be the set of all male students in the class. The complement of set A is written A' (read ''A-prime''). The Venn diagram of Figure 8.3 shows a set B. Its complement, B', is shown in color.

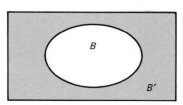

FIGURE 8.3

▶**EXAMPLE 5** Let U = {1, 2, 3, 4, 5, 6, 7}, A = {1, 3, 5, 7}, and B = {3, 4, 6}. Find the each of the following sets.

(a) A'

Set A' contains the elements of U that are not in A.

$$A' = \{2, 4, 6\}$$

(b) $B' = \{1, 2, 5, 7\}$

(c) $\emptyset' = U$ and $U' = \emptyset$ ◀ **7**

8 Find the following.

(a) {1, 2, 3, 4} ∩
{3, 5, 7, 9}

(b) {a, b, c, d, e, g} ∩
{a, c, e, g, h, j}

Answer:

(a) {3}

(b) {a, c, e, g}

9 Find the following.

(a) {a, b, c} ∪ {a, c, e}

(b) {a, c, d, e, f} ∪
{a, b, c, d, g}

Answer:

(a) {a, b, c, e}

(b) {a, b, c, d, e, f, g}

Given two sets *A* and *B*, the set of all elements belonging to *both* set *A* and set *B* is called the **intersection** of the two sets, written *A* ∩ *B*. For example, the elements that belong to both *A* = {1, 2, 4, 5, 7} and *B* = {2, 4, 5, 7, 9, 11} are 2, 4, 5, and 7, so that

$$A \cap B = \{1, 2, 4, 5, 7\} \cap \{2, 4, 5, 7, 9, 11\} = \{2, 4, 5, 7\}.$$

The Venn diagram of Figure 8.4 shows two sets *A* and *B* with their intersection, *A* ∩ *B*, shown in color.

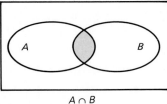

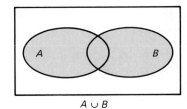

A ∩ B   A ∪ B

FIGURE 8.4   **FIGURE 8.5**

▶ **EXAMPLE 6**   **(a)** {9, 15, 25, 36} ∩ {15, 20, 25, 30, 35} = {15, 25}
The elements 15 and 25 are the only ones belonging to both sets.
(b) {−2, −3, −4, −5, −6} ∩ {−4, −3, −2, −1, 0, 1, 2} = {−4, −3, −2}
◀ **8**

Two sets that have no elements in common are called **disjoint sets.** For example, there are no elements common to both {50, 51, 54} and {52, 53, 55, 56}, so that these two sets are disjoint, and

$$\{50, 51, 54\} \cap \{52, 53, 55, 56\} = \emptyset.$$

The result of this example can be generalized: For any sets *A* and *B*,

if *A* and *B* are disjoint sets then *A* ∩ *B* = ∅.

The set of all elements belonging to set *A*, to set *B*, or to both sets is called the **union** of the two sets, written *A* ∪ *B*. For example,

$$\{1, 3, 5\} \cup \{3, 5, 7, 9\} = \{1, 3, 5, 7, 9\}.$$

The Venn diagram of Figure 8.5 shows two sets *A* and *B*, with their union *A* ∪ *B* shown in color.

▶ **EXAMPLE 7**   **(a)** Find the union of {1, 2, 5, 9, 14} and {1, 3, 4, 8}.
Begin by listing the elements of the first set, {1, 2, 5, 9, 14}. Then include any elements from the second set *that are not already listed*. Doing this gives

$$\{1, 2, 5, 9, 14\} \cup \{1, 3, 4, 8\} = \{1, 2, 3, 4, 5, 8, 9, 14\}.$$

(b) {1, 3, 5, 7} ∪ {2, 4, 6} = {1, 2, 3, 4, 5, 6, 7}. ◀ **9**

Finding the complement of a set, the intersection of two sets, or the union of two sets are examples of *set operations*. These set operations are ways of combining sets to get other sets, just as operations on numbers, such as addition, subtraction and multiplication usually produce other numbers.

The set operations are summarized below.

Operations on Sets

Let A and B be any sets with U the universal set. Then

the **complement** of A, written A', is

$$A' = \{x \mid x \notin A \text{ and } x \in U\};$$

the **intersection** of A and B is

$$A \cap B = \{x \mid x \in A \text{ and } x \in B\};$$

the **union** of A and B is

$$A \cup B = \{x \mid x \in A \text{ or } x \in B\}.$$

Caution As shown in the definitions given above, an element is in the *intersection* of sets A and B if it is in A *and* B. On the other hand, an element is in the *union* of sets A and B if it is in A *or* B (or both).

▶**EXAMPLE 8** The table below gives the current dividend, price to earnings ratio (the quotient of the price per share and the annual earnings per share), and price change at the end of a day for six companies, as listed on the New York Stock Exchange.

Stock	Dividend	Price to Earnings Ratio	Price Change
ATT	1.20	17	$-\frac{1}{2}$
GE	2.32	14	$+1\frac{3}{4}$
Hershey	1.5	13	$-1\frac{1}{8}$
IBM	4.4	16	$+3\frac{3}{8}$
Mobil	2.2	14	$-\frac{1}{2}$
PepsiCo	.64	17	$+\frac{1}{2}$

Let set A include all stocks with a dividend greater than \$2: B all stocks with a price to earnings ratio of at least 16, and C all stocks with a positive price change. Find the following.

(a) A'

Set A' contains all the listed stocks outside set A, those with a dividend less than or equal to \$2, so

$$A' = \{\text{ATT, Hershey, PepsiCo}\}.$$

10 If $U =$
$\{-3, -2, -1, 0, 1, 2, 3\}$,
$A = \{-1, -2, -3\}$, and
$B = \{-2, 0, 2\}$, find

(a) $A \cup B$;   **(b)** B';   **(c)** $A \cap B$.

Answer

(a) $\{-3, -2, -1, 0, 2\}$

(b) $\{-3, -1, 1, 3\}$ **(c)** $\{-2\}$

(b) $A \cap B$

The intersection of A and B will contain those stocks that offer a dividend greater than $2 *and* have a price to earnings ratio of at least 16.

$$A \cap B = \{\text{IBM}\}$$

(c) $A \cup C$

The union of A and C contains all stocks with a dividend greater than $2 or a positive price change (or both).

$$A \cup C = \{\text{GE, IBM, Mobil, PepsiCo}\} \quad \blacktriangleleft \quad \boxed{10}$$

8.1 EXERCISES

In each of the following, insert one of the symbols $\in$, $\notin$,
$=$, *or $\neq$ to make the statement true.*

1. 3 _____ $\{2, 5, 7, 9, 10\}$

2. 6 _____ $\{-2, 6, 9, 5\}$

3. 1 _____ $\{3, 4, 5, 1, 11\}$

4. 12 _____ $\{19, 17, 14, 13, 12\}$

5. 9 _____ $\{2, 1, 5, 8\}$

6. 3 _____ $\{7, 6, 5, 4\}$

7. $\{2, 5, 8, 9\}$ _____ $\{2, 5, 9, 8\}$

8. $\{3, 0, 9, 6, 2\}$ _____ $\{2, 9, 0, 3, 6\}$

9. $\{5, 8, 9\}$ _____ $\{5, 8, 9, 0\}$

10. $\{3, 7, 12, 14\}$ _____ $\{3, 7, 12, 14, 0\}$

11. {all counting numbers less than 6} _____ $\{1, 2, 3, 4, 5, 6\}$

12. {all whole numbers greater than 7 and less than 10} _____ $\{8, 9\}$

13. {all whole numbers not greater than 4} _____ $\{0, 1, 2, 3\}$

14. {all counting numbers not greater than 3} _____ $\{0, 1, 2\}$

15. $\{x \mid x$ is a whole number, $x \leq 5\}$ _____ $\{0, 1, 2, 3, 4, 5\}$

16. $\{x \mid x$ is an integer, $-3 \leq x < 4\}$
_____ $\{-3, -2, -1, 0, 1, 2, 3, 4\}$

17. $\{x \mid x$ is an odd integer, $6 \leq x \leq 18\}$
_____ $\{7, 9, 11, 15, 17\}$

18. $\{x \mid x$ is an even counting number, $x \leq 9\}$
_____ $\{0, 2, 4, 6, 8\}$

19. $\{5, 7, 9, 19\} \cap \{7, 9, 11, 15\}$ _____ $\{7, 9\}$

20. $\{8, 11, 15\} \cap \{8, 11, 19, 20\}$ _____ $\{8, 11\}$

21. $\{2, 1, 7\} \cup \{1, 5, 9\}$ _____ $\{1\}$

22. $\{6, 12, 14, 16\} \cup \{6, 14, 19\}$ _____ $\{6, 14\}$

23. $\{3, 2, 5, 9\} \cap \{2, 7, 8, 10\}$ _____ $\{2\}$

24. $\{8, 9, 6\} \cup \{9, 8, 6\}$ _____ $\{8, 9\}$

25. $\{3, 5, 9, 10\} \cap \emptyset$ _____ $\{3, 5, 9, 10\}$

26. $\{3, 5, 9, 10\} \cup \emptyset$ _____ $\{3, 5, 9, 10\}$

27. $\{1, 2, 4\} \cup \{1, 2, 4\}$ _____ $\{1, 2, 4\}$

28. $\{1, 2, 4\} \cap \{1, 2, 4\}$ _____ $\emptyset$

29. $\emptyset \cup \emptyset$ _____ $\emptyset$

30. $\emptyset \cap \emptyset$ _____ $\emptyset$

Let $A = \{2, 4, 6, 8, 10, 12\}$, $B = \{2, 4, 8, 10\}$, $C = \{4, 10, 12\}$, $D = \{2, 10\}$, and $U = \{2, 4, 6, 8, 10, 12, 14\}$. Insert $\subseteq$ or $\nsubseteq$ in the blank to make the statement true. (See Examples 2–3.)

31. A _____ U

32. C _____ U

33. D _____ B

34. D _____ A

35. A _____ B

36. B _____ C

37. $\emptyset$ _____ A

38. $\emptyset$ _____ $\emptyset$

39. $\{4, 8, 10\}$ _____ B

40. $\{0, 2\}$ _____ D

41. B _____ D

42. A _____ C

Insert a number in the blank that makes each statement true. (See Example 4.)

43. There are exactly _____ subsets of *A*.

44. There are exactly _____ subsets of *B*.

45. There are exactly _____ subsets of *C*.

46. There are exactly _____ subsets of *D*.

Find the number of subsets for each of the following sets. (See Examples 3 and 4.)

47. $\{4, 5, 6\}$

48. $\{3, 7, 9, 10\}$

49. $\{4, 9, 10, 15, 17\}$

50. $\{6, 9, 1, 4, 3, 2\}$

51. $\emptyset$

52. $\{0\}$

53. $\{x \mid x$ is a counting number between (not including) 6 and 12$\}$ **54.** $\{x \mid x$ is a whole number between (not including) 8 and 12$\}$

Let $U = \{2, 3, 4, 5, 7, 9\}$, $X = \{2, 3, 4, 5\}$, $Y = \{3, 5, 7, 9\}$, and $Z = \{2, 4, 5, 7, 9\}$. Find each of the following sets. (See Examples 5–7.)

55. $X \cap Y$

56. $X \cup Y$

57. $X \cup U$

58. $Y \cap U$

59. X'

60. Y'

61. $X' \cap Y'$

62. $X' \cap Z$

Let $U = \{a, b, c, d, e, f, 1, 2, 3, 4, 5, 6\}$, $A = \{a, c, e, 2, 4, 6\}$, $B = \{a, b, c, 1, 2, 3\}$, $C = \{d, e, f\}$, $D = \{c, f, 3, 5\}$. Find each of the following sets. (See Examples 5–7.)

63. (a) $C \cup D$
(b) $B \cup (C \cup D)$
(c) $B \cup C$
(d) $(B \cup C) \cup D$

64. (a) $B \cap C$
(b) $A \cap (B \cap C)$
(c) $A \cap B$
(d) $(A \cap B) \cap C$

65. (a) $B \cap (C \cup D)$
(b) $(B \cap C) \cup (B \cap D)$

66. (a) $A \cup (B \cap C)$
(b) $(A \cup B) \cap (A \cup C)$

67. (a) $(A \cup B)'$
(b) $A' \cup B'$
(c) $A' \cap B'$

68. (a) $(C \cap D)'$
(b) $C' \cap D'$
(c) $C' \cup D'$

Let $U = \{$all students in this school$\}$,
$M = \{$all students taking this course$\}$,
$N = \{$all students taking accounting$\}$,
$P = \{$all students taking zoology$\}$.

Describe each of the following sets in words.

69. M'

70. $M \cup N$

71. $N \cap P$

72. $N' \cap P'$

73. $M \cup P$

74. $P' \cup M'$

Natural Science *Work the following problems. (See Example 8.)*

75. The lists below show some symptoms of an overactive thyroid and an underactive thyroid.

Underactive Thyroid	Overactive Thyroid
Sleepiness, *s*	Insomnia, *i*
Dry hands, *d*	Moist hands, *m*
Intolerance of cold, *c*	Intolerance of heat, *h*
Goiter, *g*	Goiter, *g*

(a) Find the smallest possible universal set *U* that includes all the symptoms listed.

Let *N* be the set of symptoms for an underactive thyroid, and let *O* be the set of symptoms for an overactive thyroid. Find each of the following sets.

(b) O' **(c)** N' **(d)** $N \cap O$

(e) $N \cup O$ **(f)** $N \cap O'$

76. The following list includes the industries that produce most (93%) of the hazardous wastes in the United States.

Industry	% (in 1977)
Inorganic chemicals (1)	11
Organic chemicals (2)	34
Electroplating (3)	12
Petroleum refining (4)	5
Smelting and refining (5)	26
Textiles, dyeing and finishing (6)	5

Let x be the number associated with an industry in the table above. The universal set is $U = \{1, 2, 3, 4, 5, 6\}$. List the elements of the following subsets of U.

(a) $\{x \mid x$ is an industry that produces more than 15% of the wastes$\}$

(b) $\{x \mid x$ is an industry that produces less than 5% of the wastes$\}$

(c) Find all subsets of U containing industries that produce a total of at least 80% of the waste.

77. If A is a subset of the universal set U, then what is the set A''?

78. The elements of a set may be sets themselves, as in $\{1, \{1, 2\}, \{a, b, c\}, d\}$. Explain why the set $\{\emptyset\}$ is not the same set as $\emptyset$.

8.2 APPLICATIONS OF VENN DIAGRAMS

Venn diagrams were used in the last section to illustrate set union and intersection. The rectangular region in a Venn diagram represents the universal set, U. Including only a single set, A, inside the universal set, as in Figure 8.6, divides U into two nonoverlapping regions. Region 1 represents those elements outside set A, while region 2 represents those elements belonging to set A. (The numbering of these regions is arbitrary.)

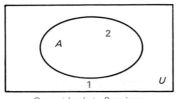

One set leads to 2 regions
(numbering is arbitrary)

FIGURE 8.6

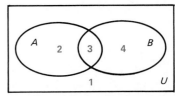

Two sets lead to 4 regions
(numbering is arbitrary)

FIGURE 8.7

The Venn diagram of Figure 8.7 shows two sets inside U. These two sets divide the universal set into four nonoverlapping regions. As labeled in Figure 8.7, region 1 includes those elements outside both set A and set B. Region 2 includes those elements belonging to A and not to B. Region 3 includes those elements belonging to both A and B. Which elements belong to region 4? (Again, the numbering is arbitrary.)

1 Draw Venn diagrams for the following.

(a) $A \cup B'$

(b) $A' \cap B'$

Answer:

(a)

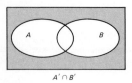

$A \cup B'$

(b)

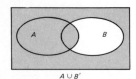

$A' \cap B'$

2 Draw a Venn diagram for the following.

(a) $B' \cap A$

(b) $(A \cup B)' \cap C$

Answer:

(a)

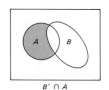

$B' \cap A$

(b)

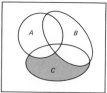

$(A \cup B)' \cap C$

▶ **EXAMPLE 1** Draw Venn diagrams similar to Figure 8.7 and shade the regions representing the following sets.

(a) $A' \cap B$

Set A' contains all the elements outside set A. As labeled in Figure 8.7, A' is represented by regions 1 and 4. Set B is represented by the elements in regions 3 and 4. The intersection of sets A' and B, the set $A' \cap B$, is given by the region common to regions 1 and 4 and regions 3 and 4. The result, region 4, is shaded in Figure 8.8.

(b) $A' \cup B'$

Again, set A' is represented by regions 1 and 4, and set B' by regions 1 and 2. To find $A' \cup B'$, identify the region that represents the set of all elements in A', B', or both. The result, which is shaded in Figure 8.9, includes regions 1, 2, and 4. ◀ **1**

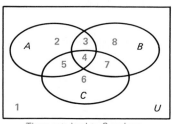

$A' \cap B$

FIGURE 8.8

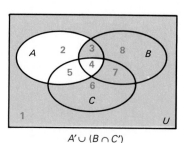

$A' \cup B'$

FIGURE 8.9

Venn diagrams also can be drawn with three sets inside U. These three sets divide the universal set into eight nonoverlapping regions which can be numbered (arbitrarily) as in Figure 8.10.

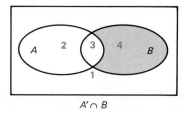

Three sets lead to 8 regions
(numbering is arbitrary)

FIGURE 8.10

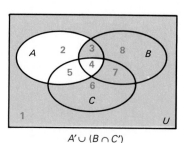

$A' \cup (B \cap C')$

FIGURE 8.11

▶ **EXAMPLE 2** Shade $A' \cup (B \cap C')$ on a Venn diagram.

First find $B \cap C'$. Set B is represented by regions 3, 4, 7, and 8, and C' by regions 1, 2, 3, and 8. The overlap of these regions, regions 3 and 8, represents the set $B \cap C'$. Set A' is represented by regions 1, 6, 7, and 8. The union of regions 3 and 8 and regions 1, 6, 7, and 8 includes regions 1, 3, 6, 7, and 8, which are shaded in Figure 8.11. ◀ **2**

3 Place numbers in the regions on a Venn diagram if the data on the 50 surfers showed

29 blondes;
17 with blue eyes;
10 blue-eyed blondes.

Answer:

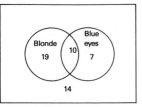

Venn diagrams can be used to solve problems that result from surveying groups of people. As an example, suppose a researcher collecting data on a group of 50 surfers found

32 were blonde,
28 had blue eyes,
15 were blue-eyed blondes.

The researcher wishes to use this data to answer the following questions.

(a) How many were not blonde?

(b) How many were neither blue-eyed nor blonde?

(c) How many were blonde but not blue-eyed?

Since some of the surfers are listed in both categories, some in only one category, and some in neither category, a Venn diagram like the one in Figure 8.12(a) can help to clarify the information. In Figure 8.12(b), place 15 in the region common to both blonde and blue eyes, since there are 15 surfers with both characteristics. Of the 32 blondes, 15 also have blue eyes, so place $32 - 15 = 17$ in the region for blonde but not blue-eyed. Since 15 of the 28 with blue eyes are also blonde, write $28 - 15 = 13$ in the region for blue-eyed but not blonde. Finally, of the 50 surfers, $50 - 17 - 15 - 13 = 5$ have neither blonde hair nor blue eyes, so put 5 in the region outside both circles. Now the questions can be answered:

(a) $13 + 5 = 18$ were not blonde;

(b) 5 were neither blue-eyed nor blonde;

(c) 17 were blonde but not blue-eyed. **3**

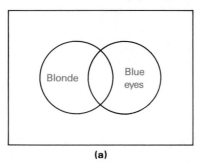

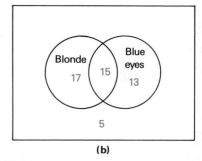

(a)   (b)

FIGURE 8.12

▶**EXAMPLE 3**   A group of 60 freshman business students at a large university was surveyed, with the following results.

19 of the students read *Business Week;*
18 read *The Wall Street Journal;*
50 read *Fortune;*
13 read *Business Week* and *The Journal;*
11 read *The Journal* and *Fortune;*
13 read *Business Week* and *Fortune;*
 9 read all three.

Use this data to answer the following questions.

(a) How many students read none of the publications?
(b) How many read only *Fortune?*
(c) How many read *Business Week* and *The Journal,* but not *Fortune?*

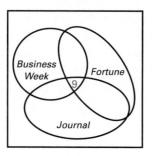

FIGURE 8.13(a)

Once again, use a Venn diagram to represent the data. Since 9 students read all three publications, begin by placing 9 in the area in Figure 8.13(a) that belongs to all three regions.

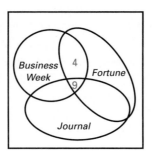

FIGURE 8.13(b)

Of the 13 students who read *Business Week* and *Fortune,* 9 also read *The Journal.* Therefore only 13 − 9 = 4 students read just *Business Week* and *Fortune.* So place a 4 in the region common only to *Business Week* and *Fortune* readers, as in Figure 8.13(b).

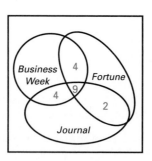

FIGURE 8.13(c)

In the same way, place a 4 in the region of Figure 8.13(c) common only to *Business Week* and *The Journal,* and 2 in the region common only to *Fortune* and *The Journal.*

4 In the example about the three publications, how many students read exactly

(a) 1;

(b) 2
of the publications?

Answer:

(a) 40

(b) 10

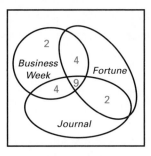

FIGURE 8.13(d)

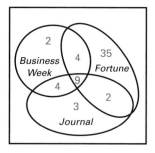

FIGURE 8.13(e)

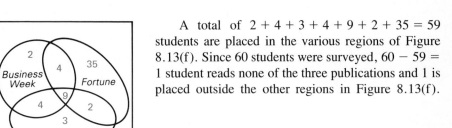

FIGURE 8.13(f)

The data shows that 19 students read *Business Week*. However, $4 + 9 + 4 = 17$ readers have already been placed in the *Business Week* region. The balance of this region in Figure 8.13(d) will contain only $19 - 17 = 2$ students. These 2 students read *Business Week* only—not *Fortune* and not *The Journal*.

In the same way, 3 students read only *The Journal* and 35 read only *Fortune,* as shown in Figure 8.13(e).

A total of $2 + 4 + 3 + 4 + 9 + 2 + 35 = 59$ students are placed in the various regions of Figure 8.13(f). Since 60 students were surveyed, $60 - 59 = 1$ student reads none of the three publications and 1 is placed outside the other regions in Figure 8.13(f).

Figure 8.13(f) can now be used to answer the questions asked above.

(a) Only 1 student reads none of the publications.

(b) There are 35 students who read only *Fortune*.

(c) The overlap of the regions representing *Business Week* and *The Journal* shows that 4 students read *Business Week* and *The Journal* but not *Fortune.* ◀ **4**

5 In Example 4, suppose 46 employees can cut tall trees. Then how many

(a) can only cut tall trees?

(b) can cut trees or climb poles?

(c) can cut trees or climb poles or splice wire?

Answer:

(a) 4

(b) 68

(c) 91

▶**EXAMPLE 4**   Jeff Friedman is a section chief for an electric utility company. The employees in his section cut down tall trees, climb poles, and splice wire. Friedman reported the following information to the management of the utility.

> Of the 100 employees in my section,
> 45 can cut tall trees;
> 50 can climb poles;
> 57 can splice wire;
> 28 can cut trees and climb poles;
> 20 can climb poles and splice wire;
> 25 can cut trees and splice wire;
> 11 can do all three;
> 9 can't do any of the three (management trainees).

The data supplied by Friedman lead to the numbers shown in Figure 8.14. Add the numbers from all the regions to get the total number of Friedman's employees.

$$9 + 3 + 14 + 23 + 11 + 9 + 17 + 13 = 99$$

Friedman claimed to have 100 employees, but his data indicates only 99. The management decided that Friedman didn't qualify as a section chief, and reassigned him as a nightshift meter reader in Guam. (Moral: he should have taken this course.) ◀ **5**

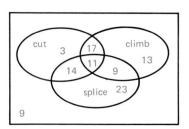

FIGURE 8.14

Caution   Note that in all the examples above, we started in the innermost region with the intersection of the three categories. This is usually the best way to begin solving problems of this type.

When dealing with a set A, we use the symbol $n(A)$ to denote the *number* of elements in A. For instance, if $A = \{w, x, y, z\}$, then $n(A) = 4$. The following useful fact is proved below.

$$n(A \cup B) = n(A) + n(B) - n(A \cap B)$$

6 If $n(A) = 10$, $n(B) = 7$, and $n(A \cap B) = 3$, find $n(A \cup B)$.

Answer:
14

For example, if $A = \{r, s, t, u, v\}$ and $B = \{r, t, w\}$, then $A \cap B = \{r, t\}$, so that $n(A) = 5$, $n(B) = 3$, and $n(A \cap B) = 2$. By the formula in the box, $n(A \cup B) = 5 + 3 - 2 = 6$, which is certainly true since $A \cup B = \{r, s, t, u, v, w\}$.

Here is a proof of the statement in the box: let x be the number of elements in A that are not in B, y the number of elements in $A \cap B$, and z the number of elements in B that are not in A, as indicated in Figure 8.15. That diagram shows that $n(A \cup B) = x + y + z$. It also shows that $n(A) = x + y$ and $n(B) = y + z$, so that

$$n(A) + n(B) - n(A \cap B) = (x + y) + (z + y) - y$$
$$= x + y + z.$$
$$= n(A \cup B).$$

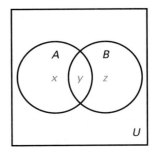

FIGURE 8.15

▶**EXAMPLE 5** If A and B are sets such that $n(A) = 5$, $n(B) = 7$, and $n(A \cup B)$ is 10, find $n(A \cap B)$.

We have

$$n(A \cup B) = n(A) + n(B) - n(A \cap B)$$
$$10 = 5 + 7 - n(A \cap B),$$

so that $n(A \cap B) = 5 + 7 - 10 = 2.$ ◀ **6**

8.2 EXERCISES

Sketch a Venn diagram like the one to the right and use shading to show each of the following sets. (See Example 1.)

1. $B \cap A'$ **2.** $A \cup B'$

3. $A' \cup B$ **4.** $A' \cap B'$

5. $B' \cup (A' \cap B')$ **6.** $(A \cap B) \cup B'$

7. U' **8.** $\emptyset'$

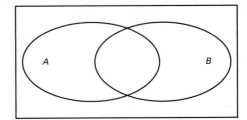

Sketch a Venn diagram like the one shown, and use shading to show each of the following sets. (See Example 2.)

9. $(A \cap B) \cap C$

10. $(A \cap C') \cup B$

11. $A \cap (B \cup C')$

12. $A' \cap (B \cap C)$

13. $(A' \cap B') \cap C$

14. $(A \cap B') \cup C$

15. $(A \cap B') \cap C$

16. $A' \cap (B' \cup C)$

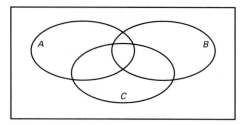

Use Venn diagrams to answer the following questions. (See Examples 3 and 4.)

17. Jeff Friedman, of Example 4 in the text, was again reassigned, this time to the home economics department of the electric utility. He interviewed 140 people in a suburban shopping center to find out some of their cooking habits. He obtained the following results. Should he be reassigned yet one more time?

> 58 use microwave ovens
> 63 use electric ranges;
> 58 use gas ranges;
> 19 use microwave ovens and electric ranges;
> 17 use microwave ovens and gas ranges;
> 4 use both gas and electric ranges;
> 1 uses all three;
> 2 cook only with solar energy.

18. Toward the middle of the harvesting season, peaches for canning come in three types, earlies, lates, and extra lates, depending on the expected date of ripening. During a certain week, the following data was recorded at a fruit delivery station.

> 34 trucks went out carrying early peaches;
> 61 had late peaches;
> 50 had extra lates;
> 25 had earlies and lates;
> 30 had lates and extra lates;
> 8 had earlies and extra lates;
> 6 had all three;
> 9 had only figs (no peaches at all).

(a) How many trucks had only late variety peaches?

(b) How many had only extra lates?

(c) How many had only one type of peach?

(d) How many trucks in all went out during the week?

19. Social Science A recent nationwide survey revealed the following data concerning unemployment in the United States. Out of the total population of people who are working or seeking work,

> 9% are unemployed;
> 11% are black;
> 30% are youths;
> 1% are unemployed black youths;
> 5% are unemployed youths;
> 7% are employed black adults;
> 23% are nonblack employed youths.

What percent are
(a) not black?
(b) employed adults?
(c) unemployed adults?
(d) unemployed black adults?
(e) unemployed non-black youths?

20. Country-western songs emphasize three basic themes: love, prison, and trucks. A survey of the local country-western radio station produced the following data.

> 12 songs were about a truck driver who was in love while in prison;
> 13 about a prisoner in love;
> 28 about a person in love;
> 18 about a truck driver in love;
> 3 about a truck driver in prison who was not in love;
> 2 about a prisoner who was not in love and did not drive a truck;
> 8 about a person out of jail who was not in love, and did not drive a truck;
> 16 about truck drivers who were not in prison.

(a) How many songs were surveyed?
Find the number of songs about
(b) truck drivers;
(c) prisoners;
(d) truck drivers in prison;
(e) people not in prison;
(f) people not in love.

21. Natural Science After a genetics experiment, the number of pea plants having certain characteristics was tallied, with the results as follows.

22 were tall;
25 had green peas;
39 had smooth peas;
9 were tall and had green peas;
17 were tall and had smooth peas;
20 had green peas and smooth peas;
6 had all three characteristics;
4 had none of the characteristics.

(a) Find the total number of plants counted.
(b) How many plants were tall and had peas that were neither smooth nor green?
(c) How many plants were not tall but had peas that were smooth and green?

22. Natural Science Human blood can contain either no antigens, the A antigen, the B antigen, or both the A and B antigens. A third antigen, called the Rh antigen, is important in human reproduction, and again may or may not be present in an individual. Blood is called type A-positive if the individual has the A and Rh, but not the B antigen. A person having only the A and B antigens is said to have type AB-negative blood. A person having only the Rh antigen has type O-positive blood. Other blood types are defined in a similar manner. Identify the blood type of the individuals in regions (a)–(g) of the Venn diagram.

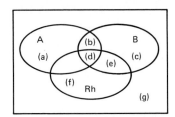

23. Natural Science (Use the diagram from Exercise 22.) In a certain hospital, the following data was recorded.

25 patients had the A antigen;
17 had the A and B antigens;
27 had the B antigen;
22 had the B and Rh antigens;
30 had the Rh antigen;
12 had none of the antigens;
16 had the A and Rh antigens;
15 had all three antigens.

How many patients
(a) were represented?
(b) had exactly one antigen?
(c) had exactly two antigens?
(d) had O-positive blood?
(e) had AB-positive blood?
(f) had B-negative blood?
(g) had O-negative blood?
(h) had A-positive blood?

24. A survey of 80 sophomores at a western college showed that

36 take English;
32 take history;
32 take political science;
16 take political science and history;
16 take history and English;
14 take political science and English;
6 take all three.

How many students
(a) take English and neither of the other two?
(b) take none of the three courses?
(c) take history, but neither of the other two?
(d) take political science and history, but not English?
(e) do not take political science?

25. The following table shows the number of people in a certain small town in Georgia who fit in the given categories.

Age	Drink Vodka (V)	Drink Bourbon (B)	Drink Gin (G)	Totals
20–25 (Y)	40	15	15	70
26–35 (M)	30	30	20	80
Over 35 (O)	10	50	10	70
Totals	80	95	45	220

Using the letters given in the table, find the number of people in each of the following sets.
(a) $Y \cap V$ ⠀⠀⠀⠀⠀**(b)** $M \cap B$
(c) $M \cup (B \cap Y)$ ⠀⠀**(d)** $Y' \cap (B \cup G)$
(e) $O' \cup G$ ⠀⠀⠀⠀⠀**(f)** $M' \cap (V' \cap G')$

26. The following table shows the results of a survey in a medium-sized town in Tennessee. The survey asked questions about the investment habits of local citizens.

Age	Stocks (S)	Bonds (B)	Savings Accounts (A)	Totals
18–29 (Y)	6	2	15	23
30–49 (M)	14	5	14	33
50 or over (O)	32	20	12	64
Totals	52	27	41	120

Using the letters given in the table, find the number of people in each of the following sets.

(a) $Y \cap B$ (b) $M \cup A$

(c) $Y \cap (S \cup B)$ (d) $O' \cup (S \cup A)$

(e) $(M' \cup O') \cap B$

Use Venn diagrams to answer the following questions. (See Example 5.)

27. If $n(A) = 5$, $n(B) = 8$, and $n(A \cap B) = 4$, what is $n(A \cup B)$?

28. If $n(A) = 12$, $n(B) = 27$, and $n(A \cup B) = 30$, what is $n(A \cap B)$?

29. Suppose $n(B) = 7$, $n(A \cap B) = 3$, and $n(A \cup B) = 20$. What is $n(A)$?

30. Suppose $n(A \cap B) = 5$, $n(A \cup B) = 35$, and $n(A) = 13$. What is $n(B)$?

Draw a Venn diagram and use the given information to fill in the number of elements for each region.

31. $n(U) = 38$, $n(A) = 16$, $n(A \cap B) = 12$, $n(B') = 20$

32. $n(A) = 26$, $n(B) = 10$, $n(A \cup B) = 30$, $n(A') = 17$

33. $n(A \cup B) = 17$, $n(A \cap B) = 3$, $n(A) = 8$, $n(A' \cup B') = 21$

34. $n(A') = 28$, $n(B) = 25$, $n(A' \cup B') = 45$, $n(A \cap B) = 12$

35. $n(A) = 28$, $n(B) = 34$, $n(C) = 25$, $n(A \cap B) = 14$, $n(B \cap C) = 15$, $n(A \cap C) = 11$, $n(A \cap B \cap C) = 9$, $n(U) = 59$

36. $n(A) = 54$, $n(A \cap B) = 22$, $n(A \cup B) = 85$, $n(A \cap B \cap C) = 4$, $n(A \cap C) = 15$, $n(B \cap C) = 16$, $n(C) = 44$, $n(B') = 63$

37. $n(A \cap B) = 6$, $n(A \cap B \cap C) = 4$, $n(A \cap C) = 7$, $n(B \cap C) = 4$, $n(A \cap C') = 11$, $n(B \cap C') = 8$, $n(C) = 15$, $n(A' \cap B' \cap C') = 5$

38. $n(A) = 13$, $n(A \cap B \cap C) = 4$, $n(A \cap C) = 6$, $n(A \cap B') = 6$, $n(B \cap C) = 6$, $n(B \cap C') = 11$, $n(B \cup C) = 22$, $n(A' \cap B' \cap C') = 5$

In Exercises 39–42, prove that the statements are true by drawing Venn diagrams and shading the regions representing the sets on each side of the equals signs.*

39. $(A \cup B)' = A' \cap B'$

40. $(A \cap B)' = A' \cup B'$

41. $A \cap (B \cup C) = (A \cap B) \cup (A \cap C)$

42. $A \cup (B \cap C) = (A \cup B) \cap (A \cup C)$

43. Use Venn diagrams to show that if U is the universal set, then $n(A') = n(U) - n(A)$.

44. Use Venn diagrams to show that $n(A \cup B \cup C) = n(A) + n(B) + n(C) - n(A \cap B) - n(A \cap C) - n(B \cap C) + n(A \cap B \cap C)$.

---

*The statements in Exercises 39 and 40 are known as De Morgan's laws. They are named for the English mathematician Augustus De Morgan (1806–71).

8.3 PROBABILITY

If you go to a supermarket and buy five pounds of peaches at 54¢ per pound, you can easily find the *exact* price of your purchase: $2.70. On the other hand, the produce manager of the market is faced with the problem of ordering peaches. The manager may have a good estimate of the number of pounds of peaches that will be sold during the day, but there is no way to know *exactly*. The number of pounds that customers will purchase during a day is **random:** it cannot be predicted exactly.

1 Find the probability that the spinner of Figure 8.16

(a) points to 3;

(b) does not point to 1.

Answer:

(a) 1/3

(b) 2/3

Many problems that come up in applications of mathematics involve random phenomena for which exact prediction is impossible. The best that we can do is to determine the *probability* of the possible outcomes. In this section and the next we introduce some of the terminology and basic premises of probability theory.

In probability, an **experiment** is an activity or occurrence with an observable result. Each repetition of an experiment is called a **trial.** The possible results of each trial are called **outcomes.** An example of a probability experiment is the tossing of a coin. Each trial of the experiment (each toss) has two possible outcomes: heads, *h*, and tails, *t*. If the outcomes *h* and *t* are equally likely to occur, then the coin is not "loaded" to favor one side over the other. Such a coin is called **fair.** For a coin that is not loaded, this "equally likely" assumption is made for each trial.

Since *two* equally likely outcomes are possible, *h* and *t*, theoretically a coin tossed many, many times should come up heads approximately 1/2 of the time. Also, the more times the coin is tossed, the closer the occurrence of heads should be to 1/2. Therefore, it is reasonable to define the probability of a head coming up to be 1/2. Similarly, we define the probability of tails coming up to be 1/2.

For now we shall concentrate on experiments, like coin tossing, in which each outcome is equally likely and use the probabilities suggested by this fact.

▶**EXAMPLE 1**   Suppose a spinner like the one shown is Figure 8.16 is spun. **(a)** Find the probability that the spinner will point to 1.

It is natural to assume that if this spinner were spun many, many times, it would point to 1 about 1/3 of the time, so that the required probability is 1/3. **(b)** Find the probability that the spinner will point to 2.

Since 2 is one of three possible outcomes, the probability is 1/3. ◀ **1**

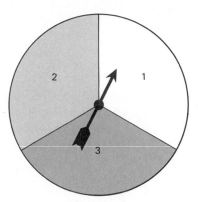

FIGURE 8.16

An ordinary die is a cube whose six faces show the following numbers of dots: 1, 2, 3, 4, 5, and 6. If the die is not "loaded" to favor certain faces over others (a fair die), then any one of the six faces is equally likely to come up when the die is rolled.

2 A jar holds a red marble, a blue marble, and a green marble. If one marble is drawn at random, what is the probability it is

(a) green?

(b) red?

Answer:

(a) 1/3

(b) 1/3

3 Write an equally likely sample space for the experiment of rolling a single fair die.

Answer:
{1, 2, 3, 4, 5, 6}

▶**EXAMPLE 2** **(a)** If a single fair die is rolled, find the probability of rolling the number 4.

Since one out of six faces is a 4, the required probability is 1/6.

(b) Using the same die, find the probability of rolling a 6.

Since one out of the six faces shows a 6, the probability is 1/6. ◀ **2**

In the general case, probabilities are defined by the same principle used in the preceding examples.

If there are n equally likely outcomes of an experiment, then the probability of any one outcome is $1/n$.

SAMPLE SPACES   The set of all possible outcomes for an experiment is the **sample space** for the experiment. A sample space for the experiment of tossing a coin has two outcomes: heads, h, and tails, t. If S represents this sample space, then

$$S = \{h, t\}.$$

▶**EXAMPLE 3**   Give the sample space for each experiment.

(a) A spinner like the one in Figure 8.16 is spun.

The three outcomes 1, 2, or 3 are equally likely, so the sample space is

$$\{1, 2, 3\}.$$

(b) For the purposes of a public opinion poll, people are classified as young, middle-aged, or older, and as male or female.

A sample space for this poll could be written as a set of ordered pairs of these types.

{(young, male), (young, female), (middle-aged, male),

(middle-aged, female), (older, male), (older, female)}

(c) An experiment consists of studying the number of boys and girls in families with exactly 3 children. Let b represent *boy* and g represent *girl*.

A 3-child family can have 3 boys, written *bbb*, 3 girls, *ggg*, or various combinations, such as *bgg*. A sample space with four outcomes (not equally likely) is

$$S_1 = \{3 \text{ boys, 2 boys and 1 girl, 1 boy and 2 girls, 3 girls}\}.$$

Another sample space that considers the ordering of the births of the 3 children with, for example, *bgg* different from *gbg* or *ggb*, is

$$S_2 = \{bbb, bbg, bgb, gbb, bgg, gbg, ggb, ggg\}.$$

The second sample space, S_2, has equally likely outcomes; S_1 does not, since there is more than one way to get a family with 2 boys and 1 girl or a family of 2 girls and 1 boy, but only one way to get 3 boys or 3 girls. ◀ **3**

4 Suppose a die is tossed. Write the following events.

(a) The number showing is less than 3.

(b) The number showing is 5.

(c) The number showing is 10.

Answer:

(a) {1, 2}

(b) {5}

(c) ∅

Caution An experiment may have more than one sample space, as shown in Example 3(c). The most useful sample spaces have equally likely outcomes, but it is not always possible to choose such a sample space.

EVENTS An **event** is a subset of a sample space. If the sample space for tossing a coin is $S = \{h, t\}$, then one event is the subset $\{h\}$, which represents the outcome "heads."

For the sample space of rolling a single fair die, $S = \{1, 2, 3, 4, 5, 6\}$, some possible events are listed below.

The die shows an even number: $E_1 = \{2, 4, 6\}$.

The die shows a 1: $E_2 = \{1\}$.

The die shows a number less than 5: $E_3 = \{1, 2, 3, 4\}$.

The die shows a multiple of 3: $E_4 = \{3, 6\}$.

▶**EXAMPLE 4** For the sample space S_2 in Example 3(c), write the following events.

(a) Event H: the family has exactly 2 girls

Families can have exactly 2 girls with either bgg, gbg, or ggb, so that event H is

$$H = \{bgg, \, gbg, \, ggb\}.$$

(b) Event K: the 3 children are the same sex

Two outcomes satisfy this condition, all boys or all girls.

$$K = \{bbb, \, ggg\}.$$

(c) Event J: the family has 3 girls

Only ggg satisfies this condition, so

$$J = \{ggg\}. \quad ◀ \quad \boxed{4}$$

In Example 4(c), event J had only one possible outcome, *ggg*. Such an event, with only one possible outcome, is a **simple event.** If an event E equals the sample space S, then E is a **certain event.** If event $E = \emptyset$, then E is an **impossible event.**

▶**EXAMPLE 5** Suppose a die is rolled. As shown above, the sample space is $\{1, 2, 3, 4, 5, 6\}$.

(a) The event "the die shows a 4," $\{4\}$, has only one possible outcome. It is a simple event.

5 Which of the events listed in problem 4 at the side is

(a) simple?

(b) certain?

(c) impossible?

Answer:

(a) Part (b)

(b) None

(c) Part (c)

6 A fair die is rolled. Find the probability of rolling

(a) an odd number;

(b) 2, 4, 5, or 6;

(c) a number greater than 5;

(d) the number 7.

Answer:

(a) 1/2

(b) 2/3

(c) 1/6

(d) 0

(b) The event "the number showing is less than 10" equals the sample space $S = \{1, 2, 3, 4, 5, 6\}$. This event is a certain event; if a die is rolled the number showing (either 1, 2, 3, 4, 5, or 6), must be less than 10.

(c) The event "the die shows a 7" is the empty set, $\emptyset$; this is an impossible event. ◀ **5**

PROBABILITY For sample spaces with *equally likely* outcomes, the *probability of an event* is defined as follows.

Basic Probability Principle

Let a sample space S have n equally likely outcomes. Let event E contain m of these outcomes. Then the **probability that event E occurs**, written $P(E)$, is

$$P(E) = \frac{m}{n}.$$

This definition means that the probability of an event is a number that indicates the relative likelihood of the event.

▶**EXAMPLE 6** Suppose a single die is rolled. Use the sample space $S = \{1, 2, 3, 4, 5, 6\}$ and give the probability of each of the following events.

(a) E: the die shows an even number

Here, $E = \{2, 4, 6\}$, a set with three elements. Since S contains six elements,

$$P(E) = \frac{3}{6} = \frac{1}{2}.$$

(b) F: the die shows a number greater than 4

Since F contains two elements, 5 and 6,

$$P(F) = \frac{2}{6} = \frac{1}{3}.$$

(c) G: the die shows a number less than 10

Event G is a certain event, with

$$G = \{1, 2, 3, 4, 5, 6\}, \quad \text{so} \quad P(G) = \frac{6}{6} = 1.$$

(d) H: the die shows an 8

This event is impossible, so

$$P(H) = 0. \quad ◀ \quad \boxed{6}$$

7 A single playing card is drawn at random from an ordinary 52-card deck. Find the probability of drawing

(a) a queen;

(b) a diamond;

(c) a red card.

Answer:

(a) 1/13

(b) 1/4

(c) 1/2

▶**EXAMPLE 7**   If a single card is drawn at random from an ordinary 52-card bridge deck (shown in Figure 8.17), find the probability of each of the following events.

(a) Drawing an ace

There are 4 aces in the deck. The event "drawing an ace" is

{heart ace, diamond ace, club ace, spade ace}.

Therefore,

$$P(\text{ace}) = \frac{4}{52} = \frac{1}{13}.$$

(b) Drawing a face card

Since there are 12 face cards,

$$P(\text{face card}) = \frac{12}{52} = \frac{3}{13}.$$

(c) Drawing a spade

The deck contains 13 spades, so

$$P(\text{spade}) = \frac{13}{52} = \frac{1}{4}.$$

(d) Drawing a spade or a heart

Besides the 13 spades, the deck contains 13 hearts, so

$$P(\text{spade or heart}) = \frac{26}{52} = \frac{1}{2}. \quad ◀ \;\; \boxed{7}$$

FIGURE 8.17

The probability of each event in the preceding examples was a number between 0 and 1. The same thing is true in the general case. If an event E consists of m of the possible n outcomes, then $0 \le m \le n$ and hence, $0 \le m/n \le 1$. Since $P(E) = m/n$, $P(E)$ is between 0 and 1.

For any event E, $0 \le P(E) \le 1.$

8.3 EXERCISES

Write sample spaces for the following experiments. (See Example 3.)

1. A month of the year is chosen for a wedding.

2. A day in April is selected for a bicycle race.

3. A student is asked how many points she earned on a recent 80-point test.

4. A person is asked the number of hours (to the nearest hour) he watched television yesterday.

5. The management of an oil company must decide whether to go ahead with a new oil shale plant or to cancel it.

6. A record is kept for 3 days about whether a particular stock goes up or down.

7. A coin is tossed and a die is rolled.

8. A box contains five balls, numbered 1, 2, 3, 4, and 5. A ball is drawn at random, the number on it recorded, and the ball replaced. The box is shaken, a second ball is drawn, and its number is recorded.

For the experiments in Exercises 9–14, write out an equally likely sample space, and then write the indicated events in set notation. (See Examples 3 and 4.)

9. A committee of 2 people is selected from 5 executives, Alam, Bartolini, Chinn, Dickson, and Ellsberg.
   (a) Chinn is on the committee.
   (b) Dickson and Ellsberg are not both on the committee.
   (c) Both Alam and Chinn are on the committee.

10. A marble is drawn at random from a bowl containing 3 yellow, 4 white, and 8 blue marbles.
    (a) A yellow marble is drawn.
    (b) A blue marble is drawn.
    (c) A white marble is drawn.
    (d) A black marble is drawn.

11. Slips of paper marked with the numbers 1, 2, 3, 4, and 5 are placed in a box. After being mixed, two slips are drawn.
    (a) Both slips are marked with even numbers.
    (b) One slip is marked with an odd numbr and the other is marked with an even number.
    (c) Both slips are marked with the same number.

12. An unprepared student takes a three-question true/false quiz in which he guesses the answers to all three questions.
    (a) The student gets three answers wrong.
    (b) The student gets exactly two answers correct.
    (c) The student gets only the first answer correct.

13. A die is tossed twice, with the tosses recorded as ordered pairs.
    (a) The first die shows a 3.
    (b) The sum of the numbers showing is 8.
    (c) The sum of the numbers showing is 13.

14. One urn contains four balls, labeled 1, 2, 3, and 4. A second urn contains five balls, labeled 1, 2, 3, 4, and 5. An experiment consists of taking one ball from the first urn, and then taking a ball from the second urn.
    (a) The number on the first ball is even.
    (b) The number on the second ball is even.
    (c) The sum of the numbers on the two balls is 5.
    (d) The sum of the numbers on the two balls is 1.

15. **Management** The management of a firm wishes a check on the opinions of its assembly line workers. Before the workers are interviewed, they are divided into various categories. Define events E, F, and G as follows.

    E: worker is female
    F: worker has worked less than 5 years
    G: worker contributes to a voluntary retirement plan

Describe each of the following events in words.
(a) E' (b) F' (c) $E \cap F$
(d) $F \cup G$ (e) $E \cup G'$ (f) $F' \cap G'$

16. Natural Science For a medical experiment, people are classified as to whether they smoke, have a family history of heart disease, or are overweight. Define events E, F, and G as follows.

E: person smokes
F: person has a family history of heart disease
G: person is overweight

Describe each of the following events in words.
(a) G' (b) $E \cup F$ (c) $F \cap G$
(d) $E' \cap F$ (e) $E \cup G'$ (f) $F' \cup G'$

A single fair die is rolled. Find the probabilities of the following events. (See Example 6.)

17. Getting a 2

18. Getting an odd number

19. Getting a number less than 5

20. Getting a number greater than 2

21. Getting a 3 or a 4

22. Getting any number except 3

A card is drawn from a well-shuffled deck of 52 cards. Find the probability of drawing each of the following. (See Example 7.)

23. A 9

24. A black card

25. A black 9

26. A heart

27. The 9 of hearts

28. A face card

29. A 2 or a queen

30. A black 7 or a red 8

31. A heart or a spade

32. A red face card

33. A 2, 3, 4, or 5

34. Anything except a jack or a 7

A jar contains 5 red, 4 black, 7 purple, and 9 green marbles. If a marble is drawn at random, what is the probability that the marble is

35. red?

36. black?

37. green?

38. purple?

39. not black?

40. not purple?

41. red or black?

42. black or purple?

A letter is chosen at random from the word ABRACADABRA. What is the probability that the chosen letter is an

43. A?

44. B?

45. C or D?

46. E?

8.4 BASIC CONCEPTS OF PROBABILITY

We discuss the probability of more complex events in this section. Since events are sets, we can use set operations to find unions, intersections, and complements of events.

▶**EXAMPLE 1** Suppose a die is tossed. Let E be the event "the die shows a number greater than 3," and let F be the event "the die shows an even number." Then

$$E = \{4, 5, 6\} \quad \text{and} \quad F = \{2, 4, 6\}.$$

1 Give the following events for the experiment of Example 1 if $E = \{1, 3\}$ and $F = \{2, 3, 4, 5\}$.

(a) $E \cap F$

(b) $E \cup F$

(c) E'

Answer:

(a) $\{3\}$

(b) $\{1, 2, 3, 4, 5\}$

(c) $\{2, 4, 5, 6\}$

Find each of the following.

(a) $E \cap F$

The event $E \cap F$ is "the die shows a number that is greater than 3 *and* is even." Event $E \cap F$ includes the outcomes common to *both* E and F.

$$E \cap F = \{4, 6\}$$

(b) $E \cup F$

The event $E \cup F$ is "the die shows a number that is greater than 3 *or* is even." Event $E \cup F$ inlcudes the outcomes of E *or* F, or both.

$$E \cup F = \{2, 4, 5, 6\}$$

(c) E'

Event E' is "the die shows a number that is *not* greater than 3." E' includes the elements of the sample space that are not in E. (Event E' is the complement of event E.)

$$E' = \{1, 2, 3\}$$

The Venn diagrams of Figure 8.18 show the events $E \cap F$, $E \cup F$, and E'. ◀ **1**

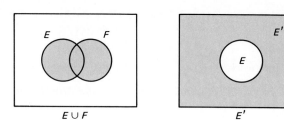

$E \cap F$ $E \cup F$ E'

FIGURE 8.18

Two events that cannot both occur at the same time, such as getting both a head and a tail on the same toss of a coin, are called **mutually exclusive events.**

Events A and B are mutually exclusive events if $A \cap B = \emptyset$.

For any event E, E and E' are mutually exclusive.

2 In Example 2, let $F =$ $\{2, 4, 6\}$ and $K = \{1, 3, 5\}$. Are the following events mutually exclusive?

(a) F and K

(b) F and G

Answer:

(a) Yes

(b) No

▶**EXAMPLE 2**   Let $S = \{1, 2, 3, 4, 5, 6\}$, the sample space for tossing a die. Let $E = \{4, 5, 6\}$, and let $G = \{1, 2\}$. Then E and G are mutually exclusive events since they have no outcomes in common; $E \cap G = \emptyset$. See Figure 8.19.   ◀ **2**

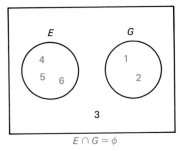

$$E \cap G = \phi$$

FIGURE 8.19

A summary of the set operations for events is given below.

Let E and F be events for a sample space S.

$E \cap F$ occurs when both E and F occur.
$E \cup F$ occurs when either E or F or both occur.
E' occurs when E does not.
E and F are mutually exclusive if $E \cap F = \emptyset$.

Recall that $n(A)$ denotes the *number* of elements in the set A and that the probability of an event E in a sample space S of equally likely outcomes is $P(E) = n(E)/n(S)$.

To determine the probability of the complement E' of an event E in a sample space S, recall that by definition, $E \cup E' = S$ and $E \cap E' = \emptyset$. Hence, $n(E \cap E') = n(\emptyset) = 0$, and by the rule on page 412,

$$n(S) = n(E \cup E') = n(E) + n(E') - n(E \cap E')$$
$$= n(E) + n(E') - 0.$$

Dividing both sides of this equation by $n(S)$ shows that

$$\frac{n(S)}{n(S)} = \frac{n(E)}{n(S)} + \frac{n(E')}{n(S)}$$
$$1 = P(E) + P(E').$$

3 **(a)** Let $P(K) = 2/3$. Find $P(K')$.

(b) If $P(X') = 3/4$, find $P(X)$.

Answer:

(a) 1/3

(b) 1/4

4 A single card is drawn from an ordinary deck. Find the probability that it is black or a 9.

Answer:
7/13

Rearranging these terms produces the following rule.

Complement Rule

For any event E,

$$P(E') = 1 - P(E) \quad \text{and} \quad P(E) = 1 - P(E').$$

▶**EXAMPLE 3** If a fair die is rolled, what is the probability that any number but 5 will come up?

If E is the event that 5 comes up, then E′ is the event that any number but 5 comes up. Since $P(E) = 1/6$, we have $P(E') = 1 - 1/6 = 5/6$. ◀ **3**

To determine the probability of the union of two events E and F in a sample space S, use the fact that

$$n(E \cup F) = n(E) + n(F) - n(E \cap F).$$

Dividing both sides by $n(S)$ shows that

$$\frac{n(E \cup F)}{n(S)} = \frac{n(E)}{n(S)} + \frac{n(F)}{n(S)} - \frac{n(E \cap F)}{n(S)}$$

$$P(E \cup F) = P(E) + P(F) - P(E \cap F).$$

This discussion is summarized below.

Union Rule

For any events E and F from a sample space S,

$$P(E \cup F) = P(E) + P(F) - P(E \cap F).$$

▶**EXAMPLE 4** If a single card is drawn from an ordinary deck of cards, find the probability that it will be red or a face card.

Let R represent the event ''red card'' and F the event ''face card.'' There are 26 red cards in the deck, so $P(R) = 26/52$. There are 12 face cards in the deck, so $P(F) = 12/52$. Since there are 6 red face cards in the deck, $P(R \cap F) = 6/52$. By the union rule, the probability that the card will be red or a face card is

$$P(R \cup F) = P(R) + P(F) - P(R \cap F)$$

$$= \frac{26}{52} + \frac{12}{52} - \frac{6}{52} = \frac{32}{52} = \frac{8}{13}. \quad ◀ \mathbf{4}$$

5 In the experiment of Example 5, find the following probabilities.

(a) The sum is 5 or the second die shows a 3.

(b) Both dice show the same number or the sum is at least 11.

Answer:

(a) 1/4

(b) 2/9

▶**EXAMPLE 5**   Suppose two fair dice are rolled. Find each of the following probabilities.

(a) The first die shows a 2 or the sum of the results is 6 or 7.

The sample space for the throw of two dice is shown in Figure 8.20, where 1—1 represents the event ''the first die shows a 1 and the second die shows a 1'', 1—2 represents ''the first die shows a 1 and the second die shows a 2,'' and so on. Let A represent the event ''the first die shows a 2'' and B represent the event ''the sum of the results is 6 or 7.'' These events are indicated in Figure 8.20. From the diagram, event A has 6 elements, B has 11 elements, and the sample space has 36 elements. Thus,

$$P(A) = \frac{6}{36}, \quad P(B) = \frac{11}{36}, \quad \text{and} \quad P(A \cap B) = \frac{2}{36}.$$

By the union rule,

$$P(A \cup B) = P(A) + P(B) - P(A \cap B),$$

$$P(A \cup B) = \frac{6}{36} + \frac{11}{36} - \frac{2}{36} = \frac{15}{36} = \frac{5}{12}.$$

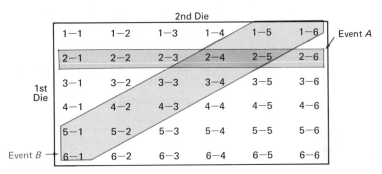

FIGURE 8.20

(b) The sum is 11 or the second die is 5.

$P(\text{sum is } 11) = 2/36$, $P(\text{second die is } 5) = 6/36$, and $P(\text{sum is } 11 \text{ and second die is } 5) = 1/36$, so

$$P(\text{sum is 11 or second die is 5}) = \frac{2}{36} + \frac{6}{36} - \frac{1}{36} = \frac{7}{36}. \quad ◀ \boxed{5}$$

When events E and F are mutually exclusive, then $E \cap F = \emptyset$ by definition, hence, $P(E \cap F) = 0$. Applying the union rule yields this useful fact.

6 In Example 6, find the probability of no more than 2 girls.

Answer:
7/8

7 In Example 7, find the probability that the sum of the numbers rolled is at least 5.

Answer:
5/6

For mutually exclusive events E and F,

$$P(E \cup F) = P(E) + P(F).$$

▶**EXAMPLE 6** Assume that the probability of a couple having a boy is the same as the probability of their having a girl. If the couple has 3 children, find the probability that at least 2 of them are girls.

The event of having at least 2 girls is the union of the mutually exclusive events E = "the family has exactly 2 girls" and F = "the family has exactly 3 girls." Using the equally likely sample space

$$\{ggg, ggb, gbg, bgg, gbb, bgb, bbg, bbb\},$$

we see that $P(2 \text{ girls}) = 3/8$ and $P(3 \text{ girls}) = 1/8$. Therefore,

$$P(\text{at least 2 girls}) = P(2 \text{ girls}) + P(3 \text{ girls})$$

$$= \frac{3}{8} + \frac{1}{8} = \frac{1}{2}. \quad \blacktriangleleft \;\; \boxed{6}$$

▶**EXAMPLE 7** If two fair dice are rolled, find the probability that the sum of the numbers showing is greater than 3.

To calculate this probability directly, we must find the probabilities that the sum is 4, 5, 6, 7, 8, 9, 10, 11, or 12 and then add them. It is much simpler to first find the probability of the complement, the event that the sum is less than or equal to 3.

$$P(\text{sum} \leq 3) = P(\text{sum is 2}) + P(\text{sum is 3})$$

$$= \frac{1}{36} + \frac{2}{36}$$

$$= \frac{3}{36} = \frac{1}{12}$$

Now use the fact that $P(E) = 1 - P(E')$ to get

$$P(\text{sum} > 3) = 1 - P(\text{sum} \leq 3)$$

$$= 1 - \frac{1}{12} = \frac{11}{12}. \quad \blacktriangleleft \;\; \boxed{7}$$

Venn diagrams can be useful in finding probabilities, as the next examples show.

8 Suppose $P(A) = .2$, $P(B) = .5$, and $P(A \cap B) = .1$. Find the following.

(a) $P(A \cup B)'$

(b) $P(A' \cup B)$

Answer:

(a) .4

(b) .9

▶**EXAMPLE 8**   Suppose E and F are events such that $P(E) = .4$, $P(E \cup F) = .8$, and $P(E \cap F') = .3$.

(a) Find $P(F')$.

Draw a Venn diagram, as in Figure 8.21. It shows that the sample space S consists of four mutually exclusive events (regions). $E \cap F'$ is region I, hence, $P(\text{I}) = .3$. Region IV is the complement of $E \cup F$, so that $P(\text{IV}) = 1 - .8 = .2$. The diagram shows that $F' = \text{I} \cup \text{IV}$. Since I and IV are mutually exclusive,

$$P(F') = P(\text{I} \cup \text{IV}) = P(\text{I}) + P(\text{IV}) = .3 + .2 = .5.$$

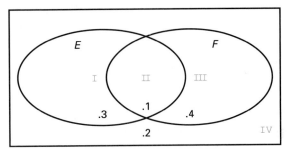

FIGURE 8.21

(b) Find $P(E \cap F)$ and $P(E' \cap F)$.

Since $E = \text{I} \cup \text{II}$, with I and II mutually exclusive, we have $P(E) = P(\text{I}) + P(\text{II})$. Hence,

$$P(E \cap F) = P(\text{II}) = P(E) - P(\text{I}) = .4 - .3 = .1.$$

Similarly,

$$P(E \cup F) = P(E \cup \text{III}) = P(E) + P(\text{III}),$$

so that

$$P(E' \cap F) = P(\text{III}) = P(E \cup F) - P(E) = .8 - .4 = .4. \quad ◀ \quad \boxed{8}$$

▶**EXAMPLE 9**   Susan is a college student who receives heavy sweaters from her aunt at the first sign of cold weather. Suppose the probability that a sweater is the wrong size is .47, the probability that it is a loud color is .59, and the probability that it is both the wrong size and a loud color is .31.

(a) Find the probability that the sweater is the correct size and not a loud color.

Let W be the event "wrong size" and L be the event "loud color." Draw a Venn diagram and determine the probabilities of the mutually exclusive events I–IV in Figure 8.22 as follows. We are given $P(\text{II}) = P(W \cap L) = .31$. Since regions I and II are mutually exclusive, $P(W) = P(\text{I}) + P(\text{II})$, so that $P(\text{I}) = P(W) - P(\text{II}) = .47 - .31 = .16$. Similarly, $P(\text{III}) = P(L) - P(\text{II}) = .59 - .31 = .28$. Finally, since IV is the complement of $W \cup L$, we have

$$P(\text{IV}) = 1 - P(W \cup L) = 1 - (.16 + .31 + .28) = .25.$$

9 Find the probability that the sweater in Example 9 is

(a) not a loud color;

(b) the correct size *and* a loud color.

Answer:

(a) .41

(b) .28

The event "correct size *and* not a loud color" is region IV, the area outside of both W and L. Therefore, its probability is .25.

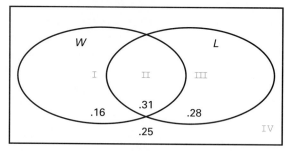

FIGURE 8.22

(b) Find the probability that the sweater is the correct size *or* not a loud color.

The event "correct size *or* not a loud color" is the region outside of region II (in II every sweater is both the wrong size and a loud color), that is, the complement of II. Hence, its probability is $1 - .31 = .69$. ◀ **9**

ODDS Sometimes probability statements are given in terms of **odds**, a comparison of $P(E)$ with $P(E')$. For example, suppose $P(E) = \frac{4}{5}$. Then $P(E') = 1 - \frac{4}{5} = \frac{1}{5}$. These probabilities predict that E will occur 4 out of 5 times and E' will occur 1 out of 5 times. Then we say the **odds in favor** of E are 4 to 1.

Odds

The **odds in favor** of an event E are defined as the ratio of $P(E)$ to $P(E')$, or

$$\frac{P(E)}{P(E')}, \quad P(E') \neq 0.$$

▶**EXAMPLE 10** Suppose the weather forecaster says that the probability of rain tomorrow is 1/3. Find the odds in favor of rain tomorrow.

Let E be the event "rain tomorrow." Then E' is the event "no rain tomorrow." Since $P(E) = 1/3$, $P(E') = 2/3$. By the definition of odds, the odds in favor of rain are

$$\frac{\frac{1}{3}}{\frac{2}{3}} = \frac{1}{2}, \quad \text{written 1 to 2} \quad \text{or} \quad 1:2.$$

10 Suppose $P(E) = 9/10$. Find the odds

(a) in favor of E;

(b) against E.

Answer:

(a) 9 to 1

(b) 1 to 9

11 If the odds in favor of event E are 1 to 5, find

(a) $P(E)$;

(b) $P(E')$.

Answer:

(a) 1/6

(b) 5/6

12 If the odds in favor of event E are 7 to 9, what is the probability of E?

Answer:
7/16

On the other hand, the odds that it will *not* rain, or the odds *against* rain, are

$$\frac{\frac{2}{3}}{\frac{1}{3}} = \frac{2}{1}, \quad \text{written 2 to 1.} \quad \blacktriangleleft \quad \boxed{10}$$

If the odds in favor of an event are, say, 3 to 5, then the probability of the event is 3/8, while the probability of the complement of the event is 5/8. (Odds of 3 to 5 indicate 3 outcomes in favor of the event out of a total of 8 outcomes.) This example suggests the following generalization.

If the odds favoring event E are m to n, then

$$P(E) = \frac{m}{m + n} \quad \text{and} \quad P(E') = \frac{n}{m + n}.$$

▶**EXAMPLE 11**   The odds that a particular bid will be the low bid are 4 to 5.
(a) Find the probability that the bid will be the low bid.
    Odds of 4 to 5 show 4 favorable chances out of $4 + 5 = 9$ chances altogether.

$$P(\text{bid will be low bid}) = \frac{4}{4 + 5} = \frac{4}{9}$$

(b) Find the odds against that bid being the low bid.
    There is a 5/9 chance that the bid will not be the low bid, so the odds against a low bid are

$$\frac{P(\text{bid will not be low})}{P(\text{bid will be low})} = \frac{5/9}{4/9} = \frac{5}{4},$$

or 5:4 $\blacktriangleleft$ $\boxed{11}$

▶**EXAMPLE 12**   If the odds in favor of a particular horse's winning a race are 5 to 7, what is the probability that the horse will win the race?
    The odds indicate chances of 5 out of 12 $(5 + 7 = 12)$ that the horse will win, so

$$P(\text{winning}) = \frac{5}{12}. \quad \blacktriangleleft \quad \boxed{12}$$

EMPIRICAL PROBABILITY   In many realistic problems, it is not possible to establish exact probabilities for events. Useful approximations, however, often can be found by using past experience as a guide to the future. The next example shows one approach to such **empirical probabilities.**

13 A traffic engineer gathered the following data on the number of accidents at four busy intersections.

Intersection	Number of Accidents
1	25
2	32
3	28
4	35

If an accident occurs at one of these four intersections, what is the probability that it occurs at intersections 1 or 3?

Answer:
.44

▶**EXAMPLE 13**   The manager of a store has decided to make a study of the amounts of money spent by people coming into the store. To begin, he chooses a day that seems fairly typical and gathers the following data. (Purchases have been rounded to the nearest dollar, with sales tax ignored.)

Amount Spent	Number of Customers
$0	158
$1–$5	94
$6–$9	203
$10–$19	126
$20–$49	47
$50–$99	38
$100 and over	53

First, the manager might add the numbers of customers to find that 719 people came into the store that day. Of these 719 people, $126/719 \approx .175$ made a purchase of at least $10 but no more than $19. Also, $53/719 \approx .074$ of the customers spent $100 or more. For this day, the probability that a customer entering the store will spend from $10 to $19 is .175, and the probability that the customer will spend $100 or more is .074. Probabilities for the other purchase amounts can be assigned in the same way, giving the results below.

Amount Spent	Probability
$0	.220
$1–$5	.131
$6–$9	.282
$10–$19	.175
$20–$49	.065
$50–$99	.053
$100 and over	.074
Total	1.000

The categories in the table are mutually exclusive simple events. Thus, by the union rule,
$$P(\text{spending} < \$10) = .220 + .131 + .282 = .633.$$

From the table, .282 of the customers spend from $6 to $9, inclusive. Since this price range attracts more than a quarter of the customers, perhaps the store's advertising should emphasize items in, or near, this price range.

The manager should use this table of probabilities to help in predicting the results on other days only if the manager is reasonably sure that the day when the measurements were made is fairly typical of the other days the store is open. For example, on the last few days before Christmas the probabilities might be quite different. ◀ **13**

A table of probabilities, as in Example 13, sets up a **probability distribution:** that is, for each possible outcome of an experiment, a number, called the probability of that outcome, is assigned. This assignment may be done in any reasonable way (on an empirical basis, as in Example 13, or by theoretical reasoning, as in Section 8.3), provided that it satisfies certain conditions.

Properties of Probability

Let S be a sample space consisting of n distinct outcomes $s_1, s_2, \ldots, s_n$. An acceptable probability assignment consists of assigning to each outcome s_i a number p_i (the probability of s_i) according to these rules.

1. The probability of each outcome is a number between 0 and 1.

$$0 \leq p_1 \leq 1, \quad 0 \leq p_2 \leq 1, \ldots, \quad 0 \leq p_n \leq 1$$

2. The sum of the probabilities of all possible outcomes is 1.

$$p_1 + p_2 + p_3 + \cdots + p_n = 1.$$

Another example of empirical probability occurs in weather forecasting. In one method, the forecaster compares atmospheric conditions with similar conditions in the past, and predicts the weather according to what happened previously under such conditions. For example, if certain kinds of atmospheric conditions have been followed by rain 80 times out of 100, the forecaster will announce an 80% chance of rain under those conditions.

One difficulty with empirical probability is that different people may assign different probabilities to the same event. Nevertheless, empirical probabilities can be used in many situations where the objective approach to probability cannot be used. In the next chapter the use of empirical probability is discussed in more detail.

8.4 EXERCISES

Decide whether the events in Exercises 1–6 are mutually exclusive. (See Example 2.)

1. Owning a car and owning a truck
2. Wearing glasses and wearing sandals
3. Being married and being over 30 years old
4. Being a teenager and being over 30 years old
5. Rolling a die once, getting a 4, and getting an odd number
6. Being a male and being a postal worker

Two dice are rolled. Find the probabilities of rolling the given sums. (See Example 5.)

7. (a) 2　　(b) 4　　(c) 5　　(d) 6
8. (a) 8　　(b) 9　　(c) 10　　(d) 13
9. (a) 9 or more　　(b) Less than 7　　(c) Between 5 and 8 (exclusive)
10. (a) Not more than 5　　(b) Not less than 8　　(c) Between 3 and 7 (exclusive)

One card is drawn from an ordinary deck of 52 cards. Find the probabilities of drawing the following cards. (See Example 4.)

11. (a) A 9 or 10
(b) A red card or a 3
(c) A 9 or a black 10
(d) A heart or a black card

12. (a) Less than a 4 (count aces as ones)
(b) A diamond or a 7
(c) A black card or an ace
(d) A heart or a jack

Ms. Elliott invites 10 relatives to a party: her mother, 2 aunts, 3 uncles, 2 brothers, 1 male cousin, and 1 female cousin. If the chances of any one guest arriving first are equally likely, find the probabilities that the first guest to arrive is as follows.

13. (a) A brother or an uncle
(b) A brother or a cousin
(c) A brother or her mother

14. (a) An uncle or a cousin
(b) A male or a cousin
(c) A female or a cousin

The numbers 1, 2, 3, 4, and 5 are written on slips of paper, and two slips are drawn at random without replacement. Find each of the probabilities in Exercises 15–16.

15. (a) The sum of the numbers is 9.
(b) The sum of the numbers is 5 or less.
(c) The first number is 2 or the sum is 6.

16. (a) Both numbers are even.
(b) One of the numbers is even or greater than 3.
(c) The sum is 5 or the second number is 2.

Use Venn diagrams to work Exercises 17–24. (See Examples 8 and 9.)

17. Suppose $P(E) = .26$, $P(F) = .41$, and $P(E \cap F) = .17$. Find each of the following.
(a) $P(E \cup F)$ (b) $P(E' \cap F)$
(c) $P(E \cap F')$ (d) $P(E' \cup F')$

18. Let $P(Z) = .42$, $P(Y) = .38$, and $P(Z \cup Y) = .61$. Find each of the following.
(a) $P(Z' \cap Y')$ (b) $P(Z' \cup Y')$
(c) $P(Z' \cup Y)$ (d) $P(Z \cap Y')$

19. Social Science   Fifty students in a Texas school were interviewed with the following results: 45 spoke Spanish, 10 spoke Vietnamese, and 8 spoke both languages. Find the probability that a randomly selected student from this group
(a) speaks both languages;
(b) speaks neither language;
(c) speaks only one of the two languages.

20. Social Science   A study on body types gave the following results: 45% were short, 25% were short and overweight, and 24% were tall and not overweight. Find the probability that a person is
(a) overweight;
(b) short, but not overweight;
(c) tall and overweight.

21. Social Science   The following data were gathered for 130 adult U.S. workers: 55 were women, 3 women earned more than $40,000, 62 men earned less than $40,000. Find the probability that an individual is
(a) a woman earning less than $40,000;
(b) a man earning more than $40,000;
(c) a man or is earning more than $40,000;
(d) a woman or is earning less than $40,000.

22. Social Science   A survey of 100 people about their music expenditures gave the following information: 38 bought rock music, 20 were teenagers who bought rock music, 26 were teenagers. Find the probability that a person
(a) is a teenager who buys nonrock music;
(b) buys rock music or is a teenager;
(c) is not a teenager;
(d) is not a teenager, but buys rock music.

23. Management   Suppose than 8% of a certain batch of calculators have a defective case, and that 11% have defective batteries. Also, 3% have both a defective case and defective batteries. A calculator is selected from the batch at random. Find the probability that the calculator has a good case and good batteries.

24. A student believes that her probability of passing accounting is .74, of passing mathematics is .39, and that her probability of passing both courses is .25. Find the probability that the student passes at least one course.

Natural Science *Color blindness is an inherited character-istic which is sex-linked, so that it is more common in males than in females. If M represents male and C repre-sents red-green color blindness, using the relative frequen-cies of the incidence of males and red-green color blind-ness as probabilities, $P(C) = .049$, $P(M \cap C) = .042$, $P(M \cup C) = .534$. Find the following. (Hint: use a Venn diagram with two circles labeled M and C.)*

25. $P(C')$ **26.** $P(M)$ **27.** $P(M')$

28. $P(M' \cap C')$ **29.** $P(C \cap M')$ **30.** $P(C \cup M')$

Work the following exercises on odds. (See Examples 10–12.)

A single fair die is rolled. Find the odds in favor of getting the following results.

31. 5   **32.** 3, 4, or 5   **33.** 1, 2, 3, or 4

34. Some number less than 2

35. Refer to the table below.

Golf Score	Probability
Below 60	.01
60–64	.08
65–69	.15
70–74	.28
75–79	.22
80–84	.08
85–89	.06
90–94	.04
95–99	.02
100 or more	.06

Find the following odds.
(a) The odds in favor of the golfer shooting *below* 75
(b) The odds against the golfer shooting in the 70s

36. A marble is drawn from a box containing 3 yellow, 4 white, and 8 blue marbles. Find the odds in favor of draw-ing the following.
(a) A yellow marble
(b) A blue marble
(c) A white marble

37. Find the odds of *not* drawing a white marble in Exer-cise 36.

38. The probability that a company will make a profit this year is .74. Find the odds against the company making a profit.

39. If the odds that it will rain are 4 to 7, what is the probabil-ity of rain?

40. If the odds that a given candidate will win an election are 3 to 2, what is the probability that the candidate will lose?

41. On page 134 of Roger Staubach's autobiography, *First Down, Lifetime to Go,* Staubach makes the following statement regarding his experience in Vietnam: "Odds against a direct hit are very low but when your life is in danger, you don't worry too much about the odds." Is this wording consistent with our definition of odds for and against? How could it have been said so as to be techni-cally correct?

An experiment is conducted for which the sample space is $S = \{s_1, s_2, s_3, s_4, s_5\}$. Which of the probability assignments in Exercises 42–47 is possible for this experiment? If an assignment is not possible, tell why.

42.

Outcomes	s_1	s_2	s_3	s_4	s_5
Probabilities	.09	.32	.21	.25	.13

43.

Outcomes	s_1	s_2	s_3	s_4	s_5
Probabilities	.92	.03	0	.02	.03

44.

Outcomes	s_1	s_2	s_3	s_4	s_5
Probabilities	$\frac{1}{3}$	$\frac{1}{4}$	$\frac{1}{6}$	$\frac{1}{8}$	$\frac{1}{10}$

45.

Outcomes	s_1	s_2	s_3	s_4	s_5
Probabilities	$\frac{1}{5}$	$\frac{1}{3}$	$\frac{1}{4}$	$\frac{1}{5}$	$\frac{1}{10}$

46.

Outcomes	s_1	s_2	s_3	s_4	s_5
Probabilities	.64	−.08	.30	.12	.02

47.

Outcomes	s_1	s_2	s_3	s_4	s_5
Probabilities	.05	.35	.5	.2	−.3

Work the following problems. (See Example 13.)

Social Science *A consumer survey randomly selects 2,000 people and asks them about their income and their TV-watching habits. The results are shown in this table.*

Annual Income	Hours of TV Watched per Week				
	0–8	9–15	16–22	23–30	More than 30
Less than $12,000	11	15	8	14	120
$12,000–$24,999	10	19	28	96	232
$25,000–$39,999	18	32	88	327	189
$40,000–$59,999	31	85	160	165	100
$60,000 or more	73	60	52	55	12

As a reward for participation, each person questioned is given a small prize. Then the names of all participants are placed in a box and one name is randomly drawn to receive the grand prize of a free vacation trip. What is the probability that the winner of the grand prize

48. has an income of $25,000–$39,999?

49. watches TV for 23–30 hours per week?

50. watches TV at least 16 hours per week?

51. has an income of at least $25,000?

52. has an income of at least $40,000 and watches TV at least 16 hours per week?

53. has an income less than $25,000 and watches TV at least 9 hours per week?

54. has an income of $12,000–$59,999 and watches TV for 9–30 hours per week?

55. has an income of $0–$24,999 and watches TV more than 30 hours per week?

56. **Management** The table below shows the probability that a customer of a department store will make a purchase in the indicated range.

Cost	Probability
Below $2	.07
$2–$4.99	.18
$5–$9.99	.21
$10–$19.99	.16
$20–$39.99	.11
$40–$69.99	.09
$70–$99.99	.07
$100–$149.99	.08
$150 or over	.03

Find the probability that a customer makes a purchase that is
(a) less than $5;
(b) $10 to $69.99;
(c) $20 or more;
(d) more than $4.99;
(e) less than $100;
(f) $100 or more.

57. **Social Science** The table shows the probability of a person accumulating credit card charges over a 12-month period.

Charges	Probability
Under $100	.31
$100–$499	.18
$500–$999	.18
$1000–$1999	.13
$2000–$2999	.08
$3000–$4999	.05
$5000–$9999	.06
$10000 or more	.01

Find the probability that a person's total charges during the period are
(a) $500 or more;   (b) less than $1000;
(c) $500 to $2999;   (d) $3000 or more.

58. **Social Science** The table below gives the probability that a person has life insurance in the indicated range.

Amount of Insurance	Probability
None	.17
Less than $10,000	.20
$10,000–$24,999	.17
$25,000–$49,999	.14
$50,000–$99,999	.15
$100,000–$199,999	.12
$200,000 or more	.05

Find the probability that an individual has the following amounts of life insurance.
(a) Less than $10,000
(b) $10,000 to $99,999
(c) $50,000 or more
(d) Less than $50,000 or $100,000 or more

59. **Natural Science** The results of a study relating blood cholesterol level to coronary disease are given below.

Cholesterol Level	Probability of Coronary Disease
Under 200	.10
200–219	.15
220–239	.20
240–259	.26
Over 259	.29

Find the probability of coronary disease if the cholesterol level is
(a) less than 240;   (b) 220 or more;
(c) from 200 to 239;   (d) from 220 to 259.

Natural Science *Gregor Mendel, an Austrian monk, was the first to use probability in the study of genetics. In an effort to understand the mechanisms of character transmittal from one generation to the next in plants, he counted the number of occurrences of various characteristics. Mendel found that the flower color in certain pea plants obeyed this scheme:*

Pure red crossed with pure white produces red.

The red offspring received from its parents genes for both red (R) and white (W), but in this case red is dominant *and white* recessive, *so the offspring exhibits the color red. However, the offspring still carries both genes, and when two such offspring are crossed, several things can happen in the third generation, as shown in the table below, which is called a* Punnet square.

		2nd Parent	
		R	W
1st parent	R W	RR WR	RW WW

This table shows the possible combinations of genes. Use the fact that red is dominant over white to find

60. *P*(red); **61.** *P*(white).

Natural Science *Mendel found no dominance in snapdragons, with one red gene and one white gene producing pink-flowering offspring. These second-generation pinks, however, still carry one red and one white gene, and when they are crossed, the next generation still yields the Punnet square above. Find*

62. *P*(red); **63.** *P*(pink); **64.** *P*(white.)

(Mendel verified these probability ratios experimentally with large numbers of observations, and did the same for many character units other than flower color. The importance of his work, published in 1866, was not recognized until 1900.)

Natural Science *In most animals and plants, it is very unusual for the number of main parts of the organism (arms, legs, toes, flower petals, etc.) to vary from generation to generation. Some species, however, have* meristic variability, *in which the number of certain body parts varies from generation to generation. One researcher* studied the*

---

*From ''An Analysis of Variability in Guinea Pigs'' by J. R. Wright, from *Genetics*, Vol. 19, 1934, pp. 506–36. Reprinted by permission.

front feet of certain guinea pigs and produced the probabilities shown below.

$$P(\text{only four toes, all perfect}) = .77$$
$$P(\text{one imperfect toe and four good ones}) = .13$$
$$P(\text{exactly five good toes}) = .10$$

Find the probability of having the following.

65. No more than four good toes

66. Five toes, whether perfect or not

Monte Carlo Method *One way to solve a probability problem is to repeat the experiment (or a simulation of the experiment) many times, keeping track of the results. Then the probability can be estimated using the relative frequency m/n, where m favorable outcomes occur in n trials of an experiment. Use the computer to generate random numbers between 0 and 1. Assign the numbers x, $0 \le x < .5$, to represent heads and the numbers $.5 \le x < 1$ to represent tails. Then a series of, say, five of these random numbers represents the results of five coin tosses. For example, the three numbers .001234, .569241, .764219 represent heads, tails, tails.*

Using random numbers to estimate probabilities is called the Monte Carlo *method of finding probabilities. Suppose a coin is tossed five times. Use the Monte Carlo method to approximate the probabilities in Exercises 67 and 68. Then calculate the theoretical probabilities using the methods of the text and compare the results.*

67. *P*(4 heads)

68. *P*(2 heads, 1 tail, 2 heads) (in the order given)

Use the Monte Carlo method to approximate the following probabilities if 4 cards are drawn from a 52-card deck.

69. *P*(any 2 cards and then 2 kings)

70. *P*(exactly 2 kings)

Use Monte Carlo methods to approximate the probabilities in Exercises 71 and 72.

71. A jeweler received 8 identical watches each in a box marked with the series number of the watch. An assistant, who does not know that the boxes are marked, is told to polish the watches and then put them back in the boxes. She puts them in the boxes at random. What is the probability that she gets at least one watch in the right box?

72. A check room attendant has 10 hats but has lost the numbers identifying them. If he gives them back randomly, what is the probability that at least 2 of the hats are given back correctly?

1 Use the data in the table to find

(a) $P(B)$,

(b) $P(A')$,

(c) $P(B')$.

Answer:

(a) .45

(b) .4

(c) .55

8.5 CONDITIONAL PROBABILITY

The training manager for a large stockbrokerage firm has noticed that some of the firm's brokers use the firm's research advice, while other brokers tend to go with their own feelings of which stocks will go up. To see whether the research department performs better than the feelings of the brokers, the manager conducted a survey of 100 brokers, with results as shown in the following table.

	Picked Stocks that Went Up	Didn't Pick Stocks that Went Up	Totals
Used research	30	15	45
Didn't use research	30	25	55
Totals	60	40	100

Letting A be the event "picked stocks that went up" and letting B be the event "used research," $P(A)$, $P(A')$, $P(B)$, and $P(B')$ can be found. For example, the chart shows that a total of 60 brokers picked stocks that went up, so $P(A) = 60/100 = .6$. **1**

Suppose we want to find the probability that a broker using research will pick stocks that go up. From the table above, of the 45 brokers who use research, there are 30 who picked stocks that went up, so

$$P(\text{broker who uses research picks stocks that go up}) = \frac{30}{45} = .667$$

This is a different number than the probability that a broker picks stocks that go up, .6, since *we have additional information* (the broker uses research) *that has reduced the sample space*. In other words, we found the probability that a broker picks stocks that go up, A, given the additional information that the broker uses research, B. This is called the *conditional probability* of event A, given that event B has occurred, written $P(A \mid B)$. In the example above,

$$P(A \mid B) = \frac{30}{45}.$$

If we divide the numerator and denominator by 100, this can be written as

$$P(A \mid B) = \frac{\dfrac{30}{100}}{\dfrac{45}{100}} = \frac{P(A \cap B)}{P(B)},$$

where $P(A \cap B)$ represents, as usual, the probability that both A and B will occur.

To generalize this result, assume that E and F are two events for a particular experiment. Assume that the sample space S for this experiment has n possible equally likely outcomes. Suppose event F has m elements, and $E \cap F$ has k elements ($k \le m$). Using the fundamental principle of probability,

$$P(F) = \frac{m}{n} \quad \text{and} \quad P(E \cap F) = \frac{k}{n}.$$

We now want $P(E \mid F)$, the probability that E occurs given that F has occurred. Since we assume F has occurred, reduce the sample space to F: look only at the m elements inside F. (See Figure 8.23). Of these m elements, there are k elements where E also occurs, since $E \cap F$ has k elements. This makes

$$P(E \mid F) = \frac{k}{m}.$$

Divide numerator and denominator by n to get

$$P(E \mid F) = \frac{\dfrac{k}{n}}{\dfrac{m}{n}} = \frac{P(E \cap F)}{P(F)}.$$

This last result gives the definition of conditional probability.

The **conditional probability** of an event E given event F, written $P(E \mid F)$, is

$$P(E \mid F) = \frac{P(E \cap F)}{P(F)}, \quad P(F) \ne 0.$$

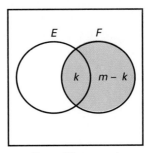

Event F has a total
of m elements.

FIGURE 8.23

▶ **EXAMPLE 1** Use the chart in the stockbroker's problem at the beginning of this section to find the following probabilities, where A is the event "picked stocks that went up" and B is the event "used research."

(a) $P(B \mid A)$

By the definition of conditional probability,

$$P(B \mid A) = \frac{P(B \cap A)}{P(A)}.$$

2 The table shows the results of a survey of a buffalo herd.

	Males	Females	Totals
Adults	500	1300	1800
Calves	520	500	1020
Totals	1020	1800	2820

Let M represent "male" and A represent "adult." Find each of the following.

(a) $P(M \mid A)$

(b) $P(M' \mid A)$

(c) $P(A \mid M')$

(d) $P(A' \mid M)$

Answer:

(a) 5/18

(b) 13/18

(c) 13/18

(d) 26/51

In the example, $P(B \cap A) = 30/100$, and $P(A) = 60/100$.

$$P(B \mid A) = \frac{\dfrac{30}{100}}{\dfrac{60}{100}} = \frac{1}{2}$$

If a broker picked stocks that went up, then the probability is 1/2 that the broker used research.

(b) $P(A' \mid B)$

In words, this is the probability that a broker picks stock that do not go up, even though he used research.

$$P(A' \mid B) = \frac{P(A' \cap B)}{P(B)} = \frac{\dfrac{15}{100}}{\dfrac{45}{100}} = \frac{1}{3}$$

(c) $P(B' \mid A')$

Here, we want the probability that a broker who picked stocks that did not go up, did not use research.

$$P(B' \mid A') = \frac{P(B' \cap A')}{P(A')} = \frac{\dfrac{25}{100}}{\dfrac{40}{100}} = \frac{5}{8} \quad \blacktriangleleft \;\; \boxed{2}$$

Venn diagrams can be used to illustrate problems in conditional probability. A Venn diagram for Example 1, in which the probabilities are used to indicate the number in the set defined by each region, is shown in Figure 8.24. In the diagram, $P(B \mid A)$ is found by *reducing the sample space to just set A*. Then $P(B \mid A)$ is the ratio of the number in that part of set B which is also in A to the number in set A, or $.3/.6 = .5$.

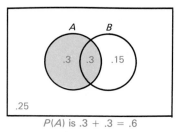

$P(A)$ is $.3 + .3 = .6$

FIGURE 8.24

▶**EXAMPLE 2** Given $P(E) = .4$, $P(F) = .5$, and $P(E \cup F) = .7$, find $P(E \mid F)$.

Find $P(E \cap F)$ first. Then use a Venn diagram to find $P(E \mid F)$. By the union rule,

$$P(E \cup F) = P(E) + P(F) - P(E \cap F)$$
$$.7 = .4 + .5 - P(E \cap F)$$
$$P(E \cap F) = .2.$$

3 Find $P(F \mid E)$ if $P(E) = .3$, $P(F) = .4$, and $P(E \cup F) = .6$.

Answer:
1/3

4 In Example 3, find the probability that exactly one coin showed a head, given that at least one was a head.

Answer:
2/3

Now use the probabilities to indicate the number in each region of the Venn diagram in Figure 8.25. $P(E \mid F)$ is the ratio of the probability of that part of E which is in F to the probability of F or

$$P(E \mid F) = \frac{P(E \cap F)}{P(F)} = \frac{.2}{.5} = \frac{2}{5} = .4. \quad \blacktriangleleft \quad \boxed{3}$$

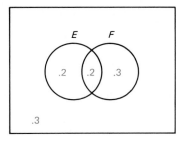

FIGURE 8.25

▶**EXAMPLE 3**   Two fair coins were tossed, and it is known that at least one was a head. Find the probability that both were heads.

The sample space has four equally likely outcomes, $S = \{hh, ht, th, tt\}$. Define two events:

$$E_1 = \text{at least 1 head} = \{hh, ht, th\}$$

and

$$E_2 = 2 \text{ heads} = \{hh\}.$$

Since there are four equally likely outcomes, $P(E_1) = 3/4$. Also, $P(E_1 \cap E_2) = 1/4$. We want the probability that both were heads, given that at least one was a head; that is, we want to find $P(E_2 \mid E_1)$. Because of the condition that at least one coin was a head, the reduced sample space is

$$\{hh, ht, th\}.$$

Since only one outcome in this reduced sample space is 2 heads,

$$P(E_2 \mid E_1) = \frac{1}{3}.$$

Alternatively, use the definition given above.

$$P(E_2 \mid E_1) = \frac{P(E_2 \cap E_1)}{P(E_1)} = \frac{1/4}{3/4} = \frac{1}{3} \quad \blacktriangleleft \quad \boxed{4}$$

PRODUCT RULE   If $P(E) \neq 0$ and $P(F) \neq 0$, then the definition of conditional probability shows that

$$P(E \mid F) = \frac{P(E \cap F)}{P(F)} \quad \text{and} \quad P(F \mid E) = \frac{P(F \cap E)}{P(E)}.$$

5 In a litter of puppies, 3 were female and 4 were male. Half the males were black. Find the probability that a puppy chosen at random from the litter would be a black male.

Answer:
2/7

Using the fact that $P(E \cap F) = P(F \cap E)$, and solving each of these equations for $P(E \cap F)$, we obtain the following rule.

Product Rule of Probability

If E and F are events, then $P(E \cap F)$ may be found by either of these formulas.

$$P(E \cap F) = P(F) \cdot P(E \mid F) \quad \text{or} \quad P(E \cap F) = P(E) \cdot P(F \mid E).$$

The product rule gives a method for finding the probability that events E and F both occur, as illustrated by the next few examples.

▶**EXAMPLE 4**   In a class with 2/5 women and 3/5 men, 25% of the women are business majors. Find the probability that a student chosen at random from the class is a female business major.

Let B and W represent the events "business major" and "women," respectively. We want to find $P(B \cap W)$. By the product rule,

$$P(B \cap W) = P(W) \cdot P(B \mid W).$$

From the given information, $P(W) = 2/5 = .4$ and the probability that a woman is a business major is $P(B \mid W) = .25$. Then

$$P(B \cap W) = .4(.25) = .10. \quad ◀ \quad \boxed{5}$$

The next examples show how a tree diagram is used to find conditional probabilities.

▶**EXAMPLE 5**   A company needs to hire a new director of advertising. It has decided to try to hire either person A or person B, who are assistant advertising directors for its major competitor. To decide between A and B, the company does research on the campaigns managed by A or B (none are managed by both), and finds that A is in charge of twice as many advertising campaigns as B. Also, A's campaigns have satisfactory results three out of four times, while B's campaigns have satisfactory results only two out of five times. Suppose one of the competitor's advertising campaigns (managed by A or B) is selected randomly.

We can represent this situation schematically as follows. Let A denote the event "Person A does the job" and B the event "Person B does the job." Let S be the event "satisfactory results" and U the event "unsatisfactory results." Then the given information can be summarized in the tree diagram in Figure 8.26. Since A does twice as many jobs as B, $P(A) = 2/3$ and $P(B) = 1/3$, as noted on the first stage branches of the tree. When A does a job, the probability of satisfactory results is 3/4 and of unsatisfactory results 1/4, as noted on the second-stage branches.

Similarly, the probabilities when B does the job are noted on the remaining second-stage branches. The composite branches labeled 1–4 represent the four mutually exclusive possibilities for the running and outcome of the campaign.

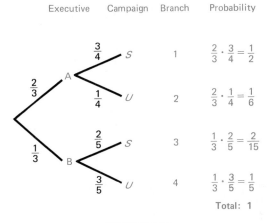

FIGURE 8.26

(a) Find the probability that A is in charge of a campaign that produces satisfactory results.

We are asked to find $P(A \cap S)$. We know that when A does the job, the probability of success is 3/4, that is, $P(S \,|\, A) = 3/4$. Hence, by the product rule,

$$P(A \cap S) = P(A) \cdot P(S \,|\, A) = \frac{2}{3} \cdot \frac{3}{4} = \frac{1}{2}.$$

The event $A \cap S$ is represented by branch 1 of the tree, and as we have just seen, its probability is the product of the probabilities of the pieces that make up that branch.

(b) Find the probability that B runs a campaign that produces satisfactory results.

We must find $P(B \cap S)$. The event is represented by branch 3 of the tree and, as before, its probability is the product of the probabilities of the pieces of that branch:

$$P(B \cap S) = P(B) \cdot P(S \,|\, B) = \frac{1}{3} \cdot \frac{2}{5} = \frac{2}{15}.$$

(c) What is the probability that the campaign is satisfactory?

The event S is the union of the mutually exclusive events $A \cap S$ and $B \cap S$, which are represented by branches 1 and 3 of the tree diagram. By the union rule,

$$P(S) = P(A \cap S) + P(B \cap S) = \frac{1}{2} + \frac{2}{15} = \frac{19}{30}.$$

6 Find each of the following probabilities for Example 5.

(a) $P(U \mid A)$

(b) $P(U \mid B)$

Answer:

(a) 1/6

(b) 1/5

7 Find the probability of drawing a green marble and then a white marble.

Answer:
1/5

8 In Example 6, find the probability of drawing 1 white and 1 red marble.

Answer:
1/5

Thus, the probability of an event that appears on several branches is the sum of the probabilities of each of these branches.

(d) What is the probability that the campaign is unsatisfactory?

$P(U)$ can be read from branches 2 and 4 of the tree.

$$P(U) = \frac{1}{6} + \frac{1}{5} = \frac{11}{30}$$

Alternatively, since U is the complement of S,

$$P(U) = 1 - P(S) = 1 - \frac{19}{30} = \frac{11}{30}.$$

(e) Find the probability that either A runs the campaign or the results are unsatisfactory (or possibly both).

Event A combines branches 1 and 2, while event S combines branches 1 and 3, so use branches 1, 2, and 3.

$$P(A \cup S) = \frac{1}{2} + \frac{1}{6} + \frac{2}{15} = \frac{4}{5} \quad \blacktriangleleft \quad \boxed{6}$$

▶**EXAMPLE 6**   From a box containing 1 red, 3 white, and 2 green marbles, two marbles are drawn one at a time without replacing the first before the second is drawn. Find the probability that one white and one green marble are drawn.

A tree diagram showing the various possible outcomes is given in Figure 8.27. In this diagram, W represents the event "drawing a white marble" and G represents "drawing a green marble." On the first draw, $P(W$ on the 1st$) = 3/6 = 1/2$ because three of the six marbles in the box are white. On the second draw, $P(G$ on the 2nd $\mid W$ on the 1st$) = 2/5$. One white marble has been removed, leaving 5, of which 2 are green.

We want to find the probability of drawing exactly one white marble and exactly one green marble. Two events satisfy this condition: drawing a white marble first and then a green one (branch 2 of the tree diagram), or drawing a green marble first and then a white one (branch 4). For branch 2,

$$P(W \text{ on 1st}) \cdot P(G \text{ on 2nd} \mid W \text{ on 1st}) = \frac{1}{2} \cdot \frac{2}{5} = \frac{1}{5}. \quad \boxed{7}$$

Since these two events are mutually exclusive, the final probability is the sum of the two probabilities.

$P(\text{one } W, \text{ one } G) = P(W \text{ on 1st}) \cdot P(G \text{ on 2nd} \mid W \text{ on 1st})$

$$+ P(G \text{ on 1st}) \cdot P(W \text{ on 2nd} \mid G \text{ on 1st}) = \frac{2}{5} \quad \blacktriangleleft \quad \boxed{8}$$

9 Find the probability of drawing a heart on the first draw and a black card on the second, if two cards are drawn without replacement.

Answer:
13/102 or .1275.

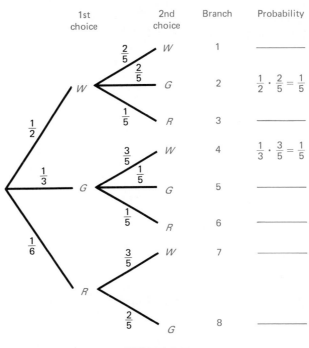

1st choice	2nd choice	Branch	Probability

FIGURE 8.27

The product rule is often used with *stochastic processes,* where the outcome of an experiment depends on the outcomes of previous experiments. For example, the outcome of a draw of a card from a deck depends on any cards previously drawn. (Stochastic processes are studied further in Section 9.4.)

▶ **EXAMPLE 7**   Two cards are drawn without replacement from an ordinary deck (52 cards). Find the probability that the first card is a heart and the second card is red.

Start with the tree diagram of Figure 8.28. (You may wish to refer to the deck of cards shown in Figure 8.17.) On the first draw, since there are 13 hearts in the 52 cards, the probability of drawing a heart first is $13/52 = 1/4$. On the second draw, since a (red) heart has been drawn already, there are 25 red cards in the remaining 51 cards. Thus the probability of drawing a red card on the second draw, given that the first is a heart, is 25/51. By the product rule of probability

$$P(\text{heart on 1st and red on 2nd})$$
$$= P(\text{heart on 1st}) \cdot P(\text{red on 2nd} \mid \text{heart on 1st})$$
$$= \frac{1}{4} \cdot \frac{25}{51} = \frac{25}{204} = .1225. \quad ◀ \quad \boxed{9}$$

10 Use the tree in Example 8 to find the probability that exactly one of the cards is red.

Answer:
$13/34 \approx .382$

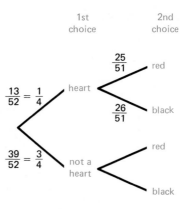

1st choice / 2nd choice

$\frac{13}{52} = \frac{1}{4}$ heart
$\frac{25}{51}$ red
$\frac{26}{51}$ black

$\frac{39}{52} = \frac{3}{4}$ not a heart
red
black

FIGURE 8.28

▶**EXAMPLE 8** Three cards are drawn, without replacement, from an ordinary deck. Find the probability that exactly 2 of the cards are red.

Here we need a tree diagram with three stages, as shown in Figure 8.29. The three branches indicated with arrows produce exactly 2 red cards from the draws. Multiply the probabilities along each of these branches and then add.

$$P(\text{exactly 2 red cards}) = \frac{26}{52} \cdot \frac{25}{51} \cdot \frac{26}{50} + \frac{26}{52} \cdot \frac{26}{51} \cdot \frac{25}{50} + \frac{26}{52} \cdot \frac{26}{51} \cdot \frac{25}{50}$$

$$= \frac{50,700}{132,600} = \frac{13}{34} \approx .382 \quad ◀ \quad \boxed{10}$$

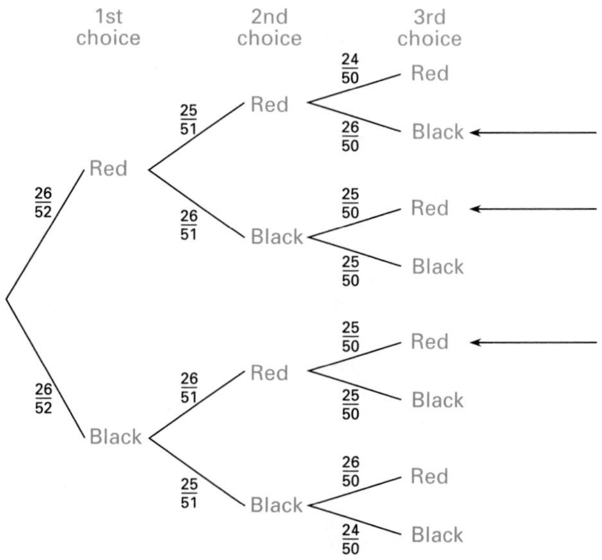

1st choice / 2nd choice / 3rd choice

$\frac{26}{52}$ Red
$\frac{25}{51}$ Red
$\frac{24}{50}$ Red
$\frac{26}{50}$ Black ←

$\frac{26}{51}$ Black
$\frac{25}{50}$ Red ←
$\frac{25}{50}$ Black

$\frac{26}{52}$ Black
$\frac{26}{51}$ Red
$\frac{25}{50}$ Red ←
$\frac{25}{50}$ Black

$\frac{25}{51}$ Black
$\frac{26}{50}$ Red
$\frac{24}{50}$ Black

FIGURE 8.29

11 Find the probability of getting 4 successive heads on 4 tosses of a fair coin.

Answer:
1/16

INDEPENDENT EVENTS Suppose a fair coin is tossed and shows heads. The probability of heads on the next toss is still 1/2; the fact that heads was the result on a given toss has no effect on the outcome of the next toss. Coin tosses are **independent events,** since the outcome of one toss does not help decide the outcome of the next toss. Rolls of a fair die are independent events; the fact that a 2 came up on one roll does not increase our knowledge of the outcome of the next roll. On the other hand, the events "today is cloudy" and "today is rainy" are **dependent events;** if it is cloudy, then there is an increased chance of rain.

If events E and F are independent, then the knowledge that E has occurred gives no (probability) information about the occurrence or nonoccurrence of event F. That is, $P(F)$ is exactly the same as $P(F \mid E)$, or

$$P(F \mid E) = P(F).$$

In fact, this is the formal definition of independent events.

E and F are **independent events** if

$$P(F \mid E) = P(F) \text{ or } P(E \mid F) = P(E).$$

When E and F are independent events, then $P(F \mid E) = P(F)$ and the product rule becomes

$$P(E \cap F) = P(E) \cdot P(F \mid E) = P(E) \cdot P(F).$$

Conversely, if this equation holds, then it follows that $P(F) = P(F \mid E)$. Consequently, we have this useful fact

Product Rule for Independent Events

E and F are independent events if and only if

$$P(E \cap F) = P(E) \cdot P(F).$$

▶ **EXAMPLE 9**   A calculator requires a key-stroke assembly and a logic circuit. Assume that 99% of the key-stroke assemblies are satisfactory and 97% of the logic circuits are satisfactory. Find the probability that a finished calculator will be satisfactory.

If the failure of a key-stroke assembly and the failure of a logic circuit are independent events, then

P (satisfactory calculator)

   $= P(\text{satisfactory key-stroke assembly}) \cdot P(\text{satisfactory logic circuit})$

   $= (.99)(.97) \approx .96.$

The probability of a defective calculator is $1 - .96 = .04.$ ◀   **11**

12 Find the probability that the mice of Example 10 will have a black-coated offspring.

Answer:
1/2

▶ **EXAMPLE 10**   When black-coated mice are crossed with brown-coated mice, a pair of genes, one from each parent, determines the coat color of the offspring. Let *b* represent the gene for brown and *B* the gene for black. If a mouse carries either one *B* gene and one *b* gene (*Bb* or *bB*) or two *B* genes (*BB*), the coat will be black. If the mouse carries two *b* genes (*bb*), the coat will be brown. Find the probability that a mouse, born to a brown-coated female and a black-coated male who is known to carry the *Bb* combination, will be brown.

To be brown-coated, the offspring must receive one *b* gene from each parent. The brown-coated parent carries two *b* genes, so that the probability of getting one *b* gene from the mother is 1. The probability of getting one *b* gene from the black-coated father is 1/2. Therefore, since these are independent events, the probability of a brown-coated offspring from these parents is $1 \cdot 1/2 = 1/2$. ◀ **12**

Caution   It is common for students to confuse the ideas of *mutually exclusive* events and *independent* events. Events *E* and *F* are mutually exclusive if $E \cap F = \emptyset$. For example, if a family has exactly one child, the only possible outcomes are $B = \{\text{boy}\}$ and $G = \{\text{girl}\}$. These two events are mutually exclusive. However, the events are *not* independent, since $P(G \mid B) = 0$ (if a family with only one child has a boy, the probability it has a girl is then 0). Since $P(G \mid B) \neq P(G)$, the events are not independent.

Of all the families with exactly *two* children, the events $G_1 = \{\text{first child is a girl}\}$ and $G_2 = \{\text{second child is a girl}\}$ are independent, since $P(G_2 \mid G_1)$ equals $P(G_2)$. However, G_1 and G_2 are not mutually exclusive, since $G_1 \cap G_2 = \{\text{both children are girls}\} \neq \emptyset$.

The only way to show that two events *E* and *F* are independent is to show that $P(F \mid E) = P(F)$ (or $P(E \mid F) = P(E)$).

8.5 EXERCISES

If a single fair die is rolled, find the probability of rolling the following.

1. 2, given that the number rolled was odd

2. 4, given that the number rolled was even

3. An even number, given that the number rolled was 6

If two fair dice are rolled (recall the 36-outcome sample space), find the probability of rolling the following.

4. A sum of 8, given the sum was greater than 7

5. A sum of 6, given the roll was a "double" (two identical numbers)

6. A double, given that the sum was 9

If two cards are drawn without replacement from an ordinary deck (see Example 7), find the following probabilities.

7. The second is a heart, given that the first is a heart.

8. The second is black, given that the first is a spade.

9. The second is a face card, given that the first is a jack.

10. The second is an ace, given that the first is not an ace.

11. A jack and a 10 are drawn.

12. An ace and a 4 are drawn.

13. Two black cards are drawn.

14. Two hearts are drawn.

Social Science *Use a tree diagram in Exercises 15–33. (See Examples 5–8.)*

Suppose 20% of the population is 65 or over, 26% of those 65 or over have loans, and 53% of those under 65 have loans. Find the probability that a person

15. is 65 or over and has a loan;

16. has a loan.

In a certain area, 15% of the population are joggers and 40% of the joggers are women. If 55% of those who do not jog are women, find the probability that an individual from that community is

17. a woman jogger;

18. not a jogger.

A survey has shown that 52% of the women in a certain community work outside the home. Of these women, 64% are married, while 86% of the women who do not work outside the home are married. Find the probability that a woman in that community is

19. married;

20. a single woman working outside the home.

Two-thirds of the population is on a diet at least occasionally. Of this group, 4/5 drink diet soft drinks, while 1/2 of the rest of the population drinks diet soft drinks. Find the probability that a person

21. drinks diet soft drinks;

22. diets but does not drink diet soft drinks.

A smooth-talking young man has a 1/3 probability of talking a policeman out of giving him a speeding ticket. The probability that he is stopped for speeding during a given weekend is 1/2. Find the following probabilities.

23. He will receive no tickets on a given weekend.

24. He will receive no tickets on three consecutive weekends.

Slips of paper marked with the digits 1, 2, 3, 4, and 5 are placed in a box and mixed well. If two slips are drawn (without replacement), find the following probabilities.

25. The first is even and the second is odd.

26. The first is a 3 and the second a number greater than 3.

27. Both are even.

28. Both are marked 3.

Two marbles are drawn without replacement from a jar with four black and three white marbles. Find the following probabilities.

29. Both are white.

30. Both are black.

31. The second is white, given that the first is black.

32. The first is black and the second is white.

33. One is black and the other is white.

The Midtown Bank has found that most customers at the tellers' windows either cash a check or make a deposit. The chart below indicates the transactions for one teller for one day.

	Cash Check	No Check	Totals
Make deposit	50	20	70
No deposit	30	10	40
Totals	80	30	110

Letting C represent "cashing a check" and D represent "making a deposit," express each of the following probabilities in words and find its value. (See Example 1.)

34. $P(C \mid D)$ **35.** $P(D' \mid C)$ **36.** $P(C' \mid D')$ **37.** $P(C' \mid D)$ **38.** $P[(C \cap D)']$

A pet shop has 10 puppies, 6 of them males. There are 3 beagles (1 male), 1 cocker spaniel (male), and 6 poodles. Construct a table similar to the one above and find the probability that one of these puppies, chosen at random, is

39. a beagle;

40. a beagle, given that it is a male;

41. a male, given that it is a beagle;

42. a cocker spaniel, given that it is a female;

43. a poodle, given that it is a male;

44. a female, given that it is a beagle.

A bicycle factory runs two assembly lines, A and B. If 95% of line A's products pass inspection, while only 90% of line B's products pass inspection, and 60% of the factory's bikes come off assembly line B (the rest off A), find the probability that one of the factory's bikes did not pass inspection and came off

45. assembly line A;

46. assembly line B.

---

47. **Natural Science** Both of a certain pea plant's parents had a gene for red and a gene for white flowers. (See Exercises 60–64 in Section 8.4.) If the offspring has red flowers, find the probability that it combined a gene for red and a gene for white (rather than two for red).

Assume that boy and girl babies are equally likely.

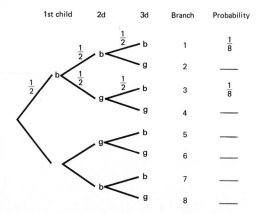

	1st child	2d	3d	Branch	Probability

Fill in the remaining probabilities on the tree diagram above and use the information to find the probability that a family with three children has all girls, given the following.

48. The first is a girl

49. The third is a girl

50. The second is a girl

51. At least two are girls

52. At least one is a girl

Natural Science *The following table shows frequencies for red-green color blindness, where M represents male and C represents color-blind.*

	M	M'	Totals
C	.042	.007	.049
C'	.485	.466	.951
Totals	.527	.473	1.000

Use this table to find the following probabilities.

53. $P(M)$ **54.** $P(C)$ **55.** $P(M \cap C)$

56. $P(M \cup C)$ **57.** $P(M \mid C)$ **58.** $P(M' \mid C)$

59. Are the events C and M described above dependent? (Recall that two events E and F are dependent if $P(E \mid F) \neq P(E)$.)

60. A scientist wishes to determine if there is any dependence between color blindness (C) and deafness (D). Given the probabilities listed in the table below, what should his findings be? (See Exercises 53–59.)

	D	D'	Totals
C	.0004	.0796	.0800
C'	.0046	.9154	.9200
Totals	.0050	.9950	1.0000

The Motor Vehicle Department has found that the probability of a person passing the test for a driver's license on the first try is .75. The probability that an individual who fails on the first test will pass on the second try is .80, and the probability that an individual who fails the first and second tests will pass the third time is .70. Find the probability that an individual

61. fails both the first and second tests;

62. will fail three times in a row;

63. will require at least two tries to pass the test.

Four different medications, C, D, E, and F, may be used to control high blood pressure. A physician usually prescribes C first because it is least likely to cause side effects. If blood pressure remains high, the patient is switched to D. If this fails to work, the patient is switched to E, and if necessary to F. The probability that C will work is .7. If C fails, the probability that D will work is .8. If D fails, the probability that E will work is .62, If E fails, the probability that F will work is .45 Find the probability that

64. a patient's blood pressure will not be reduced by any of the medications.

65. a patient will have to take at least two medications and will have his or her blood pressure reduced.

66. If medications A, B, and C fail, what is the probability that a patient's blood pressure will be reduced by medication D or E?

An insurance study produced the following information about beer drinking and driving on a typical weekend night.

	Percent of Drivers	Probability of Accident	Probability of Driver Being Injured if Accident Occurs
No drinks	29%	.003	.15
1–2 beers	37%	.004	.18
3–5 beers	23%	.011	.32
More than 5 beers	11%	.122	.54

Find the probability that a driver selected at random will

67. have an accident.

68. be injured in an accident.

69. be a nondrinker injured in an accident.

70. have drunk more than 2 beers and be injured in an accident.

71. have drunk less than 6 beers and have an accident, but escape injury.

72. be a drinker who is injured in an accident.

The probability that the first record by a singing group will be a hit is .32. If their first record is a hit, so are all their subsequent records. If their first record is not a hit, the probability of their second record and all subsequent ones being hits is .16. If the first two records are not hits, the probability that the third is a hit is .08. The probability of a hit continues to decrease by half with each successive nonhit record. Find the probability that

73. the group will have at least one hit in their first four records.

74. the group will have exactly one hit in their first three records.

75. the group will have a hit in their first six records if the first three are not hits.

According to a booklet put out by EastWest Airlines, 98% of all scheduled EastWest flights actually take place. (The other flights are canceled due to weather, equipment problems, and so on.) Assume that the event that a given flight takes place is independent of the event that another flight takes place. (See Example 9.)

76. Elizabeth Guerera plans to visit her company's branch offices; her journey requires three separate flights on East-West Airlines. What is the probability that all these flights will take place?

77. Based on the reasons we gave for a flight to be canceled, how realistic is the assumption of independence that we made?

78. In one area, 4% of the population drives a luxury car. However, 17% of the CPAs drive a luxury car. Are the events "drive a luxury car" and "person is a CPA" independent?

79. Corporations such as banks, where a computer is essential to day-to-day operations, often have a second, backup computer in case of failure by the main computer. Suppose that there is a .003 chance that the main computer will fail in a given time period and a .005 chance that the backup computer will fail while the main computer is being repaired. Assume these failures represent independent events, and find the fraction of the time that the corporation can assume it will have computer service. How realistic is our assumption of independence?

80. A key component of a space rocket will fail with a probability of .03. How many such components must be used as backups to ensure the probability that at least one of the components will work is .999999?

8.6 BAYES' FORMULA

Suppose the probability that a person smokes a pack or more of cigarettes daily, given that the person has lung cancer, is known. For a research project, it might be necessary to know the probability that a person gets lung cancer, given that the person smokes a pack or more of cigarettes daily. More generally, if $P(E \mid F)$ is known for two events E and F, can $P(F \mid E)$ be found? It turns out that it can, using the formula to be developed in this section. To find this formula, we start with the product rule:

$$P(E \cap F) = P(E) \cdot P(F \mid E) \quad \text{or} \quad P(F \cap E) = P(F) \cdot P(E \mid F).$$

From the fact that $P(E \cap F) = P(F \cap E)$, we can set the right-hand sides equal to get

$$P(E) \cdot P(F \mid E) = P(F) \cdot P(E \mid F),$$

from which

$$P(F \mid E) = \frac{P(F) \cdot P(E \mid F)}{P(E)}. \quad \boxed{1} \qquad (1)$$

Given the two events E and F, if E occurs, then either F also occurs or F' also occurs. The probabilities of $E \cap F$ and $E \cap F'$ can be expressed as follows.

$$P(E \cap F) = P(F) \cdot P(E \mid F)$$
$$P(E \cap F') = P(F') \cdot P(E \mid F')$$

Since $(E \cap F) \cup (E \cap F') = E$ (because $F \cup F'$ is the entire sample space), and $(E \cap F) \cap (E \cap F') = \emptyset$ (because $F \cap F' = \emptyset$),

$$P(E) = P(E \cap F) + P(E \cap F')$$

or

$$P(E) = P(F) \cdot P(E \mid F) + P(F') \cdot P(E \mid F').$$

Substituting this in equation (1) produces the following result, a special case of Bayes' formula, which is generalized later in this section.

Bayes' Formula (Special Case)

$$P(F \mid E) = \frac{P(F) \cdot P(E \mid F)}{P(F) \cdot P(E \mid F) + P(F') \cdot P(E \mid F')}. \quad \boxed{2} \qquad (2)$$

---

1 Use equation (1) to find $P(F \mid E)$ if $P(F) = 1/4$, $P(E) = 1/3$, and $P(E \mid F) = 1/2$.

Answer:
3/8

2 Use the special case of Bayes' formula to find $P(F \mid E)$ if $P(F) = .2$, $P(E \mid F) = .1$, and $P(E \mid F') = .3$. (Hint: $P(F') = 1 - P(F)$.)

Answer:
$1/13 \approx .077$

3 In Example 1, find $P(E' \mid A)$.

Answer:
$6/7 \approx .857$

▶**EXAMPLE 1**   For a fixed length of time, the probability of worker error on a certain production line is .1, the probability that an accident will occur when there is a worker error is .3, and the probability that an accident will occur when there is no worker error is .2. Find the probability of a worker error if there is an accident.

Let A represent "accident" and E represent "worker error." From the information above,

$$P(E) = .1, \quad P(A \mid E) = .3, \quad \text{and} \quad P(A \mid E') = .2.$$

These probabilities are shown in the tree diagram in Figure 8.30.

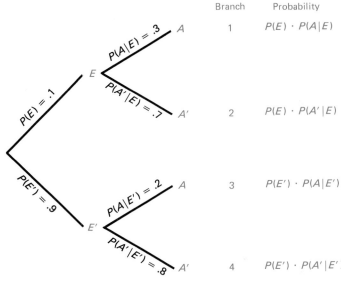

	Branch	Probability
A	1	$P(E) \cdot P(A \mid E)$
A'	2	$P(E) \cdot P(A' \mid E)$
A	3	$P(E') \cdot P(A \mid E')$
A'	4	$P(E') \cdot P(A' \mid E')$

FIGURE 8.30

Find $P(E \mid A)$ using equation (2) above:

$$P(E \mid A) = \frac{P(E) \cdot P(A \mid E)}{P(E) \cdot P(A \mid E) + P(E') \cdot P(A \mid E')}$$

$$= \frac{(.1)(.3)}{(.1)(.3) + (.9)(.2)} = \frac{1}{7} \approx .143. \quad ◀ \boxed{3}$$

Equation (2) above can be generalized to more than two possibilities with the tree diagram of Figure 8.31. This diagram shows the paths that can produce an event E. We assume that events $F_1, F_2, \ldots , F_n$ are pairwise mutually exclusive events (that is, events which, taken two at a time, are disjoint), whose union is the sample space, and E is an event that has occurred. See Figure 8.32.

The probability $P(F_i \mid E)$, where $1 \le i \le n$, can be found by dividing the probability for the branch containing $P(E \mid F_i)$ by the sum of the probabilities of all the branches producing event E.

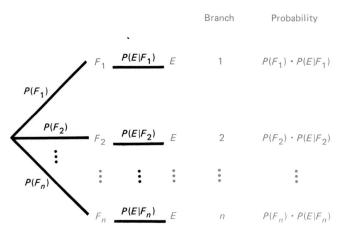

FIGURE 8.31

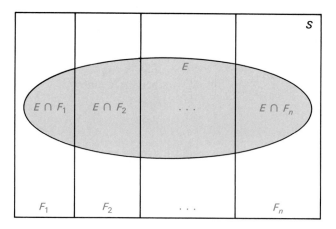

FIGURE 8.32

Bayes' Formula

$$P(F_i \mid E) = \frac{P(F_i) \cdot P(E \mid F_i)}{P(F_1) \cdot P(E \mid F_1) + \ldots + P(F_n) \cdot P(E \mid F_n)}.$$

This result is known as **Bayes' formula,** after the Reverend Thomas Bayes (1702–61), whose paper on probability was published about two hundred years ago.

The statement of Bayes' formula can be daunting. Actually, it is easier to remember the formula by thinking of the tree diagram that produced it. Go through the following steps.

4 In Example 2, find

(a) $P(F_1 \mid E)$;

(b) $P(F_2 \mid E)$.

Answer:

(a) $4/13 \approx .3075$

(b) $5/13 \approx .385$

Using Bayes' Formula

Step 1 Start a tree diagram with branches representing events F_1, F_2, . . . , F_n. Label each branch with its corresponding probability.

Step 2 From the end of each of these branches, draw a branch for event E. Label this branch with the probability of getting to it, or $P(E \mid F_i)$.

Step 3 There are now n different paths that result in event E. Next to each path, put its probability—the product of the probabilities that the first branch occurs, $P(F_i)$, and that the second branch occurs, $P(E \mid F_i)$: that is, $P(F_i) \cdot P(E \mid F_i)$.

Step 4 $P(F_i \mid E)$ is found by dividing the probability of the branch for F_i by the sum of the probabilities of all the branches producing event E.

▶**EXAMPLE 2** Based on past experience, a company knows that an experienced machine operator (one or more years of experience) will produce a defective item 1% of the time. Operators with some experience (up to one year) have a 2.5% defect rate, while new operators have a 6% defect rate. At any one time, the company has 60% experienced employees, 30% with some experience, and 10% new employees. Find the probability that a particular defective item was produced by a new operator.

Let E represent the event "item is defective," with F_1 representing "item was made by an experienced operator," F_2 "item was made by an operator with some experience," and F_3 "item was made by a new operator." Then

$$P(F_1) = .60 \qquad P(E \mid F_1) = .01$$
$$P(F_2) = .30 \qquad P(E \mid F_2) = .025$$
$$P(F_3) = .10 \qquad P(E \mid F_3) = .06.$$

We need to find $P(F_3 \mid E)$, the probability that an item was produced by a new operator, given that it is defective. First, draw a tree diagram using the given information, as in Figure 8.33. The steps leading to event E are shown in heavy type.

Find $P(F_3 \mid E)$ using the bottom branch of the tree in Figure 8.33: divide the probability for this branch by the sum of the probabilities of all the branches leading to E.

$$P(F_3 \mid E) = \frac{.10(.06)}{.60(.01) + .30(.025) + .10(.06)} = \frac{.006}{.0195} = \frac{4}{13} \approx .3075 \quad ◀ \quad \boxed{4}$$

After working Problem 4 at the side, check that $P(F_1 \mid E) + P(F_2 \mid E) + P(F_3 \mid E) = 1$. (That is, the defective item was made by *someone*.)

5 In Example 3, find the probability that the defective item came from

(a) supplier 3;

(b) supplier 6.

Answer:

(a) .340

(b) .137

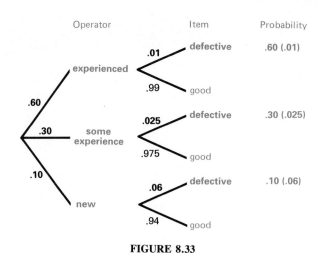

FIGURE 8.33

▶**EXAMPLE 3**   A manufacturer buys items from six different suppliers. The fraction of the total number of items obtained from each supplier, along with the probability that an item purchased from that supplier is defective, is shown in the following table.

Supplier	*Fraction of Total Supplied*	*Probability of Defective*
1	.05	.04
2	.12	.02
3	.16	.07
4	.23	.01
5	.35	.03
6	.09	.05

Find the probability that a defective item came from supplier 5.

Let F_1 be the event that an item came from supplier 1, with F_2, F_3, F_4, F_5, and F_6 defined in a similar manner. Let E be the event that an item is defective. We want to find $P(F_5 \mid E)$. Use the probabilities in the table above to prepare a tree diagram. By Bayes' formula,

$$P(F_5 \mid E) =$$
$$\frac{(.35)(.03)}{(.05)(.04) + (.12)(.02) + (.16)(.07) + (.23)(.01) + (.35)(.03) + (.09)(.05)}$$
$$= \frac{.0105}{.0329} \approx .319.$$

There is about a 32% chance that a defective item came from supplier 5. ◀ **5**

8.6 EXERCISES

For two events, M and N, P(M) = .4, P(N | M) = .3, and P(N | M') = .4. Find each of the following. (See Example 1.)

1. $P(M \mid N)$ **2.** $P(M' \mid N)$

For mutually exclusive events R_1, R_2, R_3, $P(R_1) = .05$, $P(R_2) = .6$, and $P(R_3) = .35$. Also, $P(Q \mid R_1) = .40$, $P(Q \mid R_2) = .30$, and $P(Q \mid R_3) = .60$. Find each of the following. (See Example 2.)

3. $P(R_1 \mid Q)$ **4.** $P(R_2 \mid Q)$ **5.** $P(R_3 \mid Q)$ **6.** $P(R_1' \mid Q)$

Suppose three jars have the following contents: 2 black balls and 1 white ball in the first; 1 black ball and 2 white balls in the second; 1 black ball and 1 white ball in the third. If the probability of selecting one of the three jars is 1/2, 1/3, and 1/6, respectively, find the probability that if a white ball is drawn, it came from the

7. second jar;   **8.** third jar.

Management *Of all the people applying for a certain job, 70% are qualified, and 30% are not. The personnel manager claims that she approves qualified people 85% of the time; she approves an unqualified person 20% of the time. Find each of the following probabilities.*

9. A person is qualified if he or she was approved by the manager.

10. A person is unqualified if he or she was approved by the manager.

Management *A building contractor buys 70% of his cement from supplier A, and 30% from supplier B. A total of 90% of the bags from A arrive undamaged, while 95% of the bags from B come undamaged. Give the probability that a damaged bag is from supplier*

11. A;   **12.** B.

Management *The probability that a customer of a local department store will be a "slow pay" is .02. The probability that a "slow pay" will make a large down payment when buying a refrigerator is .14. The probability that a person who is not a "slow pay" will make a large down payment when buying a refrigerator is .50. Suppose a customer makes a large down payment on a refrigerator. Find the probability that the customer is*

13. a "slow pay";   **14.** not a "slow pay."

Management *Companies A, B, and C produce 15%, 40%, and 45%, respectively, of the major appliances sold in a certain area. In that area, 1% of the Company A appliances, $1\frac{1}{2}$% of the Company B appliances, and 2% of the Company C appliances need service within the first year. Suppose an appliance that needs service within the first year is chosen at random; find the probability that it was manufactured by Company*

15. A;   **16.** B.

Management *On a given weekend in the fall, a tire company can buy television advertising time for a college football game, a baseball game, or a professional football game. If the company sponsors the college game, there is a 70% chance of a high rating, a 50% chance if they sponsor a baseball game, and a 60% chance if they sponsor a professional football game. The probability of the company sponsoring these various games is .5, .2, and .3, respectively. Suppose the company does get a high rating; find the probability that it sponsored*

17. a college game;   **18.** a professional football game.

Management *According to a business publication, there is a 50% chance of a booming economy next summer, a 20% chance of a mediocre economy, and a 30% chance of a recession. The probabilities that a particular investment strategy will produce a high profit under each of these possibilities are .1, .6, and .3, respectively. Suppose it turns out that the strategy does produce huge profits; find the probability that the economy was*

19. booming;   **20.** in recession.

21. Management A manufacturing firm finds that 70% of its new hires turn out to be good workers and 30% poor workers. All current workers are given a reasoning test. Of the good workers, 80% pass it; 40% of the poor workers pass it. Assume that these figures will hold true in the future. If the company makes the test part of its hiring procedure and only hires people who meet the previous requirements and pass the test, what percent of the new hires will turn out to be good workers?

22. Management A bank finds that the relationship between mortgage defaults and the size of the down payment is given by this table.

Down Payment (%)	5%	10%	20%	25%
Number of mortgages with this down payment	1260	700	560	280
Probability of default	.05	.03	.02	.01

If a default occurs, what is the probability that it is on a mortgage with a 5% down payment? (See Examples 2 and 3.)

Natural Science *In a test for toxemia, a disease that affects pregnant women, the woman lies on her left side and then rolls over on her back. The test is considered positive if there is a 20 mm rise in her blood pressure within one minute. The results have produced the following probabilities, where T represents having toxemia at some time during the pregnancy, and N represents a negative test.*

$$P(T' \mid N) = .90 \quad \text{and} \quad P(T \mid N') = .75$$

Assume that $P(N') = .11$, and find each of the following.

23. $P(N \mid T)$ **24.** $P(N' \mid T)$

25. Natural Science The probability that a person with certain symptoms has hepatitis is .8. The blood test used to confirm this diagnosis gives positive results for 90% of those who have the disease and 5% of those without the disease. What is the probability that an individual with the symptoms who reacts positively to the test has hepatitis?

26. Social Science In a certain county, the Democrats have 53% of the registered voters, 12% of whom are under 21. The Republicans have 47% of all registered voters, of whom 10% are under 21. If Kay is a registered voter who is under 21, what is the probability that she is a Democrat?

Social Science *The following table shows the fraction of the population in various income levels, as well as the probability that a person from that income level will take an airline flight within the next year.*

Income Level	Proportion of Population	Probability of a Flight During the Next Year
$0–$5999	12.8%	.04
$6000–$9999	14.6	.06
$10,000–$14,999	18.5	.07
$15,000–$19,999	17.8	.09
$20,000–$24,999	13.9	.12
$25,000 and over	22.4	.13

If a person is selected at random from an airline flight, find the probability that the person has the following income level. (See Example 3.)

27. $10,000-$14,999 **28.** $25,000 and over

Social Science *The following table shows the proportion of people over 18 who are in the various age categories, along with the probability that a person in a given age category will vote in a general election.*

Age	Proportion of Voting Age Population	Probability of a Person of this Age Voting
18–21	11.0%	.48
22–24	7.6	.53
25–44	37.6	.68
45–64	28.3	.64
65 or over	15.5	.74

Suppose a voter is picked at random. Find the probability that the voter is in the following age categories.

29. 18–21 **30.** 65 or over

KEY TERMS AND SYMBOLS

8.1 { } set braces
  $\in$ is an element of
  $\notin$ is not an element of
  $\subseteq$ is a subset of
  $\nsubseteq$ is not a subset of
  A' complement of set A
  $\cap$ set intersection
  $\cup$ set union
  set
  element (member) of a set
  equal sets
  set-builder notation
  universal set
  subset
  empty set
  tree diagram
  Venn diagram
  complement
  intersection
  disjoint sets
  union

8.3 $P(E)$ probability of event E
  random
  experiment
  trial

outcome
fair coin
sample space
event
simple event
certain event
impossible event
basic probability principle
probability of an event

8.4 mutually exclusive events
  Union Rule
  odds
  empirical probability
  probability distribution

8.5 $P(F \mid E)$ probability of F, given
    that E has occurred
  conditional probability
  Product Rule of Probability
  independent events
  dependent events
  Product Rule for Independent
    Events

8.6 Bayes' formula

KEY CONCEPTS

SETS

Set A is a **subset** of set B if every element of A is also an element of B.

A set of n elements has 2^n subsets.

Operations on Sets: Let A and B be any sets with universal set U.

The **complement** of A is $A' = \{x \mid x \notin A \text{ and } x \in U\}$.

The **intersection** of A and B is $A \cap B = \{x \mid x \in A \text{ and } x \in B\}$.

The **union** of A and B is $A \cup B = \{x \mid x \in A \text{ or } x \in B \text{ or both}\}$.

$n(A \cup B) = n(A) + n(B) - n(A \cap B)$

PROBABILITY

If $n(S) = n$ and $n(E) = m$, where S is the sample space, then $P(E) = \dfrac{m}{n}$.

The probability of any outcome is a number between 0 and 1, inclusive. The sum of the probabilities of all possible distinct outcomes in a sample space is 1.

Let E and F be events from a sample space S.

$P(E') = 1 - P(E)$ and $P(E) = 1 - P(E')$.

$P(E \cup F) = P(E) + P(F) - P(E \cap F)$.

$P(E \mid F) = \dfrac{P(E \cap F)}{P(F)}, \quad P(F) \neq 0$.

$P(E \cap F) = P(F) \cdot P(E \mid F) \quad \text{or} \quad P(E \cap F) = P(E) \cdot P(F \mid E)$.

Odds: The odds in favor of event E are given by the ratio of $P(E)$ to $P(E')$.

Events E and F are **mutually exclusive** if $E \cap F = \emptyset$. In that case, $P(E \cup F) = P(E) + P(F)$.

Events E and F are **independent events** if $P(F \mid E) = P(F)$ or $P(E \mid F) = P(E)$. In that case, $P(E \cap F) = P(E) \cdot P(F)$.

Bayes' Formula: $P(F_i \mid E) = \dfrac{P(F_i) \cdot P(E \mid F_i)}{P(F_i) \cdot P(E \mid F_i) + \ldots + P(F_n) \cdot P(E \mid F_n)}$.

CHAPTER 8   REVIEW EXERCISES

Write true *or* false *for each of the following.*

1. $-6 \in \{8, 4, -3, -9, 6\}$ **2.** $8 \in \{3, 9, 7\}$ **3.** $2 \notin \{0, 1, 2, 3, 4\}$ **4.** $0 \in \{0, 1, 2, 3, 4\}$

5. $\{3, 4, 5\} \subseteq \{2, 3, 4, 5, 6\}$ **6.** $\{1, 2, 5, 8\} \subseteq \{1, 2, 5, 10, 11\}$ **7.** $\{3, 6, 9, 10\} \subseteq \{3, 9, 11, 13\}$

8. $\emptyset \subseteq \{1\}$ **9.** $\{2, 8\} \not\subseteq \{2, 4, 6, 8\}$ **10.** $0 \subseteq \emptyset$

List the elements in the following sets.

11. $\{x \mid x$ is a counting number more than 8 and less than 5$\}$ **12.** $\{x \mid x$ is an integer, $-3 \leq x < 1\}$

13. $\{$all counting numbers less than 5$\}$ **14.** $\{$all whole numbers not greater than 2$\}$

Let $U = \{a, b, c, d, e, f, g\}$, $K = \{c, d, f, g\}$, *and* $R = \{a, c, d, e, g\}$. *Find the following.*

15. The number of subsets of K **16.** The number of subsets of R

17. K' **18.** R' **19.** $K \cap R$ **20.** $K \cup R$ **21.** $(K \cap R)'$

22. $(K \cup R)'$ **23.** $\emptyset'$ **24.** U'

Let $U = \{$all employees of the K.O. Brown Company$\}$

   $A = \{$employees in the accounting department$\}$

   $B = \{$employees in the sales department$\}$

   $C = \{$employees with at least 10 years' service$\}$

   $D = \{$employees with an MBA degree$\}$.

Describe the following sets in words.

25. $A \cap C$ **26.** $B \cap D$ **27.** $A \cup D$ **28.** $A' \cap D$ **29.** $B' \cap C'$ **30.** $(B \cup C)'$

Draw Venn diagrams for Exercises 31-34.

31. $B \cup A'$ **32.** $A' \cap B$ **33.** $A' \cap (B' \cap C)$ **34.** $(A \cup B)' \cap C$

Social Science *The following data apply to Exercises 35–38. A telephone survey of television viewers revealed that*

20 *watch situation comedies*
19 *watch game shows*
27 *watch movies*
 5 *watch both situation comedies and game shows*
 8 *watch both game shows and movies*
10 *watch both situation comedies and movies*
 3 *watch all three*
 6 *watch none of these.*

35. How many viewers were interviewed?

36. How many viewers watch comedies and movies but not game shows?

37. How many viewers watch only movies?

38. How many viewers watch comedies and game shows but not movies?

Write sample spaces for the following.

39. A die is rolled.

40. A card is drawn from a deck containing only the thirteen spades.

41. The weight of a person is measured to the nearest half pound; the scale will not measure more than 300 pounds.

42. A coin is tossed four times.

An urn contains five balls labeled 3, 5, 7, 9, and 11, while a second urn contains four red and two green balls. An experiment consists of pulling one ball from each urn, in turn. Write each of the following.

43. The sample space

44. Event E, the first ball is greater than 5

45. Event F, the second ball is green

46. Are the outcomes in the sample space equally likely?

Management *A company sells typewriters and copiers. Let E be the event "a customer buys a typewriter," and let F be the event "a customer buys a copier." Write each of the following using ∩, ∪, or ', as necessary.*

47. A customer buys neither.

48. A customer buys at least one.

When a single card is drawn from an ordinary deck, find the probability that it will be

49. a heart; **50.** a red queen; **51.** a face card; **52.** black or a face card;

53. red, given it is a queen; **54.** a jack, given it is a face card; **55.** a face card, given it is a king.

Find the odds in favor of a card drawn being

56. a club; **57.** a black jack; **58.** a red face card or a queen.

A sample shipment of five swimming pool filters is chosen at random. The probability of exactly 0, 1, 2, 3, 4, or 5 filters being defective is given in the following table.

Number defective	0	1	2	3	4	5
Probability	.31	.25	.18	.12	.08	.06

Find the probability that the following number of filters is defective.

59. No more than 3

60. At least 3

Natural Science *The square shows the four possible (equally likely) combinations when both parents are carriers of the sickle cell anemia trait. Each carrier parent has normal cells (N) and trait cells (T).*

		2nd Parent	
		N_2	T_2
1st parent	N_1 T_1		$N_1 T_2$

61. Complete the table.

62. If the disease occurs only when two trait cells combine, find the probability that a child born to these parents will have sickle cell anemia.

63. The child will carry the trait but not have the disease if a normal cell combines with a trait cell. Find this probability.

64. Find the probability that the child is neither a carrier nor has the disease.

Find the probability for the following sums when two fair dice are rolled.

65. 8   **66.** 0   **67.** At least 10

68. No more than 5   **69.** Odd and greater than 8

70. 12, given it is greater than 10

71. 7, given that one die is 4

72. At least 9, given that one die is 5

Suppose $P(E) = .51$, $P(F) = .37$, and $P(E \cap F) = .22$. Find each of the following probabilities.

73. $P(E \cup F)$ **74.** $P(E \cap F')$

75. $P(E' \cup F)$ **76.** $P(E' \cap F')$

Social Science *The table below shows the results of a survey of 1000 new or used car buyers of a certain model car.*

	Satisfied	Not Satisfied	Totals
New	300	100	400
Used	450	150	600
Totals	750	250	1000

Let S represent the event ''satisfied,'' and N the event ''bought a new car.'' Find each of the following.

77. $P(N \cap S)$ **78.** $P(N \cup S')$ **79.** $P(N \mid S)$

80. $P(N' \mid S)$ **81.** $P(S \mid N')$ **82.** $P(S' \mid N')$

83. For events E and F, $P(E) = .2$, $P(F \mid E) = .3$, and $P(F \mid E') = .2$. Find each of the following.
   (a) $P(E \mid F)$ **(b)** $P(E \mid F')$

Of the appliance repair shops listed in the phone book, 80% are competent and 20% are not. A competent shop can repair an appliance correctly 95% of the time; an incompetent shop can repair an appliance correctly 60% of the time. Suppose an appliance was repaired correctly. Find the probability that it was repaired by

84. a competent shop;   **85.** an incompetent shop.

Suppose an appliance was repaired incorrectly. Find the probability that it was repaired by

86. a competent shop;   **87.** an incompetent shop.

88. Box A contains 5 red balls and 1 black ball; box B contains 2 red and 3 black balls. A box is chosen at random, and a ball is selected from it. The probability of choosing box A is 3/8. If the selected ball is black, what is the probability that it came from box A?

89. Find the probability that the ball in Exercise 88 came from box B, given that it is red.

90. A manufacturer buys items from four different suppliers. The fraction of the total number of items that is obtained from each supplier, along with the probability that an item purchased from that supplier is defective, is shown in the following table.

Supplier	Fraction of Total Supplied	Probability of Defective
1	.17	.04
2	.39	.02
3	.35	.07
4	.09	.03

(a) Find the probability that a defective item came from supplier 4.
(b) Find the probability that a defective item came from supplier 2.

In searching for a new drug with commerical possibilities, drug company researchers use the ratio

$$N_S:N_A:N_D:1.$$

That is, if the company gives preliminary screening to N_S substances, it may find that N_A of them are worthy of further study, with N_D of these surviving into full-scale development. Finally, one of the substances will result in a marketable drug. Typical numbers used by Smith Kline and French Laboratories in planning research budgets might be $2000:30:8:1$.

EXERCISES

1. Suppose a compound has been chosen for preliminary study. Use the ratio 2000:30:8:1 to find the probability that the compound will survive and become a marketable drug.

2. Find the probability that the compound will not lead to a marketable drug.

3. Suppose a such compounds receive preliminary screening. Set up the probability that none of them produces a marketable drug. (Assume independence throughout these exercises.)

4. Use your results from exercise 3 to set up the probability that a least one of the drugs will prove marketable.

5. Suppose now that N scientists are employed in the preliminary screening, and that each scientist can screen c compounds per year. Set up the probability that no marketable drugs will be discovered in a year.

6. Set up the probability that at least one marketable drug will be discovered.

For the following exercises, evaluate your answer in Exercise 6, which should have been $1 - (1999/2000)^{Nc}$, for the following values of N and c. Use a calculator with a y^x key or a computer.

7. $N = 100, \quad c = 6$ **8.** $N = 25, \quad c = 10$

*From "Scientific Manpower Allocation to New Drug Screening Program" by E. B. Pyle III, B. Douglas, G. W. Ebright, W. J. Westlake, and A. D. Bender, from *Management Science*, Vol. 19, No. 12, August 1973. Copyright © 1973 The Institute of Management Sciences. Reprinted by permission.

A first down is desirable in football—it guarantees four more plays by the team making it, assuming no score or turnover occurs in the plays. After getting a first down, a team can get another by advancing the ball at least 10 yards. During the four plays given by a first down, a team's position will be indicated by a phrase such as "third and 4," which means that the team has already had two of its four plays, and that 4 more yards are needed to get the 10 yards necessary for another first down. An article in a management journal* offers the following results for 189 games of a recent National Football League season. "Trials" represents the number of times a team tried to make a first down, given that it was currently playing either a third or a fourth down. Here n represents the number of yards still needed for a first down.

n	Trials	Successes	Probability of Making First Down with n Yards to Go
1	543	388	
2	327	186	
3	356	146	
4	302	97	
5	336	91	

EXERCISES

1. Use the basic probability definition to complete the table.

2. Why isn't the sum of the answers in Exercise 1 equal to 1?

*From "Optimal Strategies of Fourth Down" by Virgil Carter and Robert Machols from *Management Science*, vol. 24, no. 16, December 1978. Copyright © 1978 The Institute of Management Sciences. Reprinted by permission.

When a patient is examined, information, typically incomplete, is obtained about his state of health. Probability theory provides a mathematical model appropriate for this situation, as well as a procedure for quantitatively interpreting such partial information to arrive at a reasonable diagnosis.*

To do this, list the states of health that can be distinguished in such a way that the patient can be in one and only one state at the time of the examination. Each state of health H is associated with a number $P(H)$ between 0 and 1 such that the sum of all these numbers is 1. This number $P(H)$ represents the probability, before examination, that a patient is in the state of health H, and $P(H)$ may be chosen subjectively from medical experience, using any information available prior to the examination. The probability may be most conveniently established from clinical records; that is, a mean probability is established for patients in general, although the number would vary from patient to patient. Of course, the more information that is brought to bear in establishing $P(H)$, the better the diagnosis.

For example, limiting the discussion to the condition of a patient's heart, suppose there are exactly 3 states of health, with probabilities as follows.

State of Health H		$P(H)$
H_1	patient has a normal heart	.8
H_2	patient has minor heart irregularities	.15
H_3	patient has a severe heart condition	.05

Having selected $P(H)$, the information of the examination is processed. First, the results of the examination must be classified. The examination itself consists of observing the state of a number of characteristics of the patient. Let us assume that the examination for a heart condition consists of a stethoscope examination and a cardiogram. The outcome of such an examination, C, might be one of the following:

C_1—stethoscope shows normal heart and cardiogram shows normal heart;

C_2—stethoscope shows normal heart and cardiogram shows minor irregularities;

and so on.

*From "Probabilistic Medical Diagnosis," Roger Wright, *Some Mathematical Models in Biology*, Robert M. Thrall, ed., (The University of Michigan, 1967), by permission of Robert M. Thrall.

It remains to assess for each state of health H the conditional probability $P(C \mid H)$ of each examination outcome C using only the knowledge that a patient is in a given state of health. (This may be based on the medical knowledge and clinical experience of the doctor.) The conditional probabilities $P(C \mid H)$ will not vary from patient to patient, so that they may be built into a diagnostic system, although they should be reviewed periodically.

Suppose the result of the examination is C_1. Let us assume the following probabilities.

$$P(C_1 \mid H_1) = .9$$
$$P(C_1 \mid H_2) = .4$$
$$P(C_1 \mid H_3) = .1$$

Now, for a given patient, the appropriate probability associated with each state of health H, after examination, is $P(H \mid C)$ where C is the outcome of the examination. This can be calculated by using Bayes' theorem. For example, to find $P(H_1 \mid C_1)$—that is, the probability that the patient has a normal heart given that the examination showed a normal stethoscope examination and a normal cardiogram—we use Bayes' theorem as follows.

$P(H_1 \mid C_1)$

$$= \frac{P(C_1 \mid H_1)P(H_1)}{P(C_1 \mid H_1)P(H_1) + P(C_1 \mid H_2)P(H_2) + P(C_1 \mid H_3)P(H_3)}$$

$$= \frac{(.9)(.8)}{(.9)(.8) + (.4)(.15) + (.1)(.05)} \approx .92$$

Hence, the probability is about .92 that the patient has a normal heart on the basis of the examination results. This means that in 8 out of 100 patients, some abnormality will be present and not be detected by the stethoscope or the cardiogram.

EXERCISES

1. Find $P(H_2 \mid C_1)$.

2. Assuming the following probabilities, find $P(H_1 \mid C_2)$:
$P(C_2 \mid H_1) = .2$, $P(C_2 \mid H_2) = .8$, $P(C_2 \mid H_3) = .3$.

3. Assuming the probabilities of Exercise 2, find $P(H_3 \mid C_2)$.

CHAPTER 9

Further Topics in Probability

This chapter introduces some important techniques and applications of probability, which illustrate its usefulness in such diverse fields as management, medicine, insurance, and the military.

9.1 PERMUTATIONS AND COMBINATIONS

Many probability problems involve counting all the possibilities. This section deals with counting methods that are useful in a wide variety of situations.

Let us begin with a simple example. If there are 3 roads from town A to town B and 2 roads from town B to town C, in how many ways can someone travel from A to C by way of B? For each of the 3 roads from A there are 2 different routes leading from B to C, making $3 \cdot 2 = 6$ different trips, as shown in Figure 9.1.

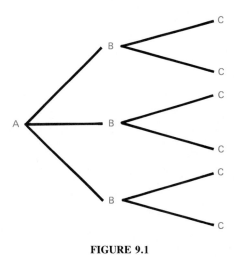

FIGURE 9.1

This example illustrates a general principle of counting called the *multiplication principle*.

Multiplication Principle

Suppose n choices must be made, with

$$m_1 \text{ ways to make choice 1,}$$

and for each of these,

$$m_2 \text{ ways to make choice 2,}$$

and so on, with

$$m_n \text{ ways to make choice } n.$$

Then there are

$$m_1 \cdot m_2 \cdots m_n$$

different ways to make the entire sequence of choices.

▶**EXAMPLE 1**   A combination lock can be set to open to any 3-letter sequence. How many such sequences are possible?

Since there are 26 letters in the alphabet, there are 26 choices for each of the 3 letters, and, by the multiplication principle, $26 \cdot 26 \cdot 26 = 17{,}576$ different sequences. ◀

▶**EXAMPLE 2**   Morse code uses a sequence of dots and dashes to represent letters and words. How many sequences are possible with at most 3 symbols?

"At most 3" means "1 or 2 or 3." Each symbol may be either a dot or a dash. Thus, the following number of sequences are possible in each case.

Number of Symbols	Number of Sequences
1	2
2	$2 \cdot 2 = 4$
3	$2 \cdot 2 \cdot 2 = 8$

Altogether, $2 + 4 + 8 = 14$ different sequences of at most 3 symbols are possible. Since there are 26 letters in the alphabet, some letters must be represented by 4 symbols in Morse code. ◀

1 (a) In how many ways can 6 business tycoons line up their golf carts at the country club?

(b) How many ways can 4 pupils be seated in a row with 4 seats?

Answer:

(a) $6 \cdot 5 \cdot 4 \cdot 3 \cdot 2 \cdot 1 = 720$

(b) $4 \cdot 3 \cdot 2 \cdot 1 = 24$

2 Evaluate:

(a) 4!

(b) 6!

(c) 7!

(d) 1!

Answer:

(a) 24

(b) 720

(c) 5040

(d) 1

3 In how many ways can 3 of 7 items be arranged?

Answer:
$7 \cdot 6 \cdot 5 = 210$

▶**EXAMPLE 3** A teacher has 5 different books to be arranged side by side. How many different arrangements are possible?

Five choices will be made, 1 for each space that will hold a book. Any of the 5 possible books could be chosen for the first space. There are 4 possible choices for the second space, since 1 book has already been placed in the first space, 3 possible choices for the third space, and so on. By the multiplication principle, the number of different possible arrangements (sequence of choices) is $5 \cdot 4 \cdot 3 \cdot 2 \cdot 1 = 120$. ◀ **1**

The use of the multiplication principle often leads to products such as $5 \cdot 4 \cdot 3 \cdot 2 \cdot 1$, the product of all the natural numbers from 5 down to 1. If n is a natural number, the symbol $n!$ (read ''*n factorial*'') denotes the product of all the natural numbers from n down to 1. If $n = 1$, this formula is understood to give $1! = 1$.

n-factorial

For any natural number n,

$$n! = n(n - 1)(n - 2) \ldots (3)(2)(1).$$

Also, 0! is defined to be the number 1.

With this symbol, the product $5 \cdot 4 \cdot 3 \cdot 2 \cdot 1$ can be written as 5!. Also, $3! = 3 \cdot 2 \cdot 1 = 6$. The definition of $n!$ could be used to show that $n! = n \cdot (n - 1)!$ for all natural numbers $n \geq 2$. It is helpful if this result also holds for $n = 1$. This can only happen if 0! equals 1, as defined above. **2**

Some calculators have an $\boxed{n!}$ key. A calculator with a 10-digit display and scientific notation capability will usually give the exact value of $n!$ for $n \leq 13$ and approximate values of $n!$ for $14 \leq n \leq 69$.

▶**EXAMPLE 4** Suppose the teacher in Example 3 wishes to place only 3 of the 5 books on his desk. How many arrangements of 3 books are possible?

The teacher again has 5 ways to fill the first space, 4 ways to fill the second space, and 3 ways to fill the third. Since he wants to use only 3 books, there are only 3 spaces to be filled giving $5 \cdot 4 \cdot 3 = 60$ arrangements. ◀ **3**

PERMUTATIONS The answer 60 in Example 4 is called the number of *permutations* of 5 things taken 3 at a time. A **permutation** of r (where $r \geq 1$) elements from a set of n elements is any arrangement, *without repetition,* of the r elements. The number of permutations of n things taken r at a time (with $r \leq n$) is written $P(n, r)$. Based on the work in Example 4,

$$P(5, 3) = 5 \cdot 4 \cdot 3 = 60.$$

4 Find the number of permutations of

(a) 5 things taken 2 at a time;

(b) 9 things taken 3 at a time.

Find each of the following.

(c) $P(3, 1)$

(d) $P(7, 3)$

(e) $P(12, 2)$

Answer:

(a) 20

(b) 504

(c) 3

(d) 210

(e) 132

5 Find the number of permutations of the letters C, O, D, and E

(a) using all the letters;

(b) using 2 of the 4 letters.

Answer:

(a) 24

(b) 12

Factorial notation can be used to express this product as follows.

$$5 \cdot 4 \cdot 3 = 5 \cdot 4 \cdot 3 \cdot \frac{2 \cdot 1}{2 \cdot 1} = \frac{5 \cdot 4 \cdot 3 \cdot 2 \cdot 1}{2 \cdot 1} = \frac{5!}{2!} = \frac{5!}{(5-3)!}$$

This example illustrates the general rule of permutations, which is stated below.

Permutations

If $P(n, r)$ (where $r \le n$) is the number of permutations of n elements taken r at a time, then

$$P(n, r) = \frac{n!}{(n-r)!}.$$

To find $P(n, r)$, we can use either the rule above or direct application of the multiplication principle, as the following example shows.

▶**EXAMPLE 5** Find the number of permutations of 8 elements taken 3 at a time.

Since there are 3 choices to be made, the multiplication principle gives $P(8, 3) = 8 \cdot 7 \cdot 6 = 336$. Alternatively, by the formula for $P(n, r)$,

$$P(8, 3) = \frac{8!}{(8-3)!} = \frac{8!}{5!} = \frac{8 \cdot 7 \cdot 6 \cdot 5 \cdot 4 \cdot 3 \cdot 2 \cdot 1}{5 \cdot 4 \cdot 3 \cdot 2 \cdot 1} = 8 \cdot 7 \cdot 6 = 336. \; ◀ \; \boxed{4}$$

▶**EXAMPLE 6** Find each of the following.

(a) The number of permutations of the letters A, B, and C

By the formula for $P(n, r)$ with both n and r equal to 3,

$$P(3, 3) = \frac{3!}{(3-3)!} = \frac{3!}{0!} = \frac{3!}{1} = 3! = 3 \cdot 2 \cdot 1 = 6.$$

The 6 permutations (or arrangements) are

$$\text{ABC, ACB, BAC, BCA, CAB, CBA.}$$

(b) The number of permutations possible using just 2 of the letters A, B, and C

Find $P(3, 2)$:

$$P(3, 2) = \frac{3!}{(3-2)!} = \frac{3!}{1!} = 3! = 6,$$

This result is exactly the same answer as in part (a). This is because, in the case of $P(3, 3)$, after the first 2 choices are made, the third is already determined, as shown in the table below.

First two letters	AB	AC	BA	BC	CA	CB
Third letter	C	B	C	A	B	A

◀ **5**

6 A collection of 3 paintings by one artist and 2 by another is to be displayed. In how many ways can the paintings be shown

(a) in a row?

(b) if the works of the artists are to be alternated?

Answer:

(a) 120

(b) 12

▶**EXAMPLE 7**   A televised talk show will include 4 women and 3 men as panelists.

(a) In how many ways can the panelists be seated in a row of 7 chairs?

Find $P(7, 7)$, the total number of ways to seat 7 panelists in 7 chairs.

$$P(7, 7) = \frac{7!}{(7 - 7)!} = \frac{7!}{0!} = \frac{7!}{1} = 7 \cdot 6 \cdot 5 \cdot 4 \cdot 3 \cdot 2 \cdot 1 = 5040$$

There are 5040 ways to seat the 7 panelists.

(b) In how many ways can the panelists be seated if the men and women are to be alternated?

In order to alternate men and women, a woman must be seated in the first chair (since there are 4 women and only 3 men), any of the men next, and so on. Thus, there are 4 ways to fill the first seat, 3 ways to fill the second seat, 3 ways to fill the third seat (with any of the 3 remaining women), and so on. This gives

$$4 \cdot 3 \cdot 3 \cdot 2 \cdot 2 \cdot 1 \cdot 1 = 144$$

ways to seat the panelists.   ◀ **6**

COMBINATIONS   In Example 4, we found that there are 60 ways that a teacher can arrange 3 of 5 different books on a desk. That is, there are 60 permutations of 5 things taken 3 at a time. Suppose now that the teacher does not wish to arrange the books on his desk, but rather wishes to choose, at random, any 3 of the 5 books to give to a book sale to raise money for his school. In how many ways can he do this?

At first glance, we might say 60 again, but this is incorrect. The number 60 counts all possible *arrangements* of 3 books chosen from 5. However, the following arrangements would all lead to the same set of 3 books being given to the book sale.

mystery–biography–textbook	biography–textbook–mystery
mystery–textbook–biography	textbook–biography–mystery
biography–mystery–textbook	textbook–mystery–biography

The list shows 6 different *arrangements* of 3 books, but only one subset of 3 books selected from the 5 books for the book sale. A subset of items selected *without regard to order* is called a **combination.** The number of combinations of 5 things taken 3 at a time is written $\binom{5}{3}$ (read "5 over 3"). Since they are subsets, combinations are *not ordered*.

To evaluate $\binom{5}{3}$, start with the $5 \cdot 4 \cdot 3$ *permutations* of 5 things taken 3 at a time. Since combinations are unordered, find the number of combinations by dividing the number of permutations by the number of ways each group of 3 can be ordered—that is, by 3!.

$$\binom{5}{3} = \frac{5 \cdot 4 \cdot 3}{3!} = \frac{5 \cdot 4 \cdot 3}{3 \cdot 2 \cdot 1} = 10$$

7 Evaluate $\dfrac{P(n, r)}{r!}$ for the following values.

(a) $n = 6$, $r = 2$

(b) $n = 8$, $r = 4$

(c) $n = 4$, $r = 1$

Answer:

(a) 15

(b) 70

(c) 4

8 Find the following.

(a) $\dbinom{6}{3}$

(b) $\dbinom{5}{1}$

(c) $\dbinom{7}{0}$

Answer:

(a) 20

(b) 5

(c) 1

There are 10 ways that the teacher can choose 3 books at random for the book sale.

Generalizing this discussion gives the formula for the number of combinations of n elements taken r at a time, written $\dbinom{n}{r}$:*

$$\binom{n}{r} = \frac{P(n, r)}{r!}. \quad \boxed{7}$$

A more useful version of this formula can be found using the result for $P(n, r)$ given above.

$$\binom{n}{r} = \frac{P(n, r)}{r!}$$

$$= \frac{n!}{(n - r)!} \cdot \frac{1}{r!}$$

$$= \frac{n!}{(n - r)!\,r!}$$

This last form is the most useful for calculation. The steps above lead to the following result.

Combinations

If $\dbinom{n}{r}$ denotes the number of combinations of n elements taken r at a time, where $r \leq n$, then

$$\binom{n}{r} = \frac{n!}{(n - r)!\,r!}.$$

Table 7 at the back of the book gives the values of $\dbinom{n}{r}$ for $n \leq 20$.

▶ **EXAMPLE 8** How many committees of 3 people can be formed from a group of 8 people?

A committee is an unordered group, so find $\dbinom{8}{3}$. By the formula for combinations,

$$\binom{8}{3} = \frac{8!}{5!3!} = \frac{8 \cdot 7 \cdot 6 \cdot 5 \cdot 4 \cdot 3 \cdot 2 \cdot 1}{5 \cdot 4 \cdot 3 \cdot 2 \cdot 1 \cdot 3 \cdot 2 \cdot 1} = \frac{8 \cdot 7 \cdot 6}{3 \cdot 2 \cdot 1} = 56. \quad ◀ \quad \boxed{8}$$

_____

*Other common notations for $\dbinom{n}{r}$ are $_nC_r$ and $C(n, r)$.

9 Five orchids from a collection of 20 are to be selected for a flower show.

(a) In how many ways can this be done?

(b) In how many different ways can the group of 5 be selected if 2 particular orchids must be included?

(c) In how many ways can at least 1 and at most 5 orchids be selected? (Hint: refer to Table 7.)

Answer:

(a) $\binom{20}{5} = 15{,}504$

(b) $\binom{18}{3} = 816$

(c) 21,699

▶**EXAMPLE 9**   Three secretaries are to be selected from a group of 30 to work on a special project.

(a) In how many different ways can the secretaries be selected?

Here we wish to know the number of 3-element combinations that can be formed from a set of 30 elements. (We want combinations and not permutations, since order within the group of 3 does not matter.)

$$\binom{30}{3} = \frac{30!}{27!3!} = \frac{30(29)(28)(27) \cdots (3)(2)(1)}{27(26)(25) \cdots (2)(1)(3)(2)(1)}$$

$$= \frac{30(29)(28)}{3(2)(1)}$$

$$= 4060$$

There are 4060 ways to select the project group.

(b) In how many ways can the group of 3 be selected if a certain secretary must work on the project?

Since 1 secretary has already been selected for the project, the problem is reduced to selecting 2 more from the remaining 29 secretaries.

$$\binom{29}{2} = \frac{29!}{27!2!} = \frac{29 \cdot 28 \cdot 27!}{27! \cdot 2 \cdot 1} = \frac{29 \cdot 28}{2 \cdot 1}$$

$$= 29 \cdot 14$$

$$= 406$$

In this case, the project group can be selected in 406 ways.

(c) In how many ways can a nonempty group of at most 3 secretaries be selected from these 30 secretaries.

Here "at most 3" means "1 or 2 or 3." Find the number of ways for each case.

Case	Number of Ways		
1	$\binom{30}{1} = \dfrac{30!}{29!1!} = \dfrac{30 \cdot 29!}{29!(1)} = 30$		
2	$\binom{30}{2} = \dfrac{30!}{28!2!} = \dfrac{30 \cdot 29 \cdot 28!}{28! \cdot 2 \cdot 1} = 435$		
3	$\binom{30}{3} = \dfrac{30!}{27!3} = \dfrac{30 \cdot 29 \cdot 28 \cdot 27!}{27! \cdot 3 \cdot 2 \cdot 1} = 4060$		

The total number of ways to select at most 3 secretaries will be the sum

$$30 + 435 + 4060 = 4525. \quad ◀ \quad \boxed{9}$$

10 Solve the problems in Example 10.

Answer:

(a) 5040

(b) 455

(c) 28

(d) 360

The formulas for permutations and combinations given in this section will be very useful in solving probability problems in later sections. Any difficulty in using these formulas usually comes from being unable to differentiate between them. Both permutations and combinations give the number of ways to choose r objects from a set of n objects. The differences between permutations and combinations are outlined below.

Permutations	Combinations
Different orderings or arrangements of the r objects are different permutations.	Each choice or subset of r objects gives 1 combination. Order within the r objects does not matter.
$$P(n, r) = \frac{n!}{(n-r)!}$$	$$\binom{n}{r} = \frac{n!}{(n-r)!r!}$$
Clue words: Arrangement, Schedule, Order	Clue words: Group, Committee, Sample

In the next examples, concentrate on recognizing which of the formulas should be applied.

▶**EXAMPLE 10** For each of the following problems, tell whether permutations or combinations should be used to solve the problem.

(a) How many 4-digit code numbers are possible if no digits are repeated?

Since changing the order of the 4 digits results in a different code, use permutations.

(b) A sample of 3 light bulbs is randomly selected from a batch of 15 items. How many different samples are possible?

The order in which the 3 light bulbs items are selected is not important. The sample is unchanged if the items are rearranged, so combinations should be used.

(c) In a basketball tournament with 8 teams, how many games must be played so that each team plays every other team exactly once?

Selection of 2 teams for a game is an *unordered* subset of 2 from the set of 8 teams. Use combinations again.

(d) In how many ways can 4 patients be assigned to 6 hospital rooms so that each patient has a private room?

The room assignments are an *ordered* selection of 4 rooms from the 6 rooms. Exchanging the rooms of any 2 patients within a selection of 4 rooms gives a different assignment, so permutations should be used. ◀ **10**

11 A salesman has the names of 6 prospects.

(a) In how many ways can he arrange his schedule if he calls on all 6?

(b) In how many ways can he do it if he decides to call on only 4 of the 6?

Answer:

(a) 720

(b) 360

12 In how many ways can 4 aces and any other card be dealt?

Answer:
48

▶ **EXAMPLE 11**   A manager must select 4 employees for promotion: 12 employees are eligible.

(a) In how many ways can the 4 be chosen?

Since there is no reason to differentiate among the 4 who are selected, use combinations.

$$\binom{12}{4} = \frac{12!}{4!8!} = 495$$

(b) In how many ways can 4 employees be chosen (from 12) to be placed in 4 different jobs?

In this case, once a group of 4 is selected, they can be assigned in many different ways (or arrangements) to the 4 jobs. Therefore, this problem requires permutations.

$$P(12, 4) = \frac{12!}{8!} = 11,880 \quad \blacktriangleleft \quad \boxed{11}$$

The following problems involve a standard deck of 52 playing cards, shown in Figure 8.17 on page 421.

▶ **EXAMPLE 12**   In how many ways can a full house of aces and eights (3 aces and 2 eights) be dealt in 5-card poker?

The arrangement of the 3 aces or the 2 eights does not matter, so use combinations and the multiplication principle. There are $\binom{4}{3}$ ways to get 3 aces from the 4 aces in the deck, and $\binom{4}{2}$ ways to get 2 eights. By the multiplication principle, the number of ways to get 3 aces and 2 eights is

$$\binom{4}{3} \cdot \binom{4}{2} = 4 \cdot 6 = 24. \quad \blacktriangleleft \quad \boxed{12}$$

▶ **EXAMPLE 13**   Five cards are dealt from a standard 52-card deck.

(a) How many such hands have all face cards?

The face cards are the king, queen, and jack of each suit. Since there are 4 suits, there are 12 face cards. The arrangement of the 5 cards is not important, so use combinations to get

$$\binom{12}{5} = \frac{12!}{5!7!} = 792.$$

(b) How many 5-card hands have all cards of the same suit?

The total number of ways that 5 cards of a particular suit of 13 cards can be dealt is $\binom{13}{5}$. The arrangement of the 5 cards is not important, so use combinations.

Since there are 4 different suits, by the multiplication principle, there are

13 In how many ways can 5 red cards be dealt?

Answer:
65,780

$$4 \cdot \binom{13}{5} = 4 \cdot 1287 = 5148$$

ways to deal 5 cards of the same suit. ◄ **13**

As Example 13 shows, often both combinations and the multiplication principle must be used in the same problem.

►**EXAMPLE 14** To illustrate the differences between permutations and combinations in another way, suppose 2 cans of soup are to be selected from 4 cans on a shelf: noodle (N), bean (B), mushroom (M), and tomato (T). As shown in Figure 9.2 (a), there are 12 ways to select 2 cans from the 4 cans if the order matters (if noodle first and bean second is considered different from bean, then noodle, for example). On the other hand, if order is unimportant, then there are 6 ways to choose 2 cans of soup from the 4, as illustrated in Figure 9.2(b). ◄

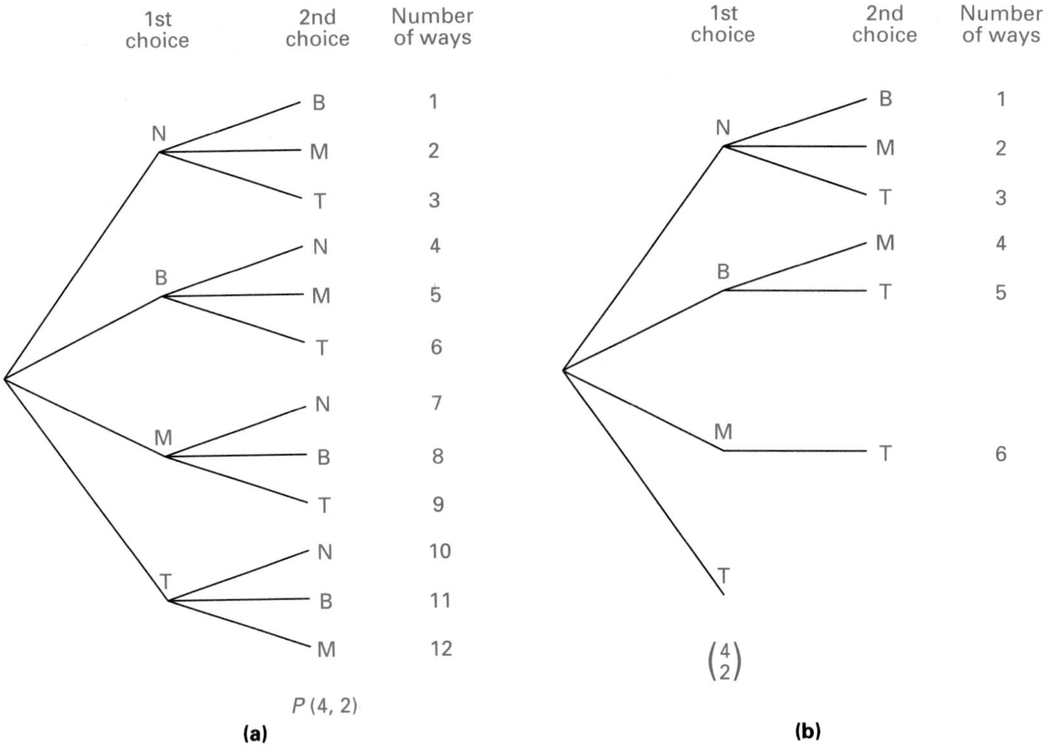

FIGURE 9.2

Caution It should be stressed that not all counting problems lend themselves to either permutations or combinations. Whenever a tree diagram or the multiplication principle can be used directly, as in the example at the beginning of this section, then use it.

9.1 EXERCISES

Evaluate the following factorials, permutations, and combinations.

1. $P(4, 2)$

2. $3!$

3. $\binom{8}{3}$

4. $7!$

5. $P(8, 1)$

6. $\binom{8}{1}$

7. $4!$

8. $P(4, 4)$

9. $\binom{12}{5}$

10. $\binom{10}{8}$

11. $P(13, 2)$

12. $P(12, 3)$

 Use a calculator to find values for Exercises 13–20.

13. $P(25, 5)$

14. $P(38, 4)$

15. $P(14, 5)$

16. $P(17, 8)$

17. $\binom{21}{10}$

18. $\binom{34}{25}$

19. $\binom{25}{16}$

20. $\binom{30}{15}$

Use the multiplication principle to solve the following problems. (See Examples 1–4.)

21. How many different types of homes are available if a builder offers a choice of 5 basic plans, 3 roof styles, and 2 exterior finishes?

22. An auto manufacturer produces 7 models, each available in 6 different colors, with 4 different upholstery fabrics, and 5 interior colors. How many varieties of the auto are available?

23. How many different 4-letter radio station call letters can be made
   (a) if the first letter must be K or W and no letter may be repeated?
   (b) if repeats are allowed (but the first letter is K or W)?
   (c) How many of the 4-letter call letters (starting with K or W) with no repeats end in R?

24. A menu offers a choice of 3 salads, 8 main dishes, and 5 desserts. How many different 3-course meals (salad, main dish, dessert) are possible?

25. A couple has narrowed down the choice of a name for their new baby to 3 first names and 5 middle names. How many different first- and middle-name arrangements are possible?

26. A concert to raise money for an economics prize is to consist of 5 works: 2 overtures, 2 sonatas, and a piano concerto. In how many ways can a program with these 5 works be arranged?

27. How many different license numbers consisting of 3 letters followed by 3 digits are possible?

28. How many different license plate numbers can be formed using 3 letters followed by 3 digits if no repeats are allowed?

29. How many license plate numbers (see Exercise 28) are possible if there are no repeats and either numbers or letters can come first?

30. How many 7-digit telephone numbers are possible if the first digit cannot be zero and
   (a) only odd digits may be used;
   (b) the telephone number must be a multiple of 10 (that is, it must end in zero);
   (c) the telephone number must be a multiple of 100;
   (d) the first 3 digits are 481;
   (e) no repetitions are allowed?

Use permutations to solve each of the following problems. (See Examples 5–7.)

31. In an experiment on social interaction, 6 people will sit in 6 seats in a row. In how many ways can this be done?

32. In how many ways can 7 of 10 monkeys be arranged in a row for a genetics experiment?

33. A business school gives courses in typing, shorthand, transcription, business English, technical writing, and accounting. In how many ways can a student arrange a schedule if 3 courses are taken?

34. If your college offers 400 courses, 20 of which are in mathematics, and your counselor arranges your schedule of 4 courses by random selection, how many schedules are possible that do not include a math course?

35. In a club with 15 members, how many ways can a slate of 3 officers consisting of president, vice-president, and secretary/treasurer be chosen?

36. A baseball team has 20 players. How many 9-player batting orders are possible?

Use combinations to solve each of the following problems. (See Examples 8–9 and 12–13.)

37. How many different 2-card hands can be dealt from an ordinary deck (52 cards)?

38. How many different 13-card bridge hands can be dealt from an ordinary deck?

39. Five cards are marked with the numbers 1, 2, 3, 4, and 5, then shuffled, and 2 cards are drawn. How many different 2-card combinations are possible?

40. Five cards are drawn from an ordinary deck. In how many ways is it possible to draw
   (a) all queens;
   (b) all face cards (face cards are the Jack, Queen, and King);
   (c) no face card;
   (d) exactly 2 face cards;
   (e) 1 heart, 2 diamonds, and 2 clubs.

41. If a baseball coach has 5 good hitters and 4 poor hitters on the bench and chooses 3 players at random, in how many ways can he choose at least 2 good hitters?

42. An economics club has 30 members. If a committee of 4 is to be selected, in how many ways can it be done?

For each of the following problems, decide if it is a permutations or combinations problem, and then solve the problem. (See Examples 10, 11, and 14.)

43. In how many ways can an employer select 2 new employees from a group of 4 applicants?

44. Hal's Hamburger Palace sells hamburgers with cheese, relish, lettuce, tomato, mustard, or catsup. How many different kinds of hamburgers can be made using any 3 of the extras?

45. In a club with 8 men and 11 women members, how many 5-member committees can be chosen that have
   (a) all men;   **(b)** all women;
   (c) 3 men and 2 women;   **(d)** no more than 3 women?

46. A contest requires matching 8 baby pictures of celebrities with their names. How many different entries are possible?

47. A group of 3 students is to be selected from a group of 12 students to take part in a class in cell biology.
   (a) In how many ways can this be done?
   (b) In how many ways can the group which will *not* take part be chosen?

48. The coach of the Morton Valley Softball Team has 6 good hitters and 8 poor hitters. He chooses 3 hitters at random.
   (a) In how many ways can he choose 2 good hitters and 1 poor hitter?
   (b) In how many ways can he choose all good hitters?
   (c) In how many ways can he choose at least 2 good hitters?

49. In a game of musical chairs, 12 children will sit in 11 chairs (1 will be left out). How many seatings are possible?

50. Marbles are drawn without replacement from a bag containing 15 marbles.
   (a) How many samples of 2 marbles can be drawn?
   (b) How many samples of 4 marbles can be drawn?
   (c) If the bag contains 3 yellow, 4 white, and 8 blue marbles, how many samples of 2 marbles can be drawn in which both marbles are blue?

51. A city council is composed of 5 liberals and 4 conservatives. A delegation of 3 is to be selected to attend a convention.
   (a) How many delegations are possible?
   (b) How many delegations could have all liberals?
   (c) How many delegations could have 2 liberals and 1 conservative?
   (d) If 1 member of the council serves as mayor, how many delegations which include the mayor are possible?

52. In an experiment on plant hardiness, a researcher gathers 6 wheat plants, 3 barley plants, and 2 rye plants. She wishes to select 4 plants at random.
   (a) In how many ways can this be done?
   (b) In how many ways can this be done if 2 wheat plants must be included?

53. A group of 7 workers decides to send a delegation of 2 to their supervisor to discuss their grievances.
   (a) How many delegations are possible?
   (b) If it is decided that a particular employee must be in the delegation, how many different delegations are possible?
   (c) If there are 2 women and 5 men in the group, how many delegations would include at least 1 woman?

54. From a pool of 7 secretaries, 3 are selected to be assigned to 3 managers. In how many ways can this be done?

55. From 10 names on a ballot, 4 will be elected to a political party committee. In how many ways can the committee of 4 be formed if each person will have a different responsibility?

56. There are 5 rotten apples in a crate of 25 apples.
   (a) How many samples of 3 apples can be drawn from the crate?
   (b) How many samples of 3 could be drawn in which all 3 are rotten?
   (c) How many samples of 3 could be drawn in which there are 2 good apples and 1 rotten one?

57. A bag contains 5 black, 1 red, and 3 yellow jelly beans; you take 3 at random. How many samples are possible in which the jelly beans are
   (a) all black;   **(b)** all red;
   (c) all yellow;   **(d)** 2 black, 1 red;
   (e) 2 black, 1 yellow; **(f)** 2 yellow, 1 black;
   (g) 2 red, 1 yellow.

58. In how many ways can 5 out of 9 plants be arranged in a row on a window sill?

*If the n objects in a permutations problem are not all distinguishable, that is, there are n_1 of type 1, n_2 of type 2, and so on for r different types, then the number of **distinguishable permutations** is*

$$\frac{n!}{n_1!n_2! \ldots n_r!}.$$

Example *In how many ways can you arrange the letters in the word Mississippi?*

This word contains 1 m, 4 i's, 4 s's, and 2 p's. To use the formula, let $n = 11$, $n_1 = 1$, $n_2 = 4$, $n_3 = 4$, $n_4 = 2$ to get

$$\frac{11!}{1!4!4!2!} = 34,650$$

arrangements. The letters in a word with 11 different letters can be arranged in 11! = 39,916,800 ways. ◀

59. Find the number of distinguishable permutations of the letters in each of the following words.
   (a) initial   **(b)** little   **(c)** decreed

60. A printer has 5 A's, 4 B's, 2 C's, and 2 D's. How many different "words" are possible which use all these letters? (A "word" does not have to have any meaning here.)

61. Mike has 4 blue, 3 green, and 2 red books to arrange on a shelf.
   (a) In how many ways can this be done if they can be arranged in any order?
   (b) In how many ways if books of the same color are identical and must be grouped together?
   (c) In how many distinguishable ways if books of the same color are identical but need not be grouped together?

62. A child has a set of different shaped plastic objects. There are 3 pyramids, 4 cubes, and 7 spheres.
   (a) In how many ways can she arrange them in a row if they are all different colors?
   (b) In how many ways if the same shapes must be grouped?
   (c) In how many distinguishable ways can they be arranged in a row if objects of the same shape are also the same color, but need not be grouped?

Use a calculator to solve the following exercises.

63. How many 5-card poker hands are possible with a regular deck of 52 cards?

64. How many 13-card bridge hands with 4 aces are possible?

65. How many 13-card bridge hands are possible with a regular deck of 52 cards?

66. How many 13-card bridge hands with exactly 3 aces are possible?

67. How many 13-card bridge hands with at least 3 aces are possible?

68. Natural Science Eleven drugs have been found to be effective in the treatment of a disease. It is believed that the sequence in which the drugs are administered is important in the effectiveness of the treatment. In how many orders can 5 of the 11 drugs be administered?

69. Natural Science A biologist is attempting to classify 52,000 species of insects by assigning 3 initials to each species. Is it possible to classify all the species in this way? If not, how many initials should be used?

70. One play in a state lottery consists of choosing 6 numbers from 1 to 44. If your 6 numbers are drawn (in any order), you win the jackpot.
   (a) How many possible ways are there to draw the 6 numbers?
   (b) If you get 2 plays for a dollar, how much would it cost to guarantee that 1 of your choices would be drawn?

(c) Assuming that you work alone and can fill out a betting ticket (for 2 plays) every second and the lotto drawing will take place 3 days from now, can you place enough bets to guarantee that 1 of your choices will be drawn?

71. In the State of Confusion, each automobile license plate consists of 3 numerals (0 through 9) followed by 3 letters.
(a) How many possible license plates are there?
(b) There are currently 12 million cars licensed in the state. The number of cars is expected to increase by 10% each year in the coming decade. How long will it take before the state no longer has enough license plates to go around?

72. In how many different orders could King Arthur and 12 knights be seated at the Round Table? (It may help to consider a similar problem with smaller numbers. Note that if 3 people sit at a straight table, ABC, BCA, and CAB are different orders; but at a round table, they produce the same order.)

9.2 APPLICATIONS OF COUNTING

Many of the probability problems involving *dependent* events that were solved by tree diagrams in Chapter 8 can also be solved by using combinations. Combinations are especially helpful when the numbers involved would require a tree with a large number of branches.

The use of combinations to solve probability problems depends on the basic probability principle introduced in Section 8.3 and repeated here.

Let S be a sample space with n equally likely outcomes. Let event E contain m of these outcomes. Then the probability that event E occurs, written $P(E)$, is

$$P(E) = \frac{m}{n}.$$

To compare the method of using combinations with the method of tree diagrams used in Section 8.5, the first example repeats Example 6 from that section.

▶**EXAMPLE 1**   From a box containing 3 white, 2 green, and 1 red marble, 2 marbles are drawn one at a time without replacement. Find the probability that 1 white and 1 green marble are drawn.

In Example 6 of Section 8.5, it was necessary to consider the order in which the marbles were drawn. With combinations, it is not necessary. Simply count the number of ways in which 1 white and 1 green marble can be drawn from the given selection. The white marble can be drawn from the 3 white marbles in $\binom{3}{1}$ ways, and the green marble can be drawn from the 2 green marbles in $\binom{2}{1}$ ways. By the multiplication principle, both results can occur in

$$\binom{3}{1} \cdot \binom{2}{1} \text{ ways,}$$

1 A jar contains 1 white and 4 red jelly beans.

(a) What is the probability that out of 2 jelly beans selected at random from the jar, 1 will be white?

(b) What is the probability of choosing 3 red jelly beans?

Answer:

(a) 2/5

(b) 2/5

giving the numerator of the probability fraction, $P(E) = m/n$. For the denominator, 2 marbles are to be drawn from a total of 6 marbles. This can occur in $\binom{6}{2}$ ways. The required probability is

$$P(1 \text{ white and } 1 \text{ green}) = \frac{\binom{3}{1}\binom{2}{1}}{\binom{6}{2}} = \frac{\dfrac{3!}{2!1!} \cdot \dfrac{2!}{1!1!}}{\dfrac{6!}{4!2!}} = \frac{6}{15} = \frac{2}{5}.$$

This agrees with the answer found earlier. ◀

▶**EXAMPLE 2**　From a group of 22 nurses, 4 are to be selected to present a list of grievances to management.
(a) In how many ways can this be done?

Four nurses from a group of 22 can be selected in $\binom{22}{4}$ ways. (Use combinations, since the group of 4 is an unordered set.)

$$\binom{22}{4} = \frac{22!}{4!18!} = \frac{22(21)(20)(19)}{4(3)(2)(1)} = 7315.$$

There are 7315 ways to choose 4 people from 22.
(b) One of the nurses is Michael Branson. Find the probability that Branson will be among the 4 selected.

The probability that Branson will be selected is given by m/n, where m is the number of ways the chosen group includes him, and n is the total number of ways the group of 4 can be chosen. If Branson must be one of the 4 selected, the problem reduces to finding the number of ways that the 3 additional nurses can be chosen. The 3 are chosen from 21 nurses; this can be done in

$$\binom{21}{3} = \frac{21!}{3!18!} = 1330$$

ways, so $m = 1330$. Since n is the number of ways 4 nurses can be selected from 22,

$$n = \binom{22}{4} = 7315.$$

The probability that Branson will be one of the 4 chosen is

$$P(\text{Branson is chosen}) = \frac{1330}{7315} \approx .182.$$

(b) Find the probability that Branson will not be selected.
The probability that he will not be chosen is $1 - .182 = .818.$ ◀ **1**

2 Calculate $\binom{2}{1}\binom{10}{2}$.

Answer:
90

3 Calculate $\binom{2}{2}\binom{10}{1}$.

Answer:
10

▶**EXAMPLE 3** When shipping diesel engines abroad, it is common to pack 12 engines in 1 container which is then loaded on a railcar and sent to a port. Suppose that a company has received complaints from its customers that many of the engines arrive in nonworking condition. To help solve this problem, the company decides to make a spot check of containers after loading—the company will test 3 engines from a container at random; if any of the 3 is nonworking, the container will not be shipped until each engine in it is checked. Suppose a given container has 2 non-working engines. Find the probability that the container will not be shipped.

The container will not be shipped if the sample of 3 engines contains 1 or 2 defective engines. Thus, letting $P(1 \text{ defective})$ represent the probability of exactly 1 defective engine in the sample,

$$P(\text{not shipping}) = P(1 \text{ defective}) + P(2 \text{ defectives}).$$

There are $\binom{12}{3}$ ways to choose the 3 engines for testing:

$$\binom{12}{3} = \frac{12!}{3!9!} = \frac{12(11)(10)}{3(2)(1)} = 220.$$

There are $\binom{2}{1}$ ways of choosing 1 defective engine from the 2 in the container, and for each of these ways, there are $\binom{10}{2}$ ways of choosing 2 good engines from among the 10 in the container. By the multiplication principle, there are

$$\binom{2}{1}\binom{10}{2}$$

ways of choosing a sample of 3 engines containing 1 defective. **2**

Using the result from Problem 2 at the side,

$$P(1 \text{ defective}) = \frac{90}{220}.$$

There are $\binom{2}{2}$ ways of choosing 2 defective engines from the 2 defective engines in the container, and $\binom{10}{1}$ ways of choosing 1 good engine from among the 10 good engines, giving

$$\binom{2}{2}\binom{10}{1}$$

ways of choosing a sample of 3 engines containing 2 defectives. **3**

Then, using the result from Problem 3 at the side,

$$P(2 \text{ defectives}) = \frac{10}{220}$$

and

$$P(\text{not shipping}) = P(1 \text{ defective}) + P(2 \text{ defectives})$$

$$= \frac{90}{220} + \frac{10}{220} = \frac{100}{220} = \approx .455$$

4 In Example 3, if a sample of 2 engines is tested, what is the probability that the container will not be shipped?

Answer:
.318

The probability is $1 - .455 = .545$ that the container *will* be shipped, even though it has 2 defective engines. The management must decide if this probability is acceptable; if not, if may be necessary to test more than 3 engines from a container. ◄

Instead of finding the sum $P(1 \text{ defective}) + P(2 \text{ defectives})$, the result in Example 3 could be found by calculating $1 - P(\text{no defectives})$.

$$P(\text{not shipping}) = 1 - P(\text{no defectives in sample})$$

$$= 1 - \frac{\binom{2}{0}\binom{10}{3}}{\binom{12}{3}} = 1 - \frac{1(120)}{220}$$

$$= 1 - \frac{120}{220} = \frac{100}{220} \approx .455 \quad \boxed{4}$$

▶ **EXAMPLE 4**   In a common form of the card game *poker,* a hand of 5 cards is dealt to each player from a deck of 52 cards. There are a total of

$$\binom{52}{5} = \frac{52!}{5!47!} = 2{,}598{,}960$$

such hands possible. Find each of the following probabilities.

(a) A hand containing only hearts, called a *heart flush*

There are 13 hearts in a deck; there are

$$\binom{13}{5} = \frac{13!}{5!8!} = \frac{13(12)(11)(10)(9)}{5(4)(3)(2)(1)} = 1287$$

different hands containing only hearts. The probability of a heart flush is

$$P(\text{heart flush}) = \frac{1287}{2{,}598{,}960} = \frac{33}{66{,}640} \approx .000495.$$

(b) A flush of any suit (5 cards, all from 1 suit)

There are 4 suits to a deck, so

$$P(\text{flush}) = 4 \cdot P(\text{heart flush}) = 4 \cdot \frac{33}{66{,}640} \approx .00198.$$

(c) A full house of aces and eights (3 aces and 2 eights)

There are $\binom{4}{3}$ ways to choose 3 aces from among the 4 in the deck, and $\binom{4}{2}$ ways to choose 2 eights.

$$P(3 \text{ aces, 2 eights}) = \frac{\binom{4}{3} \cdot \binom{4}{2}}{2{,}598{,}960} = \frac{1}{108{,}290} = \approx .00000923$$

5 Find the probability of being dealt a poker hand (5 cards) with 4 kings.

Answer:
.00001847

6 An office manager must select 4 employees, A, B, C, and D, to work on a project. Each employee will be responsible for a specific task and 1 will be the coordinator.

(a) If the manager assigns the tasks randomly, what is the probability that B is the coordinator?

(b) What is the probability of one specific assignment of tasks?

Answer:

(a) 1/4

(b) 1/24

(d) Any full house (3 cards of one value, 2 of another)

There are 13 values in a deck, so there are 13 choices for the first value mentioned, leaving 12 choices for the second value (order *is* important here, since a full house of aces and eights, for example, is not the same as a full house of eights and aces). Since, from part (c), the probability of any particular full house is 1/108,290,

$$P(\text{full house}) = 13 \cdot 12 \cdot \left(\frac{1}{108,290} \right) = \frac{156}{108,290} \approx .00144. \quad \blacktriangleleft \quad \boxed{5}$$

▶**EXAMPLE 5** A music teacher has 3 violin pupils, Fred, Carl, and Helen. For a recital, the teacher selects a first violinist and a second violinist. The third pupil will play with the others, but not solo. If the teacher selects randomly, what is the probability that Helen is first violinist, Carol is second violinist, and Fred does not solo?

Use *permutations* to find the number of arrangements in the sample space.

$$P(3, 3) = 3! = 6$$

The 6 arrangements are equally likely, since the teacher will select randomly. Thus, the required probability is 1/6. ◀ $\boxed{6}$

▶**EXAMPLE 6** Suppose a group of n people is in a room. Find the probability that at least 2 of the people have the same birthday.

''Same birthday'' refers to the month and the day, not necessarily the same year. Also, ignore leap years, and assume that each day in the year is equally likely as a birthday. First find the probability that *no 2 people* among 5 people have the same birthday. There are 365 different birthdays possible for the first of the 5 people, 364 for the second (so that the people have different birthdays), 363 for the third, and so on. The number of ways the 5 people can have different birthdays is thus the number of permutations of 365 things (days) taken 5 at a time or

$$P(365, 5) = 365 \cdot 364 \cdot 363 \cdot 362 \cdot 361.$$

The number of ways that the 5 people can have the same or different birthdays is

$$365 \cdot 365 \cdot 365 \cdot 365 \cdot 365 = (365)^5.$$

Finally, the *probability* that none of the 5 people have the same birthday is

$$\frac{P(365, 5)}{(365)^5} = \frac{365 \cdot 364 \cdot 363 \cdot 362 \cdot 361}{365 \cdot 365 \cdot 365 \cdot 365 \cdot 365} \approx .973.$$

The probability that at least 2 of the 5 people *do* have the same birthday is $1 - .973 = .027$.

7 Evaluate

$$1 - \frac{P(365, n)}{(365)^n}$$

for

(a) $n = 3$;

(b) $n = 6$.

Answer:

(a) .008

(b) .040

8 Set up (do not calculate) the probability that at least 2 of the 7 astronauts in the Mercury project have the same birthday.

Answer:
$1 - P(365, 7)/365^7$

This result can be extended for more than 5 people. In general, the probability that no 2 people among n people have the same birthday is

$$\frac{P(365, n)}{(365)^n}.$$

The probability that at least 2 of the n people *do* have the same birthday is

$$1 - \frac{P(365, n)}{(365)^n}. \blacktriangleleft \quad \boxed{7}$$

The following table shows this probability for various values of n.

Number of People, n	Probability that 2 Have the Same Birthday
5	.027
10	.117
15	.253
20	.411
22	.476
23	.507
25	.569
30	.706
35	.814
40	.891
50	.970
366	1

The probability that 2 people among 23 have the same birthday is .507, a little more than half. Many people are surprised at this result—somehow it seems that a larger number of people should be required. **8**

9.2 EXERCISES

Management *A shipment of 9 typewriters contains 2 defectives. Find the probability that a sample of the following size, drawn from the 9, will not contain a defective. (See Example 3.)*

1. 1   **2.** 2   **3.** 3   **4.** 4

Management *Refer to Example 3. The management feels that the probability of .545 that a container will be shipped even though it contains 2 defectives is too high. They decide to increase the sample size chosen. Find the probability that a container will be shipped even though it contains 2 defectives if the sample size is increased to*

5. 4;   **6.** 5.

A basket contains 6 red apples and 4 yellow apples. A sample of 3 apples is drawn. Find the probability that the sample contains each of the following. (See Examples 1 and 2.)

7. All red apples

8. All yellow apples

9. 2 yellow and 1 red apple

10. More red than yellow apples

Two cards are drawn at random from an ordinary deck of 52 cards. (See Example 4.)

11. How many 2-card hands are possible?

Find the probability that the 2-card hand in Exercise 11 contains

12. 2 aces

13. at least 1 ace

14. all spades

15. 2 cards of the same suit

16. only face cards

17. no face cards

18. no card higher than 8 (count ace as 1)

Twenty-six slips of paper are each marked with a different letter of the alphabet and placed in a basket. A slip is pulled out, its letter recorded (in the order in which the slip was drawn), and the slip replaced. This is done 5 times. Find the probabilities that the "word" formed

19. is "chuck";

20. starts with p;

21. has all different letters;

22. contains no x, y, or z.

Find the probability of the following hands at poker. Assume aces are either high or low. (See Example 4.)

23. Royal flush (5 highest cards of a single suit)

24. Straight flush (5 in a row in a single suit, but not a royal flush)

25. Four of a kind (4 cards of the same value)

26. Straight (5 cards in a row, not all of the same suit with ace either high or low)

A bridge hand is made up of 13 cards from a deck of 52. Set up the probability that a hand chosen at random

27. contains only hearts;

28. has 4 aces;

29. contains exactly 3 aces and exactly 3 kings;

30. has 6 of 1 suit, 5 of another, and 2 of another.

31. At a conference of black writers in Detroit, special-edition books were selected to be given away in contests. There were 9 books written by Langston Hughes, 5 books by James Baldwin, and 7 books by Toni Morrison. The judge of one contest selected 6 books at random for prizes. Find the probabilities that the selection consisted of the following.
   (a) Three Hughes and 3 Morrison books
   (b) Exactly 4 Baldwin books
   (c) Two Hughes, 3 Baldwin, and 1 Morrison book
   (d) At least 4 Hughes books
   (e) Exactly 4 books written by males (Morrison is female)
   (f) No more than 2 books written by Baldwin

32. At the first meeting of a committee to plan a Northern California pow-wow, there were 3 women and 3 men from the Miwok tribe, 2 men and 3 women from the Hoopa tribe, and 4 women and 5 men from the Pomo tribe. If the ceremony subcouncil consists of 5 people and is randomly selected, find the probabilities that the subcouncil contains the following.
   (a) 3 men and 2 women
   (b) Exactly 3 Miwoks and 2 Pomos
   (c) 2 Miwoks, 2 Hoopas, and a Pomo
   (d) 2 Miwoks, 2 Hoopas, and 2 Pomos
   (e) More women than men
   (f) Exactly 3 Hoopas
   (g) At least 2 Pomos

Work the following exercises. (See Examples 4 and 5.)

33. A box of 14 smoke detectors contains 4 defective ones.
   (a) If 3 are randomly taken from the box, what is the probability that all 3 work?
   (b) If 2 are randomly taken from the box, what is the probability that 1 is defective and 1 is not?
   (c) If 2 defective detectors are removed from the box and then 3 are randomly chosen from the remaining ones, what is the probability that all 3 work?

34. A computer manufacturing company randomly selects 5 computers from each batch of 25 coming off the assembly line and tests them. If at least 4 of the 5 pass inspection, the batch of 25 is considered acceptable. Find the probability that a batch will be acceptable if it contains
(a) no defective computers;
(b) 2 defective computers;
(c) 5 defective computers.

35. The State Senate has 25 members, 10 Democrats and 15 Republicans. Four members are randomly selected to serve on a special committee. Find the probability that
(a) all 4 are Democrats
(b) all 4 are Republicans;
(c) 2 are Democrats and 2 are Republicans;
(d) 1 is a Democrat and 3 are Republicans;
(e) Rushby and McCullar are Democrats. If they are not chosen, what is the probability that the committee consists of 1 Democrat and 3 Republicans?

36. Show that the probability that in a group of *n* people *exactly 1* pair have the same birthday is
$$\binom{n}{2} \cdot \frac{P(365, \, n-1)}{(365)^n}.$$

37. To win a contest, a player must match 4 movies stars with their baby pictures. Suppose this is done at random. Find the probability of getting no matches correct; of getting exactly 2 correct.

38. A contractor has hired a decorator to send 3 different sofas out to the contractor's model homes each week for a year. The contractor does not want exactly the same 3 sofas sent out twice. Find the minimum number of sofas that the decorator will need.

39. An elevator has 4 passengers and stops at 7 floors. It is equally likely that a person will get off at any 1 of the 7 floors. Find the probability that no 2 passengers leave at the same floor.

40. A car dealer has 8 red, 11 gray, and 9 blue cars in stock. Ten cars are randomly chosen to be displayed in front of the dealership. Find the probability that
(a) 4 are red and the others are blue.
(b) 3 are red, 3 are blue, and 4 are gray.

(c) exactly 5 are gray and none are blue.
(d) all 10 are gray.

The rest of these exercises involve the idea of a circular permutation: the number of ways of arranging distinct objects in a circle. The number of ways of arranging n distinct objects in a line is n!, but there are (n − 1)! ways for arranging the n items in a circle (see Exercise 47). Find the number of ways of arranging the following number of distinct items in a circle.

41. 4   **42.** 7   **43.** 10

44. Suppose that 8 people sit at a circular table. Find the probability that 2 particular people are sitting next to each other.

45. A key ring contains 7 keys: 1 black, 1 gold, and 5 silver. If the keys are arranged at random on the ring, find the probability that the black key is next to the gold key.

46. A circular table for a board of directors has 10 seats for the 10 attending members of the board. The chairman of the board always sits closest to the window. The vice president for sales, who is currently out of favor, will sit 3 positions to the chairman's left, since the chairman doesn't see so well out of his left eye. The chairman's daughter-in-law will sit opposite him. All other members take seats at random. Find the probability that a particular other member will sit next to the chairman.

47. Show that the number of ways of arranging *n* distinct items in a circle is (*n* − 1)!.

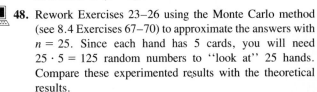

48. Rework Exercises 23–26 using the Monte Carlo method (see 8.4 Exercises 67–70) to approximate the answers with *n* = 25. Since each hand has 5 cards, you will need 25 · 5 = 125 random numbers to "look at" 25 hands. Compare these experimented results with the theoretical results.

49. Rework Exercises 27–30 using the Monte Carlo method (see 8.4 Exercises 67–70) to approximate the answers with *n* = 20. Since each hand has 13 cards, you will need 20 · 13 = 260 random numbers to "look at" 20 hands.

9.3 BINOMIAL TRIALS

Many probability problems are concerned with experiments in which an event is repeated many times. Some examples include finding the probability of getting 7 heads in 8 tosses of a coin, of hitting a target 6 times out of 6, and of finding 1 defective item in a sample of 15 items. Probability problems of this kind are called **binomial trials** problems, or **Bernoulli processes,** named after the Swiss mathematician Jakob Bernoulli (1654–1705). In each case, some outcome is designated a success, and any other outcome is considered a failure. Thus, if the probability of a success in a single trial is p, the probability of failure will be $1 - p$. Binomial trials problems must satisfy the following conditions.

Binomial Trials

1. The same experiment is repeated several times.
2. There are only two possible outcomes, success or failure.
3. The repeated trials are independent so that the probability of each outcome remains the same for each trial.

Consider a typical problem of this type: find the probability of getting 5 1's on 5 rolls of a die. Here, the experiment, rolling a die, is repeated 5 times. If getting a 1 is designated as a success, then getting any other outcome is a failure. The 5 trials are independent; the probability of success (getting a 1) is 1/6 on each trial, while the probability of a failure (any other result) is 5/6. Thus, the required probability is

$$P(5 \text{ 1's on 5 rolls}) = P(1) \cdot P(1) \cdot P(1) \cdot P(1) \cdot P(1) = \left(\frac{1}{6}\right)^5$$

$$\approx .00013.$$

Now suppose the problem is changed to that of finding the probability of getting a 1 exactly 4 times in 5 rolls of the die. Again, a success on any roll of the die is defined as getting a 1, and the trials are independent, with the same probability of success each time. The desired outcome for this experiment can occur in more than one way, as shown below, where s represents a success (getting a 1), and f represents a failure (getting any other result), and each row represents a different outcome.

$$s \quad s \quad s \quad s \quad f$$
$$s \quad s \quad s \quad f \quad s$$
$$s \quad s \quad f \quad s \quad s$$
$$s \quad f \quad s \quad s \quad s$$
$$f \quad s \quad s \quad s \quad s$$

1 Find the probability of rolling a 1

(a) exactly twice in 5 rolls of a die;

(b) exactly once in 5 rolls of a die.

Answer:

(a) .161

(b) .402

The probability of each of these 5 outcomes is

$$\left(\frac{1}{6}\right)^4\left(\frac{5}{6}\right).$$

Since the 5 outcomes represent mutually exclusive alternative events, add the 5 probabilities.

$$P(4 \text{ 1's in 5 rolls}) = 5\left(\frac{1}{6}\right)^4\left(\frac{5}{6}\right) = \frac{5^2}{6^5} \approx .0032$$

The probability of rolling a 1 exactly 3 times in 5 rolls of a die can be computed in the same way. The probability of any one way of achieving 3 successes and 2 failures will be

$$\left(\frac{1}{6}\right)^3\left(\frac{5}{6}\right)^2.$$

Again the desired outcome can occur in more than one way. To see this, let the set $\{1, 2, 3, 4, 5\}$ represent the first, second, third, fourth, and fifth tosses. The number of 3-element subsets of this set will correspond to the number of ways in which 3 successes and 2 failures can occur. Using combinations, there are $\binom{5}{3}$ such subsets.

Since $\binom{5}{3} = 5!/(3!2!) = 10,$

$$P(3 \text{ ones in 5 rolls}) = 10\left(\frac{1}{6}\right)^3\left(\frac{5}{6}\right)^2 = \frac{250}{6^5} \approx .032. \quad \boxed{1}$$

A similar argument works in the general case.

Probability in a Binomial Experiment

If p is the probability of success in a single trial of a binomial experiment, the probability of x successes and $n - x$ failures in n independent repeated trials of the experiment is

$$\binom{n}{x}p^x(1 - p)^{n-x}.$$

This formula plays an important role in the study of *binomial distributions*, which are considered in Section 10.5.

▶**EXAMPLE 1** The advertising agency that handles the Diet Supercola account believes that 40% of all consumers prefer this product over its competitors. Suppose a random sample of 6 people is chosen. Assume that all responses are independent of each other. Find the probability of the following.

2 Eighty percent of all students at a certain school ski. If a sample of 5 students at this school is selected, and if their responses are independent, find the probability that exactly

(a) 1 of the 5 students skis;

(b) 4 of the 5 students ski.

Answer:

(a) $\binom{5}{1}(.8)^1(.2)^4 = .0064$

(b) $\binom{5}{4}(.8)^4(.2)^1 = .4096$

3 Find the probability of getting 2 fours in 8 tosses of a die.

Answer:

$\binom{8}{2}\left(\frac{1}{6}\right)^2\left(\frac{5}{6}\right)^6 \approx .2605$

(a) Exactly 3 of the 6 people prefer Diet Supercola.

Think of the 6 responses as 6 independent trials. A success occurs if a person prefers Diet Supercola. Then this is a binomial experiment with $p = P(\text{success}) = P(\text{prefer Diet Supercola}) = .4$. The sample is made up of 6 people, so $n = 6$. To find the probability that exactly 3 people prefer this drink, let $x = 3$ and use the result in the box.

$$P(\text{exactly } 3) = \binom{6}{3}(.4)^3(1 - .4)^{6-3}$$
$$= 20(.4)^3(.6)^3$$
$$= 20(.064)(.216) \qquad \text{Use a calculator}$$
$$= .27648$$

(b) None of the 6 people prefers Diet Supercola.

Let $x = 0$.

$$P(\text{exactly } 0) = \binom{6}{0}(.4)^0(1 - .4)^6$$
$$= 1(1)(.6)^6$$
$$\approx .0467 \qquad \blacktriangleleft \quad \boxed{2}$$

▶**EXAMPLE 2** Find the probability of getting exactly 7 heads in 8 tosses of a fair coin.

The probability of success, getting a head in a single toss, is 1/2, so the probability of failure, getting a tail, is 1/2.

$$P(\text{7 heads in 8 tosses}) = \binom{8}{7}\left(\frac{1}{2}\right)^7\left(\frac{1}{2}\right)^1 = 8\left(\frac{1}{2}\right)^8 = .03125 \qquad \blacktriangleleft \quad \boxed{3}$$

▶**EXAMPLE 3** Assuming that selection of items for a sample can be treated as independent trials, and that the probability that any 1 item is defective is .01, find the following.

(a) The probability of 1 defective item in a random sample of 15 items from a production line

Here, a "success" is a defective item. Since selecting each item for the sample is assumed to be an independent trial, the Bernoulli trials formula applies. The probability of success (a defective item) is .01, while the probability of failure (an acceptable item) is .99. This makes

$$P(\text{1 defective in 15 items}) = \binom{15}{1}(.01)^1(.99)^{14}$$
$$= 15(.01)(.99)^{14}$$
$$\approx .130.$$

4 Five percent of the clay pots fired in a certain way are defective. Find the probability of getting exactly 2 defective pots in a sample of 12.

Answer:

$\binom{12}{2}(.05)^2(.95)^{10} \approx .0988$

5 In Example 4, find the probability that out of 10 customers

(a) 6 or more buy the shoe;

(b) no more than 5 buy the shoe.

Answer:

(a) .0473490

(b) .9526510

(b) The probability of at most 1 defective item in a random sample of 15 items from a production line

"At most 1" means 0 defective items or 1 defective item. Since 0 defective items is equivalent to 15 acceptable items,

$$P(0 \text{ defective}) = (.99)^{15} \approx .860.$$

Use the union rule, noting that 0 defective and 1 defective are mutually exclusive events, to get

$$P(\text{at most 1 defective}) = P(0 \text{ defective}) + P(1 \text{ defective})$$
$$\approx .860 + .130$$
$$= .990. \blacktriangleleft \boxed{4}$$

▶**EXAMPLE 4**   A new style of shoe is sweeping the country. In one area, 30% of all the shoes sold are of this new style. Assume that these sales are independent events. Find the following probabilities.

(a) Out of 10 customers in a shoe store, at least 8 buy the new shoe style.

Let success be "buy the new style," so that $P(\text{success}) = .3$. For at least 8 people out of 10 to buy the shoe, it must be sold to 8, 9, or 10 people. This means

$$P(\text{at least 8}) = P(8) + P(9) + P(10)$$
$$= \binom{10}{8}(.3)^8(.7)^2 + \binom{10}{9}(.3)^9(.7)^1 + \binom{10}{10}(.3)^{10}(.7)^0$$
$$\approx .0014467 + .0001378 + .0000059$$
$$= .0015904.$$

(b) Out of 10 customers in a shoe store, no more than 7 buy the new shoe style.

"No more than 7" means 0, 1, 2, 3, 4, 5, 6, or 7 people buy the shoe. We could add $P(0)$, $P(1)$, and so on, but it is easier to use the formula $P(E) = 1 - P(E')$. The complement of "no more than 7" is "8 or more," so

$$P(\text{no more than 7}) = 1 - P(8 \text{ or more})$$
$$\approx 1 - .0015904 \qquad \text{Answer from part (a)}$$
$$= .9984096. \blacktriangleleft \boxed{5}$$

9.3 EXERCISES

Suppose that a family has 5 children and that the probability of having a girl is 1/2. (See Examples 1–4.) Find the probability that the family will have

1. exactly 2 girls;

2. exactly 3 girls;

3. no girls;

4. no boys;

5. at least 4 girls;

6. at least 3 boys;

7. no more than 3 boys;

8. no more than 4 girls.

A die is rolled 12 *times. Find the probability of rolling*

9. exactly 12 ones;

10. exactly 6 ones;

11. exactly 1 one;

12. exactly 2 ones;

13. no more than 3 ones;

14. no more than 1 one.

A coin is tossed 5 *times. Find the probability of getting*

15. all heads;

16. exactly 3 heads;

17. no more than 3 heads;

18. at least 3 heads.

Management   *A factory tests a random sample of* 20 *transistors for defectives. The probability that a particular transistor will be defective has been established by past experience to be* .05. *(See Example 3.)*

19. What is the probability that there are no defectives in the sample?

20. What is the probability that the number of defectives in the sample is at most 2?

Management   *A company gives prospective employees a 6-question multiple-choice test. Each question has 5 possible answers, so that there is a 1/5 or 20% chance of an-*

swering a question correctly just by guessing. Find the probability of guessing the answers to

21. exactly 2 questions correctly;

22. no questions correctly;

23. at least 4 correctly;

24. no more than 3 correctly.

25. Five out of the 50 clients of a certain stockbroker will lose their life savings as a result of his advice. Find the probability that in a sample of 3 clients, exactly 1 loses all his money. (Assume independence.)

Management   *According to a recent article in a business publication, only* 20% *of the population of the United States has never had a Big Mac hamburger at McDonald's. Assume independence and find the probability that in a random sample of* 10 *people*

26. exactly 2 never had a Big Mac;

27. exactly 5 never had a Big Mac;

28. 3 or fewer never had a Big Mac;

29. 4 or more *have* had a Big Mac.

Natural Science   *A new drug cures* 70% *of the people taking it. Suppose* 20 *people take the drug; find the probability that*

30. exactly 18 are cured;

31. exactly 17 are cured;

32. at least 17 are cured;

33. at least 18 are cured.

In a 10-question multiple-choice biology test with 5 *choices for each question, an unprepared student guesses on each item. Find the probability that he answers*

34. exactly 6 questions correctly;

35. exactly 7 correctly;

36. at least 8 correctly;

37. fewer than 8 correctly.

Natural Science   *Assume that the probability that a person will die within a month after a certain operation is* 20%. *Find the probability that in* 3 *such operations*

38. all 3 people survive;

39. exactly 1 person survives;

40. at least 2 people survive;

41. no more than 1 person survives.

Natural Science   *Six mice from the same litter, all suffering from a vitamin A deficiency, are fed a certain dose of carrots. If the probability of recovery under such treatment is* .70, *find the probability that*

42. none recover;

43. exactly 3 of the 6 recover;

44. all recover;

45. no more than 3 recover.

46. Natural Science In an experiment on the effects of a radiation dose on cells, a beam of radioactive particles is aimed at a group of 10 cells. Find the probability that 8 of the cells will be hit by the beam if the probability that any single cell will be hit is .6. (Assume independence.)

47. Natural Science The probability of a mutation of a given gene under a dose of 1 roentgen of radiation is approximately 2.5×10^{-7}. What is the probability that in 10,000 genes, at least 1 mutation occurs?

48. Natural Science A new drug being tested causes a serious side effect in 5 out of 100 patients. What is the probability that no side effects occur in a sample of 10 patients taking the drug?

49. Children in a certain school are monitored for a year and it is found that 30% of those who do not brush their teeth regularly don't have a cavity during the year. A child who brushes regularly has a 70% chance of having no cavities during the year.
   (a) If 10 children brush regularly, what is the probability that at least 9 will have no cavities during the year?
   (b) 150 children who brush regularly are divided into 15 groups of 10 each. What is the probability that in at least one group of 10, at least 9 will have no cavities during the year?

50. A building has a ground floor and floors 2–12 above it. Ten people get in an elevator on the ground floor. Find the probability that
   (a) all of them get off on the fifth floor.
   (b) all of them get off on the same floor.

(c) exactly 4 get off on the ninth floor.
   (d) 4 get off on the fourth floor and 6 get off on the sixth floor.
   (e) everyone gets off on a different floor.

Use a calculator or computer to calculate each of the following probabilities.

51. A flu vaccine has a probability of 80% of preventing a person who is inoculated from getting flu. A county health office inoculates 134 people. What is the probablility that
   (a) exactly 10 of them get the flu?
   (b) no more than 10 get the flu?
   (c) none of them get the flu?

52. The probability that a male will be color-blind is .042. What is the probability that in a group of 53 men
   (a) exactly 5 are color-blind?
   (b) no more than 5 are color-blind?
   (c) at least 1 is color-blind?

53. The probability that a certain machine turns out a defective item is .05. What is the probability that in a run of 75 items
   (a) exactly 5 defectives are produced?
   (b) no defectives are produced?
   (c) at least 1 defective is produced?

54. A company is taking a survey to find out if people like their product. Their last survey indicated that 70% of the population like their product. Based on that, of a sample of 58 people, what is the probability that
   (a) all 58 like the product?
   (b) from 28 to 30 (inclusive) like the product?

9.4 MARKOV CHAINS

In Section 8.5, we touched on **stochastic processes,** experiments in which the outcome of a trial depends on the outcomes of previous trials. In this section we introduce a special type of stochastic process, called a **Markov chain,** where the outcome of a trial depends only on the outcome of the previous trial. That is, the next state of the experiment depends only on the present state, not on preceding states. Such experiments are common enough in applications to make their study worthwhile. Markov chains are named after the Russian mathematician A. A. Markov, (1856–1922) who started the theory of such processes. To see how Markov chains work, we look at an example.

1 Given the transition matrix

$$\begin{array}{c} \quad State \\ \quad 1 \quad 2 \\ State \begin{array}{c} 1 \\ 2 \end{array} \left[\begin{array}{cc} .3 & .7 \\ .1 & .9 \end{array} \right], \end{array}$$

(a) what is the probability of changing from state 1 to state 2?

(b) what does the number .1 represent?

Answer:

(a) .7

(b) The probability of changing from state 2 to state 1

▶**EXAMPLE 1**   A small town has only two dry cleaners, Johnson and NorthClean. Johnson's manager hopes to increase the firm's market share by an extensive advertising campaign. After the campaign, a market research firm finds that there is a probability of .8 that a Johnson customer will bring his next batch of dirty items to Johnson, and a .35 chance that a NorthClean customer will switch to Johnson for his next batch. Assume that the probability that a customer comes to a given cleaners depends only on where the last load of clothes was taken. If there is an .8 chance that a Johnson customer will return to Johnson, then there must be a $1 - .8 = .2$ chance that the customer will switch to NorthClean. In the same way, there is a $1 - .35 = .65$ chance that a NorthClean customer will return to NorthClean. If an individual bringing a load to Johnson is said to be in state 1 and an individual bringing a load to NorthClean is said to be in state 2, then these probabilities of change from one cleaner to the other are as shown in the following table.

		Second Load	
	State	*1*	*2*
First	1	.8	.2
Load	2	.35	.65

◀

The information from the table can be written as a *transition matrix,* where the states are indicated at the side and top, as follows.

$$\begin{array}{c} \qquad\qquad\qquad\qquad \textit{Second Load} \\ \qquad\qquad\qquad \text{Johnson} \quad \text{NorthClean} \\ \textit{First Load} \begin{array}{c} \text{Johnson} \\ \text{NorthClean} \end{array} \left[\begin{array}{cc} .8 & .2 \\ .35 & .65 \end{array} \right] \end{array}$$

A **transition matrix** has the following features.

1. It is square, since all possible states must be used both as rows and as columns.
2. All entries are between 0 and 1, inclusive, because all entries represent probabilities.
3. The sum of the entries in any row must be 1, since the numbers in the row give the probability of changing from the state at the left to one of the states indicated across the top.

▶**EXAMPLE 2**   Suppose that when the new promotional campaign began, Johnson had 40% of the market and NorthClean had 60%. Use the tree diagram in Figure 9.3 to find how these proportions would change after another week of advertising.

2 Find the product

$$[.53 \quad .47]\begin{bmatrix} .8 & .2 \\ .35 & .65 \end{bmatrix}.$$

Answer:
[.59   .41]   (rounded)

3 Find each cleaner's market share after 3 weeks.

Answer:
[.62   .38]

Add the numbers indicated with arrows to find the proportion of people taking their cleaning to Johnson after 1 week.

$$.32 + .21 = .53$$

Similarly, the proportion taking their cleaning to NorthClean is

$$.08 + .39 = .47.$$

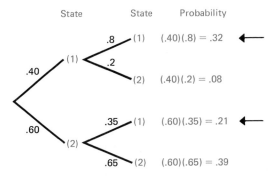

State 1: taking cleaning to Johnson
State 2: taking cleaning to NorthClean

FIGURE 9.3

The initial distribution of 40% and 60% becomes, after 1 week, 53% and 47%. ◀

These distributions can be written as the probability vectors

$$[.40 \quad .60] \quad \text{and} \quad [.53 \quad .47].$$

A **probability vector** is a matrix of only one row, having nonnegative entries with the sum of the entries equal to 1.

The results from the tree diagram above are exactly the same as the result of multiplying the initial probability vector by the transition matrix. (Multiplication of matrices was discussed in Section 6.4.)

$$[.4 \quad .6]\begin{bmatrix} .8 & .2 \\ .35 & .65 \end{bmatrix} = [.53 \quad .47]$$

Find the market share after 2 weeks by multiplying the vector [.53   .47] and the transition matrix. **2**

The product from Problem 2 at the side (with the numbers rounded to the nearest hundredth) shows that after 2 weeks of advertising, Johnson's share of the market has increased to 59%. To get the share after 3 weeks, multiply the probability vector [.59   .41] and the transition matrix. Continuing this process gives each cleaner's share of the market after any number of weeks. **3**

The following table gives the market share for each cleaner at the end of week n for several values of n.

Week	Johnson	NorthClean
Start	.4	.6
1	.53	.47
2	.59	.41
3	.62	.38
4	.63	.37
5	.63	.37
12	.64	.36

The results seem to approach the probability vector [.64 .36]. This vector is called the **equilibrium vector** or **fixed vector** for the given transition matrix. The equilibrium vector gives a long-range prediction—the shares of the market will stabilize (under the same conditions) at 64% for Johnson and 36% for NorthClean.

Starting with some other initial probability vector and going through the steps above would give the same equilibrium vector. In fact, the long-range trend is the same no matter what the initial vector is. The long-range trend depends only on the transition matrix, not on the initial distribution.

One of the many applications of Markov chains is in finding these long-range predictions. It is not possible to make long-range predictions with all transition matrices, but there is a large set of transition matrices for which long-range predictions *are* possible. Such predictions are always possible with *regular transition matrices*. A transition matrix is **regular** if some power of the matrix contains all positive entries. A Markov chain is a **regular Markov chain** if its transition matrix is regular. A regular Markov chain always has an equilibrium vector.

▶**EXAMPLE 3** Decide if the following transition matrices are regular.

(a) $A = \begin{bmatrix} .75 & .25 & 0 \\ 0 & .5 & .5 \\ .6 & .4 & 0 \end{bmatrix}$

Square the matrix A by multiplying it by itself.

$$A^2 = \begin{bmatrix} .5625 & .3125 & .125 \\ .3 & .45 & .25 \\ .45 & .35 & .2 \end{bmatrix}$$

All entries in A^2 are positive, so that matrix A is regular.

4 Decide if the following transition matrices are regular.

(a) $\begin{bmatrix} 0 & 1 \\ 1 & 0 \end{bmatrix}$

(b) $\begin{bmatrix} .3 & .7 \\ 1 & 0 \end{bmatrix}$

Answer:

(a) No

(b) Yes

5 Solve the equation for v_2. Round to the nearest thousandth.

Answer:

.364

(b) $B = \begin{bmatrix} .5 & 0 & .5 \\ 0 & 1 & 0 \\ 0 & 0 & 1 \end{bmatrix}$

Find various powers of B.

$$B^2 = \begin{bmatrix} .25 & 0 & .75 \\ 0 & 1 & 0 \\ 0 & 0 & 1 \end{bmatrix} \quad B^3 = \begin{bmatrix} .125 & 0 & .875 \\ 0 & 1 & 0 \\ 0 & 0 & 1 \end{bmatrix} \quad B^4 = \begin{bmatrix} .0625 & 0 & .9375 \\ 0 & 1 & 0 \\ 0 & 0 & 1 \end{bmatrix}$$

Notice that all of the powers of B shown here have zeros in the same locations. Thus, further powers of B will still give the same zero entries, so that no power of matrix B contains all positive entries. For this reason, B is not regular. ◀ **4**

The equilibrium vector can be found without doing all the work shown above. Let $V = [v_1 \quad v_2]$ represent the desired vector. Let P represent the transition matrix. By definition, V is the fixed probability vector if $VP = V$. In our example,

$$[v_1 \quad v_2] \begin{bmatrix} .8 & .2 \\ .35 & .65 \end{bmatrix} = [v_1 \quad v_2].$$

Since $[v_1 \quad v_2]$ is a probability vector, $v_1 + v_2 = 1$. Multiply on the left to get

$$[.8v_1 + .35v_2 \quad .2v_1 + .65v_2] = [v_1 \quad v_2].$$

Set corresponding entries from the two matrices equal to get

$$.8v_1 + .35v_2 = v_1 \quad .2v_1 + .65v_2 = v_2.$$

Simplify each of these equations.

$$-.2v_1 + .35v_2 = 0 \quad .2v_1 - .35v_2 = 0$$

These last two equations are really the same. (The equations in the system obtained from $VP = V$ are always dependent.) To find the values of v_1 and v_2, recall that $V = [v_1 \quad v_2]$ is a probability vector, so that

$$v_1 + v_2 = 1.$$

Find v_1 and v_2 by solving the system

$$-.2v_1 + .35v_2 = 0$$
$$v_1 + v_2 = 1.$$

From the second equation, $v_1 = 1 - v_2$. Substitute for v_1 in the first equation:

$$-.2(1 - v_2) + .35v_2 = 0. \quad \textbf{5}$$

Since $v_2 = .364$ (from Problem 5 at the side) and $v_1 = 1 - v_2$, then $v_1 = 1 - .364 = .636$ and the equilibrium vector is $[.636 \quad .364]$.

6 Find the equilibrium vector for the transition matrix

$$P = \begin{bmatrix} .3 & .7 \\ .5 & .5 \end{bmatrix}.$$

Answer:

[5/12 7/12]

▶ **EXAMPLE 4** Find the equilibrium vector for the transition matrix

$$P = \begin{bmatrix} \frac{1}{3} & \frac{2}{3} \\ \frac{3}{4} & \frac{1}{4} \end{bmatrix}.$$

Look for a vector $[v_1 \quad v_2]$ such that

$$[v_1 \quad v_2]\begin{bmatrix} \frac{1}{3} & \frac{2}{3} \\ \frac{3}{4} & \frac{1}{4} \end{bmatrix} = [v_1 \quad v_2].$$

Multiply on the left, obtaining the two equations

$$\frac{1}{3}v_1 + \frac{3}{4}v_2 = v_1 \qquad \frac{2}{3}v_1 + \frac{1}{4}v_2 = v_2.$$

Simplify each of these equations.

$$-\frac{2}{3}v_1 + \frac{3}{4}v_2 = 0 \qquad \frac{2}{3}v_1 - \frac{3}{4}v_2 = 0$$

Since $v_1 + v_2 = 1$, then $v_1 = 1 - v_2$. Substitute this into either of the two given equations. Using the first equation gives

$$-\frac{2}{3}(1 - v_2) + \frac{3}{4}v_2 = 0$$

$$-\frac{2}{3} + \frac{2}{3}v_2 + \frac{3}{4}v_2 = 0$$

$$-\frac{2}{3} + \frac{17}{12}v_2 = 0$$

$$\frac{17}{12}v_2 = \frac{2}{3}$$

$$v_2 = \frac{8}{17},$$

and $v_1 = 1 - 8/17 = 9/17$. The equilibrium vector is [9/17 8/17]. ◀ **6**

▶ **EXAMPLE 5** The probability that a complex assembly line works correctly depends on whether or not the line worked correctly the last time it was used. The various probabilities are as given in the following transition matrix.

	Works Properly Now	Does Not
Worked Properly Before	.9	.1
Did Not	.7	.3

Find the long-range probability that the assembly line will work properly.

7 In Example 5, suppose the company modifies the line so that the transition matrix becomes

$$\begin{bmatrix} .95 & .05 \\ .8 & .2 \end{bmatrix}.$$

Find the long-range probability that the assembly line will work properly.

Answer:
16/17

Begin by finding the equilibrium vector $[v_1 \quad v_2]$, where

$$[v_1 \quad v_2]\begin{bmatrix} .9 & .1 \\ .7 & .3 \end{bmatrix} = [v_1 \quad v_2].$$

Multiplying on the left and setting corresponding entries equal gives the equations

$$.9v_1 + .7v_2 = v_1 \quad \text{and} \quad .1v_1 + .3v_2 = v_2$$

or

$$-.1v_1 + .7v_2 = 0 \quad \text{and} \quad .1v_1 - .7v_2 = 0.$$

Substitute $v_1 = 1 - v_2$ in the first of these equations to get

$$-.1(1 - v_2) + .7v_2 = 0$$
$$-.1 + .1v_2 + .7v_2 = 0$$
$$-.1 + .8v_2 = 0$$
$$.8v_2 = .1$$
$$v_2 = \frac{1}{8},$$

and $v_1 = 1 - 1/8 = 7/8$. The equilibrium vector is $[7/8 \quad 1/8]$. In the long run, the company can expect the assembly line to run properly 7/8 of the time. ◀ **7**

The examples suggest the following generalization. If a regular Markov chain has transition matrix P and initial probability vector v, the products

$$vP$$
$$vPP = vP^2$$
$$vP^2P = vP^3$$
$$\vdots$$
$$vP^n$$

lead to the equilibrium vector V. The results of this section can be summarized as follows.

Suppose a regular Markov chain has a transition matrix P.

1. For any initial probability vector v, the products vP^n approach a unique vector V as n gets larger and larger. Vector V is called the **equilibrium** or **fixed vector.**
2. Vector V has the property that $VP = V$.
3. Vector V is found by solving a system of equations obtained from the matrix equation $VP = V$ and from the fact that the sum of the entries of V is 1.

9.4 EXERCISES

Which of the following could be a probability vector?

1. $\begin{bmatrix} \frac{2}{3} & \frac{1}{3} \end{bmatrix}$　　**2.** $\begin{bmatrix} \frac{1}{2} & 1 \end{bmatrix}$　　**3.** $\begin{bmatrix} 0 & 1 \end{bmatrix}$　　**4.** $\begin{bmatrix} .1 & .1 \end{bmatrix}$　　**5.** $\begin{bmatrix} .4 & .2 & 0 \end{bmatrix}$　　**6.** $\begin{bmatrix} \frac{1}{4} & \frac{1}{8} & \frac{5}{8} \end{bmatrix}$

Which of the following could be a transition matrix, by definition?

7. $\begin{bmatrix} .5 & 0 \\ 0 & .5 \end{bmatrix}$　　**8.** $\begin{bmatrix} \frac{2}{3} & \frac{1}{3} \\ 1 & 0 \end{bmatrix}$　　**9.** $\begin{bmatrix} \frac{1}{4} & \frac{3}{4} \\ \frac{1}{2} & \frac{1}{2} \end{bmatrix}$　　**10.** $\begin{bmatrix} \frac{1}{4} & \frac{3}{4} & 0 \\ 2 & 0 & 1 \\ 1 & \frac{2}{3} & 3 \end{bmatrix}$

11. $\begin{bmatrix} \frac{1}{3} & \frac{1}{3} & \frac{1}{3} \\ 0 & 1 & 0 \\ \frac{1}{2} & 0 & \frac{1}{2} \end{bmatrix}$　　**12.** $\begin{bmatrix} \frac{1}{3} & \frac{1}{2} & 1 \\ 0 & 1 & 0 \\ \frac{1}{2} & 0 & \frac{1}{2} \end{bmatrix}$　　**13.** $\begin{bmatrix} \frac{1}{3} & \frac{1}{2} & 1 \\ \frac{1}{3} & 0 & 0 \\ \frac{1}{3} & \frac{1}{2} & 0 \end{bmatrix}$　　**14.** $\begin{bmatrix} .9 & .1 & 0 \\ .1 & .6 & .3 \\ 0 & .3 & .7 \end{bmatrix}$

Which of the following transition matrices are regular? (See Example 3.)

15. $\begin{bmatrix} .2 & .8 \\ .9 & .1 \end{bmatrix}$　　**16.** $\begin{bmatrix} .22 & .78 \\ .43 & .57 \end{bmatrix}$　　**17.** $\begin{bmatrix} 1 & 0 \\ .6 & .4 \end{bmatrix}$　　**18.** $\begin{bmatrix} .55 & .45 \\ 0 & 1 \end{bmatrix}$

19. $\begin{bmatrix} 0 & 1 & 0 \\ .4 & .2 & .4 \\ 1 & 0 & 0 \end{bmatrix}$　　**20.** $\begin{bmatrix} .3 & .5 & .2 \\ 1 & 0 & 0 \\ .5 & .1 & .4 \end{bmatrix}$

Find the equilibrium vector for each of the following transition matrices. (See Examples 4 and 5.)

21. $\begin{bmatrix} \frac{1}{4} & \frac{3}{4} \\ \frac{1}{2} & \frac{1}{2} \end{bmatrix}$　　**22.** $\begin{bmatrix} \frac{2}{3} & \frac{1}{3} \\ \frac{1}{8} & \frac{7}{8} \end{bmatrix}$　　**23.** $\begin{bmatrix} .3 & .7 \\ .4 & .6 \end{bmatrix}$　　**24.** $\begin{bmatrix} .8 & .2 \\ .1 & .9 \end{bmatrix}$

25. $\begin{bmatrix} .1 & .1 & .8 \\ .4 & .4 & .2 \\ .1 & .2 & .7 \end{bmatrix}$　　**26.** $\begin{bmatrix} .5 & .2 & .3 \\ .1 & .4 & .5 \\ .2 & .2 & .6 \end{bmatrix}$　　**27.** $\begin{bmatrix} .25 & .35 & .4 \\ .1 & .3 & .6 \\ .55 & .4 & .05 \end{bmatrix}$　　**28.** $\begin{bmatrix} .16 & .28 & .56 \\ .43 & .12 & .45 \\ .86 & .05 & .09 \end{bmatrix}$

29. Management The probability that a complex assembly line works correctly depends on whether the line worked correctly the last time it was used. There is a .95 chance that the line will work correctly if it worked correctly the time before, and a .7 chance that it will work correctly if it did *not* work correctly the time before. Set up a transition matrix with this information and find the long-run probability that the line will work correctly. (See Example 5.)

30. Management Suppose improvements are made in the assembly line of Exercise 29, so that the transition matrix becomes

	Works	Doesn't Work
Works	.95	.05
Doesn't Work	.85	.15

Find the new long-run probability that the line will work properly.

31. Natural Science In Exercises 60–64 of Section 8.4 we discussed the effect on flower color of cross-pollinating pea plants. As shown in Exercises 60–61, since the gene for red is dominant and the gene for white is recessive, 75% of these pea plants have red flowers and 25% have white flowers, because plants with 1 red and 1 white gene appear red. If a red-flowered plant is crossed with a red-flowered plant known to have 1 red and 1 white gene, then 75% of the offspring will be red and 25% will be white. Crossing a red-flowered plant that has 1 red and 1 white gene with a white-flowered plant produces 50% red-flowered offspring and 50% white-flowered offspring.

(a) Write a transition matrix using this information.

(b) Write a probability vector for the initial distribution of colors.

(c) Find the distribution of colors after 4 generations.

(d) Find the long-range distribution of colors.

32. Natural Science Snapdragons with 1 red gene and 1 white gene produce pink-flowered offspring. If a red snapdragon is crossed with a pink snapdragon, the probabilities that the offspring will be red, pink, or white are 1/2, 1/2, and 0, respectively. If 2 pink snapdragons are crossed, the probabilities of red, pink, or white offspring are 1/4, 1/2, and 1/4, respectively. For a cross between a white and a pink snapdragon, the corresponding probabilities are 0, 1/2, and 1/2. Find the long-range prediction for the fraction of red, pink, and white snapdragons.

33. Management Each month, a sales manager classifies her salespeople as low, medium, or high producers. There is a .4 chance that a low producer one month will become a medium producer the following month, and a .1 chance that a low producer will become a high producer. A medium producer will become a low or high producer, respectively, with probabilities .25 and .3. A high producer will become a low or medium producer, respectively, with probabilities .05 and .4. Find the long-range trend for the proportion of low, medium, and high producers.

34. Management An insurance company classifies its drivers into three groups: G_0 (no accidents), G_1 (one accident), and G_2 (more than one accident). The probability that a driver in G_0 will stay in G_0 after 1 year is .85, that he will become a G_1 is .10, and that he will become a G_2 is .05. A driver in G_1 cannot move to G_0 (this insurance company has a long memory). There is an .80 probability that a G_1 driver will stay in G_1 and a .20 probability that he will become a G_2. A driver in G_2 must stay in G_2.

(a) Write a transition matrix using this information. Suppose that the company accepts 50,000 new policyholders, all of whom are in group G_0. Find the number in each group

(b) after 1 year;   **(c)** after 2 years;

(d) after 3 years;   **(e)** after 4 years.

(f) Find the equilibrium vector here. Interpret your result.

35. Management The difficulty with the mathematical model of Exercise 34 is that no "grace period" is provided; there should be a certain probability of moving from

G_1 or G_2 back to G_0 (say, after 4 years with no accidents). A new system with this feature might produce the following transition matrix.

$$\begin{bmatrix} .85 & .10 & .05 \\ .15 & .75 & .10 \\ .10 & .30 & .60 \end{bmatrix}$$

Suppose that when this new policy is adopted, the company has 50,000 policyholders in group G_0. Find the number of these in each group

(a) after 1 year;   **(b)** after 2 years;

(c) after 3 years.

(d) Find the equilibrium vector here. Interpret your result.

36. Management Research done by the Gulf Oil Corporation* produced the following transition matrix for the probability that a person with one form of home heating would switch to another.

		Will Switch to		
		Oil	Gas	Electric
Now Has	Oil	.825	.175	0
	Gas	.060	.919	.021
	Electric	.049	0	.951

The current share of the market held by these three types of heat is given by [.26 .60 .14]. Find the share of the market held by each type of heat after

(a) 1 year;   **(b)** 2 years;   **(c)** 3 years.

(d) What is the long-range prediction?

37. Natural Science A certain genetic defect is carried only by males. Suppose the probability is .95 that a male offspring will have the defect if his father did, and .10 that a male offspring will have it if his father did not. Find the long-range prediction for the fraction of the males in the population who will have this defect.

---

*From "Forecasting Market Shares of Alternative Home-Heating Units by Markov Process Using Transition Probabilities Estimates from Aggregate Time Series Data" by Ali Ezzati, from *Management Science,* Vol. 21, No. 4, December 1974. Copyright © 1974 The Institute of Management Sciences. Reprinted by permission.

38. Natural Science A large group of mice is kept in a cage having connected compartments A, B, and C. Mice in compartment A move to B with probability .3 and to C with probability .4. Mice in B move to A or C with probabilities of .15 and .55, respectively. Mice in C move to A and B with probabilities of .3 and .6, respectively. Find the long-range prediction for the fraction of mice in each of the compartments.

39. Social Science In a survey investigating change in housing patterns in one urban area, it was found that 75% of the population lived in single-family dwellings and 25% in multiple housing of some kind. Five years later, in a follow-up survey, of those who had been living in single-family dwellings, 90% still did so, but 10% had moved to multiple-family dwellings. Of those in multiple housing, 95% were still living in that type of housing, while 5% had moved to single-family dwellings. Assume that these trends continue.

(a) Write a transition matrix for this information.

(b) Write a probability vector for the initial distribution of housing.

(c) What percent of the population can be expected in each category 5 years later?

(d) What is the long-range prediction?

40. Social Science In one state, a land-use survey showed that 35% of the land was used for agricultural purposes, while 10% was urban. Ten years later, of the agricultural land, 15% had become urban and 80% had remained agricultural. (The remainder lay idle.) Of the idle land, 20% had become urbanized and 10% had been converted for agricultural use. Of the urban land, 90% remained urban and 10% was idle. Assume that these trends continue.

(a) Write a transition matrix for this information.

(b) Write a probability vector for the initial distribution of land.

(c) What is the land-use pattern after 20 years?

(d) What is the long-range prediction for land use in this state?

41. Social Science At one liberal arts college, students are classified as humanities majors, science majors, or undecideds. There is a 20% chance that a humanities major will change to a science major from one year to the next, and a 45% chance that a humanities major will change to undecided. A science major will change to humanities with probability .15, and to undecided with probability .35. An undecided will switch to humanities or science with probabilities of .5 and .3, respectively. Find the long-

range prediction for the fraction of students in each of these three majors.

42. Natural Science The weather in a certain spot is classified as fair, cloudy without rain, or rainy. A fair day is followed by a fair day 60% of the time, and by a cloudy day 25% of the time. A cloudy day is followed by a cloudy day 35% of the time, and by a rainy day 25% of the time. A rainy day is followed by a cloudy day 40% of the time, and by another rainy day 25% of the time. Find the long-range prediction for the proportion of fair, cloudy, and rainy days.

Exercises 43–48 are computer problems. For each of the following transition matrices, find the first five powers of the matrix. Then find the probability that state 2 changes to state 4 after 5 repetitions of the experiment.

43.
$$\begin{bmatrix} .1 & .2 & .2 & .3 & .2 \\ .2 & .1 & .1 & .2 & .4 \\ .2 & .1 & .4 & .2 & .1 \\ .3 & .1 & .1 & .2 & .3 \\ .1 & .3 & .1 & .1 & .4 \end{bmatrix}$$

44.
$$\begin{bmatrix} .3 & .2 & .3 & .1 & .1 \\ .4 & .2 & .1 & .2 & .1 \\ .1 & .3 & .2 & .2 & .2 \\ .2 & .1 & .3 & .2 & .2 \\ .1 & .1 & .4 & .2 & .2 \end{bmatrix}$$

45. Management A company with a new training program classified each employee in one of the four states: s_1, never in the program; s_2, currently in the program; s_3, discharged; s_4, completed the program. The transition matrix for this company is given below

$$\begin{array}{c c} & \begin{array}{cccc} s_1 & s_2 & s_3 & s_4 \end{array} \\ \begin{array}{c} s_1 \\ s_2 \\ s_3 \\ s_4 \end{array} & \begin{bmatrix} .4 & .2 & .05 & .35 \\ 0 & .45 & .05 & .5 \\ 0 & 0 & 1.0 & 0 \\ 0 & 0 & 0 & 1.0 \end{bmatrix} \end{array}$$

(a) What percent of employees who had never been in the program (state s_1) completed the program (state s_4) after the program had been offered five times?

(b) If the initial percent of employees in each state was [.5 .5 0 0], find the corresponding percents after the program had been offered four times.

Find the equilibrium vector for the transition matrices in Exercises 46 and 47 by taking powers of the matrix.

46. $\begin{bmatrix} .1 & .2 & .2 & .3 & .2 \\ .2 & .1 & .1 & .2 & .4 \\ .2 & .1 & .4 & .2 & .1 \\ .3 & .1 & .1 & .2 & .3 \\ .1 & .3 & .1 & .1 & .4 \end{bmatrix}$

47. $\begin{bmatrix} .3 & .2 & .3 & .1 & .1 \\ .4 & .2 & .1 & .2 & .1 \\ .1 & .3 & .2 & .2 & .2 \\ .2 & .1 & .3 & .2 & .2 \\ .1 & .1 & .4 & .2 & .2 \end{bmatrix}$

48. Find the long-range prediction for the percent of employees in each state for the company training program from Exercise 45.

9.5 PROBABILITY DISTRIBUTIONS; EXPECTED VALUE

The concept of *probability distributions* was discussed briefly in Chapter 8. A probability distribution depends on the idea of a *random variable*.

RANDOM VARIABLES Suppose a bank is interested in improving its services to the public. The manager decides to begin by finding the amount of time tellers spend on each transaction, rounded to the nearest minute. To each transaction, then, will be assigned one of the numbers 0, 1, 2, 3, 4, ... That is, if t represents an outcome of the experiment of timing a transaction, then t may take on any of the values from the list 0, 1, 2, 3, 4, ... Since the value that t takes on for a particular transaction is random, t is called a random variable.

A **random variable** is a function that assigns a real number to each outcome of an experiment.

PROBABILITY DISTRIBUTION Suppose that the bank manager in our example records the times for 75 different transactions, with results as shown in Table 1. As the table shows, the shortest transaction time was 1 minute, with 3 transactions of 1-minute duration. The longest time was 10 minutes. Only 1 transaction took that long.

In Table 1, the first column gives the 10 values assumed by the random variable t, and the second column shows the number of occurrences corresponding to each of these values, the **frequency** of that value. Table 1 is an example of a **frequency distribution,** a table listing the frequencies for each value a random variable may assume.

Table 1

Time	Frequency
1	3
2	5
3	9
4	12
5	15
6	11
7	10
8	6
9	3
10	1
Total:	75

Now suppose that several weeks after starting new procedures to speed up transactions, the manager takes another survey. This time she observes 57 transactions, and she records their times as shown in the frequency distribution in Table 2. (Table 1 is repeated to allow comparison of the two tables.)

Table 1

Time	Frequency
1	3
2	5
3	9
4	12
5	15
6	11
7	10
8	6
9	3
10	1
Total:	75

Table 2

Time	Frequency
1	4
2	5
3	8
4	10
5	12
6	17
7	0
8	1
9	0
10	0
Total:	57

Do the results in Table 2 indicate an improvement? It is hard to compare the two tables, since one is based on 75 transactions and the other on 57. To make them comparable, we can add a column to each table that gives the *relative frequency* of each transaction time. These results are shown in Tables 3 and 4. Where necessary, decimals are rounded to the nearest hundredth. To find a **relative frequency,** divide

1 Complete the probability distribution below. (f represents frequency and p represents probability.)

x	f	p
1	3	
2	7	
3	9	
4	3	
5	2	

Answer:

x	f	p
1	3	.13
2	7	.29
3	9	.38
4	3	.13
5	2	.08
Total:	24	

each frequency by the total of the frequencies. Here the individual frequencies are divided by 75 or 57.

Table 3

Time	Frequency	Relative Frequency
1	3	$3/75 = .04$
2	5	$5/75 \approx .07$
3	9	$9/75 = .12$
4	12	$12/75 = .16$
5	15	$15/75 = .20$
6	11	$11/75 \approx .15$
7	10	$10/75 \approx .13$
8	6	$6/75 = .08$
9	3	$3/75 = .04$
10	1	$1/75 \approx .01$

Table 4

Time	Frequency	Relative Frequency
1	4	$4/57 \approx .07$
2	5	$5/57 \approx .09$
3	8	$8/57 \approx .14$
4	10	$10/57 \approx .18$
5	12	$12/57 \approx .21$
6	17	$17/57 \approx .30$
7	0	$0/57 = 0$
8	1	$1/57 \approx .02$
9	0	$0/57 = 0$
10	0	$0/57 = 0$

Whether the differences in relative frequency between the distributions in Tables 3 and 4 are interpreted as desirable or undesirable depends on management goals. If the manager wanted to eliminate the most time-consuming transactions, the results appear to be desirable. However, before the new procedures were followed, the largest relative frequency of transactions, .20, was for transactions of 5 minutes. With the new procedures, the largest relative frequency, .30, corresponds to a transaction of 6 minutes. At any rate, the results shown in the two tables are easier to compare using relative frequencies.

The relative frequencies in Tables 3 and 4 can be considered as probabilities. Such a table that lists all the possible values of a random variable, together with the corresponding probabilities, is called a **probability distribution.** The sum of the probabilities in a probability distribution must always be 1. (The sum in an actual distribution may vary slightly from 1 because of rounding.)   **1**

▶**EXAMPLE 1**   Many plants have seed pods with variable numbers of seeds. One variety of green beans has no more than 6 seeds per pod. Suppose that examination of 30 such bean pods gives the results shown in Table 5. Here the random variable x tells the number of seeds per pod. Give a probability distribution for these results.

The probabilities are found by computing the relative frequencies. A total of 30 bean pods were examined, so each frequency should be divided by 30 to get the probability distribution shown as Table 6. Some of the results have been rounded to the nearest hundredth.

2 In Example 1, what is

(a) $P(x = 0)$?

(b) $P(x = 1)$?

Answer:

(a) .10

(b) .13

Table 5

x	Frequency
0	3
1	4
2	6
3	8
4	5
5	3
6	1
	Total: 30

Table 6

x	Frequency	Probability
0	3	3/30 = .10
1	4	4/30 ≈ .13
2	6	6/30 = .20
3	8	8/30 ≈ .27
4	5	5/30 ≈ .17
5	3	3/30 = .10
6	1	1/30 ≈ .03
	Total: 30	

As shown in Table 6, the probability that the random variable x takes on the value 2 is 6/30, or .20. This is often written as

$$P(x = 2) = .20.$$

Also, $P(x = 5) = .10$, and $P(x = 6) \approx .03$. ◀ **2**

Instead of writing the probability distribution in Example 1 as a table, we could write the same information as a set of ordered pairs:

$$\{(0, .10), (1, .13), (2, .20), (3, .27), (4, .17), (5, .10), (6, .03)\}.$$

There is just one probability for each value of the random variable. Thus, a probability distribution defines a function, called a **probability distribution function** or simply a **probability function.** We shall use the terms "probability distribution" and "probability function" interchangeably.

The information in a probability distribution is often displayed graphically as a special kind of bar graph called a **histogram.** The bars of a histogram all have the same width, usually 1. The heights of the bars are determined by the frequencies. A histogram for the data in Table 3 is given in Figure 9.4. A histogram shows important characteristics of a distribution that may not be readily apparent in tabular form, such as the relative sizes of the probabilities and any symmetry in the distribution.

The area of the bar above $t = 1$ in Figure 9.4 is the product of 1 and .04, or $.04 \times 1 = .04$. Since each bar has a width of 1, its area is equal to the probability that corresponds to that value of t. The probability that a particular value will occur is thus given by the area of the appropriate bar of the graph. For example, the probability that a transaction will take less than 4 minutes is the sum of the areas for $t = 1$, $t = 2$, and $t = 3$. This area, shown in color in Figure 9.5, corresponds to 23% of the total area, since

$$P(t < 4) = P(t = 1) + P(t = 2) + P(t = 3)$$
$$= .04 + .07 + .12 = .23.$$

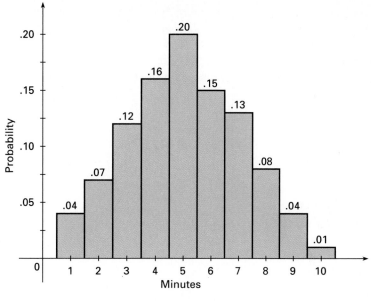

FIGURE 9.4

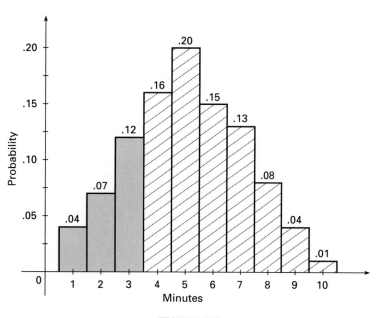

FIGURE 9.5

3 **(a)** Prepare a histogram for the probability distribution in Problem 1 at the side.

(b) Find $P(x \leq 2)$.

Answer:

(a)

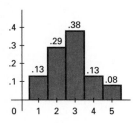

(b) .42

▶**EXAMPLE 2** Construct a histogram for the probability distribution in Example 1. Find the area that gives the probability that the number of seeds will be more than 4.

A histogram for this distribution is shown in Figure 9.6. The portion of the histogram in color represents

$$P(x > 4) = P(x = 5) + P(x = 6)$$
$$= .10 + .03 = .13,$$

or 13% of the total area. ◀ **3**

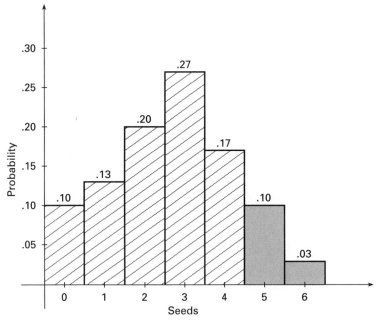

FIGURE 9.6

▶**EXAMPLE 3** **(a)** Give the probability distribution for the number of heads showing when 2 coins are tossed.

Let x represent the random variable "number of heads," Then x can take on the values 0, 1, or 2. Now find the probability of each outcome. The results are shown in Table 7.

Table 7

x	$P(x)$
0	1/4
1	1/2
2	1/4

4 **(a)** Give the probability distribution for the number of heads showing when 3 coins are tossed.

(b) Find the probability that no more than 1 coin shows heads.

Answer:

(a)

x	$P(x)$
0	1/8
1	3/8
2	3/8
3	1/8

(b) 1/2

(b) Draw a histogram for the distribution in Table 7. Find the probability that at least 1 coin comes up heads.

The histogram is shown in Figure 9.7. The portion in color represents

$$P(x \geq 1) = P(x = 1) + P(x = 2)$$
$$= \frac{3}{4}. \quad \blacktriangleleft \quad \boxed{4}$$

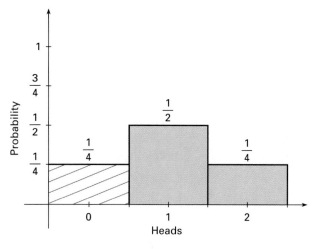

FIGURE 9.7

In working with experimental data, it is often useful to have a typical or "average" number that represents the entire set of data. For example, we compare our heights and weights to those of the typical or "average" person on weight charts. Students are familiar with the "class average" and their own "average" in a given course.

In a recent year, a citizen of the United States could expect to complete about 12 years of school, to be a member of a household earning $27,700 per year, and to live in a household of 2.7 people. What do we mean here by "expect"? Many people have completed less than 12 years of school; many others have completed more. Many households have less income than $27,700 per year; many others have more. The idea of a household of 2.7 people is a little hard to swallow. The numbers all refer to *averages*. When the term "expect" is used in this way, it refers to *mathematical expectation,* which is a kind of average.

EXPECTED VALUE What about an average value for a random variable? As an example, let us find the average number of offspring for a certain species of pheasant, given the probability distribution in Table 8.

Table 8

Number of Offspring	Frequency	Probability
0	8	.08
1	14	.14
2	29	.29
3	32	.32
4	17	.17
Total:	100	

We might be tempted to find the typical number of offspring by averaging the numbers 0, 1, 2, 3, and 4, which represent the numbers of offspring possible. This won't work, however, since the various numbers of offspring do not occur with equal probability: for example, a brood with 3 offspring is much more common than one with 0 or 1 offspring. The differing probabilities of occurrence can be taken into account with a **weighted average,** found by multiplying each of the possible numbers of offspring by its corresponding probability, as follows.

$$\text{Typical number of offspring} = 0(.08) + 1(.14) + 2(.29) + 3(.32) + 4(.17)$$
$$= 0 + .14 + .58 + .96 + .68$$
$$= 2.36$$

Based on the data given above, a typical brood of pheasants includes 2.36 offspring.

It is certainly not possible for a pair of pheasants to produce 2.36 offspring. If the numbers of offspring produced by many different pairs of pheasants are found, however, the average (or the mean) of these numbers will be about 2.36.

We can use the results of this example to define the mean, or *expected value,* of a probability distribution as follows.

Expected Value

Suppose the random variable x can take on the n values $x_1, x_2, x_3, \ldots, x_n$. Also, suppose the probabilities that these values occur are respectively $p_1, p_2, p_3, \ldots, p_n$. Then the **expected value** of the random variable is

$$E(x) = x_1p_1 + x_2p_2 + x_3p_3 + \ldots + x_np_n.$$

As in the example above, the expected value of a random variable may be a number that can never occur in any one trial of the experiment.

Physically, the expected value of a probability distribution represents a balance point. Figure 9.8 shows a histogram for the distribution of the pheasant offspring. If the histogram is thought of as a series of weights with magnitudes represented by the heights of the bars, then the system would balance if supported at the point corresponding to the expected value.

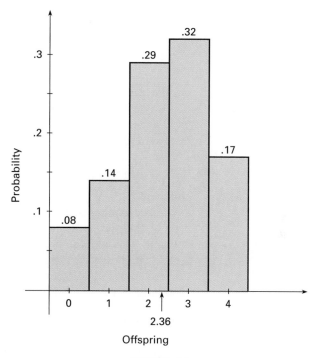

FIGURE 9.8

▶**EXAMPLE 4** Suppose a local church decides to raise money by raffling a microwave oven worth $400. A total of 2000 tickets are sold at $1 each. Find the expected value of winning for a person who buys 1 ticket in the raffle.

Here the random variable represents the possible amounts of net winnings, where net winnings = amount won − cost of ticket. The net winnings of the person winning the oven are $400 (amount won) − $1 (cost of ticket) = $399. The net winnings for each losing ticket are $0 − $1 = −$1.

The probability of winning is 1 in 2000, or 1/2000, while the probability of losing is 1999/2000. See Table 9.

Table 9

Outcome (net winnings)	Probability
$399	$\dfrac{1}{2000}$
−$1	$\dfrac{1999}{2000}$

5 Suppose you buy 1 of 1000 tickets at 10¢ each in a lottery where the first prize is $50. What are your expected net winnings?

Answer:
−5¢

The expected winnings for a person buying 1 ticket are

$$399\left(\frac{1}{2000}\right) + (-1)\left(\frac{1999}{2000}\right) = \frac{399}{2000} - \frac{1999}{2000}$$

$$= -\frac{1600}{2000}$$

$$= -.80.$$

On the average, a person buying 1 ticket in the raffle will lose $.80, or 80¢.

It is not possible to lose 80¢ in this raffle—either you lose $1, or you win a $400 prize. If you bought tickets in many such raffles over a long period of time, however, you would lose 80¢ per ticket, on the average. ◀ **5**

▶**EXAMPLE 5**   What is the expected number of girls in a family having exactly 3 children?

Some families with 3 children have 0 girls, others have 1 girl, and so on. We need to find the probabilities associated with 0, 1, 2 or 3 girls in a family of 3 children. To find these probabilities, first write the sample space S of all possible 3-child families: $S = \{ggg, ggb, bgg, gbg, gbb, bgb, bbg, bbb\}$. This sample space gives the probabilities shown in Table 10, assuming that the probability of a girl at each birth is 1/2.

Table 10

Outcome (number of girls)	Probability
0	1/8
1	3/8
2	3/8
3	1/8

The expected number of girls can now be found by multiplying each outcome (number of girls) by its corresponding probability and finding the sum of these values.

$$\text{Expected number of girls} = 0 \cdot \frac{1}{8} + 1 \cdot \frac{3}{8} + 2 \cdot \frac{3}{8} + 3 \cdot \frac{1}{8}$$

$$= \frac{3}{8} + \frac{6}{8} + \frac{3}{8}$$

$$= \frac{12}{8} = \frac{3}{2} = 1.5$$

6 Find the expected outcome for throwing 1 die. Begin by completing this table.

Outcome	1	2	3	4	5	6
Probability	$\frac{1}{6}$					

Answer:
All probabilities are 1/6; expected outcome is 3.5.

On the average, a 3-child family will have 1.5 girls. This result agrees with our intuition that, on the average, half the children born will be girls. ◀ **6**

▶**EXAMPLE 6**   Each day Donna and Mary toss a coin to see who buys the coffee (40¢ a cup). One tosses and the other calls the outcome. If the person who calls the outcome is correct, the other buys the coffee; otherwise the caller pays. Find Donna's expected winnings.

Assume that an honest coin is used, that Mary tosses the coin, and that Donna calls the outcome. The possible results and corresponding probabilities are shown below.

	Possible Results			
Result of toss	Heads	Heads	Tails	Tails
Call	Heads	Tails	Heads	Tails
Caller wins?	Yes	No	No	Yes
Probability	1/4	1/4	1/4	1/4

Donna wins a 40¢ cup of coffee whenever the results and calls match, and she loses a 40¢ cup when there is no match. Here expected winnings are

$$(.40)\left(\frac{1}{4}\right) + (-.40)\left(\frac{1}{4}\right) + (-.40)\left(\frac{1}{4}\right) + (.40)\left(\frac{1}{4}\right) = 0.$$

On the average, over the long run. Donna neither wins nor loses. ◀

A game with an expected value of 0 (such as the one in Example 6) is called a **fair game.** Casinos do not offer fair games. If they did, they would win (on the average) $0, and have a hard time paying the help! Casino games have expected winnings for the house that vary from 1.5¢ per dollar to 60¢ per dollar. Exercises 36–39 at the end of the section ask you to find the expected winnings for certain games of chance.

The idea of expected value can be very useful in decision making, as shown by the next example.

▶**EXAMPLE 7**   At age 50, you receive a letter from the Mutual of Mauritania Insurance Company. According to the letter, you must tell the company immediately which of the following two options you will choose: take $20,000 at age 60 (if

7 A person can take one of two jobs. With job A, there is a 50% chance of making $60,000 per year after 5 years and a 50% chance of making $30,000. With job B, there is a 30% chance of making $90,000 per year after 5 years and a 70% chance of making $20,000. Based strictly on expected value, which job should be taken?

Answer:
Job A has an expected salary of $45,000; job B, $41,000. Take job A.

you are alive, $0 otherwise) or $30,000 at age 70 (again, if you are alive, $0 otherwise). Based *only* on the idea of expected value, which should you choose?*

Life insurance companies have constructed elaborate tables showing the probability of a person living a given number of years in the future. From a recent such table, the probability of living from age 50 to age 60 is .88, while the probability of living from age 50 to 70 is .64. The expected values of the two options are given below.

First option: $(20,000)(.88) + (0)(.12) = 17,600$
Second option: $(30,000)(.64) + (0)(.36) = 19,200$

Based strictly on expected values, choose the second option ◀ **7**

9.5 EXERCISES

In Exercises 1–6, (a) give the probability distribution, and (b) sketch its histogram. (See Examples 1–3.)

1. In a seed viability test, 50 seeds were placed in 10 rows of 5 seeds each. After a period of time, the number that germinated in each row were counted, giving the results in the table.

Number Germinated	Frequency
0	0
1	0
2	1
3	3
4	4
5	2
	Total: 10

2. At a large supermarket during the 5 P.M. rush, the manager counted the number of customers waiting in each of 10 checkout lines. The results are shown in the table.

Number Waiting	Frequency
2	1
3	2
4	4
5	2
6	0
7	1
	Total: 10

*Other considerations might affect your decision, such as the rate at which you might invest the $20,000 at age 60 for 10 years.

3. At a training program for police officers, each member of a class of 25 took 6 shots at a target. The numbers of bullseyes shot are shown in the table.

Number of Bullseyes	Frequency
0	0
1	1
2	0
3	4
4	10
5	8
6	2
Total:	25

4. A class of 42 students took a 10-point quiz. The frequency of scores is given in the table.

Number of Points	Frequency
5	2
6	5
7	10
8	15
9	7
10	3
Total:	42

5. Five people are given a pain reliever. After a certain time period, the number who experience relief is noted. The experiment is repeated 20 times with the results shown in the table.

Number Who Experience Relief	Frequency
0	3
1	5
2	6
3	3
4	2
5	1
Total:	20

6. The telephone company kept track of the calls for the correct time during a 24-hour period for 2 weeks. The results are shown in the table.

Number of Calls	Frequency
28	1
29	1
30	2
31	3
32	2
33	2
34	2
35	1
Total:	14

For each of the experiments described below, let x represent a random variable, and use your knowledge of probability to prepare a probability distribution.

7. Four coins are tossed, and the number of heads is noted.

8. Two dice are rolled, and the total number of points is recorded.

9. Three cards are drawn from a deck. The number of aces is counted.

10. Two balls are drawn from a bag in which there are 4 white balls and 2 black balls. The number of black balls is counted.

11. A ballplayer with a batting average of .290 comes to bat 4 times in a game. The number of hits is counted.

12. Five cards are drawn from a deck. The number of black threes is counted.

Draw a histogram for each of the following, and shade the region that gives the indicated probability.

13. Exercise 7; $P(x \leq 2)$

14. Exercise 8; $P(x \geq 11)$

15. Exercise 9; P(at least one ace)

16. Exercise 10; P(at least one black ball)

17. Exercise 11; $P(x = 2 \text{ or } x = 3)$

18. Exercise 12; $P(1 \leq x \leq 2)$

Find the expected value for each random variable. (See Example 5.)

19.

x	2	3	4	5
P(x)	.1	.4	.3	.2

20.

y	4	6	8	10
P(y)	.4	.4	.05	.15

21.

z	9	12	15	18	21
P(z)	.14	.22	.36	.18	.10

22.

x	30	32	36	38	44
P(x)	.31	.30	.29	.06	.04

Find the expected values for the random variables x having the probability functions graphed below.

23.

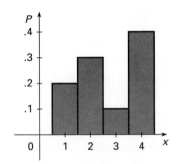

24.

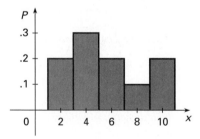

25.

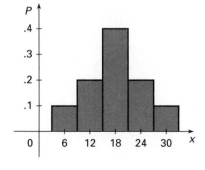

26.

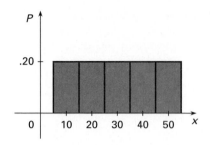

27. A raffle offers a first prize of $100, and two second prizes of $40 each. One ticket costs $1, and 500 tickets are sold. Find the expected winnings for a person who buys 1 ticket. (See Example 4.)

28. A raffle offers a first prize of $1000, two second prizes of $300 each, and twenty prizes of $10 each. If 10,000 tickets are sold at 50¢ each, find the expected winnings for a person buying 1 ticket.

Many of the following exercises use the idea of combinations discussed in Section 9.1.

29. If 3 marbles are drawn from a bag containing 3 yellow and 4 white marbles, what is the expected number of yellow marbles in the sample?

30. If 5 apples in a barrel of 25 apples are known to be rotten, what is the expected number of rotten apples in a sample of 2 apples?

31. A delegation of 3 is selected from a city council made up of 5 liberals and 4 conservatives.
   (a) What is the expected number of liberals on the committee?
   (b) What is the expected number of conservatives?

32. From a group of 2 women and 5 men, a delegation of 2 is selected. Find the expected number of women in the delegation.

33. In a club with 20 senior and 10 junior members, what is the expected number of junior members on a 3-member committee?

34. If 2 cards are drawn at one time from a deck of 52 cards, what is the expected number of diamonds?

35. Suppose someone offers to pay you $5 if you draw 2 diamonds in the game of Exercise 34. He says that you should pay 50¢ for the chance to play. Is this a fair game?

Find the expected winnings for each of the following games of chance.

36. In one form of roulette, you bet $1 on "even." If 1 of the 18 even numbers comes up, you get your dollar back, plus another one. If 1 of the 20 noneven (18 odd, 0, and 00) numbers comes up, you lose.

37. In another form of roulette, there are only 19 noneven numbers (no 00).

38. Numbers is an illegal game in which you bet $1 on any 3-digit number from 000 to 999. If your number comes up, you get $500.

39. In Keno, the house has a pot containing 80 balls, each marked with a different number from 1 to 80. You buy a ticket for $1 and mark 1 of the 80 numbers on it. The house then selects 20 numbers at random. If your number is among the 20, you get $3.20 (for a net winning of $2.20).

40. Use the assumptions of Example 6 to find Mary's expected winnings. If Mary tosses and Donna calls, is it still a fair game?

41. Suppose one day Mary brings a two-headed coin and uses it to toss for the coffee. Since Mary tosses, Donna calls.
(a) Is this still a fair game?
(b) What is Donna's expected gain if she calls heads?
(c) What is it if she calls tails?

42. Jack must choose at age 40 to inherit $25,000 at age 50 (if he is still alive) or $30,000 at age 55 (if he is still alive). If the probabilities for a person of age 40 to live to be 50 and 55 are .90 and .85, respectively, which choice gives him the larger expected inheritance? (See Example 7.)

43. Management An insurance company has written 100 policies of $10,000, 500 of $5000, and 1000 policies of $1000 on people of age 20. If experience shows that the probability of dying during the 20th year of life is .001, how much can the company expect to pay out during the year the policies were written?

44. Management A builder is considering a job which promises a profit of $30,000 with a probability of .7 or a loss (due to bad weather, strikes, and such) of $10,000 with a probability of .3. What is her expected profit?

45. Management Experience has shown that a ski lodge will be full (160 guests) during the Christmas holidays if there is a heavy snow pack in December, while a light snowfall in December means that it will have only 90 guests. What is the expected number of guests if the probability for a heavy snow in December is .40? (Assume that there must either be a light snowfall or a heavy snowfall.)

46. Management A magazine distributor offers a first prize of $100,000, two second prizes of $40,000 each, and two third prizes of $10,000 each. A total of 2,000,000 entries are received in the contest. Find the expected winnings if you submit 1 entry to the contest. If it would cost you 25¢ in time, paper, and stamps to enter, would it be worth it?

47. Management A local car dealer gets complaints about his cars, as shown in the following table.

Number of complaints per day	0	1	2	3	4	5	6
Probability	.01	.05	.15	.26	.33	.14	.06

Find the expected number of complaints per day.

48. Management Levi Strauss and Company* use expected value to help its salespeople rate their accounts. For each account, a salesperson estimates potential additional volume and the probability of getting it. The product of these gives the expected value of the potential, which is added to the existing volume. The totals are then classified as A, B, or C as follows: below $40,000, class C; between $40,000 and $55,000, class B; above $55,000, class A. Complete the chart below for one of its salespeople.

---

*This example was supplied by James McDonald, Levi Strauss and Company, San Francisco.

Account Number	Existing Volume	Potential Additional Volume	Probability of Additional Volume	Expected Value of Potential	Existing Volume + Expected Volume of Potential	Class
1	$15,000	$10,000	.25	$2,500	$17,500	C
2.	40,000	0	—	—	40,000	C
3	20,000	10,000	.20			
4	50,000	10,000	.10			
5	5,000	50,000	.50			
6	0	100,000	.60			
7	30,000	20,000	.80			

49. At the end of play in a major golf tournament, two players, an ''old pro'' and a ''new kid,'' are tied. Suppose first prize is $80,000 and second prize is $20,000. Find the expected winnings for the old pro if
(a) both players are of equal ability;
(b) the new kid will freeze up, giving the old pro a 3/4 chance of winning.

50. Natural Science In a certain animal species, the probability that a healthy adult female will have no offspring in a given year is .31, while the probability of 1, 2, 3, or 4 offspring are, respectively, .21, .19, .17, and .12. Find the expected number of offspring.

51. A recent McDonald's contest offered cash prizes and probabilities of winning on one visit, as shown in the following table. Suppose you spend $1 to buy a bus pass that lets you go to 25 different McDonald's and pick up entry forms. Find the expected value of winning a prize.

Prize	Probability
$100,000	$\dfrac{1}{176,402,500}$
25,000	$\dfrac{1}{39,200,556}$
5000	$\dfrac{1}{17,640,250}$
1000	$\dfrac{1}{1,568,022}$
100	$\dfrac{1}{282,244}$
5	$\dfrac{1}{7056}$
1	$\dfrac{1}{588}$

52. Natural Science One of the few methods that can be used in an attempt to cut the severity of a hurricane is to *seed* the storm. In this process, silver iodide crystals are dropped into the storm. Unfortunately, silver iodide crystals some-

times cause the storm to *increase* its speed. Wind speeds may also increase or decrease even with no seeding. The probabilities and amounts of property damage shown in the tree diagram below are from an article by R. A. Howard, J. E. Matheson, and D. W. North, ''The Decision to Seed Hurricanes.''*

(a) Find the expected amount of damage under each option, ''seed'' and ''do not seed.''
(b) To minimize total expected damage, what option should be chosen?

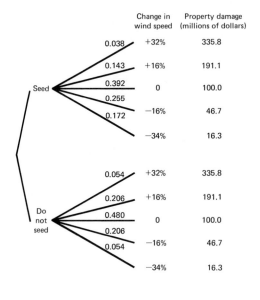

*From ''The Decision to Seed Hurricanes,'' Howard R. A., *Science,* Vol. 176, pp. 1191–1202, Fig. 7, 16 June 1972. Copyright 1972 by the American Association for the Advancement of Science. Reprinted by permission.

9.6 DECISION MAKING

John F. Kennedy once remarked he had assumed that as president it would be difficult to choose between distinct, opposite alternatives when a decision needed to be made. Actually, however, he found that such decisions were easy to make; the hard decisions came when he was faced with choices that were not as clear-cut. Most decisions fall in this last category—decisions which must be made under conditions of uncertainty. In the previous section we saw how to use expected values to help make a decision. These ideas are extended in this section, where we consider decision making in the face of uncertainty. Let us begin with an example.

Freezing temperatures are endangering the orange crop in central California. A farmer can protect his crop by burning smudge pots—the heat from the pots keeps the oranges from freezing. However, burning the pots is expensive; the cost is $4000. The farmer knows that if he burns smudge pots he will be able to sell his crop for a net profit (after smudge pot costs are deducted) of $10,000, provided that the freeze does develop and wipes out other orange crops in California. If he does nothing he will either lose $2000 already invested in the crop if it does freeze, or make a profit of $9600 if it does not freeze. (If it does not freeze, there will be a large supply of oranges, and thus his profit will be lower than if there was a small supply.)

What should the farmer do? He should begin by carefully defining the problem. First, he must decide on the **states of nature,** the possible alternatives over which he has no control. Here there are two: freezing temperatures, or no freezing temperatures. Next, the farmer should list the things he can control—his actions or **strategies.** He has two possible strategies: to use smudge pots or not. The consequences of each action under each state of nature, called **payoffs,** are summarized in a **payoff matrix,** as shown below. The payoffs in this case are the profits for each possible combination of events.

$$
\begin{array}{c}
 \\
\textit{Strategies of Farmer}
\end{array}
\quad
\begin{array}{c}
 \\
\text{Use Smudge Pots} \\
\text{Do not Use Pots}
\end{array}
\quad
\overset{\textstyle\textit{States of Nature}}{\overset{\textstyle\text{Freeze}\quad\text{No Freeze}}{\begin{bmatrix} \$10{,}000 & \$5600 \\ -\$2000 & \$9600 \end{bmatrix}}}
$$

To get the $5600 entry in the payoff matrix, use the profit if there is no freeze, $9600, and subtract the $4000 cost of using the pots. **1**

Once the farmer makes the payoff matrix, what then? The farmer might be an optimist (some might call him a gambler); in this case he might assume that the best will happen and go for the biggest number of the matrix ($10,000). For this profit, he must adopt the strategy "use smudge pots."

On the other hand, if the farmer is a pessimist, he would want to minimize the worst thing that could happen. If he uses smudge pots, the worst thing that could happen to him would be profit of $5600, which will result if there is no freeze. If

2 What should the farmer do if the probability of a freeze is .6? What is his expected profit?

Answer:
Use smudge pots; $8240

he does not use smudge pots, he might face a loss of $2000. To minimize the worst, he once again should adopt the strategy "use smudge pots."

Suppose the farmer decides that he is neither an optimist nor a pessimist, but would like further information before choosing a strategy. For example, he might call the weather forecaster and ask for the probability of a freeze. Suppose the forecaster says that this probability is only .1. What should the farmer do? He should recall the discussion of expected value from the previous section and work out the expected profit for each of his two possible strategies. If the probability of a freeze is .1, then the probability that there is no freeze is .9. This information leads to the following expected values.

$$\text{If smudge pots are used:} \qquad 10{,}000(.1) + 5600(.9) = 6040$$
$$\text{If no smudge pots are used:} \quad -2000(.1) + 9600(.9) = 8440$$

Here the maximum expected profit, $8440, is obtained if smudge pots are *not* used. **2**

As the example shows, the farmer's beliefs about the probabilities of a freeze affect his choice of strategies.

▶ **EXAMPLE 1**   A small manufacturer of Christmas cards must decide in February about the type of cards she should emphasize in her fall line. She has three possible strategies: emphasize modern cards, emphasize old-fashioned cards, or a mixture of the two. Her success is dependent on the state of the economy in December. If the economy is strong, she will do well with her modern cards, while in a weak economy people long for the old days and buy old-fashioned cards. In an in-between economy, her mixture of lines would do the best. She first prepares a payoff matrix for all three possibilities. The numbers in the matrix represent her profits in thousands of dollars.

		States of Nature		
		Weak Economy	In-Between	Strong Economy
	Modern	40	85	120
Strategies	Old-fashioned	106	46	83
	Mixture	72	90	68

(a) If the manufacturer is an optimist, she should aim for the biggest number on the matrix, 120 (representing $120,000 in profit). Her strategy in this case would be to produce modern cards.

(b) A pessimistic manufacturer wants to find the best of the worst of all bad things that can happen. If she produces modern cards, the worst that can happen is a profit of $40,000. For old-fashioned cards, the worst is a profit of $46,000, while the worst that can happen from a mixture is a profit of $68,000. Her strategy here is to use a mixture.

3 Suppose the manufacturer reads another article which gives the following predictions: 35% chance of a weak economy, 25% chance of an in-between economy, and a 40% chance of a strong economy. What is the best strategy now? What is the expected profit?

Answer:
Modern; $83,250

(c) Suppose the manufacturer reads an article in *The Wall Street Journal* that claims that leading experts believe there is a 50% chance of a weak economy at Christmas, a 20% chance of an in-between economy, and a 30% chance of a strong economy. The manufacturer can now use this information to find her expected profit for each possible strategy.

Modern: $40(.50) + 85(.20) + 120(.30) = 73$

Old-fashioned: $106(.50) + 46(.20) + 83(.30) = 87.1$

Mixture: $72(.50) + 90(.20) + 68(.30) = 74.4$

Here the best strategy is old-fashioned cards; the expected profit is 87.1, or $87,100. ◄ **3**

9.6 EXERCISES

1. An investor has $20,000 to invest in stocks. She has two possible strategies: buy conservative blue-chip stocks or buy highly speculative stocks. There are two states of nature: the market goes up or the market goes down. The following payoff matrix shows the net amounts she will have under the various circumstances.

	Market Up	Market Down
Buy Blue-Chip	$25,000	$18,000
Buy Speculative	$30,000	$11,000

What should the investor do if she is
(a) an optimist?   **(b)** a pessimist?
(c) Suppose there is a .7 probability of the market going up. What is the best strategy? What is the expected profit?
(d) What is the best strategy if the probability of a market rise is .2?

2. Management   A developer has $100,000 to invest in land. He has a choice of two parcels (at the same price), one on the highway and one on the coast. With both parcels, his ultimate profit depends on whether he faces light opposition from environmental groups or heavy opposition. He estimates that the payoff matrix is as follows (the numbers represent his profit).

	Opposition	
	Light	Heavy
Highway	$70,000	$30,000
Coast	$150,000	−$40,000

What should the developer do if he is
(a) an optimist?   **(b)** a pessimist?

(c) Suppose the probability of heavy opposition is .8. What is his best strategy? What is the expected profit?
(d) What is the best strategy if the probability of heavy opposition is only .4?

3. Hillsdale College has sold out all tickets for a jazz concert to be held in the stadium. If it rains, the show will have to be moved to the gym, which has a much smaller seating capacity. The dean must decide in advance whether to set up the seats and the stage in the gym or in the stadium, or both, just in case. The payoff matrix below shows the net profit in each case.

		States of Nature	
		Rain	No Rain
	Set up in Stadium	−$1550	$1500
Strategies	Set up in Gym	$1000	$1000
	Set up Both	$750	$1400

What strategy should the dean choose if she is
(a) an optimist?   **(b)** a pessimist?
(c) If the weather forecaster predicts rain with a probability of .6, what strategy should she choose to maximize expected profit? What is the maximum expected profit?

4. Management   An analyst must decide what fraction of the items produced by a certain machine are defective. He has already decided that there are three possibilities for the fraction of defective items: .01, .10, and .20. He may recommend two courses of action: repair the machine or make no repairs. The payoff matrix below represents the *costs* to the company in each case, in hundreds of dollars.

States of Nature

		.01	.10	.20
Strategies	Repair	130	130	130
	No Repair	25	200	500

What strategy should the analyst recommend if he is
(a) an optimist? (b) a pessimist?
(c) Suppose the analyst is able to estimate probabilities for the three states of nature as follows.

Fraction of Defectives	Probability
.01	.70
.10	.20
.20	.10

Which strategy should he recommend? Find the expected cost to the company if this strategy is chosen.

5. **Management** The research department of the Allied Manufacturing Company has developed a new process which it believes will result in an improved product. Management must decide whether or not to go ahead and market the new product. The new product may be better than the old or it may not be better. If the new product is better and the company decides to market it, sales should increase by $50,000. If it is not better and they replace the old product with the new product on the market, they will lose $25,000 to competitors. If they decide not to market the new product they will lose $40,000 if it is better, and research costs of $10,000 if it is not.
(a) Prepare a payoff matrix.
(b) If management believes the probability that the new product is better to be .4, find the expected profits under each strategy and determine the best action.

6. **Management** A businessman is planning to ship a used machine to his plant in Nigeria. He would like to use it there for the next 4 years. He must decide whether or not to overhaul the machine before sending it. The cost of overhaul is $2600. If the machine fails when in operation in Nigeria, it will cost him $6000 in lost production and repairs. He estimates the probability that it will fail at .3 if he does not overhaul it, and .1 if he does overhaul it. Neglect the possibility that the machine might fail more than once in the 4 years.
(a) Prepare a payoff matrix.
(b) What should the businessman do to minimize his expected costs?

7. **Management** A contractor prepares to bid on a job. If all goes well, his bid should be $30,000, which will cover his costs plus his usual profit margin of $4500. However, if a threatened labor strike actually occurs, his bid should be $40,000 to give him the same profit. If there is a strike and he bids $30,000, he will lose $5500. If his bid is too high, he may lose the job entirely, while if it is too low, he may lose money.
(a) Prepare a payoff matrix.
(b) If the contractor believes that the probability of a strike is .6, how much should he bid?

8. **Natural Science** A community is considering an anti-smoking campaign.* The city council will choose one of three possible strategies: a campaign for everyone over age 10 in the community, a campaign for youths only, or no campaign at all. The two states of nature are a true cause-effect relationship between smoking and cancer and no cause-effect relationship. The costs to the community (including loss of life and productivity) in each case are as shown below.

States of Nature

Strategies	Cause-Effect Relationship	No Cause-Effect Relationship
Campaign for All	$100,000	$800,000
Campaign for Youth	$2,820,000	$20,000
No Campaign	$3,100,100	$0

What action should the city council choose if it is
(a) optimistic? (b) pessimistic?
(c) If the Director of Public Health estimates that the probability of a true cause-effect relationship is .8, which strategy should the city council choose?

*Sometimes the numbers (or payoffs) in a payoff matrix do not represent money (profits or costs, for example), but utility. A **utility** is a number that measures the satisfaction (or lack of it) that results from a certain action. The numbers must be assigned by each individual, depending on how he or she feels about a situation. For example, one person might assign a utility of +20 for a week's vacation in San Francisco, with −6 being assigned if the vacation were moved to Sacramento. Work the following problems in the same way as those above.*

---

*This problem is based on an article by B. G. Greenberg in the September 1969 issue of the *Journal of the American Statistical Association*.

9. Political Science A politician must plan her reelection strategy. She can emphasize jobs or she can emphasize the environment. The voters can be concerned about jobs or about the environment. A payoff matrix showing the utility of each possible outcome is shown below.

$$\begin{array}{c} & & \textit{Voters} \\ & & \text{Jobs} \quad \text{Environment} \\ \textit{Candidate} & \begin{array}{c} \text{Jobs} \\ \text{Environment} \end{array} & \begin{bmatrix} +25 & -10 \\ -15 & +30 \end{bmatrix} \end{array}$$

The political analysts feel that there is a .35 chance that the voters will emphasize jobs. What strategy should the candidate adopt? What is its expected utility?

10. In an accounting class, the instructor permits the students to bring a calculator or a reference book (but not both) to an examination. The examination itself can emphasize either numerical problems or definitions. In trying to decide which aid to take to an examination, a student first decides on the utilities shown in the following payoff matrix.

$$\begin{array}{c} & & \textit{Exam Emphasizes} \\ & & \text{Numbers} \quad \text{Definitions} \\ \textit{Student Chooses} & \begin{array}{c} \text{Calculator} \\ \text{Book} \end{array} & \begin{bmatrix} +50 & 0 \\ +10 & +40 \end{bmatrix} \end{array}$$

(a) What strategy should the student choose if the probability that the examination will emphasize numbers is .6? What is the expected utility in this case?

(b) Suppose the probability that the examination emphasizes numbers is .4. What strategy should be chosen by the student?

CHAPTER 9 SUMMARY

KEY TERMS AND SYMBOLS

9.1 $n!$ n factorial
multiplication principle
permutations
combinations

9.3 binomial trials

9.4 stochastic processes
Markov chain
probability vector
equilibrium vector (fixed vector)
regular transition matrix
regular Markov chain

9.5 random variable
frequency distribution
relative frequency
probability distribution
probability function (probability
   distribution function)
histogram
weighted average
expected value
fair game

9.6 states of nature
strategies
payoffs
payoff matrix

KEY CONCEPTS

Multiplication Principle: If there are m_1 ways to make a first choice, m_2 ways to make a second choice, and so on, where none of the choices depend on any of the others, then there are $m_1 m_2 \ldots m_n$ different ways to make the entire sequence of m choices.

The number of **permutations** of n elements taken r at a time is $P(n,r) = \dfrac{n!}{(n-r)!}$.

The number of **combinations** of n elements taken r at a time is

$$\binom{n}{r} = \frac{n!}{(n-r)!\, r!}.$$

Binomial trials experiments have the following characteristics: The same experiment is repeated several times. There are only *two* outcomes, labeled success and failure. The trials are independent so that the probability of success is the same for each trial. If the probability of success in a single trial is p, then the probability of x successes in n trials is

$$\binom{n}{x} p^x (1-p)^{n-x}.$$

Markov Chains: A **transition matrix** must be square, with all entries between 0 and 1 inclusive, and the sum of the entries in any row must be 1. A Markov chain is *regular* if some power of its transition matrix P contains all positive entries. The long-range probabilities for a regular Markov chain are given by the **equilibrium or fixed vector** V, where for any initial probability vector v, the products vP^n approach V as n gets larger and larger, and $VP = V$. To find V, solve the system of equations formed by $VP = V$ and the fact that the sum of the entries of V is 1.

Expected Value of a Probability Distribution: For a random variable x with values $x_1, x_2,$ . . . , x_n and probabilities $p_1, p_2,$ . . . , p_n, respectively, the expected value is

$$E(x) = x_1 p_1 + x_2 p_2 + \ \ldots \ + x_n p_n.$$

Decision Making: A **payoff matrix** which includes all available strategies and states of nature is used in decision making to define the problem and the possible solutions. The expected value of each strategy can help to determine the best course of action.

CHAPTER 9   REVIEW EXERCISES

1. In how many ways can 6 taxicabs line up at the train station?

2. How many possible finishes (first, second, third place) are there in an 8-horse race?

3. In how many ways can a sample of 3 oranges be taken from a bag of a dozen oranges?

4. In how many ways can a selection of 2 pictures from a group of 5 pictures be arranged in a row on a wall?

5. In how many ways can all 5 pictures of Exercise 4 be arranged if 1 must be first?

6. In a Chinese restaurant the menu lists 8 items in column A and 6 items in column B. To order a dinner, the diner is told to select 3 from column A and 2 from column B. How many dinners are possible?

7. A spokesperson is to be selected from each of 3 departments in a small college. If there are 7 people in the first department, 5 in the second department, and 4 in the third department, how many different groups of 3 representatives are possible?

8. How many different committees consisting of 2 men and 2 women can be formed from a group of 8 women and 5 men?

A basket contains 4 black, 2 blue, and 5 green balls. A sample of 3 balls is drawn. Find the probability that the sample contains

9. all black balls;

10. all blue balls;

11. 2 black balls and 1 green ball;

12. 2 black balls;

13. 2 green and 1 blue ball;

14. 1 blue ball.

Suppose 2 cards are drawn without replacement from an ordinary deck of 52 cards. Find the probability that

15. both cards are red;

16. both cards are spades;

17. at least 1 card is a spade;

18. the cards are a king and an ace.

Suppose a family plans 6 children, and the probability that a particular child is a girl is 1/2. Find the probability that the family will have

19. exactly 3 girls

20. all girls;

21. at least 4 girls;

22. no more than 2 boys.

A certain machine used to manufacture screws produces a defective rate of .01. A random sample of 20 screws is selected. Find the probability that the sample contains

23. exactly 4 defective screws;

24. exactly 3 defective screws;

25. no more than 4 defective screws.

26. Set up the probability that the sample has 12 or more defective screws. (Do not evaluate.)

Which of the following are transition matrices?

27. $\begin{bmatrix} .4 & .6 \\ 1 & 0 \end{bmatrix}$

28. $\begin{bmatrix} -.2 & 1.2 \\ .8 & .2 \end{bmatrix}$

29. $\begin{bmatrix} .8 & .2 & 0 \\ 0 & 1 & 0 \\ .1 & .4 & .5 \end{bmatrix}$

30. $\begin{bmatrix} .6 & .2 & .3 \\ .1 & .5 & .4 \\ .3 & .3 & .4 \end{bmatrix}$

31. Currently, 35% of all hot dogs sold in one area are made by Dogkins, while 65% are made by Long Dog. Suppose that Dogkins starts a heavy advertising campaign, with the campaign producing the following transition matrix.

After Campaign

$\text{Before Campaign} \quad \begin{array}{c} \text{Dogkins} \\ \text{Long Dog} \end{array} \begin{array}{cc} \text{Dogkins} & \text{Long Dog} \\ \begin{bmatrix} .8 & .2 \\ .4 & .6 \end{bmatrix} \end{array}$

(a) Find the share of the market for each company after the campaign.

(b) Find the share of the market for each company after three such campaigns.

(c) Predict the long-range market share for Dogkins.

32. A credit card company classifies its customers in three groups: nonusers in a given month, light users, and heavy users. The transition matrix for these states is

	Nonuser	Light	Heavy
Nonuser	.8	.15	.05
Light	.25	.55	.2
Heavy	.04	.21	.75

Suppose the initial distribution for the three states is [.4 .4 .2]. Find the distribution after

(a) 1 month;   (b) 2 months.

(c) What is the long-range prediction for the distribution of users?

33. A medical researcher is studying the risk of heart attack in men. She first divides men into three weight categories: thin, normal, and overweight. By studying the ancestors, children, and grandchildren of these men, the researcher comes up with the following transition matrix.

$$
\begin{array}{c}
\\
\text{Thin} \\
\text{Normal} \\
\text{Overweight}
\end{array}
\begin{array}{ccc}
\text{Thin} & \text{Normal} & \text{Overweight} \\
\left[\begin{array}{ccc}
.3 & .5 & .2 \\
.2 & .6 & .2 \\
.1 & .5 & .4
\end{array} \right]
\end{array}
$$

Suppose that the distribution of men by weight is initially given by [.2 .55 .25]. Find the distribution after
(a) 1 generation; **(b)** 2 generations.
(c) Find the long-range prediction for the distribution of weights.

*In Exercises 34–38, **(a)** give a probability distribution, and **(b)** sketch its histogram.*

34.

x	1	2	3	4	5
Frequency	3	7	9	3	2

35.

x	8	9	10	11	12	13	14
Frequency	1	0	2	5	8	4	3

36. A coin is tossed 3 times and the number of heads is recorded.

37. A pair of dice are rolled and the sum of the results for each roll is recorded.

38. Patients in groups of 5 are given a new treatment for a fatal disease. The experiment is repeated 10 times with the following results.

Number Who Survived	Frequency
0	1
1	1
2	2
3	3
4	3
5	0
	Total: 10

In Exercises 39 and 40, give the probability that corresponds to the shaded region of each histogram.

39.

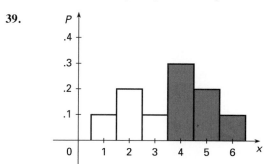

40.

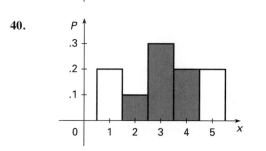

Solve the following problems.

41. You pay $6 to play in a game where you will roll a die, with payoffs as follows: $8 for a 6, $7 for a 5, and $4 for any other results. What are your expected winnings? Is the game fair?

42. A lottery has a first prize of $5000, two second prizes of $1000 each, and two $100 third prizes. A total of 10,000 tickets is sold, at $1 each. Find the expected winnings of a person buying 1 ticket.

43. Find the expected number of girls in a family of 5 children.

44. A developer can buy a piece of property that will produce a profit of $16,000 with probability .7, or a loss of $9000 with probability .3. What is the expected profit?

45. Game boards for a recent United Airlines contest could be obtained by sending a self-addressed stamped envelope to a certain address. The prize was a ticket for any city to which United flies. Assume that the value of the ticket was $1000 (we might as well go first class), and that the probability that a particular game board would win was 1/4000. If the stamps to enter the contest cost 30¢ and the envelopes are 1¢ each, find the expected winnings for a person ordering one game board.

46. Three cards are drawn from a standard deck of 52 cards.
(a) What is the expected number of aces?
(b) What is the expected number of clubs?

47. Suppose someone offers to pay you $100 if you draw 3 cards from a standard deck of 52 cards and all the cards are clubs. What should you pay for the chance to win if it is a fair game?

48. In labor-management relations, both labor and management can adopt either a friendly or a hostile attitude. The results are shown in the following payoff matrix. The numbers give the wage gains made by an average worker.

$$\begin{array}{c} \qquad\qquad\text{\textit{Management}} \\ \qquad\qquad \text{Friendly}\quad\text{Hostile} \\ \textit{Labor}\;\begin{array}{c}\text{Friendly}\\ \text{Hostile}\end{array}\left[\begin{array}{cc} \$600 & \$800 \\ \$400 & \$950 \end{array}\right] \end{array}$$

(a) Suppose the chief negotiator for labor is an optimist. What strategy should he choose?
(b) What strategy should he choose if he is a pessimist?
(c) The chief negotiator for labor feels that there is a 70% chance that the company will be hostile. What strategy should he adopt? What is the expected payoff?

(d) Just before negotiations begin, a new management is installed in the company. There is only a 40% chance that the new management will be hostile. What strategy should be adopted by labor?

49. A candidate for city council can come out in favor of a new factory, be opposed to it, or waffle on the issue. The change in votes for the candidate depends on what her opponent does, with payoffs as shown.

$$\begin{array}{c} \qquad\qquad\qquad\text{\textit{Opponent}} \\ \qquad\qquad \text{Favors}\quad\text{Waffles}\quad\text{Opposes} \\ \textit{Candidate}\;\begin{array}{c}\text{Favors}\\ \text{Waffles}\\ \text{Opposes}\end{array}\left[\begin{array}{ccc} 0 & -1000 & -4000 \\ 1000 & 0 & -500 \\ 5000 & 2000 & 0 \end{array}\right] \end{array}$$

(a) What should the candidate do if she is an optimist?
(b) What should she do if she is a pessimist?
(c) Suppose the candidate's campaign manager feels there is a 40% chance that the opponent will favor the plant, and a 35% chance that he will waffle. What strategy should the candidate adopt? What is the expected change in the number of votes?
(d) The opponent conducts a new poll which shows strong opposition to the new factory. This changes the probability he will favor the factory to 0 and the probability he will waffle to .7. What strategy should our candidate adopt? What is the expected change in the number of votes now?

For many different items it is difficult or impossible to take the item to a central repair facility when service for the item is required. Washing machines, large television sets, office copiers, and computers are only a few examples of such items. Service for items of this type is commonly performed by sending a repair person to the item, with the person driving to the item in a truck containing various parts that might be required in repairing the item. Ideally, the truck should contain all the parts that might be required in repairing the item. However, most parts would be needed only infrequently, so that inventory costs for the parts would be high.

An optimum policy for deciding on the parts to stock on a truck would require that the probability of not being able to repair an item without a trip back to the warehouse for needed parts be as low as possible, consistent with minimum inventory costs. An analysis similar to the one below was developed at the Xerox Corporation.*

To setup a mathematical model for deciding on the optimum truck stocking policy, let us assume that a broken machine might require 1 of 5 different parts (we could assume any number of different parts—we use 5 to simplify the notation). Suppose also that the probability that a particular machine requires part 1 is p_1, that it requires part 2 is p_2, and so on. Assume also that the failure of different part types are independent, and that at most 1 part of each type is used on a given job.

Suppose that, on the average, a repair person makes N service calls per time period. If the repair person is unable to make a repair because at least one of the parts is unavailable, there is a penalty cost, L, corresponding to wasted time for the repair person, an extra trip to the parts depot, customer unhappiness, and so on. For each of the parts carried on the truck, an average inventory cost is incurred. Let H_i be the average inventory cost for part i, where $1 \leq i \leq 5$.

Let M_1 represent a policy of carrying only part 1 on the repair truck, M_{24} represent a policy of carrying only parts 2 and

*From "Optimal Inventories Based on Job Completion Rate for Repairs Requiring Multiple Items" by Stephen Smith, John Chambers, and Eli Shlifer from *Management Science,* vol. 26, no. 8, August 1980. Copyright © 1980 The Institute of Management Sciences. Reprinted by permission.

4, with M_{12345} and M_0 representing policies of carrying all parts and no parts, respectively.

For policy M_{35}, carrying parts 3 and 5 only, the expected cost per time period per repair person, written $C(M_{35})$, is

$$C(M_{35}) = (H_3 + H_5) + NL[1 - (1 - p_1)(1 - p_2)(1 - p_4)].$$

(The expression in brackets represents the probability of needing at least one of the parts not carried, 1, 2, or 4 here.) As further examples, $C(M_{125})$ is

$$C(M_{125}) = (H_1 + H_2 + H_5) + NL[1 - (1 - p_3)(1 - p_4)],$$

while

$$C(M_{12345}) = (H_1 + H_2 + H_3 + H_4 + H_5) + NL(1 - 1)$$
$$= H_1 + H_2 + H_3 + H_4 + H_5,$$

and

$$C(M_0) = NL[1 - (1 - p_1)(1 - p_2)(1 - p_3)(1 - p_4)(1 - p_5)].$$

To find the best policy, evaluate $C(M_0)$, $C(M_1)$, . . . , $C(M_{12345})$ and choose the smallest result. (A general solution method is in the *Management Science* paper.)

Example Suppose that, for a particular item, only 3 possible parts might need to be replaced. By studying past records of failures of the item and finding necessary inventory costs, suppose that the following values have been found.

p_1	p_2	p_3	H_1	H_2	H_3
.09	.24	.17	$15	$40	$9

Suppose $N = 3$ and L is $54. Then, as an example,

$$C(M_1) = H_1 + NL[1 - (1 - p_2)(1 - p_3)]$$
$$= 15 + 3(54)[1 - (1 - .24)(1 - .17)]$$
$$= 15 + 3(54)[1 - (.76)(.83)]$$
$$= 15 + 59.81$$
$$= 74.81$$

Thus, if policy M_1 is followed (carrying only part 1 on the truck), the expected cost per repair person per time period is $74.81. Also,

$$C(M_{23}) = H_2 + H_3 + NL[1 - (1 - p_1)]$$
$$= 40 + 9 + 3(54)(.09)$$
$$= 63.58,$$

so that M_{23} is a better policy than M_1. By finding the expected values for all other possible policies (see the exercises below), the optimum policy may be chosen. ◀

EXERCISES

1. Refer to the example above and find each of the following.
   (a) $C(M_0)$ (b) $C(M_2)$ (c) $C(M_3)$
   (d) $C(M_{12})$ (e) $C(M_{13})$ (f) $C(M_{123})$.

2. Which policy leads to lowest expected cost?

3. In the example above, $p_1 + p_2 + p_3 = .09 + .24 + .17 = .50$. Why is it not necessary that the probabilities add to 1?

4. Suppose an item to be repaired might need 1 of n different parts. How many different policies would then need to be evaluated?

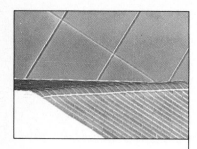

CHAPTER 10

Introduction to Statistics

Statistics deals with the collection and organization of data, and methods of drawing conclusions about a population based on data from a sample of the population. Statistical models have become increasingly useful in a variety of fields—for example, manufacturing, government, agriculture, medicine, the social sciences, and in all types of research. This chapter gives a brief introduction to some of the key topics from statistical theory.

Often, a researcher wishes to learn something about a characteristic of a population, but because the population is very large or mobile, it is not possible to examine all of its elements. Instead, a limited sample drawn from the population is studied to determine the characteristics of the population. For these inferences to be correct, the sample chosen must be a **random sample.** Random samples are representative of the population because they are chosen so that every element of the population is equally likely to be selected. A hand dealt from a well-shuffled deck of cards is a random sample. In other situations, a random sample can be selected by assigning a number to each element in the population and then by using some chance method to select those numbers that will be included in the sample. Random sampling typically relies on a chance device.

10.1 FREQUENCY DISTRIBUTIONS

Once a sample has been chosen and all data of interest are collected, the data must be organized so that conclusions may be more easily drawn. One method of organization is to group the data into intervals; equal intervals are usually chosen. For example, suppose the data are the scores of 30 students on a mathematics test, as shown below.

60	90	76	81	59	46
48	61	57	78	86	65
63	54	68	93	71	78
79	67	75	87	76	74
86	57	62	95	80	70

The intervals must be chosen so that each score will fall into exactly one interval. The highest score on the list is 95 and the lowest is 46. This is a difference of about 50. Using intervals of width 10 will give six intervals, starting with 40–49 and ending with 90–99. This gives an interval for each score in the data and results in six equal intervals of a convenient size. Too many intervals of smaller size would not simplify the data enough, while too few intervals of larger size would conceal information that the data might provide. A rule of thumb is to use from six to fifteen intervals.

In Table 1 below, the six intervals, or **classes,** chosen above are listed in the first column, the number of grades falling into each class are tallied in the second column, and the tallies are totaled and their frequency is then entered in the third column. Table 1 is called a **grouped frequency distribution.**

Table 1

Class	Tally	Frequency
40–49	||	2
50–59	||||	4
60–69	ⅢⅡ ||	7
70–79	ⅢⅡ ||||	9
80–89	ⅢⅡ	5
90–99	|||	3
		Total: 30

The frequency distribution in Table 1 shows information about the data that might not have been noticed before. For example, the class with the largest number of scores is 70–79, and 16 scores (more than half) were between 59 and 80. Also, the number of scores in each class increases rather evenly (up to 9) and then decreases at about the same pace. However, some information is lost; for example, we no longer know how many students had a score of 65.

▶ **EXAMPLE 1**   Suppose 30 business executives give their recommendations as to the number of college units in management that a business major should have. Use the following results to make a grouped frequency distribution for this data. Use classes of 0–4, 5–9, and so on.

3	25	22	16	0	9	14	8	34	21
15	12	9	3	8	15	20	12	28	19
17	16	23	19	12	14	29	13	24	18

First tally the number of college units falling into each class. Then total the tallies in each class, as in Table 2.

1 An accounting firm selected 24 complex tax returns prepared by a certain tax preparer. The number of errors per return were as follows.

8	12	0	6	10	8	0	14
8	12	14	16	4	14	7	11
9	12	7	5	11	21	22	19

Prepare a grouped frequency distribution for this data. Use classes 0–4, 5–9, and so on.

Answer:

Interval	Frequency
0–4	3
5–9	7
10–14	9
15–19	3
20–24	2
	Total: 24

Table 2

Class	Tally	Frequency
0–4	|||	3
5–9	||||	4
10–14	⧄|	6
15–19	⧄ |||	8
20–24	⧄	5
25–29	|||	3
30–34	|	1
	Total:	30

◄ **1**

The information in a grouped frequency distribution can be displayed in a special kind of bar graph called a **histogram.** In a histogram, class interval sizes correspond to widths of the bars, and since equal intervals are used, the bars in a histogram are all the same width. The heights of the bars are determined by the frequencies. (Histograms were discussed in Section 9.5.) A histogram for the data in Table 1 is shown in Figure 10.1. The break in the horizontal axis indicates that the distance from 0 is not 40 units. It allows us to place the bars close to the vertical axis.

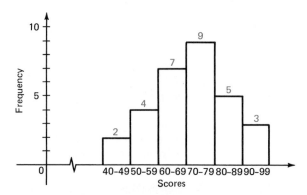

FIGURE 10.1

A **frequency polygon** is another form of graph that illustrates a grouped frequency distribution. As shown in Figure 10.2, the values of the endpoints of the class intervals, called the **class boundaries,** are halfway between the end of one interval and the beginning of the next interval. For example, the interval 40–49 has class boundaries of 39.5–49.5. The polygon is formed by joining consecutive midpoints of the tops of the histogram bars with straight-line segments. The midpoint of the first class interval is found by adding the class boundaries and dividing by 2.

$$\frac{39.5 + 49.5}{2} = \frac{89.0}{2} = 44.5$$

2 In Problem 1 at the side, the following grouped frequency distribution was found.

Interval	Frequency
0–4	3
5–9	7
10–14	9
15–19	3
20–24	2

Make a histogram and a frequency polygon for this distribution.

Answer:

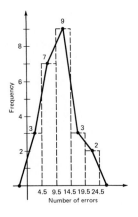

Add the class width, 10 in the example, to the first midpoint to get the succeeding midpoints, 54.5, 64.5, and so on. The midpoints of the first and last bars are joined to endpoints on the horizontal axis where the next midpoint would appear. A frequency polygon for the data of Table 1 is illustrated in Figure 10.2

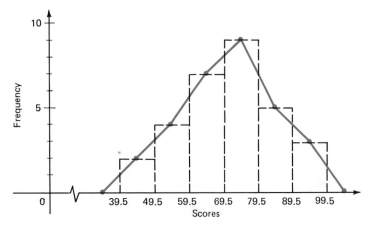

FIGURE 10.2

▶**EXAMPLE 2**   The grouped frequency distribution of college units shown in Table 3 was found in Example 1. Draw a histogram and a frequency polygon for this distribution.

Table 3

Interval	Frequency
0–4	3
5–9	4
10–14	6
15–19	8
20–24	5
25–29	3
30–34	1

First draw a histogram, shown in black in Figure 10.3. To get a frequency polygon, connect consecutive midpoints of the tops of the bars. The frequency polygon is shown in color in Figure 10.3 ◀ **2**

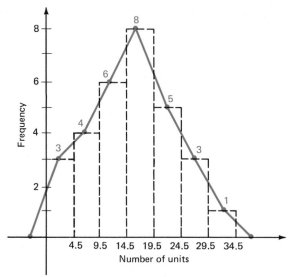

FIGURE 10.3

We can obtain further information from Table 1 by adding a new column, called *cumulative frequency,* as shown in Table 4. The entries in this new column give the cumulative number of scores up to and including those in a given interval. For example, Table 4 shows that 13 of the scores were below 70, 22 were below 80, and 30 (all scores) were below 100. Table 4 is called a **cumulative frequency distribution.***

Table 4

Interval	Frequency	Cumulative Frequency
40–49	2	2
50–59	4	6
60–69	7	13
70–79	9	22
80–89	5	27
90–99	3	30

A graph that illustrates the cumulative frequencies, called a **cumulative frequency polygon,** or **ogive,** is shown in Figure 10.4. Note that the points are not at the midpoints of the intervals, as in a frequency polygon, but at the end of each interval, to correspond with the information that the cumulative frequency provides.

---

*In some situations it might be more useful to accumulate the scores in the other direction, so the cumulative frequency would tell how many scores are above 80, for example.

3 Make a cumulative frequency distribution for the data of Problem 2 at the side. Draw a cumulative frequency polygon.

Answer:

Interval	Frequency	Cumulative Frequency
0–4	3	3
5–9	7	10
10–14	9	19
15–19	3	22
20–24	2	24

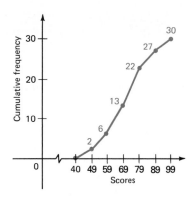

FIGURE 10.4

▶**EXAMPLE 3** Complete a cumulative frequency distribution for the data of Example 2. Draw a cumulative frequency polygon.

The first number in the cumulative frequency column is the same as the first frequency, the second number is the same as the sum of the first two frequencies, and so on.

Table 5

Interval	Frequency	Cumulative Frequency
0–4	3	3
5–9	4	3 + 4 = 7
10–14	6	7 + 6 = 13
15–19	8	13 + 8 = 21
20–24	5	21 + 5 = 26
25–29	3	26 + 3 = 29
30–34	1	29 + 1 = 30

The cumulative frequency polygon is shown in Figure 10.5. ◀ **3**

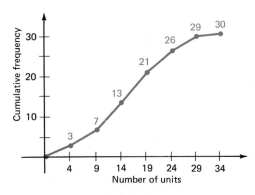

FIGURE 10.5

10.1 EXERCISES

1. Social Science The histogram below shows the percent of the U.S. population in each age group in a recent year.* What percent of the population was in each of the following age groups?
(a) 10–19   **(b)** 60–69
(c) What age group has the largest percent of the population?

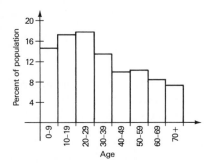

2. Social Science The histogram below shows the estimated percent of the U.S. population in each age group in the year 2000.* What percent of the population is estimated to be in each of the following age groups then?
(a) 20–29   **(b)** 70+
(c) What age group has the largest percent of the population?
(d) Compare the histogram in Exercise 1 with the histogram below. What seems to be true of the U.S. population?

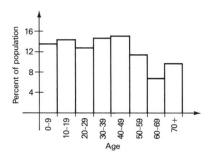

*Source: U.S. Census Bureau.

For Exercises 3–6:
(a) *group the data as indicated;*
(b) *prepare a frequency distribution with a column for intervals, frequencies, and cumulative frequencies;*
(c) *construct a histogram;*
(d) *construct a frequency polygon;*
(e) *construct a cumulative frequency polygon.*

(See Examples 1–3.)

3. The data below give the number of college units completed by 30 of the EZ Life Company's employees. Group the data into six intervals, starting with 0–24.

74	133	4	127	20	30
103	27	139	118	138	121
149	132	64	141	130	76
42	50	95	56	65	104
4	140	12	88	119	64

4. The scores of 80 students on a business law test were as follows. Group the data into seven intervals, starting with 30–39.

79	71	78	87	69	50	63	51
60	46	65	65	56	88	94	56
74	63	87	62	84	76	82	67
59	66	57	81	93	93	54	88
55	69	78	63	63	48	89	81
98	42	91	66	60	70	64	70
61	75	82	65	68	39	77	81
67	62	73	49	51	76	94	54
83	71	94	45	73	95	72	66
71	77	48	51	54	57	69	87

5. Natural Science The daily high temperatures in Tucson for the month of July for one year were as follows. Group the data using 70–74 as the first interval.

79	84	88	96	102	104	110	108	106	106
104	99	97	92	94	90	82	74	72	83
85	92	100	99	101	107	111	102	97	94	92

6. Natural Science The heights of 40 women (in centimeters) are as follows. Group the data using 140–149 as the first interval.

174	190	172	182	179	186	171	152	174	185
180	170	160	173	163	177	165	157	149	167
169	182	178	158	182	169	181	173	183	176
170	162	159	147	150	192	179	165	148	188

7. Management   A firm took a random sample of days absent in a year for 40 employees with results as shown below. Sketch a histogram and a frequency polygon for this data.

Days Absent	Frequency
0–2	23
3–5	11
6–8	5
9–11	0
12–14	1

8. Natural Science   The home range (in square kilometers) of several pandas were surveyed over a year's time with the following results. Sketch a histogram and frequency polygon for the data.

Home Range	Frequency
.1–.5	11
.6–1.0	12
1.1–1.5	7
1.6–2.0	6
2.1–2.5	2
2.6–3.0	1
3.1–3.5	1

9. Get 5 coins and toss them all at once. Keep track of the number of heads. Repeat this experiment 64 times. Make a histogram of your results.

10. Use the results of Exercise 9 to make a frequency polygon.

11. The frequency distribution below gives the theoretical results of tossing 5 coins 64 times. Use this distribution to make a frequency polygon. Compare it to the one you got in Exercise 10.

Number of Heads	Frequency
0	2
1	10
2	20
3	20
4	10
5	2

12. Toss 6 coins a total of 64 times. Keep track of the number of heads. Make a histogram of your results.

13. Make a frequency polygon using the results of Exercise 12.

14. The frequency distribution below gives the theoretical results of tossing 6 coins 64 times. Use the distribution to make a frequency polygon. Compare it to the one you made in Exercise 13.

Number of Heads	Frequency
0	1
1	6
2	15
3	20
4	15
5	6
6	1

15. An experiment consists of drawing 5 cards and observing the number of red cards drawn. The theoretical results from repeating this experiment 1000 times are shown below. Sketch a frequency polygon for this data.

Red Cards	Frequency
0	25
1	150
2	325
3	325
4	150
5	25

16. Four cards are drawn from a well-shuffled deck of 52 cards and the number of aces drawn is observed. The theoretical results if this experiment is repeated 1000 times are shown below. Sketch a frequency polygon for the data.

Aces	Frequency
0	719
1	256
2	25
3	1
4	0

17. Social Science   The frequency with which letters occur in a large sample of any written language does not vary much. Therefore, determining the frequency of each letter in a coded message is usually the first step in deciphering it. The percent frequencies of the letters in the English language are as follows.

Letter	%	Letter	%
E	13	L	3.5
T	9	C, M, U	3
A, O	8	F, P, Y	2
N	7	B, G, W	1.5
I, R	6.5	V	1
H, S	6	J, K, X	0.5
D	4	Q, Z	0.2

Use the introductory paragraph of this exercise as a sample of the English language. Find the percent frequency for each letter in the sample. Compare your results with the frequencies given above.

18. **Social Science** The following message is written in a code in which the frequency of the symbols is the main key to the solution.

)?−−8))6*+8506*3×6;4?*7*&×*−6.48()6)985)?
(8+2:;48)81&?(;46*3)6*;48&(+8(*598+.8()8 = 8
(5*−8−5(81?098;4&+)&15*50:)6)6*;?6;6&*0?−7

(a) Find the frequency of each symbol.
(b) By comparing the high-frequency symbols with the high-frequency letters in English, and the low-frequency symbols with the low-frequency letters, try to decipher the message. (Hint: look for repeated two-symbol combinations and double letters for added clues. Try to identify vowels first.)

10.2 MEASURES OF CENTRAL TENDENCY

In summarizing data, it is often useful to have a single number, a typical or "average" number, that is representative of the entire collection of data. Such numbers are referred to as **measures of central tendency.** As mentioned in Section 9.5, everyone is familiar with some type of average. For example, we compare our heights and weights to those of the typical or "average" person on weight charts. Students are quite familiar with the "class average" and their own "average" at any time in a given course.

MEAN The **arithmetic average** or **mean** of a sample of n numbers is the sum of the numbers, divided by n. The sum of the n numbers $x_1, x_2, x_3, \ldots, x_n$ is written in a compact way by using **summation notation,** also called **sigma notation.** This notation uses the symbol Σ (sigma), the Greek capital S (for sum). Using Σ, the sum $x_1 + x_2 + x_3 + \cdots + x_n$ is written

$$x_1 + x_2 + x_3 + \cdots + x_n = \sum_{i=1}^{n} x_i.$$

For instance, the mean of the sample consisting of the six numbers 12, 15, 10, 8, 14, 20 is

$$\sum_{i=1}^{6} x_i = \frac{12 + 15 + 10 + 8 + 14 + 20}{6} = 13.167.$$

In this example, $x_1 = 12$, $x_2 = 15$, $x_3 = 10$, and so on. **1**

1 Let $x_1 = 2$, $x_2 = 3$, $x_3 = 7$, and $x_4 = 9$. Find

(a) $\displaystyle\sum_{i=1}^{4} x_i$;

(b) $\displaystyle\sum_{i=1}^{4} x_i^2$

Answer:

(a) 21

(b) 143

2 Find the mean of the following list of numbers.

$25.12	$42.58
$76.19	$32
$81.11	$26.41
$19.76	$59.32
$71.18	$21.03

Answer:

$45.47

In statistics, $\sum_{i=1}^{n} x_i$ is often abbreviated as just $\Sigma(x)$. The symbol $\bar{x}$ (read "x-bar") is used to represent the sample mean.

Mean

The mean of the n numbers, $x_1, x_2, x_3, \ldots, x_n$, is

$$\bar{x} = \frac{\Sigma(x)}{n}.$$

▶**EXAMPLE 1**   Daily sales at Lupe's Laundry last week were $86, $103, $118, $117, $126, $158, and $149. Find the mean, $\bar{x}$, of these sales.

Let $x_1 = 86$, $x_2 = 103$, and so on. ($n = 7$ since there are 7 numbers.) Then

$$\bar{x} = \text{mean} = \frac{\Sigma(x)}{n} = \frac{86 + 103 + 118 + 117 + 126 + 158 + 149}{7}$$

$$= \frac{857}{7}$$

$$\bar{x} \approx 122.43.$$

The mean sales at the laundry were $122.43 per day, rounded to the nearest cent. ◀ **2**

The mean of data that have been arranged in a frequency distribution is found in a similar way. For example, suppose the following data are collected.

Value	Frequency
84	2
87	4
88	7
93	4
99	3
Total:	20

The value 84 appears twice, 87 four times, and so on. To find the mean, first add 84 two times, 87 four times, and so on; or get the same result faster by multiplying 84 by 2, 87 by 4, and so on, and then by adding the results. Dividing the sum by 20, the total of the frequencies, gives the mean.

3 Find $\bar{x}$ for the following frequency distribution.

Value	Frequency
7	2
9	3
11	6
13	4
15	1
17	4

Answer:
$\bar{x} = 12.1$

$$\bar{x} = \frac{(84 \cdot 2) + (87 \cdot 4) + (88 \cdot 7) + (93 \cdot 4) + (99 \cdot 3)}{20}$$

$$= \frac{168 + 348 + 616 + 372 + 297}{20} = \frac{1801}{20}$$

$$\bar{x} = 90.05$$

▶**EXAMPLE 2**   Find the mean for the data shown in the following frequency distribution.

Value	Frequency	Value × Frequency
30	6	$30 \cdot 6 =$ 180
32	9	$32 \cdot 9 =$ 288
33	7	$33 \cdot 7 =$ 231
37	12	$37 \cdot 12 =$ 444
42	6	$42 \cdot 6 =$ 252
	Total: 40	Total: 1395

A new column, ''Value × Frequency,'' has been added to the frequency distribution. Adding the products from this column gives a total of 1395. The total from the frequency column is 40. The mean is

$$\bar{x} = \frac{1395}{40} = 34.875. \quad ◀ \quad \textbf{3}$$

The mean of grouped data is found in a similar way. For grouped data, intervals are used rather than single values. To calculate the mean, it is assumed that all these values are located at the midpoint of the interval. The letter x is used to represent the midpoints and f represents the frequencies, as shown in the next example.

▶**EXAMPLE 3**   Find the mean for the following grouped frequency distribution (Table 1 of the previous section).

Interval	Midpoint, x	Frequency, f	Product, xf
40–49	44.5	2	89
50–59	54.5	4	218
60–69	64.5	7	451.5
70–79	74.5	9	670.5
80–89	84.5	5	422.5
90–99	94.5	3	283.5
		Total: 30	Total: 2135

4 Find the mean of the following grouped frequency distribution.

Interval	Frequency
0–4	6
5–9	4
10–14	7
15–19	3

Answer:

8.75

An additional column for the midpoint of each interval has been included. Recall that we get the numbers in this column by adding the endpoints of each interval and dividing by 2. For the interval 40–49, the midpoint is (39.5 + 49.5)/2 = 44.5 (or (40 + 49)/2 = 44.5). The numbers in the product column on the right are found by multiplying frequencies and corresponding midpoints. (A calculator with a Σ key can speed up the work in finding the total of the *xf* column.) Finally, divide the total of the product column by the total of the frequency column to get

$$\bar{x} = \frac{2135}{30} = 71.2 \quad \text{(to the nearest tenth).} \quad \blacktriangleleft$$

The formula for the **mean of a grouped frequency distribution** is given below.

Mean of a Grouped Distribution

The mean of a distribution where x represents the midpoints, f the frequencies, and $n = \Sigma(f)$, is

$$\bar{x} = \frac{\Sigma\,(xf)}{n}.$$

Caution Note that in the formula above, n is the sum of the frequencies in the entire data set, not the number of classes. **4**

MEDIAN Asked by a reporter to give the average height of the players on his team the Little League coach lined up his 15 players by increasing height. He picked out the player in the middle and pronounced this player to be of average height. This kind of average, called the **median,** is defined as the middle entry in a set of data arranged in either increasing or decreasing order. If the number of entries, n, is odd, the location of the median can be found by dividing n by 2 and adding .5. Count down that many entries to identify the median. If n is even, the median is defined to be the mean of the two center entries. In that case, divide n by 2 and count down $n/2$ entries. Add that entry and the next one and divide by 2 to get the median.

Odd Number of Entries	Even Number of Entries
8	2
7	3
Median = 4	4 ⎫
3	7 ⎬ Median = $\frac{4 + 7}{2}$ = 5.5
1	9
	12

Note As shown in the table above, when there are an even number of entries, the median is not always equal to one of the data entries.

5 Find the median for each of the following lists of numbers.

(a) 12, 15, 17, 19, 35, 42, 58

(b) 28, 68, 7, 15, 47, 59, 13, 74, 32, 25

Answer:

(a) 19

(b) 30

The procedure for finding the median of a grouped frequency distribution is more complicated. We omit it here because it is more common to find the mean when working with grouped frequency distributions.

▶**EXAMPLE 4** Find the median for the following lists of numbers.

(a) 11, 12, 17, 20, 23, 28, 29

The median is the middle number, in this case, 20. (Note that the numbers are aleady arranged in numerical order.) In this list, three numbers are smaller than 20 and three are larger.

(b) 15, 13, 7, 11, 19, 30, 39, 5, 10

First arrange the numbers in numerical order, from smallest to largest.

$$5, \ 7, \ 10, \ 11, \ 13, \ 15, \ 19, \ 30, \ 39$$

The middle number can now be determined: the median is 13.

(c) 47, 59, 32, 81, 74, 153

Write the numbers in numerical order.

$$32, \ 47, \ 59, \ 74, \ 81, \ 153$$

There are six numbers here; the median is the mean of the two middle numbers, or

$$\text{Median} = \frac{59 + 74}{2} = \frac{133}{2} = 66\frac{1}{2}. \ \blacktriangleleft$$

Caution Remember, the data must be arranged in numerical order before locating the median. **5**

In some situations, the median gives a truer representative or typical element of the data than the mean. For example, suppose in an office there are 10 salespersons, 4 secretaries, the sales manager, and Ms. Daly, who owns the business. Their annual salaries are as follows: secretaries, $15,000 each; salespersons, $25,000 each; manager, $35,000; and owner, $200,000. The mean salary is

$$\bar{x} = \frac{(15,000)4 + (25,000)10 + 35,000 + 200,000}{16} = \$34,062.50.$$

However, since 14 people earn less than $34,062.50 and only 2 earn more, this does not seem very representative. The median salary is found by ranking the salaries by size: $15,000, $15,000, $15,000, $15,000, $25,000, $25,000, . . . , $200,000. Since there are 16 salaries (an even number) in the list, the mean of the 8th and 9th entries will give the value of the median. The 8th and 9th entries are both $25,000, so the median is $25,000. In this example, the median is more representative of the distribution than the mean.

6 Find the mode for each of the following lists of numbers.

(a) 29, 35, 29, 18, 29, 56, 48

(b) 13, 17, 19, 20, 20, 13, 25, 27, 13, 20

(c) 512, 546, 318, 729, 854, 253

Answer:

(a) 29

(b) 13 and 20

(c) No mode

MODE   Sue's scores on ten class quizzes include one 7, two 8's, six 9's, and one 10. She claims that her average grade on quizzes is 9, because most of her scores are 9's. This kind of "average," found by selecting the most frequent entry, is called the **mode.** Similarly, in a grouped frequency distribution, the interval with the greatest frequency is the **modal class.**

▶**EXAMPLE 5**   Find the mode for each list of numbers.
(a) 57, 38, 55, 55, 80, 87, 98
   The number 55 occurs more often than any other, and is the mode. It is not necessary to place the numbers in numerical order when looking for the mode.
(b) 182, 185, 183, 185, 187, 187, 189
   Both 185 and 187 occur twice. Since this list has two modes, it is called **bimodal.**
(c) 10,708   11,519,   10,972   17,546,   13,905,   12,182
   No number occurs more than once. This list has no mode.   ◀   **6**

The mode is often used to represent data that are not numerical.

▶**EXAMPLE 6**   **(a)** Refer to the table in Section 10.1, Exercise 17 on the percent of frequency of the letters in the English language. Find the mode.
   The table shows that the mode is the letter E, since it occurs most frequently.
(b) A college class has 4 freshmen, 15 sophomores, and 6 juniors. What is the mode?
   The mode is sophomore, the most frequent group.   ◀

In the examples given in Example 6, the mode is the most meaningful measure of central tendency. The mode has the advantages of being easily found and not being influenced by extremes—data which are very large or very small compared to the rest of the data. It is often used in samples where the data to be "averaged" are not numerical. A major disadvantage to the mode is that there may be more than one, in case of ties, or there may be no mode at all when all entries occur with the same frequency.

The mean is the most commonly used measure of central tendency. Its advantages are that it is easy to compute, it takes all the data into consideration, and it is reliable—that is, repeated samples are likely to give very similar means. A disadvantage of the mean is that it is influenced by extreme values, as illustrated in the salary example above.

The median can be easy to compute and is influenced very little by extremes. Like the mode, the median can be found in situations where the data are not numerical. A disadvantage of the median is the need to rank the data in order; this can be tedious when the number of items is large.

10.2 EXERCISES

A calculator will be helpful with many of the exercises in this set.

Find the mean for each list of numbers. Round to the nearest tenth. (See Example 1.)

1. 8, 10, 16, 21, 25

2. 44, 41, 25, 36, 67, 51

3. 130, 141, 149, 152, 158, 163, 139, 170

4. 42, 48, 54, 62, 69, 75, 90, 94

5. 21,900, 22,850, 24,930, 29,710, 28,340, 40,000

6. 38,500, 39,720, 42,183, 21,982, 43,250

7. 9.4, 11.3, 10.5, 7.4, 9.1, 8.4, 9.7, 5.2, 1.1, 4.7

8. 30.1, 42.8, 91.6, 51.2, 88.3, 21.9, 43.7, 51.2

9. .06, .04, .05, .08, .03, .14, .18, .29, .07, .01

10. .31, .09, .08, .22, .46, .51, .48, .42, .53, .42

Find the mean for each of the following. Round to the nearest tenth. (See Example 2.)

11.

Value	Frequency
3	4
5	2
9	1
12	3

12.

Value	Frequency
9	3
12	5
15	1
18	1

13.

Value	Frequency
12	4
13	2
15	5
19	3
22	1
23	5

14.

Value	Frequency
25	1
26	2
29	5
30	4
32	3
33	5

15.

Value	Frequency
104	6
112	14
115	21
119	13
123	22
127	6
132	9

16.

Value	Frequency
246	2
291	4
295	3
304	8
307	9
319	2

Find the median for each of the following lists of numbers. Don't forget to first place the numbers in numerical order if necessary. (See Example 4.)

17. 12, 18, 32, 51, 58, 92, 106

18. 596, 604, 612, 683, 719

19. 100, 114, 125, 135, 150, 172

20. 298, 346, 412, 501, 515, 521, 528, 621

21. 32, 58, 97, 21, 49, 38, 72, 46, 53

22. 1072, 1068, 1093, 1042, 1056, 1005, 1009

23. 576, 578, 542, 551, 565, 525, 590, 559

24. 7, 15, 28, 3, 14, 18, 46, 59, 1, 2, 9, 21

25. 28.4, 9.1, 3.4, 27.6, 59.8, 32.1, 47.6, 29.8

26. .6, .4, .9, 1.2, .3, 4.1, 2.2, .4, .7, .1

Find the mode or modes for each of the following lists of numbers. (See Example 5.)

27. 4, 9, 8, 6, 9, 2, 1, 3

28. 21, 32, 46, 32, 49, 32, 49

29. 80, 72, 64, 64, 72, 53, 64

30. 97, 95, 94, 95, 94, 97, 97

31. 74, 68, 68, 68, 75, 75, 74, 74, 70

32. 158, 162, 165, 162, 165, 157, 163

33. 5, 9, 17, 3, 2, 8, 19, 1, 4, 20

34. 12, 15, 17, 18, 21, 29, 32, 74, 80

35. 6.1, 6.8, 6.3, 6.3, 6.9, 6.7, 6.4, 6.1, 6.0

36. 12.75, 18.32, 19.41, 12.75, 18.30, 19.45, 18.33

For grouped data, the modal class is the interval containing the most data values. Give the mean and modal class for each of the following collections of grouped data. (See Example 3.)

37.

College Units	Frequency	College Units	Frequency
0–24	4	75–99	3
25–49	3	100–124	5
50–74	6	125–149	9

38. (You will need to use Exercise 5, Section 10.1, to obtain the frequencies for this distribution before finding the mean and modal class.)

Temperature	Frequency	Temperature	Frequency
70–74		95–99	
75–79		100–104	
80–84		105–109	
85–89		110–114	
90–94			

39. Use the distribution of Exercise 4, Section 10.1.

40. Use the distribution of Exercise 6, Section 10.1.

The table below gives the average monthly temperatures, in degrees Fahrenheit for a certain area.

Month	Maximum	Minimum
January	39	16
February	39	18
March	44	21
April	50	26
May	60	32
June	69	37
July	79	43
August	78	42
September	70	37
October	51	31
November	47	24
December	40	20

Find the mean for each of the following.

41. The maximum temperature

42. The minimum temperature

U.S. wheat prices and production figures for a recent 10-year period are given below.*

Year	Price ($ per bushel)	Production (millions of bushels)
1976	2.70	2200
1977	2.30	2000
1978	2.95	1750
1979	3.80	2200
1980	3.90	2400
1981	3.60	2800
1982	3.55	2800
1983	3.50	2450
1984	3.35	2600
1985	3.20	2750

Find the mean for each of the following.

43. Price per bushel of wheat

44. Wheat production

Find the mean, median, and mode for each set of ungrouped data.

45. The weight gains of 10 experimental rats fed on a special diet were −1, 0, −3, 7, 1, 1, 5, 4, 1, 4.

46. A sample of 7 measurements of the thickness of a copper wire were .010, .010, .009, .008, .007, .009, .008.

47. The times in minutes that 12 patients spent in a doctor's office were 20, 15, 18, 22, 10, 12, 16, 17, 19, 21, 23, 13.

48. The scores on a 10-point botany quiz were 6, 6, 8, 10, 9, 7, 6, 5, 6, 8, 3.

---

*Source: U.S. Department of Agriculture.

10.3 MEASURES OF VARIATION

The mean gives a measure of central tendency of a list of numbers, but tells nothing about the *spread* of the numbers in the list. For example, look at the following three samples.

I	3	5	6	3	3
II	4	4	4	4	4
III	10	1	0	0	9

1 Find the range for the numbers

159, 283, 490, 390, 375, 297.

Answer:
331

2 Find the deviations from the mean for each set of numbers.

(a) 19, 25, 36, 41, 52, 61

(b) 6, 9, 5, 11, 3, 2

Answer:

(a) Mean is 39; deviations are -20, -14, -3, 2, 13, 22

(b) Mean is 6; deviations from the mean are 0, 3, -1, 5, -3, -4

Each of these three samples has a mean of 4, and yet they are quite different; the amount of dispersion or variation within the samples is different. Therefore, in addition to a measure of central tendency, another kind of measure is needed that describes how much the numbers vary.

The largest number in sample I is 6, while the smallest is 3, a difference of 3. In sample II this difference is 0; in sample III it is 10. The difference between the largest and smallest number in a sample is called the **range,** one example of a measure of variation. The range of sample I is 3, of sample II is 0, and of sample III is 10. The range has the advantage of being very easy to compute and gives a rough estimate of the variation among the data in the sample. However, it depends only on the two extremes and tells nothing about how the other data are distributed between the extremes.

▶ **EXAMPLE 1**   Find the range for each list of numbers.
(a) 12, 27, 6, 19, 38, 9, 42, 15

The highest number here is 42; the lowest is 6. The range is the difference of these numbers, or
$$42 - 6 = 36.$$
(b) 74, 112, 59, 88, 200, 73, 92, 175
$$\text{Range} = 200 - 59 = 141 \quad ◀ \quad \boxed{1}$$

The most useful measure of variation is the *standard deviation*. Before defining it, however, we must find the **deviations from the mean,** the differences found by subtracting the mean from each number in a distribution.

▶ **EXAMPLE 2**   Find the deviations from the mean for the numbers
$$32, \quad 41, \quad 47, \quad 53, \quad 57.$$

Adding these numbers and dividing by 5 gives a mean of 46. To find the deviations from the mean, subtract 46 from each number in the list. For example, the first deviation from the mean is $32 - 46 = -14$; the last is $57 - 46 = 11$.

Number	Deviations from Mean
32	-14
41	-5
47	1
53	7
57	11

To check your work, find the sum of these deviations. It should always equal 0. (The answer is always 0 because the positive and negative numbers cancel each other.)   ◀   **2**

To find a measure of variation, we might be tempted to use the mean of the deviations. However, as mentioned above, this number is always 0, no matter how widely the data are dispersed. One way to solve this problem is to use absolute value and find the mean of the absolute values of the deviations from the mean.

$$\frac{|-14| + |-5| + |1| + |7| + |11|}{5} = \frac{38}{5} = 7.6$$

Absolute value is awkward to work with algebraically, however.

Another way to get a list of positive numbers is to square each deviation as shown in the table below and then find the mean.

Number	Deviation from Mean	Square of Deviation
32	−14	196
41	−5	25
47	1	1
53	7	49
57	11	121

To find the mean of the squared deviations, statisticians prefer to divide by $n - 1$, rather than n, since a sample usually has less variation than the entire population.* For the distribution above, this gives

$$\frac{196 + 25 + 1 + 49 + 121}{5 - 1} = \frac{392}{4} = 98.$$

This number 98 is called the **variance** of the distribution. Since it is found by averaging a list of squares, the variance of a sample is represented by s^2. The Greek letter σ (sigma) is used for the population variance, which is σ^2.

For a sample of n numbers, the variance is found with the following formula.

Variance

The variance of a sample of n numbers, $x_1, x_2, x_3, \ldots, x_n$, with mean $\bar{x}$, is

$$s^2 = \frac{\Sigma (x - \bar{x})^2}{n - 1}.$$

To find the variance, we square the deviations from the mean, so the variance is in squared units. To return to the same units as the data, we use the *square root* of the variance, called the **standard deviation.**

*Dividing by the smaller number $n - 1$ instead of n gives a larger quotient.

3 Find the standard deviation of each set of numbers. The deviations from the mean were found in Problem 2 at the side.

(a) 19, 25, 36, 41, 52, 61

(b) 6, 9, 5, 11, 3, 2

Answer:

(a) 15.9

(b) 3.5

Standard Deviation

The standard deviation of the n numbers, $x_1, x_2, x_3, \ldots, x_n$, with mean $\bar{x}$, is

$$s = \sqrt{\frac{\Sigma (x - \bar{x})^2}{n - 1}}.$$

As its name indicates, the standard deviation is the most commonly used measure of variation. The standard deviation is a measure of the variation from the mean. The size of the standard deviation tells us something about how spread out the data are from the mean.

▶**EXAMPLE 3** Find the standard deviation of the numbers

7, 9, 18, 22, 27, 29, 32, 40.

The mean of the numbers is

$$\frac{7 + 9 + 18 + 22 + 27 + 29 + 32 + 40}{8} = 23.$$

Arrange the work in columns as follows.

Number	Deviation from Mean	Square of Deviation
7	−16	256
9	−14	196
18	−5	25
22	−1	1
27	4	16
29	6	36
32	9	81
40	17	289
	Total:	900

Divide the total of the last column by $n - 1$, where n is the number of squares in the list. In this case $n = 8$, so the quotient is

$$\frac{900}{7} = 128.6,$$

(rounded), and the standard deviation is

$$\sqrt{128.6} = 11.3 \quad ◀$$

Caution Be careful to divide by $n - 1$, not n, when calculating the standard deviation of a sample. **3**

4 Find the standard deviation for the grouped data that follows. (Hint: $\bar{x} = 28.5$)

Value	Frequency
20–24	3
25–29	2
30–34	4
35–39	1

Answer:
$\sqrt{28.06} = 5.3$

For data in a grouped frequency distribution, a slightly different formula for the standard deviation is used.

Standard Deviation for a Grouped Distribution

The standard deviation for a distribution with mean $\bar{x}$, where x is an interval midpoint with frequency f, and $n = \Sigma f$, is

$$s = \sqrt{\frac{\Sigma\,(x - \bar{x})^2 f}{n - 1}}.$$

The formula indicates that the product $(x - \bar{x})^2 f$ is to be found for each interval. Then these products are summed, and the sum is divided by one less than the total frequency, that is by $n - 1$. The square root of this result is s, the standard deviation.

▶ **EXAMPLE 4** Find s for the grouped data of Example 3, Section 10.2.
Begin by including columns for $x - \bar{x}$, $(x - \bar{x})^2$, and $(x - \bar{x})^2 \cdot f$. Recall from Section 10.2. that $\bar{x} = 71.2$.

Interval	f	x	$x - \bar{x}$	$(x - \bar{x})^2$	$(x - \bar{x})^2 \cdot f$
40–49	2	44.5	−26.7	712.89	1425.78
50–59	4	54.5	−16.7	278.89	1115.56
60–69	7	64.5	−6.7	44.89	314.23
70–79	9	74.5	3.3	10.89	98.01
80–89	5	84.5	13.3	176.89	884.45
90–99	3	94.5	23.3	542.89	1628.67
Total:	30				Total: 5466.70

Use the formula above with $n = 30$ to find s.

$$s = \sqrt{\frac{\Sigma\,(x - \bar{x})^2 \cdot f}{n - 1}} = \frac{\sqrt{5466.70}}{29} \approx \sqrt{188.5} \approx 13.7 \quad ◀ \;\; \boxed{4}$$

A calculator is almost a necessity for finding a standard deviation. A good procedure to follow is to first calculate $\bar{x}$. Then for each x, find $x - \bar{x}$, square that answer, then multiply the result by the appropriate frequency. If your calculator has a key that accumulates a sum, use it to accumulate the total in the last column of the table. Divide that total by $n - 1$ and take the square root.*

---

*Some calculators are equipped with statistical keys that compute the variance and standard deviation of a list of numbers directly, without your having to do any arithmetic or taking square roots. Some of these calculators use $n - 1$ and others use n for these computations. Check the instruction book before using a statistical calculator for the exercises.

10.3 EXERCISES

A calculator will be helpful with many of the exercises in this set.

Find the range and standard deviation for each of the following sets of numbers. (See Examples 1 and 3.)

1. 6, 8, 9, 10, 12

2. 12, 15, 19, 23, 26

3. 7, 6, 12, 14, 18, 15

4. 4, 3, 8, 9, 7, 10, 1

5. 42, 38, 29, 74, 82, 71, 35

6. 122, 132, 141, 158, 162, 169, 180

7. 241, 248, 251, 257, 252, 287

8. 51, 58, 62, 64, 67, 71, 74, 78, 82, 93

9. 3, 7, 4, 12, 15, 18, 19, 27, 24, 11

10. 15, 42, 53, 7, 9, 12, 28, 47, 63, 14

11. 21, 28, 32, 42, 51

12. 76, 78, 92, 104, 111

Find the standard deviation for the following grouped data. (See Example 4.)

13. (From Exercise 3, Section 10.1)

College Units	Frequency
0–24	4
25–49	3
50–74	6
75–99	3
100–124	5
125–149	9

14. (From Exercise 4, Section 10.1)

Scores	Frequency
30–39	1
40–49	6
50–59	13
60–69	22
70–79	17
80–89	13
90–99	8

15. The data of Exercise 5, Section 10.1

16. The data of Exercise 6, Section 10.1

An application of standard deviation is given by **Chebyshev's theorem.** *(P. L. Chebyshev was a Russian mathematician who lived from 1821 to 1894.) This theorem applies to* any *distribution of numbers. It states:*

> For any distribution of numbers, at least $1 - 1/k^2$ of the numbers lie within k standard deviations of the mean.

Example *For any distribution, at least*

$$1 - \frac{1}{3^2} = 1 - \frac{1}{9} = \frac{8}{9}$$

of the numbers lie within 3 standard deviations of the mean. ◄

Find the fraction of all the numbers of a data set lying within the following numbers of standard deviations from the mean.

17. 2   **18.** 4   **19.** 5

In a certain distribution of numbers, the mean is 50 with a standard deviation of 6. Use Chebyshev's theorem to tell what percent of the numbers are

20. between 38 and 62;   **21.** between 32 and 68;

22. between 26 and 74;   **23.** between 20 and 80;

24. less than 38 or more than 62;

25. less than 32 or more than 68;

26. less than 26 or more than 74.

27. **Management** The Forever Power Company conducted tests on the life of its batteries and those of a competitor (Brand X) with the following results for samples of 10 batteries of each brand.

	Hours of Use									
Forever Power	20	22	22	25	26	27	27	28	30	35
Brand X	15	18	19	23	25	25	28	30	34	38

Compute the mean and standard deviation for each sample. Compare the means and standard deviations of the two brands and then answer the questions below.

(a) Which batteries have a more uniform life in hours?

(b) Which batteries have the highest average life in hours?

28. Management The weekly wages of the six employees of Harold's Hardware Store are $300, $320, $380, $420, $500, and $2000.

(a) Find the mean and standard deviation of this distribution.

(b) How many of the employees earn within one standard deviation of the mean? How many earn within two standard deviations of the mean?

29. Management The Quaker Oats Company conducted a survey to determine if a proposed premium, to be included in their cereal, was appealing enough to generate new sales.* Four cities were used as test markets, where the cereal was distributed with the premium, and four cities as control markets, where the cereal was distributed without the premium. The eight cities were chosen on the basis of their similarity in terms of population, per capita income, and total cereal purchase volume. The results were as follows.

		Percent Change in Average Market Shares per Month
Test Cities	1	+18
	2	+15
	3	+7
	4	+10
Control Cities	1	+1
	2	−8
	3	−5
	4	0

(a) Find the mean of the change in market share for the four test cities.

(b) Find the mean of the change in market share for the four control cities.

(c) Find the standard deviation of the change in market share for the test cities.

(d) Find the standard deviation of the change in market share for the control cities.

(e) Find the difference between the mean of part (a) and the mean of part (b). This represents the estimate of the percent change in sales due to the premium.

---

*This example was supplied by Jeffery S. Berman, Senior Analyst, Marketing Information, Quaker Oats Company.

(f) The two standard deviations from part (c) and part (d) were used to calculate an "error" of ±7.95 for the estimate in part (e). With this amount of error what is the smallest and largest estimate of the increase in sales? (Hint: use the answer to part (e).)

On the basis of the results of Exercise 29, the company decided to mass produce the premium and distribute it nationally.

Use a calculator or computer to solve the problems in Exercises 30–33.

30. Natural Science Twenty-five laboratory rats, used in an experiment to test the food value of a new product, made the following weight gains in grams.

5.25	5.03	4.90	4.97	5.03
5.12	5.08	5.15	5.20	4.95
4.90	5.00	5.13	5.18	5.18
5.22	5.04	5.09	5.10	5.11
5.23	5.22	5.19	4.99	4.93

Find the mean gain and the standard deviation of the gains.

31. Management An artisan turns out hand-crafted bowls with the following thicknesses in mm.

1.20	1.01	1.25	2.20	2.58	2.19
1.29	1.15	2.05	1.46	1.90	2.03
2.13	1.86	1.65	2.27	1.64	2.19
2.25	2.08	1.96	1.83	1.17	2.24

Find the mean and standard deviation of these thicknesses.

32. Management The prices of pork bellies futures on the Chicago Mercantile Exchange over a period of several weeks were as follows.

48.25	48.50	47.75	48.45	46.85
47.10	46.50	46.90	46.60	47.00
46.35	46.65	46.85	47.20	46.60
47.00	45.00	45.15	44.65	45.15
46.25	45.90	46.10	45.82	45.70
47.05	46.95	46.90	47.15	47.10

Find the mean and standard deviation.

33. Natural Science A medical laboratory tested 21 samples of human blood for acidity on the pH scale with the following results.

7.1	7.5	7.3	7.4	7.6	7.2	7.3
7.4	7.5	7.3	7.2	7.4	7.3	7.5
7.5	7.4	7.4	7.1	7.3	7.4	7.4

Find the mean and standard deviation.

1 Give a probability distribution for the table below.

Number	Frequency
0	2
1	4
2	6
3	4
4	4
Total:	20

Answer:

Number	Probability
0	.1
1	.2
2	.3
3	.2
4	.2

10.4 NORMAL DISTRIBUTIONS

A **probability distribution** assigns to each outcome of an experiment the probability of the occurrence of that outcome. The probabilities are determined theoretically or, in an experimental situation, by using frequencies. The following example was first discussed in Section 9.5. As shown there, probability distributions are usually presented as tables.

▶**EXAMPLE 1**   A bank wishes to improve its services to the public. The manager decides to begin by finding the number of minutes the tellers spend on each transaction. The transaction times then may take any of the values 0, 1, 2, 3, ... Suppose the manager found the times for 75 transactions, with results as shown in Table 1 below. Give a probability distribution for these results.

Table 1

Minutes	Frequency
1	3
2	5
3	9
4	12
5	15
6	11
7	10
8	6
9	3
10	1
Total:	75

Table 2

Minutes	Probability
1	.04
2	.07
3	.12
4	.16
5	.20
6	.15
7	.13
8	.08
9	.04
10	.01
Total:	1.00

To get a probability distribution, use the relative frequencies as probabilities. A total of 75 transaction times were found, so each frequency should be divided by 75 to get the probabilities shown in Table 2. Some of the results are rounded to the nearest hundredth. The total of the probabilities should be 1, since all possible outcomes are included.   ◀   **1**

Figure 10.6(a) shows a histogram and frequency polygon for the information in Table 2. The heights of the bars are the probabilities. The transaction times in Example 1 were given to the nearest minute. Theoretically at least, they could have been timed to the nearest tenth of a minute, or hundredth of a minute, or even more accurately. In each case, a histogram and frequency polygon could be drawn. If the times are measured with smaller and smaller units, there are more bars in the histogram, and the frequency polygon begins to look more and more like the curve in Figure 10.6(b) instead of a polygon. Actually it is possible for the transaction times to take on any real number value greater than 0. A distribution in which the outcomes can take any real number value within some interval is a **continuous distribution.** The graph of a continuous distribution is a curve.

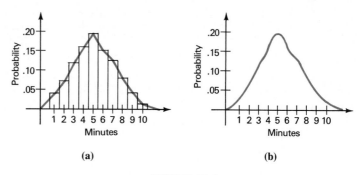

FIGURE 10.6

The distribution of heights (in inches) of college women is another example of a continuous distribution, since these heights include infinitely many possible measurements, such as 53, 58.5, 66.3, 72.666, . . . , and so on. Figure 10.7 shows the continuous distribution of heights of college women. Here the most frequent heights occur near the center of the interval shown.

Another continuous curve, which approximates the distribution of yearly incomes in the United States, is given in Figure 10.8. The graph shows that the most frequent incomes are grouped near the low end of the interval. This kind of distribution, where the peak is not at the center, is called **skewed.**

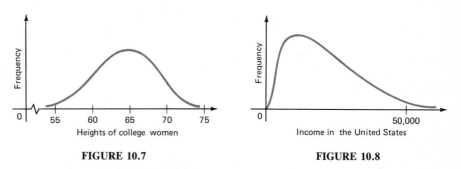

FIGURE 10.7   **FIGURE 10.8**

Many natural and social phenomena produce continuous probability distributions whose graphs are bell-shaped curves, such as those shown in Figure 10.9. Such distributions are called **normal distributions** and their graphs are called **normal curves.** Examples of normal distributions are the heights of college women and the errors made in filling 1-pound cereal boxes. We use the Greek letters μ (mu) to denote the mean and σ (sigma) to denote the standard deviation of a normal distribution.

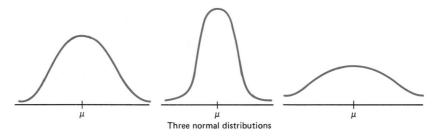

Three normal distributions

FIGURE 10.9

There are many normal distributions. Some of the corresponding normal curves are tall and thin and others short and wide, as shown in Figure 10.9. But every normal curve has the following properties.

1. Its peak occurs directly above the mean μ.
2. The curve is symmetric about the vertical line through the mean (that is, if you fold the page along this line, the left half of the graph will fit exactly on the right half).
3. The area under the curve (and above the horizontal axis) is 1. (As is shown in calculus, this is a consequence of the fact that the sum of the probabilities in any distribution is 1.)

It can be shown that a normal distribution is completely determined by its mean μ and standard deviation σ.* A small standard deviation leads to a tall, narrow curve like the one in the center of Figure 10.9. A large standard deviation produces a flat, wide curve like the one on the right in Figure 10.9.

Since the area under a normal curve is 1, parts of this area can be used to determine certain probabilities. For instance, Figure 10.10(a) is the probability distribution of the annual rainfall in a certain region. Calculus can be used to show that the probability that the annual rainfall will be between 25 and 35 inches is the area under the curve from 25 to 35. The general case, shown in Figure 10.10(b), can be stated as follows.

The area of the shaded region under the normal curve from a to b is the probability that an observed data value will be between a and b.

*As is shown in more advanced courses, its graph is the graph of the function

$$f(x) = \frac{1}{\sigma\sqrt{2\pi}}e^{-(x-\mu)^2/(2\sigma^2)},$$

where $e \approx 2.71828$ is the real number discussed in Section 4.2.

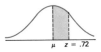

2 Find the percent of area between the mean and

(a) $z = 1.51$;

(b) $z = -2.04$.

In (c) and (d) find the percent of area in the shaded region.

(c)

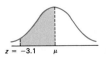

(d)

Answer:

(a) 43.45%

(b) 47.93%

(c) 26.42%

(d) 49.9%

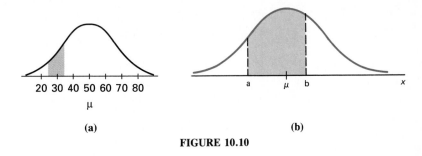

(a)   **(b)**

FIGURE 10.10

To use normal curves effectively we must be able to calculate areas under portions of these curves. These calculations have already been done for the normal curve with mean $\mu = 0$ and standard deviation $\sigma = 1$ (which is called the **standard normal curve**) and are available in Table 8 at the back of the book. The following examples demonstrate how to use Table 8 to find such areas. Later we shall see how the standard normal curve may be used to find areas under any normal curve.

▶**EXAMPLE 2**   The horizontal axis of the standard normal curve is usually labeled z. Find the following areas under the standard normal curve.
(a) The area between $z = 0$ and $z = 1$ (the shaded region in Figure 10.11)
    Find the entry 1 in the z-column of Table 8. The entry next to it in the A-column is .3413, which means that the area between $z = 0$ and $z = 1$ is .3413. Since the total area under the curve is 1, the shaded area in Figure 10.11 is 34.13% of the total area under the normal curve.
(b) The area between $z = -2.43$ and $z = 0$.
    Table 8 lists only positive values of z. But the normal curve is symmetric around the mean $z = 0$, so the area between $z = 0$ and $z = -2.43$ is the same as the area between $z = 0$ and $z = 2.43$. Find 2.43 in the z-column of Table 8. The entry next to it in the A-column shows that the area is .4925. Hence, the shaded area in Figure 10.12 is 49.25% of the total area under the curve.   ◀ **2**

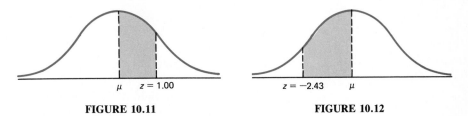

FIGURE 10.11   **FIGURE 10.12**

Since the standard deviation of the standard normal curve is 1, the numbers along the horizontal axis (the z-values) measure the number of standard deviations above or below the mean $z = 0$.

▶**EXAMPLE 3** Find the percent of the total area for the following areas under the standard normal curve.

(a) The area between 1.41 standard deviations *below* the mean and 2.25 standard deviations *above* the mean (that is, between $z = -1.41$ and $z = 2.25$)

First, draw a sketch showing the desired area, as in Figure 10.13. From Table 8, the area between the mean and 1.41 standard deviations below the mean is .4207. Also, the area from the mean to 2.25 standard deviations above the mean is .4878. As the figure shows, the total desired area can be found by *adding* these numbers.

$$\begin{array}{r} .4207 \\ +.4878 \\ \hline .9085 \end{array}$$

The shaded area in Figure 10.13 represents 90.85% of the total area under the normal curve.

(b) The area between .58 standard deviations above the mean and 1.94 standard deviations above the mean

Figure 10.14 shows the desired area. The area between the mean and .58 standard deviations above the mean is .2190. The area between the mean and 1.94 standard deviations above the mean is .4738. As the figure shows, the desired area is found by *subtracting* the two areas.

$$\begin{array}{r} .4738 \\ -.2190 \\ \hline .2548 \end{array}$$

The shaded area of Figure 10.14 represents 25.48% of the total area under the normal curve.

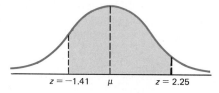

$z = -1.41 \quad \mu \quad z = 2.25$

FIGURE 10.13

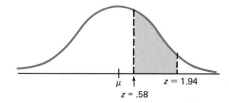

$\mu \quad z = 1.94$
$z = .58$

FIGURE 10.14

(c) The area to the right of 2.09 standard deviations above the mean

The total area under a normal curve is 1. Thus, the total area to the right of the mean is 1/2, or .5000. From Table 8, the area from the mean to 2.09 standard deviations above the mean is .4817. The area to the right of 2.09 standard deviations is found by subtracting .4817 from .5000.

$$\begin{array}{r} .5000 \\ -.4817 \\ \hline .0183 \end{array}$$

3 Find the following standard normal curve areas as percents of the total area.

(a) Between .31 standard deviations below the mean and 1.01 standard deviations above the mean

(b) Between .38 and 1.98 standard deviations below the mean

(c) To the right of 1.49 standard deviations above the mean

Answer:

(a) 46.55%

(b) 32.82%

(c) 6.81%

4 Find each z-score using the information in Example 4.

(a) $x = 36$

(b) $x = 55$

Answer:

(a) -3.5

(b) 1.25

A total of 1.83% of the total area is to the right of 2.09 standard deviations above the mean. Figure 10.15 shows the desired area. ◄ **3**

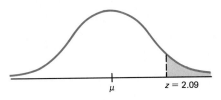

FIGURE 10.15

The key to finding areas under *any* normal curve is to express each number x on the horizontal axis in terms of standard deviations above or below the mean. The **z-score** for x is the number of standard deviations that x lies from the mean (positive if x is above the mean, negative if x is below the mean).

▶**EXAMPLE 4** If a normal distribution has mean 50 and standard deviation 4, find the following z-scores.

(a) The z-score for $x = 46$

Since 46 is 4 units below 50 and the standard deviation is 4, 46 is 1 standard deviation below the mean. So, its z-score is -1.

(b) The z-score for $x = 60$

The z-score is 2.5 because 60 is 10 units above the mean (since $60 - 50 = 10$) and 10 units is 2.5 standard deviations (since $10/4 = 2.5$). ◄ **4**

In Example 4(b) we found the z-score by taking the difference between 60 and the mean and dividing this difference by the standard deviation. The same procedure works in the general case.

If a normal distribution has mean μ and standard deviation σ, then the z-score for the number x is

$$z = \frac{x - \mu}{\sigma}.$$

The importance of z-scores is the following fact, whose proof is omitted.

Area Under a Normal Curve

The area under a normal curve between $x = a$ and $x = b$ is the same as the area under the standard normal curve between the z-score for a and the z-score for b.

Therefore, by converting to z-scores and using Table 8 for the standard normal curve, we can find areas under any normal curve. Since these areas are probabilities (as explained on page 552), we can now handle a variety of applications.

▶**EXAMPLE 5**   Dixie Office Supplies finds that its sales force drives an average of 1200 miles per month per person, with a standard deviation of 150 miles. Assume that the number of miles driven by a salesperson is closely approximated by a normal distribution.

(a) Find the probability that a salesperson drives between 1200 and 1600 miles per month.

Here $\mu = 1200$ and $\sigma = 150$, and we must find the area under the normal distribution curve between $x = 1200$ and $x = 1600$. We begin by finding the z-score for $x = 1200$.*

$$z = \frac{x - \mu}{\sigma} = \frac{1200 - 1200}{150} = \frac{0}{150} = 0.$$

The z-score for $x = 1600$ is

$$z = \frac{x - \mu}{\sigma} = \frac{1600 - 1200}{150} = \frac{400}{150} = 2.67.$$

So the area under the curve from $x = 1200$ to $x = 1600$ is the same as the area under the standard normal curve from $z = 0$ to $z = 2.67$, as indicated in Figure 10.16. Table 8 shows that this area is .4962. Therefore, the probability that a salesperson drives between 1200 and 1600 miles per month is .4962.

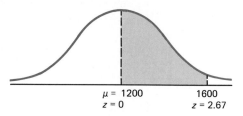

μ = 1200   1600
z = 0   z = 2.67

FIGURE 10.16

---

*All z-scores here are rounded to two decimal places.

5 The heights of female sopho-more college students at one school have $\mu = 172$ centimeters, with $\sigma = 10$ centimeters. Find the probability that the height of such a student is

(a) between 172 cm and 185 cm;

(b) between 160 cm and 180 cm;

(c) less than 165 cm.

Answer:

(a) .4032

(b) .6730

(c) .2420

(b) Find the probability that a salesperson drives between 1000 and 1500 miles per month.

As shown in Figure 10.17, z-scores for both $x = 1000$ and $x = 1500$ are needed.

For $x = 1000$,

$$z = \frac{1000 - 1200}{150}$$

$$= \frac{-200}{150}$$

$$= -1.33$$

For $x = 1500$,

$$z = \frac{1500 - 1200}{150}$$

$$= \frac{300}{150}$$

$$= 2.00$$

From the table, $z = 1.33$ leads to an area of .4082, while $z = 2.00$ corresponds to .4773. A total of $.4082 + .4773 = .8855$, or 88.55%, of all drivers travel between 1000 and 1500 miles per month. From this, the probability that a driver travels between 1000 and 1500 miles per month is .8855. ◀ **5**

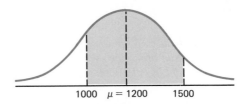

1000 $\mu = 1200$ 1500

FIGURE 10.17

▶**EXAMPLE 6**   A tire store finds that the tread life of its tires is normally distributed, with a mean of 26,640 miles and a standard deviation of 4000 miles. The store sold 9000 tires this month. How many of them can be expected to last more than 35,000 miles?

Here $\mu = 26,640$ and $\sigma = 4000$. The probability that a tire will last more than 35,000 miles is the area under the normal curve to the right of $x = 35,000$. The $z =$ score for $x = 35,000$ is

$$z = \frac{x - \mu}{\sigma} = \frac{35,000 - 26,640}{4000} = \frac{8360}{4000} = 2.09.$$

Example 3(c) and Figure 10.15 show that the area to the right of $z = 2.09$ is .0183, which is 1.83% of the total area under the curve. Therefore, 1.83% of the tires can be expected to last more than 35,000 miles. Since

$$1.83\% \text{ of } 9000 = .0183 \cdot 9000 = 164.7,$$

approximately 165 tires can be expected to last more than 35,000 miles. ◀

10.4 EXERCISES

Find the percent of area under a normal curve between the mean and the following number of standard deviations from the mean. (See Example 3.)

1. 2.50   **2.** 1.68   **3.** .45   **4.** .81   **5.** −1.71   **6.** −2.04

7. 3.11   **8.** 2.80

Find the percent of the total area under the standard normal curve between the following z-scores. (See Example 2.)

9. $z = 1.41$ and $z = 2.83$ **10.** $z = .64$ and $z = 2.11$ **11.** $z = -2.48$ and $z = -.05$

12. $z = -1.74$ and $z = -1.02$ **13.** $z = -3.11$ and $z = 1.44$ **14.** $z = -2.94$ and $z = -.43$

15. $z = -.42$ and $z = .42$ **16.** $z = -1.98$ and $z = 1.98$

Find a z-score satisfying each of the following conditions. (Hint: use Table 8 backwards.)

17. 5% of the total area is to the right of z.   **18.** 1% of the total area is to the left of z.

19. 15% of the total area is to the left of z.   **20.** 25% of the total area is to the right of z.

Work the following applied problems. (See Examples 5 and 6.)

Management *A certain type of light bulb has an average life of* 500 *hours, with a standard deviation of* 100 *hours. The length of life of the bulb can be closely approximated by a normal curve. An amusement park buys and installs* 10,000 *such bulbs. Find the total number that can be expected to last*

21. at least 500 hours;   **22.** less than 500 hours;   **23.** between 500 and 650 hours;

24. between 300 and 500 hours;   **25.** between 650 and 780 hours;   **26.** between 290 and 540 hours;

27. less than 740 hours;   **28.** more than 300 hours;   **29.** more than 790 hours;

30. less than 410 hours.

Management *A box of oatmeal must contain* 16 *ounces. The machine that fills the oatmeal boxes is set so that, on the average, a box contains* 16.5 *ounces. The boxes filled by the machine have weights that can be closely approximated by a normal curve. What percent of the boxes filled by the machine are underweight if the standard deviation is*

31. .5 ounce;   **32.** .3 ounce;   **33.** .2 ounce;   **34.** .1 ounce?

The chickens at Colonel Thompson's Ranch have a mean weight of 1850 *grams with a standard deviation of* 150 *grams. The weights of the chickens are closely approximated by a normal curve. Find the percent of all chickens weighing*

35. more than 1700 grams;   **36.** less than 1800 grams;

37. between 1750 grams and 1900 grams;   **38.** between 1600 grams and 2000 grams;

39. less than 1550 grams;   **40.** more than 2100 grams.

Natural Science *In nutrition, the recommended daily allowance of vitamins is a number set by the government as a guide to an individual's daily vitamin intake. Actually, vitamin needs vary drastically from person to person, but the needs are very closely approximated by a normal curve. To calculate the recommended daily allowance, the government first finds the average need for vitamins among people in the population and then the standard deviation. The recommended daily allowance is defined as the mean plus 2.5 times the standard deviation.*

41. What percentage of the population will receive adequate amounts of vitamins under this plan?

Find the recommended daily allowance for the following vitamins.

42. Mean = 1800 units, standard deviation = 140 units

43. Mean = 159 units, standard deviation = 12 units

44. Mean = 1200 units, standard deviation = 92 units

Assume the following distributions are all normal, and use the areas under the normal curve given in Table 8 to answer the questions.

45. Management A machine produces bolts with an average diameter of .25 inches and a standard deviation of .02 inches. What is the probability that a bolt will be produced with a diameter greater than .3 inches?

46. Management The mean monthly income of the trainees of an engineering firm is $1200 with a standard deviation of $200. Find the probability that an individual trainee earns less than $1000 per month.

47. Management A machine that fills quart milk cartons is set up to average 32.2 ounces per carton, with a standard deviation of 1.2 ounces. What is the probability that a filled carton will contain less than 32 ounces of milk?

48. The average contribution to the campaign of Polly Potter, a candidate for city council, was $50 with a standard deviation of $15. How many of the 200 people who contributed to Potter's campaign gave between $30 and $100?

49. Management At the Discount Market, the average weekly grocery bill is $52.25 with a standard deviation of $15.50. What are the largest and smallest amounts spent by the middle 50% of this market's customers?

50. Natural Science The mean clotting time of blood is 7.45 seconds with a standard deviation of 3.6 seconds. What is the probability that an individual's blood clotting time will be less than 7 seconds or greater than 8 seconds?

51. The average size of the fish in Lake Amotan is 12.3 inches with a standard deviation of 4.1 inches. Find the probability of catching a fish there that is longer than 18 inches.

52. To be graded extra large, an egg must weigh at least 2.2 ounces. If the average weight for an egg is 1.5 ounces with a standard deviation of .4 ounces, how many of five dozen eggs would you expect to grade extra large?

A teacher gives a test to a large group of students. The results are closely approximated by a normal curve. The mean is 74, with a standard deviation of 6. The teacher wishes to give A's to the top 8% of the students and F's to the bottom 8%. A grade of B is given to the next 15%, with D's given similarly. All other students get C's. Find the bottom cutoff (rounded to the nearest whole number) for the following grades. (Hint: use Table 8 backwards.)

53. A   **54.** B   **55.** C   **56.** D

57. The birth weights of babies born at Metropolitan Hospital are normally distributed with a mean of 7.5 pounds and a standard deviation of 21 ounces. If a newborn baby is randomly chosen, what is the probability that he or she weighs less than 4 pounds 15 ounces? (Hint: convert all weights to ounces.)

58. The number of hours that a 10-year-old child in California watches TV per week is normally distributed, with a mean of 12 hours and a standard deviation of 1.5 hours. What is the probability that a randomly selected child watches TV between 9 and 14 hours per week?

59. A commuter finds that her driving time to work each day is approximated by a normal distribution. If she averages 35 minutes per day, with a standard deviation of 7 minutes, when should she leave home so that she has a 95% chance of arriving at work by 8 A.M.?

60. The life of a certain brand of color television tubes is approximated by a normal distribution with a standard deviation of 1.53 years. If 7% of these tubes last more than 6.9944 years, what is the mean life of a tube?

10.5 THE BINOMIAL DISTRIBUTION

Many practical experiments have only two possible outcomes, *success* or *failure*. Such experiments are called *binomial trials* or *Bernoulli trials* and were first studied in Section 9.3. Examples of binomial trials include flipping a coin (with heads being a success, for instance, and tails a failure) or testing TVs coming off the assembly line to see whether or not they are defective.

A **binomial distribution** is a probability distribution that satisfies the following conditions.

1. The experiment is a series of independent trials.
2. There are only two possible outcomes for each trial, success or failure.
3. The probability of each outcome is constant from trial to trial.

For example, suppose a fair die is tossed 5 times, with 1 or 2 considered a success and 3, 4, 5, or 6 a failure. Each trial is independent of the others and the probability of success in each trial is $p = 2/6 = 1/3$. In 5 tosses there can be any number of successes from 0 through 5. But these 6 possible results are not equally likely, as shown in the chart at the side, which was obtained by using the following formula from Section 9.3.

$$P(x) = \binom{n}{x} p^x (1 - p)^{n-x},$$

where n is the number of trials (here, $n = 5$), x is the number of successes, p is the probability of success in a single trial (here, $p = 1/3$), and $P(x)$ is the probability that exactly x of the n trials result in success.

As we saw in Section 9.5, the expected value of this experiment (that is, the expected number of successes in 5 trials) is the sum

$$0 \cdot P(0) + 1 \cdot P(1) + 2 \cdot P(2) + 3 \cdot P(3) + 4 \cdot P(4) + 5 \cdot P(5)$$

$$= 0\left(\frac{32}{243}\right) + 1\left(\frac{80}{243}\right) + 2\left(\frac{80}{243}\right) + 3\left(\frac{40}{243}\right) + 4\left(\frac{10}{243}\right) + 5\left(\frac{1}{243}\right)$$

$$= \frac{405}{243} = 1\frac{2}{3} \quad \text{or} \quad \frac{5}{3}.$$

This result agrees with our intuition since there is, on average, 1 success in 3 trials; so we would expect 2 successes in 6 trials, and a bit less than 2 in 5 trials. From another point of view, since the probability of success is 1/3 in each trial and 5 trials were performed, the expected number of successes should be $5 \cdot (1/3) = 5/3$. More generally, we can show the following.

x	$P(x)$
0	$\binom{5}{0}\left(\frac{1}{3}\right)^0\left(\frac{2}{3}\right)^5 = \frac{32}{243}$
1	$\binom{5}{1}\left(\frac{1}{3}\right)^1\left(\frac{2}{3}\right)^4 = \frac{80}{243}$
2	$\binom{5}{2}\left(\frac{1}{3}\right)^2\left(\frac{2}{3}\right)^3 = \frac{80}{243}$
3	$\binom{5}{3}\left(\frac{1}{3}\right)^3\left(\frac{2}{3}\right)^2 = \frac{40}{243}$
4	$\binom{5}{4}\left(\frac{1}{3}\right)^4\left(\frac{2}{3}\right)^1 = \frac{10}{243}$
5	$\binom{5}{5}\left(\frac{1}{3}\right)^5\left(\frac{2}{3}\right)^0 = \frac{1}{243}$

The expected number of successes in n binomial trials is np, where p is the probability of success in a single trial.

1 Find μ and σ for a binomial distribution having $n = 120$ and $p = 1/6$.

Answer:
$\mu = 20$; $\sigma = 4.08$

The expected value of any probability distribution is actually its *mean,* as we now show. Suppose an experiment has possible numerical outcomes $x_1, x_2, \ldots,$ x_k, which occur with frequencies $f_1, f_2, \ldots, f_k$, respectively. If n is the total number of items, then (as shown in Section 10.2) the mean of this grouped distribution is

$$\mu = \frac{x_1 f_1 + x_2 f_2 + \cdots + x_k f_k}{n} = \frac{x_1 f_1}{n} + \frac{x_2 f_2}{n} + \cdots + \frac{x_k f_k}{n}$$

$$= x_1\left(\frac{f_1}{n}\right) + x_2\left(\frac{f_2}{n}\right) + \cdots + x_k\left(\frac{f_k}{n}\right).$$

But in each case, the fraction f_i/n (frequency over total number) is the probability p_i of outcome x_i. Hence,

$$\mu = x_1 p_1 + x_2 p_2 + \cdots + x_k p_k,$$

which is the expected value.

We saw above that the expected value (mean) of a binomial distribution is np. It can be shown that the variance and standard deviation of a binomial distribution are also given by convenient formulas.

$$\sigma^2 = np(1 - p) \quad \text{and} \quad \sigma = \sqrt{np(1 - p)}$$

Substituting the appropriate values for n and p from the example into this new formula gives

$$\sigma^2 = 5\left(\frac{1}{3}\right)\left(\frac{2}{3}\right) = \frac{10}{9}. \quad \text{■ 1}$$

A summary of these results follows.

Binomial Distribution

Suppose an experiment is a series of n independent repeated trials, where the probability of a success in a single trial is always p. Let x be the number of successes in the n trials. Then the probability that exactly x successes will occur in n trials is given by

$$\binom{n}{x} p^x (1 - p)^{n-x}.$$

The mean μ and variance σ^2 of a binomial distribution are, respectively,

$$\mu = np \quad \text{and} \quad \sigma^2 = np(1 - p).$$

The standard deviation is

$$\sigma = \sqrt{np(1 - p)}.$$

2 The probability that a can of beer from a certain plant is defective is .005. A sample of 4 cans is selected at random. Write a distribution for the number of defective cans in the sample, and give its mean and standard deviation.

Answer:

x	$P(x)$
0	.961
1	.020
2	.015
3	.00002
4	.00000000

$\mu = .02$, $\sigma = .14$

▶**EXAMPLE 1**   The probability that a plate picked at random from the assembly line in a china factory will be defective is .01. A sample of 3 is to be selected. Write the distribution for the number of defective plates in the sample, and give its mean and standard deviation.

Since 3 plates will be selected, the possible number of defective plates ranges from 0 to 3. Here, n (the number of trials) is 3; p (the probability of selecting a defective on a single trial) is .01. The distribution and the probability of each outcome are shown below.

x	$P(x)$
0	$\binom{3}{0}(.01)^0(.99)^3 = .970$
1	$\binom{3}{1}(.01)(.99)^2 = .029$
2	$\binom{3}{2}(.01)^2(.99) = .0003$
3	$\binom{3}{3}(.01)^3(.99)^0 = .000001$

The mean of the distribution is

$$\mu = np = 3(.01) = .03.$$

The standard deviation is

$$\sigma = \sqrt{np(1-p)} = \sqrt{3(.01)(.99)} = \sqrt{.0297} \approx .17. \quad \blacktriangleleft \quad \boxed{2}$$

Binomial distributions are very useful, but the probability calculations become difficult if n is large. However, the standard normal distribution of the previous section (which shall be referred to as the normal distribution from now on) can be used to get a good approximation of the binomial distributions.

As an example, to show how the normal distribution is used, consider the distribution for the expected number of heads if 1 coin is tossed 15 times. This is a binomial distribution with $p = 1/2$ and $n = 15$. The distribution is shown with Figure 10.18. The mean of the distribution is

$$\mu = np = 15\left(\frac{1}{2}\right) = 7.5.$$

The standard deviation is

$$\sigma = \sqrt{np(1-p)} = \sqrt{15\left(\frac{1}{2}\right)\left(1 - \frac{1}{2}\right)}$$

$$= \sqrt{15\left(\frac{1}{2}\right)\left(\frac{1}{2}\right)} = \sqrt{3.75} \approx 1.94.$$

In Figure 10.18 the normal curve with $\mu = 7.5$ and $\sigma = 1.94$ has been superimposed over the histogram of the distribution.

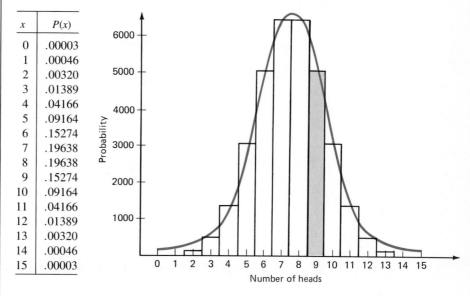

x	$P(x)$
0	.00003
1	.00046
2	.00320
3	.01389
4	.04166
5	.09164
6	.15274
7	.19638
8	.19638
9	.15274
10	.09164
11	.04166
12	.01389
13	.00320
14	.00046
15	.00003

FIGURE 10.18

As shown in the distribution, the probability of getting exactly 9 heads in the 15 tosses is approximately .153. This answer is about the same fraction that would be found by dividing the area of the bar in color in Figure 10.18 by the total area of all 16 bars in the graph. (Some of the bars at the extreme left and right ends of the graph are too short to show up.)

As the graph suggests, the area in color is approximately equal to the area under the normal curve from $x = 8.5$ to $x = 9.5$. The normal curve is higher than the top of the bar in the left half but lower in the right half.

To find the area under the normal curve from $x = 8.5$ to $x = 9.5$, first find z-scores, as in the previous section. Use the mean and the standard deviation for the distribution, which we have already calculated, to get z-scores for $x = 8.5$ and $x = 9.5$.

For $x = 8.5$,
$$z = \frac{8.5 - 7.5}{1.94}$$
$$= \frac{1.00}{1.94}$$
$$z \approx .52$$

For $x = 9.5$,
$$z = \frac{9.5 - 7.5}{1.94}$$
$$= \frac{2.00}{1.94}$$
$$z \approx 1.03$$

3 Use the normal distribution to find the probability of getting exactly the following number of heads in 15 tosses of a coin.

(a) 7

(b) 10

Answer:

(a) .1985

(b) .0909

From Table 8, $z = .52$ gives an area of .1985, while $z = 1.03$ gives .3485. Find the required result by subtracting these two numbers.

$$.3485 - .1985 = .1500$$

This answer, .1500, is close to the exact answer, .153, found above. **3**

▶**EXAMPLE 2**   About 6% of the bolts produced by a certain machine are defective.

(a) Find the probability that in a sample of 100 bolts, 3 or fewer are defective.

This problem satisfies the conditions of the definition of a binomial distribution, so the normal curve approximation can be used. First find the mean and the standard deviation using $n = 100$ and $p = 6\% = .06$.

$$\mu = 100(.06) \qquad \sigma = \sqrt{100(.06)(1 - .06)}$$
$$\mu = 6 \qquad\qquad = \sqrt{100(.06)(.94)}$$
$$= \sqrt{5.64}$$
$$\sigma \approx 2.37$$

As the graph of Figure 10.19 shows, the area to the left of $x = 3.5$ (since we want 3 or fewer defective bolts) must be found. For $x = 3.5$,

$$z = \frac{3.5 - 6}{2.37} = \frac{-2.5}{2.37} \approx -1.05.$$

From Table 8, $z = -1.05$ corresponds to an area of .3531. Find the area to the left of z by subtracting .3531 from .5000 to get .1469. The probability of 3 or fewer defective bolts in a set of 100 bolts is .1469, or 14.69%.

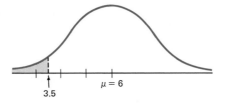

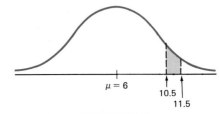

| FIGURE 10.19 | FIGURE 10.20 |

(b) Find the probability of exactly 11 defective bolts in a sample of 100 bolts.

As Figure 10.20 shows, the area between $x = 10.5$ and $x = 11.5$ must be found.

$$\text{If } x = 10.5, \text{ then } z = \frac{10.5 - 6}{2.37} = 1.90.$$

$$\text{If } x = 11.5, \text{ then } z = \frac{11.5 - 6}{2.37} = 2.32.$$

4 About 9% of the transistors produced by a certain factory are defective. Find the probability that in a sample of 200, the following numbers of transistors will be defective. (Hint: $\sigma = 4.05$)

(a) Exactly 11   **(b)** 16 or fewer

(c) More than 14

Answer:

(a) .0226   **(b)** .3557

(c) .8051

Look in Table 8; $z = 1.90$ gives an area of .4713, while $z = 2.32$ yields .4898. The final answer is the difference of these numbers, or

$$.4898 - .4713 = .0185.$$

There is about a 1.85% chance of having exactly 11 defective bolts. ◄ **4**

Caution The normal curve approximation to a binomial distribution is quite accurate *provided that n is large and p is not close to 0 or 1.* As a rule of thumb, the normal curve approximation can be used as long as both np and $n(1 - p)$ are at least 5.

10.5 EXERCISES

In Exercises 1–6, several binomial experiments are described. For each one, (a) write the distribution; (b) find the mean; (c) find the standard deviation. (See Example 1.)

1. A die is rolled six times and the number of 1's that come up is tallied.

2. A 6-item multiple choice test has four possible answers for each item. A student selects all his answers randomly. Write the distribution for the number of correct answers.

3. **Management** To maintain quality control on the production line, the Bright Lite Company randomly selects three light bulbs each day for testing. Experience has shown a defective rate of .02. Write the distribution for the number of defectives in the daily samples.

4. In a taste test, each member of a panel of four is given two glasses of Super Cola, one made using the old formula and one with the new formula, and asked to identify the new formula. Assuming the panelists operate independently, write the distribution of the number of successful identifications, if each judge actually guesses.

5. The probability that a radish seed will germinate is .7. Joe's mother gives him four seeds to plant. Write the distribution for the number of seeds that he can expect to germinate.

6. **Natural Science** Five patients in Ward 8 of Memorial Hospital have a disease with a known mortality rate of .1.

Write the distribution of the number who can be expected to survive.

Natural Science *Work the following exercises involving binomial experiments.*

7. The probability that an infant will die in the first year of life is about .025. In a group of 500 babies, what are the mean and standard deviation of the number of babies who can be expected to die in their first year of life?

8. The probability that a particular kind of mouse will have a brown coat is 1/4. In a litter of 8, assuming independence, how many could be expected to have a brown coat? With what standard deviation?

9. A certain drug is effective 80% of the time. Give the mean and standard deviation of the number of patients using the drug who recover, out of a group of 64 patients.

10. The probability that a newborn infant will be a girl is .49. If 50 infants are born on Susan B. Anthony's birthday, how many can be expected to be girls? With what standard deviation?

For the remaining exercises, use the normal curve approximation to a binomial distribution. (See Example 2.)

Suppose 16 coins are tossed. Find the probability of getting exactly

11. 8 heads;   **12.** 7 heads;   **13.** 10 tails;   **14.** 12 tails.

Suppose 1000 *coins are tossed. Find the probability of getting each of the following. (Hint:* $\sqrt{250} = 15.8$*)*

15. Exactly 500 heads

16. Exactly 510 heads

17. 480 heads or more

18. Fewer than 470 tails

19. Fewer than 518 heads

20. More than 550 tails

A die is tossed 120 *times. Find the probability of getting each of the following. (Hint:* $\sigma = 4.08$*)*

21. Exactly 20 fives

22. Exactly 24 sixes

23. Exactly 17 threes

24. Exactly 22 twos

25. More than 18 threes

26. Fewer than 22 sixes

Management *Two percent of the hamburgers sold at Tom's Burger Queen are defective. Tom sold* 10,000 *burgers last week. Find the probability that among these burgers*

27. fewer than 170 were defective;

28. more than 222 were defective.

Natural Science *A new drug cures* 80% *of the patients to whom it is administered. It is given to* 25 *patients. Find the probability that among these patients*

29. exactly 20 are cured;

30. exactly 23 are cured;

31. all are cured;

32. no one is cured;

33. 12 or fewer are cured;

34. between 17 and 23 (inclusive) are cured.

 Use a calculator to work the following problems.

35. Rework Exercises 51–54 of Section 9.3 using the normal curve to approximate the binomial probabilities. Compare your answers for the binomial distribution with the results found in Section 9.3.

36. A coin is tossed 100 times. Find the probability of **(a)** exactly 50 heads; **(b)** at least 55 heads; **(c)** no more than 40 heads.

CHAPTER 10   SUMMARY

KEY TERMS AND SYMBOLS

	random sample	
10.1	class	
	grouped frequency distribution	
	histogram	
	frequency polygon	
	class boundary	
	cumulative frequency distribution	
	cumulative frequency polygon (ogive)	
10.2	Σ summation (sigma) notation	
	$\bar{x}$ sample mean	
	measures of central tendency	
	mean (arithmetic average)	
	median	
	mode	
	modal class	
	bimodal	

10.3	s^2 sample variance
	s sample standard deviation
	σ^2 population variance
	σ population standard deviation
	range
	deviations from the mean
10.4	μ population mean
	probability distribution
	continuous distribution
	skewed distribution
	normal distributions
	normal curves
	standard normal curve
	z-score
10.5	binomial trials (Bernoulli trials)
	binomial distribution.

KEY CONCEPTS

To organize the data from a sample, we use a **grouped frequency distribution,** a set of intervals with their corresponding frequencies. The same information can be displayed with a **histogram,** a bar graph with a bar for each interval. Each bar has width 1 and height equal to the probability of the corresponding interval. Another way to display this information is with a **frequency polygon,** which is formed by connecting the midpoints of consecutive bars of the histogram. A **cumulative frequency distribution** gives the total frequency up to and including each interval.

The **mean** $\bar{x}$ of a frequency distribution is the expected value.

For n numbers $x_1, x_2, \ldots, x_n$ For a grouped distribution

$$\bar{x} = \frac{\Sigma(x)}{n}. \qquad\qquad \bar{x} = \frac{\Sigma(xf)}{n}.$$

The **median** is the middle entry in a set of data arranged in either increasing or decreasing order.

The **mode** is the most frequent entry in a set of numbers. The **modal class** is the interval in a grouped distribution with the greatest frequency.

The **range** of a distribution is the difference between the largest and smallest numbers in the distribution.

The **standard deviation** s is the square root of the **variance.**

For n numbers For a grouped distribution

$$s = \sqrt{\frac{\Sigma(x - \bar{x})^2}{n - 1}}. \qquad\qquad s = \sqrt{\frac{\Sigma(x - \bar{x})^2 f}{n - 1}}.$$

A **normal distribution** is a continuous distribution with the following properties: The highest frequency is at the mean; the graph is symmetric about a vertical line through the mean; the total area under the curve, above the x-axis, is 1. If a normal distribution has mean μ and standard deviation σ, then the z-score for the number x is $z = \dfrac{x - \mu}{\sigma}$.

Area Under a Normal Curve The area under a normal curve between $x = a$ and $x = b$ gives the probability that an observed data value will be between a and b.

The **binomial distribution** is a distribution with the following properties: For n independent repeated trials, where the probability of success in a single trial is p, the probability of x successes is $\binom{n}{x} p^x (1 - p)^{n-x}$. The mean is $\mu = np$ and the standard deviation is

$$\sigma = \sqrt{np(1 - p)}.$$

*In Exercises 1 and 2, (**a**) write a frequency distribution; (**b**) draw a histogram; (**c**) draw a frequency polygon.*

1. The following numbers give the sales in dollars for the lunch hour at a local hamburger store for the last twenty Fridays. (Use intervals 450–474, 475–499, and so on.)

| 480 | 451 | 501 | 478 | 512 | 473 | 509 | 515 | 458 | 566 |
| 516 | 535 | 492 | 558 | 488 | 547 | 461 | 475 | 492 | 471 |

2. The number of units carried in one semester by the students in a business mathematics class was as follows. (Use intervals of 9–10, 11–12, 13–14, 15–16.)

| 10 | 9 | 16 | 12 | 13 | 15 | 13 | 16 | 15 | 11 | 13 |
| 12 | 12 | 15 | 12 | 14 | 10 | 12 | 14 | 15 | 15 | 13 |

Find the mean for each of the following.

3. 41, 60, 67, 68, 72, 74, 78, 83, 90, 97

4. 105, 108, 110, 115, 106, 110, 104, 113, 117

5.

Interval	Frequency
10–19	6
20–29	12
30–39	14
40–49	10
50–59	8

6.

Interval	Frequency
40–44	2
45–49	5
50–54	7
55–59	10
60–64	4
65–69	1

Find the median and the mode (or modes) for each of the following.

7. 32, 35, 36, 44, 46, 46, 59

8. 38, 36, 42, 44, 38, 36, 48, 35

Find the modal class for the distributions of

9. Exercise 5 above;

10. Exercise 6 above.

Find the range and standard deviation for each of the following distributions.

11. 14, 17, 18, 19, 32

12. 26, 43, 51, 29, 37, 56, 29, 82, 74, 93

Find the standard deviation for the following.

13. Exercise 5 above

14. Exercise 6 above

15. The annual returns of two stocks for three years are given below.

	1988	1989	1990
Stock I	11%	−1%	14%
Stock II	9%	5%	10%

(**a**) Find the mean and standard deviation for each stock over the three-year period.

(**b**) If you are looking for security with an 8% return, which of these two stocks would you choose?

16. The weight gains of 2 groups of 10 rats fed on two different experimental diets were as follows.

	Weight Gains									
Diet A	1	0	3	7	1	1	5	4	1	4
Diet B	2	1	1	2	3	2	1	0	1	0

Compute the mean and standard deviation for each group and compare them to answer the questions below.

(**a**) Which diet produced the greatest mean gain?

(**b**) Which diet produced the most consistent gain?

17. A probability distribution has an expected value of 28 and a standard deviation of 4. Use Chebyshev's theorem (Section 10.3, Exercises 17–26) to decide what percent of the distribution is
 (a) between 20 and 36;
 (b) less than 23.2 or greater than 32.8.

18. (a) Find the percent of the area under a normal curve within 2.5 standard deviations of the mean.
 (b) Compare your answer to part (a) with the result using Chebyshev's theorem.

Find the following areas under the standard normal curve.

19. Between $z = 0$ and $z = 1.27$

20. Between $z = -1.88$ and $z = 2.10$

21. To the left of $z = -.41$

22. To the right of $z = 2.50$

On standard IQ tests, the mean is 100, *with a standard deviation of* 15. *The results are very close to fitting a normal curve. Suppose an IQ test is given to a very large group of people. Find the percentage of those people whose IQ score is*

23. more than 130;   24. less than 85;

25. between 85 and 115.

26. A machine that fills quart milk cartons is set to fill them with 32.1 oz of milk. If the actual contents of the cartons vary normally, with a standard deviation of .1 oz, what percent of the cartons contain less than a quart (32 oz)?

27. The probability that a can of peaches from a certain cannery is defective is .005. A sample of 4 cans is selected at random. Write a binomial distribution for the number of defective cans in the sample, and give its mean and standard deviation.

28. Use the normal curve to approximate the probability of getting the following number of heads in 15 tosses of a coin.
 (a) Exactly 7   (b) Between 7 and 10 (exclusive)
 (c) At least 10

29. The probability that a small business will go bankrupt in its first year is .21. For 50 such small businesses, use the normal curve to approximate the following probabilities.
 (a) Exactly 8 go bankrupt
 (b) No more than 2 go bankrupt

The average resident of a certain Eastern suburb spends 42 *minutes per day commuting, with a standard deviation of* 12 *minutes. Assume a normal distribution. Find the percent of all residents of this suburb who commute*

30. at least 50 minutes per day;

31. no more than 35 minutes per day;

32. between 32 and 40 minutes per day;

33. between 38 and 60 minutes per day.

About 6% *of the frankfurters produced by a certain machine are overstuffed. Assume a binomial distribution and use the normal distribution to find the probability that, in a sample of* 500 *frankfurters,*

34. 25 or fewer are overstuffed (Hint: $\sigma = 5.3$);

35. exactly 30 are overstuffed.

36. Find a z-score such that 8% of the area under the curve is to the right of z.

An area infested with fruit flies is to be sprayed with a chemical which is known to be 98% *effective for each application. Assume a sample of* 100 *flies is checked. Set up the calculation; don't solve.*

37. Find the probability that 95% of the flies are killed in one application.

38. Find the probability that at least 95% of the flies are killed in one application.

39. Find the probability that at least 90% of the flies are killed in one application.

40. Find the probability that all the flies are killed in one application.

A department store must control its inventory carefully.* It should not reorder too often, because it then builds up a large warehouse full of merchandise, which is expensive to hold. On the other hand, it must reorder sufficiently often to be sure of having sufficient stock to meet customer demand. The company desires a simple chart that can be used by its employees to determine the best possible time to reorder merchandise. The merchandise level on hand will be checked periodically. At the end of each period, if the level on hand is less than some predetermined level given in the chart, which considers sales rate and waiting time for orders, the item will be reordered. The example uses the following variables.

F = frequency of stock review (in weeks)

r = acceptable risk of being out of stock (in percent)

P = level of inventory at which reordering should occur

L = waiting time for order to arrive (in weeks)

S = sales rate (in units per week)

M = minimum level of merchandise to guarantee that the probability of being out of stock is no higher than r

z = z-score (from Table 8 at back of book) corresponding to r

---

*Example supplied by Leonard W. Cooper, Operations Research Project Director, Federated Department Stores.

Goods should not be reordered until inventory on hand has declined to a level less than the rate of sales per week, S, times the sum of the number of weeks until the next stock review, F, and the expected waiting time in weeks, L, plus a minimum level of merchandise, M, necessary to guarantee that the probability of being out of stock is no higher than r. That is, $P = S(F + L) + M$.

To find M, which depends on S, F, and L, assume that both sales and waiting time are normally distributed. With this assumption, using a formula from more advanced statistics courses gives $M = z\sqrt{2S(F + L)}$, where z is the z-score (from Table 8) corresponding to r, and $\sqrt{2S(F + L)}$ is the standard deviation of normally distributed deviations in sales rate and waiting time.

Combining these two formulas,

$$P = S(F + L) + z\sqrt{2S(F + L)}.$$

Suppose the firm wishes to be 95% sure of having goods to sell, so that $r = 5\%$. From Table 8, $z = 1.64$. Hence, for $r = 5\%$,

$$P = S(F + L) + 1.64\sqrt{2S(F + L)}.$$

Based on this formula, the chart on the facing page was prepared.

To use the chart, for example, if an item is reviewed every 4 weeks ($F = 4$), the waiting time for a reorder is 3 weeks ($L = 3$), and the rate of sales is 2 per week ($S = 2$), then the item should be reordered when the number on hand falls below 23.

Reorder Levels (*P*)

Reorder merchandise when inventory on hand falls below the levels given in the chart.

$r = 5\%$, $F = 1$ week $r = 5\%$, $F = 4$ weeks

Rate of Sales (S) (units/week)	Waiting Time in Weeks for Order (L)						Rate of Sales(S) (units/week)	Waiting Time in Weeks for Order (L)					
	1	2	3	4	5	6		1	2	3	4	5	6
9	28	39	50	61	71	81	1	10	12	13	15	16	17
10	30	43	55	66	78	89	2	17	20	23	25	28	30
11	33	46	59	72	85	97	3	24	28	32	35	39	43
12	35	50	64	78	92	105	4	30	35	40	45	50	55
13	38	54	69	84	99	113							
14	40	57	83	89	105	121							
15	43	61	78	95	112	129							

EXERCISES

1. Suppose an item is reviewed weekly, and the waiting time for a reorder is 4 weeks. If the average sales per week of the item is 12 units and the current inventory level is 85 units, should it be reordered? What if the inventory level is 50 units?

2. Suppose an item is reviewed every 4 weeks. If orders require a 5-week waiting time and if sales average 3 units per week, should the item be reordered if current inventory is 50 units? What if current inventory is 30?

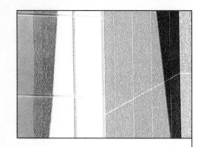

CHAPTER 11

Differential Calculus

The algebraic problems considered in earlier chapters dealt with *static* situations.

> What is the revenue when x items are sold?
>
> How much interest is earned in 2 years?
>
> What is the equilibrium price?

Calculus, on the other hand, deals with *dynamic* situations.

> At what rate is the economy growing?
>
> How fast is a rocket going at any instant after lift-off?
>
> How quickly can production be increased without adversely
>
> affecting profits?

The techniques of calculus will allow us to answer many questions like these that deal with rates of change.

The key idea underlying the development of calculus is the concept of limit. So we begin by studying limits.

11.1 LIMITS

It is easy to see that the value of the function

$$f(x) = \frac{x^2 - 4}{x - 2}$$

when $x = 1$ is the number $f(1) = -3/-1 = 3$. It is also true that when x is a number *very close* to 1 (on either side of 1), then $f(x)$ is a number *very close* to 3 as shown in the following table.

x	.99	.9999	1	1.0001	1.01
$f(x)$	2.99	2.9999	3	3.0001	3.01

1 Find $\lim\limits_{x \to 4} \dfrac{x^2 - 4}{x - 2}$ by completing the following table. A calculator would be helpful.

x	$f(x)$
3.8	
3.9	
3.99	
3.999	
4.001	
4.1	
4.2	

Answer:
5.8; 5.9; 5.99; 5.999; 6.001; 6.1; 6.2; limit is 6

At $x = 2$ the situation is different: $f(2)$ is *not defined* because substituting 2 in the rule of the function makes the denominator 0. But we can still ask, what happens to $f(x)$ when x is a number *very close* to (but not equal to) 2? The following table provides an answer.

x	1.9	1.99	1.9999	2	2.00001	2.001	2.05	2.1
$f(x)$	3.9	3.99	3.9999		4.00001	4.001	4.05	4.1

The table suggests that

as x gets closer and closer to 2 from either direction, the corresponding value of $f(x)$ gets closer and closer to 4.

In fact, by experimenting with a calculator you can convince yourself that the values of $f(x)$ can be made *as close as you want* to 4 by taking values of x close enough to 2. This situation is usually described by saying that "the *limit* of $f(x)$ as x approaches 2 is the number 4," which is written symbolically as

$$\lim_{x \to 2} f(x) = 4.$$

Similarly, we write $\qquad \lim\limits_{x \to 1} f(x) = 3,$

which is read "the limit of $f(x)$ as x approaches 1 is 3." This means that

as x takes values closer and closer to 1 from either direction, the corresponding values of $f(x)$ gets closer and closer to 3

and that

the values of $f(x)$ can be made *arbitrarily close* (as close as you want) to 3 by taking values of x close enough to 1. **1**

In the general case, we have this informal definition.

Limit of a Function

Let f be a function and let a and L be real numbers. Suppose that

as x takes values closer and closer (but not equal) to a (on both sides of a), the corresponding values of $f(x)$ get closer and closer (and possibly are equal) to L;

and that

the values of $f(x)$ can be made arbitrarily close to L by taking values of x close enough to a.

Then L is the **limit** of the function $f(x)$ as x approaches a, written:

$$\lim_{x \to a} f(x) = L.$$

This definition is *informal* because the expressions "closer and closer" and "arbitrarily close" have not been precisely defined. In particular, the charts used above provide strong intuitive confirmation, but not a rigorous proof, that the limits are what was claimed.

Several observations about limits are in order. First, the definition of limit deals with the behavior of a function f *near* a number a, without regard to whether or not $f(a)$ is defined. Second, the definition of limit implies that *if* a limit exists, it is necessarily unique (because the values of $f(x)$ cannot be arbitrarily close to two different numbers).

▶**EXAMPLE 1**   If $f(x) = x^2 + x + 1$, then $\lim\limits_{x \to 3} f(x)$ can be written $\lim\limits_{x \to 3} (x^2 + x + 1)$. Find this limit.

To find this limit, we make a chart showing the values of the function at numbers very close to 3.

	x approaches 3 from the left →			3	← *x* approaches 3 from the right		
x	2.9	2.99	2.9999	3	3.0001	3.01	3.1
$f(x)$	12.31	12.9301	12.9993 . . .		13.0007 . . .	13.0701	13.71

The chart strongly suggests that as x approaches 3 from either direction, $f(x)$ gets closer and closer to 13 and, hence, that $\lim\limits_{x \to 3} (x^2 + x + 1) = 13$. ◀

The function in Example 1 is defined at 3 and $f(3) = 3^2 + 3 + 1 = 13$. So the limit of $f(x)$ as x approaches 3 is $f(3)$, the value of the function at 3 that is

$$\lim_{x \to 3} f(x) = f(3).$$

It can be shown that what happened in Example 1 always occurs for *polynomial* functions.

If $f(x)$ is a polynomial function and a is a real number, then

$$\lim_{x \to a} f(x) = f(a).$$

In cases such as this, where the limit as x approaches a is the value of the function at a, the function f is said to be *continuous* at a. Continuous functions are discussed in Section 11.8.

2 If $f(x) = 2x^4 - 4x^3 + 3x$, find

(a) $\lim\limits_{x \to 2} f(x)$;

(b) $\lim\limits_{x \to -1} f(x)$.

Answer:

(a) 6

(b) 3

▶**EXAMPLE 2**   According to the fact in the box,

$$\lim_{x \to -1} (x^3 - 2x^2 - 8x - 3) = (-1)^3 - 2(-1)^2 - 8(-1) - 3 = 2.$$ ◀ **2**

Similarly, the limit of any constant polynomial as x approaches any constant a is the value of that constant.

$$\lim_{x \to a} 5 = 5 \quad \text{and} \quad \lim_{x \to a} (-7) = -7$$

When $f(x)$ is not a polynomial function and $f(a)$ is defined, it is possible that the limit of $f(x)$ as x approaches a may not be $f(a)$.

▶**EXAMPLE 3**   Let f be the function whose rule is

$$f(x) = \begin{cases} 0 \text{ if } x \text{ is an integer} \\ 1 \text{ if } x \text{ is not an integer} \end{cases}$$

and whose graph is shown in Figure 11.1.

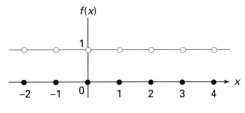

FIGURE 11.1

The graph shows that whenever x is a number very close to (but not equal to) 3, then $f(x) = 1$. Therefore, $\lim\limits_{x \to 3} f(x) = 1$. But since 3 is an integer, $f(3) = 0$. Therefore, $\lim\limits_{x \to 3} f(x) \neq f(3)$. ◀

It is possible that $\lim\limits_{x \to a} f(x)$ may not exist, that is, there may be no number L satisfying the definition. This can happen in several ways.

▶**EXAMPLE 4**   Let $g(x) = \dfrac{x^2 + 4}{x - 2}$ and find $\lim\limits_{x \to 2} g(x)$.

Make a table of values.

x approaches 2 from the left $\to 2 \leftarrow$ *x approaches 2 from the right*

x	1.8	1.9	1.99	1.999	2	2.001	2.01	2.05
$g(x)$	-36.2	-76.1	-796	-7996		8004	804	164
	g(x) gets smaller and smaller					*g(x) gets larger and larger*		

3 Let $f(x) = \dfrac{x^2 + 9}{x - 3}$.

Find the following.

(a) $\lim\limits_{x \to 3} f(x)$

(b) $\lim\limits_{x \to 0} f(x)$

Answer:

(a) Does not exist

(b) -3

The table above and the graph of $g(x)$ in Figure 11.2 show that as x approaches 2 from the left, $g(x)$ gets smaller and smaller, but as x approaches 2 from the right, $g(x)$ gets larger and larger. Since $g(x)$ does not get closer and closer to a single real number as x approaches 2 from either side,

$$\lim_{x \to 2} \frac{x^2 + 4}{x - 2} \text{ does not exist.} \qquad \blacktriangleleft \quad \boxed{3}$$

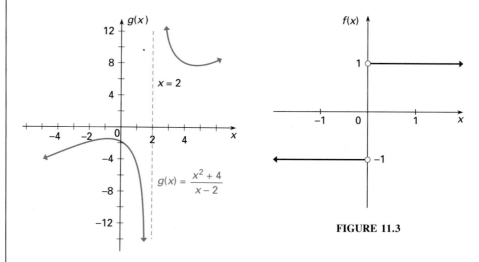

FIGURE 11.2

FIGURE 11.3

▶ **EXAMPLE 5**   What is $\lim\limits_{x \to 0} \dfrac{|x|}{x}$?

The function $f(x) = |x|/x$ is not defined when $x = 0$. When $x > 0$, then the definition of absolute value shows that $f(x) = |x|/x = x/x = 1$. When $x < 0$, then $|x| = -x$ and $f(x) = -x/x = -1$. The graph of f is shown in Figure 11.3. As x approaches 0 from the right, x is always positive and the corresponding value of $f(x)$ is 1. But as x approaches 0 from the left, x is always negative and the corresponding value of $f(x)$ is -1. Thus, as x approaches 0 from *both* sides, the corresponding values of $f(x)$ do not get closer and closer to a *single* real number. Therefore, the limit does not exist.* ◀

---

*In a situation like this, one sometimes says that -1 is the *limit of f(x) from the left* and that 1 is the *limit of f(x) from the right*. The concept of limit as we have defined it is sometimes called a *two-sided limit*.

The preceding examples illustrate the following facts.

Existence of Limits

1. The limit of a function f as x approaches a may fail to exist.
   (a) If $f(x)$ becomes infinitely large or infinitely small as x approaches a from either side, then the limit does not exist.
   (b) If $f(x)$ gets closer and closer to L as x approaches a from the left and $f(x)$ gets closer and closer to M as x approaches a from the right and $L \neq M$, then the limit does not exist.
2. If the limit does exist, it may not be equal to $f(a)$. In fact, $f(a)$ may not even be defined.
3. But for a polynomial function $p(x)$, the limit as x approaches a is always defined and is equal to $p(a)$.

The function f whose graph is shown in Figure 11.4 illustrates the facts listed in the box above.

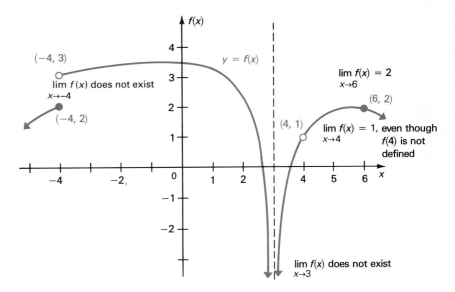

FIGURE 11.4

RULES FOR LIMITS   As shown by the examples above, tables and graphs can be used to find limits. However, it is usually more efficient to find limits algebraically by using the following rules for limits. The proofs of these rules are omitted since they require a rigorous definition of limit, which we have not given.

Rules for Limits

Let a, k, n, A, and B be real numbers, and let f and g be functions such that

$$\lim_{x \to a} f(x) = A \quad \text{and} \quad \lim_{x \to a} g(x) = B.$$

1. If k is a constant, then (a) $\lim_{x \to a} k = k$ and (b) $\lim_{x \to a} k \cdot f(x) = k \cdot \lim_{x \to a} f(x)$.

2. $\lim_{x \to a} [f(x) \pm g(x)] = \lim_{x \to a} f(x) \pm \lim_{x \to a} g(x) = A \pm B$

   (The limit of a sum or difference is the sum or difference of the limits.)

3. If $p(x)$ is a polynomial, then $\lim_{x \to a} p(x) = p(a)$.

4. $\lim_{x \to a} [f(x) \cdot g(x)] = [\lim_{x \to a} f(x)] \cdot [\lim_{x \to a} g(x)] = A \cdot B$

   (The limit of a product is the product of the limits.)

5. $\lim_{x \to a} \dfrac{f(x)}{g(x)} = \dfrac{\lim_{x \to a} f(x)}{\lim_{x \to a} g(x)} = \dfrac{A}{B}$ if $B \neq 0$.

   (The limit of a quotient is the quotient of the limits, provided the limit of the denominator is not zero.)

6. For any real number n for which A^n exists, $\lim_{x \to a} [f(x)]^n = [\lim_{x \to a} f(x)]^n = A^n$.

7. $\lim_{x \to a} f(x) = \lim_{x \to a} g(x)$ if $f(x) = g(x)$ for all $x \neq a$.

This list may seem imposing, but most limit problems have solutions that agree with your common sense. Algebraic techniques and tables of values can also help resolve questions about limits, as some of the following examples will show.

▶**EXAMPLE 6** Find each limit.

(a) $\lim_{x \to 2} [(x^2 + 1) + (x^3 - x + 3)]$

$$\lim_{x \to 2} [(x^2 + 1) + (x^3 - x + 3)]$$

$$= \lim_{x \to 2} (x^2 + 1) + \lim_{x \to 2} (x^3 - x + 3) \qquad \text{Rule 2}$$

$$= (2^2 + 1) + (2^3 - 2 + 3) = 5 + 9 = 14 \qquad \text{Rule 3}$$

4 Use the limit rules to find the following.

(a) $\lim\limits_{x \to 4} (3x - 9)$

(b) $\lim\limits_{x \to -1} (2x^2 - 4x + 1)$

(c) $\lim\limits_{x \to 2} \dfrac{x - 1}{3x + 2}$

(d) $\lim\limits_{x \to 2} \sqrt{3x + 3}$

Answer:

(a) 3   **(b)** 7

(c) 1/8   **(d)** 3

(b) $\lim\limits_{x \to -1} (x^3 + 4x)(2x^2 - 3x)$

$$\lim\limits_{x \to -1} (x^3 + 4x)(2x^2 - 3x)$$

$$= \lim\limits_{x \to -1} (x^3 + 4x) \cdot \lim\limits_{x \to -1} (2x^2 - 3x) \qquad \text{Rule 4}$$

$$= [(-1)^3 + 4(-1)] \cdot [2(-1)^2 - 3(-1)] \qquad \text{Rule 3}$$

$$= (-1 - 4)(2 + 3) = -25$$

(c) $\lim\limits_{x \to -1} 5(3x^2 + 2)$

$$\lim\limits_{x \to -1} 5(3x^2 + 2) = 5 \cdot \lim\limits_{x \to -1} (3x^2 + 2) \qquad \text{Rule 1}$$

$$= 5[3(-1)^2 + 2] \qquad \text{Rule 3}$$

$$= 25$$

(d) $\lim\limits_{x \to 4} \dfrac{x}{x + 2}$

$$\lim\limits_{x \to 4} \dfrac{x}{x + 2} = \dfrac{\lim\limits_{x \to 4} x}{\lim\limits_{x \to 4} (x + 2)} \qquad \text{Rule 5}$$

$$= \dfrac{4}{4 + 2} = \dfrac{2}{3} \qquad \text{Rule 3}$$

(e) $\lim\limits_{x \to 9} \sqrt{4x - 11}$

As $x \to 9$, the expression $4x - 11$ approaches $4 \cdot 9 - 11 = 25$. Using Rule 6, with $n = 1/2$, gives

$$\lim\limits_{x \to 9} (4x - 11)^{1/2} = [\lim\limits_{x \to 9} (4x - 11)]^{1/2} = \sqrt{25} = 5. \quad \blacktriangleleft \quad \boxed{4}$$

Rule 7 for limits reflects the fact that the limit as x approaches a involves only the values of the function *near a*, but not *at a*. So two functions that agree for all values of x except $x = a$ will necessarily have the same limit at a. This fact and some algebraic manipulation can be used to evaluate many limits involving quotients.

▶ **EXAMPLE 7**   Find $\lim\limits_{x \to 2} \dfrac{x^2 + x - 6}{x - 2}$.

Rule 5 cannot be used here, since

$$\lim\limits_{x \to 2} (x - 2) = 0.$$

We can, however, simplify the function by rewriting the fraction as

$$\dfrac{x^2 + x - 6}{x - 2} = \dfrac{(x + 3)(x - 2)}{x - 2}.$$

5 Use the limit rules to find

$$\lim_{x\to 1} \frac{2x^2 + x - 3}{x - 1}.$$

Answer:
5

6 Find the following.

(a) $\lim\limits_{x\to 1} \dfrac{\sqrt{x} - 1}{x - 1}$

(b) $\lim\limits_{x\to 9} \dfrac{\sqrt{x} - 3}{x - 9}$

Answer:

(a) 1/2

(b) 1/6

When $x \neq 2$, the quantity $x - 2$ is nonzero and may be cancelled, so that

$$\frac{x^2 + x - 6}{x - 2} = x + 3 \quad \text{for all } x \neq 2.$$

Now Rule 7 can be used.

$$\lim_{x\to 2} \frac{x^2 + x - 6}{x - 2} = \lim_{x\to 2} (x + 3) = 2 + 3 = 5 \quad \blacktriangleleft \;\boxed{5}$$

▶**EXAMPLE 8** Find $\lim\limits_{x\to 4} \dfrac{\sqrt{x} - 2}{x - 4}$.

As $x \to 4$, the numerator approaches 0 and the denominator also approaches 0, giving the meaningless expression 0/0. To change the form of the expression, algebra can be used to rationalize the numerator by multiplying both the numerator and the denominator by $\sqrt{x} + 2$. This gives

$$\frac{\sqrt{x} - 2}{x - 4} = \frac{\sqrt{x} - 2}{x - 4} \cdot \frac{\sqrt{x} + 2}{\sqrt{x} + 2} = \frac{\sqrt{x} \cdot \sqrt{x} - 2\sqrt{x} + 2\sqrt{x} - 4}{(x - 4)(\sqrt{x} + 2)}$$

$$= \frac{x - 4}{(x - 4)(\sqrt{x} + 2)} = \frac{1}{\sqrt{x} + 2}$$

for all $x \neq 4$. Now use rules for limits.

$$\lim_{x\to 4} \frac{\sqrt{x} - 2}{x - 4} = \lim_{x\to 4} \frac{1}{\sqrt{x} + 2} = \frac{1}{\sqrt{4} + 2} = \frac{1}{2 + 2} = \frac{1}{4} \quad \blacktriangleleft \;\boxed{6}$$

▶**EXAMPLE 9** Find $\lim\limits_{x\to 1} \dfrac{x + 1}{x^2 - 1}$.

Again, Rule 5 cannot be used since $\lim x^2 - 1 = 0$. However,

$$\frac{x + 1}{x^2 - 1} = \frac{x + 1}{(x + 1)(x - 1)} = \frac{1}{x - 1} \quad \text{for all } x \neq 1.$$

Hence,

$$\lim_{x\to 1} \frac{x + 1}{x^2 - 1} = \lim_{x\to 1} \frac{1}{x - 1}$$

by Rule 7. None of the rules can be used to find

$$\lim_{x\to 1} \frac{1}{x - 1}.$$

7 Find $\lim\limits_{x \to 2} \dfrac{2}{3x - 6}$.

Answer:
Does not exist

A calculator shows that the values of $1/(x - 1)$ are increasing positive numbers as x approaches 1 from the right; however, the values of $1/(x - 1)$ are decreasing negative numbers as x approaches 1 from the left. Therefore,

$$\lim_{x \to 1} \frac{1}{x - 1} \text{ does not exist.} \quad \blacktriangleleft \quad \boxed{7}$$

LIMITS AT INFINITY Sometimes it is useful to examine the behavior of the values of $f(x)$ as x gets larger and larger (or smaller and smaller). For example, suppose a small pond normally contains 12 units of dissolved oxygen in a fixed volume of water. Suppose also that at time $t = 0$ a quantity of organic waste is introduced into the pond, with oxygen concentration t weeks later given by

$$f(t) = \frac{12t^2 - 15t + 12}{t^2 + 1}.$$

As time goes on, what will be the ultimate concentration of oxygen? Will it return to 12 units? After 2 weeks, the pond contains

$$f(2) = \frac{12 \cdot 2^2 - 15 \cdot 2 + 12}{2^2 + 1} = \frac{30}{5} = 6$$

units of oxygen, and after 4 weeks, it contains

$$f(4) = \frac{12 \cdot 4^2 - 15 \cdot 4 + 12}{4^2 + 1} \approx 8.5$$

units. Choosing several values of t and finding the corresponding values of $f(t)$ leads to the graph in Figure 11.5.

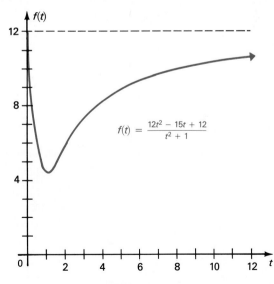

$$f(t) = \frac{12t^2 - 15t + 12}{t^2 + 1}$$

FIGURE 11.5

The graph suggests that as time goes on, the oxygen level gets closer and closer to the original 12 units. Consider a table of values as t gets larger and larger.

t	10	100	1000	10,000	100,000
$f(t)$	10.5	11.85	11.985	11.9985	11.9999

The table suggests that

as t takes larger and larger values, beyond all bounds, the corresponding values of $f(t)$ get closer and closer to 12,

and that

the values of $f(t)$ can be made arbitrarily close to 12 by taking large enough values of t.

We express these facts by writing

$$\lim_{x \to \infty} f(t) = 12,$$

which is read "the limit of $f(t)$ as t approaches infinity is 12." In terms of the pond, this means that the oxygen concentration will approach 12, but never be exactly 12.

The preceding discussion is an example of a **limit at infinity.** The phrase "t approaches infinity" (symbolically, $t \to \infty$) is simply convenient shorthand to express the fact that t takes larger and larger values without bound. Similarly, the phrase "t approaches negative infinity" (symbolically, $t \to -\infty$) means that t takes smaller and smaller values without bound (such as -100, -1000, $-10,000$, etc.), in which case, the function *may* have a limit.

The graphs of $f(x) = 1/x$ (in black) and $f(x) = 1/x^2$ (in color) and the table in Figure 11.6 suggest that

$$\lim_{x \to \infty} \frac{1}{x} = 0, \quad \lim_{x \to -\infty} \frac{1}{x} = 0, \quad \lim_{x \to \infty} \frac{1}{x^2} = 0, \quad \lim_{x \to -\infty} \frac{1}{x^2} = 0.$$

These are the cases $n = 1$ and $n = 2$ of the following rule.

Limits at Infinity

For any positive integer n,

$$\lim_{x \to \infty} \frac{1}{x^n} = 0 \quad \text{and} \quad \lim_{x \to -\infty} \frac{1}{x^n} = 0.$$

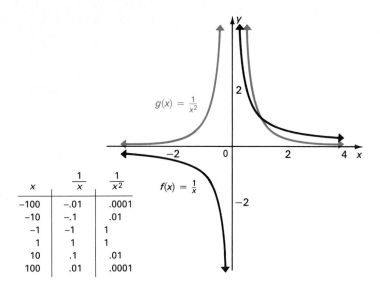

FIGURE 11.6

All the rules for limits given earlier in this section (except Rule 3) remain valid when a is replaced by ∞ or $-\infty$.

▶**EXAMPLE 10** Find each limit.

(a) $\lim\limits_{x\to\infty} \dfrac{8x + 6}{3x - 1}$

First, we divide both numerator and denominator by x, which does not change the value of the quotient.

$$\lim_{x\to\infty} \frac{8x + 6}{3x - 1} = \lim_{x\to\infty} \frac{\dfrac{8x}{x} + \dfrac{6}{x}}{\dfrac{3x}{x} - \dfrac{1}{x}}$$

$$= \lim_{x\to\infty} \frac{8 + 6 \cdot \dfrac{1}{x}}{3 - \dfrac{1}{x}}$$

Now apply the limit rules and the fact that $\lim\limits_{x\to\infty} 1/x^n = 0$.

8 Find the following limits.

(a) $\displaystyle\lim_{x \to \infty} \frac{4x^2 + 6x}{9x^2 - 3}$

(b) $\displaystyle\lim_{x \to \infty} \frac{3x^2 + 4x + 5}{8x^3 + 2x + 1}$

(c) $\displaystyle\lim_{x \to \infty} \frac{4x^2 + 9x + 1}{2x + 3}$

Answer:

(a) 4/9

(b) 0

(c) Does not exist

$$\frac{\displaystyle\lim_{x \to \infty}\left(8 + 6 \cdot \frac{1}{x}\right)}{\displaystyle\lim_{x \to \infty}\left(3 - \frac{1}{x}\right)} = \frac{\displaystyle\lim_{x \to \infty} 8 + \lim_{x \to \infty} 6 \cdot \frac{1}{x}}{\displaystyle\lim_{x \to \infty} 3 - \lim_{x \to \infty} \frac{1}{x}} \qquad \text{Rules 5 and 2}$$

$$= \frac{8 + 6\left(\displaystyle\lim_{x \to \infty} \frac{1}{x}\right)}{3 - \displaystyle\lim_{x \to \infty} \frac{1}{x}} = \frac{8 + 6(0)}{3 - 0} = \frac{8}{3} \qquad \begin{array}{l}\text{Rule 1, limits} \\ \text{at infinity}\end{array}$$

(b) $\displaystyle\lim_{x \to -\infty} \frac{4x^2 - 6x + 3}{2x^2 - x + 4}$

Divide each term of the numerator and denominator by x^2, the highest power of x, and proceed as in part (a).

$$\lim_{x \to -\infty} \frac{4x^2 - 6x + 3}{2x^2 - x + 4} = \lim_{x \to -\infty} \frac{4 - 6 \cdot \dfrac{1}{x} + 3 \cdot \dfrac{1}{x^2}}{2 - \dfrac{1}{x} + 4 \cdot \dfrac{1}{x^2}}$$

$$= \frac{4 - 0 + 0}{2 - 0 + 0} = \frac{4}{2} = 2$$

(c) $\displaystyle\lim_{x \to \infty} \frac{3x + 2}{4x^3 - 1} = \lim_{x \to \infty} \frac{3 \cdot \dfrac{1}{x^2} + 2 \cdot \dfrac{1}{x^3}}{4 - \dfrac{1}{x^3}} = \frac{0 + 0}{4 - 0} = \frac{0}{4} = 0$$

Here, the highest power of x is x^3, which is used to divide each term in numerator and denominator.

(d) $\displaystyle\lim_{x \to \infty} \frac{3x^2 + 2}{4x - 3} = \lim_{x \to \infty} \frac{3 + \dfrac{2}{x^2}}{\dfrac{4}{x} - \dfrac{3}{x^2}} = \frac{3 + 0}{0 - 0} = \frac{3}{0}$$

Division by 0 is undefined, so the limit does not exist. ◄ **8**

The method used in Example 10 is a useful way to rewrite expressions with fractions so that the rules for limits at infinity can be used.

Finding Limits at Infinity

If $f(x) = \dfrac{p(x)}{q(x)}$, for polynomials $p(x)$ and $q(x)$, $q(x) \neq 0$, $\lim\limits_{x \to -\infty} f(x)$ and $\lim\limits_{x \to \infty} f(x)$ can be found as follows.

1. Divide $p(x)$ and $q(x)$ by the highest power of x in either polynomial.
2. Use the rules for limits, including the rules for limits at infinity,

$$\lim_{x \to \infty} \frac{1}{x^n} = 0 \quad \text{and} \quad \lim_{x \to -\infty} \frac{1}{x^n} = 0,$$

to find the limit of the result from step 1.

Limits at infinity can be used to find horizontal asymptotes when graphing rational functions.

▶ **EXAMPLE 11**   Graph $f(x) = \dfrac{3x^2}{x^2 + 5}$.

There is no vertical asymptote, because $x^2 + 5 \neq 0$ for all values of x. Find any horizontal asymptote by calculating $\lim\limits_{x \to \infty} f(x)$ and $\lim\limits_{x \to -\infty} f(x)$. First, divide both the numerator and the denominator of $f(x)$ by x^2.

$$\lim_{x \to \infty} \frac{3x^2}{x^2 + 5} = \lim_{x \to \infty} \frac{\dfrac{3x^2}{x^2}}{\dfrac{x^2}{x^2} + \dfrac{5}{x^2}} = \frac{3}{1 + 0} = 3$$

Verify that the limit of $f(x)$ as $x \to -\infty$ is also 3. Thus, the horizontal asymptote is $y = 3$. Plot a few points (several are needed near the origin) including the intercept at the origin, and use the fact that the graph approaches the line $y = 3$ as $x \to \infty$ and also as $x \to -\infty$, to get the graph that is shown in Figure 11.7.   ◀

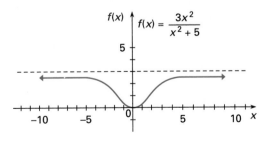

FIGURE 11.7

11.1 EXERCISES

In each of the following, use the graph to determine the value of the indicated limits. (See Figure 11.4).

1. (a) $\lim\limits_{x \to 3} f(x)$;   **(b)** $\lim\limits_{x \to -1.5} f(x)$

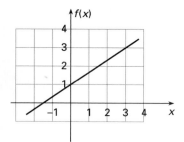

2. (a) $\lim\limits_{x \to 2} F(x)$ **(b)** $\lim\limits_{x \to -1} F(x)$

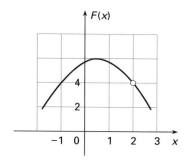

3. (a) $\lim\limits_{x \to -2} f(x)$ **(b)** $\lim\limits_{x \to 1} f(x)$

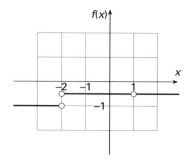

4. (a) $\lim\limits_{x \to -1} g(x)$ **(b)** $\lim\limits_{x \to 3} g(x)$

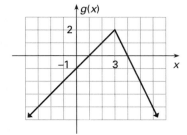

5. (a) $\lim\limits_{x \to 0} f(x)$ **(b)** $\lim\limits_{x \to -1} f(x)$

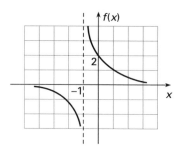

6. (a) $\lim\limits_{x \to 1} h(x)$ **(b)** $\lim\limits_{x \to 2} h(x)$

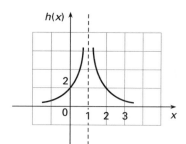

7. (a) $\lim\limits_{x \to 1} g(x)$ **(b)** $\lim\limits_{x \to -1} g(x)$

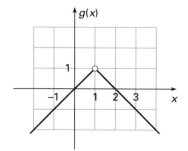

8. (a) $\lim\limits_{x \to \infty} f(x)$ **(b)** $\lim\limits_{x \to 0} f(x)$

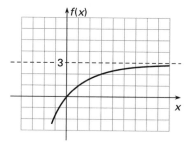

9. (a) $\lim\limits_{x \to \infty} g(x)$ **(b)** $\lim\limits_{x \to 0} g(x)$

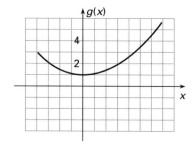

In each of the following, use a calculator to complete the table and estimate the limit (See Examples 1–4).

10. $f(x) = \ln x$; find $\lim\limits_{x \to 6} f(x)$.

x	5.9	5.99	5.999	5.9999	6.0001	6.001	6.01	6.1
$f(x)$								

11. $f(x) = \ln \sqrt{x + 1}$; find $\lim\limits_{x \to 3} f(x)$.

x	2.9	2.99	2.999	2.9999	3.0001	3.001	3.01	3.1
$f(x)$								

12. $f(x) = e^x$; find $\lim\limits_{x \to 2} f(x)$.

x	1.9	1.99	1.999	1.9999	2.0001	2.001	2.01	2.1
$f(x)$								

13. $f(x) = e^{3x+2}$; find $\lim\limits_{x \to -1} f(x)$.

x	-1.1	-1.01	-1.001	-1.0001	$-.9999$	$-.999$	$-.99$	$-.9$
$f(x)$								

Use algebra and the rules for limits as needed to find the following limits. If the limit does not exist, say so. (See Examples 5–9.)

14. $\lim\limits_{x \to 2} (2x^3 + 5x^2 + 2x + 1)$

15. $\lim\limits_{x \to -1} (4x^3 - x^2 + 3x - 1)$

16. $\lim\limits_{x \to 3} \dfrac{5x - 6}{2x + 1}$

17. $\lim\limits_{x \to -2} \dfrac{2x + 1}{3x - 4}$

18. $\lim\limits_{x \to 1} \dfrac{2x^2 - 6x + 3}{3x^2 - 4x + 2}$

19. $\lim\limits_{x \to 2} \dfrac{-4x^2 + 6x - 8}{3x^2 + 7x - 2}$

20. $\lim\limits_{x \to 3} \dfrac{x^2 - 9}{x - 3}$

21. $\lim\limits_{x \to -2} \dfrac{x^2 - 4}{x + 2}$

22. $\lim\limits_{x \to -2} \dfrac{x^2 - x - 6}{x + 2}$

23. $\lim\limits_{x \to 5} \dfrac{x^2 - 3x - 10}{x - 5}$

24. $\lim\limits_{x \to 2} \dfrac{x^2 - 5x + 6}{x^2 - 6x + 8}$

25. $\lim\limits_{x \to -2} \dfrac{x^2 + 3x + 2}{x^2 - x - 6}$

26. $\lim\limits_{x \to 4} \dfrac{x^2 - 7x + 12}{x^2 - 8x + 16}$

27. $\lim\limits_{x \to -3} \dfrac{x^2 + 5x + 6}{x^2 + 2x - 3}$

28. $\lim\limits_{x \to 4} \dfrac{(x + 4)^2(x - 5)}{(x - 4)(x + 4)^2}$

29. $\lim\limits_{x \to -3} \dfrac{(x + 3)(x - 3)(x + 4)}{(x + 8)(x + 3)(x - 4)}$

30. $\lim\limits_{x \to 3} \sqrt{x^2 - 4}$

31. $\lim\limits_{x \to 3} \sqrt{x^2 - 5}$

32. $\lim\limits_{x \to 4} \dfrac{-6}{(x - 4)^2}$

33. $\lim\limits_{x \to -2} \dfrac{3x}{(x + 2)^3}$

34. $\lim\limits_{x \to 0} \dfrac{x^3 - 4x^2 + 8x}{2x}$

35. $\lim\limits_{x \to 0} \dfrac{-x^5 - 9x^3 + 8x^2}{5x}$

36. $\lim\limits_{x \to 0} \dfrac{[1/(x + 3)] - 1/3}{x}$

37. $\lim\limits_{x \to 0} \dfrac{[-1/(x + 2)] + 1/2}{x}$

38. $\lim\limits_{x \to 25} \dfrac{\sqrt{x} - 5}{x - 25}$

39. $\lim\limits_{x \to 36} \dfrac{\sqrt{x} - 6}{x - 36}$

40. $\lim\limits_{x \to 5} \dfrac{\sqrt{x} - \sqrt{5}}{x - 5}$

41. $\lim\limits_{x \to 8} \dfrac{\sqrt{x} - \sqrt{8}}{x - 8}$

42. $\lim\limits_{x \to 0} \left(\dfrac{|x|}{x} - \dfrac{x}{|x|} \right)$

43. $\lim\limits_{x \to 2} \dfrac{(x - 2)^2}{|x - 2|}$

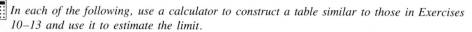

In each of the following, use a calculator to construct a table similar to those in Exercises 10–13 and use it to estimate the limit.

44. $\lim\limits_{x \to 1} \dfrac{\ln x}{x - 1}$

45. $\lim\limits_{x \to 3} \dfrac{\ln x - \ln 3}{x - 3}$

46. $\lim\limits_{x \to 0} \dfrac{e^{2x} - 1}{x}$

47. $\lim\limits_{x \to 0} (x \cdot \ln |x|)$

Find each of the following limits that exist. (See Example 10.)

48. $\lim\limits_{x \to \infty} \dfrac{3x}{5x - 1}$

49. $\lim\limits_{x \to \infty} \dfrac{5x}{3x - 1}$

50. $\lim\limits_{x \to \infty} \dfrac{2x + 3}{4x - 7}$

51. $\lim\limits_{x \to \infty} \dfrac{8x + 2}{2x - 5}$

52. $\lim\limits_{x \to \infty} \dfrac{x^2 + 2x}{2x^2 - 2x + 1}$

53. $\lim\limits_{x \to \infty} \dfrac{x^2 + 2x - 5}{3x^2 + 2}$

54. $\lim\limits_{x \to \infty} \dfrac{3x^3 + 2x - 1}{2x^4 - 3x^3 - 2}$

55. $\lim\limits_{x \to \infty} \dfrac{2x^2 + 11x - 10}{5x^3 + 3x^2 + 2x}$

56. $\lim\limits_{x \to \infty} \dfrac{2x^4 + 3x + 4}{x^2 - 4x + 1}$

57. $\lim\limits_{x \to \infty} \dfrac{2x^2 - 1}{3x^4 + 2}$

---

58. Management The graph below shows the profit from the daily production of x thousand kilograms of an industrial chemical. Use the graph to find the following limits.

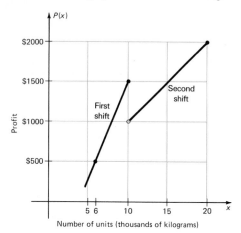

(a) $\lim\limits_{x \to 6} P(x)$ **(b)** $\lim\limits_{x \to 10} P(x)$ **(c)** $\lim\limits_{x \to 15} P(x)$

(d) Use the graph to estimate the number of units of the chemical that must be produced before the second shift is beneficial.

59. Management The cost for manufacturing a particular videotape is

$$c(x) = 15{,}000 + 6x,$$

where x is the number of tapes produced. The average cost per tape, denoted by $\bar{c}(x)$, is found by dividing $c(x)$ by x. Find the following.

(a) $\bar{c}(1000)$ **(b)** $\bar{c}(100{,}000)$

(c) $\lim\limits_{x \to 10{,}000} \bar{c}(x)$ **(d)** $\lim\limits_{x \to \infty} \bar{c}(x)$

60. Management A company training program has determined that, on the average, a new employee can do $P(s)$ pieces of work per day after s days of on-the-job training, where

$$P(s) = \frac{75s}{s + 8}.$$

Find the following.

(a) $P(1)$ **(b)** $P(11)$ **(c)** $\lim\limits_{s \to 11} P(s)$ **(d)** $\lim\limits_{s \to \infty} P(s)$

61. Management When the price of an essential commodity (such as gasoline) rises rapidly, consumption drops slowly at first. If the price continues to rise, however, a "tipping" point may be reached, at which consumption takes a sudden, substantial drop. Suppose the accompanying graph shows the consumption of gasoline, $G(t)$, in millions of gallons, in a certain area. We assume that the price is rising rapidly. Here t is time in months after the price began rising. Use the graph to find the following.

(a) $\lim\limits_{t \to 12} G(t)$ **(b)** $\lim\limits_{t \to 16} G(t)$ **(c)** $G(16)$

(d) The tipping point (in months)

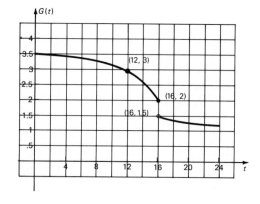

62. Natural Science The concentration of a drug in a patient's bloodstream h hours after it was injected is given by

$$A(h) = \frac{.17h}{h^2 + 2}.$$

Find the following.
(a) $A(.5)$ **(b)** $A(1)$ **(c)** $\lim_{h \to 1} A(h)$ **(d)** $\lim_{h \to \infty} A(h)$

63. Political Science Members of a legislature must often vote repeatedly on the same issue. As time goes on, the members may change their vote. A formula* for the chance that a legislator will vote *yes* on the nth roll call vote on the same issue is

$$p_n = \frac{1}{2} + \left(p_0 - \frac{1}{2}\right)(1 - 2p)^n,$$

---

*See John W. Bishir and Donald W. Drewes, *Mathematics in the Behavioral and Social Sciences* (New York: Harcourt Brace Jovanovich, 1970), p. 538.

where n is the number of roll calls taken, p_n is the chance that the member will vote *yes* on the nth roll call vote, p is the chance that the legislator will change his or her position on successive roll calls, and p_0 is the chance that the member favors the issue at the beginning (p_n, p, and p_0 are always between 0 and 1, inclusive). Suppose that $p_0 = .7$ and $p = .2$. Find p_n for
(a) $n = 2$; **(b)** $n = 4$; **(c)** $n = 8$.
(d) Find $\lim_{n \to \infty} p_n$.

Use the limit rules and a calculator (where necessary) to find the following limits.

64. $\lim_{x \to 1.3} \dfrac{x^3 - 2x^2 + 5x - 8}{2x^4 - 3x^2}$

65. $\lim_{x \to 5.2} \dfrac{x^4 - 6x^3 - 10x + 5}{4x^3 + 6x^2 - 12x - 6}$

66. $\lim_{x \to 2.5} \dfrac{x^3 - 4x^2 + 2x}{x^4 + 5x^3 + 4x}$

67. $\lim_{x \to 1.89} \dfrac{3.1x^3 + 5.2\sqrt{x}}{-2x^3 + 12.3x - 2\sqrt{x}}$

11.2 RATES OF CHANGE

One of the main applications of calculus is determining how one variable changes in relation to another. A person in business wants to know how profit changes with respect to advertising, while a person in medicine wants to know how a patient's reaction to a drug changes with respect to the dose.

We begin the discussion with a familiar situation. A driver makes the 168-mile trip from Cleveland to Columbus, Ohio, in 3 hours. The following table shows how far the driver has traveled from Cleveland at various times.

Time (in hours)	0	.5	1	1.5	2	2.5	3
Distance (in miles)	0	22	52	86	118	148	168

If f is the function whose rule is

$$f(x) = \text{distance from Cleveland at time } x,$$

then the table shows, for example, that $f(2) = 118$ and $f(3) = 168$. So the distance traveled from time $x = 2$ to $x = 3$ is $168 - 118$, that is, $f(3) - f(2)$. In a similar fashion, we obtain the other entries in this chart.

1 Find the average speed

(a) from $t = 1.5$ to $t = 2$;

(b) from $t = s$ to $t = r$.

Answer:

(a) 64 mph

(b) $\dfrac{f(r) - f(s)}{r - s}$

Time Interval	Distance Traveled
$x = 2$ to $x = 3$	$f(3) - f(2) = 168 - 118 = 50$
$x = 1$ to $x = 3$	$f(3) - f(1) = 168 - 52 = 116$
$x = 0$ to $x = 2.5$	$f(2.5) - f(0) = 148 - 0 = 148$
$x = .5$ to $x = 2$	$f(2) - f(.5) = 118 - 22 = 96$
$x = a$ to $x = b$	$f(b) - f(a)$

The last line of the chart shows how to find the distance traveled in any time interval $(0 \leq a < b \leq 3)$.

Since distance = average speed × time,

$$\text{Average speed} = \frac{\text{distance traveled}}{\text{time interval}}.$$

In the chart above, you can compute the length of each time interval by taking the difference between the two times. Thus, for example, from $x = 1$ to $x = 3$ is a time interval of length $3 - 1 = 2$ hours and, hence, the average speed over this interval is $116/2 = 58$ mph. Similarly, we have the following information.

Time Interval	Average Speed $= \dfrac{\text{Distance Traveled}}{\text{Time Interval}}$
$x = 2$ to $x = 3$	$\dfrac{f(3) - f(2)}{3 - 2} = \dfrac{168 - 118}{3 - 2} = \dfrac{50}{1} = 50$ mph
$x = 1$ to $x = 3$	$\dfrac{f(3) - f(1)}{3 - 1} = \dfrac{168 - 52}{3 - 1} = \dfrac{116}{2} = 58$ mph
$x = 0$ to $x = 2.5$	$\dfrac{f(2.5) - f(0)}{2.5 - 0} = \dfrac{148 - 0}{2.5 - 0} = \dfrac{148}{2.5} = 59.2$ mph
$x = .5$ to $x = 2$	$\dfrac{f(2) - f(.5)}{2 - .5} = \dfrac{118 - 22}{2 - .5} = \dfrac{96}{1.5} = 64$ mph
$x = a$ to $x = b$	$\dfrac{f(b) - f(a)}{b - a}$ mph

The last line of the chart shows how to compute the average speed over any time interval $(0 \leq a < b \leq 3)$. **1**

Now speed (miles per hour) is simply the *rate of change* of distance with respect to time and what was done for the distance function f in the preceding discussion can be done with any function.

2 Find the average rate of change of $f(x)$ in Example 1 when x changes from

(a) 0 to 4;

(b) 2 to 7.

Answer:

(a) 8

(b) 13

Quantity	Meaning for the Distance Function	Meaning for an Arbitrary Function f
$b - a$	time interval = change in time from $x = a$ to $x = b$	change in x from $x = a$ to $x = b$
$f(b) - f(a)$	distance traveled = corresponding change in distance as time changes from a to b	corresponding change in $f(x)$ as x changes from a to b
$\dfrac{f(b) - f(a)}{b - a}$	average speed = average rate of change of distance with respect to time as time changes from a to b	**average rate of change** of $f(x)$ with respect to x as x changes from a to b (where $a < b$)

▶**EXAMPLE 1**   If $f(x) = x^2 + 4x + 5$, find the average rate of change of $f(x)$ with respect to x as x changes from -2 to 3.

This is the situation described in the last line of the chart above, with $a = -2$ and $b = 3$. The average rate of change is

$$\frac{f(3) - f(-2)}{3 - (-2)} = \frac{26 - 1}{5} = \frac{25}{5} = 5. \quad ◀ \quad \boxed{2}$$

▶**EXAMPLE 2**   The graph in Figure 11.8 shows the profit $P(x)$, in hundreds of thousands of dollars, from a popular new computer program x months after its introduction on the market.

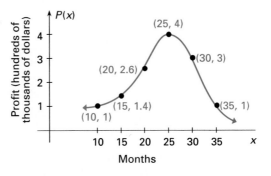

FIGURE 11.8

3 Use Figure 11.8 and find the average rate of change of profit with respect to time if time changes from

(a) $x = 15$ to $x = 20$;

(b) $x = 20$ to $x = 25$;

(c) $x = 25$ to $x = 35$;

(d) $x = 30$ to $x = 35$.

Answer:

(a) .24 hundred thousand, or $24,000

(b) .28 hundred thousand, or $28,000

(c) $-.3$ hundred thousand, or $-$30,000

(d) $-.4$ hundred thousand, or $-$40,000

(a) The graph shows that $P(25) = 4$ and $P(15) = 1.4$. So the average rate of change of profit from the 15th to the 25th month is

$$\frac{P(25) - P(15)}{25 - 15} = \frac{4 - 1.4}{10} = \frac{2.6}{10} = .26.$$

Thus, the average rate of change is .26 hundred thousand dollars, that is, $26,000 per month.

(b) From 25 months to 30 months, the average rate of change is

$$\frac{P(30) - P(25)}{30 - 25} = \frac{3 - 4}{5} = \frac{-1}{5} = -.20,$$

with each month in this interval showing an average *decline* in profits of .20 hundred thousand dollars, or $20,000.

(c) From 10 to 35 months, the average rate of change of profit is

$$\frac{P(35) - P(10)}{35 - 10} = \frac{1 - 1}{25} = \frac{0}{25} = 0. \quad \blacktriangleleft \quad \boxed{3}$$

INSTANTANEOUS RATE OF CHANGE   Suppose a car is stopped at a traffic light. When the light turns green, the car begins to move along a straight road. Assume that the distance traveled by the car is given by the function

$$s(t) = 2t^2 \quad (0 \le t \le 30),$$

where time t is measured in seconds and the distance $s(t)$ at time t is measured in feet.

It is easy to see that the speed of the car must be steadily increasing. For instance, in the first 5 seconds it travels only $2 \cdot 5^2 = 50$ feet, but in the first 10 seconds it travels $2 \cdot 10^2 = 200$ feet.

We know how to find the *average* speed of the car over any time interval, so we now turn to a different problem: determining the *exact* speed of the car at a particular instant, say $t = 10$.*

The intuitive idea is that the exact speed at $t = 10$ is very close to the average speed over a very short time interval near $t = 10$. If we take shorter and shorter time intervals near $t = 10$, the average speeds over these intervals should get closer and closer to the exact speed at $t = 10$. In other words,

> the exact speed at $t = 10$ is the limit of the average speeds over shorter and shorter time intervals near $t = 10$.

The following chart illustrates this idea.

---

*Since distance is measured in feet and time in seconds here, speed is measured in feet per second. It may help to know that 15 mph is equivalent to 22 ft/sec and 60 mph to 88 ft/sec.

Interval	Average Speed
$t = 10$ to $t = 10.1$	$\dfrac{s(10.1) - s(10)}{10.1 - 10} = \dfrac{204.02 - 200}{.1} = 40.2$
$t = 10$ to $t = 10.01$	$\dfrac{s(10.01) - s(10)}{10.1 - 10} = \dfrac{200.4002 - 200}{.01} = 40.02$
$t = 10$ to $t = 10.001$	$\dfrac{s(10.001) - s(10)}{10.001 - 10} = \dfrac{200.040002 - 200}{.001} = 40.002$

The chart suggests that the exact speed at $t = 10$ is 40 ft/sec. We can confirm this intuition by computing the average speed from $t = 10$ to $t = 10 + h$, where h is any very small nonzero number. (The chart does this for $h = .1$, $h = .01$, and $h = .001$). The average speed from $t = 10$ to $t = 10 + h$ is

$$\frac{s(10 + h) - s(10)}{(10 + h) - 10} = \frac{s(10 + h) - s(10)}{h}$$

$$= \frac{2(10 + h)^2 - 2 \cdot 10^2}{h}$$

$$= \frac{2(100 + 20h + h^2) - 200}{h}$$

$$= \frac{200 + 40h + 2h^2 - 200}{h}$$

$$= \frac{40h + 2h^2}{h} = \frac{h(40 + 2h)}{h}$$

$$= 40 + 2h.$$

Saying that the time interval gets shorter and shorter is equivalent to saying h gets closer and closer to 0. Hence, the exact speed at $t = 10$ is the limit as h approaches 0 of the average speeds over the intervals from $t = 10$ to $t = 10 + h$; that is,

$$\lim_{h \to 0} \frac{s(10 + h) - s(10)}{h} = \lim_{h \to 0} (40 + 2h)$$

$$= 40 \text{ ft/sec.}$$

The preceding example can easily be generalized. Suppose an object is moving in a straight line with its position (distance from some fixed point) at time t given by the function $s(t)$. The speed of the object is called its **velocity** and its exact speed at time t is called the **instantaneous velocity at time t** (or just velocity at time t).

Let t_0 be a constant. By replacing 10 by t_0 in the discussion above, we see that the average velocity of the object from time $t = t_0$ to time $t = t_0 + h$ is the quotient

$$\frac{s(t_0 + h) - s(t_0)}{(t_0 + h) - t_0} = \frac{s(t_0 + h) - s(t_0)}{h}.$$

4 In Example 3, if $s(t) = s^2 + 3$, find

(a) the average velocity from 1 second to 5 seconds;

(b) the instantaneous velocity at 5 seconds.

Answer:

(a) 6 ft per second

(b) 10 ft per second

The instantaneous velocity at time t is the limit of this quotient as h approaches 0.

Velocity

If an object moves along a straight line, with position $s(t)$ at time t, then the velocity of the object at $t = t_0$ is

$$\lim_{h \to 0} \frac{s(t_0 + h) - s(t_0)}{h},$$

provided this limit exists.

▶ **EXAMPLE 3**   The distance in feet of an object from a starting point is given by $s(t) = 2t^2 - 5t + 40$, where t is time in seconds.

(a) Find the average velocity of the object from 2 seconds to 4 seconds.

The average velocity is

$$\frac{s(4) - s(2)}{4 - 2} = \frac{52 - 38}{2} = \frac{14}{2} = 7$$

feet per second.

(b) Find the instantaneous velocity at 4 seconds.

For $t = 4$, the instantaneous velocity is

$$\lim_{h \to 0} \frac{s(4 + h) - s(4)}{h}$$

feet per second. We have

$$
\begin{aligned}
s(4 + h) &= 2(4 + h)^2 - 5(4 + h) + 40 \\
&= 2(16 + 8h + h^2) - 20 - 5h + 40 \\
&= 32 + 16h + 2h^2 - 20 - 5h + 40 \\
&= 2h^2 + 11h + 52
\end{aligned}
$$

and

$$s(4) = 2(4)^2 - 5(4) + 40 = 52.$$

Thus,

$$s(4 + h) - s(4) = (2h^2 + 11h + 52) - 52 = 2h^2 + 11h$$

and the instantaneous velocity at $t = 4$ is

$$
\lim_{h \to 0} \frac{2h^2 + 11h}{h} = \lim_{h \to 0} \frac{h(2h + 11)}{h}
$$

$$
= \lim_{h \to 0} (2h + 11) = 11. \quad ◀ \boxed{4}
$$

5 Repeat Example 4 with $s(t) = .3t - 2$.

Answer:
The velocity is .3 millimeters per second.

▶**EXAMPLE 4** The velocity of blood cells is of interest to physicians; a slower velocity than normal might indicate a constriction, for example. Suppose the position of a red blood cell in a capillary is given by

$$s(t) = 1.2t + 5,$$

where $s(t)$ gives the position of a cell in millimeters from some reference point and t is time in seconds. Find the velocity of this cell at time t.

Evaluate the limit given above. To find $s(t + h)$, substitute $t + h$ for the variable t in $s(t) = 1.2t + 5$.

$$s(t + h) = 1.2(t + h) + 5$$

Now use the definition of velocity.

$$v(t) = \lim_{h \to 0} \frac{s(t + h) - s(t)}{h}$$

$$= \lim_{h \to 0} \frac{1.2(t + h) + 5 - (1.2t + 5)}{h}$$

$$= \lim_{h \to 0} \frac{1.2t + 1.2h + 5 - 1.2t - 5}{h} = \lim_{h \to 0} \frac{1.2h}{h} = 1.2$$

The velocity of the blood cell is a constant 1.2 millimeters per second. ◀ **5**

The ideas underlying the concept of the velocity of a moving object can be extended to any function $f(x)$. In place of average velocity at time t, we have the average rate of change of $f(x)$ with respect to x as x changes from one value to another.

The **instantaneous rate of change** for a function f when $x = x_0$ is

$$\lim_{h \to 0} \frac{f(x_0 + h) - f(x_0)}{h},$$

provided this limit exists.

▶**EXAMPLE 5** A company determines that the cost in dollars of manufacturing x units of a certain item is

$$C(x) = 100 + 5x - x^2.$$

(a) Find the average rate of change of cost per item for manufacturing between 1 and 5 items.

Use the formula for average rate of change. The cost to manufacture 1 item is

$$C(1) = 100 + 5(1) - (1)^2 = 104,$$

or $104. The cost to manufacture 5 items is

$$C(5) = 100 + 5(5) - (5)^2 = 100,$$

or $100. The average rate of change of cost is

$$\frac{C(5) - C(1)}{5 - 1} = \frac{100 - 104}{4}$$

$$= -1.$$

Thus, on the average, cost is decreasing at the rate of $1 per item when production is increased from 1 to 5 items.

(b) Find the instantaneous rate of change of cost with respect to the number of items produced when just 1 item is produced.

The instantaneous rate of change for $x = 1$ is given by

$$\lim_{h \to 0} \frac{C(1 + h) - C(1)}{h}$$

$$= \lim_{h \to 0} \frac{[100 + 5(1 + h) - (1 + h)^2] - [100 + 5(1) - 1^2]}{h}$$

$$= \lim_{h \to 0} \frac{[100 + 5 + 5h - 1 - 2h - h^2 - 104]}{h}$$

$$= \lim_{h \to 0} \frac{3h - h^2}{h} \qquad \text{Combine terms}$$

$$= \lim_{h \to 0} (3 - h) \qquad \text{Divide by } h$$

$$= 3. \qquad \text{Calculate the limit}$$

When 1 item is manufactured, the cost is increasing at the rate of $3 per item.

(c) Find the instantaneous rate of change of cost when 5 items are made.

The instantaneous rate of change for $x = 5$ is given by

$$\lim_{h \to 0} \frac{C(5 + h) - C(5)}{h}$$

$$= \lim_{h \to 0} \frac{[100 + 5(5 + h) - (5 + h)^2] - [100 + 5(5) - 5^2]}{h}$$

$$= \lim_{h \to 0} \frac{[100 + 25 + 5h - 25 - 10h - h^2] - 100}{h}$$

$$= \lim_{h \to 0} \frac{-5h - h^2}{h}$$

$$= \lim_{h \to 0} (-5 - h)$$

$$= -5.$$

When 5 items are manufactured, the cost is *decreasing* at the rate of $5 per item. ◀

The rate of change of cost function is called the **marginal cost.*** Similarly, **marginal revenue** and **marginal profit** are the rates of change of the revenue and profit functions, respectively.

Part (b) of Example 5 shows that the marginal cost when 1 item is manufactured is $3 and part (c) shows that the marginal cost when 5 items are manufactured is −$5. Notice that as the number of items produced goes up, the marginal cost goes down.

---

*Marginal cost for linear cost functions was discussed in Section 2.5.

11.2 EXERCISES

Find the average rate of change for the functions defined as follows over the given closed intervals. (See Example 1.)

1. $f(x) = x^2 + 2x$ between $x = 0$ and $x = 3$

2. $f(x) = -4x^2 - 6$ between $x = 2$ and $x = 5$

3. $f(x) = 2x^3 - 4x^2 + 6x$ between $x = -1$ and $x = 1$

4. $f(x) = -3x^3 + 2x^2 - 4x + 1$ between $x = 0$ and $x = 1$

5. $f(x) = \sqrt{x}$ between $x = 1$ and $x = 4$

6. $f(x) = \sqrt{3x - 2}$ between $x = 1$ and $x = 2$

7. $f(x) = \dfrac{1}{x - 1}$ between $x = -2$ and $x = 0$

8. $f(x) = \dfrac{-5}{2x - 3}$ between $x = 2$ and $x = 4$

Exercises 9–13 deal with a car moving along a straight road, as discussed on pages 592–94. At time t seconds the distance of the car (in feet) from the starting point is $s(t) = 2t^2$. Find the instantaneous velocity (speed) of the car at

9. $t = 5$;

10. $t = 20$;

11. $t = 25$;

12. $t = 30$.

13. What was the average speed of the car during the first 30 seconds?

An object moves along a straight line; its distance (in feet) from a fixed point at time t seconds is $s(t) = t^2 + 5t + 2$. Find the instantaneous velocity of the object at the following times. (See Example 3.)

14. $t = 6$

15. $t = 1$

16. $t = 10$

---

17. Management The graph shows the total sales in thousands of dollars from the distribution of x thousand catalogs. Find the average rate of change of sales with respect to the number of catalogs distributed for the following changes in x. (See Example 2.)

(a) 10 to 20 **(b)** 10 to 40

(c) 20 to 30 **(d)** 30 to 40

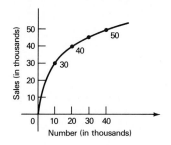

18. Management The graph shows annual sales (in units) of a typical product. Sales increase slowly at first to some peak, hold steady for a while, and then decline as the product goes out of style. Find the average annual rate of change in sales for the following changes in years.

(a) 1 to 3   **(b)** 2 to 4   **(c)** 3 to 6   **(d)** 5 to 7
(e) 7 to 9   **(f)** 8 to 11   **(g)** 9 to 10   **(h)** 10 to 12

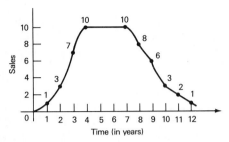

19. Natural Science The graph below shows the number of protected areas established worldwide from 1870 to 1985.* Estimate the average rate of change for nature reserves, and so on, over the following intervals.

(a) 1870 to 1935   **(b)** 1945 to 1975
(c) 1975 to 1985

Estimate the average rate of change for national parks over the following intervals.

(d) 1945 to 1970   **(e)** 1970 to 1980
(f) 1980 to 1985

FIGURE 1. Number of protected areas[a] established worldwide, 1870–1985.

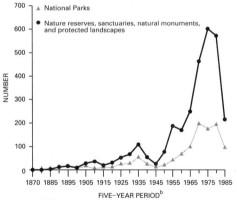

[a]Sites larger than 1,000 hectares.
[b]From the date shown, for example, 1970 is 1970–1975.
SOURCE: World Conservation Monitoring Centre, (Cambridge, England).

20. Natural Science The graph shows the temperature in degrees Celsius as a function of the altitude h in feet when an inversion layer is over southern California. (An inversion layer is formed when air at a higher altitude, say 3000 feet, is warmer than air at sea level, even though air nor-

mally is cooler with increasing altitude.) Find the average rate of change in temperature as the altitude changes from

(a) 1000 to 3000 feet;   **(b)** 1000 to 5000 feet;
(c) 3000 to 9000 feet;   **(d)** 1000 to 9000 feet.

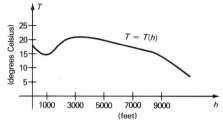

21. Physical Science A car is moving along a straight test track. The position in feet of the car, $s(t)$, at various times t is measured, with the following results.

t (seconds)	0	2	6	10	12	18	20
$s(t)$ (feet)	0	10	14	18	30	36	40

Find the average velocity (velocity equals distance divided by time) as the time in seconds changes from

(a) 0 to 6;   **(b)** 2 to 10;   **(c)** 2 to 12;
(d) 6 to 10;   **(e)** 6 to 18;   **(f)** 12 to 20.

22. Management The graph shows the annual net income in millions of dollars for Orbital Sciences, a small-satellite company, from 1984 through 1989.* Estimate and interpret the average rate of change of income between the following years.

(a) 1984 and 1986   **(b)** 1986 and 1989
(c) 1984 and 1989   **(d)** 1988 and 1989

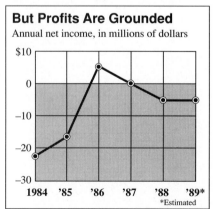

*"Number of Protected Areas Established Worldwide" from "the Future of National Parks" by Jeffrey A. McNeely in *Environment*, January/February 1990. Reprinted by permission of the author.

*Figure "But Profits are Grounded" from "Space Gamble: Start-Up Firm is Facing High Risks in Plans to Launch Rocket from Plane Next Month" in *The Wall Street Journal*, March 23, 1990. Copyright © 1990 by Dow, Jones & Company, Inc. Reprinted by permission of The Wall Street Journal. All Rights Reserved Worldwide.

Management *The graph shows the U.S. market share in percent for Japan's top three carmakers from 1985 through 1989.* Estimate and interpret the average rate of change of percent of the market for the following.*

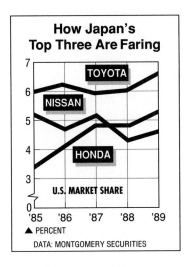

How Japan's Top Three Are Faring

TOYOTA
NISSAN
HONDA

U.S. MARKET SHARE

'85 '86 '87 '88 '89

▲ PERCENT

DATA: MONTGOMERY SECURITIES

23. Toyota from **(a)** 1985 to 1989   **(b)** 1986 to 1989
(c) 1987 to 1989

24. Nissan from **(a)** 1985 to 1987   **(b)** 1987 to 1989;
Honda from **(c)** 1985 to 1989   **(d)** 1987 to 1989

25. Management Profit figures for a small firm over a 7-year period are shown in the following table.

Year	1981	1982	1983	1984	1985	1986	1987
Profit (in dollars)	5000	6000	6500	6800	7200	7500	8000

(a) Find the average rate of change of profit from 1981 to 1987.

(b) Find the average rate of change of profit from 1982 to 1986.

---

*From "The American Drivers Steering Japan Through the States" by Larry Armstrong. Reprinted from March 12, 1990 issue of *Business Week* by special permission. Copyright © 1990 by McGraw-Hill, Inc.

26. Management Suppose that the total profit in hundreds of dollars from selling x items is given by

$$P(x) = 2x^2 - 5x + 6.$$

Find the average rate of change of profit for the following changes in x.

(a) 2 to 4   **(b)** 2 to 3

(c) Use $P(x) = 2x^2 - 5x + 6$ to complete the following table.

h	1	.1	.01	.001	.0001
$2 + h$	3	2.1	2.01	2.001	2.0001
$P(2 + h)$	9	4.32	4.0302	4.003002	4.00030002
$P(2)$	4	4	4	4	4
$P(2 + h) - P(2)$	5	.32	.0302	.003002	.00030002
$\dfrac{P(2 + h) - P(2)}{h}$	5	3.2			

(d) Use the bottom row of the chart to find

$$\lim_{h \to 0} \frac{P(2 + h) - P(2)}{h}.$$

(e) Use the Rules for Limits from Section 11.1 to find

$$\lim_{h \to 0} \frac{P(2 + h) - P(2)}{h}.$$

(f) What is the instantaneous rate of change of profit with respect to the number of items produced when $x = 2$?

27. Management Redo the table in Exercise 26, replacing $2 + h$ with $4 + h$. Then find

$$\lim_{h \to 0} \frac{P(4 + h) - P(4)}{h}.$$

28. Management The revenue from producing x units of an item is $R = 10x - .002x^2$.

(a) Find the marginal revenue when 1000 units are produced. (See Example 5.)

(b) Find the additional revenue if production is increased from 1000 to 1001 units.

(c) Find the average rate of change of revenue when production is increased from 1000 to 3000 units.

29. Management Suppose customers in a hardware store are willing to buy $N(p)$ boxes of nails at p dollars per box, as given by $N(p) = 80 - 5p^2$, $1 \leq p \leq 4$. (See Example 5.)

(a) Find the average rate of change of demand for a change in price from $2 to $3.

(b) Find the instantaneous rate of change of demand when the price is $2.

(c) Find the instantaneous rate of change of demand when the price is $3.

30. Natural Science The graph shows the population in millions of bacteria t minutes after a bactericide is introduced into a culture. Find and interpret the average rate of change of population with respect to time for the following time intervals.

(a) 1 to 2   (b) 1 to 3   (c) 2 to 3
(d) 2 to 5   (e) 3 to 4   (f) 4 to 5

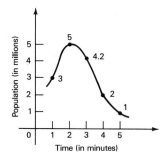

31. Social Science In an experiment, teenagers memorize a list of 15 words and then listen to rock music for certain periods of time. After t minutes of listening to rock music, the typical subject can remember approximately

$$R(t) = -.03t^2 + 15$$

of the words, where $0 \leq t \leq 20$. Find the instantaneous rate at which the typical student remembers the words at the following times.

(a) $t = 5$ (b) $t = 15$

32. Physical Science The distance of a particle from some fixed point is given by

$$s(t) = t^2 + 5t + 2,$$

where t is time measured in seconds. Find the average velocity of the particle over each of the following intervals.

(a) 4 to 6 sec   (b) 4 to 5 sec

(c) Complete the following table.

h	1	.1	.01	.001	.0001
$4 + h$	5	4.1	4.01	4.001	4.0001
$s(4 + h)$	52	39.31	38.1301	38.013001	38.00130001
$s(4)$	38	38	38	38	38
$s(4 + h) - s(4)$	14	1.31	.1301	.013001	.00130001
$\dfrac{s(4 + h) - s(4)}{h}$	14	13.1			

(d) Find $\lim\limits_{h \to 0} \dfrac{s(4 + h) - s(4)}{h}$, and give the instantaneous velocity of the particle when $x = 4$.

11.3 DEFINITION OF THE DERIVATIVE

In the previous section, the formula

$$\lim_{h \to 0} \frac{f(x_0 + h) - f(x_0)}{h}$$

was used to calculate the instantaneous rate of change of a function f at the point where $x = x_0$. Now we will give a geometric interpretation of this limit in terms of tangent lines.

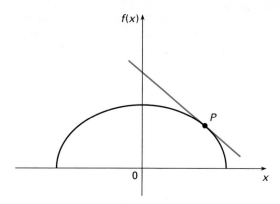

FIGURE 11.9

In geometry, a tangent line to a circle is defined as a line that touches the circle at only one point, as at the point P in Figure 11.9 (which shows only the top half of the circle). If you think of this half-circle as a road on which you are driving at night, then the tangent line indicates the direction of the light beam from your headlights as you pass through the point P. Intuitively, the tangent line to an arbitrary curve at a point P on the curve ought to have these same properties if possible.

1. It should touch the curve at P, but not at any nearby points;
2. It should indicate the "direction" of the curve as you move left to right through the point P.

For example, the lines through P_1 and P_3 in Figure 11.10 meet these intuitive criteria for tangent lines, but the lines through P_2 and P_4 do not.

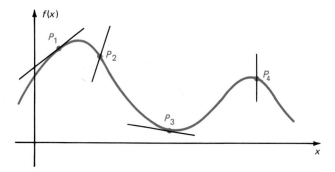

FIGURE 11.10

We can use these ideas to develop a precise definition of the tangent line to the graph of a function f at the point R. Let R have coordinates $(x_0, f(x_0))$ as in Figure 11.11. Choose a different point S on the graph and draw the line through R and S:

this line is called a **secant line.** If S has coordinates $(x_0 + h, f(x_0 + h))$, then by the definition of slope, the slope of the secant line RS is

$$\text{Slope of secant} = \frac{f(x_0 + h) - f(x_0)}{x_0 + h - x_0} = \frac{f(x_0 + h) - f(x_0)}{h}.$$

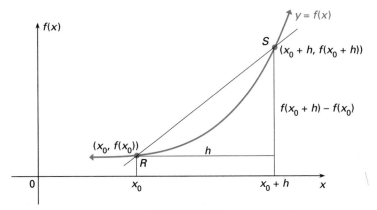

FIGURE 11.11

The slope corresponds to the average rate of change of y with respect to x over the interval from x_0 to $x_0 + h$. As h approaches 0, point S will slide along the curve, getting closer and closer to the fixed point R. See Figure 11.12, which shows successive positions S_1, S_2, S_3, S_4 of the point S. If the slopes of the corresponding secant lines approach a limit as h approaches 0, then this limit is defined to be the slope of the **tangent line** at point R.

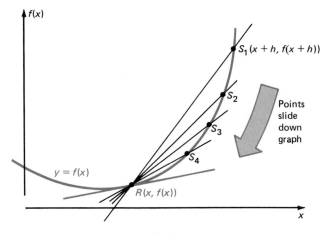

FIGURE 11.12

1 Rework Example 1 for $x = 2$.

Answer:
Slope is 4; equation is $y = 4x - 2$.

Tangent Line

The tangent line to the graph of $y = f(x)$ at the point $(x_0, f(x_0))$ is the line through this point having slope

$$\lim_{h \to 0} \frac{f(x_0 + h) - f(x_0)}{h},$$

provided this limit exists. If this limit does not exist, then there is no tangent at the point.

The slope of this line at a point is also called the **slope of the curve** at the point and corresponds to the instantaneous rate of change of y with respect to x at the point.

In certain applications of mathematics it is necessary to determine the equation of a line tangent to the graph of a function at a given point, as in the next example.

▶**EXAMPLE 1** Find the slope of the tangent line to the graph of $y = x^2 + 2$ when $x = -1$. Find the equation of the tangent line.

Use the definition above, with $f(x) = x^2 + 2$ and $x_0 = -1$. The slope of the tangent line is

$$\text{Slope of tangent} = \lim_{h \to 0} \frac{f(x_0 + h) - f(x_0)}{h}$$

$$= \lim_{h \to 0} \frac{[(-1 + h)^2 + 2] - [(-1)^2 + 2]}{h}$$

$$= \lim_{h \to 0} \frac{[1 - 2h + h^2 + 2] - [1 + 2]}{h}$$

$$= \lim_{h \to 0} \frac{-2h + h^2}{h} = \lim_{h \to 0} (-2 + h) = -2.$$

The slope of the tangent line at $(-1, f(-1)) = (-1, 3)$ is -2. The equation of the tangent line can be found with the point-slope form of the equation of a line from Chapter 2.

$$y - y_1 = m(x - x_1)$$
$$y - 3 = -2[x - (-1)]$$
$$y - 3 = -2(x + 1)$$
$$y - 3 = -2x - 2$$
$$y = -2x + 1$$

Figure 11.13 shows a graph of $f(x) = x^2 + 2$, along with a graph of the tangent line at $x = -1$. ◀ **1**

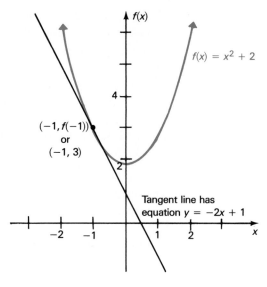

FIGURE 11.13

THE DERIVATIVE If $y = f(x)$ is a function and x_0 is a number in its domain, then we shall use the symbol $f'(x_0)$ to denote the special limit

$$\lim_{h \to 0} \frac{f(x_0 + h) - f(x_0)}{h},$$

provided that it exists. In other words, to each number x_0, we can assign the number $f'(x_0)$ obtained by calculating this limit. This process defines an important new function.

The **derivative** of the function f is the function denoted f' whose value at the number x is defined to be the number

$$f'(x) = \lim_{h \to 0} \frac{f(x + h) - f(x)}{h},$$

provided this limit exists.

The derivative function f' has as its domain all the points at which the specified limit exists, and the value of the derivative function at the number x is the number $f'(x)$. Using x instead of x_0 here is similar to the way that $g(x) = 2x$ denotes the function that assigns to each number x_0 the number $2x_0$.

If $y = f(x)$ is a function, then its derivative is denoted either by f' or by y'. If x is a number in the domain of $y = f(x)$ such that $y' = f'(x)$ is defined, then the function f is said to be **differentiable** at x. The process that produces the function f' from the function f is called **differentiation.**

The derivative function may be interpreted in many ways, two of which were discussed above.

1. The derivative function f' gives the *slope* of the graph of f at any point. If the derivative is evaluated at $x = x_0$, then $f'(x_0)$ is the slope of the tangent line to the curve at the point $(x_0, f(x_0))$.
2. The derivative function f' gives the *instantaneous rate of change* of $y = f(x)$ with respect to x. This instantaneous rate of change can be interpreted as marginal cost, revenue, or profit (if the original function represents cost, revenue, or profit) or as velocity (if the original function represents displacement along a line). From now on we will use ''rate of change'' to mean ''instantaneous rate of change.''

The next few examples show how to use the definition to find the derivative of a function by means of a four-step procedure.

▶**EXAMPLE 2** Let $f(x) = x^2$.

(a) Find the derivative.

Let x be a fixed number for which $f'(x)$ is defined. Then by definition,

$$f'(x) = \lim_{h \to 0} \frac{f(x + h) - f(x)}{h}.$$

In evaluating this limit, x is fixed and h is the variable. Use the following sequence of steps to evaluate the limit.

Step 1 Find $f(x + h)$.
   Replace x with $x + h$ in the equation for $f(x)$. Simplify the result.

$$f(x) = x^2$$
$$f(x + h) = (x + h)^2$$
$$= x^2 + 2xh + h^2$$

Step 2 Find $f(x + h) - f(x)$.
   Since $f(x) = x^2$,

$$f(x + h) - f(x) = x^2 + 2xh + h^2 - x^2$$
$$= 2xh + h^2.$$

Step 3 Form and simplify the quotient $\dfrac{f(x + h) - f(x)}{h}$.

$$\frac{f(x + h) - f(x)}{h} = \frac{2xh + h^2}{h} = \frac{h(2x + h)}{h} = 2x + h$$

2 Let $f(x) = -2x^2 + 7$. Find the following.

(a) $f(x + h)$

(b) $f(x + h) - f(x)$

(c) $\dfrac{f(x + h) - f(x)}{h}$

(d) $f'(x)$

(e) $f'(4)$

(f) $f'(0)$

Answer:

(a) $-2x^2 - 4xh - 2h^2 + 7$

(b) $-4xh - 2h^2$

(c) $-4x - 2h$

(d) $-4x$

(e) -16

(f) 0

Step 4   Find the limit of the result in step 3 as h approaches 0.

$$f'(x) = \lim_{h \to 0} \frac{f(x + h) - f(x)}{h} = \lim_{h \to 0} (2x + h) = 2x + 0 = 2x$$

From the last step, the derivative of $f(x) = x^2$ is $f'(x) = 2x$.

(b) Calculate and interpret $f'(3)$.

$$f'(3) = 2(3) = 6$$

The number 6 is the slope of the tangent line to the graph of $f(x) = x^2$ at the point where $x = 3$, that is, at $(3, f(3)) = (3, 9)$. See Figure 11.14.   ◀ **2**

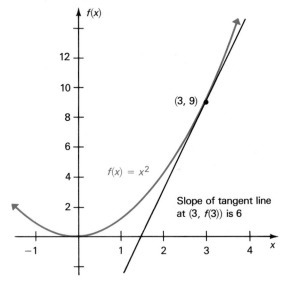

FIGURE 11.14

Caution   In Example 2(a), note that $f(x + h)$ is *not* equal to $f(x) + h$ because

$$f(x + h) = (x + h)^2 = x^2 + 2xh + h^2,$$

but

$$f(x) + h = x^2 + h.$$

In Example 2(b), do not confuse $f(3)$ and $f'(3)$. The value $f(3)$ is found by substituting 3 for x in the rule of the original function $f(x) = x^2$.

$$f(3) = 3^2 = 9$$

On the other hand, $f'(3)$ is the slope of the tangent line to the curve $y = f(x)$ at $x = 3$. It is found by substituting 3 for x in the rule of the derivative function $f'(x) = 2x$.

$$f'(3) = 2 \cdot 3 = 6$$

3 Let $f(x) = -3x^3 + 2x$. Find the following.

(a) $f(x + h)$

(b) $f(x + h) - f(x)$

(c) $\dfrac{f(x + h) - f(x)}{h}$

(d) $f'(x)$

(e) $f'(-1)$

Answer:

(a) $-3x^3 - 9x^2h - 9xh^2 - 3h^3 + 2x + 2h$

(b) $-9x^2h - 9xh^2 - 3h^3 + 2h$

(c) $-9x^2 - 9xh - 3h^2 + 2$

(d) $-9x^2 + 2$

(e) -7

The four steps used when finding the derivative of $f(x)$ are summarized below.

Finding $f'(x)$ from the Definition of the Derivative

Step 1 Find $f(x + h)$.

Step 2 Find $f(x + h) - f(x)$.

Step 3 Divide by h to get $\dfrac{f(x + h) - f(x)}{h}$.

Step 4 Let $h \to 0$; $f'(x) = \lim\limits_{h \to 0} \dfrac{f(x + h) - f(x)}{h}$ if this limit exists.

▶**EXAMPLE 3**   Let $f(x) = 2x^3 + 4x$. Find $f'(x)$. Then find $f'(2)$ and $f'(-3)$. Go through the four steps used above.

Step 1 Find $f(x + h)$ by replacing x with $x + h$.

$$f(x + h) = 2(x + h)^3 + 4(x + h)$$
$$= 2(x^3 + 3x^2h + 3xh^2 + h^3) + 4(x + h)$$
$$= 2x^3 + 6x^2h + 6xh^2 + 2h^3 + 4x + 4h.$$

Step 2 $f(x + h) - f(x) = 2x^3 + 6x^2h + 6xh^2 + 2h^3 + 4x + 4h - 2x^3 - 4x$
$$= 6x^2h + 6xh^2 + 2h^3 + 4h.$$

Step 3 $\dfrac{f(x + h) - f(x)}{h} = \dfrac{6x^2h + 6xh^2 + 2h^3 + 4h}{h}$

$$= \dfrac{h(6x^2 + 6xh + 2h^2 + 4)}{h}$$

$$= 6x^2 + 6xh + 2h^2 + 4.$$

Step 4 Now use the rules for limits to get

$$f'(x) = \lim_{h \to 0} \dfrac{f(x + h) - f(x)}{h} = \lim_{h \to 0} (6x^2 + 6xh + 2h^2 + 4)$$

$$= 6x^2 + 6x(0) + 2(0)^2 + 4 = 6x^2 + 4.$$

Use this result to find $f'(2)$ and $f'(-3)$.

$$f'(2) = 6 \cdot 2^2 + 4 = 28$$
$$f'(-3) = 6 \cdot (-3)^2 + 4 = 58 \quad ◀ \; \boxed{3}$$

4 Let $f(x) = -5/x$. Find the following.

(a) $f(x + h)$

(b) $f(x + h) - f(x)$

(c) $\dfrac{f(x + h) - f(x)}{h}$

(d) $f'(x)$

(e) $f'(-1)$

Answer:

(a) $\dfrac{-5}{x + h}$

(b) $\dfrac{5h}{x(x + h)}$

(c) $\dfrac{5}{x(x + h)}$

(d) $\dfrac{5}{x^2}$

(e) 5

▶**EXAMPLE 4** Let $f(x) = 4/x$. Find $f'(x)$.

Step 1 $f(x + h) = \dfrac{4}{x + h}$

Step 2 $f(x + h) - f(x) = \dfrac{4}{x + h} - \dfrac{4}{x}$

$= \dfrac{4x - 4(x + h)}{x(x + h)}$ **Find a common denominator**

$= \dfrac{4x - 4x - 4h}{x(x + h)}$ **Simplify the numerator**

$= \dfrac{-4h}{x(x + h)}$

Step 3 $\dfrac{f(x + h) - f(x)}{h} = \dfrac{\dfrac{-4h}{x(x + h)}}{h}$

$= \dfrac{-4h}{x(x + h)} \cdot \dfrac{1}{h}$ **Invert and multiply**

$= \dfrac{-4}{x(x + h)}$

Step 4 $f'(x) = \lim\limits_{h \to 0} \dfrac{f(x + h) - f(x)}{h} = \lim\limits_{h \to 0} \dfrac{-4}{x(x + h)}$

$= \dfrac{-4}{x(x + 0)} = \dfrac{-4}{x(x)} = \dfrac{-4}{x^2}$ ◀ **4**

▶**EXAMPLE 5** Let $f(x) = \sqrt{x}$. Find $f'(x)$.

Step 1 $f(x + h) = \sqrt{x + h}$

Step 2 $f(x + h) - f(x) = \sqrt{x + h} - \sqrt{x}$

Step 3 $\dfrac{f(x + h) - f(x)}{h} = \dfrac{\sqrt{x + h} - \sqrt{x}}{h}$

At this point, in order to be able to divide by h, multiply both numerator and denominator by $\sqrt{x + h} + \sqrt{x}$; that is, rationalize the *numerator*.

$$\frac{f(x + h) - f(x)}{h} = \frac{\sqrt{x + h} - \sqrt{x}}{h} \cdot \frac{\sqrt{x + h} + \sqrt{x}}{\sqrt{x + h} + \sqrt{x}} = \frac{(\sqrt{x + h})^2 - (\sqrt{x})^2}{h(\sqrt{x + h} + \sqrt{x})}$$

$$= \frac{x + h - x}{h(\sqrt{x + h} + \sqrt{x})} = \frac{1}{\sqrt{x + h} + \sqrt{x}}$$

Step 4 $f'(x) = \lim\limits_{h \to 0} \dfrac{1}{\sqrt{x + h} + \sqrt{x}} = \dfrac{1}{\sqrt{x} + \sqrt{x}} = \dfrac{1}{2\sqrt{x}}$ ◀

5 Go through the four steps to find $s'(t)$, the velocity of the car at any time t.

Answer:
$s'(t) = -15t^2 + 60t$

▶**EXAMPLE 6**   The cost in dollars to manufacture x graphic calculators is given by $C(x) = -.005x^2 + 20x + 150$. Find the rate of change of cost with respect to the number manufactured when 100 calculators are made and when 1000 calculators are made.

The rate of change of cost is given by the derivative of the cost function,

$$C'(x) = \lim_{h \to 0} \frac{C(x + h) - C(x)}{h}.$$

Going through the steps for finding $C'(x)$ gives

$$C'(x) = -.01x + 20.$$

When $x = 100$, $C'(100) = -.01(100) + 20 = 19.$

When 1000 calculators are made, the marginal cost is

$$C'(1000) = -.01(1000) + 20 = 10. \quad ◀$$

▶**EXAMPLE 7**   A sales representative for a textbook publishing company frequently makes a 4-hour drive from her home in a large city to a university in another city. If $s(t)$ represents her distance (in miles) from home t hours into the trip, then $s(t)$ is given by
$$s(t) = -5t^3 + 30t^2.$$

(a) How far from home will she be after 1 hour? after $1\frac{1}{2}$ hours?
   Her distance from home after 1 hour is

$$s(1) = -5(1)^3 + 30(1)^2 = 25,$$

or 25 miles. After $1\frac{1}{2}$ (or $\frac{3}{2}$) hours, her distance from home is

$$s\left(\frac{3}{2}\right) = -5\left(\frac{3}{2}\right)^3 + 30\left(\frac{3}{2}\right)^2 = \frac{405}{8} = 50.625,$$

or 50.625 miles.
(b) How far apart are the two cities?
   Since the trip takes 4 hours and the distance is given by $s(t)$, the university city is $s(4) = 160$ miles from her home.
(c) How fast is she driving 1 hour into the trip? $1\frac{1}{2}$ hours into the trip?
   Velocity (or speed) is the instantaneous rate of change in position with respect to time. We need to find the value of the derivative $s'(t)$ at $t = 1$ and $t = 1\frac{1}{2}$. **5**
   From Problem 5 at the side, $s'(t) = -15t^2 + 60t$. At $t = 1$, the velocity is

$$s'(1) = -15(1)^2 + 60(1) = 45, \qquad \bullet$$

or 45 miles per hour. At $t = 1\frac{1}{2}$, the velocity is

$$s'\left(\frac{3}{2}\right) = -15\left(\frac{3}{2}\right)^2 + 60\left(\frac{3}{2}\right) \approx 56,$$

about 56 miles per hour.
(d) Does she ever exceed the speed limit of 65 miles per hour on the trip?

To find the maximum velocity, notice that the graph of the velocity function $s'(t) = -15t^2 + 60t$ is a parabola opening downward. The maximum velocity will occur at the vertex. Verify that the vertex of the parabola is (2, 60). Thus, her maximum velocity during the trip is 60 miles per hour, so she never exceeds the speed limit. ◄

EXISTENCE OF THE DERIVATIVE The definition of the derivative included the phrase "provided this limit exists." If the limit used to define $f'(x)$ does not exist, then of course the derivative does not exist at that x. For example, a derivative cannot exist at a point where the function itself is not defined. If there is no function value for a particular value of x, there can be no tangent line for that value. This was the case in Example 4—there was no tangent line (and no derivative) when $x = 0$.

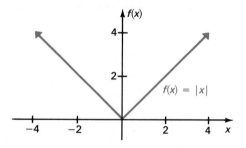

FIGURE 11.15

Derivatives also do not exist at "corners" or "sharp points" on a graph. For example, the function graphed in Figure 11.15 is the *absolute value function*, defined as

$$f(x) = \begin{cases} x & \text{if } x \geq 0 \\ -x & \text{if } x < 0, \end{cases}$$

and written $f(x) = |x|$. By the definition of derivative, the derivative at any value of x is given by

$$\lim_{h \to 0} \frac{f(x+h) - f(x)}{h},$$

provided this limit exists. To find the derivative at 0 for $f(x) = |x|$, replace x with 0 and $f(x)$ with $|0|$ to get

$$\lim_{h \to 0} \frac{|0 + h| - |0|}{h} = \lim_{h \to 0} \frac{|h|}{h}.$$

In Example 5 of Section 11.1 (with x in place of h) we showed that

$$\lim_{h \to 0} \frac{|h|}{h} \text{ does not exist.}$$

Therefore, there is no derivative at 0. However, the derivative of $f(x) = |x|$ *does* exist for all values of x other than 0.

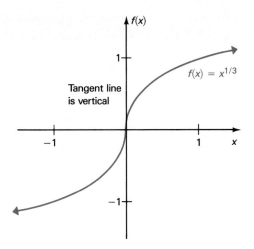

FIGURE 11.16

A graph of $f(x) = x^{1/3}$ is shown in Figure 11.16. As the graph suggests, the tangent line is vertical when $x = 0$. Since a vertical line has an undefined slope, the derivative of $f(x) = x^{1/3}$ cannot exist when $x = 0$.

Figure 11.17 summarizes the various ways that a derivative can fail to exist.

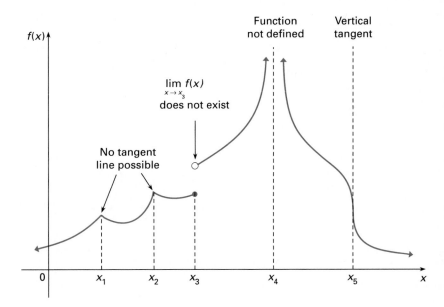

FIGURE 11.17

11.3 EXERCISES

Find the slope of the tangent line to each of the following curves when x has the given value.
(Hint for Exercise 5: in step 3, multiply numerator and denominator by $\sqrt{16 + h} + \sqrt{16}$.)
(See Example 1.)

1. $f(x) = -4x^2 + 11x$; $x = -2$

2. $f(x) = 6x^2 - 4x$; $x = -1$

3. $f(x) = -\dfrac{2}{x}$; $x = 4$

4. $f(x) = \dfrac{6}{x}$; $x = -1$

5. $f(x) = \sqrt{x}$; $x = 16$

6. $f(x) = -3\sqrt{x}$; $x = 1$

Find the equation of the tangent line to each of the following curves when x has the given value. (See Example 1.)

7. $f(x) = x^2 + 2x$; $x = 3$

8. $f(x) = 6 - x^2$; $x = -1$

9. $f(x) = \dfrac{5}{x}$; $x = 2$

10. $f(x) = -\dfrac{3}{x + 1}$; $x = 1$

11. $f(x) = 4\sqrt{x}$; $x = 9$

12. $f(x) = \sqrt{x}$; $x = 25$

Estimate the slope of the tangent line at the given point (x, y) in each of the following graphs.

13.

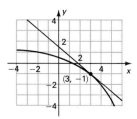

14.

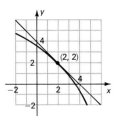

15.

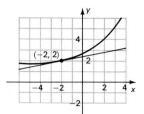

16.

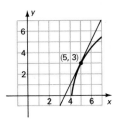

17.

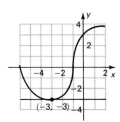

18.

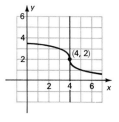

Find f′(x) for each function defined as follows. (Many of these derivatives were found in Exercises 1–6.) Then find f′(2), f′(0), and f′(−3). (See Examples 2–5.)

19. $f(x) = -4x^2 + 11x$

20. $f(x) = 6x^2 - 4x$

21. $f(x) = 8x + 6$

22. $f(x) = -9x - 5$

23. $f(x) = x^3 + 3x$

24. $f(x) = 2x^3 - 14x$

25. $f(x) = -\dfrac{2}{x}$

26. $f(x) = \dfrac{6}{x}$

27. $f(x) = \dfrac{4}{x - 1}$

28. $f(x) = \dfrac{3}{x + 2}$

29. $f(x) = \sqrt{x}$

30. $f(x) = -3\sqrt{x}$

Find all points where the functions graphed as shown do not have derivatives.

31.

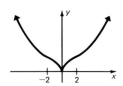

32.

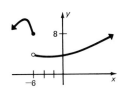

33.

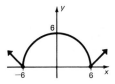

34.

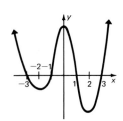

35.

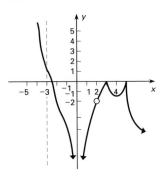

36.

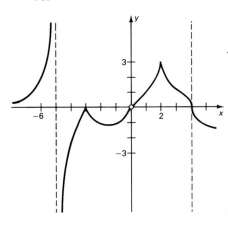

Work the following exercises. (See Examples 6 and 7.)

37. What do each of the following values of $P'(x)$ tell you about the profit $P(x)$?
 (a) $P'(10) = 8$ **(b)** $P'(12) = 0$
 (c) $P'(15) = -4$

38. What can you say about the average cost $\overline{C}$ if $\overline{C}'$ has the following values?
 (a) $\overline{C}'(10) = -8$ **(b)** $\overline{C}'(12) = 0$
 (c) $\overline{C}'(15) = 5$

39. Management The revenue generated from the sale of x picnic tables is given by

$$R(x) = 20x - \frac{x^2}{500}.$$

 (a) Find the marginal revenue when $x = 1000$ units.
 (b) Determine the actual revenue from the sale of the 1001st item.

(c) Compare the answers to parts (a) and (b). How are they related?

40. Management The cost of producing x tacos is $C(x) = 1000 + .24x^2$, $0 \le x \le 30{,}000$.
 (a) Find the marginal cost, $C'(x)$.
 (b) Find and interpret $C'(100)$.
 (c) Find the exact cost to produce the 101st taco.
 (d) Compare the answers to parts (b) and (c). How are they related?

41. Management Suppose the demand for a certain item is given by

$$D(x) = -2x^2 + 4x + 6,$$

where x represents the price of the item. Find the rate of change of demand with respect to price for the following values of x. (See Example 6.)
 (a) $x = 3$ **(b)** $x = 6$

42. Management Suppose the profit from the expenditure of x thousand dollars on advertising is given by

$$P(x) = 1000 + 32x - 2x^2.$$

Find the rate of change of profit with respect to the expenditure on advertising for the following amounts. In each case, decide if the firm should increase the expenditure.
(a) $x = 8$ **(b)** $x = 6$ **(c)** $x = 12$ **(d)** $x = 20$

43. Natural Science A biologist estimates that when a bactericide is introduced into a culture of bacteria, the number of bacteria present is given by

$$B(t) = 1000 + 50t - 5t^2,$$

where $B(t)$ is the number of bacteria, in millions, present at time t, measured in hours. Find the rate of change of the number of bacteria with respect to time for the following values of t.
(a) $t = 2$ **(b)** $t = 3$ **(c)** $t = 4$
(d) $t = 5$ **(e)** $t = 6$ **(f)** $t = 8$

44. Natural Science When does the population of bacteria in Exercise 43 start to decline?

45. Natural Science (Continuation of Exercise 20 from the previous section.) The graph below shows the temperature in degrees Celsius as a function of the altitude h in feet when an inversion layer is over southern California. Estimate the derivatives of $T(h)$ at the marked points.

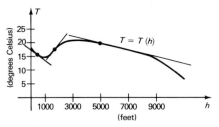

46. Natural Science In a research study, the population of a certain shellfish in an area at time t was closely approximated by the graph below. Estimate the derivative at the marked points.

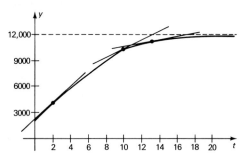

Use a calculator or a computer to find the slope of the tangent line to the graph of each function defined as follows.

47. $f(x) = \dfrac{x^2 + 4.9}{3x^2 - 1.7x}$ at $x = 4.9$

48. $f(x) = \dfrac{3x - 7.2}{x^2 + 5x + 2.3}$ at $x = -6.2$

49. $f(x) = \sqrt{x^3 + 4.8}$ at $x = 8.17$

50. $f(x) = \sqrt{6x^2 + 3.5}$ at $x = 3.49$

51. $f(x) = \dfrac{2.8}{\sqrt{x} + 1.2}$ at $x = 2.65$

52. $f(x) = \dfrac{-3.7}{4.1 + 2\sqrt{x}}$ at $x = 1.23$

11.4 TECHNIQUES FOR FINDING DERIVATIVES

In the previous section, the derivative of a function was defined as a special limit. The mathematical process of finding this limit, called *differentiation*, resulted in a new function that was interpreted in several different ways. Using the definition to calculate the derivative of a function is a very involved process even for simple

1 Use the results of some of Exercises 23–30 in the previous section to find each of the following.

(a) $\dfrac{d}{dx}(x^3 + 3x)$

(b) $\dfrac{d}{dx}\left(\dfrac{-2}{x}\right)$

(c) $D_x\left(\dfrac{4}{x-1}\right)$

(d) $D_x(-3\sqrt{x})$

Answer:

(a) $3x^2 + 3$

(b) $\dfrac{2}{x^2}$

(c) $\dfrac{-4}{(x-1)^2}$

(d) $\dfrac{-3}{2\sqrt{x}}$

functions. In this section we develop rules that make the calculation of derivatives much easier. Keep in mind that even though the process of finding a derivative will be greatly simplified with these rules, *the interpretation of the derivative will not change*.

In addition to y' and $f'(x)$, there are several other commonly used notations for the derivative.

Notations for the Derivative

The derivative of the function $y = f(x)$ may be denoted in any of the following ways.

$$f'(x), \quad y', \quad \frac{dy}{dx}, \quad \frac{d}{dx}[f(x)], \quad D_x y, \quad D_x[f(x)]$$

The dy/dx notation for the derivative is sometimes referred to as Leibniz notation, named after one of the co-inventors of calculus, Gottfried Wilhelm Leibniz (1646–1716). (The other was Sir Isaac Newton (1642–1727).)

For example, the derivative of $y = 2x^3 + 4x$, which we found in Example 2 of the last section to be $y' = 6x^2 + 4$, can also be written

$$\frac{dy}{dx} = 6x^2 + 4$$

$$\frac{d}{dx}(2x^3 + 4x) = 6x^2 + 4$$

$$D_x(2x^3 + 4x) = 6x^2 + 4. \quad \boxed{1}$$

A variable other than x often may be used as the independent variable. For example, if $y = f(t)$ gives population growth as function of time, then the derivative of y with respect to t could be written

$$f'(t), \quad \frac{dy}{dt}, \quad \frac{d}{dt}[f(t)], \quad \text{or} \quad D_t[f(t)].$$

In this section the definition of the derivative,

$$f'(x) = \lim_{h \to 0} \frac{f(x + h) - f(x)}{h},$$

is used to develop some rules for finding derivatives more easily than by the four-step process of the previous section.

The first rule tells how to find the derivative of a constant function $f(x) = k$, where k is a constant real number. Since $f(x + h)$ is also k, by definition $f'(x)$ is

2 Find the derivatives of the following.

(a) $y = -4$

(b) $f(x) = \pi^3$

(c) $y = 0$

Answer:

(a) 0

(b) 0

(c) 0

$$f'(x) = \lim_{h \to 0} \frac{f(x + h) - f(x)}{h}$$

$$= \lim_{h \to 0} \frac{k - k}{h} = \lim_{h \to 0} \frac{0}{h} = \lim_{h \to 0} 0 = 0,$$

establishing the following rule.

Constant Rule

If $f(x) = k$ where k is any real number, then

$$f'(x) = 0.$$

(The derivative of a constant function is 0.)

(For reference, a summary of all the formulas for derivatives is given at the end of this chapter.)

Figure 11.18 illustrates the constant rule; it shows a graph of the horizontal line $y = k$. At any point P on this line, the tangent line at P is the line itself. Since a horizontal line has a slope of 0, the slope of the tangent line is 0. This agrees with the result above: the derivative of a constant is 0.

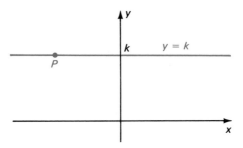

FIGURE 11.18

▶ **EXAMPLE 1**

(a) If $f(x) = 9$, then $f'(x) = 0$.

(b) If $y = \pi$, then $y' = 0$.

(c) If $y = 2^3$, then $dy/dx = 0$. ◀ **2**

Functions of the form $y = x^n$, where n is a real number, are very common in applications. To get a rule for finding the derivative of such a function, we can use the definition to work out the derivatives for various special values of n. This was done in the previous section in Example 2 to show that for $f(x) = x^2$, $f'(x) = 2x$.

For $f(x) = x^3$, the derivative is found as follows.

$$f'(x) = \lim_{h \to 0} \frac{f(x + h) - f(x)}{h}$$

$$= \lim_{h \to 0} \frac{(x + h)^3 - x^3}{h}$$

$$= \lim_{h \to 0} \frac{(x^3 + 3x^2h + 3xh^2 + h^3) - x^3}{h}$$

$$= \lim_{h \to 0} \frac{3x^2h + 3xh^2 + h^3}{h}$$

$$= \lim_{h \to 0} (3x^2 + 3xh + h^2)$$

$$= 3x^2$$

The results in the table below were found in a similar way, using the definition of the derivative. (These results are modifications of some of the examples and exercises from the previous section.)

Function	n	Derivative
$y = x^2$	2	$y' = 2x = 2x^1$
$y = x^3$	3	$y' = 3x^2$
$y = x^4$	4	$y' = 4x^3$
$y = x^{-1}$	-1	$y' = -1 \cdot x^{-2} = \dfrac{-1}{x^2}$
$y = x^{1/2}$	$\dfrac{1}{2}$	$y' = \dfrac{1}{2}x^{-1/2} = \dfrac{1}{2x^{1/2}}$

These results suggest the following rule.

POWER RULE

If $f(x) = x^n$ for any nonzero real number n, then

$$f'(x) = nx^{n-1}.$$

(The derivative of $f(x) = x^n$ is found by multiplying by the exponent n and decreasing the exponent on x by 1.)

For a proof of the power rule, see Exercise 67 at the end of this section.

3 Find the following.

(a) $y = x^4$; find y'

(b) $y = x^{17}$; find y'

(c) $y = x^{-2}$; find dy/dx

(d) $y = t^{-5}$; find dy/dt

(e) $D_t(t^{3/2})$

Answer:

(a) $y' = 4x^3$

(b) $y' = 17x^{16}$

(c) $dy/dx = -2/x^3$

(d) $dy/dt = -5/t^6$

(e) $D_t(t^{3/2}) = (3/2)t^{1/2}$

▶**EXAMPLE 2** Find each of the following.

(a) If $y = x^6$, find y'.

Multiply x^{6-1} by 6.

$$y' = 6 \cdot x^{6-1} = 6x^5$$

(b) If $y = x = x^1$, find y'.

$$y' = 1 \cdot x^{1-1} = 1 \cdot x^0 = 1 \cdot 1 = 1.$$

(Recall that $x^0 = 1$ if $x \neq 0$.)

(c) If $y = t^{-3}$, find dy/dt.

$$\frac{dy}{dt} = -3 \cdot t^{-3-1} = -3 \cdot t^{-4} = -\frac{3}{t^4}.$$

(d) Find $D_t(t^{4/3})$.

$$D_t(t^{4/3}) = \frac{4}{3}t^{4/3-1} = \frac{4}{3}t^{1/3}.$$

(e) If $y = \sqrt{x}$, find dy/dx.

Replace $\sqrt{x}$ by the equivalent expression $x^{1/2}$. Then

$$\frac{dy}{dx} = \frac{1}{2} \cdot x^{(1/2)-1} = \frac{1}{2}x^{-1/2} = \frac{1}{2x^{1/2}} = \frac{1}{2\sqrt{x}}. \quad ◀ \boxed{3}$$

The next rule shows how to find the derivative of the product of a constant and a function.

Constant Times a Function

Let k be a real number. If $g'(x)$ exists, then the derivative of $f(x) = k \cdot g(x)$ is

$$f'(x) = k \cdot g'(x).$$

(The derivative of a constant times a function is the constant times the derivative of the function.)

▶**EXAMPLE 3** Find the derivatives of the following functions.

(a) $y = 8x^4$; find y'.

Since the derivative of $g(x) = x^4$ is $g'(x) = 4x^3$ and $y = 8x^4 = 8g(x)$,

$$y' = 8g'(x) = 8(4x^3) = 32x^3.$$

4 Find the derivatives of the following.

(a) $y = 12x^3$

(b) $f(x) = 30t^7$

(c) $y = -35t$

(d) $y = 5\sqrt{x}$

(e) $y = -10/t$

Answer:

(a) $36x^2$

(b) $210t^6$

(c) -35

(d) $(5/2)x^{-1/2}$ or $5/(2\sqrt{x})$

(e) $10t^{-2}$ or $10/t^2$

(b) $y = -\dfrac{3}{4}t^{12}$; find dy/dt.

$$\frac{dy}{dt} = -\frac{3}{4}\left[\frac{dy}{dt}(t^{12})\right] = -\frac{3}{4}(12t^{11}) = -9t^{11}$$

(c) Find $D_x(15x)$.

$$D_x(15x) = 15 \cdot D_x(x) = 15(1) = 15$$

(d) Find $D_x(10x^{3/2})$.

$$D_x(10x^{3/2}) = 10\left(\frac{3}{2}x^{1/2}\right) = 15x^{1/2}$$

(e) $y = \dfrac{6}{x}$; find y'.

Replace $\dfrac{6}{x}$ with $6 \cdot \dfrac{1}{x}$, or $6x^{-1}$. Then

$$y' = 6(-1x^{-2}) = -6x^{-2} = -\frac{6}{x^2}. \quad \blacktriangleleft \quad \boxed{4}$$

The final rule in this section is for the derivative of a function that is a sum or difference of functions.

Sum or Difference Rule

If $f(x) = u(x) + v(x)$, and if $u'(x)$ and $v'(x)$ exist, then

$$f'(x) = u'(x) + v'(x).$$

If $f(x) = u(x) - v(x)$, then

$$f'(x) = u'(x) - v'(x).$$

(The derivative of a sum or difference of two functions is the sum or difference of the derivatives of the functions.)

For a proof of this rule, see Exercise 68 at the end of this section. This rule can be generalized to sums and differences with more than two terms.

▶**EXAMPLE 4** Find the derivatives of the following functions.
(a) $y = 6x^3 + 15x^2$

Let $u(x) = 6x^3$ and $v(x) = 15x^2$; then $y = u(x) + v(x)$. Since $u'(x) = 18x^2$ and $v'(x) = 30x$,

$$\frac{dy}{dx} = 18x^2 + 30x.$$

5 Find the derivatives of the following.

(a) $y = -4x^5 - 8x + 6$

(b) $y = 32x^5 - 100x^2 + 12x$

(c) $y = 8t^{3/2} + 2t^{1/2}$

(d) $f(t) = -\sqrt{t} + 6/t$

Answer:

(a) $y' = -20x^4 - 8$

(b) $y' = 160x^4 - 200x + 12$

(c) $y' = 12t^{1/2} + t^{-1/2}$
or $12t^{1/2} + 1/t^{1/2}$

(d) $f'(t) = -1/(2\sqrt{t}) - 6/t^2$

(b) $p(t) = 8t^4 - 6\sqrt{t} + \dfrac{5}{t}$

Rewrite $p(t)$ as $p(t) = 8t^4 - 6t^{1/2} + 5t^{-1}$; then $p'(t) = 32t^3 - 3t^{-1/2} - 5t^{-2}$, which also may be written as $p'(t) = 32t^3 - \dfrac{3}{\sqrt{t}} - \dfrac{5}{t^2}$.

(c) $f(x) = 5\sqrt[3]{x^2} + 4x^{-2} + 7$

Rewrite $f(x)$ as $f(x) = 5x^{2/3} + 4x^{-2} + 7$. Then

$$D_x[f(x)] = \frac{10}{3}(x^{-1/3}) - 8x^{-3},$$

or

$$D_x[f(x)] = \frac{10}{3\sqrt[3]{x}} - \frac{8}{x^3}. \blacktriangleleft \boxed{5}$$

The rules developed in this section make it possible to find the derivative of a function more directly, so that applications of the derivative can be dealt with more effectively. The following examples illustrate some business applications.

MARGINAL ANALYSIS In business and economics the rates of change of such variables as cost, revenue, and profit are important considerations. Economists use the word marginal to refer to rates of change: for example, *marginal cost* refers to the rate of change of cost. Since the derivative of a function gives the rate of change of the function, a marginal cost (or revenue, or profit) function is found by taking the derivative of the cost (or revenue, or profit) function. Roughly speaking, the marginal cost at some level of production x is the cost to produce the $(x + 1)$st item, as we now show. (Similar statements could be made for revenue or profit.)

Look at Figure 11.19, where $C(x)$ represents the cost of producing x units of some item. Then the cost of producing $x + 1$ units is $C(x + 1)$. The cost of the $(x + 1)$st unit is, therefore, $C(x + 1) - C(x)$. This quantity is shown on the graph in Figure 11.19.

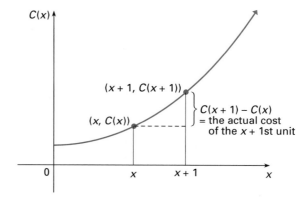

FIGURE 11.19

Now if $C(x)$ is the cost function, then the marginal cost $C'(x)$ represents the slope of the tangent line at any point $(x, C(x))$. The graph in Figure 11.20 shows the cost function $C(x)$ and the tangent line at point $P = (x, C(x))$. We know that the slope of the tangent line is $C'(x)$ and that the slope can be computed using the triangle PQR in Figure 11.20.

$$C'(x) = \text{slope} = \frac{\text{rise}}{\text{run}} = \frac{QR}{PR} = \frac{QR}{1} = QR$$

So the length of the line segment QR is the number $C'(x)$.

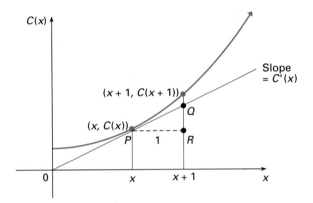

FIGURE 11.20

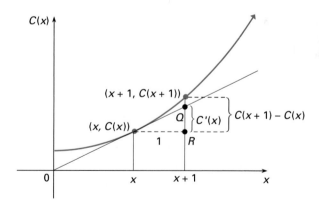

FIGURE 11.21

Superimposing the graphs from Figures 11.19 and 11.20 as in Figure 11.21 shows that $C'(x)$ is indeed very close to $C(x + 1) - C(x)$. The two values are closest when $C'(x)$ is very large, so that 1 unit is relatively small.

6 The cost in dollars to produce x units of wheat is given by $C(x) = 5000 + 20x + 10\sqrt{x}$. Find the marginal cost when

(a) $x = 9$;

(b) $x = 16$;

(c) $x = 25$.

(d) As more wheat is produced, what happens to the marginal cost?

Answer:

(a) $\$65/3 \approx \$21.67$

(b) $\$85/4 = \$21.25$

(c) $\$21$

(d) It decreases and approaches $\$20$.

▶ **EXAMPLE 5** Suppose that the total cost in hundreds of dollars to produce x thousand barrels of a beverage is given by

$$C(x) = 4x^2 + 100x + 500.$$

Find the marginal cost for the following values of x.
(a) $x = 5$

To find the marginal cost, first find $C'(x)$, the derivative of the total cost function.

$$C'(x) = 8x + 100$$

When $x = 5$,

$$C'(5) = 8(5) + 100 = 140.$$

After 5 thousand barrels of the beverage have been produced, the cost to produce 1 thousand more barrels will be *approximately* 140 hundred dollars, or $14,000.
The *actual* cost to produce 1 thousand more barrels is $C(6) - C(5)$.

$$C(6) - C(5) = (4 \cdot 6^2 + 100 \cdot 6 + 500) - (4 \cdot 5^2 + 100 \cdot 5 + 500)$$
$$= 1244 - 1100$$
$$= 144$$

The actual cost is 144 hundred dollars, or $14,400.
(b) $x = 30$

After 30 thousand barrels have been produced, the cost to produce 1 thousand more barrels will be approximately

$$C'(30) = 8(30) + 100 = 340,$$

or $34,000. Notice that the cost to produce an additional thousand barrels of beverage has increased by approximately $20,000 at a production level of 30 thousand barrels compared to a production level of 5 thousand barrels. Management must be careful to keep track of marginal costs. If the marginal cost of producing an extra unit exceeds the revenue received from selling it, then the company will lose money on that unit. ◀ **6**

DEMAND FUNCTIONS The **demand function,** defined by $p = f(x)$, relates the number of units x of an item that consumers are willing to purchase at the price p. (Demand functions were also discussed in Chapter 1.) The total revenue $R(x)$ is related to price per unit and the amount demanded (or sold) by the equation

$$R(x) = xp = x \cdot f(x).$$

7 Suppose the demand function for x units of an item is

$$p = 5 - \frac{x}{1000},$$

where x is the price in dollars. Find

(a) the marginal revenue;

(b) marginal revenue at $x = 500$;

(c) marginal revenue at $x = 1000$.

Answer:

(a) $R'(x) = 5 - \dfrac{x}{500}$

(b) $4

(c) $3

▶ **EXAMPLE 6** The demand function for a certain product is given by

$$p = \frac{50,000 - x}{25,000}.$$

Find the marginal revenue when $x = 10,000$ units and p is in dollars.

From the given function for p, the revenue function is given by

$$R(x) = xp$$

$$= x\left(\frac{50,000 - x}{25,000}\right)$$

$$= \frac{50,000x - x^2}{25,000}$$

$$= 2x - \frac{1}{25,000}x^2.$$

The marginal revenue is

$$R'(x) = 2 - \frac{2}{25,000}x.$$

When $x = 10,000$, the marginal revenue is

$$R'(10,000) = 2 - \frac{2}{25,000}(10,000) = 1.2,$$

or $1.20 per unit. Thus, the next unit sold (at sales of 10,000) will produce additional revenue of about $1.20 per unit. ◀ **7**

In economics, the demand function is written in the form $p = f(x)$, as shown above. From the perspective of a consumer, it is probably more reasonable to think of the quantity demanded as a function of price. Mathematically, these two viewpoints are equivalent. In Example 6, the demand function could have been written from the consumer's viewpoint as

$$x = 50,000 - 25,000p.$$

▶ **EXAMPLE 7** Suppose that the cost function for the product in Example 6 is given by

$$C(x) = 2100 + .25x \quad \text{where } 0 \le x \le 30,000.$$

Find the marginal profit from the production of the following numbers of units.
(a) 15,000

From Example 6, the revenue from the sale of x units is

$$R(x) = 2x - \frac{1}{25,000}x^2.$$

8 For a certain product, the cost $C(x) = 1250 + .75x$ and the revenue $R(x) = 5x - \dfrac{x^2}{10,000}$ for x units.

(a) Find the profit $P(x)$.

(b) Find $P'(20,000)$.

(c) Find $P'(30,000)$.

(d) Interpret the results of (b) and (c).

Answer:

(a) $P(x) = 4.25x - x^2/10,000 - 1250$

(b) .25

(c) -1.75

(d) Profit is increasing by \$.25 per unit at 20,000 units in (b) and decreasing by \$1.75 per unit at 30,000 units in (c).

Since profit, P, is given by $P = R - C$,

$$P(x) = R(x) - C(x)$$

$$= \left(2x - \frac{1}{25,000}x^2\right) - (2100 + .25x)$$

$$= 2x - \frac{1}{25,000}x^2 - 2100 - .25x$$

$$= 1.75x - \frac{1}{25,000}x^2 - 2100.$$

The marginal profit from the sale of x units is

$$P'(x) = 1.75 - \frac{2}{25,000}x = 1.75 - \frac{1}{12,500}x.$$

At $x = 15,000$, the marginal profit is

$$P'(15,000) = 1.75 - \frac{1}{12,500}(15,000) = .55,$$

or \$.55 per unit.

(b) 21,875

When $x = 21,875$, the marginal profit is

$$P'(21,875) = 1.75 - \frac{1}{12,500}(21,875) = 0.$$

(c) 25,000

When $x = 25,000$, the marginal profit is

$$P'(25,000) = 1.75 - \frac{1}{12,500}(25,000) = -.25,$$

or $-\$.25$ per unit.

As shown by parts (b) and (c), if more than 21,875 units are sold, the marginal profit is negative. This indicates that increasing production beyond that level will *reduce* profit. ◀ **8**

The final example shows a medical application of the derivative as the rate of change of a function.

▶**EXAMPLE 8**   A tumor has the approximate shape of a cone. See Figure 11.22. The radius of the tumor is fixed by the bone structure at 2 centimeters, but the tumor is growing along the height of the cone. The formula for the volume of a cone is $V = \frac{1}{3}\pi r^2 h$, where r is the radius of the base and h is the height of the cone. Find the rate of change in the volume of the tumor with respect to the height.

9 A balloon is spherical. The formula for the volume of a sphere is $V = (4/3)\pi r^3$, where r is the radius of the sphere. Find the following.

(a) dV/dr

(b) The rate of change of the volume when $r = 3$ inches

Answer:

(a) $4\pi r^2$

(b) 36π cubic inches per inch

To emphasize that the rate of change of the volume is found with respect to the height, use the symbol dV/dh for the derivative. For this tumor, r is fixed at 2 cm. By substituting 2 for r,

$$V = \frac{1}{3}\pi r^2 h \quad \text{becomes} \quad V = \frac{1}{3}\pi \cdot 2^2 \cdot h \quad \text{or} \quad V = \frac{4}{3}\pi h.$$

Since $4\pi/3$ is constant,

$$\frac{dV}{dh} = \frac{4\pi}{3} \approx 4.2 \text{ cu cm per cm.}$$

For each additional centimeter that the tumor grows in height, its volume will increase approximately 4.2 cubic centimeters. ◀ **9**

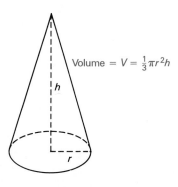

Volume $= V = \frac{1}{3}\pi r^2 h$

FIGURE 11.22

11.4 EXERCISES

Find the derivatives of the functions defined as follows. (See Examples 1–4.)

1. $f(x) = 9x^2 - 8x + 4$

2. $f(x) = 10x^2 + 4x - 9$

3. $y = 10x^3 - 9x^2 + 6x$

4. $y = 3x^3 - x^2 - 12x$

5. $y = x^4 - 5x^3 + 9x^2 + 5$

6. $y = 3x^4 + 11x^3 + 2x^2 - 4x$

7. $f(x) = 6x^{1.5} - 4x^{.5}$

8. $f(x) = -2x^{2.5} + 8x^{.5}$

9. $y = -15x^{3.2} + 2x^{1.9}$

10. $y = 18x^{1.6} - 4x^{3.1}$

11. $y = 24t^{3/2} + 4t^{1/2}$

12. $y = -24t^{5/2} - 6t^{1/2}$

13. $y = 8\sqrt{x} + 6x^{3/4}$

14. $y = -100\sqrt{x} - 11x^{2/3}$

15. $g(x) = 6x^{-5} - x^{-1}$

16. $y = -2x^{-4} + x^{-1}$

17. $y = -4x^{-2} - 3x^{-3}$

18. $g(x) = 8x^{-4} - 9x^{-2}$

19. $y = 10x^{-2} + 3x^{-4} - 6x$

20. $y = x^{-5} - x^{-2} + 5x^{-1}$

21. $f(t) = \dfrac{6}{t} - \dfrac{8}{t^2}$

22. $f(t) = \dfrac{4}{t} + \dfrac{2}{t^3}$

23. $y = \dfrac{9 - 8x + 2x^3}{x^4}$

24. $y = \dfrac{3 + x - 7x^4}{x^6}$

25. $f(x) = 12x^{-1/2} - 3x^{1/2}$

26. $r(x) = -30x^{-1/2} + 5x^{1/2}$

27. $p(x) = -10x^{-1/2} + 8x^{-3/2}$

28. $h(x) = x^{-1/2} - 14x^{-3/2}$

29. $y = \dfrac{6}{\sqrt[4]{x}}$

30. $y = \dfrac{-2}{\sqrt[3]{x}}$

31. $y = \dfrac{-5t}{\sqrt[3]{t^2}}$

32. $g(t) = \dfrac{9t}{\sqrt{t^3}}$

Find each of the following.

33. $\dfrac{dy}{dx}$ if $y = 8x^{-5} - 9x^{-4}$

34. $\dfrac{dy}{dx}$ if $y = -3x^{-2} - 4x^{-5}$

35. $D_x\left(9x^{-1/2} + \dfrac{2}{x^{3/2}}\right)$

36. $D_x\left(\dfrac{8}{\sqrt[4]{x}} - \dfrac{3}{\sqrt{x^3}}\right)$

37. $f'(-2)$ if $f(x) = 6x^2 - 4x$

38. $f'(3)$ if $f(x) = 9x^3 - 8x^2$

39. $f'(4)$ if $f(t) = 2\sqrt{t} - \dfrac{3}{\sqrt{t}}$

40. $f'(8)$ if $f(t) = -5\sqrt[3]{t} + \dfrac{6}{\sqrt[3]{t}}$

Find the slope of the tangent line to the graph of each function defined as follows at the given value of x. Find the equation of the tangent line in Exercises 41 and 42.

41. $y = x^4 - 5x^3 + 2$; $x = 2$

42. $y = -2x^5 - 7x^3 + 8x^2$; $x = 1$

43. $y = 3x^{3/2} - 2x^{1/2}$; $x = 9$

44. $y = -2x^{1/2} + x^{3/2}$; $x = 4$

45. $y = 5x^{-1} - 2x^{-2}$; $x = 3$

46. $y = -x^{-3} + x^{-2}$; $x = 1$

Work the following exercises. (See Examples 5–8.)

47. Management   Assume that a demand equation is given by $x = 5000 - 100p$. Find the marginal revenue for the following production levels (values of x). (Hint: solve the demand equation for p and use $R(x) = xp$.)
(a) 1000 units   (b) 2500 units   (c) 3000 units

48. Management   Suppose that for the situation in Exercise 47, the cost of producing x units is given by $C(x) = 3000 - 20x + .03x^2$. Find the marginal profit for each of the following production levels.
(a) 500 units   (b) 815 units   (c) 1000 units

49. Management   If the price of a product is given by

$$P(x) = \frac{-1000}{x} + 1000,$$

where x represents the demand for the product, find the rate of change of price when the demand is 10.

50. Management   Often sales of a new product grow rapidly at first and then level off with time. This is the case with the sales represented by the function

$$S(t) = 100 - 100t^{-1},$$

where t represents time in years. Find the rate of change of sales for the following values of t.
(a) $t = 1$ (b) $t = 10$

51. Natural Science   Suppose $P(t) = 100/t$ represents the percent of acid in a chemical solution after t days of exposure to an ultraviolet light source. Find the percent of acid in the solution after the following number of days.
(a) 1 day   (b) 100 days
(c) Find and interpret $P'(100)$.

52. Natural Science   Insulin affects the glucose, or blood sugar, level of some diabetics according to the function

$$G(x) = -.2x^2 + 450,$$

where $G(x)$ is the blood sugar level 1 hour after x units of insulin are injected. (This mathematical model is only approximate and valid only for values of x less than about 40.)
(a) Find $G(0)$.   (b) Find $G(25)$.

53. Use the function G of Exercise 52 and find dG/dx for the following values of x. Interpret your answers.
(a) $x = 10$ (b) $x = 25$

54. Natural Science   A short length of blood vessel has a cylindrical shape. The volume of a cylinder is given by $V = \pi r^2 h$. Suppose an experimental device is set up to measure the volume of blood in a blood vessel of fixed length 80 mm as the radius changes.
(a) Find dV/dr.

Suppose a drug is administered which causes the blood vessel to expand. Evaluate dV/dr for the following values of r and interpret your answers.

(b) 4 mm   **(c)** 6 mm   **(d)** 8 mm

55. Natural Science   In an experiment testing methods of sexually attracting male insects to sterile females, equal numbers of males and females of a certain species are permitted to intermingle. Assume that

$$M(t) = 4t^{3/2} + 2t^{1/2}$$

approximates the number of matings observed among the insects in an hour, where t is the temperature in degrees Celsius. (This formula is only valid for certain temperature ranges.) Find each of the following.

(a) $M(16)$ **(b)** $M(25)$
(c) The rate of change of M when $t = 16$
(d) Interpret your answer for part (c).

56. Management   The profit in dollars from the sale of x expensive cassette recorders is

$$P(x) = x^3 - 5x^2 + 7x + 10.$$

Find the marginal profit for the following values of x.

(a) $x = 4$ **(b)** $x = 8$ **(c)** $x = 10$ **(d)** $x = 12$

57. Management   The total cost to produce x handcrafted weathervanes is

$$C(x) = 100 + 8x - x^2 + 4x^3.$$

Find the marginal cost for the following values of x.

(a) $x = 0$ **(b)** $x = 4$ **(c)** $x = 6$ **(d)** $x = 8$

58. Management   An analyst has found that a company's costs and revenues for one product are given by

$$C(x) = 2x \quad \text{and} \quad R(x) = 6x - \frac{x^2}{1000},$$

respectively, where x is the number of items produced.
(a) Find the marginal cost function.
(b) Find the marginal revenue function.
(c) Using the fact that profit is the difference between revenue and costs, find the marginal profit function.
(d) What value of x makes marginal profit equal 0?
(e) Find the profit when the marginal profit is 0.
(As we shall see in the next chapter, this process is used to find *maximum* profit.)

Physical Science   *We saw earlier that the velocity of a particle moving in a straight line is given by*

$$\lim_{h \to 0} \frac{s(t + h) - s(t)}{h},$$

where $s(t)$ gives the position of the particle at time t. This limit is the derivative of $s(t)$, so the velocity of a particle is given by $s'(t)$. If $v(t)$ represents velocity at time t, then $v(t) = s'(t)$. For each of the following position functions, find (a) $v(t)$; (b) the velocity when $t = 0$, $t = 5$, and $t = 10$.

59. $s(t) = 6t + 5$

60. $s(t) = 9 - 2t$

61. $s(t) = 11t^2 + 4t + 2$

62. $s(t) = 25t^2 - 9t + 8$

63. $s(t) = 4t^3 + 8t^2$

64. $s(t) = -2t^3 + 4t^2 - 1$

65. Physical Science   If a rock is dropped from a 144-ft building, its position (in feet above the ground) is given by $s(t) = -16t^2 + 144$, where t is the time in seconds since it was dropped.
(a) What is its velocity 1 second after being dropped? 2 seconds after being dropped?
(b) When will it hit the ground?
(c) What is its velocity upon impact?

66. Physical Science   A ball is thrown vertically upward from the ground at a velocity of 64 feet per second. Its distance from the ground at t seconds is given by $s(t) = -16t^2 + 64t$.
(a) How fast is the ball moving 2 seconds after being thrown? 3 seconds after being thrown?
(b) How long after the ball is thrown does it reach its maximum height?
(c) How high will it go?

67. Perform each step and give reasons for your results in the following proof that the derivative of $y = x^n$ is $y' = n \cdot x^{n-1}$. (We prove this result only for positive integer values of n, but it is valid for all values of n.)
(a) Recall the binomial theorem from algebra:

$$(p + q)^n = p^n + n \cdot p^{n-1}q$$
$$+ \frac{n(n-1)}{2}p^{n-2}q^2 + \cdots + q^n.$$

Let $y = (x + h)^n$, and evaluate $(x + h)^n$.
(b) Find the quotient $\dfrac{(x + h)^n - x^n}{h}$.
(c) Use the definition of derivative to find y'.

68. Perform each step and give reasons for your result in the proof that the derivative of $y = f(x) + g(x)$ is $y' = f'(x) + g'(x)$.

(a) Let $s(x) = f(x) + g(x)$. Show that

$$s'(x) = \lim_{h \to 0} \frac{[f(x + h) + g(x + h)] - [f(x) + g(x)]}{h}.$$

(b) Show that $s'(x)$ equals

$$\lim_{h \to 0} \left[\frac{f(x + h) - f(x)}{h} + \frac{g(x + h) - g(x)}{h} \right].$$

(c) Finally, show that $s'(x) = f'(x) + g'(x)$.

11.5 DERIVATIVES OF PRODUCTS AND QUOTIENTS

In the previous section we developed several rules for finding derivatives. We develop two additional rules in this section, again using the definition of the derivative.

The derivative of a sum of two functions is found from the sum of the derivatives. What about products? Is the derivative of a product equal to the product of the derivatives? For example, if

$$u(x) = 2x + 3 \quad \text{and} \quad v(x) = 3x^2;$$

then $\qquad u'(x) = 2 \qquad$ and $\quad v'(x) = 6x.$

Let $f(x)$ be the product of u and v; that is, $f(x) = (2x + 3)(3x^2) = 6x^3 + 9x^2$. By the rules of the preceding section, $f'(x) = 18x^2 + 18x$. On the other hand, $u'(x) = 2$ and $v'(x) = 6x$, with the product $u'(x) \cdot v'(x) = 2(6x) = 12x \neq f'(x)$. In this example, the derivative of a product is *not* equal to the product of the derivatives, nor is this usually the case.

The rule for finding derivatives of products is given below.

Product Rule

If $f(x) = u(x) \cdot v(x)$, and if both $u'(x)$ and $v'(x)$ exist, then

$$f'(x) = u(x) \cdot v'(x) + v(x) \cdot u'(x).$$

(The derivative of a product of two functions is the first function times the derivative of the second, plus the second function times the derivative of the first.)

To sketch the method used to prove the product rule, let

$$f(x) = u(x) \cdot v(x).$$

1 Use the product rule to find the derivatives of the following.

(a) $f(x) = (5x^2 + 6)(3x)$

(b) $g(x) = (8x)(4x^2 + 5x)$

Answer:

(a) $45x^2 + 18$

(b) $96x^2 + 80x$

Then $f(x + h) = u(x + h) \cdot v(x + h)$, and, by definition, $f'(x)$ is given by

$$f'(x) = \lim_{h \to 0} \frac{f(x + h) - f(x)}{h}$$

$$= \lim_{h \to 0} \frac{u(x + h) \cdot v(x + h) - u(x) \cdot v(x)}{h}.$$

Now subtract and add $u(x + h) \cdot v(x)$ in the numerator, giving

$$f'(x) = \lim_{h \to 0} \frac{u(x + h) \cdot v(x + h) - u(x + h) \cdot v(x) + u(x + h) \cdot v(x) - u(x) \cdot v(x)}{h}$$

$$= \lim_{h \to 0} \frac{u(x + h)[v(x + h) - v(x)] + v(x)[u(x + h) - u(x)]}{h}$$

$$= \lim_{h \to 0} u(x + h)\left[\frac{v(x + h) - v(x)}{h}\right] + \lim_{h \to 0} v(x)\left[\frac{u(x + h) - u(x)}{h}\right]$$

$$= \lim_{h \to 0} u(x + h) \cdot \lim_{h \to 0} \frac{v(x + h) - v(x)}{h} + \lim_{h \to 0} v(x) \cdot \lim_{h \to 0} \frac{u(x + h) - u(x)}{h} \text{(*)}$$

If u' and v' both exist, then

$$\lim_{h \to 0} \frac{u(x + h) - u(x)}{h} = u'(x) \quad \text{and} \quad \lim_{h \to 0} \frac{v(x + h) - v(x)}{h} = v'(x).$$

The fact that u' exists can be used to prove

$$\lim_{h \to 0} u(x + h) = u(x),$$

and since no h is involved in $v(x)$,

$$\lim_{h \to 0} v(x) = v(x).$$

Substituting these results into equation (*) gives

$$f'(x) = u(x) \cdot v'(x) + v(x) \cdot u'(x),$$

the desired result.

▶ **EXAMPLE 1** Let $f(x) = (2x + 3)(3x^2)$. Use the product rule to find $f'(x)$.

Here f is given as the product of $u(x) = 2x + 3$ and $v(x) = 3x^2$. By the product rule and the fact that $u'(x) = 2$ and $v'(x) = 6x$,

$$f'(x) = u(x) \cdot v'(x) + v(x) \cdot u'(x)$$
$$= (2x + 3)(6x) + (3x^2)(2)$$
$$= 12x^2 + 18x + 6x^2 = 18x^2 + 18x.$$

This result is the same as that found at the beginning of the section. ◀ **1**

2 Find the derivatives of the following.

(a) $f(x) = (x^2 - 3)(\sqrt{x} + 5)$

(b) $g(x) = (\sqrt{x} + 4)(5x^2 + x)$

Answer:

(a) $\frac{5}{2}x^{3/2} + 10x - \frac{3}{2}x^{-1/2}$

(b) $\frac{25}{2}x^{3/2} + 40x$

$\quad + \frac{3}{2}x^{1/2} + 4$

3 Find the derivatives of the following.

(a) $f(x) = \dfrac{3x + 7}{5x + 8}$

(b) $g(x) = \dfrac{2x + 11}{5x - 1}$

Answer:

(a) $\dfrac{-11}{(5x + 8)^2}$

(b) $\dfrac{-57}{(5x - 1)^2}$

▶ **EXAMPLE 2** Find the derivative of $y = (\sqrt{x} + 3)(x^2 - 5x)$.

Let $u(x) = \sqrt{x} + 3 = x^{1/2} + 3$, and $v(x) = x^2 - 5x$. Then

$$y' = u(x) \cdot v'(x) + v(x) \cdot u'(x)$$

$$= (x^{1/2} + 3)(2x - 5) + (x^2 - 5x)\left(\frac{1}{2}x^{-1/2}\right).$$

$$= 2x^{3/2} + 6x - 5x^{1/2} - 15 + \frac{1}{2}x^{3/2} - \frac{5}{2}x^{1/2}$$

$$= \frac{5}{2}x^{3/2} + 6x - \frac{15}{2}x^{1/2} - 15. \quad ◀ \quad \boxed{2}$$

We could have found the derivatives above by multiplying out the original functions. The product rule then would not have been needed. In the next section, however, we shall see products of functions where the product rule is essential.

What about *quotients* of functions? To find the derivative of the quotient of two functions, use the next rule.

QUOTIENT RULE

If $f(x) = \dfrac{u(x)}{v(x)}$, if all indicated derivatives exist, and if $v(x) \neq 0$, then

$$f'(x) = \frac{v(x) \cdot u'(x) - u(x) \cdot v'(x)}{[v(x)]^2}.$$

(The derivative of a quotient is the denominator times the derivative of the numerator, minus the numerator times the derivative of the denominator, all divided by the square of the denominator.)

The proof of the quotient rule is similar to that of the product rule and is omitted here.

▶ **EXAMPLE 3** Find $f'(x)$ if $f(x) = \dfrac{2x - 1}{4x + 3}$.

Let $u(x) = 2x - 1$, with $u'(x) = 2$. Also, let $v(x) = 4x + 3$, with $v'(x) = 4$. Then, by the quotient rule,

$$f'(x) = \frac{v(x) \cdot u'(x) - u(x) \cdot v'(x)}{[v(x)]^2}$$

$$= \frac{(4x + 3)(2) - (2x - 1)(4)}{(4x + 3)^2}$$

$$= \frac{8x + 6 - 8x + 4}{(4x + 3)^2}$$

$$f'(x) = \frac{10}{(4x + 3)^2}. \quad ◀ \quad \boxed{3}$$

4 Find each derivative. Write answers with positive exponents.

(a) $D_x\left(\dfrac{x^{-2} - 1}{x^{-1} + 2}\right)$

(b) $D_x\left(\dfrac{2 + x^{-1}}{x^{-3} + 1}\right)$

Answer:

(a) $\dfrac{-1 - 4x - x^2}{x^2 + 4x^3 + 4x^4}$

(b) $\dfrac{2x + 6x^2 - x^4}{1 + 2x^3 + x^6}$

5 Find each derivative.

(a) $D_x\left(\dfrac{(3x - 1)(4x + 2)}{2x}\right)$

(b) $D_x\left(\dfrac{5x^2}{(2x + 1)(x - 1)}\right)$

Answer:

(a) $\dfrac{6x^2 + 1}{x^2}$

(b) $\dfrac{-5x^2 - 10x}{(2x + 1)^2(x - 1)^2}$

▶**EXAMPLE 4** Find $D_x\left(\dfrac{x^{-1} - 2}{x^{-2} + 4}\right)$.

Use the quotient rule.

$$D_x\left(\frac{x^{-1} - 2}{x^{-2} + 4}\right) = \frac{(x^{-2} + 4) \cdot D_x(x^{-1} - 2) - (x^{-1} - 2) \cdot D_x(x^{-2} + 4)}{(x^{-2} + 4)^2}$$

$$= \frac{(x^{-2} + 4)(-x^{-2}) - (x^{-1} - 2)(-2x^{-3})}{(x^{-2} + 4)^2}$$

$$= \frac{-x^{-4} - 4x^{-2} + 2x^{-4} - 4x^{-3}}{(x^{-2} + 4)^2} = \frac{x^{-4} - 4x^{-3} - 4x^{-2}}{(x^{-2} + 4)^2}$$

When derivatives are used to solve practical problems, the work is easier if the derivatives are first simplified. This derivative, for example, can be simplified if the negative exponents are removed by multiplying numerator and denominator by x^4, and using $(x^{-2} + 4)^2 = x^{-4} + 8x^{-2} + 16$.

$$\frac{x^{-4} - 4x^{-3} - 4x^{-2}}{(x^{-2} + 4)^2} = \frac{1 - 4x - 4x^2}{x^4(x^{-4} + 8x^{-2} + 16)} = \frac{1 - 4x - 4x^2}{1 + 8x^2 + 16x^4} \quad ◀ \boxed{4}$$

▶**EXAMPLE 5** Find $Dx\left(\dfrac{(3 - 4x)(5x + 1)}{7x - 9}\right)$.

This function has a product within a quotient. Instead of multiplying the factors in the numerator first (which is an option), we can use the quotient rule together with the product rule, as follows. Use the quotient rule first to get

$$D_x\left(\frac{(3 - 4x)(5x + 1)}{7x - 9}\right)$$

$$= \frac{(7x - 9)[D_x(3 - 4x)(5x + 1)] - [(3 - 4x)(5x + 1)D_x(7x - 9)]}{(7x - 9)^2}$$

Now use the product rule to find $D_x(3 - 4x)(5x + 1)$ in the numerator.

$$= \frac{(7x - 9)[(3 - 4x)5 + (5x + 1)(-4)] - (3 + 11x - 20x^2)(7)}{(7x - 9)^2}$$

$$= \frac{(7x - 9)(15 - 20x - 20x - 4) - (21 + 77x - 140x^2)}{(7x - 9)^2}$$

$$= \frac{(7x - 9)(11 - 40x) - 21 - 77x + 140x^2}{(7x - 9)^2}$$

$$= \frac{-280x^2 + 437x - 99 - 21 - 77x + 140x^2}{(7x - 9)^2}$$

$$= \frac{-140x^2 + 360x - 120}{(7x - 9)^2} \quad ◀ \boxed{5}$$

6 The total revenue in thousands of dollars from the sale of x dozen CB radios is given by

$$R(x) = 32x^2 + 7x + 80.$$

(a) Find the average revenue.

(b) Find the marginal average revenue.

Answer:

(a) $\dfrac{32x^2 + 7x + 80}{x}$

(b) $\dfrac{32x^2 - 80}{x^2}$

AVERAGE COST Suppose $y = C(x)$ gives the total cost to manufacture x items. As mentioned earlier, the average cost per item is found by dividing the total cost by the number of items. The rate of change of average cost, called the *marginal average cost,* is the derivative of the average cost.

Average Cost

If the total cost to manufacture x items is given by $C(x)$, then the **average cost per item** is

$$\overline{C}(x) = \frac{C(x)}{x}.$$

The **marginal average cost** is the derivative of the average cost function, $\overline{C}'(x)$.

A company naturally would be interested in making the average cost as small as possible. We will see in the next chapter that this can be done by using the derivative of $C(x)/x$. This derivative often can be found with the quotient rule, as in the next example.

▶**EXAMPLE 6** The total cost in thousands of dollars to manufacture x electrical generators is given by $C(x)$, where

$$C(x) = -x^3 + 15x^2 + 1000.$$

(a) Find the average cost per generator.

The average cost is given by the total cost divided by the number of items, or

$$\frac{C(x)}{x} = \frac{-x^3 + 15x^2 + 1000}{x}.$$

(b) Find the marginal average cost.

The marginal average cost is the derivative of the average cost function. Using the quotient rule,

$$\frac{d}{dx}\left(\frac{C(x)}{x}\right) = \frac{x(-3x^2 + 30x) - (-x^3 + 15x^2 + 1000)(1)}{x^2}$$

$$= \frac{-3x^3 + 30x^2 + x^3 - 15x^2 - 1000}{x^2}$$

$$= \frac{-2x^3 + 15x^2 - 1000}{x^2}. \quad ◀ \boxed{6}$$

7 If the cost in Example 7 is given by

$$C(x) = x^2 + 10x + 16,$$

find the production level at which the marginal average cost is zero.

Answer:
400 items

▶**EXAMPLE 7** Suppose the cost in dollars of manufacturing x hundred items is given by

$$C(x) = 3x^2 + 7x + 12.$$

(a) Find the average cost.
The average cost is

$$\overline{C}(x) = \frac{C(x)}{x} = \frac{3x^2 + 7x + 12}{x} = 3x + 7 + \frac{12}{x}.$$

(b) Find the marginal average cost.
The marginal average cost is

$$\frac{d}{dx}(\overline{C}(x)) = \frac{d}{dx}\left(3x + 7 + \frac{12}{x}\right) = 3 - \frac{12}{x^2}.$$

(c) Find the marginal cost.
The marginal cost is

$$\frac{d}{dx}(C(x)) = \frac{d}{dx}(3x^2 + 7x + 12) = 6x + 7.$$

(d) Find the level of production at which the marginal average cost is zero.
Set the derivative $\overline{C}'(x) = 0$ and solve for x.

$$3 - \frac{12}{x^2} = 0$$

$$\frac{3x^2 - 12}{x^2} = 0$$

$$3x^2 - 12 = 0$$

$$x^2 = 4$$

$$x = 2$$

Since x is in hundreds, production of 200 items will produce a marginal average cost of zero dollars. ◀ **7**

11.5 EXERCISES

Use the product rule to find the derivative of the functions defined as follows. (See Examples 1 and 2.) (In Exercises 11–14, use the fact that $p^2 = p \cdot p$.)

1. $y = (2x - 5)(x + 4)$

2. $y = (3x + 7)(x - 1)$

3. $y = (8x - 2)(3x + 9)$

4. $y = (4x + 1)(7x + 12)$

5. $y = (3x^2 + 2)(2x - 1)$

6. $y = (5x^4 - 1)(4x + 3)$

7. $y = (x^2 + x)(3x - 5)$

8. $y = (2x^2 - 6x)(x + 2)$

9. $y = (9x^4 + 7x)(x^2 - 1)$

10. $y = (2x^2 - 4x)(5x^3 + 4)$

11. $y = (2x - 5)^2$

12. $y = (7x - 6)^2$

13. $y = (x^2 - 1)^2$

14. $y = (3x^3 + 2)^2$

15. $y = (x + 1)(\sqrt{x} + 2)$

16. $y = (2x - 3)(\sqrt{x} - 1)$

17. $y = (5\sqrt{x} - 1)(2\sqrt{x} + 1)$

18. $y = (-3\sqrt{x} + 6)(4\sqrt{x} - 2)$

Use the quotient rule to find the derivatives of the functions defined as follows. (See Examples 3 and 4.)

19. $y = \dfrac{x + 1}{2x - 1}$

20. $y = \dfrac{3x - 5}{x - 4}$

21. $f(x) = \dfrac{7x + 1}{3x + 8}$

22. $f(x) = \dfrac{6x - 11}{8x + 1}$

23. $y = \dfrac{2}{3x - 5}$

24. $y = \dfrac{-4}{2x - 11}$

25. $y = \dfrac{5 - 3x}{4 + x}$

26. $y = \dfrac{9 - 7x}{1 - x}$

27. $f(t) = \dfrac{t^2 + t}{t - 1}$

28. $f(t) = \dfrac{t^2 - 4t}{t + 3}$

29. $y = \dfrac{x - 2}{x^2 + 1}$

30. $y = \dfrac{4x + 11}{x^2 - 3}$

31. $g(x) = \dfrac{3x^2 + x}{2x^3 - 1}$

32. $k(x) = \dfrac{-x^2 + 6x}{4x^3 + 1}$

33. $y = \dfrac{x^2 - 4x + 2}{x + 3}$

34. $y = \dfrac{x^2 + 7x - 2}{x - 2}$

35. $p(t) = \dfrac{\sqrt{t}}{t - 1}$

36. $r(t) = \dfrac{\sqrt{t}}{2t + 3}$

37. $y = \dfrac{5x + 6}{\sqrt{x}}$

38. $y = \dfrac{9x - 8}{\sqrt{x}}$

Find the derivative of each of the following. (See Example 5.)

39. $f(p) = \dfrac{(2p + 3)(4p - 1)}{3p + 2}$

40. $g(t) = \dfrac{(5t - 2)(2t + 3)}{t - 4}$

41. $g(x) = \dfrac{x^3 + 1}{(2x + 1)(5x + 2)}$

42. $f(x) = \dfrac{x^3 - 4}{(2x + 1)(3x - 2)}$

Work the following exercises. (See Examples 6 and 7.)

43. Management The total cost to produce x units of perfume is given by

$$C(x) = (3x + 2)(3x + 4).$$

Find the average cost per unit to produce the following.
(a) 10 units   **(b)** 20 units   **(c)** x units
(d) Find the derivative of the average cost function.

44. Management The total profit from selling x units of mathematics textbooks is given by

$$P(x) = (5x - 6)(2x + 3).$$

Find the average profit from selling
(a) 8 units;   **(b)** 15 units;   **(c)** x units.
(d) Find the marginal average profit.

45. Management A company that manufactures bicycles has determined that a new employee can assemble $M(d)$ bicycles per day after d days of on-the-job training, where

$$M(d) = \frac{200d}{3d + 10}.$$

(a) Find $M'(d)$.
(b) Find and interpret $M'(2)$ and $M'(5)$.

46. Management Suppose you are the manager of a trucking firm, and one of your drivers reports that, according to her calculations, her truck burns fuel at the rate of

$$G(x) = \frac{1}{200}\left(\frac{800}{x} + x\right)$$

gallons per mile when traveling at x miles per hour on a smooth, dry road.

(a) If the driver tells you that she wants to travel 20 miles per hour, what should you tell her? (Hint: take the derivative of G and evaluate it for $x = 20$. Then interpret your results.)
(b) If the driver wants to go 40 miles per hour, what should you say? (Hint: find $G'(40)$.)

47. Natural Science When a certain drug is introduced into a muscle, the muscle responds by contracting. The amount of contraction, s, in millimeters, is related to the concentration of the drug, x, in milliliters, by

$$s(x) = \frac{x}{m + nx},$$

where m and n are constants.
(a) Find $s'(x)$.
(b) Evaluate $s'(x)$ when $x = 50$, $m = 10$, and $n = 3$.
(c) Interpret your results in part (b).

48. Social Science According to the psychologist L. L. Thurstone, the number of facts of a certain type that are remembered after t hours is given by

$$f(t) = \frac{kt}{at - m},$$

where k, m, and a are constants. Find $f'(t)$ if $a = 99$, $k = 90$, $m = 90$, and
(a) $t = 1$; **(b)** $t = 10$.
(c) Interpret the results.

11.6 THE CHAIN RULE

Many of the most useful functions for applications are created by combining simpler functions. Viewing complex functions as combinations of simpler functions often makes them easier to understand and use.

COMPOSITION OF FUNCTIONS Suppose a function f assigns to each element x in set X some element $y = f(x)$ in set Y. Suppose also that a function g takes each element in set Y and assigns to it a value $z = g(f(x))$ in set Z. By using both f and g, an element x in X is assigned to an element z in Z, as illustrated in Figure 11.23.

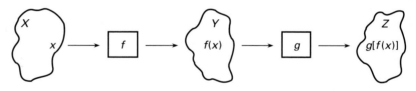

FIGURE 11.23

The result of this process is a new function called the *composition* of functions g and f and defined as follows.

COMPOSITE FUNCTION

Let f and g be functions. A **composite function,** or **composition,** of g and f is the function whose values are given by $g[f(x)]$ for all x in the domain of f such that $f(x)$ is in the domain of g.

▶**EXAMPLE 1** Let $f(x) = 2x - 1$ and $g(x) = \sqrt{3x + 5}$. Find each of the following.
(a) $g[f(4)]$
   Find $f(4)$ first.

$$f(4) = 2 \cdot 4 - 1 = 8 - 1 = 7$$

1 Let $f(x) = 3x - 2$ and $g(x) = (x - 1)^5$. Find the following.

(a) $g[f(2)]$

(b) $f[g(2)]$

Answer:

(a) 243

(b) 1

2 Let $f(x) = \sqrt{x + 4}$ and $g(x) = x^2 + 5x + 1$. Find $f[g(x)]$ and $g[f(x)]$.

Answer:
$f[g(x)] = \sqrt{x^2 + 5x + 5}$,
$g[f(x)] = (\sqrt{x + 4})^2$
$\qquad + 5\sqrt{x + 4} + 1$
$\qquad = x + 5 + 5\sqrt{x + 4}$

Then

$$g[f(4)] = g[7] = \sqrt{3 \cdot 7 + 5} = \sqrt{26}.$$

(b) $f[g(4)]$

Since $g(4) = \sqrt{3 \cdot 4 + 5} = \sqrt{17}$.

$$f[g(4)] = 2 \cdot \sqrt{17} - 1 = 2\sqrt{17} - 1.$$

(c) $f[g(-2)]$ does not exist since -2 is not in the domain of g. ◀ **1**

▶**EXAMPLE 2** Let $f(x) = 4x + 1$ and $g(x) = 2x^2 + 5x$. Find each of the following.

(a) $g[f(x)]$

Using the given functions,

$$g[f(x)] = g[4x + 1]$$
$$= 2(4x + 1)^2 + 5(4x + 1)$$
$$= 2(16x^2 + 8x + 1) + 20x + 5$$
$$= 32x^2 + 16x + 2 + 20x + 5$$
$$= 32x^2 + 36x + 7.$$

(b) $f[g(x)]$

By the definition above, with f and g interchanged,

$$f[g(x)] = f[2x^2 + 5x]$$
$$= 4(2x^2 + 5x) + 1$$
$$= 8x^2 + 20x + 1. \quad ◀ \; \boxed{2}$$

As Example 2 shows, it is not always true that $f[g(x)] = g[f(x)]$. In fact, it is rare to find two functions f and g such that $f[g(x)] = g[f(x)]$.

THE CHAIN RULE A leaking oil well off the Gulf Coast is spreading a circular film of oil over the water surface. At any time t, in minutes, after the beginning of the leak, the radius of the circular oil slick is given by

$$r(t) = 4t, \quad \text{with} \quad \frac{dr}{dt} = 4,$$

where dr/dt is the rate of change in radius over time. The area of the oil slick is given by

$$A(r) = \pi r^2, \quad \text{with} \quad \frac{dA}{dr} = 2\pi r,$$

where dA/dr is the rate of change in area per unit change in radius.

3 Use the chain rule to find dy/dx if $y = 10(2x^2 + 1)^4$.

Answer:
$160x(2x^2 + 1)^3$

As these derivatives show, the radius is increasing 4 times as fast as the time t, and the area is increasing $2\pi r$ times as fast as the radius r. It seems reasonable, then, that the area is increasing $2\pi r \cdot 4 = 8\pi r$ times as fast as time. That is,

$$\frac{dA}{dt} = \frac{dA}{dr} \cdot \frac{dr}{dt} = 2\pi r \cdot 4 = 8\pi r.$$

Since $r = 4t$,

$$\frac{dA}{dt} = 8\pi(4t) = 32\pi t.$$

To check this, use the fact that $r = 4t$ and $A = \pi r^2$ to get the same result:

$$A = \pi(4t)^2 = 16\pi t^2, \quad \text{with} \quad \frac{dA}{dt} = 32\pi t.$$

The product used above, $$\frac{dA}{dt} = \frac{dA}{dr} \cdot \frac{dr}{dt},$$

is an example of the **chain rule,** which is used to find the derivative of a composite function.

Chain Rule

If y is a function of u, say $y = f(u)$, and if u is a function of x, say $u = g(x)$, then $y = f(u) = f[g(x)]$, and

$$\frac{dy}{dx} = \frac{dy}{du} \cdot \frac{du}{dx}.$$

One way to remember the chain rule is to *pretend* that dy/du and du/dx are fractions, with du "canceling out." The proof of the chain rule requires advanced concepts, and is not given here.

▶**EXAMPLE 3** Find dy/dx if $y = (3x^2 - 5x)^{1/2}$.

Let $y = u^{1/2}$, and $u = 3x^2 - 5x$. Then

$$\frac{dy}{dx} = \frac{dy}{du} \cdot \frac{du}{dx}$$

$$= \frac{1}{2}u^{-1/2} \cdot (6x - 5).$$

Replacing u with $3x^2 - 5x$ gives

$$\frac{dy}{dx} = \frac{1}{2}(3x^2 - 5x)^{-1/2}(6x - 5) = \frac{6x - 5}{2(3x^2 - 5x)^{1/2}} \quad ◀ \boxed{3}$$

4 Use the alternative form of the chain rule to find $D_x\sqrt{14x + 1}$.

Answer:

$$\frac{7}{\sqrt{14x + 1}}$$

The following alternative version of the chain rule is stated in terms of composite functions.

The Chain Rule (alternative form)

If $y = f[g(x)]$, then

$$y' = f'[g(x)] \cdot g'(x).$$

(To find the derivative of $f[g(x)]$, find the derivative of $f(x)$, replace each x with $g(x)$, and then multiply the result by the derivative of $g(x)$.)

▶ **EXAMPLE 4** Use the chain rule to find $D_x\sqrt{15x^2 + 1}$.

Write $\sqrt{15x^2 + 1}$ as $(15x^2 + 1)^{1/2}$. Let $f(x) = x^{1/2}$ and $g(x) = 15x^2 + 1$. Then $\sqrt{15x^2 - 1} = f[g(x)]$ and

$$D_x(15x^2 + 1)^{1/2} = f'[g(x)] \cdot g'(x).$$

Here $f'(x) = \frac{1}{2}x^{-1/2}$, with $f'[g(x)] = \frac{1}{2}[g(x)]^{-1/2} = \frac{1}{2}(15x^2 + 1)^{-1/2}$, and

$$D_x\sqrt{15x^2 + 1} = \frac{1}{2}[g(x)]^{-1/2} \cdot g'(x)$$

$$= \frac{1}{2}(15x^2 + 1)^{-1/2} \cdot (30x)$$

$$= \frac{15x}{(15x^2 + 1)^{1/2}}. \quad ◀ \quad \boxed{4}$$

While the chain rule is essential for finding the derivatives of some of the functions discussed later, the derivatives of the algebraic functions discussed so far can be found by the following *generalized power rule,* a special case of the chain rule.

Generalized Power Rule

Let u be a function of x, and let $y = u^n$, for any real number n. Then

$$y' = n \cdot u^{n-1} \cdot u'.$$

(The derivative of $y = u^n$ is found by decreasing the exponent on u by 1 and multiplying the result by the exponent n and by the derivative of u with respect to x.)

5 Find dy/dx for the following.

(a) $y = (2x + 5)^6$

(b) $y = (4x^2 - 7)^3$

(c) $f(x) = \sqrt{3x^2 - x}$

(d) $g(x) = (2 - x^4)^{-3}$

Answer:

(a) $12(2x + 5)^5$

(b) $24x(4x^2 - 7)^2$

(c) $\dfrac{6x - 1}{2\sqrt{3x^2 - x}}$

(d) $\dfrac{12x^3}{(2 - x^4)^4}$

6 Find dy/dx for the following.

(a) $y = 12(x^2 + 6)^5$

(b) $y = 8(4x^2 + 2)^{3/2}$

Answer:

(a) $120x(x^2 + 6)^4$

(b) $96x(4x^2 + 2)^{1/2}$

▶**EXAMPLE 5** **(a)** Use the generalized power rule to find the derivative of $y = (3 + 5x)^2$.

Let $u = 3 + 5x$, and $n = 2$. Then $u' = 5$. By the generalized power rule,

$$y' = \frac{dy}{dx} = n \cdot u^{n-1} \cdot u'$$

$$= \overset{n}{2} \cdot \overset{u}{(3 + 5x)}^{\overset{n-1}{2-1}} \cdot \overset{u'}{\frac{d}{dx}(3 + 5x)}$$

$$= 2(3 + 5x)^{2-1} \cdot 5 = 10(3 + 5x)$$

$$= 30 + 50x.$$

(b) Find y' if $y = (3 + 5x)^{-3/4}$.

Use the generalized power rule with $n = -\dfrac{3}{4}$, $u = 3 + 5x$, and $u' = 5$.

$$y' = -\frac{3}{4}(3 + 5x)^{(-3/4)-1}(5)$$

$$= -\frac{15}{4}(3 + 5x)^{-7/4}$$

This result could not have been found by any of the rules given earlier. ◀ **5**

▶**EXAMPLE 6** Find the derivative of the following.

(a) $y = 2(7x^2 + 5)^4$

Let $u = 7x^2 + 5$. Then $u' = 14x$, and

$$y' = \overset{n}{2} \cdot \overset{u}{4(7x^2 + 5)}^{\overset{n-1}{4-1}} \cdot \overset{u'}{\frac{d}{dx}(7x^2 + 5)}$$

$$= 2 \cdot 4(7x^2 + 5)^3(14x)$$

$$= 112x(7x^2 + 5)^3.$$

(b) $y = \sqrt{9x + 2}$

Write $y = \sqrt{9x + 2}$ as $y = (9x + 2)^{1/2}$. Then

$$y' = \frac{1}{2}(9x + 2)^{-1/2}(9) = \frac{9}{2}(9x + 2)^{-1/2}.$$

The derivative also can be written as

$$y' = \frac{9}{2(9x + 2)^{1/2}} \quad \text{or} \quad y' = \frac{9}{2\sqrt{9x + 2}}. \quad ◀ \boxed{6}$$

7 Find the derivatives of the following.

(a) $y = 6x(x + 2)^2$

(b) $y = -9x(2x^2 + 1)^3$

Answer:

(a) $6(x + 2)(3x + 2)$

(b) $-9(2x^2 + 1)^2(14x^2 + 1)$

8 Find the derivatives of the following.

(a) $y = \dfrac{(2x + 1)^3}{3x}$

(b) $y = \dfrac{(x - 6)^5}{3x - 5}$

Answer:

(a) $\dfrac{(2x + 1)^2(4x - 1)}{3x^2}$

(b) $\dfrac{(x - 6)^4(12x - 7)}{(3x - 5)^2}$

Sometimes both the generalized power rule and either the product or quotient rule are needed to find a derivative, as the next examples show.

▶**EXAMPLE 7**   Find the derivative of $y = 4x(3x + 5)^5$.
Write $4x(3x + 5)^5$ as the product

$$4x \cdot (3x + 5)^5.$$

To find the derivative of $(3x + 5)^5$, let $u = 3x + 5$ with $u' = 3$. Now use the product rule and the generalized power rule.

$$y' = 4x\overbrace{[5(3x + 5)^4 \cdot 3]}^{\text{derivative of }(3x + 5)^5} + (3x + 5)^5\overbrace{(4)}^{\text{derivative of }4x}$$
$$= 60x(3x + 5)^4 + 4(3x + 5)^5$$
$$= 4(3x + 5)^4[15x + (3x + 5)^1] \qquad \text{Factor out the greatest}$$
$$\text{common factor, }4(3x + 5)^4$$
$$= 4(3x + 5)^4(18x + 5). \quad ◀ \boxed{7}$$

▶**EXAMPLE 8**   Find the derivative of $y = \dfrac{(3x + 2)^7}{x - 1}$.
Use the quotient rule and the generalized power rule.

$$\frac{dy}{dx} = \frac{(x - 1)[7(3x + 2)^6 \cdot 3] - (3x + 2)^7(1)}{(x - 1)^2}$$
$$= \frac{21(x - 1)(3x + 2)^6 - (3x + 2)^7}{(x - 1)^2}$$
$$= \frac{(3x + 2)^6[21(x - 1) - (3x + 2)]}{(x - 1)^2} \qquad \text{Factor out the greatest}$$
$$\text{common factor, }(3x + 2)^6$$
$$= \frac{(3x + 2)^6[21x - 21 - 3x - 2]}{(x - 1)^2} \qquad \text{Simplify inside brackets.}$$
$$\frac{dy}{dx} = \frac{(3x + 2)^6(18x - 23)}{(x - 1)^2} \quad ◀ \boxed{8}$$

Some applications requiring the use of the chain rule or the generalized power rule are illustrated in the next examples.

▶**EXAMPLE 9**   The revenue realized by a small city from the collection of fines from parking tickets is given by

$$R(n) = \frac{8000n}{n + 2},$$

where n is the number of work hours each day that can be devoted to parking patrol. At the outbreak of a flu epidemic, 30 work hours are used daily in parking patrol,

9 Suppose the revenue in Example 9 is given by

$$R(n) = \frac{4500n}{n + 5}$$

and the work hours are decreasing at the rate of 4 per day. How fast is the revenue decreasing?

Answer:
About $73.47 per day

but during the epidemic that number is decreasing at the rate of 6 work hours per day. How fast is revenue from parking fines decreasing during the epidemic?

We want to find dR/dt, the change in revenue with respect to time. By the chain rule,

$$\frac{dR}{dt} = \frac{dR}{dn} \cdot \frac{dn}{dt}.$$

First find dR/dn, as follows.

$$\frac{dR}{dn} = \frac{(n + 2)(8000) - 8000n(1)}{(n + 2)^2} = \frac{16000}{(n + 2)^2}$$

Since $n = 30$, $dR/dn = 15.625$. Also, $dn/dt = -6$. Thus,

$$\frac{dR}{dt} = (15.625)(-6) = -93.75.$$

Revenue is being lost at the rate of approximately $94 per day. ◀ **9**

▶**EXAMPLE 10**   Suppose a sum of $500 is deposited in an account with an interest rate of r percent per year compounded monthly. At the end of 10 years, the balance in the account is given by

$$A = 500\left(1 + \frac{r}{1200}\right)^{120}.$$

Find the rate of change of A with respect to r if $r = 5$, 7, or 9.

First find dA/dr using the generalized power rule.

$$\frac{dA}{dr} = (120)(500)\left(1 + \frac{r}{1200}\right)^{119}\left(\frac{1}{1200}\right) = 50\left(1 + \frac{r}{1200}\right)^{119}$$

If $r = 5$,

$$\frac{dA}{dr} = 50\left(1 + \frac{5}{1200}\right)^{119} \approx 82.01,$$

or $82.01 per percentage point. If $r = 7$,

$$\frac{dA}{dr} = 50\left(1 + \frac{7}{1200}\right)^{119} \approx 99.90,$$

or $99.90 per percentage point. If $r = 9$,

$$\frac{dA}{dr} = 50\left(1 + \frac{9}{1200}\right)^{119} \approx 121.66.$$

or $121.66 per percentage point. ◀

The chain rule can be used to develop the formula for **marginal revenue product,** an economic concept that approximates the change in revenue when a manufacturer hires an additional employee. Start with $R = px$, where R is total revenue from the daily production of x units and p is the price per unit. The demand function is $p = f(x)$, as before. Also, x can be considered a function of the number of employees, n. Since $R = px$, and x and therefore p depend on n, R can also be considered a function of n. To find an expression for dR/dn, use the product rule for derivatives on the function $R = px$ to get

$$\frac{dR}{dn} = p \cdot \frac{dx}{dn} + x \cdot \frac{dp}{dn}. \tag{1}$$

By the chain rule,

$$\frac{dp}{dn} = \frac{dp}{dx} \cdot \frac{dx}{dn}.$$

Substituting for dp/dn in equation (1) gives

$$\frac{dR}{dn} = p \cdot \frac{dx}{dn} + x\left(\frac{dp}{dx} \cdot \frac{dx}{dn}\right)$$

$$= \left(p + x \cdot \frac{dp}{dx}\right)\frac{dx}{dn}. \qquad \text{Factor out } \frac{dx}{dn}$$

The equation for dR/dn gives the marginal revenue product.

▶**EXAMPLE 11** Find the marginal revenue product dR/dn (in dollars) when $n = 20$ if the demand function is $p = 600/\sqrt{x}$ and $x = 5n$.

As shown above,

$$\frac{dR}{dn} = \left(p + x \cdot \frac{dp}{dx}\right)\frac{dx}{dn}.$$

Find dp/dx and dx/dn. From

$$p = \frac{600}{\sqrt{x}} = 600x^{-1/2},$$

we have the derivative

$$\frac{dp}{dx} = -300x^{-3/2}.$$

Also, from $x = 5n$,

$$\frac{dx}{dn} = 5.$$

Then, by substitution,

$$\frac{dR}{dn} = \left[\frac{600}{\sqrt{x}} + x(-300x^{-3/2})\right]5 = \frac{1500}{\sqrt{x}}.$$

10 Find the marginal revenue product at $n = 10$ if the demand function is $p = 1000/x^2$ and $x = 8n$. Interpret your answer.

Answer:
$-$\$1.25; hiring an additional employee will produce a decrease in revenue of \$1.25.

If $n = 20$, then $x = 100$ and

$$\frac{dR}{dn} = \frac{1500}{\sqrt{100}}$$

$$= 150.$$

This means that hiring an additional employee when production is at a level of 20 items will produce an increase in revenue of \$150. ◀ **10**

11.6 EXERCISES

Let $f(x) = 4x^2 - 2x$ and $g(x) = 8x + 1$. Find each of the following. (See Example 1.)

1. $f[g(2)]$ **2.** $f[g(-5)]$ **3.** $g[f(2)]$ **4.** $g[f(-5)]$ **5.** $f[g(k)]$ **6.** $g[f(5z)]$

Find $f[g(x)]$ and $g[f(x)]$ in each of the following. (See Example 2.)

7. $f(x) = 8x + 12$; $g(x) = 3x - 1$ **8.** $f(x) = -6x + 9$; $g(x) = 5x + 7$ **9.** $f(x) = -x^3 + 2$; $g(x) = 4x$

10. $f(x) = 2x$; $g(x) = 6x^2 - x^3$ **11.** $f(x) = \dfrac{1}{x}$; $g(x) = x^2$ **12.** $f(x) = \dfrac{2}{x^4}$; $g(x) = 2 - x$

13. $f(x) = \sqrt{x + 2}$; $g(x) = 8x^2 - 6$ **14.** $f(x) = 9x^2 - 11x$; $g(x) = 2\sqrt{x + 2}$

Write each function as a composition of two functions. (There may be more than one way to do this.)

15. $y = (3x - 7)^{1/3}$ **16.** $y = (5 - x)^{2/5}$ **17.** $y = \sqrt{9 - 4x}$

18. $y = -\sqrt{13 + 7x}$ **19.** $y = \dfrac{\sqrt{x} + 3}{\sqrt{x} - 3}$ **20.** $y = \dfrac{2}{\sqrt{x} + 5}$

21. $y = (x^{1/2} - 3)^2 + (x^{1/2} - 3) + 5$ **22.** $y = (x^2 + 5x)^{1/3} - 2(x^2 + 5x)^{2/3} + 7$

Find the derivatives of the functions defined as follows. (See Examples 3–6.)

23. $y = (2x + 9)^2$ **24.** $y = (8x - 3)^2$ **25.** $y = 6(5x - 1)^3$ **26.** $y = -8(3x + 2)^3$

27. $y = -2(12x^2 + 4)^3$ **28.** $y = 5(3x^2 - 5)^3$ **29.** $y = 9(x^2 + 5x)^4$ **30.** $y = -3(x^2 - 5x)^4$

31. $y = 12(2x + 5)^{3/2}$ **32.** $y = 45(3x - 8)^{3/2}$ **33.** $y = -7(4x^2 + 9x)^{3/2}$ **34.** $y = 11(5x^2 + 6x)^{3/2}$

35. $y = 8\sqrt{4x + 7}$ **36.** $y = -3\sqrt{7x - 1}$ **37.** $y = -2\sqrt{x^2 + 4x}$ **38.** $y = 4\sqrt{2x^2 + 3}$

Use the product or quotient rule to find the derivatives of the functions defined as follows. (See Examples 7–8.)

39. $y = 4x(2x + 3)^2$ **40.** $y = -6x(5x - 1)^2$ **41.** $y = (x + 2)(x - 1)^2$ **42.** $y = (3x + 1)^2(x + 4)$

43. $y = 5(x + 3)^2(2x - 1)^5$ **44.** $y = -9(x + 4)^2(2x - 3)^2$ **45.** $y = (3x + 1)^3\sqrt{x}$ **46.** $y = (3x + 5)^2\sqrt{x}$

47. $y = \dfrac{1}{(x - 4)^2}$ **48.** $y = \dfrac{-5}{(2x + 1)^2}$ **49.** $y = \dfrac{(4x + 3)^2}{2x - 1}$ **50.** $y = \dfrac{(x - 6)^2}{3x + 4}$

51. $y = \dfrac{x^2 + 4x}{(5x + 2)^3}$ **52.** $y = \dfrac{3x^2 - x}{(x - 1)^2}$ **53.** $y = (x^{1/2} + 1)(x^{1/2} - 1)^{1/2}$ **54.** $y = (3 - x^{2/3})(x^{2/3} + 2)^{1/2}$

Work the following exercises. (See Examples 9–11.)

55. Management Suppose the demand for a certain brand of vacuum cleaner is given by

$$D(p) = \frac{-p^2}{100} + 500,$$

where p is the price in dollars. If the price, in terms of the cost c, is expressed as

$$p(c) = 2c - 10,$$

find the demand in terms of the cost.

56. Management Assume that the total revenue from the sale of x television sets is given by

$$R(x) = 1000\left(1 - \frac{x}{500}\right)^2.$$

Find the marginal revenue for the following values of x.
(a) $x = 100$ **(b)** $x = 150$
(c) $x = 200$ **(d)** $x = 400$
(e) Find the average revenue from the sale of x sets.
(f) Find the marginal average revenue.

57. Management A sum of $1500 is deposited in an account with an interest rate of r percent per year compounded daily. At the end of 5 years, the balance in the account is given by

$$A = 1500\left(1 + \frac{r}{36500}\right)^{1825}.$$

Find the rate of change of A with respect to r for the following interest rates.
(a) $r = 6$ **(b)** $r = 8$ **(c)** $r = 9$

58. Management Suppose a demand function is given by

$$x = 30\left(5 - \frac{p}{\sqrt{p^2 + 1}}\right),$$

where x is the demand for a product and p is the price per item in dollars. Find the rate of change in the demand for the product (i.e., find dx/dp).

59. Management A certain truck depreciates according to the formula

$$V = \frac{6000}{1 + .3t + .1t^2},$$

where t is measured in years and $t = 0$ represents the time of purchase (in years). Find the rate at which the value of the truck is changing at the following times.
(a) $t = 2$ **(b)** $t = 4$

60. Management Suppose the cost in dollars of manufacturing x items is given by

$$C = 2000x + 3500,$$

and the demand equation is given by

$$x = \sqrt{15,000 - 1.5p}.$$

(a) Find an expression for the revenue R.
(b) Find an expression for the profit P.
(c) Find an expression for the marginal profit.
(d) Determine the value of the marginal profit for $p = $25.

61. Management Find the marginal revenue product for a manufacturer with 8 workers if the demand function is $p = 300/x^{1/3}$ and if $x = 8n$.

62. Management Suppose the demand function for a product is $p = 200/x^{1/2}$. Find the marginal revenue product if there are 25 employees and if $x = 15n$.

63. Natural Science Suppose the population P of a certain species of fish depends on the number x (in hundreds) of a smaller fish that serves as its food supply, so that

$$P(x) = 2x^2 + 1.$$

Suppose, also, that the number x (in hundreds) of the smaller species of fish depends upon the amount a (in appropriate units) of its food supply, a kind of plankton. Suppose

$$x = f(a) = 3a + 2.$$

Find $P[f(a)]$, the relationship between the population P of the large fish and the amount a of plankton available.

64. Natural Science An oil well off the Gulf Coast is leaking, with the leak spreading oil over the surface as a circle. At any time t, in minutes, after the beginning of the leak, the radius of the circular oil slick on the surface is $r(t) = t^2$ feet. Let $A(r) = \pi r^2$ represent the area of a circle of radius r. Find and interpret $A[r(t)]$.

65. Natural Science When there is a thermal inversion layer over a city (as happens often in Los Angeles), pollutants cannot rise vertically but are trapped below the layer and must disperse horizontally. Assume that a factory smokestack begins emitting a pollutant at 8 A.M. Assume that the pollutant disperses horizontally, forming a circle. If t represents the time, in hours, since the factory began emitting pollutants ($t = 0$ represents 8 A.M.), assume that the radius of the circle of pollution is $r(t) = 2t$ miles. Let $A(r) = \pi r^2$ represent the area of a circle of radius r. Find and interpret $A[r(t)]$.

66. Natural Science The total number of bacteria (in millions) present in a culture is given by

$$N(t) = 2t(5t + 9)^{1/2} + 12,$$

where t represents time in hours after the beginning of an experiment. Find the rate of change of the population of bacteria with respect to time for each of the following.
(a) $t = 0$ (b) $t = 7/5$ (c) $t = 8$ (d) $t = 11$.

67. Natural Science To test an individual's use of calcium, a researcher injects a small amount of radioactive calcium into the person's bloodstream. The calcium remaining in the bloodstream is measured each day for several days. Suppose the amount of the calcium remaining in the bloodstream in milligrams per cubic centimeter t days after the initial injection is approximated by

$$C(t) = \frac{1}{2}(2t + 1)^{-1/2}.$$

Find the rate of change of C with respect to time for each of the following.
(a) $t = 0$ (b) $t = 4$ (c) $t = 6$ (Use a calculator.)
(d) $t = 7.5$

68. Natural Science The strength of a person's reaction to a certain drug is given by

$$R(Q) = Q\left(C - \frac{Q}{3}\right)^{1/2},$$

where Q represents the quantity of the drug given to the patient and C is a constant.
(a) The derivative $R'(Q)$ is called the *sensitivity* to the drug. Find $R'(Q)$.
(b) Find $R'(Q)$ if $Q = 87$ and $C = 59$.

11.7 DERIVATIVES OF EXPONENTIAL AND LOGARITHMIC FUNCTIONS

Exponential and logarithmic functions to the base e were examined in detail in Chapter 4. (Recall that to seven decimal places $e = 2.7182818$.) In this section formulas will be developed for the derivatives of $f(x) = e^x$ and $g(x) = \ln x$.

To find the derivative of $f(x) = e^x$, we use the definition of the derivative function:

$$f'(x) = \lim_{h \to 0} \frac{f(x + h) - f(x)}{h},$$

provided this limit exists. (Remember that h is the variable here and x is treated as a constant.)

Since $f(x) = e^x$, we see that

$$f'(x) = \lim_{h \to 0} \frac{e^{x+h} - e^x}{h}$$

$$= \lim_{h \to 0} \frac{e^x e^h - e^x}{h} \qquad \text{Product property of exponents}$$

$$= \lim_{h \to 0} \frac{e^x(e^h - 1)}{h}$$

$$= \lim_{h \to 0} e^x \cdot \lim_{h \to 0} \frac{e^h - 1}{h}, \qquad \text{Product property of limits}$$

1 Differentiate the following.

(a) $(2x^2 - 1)e^x$

(b) $(1 - e^x)^{1/2}$

Answer:

(a) $(2x^2 - 1)e^x + 4xe^x$

(b) $\dfrac{-e^x}{2(1 - e^x)^{1/2}}$

provided that both of these last limits exist. But h is the variable here and x is constant and therefore,

$$\lim_{h \to 0} e^x = e^x.$$

We claim that

$$\lim_{h \to 0} \frac{e^h - 1}{h} = 1.$$

Although a rigorous proof of this fact is beyond the scope of this book, the following chart (constructed with a calculator) makes it highly plausible.

h approaches 0 from the left $\to 0 \leftarrow h$ approaches 0 from the right

h	$-.001$	$-.0001$	$-.00001$	0	.00001	.0001	.001
$\dfrac{e^h - 1}{h}$	.999500	.999950	.999995		1.000005	1.000050	1.000500

Therefore,

$$f'(x) = \lim_{h \to 0} e^x \cdot \lim_{h \to 0} \frac{e^h - 1}{h} = e^x \cdot 1 = e^x.$$

In other words, *the exponential function $f(x) = e^x$ is its own derivative.*

▶**EXAMPLE 1** Find each derivative.

(a) $y = x^3 e^x$

The fact just proved and the product rule show that

$$y' = x^3 \cdot D_x(e^x) + D_x(x^3) \cdot e^x$$
$$= x^3 \cdot e^x + 3x^2 \cdot e^x = e^x(x^3 + 3x^2).$$

(b) $y = (2e^x + x)^5$

By the generalized power, sum, and constant rules,

$$y' = 5(2e^x + x)^4 \cdot D_x(2e^x + x)$$
$$= 5(2e^x + x)^4 \cdot [D_x(2e^x) + D_x(x)]$$
$$= 5(2e^x + x)^4 \cdot [2D_x(e^x) + 1]$$
$$= 5(2e^x + x)^4(2e^x + 1). \quad ◀ \quad \boxed{1}$$

▶**EXAMPLE 2** Find the derivative of $y = e^{x^2 - 3x}$.

Let $f(x) = e^x$ and $g(x) = x^2 - 3x$. Then

$$y = e^{x^2 - 3x} = e^{g(x)} = f[g(x)]$$

and $f'(x) = e^x$ and $g'(x) = 2x - 3$. By the alternative form of the chain rule,

$$y' = f'[g(x)] \cdot g'(x)$$
$$= e^{g(x)} \cdot (2x - 3)$$
$$= e^{x^2 - 3x} \cdot (2x - 3) = (2x - 3)e^{x^2 - 3x}. \quad ◀$$

2 Find each derivative.

(a) $y = 3e^{12x}$

(b) $y = -6e^{(-10x+1)}$

(c) $y = e^{-x^2}$

Answer:

(a) $y' = 36e^{12x}$

(b) $y' = 60e^{(-10x+1)}$

(c) $y' = -2xe^{-x^2}$

3 Find each derivative.

(a) $y = \dfrac{e^x}{1 + x}$

(b) $y = \dfrac{10{,}000}{1 + 2e^x}$

Answer:

(a) $y' = \dfrac{xe^x}{(1 + x)^2}$

(b) $y' = \dfrac{-20{,}000e^x}{(1 + 2e^x)^2}$

The argument used in Example 2 can be used to find the derivative of $y = e^{g(x)}$ for any differentiable function g. By the chain rule,

$$y' = f'[g(x)] \cdot g'(x) = e^{g(x)} \cdot g'(x) = g'(x)e^{g(x)}.$$

We can summarize these results as follows.

Derivative of e^x and $e^{g(x)}$

If $y = e^x$, then $y' = e^x$.
If $y = e^{g(x)}$, then $y' = g'(x) \cdot e^{g(x)}$.

▶**EXAMPLE 3** Find derivatives of the functions defined as follows.

(a) $y = 4e^{5x}$

Let $g(x) = 5x$, with $g'(x) = 5$. Then

$$y' = 4 \cdot 5e^{5x} = 20e^{5x}.$$

(b) $y = 3e^{-4x}$

$$y' = 3(-4e^{-4x}) = -12e^{-4x}$$

(c) $y = 10e^{3x^2}$

$$y' = 6x(10e^{3x^2}) = 60xe^{3x^2} \quad ◀ \; \boxed{2}$$

▶**EXAMPLE 4** Let $y = \dfrac{100{,}000}{1 + 100e^{-.3x}}$. Find y'.

Use the quotient rule.

$$y' = \frac{(1 + 100e^{-.3x})(0) - 100{,}000(-30e^{-.3x})}{(1 + 100e^{-.3x})^2}$$

$$= \frac{3{,}000{,}000e^{-.3x}}{(1 + 100e^{-.3x})^2} \quad ◀ \; \boxed{3}$$

To find the derivative of $g(x) = \ln x$, we use the definition of a logarithm (see Section 4.3):

$$g(x) = \ln x \quad \text{means} \quad e^{g(x)} = x.$$

Note that $x > 0$ since logarithms of negative numbers are not defined. Differentiating with respect to x on each side of $e^{g(x)} = x$ shows that

$$D_x(e^{g(x)}) = D_x(x)$$

$$e^{g(x)} \cdot g'(x) = 1.$$

Since $e^{g(x)} = x$, this last equation becomes

$$x \cdot g'(x) = 1$$

$$g'(x) = \frac{1}{x} \quad \text{for all } x > 0.$$

▶ **EXAMPLE 5**   Find the derivative of $y = \ln 6x$. Assume $x > 0$.

Use the properties of logarithms and the rules for derivatives.

$$y' = \frac{d}{dx}(\ln 6x)$$

$$= \frac{d}{dx}(\ln 6 + \ln x) \qquad\qquad \text{Product rule for logarithms}$$

$$= \frac{d}{dx}(\ln 6) + \frac{d}{dx}(\ln x) = 0 + \frac{1}{x} = \frac{1}{x} \qquad \text{ln 6 is a constant.} \quad ◀$$

The derivative of $y = \ln g(x)$, where $g(x) > 0$, can be found by letting $f(x) = \ln x$, so that $y = f[g(x)]$, and applying the chain rule:

$$y' = f'[g(x)] \cdot g'(x) = \frac{1}{g(x)} \cdot g'(x) = \frac{g'(x)}{g(x)}.$$

We can summarize these results as follows.

Derivative of ln x and ln $g(x)$

If $y = \ln x$, then $y' = \dfrac{1}{x}$ $(x > 0)$.

If $y = \ln g(x)$, then $y' = \dfrac{g'(x)}{g(x)}$ $(g(x) > 0)$.

▶ **EXAMPLE 6**   Find the derivative of the functions defined as follows.

(a) $y = \ln 5x$

Let $g(x) = 5x$, so that $g'(x) = 5$. From the formula above,

$$y' = \frac{g'(x)}{g(x)} = \frac{5}{5x} = \frac{1}{x}.$$

(b) $y = \ln (3x^2 - 4x)$

$$y' = \frac{6x - 4}{3x^2 - 4x}$$

4 Find y' for the following.

(a) $y = \ln (7 + x)$

(b) $y = \ln (4x^2)$

(c) $y = \ln (8x^3 - 3x)$

(d) $y = x^2 \ln x$

Answer:

(a) $y' = \dfrac{1}{7 + x}$

(b) $y' = \dfrac{2}{x}$

(c) $y' = \dfrac{24x^2 - 3}{8x^3 - 3x}$

(d) $y' = x(1 + 2 \ln x)$

5 Find each derivative.

(a) $y = e^{x^2} \ln |x|$

(b) $y = x^2/\ln |x|$

Answer:

(a) $e^{x^2}\left(\dfrac{1}{x} + 2x \ln |x|\right)$

(b) $\dfrac{2x \ln |x| - x}{(\ln |x|)^2}$

(c) $y = 3x \ln x^2$

Since $3x \ln x^2$ is the product of $3x$ and $\ln x^2$, use the product rule.

$$y' = (3x)\left(\frac{d}{dx} \ln x^2\right) + (\ln x^2)\left(\frac{d}{dx} 3x\right)$$

$$= 3x\left(\frac{2x}{x^2}\right) + (\ln x^2)(3) \qquad \text{Take derivatives}$$

$$= 6 + 3 \ln x^2$$

$$= 6 + \ln (x^2)^3 \qquad \text{Property of logarithms}$$

$$y' = 6 + \ln x^6 \qquad \text{Property of exponents}$$

Alternatively, write the answer as $y' = 6 + 6 \ln x$. ◀ **4**

The function $y = \ln (-x)$ is defined for all $x < 0$ (since $-x > 0$ when $x < 0$). Its derivative can be found by applying the derivative rule for $\ln g(x)$ with $g(x) = -x$.

$$y' = \frac{g'(x)}{g(x)} = \frac{-1}{-x} = \frac{1}{x}$$

This is the same as the derivative of $y = \ln x$, with $x > 0$. Since

$$|x| = \begin{cases} x & \text{if } x > 0 \\ -x & \text{if } x < 0, \end{cases}$$

we can combine two rules into one, as follows.

If $y = \ln |x|$, then $y' = \dfrac{1}{x}$.

▶ **EXAMPLE 7** Let $y = e^x \cdot \ln |x|$. Find y'.

Use the product rule.

$$y' = e^x \cdot \frac{1}{x} + \ln |x| \cdot e^x$$

$$y' = e^x\left(\frac{1}{x} + \ln |x|\right) \qquad ◀ \; \boxed{5}$$

Often a population, or the sales of a certain product, will start growing slowly, then grow more rapidly, and then gradually level off. Such growth can often be approximated by a mathematical model of the form

$$f(x) = \frac{b}{1 + ae^{kx}}$$

for appropriate constants a, b, and k.

6 Suppose a deer population is given by

$$f(x) = \frac{10,000}{1 + 2e^x},$$

where x is time in years. (See Problem 3(b) at the side.) Find the rate of change of the population when

(a) $x = 0$;

(b) $x = 5$.

(c) Is the population increasing or decreasing?

Answer:

(a) About -2200

(b) About -33

(c) Decreasing

▶**EXAMPLE 8**   Suppose that the sales of a new product can be approximated for its first few years on the market by

$$S(x) = \frac{100,000}{1 + 100e^{-.3x}},$$

where x is time in years since the introduction of the product. Find the rate of change of the sales when $x = 4$.

The derivative was given in Example 4. Using this derivative,

$$S'(4) = \frac{3,000,000e^{-.3(4)}}{(1 + 100e^{-.3(4)})^2}$$

$$= \frac{3,000,000e^{-1.2}}{(1 + 100e^{-1.2})^2}.$$

Using a calculator, $e^{-1.2} \approx .301$, with

$$S'(4) \approx \frac{3,000,000(.301)}{[1 + 100(.301)]^2} = \frac{903,000}{(1 + 30.1)^2} = \frac{903,000}{967.21} \approx 934.$$

The rate of change of sales at time $x = 4$ is an increase of about 934 units per year. ◀ **6**

11.7   EXERCISES

Find derivatives of the functions defined as follows. (See Examples 1–7.)

1. $y = e^{4x}$

2. $y = e^{-2x}$

3. $y = -6e^{-2x}$

4. $y = 8e^{4x}$

5. $y = -8e^{2x}$

6. $y = .2e^{5x}$

7. $y = -16e^{x+1}$

8. $y = -4e^{-.1x}$

9. $y = e^{x^2}$

10. $y = e^{-x^2}$

11. $y = 3e^{2x^2}$

12. $y = -5e^{4x^3}$

13. $y = 4e^{2x^2-4}$

14. $y = -3e^{3x^2+5}$

15. $y = xe^x$

16. $y = x^2e^{-2x}$

17. $y = (x - 3)^2e^{2x}$

18. $y = (3x^2 - 4x)e^{-3x}$

19. $y = \ln(3 - x)$

20. $y = \ln(1 + x^2)$

21. $y = \ln(2x^2 - 7x)$

22. $y = \ln(-8x^2 + 6x)$

23. $y = \ln\sqrt{x + 5}$

24. $y = \ln\sqrt{2x + 1}$

25. $y = \ln[(2x + 7)(3x - 2)]$

26. $y = \ln[(x^2 - 1)(3x + 8)]$

27. $y = \ln\left(\frac{5x - 1}{2x + 4}\right)$

28. $y = \ln\left(\frac{6 - x}{3x + 5}\right)$

29. $y = \ln(x^4 + 5x^2)^{3/2}$

30. $y = \ln(5x^3 - 2x)^{3/2}$

31. $y = -3x \ln(x + 2)$

32. $y = (3x + 1)\ln(x - 1)$

33. $y = x^2 \ln|x|$

34. $y = x \ln(2 - x^2)$

35. $y = \dfrac{x^2}{e^x}$

36. $y = \dfrac{e^x}{2x + 1}$

37. $y = (2x^3 - 1)\ln|x|$

38. $y = \dfrac{\ln|x|}{x^3}$

39. $y = \dfrac{\ln|x|}{4x + 7}$

40. $y = \dfrac{-2\ln|x|}{3x - 1}$

41. $y = \dfrac{\ln|x|}{e^x}, x > 0$

42. $y = \dfrac{x^3}{e^{2x}}$

43. $y = \dfrac{3x^2}{\ln |x|}$

44. $y = \dfrac{x^3 - 1}{2 \ln |x|}$

45. $y = [\ln (x + 1)]^4$

46. $y = \sqrt{\ln (x - 3)}$

47. $y = \dfrac{e^x}{\ln |x|}$

48. $y = \dfrac{e^x - 1}{\ln |x|}$

49. $y = \dfrac{e^x + e^{-x}}{x}$

50. $y = \dfrac{e^x - e^{-x}}{x}$

51. $y = e^{x^2} \ln |x|$

52. $y = e^{2x-1} \ln (2x - 1)$

53. $y = \dfrac{5000}{1 + 10e^{.4x}}$

54. $y = \dfrac{600}{1 - 50e^{.2x}}$

55. $y = \dfrac{10,000}{9 + 4e^{-.2x}}$

56. $y = \dfrac{500}{12 + 5e^{-.5x}}$

57. $y = \ln (\ln |x|)$

58. $y = (\ln 4)[\ln (3x)]$

Work the following exercises. (See Example 8.)

59. Management Suppose the demand function for x thousand of a certain item is

$$p = 100 + \frac{50}{\ln x}, \quad x > 1,$$

where p is in dollars.
(a) Find the marginal revenue.
(b) Find the revenue from the next thousand items at a demand of 8000 ($x = 8$).

60. Management Assume that the total revenue received from the sale of x items is given by

$$R(x) = 30 \ln (2x + 1),$$

while the total cost to produce x items is $C(x) = x/2$. Find the number of items that should be manufactured so that marginal profit is 0.

61. Management If the cost function in dollars for x thousand of the item in Exercise 59 is $C(x) = 100x + 100$, find the following.
(a) The marginal cost
(b) The profit function $P(x)$
(c) The profit from the next thousand items at a demand of 8000 ($x = 8$).

62. Management The demand function for x units of a product is

$$p = 100 - 10 \ln x, \quad 1 < x < 20,000,$$

where $x = 6n$ and n is the number of employees producing the product.
(a) Find the revenue function $R(x)$.
(b) Find the marginal revenue product function. (See Example 11 in Section 11.6.)
(c) Evaluate and interpret the marginal revenue product when $x = 20$.

63. Management Often, sales of a new product grow rapidly and then level off with time. Such a situation can be represented by an equation of the form

$$S(t) = 100 - 90e^{-.3t},$$

where t represents time in years and $S(t)$ represents sales. Find the rate of change of sales when
(a) $t = 1$; (b) $t = 10$.

64. Management Suppose $P(x) = e^{-.02x}$ represents the proportion of cars manufactured by a given company that are still free of defects after x months of use. Find the proportion of cars free of defects after
(a) 1 month; (b) 10 months; (c) 100 months.
(d) Calculate and interpret $P'(100)$.

65. Management Based on data in a car magazine, we constructed the mathematical model

$$y = 100e^{-.03045t}$$

for the percent of cars of a certain type still on the road after t years. Find the percent of cars on the road after the following number of years.
(a) 0 (b) 2 (c) 4 (d) 6
Find y' for the following values of t.
(e) 0 (f) 2 (g) 4 (h) 6

66. Recall from Chapter 5 the formula for interest compounded continuously:

$$A = Pe^{rt},$$

where A is the final amount on deposit when P dollars are deposited at a rate of interest r compounded continuously for t years. Assume $P = 100$, and $t = 5$. Find A' for the following values of r.
(a) .04 (b) .08 (c) .12 (d) .16

67. Natural Science Suppose that the population of a certain collection of rare Brazilian ants is given by

$$P(t) = 1000e^{.2t},$$

where t represents the time in days. Find the rate of change of the population when $t = 2$; when $t = 8$.

68. Natural Science Assume that the amount of a radioactive substance present at time t is given by

$$A(t) = 500e^{-.25t}$$

grams. Find the rate of change of the quantity present when

(a) $t = 0$; (b) $t = 4$; (c) $t = 6$; (d) $t = 10$.

69. Natural Science Consider an experiment in which equal numbers of male and female insects of a certain species are permitted to intermingle. Assume that

$$M(t) = (e^{.1t} + 1) \ln \sqrt{t}$$

represents the number of matings observed among the insects in an hour, where t is the temperature in degrees Celsius. (Note: The formula is an approximation at best and holds only for specific temperature intervals.) Find

(a) $M(15)$; (b) $M(25)$.
(c) Find the rate of change of $M(t)$ when $t = 15$.

70. Natural Science Suppose that the population of a certain colony of honey bees is given by

$$P(t) = (t + 100) \ln (t + 2),$$

where t represents the time in days. Find the rates of change of the population when $t = 2$ and when $t = 8$.

71. Natural Science The concentration of pollutants, in grams per liter, in the east fork of the Big Weasel River is approximated by

$$P(x) = .04e^{-4x},$$

where x is the number of miles downstream from a paper mill that the measurement is taken. Find

(a) $P(.5)$; (b) $P(1)$; (c) $P(2)$.

Find the rate of change of the concentration with respect to distance at

(d) $x = .5$; (e) $x = 1$; (f) $x = 2$.

72. Social Science According to work by the psychologist C. L. Hull, the strength of a habit is a function of the number of times the habit is repeated. If N is the number of repetitions and $H(N)$ is the strength of the habit, then

$$H(N) = 1000(1 - e^{-kN}),$$

where k is a constant. Find $H'(N)$ if $k = .1$ and

(a) $N = 10$; (b) $N = 100$; (c) $N = 1000$.
(d) Show that $H'(N)$ is always positive. What does this mean?

11.8 CONTINUITY AND DIFFERENTIABILITY

Intuitively speaking, a function is **continuous** at a point if you can draw the graph of the function near that point without lifting your pencil from the paper. Conversely, a function is **discontinuous** at a point if the pencil *must* be lifted from the paper in order to draw the graph on both sides of the point.

Looking first at graphs having points of discontinuity will clarify the idea of continuity at a point. For example, the function shown in Figure 11.24(a) is discontinuous at $x = 2$ because of the "hole" in the graph at $(2, 3)$. To draw the graph from $x = 1$ to $x = 3$, you must lift the pencil for an instant as you pass through $(2, 3)$.

1 Find any points of discontinuity for the following functions.

(a)

(b)

(c)

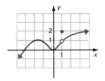

Answer:

(a) −1, 1

(b) −2

(c) 1

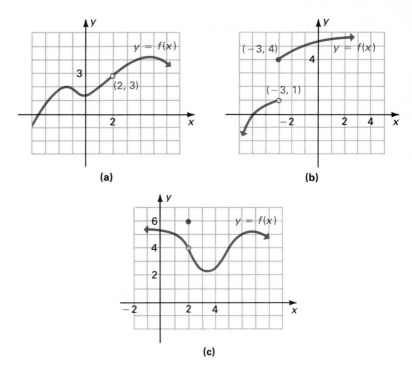

(a)

(b)

(c)

FIGURE 11.24

The function in Figure 11.24(b) is discontinuous at $x = -3$ because of the "jump" in the graph there (which necessitates lifting the pencil to draw the graph on both sides of $x = -3$). Although the function is not continuous at $x = -3$, it *is* continuous at $x = -1$ because you can draw the graph from $x = -2$ to $x = 0$, say, without lifting pencil from paper.

Finally, the function in Figure 11.24(c) is discontinuous at $x = 2$. When x is near (but not equal to) 2, all of the corresponding values of $f(x)$ are very near 4, so that on either side of $x = 2$, the graph is very near the point (2, 4) and can be drawn without lifting pencil from paper. But the instant $x = 2$, you must lift the pencil to the point $(2, f(2)) = (2, 6)$. Looked at from another point of view, as x gets closer and closer to 2, $f(x)$ gets closer and closer to 4, that is, $\lim_{x \to 2} f(x) = 4$. But $f(2) = 6$ and, hence,

$$\lim_{x \to 2} f(x) \neq f(2). \quad \boxed{1}$$

Now let's consider what it means for a function f to be continuous at $x = c$. If you *can* draw the graph of f around $x = c$ without lifting pencil from paper, then at the very least, $f(c)$ must be defined (otherwise there would be a hole in the graph). But the last example shows that this is not enough to guarantee continuity: as x gets very close to c, $f(x)$ must get very close to $f(c)$ (otherwise you have to lift the pencil at $x = c$). These considerations lead to this definition.

Definition of Continuity at a Point

A function f is **continuous** at $x = c$ if

(a) $f(c)$ is defined;
(b) $\lim\limits_{x \to c} f(x)$ exists;
(c) $\lim\limits_{x \to c} f(x) = f(c)$.

If f is not continuous at $x = c$, it is **discontinuous** there.

▶ **EXAMPLE 1**   Tell why the following functions are discontinuous at the indicated points.

(a) $f(x)$ in Figure 11.25 at $x = 3$

The open circle on the graph of Figure 11.25 at the point where $x = 3$ means that $f(3)$ does not exist. Because of this, part (a) of the definition fails.

(b) $h(x)$ in Figure 11.26 at $x = 0$

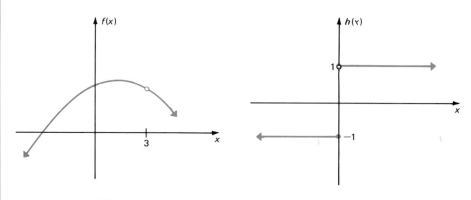

| FIGURE 11.25 | FIGURE 11.26 |

The graph of Figure 11.26 shows that $h(0) = -1$. Also, as x approaches 0 from the left, $h(x)$ is -1. However, as x approaches 0 from the right, $h(x)$ is 1. As mentioned in Section 11.1, for a limit to exist at a particular value of x, the values of $h(x)$ must approach a single number. Since no single number is approached by the values of $h(x)$ as x approaches 0, $\lim\limits_{x \to 0} h(x)$ does not exist, and part (b) of the definition fails.

(c) $g(x)$ at $x = 4$ in Figure 11.27

2 Tell why the following functions are discontinuous at the indicated points.

(a)

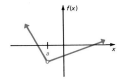

(b)

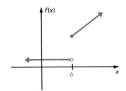

Answer:

(a) $f(a)$ does not exist

(b) $\lim\limits_{x \to b} f(x)$ does not exist

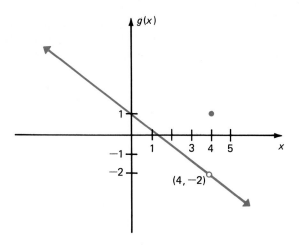

FIGURE 11.27

In Figure 11.27, the heavy dot above 4 shows that $g(4)$ is defined. In fact, $g(4) = 1$. However, the graph also shows that

$$\lim_{x \to 4} g(x) = -2,$$

so $\lim\limits_{x \to 4} g(x) \neq g(4)$, and part (c) of the definition fails.

(d) $f(x)$ in Figure 11.28 at $x = -2$

The function f graphed in Figure 11.28 is not defined at -2, and $\lim\limits_{x \to -2} f(x)$ does not exist. Either of these reasons is sufficient to show that f is not continuous at -2. (Function f *is* continuous at any value of x greater than -2, however.) ◀ **2**

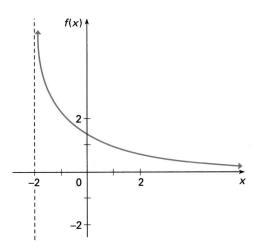

FIGURE 11.28

3 Write each of the following using interval notation.

(a)

(b)

(c)

Answer:

(a) $(-5, 3)$

(b) $[4, 7]$

(c) $(-\infty, -1]$

4 Are the functions with graphs as shown continuous on the indicated intervals?

(a) $(-4, -2)$; $(-3, 0)$

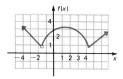

(b) $(-1, 1)$; $(0, 2)$

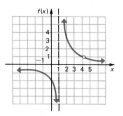

Answer:

(a) Yes; no

(b) Yes; no

When discussing the continuity of a function, it is often helpful to use interval notation, which was introduced in Chapter 1. The following chart should help you recall how it is used.

Interval	Name	Description	Interval Notation
	Open interval	$-2 < x < 3$	$(-2, 3)$
	Closed interval	$-2 \le x \le 3$	$[-2, 3]$
	Open interval	$x < 3$	$(-\infty, 3)$
	Open interval	$x > -5$	$(-5, \infty)$

Remember, the symbol ∞ does not represent a number; ∞ is used for convenience in interval notation to indicate that the interval extends without bound in the positive direction. Also, $-\infty$ indicates no bound in the negative direction. **3**

Continuity at a point was defined above; *continuity on an open interval* is defined as follows.

If a function is continuous at each point of an open interval, it is said to be **continuous on the open interval.**

Intuitively, the function f is continuous on the interval (a, b) if you can draw the graph between $x = a$ and $x = b$ without lifting your pencil from the paper.

▶**EXAMPLE 2** Is the function of Figure 11.29 continuous on the following x-intervals?
(a) $(-2, -1)$

The function is discontinuous only at $x = -2, 0,$ and 2. Thus, it is continuous at every point of the open interval $(-2, -1)$, and, therefore, is continuous on the open interval.
(b) $(1, 3)$

Since this interval includes the point of discontinuity $x = 2$, the function is not continuous on the open interval $(1, 3)$. ◀ **4**

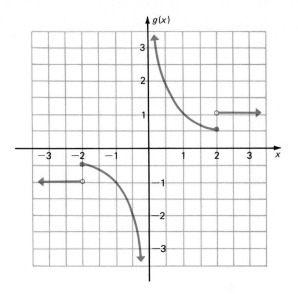

FIGURE 11.29

▶**EXAMPLE 3** A trailer rental firm charges a flat $4 to rent a hitch. The trailer itself is rented for $11 per day or fraction of a day. Let $C(x)$ represent the cost of renting a hitch and trailer for x days.

(a) Graph C.

The charge for 1 day is $4 for the hitch and $11 for the trailer, or $15. In fact, in the interval (0, 1], $C(x) = 15$. To rent the trailer for more than 1 day, but not more than 2 days, the charge is $4 + 2 \cdot 11 = 26$ dollars. For any value of x in the interval (1, 2], $C(x) = 26$. Also, in (2, 3] $C(x) = 37$. These results lead to the graph of Figure 11.30.

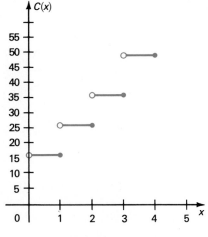

FIGURE 11.30

5 Suppose the cost is $2.25 to mail a package weighing up to 1 pound plus $.50 for each additional pound or fraction of a pound. Let $P(x)$ represent the cost of mailing a package weighing x pounds. Find any points of discontinuity for P.

Answer:

$x = 2.25, 2.75, 3.25, 3.75, \ldots$

(b) Find any points of discontinuity for C.

As the graph suggests, C is discontinuous at 1, 2, 3, 4, and all other positive integers. ◄ **5**

CONTINUITY AND DIFFERENTIABILITY As shown earlier in this chapter, a function fails to have a derivative at a point where the function is not defined, where the graph of the function has a "sharp point," or where the graph has a vertical tangent line. (See Figure 11.31.)

The function graphed in Figure 11.31 is continuous on the interval (x_1, x_2) and has a derivative at each point on this interval. On the other hand, the function is also continuous on the interval $(0, x_2)$ but does *not* have a derivative at each point on the interval (see x_1 on the graph).

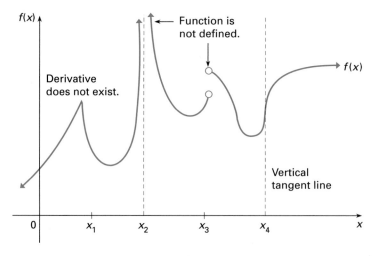

FIGURE 11.31

If the derivative of a function exists at a point, then the function is continuous at that point.

Intuitively, this means that a graph can have a derivative at a point only if the graph is "smooth" in the vicinity of the point.

► **EXAMPLE 4** A nova is a star whose brightness suddenly increases and then gradually fades. The cause of the sudden increase in brightness is thought to be an explosion of some kind. The intensity of light emitted by a nova as a function of time is shown in Figure 11.32.* Notice that although the graph is a continuous curve, it is not differentiable at the point of the explosion. ◀

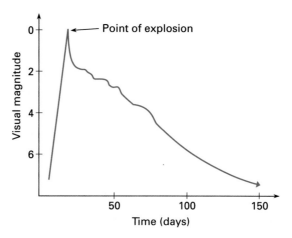

FIGURE 11.32

---

*Reprinted with permission of Macmillan Publishing Company from *Astronomy: The Structure of the Universe* by William J. Kaufmann, III. Copyright © 1977 by William J. Kaufmann, III.

11.8 EXERCISES

Find all points of discontinuity for the following. (See Example 1.)

1.

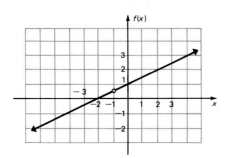

2.

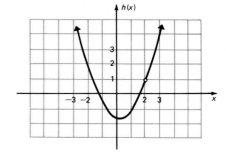

3.

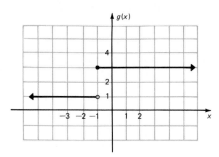

4.

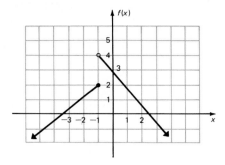

5.

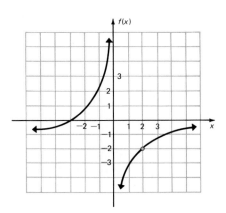

6.

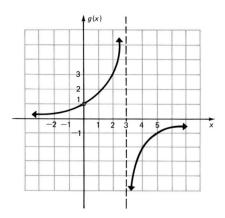

7.

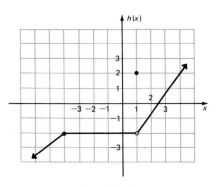

8.

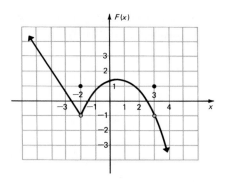

Are the functions defined as follows continuous at the given values of x? (See Example 1.)

9. $f(x) = \dfrac{2}{x-3}$; $x = 0$, $x = 3$

10. $f(x) = \dfrac{6}{x}$; $x = 0$, $x = -1$

11. $g(x) = \dfrac{1}{x(x-2)}$; $x = 0$, $x = 2$, $x = 4$

12. $h(x) = \dfrac{-2}{3x(x+5)}$; $x = 0$, $x = 3$, $x = -5$

13. $h(x) = \dfrac{1+x}{(x-3)(x+1)}$; $x = 0$, $x = 3$, $x = -1$

14. $g(x) = \dfrac{-2x}{(2x+1)(3x+6)}$; $x = 0$, $x = -\dfrac{1}{2}$, $x = -2$

15. $k(x) = \dfrac{5+x}{2+x}$; $x = 0$, $x = -2$, $x = -5$

16. $f(x) = \dfrac{4-x}{x-9}$; $x = 0$, $x = 4$, $x = 9$

17. $g(x) = \dfrac{x^2-4}{x-2}$; $x = 0$, $x = 2$, $x = -2$

18. $h(x) = \dfrac{x^2-25}{x+5}$; $x = 0$, $x = 5$, $x = -5$

19. $p(x) = x^2 - 4x + 11$; $x = 0$, $x = 2$, $x = -1$

20. $q(x) = -3x^3 + 2x^2 - 4x + 1$; $x = -2$, $x = 3$, $x = 1$

21. $p(x) = \dfrac{|x+2|}{x+2}$; $x = -2$, $x = 0$, $x = 2$

22. $r(x) = \dfrac{|5-x|}{x-5}$; $x = -5$, $x = 0$, $x = 5$

Work the following problems. (See Example 3.)

23. Management A company charges $1.20 per pound for a certain fertilizer on all orders not over 100 pounds, and $1 per pound for orders over 100 pounds. Let $F(x)$ represent the cost for buying x pounds of the fertilizer.

(a) Find $F(80)$.   (b) Find $F(150)$.

(c) Graph $y = F(x)$.

(d) Where is F discontinuous?

24. Management The cost to transport a mobile home depends on the distance, x, in miles that the home is moved. Let $C(x)$ represent the cost to move a mobile home x miles. One firm charges as follows.

Cost per Mile	Distance in Miles
$2	if $0 < x \le 150$
1.50	if $150 < x \le 400$
1.25	if $400 < x$

(a) Find $C(130)$.   (b) Find $C(210)$.

(c) Find $C(350)$.   (d) Find $C(500)$.

(e) Graph $y = C(x)$.

(f) For what positive values of x is C discontinuous?

25. Management Recently, a car rental firm charged $30 a day to rent a car for a rental period of 1 through 5 days. Days 6 and 7 were then "free," with days 8 through 12 again $30 each. Let $A(t)$ represent the average cost per day to rent the car for t days, where $0 < t \le 12$. (The average cost for t days is the total cost for t days, divided by the number of days.) Find each of the following.

(a) $A(4)$ (b) $A(5)$ (c) $A(6)$ (d) $A(7)$

(e) $A(8)$ (f) $\lim\limits_{t\to 5} A(t)$ (g) $\lim\limits_{t\to 6} A(t)$

26. Social Science With certain skills (such as music) learning is rapid at first and then levels off. Sudden insights may cause learning to speed up sharply. A typical graph of such learning is shown in the figure. Where is the function discontinuous?

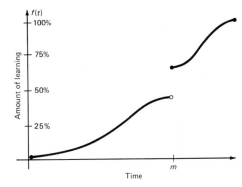

27. Management The graph below shows the interest rate for business loans for the business days of a recent month. Where is the graph discontinuous?

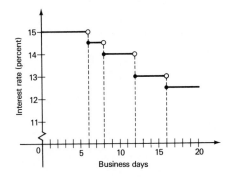

28. Natural Science Suppose a gram of ice is at a temperature of $-100°C$. The graph at the right shows the temperature of the ice as an increasing number of calories of heat are applied. It takes 80 calories to melt 1 gram of ice at 0°C into water, and 539 calories to boil 1 gram of water at 100°C into steam. Where is this graph discontinuous?

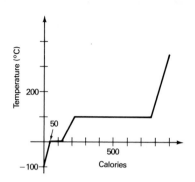

Write each of the following in interval notation.

29.

30.

31.

32.

33. $\{x \mid -11 \le x \le 9\}$

34. $\{x \mid -8 \le x \le 0\}$

35. $\{x \mid -6 < x < -2\}$

36. $\{x \mid 12 < x < 20\}$

37. $\{x \mid x > -4\}$

38. $\{x \mid x < 3\}$

39. $\{x \mid x < 0\}$

40. $\{x \mid x > -10\}$

Are the functions graphed as shown continuous on the indicated intervals? (See Example 2.)

41. $(-3, 0)$; $(0, 3)$; $(0, 4)$

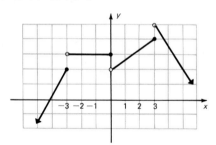

42. $(-6, 0)$; $(0, 6)$; $(4, 8)$

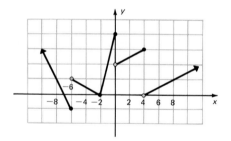

43. $(-12, 6)$; $(0, 6)$; $(6, 12)$

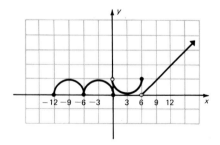

44. $(-4, 0)$; $(0, 3)$; $(0, 5)$

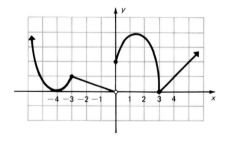

KEY TERMS AND SYMBOLS

11.1 $\lim\limits_{x \to a} f(x)$ limit of a function

limits at infinity

11.2   average rate of change
velocity
instantaneous rate of change
marginal cost, revenue, profit

11.3 y' derivative of y
$f'(x)$ derivative of $f(x)$
secant line
tangent line
slope of a curve
derivative
differentiation

11.4 $\dfrac{dy}{dx}$ derivative of $y = f(x)$

$D_x[f(x)]$ derivative of $f(x)$

$\dfrac{d}{dx}[f(x)]$ derivative of $f(x)$

demand function

11.5 $\overline{C}(x)$ average cost per item

11.6 $g[f(x)]$ composite function
marginal revenue product

11.8   continuous at a point
discontinuous
interval notation
continuous on an open interval

KEY CONCEPTS

Limit of a Function

Let f be a function and let a and L be real numbers. Suppose that as x takes values closer and closer (but not equal) to a (on both sides of a), the corresponding values of $f(x)$ get closer and closer (and possibly are equal) to L; and that the values of $f(x)$ can be made arbitrarily close to L by taking values of x close enough to a. Then L is the **limit** of f as x approaches a, written $\lim\limits_{x \to a} f(x) = L$.

Rules for Limits

Let a, k, n, A, and B be real numbers, and let f and g be functions such that $\lim\limits_{x \to a} f(x) = A$ and $\lim\limits_{x \to a} g(x) = B$.

1. If k is a constant, then (a) $\lim\limits_{x \to a} k = k$ and (b) $\lim\limits_{x \to a} k \cdot f(x) = k \cdot \lim\limits_{x \to a} f(x)$.

2. $\lim\limits_{x \to a} [f(x) \pm g(x)] = \lim\limits_{x \to a} f(x) \pm \lim\limits_{x \to a} g(x) = A \pm B$

3. If $p(x)$ is a polynomial, then $\lim\limits_{x \to a} p(x) = p(a)$.

4. $\lim\limits_{x \to a} [f(x) \cdot g(x)] = [\lim\limits_{x \to a} f(x)] \cdot [\lim\limits_{x \to a} g(x)] = A \cdot B$

5. $\lim\limits_{x \to a} \dfrac{f(x)}{g(x)} = \dfrac{\lim\limits_{x \to a} f(x)}{\lim\limits_{x \to a} g(x)} = \dfrac{A}{B}$ if $B \neq 0$

6. For any real number n for which A^n exists, $\lim\limits_{x \to a} [f(x)]^n = [\lim\limits_{x \to a} f(x)]^n = A^n$.

7. $\lim\limits_{x \to a} f(x) = \lim\limits_{x \to a} g(x)$ if $f(x) = g(x)$ for all $x \neq a$.

8. For any positive integer n, $\lim\limits_{x \to \infty} \dfrac{1}{x^n} = 0$ and $\lim\limits_{x \to -\infty} \dfrac{1}{x^n} = 0$.

The **instantaneous rate of change** for a function f when $x = x_0$ is $\lim\limits_{h \to 0} \dfrac{f(x_0 + h) - f(x_0)}{h}$, provided this limit exists.

The **tangent line** to the graph of $y = f(x)$ at the point $(x_0, f(x_0))$ is the line through this point having slope $\lim\limits_{h \to 0} \dfrac{f(x_0 + h) - f(x_0)}{h}$, provided this limit exists.

The **derivative** of the function f is the function denoted f' whose value at the number x is $f'(x) = \lim\limits_{h \to 0} \dfrac{f(x + h) - f(x)}{h}$, provided this limit exists.

Rules for Derivatives

(Assume all indicated derivatives exist.)

Constant Function

If $f(x) = k$, where k is any real number, then $\quad f'(x) = 0.$

Power Rule

If $f(x) = x^n$, for any real number n, then $\quad f'(x) = n \cdot x^{n-1}.$

Constant Times a Function

Let k be a real number. Then the derivative of $y = k \cdot f(x)$ is $\quad y' = k \cdot f'(x).$

Sum or Difference Rule

If $y = f(x) \pm g(x)$, then $\quad y' = f'(x) \pm g'(x).$

Product Rule

If $f(x) = g(x) \cdot k(x)$, then $\quad f'(x) = g(x) \cdot k'(x) + k(x) \cdot g'(x).$

Quotient Rule

If $f(x) = \dfrac{g(x)}{k(x)}$, and $k(x) \neq 0$, then $\quad f'(x) = \dfrac{k(x) \cdot g'(x) - g(x) \cdot k'(x)}{[k(x)]^2}.$

Chain Rule

If y is a function of u, say $y = f(u)$, and if u is a function of x, say $u = g(x)$, then $y = f(u) = f[g(x)]$, and

$$\frac{dy}{dx} = \frac{dy}{du} \cdot \frac{du}{dx}.$$

Chain Rule (alternate form)

Let $y = f[g(x)]$. Then $\quad y' = f'[g(x)] \cdot g'(x).$

Generalized Power Rule

Let u be a function of x, and let $y = u^n$ for any real number n. Then

$$y' = n \cdot u^{n-1} \cdot u'.$$

Exponential Function

If $y = e^{g(x)}$, then $\quad y' = g'(x) \cdot e^{g(x)}.$

Natural Logarithmic Function

If $y = \ln [g(x)]$, then $\quad y' = \dfrac{g'(x)}{g(x)}.$

If $y = \ln |x|$, then $y' = \dfrac{1}{x}.$

A function f is **continuous** at $x = c$ if $f(c)$ is defined, $\lim\limits_{x \to c} f(x)$ exists, and $\lim\limits_{x \to c} f(x) = f(c).$

Decide if the limits in Exercises 1–18 exist. If a limit exists, find its value.

1. $\lim_{x \to -3} f(x)$

2. $\lim_{x \to -1} g(x)$

3. $\lim_{x \to \infty} f(x)$

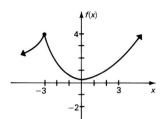

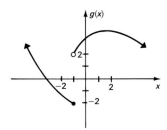

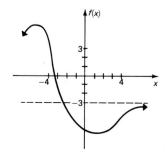

4. $\lim_{x \to 0} \dfrac{x^2 - 5}{2x}$

5. $\lim_{x \to -1} (2x^2 + 3x + 5)$

6. $\lim_{x \to 2} (-x^2 + 4x + 1)$

7. $\lim_{x \to 6} \dfrac{2x + 5}{x - 3}$

8. $\lim_{x \to 3} \dfrac{2x + 5}{x - 3}$

9. $\lim_{x \to 4} \dfrac{x^2 - 16}{x - 4}$

10. $\lim_{x \to 2} \dfrac{x^2 + 3x - 10}{x - 2}$

11. $\lim_{x \to -4} \dfrac{2x^2 + 3x - 20}{x + 4}$

12. $\lim_{x \to 3} \dfrac{3x^2 - 2x - 21}{x - 3}$

13. $\lim_{x \to 9} \dfrac{\sqrt{x} - 3}{x - 9}$

14. $\lim_{x \to 16} \dfrac{\sqrt{x} - 4}{x - 16}$

15. $\lim_{x \to \infty} \dfrac{x^2 + 5}{5x^2 - 1}$

16. $\lim_{x \to \infty} \dfrac{x^2 + 6x + 8}{x^3 + 2x + 1}$

17. $\lim_{x \to \infty} \left(\dfrac{3}{4} + \dfrac{2}{x} - \dfrac{5}{x^2} \right)$

18. $\lim_{x \to \infty} \left(\dfrac{9}{x^4} + \dfrac{1}{x^2} - 3 \right)$

Use the graph to find the average rate of change of f on the following intervals.

19. $x = 0$ to $x = 4$
20. $x = 2$ to $x = 8$
21. $x = 2$ to $x = 4$
22. $x = 0$ to $x = 6$

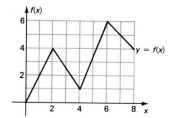

Find the average rate of change for the functions defined as follows.

23. $y = 6x^2 + 2$, from $x = 1$ to $x = 4$

24. $y = -2x^3 - x^2 + 5$, from $x = -2$ to $x = 6$

25. $y = \dfrac{-6}{3x - 5}$, from $x = 4$ to $x = 9$

26. $y = \dfrac{x + 4}{x - 1}$, from $x = 2$ to $x = 5$

Use the definition of the derivative to find the derivative of the functions defined as follows.

27. $y = 4x + 3$

28. $y = 5x^2 + 6x$

29. $y = -x^3 + 7x$

30. $y = 11x^2 - x^3$

Find the slope of the tangent line to the given curve at the given value of x. Find the equation of each tangent line.

31. $y = x^2 - 6x$; at $x = 2$

32. $y = 8 - x^2$; at $x = 1$

33. $y = \dfrac{3}{x - 1}$; at $x = -1$

34. $y = \dfrac{-2}{x + 5}$; at $x = -2$

35. $y = (3x^2 - 5x)(2x)$; at $x = -1$

36. $y = \dfrac{3}{x^2 - 1}$; at $x = 2$

37. $y = \sqrt{6x - 2}$; at $x = 3$

38. $y = -\sqrt{8x + 1}$; at $x = 3$

---

39. A company charges \$1.50 per pound when a certain chemical is bought in lots of 125 pounds or less, with a price per pound of \$1.35 if more than 125 pounds are purchased. Let $C(x)$ represent the cost of x pounds. Find each of the following.
 (a) $C(100)$ **(b)** $C(125)$ **(c)** $C(140)$

 (d) Graph $y = C(x)$.
 (e) Where is C discontinuous?

40. Use the information in Exercise 39 to find the average cost per pound if the following number of pounds are bought.
 (a) 100 **(b)** 125 **(c)** 140 **(d)** 200

Find the derivative of each of the following.

41. $y = 5x^2 - 7x - 9$

42. $y = x^3 - 4x^2$

43. $y = 6x^{7/3}$

44. $y = -3x^{-2}$

45. $f(x) = x^{-3} + \sqrt{x}$

46. $f(x) = 6x^{-1} - 2\sqrt{x}$

47. $y = (3t^2 + 7)(t^3 - t)$

48. $y = (-5t + 4)(t^3 - 2t^2)$

49. $y = 4\sqrt{x}(2x - 3)$

50. $y = -3\sqrt{x}(8 - 5x)$

51. $g(t) = -3t^{1/3}(5t + 7)$

52. $p(t) = 8t^{3/4}(7t - 2)$

53. $y = 12x^{-3/4}(3x + 5)$

54. $y = 15x^{-3/5}(x + 6)$

55. $k(x) = \dfrac{3x}{x + 5}$

56. $r(x) = \dfrac{-8}{2x + 1}$

57. $y = \dfrac{\sqrt{x} - 1}{x + 2}$

58. $y = \dfrac{\sqrt{x} + 6}{x - 3}$

59. $y = \dfrac{x^2 - x + 1}{x - 1}$

60. $y = \dfrac{2x^3 - 5x^2}{x + 2}$

61. $f(x) = (3x - 2)^4$

62. $k(x) = (5x - 1)^6$

63. $y = \sqrt{2t - 5}$

64. $y = -3\sqrt{8t - 1}$

65. $y = 3x(2x + 1)^3$

66. $y = 4x^2(3x - 2)^5$

67. $r(t) = \dfrac{5t^2 - 7t}{(3t + 1)^3}$

68. $s(t) = \dfrac{t^3 - 2t}{(4t - 3)^4}$

69. $y = \dfrac{x^2 + 3x - 10}{x - 2}$

70. $y = \dfrac{x^2 - x - 6}{x - 3}$

71. $y = -6e^{2x}$

72. $y = 8e^{.5x}$

73. $y = e^{-2x^3}$

74. $y = -4e^{x^2}$

75. $y = 5x \cdot e^{2x}$

76. $y = -7x^2 \cdot e^{-3x}$

77. $y = \ln(2 + x^2)$

78. $y = \ln(5x + 3)$

79. $y = \dfrac{\ln 3x}{x - 3}$

80. $y = \dfrac{\ln(2x - 1)}{x + 3}$

Find each of the following.

81. $D_x\!\left(\dfrac{\sqrt{x} + 1}{\sqrt{x} - 1}\right)$

82. $D_x\!\left(\dfrac{2x + \sqrt{x}}{1 - x}\right)$

83. $\dfrac{dy}{dt}$ if $y = \sqrt{t^{1/2} + t}$

84. $\dfrac{dy}{dx}$ if $y = \dfrac{\sqrt{x} - 1}{x}$

85. $f'(1)$ if $f(x) = \dfrac{\sqrt{8 + x}}{x + 1}$

86. $f'(-2)$ if $f(t) = \dfrac{2 - 3t}{\sqrt{2 + t}}$

Find all points of discontinuity for the following.

87.

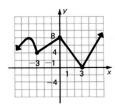

88.

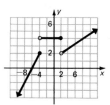

Are the functions defined as follows continuous at the given points?

89. $f(x) = \dfrac{6x + 1}{2x - 3}$; $x = \dfrac{3}{2}$, $x = -\dfrac{1}{6}$, $x = 0$

90. $f(x) = \dfrac{2}{x(x + 4)}$; $x = 2$, $x = 0$, $x = -4$

91. $f(x) = \dfrac{-5}{3x(2x - 1)}$; $x = -5$, $x = 0$, $x = -\dfrac{1}{3}$, $x = \dfrac{1}{2}$

92. $f(x) = \dfrac{2 - 3x}{(1 + x)(2 - x)}$; $x = \dfrac{2}{3}$, $x = -1$, $x = 2$, $x = 0$

93. $f(x) = \dfrac{x - 6}{x + 5}$; $x = 6$, $x = -5$, $x = 0$

94. $f(x) = \dfrac{x^2 - 9}{x + 3}$; $x = 3$, $x = -3$, $x = 0$

95. $f(x) = x^2 + 3x - 4$; $x = 1$, $x = -4$, $x = 0$

96. $f(x) = 2x^2 - 5x - 3$; $x = -\dfrac{1}{2}$, $x = 3$, $x = 0$

Management   *Work the following exercises.*

97. Suppose that the profit in cents from selling x pounds of potatoes is given by

$$P(x) = 15x + 25x^2.$$

Find the marginal profit when
(a) $x = 6$;   **(b)** $x = 20$;   **(c)** $x = 30$.

98. In Exercise 97, find the average profit function and the marginal average profit function.

99. The sales of a company are related to its expenditure on research by

$$S(x) = 1000 + 50\sqrt{x} + 10x,$$

where $S(x)$ gives sales in millions when x thousand dollars are spent on research. Find the rate of change of sales, dS/dx, when
(a) $x = 9$;   **(b)** $x = 16$;   **(c)** $x = 25$.

100. Suppose that the profit in hundreds of dollars from selling x units of a product is given by $P(x)$, where

$$P(x) = \dfrac{x^2}{x - 1}.$$

Find the marginal profit when
(a) $x = 4$;   **(b)** $x = 12$;   **(c)** $x = 20$.

101. A company finds its sales are related to the amount spent on training programs by

$$T(x) = \dfrac{1000 + 50x}{x + 1},$$

where $T(x)$ is sales in hundreds of thousands of dollars when x thousand dollars are spent on training. Find the rate of change of sales, $T'(x)$, when
(a) $x = 9$;   **(b)** $x = 19$.

102. Waverly Products has determined that its profits are related to advertising expenditures by the function

$$P(x) = 5000 + 16x - 3x^2,$$

where $P(x)$ is the profit when x hundred dollars are spent on advertising.
(a) Find the marginal profit function.
(b) At an expenditure of \$1000 ($x = 10$), is the profit increasing or decreasing?

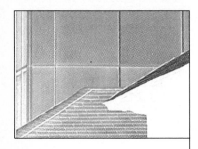

CHAPTER 12

Applications of the Derivative

If the graph of a function is given, its maximum and minimum values can usually be read directly. The graph also shows where the function is increasing or decreasing in value. We saw earlier how to determine these facts algebraically for a quadratic function (whose graph is a parabola) by finding the vertex. For more complicated functions, whose graphs may be difficult to draw, other techniques are needed. In this chapter we shall see how derivatives can be used to determine the maximum and minimum values of a function, as well as the intervals where the function is increasing or decreasing in value.

12.1 MAXIMA AND MINIMA

The graph of a typical function may have "peaks" and "valleys," as shown in Figure 12.1. A peak is not necessarily the highest point on the graph, but it is higher than all the points that are near it. This means that if the peak is the point $(c, f,(c))$, then $f(x) \leq f(c)$ for all values of x *near* c. Analogous remarks apply to valleys and motivate this formal definition.

Let c be a number in the domain of a function f.

1. $f(c)$ is a **relative maximum** for f if there exists an open interval (a, b) containing c such that
$$f(x) \leq f(c)$$
for all x in (a, b).
2. $f(c)$ is a **relative minimum** for f if there exists an open interval (a, b) containing c such that
$$f(x) \geq f(c)$$
for all x in (a, b).

The function f is said to have a **relative extremum** at c if it has a relative maximum or minimum there.

1 Identify the x-values of all points where these graphs have relative maxima or relative minima.

(a)

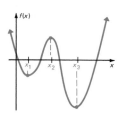

(b)

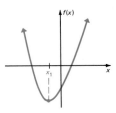

(c)

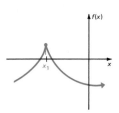

Answer:

(a) Relative maximum at x_2; relative minima at x_1 and x_3

(b) No relative maximum; relative minimum at x_1

(c) Relative maximum at x_1; no relative minimum

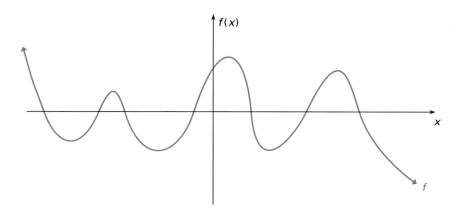

FIGURE 12.1

Note The plurals of maximum, minimum, and extremum, respectively, are maxima, minima, and extrema.

▶**EXAMPLE 1** Identify the relative extrema of the function whose graph is shown in Figure 12.2.

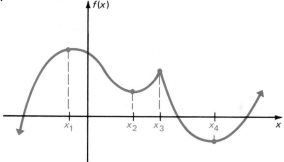

FIGURE 12.2

The function has relative maxima at x_1 and x_3 and relative minima at x_2 and x_4. ◀ **1**

When the graph of a function is given, as in the preceding example, determining the relative maxima and minima is easy. We now develop methods for doing this when the graph isn't known.

Let f be a function and think of the graph of f as a roller coaster track, with the roller coaster cars moving from left to right along the graph, as shown in Figure 12.3. Now picture one of the cars on the roller coaster. As the car moves up towards a peak, its floor tilts upward. At the instant the car reaches the peak, its floor is level, but then it begins to tilt downward (perhaps very steeply) as the car rolls down toward a valley.

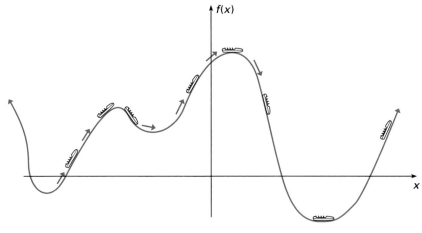

FIGURE 12.3

At any point along the graph, the floor of the car (a straight-line segment in the figure) represents the tangent line to the graph at that point. Using this analogy, we see that as the car passes through the peaks and valleys at A, B, C in Figure 12.4, the tangent line is horizontal and has slope 0. At peak D and valley E, however, a real roller coaster car would have trouble. It would fly off the track at peak D and be unable to make the 90° change of direction at valley E. This corresponds to the fact that although the graph is continuous at D and E, there is no tangent line at D or E because of the sharp corners (see the discussion on the existence of the derivative in Section 11.3).

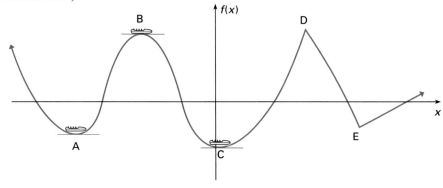

FIGURE 12.4

Thus, the points where a peak or a valley occurs have this property: the tangent line is horizontal and has slope 0 there *or* no tangent line is defined there. The slope of the tangent line to the graph of the function f at $(x, f(x))$ is the value of the derivative, $f'(x)$. Consequently, the following result, whose proof is omitted, should be quite plausible.

Let c be a number in the domain of a function f. If f has a relative extremum at c, then either

$$f'(c) = 0 \quad \text{or} \quad f'(c) \text{ does not exist.}$$

Because of this fact, a number c *in the domain of f* for which $f'(c) = 0$ or $f'(c)$ does not exist is called a **critical number** of f. The corresponding point $(c, f(c))$ on the graph of f is called a **critical point.**

Caution Be very careful not to get the result in the box above backwards. It says every relative extremum is a critical point, but *not* that every critical point is a relative extremum. For example, Figure 12.5 shows the graph of $f(x) = x^3$. The derivative, $f'(x) = 3x^2$, is 0 when $x = 0$, so that 0 is a critical number for f. But the graph shows that $f(x) = x^3$ does *not* have a relative maximum or minimum at $x = 0$ (or anywhere else, for that matter).

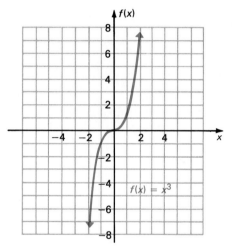

FIGURE 12.5

▶**EXAMPLE 2**   Find the critical numbers of the following functions.
(a) $f(x) = 2x^3 - 3x^2 - 72x + 15$

We have $f'(x) = 6x^2 - 6x - 72$, so $f'(x)$ exists for every x. Setting $f'(x) = 0$ shows that

$$6x^2 - 6x - 72 = 0$$
$$6(x^2 - x - 12) = 0$$
$$x^2 - x - 12 = 0$$
$$(x + 3)(x - 4) = 0$$
$$x + 3 = 0 \quad \text{or} \quad x - 4 = 0$$
$$x = -3 \quad \text{or} \qquad x = 4.$$

2 Find the critical numbers for each of the following.

(a) $\dfrac{1}{3}x^3 - x^2 - 15x + 6$

(b) $6x^{2/3} - 4x$

Answer:

(a) $-3, 5$

(b) $0, 1$

Therefore, -3 and 4 are the critical numbers of f, so $f(-3) = 150$ and $f(4) = -193$ are the only *possibilities* for the relative maxima and minima of the function.

(b) $f(x) = 3x^{4/3} - 12x^{1/3}$.

We first compute the derivative.

$$f'(x) = 3 \cdot \frac{4}{3}x^{1/3} - 12 \cdot \frac{1}{3}x^{-2/3} = 4x^{1/3} - \frac{4}{x^{2/3}}$$

$$= \frac{4x^{1/3}x^{2/3}}{x^{2/3}} - \frac{4}{x^{2/3}} = \frac{4x - 4}{x^{2/3}}$$

The derivative fails to exist when $x = 0$. Since the original function f is defined when $x = 0$, 0 is a critical number of f. If $x \neq 0$, then $f'(x)$ is 0 only when the numerator $4x - 4 = 0$; that is, when $x = 1$. So the critical numbers of f are 0 and 1; $f(0) = 0$ and $f(1) = -9$ are the *possible* relative maxima and minima. ◀ **2**

Once all the critical numbers of a function have been found, as in Example 2, they must be tested to see which, if any, lead to relative extrema. One method of doing this is suggested by thinking again of the graph of f as a roller coaster track, with the cars moving from left to right, as in Figure 12.6. As the roller coaster car moves uphill from left to right toward a relative maximum, its floor (the tangent line to the graph at that point) is always tilted upward, so the slope of the tangent line is positive on the left side of a relative maximum. As the car moves downhill from the relative maximum, its floor is tilted downward, so the slope of the tangent line is negative on the right side of a relative maximum. Near a relative minimum, the opposite occurs. The tangent line has negative slope on the left side (as the car moves downhill) and positive slope on the right side (as the car moves uphill again).

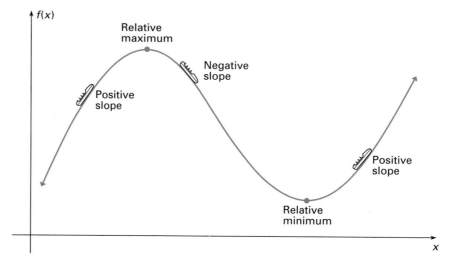

FIGURE 12.6

When this intuitive idea is given a rigorous mathematical proof, we get the following test for relative extrema. Since the derivative $f'(x)$ gives the slope of f at x, our conclusions are restated in terms of $f'(x)$.

First Derivative Test

Assume that $a < c < b$ and that c is the only critical number for a function f in $[a, b]$. Assume that f is differentiable for all x in $[a, b]$ except possibly at $x = c$.

1. If $f'(a) > 0$ and $f'(b) < 0$, then $f(c)$ is a relative maximum.
2. If $f'(a) < 0$ and $f'(b) > 0$, then $f(c)$ is a relative minimum.
3. If $f'(a)$ and $f'(b)$ are both positive or both negative, then $f(c)$ is not a relative extremum.

The sketches in the following table show how the first derivative test works. Assume the same conditions on a, b, c as those stated in the box.

$f(x)$ Has:	Sign of $f'(a)$	Sign of $f'(b)$	Sketches
Relative maximum	$+$	$-$	
Relative minimum	$-$	$+$	
No relative extrema	$+$	$+$	
No relative extrema	$-$	$-$	

3 Find all relative extrema for the functions defined as follows.

(a) $f(x) = 2x^2 - 8x + 1$

(b) $f(x) = 3 + 4x - 3x^2$

Answer:

(a) Relative minimum at $x = 2$; minimum is $f(2) = -7$

(b) Relative maximum at $x = 2/3$; maximum is $f(2/3) = 13/3$

▶**EXAMPLE 3**   In Example 2(a) we found that the critical points of $f(x) = 2x^3 - 3x^2 - 72x + 15$ are -3 and 4. To test $c = -3$, we can use $a = -4$ and $b = 0$ since -3 is the only critical number between -4 and 0. Many other choices of a and b are possible, but we try to select numbers that will make the computations easy. Since

$$f'(x) = 6x^2 - 6x - 72 = 6(x^2 - x - 12)$$
$$= 6(x + 3)(x - 4),$$

we see that

$$f'(-4) = 6(-4 + 3)(-4 - 4) = 6(-1)(-8) > 0;$$

and

$$f'(0) = 6(0 + 3)(0 - 4) = 6(3)(-4) < 0.$$

(Note that it's not necessary to finish calculating the exact value of $f'(x)$ in order to determine its sign.) Thus, the value of the derivative is positive to the left of -3 and negative to the right of -3, as shown in Figure 12.7. By part 1 of the first derivative test, $f(-3) = 150$ is a relative maximum.

Similarly, we can use $a = 0$ and $b = 5$ to test the critical number $c = 4$. We just saw that $f'(0) < 0$; and

$$f'(5) = 6(5 + 3)(5 - 4) = 6(8)(1) > 0.$$

Hence, by part 2 of the first derivative test, $f(4) = -193$ is a relative minimum. ◀ **3**

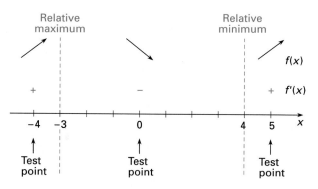

FIGURE 12.7

4 Find all relative extrema for $f(x) = x^3 + 4x^2 - 3x + 5$.

Answer:
Relative maximum at -3 of $f(-3) = 23$; relative minimum at $1/3$ of $f(1/3) = 121/27$

▶**EXAMPLE 4** In Example 2(b) we found that 0 and 1 are the critical numbers for $f(x) = 3x^{4/3} - 12x^{1/3}$. We can use -1, $1/2$, and 2 to apply the first derivative test.

$$f'(x) = \frac{4x - 4}{x^{2/3}} = \frac{4x - 4}{\sqrt[3]{x^2}}$$

$$f'(-1) = \frac{4(-1) - 4}{\sqrt[3]{(-1)^2}} = \frac{-4 - 4}{1} < 0$$

$$f'\left(\frac{1}{2}\right) = \frac{4(1/2) - 4}{\sqrt[3]{(1/2)^2}} = \frac{-2}{\sqrt[3]{1/4}} < 0$$

$$f'(2) = \frac{4(2) - 4}{\sqrt[3]{2^2}} = \frac{4}{\sqrt[3]{4}} > 0$$

These results are shown in Figure 12.8.

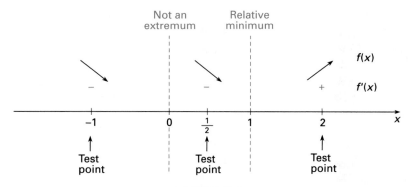

FIGURE 12.8

Since 0 is the only critical number between -1 and $1/2$ and the derivative is negative at both -1 and $1/2$, part 3 of the first derivative test shows that there is no relative extremum at $x = 0$. The only critical number between $1/2$ and 2 is $x = 1$. Since $f'(1/2) < 0$ and $f'(2) > 0$, $f(1) = -9$ is a relative minimum. ◀ **4**

▶**EXAMPLE 5** Find all relative maxima or relative minima for $f(x) = (2x + 1)e^{-x}$.

First take the derivative, using the product rule.

$$\begin{aligned} f'(x) &= (2x + 1)(-e^{-x}) + e^{-x}(2) \\ &= -2x \cdot e^{-x} - e^{-x} + 2e^{-x} \\ &= -2x \cdot e^{-x} + e^{-x} \\ &= e^{-x}(-2x + 1) \end{aligned}$$

Set the derivative equal to 0.

$$e^{-x}(-2x + 1) = 0$$

5 Find any relative extrema for $f(x) = x^2 e^{5x}$.

Answer:
Relative minimum of 0 at $x = 0$; relative maximum of about .02 at $x = -2/5$

6 Identify the location of any absolute extrema for these graphs.

(a)

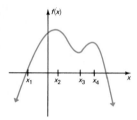

(b)

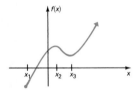

Answer:

(a) Absolute maximum at x_2; no absolute minimum

(b) Absolute minimum at x_1; no absolute maximum

Since e^{-x} is never 0, this derivative can equal 0 only when $-2x + 1 = 0$, or when $x = 1/2$.

To decide if there is a maximum or a minimum at $x = 1/2$, use the first derivative test. Since $f'(0) > 0$ and $f'(1) < 0$, there is a relative maximum of $f(1/2) = 2/e^{1/2} \approx 1.2$ at $x = 1/2$. ◀ **5**

We can now summarize the procedure for finding relative extrema.

To find the relative extrema of a function f:

Step 1 Find the derivative $f'(x)$;

Step 2 Find every critical number c (those numbers in the domain of f where $f'(c) = 0$ or $f'(c)$ does not exist);

Step 3 Use the first derivative test to identify relative maxima or minima.

ABSOLUTE MAXIMA AND MINIMA If a function f has a *relative* maximum at c, then $f(x) \le f(c)$ for all x *near* c (the graph has a peak at $(c, f(c))$), but it may not be the highest point on the graph. We say that f has an **absolute maximum** at c if $f(x) \le f(c)$ for *every* x in the *domain* of f, so that the point $(c, f(c))$ is the highest point on the graph. Similarly, f has an **absolute minimum** at c if $f(x) \ge f(c)$ for every x in the domain of f. The function graphed in Figure 12.9 has an absolute maximum at x_1 and an absolute minimum at x_2. The function graphed in Figure 12.10 has neither an absolute maximum nor an absolute minimum. **6**

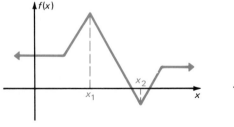

FIGURE 12.9

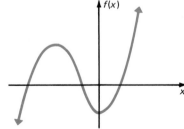

FIGURE 12.10

In many applications, we are interested only in the largest or smallest values of a function on a particular interval.

Let f be a function defined on some interval and let c be a number in the interval.

1. $f(c)$ is the **absolute maximum of f on the interval** if
$$f(x) \leq f(c)$$
for every x in the interval.
2. $f(c)$ is the **absolute minimum of f on the interval** if
$$f(x) \geq f(c)$$
for every x in the interval.

▶**EXAMPLE 6**   Consider the function graphed in Figure 12.11. The absolute maximum of f on the interval $[-2, 6]$ is $f(6)$, even though it is not a relative maximum nor is it the absolute maximum of the entire function f. Similarly, $f(3)$ is the absolute minimum of f on the interval $[-2, 6]$. $f(3)$ is also a relative minimum of f, but not an absolute minimum of the entire function.   ◀

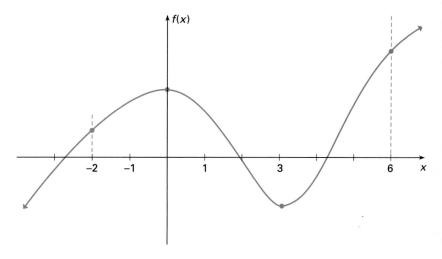

FIGURE 12.11

Example 6 suggests the truth of the following result, which is proved in more advanced courses.

Extreme Value Theorem

A function f continuous on a closed interval $[a, b]$ will have both an absolute maximum and an absolute minimum on the interval.

7 Find the absolute maximum and absolute minimum values of

(a) $f(x) = -x^2 + 4x - 8$ on $[-4, 4]$;

(b) $f(x) = 2x^2 - 6x + 6$ on $[0, 5]$.

Answer:

(a) Absolute maximum of -4 at $x = 2$; absolute minimum of -40 at $x = -4$

(b) Absolute maximum of 26 at $x = 5$; absolute minimum of 3/2 at $x = 3/2$

This extreme value theorem guarantees the existence of absolute extrema. Use the following steps to find these extrema.

To find absolute extrema for a function f continuous on an interval $[a, b]$:

Step 1 Find f' and solve $f'(x) = 0$ to find all critical points $c_1, c_2, \ldots$ in (a, b). Also, identify the critical points where f' does not exist;

Step 2 Evaluate $f(c_1), f(c_2), \ldots$ for all critical points in (a, b);

Step 3 Evaluate $f(a)$ and $f(b)$ for the endpoints a and b of the interval $[a, b]$;

Step 4 The largest value found in steps 2 and 3 is the absolute maximum for f on $[a, b]$; the smallest value found is the absolute minimum.

▶**EXAMPLE 7** Find the absolute extrema of the function given by

$$f(x) = 4x + \frac{36}{x},$$

on the interval $[1, 6]$.

First determine any critical points in $[1, 6]$.

$$f'(x) = 4 - \frac{36}{x^2} = 0$$

$$\frac{4x^2 - 36}{x^2} = 0$$

Although $f'(x)$ does not exist at $x = 0$, f is not defined at $x = 0$, so $x = 0$ is not a critical number. When $f'(x) = 0$ and $x \neq 0$, then

$$4x^2 - 36 = 0$$
$$4x^2 = 36$$
$$x^2 = 9$$
$$x = -3 \quad \text{or} \quad x = 3$$

Since -3 is not in the interval $[1, 6]$, 3 is the only critical number of interest. To find the absolute extrema, evaluate $f(x)$ at $x = 3$ and at the endpoints, where $x = 1$ and $x = 6$.

x-value	Value of Function	
1	40	← Absolute maximum
3	24	← Absolute minimum
6	30	

◀ **7**

8 For a certain firm, the cost to produce x units of an item is given by

$$C(x) = x^3 - 15x^2 + 48x + 50.$$

Because of production problems, the function is defined only for the domain [1, 6]. Find the absolute maximum and minimum costs that the firm will face.

Answer:
Absolute maximum of 94 when $x = 2$; absolute minimum of 14 when $x = 6$

Absolute extrema are particularly important in applications of mathematics. In most applications the domain is limited in some natural way that ensures the occurrence of absolute extrema.

▶**EXAMPLE 8** A company has found that its weekly profit from the sale of x units of an auto part is given by

$$P(x) = -.02x^3 + 600x - 20{,}000.$$

Production bottlenecks limit the number of units that can be made per week to no more than 150, while a long-term contract requires that at least 50 units be made each week. Find the maximum possible weekly profit that the firm can make.

Because of the restrictions, the profit function is defined only for the domain [50, 150]. Look first for critical numbers of the function in the open interval (50, 150). Here $P'(x) = -.06x^2 + 600$. Set this derivative equal to 0 and solve for x.

$$-.06x^2 + 600 = 0$$
$$-.06x^2 = -600$$
$$x^2 = 10{,}000$$
$$x = 100 \quad \text{or} \quad x = -100$$

Since $x = -100$ is not in the interval [50, 150], disregard it. Now evaluate the function at the remaining critical number 100 and at the endpoints of the domain, 50 and 150.

x-value	Value of Function	
50	7500	
100	20,000	← Absolute maximum
150	2500	

Maximum profit of $20,000 occurs when 100 units are made per week. ◀ **8**

Find the location and value of all relative maxima and relative minima for the functions graphed as follows. (See Example 1.)

1.

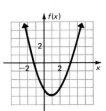

2.

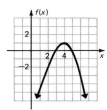

3.

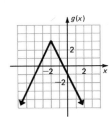

4.

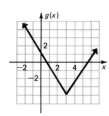

5.

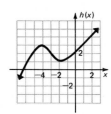

6.

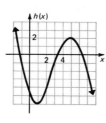

7.

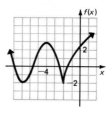

8.

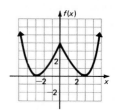

Find the x-values of all points where the functions defined as follows have any relative extrema. Find the value of any relative extrema. (See Examples 2–5.)

9. $f(x) = x^2 + 12x - 8$

10. $f(x) = 3 - 4x - 2x^2$

11. $f(x) = x^3 + 6x^2 + 9x - 8$

12. $f(x) = x^3 + 3x^2 - 24x + 2$

13. $f(x) = -\dfrac{4}{3}x^3 - \dfrac{21}{2}x^2 - 5x + 8$

14. $f(x) = -\dfrac{2}{3}x^3 - \dfrac{1}{2}x^2 + 3x - 4$

15. $f(x) = 2x^3 - 21x^2 + 60x + 5$

16. $f(x) = 2x^3 + 15x^2 + 36x - 4$

17. $f(x) = x^4 - 18x^2 - 4$

18. $f(x) = x^4 - 8x^2 + 9$

19. $f(x) = -(8 - 5x)^{2/3}$

20. $f(x) = (2 - 9x)^{2/3}$

21. $f(x) = x - \dfrac{1}{x}$

22. $f(x) = x^2 + \dfrac{1}{x}$

23. $f(x) = \dfrac{x^2}{x^2 + 1}$

24. $f(x) = \dfrac{x^2 - 2x + 1}{x - 3}$

25. $f(x) = -xe^x$

26. $f(x) = xe^{-x}$

27. $f(x) = e^x + e^{-x}$

28. $f(x) = -x^2e^x$

29. $f(x) = x \cdot \ln |x|$

30. $f(x) = x - \ln |x|$

Find the location of all absolute maxima and absolute minima (if any) for the functions graphed as follows. (See Example 6.)

31.

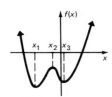

32.

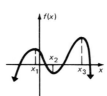

33.

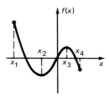

34.

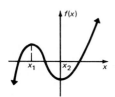

35.

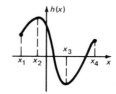

36.

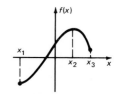

Find the location of all absolute extrema for the functions having domains as specified in Exercises 37–52. A calculator will be helpful for many of these problems. (See Example 7.)

37. $f(x) = 5 - 8x - 4x^2$; $[-5, 1]$

38. $f(x) = 9 - 6x - 3x^2$; $[-4, 3]$

39. $f(x) = x^3 - 3x^2 - 24x + 5$; $[-3, 6]$

40. $f(x) = x^3 - 6x^2 + 9x - 8$; $[0, 5]$

41. $f(x) = \dfrac{1}{3}x^3 - \dfrac{1}{2}x^2 - 6x + 3$; $[-4, 4]$

42. $f(x) = \dfrac{1}{3}x^3 + \dfrac{3}{2}x^2 - 4x + 1$; $[-5, 2]$

43. $f(x) = x^4 - 32x^2 - 7$; $[-5, 6]$

44. $f(x) = x^4 - 18x^2 + 1$; $[-4, 4]$

45. $f(x) = \dfrac{8 + x}{8 - x}$; $[4, 6]$

46. $f(x) = \dfrac{1 - x}{3 + x}$; $[0, 3]$

47. $f(x) = \dfrac{x}{x^2 + 2}$; $[0, 4]$

48. $f(x) = \dfrac{x - 1}{x^2 + 1}$; $[1, 5]$

49. $f(x) = (x^2 + 18)^{2/3}$; $[-3, 3]$

50. $f(x) = (x^2 + 4)^{1/3}$; $[-2, 2]$

51. $f(x) = \dfrac{1}{\sqrt{x^2 + 1}}$; $[-1, 1]$

52. $f(x) = \dfrac{3}{\sqrt{x^2 + 4}}$; $[-2, 2]$

Work the following exercises. (See Example 8.)

53. Management A company has found through experience that increasing its advertising also increases its sales, up to a point. The company believes that the mathematical model connecting profit in dollars, $P(x)$, and expenditures on advertising in dollars, x, is

$$P(x) = 80 + 108x - x^3.$$

(a) Find the expenditure on advertising that leads to maximum profit.

(b) Find the maximum profit.

54. The number of people visiting Timberline Ski Lodge on Washington's Birthday is approximated by

$$W(x) = -x^2 + 60x + 180,$$

where x is the total snowfall in inches for the previous week. Find the snowfall that will produce the maximum number of visitors. Find the maximum number of visitors.

55. Management An enterprising (although unscrupulous) business student has managed to get his hands on a copy of the out-of-print solutions manual for an applied calculus text. He plans to duplicate it and sell copies in the dormitories. He figures that demand will be between 100 and 1200 copies, and he wants to minimize his average cost of production. After checking into the cost of paper, duplicating, and the rental of a small van, he estimates that the cost in dollars to produce x hundred manuals is given by $C(x) = x^2 + 200x + 100$. How many should he produce in order to make the *average cost* per unit as small as possible? How much will he have to charge to make a profit?

56. Management The total profit $P(x)$ (in thousands of dollars) from the sale of x hundred thousands of automobile tires is approximated by

$$P(x) = -x^3 + 9x^2 + 120x - 400, \quad x \geq 5.$$

Find the number of hundred thousands of tires that must be sold to maximize profit. Find the maximum profit.

57. Natural Science A marshy region used for agricultural drainage has become contaminated with selenium. It has been determined that flushing the area with clean water will reduce the selenium for a while, but it will then begin to build up again. A biologist has found that the percent of selenium in the soil x months after the flushing begins is given by

$$f(x) = \frac{x^2 + 36}{2x}, \quad 1 \leq x \leq 12.$$

When will the selenium be reduced to a minimum? What is the minimum percent?

58. Natural Science The number of salmon swimming upstream to spawn is approximated by

$$S(x) = -x^3 + 3x^2 + 360x + 5000, \quad 6 \leq x \leq 20,$$

where x represents the temperature of the water in degrees Celsius. Find the water temperature that produces the maximum number of salmon swimming upstream.

59. From information given in a recent business publication we constructed the mathematical model

$$M(x) = -\frac{1}{45}x^2 + 2x - 20, \quad 30 \le x \le 65$$

to represent the miles per gallon used by a certain car at a speed of x miles per hour. Find the absolute maximum miles per gallon and the absolute minimum.

60. For a certain compact car,

$$M(x) = -.018x^2 + 1.24x + 6.2, \quad 30 \le x \le 60$$

represents the miles per gallon obtained at a speed of x miles per hour. Find the speed that produces the absolute maximum miles per gallon and the absolute minimum.

61. **Natural Science**   The microbe concentration, $B(x)$, in appropriate units, of Lake Tom depends approximately on the oxygen concentration, x, again in appropriate units, according to the function

$$B(x) = x^3 - 7x^2 - 160x + 1800.$$

(a) Find the oxygen concentration that will lead to the minimum microbe concentration.
(b) What is the minimum concentration?

62. Suppose dots and dashes are transmitted over a telegraph line so that dots occur a fraction p of the time (where $0 \le p \le 1$) and dashes occur a fraction $1 - p$ of the time. The *information content* of the telegraph line is given by $I(p)$, where

$$I(p) = -p \cdot \ln p - (1 - p) \cdot \ln (1 - p).$$

(a) Show that $I'(p) = -\ln p + \ln (1 - p)$.
(b) Let $I'(p) = 0$ and find the value of p that maximizes $I(p)$. In the long run, what fraction of the dots and dashes should be dots?

In many practical applications, only approximate locations of maxima or minima are needed (or, in fact, can even be found). To find the approximate maximum and minimum of each function on the given domains, use a calculator to evaluate the function at intervals of .1. Then estimate the absolute maximum and absolute minimum values of the function.

63. $f(x) = x^3 - 9x^2 + 18x - 7$; $[0, 1.5]$

64. $f(x) = x^3 - 3x^2 - 12x + 4$; $[-1.5, 0]$ and $[3, 4]$

65. $f(x) = x^3 + 6x^2 - 6x + 3$; $[-5, -4]$ and $[0, 1]$

66. $f(x) = x^3 - 3x^2 - 12x + 7$; $[-2, -1]$ and $[2, 3]$

12.2 THE SECOND DERIVATIVE TEST

To understand the behavior of a function on an interval, it is important to know the *rate* at which the function is increasing or decreasing. For example, suppose that your friend, a finance major, has made a study of a young company and is trying to get you to invest in its stock. He shows you the following function, which represents the price $P(t)$ of the company's stock since it became available in January 2 years ago:

$$P(t) = 17 + t^{1/2},$$

where t is the number of months since the stock became available. He points out that the derivative of the function is always positive, so the price of the stock is always increasing. He claims that you cannot help but make a fortune on it. Should you take his advice and invest?

It is true that the price function increases for all t. The derivative is

$$P'(t) = \frac{1}{2}t^{-1/2} = \frac{1}{2\sqrt{t}},$$

which is always positive because $\sqrt{t}$ is positive for $t > 0$. The catch lies in *how fast* the function is increasing. The derivative

$$P'(t) = \frac{1}{2\sqrt{t}}$$

tells how fast the price is increasing at any number of months, t, since the stock became available. For example, when $t = 1$, $P'(t) = 1/2$, and the price is increasing at the rate of 1/2 dollar, or 50¢, per month. When $t = 4$, $P'(t) = 1/4$; the stock is increasing at 25¢ per month. At $t = 9$ months, $P'(t) = 1/6$, or about 17¢ per month. By the time you could buy in at $t = 24$ months, the price is increasing at 10¢ per month, and the *rate of increase* looks as though it will continue to decrease.

The rate of increase in P' is given by the derivative of $P'(t)$, called the **second derivative** of P and denoted by $P''(t)$. Since $P'(t) = (1/2)t^{-1/2}$,

$$P''(t) = -\frac{1}{4}t^{-3/2} = -\frac{1}{4\sqrt{t^3}}.$$

$P''(t)$ is negative for all positive values of t and therefore confirms the suspicion that the *rate* of increase in price does indeed decrease for all $t \geq 0$. The price of the company's stock will not drop, but the amount of return will certainly not be the "fortune" that your friend predicts. For example, at $t = 24$ months, when you would buy, the price would be $21.90. A year later, it would be $23.00 a share. If you were rich enough to buy 100 shares for $2190, they would be worth $2300 in a year. The increase of $110 is about 5% of the investment—similar to the return that you could get in most savings accounts. The only investors to make a lot of money on this stock would be those who bought early, when the rate of increase was much greater.

As mentioned earlier, the second derivative of a function f, written f'', gives the rate of change of the *derivative* of f. Before continuing to discuss applications of the second derivative, we need to introduce some additional terminology and notation.

HIGHER DERIVATIVES   If a function f has a derivative f', then the derivative of f', if it exists, is the second derivative of f, written $f''(x)$. The derivative of $f''(x)$, if it exists, is called the **third derivative** of f, and so on. By continuing this process, we can find **fourth derivatives** and other higher derivatives. For example, if $f(x) = x^4 + 2x^3 + 3x^2 - 5x + 7$, then

$$f'(x) = 4x^3 + 6x^2 + 6x - 5, \qquad \text{First derivative of } f$$
$$f''(x) = 12x^2 + 12x + 6, \qquad \text{Second derivative of } f$$
$$f'''(x) = 24x + 12, \qquad \text{Third derivative of } f$$
and $\qquad f^{(4)}(x) = 24. \qquad \text{Fourth derivative of } f$

1 Let $f(x) = 4x^3 - 12x^2 + x - 1$. Find

(a) $f''(0)$;

(b) $f''(4)$;

(c) $f''(-2)$.

Answer:

(a) -24

(b) 72

(c) -72

2 Find the second derivatives of the following.

(a) $y = -9x^3 + 8x^2 + 11x - 6$

(b) $y = -2x^4 + 6x^2$

(c) $y = \dfrac{x + 2}{5x - 1}$

(d) $y = e^x + \ln x$

Answer:

(a) $y'' = -54x + 16$

(b) $y'' = -24x^2 + 12$

(c) $y'' = \dfrac{110}{(5x - 1)^3}$

(d) $y'' = e^x - \dfrac{1}{x^2}$

The second derivative of $y = f(x)$ can be written with any of the following notations:

$$f''(x), \quad y'', \quad \frac{d^2y}{dx^2}, \quad \text{or} \quad D_x^2[f(x)].$$

The third derivative can be written in a similar way. For $n \geq 4$, the nth derivative is written $f^{(n)}(x)$.

▶**EXAMPLE 1** Let $f(x) = x^3 + 6x^2 - 9x + 8$. Find the following.

(a) $f''(0)$

Here $f'(x) = 3x^2 + 12x - 9$, so that $f''(x) = 6x + 12$. Then

$$f''(0) = 6(0) + 12 = 12.$$

(b) $f''(-3) = 6(-3) + 12 = -6$ ◀ **1**

▶**EXAMPLE 2** Find the second derivative of the functions defined as follows.

(a) $y = 8x^3 - 9x^2 + 6x + 4$

Here $y' = 24x^2 - 18x + 6$. The second derivative is the derivative of y', or

$$y'' = 48x - 18.$$

(b) $y = \dfrac{4x + 2}{3x - 1}$

Use the quotient rule to find y'.

$$y' = \frac{(3x - 1)(4) - (4x + 2)(3)}{(3x - 1)^2} = \frac{12x - 4 - 12x - 6}{(3x - 1)^2} = \frac{-10}{(3x - 1)^2}$$

Use the quotient rule again to find y''.

$$y'' = \frac{(3x - 1)^2(0) - (-10)(2)(3x - 1)(3)}{[(3x - 1)^2]^2}$$

$$= \frac{60(3x - 1)}{(3x - 1)^4}$$

$$y'' = \frac{60}{(3x - 1)^3}$$

(c) $y = xe^x$

Using the product rule gives

$$y' = x \cdot e^x + e^x \cdot 1 = xe^x + e^x.$$

Differentiate this result to get y''.

$$y'' = (xe^x + e^x) + e^x$$

$$= xe^x + 2e^x$$

$$y'' = (x + 2)e^x \quad ◀ \quad \mathbf{2}$$

3 Rework Example 3 if $s(t) = t^4 - t^3 + 10$.

Answer:

(a) $v(t) = 4t^3 - 3t^2$

(b) $a(t) = 12t^2 - 6t$

(c) At 0 and 1/2 second

In the previous chapter we saw that the first derivative of a function represents the rate of change of the function. The second derivative, then, represents the rate of change of the first derivative. If a function describes the position of a moving object (along a straight line) at time t, then the first derivative gives the velocity of the object. That is, if $y = s(t)$ describes the position (along a straight line) of the object at time t, then $v(t) = s'(t)$ gives the velocity at time t.

The rate of change of velocity is called **acceleration.** Since the second derivative gives the rate of change of the first derivative, the acceleration is the derivative of the velocity. Thus, if $a(t)$ represents the acceleration at time t, then

$$a(t) = \frac{d}{dt}v(t) = s''(t).$$

▶**EXAMPLE 3**　Suppose that an object is moving along a straight line, with its position in feet at time t in seconds given by

$$s(t) = t^3 - 2t^2 - 7t + 9.$$

Find the following.
(a) The velocity at any time t
　　The velocity is given by

$$v(t) = s'(t) = 3t^2 - 4t - 7.$$

(b) The acceleration at any time t
　　Acceleration is given by

$$a(t) = v'(t) = s''(t) = 6t - 4.$$

(c) The object stops when velocity is zero. For $t \geq 0$, when does that occur?
　　Set $v(t) = 0$.

$$3t^2 - 4t - 7 = 0$$
$$(3t - 7)(t + 1) = 0$$
$$3t - 7 = 0 \quad \text{or} \quad t + 1 = 0$$
$$t = \frac{7}{3} \qquad\qquad t = -1$$

Since we want $t \geq 0$, only $t = 7/3$ is acceptable here. The object will stop at 7/3 seconds.　◀　**3**

The first derivative of a function gives the slope (rate of change of y with respect to x) of a line tangent to the graph of f at any value x. The second derivative (the derivative of the derivative) gives the rate of change of y' (the slope) with respect to x. As we have seen, a point on the graph just to the left of a relative minimum has a tangent line with a negative slope. Moving from left to right along the graph in a neighborhood of a relative minimum, the slope of the tangent line is *increasing*, from negative to zero to positive. See Figure 12.12. Hence, the rate of change of the slope (the derivative of the slope function) is positive. An argument based on this discussion leads to the *second derivative test* for maxima and minima.

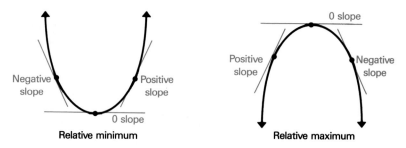

FIGURE 12.12

Second Derivative Test

Let c be a critical number of the function f such that $f'(c) = 0$ and $f''(x)$ exists for all x in some open interval containing c.

1. If $f''(c) > 0$, then $f(c)$ is a relative minimum.
2. If $f''(c) < 0$, then $f(c)$ is a relative maximum.
3. If $f''(c) = 0$, then the test gives no information about extrema.

▶**EXAMPLE 4**   Find all relative extrema for

$$f(x) = 4x^3 + 7x^2 - 10x + 8.$$

First, find the critical points. Here $f'(x) = 12x^2 + 14x - 10$. Solve the equation $f'(x) = 0$.

$$12x^2 + 14x - 10 = 0$$
$$2(6x^2 + 7x - 5) = 0$$
$$2(3x + 5)(2x - 1) = 0$$

$$3x + 5 = 0 \quad \text{or} \quad 2x - 1 = 0$$
$$3x = -5 \qquad\qquad 2x = 1$$
$$x = -\frac{5}{3} \qquad\qquad x = \frac{1}{2}$$

4 Find all relative maxima and relative minima for the following functions. Use the second derivative test.

(a) $f(x) = 6x^2 + 12x + 1$

(b) $f(x) = x^3 - 3x^2 - 9x + 8$

Answer:

(a) Relative minimum of -5 at $x = -1$

(b) Relative maximum of 13 at $x = -1$; relative minimum of -19 at $x = 3$

Now use the second derivative test. The second derivative is $f''(x) = 24x + 14$. The critical points are $-5/3$ and $1/2$. Since

$$f''\left(-\frac{5}{3}\right) = 24\left(-\frac{5}{3}\right) + 14 = -40 + 14 = -26 < 0,$$

$-5/3$ leads to a relative maximum of $f(-5/3) = 691/27 \approx 25.593$. Also,

$$f''\left(\frac{1}{2}\right) = 24\left(\frac{1}{2}\right) + 14 = 12 + 14 = 26 > 0,$$

so $1/2$ gives a relative minimum of $f(1/2) = 21/4 = 5.25$. ◀ **4**

The second derivative test works only for those critical points c that make $f'(c) = 0$. This test does not work for those critical points c for which $f'(c)$ does not exist (since $f''(c)$ would not exist either). Also, the second derivative test does not work for critical points c that make $f''(c) = 0$. In both of these cases, use the first derivative test.

12.2 EXERCISES

For each of the following, find f''. Then find $f''(0)$, $f''(2)$, and $f''(-3)$. (See Examples 1 and 2.)

1. $f(x) = 3x^3 - 4x + 5$

2. $f(x) = x^3 + 4x^2 + 2$

3. $f(x) = 3x^4 - 5x^3 + 2x^2$

4. $f(x) = -x^4 + 2x^3 - x^2$

5. $f(x) = (x + 4)^3$

6. $f(x) = (x - 2)^3$

7. $f(x) = \dfrac{2x + 1}{x - 2}$

8. $f(x) = \dfrac{x + 1}{x - 1}$

9. $f(x) = \dfrac{x^2}{1 + x}$

10. $f(x) = \dfrac{-x}{1 - x^2}$

11. $f(x) = \sqrt{x + 4}$

12. $f(x) = \sqrt{2x + 9}$

13. $f(x) = 5x^{3/5}$

14. $f(x) = -2x^{2/3}$

15. $f(x) = 2e^x$

16. $f(x) = \ln(2x - 3)$

For each of the following find $f'''(x)$, the third derivative of f, and $f^{(4)}(x)$, the fourth derivative.

17. $f(x) = -x^4 + 2x^2 + 8$

18. $f(x) = 2x^4 - 3x^3 + x^2$

19. $f(x) = 4x^5 + 6x^4 - x^2 + 2$

20. $f(x) = 3x^5 - x^4 + 2x^3 - 7x$

21. $f(x) = \dfrac{x - 1}{x + 2}$

22. $f(x) = \dfrac{x + 1}{x}$

23. $f(x) = 5e^{2x}$

24. $f(x) = 2 + e^{-x}$

25. $f(x) = \ln |x|$

26. $f(x) = \dfrac{1}{x}$

27. $f(x) = x \ln |x|$

28. $f(x) = \dfrac{\ln |x|}{x}$

Find any critical numbers for the functions defined as follows, and use the second derivative test to decide if the critical numbers lead to relative maxima or relative minima. (See Example 4.)

29. $f(x) = x^2 + 6x - 3$

30. $f(x) = -x^2 - 4x + 5$

31. $f(x) = 12 - 8x + 4x^2$

32. $f(x) = 6 + 4x + x^2$

33. $f(x) = x^3 + 2x^2 - 5$

34. $f(x) = 8x^3 - 9x^2 + 3$

35. $f(x) = -4x^3 - 15x^2 + 18x + 5$

36. $f(x) = -2x^3 - 3x^2 - 72x + 1$

37. $f(x) = \frac{2}{3}x^3 + \frac{1}{2}x^2 - x - \frac{1}{4}$

38. $f(x) = \frac{2}{3}x^3 + \frac{3}{2}x^2 - 2x + 1$

39. $f(x) = -2x^3 + 1$

40. $f(x) = (x - 1)^4$

41. $f(x) = x^4 - 8x^2$

42. $f(x) = x^4 - 32x^2 + 7$

43. $f(x) = x + \frac{3}{x}$

44. $f(x) = x - \frac{1}{x}$

45. $f(x) = \frac{x^2 + 9}{2x}$

46. $f(x) = \frac{x^2 + 16}{2x}$

47. $f(x) = \frac{2 - x}{2 + x}$

48. $f(x) = \frac{x + 2}{x - 1}$

Physical Science *Each of the functions in Exercises 49–54 gives the distance from a starting point at time t of a particle moving along a line. Find the velocity and acceleration functions. Then find the velocity and acceleration at $t = 0$ and $t = 4$. Assume that time is measured in seconds and distance is measured in centimeters. Velocity will be in centimeters per second (cm/sec) and acceleration in centimeters per second per second (cm/sec²). (See Example 3.)*

49. $s(t) = 8t^2 + 4t$

50. $s(t) = -3t^2 - 6t + 2$

51. $s(t) = -5t^3 - 8t^2 + 6t - 3$

52. $s(t) = 3t^3 - 4t^2 + 8t - 9$

53. $s(t) = \frac{-2}{3t + 4}$

54. $s(t) = \frac{1}{t + 3}$

55. Physical Science When an object is dropped straight down, the distance in feet that it travels in t seconds is given by

$$s(t) = -16t^2.$$

Find the velocity at each of the following times.
(a) After 3 seconds   **(b)** after 5 seconds
(c) After 8 seconds
(d) Find the acceleration. (The answer here is a constant, the acceleration due to the influence of gravity alone.)

56. Physical Science If an object is thrown directly upward with a velocity of 256 ft/sec, its height above the ground after t seconds is given by $s(t) = 256t - 16t^2$. Find the velocity and the acceleration after t seconds. What is the maximum height the object reaches? When does it hit the ground?

57. Management The profit from the sale of x units of a product is

$$p(x) = 1000 - 12x + 10x^2 - x^3.$$

Find the number of units that leads to a maximum profit. What is the maximum profit?

58. Management Because of raw material shortages, it is increasingly expensive to produce fine cigars. In fact, the profit in thousands of dollars from producing x hundred thousand cigars is approximated by

$$P(x) = x^3 - 23x^2 + 120x + 60.$$

Find the levels of production ($0 \le x \le 16$) that will lead to maximum profit and to any minimum profit.

59. Natural Science The number of mosquitoes, $M(x)$, in millions, in a certain area depends on the April rainfall, x, measured in inches, approximately as follows:

$$M(x) = 50 - 32x + 14x^2 - x^3.$$

Find the amount of rainfall that will produce the maximum number of mosquitoes.

60. Natural Science An autocatalytic chemical reaction is one in which the product being formed causes the rate of formation to increase. The rate of a certain autocatalytic reaction is given by

$$V(x) = 12x(100 - x),$$

where x is the quantity of the product present and 100 represents the quantity of chemical present initially. For what value of x is the rate of the reaction a maximum?

61. Natural Science The percent of concentration of a certain drug in the bloodstream x hours after the drug is administered is given by

$$K(x) = \frac{3x}{x^2 + 4}.$$

For example, after 1 hour the concentration is given by $K(1) = 3(1)/(1^2 + 4) = (3/5)\% = .6\% = .006$.

(a) Find the time at which concentration is a maximum.

(b) Find the maximum concentration.

62. Natural Science The percent of concentration of another drug in the bloodstream x hours after the drug is administered is given by

$$K(x) = \frac{4x}{3x^2 + 27}.$$

(a) Find the time at which the concentration is a maximum.

(b) Find the maximum concentration.

63. Natural Science Assume that the number of bacteria, in millions, present in a certain culture at time t is given by

$$R(t) = t^2(t - 18) + 96t + 1000.$$

(a) At what time before 8 hours will the population be maximized?

(b) Find the maximum population.

64. Management The revenue of a small firm, in thousands of dollars, is given by

$$R(x) = 2x^3 - 8x^2 + 4x,$$

where x is the number of units of product that the firm sells. Find the following. (See Example 3.)

(a) Velocity of revenue when $x = 8$

(b) When $x = 12$

(c) Acceleration of revenue when $x = 4$

(d) When $x = 8$

12.3 APPLICATIONS OF MAXIMA AND MINIMA

Finding extrema for realistic problems requires finding a function that accurately describes the situation. For example, suppose the number of units that can be produced on a production line can be closely approximated by

$$T(x) = 8x^{1/2} + 2x^{3/2} + 50,$$

where x is the number of employees on the line. Once this function has been established, it can be used to produce information about the production line. For example, the derivative $T'(x)$ could be used to estimate the change in production resulting from the addition of an extra worker to the line.

In writing the equation itself, it is important to keep track of any restrictions on the values of the variables. For example, since x represents the number of employees on a production line, x must certainly be restricted to the positive integers, or perhaps to a few common fractional values (we can conceive of halftime employees, but probably not 1/32-time employees).

On the other hand, to apply the tools of calculus to obtain an extremum for some function, the function must be defined and be meaningful at every real-number point in some interval. Because of this, the answer obtained by using a mathematical model of a practical problem might be a number that is not feasible in the setting of the problem.

Usually, the requirement that a function be continuous is of theoretical interest only. In most cases, calculus gives results that *are* acceptable in a given situation. And if the methods of calculus are used on a function f and lead to the conclusion that $80\sqrt{2}$ units should be produced in order to get the lowest possible cost, it is usually only necessary to calculate $f(80\sqrt{2})$ and then compare this result to various values of $f(x)$, where x is an acceptable number close to $80\sqrt{2}$. The lowest of these values of $f(x)$ then gives minimum cost. In most cases, the result obtained will be very close to the theoretical minimum.

This section gives several examples showing applications of calculus to extrema problems. Solve these examples with the following steps.

Solving Applied Problems

Step 1 Read the problem carefully. Make sure you understand what is given and what is asked for.

Step 2 If possible, sketch a diagram. Label the various parts.

Step 3 Decide on the variable whose values must be maximized or minimized. Express that variable as a function of *one* other variable.

Step 4 Find the critical points for the function of step 3. Test each of these for maxima or minima.

Step 5 Check the extrema at any endpoints of the domain of the function of step 3.

Step 6 Check to be sure the answer is reasonable.

▶**EXAMPLE 1**   Find two numbers x and y for which $2x + y = 30$, such that xy^2 is maximized.

First we must decide what is to be maximized and give it a variable. Here, xy^2 is to be maximized, so let

$$M = xy^2.$$

Now, express M in terms of just *one* variable. Use the equation $2x + y = 30$ to do that. Solve $2x + y = 30$ for either x or y. Solving for y gives

$$2x + y = 30$$
$$y = 30 - 2x.$$

Substitute for y in the expression for M to get

$$M = x(30 - 2x)^2$$
$$= x(900 - 120x + 4x^2)$$
$$= 900x - 120x^2 + 4x^3.$$

1 Find two numbers x and y such that $x + 2y = 60$ and xy is maximized.

Answer:
$x = 30$, $y = 15$

Find the critical points for M by finding M', then solving the equation $M' = 0$ for x.

$$M' = 900 - 240x + 12x^2 = 0$$
$$12(75 - 20x + x^2) = 0$$
$$(5 - x)(15 - x) = 0$$
$$x = 5 \quad \text{or} \quad x = 15$$

There are two critical numbers, 5 and 15. Use the second derivative test to decide which of these x-values will *maximize M*.

$$M''(x) = -240 + 24x$$
$$M''(5) = -240 + 24 \cdot 5 < 0$$
$$M''(15) = -240 + 24 \cdot 15 > 0$$

The test shows that $M = xy^2$ is maximized when $x = 5$. Since $y = 30 - 2x = 20$, the values that maximize xy^2 are $x = 5$ and $y = 20$. (There are no endpoints to check here.) ◀ **1**

▶ **EXAMPLE 2** When Power and Money, Inc., charges \$600 for a seminar on management techniques, it attracts 1000 people. For each \$20 decrease in the charge, an additional 100 people will attend the seminar.

(a) Find an expression for the total revenue if there are x \$20 decreases in the price.

The price charged will be

$$\text{Price per person} = 600 - 20x,$$

and the number of people in the seminar will be

$$\text{Number of people} = 1000 + 100x.$$

The total revenue, $R(x)$, is given by the product of the price and the number of people attending, or

$$R(x) = (600 - 20x)(1000 + 100x)$$
$$= 600{,}000 + 40{,}000x - 2000x^2.$$

(b) Find the value of x that maximizes revenue.

Set the derivative, $R'(x) = 40{,}000 - 4000x$, equal to 0.

$$40{,}000 - 4000x = 0$$
$$-4000x = -40{,}000$$
$$x = 10$$

Since $R''(x) = -4000$, $x = 10$ leads to maximum revenue.

2 An investor has built a series of self-storage units near a group of apartment houses. She must now decide on the monthly rental. From past experience, she feels that 200 units will be rented for $15 per month, with 5 additional rentals for each $.25 reduction in the rental price. Let x be the number of $.25 reductions in the price and find

(a) an expression for the number of units rented;

(b) an expression for the price per unit;

(c) an expression for the total revenue;

(d) the value of x leading to maximum revenue;

(e) the maximum revenue.

Answer:

(a) $200 + 5x$

(b) $15 - .25x$

(c) $(200 + 5x)(15 - .25x) = 3000 + 25x - 1.25x^2$

(d) $x = 10$

(e) $3125

(c) Find the maximum revenue.

Use $R(x)$ from part (a); let $x = 10$.

$$R(10) = 600{,}000 + 40{,}000(10) - 2000(10)^2 = 800{,}000$$

dollars. Since we want $R(x) \geq 0$, the domain is $[0, 30]$. At the endpoints, $R(0) = 600{,}000$ and $R(30) = 0$, so $R(10) = 800{,}000$ is the maximum revenue. For this level of revenue, $1000 + 100(10) = 2000$ people must attend the seminar; each person will pay $600 - 20(10) = \$400$. ◄ **2**

▶**EXAMPLE 3** A truck burns fuel at the rate of

$$G(x) = \frac{1}{200}\left(\frac{800 + x^2}{x}\right) \quad (x > 0)$$

gallons per mile when traveling x miles per hour on a straight, level road. If fuel costs $2 per gallon, find the speed that will produce the minimum total cost for a 1000-mile trip. Find the minimum total cost. (This is an extension of Exercise 46 in Section 11.5.)

The total cost of the trip, in dollars, is the product of the number of gallons per mile, the number of miles, and the cost per gallon. If $C(x)$ represents this cost, then

$$C(x) = \left[\frac{1}{200}\left(\frac{800 + x^2}{x}\right)\right](1000)(2)$$

$$= \frac{2000}{200}\left(\frac{800 + x^2}{x}\right)$$

$$C(x) = \frac{8000 + 10x^2}{x}.$$

To find the value of x that will minimize cost, first find the derivative.

$$C'(x) = \frac{10x^2 - 8000}{x^2}$$

Set this derivative equal to 0 (and remember $x > 0$).

$$\frac{10x^2 - 8000}{x^2} = 0$$

$$10x^2 - 8000 = 0$$

$$10x^2 = 8000$$

$$x^2 = 800$$

Take the square root on both sides to get

$$x = \pm\sqrt{800} \approx \pm 28.3 \text{ mph.}$$

3 A diesel generator burns fuel at the rate of

$$G(x) = \frac{1}{48}\left(\frac{300}{x} + 2x\right)$$

gallons per hour when producing x thousand kilowatt hours of electricity. Suppose that fuel costs \$2.25 a gallon and find the value of x that leads to minimum total cost if the generator is operated for 32 hours. Find the minimum cost.

Answer:
$x = \sqrt{150} \approx 12.2$; minimum cost is \$73.50

Reject -28.3 as a speed, leaving $x = 28.3$ as the only critical value. Find $C''(x)$ to see if 28.3 leads to a minimum cost.

$$C''(x) = -8000(-2)x^{-3} = \frac{16,000}{x^3}$$

Since $C''(28.3) > 0$, the critical value 28.3 leads to a minimum. (It is not necessary to actually calculate $C''(28.3)$; just check that $C''(28.3) > 0$.)

The minimum total cost is

$$C(28.3) = \frac{8000 + 10(28.3)^2}{28.3} = 565.69 \text{ dollars.}$$

The domain of $C(x)$ is $(0, \infty)$, so there are no endpoints to check. ◀ **3**

▶**EXAMPLE 4**   An open box is to be made by cutting a square from each corner of a 12-inch by 12-inch piece of metal and then folding up the sides. The finished box must be at least 1.5 inches deep, but not deeper than 3 inches. What size square should be cut from each corner in order to produce a box of maximum volume?

Let x represent the length of a side of the square that is cut from each corner, as shown in Figure 12.13. The width of the box is $12 - 2x$, while the length is also $12 - 2x$. As shown in Figure 12.13(b), the depth of the box will be x inches, where

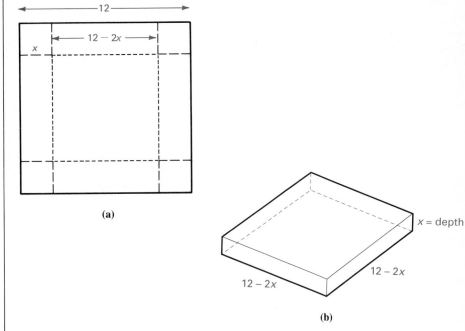

(a)

(b)

FIGURE 12.13

4 An open box is to be made by cutting squares from each corner of a 20 cm by 32 cm piece of metal and folding up the sides. Let x represent the length of the side of the square to be cut out. Find

(a) an expression for the volume of the box, $V(x)$;

(b) $V'(x)$;

(c) the value of x that leads to maximum volume. (Hint: the solutions of the equation $V'(x) = 0$ are 4 and 40/3);

(d) the maximum volume.

Answer:

(a) $V(x) = 640x - 104x^2 + 4x^3$

(b) $V'(x) = 640 - 208x + 12x^2$

(c) $x = 4$

(d) $V(4) = 1152$ cubic centimeters

x is in the interval [1.5, 3]. The volume of the box is given by the product of the length, width, and height. In this example, the volume, $V(x)$, depends on x.

$$V(x) = x(12 - 2x)(12 - 2x)$$
$$= 144x - 48x^2 + 4x^3$$

We must maximize $V(x)$ on the interval [1.5, 3].

The derivative is $V'(x) = 144 - 96x + 12x^2$. Set this derivative equal to 0.

$$12x^2 - 96x + 144 = 0$$
$$12(x^2 - 8x + 12) = 0$$
$$12(x - 2)(x - 6) = 0$$
$$x - 2 = 0 \quad \text{or} \quad x - 6 = 0$$
$$x = 2 \quad \text{or} \quad x = 6$$

To find the depth that will maximize the volume, evaluate $V(x)$ at the critical number 2 and at 1.5 and 3, the endpoints of the interval. The other critical number, $x = 6$, is outside the domain.

x	$V(x)$	
1.5	121.5	
2	128	← **Maximum**
3	108	

The chart indicates that the box will have maximum volume when $x = 2$ and that the maximum volume will be 128 cubic inches. ◀ **4**

ECONOMIC LOT SIZE Suppose that a company manufactures a constant number of units of a product per year and that the product can be manufactured in serveral batches of equal size during the year. If the company were to manufacture the item only once per year, it would minimize setup costs but incur high warehouse costs. On the other hand, if it were to make many small batches, this would increase setup costs. Calculus can be used to find the number of batches per year that should be manufactured in order to minimze total cost. This number is called the **economic lot size.**

Figure 12.14 shows several of the possibilities for a product having an annual demand of 12,000 units. The top graph shows the results if only one batch of the product is made annually: in this case, an average of 6000 items will be held in a warehouse. If four batches (of 3000 each) are made at equal time intervals during a year, the average number of units in the warehouse falls to only 1500. If twelve batches are made, an average of 500 items will be in the warehouse.

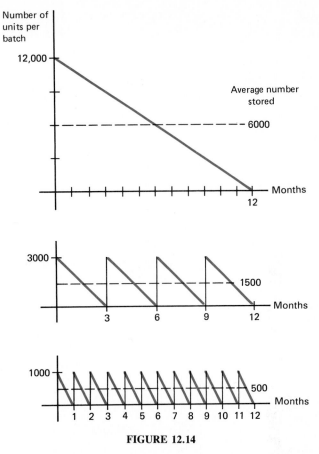

FIGURE 12.14

The following variables will be used in our discussion of economic lot size.

x = number of batches to be manufactured annually

k = cost of storing 1 unit of the product for 1 year

a = fixed setup cost to manufacture the product

b = variable cost of manufacturing a single unit of the product

M = total number of units produced annually.

The company has two types of costs associated with the production of its product: a cost associated with manufacturing the item, and a cost associated with storing the finished product.

During a year the company will produce x batches of the product, with M/x units of the product produced per batch. Each batch has a fixed cost a and a variable cost b per unit, so that the manufacturing cost per batch is

$$a + b\left(\frac{M}{x}\right).$$

There are x batches per year, so the total annual manufacturing cost is

$$\left[a + b\left(\frac{M}{x}\right)\right]x. \tag{1}$$

Since each batch consists of M/x units, and since demand is constant, it is common to assume an average inventory of

$$\frac{1}{2}\left(\frac{M}{x}\right) = \frac{M}{2x}$$

units per year. The cost to store 1 unit of the product for a year is k, so the total storage cost is

$$k\left(\frac{M}{2x}\right) = \frac{kM}{2x}. \tag{2}$$

The total production cost is the sum of the manufacturing and storage costs, or the sum of expressions (1) and (2). If $T(x)$ is the total cost of producing x batches,

$$T(x) = \left[a + b\left(\frac{M}{x}\right)\right]x + \frac{kM}{2x} = ax + bM + \left(\frac{kM}{2}\right)x^{-1}.$$

Now find the value of x that will minimize $T(x)$. (Remember that a, b, k, and M are constants.) Find $T'(x)$.

$$T'(x) = a - \frac{kM}{2}x^{-2}$$

Set this derivative equal to 0.

$$a - \frac{kM}{2}x^{-2} = 0$$

$$a = \frac{kM}{2x^2}$$

$$2ax^2 = kM$$

$$x^2 = \frac{kM}{2a}$$

$$x = \sqrt{\frac{kM}{2a}} \tag{3}$$

The second derivative test can be used to show that $\sqrt{kM/(2a)}$ is the annual number of batches that gives minimum total production cost.

5 A manufacturer of business forms has an annual demand for 30,720 units of form letters to people delinquent in their payments of installment debt. It costs $5 per year to store 1 unit of the letters and $1200 to set up the machines to produce them. Find the number of batches that should be made annually to minimize total cost.

Answer:
8 batches

6 An office uses 576 cases of copy-machine paper during the year. It costs $3 per year to store 1 case. Each reorder costs $24. Find the number of orders that should be placed annually.

Answer:
6

▶**EXAMPLE 5**   A paint company has a steady annual demand for 24,500 cans of automobile primer. The cost accountant for the company says that it costs $2 to store 1 can of paint for 1 year and $500 to set up the plant for the production of the primer. Find the number of batches of primer that should be produced for the minimum total production cost.

Use equation (3) above.

$$x = \sqrt{\frac{kM}{2a}}$$

$$x = \sqrt{\frac{2(24,500)}{2(500)}} \qquad \text{Let } k = 2, M = 24,500, a = 500$$

$$x = \sqrt{49} = 7$$

Seven batches of primer per year will lead to minimum production costs.   ◀ **5** **6**

12.3 EXERCISES

Exercises 1–8 involve maximizing and minimizing products and sums of numbers. Work all these problems using the steps shown in Exercise 1. Find only nonnegative solutions. (See Example 1.)

1. We want to find two numbers x and y such that $x + y = 100$ and the product $P = xy$ is as large as possible.
   (a) We know $x + y = 100$. Solve this equation for y.
   (b) Substitute this result for y into $P = xy$.
   (c) Find P'. Solve the equation $P' = 0$.
   (d) What are the two numbers?
   (e) What is the maximum value of P?

2. Find two numbers whose sum is 250 and whose product is as large as possible.

3. Find two numbers whose sum is 200 such that the sum of the squares of the two numbers is minimized.

4. Find two numbers whose sum is 30 such that the sum of the squares of the numbers is minimized.

5. Find positive numbers x and y such that $x + y = 150$ and x^2y is maximized.

6. Find positive numbers x and y such that $x + y = 45$ and xy^2 is maximized.

7. Find positive numbers x and y such that $x - y = 10$ and xy is minimized.

8. Find numbers x and y such that $x - y = 3$ and xy is minimized.

9. **Management** If the price charged for a candy bar is $p(x)$ cents, then x thousand candy bars will be sold in a certain city, where

$$p(x) = 100 - \frac{x}{10}.$$

   (a) find an expression for the total revenue from the sale of x thousand candy bars. (Hint: find the product of $p(x)$, x, and 1000.)

(b) Find the value of x that leads to maximum revenue.
(c) Find the maximum revenue.

10. **Management** The sale of cassette tapes of "lesser" performers is very sensitive to price. If a tape manufacturer chargers $p(x)$ dollars per tape, where

$$p(x) = 6 - \frac{x}{8},$$

then x thousand tapes will be sold.
(a) Find an expression for the total revenue from the sale of x thousand tapes. (Hint: find the product of $p(x)$, x, and 1000.)
(b) Find the value of x that leads to maximum revenue.
(c) Find the maximum revenue.

11. **Management** A truck burns fuel at the rate of $G(x)$ gallons per mile, where

$$G(x) = \frac{1}{32}\left(\frac{64}{x} + \frac{x}{50}\right),$$

while traveling x miles per hour.
(a) If fuel costs \$1.60 per gallon, find the speed that will produce minimum total cost for a 400-mile trip. (See Example 3.)
(b) Find the minimum total cost.

12. **Management** A rock-and-roll band travels from engagement to engagement in a large bus. This bus burns fuel at the rate of $G(x)$ gallons per mile, where

$$G(x) = \frac{1}{50}\left(\frac{200}{x} + \frac{x}{15}\right),$$

while traveling x miles per hour.
(a) If fuel costs \$2 per gallon, find the speed that will produce minimum total cost for a 250-mile trip.
(b) Find the minimum total cost.

13. A farmer has 1200 meters of fencing. He wants to enclose a rectangular field bordering a river, with no fencing needed along the river. (See the sketch.) Let x represent the width of the field.

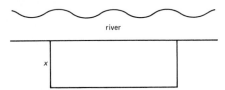

(a) Write an expression for the length of the field.
(b) Find the area of the field (area = length × width).
(c) Find the value of x leading to the maximum area.
(d) Find the maximum area.

14. Find the dimensions of the rectangular field of maximum area that can be made from 200 meters of fencing material. (This fence has four sides.)

15. **Natural Science** An ecologist is conducting a research project on breeding pheasants in captivity. She first must construct suitable pens. She wants a rectangular area with two additional fences down the center, as shown in the sketch. Find the maximum area she can enclose with 3600 meters of fencing.

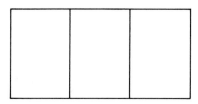

16. A rectangular field is to be enclosed with a fence. One side of the field is against an existing fence, so that no fence is needed on that side. If material for the fence costs \$2 per foot for the two ends and \$4 per foot for the side parallel to the existing fence, find the dimensions of the field of largest area that can be enclosed for \$1000.

17. A rectangular field is to be enclosed on all four sides with a fence. Fencing material costs \$3 per foot for two opposite sides, and \$6 per foot for the other two sides. Find the maximum area that can be enclosed for \$2400.

18. **Management** The manager of an 80-unit apartment complex is trying to decide on the rent to charge. It is known from experience that at a rent of \$200, all the units will be full. However, on the average, 1 additional unit will remain vacant for each \$20 increase in rent.
(a) Let x represent the number of \$20 increases. Find an expression for the rent for each apartment. (See Example 2.)
(b) Find an expression for the number of apartments rented.
(c) Find an expression for the total revenue from all rented apartments.
(d) What value of x leads to maximum revenue?
(e) What is the maximum revenue?

19. Management The manager of a peach orchard is trying to decide when to have the peaches picked. If they are picked now, the average yield per tree will be 100 pounds, which can be sold for 40¢ per pound. Past experience shows that the yield per tree will increase about 5 pounds per week, while the price will decrease about 2¢ per pound per week.

(a) Let x represent the number of weeks that the manager should wait. Find the revenue per pound.

(b) Find the number of pounds per tree.

(c) Find the total revenue from a tree.

(d) When should the peaches be picked in order to produce maximum revenue?

(e) What is the maximum revenue?

20. A local group of scouts has been collecting old aluminum beer cans for recycling. The group has already collected 12,000 pounds of cans, for which they could currently receive $4 per hundred pounds. The group can continue to collect cans at the rate of 400 pounds per day. However, a glut in the old-beer-can market has caused the recycling company to announce that it will lower its price, starting immediately, by $.10 per hundred pounds per day. The scouts can make only one trip to the recycling center. Find the best time for that single trip. What total income will be received?

21. Management In planning a small restaurant, it is estimated that a profit of $5 per seat will be made if the number of seats is between 60 and 80, inclusive. On the other hand, the profit on each seat will decrease by 5¢ for each seat above 80.

(a) Find the number of seats that will produce the maximum profit.

(b) What is the maximum profit?

22. A local club is arranging a charter flight to Hawaii. The cost of the trip is $425 each for 75 passengers, with a refund of $5 per passenger for each passenger in excess of 75.

(a) Find the number of passengers that will maximize the revenue received from the flight.

(b) Find the maximum revenue.

23. Management A television manufacturing firm needs to design an open-topped box with a square base. The box must hold 32 cubic inches. Find the dimensions of the box that can be built with the minimum amount of materials.

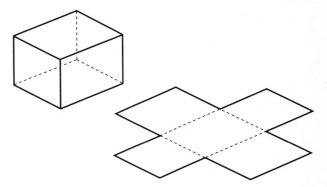

24. A closed box with a square base is to have a volume of 16,000 cubic centimeters. The material for the top and bottom of the box costs $3 per square centimeter, while the material for the sides costs $1.50 per square centimeter. Find the dimensions of the box that will lead to minimum total cost. What is the minimum total cost?

25. Management A company wishes to manufacture a box with a volume of 36 cubic feet that is open on top and that is twice as long as it is wide. Find the dimensions of the box produced from the minimum amount of material.

26. Management A mathematics book is to contain 36 square inches of printed matter per page, with margins of 1 inch along the sides, and $1\frac{1}{2}$ inches along the top and bottom. Find the dimensions of the page that will lead to the minimum amount of paper being used for a page.

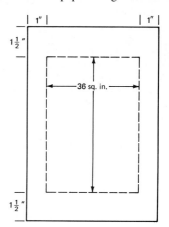

27. **Management** In Example 3 we found the speed in miles per hour that minimized cost when we considered only the cost of the fuel. Rework the problem taking into account the driver's salary of $8 per hour. (Hint: if the trip is 1000 miles at x miles per hour, the driver will be paid for $1000/x$ hours.)

28. **Management** Decide what you would do if your assistant brought you the following contract to sign:

> Your firm offers to deliver 300 tables to a dealer, at $90 per table, and to reduce the price per table on the entire order by 25¢ for each additional table over 300.

Find the dollar total involved in the largest possible transaction between you and the dealer; find the smallest possible dollar amount.

29. **Management** A company wishes to run a utility cable from point A on the shore (see the figure) to an installation at point B on the island. The island is 6 miles from the shore (at point C) and point A is 9 miles from C. It costs $400 per mile to run the cable on land and $500 per mile underwater. Assume that the cable starts at A and runs along the shoreline, then angles and runs underwater to the island. Find the point at which the line should begin to angle in order to yield the minimum total cost. (Hint: the length of the line underwater is $\sqrt{x^2 + 36}$.)

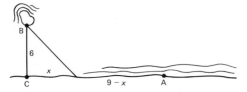

30. A hunter is at a point on a river bank. He wants to get to his cabin, located 3 miles north and 8 miles west. (See the figure.) He can travel 5 miles per hour on the river but only 2 miles per hour on this very rocky land. How far up river should he go in order to reach the cabin in minimum time? (Hint: distance = rate × time.)

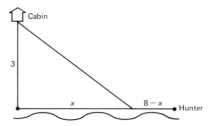

Management *The following exercises refer to economic lot size. (See Example 5.)*

31. Find the approximate number of batches that should be produced annually if 100,000 units are to be manufactured. It costs $1 to store a unit for 1 year and it costs $500 to set up the factory to produce each batch.

32. How many units per batch will be manufactured in Exercise 31?

33. A market has a steady annual demand for 16,800 cases of sugar. It costs $3 to store 1 case for 1 year. The market pays $7 for each order that is placed. Find the number of orders for sugar that should be placed each year.

34. Find the number of cases per order in Exercise 33.

35. The publisher of a best-selling book has an annual demand for 100,000 copies. It costs 50¢ to store 1 copy for 1 year. Setup of the press for a new printing costs $1000. Find the number of batches that should be printed annually.

36. A restaurant has an annual demand for 810 bottles of a California wine. It costs $1 to store 1 bottle for 1 year, and it costs $5 to place a reorder. Find the number of orders that should be placed annually.

12.4 CURVE SKETCHING

Informally, we say that a function is *increasing* in an interval if its graph is *rising* from left to right over the interval and that a function is *decreasing* in an interval if its graph is *falling* from left to right over the interval. For example, the function whose graph is shown in Figure 12.15 is rising on the intervals (a, b) and (c, ∞) and falling on the interval (b, c).

1 For what values of x is the function graphed as follows increasing? decreasing?

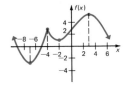

Answer:
Increasing on $(-7, -4)$ and $(-2, 3)$; decreasing on $(-\infty, -7)$, $(-4, -2)$, and $(3, \infty)$

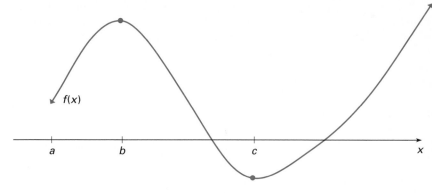

FIGURE 12.15

We can express the idea of the graph of a function rising and falling algebraically as follows.

Let f be a function defined on some interval. Then

1. f is **increasing** in the interval if for any two numbers x_1 and x_2 in the interval,

$$f(x_1) < f(x_2) \quad \text{whenever } x_1 < x_2;$$

2. f is **decreasing** in the interval if for any two numbers x_1 and x_2 in the interval,

$$f(x_1) > f(x_2) \quad \text{whenever } x_1 < x_2.$$

▶**EXAMPLE 1** For which x-intervals is the function graphed in Figure 12.16 increasing? For which intervals is it decreasing?

Moving from left to right, the function is increasing up to -4, then decreasing from -4 to 0, constant (neither increasing nor decreasing) from 0 to 4, increasing from 4 to 6, and finally, decreasing from 6 on. In interval notation, the function is increasing on $(-\infty, -4)$ and $(4, 6)$, decreasing on $(-4, 0)$ and $(6, \infty)$, and constant on $(0, 4)$. ◀ **1**

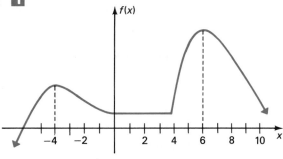

FIGURE 12.16

If f is a function, think of the graph of f as a roller coaster track, with the car moving from left to right, and the floor of the car representing the tangent line to the graph at that point as in Figure 12.17. Using this analogy, we see that the slope of the tangent line will be *positive* when the car travels uphill (the function is *increasing*) and the slope of the tangent will be *negative* when the car travels downhill (the function is *decreasing*). Since the slope of the tangent line at the point $(x, f(x))$ is given by the derivative $f'(x)$, we have this useful test.

Suppose a function f has a derivative at each point in an open interval:

1. If $f'(x) > 0$ for each x in the interval, f is *increasing* on the interval.
2. If $f'(x) < 0$ for each x in the interval, f is *decreasing* on the interval.
3. If $f'(x) = 0$ for each x in the interval, f is *constant* on the interval.

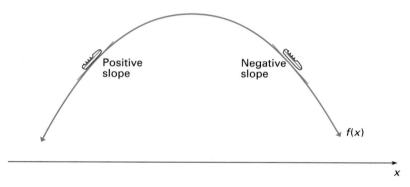

FIGURE 12.17

The derivative $f'(x)$ can change signs from positive to negative (or negative to positive) at points where $f'(x) = 0$, and also at points where $f'(x)$ does not exist. The values of x where this occurs are, by definition, the critical numbers for the function f.

It is shown in more advanced classes that if the critical numbers of a polynomial function are used to determine open intervals on a number line, then the sign of the derivative at any point in the interval will be the same as the sign of the derivative at any other point in the interval. This suggests that the test for increasing and decreasing functions be applied as follows (assuming that no open intervals exist where the function is constant).

Step 1 Locate on a number line those values of x for which $f'(x) = 0$ or $f'(x)$ does not exist. These points determine several open intervals.

Step 2 Choose a value of x in each of the intervals determined in step 1. Use these values to decide whether $f'(x) > 0$ or $f'(x) < 0$ on that interval.

Step 3 Use the test above to decide whether f is increasing or decreasing on each interval.

►**EXAMPLE 2** Find the intervals in which the function defined by $f(x) = x^3 + 3x^2 - 9x + 4$ is increasing or decreasing. Locate all points where the tangent line is horizontal. Graph the function.

Here $f'(x) = 3x^2 + 6x - 9$. To find the critical numbers, set this derivative equal to 0, and solve the resulting equation by factoring.

$$3x^2 + 6x - 9 = 0$$
$$3(x^2 + 2x - 3) = 0$$
$$3(x + 3)(x - 1) = 0$$
$$x = -3 \quad \text{or} \quad x = 1$$

The tangent line is horizontal at $x = -3$ or $x = 1$. There are no values of x where $f'(x)$ fails to exist. The critical numbers are -3 and 1. To determine where the function is increasing or decreasing, locate -3 and 1 on a number line, as in Figure 12.18. (Be sure to place the values on the number line in numerical order.) Choosing the test point -4 from the interval on the left gives

$$f'(-4) = 3(-4)^2 + 6(-4) - 9 = 15,$$

which is positive, so f is increasing on $(-\infty, -3)$. Selecting 0 from the middle interval gives $f'(0) = -9$, so f is decreasing on $(-3, 1)$. Finally, choosing 2 in the right-hand region gives $f'(2) = 15$, with f increasing on $(1, \infty)$. The intervals where f is increasing or decreasing are shown with arrows in Figure 12.18.

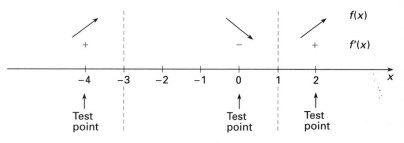

FIGURE 12.18

2 Find all intervals where the following functions are increasing or decreasing.

(a) $f(x) = 2x^2 + 5x - 4$

(b) $f(x) = 6 + 3x - x^2$

(c) $f(x) = 4x^3 + 3x^2 - 18x + 1$

Answer:

(a) Increasing on $(-5/4, \infty)$; decreasing on $(-\infty, -5/4)$

(b) Increasing on $(-\infty, 3/2)$; decreasing on $(3/2, \infty)$

(c) Increasing on $(-\infty, -3/2)$ and $(1, \infty)$; decreasing on $(-3/2, 1)$

To graph the function, plot a point in each of the intervals determined by the critical numbers, as well as the points at the critical numbers. Use these points along with the information about where the function is increasing and decreasing to draw the graph in Figure 12.19. ◀ **2**

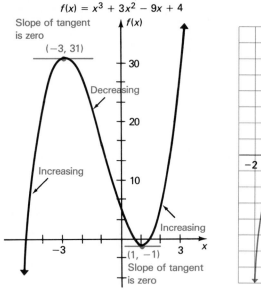

FIGURE 12.19

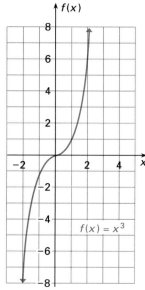

FIGURE 12.20

It is important to note that the reverse of the test for increasing and decreasing functions is not true—it is possible for a function to be increasing on an interval even though the derivative is not positive at every point in the interval. A good example is given by $f(x) = x^3$, which is increasing on every interval, even though $f'(x) = 0$ when $x = 0$. See Figure 12.20.

Knowing the intervals where a function is increasing or decreasing can be important in applications, as shown by the next examples.

▶**EXAMPLE 3** A company sells an item with the following cost and revenue functions.

$$C(x) = 1.2x - .0006x^2, \quad 0 \le x \le 1000$$
$$R(x) = 3.6x - .003x^2, \quad 0 \le x \le 1000$$

Determine any intervals on which the profit function is increasing.

First find the profit function $P(x)$.

$$P(x) = R(x) - C(x)$$
$$= (3.6x - .003x^2) - (1.2x - .0006x^2)$$
$$= 2.4x - .0024x^2$$

To find the interval where this function is increasing, set $P'(x) = 0$.

$$P'(x) = 2.4 - .0048x = 0$$
$$x = 500$$

Use $x = 500$ to determine two intervals on a number line, as shown in Figure 12.21. Choose $x = 0$ and $x = 1000$ as test points.

$$P'(0) = 2.4 - .0048(0) = 2.4$$
$$P'(1000) = 2.4 - .0048(1000) = -2.4$$

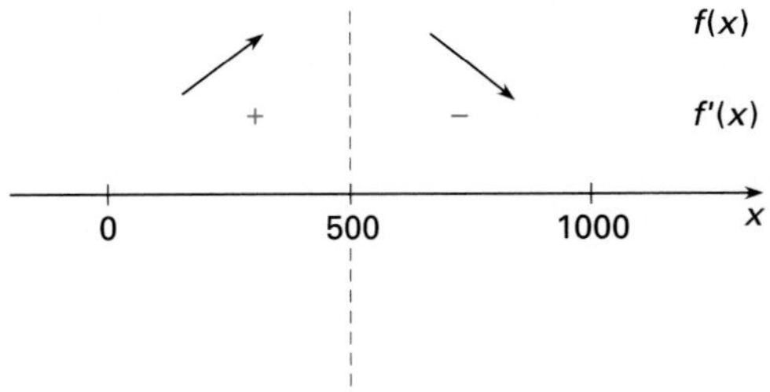

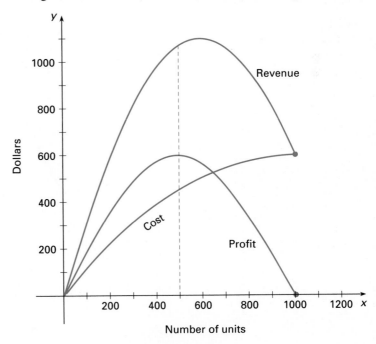

FIGURE 12.21

The test points show that the function increases on $(-\infty, 500)$ and decreases on $(500, \infty)$. See Figure 12.21. Thus, the profit is increasing on the interval $(0, 500)$ and decreasing on the interval $(500, 1000)$, as shown in Figure 12.22.

FIGURE 12.22

3 Find the intervals where profit is increasing and $P(x) > 0$ if the profit function is defined as $P(x) = 1000 + 90x - x^2$.

Answer:
$(0, 45)$

As the graph in Figure 12.22 shows, the profit will increase as long as the revenue function increases faster than the cost function. That is, increasing production will produce more profit as long as the marginal revenue is greater than the marginal cost. ◀ **3**

▶**EXAMPLE 4**   In the exercises in the previous chapter, the function

$$f(t) = \frac{90t}{99t - 90}$$

gave the number of facts recalled after t hours for $t > 10/11$. Find the intervals in which $f(t)$ is increasing or decreasing.

First find the derivative, $f'(t)$.

$$f'(t) = \frac{(99t - 90)(90) - 90t(99)}{(99t - 90)^2}$$

$$= \frac{8910t - 8100 - 8910t}{(99t - 90)^2} = \frac{-8100}{(99t - 90)^2}$$

Since $(99t - 90)^2$ is positive everywhere in the domain of the function and since the numerator is a negative constant, $f'(t) < 0$ for all t in the domain of $f(t)$. Thus, $f(t)$ always decreases and, as expected, the number of words recalled decreases steadily over time. ◀

CONCAVITY   The first derivative has been used to show where a function is increasing or decreasing and where the extrema occur. The second derivative gives the rate of change of the first derivative; it indicates *how fast* the function is increasing or decreasing. The rate of change of the derivative (the second derivative) affects the *shape* of the graph. Intuitively, we say that a graph is *concave upward* on an interval if it "holds water" and *concave downward* if it "spills water." See Figure 12.23.

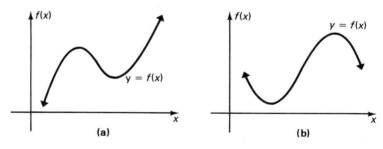

(a)   **(b)**

FIGURE 12.23

More precisely, a function is **concave upward** on an interval (a, b) if the graph of the function lies above its tangent line at each point of (a, b); a function is **concave downward** on (a, b) if the graph of the function lies below its tangent line at each point of (a, b). A point where a graph changes concavity is called a **point of inflection.** See Figure 12.24.

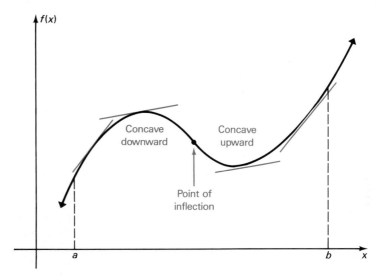

FIGURE 12.24

Just as a function can be either increasing or decreasing on an interval, it can be either concave upward or concave downward on an interval. Examples of various combinations are shown in Figure 12.25.

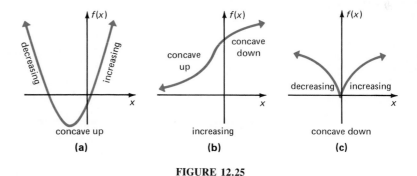

FIGURE 12.25

Figure 12.26 shows the graph of two functions that are concave upward on an interval (a, b). Several tangent lines are also shown. In Figure 12.26(a), the slopes of these tangent lines (moving from left to right) are first negative, then 0, and then positive. In Figure 12.26(b), the slopes are all positive, but getting larger.

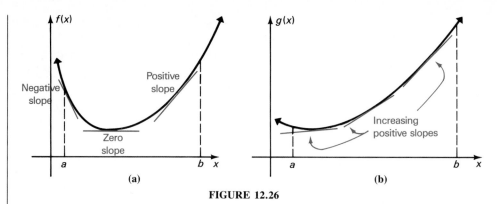

FIGURE 12.26

In both cases, the slopes are *increasing*. The slope at a point on a curve is given by the derivative. Since a function is increasing if its derivative is positive, the slope is increasing if the derivative of the function giving the slopes is positive. Since the derivative of a derivative is the second derivative, a function is concave upward on an interval if its second derivative is positive at each point of the interval.

A similar result is suggested by Figure 12.27 for functions whose graphs are concave downward. In both graphs, the slopes of the tangent lines are *decreasing* as we move from left to right. Since a function is decreasing if its derivative is negative, a function is concave downward on an interval if its second derivative is negative at each point of the interval. These observations suggest the following test.

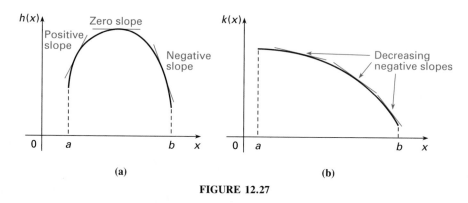

FIGURE 12.27

Let f be a function with derivatives f' and f'' existing at all points in an interval (a, b).

1. The function f is **concave upward** on (a, b) if $f''(x) > 0$ for all x in (a, b).
2. The function f is **concave downward** on (a, b) if $f''(x) < 0$ for all x in (a, b).

▶**EXAMPLE 5** Find all intervals where $f(x) = x^3 - 3x^2 + 5x - 4$ is concave upward or downward.

The first derivative is $f'(x) = 3x^2 - 6x + 5$, and the second derivative is $f''(x) = 6x - 6$. The function f is concave upward whenever $f''(x) > 0$, or

$$6x - 6 > 0$$
$$6x > 6$$
$$x > 1.$$

Also, f is concave downward if $f''(x) < 0$, or $x < 1$. In interval notation, f is concave upward on $(1, \infty)$ and concave downward on $(-\infty, 1)$, with a point of inflection at $(1, -1)$. A graph of f is shown in Figure 12.28.   ◀

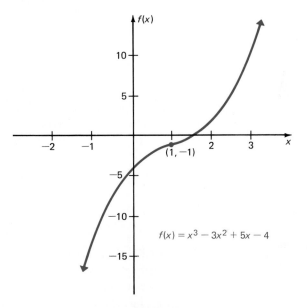

FIGURE 12.28

The graph of the function f in Figure 12.28 changes from concave downward to concave upward at $x = 1$. As mentioned earlier, a point where the direction of concavity changes is called a point of inflection. This means that the point $(1, f(1))$ or $(1, -1)$ in Figure 12.28 is a point of inflection. We can locate this point by finding values of x where the second derivative changes from negative to positive; that is, where the second derivative is 0. Setting the second derivative in Example 5 equal to 0 gives

$$6x - 6 = 0$$
$$x = 1.$$

As before, the point of inflection is $(1, f(1))$ or $(1, -1)$.

4 Find the intervals where the following are concave upward. Identify any inflection points.

(a) $f(x) = 6x^3 - 24x^2 + 9x - 3$

(b) $f(x) = 2x^2 - 4x + 8$

Answer:

(a) Concave downward on $(-\infty, 4/3)$, concave upward on $(4/3, \infty)$; point of inflection is $(4/3, -175/9)$

(b) $f''(x) = 4$, which is always positive; function is always concave upward, no inflection point

Example 5 suggests the following result.

> At a point of inflection for a function f, the second derivative is 0 or does not exist.

Caution Finding a value $f''(x_0) = 0$ does not mean that a point of inflection has been located. For example, if $f(x) = (x - 1)^4$, then $f''(x) = 12(x - 1)^2$, which is 0 at $x = 1$. The graph of $f(x) = (x - 1)^4$ is always concave upward, however, so it has no point of inflection. See Figure 12.29. **4**

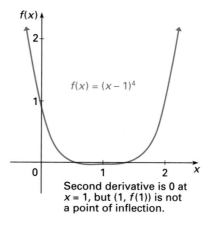

$f(x) = (x - 1)^4$

Second derivative is 0 at $x = 1$, but $(1, f(1))$ is not a point of inflection.

FIGURE 12.29

▶ **EXAMPLE 6** The graph of Figure 12.30 shows the population of catfish in a commercial catfish farm as a function of time. As the graph shows, the population increases rapidly up to a point, and then increases at a slower rate. The horizontal dashed line shows that the population will approach some upper limit determined by the capacity of the farm. The point at which the rate of population growth starts to slow is the point of inflection for the graph.

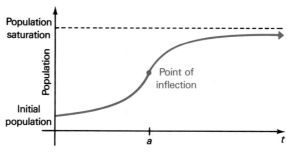

Point of inflection

FIGURE 12.30

5 Find the intervals where the function of Figure 12.30 is concave upward and concave downward.

Answer:
Concave upward on $(0, a)$ and concave downward on (a, ∞)

To produce maximum yield of catfish, harvesting should take place at the point of fastest possible growth of the population—here, this is at the point of inflection. The rate of change of the population, given by the first derivative, is increasing up to the inflection point (on the interval where the second derivative is positive) and decreasing past the inflection point (on the interval where the second derivative is negative). ◀ **5**

The *law of diminishing returns* in economics is related to the idea of concavity. The function graphed in Figure 12.31 gives the output y from a given input x. If the input were advertising costs for some product, for example, the output might be the corresponding revenue from sales.

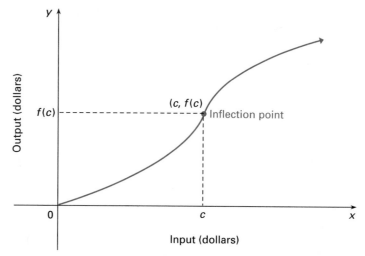

FIGURE 12.31

The graph in Figure 12.31 shows an inflection point at $(c, f(c))$. For $x < c$, the graph is concave upward, so the rate of change of the slope is increasing. This indicates that the output y is increasing at a faster rate with each additional dollar spent. When $x > c$, however, the graph is concave downward, the rate of change of the slope is decreasing, and the increase in y is smaller with each additional dollar spent. Thus, further input beyond c dollars produces diminishing returns. The point of inflection at $(c, f(c))$ is called the **point of diminishing returns.** Any investment beyond the value c is not considered a good use of capital.

▶**EXAMPLE 7**   The revenue $R(x)$ generated from sales of a certain product is related to the amount x spent on advertising by

$$R(x) = \frac{1}{150,000} (600x^2 - x^3), \quad 0 \leq x \leq 600,$$

where x and $R(x)$ are in thousands of dollars. Is there a point of diminishing returns for this function? If so, what is it?

6 In Example 7, $R(x) = \dfrac{x^3}{600} - \dfrac{x^4}{12000}$ for another product. What is the point of diminishing returns?

Answer:
(10, .833)

Since a point of diminishing returns occurs at an inflection point, look for an x-value that makes $R''(x) = 0$. Write the function as

$$R(x) = \frac{600}{150,000}x^2 - \frac{1}{150,000}x^3 = \frac{1}{250}x^2 - \frac{1}{150,000}x^3.$$

Now find $R'(x)$ and then $R''(x)$.

$$R'(x) = \frac{2x}{250} - \frac{3x^2}{150,000} = \frac{1}{125}x - \frac{1}{50,000}x^2$$

$$R''(x) = \frac{1}{125} - \frac{1}{25,000}x$$

Set $R''(x)$ equal to 0 and solve for x.

$$\frac{1}{125} - \frac{1}{25,000}x = 0$$

$$-\frac{1}{25,000}x = -\frac{1}{125}$$

$$x = \frac{25,000}{125} = 200$$

Test a number in the interval (0, 200) to see that $R''(x)$ is positive there. Then test a number in the interval (200, 600) to find $R''(x)$ negative in that interval. Since the sign of $R''(x)$ changes from positive to negative at $x = 200$, the graph changes from concave upward to concave downward at that point, and there is a point of diminishing returns at the inflection point $(200, 106\frac{2}{3})$. Any investment in advertising beyond $200,000 would not pay off. ◀ 6

CURVE SKETCHING The test for concavity and the test for increasing and decreasing functions help in sketching the graphs of a variety of functions. This process, called **curve sketching,** uses the following steps.

To sketch the graph of a function f:

1. Find f'. Locate any critical points by solving the equation $f'(x) = 0$ and determining where f' does not exist (but f does). Find any relative extrema and determine where f is increasing or decreasing.
2. Find f''. Locate potential points of inflection by solving the equation $f''(x) = 0$ and determining where f'' does not exist. Determine where f'' is concave upward or concave downward.
3. Plot the points at the critical numbers, the inflection points, and other points as needed.

▶**EXAMPLE 8**   Graph each function.

(a) $f(x) = 2x^3 - 3x^2 - 12x + 1$

   To find the intervals where the function is increasing or decreasing, find the first derivative.

$$f'(x) = 6x^2 - 6x - 12$$

This derivative is 0 when

$$6(x^2 - x - 2) = 0$$
$$6(x - 2)(x + 1) = 0$$
$$x = 2 \quad \text{or} \quad x = -1.$$

These critical numbers divide the number line in Figure 12.32 into three regions. Testing a number from each region in $f'(x)$ shows that f is increasing on $(-\infty, -1)$ and $(2, \infty)$ and decreasing on $(-1, 2)$. This is shown with the arrows in Figure 12.32. By the first derivative test, f has a relative maximum when $x = -1$ and a relative minimum when $x = 2$. The relative maximum is $f(-1) = 8$, while the relative minimum is $f(2) = -19$.

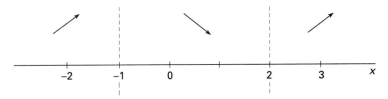

FIGURE 12.32

   Now use the second derivative to find the intervals where the function is concave upward or downward. Here

$$f''(x) = 12x - 6,$$

which is 0 when $x = 1/2$. Testing a point with x less than $1/2$, and one with x greater than $1/2$, shows that f is concave downward on $(-\infty, 1/2)$ and concave upward on $(1/2, \infty)$. The graph has an inflection point at $(1/2, f(1/2)) = (1/2, -11/2)$. This information is summarized in the chart below.

Interval	$(-\infty, -1)$	$(-1, 1/2)$	$(1/2, 2)$	$(2, \infty)$
sign of f'	+	−	−	+
sign of f''	−	−	+	+
f increasing or decreasing	increasing	decreasing	decreasing	increasing
concavity of f	downward	downward	upward	upward
shape of graph	⌒	⌒	⌣	⌣

Use this information and plot a few points to get the graph in Figure 12.33.

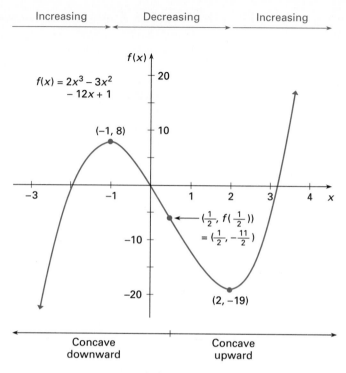

Increasing ⟶ Decreasing ⟶ Increasing ⟶

$f(x) = 2x^3 - 3x^2 - 12x + 1$

$(-1, 8)$

$(\frac{1}{2}, f(\frac{1}{2}))$
$= (\frac{1}{2}, -\frac{11}{2})$

$(2, -19)$

Concave downward   Concave upward

FIGURE 12.33

(b) $f(x) = x + 1/x$

Here $f'(x) = 1 - (1/x^2)$, which is 0 when

$$\frac{1}{x^2} = 1$$

$$x^2 = 1$$

$$x = 1 \quad \text{or} \quad x = -1.$$

The derivative fails to exist at 0. In fact, the function is undefined at 0, and $x = 0$ is a vertical asymptote. The direction of the graph may change from one side of the asymptote to the other. Evaluating $f'(x)$ in each of the regions determined by the critical numbers and the asymptote shows that f is increasing on $(-\infty, -1)$ and $(1, \infty)$ and decreasing on $(-1, 0)$ and $(0, 1)$. By the first derivative test, f has a relative maximum when $x = -1$ and a relative minimum when $x = 1$.

The second derivative is

$$f''(x) = \frac{2}{x^3},$$

which is never equal to 0 and does not exist when $x = 0$. (The function itself also does not exist at 0.) Because of this, there may be a change of concavity, but not an

7 Sketch the graphs.

(a) $f(x) = x^3 - 3x^2$

(b) $f(x) = x - 1/x$

Answer:

(a)

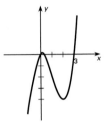

(b)

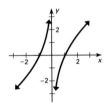

inflection point, when $x = 0$. The second derivative is negative when x is negative, making f concave downward on $(-\infty, 0)$. Also, $f''(x) > 0$ when $x > 0$, making f concave upward on $(0, \infty)$.

Interval	$(-\infty, -1)$	$(-1, 0)$	$(0, 1)$	$(1, \infty)$
sign of f'	+	−	−	+
sign of f''	−	−	+	+
f increasing or decreasing	increasing	decreasing	decreasing	increasing
concavity of f	downward	downward	upward	upward
shape of graph	⌒	⌒	⌣	⌣

Use this information and plot points as necessary to get the graph shown in Figure 12.34. ◄ **7**

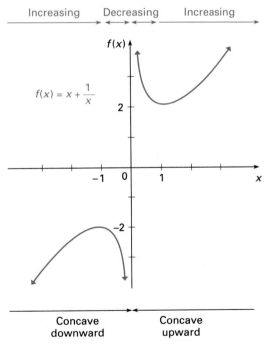

FIGURE 12.34

12.4 EXERCISES

Find the largest open intervals where the functions graphed as follows are increasing or decreasing. (See Example 1.)

1.

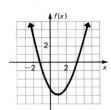

2.

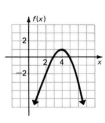

3.

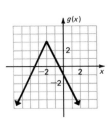

4.

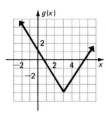

5.

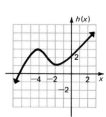

6.

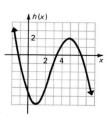

7.

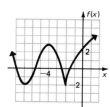

8.

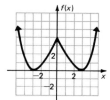

Find the largest open intervals where the functions defined as follows are increasing or decreasing. Graph each function. (See Example 2.)

9. $f(x) = x^2 + 12x - 6$

10. $f(x) = x^2 - 9x + 4$

11. $y = 5 + 9x - 3x^2$

12. $y = 3 + 4x - 2x^2$

13. $f(x) = 2x^3 - 3x^2 - 72x - 4$

14. $f(x) = 2x^3 - 3x^2 - 12x + 2$

15. $f(x) = 4x^3 - 15x^2 - 72x + 5$

16. $f(x) = 4x^3 - 9x^2 - 30x + 6$

17. $y = -3x + 6$

18. $y = 6x - 9$

19. $f(x) = \dfrac{x + 2}{x + 1}$

20. $f(x) = \dfrac{x + 3}{x - 4}$

21. $y = |x + 4|$

22. $y = -|x - 3|$

23. $f(x) = -\sqrt{x - 1}$

24. $f(x) = \sqrt{5 - x}$

25. $y = \sqrt{x^2 + 1}$

26. $y = x\sqrt{9 - x^2}$

Find the largest open intervals where the functions graphed or defined as follows are concave upward or concave downward. Find the location of any points of inflection. (See Example 5.)

27.

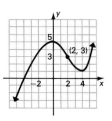

28.

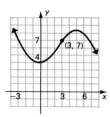

29.

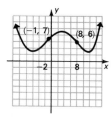

30.

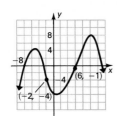

31.

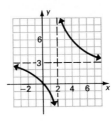

32.

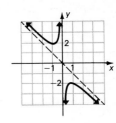

33. $f(x) = x^2 + 10x - 9$

34. $f(x) = x^2 - 4x + 3$

35. $f(x) = -3 + 8x - x^2$

36. $f(x) = 8 - 6x - x^2$

37. $f(x) = x^3 + 3x^2 - 45x - 3$

38. $f(x) = 2x^3 - 3x^2 - 12x + 1$

39. $f(x) = -2x^3 + 9x^2 + 168x - 3$

40. $f(x) = -x^3 - 12x^2 - 45x + 2$

41. $f(x) = \dfrac{3}{x - 5}$

42. $f(x) = \dfrac{-2}{x + 1}$

43. $f(x) = x(x + 5)^2$

44. $f(x) = -x(x - 3)^2$

Sketch the graphs of the following. Identify any points of inflection. (See Example 8.)

45. $f(x) = -x^2 - 10x - 25$

46. $f(x) = x^2 - 12x + 36$

47. $f(x) = 3x^3 - 3x^2 + 1$

48. $f(x) = 2x^3 - 4x^2 + 2$

49. $f(x) = -2x^3 - 9x^2 + 108x - 10$

50. $f(x) = -2x^3 - 9x^2 + 60x - 8$

51. $f(x) = 2x^3 + \dfrac{7}{2}x^2 - 5x + 3$

52. $f(x) = x^3 - \dfrac{15}{2}x^2 - 18x - 1$

53. $f(x) = (x + 3)^4$

54. $f(x) = x^3$

55. $f(x) = x^4 - 18x^2 + 5$

56. $f(x) = x^4 - 8x^2$

57. $f(x) = x + \dfrac{1}{x}$

58. $f(x) = 2x + \dfrac{8}{x}$

59. $f(x) = \dfrac{x^2 + 25}{x}$

60. $f(x) = \dfrac{x^2 + 4}{x}$

61. $f(x) = \dfrac{x - 1}{x + 1}$

62. $f(x) = \dfrac{x}{1 + x}$

Work the following problems. (See Examples 3, 4, 6, and 7.)

63. Management Suppose the total cost $C(x)$ (in dollars) to manufacture a quantity x of weed killer (in hundreds of liters) is given by

$$C(x) = x^3 - 2x^2 + 8x + 50.$$

(a) Where is $C(x)$ decreasing?

(b) Where is $C(x)$ increasing?

64. Management A county realty group estimates that the number of housing starts per year over the next 3 years will be

$$H(r) = \dfrac{300}{1 + .03r^2},$$

where r is the mortgage rate (in percent).

(a) Where is $H(r)$ increasing?

(b) Where is $H(r)$ decreasing?

65. Management A manufacturer sells video games with the following cost and revenue functions, where x is the number of games sold.

$$C(x) = 4.8x - .0004x^2, \quad 0 \le x \le 2250$$
$$R(x) = 8.4x - .002x^2, \quad 0 \le x \le 2250$$

Determine the intervals on which the profit function is increasing.

66. Management A manufacturer of compact disc players has determined that the profit $P(x)$ (in thousands of dollars) is related to the quantity x of players produced (in hundreds) per month by

$$P(x) = \dfrac{1}{3}x^3 - \dfrac{7}{2}x^2 + 10x - 2,$$

as long as the number of units produced is fewer than 800 per month.

(a) At what production levels is the profit increasing?

(b) At what levels is it decreasing?

67. Natural Science The number of people $P(t)$ (in hundreds) infected t days after an epidemic begins is approximated by

$$P(t) = 2 + 50t - \frac{5}{2}t^2.$$

When will the number of people infected start to decline?

68. Natural Science In Chapter 3 we gave the function

$$A(x) = -.015x^3 + 1.058x$$

as the approximate alcohol concentration (in tenths of a percent) in an average person's bloodstream x hours after drinking 8 ounces of 100-proof whiskey. The function applies only for the interval $[0, 8]$.

(a) On what time intervals is the alcohol concentration increasing?

(b) On what intervals is it decreasing?

69. Natural Science The percent of concentration of a drug in the bloodstream x hours after the drug is administered is given by

$$K(x) = \frac{4x}{3x^2 + 27}.$$

(a) On what time intervals is the concentration of the drug increasing?

(b) On what intervals is it decreasing?

70. Natural Science Suppose a certain drug is administered to a patient, with the percent of concentration of the drug in the bloodstream t hours later given by

$$K(t) = \frac{5t}{t^2 + 1}.$$

(a) On what time intervals is the concentration of the drug increasing?

(b) On what intervals is it decreasing?

Work the following exercises. (See Examples 6 and 7.)

71. Management The accompanying figure shows the *product life cycle* graph, with typical products marked on it.*

---

*Based on "The Product Life Cycle: A Key to Strategic Marketing Planning" in *MSU Business Topics* (Winter 1973), p. 30. Reprinted by permission of the publisher, Graduate School of Business Administration, Michigan State University.

(a) Where would you place home videotape recorders on this graph?

(b) Where would you place rotary-dial telephones?

(c) Which products on the left side of the graph are closest to the left-hand point of inflection? What does the point of inflection mean here?

(d) Which product on the right side of the graph is closest to the right-hand point of inflection? What does the point of inflection mean here?

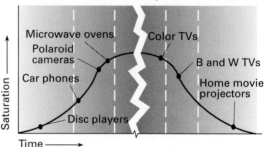

Management *In Exercises 72 and 73, find the point of diminishing returns for the given functions, where $R(x)$ represents revenue in thousands of dollars and x represents the amount spent on advertising in thousands of dollars. (See Example 7.)*

72. $R(x) = 10,000 - x^3 + 42x^2 + 800x; \ 0 \le x \le 20$

73. $R(x) = \frac{4}{27}(-x^3 + 66x^2 + 1050x - 400); \ 0 \le x \le 25$

74. Natural Science When a hardy new species is introduced into an area, the population often increases as shown. Explain the significance of the following points on the graph.

(a) f_0 **(b)** $(a, f(a))$ **(c)** f_M

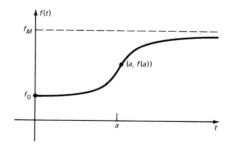

KEY TERMS AND SYMBOLS

12.1
relative maximum (maxima)
relative minimum (minima)
relative extremum (extrema)
critical number
critical point
first derivative test
absolute maximum
absolute minimum
extreme value theorem

12.2 $f''(x)$ or y'' or $\dfrac{d^2y}{dx^2}$ or $D_x^2[f(x)]$
second derivative of f

$f'''(x)$ third derivative of f
$f^{(n)}(x)$ nth derivative of f
acceleration

12.3   economic lot size

12.4
increasing function
decreasing function
concave upward
concave downward
point of inflection
point of diminishing returns
curve sketching

KEY CONCEPTS

Relative Extremum

Let c be in the interval (a, b) in the domain of function f. Then $f(c)$ is a **relative maximum** for f if $f(x) \leq f(c)$ for all x in (a, b). Also, $f(c)$ is a **relative minimum** if $f(x) \geq f(c)$ for all x in (a, b). If f has a relative extremum at c, then either $f'(c) = 0$ or $f'(c)$ does not exist.

First Derivative Test

Let f be a differentiable function for all x in $[a, b]$, except possibly at $x = c$. Assume $a < c < b$ and that c is the only critical number for f in $[a, b]$. If $f'(a) > 0$ and $f'(b) < 0$, then $f(c)$ is a relative maximum. If $f'(a) < 0$ and $f'(b) > 0$, then $f(c)$ is a relative minimum.

Absolute Extremum

Let c be in an interval where f is defined. $f(c)$ is the **absolute maximum** of f on the interval if $f(x) \leq f(c)$ for every x in the interval. $f(c)$ is the **absolute minimum** of f on the interval if $f(x) \geq f(c)$ for every x in the interval.

Second Derivative Test

Let c be a critical number of f such that $f'(c) = 0$ and $f''(x)$ exists for all x in some open interval containing c. If $f''(c) > 0$, then $f(c)$ is a relative minimum. If $f''(c) < 0$, then $f(c)$ is a relative maximum. If $f''(c) = 0$, then the test gives no information.

If $f'(x) > 0$ for each x in an interval, then f is **increasing** on the interval; if $f'(x) < 0$ for each x in the interval, then f is **decreasing** on the interval; if $f'(x) = 0$ for each x in the interval, then f is **constant** on the interval.

Let f have derivatives f' and f'' for all x in (a, b). f is **concave upward** on (a, b) if $f''(x) > 0$ for all x in (a, b). f is **concave downward** on (a, b) if $f''(x) < 0$ for all x in (a, b). f has a **point of inflection** at c if $f''(c) = 0$.

Find the largest open intervals where the functions defined as follows are increasing or decreasing.

1. $f(x) = x^2 - 5x + 3$

2. $f(x) = -2x^2 - 3x + 4$

3. $f(x) = -x^3 - 5x^2 + 8x - 6$

4. $f(x) = 4x^3 + 3x^2 - 18x + 1$

5. $f(x) = \dfrac{6}{x - 4}$

6. $f(x) = \dfrac{5}{2x + 1}$

Find the locations and the values of all relative maxima and relative minima for each of the following.

7. $f(x) = -x^2 + 4x - 8$

8. $f(x) = x^2 - 6x + 4$

9. $f(x) = 2x^2 - 8x + 1$

10. $f(x) = -3x^2 + 2x - 5$

11. $f(x) = 2x^3 + 3x^2 - 36x + 20$

12. $f(x) = 2x^3 + 3x^2 - 12x + 5$

13. $f(x) = -2x^3 - \dfrac{1}{2}x^2 + x - 3$

14. $f(x) = -\dfrac{4}{3}x^3 + x^2 + 30x - 7$

15. $f(x) = x^4 - \dfrac{4}{3}x^3 - 4x^2 + 1$

16. $f(x) = x^4 + \dfrac{8}{3}x^3 - 6x^2 + 1$

17. $f(x) = \dfrac{x - 1}{2x + 1}$

18. $f(x) = \dfrac{2x - 5}{x + 3}$

19. $f(x) = x \cdot e^x$

20. $f(x) = 3x \cdot e^{-x}$

21. $f(x) = \dfrac{e^x}{x - 1}$

22. $f(x) = (5x + 3)e^{-2x}$

Find the locations of all absolute maxima and absolute minima for the following functions on the given intervals.

23. $f(x) = -x^2 + 5x + 1;\ [1, 4]$

24. $f(x) = 4x^2 - 8x - 3;\ [-1, 2]$

25. $f(x) = x^3 + 2x^2 - 15x + 3;\ [-4, 2]$

26. $f(x) = -2x^3 - x^2 + 4x - 1;\ [-3, 1]$

Find the second derivatives of the following; then find $f''(1)$ and $f''(-3)$.

27. $f(x) = 3x^4 - 5x^2 - 11x$

28. $f(x) = \dfrac{5x - 1}{2x + 3}$

29. $f(x) = \dfrac{4 - 3x}{x + 1}$

30. $f(x) = -3e^{4x}$

31. $f(x) = e^{-x^2}$

33. $f(x) = \ln |6x + 1|$

Find the largest open intervals where the graph of f is concave upward or concave downward. Find the location of any points of inflection. Graph each function.

33. $f(x) = -4x^3 - x^2 + 4x + 5$

34. $f(x) = x^3 + \dfrac{5}{2}x^2 - 2x - 3$

35. $f(x) = x^4 + 2x^2$

36. $f(x) = 6x^3 - x^4$

37. $f(x) = \dfrac{x^2 - 4}{x}$

38. $f(x) = x + \dfrac{8}{x}$

39. $f(x) = \dfrac{2x}{3 - x}$

40. $f(x) = \dfrac{-4x}{1 + 2x}$

Work the following exercises.

41. Find two numbers whose sum is 25 and whose product is a maximum.

42. Find nonnegative x and y such that $x = 2 + y$, and xy^2 is minimized.

43. Suppose the profit from a product is $P(x) = 40x - x^2$, where x is the price in hundreds of dollars.
   (a) At what price will the maximum profit occur?
   (b) What is the maximum profit?

44. The total profit in dollars from the sale of x hundred boxes of candy is given by

$$P(x) = -x^3 + 10x^2 - 12x.$$

   (a) Find the number of boxes of candy that should be sold in order to produce maximum profit.
   (b) Find the maximum profit.

45. The packaging department of a corporation is designing a box with a square base and no top. The volume is to be 32 cubic meters. To reduce cost, the box is to have minimum surface area. What dimensions (height, length, and width) should the box have?

46. Another product (see Exercise 45) will be packaged in a closed cylindrical tin can with a volume of 54π cubic inches. Find the radius and height of the can if it is to have minimum surface area.

47. The city park department is planning an enclosed play area in a new park. One side of the area will be against an existing building, with no fence needed there. Find the dimensions of the rectangular field of maximum area that can be enclosed with 900 meters of fence.

48. A company plans to package its product in a cylinder which is open at one end. The cylinder is to have a volume of 27π cubic inches. What radius should the circular bottom of the cylinder have to minimize the cost of the material? (Hint: the volume of a circular cylinder is $\pi r^2 h$ where r is the radius of the circular base and h is the height; the surface area of an open circular cylinder is $2\pi rh + \pi r^2$.)

49. In 1 year, a health food manufacturer produces and sells 240,000 cases of vitamins. It costs $2 to store a case for 1 year and $15 to produce each batch. Find the number of batches that should be produced annually.

50. A company produces 128,000 cases of soft drink annually. It costs $1 to store a case for 1 year and $10 to produce one lot. Find the number of lots that should be produced annually.

51. If the play area referred to in Exercise 47 needs fencing on all four sides, find the dimensions of the maximum rectangular area that can be made with 900 meters of fence.

With the major increases in fuel prices recently, choosing an optimum speed for a large ocean-going ship has become a major concern of ship operators. Sailing slowly uses less fuel, but causes increased costs for salaries and interest on the value of the cargo.

The speed of sailing has a great effect on the amount of fuel used, the major expense in operating a ship. Past empirical studies have shown that the amount of fuel used by a ship is approximately proportional to the cube of the speed. This means that a 20% reduction in cruising speed produces about a 50% reduction in fuel consumption, since $(80\%)^3 \approx 50\%$. As an example, a medium sized ship that burns 40 tons of fuel per day at a certain speed could save 50% of this amount, or 20 tons, with a 20% reduction in speed. At a cost of $250 per ton, this 20% speed reduction would save $5000 per day in fuel costs.

To find the optimal sailing speed for one leg of a journey (a one-way trip with a given cargo between two points), let

R = income from the leg

P = profit from the leg

D_s = number of days at sea in the leg

D_p = number of days in port in the leg

L = distance of the leg in nautical miles

V = actual cruising speed

V_0 = normal cruising speed

C = daily cost of the vessel
    (excluding fuel for main engines)

F = actual daily fuel consumption in tons per day

F_0 = daily fuel consumption at normal cruising speed

F_c = cost of fuel in dollars per ton

V_m = minimum cruising speed.

The duration of the leg, D, is

$$D = D_s + D_p, \tag{1}$$

where

$$D_s = \frac{L}{24\,V}. \tag{2}$$

*Based on "The Effect of Oil Price on the Optimal Speed of Ships" by David Ronen from *Journal of the Operations Research Society,* Vol. 33, 1982. Copyright © 1982 Operational Research Society Ltd. Reprinted by permission.

The profit from the leg is

$$P = R - (DC + FF_cD_s). \tag{3}$$

As we mentioned above, the amount of fuel used is proportional to the cube of the speed, or

$$F = \left(\frac{V}{V_0}\right)^3 \cdot F_0. \tag{4}$$

Our goal is to maximize the daily profit, Z, given by

$$Z = \frac{P}{D}.$$

Substituting equations (1), (2), and (4) into the equation for profit, (3), and dividing by D as given by equations (1) and (2) yields

$$Z = \frac{R - D_pC - \dfrac{LC}{24\,V} - \left(\dfrac{V}{V_0}\right)^3 F_0F_c\left(\dfrac{L}{24V}\right)}{D_p + \dfrac{L}{24\,V}}. \tag{5}$$

It is not possible for a ship to sail at just any speed between 0 and its normal cruising speed, V_0. The design of the ship usually imposes some minimum cruising speed, V_m. This restriction forces V to be in the interval $[V_m, V_0]$.

To find the optimum value of V, set dZ/dV, from equation (5), equal to 0. This gives the cubic equation

$$V^3 + V^2\frac{L}{16D_p} - \frac{RV_0{}^3}{2F_0F_cD_p} = 0. \tag{6}$$

Any real solutions of this equation are potential critical values. As mentioned above, these critical values must lie in the interval $[V_m, V_0]$ to be useful. To maximize Z, therefore, we must use our work with absolute extrema and check all critical values in $[V_m, V_0]$, as well as both endpoints, V_m and V_0.

EXERCISES

1. Find the optimum sailing speed for a voyage of 1536 miles, with 8 days in port, a normal cruising speed of 20 knots, a minimum speed of 10 knots, income from the leg of $86,400, fuel costs of $250 per ton, 50 tons of fuel consumed daily at normal cruising speed, and a daily cost for the vessel of $1000. (Hint: try $V = 12$.)

2. Why does the variable C not appear in equation (6) above?

In this application, we set up a mathematical model for determining the total costs in setting up a training program. Then we use calculus to find the time between training programs that produces the minimum total cost. The model assumes that the demand for trainees is constant and that the fixed cost of training a batch of trainees is known. Also, it is assumed that people who are trained, but for whom no job is readily available, will be paid a fixed amount per month while waiting for a job to open up.

The model uses the following variables.

D = demand for trainees per month
N = number of trainees per batch
C_1 = fixed cost of training a batch of trainees
C_2 = variable cost of training per trainee per month
C_3 = salary paid monthly to a trainee who has not yet been given a job after training
m = time interval in months between successive batches of trainees
t = length of training program in months
$Z(m)$ = total monthly cost of program

The total cost of training a batch of trainees is given by $C_1 + NtC_2$. However, $N = mD$, so that the total cost per batch is $C_1 + mDtC_2$.

After training, personnel are given jobs at the rate of D per month. Thus, $N - D$ of the trainees will not get a job the first month, $N - 2D$ will not get a job the second month, and so on. The $N - D$ trainees who do not get a job the first month produce total costs of $(N - D)C_3$, those not getting jobs during the second month produce costs of $(N - 2D)C_3$, and so on. Since $N = mD$, the costs during the first month can be written as

$$(N - D)C_3 = (mD - D)C_3 = (m - 1)DC_3,$$

while the costs during the second month are $(m - 2)DC_3$, and so on. The total cost for keeping the trainees without a job is thus
$$(m - 1)DC_3 + (m - 2)DC_3$$
$$+ (m - 3)DC_3 + \cdots + 2DC_3 + DC_3,$$

which can be factored to give

$$DC_3[(m - 1) + (m - 2) + (m - 3) + \cdots + 2 + 1].$$

*Based on "A Total Cost Model for a Training Program" by P. L. Goyal and S. K. Goyal, Department of Mathematics and Computer Science, The Polytechnic of Wales, Treforest, Pontypridd. Used with permission.

The expression in brackets is the sum of the terms of an arithmetic sequence. Using formulas for arithmetic sequences, the expression in brackets can be shown to equal $m(m - 1)/2$, so that we have

$$DC_3\left[\frac{m(m - 1)}{2}\right] \tag{1}$$

as the total cost for keeping jobless trainees.

The total cost per batch is the sum of the training cost per batch, $C_1 + mDtC_2$, and the cost of keeping trainees without a proper job, given by (1). Since we assume that a batch of trainees is trained every m months, the total cost per month, $Z(m)$, is given by

$$Z(m) = \frac{C_1 + mDtC_2}{m} + \frac{DC_3\left[\frac{m(m - 1)}{2}\right]}{m}$$

$$= \frac{C_1}{m} + DtC_2 + DC_3\left(\frac{m - 1}{2}\right).$$

EXERCISES

1. Find $Z'(m)$.

2. Solve the equation $Z'(m) = 0$.
   As a practical matter, it is usually required that m be a whole number. If m does not come out to be a whole number in Exercise 1, then m^+ and m^-, the two whole numbers closest to m, must be chosen. Calculate both $Z(m^+)$ and $Z(m^-)$; the smaller of the two provides the optimum value of Z.

3. Suppose a company finds that its demand for trainees is 3 per month, that a training program requires 12 months, that the fixed cost of training a batch of trainees is $15,000, that the variable cost per trainee per month is $100, and that trainees are paid $900 per month after training but before going to work. Use your result from Exercise 2 and find m.

4. Since m is not a whole number, find m^+ and m^-.

5. Calculate $Z(m^+)$ and $Z(m^-)$.

6. What is the optimum time interval between successive batches of trainees? How many trainees should be in a batch?

7. Write a brief essay describing other considerations, perhaps not quantifiable, that a manager might want to consider in this situation.

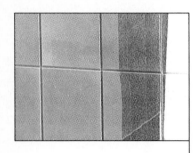

CHAPTER 13

Integral Calculus

In the previous two chapters, we studied the derivative of a function and various applications of derivatives. That material belongs to the branch of calculus called *differential calculus*. In this chapter we will study another branch of calculus, called *integral calculus*. Like the derivative of a function, the indefinite integral of a function is a special limit with many diverse applications. Geometrically, the derivative is related to the slope of the tangent line to a curve, while the definite integral is related to the area under a curve.

13.1 ANTIDERIVATIVES; INDEFINITE INTEGRALS

Functions used in applications in previous chapters have provided information about a *total amount* of a quantity, such as cost, revenue, profit, temperature, gallons of oil, or distance. Working with these functions provided information about the rate of change of these quantities and allowed us to answer important questions about the extrema of the functions. It is not always possible to find ready-made functions that provide information about the total amount of a quantity, but it is often possible to collect enough data to come up with a function that gives the *rate* of *change* of a quantity. Since we know that derivatives give the rate of change when the total amount is known, is it possible to reverse the process and use a known rate of change to get a function that gives the total amount of a quantity? The answer is yes; this reverse process, called *antidifferentiation,* is the topic of this section. The *antiderivative* of a function is defined as follows.

If $F'(x) = f(x)$, then $F(x)$ is an **antiderivative** of $f(x)$.

1 Find an antiderivative for each of the following.

(a) $3x^2$

(b) 5

(c) $8x^7$

(d) $\dfrac{1}{2}x^3$

Answer:
Only one possible antiderivative is given for each.

(a) x^3

(b) $5x$

(c) x^8

(d) $\dfrac{1}{8}x^4$

▶**EXAMPLE 1**   **(a)** If $F(x) = 10x$, then $F'(x) = 10$, so $F(x) = 10x$ is an antiderivative of $f(x) = 10$.
(b) For $F(x) = x^2$, $F'(x) = 2x$, making $F(x) = x^2$ an antiderivative of $f(x) = 2x$. ◀

▶**EXAMPLE 2**   Find an antiderivative of $f(x) = 5x^4$.
To find a function $F(x)$ whose derivative is $5x^4$, work backwards. Recall that the derivative of x^n is nx^{n-1}. If

$$nx^{n-1} \quad \text{is} \quad 5x^4,$$

then $n - 1 = 4$ and $n = 5$, so x^5 is an antiderivative of $5x^4$. ◀ **1**

▶**EXAMPLE 3**   Suppose a population is growing at a rate given by $f(x) = e^x$, where x is time in years from some initial date. Find a function giving the population at time x.
Let the population function be $F(x)$. Then

$$f(x) = F'(x) = e^x.$$

The derivative of the function defined by $F(x) = e^x$ is $F'(x) = e^x$, so a population function with the given growth rate is $F(x) = e^x$. ◀

The function from Example 1(b), defined by $F(x) = x^2$, is not the only function whose derivative is $f(x) = 2x$; for example, both $G(x) = x^2 + 2$ and $H(x) = x^2 - 4$ have $f(x) = 2x$ as a derivative. In fact, for any real number C, the function $F(x) = x^2 + C$ has $f(x) = 2x$ as its derivative. This means that there is a *family* of functions having $2x$ as an antiderivative. As the next theorem states, if two functions $F(x)$ and $G(x)$ are antiderivatives of $f(x)$, then $F(x)$ and $G(x)$ can differ only by a constant.

If $F(x)$ and $G(x)$ are both antiderivatives of $f(x)$, then there is a constant C such that

$$F(x) - G(x) = C.$$

(Two antiderivatives of a function can differ only by a constant.)

For example,

$$F(x) = x^2, \quad G(x) = x^2 + 2, \quad \text{and} \quad H(x) = x^2 - 4$$

are all antiderivatives of $f(x) = 2x$, and any two of them differ only by a constant. The derivative of a function gives the slope of the tangent line at any x-value. The fact that these three functions have the same derivative, $f(x) = 2x$, means that their slopes at any particular value of x are the same, as shown in Figure 13.1.

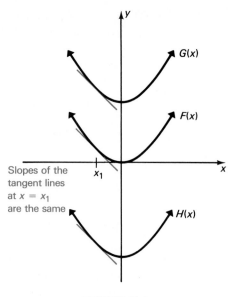

FIGURE 13.1

The family of all antiderivatives of the function f is indicated by

$$\int f(x) \; dx.$$

The symbol $\int$ is the **integral sign,** $f(x)$ is the **integrand,** and $\int f(x) \; dx$ is called an **indefinite integral,** the most general antiderivative of f. The process of finding $\int f(x) \; dx$ is called **integration.**

Indefinite Integral

If $F'(x) = f(x)$, then

$$\int f(x) \; dx = F(x) + C,$$

for any real number C.

For example, using this notation,

$$\int (2x) \; dx = x^2 + C.$$

Note The *dx* in the indefinite integral indicates that $\int f(x)\ dx$ is the "integral of *f(x) with respect to x*" just as the symbol *dy/dx* denotes the "derivative of *y* with respect to *x*." For example, in the indefinite integral $\int 2ax\ dx$, *dx* indicates that *a* is to be treated as a constant and *x* as the variable, so that

$$\int 2ax\ dx = 2a \cdot \frac{x^2}{2} = ax^2 + C.$$

On the other hand,

$$\int 2ax\ da = 2x \cdot \frac{a^2}{2} = xa^2 + C.$$

A more complete interpretation of *dx* will be discussed later.

The symbol $\int f(x)\ dx$ was created by G.W. Leibniz (1646–1716) in the latter part of the seventeenth century. The $\int$ is an elongated S from *summa,* the Latin word for *sum.* The word *integral* as a term in the calculus was coined by Jakob Bernoulli (1654–1705), a Swiss mathematician who corresponded frequently with Leibniz. The relationship between sums and integrals will be clarified in Section 13.3.

Since finding an indefinite integral is the inverse of finding a derivative, each formula for derivatives leads to a rule for indefinite integrals. As mentioned above, the derivative of x^n is found by multiplying *x* by *n* and reducing the exponent on *x* by 1. To find an indefinite integral—that is, to undo what was done—*increase* the exponent by 1 and *divide* by the new exponent, *n* + 1.

Power Rule for Indefinite Integrals

For any real number $n \neq -1$,

$$\int x^n\ dx = \frac{1}{n+1} x^{n+1} + C.$$

This result can be verified by differentiating the expression on the right above.

$$\frac{d}{dx}\left(\frac{1}{n+1} x^{n+1} + C \right) = \frac{n+1}{n+1} x^{(n+1)-1} + 0 = x^n$$

(If $n = -1$, the expression in the denominator is 0, and the above rule cannot be used. We will see later how to find an antiderivative in this case.)

2 Find each of the following.

(a) $\int x^5 \, dx$

(b) $\int \sqrt[3]{x} \, dx$

(c) $\int 5 \, dx$

Answer:

(a) $\dfrac{1}{6}x^6 + C$

(b) $\dfrac{3}{4}x^{4/3} + C$

(c) $5x + C$

▶ **EXAMPLE 4** Find each indefinite integral.

(a) $\int x^3 \, dx$

Use the power rule with $n = 3$.

$$\int x^3 \, dx = \frac{1}{3+1}x^{3+1} + C = \frac{1}{4}x^4 + C$$

(b) $\int \dfrac{1}{t^2} \, dt$

First, write $1/t^2$ as t^{-2}. Then

$$\int \frac{1}{t^2} \, dt = \int t^{-2} \, dt = \frac{1}{-2+1}t^{-2+1} = \frac{t^{-1}}{-1} + C = \frac{-1}{t} + C.$$

(c) $\int \sqrt{u} \, du$

Since $\sqrt{u} = u^{1/2}$,

$$\int \sqrt{u} \, du = \int u^{1/2} \, du = \frac{1}{1/2 + 1}u^{1/2+1} + C = \frac{2}{3}u^{3/2} + C.$$

Check this by differentiating $(2/3)u^{3/2} + C$; the derivative is $u^{1/2}$, the original function.

(d) $\int dx$

Writing dx as $1 \cdot dx$, and using the fact that $x^0 = 1$ for any nonzero number x,

$$\int dx = \int 1 \, dx = \int x^0 \, dx = \frac{1}{1}x^1 + C = x + C. \quad ◀ \;\; \boxed{2}$$

As shown in Chapter 11, the derivative of the product of a constant and a function is the product of the constant and the derivative of the function. A similar rule applies to indefinite integrals. Also, since derivatives of sums or differences are found term by term, indefinite integrals can also be found term by term.

Properties of Indefinite Integrals: Constant Multiple Rule; Sum or Difference Rule

If all indicated integrals exist,

$$\int k \cdot f(x) \, dx = k \int f(x) \, dx, \text{ for any real number } k;$$

$$\int [f(x) \pm g(x)] \, dx = \int f(x) \, dx \pm \int g(x) \, dx.$$

3 Find each of the following.

(a) $\int (-6x^4) \, dx$

(b) $\int 9x^{2/3} \, dx$

(c) $\int \dfrac{8}{x^3} \, dx$

(d) $\int (5x^4 - 3x^2 + 6) \, dx$

(e) $\int (-2x^3 + 6x^2 - 3) \, dx$

(f) $\int \left(3\sqrt{x} + \dfrac{2}{x^2}\right) dx$

Answer:

(a) $-\dfrac{6}{5}x^5 + C$

(b) $\dfrac{27}{5}x^{5/3} + C$

(c) $-4x^{-2} + C$ or $-\dfrac{4}{x^2} + C$

(d) $x^5 - x^3 + 6x + C$

(e) $-\dfrac{1}{2}x^4 + 2x^3 - 3x + C$

(f) $2x^{3/2} - \dfrac{2}{x} + C$

Caution The constant multiple rule requires that k be a *number*. The rule does not apply to a *variable*. For example,

$$\int x\sqrt{x-1} \, dx \neq x \int \sqrt{x-1} \, dx.$$

▶ **EXAMPLE 5** Find each of the following.

(a) $\int 2x^3 \, dx$

By the constant multiple rule and the power rule,

$$\int 2x^3 \, dx = 2 \int x^3 \, dx = 2\left(\frac{1}{4}x^4\right) + C = \frac{1}{2}x^4 + C.$$

Since C represents any real number, it is not necessary to multiply it by 2 in the next-to-last step.

(b) $\int \dfrac{12}{z^5} dz$

Use negative exponents.

$$\int \frac{12}{z^5} dz = \int 12z^{-5} \, dz$$

$$= 12 \int z^{-5} \, dz \qquad \text{Constant multiple rule}$$

$$= 12\left(\frac{z^{-4}}{-4}\right) + C \qquad \text{Power rule}$$

$$= -3z^{-4} + C$$

$$= \frac{-3}{z^4} + C.$$

(c) $\int (3z^2 - 4z + 5) dz$

By extending the sum or difference property given above to more than two terms,

$$\int (3z^2 - 4z + 5) dz = 3 \int z^2 \, dz - 4 \int z \, dz + 5 \int dz$$

$$= 3\left(\frac{1}{3}z^3\right) - 4\left(\frac{1}{2}z^2\right) + 5z + C$$

$$= z^3 - 2z^2 + 5z + C.$$

Only one constant C is needed in the answer: the three constants from integrating term by term are combined. ◀ **3**

4 Find each of the following.

(a) $\displaystyle \int \frac{\sqrt{x} + 1}{x^2}\, dx$

(b) $\displaystyle \int (\sqrt{x} + 2)^2$

Answer:

(a) $-\dfrac{2}{\sqrt{x}} - \dfrac{1}{x} + C$

(b) $\dfrac{x^2}{2} + \dfrac{8}{3} x^{3/2} + 4x + C$

Integration can always be checked by taking the derivative of the result. For instance, in Example 5(c) check that $z^3 - 2z^2 + 5z + C$ is the required indefinite integral by taking the derivative:

$$\frac{d}{dz}(z^3 - 2z^2 + 5z + C) = 3z^2 - 4z + 5.$$

The result is the original function to be integrated, so the work checks.

▶ **EXAMPLE 6**   Find each of the following.

(a) $\displaystyle \int \frac{x^2 + 1}{\sqrt{x}}\, dx$

First rewrite the integrand as follows.

$$\int \frac{x^2 + 1}{\sqrt{x}}\, dx = \int \left(\frac{x^2}{\sqrt{x}} + \frac{1}{\sqrt{x}} \right) dx$$

$$= \int \left(\frac{x^2}{x^{1/2}} + \frac{1}{x^{1/2}} \right) dx$$

$$= \int (x^{3/2} + x^{-1/2})\, dx \qquad \text{\textbf{Quotient rule for exponents}}$$

Now find the antiderivative.

$$\int (x^{3/2} + x^{-1/2})\, dx = \frac{x^{5/2}}{5/2} + \frac{x^{1/2}}{1/2} + C$$

$$= \frac{2}{5} x^{5/2} + 2x^{1/2} + C$$

(b) $\displaystyle \int (x^2 - 1)^2\, dx$

Square the binomial first, and then find the antiderivative.

$$\int (x^2 - 1)^2\, dx = \int (x^4 - 2x^2 + 1)\, dx$$

$$= \frac{x^5}{5} - \frac{2x^3}{3} + x + C \quad \blacktriangleleft \enspace \boxed{4}$$

As shown in Chapter 11, the derivative of $f(x) = e^x$ is $f'(x) = e^x$. Also, the derivative of $f(x) = e^{kx}$ is $f'(x) = k \cdot e^{kx}$. These results lead to the following formulas for indefinite integrals of exponential functions.

5 Find each of the following.

(a) $\displaystyle\int (-4e^x)\, dx$

(b) $\displaystyle\int e^{3x}\, dx$

(c) $\displaystyle\int (e^{2x} - 2e^x)\, dx$

(d) $\displaystyle\int (-11e^{-x})\, dx$

Answer:

(a) $-4e^x + C$

(b) $\dfrac{1}{3}e^{3x} + C$

(c) $\dfrac{1}{2}e^{2x} - 2e^x + C$

(d) $11e^{-x} + C$

6 Find each of the following.

(a) $\displaystyle\int (-9/x)\, dx$

(b) $\displaystyle\int 12x^{-1}\, dx$

(c) $\displaystyle\int (8e^{4x} - 3x^{-1})\, dx$

Answer:

(a) $-9 \cdot \ln |x| + C$

(b) $12 \cdot \ln |x| + C$

(c) $2e^{4x} - 3 \cdot \ln |x| + C$

Indefinite Integrals of Exponential Functions

If k is a real number, $k \neq 0$, then

$$\int e^x\, dx = e^x + C;$$

$$\int e^{kx}\, dx = \frac{1}{k} \cdot e^{kx} + C.$$

▶ **EXAMPLE 7** Find each of the following.

(a) $\displaystyle\int 9e^x\, dx = 9 \int e^x\, dx = 9e^x + C$

(b) $\displaystyle\int e^{9t}\, dt = \frac{1}{9}e^{9t} + C$

(c) $\displaystyle\int 3e^{(5/4)u}\, du = 3\left(\frac{1}{5/4}e^{(5/4)u}\right) + C = 3\left(\frac{4}{5}\right)e^{(5/4)u} + C$

$$= \frac{12}{5}e^{(5/4)u} + C \quad \blacktriangleleft \;\; \boxed{5}$$

The restriction $n \neq -1$ was necessary in the formula for $\int x^n\, dx$ since $n = -1$ made the denominator of $1/(n + 1)$ equal to 0. To find $\int x^n\, dx$ when $n = -1$, that is, to find $\int x^{-1}\, dx$, recall the differentiation formula for the logarithmic function: the derivative of $f(x) = \ln |x|$, where $x \neq 0$, is $f'(x) = 1/x = x^{-1}$. This formula for the derivative of $f(x) = \ln |x|$ gives a formula for $\int x^{-1}\, dx$.

Indefinite Integral of x^{-1}

$$\int x^{-1}\, dx = \int \frac{1}{x}\, dx = \ln |x| + C, \quad \text{where } x \neq 0.$$

Caution The domain of the logarithmic function is the set of positive real numbers. However, $y = x^{-1} = 1/x$ has as domain the set of all nonzero real numbers, so absolute value bars *must* be used in the antiderivative.

▶ **EXAMPLE 8** Find each of the following.

(a) $\displaystyle\int \frac{4}{x}\, dx = 4 \int \frac{1}{x}\, dx = 4 \cdot \ln |x| + C$

(b) $\displaystyle\int \left(-\frac{5}{x} + e^{-2x}\right) dx = -5 \cdot \ln |x| - \frac{1}{2}e^{-2x} + C \quad \blacktriangleleft \;\; \boxed{6}$

7 The marginal cost at a level of production of x items is

$$C'(x) = 2x^3 + 6x - 5.$$

The fixed cost is $800. Find the cost function $C(x)$.

Answer:

$C(x) = \dfrac{1}{2}x^4 + 3x^2 - 5x + 800$

In all the examples above, the antiderivative family of functions was found. In many applications, however, the given information allows us to determine the value of the integration constant C. The next examples illustrate this idea.

▶**EXAMPLE 9** Suppose a publishing company has found that the marginal cost at a level of production of x thousand books is given by

$$C'(x) = \frac{50}{\sqrt{x}}$$

and that the fixed cost (the costs before the first book can be produced) is $25,000. Find the cost function $C(x)$.

Write $50/\sqrt{x}$ as $50/x^{1/2}$ or $50x^{-1/2}$, and then use the indefinite integral rules to integrate the function.

$$\int \frac{50}{\sqrt{x}} \, dx = \int 50x^{-1/2} \, dx = 50(2x^{1/2}) + k = 100x^{1/2} + k$$

(Here k is used instead of C to avoid confusion with the cost function $C(x)$.) To find the value of k, use the fact that $C(0)$ is $25,000$.

$$C(x) = 100x^{1/2} + k$$
$$25,000 = 100 \cdot 0 + k$$
$$k = 25,000$$

With this result, the cost function is $C(x) = 100x^{1/2} + 25,000.$ ◀ **7**

▶**EXAMPLE 10** Suppose the marginal revenue from a product is given by $50 - 3x - x^2$. Find the demand function for the product.

The marginal revenue is the derivative of the revenue function

$$\frac{dR}{dx} = 50 - 3x - x^2,$$

so

$$R = \int (50 - 3x - x^2) \, dx$$

$$= 50x - \frac{3x^2}{2} - \frac{x^3}{3} + C.$$

If $x = 0$, then $R = 0$ (no items sold means no revenue), and

$$0 = 50(0) - \frac{3(0)^2}{2} - \frac{(0)^3}{3} + C$$

$$0 = C.$$

Thus,

$$R = 50x - \frac{3x^2}{2} - \frac{x^3}{3}$$

gives the revenue function. Now, recall that $R = xp$, where p is the demand function. Then

$$50x - \frac{3}{2}x^2 - \frac{1}{3}x^3 = xp$$

$$50 - \frac{3}{2}x - \frac{1}{3}x^2 = p,$$

which gives the demand function. ◀

13.1 EXERCISES

Find each of the following. (See Examples 4–8.)

1. $\int 4x\, dx$

2. $\int 8x\, dx$

3. $\int 5t^2\, dt$

4. $\int 6x^3\, dx$

5. $\int 6\, dk$

6. $\int 2\, dy$

7. $\int (2z + 3)\, dz$

8. $\int (3x - 5)\, dx$

9. $\int (x^2 + 6x)\, dx$

10. $\int (t^2 - 2t)\, dt$

11. $\int (t^2 - 4t + 5)\, dt$

12. $\int (5x^2 - 6x + 3)\, dx$

13. $\int (4z^3 + 3z^2 + 2z - 6)\, dz$

14. $\int (12y^3 + 6y^2 - 8y + 5)\, dy$

15. $\int 5\sqrt{z}\, dz$

16. $\int t^{1/4}\, dt$

17. $\int (u^{1/2} + u^{3/2})\, du$

18. $\int (4\sqrt{v} - 3v^{3/2})\, dv$

19. $\int (15x\sqrt{x} + 2\sqrt{x})\, dx$

20. $\int (x^{1/2} - x^{-1/2})\, dx$

21. $\int (10u^{3/2} - 14u^{5/2})\, du$

22. $\int (56t^{5/2} + 18t^{7/2})\, dt$

23. $\int \left(\frac{1}{z^2}\right) dz$

24. $\int \left(\frac{4}{x^3}\right) dx$

25. $\int \left(\frac{1}{y^3} - \frac{1}{\sqrt{y}}\right) dy$

26. $\int \left(\sqrt{u} + \frac{1}{u^2}\right) du$

27. $\int (-9t^{-2} - 2t^{-1})\, dt$

28. $\int (8x^{-3} + 4x^{-1})\, dx$

29. $\int e^{2t}\, dt$

30. $\int e^{-3y}\, dy$

31. $\int 3e^{-.2x}$

32. $\int -4e^{.2v}\, dv$

33. $\int \left(\frac{3}{x} + 4e^{-.5x}\right) dx$

34. $\int \left(\frac{9}{x} - 3e^{-.4x}\right) dx$

35. $\int \frac{1 + 2t^3}{t}\, dt$

36. $\int \frac{2y^{1/2} - 3y^2}{y}\, dy$

37. $\int \left(e^{2u} + \frac{u}{4}\right) du$

38. $\int \left(\frac{2}{v} - e^{3v}\right) dv$

39. $\int (x + 1)^2\, dx$

40. $\int (2y - 1)^2\, dy$

41. $\int \frac{\sqrt{x} + 1}{\sqrt[3]{x}}\, dx$

42. $\int \frac{1 - 2\sqrt[3]{z}}{\sqrt[3]{z}}\, dz$

Find the cost function for each of the following marginal cost functions. (See Example 9.)

43. $C'(x) = 4x - 5$; fixed cost is $8

44. $C'(x) = 2x + 3x^2$; fixed cost is $15

45. $C'(x) = .2x^2 + 5x$; fixed cost is $10

46. $C'(x) = .8x^2 - x$; fixed cost is $5

47. $C'(x) = x^{1/2}$; 16 units cost $40

48. $C'(x) = x^{2/3} + 2$; 8 units cost $58

49. $C'(x) = x^2 - 2x + 3$; 3 units cost $15

50. $C'(x) = x + \dfrac{1}{x^2}$; 2 units cost $5.50

51. $C'(x) = \dfrac{1}{x} + 2x$; 7 units cost $58.40

52. $C'(x) = 5x - \dfrac{1}{x}$; 10 units cost $94.20

Work the following exercises. (See Examples 9 and 10.)

53. **Management** The marginal profit of a small fast food stand is

$$P'(x) = -2x + 20,$$

where x is the sales volume in thousands of hamburgers. The "profit" is −$50 when no hamburgers are sold. Find the profit function.

54. **Management** Suppose the marginal profit from the sale of x hundred items is

$$P'(x) = 4 - 6x + 3x^2,$$

and the profit on 0 items is −$40. Find the profit function.

55. **Management** A mine begins producing at time $t = 0$. After t years, the mine is producing at the rate of

$$P'(t) = 10t - \frac{15}{\sqrt{t}}$$

tons per year. Find the total output of the mine after t years. (Assume no output when $t = 0$.)

56. The slope of the tangent line to a curve is given by

$$f'(x) = 6x^2 - 4x + 3.$$

If the point $(0, 1)$ is on the curve, find the equation of the curve.

57. Find the equation of the curve whose tangent line has a slope of

$$f'(x) = x^{2/3},$$

if the point $(1, 3/5)$ is on the curve.

58. **Natural Science** According to Fick's law, the diffusion of a solute across a cell membrane is given by

$$c'(t) = \frac{kA}{V}[C - c(t)], \tag{1}$$

where A is the area of the cell membrane, V is the volume of the cell, $c(t)$ is the concentration inside the cell at time t, C is the concentration outside the cell, and k is a constant. If c_0 represents the concentration of the solute inside the cell when $t = 0$, then it can be shown that

$$c(t) = (c_0 - C)e^{-kAt/V} + C. \tag{2}$$

(a) Use this last result to find $c'(t)$.

(b) Substitute back into equation (1) to show that (2) is indeed the correct antiderivative of (1).

---

1 Find du for the following.

(a) $u = 9x$

(b) $u = 5x^3 + 2x^2$ **(c)** $u = e^{-2x}$

Answer:

(a) $du = 9\,dx$

(b) $du = (15x^2 + 4x)\,dx$

(c) $du = -2e^{-2x}\,dx$

13.2 INTEGRATION BY SUBSTITUTION

In Section 13.1 we saw how to integrate a few simple functions. More complicated functions can sometimes be integrated by *substitution*. The technique depends on the idea of a differential. If $u = f(x)$, the **differential** of u, written du, is defined as

$$du = f'(x)\,dx.$$

For example, if $u = 6x^4$, then $du = 24x^3\,dx$. **1**

Differentials have many useful interpretations which are studied in more advanced courses. We shall only use them as a convenient notational device when finding an antiderivative such as

$$\int (3x^2 + 4)^4 6x \ dx.$$

The function $(3x^2 + 4)^4 6x$ is reminiscent of the chain rule and so we shall try to use differentials and the chain rule *in reverse* to find the antiderivative. Let $u = 3x^2 + 4$; then $du = 6x \ dx$. Now substitute u for $3x^2 + 4$ and du for $6x \ dx$ in the indefinite integral above.

$$\int (3x^2 + 4)^4 6x \ dx = \int \overbrace{(3x^2 + 4)}^{u}{}^4 \overbrace{(6x \ dx)}^{du}$$

$$= \int u^4 \ du$$

This last integral can now be found by the power rule.

$$\int u^4 \ du = \frac{u^5}{5} + C$$

Finally, substitute $3x^2 + 4$ for u.

$$\int (3x^2 + 4)^4 6x \ dx = \frac{u^5}{5} + C = \frac{(3x^2 + 4)^5}{5} + C$$

We can check the accuracy of this result by using the chain rule to take the derivative.

$$\frac{d}{dx}\left[\frac{(3x^2 + 4)^5}{5} + C \right] = \frac{1}{5} \cdot 5(3x^2 + 4)^4(6x) + 0$$

$$= (3x^2 + 4)^4 6x,$$

which is the original function.

This method of integration is called **integration by substitution.** As shown above, it is simply the chain rule for derivatives in reverse. The results can always be verified by differentiation.

This discussion leads to the following integration formula.

General Power Rule

For $u = f(x)$ and $du = f'(x) \ dx$,

$$\int u^n \ du = \frac{u^{n+1}}{n + 1} + C.$$

2 Find the following.

(a) $\displaystyle\int 8x(4x^2 - 1)^5 \, dx$

(b) $\displaystyle\int 18x^2(6x^3 - 5)^{3/2} \, dx$

Answer:

(a) $\dfrac{(4x^2 - 1)^6}{6} + C$

(b) $\dfrac{2(6x^3 - 5)^{5/2}}{5} + C$

When using integration by substitution, a certain amount of trial and error may be needed to decide on the expression to set equal to u, and even then, some rearrangement may be necessary. The integrand must be written as two factors, one of which is the derivative of the other.

▶ **EXAMPLE 1** Find $\displaystyle\int 10x(x^2 - 1)^4 \, dx$.

We try setting $u = x^2 - 1$, so that $du = 2x \, dx$. This doesn't quite match the given integrand, which has $10x$ instead of $2x$ in it. However, $10x = 5 \cdot 2x$, so we can rewrite and substitute as follows.

$$\int 10x(x^2 - 1)^4 \, dx = \int 5 \cdot 2x(x^2 - 1)^4 \, dx$$

$$= 5 \int \overbrace{(x^2 - 1)^4}^{u}\overbrace{(2x \, dx)}^{du} \qquad \text{**Constant multiple rule**}$$

$$= 5 \int u^4 \, du \qquad \text{**Substitute**}$$

This last integral can be found by the general power rule.

$$5 \int u^4 \, du = 5 \cdot \frac{u^5}{5} + C = u^5 + C$$

Now replace u with $x^2 - 1$.

$$\int 10x(x^2 - 1)^4 \, dx = u^5 + C = (x^2 - 1)^5 + C$$

You can check this answer by taking the derivative of

$$y = (x^2 - 1)^5 + C.$$
$$y' = 5(x^2 - 1)^4(2x) + 0 = 10x(x^2 - 1)^4 \quad ◀ \boxed{2}$$

▶ **EXAMPLE 2** Find $\displaystyle\int x^2\sqrt{x^3 + 1} \, dx$.

An expression raised to a power is usually a good choice for u, so because of the square root or 1/2 power, let $u = x^3 + 1$; then $du = 3x^2 \, dx$. The integrand does not contain the constant 3, which is needed for du. To take care of this, solve $du = 3x^2 \, dx$ for $x^2 \, dx$.

$$du = 3x^2 \, dx$$

$$\frac{1}{3} \, du = x^2 \, dx$$

3 Find the following.

(a) $\int x(5x^2 + 6)^4 \, dx$

(b) $\int x^2(x^3 - 2)^5 \, dx$

(c) $\int x\sqrt{x^2 + 16} \, dx$

Answer:

(a) $\dfrac{1}{50}(5x^2 + 6)^5 + C$

(b) $\dfrac{1}{18}(x^3 - 2)^6 + C$

(c) $\dfrac{1}{3}(x^2 + 16)^{3/2} + C$

4 Find the following.

(a) $\int z(z^2 + 1)^2 \, dz$

(b) $\int (r + 2)\sqrt{r^2 + 4r} \, dr$

Answer:

(a) $\dfrac{(z^2 + 1)^3}{6} + C$

(b) $\dfrac{(r^2 + 4r)^{3/2}}{3} + C$

Substitute $(1/3) \, du$ for $x^2 \, dx$.

$$\int x^2\sqrt{x^3 + 1} \, dx = \int \sqrt{x^3 + 1}(x^2 \, dx) = \int \sqrt{u} \cdot \frac{1}{3} \, du$$

Now use the constant multiple rule to bring the 1/3 outside the integral sign.

$$\int x^2\sqrt{x^3 + 1} \, dx = \int \sqrt{u} \cdot \frac{1}{3} \, du = \frac{1}{3}\int u^{1/2} \, du$$

$$= \frac{1}{3} \cdot \frac{u^{3/2}}{3/2} + C = \frac{2}{9}u^{3/2} + C$$

Since $u = x^3 + 1$,

$$\int x^2\sqrt{x^3 + 1} \, dx = \frac{2}{9}(x^3 + 1)^{3/2} + C. \blacktriangleleft \;\boxed{3}$$

Caution The substitution method given in the examples above will not always work. For example, we might try to find

$$\int x^3\sqrt{x^3 + 1} \, dx$$

by substituting $u = x^3 + 1$, so that $du = 3x^2 \, dx$. Since du must equal $x^3 \, dx$ here, the substitution method shown above will not work. This integral, and a great many others, cannot be evaluated by substitution.

▶ **EXAMPLE 3** Find $\int \dfrac{2x + 5}{(x^2 + 5x)^2} \, dx$.

Let $u = x^2 + 5x$, so that $du = (2x + 5) \, dx$. This gives

$$\int \frac{2x + 5}{(x^2 + 5x)^2} \, dx = \int \frac{du}{u^2} = \int u^{-2} \, du$$

$$= \frac{u^{-1}}{-1} + C = \frac{-1}{u} + C.$$

Substituting $x^2 + 5x$ for u gives

$$\int \frac{2x + 5}{(x^2 + 5x)^2} \, dx = \frac{-1}{x^2 + 5x} + C. \blacktriangleleft \;\boxed{4}$$

Recall the formula for $\dfrac{d}{dx}(e^u)$, where $u = f(x)$.

$$\frac{d}{dx}(e^u) = e^u\frac{d}{dx}(u)$$

5 Find the following.

(a) $\displaystyle\int 8xe^{3x^2}\,dx$

(b) $\displaystyle\int x^2 e^{x^3}\,dx$

Answer:

(a) $\dfrac{4}{3}e^{3x^2} + C$

(b) $\dfrac{1}{3}e^{x^3} + C$

For example, if $u = x^2$ then $\dfrac{d}{dx}(u) = \dfrac{d}{dx}(x^2) = 2x$, and

$$\frac{d}{dx}(e^{x^2}) = e^{x^2} \cdot 2x.$$

Working backwards, if $u = x^2$, then $du = 2x\,dx$, so

$$\int e^{x^2} \cdot 2x\,dx = \int e^u\,du = e^u + C$$

$$= e^{x^2} + C.$$

The work above suggests the following rule for the indefinite integral of e^u, where $u = f(x)$.

Indefinite Integral of e^u

If $u = f(x)$, then

$$\int e^u\,du = e^u + C.$$

▶ **EXAMPLE 4**   Find $\displaystyle\int x^2 \cdot e^{x^3}\,dx$.

Let $u = x^3$, the exponent on e. Then $du = 3x^2\,dx$. Since $(1/3)\,du = x^2\,dx$,

$$\int x^2 \cdot e^{x^3}\,dx = \int e^{x^3}(x^2\,dx)$$

$$= \int e^u\!\left(\frac{1}{3}\,du\right) \qquad \text{Substitute}$$

$$= \frac{1}{3}\int e^u\,du \qquad \text{Constant multiple rule}$$

$$= \frac{1}{3}e^u + C \qquad \text{Integrate}$$

$$= \frac{1}{3}e^{x^3} + C. \qquad \text{Substitute} \quad \blacktriangleleft \;\; \boxed{5}$$

Recall that the antiderivative of $f(x) = 1/x$ is $\ln |x|$. The next example uses $\int x^{-1}\,dx = \ln |x| + C$, and the method of substitution.

6 Find the following.

(a) $\displaystyle\int \frac{4\,dx}{x-3}$

(b) $\displaystyle\int \frac{(2x-9)\,dx}{x^2-9x}$

(c) $\displaystyle\int \frac{(3x^2+8)\,dx}{x^3+8x+5}$

Answer:

(a) $4\ln|x-3| + C$

(b) $\ln|x^2-9x| + C$

(c) $\ln|x^3+8x+5| + C$

7 Find

$$\int x(x+1)^{2/3}\,dx.$$

Answer:

$\dfrac{3}{8}(x+1)^{8/3} - \dfrac{3}{5}(x+1)^{5/3} + C$

▶**EXAMPLE 5** Find $\displaystyle\int \frac{(2x-3)\,dx}{x^2-3x}$.

Let $u = x^2 - 3x$, so that $du = (2x-3)\,dx$. Then

$$\int \frac{(2x-3)\,dx}{x^2-3x} = \int \frac{du}{u} = \ln|u| + C = \ln|x^2-3x| + C. \quad ◀ \boxed{6}$$

Generalizing from Example 5 suggests the following rule for finding the indefinite integral of u^{-1}, where $u = f(x)$.

Indefinite Integral of u^{-1}

If $u = f(x)$, then

$$\int u^{-1}\,du = \int \frac{du}{u} = \ln|u| + C.$$

▶**EXAMPLE 6** Find $\displaystyle\int x\sqrt{1-x}\,dx$.

Let $u = 1 - x$. Then $x = 1 - u$ and $dx = -du$. Now substitute.

$$\int x\sqrt{1-x}\,dx = \int (1-u)\sqrt{u}(-du) = \int (u-1)u^{1/2}\,du$$

$$= \int (u^{3/2} - u^{1/2})\,du = \frac{2}{5}u^{5/2} - \frac{2}{3}u^{3/2} + C$$

$$= \frac{2}{5}(1-x)^{5/2} - \frac{2}{3}(1-x)^{3/2} + C. \quad ◀ \boxed{7}$$

The substitution method is useful if the integral can be written in one of the following forms, where $u(x)$ is some function of x.

Substitution Method

Form of the Integral	*Form of the Antiderivative*
1. $\displaystyle\int [u(x)]^n \cdot u'(x)\,dx, \quad n \neq -1$	$\dfrac{[u(x)]^{n+1}}{n+1} + C$
2. $\displaystyle\int e^{u(x)} \cdot u'(x)\,dx$	$e^{u(x)} + C$
3. $\displaystyle\int \frac{u'(x)\,dx}{u(x)}$	$\ln|u(x)| + C$

▶**EXAMPLE 7** The research department for a hardware chain has determined that at one store the marginal price of x boxes per week of a particular type of nails is

$$p'(x) = \frac{-4000}{(2x + 15)^3}.$$

Find the demand equation if the weekly demand for this type of nails is 10 boxes when the price of a box of nails is \$4.

To find the demand function $p(x)$, first integrate $p'(x)$ as follows.

$$p(x) = \int p'(x)\, dx$$

$$= \int \frac{-4000}{(2x + 15)^3}\, dx$$

Let $u = 2x + 15$. Then $du = 2\, dx$, and

$$p(x) = -2000 \int (2x + 15)^{-3}\, 2\, dx$$

$$= -2000 \int u^{-3}\, du \qquad\qquad \textbf{Substitute}$$

$$= (-2000)\frac{u^{-2}}{-2} + C \qquad\qquad \textbf{Integrate}$$

$$= \frac{1000}{u^2} + C \qquad\qquad\qquad \textbf{Simplify}$$

$$p(x) = \frac{1000}{(2x + 15)^2} + C. \qquad\qquad \textbf{Substitute} \qquad\qquad \textbf{(1)}$$

Find the value of C by using the given information that $p = 4$ when $x = 10$.

$$4 = \frac{1000}{(2 \cdot 10 + 15)^2} + C$$

$$4 = \frac{1000}{35^2} + C$$

$$4 = .82 + C$$

$$3.18 = C$$

Replacing C with 3.18 in equation (1) gives the demand function,

$$p(x) = \frac{1000}{(2x + 15)^2} + 3.18. \quad ◀$$

▶**EXAMPLE 8**   To determine the top 100 popular songs of each year since 1956, Jim Quirin and Barry Cohen developed a function that represents the rate of change on the charts of *Billboard* magazine required for a song to earn a "star" on the *Billboard* "Hot 100" survey.* They developed the function

$$f(x) = \frac{A}{B + x},$$

where $f(x)$ represents the rate of change in position on the charts, x is the position on the "Hot 100" survey, and A and B are appropriate constants. The function

$$F(x) = \int f(x) \, dx$$

is defined as the "Popularity Index." Find $F(x)$.

Integrating $f(x)$ gives

$$F(x) = \int f(x) \, dx$$

$$= \int \frac{A}{B + x} \, dx$$

$$= A \int \frac{1}{B + x} \, dx \qquad \textbf{Constant multiple rule}$$

Let $u = B + x$, so that $du = dx$. Then

$$F(x) = A \int \frac{1}{u} \, du = A \ln u + C$$

$$= A \ln (B + x) + C.$$

(The absolute value bars are not necessary, since $B + x$ is always positive here.) ◀

---

*Formula for the "Popularity Index" from *Chartmasters' Rock 100,* Fourth Edition, by Jim Quirin and Barry Cohen. Copyright © 1987 by Chartmasters. Reprinted by permission.

13.2 EXERCISES

Use substitution to find the following indefinite integrals. (See Examples 1–6.)

1. $\int 4(2x + 3)^4 \, dx$

2. $\int (-4t + 1)^3 \, dt$

3. $\int \dfrac{4}{(y - 2)^3} \, dy$

4. $\int \dfrac{-3}{(x + 1)^4} \, dx$

5. $\int \dfrac{2 \, dm}{(2m + 1)^3}$

6. $\int \dfrac{3 \, du}{\sqrt{3u - 5}}$

7. $\int \dfrac{2x + 2}{(x^2 + 2x - 4)^4} \, dx$

8. $\int \dfrac{6x^2 \, dx}{(2x^3 + 7)^{3/2}}$

9. $\int z\sqrt{z^2 - 5} \, dz$

10. $\int r\sqrt{r^2 + 2} \, dr$

11. $\int (-4e^{2p}) \, dp$

12. $\int 5e^{-.3g} \, dg$

13. $\int 3x^2 e^{2x^3} \, dx$

14. $\int re^{-r^2} \, dr$

15. $\int (1 - t)e^{2t - t^2} \, dt$

16. $\int (x^2 - 1)e^{x^3 - 3x} \, dx$

17. $\int \dfrac{e^{1/z}}{z^2} \, dz$

18. $\int \dfrac{e^{\sqrt{y}}}{2\sqrt{y}} \, dy$

19. $\int \dfrac{-8}{1 + 3x} \, dx$

20. $\int \dfrac{9}{2 + 5t} \, dt$

21. $\int \dfrac{dt}{2t + 1}$

22. $\int \dfrac{dw}{5w - 2}$

23. $\int \dfrac{v \, dv}{(3v^2 + 2)^4}$

24. $\int \dfrac{x \, dx}{(2x^2 - 5)^3}$

25. $\int \dfrac{x - 1}{(2x^2 - 4x)^2} \, dx$

26. $\int \dfrac{2x + 1}{(x^2 + x)^3} \, dx$

27. $\int \left(\dfrac{1}{r} + r\right)\left(1 - \dfrac{1}{r^2}\right) \, dr$

28. $\int \left(\dfrac{2}{A} - A\right)\left(\dfrac{-2}{A^2} - 1\right) \, dA$

29. $\int \dfrac{x^2 + 1}{(x^3 + 3x)^{2/3}} \, dx$

30. $\int \dfrac{B^3 - 1}{(2B^4 - 8B)^{3/2}} \, dB$

31. $\int p(p + 1)^5 \, dp$

32. $\int x^3(1 + x^2)^{1/4} \, dx$

33. $\int t\sqrt{5t - 1} \, dt$

34. $\int 4r\sqrt{8 - r} \, dr$

35. $\int \dfrac{u}{\sqrt{u - 1}} \, du$

36. $\int \dfrac{2x}{(x + 5)^6} \, dx$

37. $\int 2x(x^2 + 1)^3 \, dx$

38. $\int y^2(y^3 - 4)^3 \, dy$

39. $\int \left(\sqrt{x^2 + 12x}\right)(x + 6) \, dx$

40. $\int \left(\sqrt{x^2 - 6x}\right)(x - 3) \, dx$

41. $\int \dfrac{t}{t^2 + 2} \, dt$

42. $\int \dfrac{-4x}{x^2 + 3} \, dx$

43. $\int ze^{2z^2} \, dz$

44. $\int x^2 e^{-x^3} \, dx$

Work the following exercises. (See Examples 7–8.)

45. Management Suppose the marginal revenue in dollars per plane from the sale of x jet planes is

$$R(x) = 2x(x^2 + 50)^3.$$

Find the total revenue from the sale of 3 planes.

46. Management The rate of expenditure in dollars per machine for maintenance for a particular machine is given by

$$M(x) = \sqrt{x^2 + 12x}(2x + 12),$$

where x is time measured in years. Find the total maintenance charge for the first 5 years of the machine's life.

47. Management The rate of growth of the profit (in millions of dollars per year) from a new technology is approximated by

$$p(x) = xe^{-x^2},$$

where x represents time measured in years. Find the total profit from the first 3 years that the new technology is in operation.

48. Natural Science If x milligrams of a certain drug are administered to a person, the rate of change in the person's temperature, in degrees Celsius per milligram, with respect to the dosage (the person's *sensitivity* to the drug) is given by

$$D(x) = \dfrac{2}{x + 9}.$$

Find the total change in body temperature if 3 milligrams of the drug are administered.

13.3 AREA AND THE DEFINITE INTEGRAL

Finding the area under a curve is important in a great many applications. For example, the graph and caption in Figure 13.2 is from a test report in *Modern Photography** magazine, June 1982, page 111, on the Minolta X-700 camera. The area under the curve represents the total noise made by the camera. This section shows how to find the area under a curve.

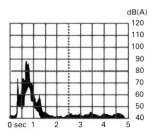

On the quiet side for a 35mm SLR, the X-700 measured a relatively low 70 dB(A) at 1/125 sec. despite having one sound peak at around 88 dB(A). Total area under curve is small due to relatively short sound duration.

FIGURE 13.2

Figure 13.3 shows the region bounded by the lines $x = 0$, the x-axis, and the graph of

$$f(x) = \sqrt{4 - x^2}.$$

A very rough approximation of the area of this region can be found by using two rectangles, as in Figure 13.4. The height of the rectangle on the left is $f(0) = 2$ and the height of the rectangle on the right is $f(1) = \sqrt{3}$. The width of each rectangle is 1, making the total area of the two rectangles

$$1 \cdot f(0) + 1 \cdot f(1) = 2 + \sqrt{3} \approx 3.7321 \text{ square units.}$$

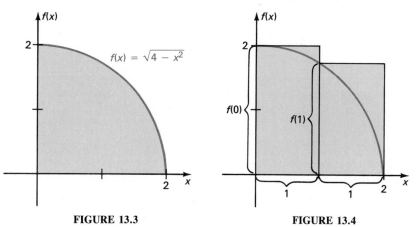

FIGURE 13.3   **FIGURE 13.4**

*From ''Sounding out the Minolta X-700'' in *Modern Photography*, June 1982. Copyright © 1981 by ABC Leisure Magazines, Inc. Reprinted by permission.

1 Calculate the sum
$$\frac{1}{4} \cdot f(0) + \frac{1}{4} \cdot f\left(\frac{1}{4}\right) + \cdots$$
$$+ \frac{1}{4} \cdot f\left(\frac{7}{4}\right), \text{ using this information.}$$

x	$f(x)$
0	2
1/4	1.98431
1/2	1.93649
3/4	1.85405
1	1.73205
5/4	1.56125
3/2	1.32288
7/4	.96825

Answer:
3.33982 square units

As Figure 13.4 suggests, this approximation is greater than the actual area. To improve the accuracy of the approximation, we could divide the interval from $x = 0$ to $x = 2$ into four equal parts, each of width 1/2, as shown in Figure 13.5. As before, the height of each rectangle is given by the value of f at the left-hand side of the rectangle, and its area is the width, 1/2, multiplied by the height. The total area of the four rectangles is

$$\frac{1}{2} \cdot f(0) + \frac{1}{2} \cdot f\left(\frac{1}{2}\right) + \frac{1}{2} \cdot f(1) + \frac{1}{2} \cdot f\left(\frac{3}{2}\right)$$

$$= \frac{1}{2}(2) + \frac{1}{2}\left(\frac{\sqrt{15}}{2}\right) + \frac{1}{2}(\sqrt{3}) + \frac{1}{2}\left(\frac{\sqrt{7}}{2}\right)$$

$$= 1 + \frac{\sqrt{15}}{4} + \frac{\sqrt{3}}{2} + \frac{\sqrt{7}}{4} \approx 3.4957 \text{ square units.}$$

This approximation looks better, but it is still greater than the actual area desired. To improve the approximation, divide the interval from $x = 0$ to $x = 2$ into eight parts with equal widths of 1/4. (See Figure 13.6) The total area of all these rectangles is

$$\frac{1}{4} \cdot f(0) + \frac{1}{4} \cdot f\left(\frac{1}{4}\right) + \frac{1}{4} \cdot f\left(\frac{1}{2}\right) + \frac{1}{4} \cdot f\left(\frac{3}{4}\right) + \frac{1}{4} \cdot f(1) + \frac{1}{4} \cdot f\left(\frac{5}{4}\right)$$

$$+ \frac{1}{4} \cdot f\left(\frac{3}{2}\right) + \frac{1}{4} \cdot f\left(\frac{7}{4}\right). \quad \boxed{1}$$

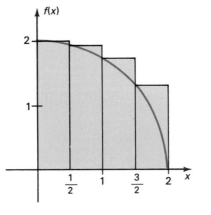

FIGURE 13.5

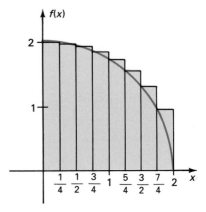

FIGURE 13.6

This process of approximating the area under a curve by using more and more rectangles to get a better and better approximation can be generalized. To do this, divide the interval from $x = 0$ to $x = 2$ into n equal parts. Each of these n intervals has width

$$\frac{2 - 0}{n} = \frac{2}{n},$$

so each rectangle has width $2/n$ and height determined by the function-value at the left side of the rectangle. A computer was used to find approximations to the area for several values of n given in the table at the side.

The areas in the last column in the table are approximations of the area under the curve, above the x-axis, and between the lines $x = 0$ and $x = 2$. As n becomes larger and larger, the approximation is better and better, getting closer to the actual area. In this example, the exact area can be found by a formula from plane geometry. Write the given function as

$$y = \sqrt{4 - x^2},$$

then square both sides to get

$$y^2 = 4 - x^2$$
$$x^2 + y^2 = 4,$$

the equation of a circle centered at the origin with radius 2. The region in Figure 13.3 is the quarter of this circle that lies in the first quadrant. The actual area of this region is one quarter of the area of the entire circle, or

$$\frac{1}{4}\pi(2)^2 = \pi \approx 3.1416.$$

As the number of rectangles increases without bound, the sum of the areas of these rectangles gets closer and closer to the actual area of the region, π. This can be written as

$$\lim_{n \to \infty} (\text{sum of areas of } n \text{ rectangles}) = \pi.$$

(The value of π was originally approximated by a process similar to this.)*

This approach could be used to find the area of any region bounded by the x-axis, the lines $x = a$, and $x = b$, and the graph of a function $y = f(x)$. At this point, we need some new notation and terminology to write sums concisely. We will indicate addition (or summation) by using the Greek letter sigma, Σ, as shown in the next example.

n	Area
2	3.7321
4	3.4957
8	3.3398
10	3.3045
20	3.2285
50	3.1783
100	3.1512
500	3.1455

*The number π is the ratio of the circumference of a circle to its diameter. It is an example of an *irrational number* and as such it cannot be expressed as a terminating or repeating decimal. Many approximations have been used for π over the years. A passage in the Bible (I Kings 7:23) indicates a value of 3. The Egyptians used the value 3.16, and Archimedes showed that its value must be between 22/7 and 223/71. A Hindu writer, Brahmagupta, used $\sqrt{10}$ as its value in the seventh century. The search for the digits of π has continued throughout the years. Two mathematicians at Columbia University have computed the value of π to more than a billion decimal places!

2 Find each sum.

(a) $\displaystyle\sum_{i=0}^{6} i$

(b) $\displaystyle\sum_{i=2}^{7}(3i + 1)$

(c) $\displaystyle\sum_{i=1}^{4}(5x_i - 1)$, if $x_1 = 9$,

$x_2 = 8$, $x_3 = 2$, $x_4 = 7$

(d) $\displaystyle\sum_{i=3}^{6}f(x_i)\,\Delta x$, if $f(x) = 3x^2 - 1$,

$x_3 = 2$, $x_4 = 5$, $x_5 = 1$, $x_6 = 5$,

and $\Delta x = \dfrac{1}{5}$

Answer:

(a) 21

(b) 87

(c) 126

(d) 32.2

▶**EXAMPLE 1**   Find the following sums.

(a) $\displaystyle\sum_{i=1}^{5} i$

Replace i in turn with the integers 1 through 5 and add the resulting terms.

$$\sum_{i=1}^{5} i = 1 + 2 + 3 + 4 + 5 = 15$$

(b) $\displaystyle\sum_{i=1}^{4} a_i = a_1 + a_2 + a_3 + a_4$

(c) $\displaystyle\sum_{i=1}^{3}(6x_i - 2)$, if $x_1 = 2$, $x_2 = 4$, $x_3 = 6$

Letting $i = 1$, 2, and 3, respectively, gives

$$\sum_{i=1}^{3}(6x_i - 2) = (6x_1 - 2) + (6x_2 - 2) + (6x_3 - 2).$$

Now substitute the given values for x_1, x_2, and x_3.

$$\sum_{i=1}^{3}(6x_i - 2) = (6 \cdot 2 - 2) + (6 \cdot 4 - 2) + (6 \cdot 6 - 2)$$
$$= 10 + 22 + 34$$
$$= 66$$

(d) $\displaystyle\sum_{i=1}^{4} f(x_i)\,\Delta x$ if $f(x) = x^2$, $x_1 = 0$, $x_2 = 2$, $x_3 = 4$, $x_4 = 6$, and $\Delta x = 2$.

$$\sum_{i=1}^{4} f(x_i)\,\Delta x = f(x_1)\,\Delta x + f(x_2)\,\Delta x + f(x_3)\,\Delta x + f(x_4)\,\Delta x$$
$$= x_1^2\,\Delta x + x_2^2\,\Delta x + x_3^2\,\Delta x + x_4^2\,\Delta x$$
$$= 0^2(2) + 2^2(2) + 4^2(2) + 6^2(2)$$
$$= 0 + 8 + 32 + 72 = 112 \quad ◀ \quad \boxed{2}$$

Now we can generalize to get a method of finding the area bounded by the curve $y = f(x)$, the x-axis, and the vertical lines $x = a$ and $x = b$, as shown in Figure 13.7. To approximate this area, we could divide the region under the curve first into ten rectangles (Figure 13.7(a)) and then into twenty rectangles (Figure 13.7(b)). The sums of the areas of the rectangles give approximations to the area under the curve.

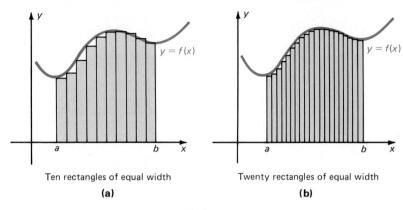

Ten rectangles of equal width

(a)

Twenty rectangles of equal width

(b)

FIGURE 13.7

To get a number that can be defined as the *exact* area, begin by dividing the interval from a to b into n pieces of equal width, using each of these n pieces as the base of a rectangle. (See Figure 13.8.) The endpoints of the n intervals are labeled $x_1, x_2, x_3, \ldots, x_{n+1}$, where $a = x_1$ and $b = x_{n+1}$. In the graph of Figure 13.8, the symbol Δx is used to represent the width of each of the intervals. The darker rectangle is an arbitrary rectangle called the ith rectangle. Its area is the product of its length and width. Since the width of the ith rectangle is Δx and the length of the ith rectangle is given by the height $f(x_i)$,

$$\text{Area of } i\text{th rectangle} = f(x_i) \cdot \Delta x.$$

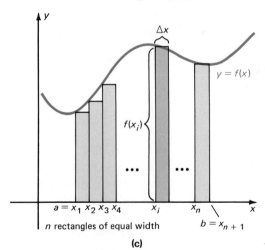

n rectangles of equal width

(c)

FIGURE 13.8

The total area under the curve is approximated by the sum of the areas of all n of the rectangles. Using the Σ symbol, the approximation to the total area becomes

$$\text{Area of all } n \text{ rectangles} = \sum_{i=1}^{n} f(x_i) \cdot \Delta x.$$

The exact area is defined to be the limit of this sum (if the limit exists) as the number of rectangles increases without bound.

$$\text{Exact area} = \lim_{n \to \infty} \sum_{i=1}^{n} f(x_i) \, \Delta x$$

This limit is called the *definite integral* of $f(x)$ from a to b and is written as follows.

The Definite Integral

If f is continuous on the interval $[a, b]$, the **definite integral** of f from a to b is given by

$$\int_a^b f(x) \, dx = \lim_{n \to \infty} \sum_{i=1}^{n} f(x_i) \, \Delta x,$$

provided the limit exists, where $\Delta x = (b - a)/n$ and x_i is *any* value of x in the ith interval.

As indicated in this definition, although the left endpoint of the ith interval has been used to find the height of the ith rectangle, any number in the ith interval can be used. (A more general definition is possible in which the rectangles do not necessarily all have the same width.) The b above the integral sign is called the **upper limit** of integration, and the a is the **lower limit** of integration. This use of the word ''limit'' has nothing to do with the limit of the sum; it refers to the limits, or boundaries, on x.

In the example at the beginning of this section, the area bounded by the x-axis, the curve $y = \sqrt{4 - x^2}$, the lines $x = 0$ and $x = 2$ could be written as the definite integral

$$\int_0^2 \sqrt{4 - x^2} \, dx = \pi.$$

Notice that unlike the indefinite integral, which is a set of *functions,* the definite integral represents a *number*. The next section will show how antiderivatives are used in finding the definite integral and, thus, the area under a curve.

Caution Keep in mind that finding the definite integral of a function can be thought of as a mathematical process that gives the sum of an infinite number of individual parts (within certain limits). The definite integral represents area only if the function involved is *nonnegative* ($f(x) \geq 0$) at every x-value in the interval $[a, b]$. There are many other interpretations of the definite integral, and all of them involve this idea of approximation by appropriate sums.

3 Divide the region of Figure 13.9 into eight rectangles of equal width whose heights are the values of the function at the midpoint of each rectangle.

(a) Complete this table.

i	x_i	$f(x)$
1	.25	
2	.75	
3	1.25	
4	1.75	
5		
6		
7		
8		

(b) Use the results from the table to approximate $\int_0^4 2x\, dx$.

Answer:

(a)

i	x_i	$f(x)$
1	.25	.50
2	.75	1.50
3	1.25	2.50
4	1.75	3.50
5	2.25	4.50
6	2.75	5.50
7	3.25	6.50
8	3.75	7.50

(b) 16

▶ **EXAMPLE 2** Approximate $\int_0^4 2x\, dx$, the area of the region under the graph of $f(x) = 2x$, above the x-axis, and between $x = 0$ and $x = 4$, by using four rectangles of equal width whose heights are the values of the function at the midpoint of each rectangle.

We want to find the area of the shaded region in Figure 13.9. The heights of the four rectangles given by $f(x_i)$ for $i = 1, 2, 3,$ and 4 are as follows.

i	x_i	$f(x_i)$
1	$x_1 = .5$	$f(.5) = 1.0$
2	$x_2 = 1.5$	$f(1.5) = 3.0$
3	$x_3 = 2.5$	$f(2.5) = 5.0$
4	$x_4 = 3.5$	$f(3.5) = 7.0$

The width of each rectangle is $\Delta x = \dfrac{4 - 0}{4} = 1$. The sum of the areas of the four rectangles is

$$\sum_{i=1}^{4} f(x_i)\, \Delta x = f(x_1)\, \Delta x + f(x_2)\, \Delta x + f(x_3)\, \Delta x + f(x_4)\, \Delta x$$
$$= f(.5)\, \Delta x + f(1.5)\, \Delta x + f(2.5)\, \Delta x + f(3.5)\, \Delta x$$
$$= (1)(1) + (3)(1) + (5)(1) + (7)(1)$$
$$= 16. \quad \boxed{3}$$

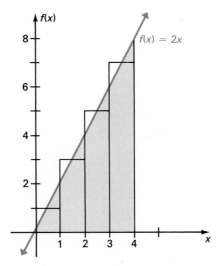

FIGURE 13.9

Using the formula for the area of a triangle, $A = (1/2)bh$, with b, the length of the base, equal to 4 and h, the height, equal to 8, gives

$$A = \frac{1}{2}bh = \frac{1}{2}(4)(8) = 16.$$

the exact value of the area. The approximation equals the exact area in this case because our use of the midpoints of each subinterval distributed the error evenly above and below the graph. ◀

We now wish to connect the area under the graph of a function with rates of change. The basic idea can be seen in a simple example. Suppose a car travels along a straight road at a constant speed of 50 mph. Then the function $v(t) = 50$ is the function that gives the speed of the car at time t. The graph of this function is the straight line shown in Figure 13.10(a).

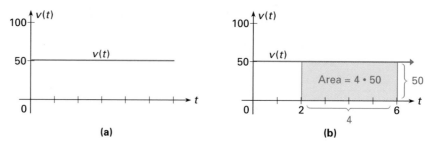

FIGURE 13.10

How far does the car travel from time $t = 2$ to $t = 6$? Since this is a 4-hour period, the answer, of course, is $4 \cdot 50 = 200$ miles. Note that this is precisely the area under the graph of the speed function $v(t)$ from $t = 2$ to $t = 6$, as shown in Figure 13.10(b) above.

As we saw in Chapter 12, the speed (or velocity) function $v(t)$ is just the rate of change of distance with respect to time, that is, the rate of change of the distance function $s(t)$ (which gives the position of the car at time t). The distance traveled from time $t = a$ to $t = b$ is the total change in the function $s(t)$ from $t = a$ to $t = b$. In this example, therefore, the total change in $s(t)$ as t goes from $t = a$ to $t = b$ is the area under $v(t)$ from $t = a$ to $t = b$, that is, the definite integral $\int_a^b v(t)\, dt$. A more complicated argument (that is omitted here) works in the general case and leads to this useful result.

4 Use Figure 13.11 to estimate the maintenance change during

(a) the first 6 years of the machine's life;

(b) the first 8 years.

Answer:

(a) $7850

(b) $14,250

Total Change in $F(x)$

Let f be a function such that f is continuous on $[a, b]$ and $f(x) \geq 0$ for all x in $[a, b]$. If $f(x)$ is the rate of change of the function $F(x)$ for x in $[a, b]$, then the **total change** in $F(x)$ as x goes from a to b is given by

$$\int_a^b f(x) \, dx.$$

In other words, the total change in a quantity can be found from the function that gives the rate of change of the quantity, using the same methods used to approximate the area under a curve.

▶**EXAMPLE 3**　Figure 13.11 shows the rate of change of the annual maintenance charges for a certain machine. Approximate the total maintenance charges over the 10-year life of the machine by using rectangles, dividing the interval from 0 to 10 into ten equal subdivisions. Each rectangle has width 1; if we use the left endpoint of each rectangle to determine the height of the rectangle, the approximation becomes

$$1 \cdot 0 + 1 \cdot 500 + 1 \cdot 750 + 1 \cdot 1800 + 1 \cdot 1800 + 1 \cdot 3000$$
$$+ 1 \cdot 3000 + 1 \cdot 3400 + 1 \cdot 4200 + 1 \cdot 5100 = 23{,}550.$$

About $23,550 will be spent on maintenance over the 10-year life of the machine.　◀　**4**

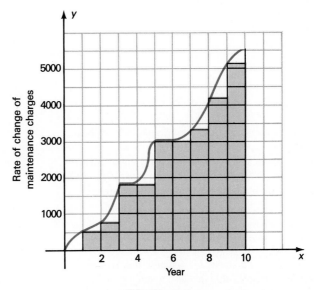

FIGURE 13.11

13.3   EXERCISES

Evaluate the following sums. (See Example 1.)

1. $\sum_{i=1}^{3} 3i$

2. $\sum_{i=1}^{6} (-5i)$

3. $\sum_{i=1}^{5} (2i + 7)$

4. $\sum_{i=1}^{10} (5i - 8)$

5. Let $x_1 = -5$,
$x_2 = 8$,
$x_3 = 7$,
$x_4 = 10$.
Find $\sum_{i=1}^{4} x_i$.

6. Let $x_1 = 10$,
$x_2 = 15$,
$x_3 = -8$,
$x_4 = -12$,
$x_5 = 0$.
Find $\sum_{i=1}^{5} x_i$.

7. Let $f(x) = x - 3$,
$x_1 = 4$,
$x_2 = 6$,
$x_3 = 7$.
Find $\sum_{i=1}^{3} f(x_i)$.

8. Let $f(x) = x^2 + 1$,
$x_1 = -2$,
$x_2 = 0$,
$x_3 = 2$,
$x_4 = 4$.
Find $\sum_{i=1}^{4} f(x_i)$.

9. Let $f(x) = 2x + 1$,
$x_1 = 0$,
$x_2 = 2$,
$x_3 = 4$,
$x_4 = 6$,
$\Delta x = 2$.
Find $\sum_{i=1}^{4} f(x_i)\, \Delta x$.

10. Let $f(x) = 1/x$,
$x_1 = 1/2$,
$x_2 = 1$,
$x_3 = 3/2$,
$x_4 = 2$,
$\Delta x = 1/2$.
Find $\sum_{i=1}^{4} f(x_i)\, \Delta x$.

Approximate the area under each given curve and above the x-axis on the given interval by using two rectangles. Let the height of the rectangle be given by the value of the function at the left side of the rectangle. Then repeat the process and approximate the area with four rectangles. (See Example 2.)

11. $f(x) = 3x + 2;\ [1, 5]$

12. $f(x) = -2x + 1;\ [-4, 0]$

13. $f(x) = x + 5;\ [2, 4]$

14. $f(x) = 3 + x;\ [1, 3]$

15. $f(x) = x^2;\ [1, 5]$

16. $f(x) = x^2;\ [0, 4]$

17. $f(x) = x^2 + 2;\ [-2, 2]$

18. $f(x) = -x^2 + 4;\ [-2, 2]$

19. $f(x) = e^x - 1;\ [0, 4]$

20. $f(x) = e^x + 1;\ [-2, 2]$

21. $f(x) = \dfrac{1}{x};\ [1, 5]$

22. $f(x) = \dfrac{2}{x};\ [1, 9]$

Work the following exercises. (See Example 2.)

23. Consider the region below $f(x) = x/2$, above the x-axis, between $x = 0$ and $x = 4$. Let x_i be the left endpoint of the ith subinterval.
   (a) Approximate the area of the region using four rectangles.
   (b) Approximate the area of the region using eight rectangles.
   (c) Find $\int_0^4 f(x)\, dx$ by using the formula for the area of a triangle.

24. Find $\int_0^5 (5 - x)\, dx$ by using the formula for the area of a triangle.

Estimate the area under the curve by summing the area of rectangles. Let the function value at the left side of each rectangle give the height of the rectangle. (See Example 3.)

25. Management   The graph below shows the rate of sales of new cars in a recent year. Estimate the total sales during that year. Use rectangles with a width of 1.

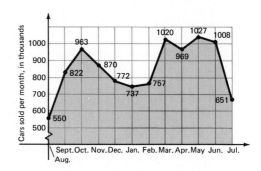

26. Management   The graph below shows the rate of use of electrical energy (in kilowatt hours) in a certain city on a very hot day. Estimate the total usage of electricity on that day. Let the width of each rectangle be 2 hours.

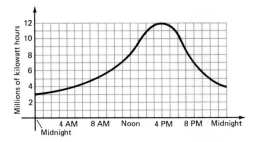

27. Natural Science   The graph below shows the approximate concentration of alcohol in a person's bloodstream t hours after drinking 2 ounces of alcohol. Estimate the total amount of alcohol in the bloodstream by estimating the area under the curve. Use rectangles of width 1 hour.

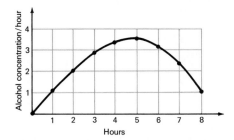

28. Natural Science   The graph below shows the rate of inhalation of oxygen by a person riding a bicycle very rapidly for 10 minutes. Estimate the total volume of oxygen inhaled in the first 20 minutes after the beginning of the ride. Use rectangles of width 1 minute.

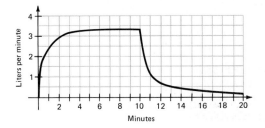

The next two graphs are from Road and Track *magazine.* The curve shows the velocity at time t, in seconds, when the car accelerates from a dead stop. To find the total distance traveled by the car in reaching 100 miles per hour, we must estimate the definite integral*

$$\int_0^T v(t)\, dt,$$

where T represents the number of seconds it takes for the car to reach 100 mph.

   Use the graphs to estimate this distance by adding the areas of rectangles with widths of 5 seconds. The last rectangle has a width of 3. To adjust your answer to miles per hour, divide by 3600 (the number of seconds in an hour). You then have the number of miles that the car traveled in reaching 100 mph. Finally, multiply by 5280 feet per mile to convert the answers to feet.

29. Estimate the distance traveled by the Porsche 928, using the graph below.

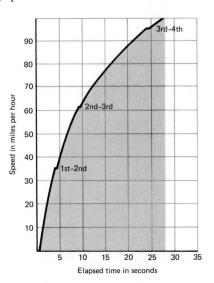

---

*From *Road & Track,* April and May, 1978. Reprinted with permission of *Road & Track.*

30. Estimate the distance traveled by the BMW 733i, using the graph below.

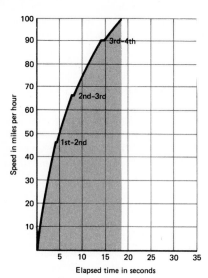

31.

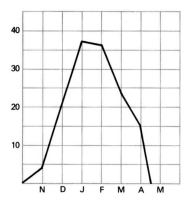

32.

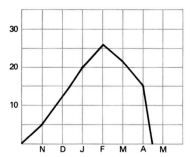

Natural Science *When doing research on the wolf-moose relationships in Michigan's Isle Royale National Park, the biologist Rolf Peterson needed a snow-depth index to help correlate populations with snow levels. The graph below is given a snow depth index of 1.0. The vertical scale gives snow depth in inches, and the horizontal scale gives months.**

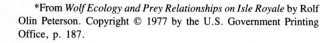

Find the snow depth index for the following years by forming the ratio of the areas under the given curves to the area under the curve above.

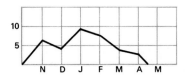 *Use a calculator and the method from the text with eight rectangles and left endpoints to approximate the area between the x-axis and the graph of each function on the given interval.*

33. $f(x) = x \ln x$; $[1, 5]$

34. $f(x) = x^2 e^{-x}$; $[-1, 3]$

35. $f(x) = \dfrac{\ln x}{x}$; $[1, 5]$

36. $f(x) = \dfrac{e^x - e^{-x}}{2}$; $[0, 4]$

---

*From *Wolf Ecology and Prey Relationships on Isle Royale* by Rolf Olin Peterson. Copyright © 1977 by the U.S. Government Printing Office, p. 187.

1 Let $C(x) = x^3 + 4x^2 - x + 3$. Find the following.

(a) $C(x)]_1^5$

(b) $C(x)]_3^4$

(c) $C(x)]_{10}^{20}$

Answer:

(a) 216

(b) 64

(c) 8190

13.4 THE FUNDAMENTAL THEOREM OF CALCULUS

In this section we shall develop the connection between definite integrals and anti-derivatives. In the last section we saw that if $f(x) \geq 0$ for all x in $[a, b]$ and if $f(x)$ is the rate of change of the function $F(x)$, then the definite integral

$$\int_a^b f(x)\, dx \text{ is the total change of } F(x) \qquad (*)$$

as x changes from a to b.

Now to say that $f(x)$ is the rate of change of $F(x)$ means that $f(x)$ is the derivative of $F(x)$, or equivalently, $F(x)$ is an antiderivative of $f(x)$. The total change in $F(x)$ from $x = a$ to $x = b$ is the difference $F(b) - F(a)$. So we can restate (*) by saying that

$$\int_a^b f(x)\, dx = F(b) - F(a).$$

It can be proved that this relationship holds in more general situations (where $f(x)$ may not always be ≥ 0).

Fundamental Theorem of Calculus

Let f be continuous on the interval $[a, b]$, and let F be *any* antiderivative of f. Then

$$\int_a^b f(x)\, dx = F(b) - F(a).$$

Caution It is important to note that the fundamental theorem does not require $f(x) \geq 0$. The condition $f(x) \geq 0$ is necessary only when using the fundamental theorem to find area. Also, note that the fundamental theorem does not *define* the definite integral; it just provides a method for evaluating it. **1**

When evaluating definite integrals, the symbol $F(x)]_a^b$ is used to denote the number $F(b) - F(a)$. For example, if $F(x) = x^4$, then $x^4]_1^2$ means $F(2) - F(1) = 2^4 - 1^4$.

2 Find each of the following.

(a) $\displaystyle\int_4^6 5z \, dz$

(b) $\displaystyle\int_2^5 8t^3 \, dt$

(c) $\displaystyle\int_1^9 \sqrt{z} \, dz$

Answer:

(a) 50

(b) 1218

(c) 52/3

▶**EXAMPLE 1** Find $\displaystyle\int_1^2 4x^3 \, dx$.

Choosing x^4 as an antiderivative gives

$$\int_1^2 4x^3 \, dx = x^4 \Big]_1^2 = 2^4 - 1^4 = 16 - 1 = 15.$$

There are infinitely many antiderivatives of $4x^3$, but all of these have the form $x^4 + C$, for a real number C. It doesn't matter which value of C is used.

$$\int_1^2 4x^3 \, dx = (x^4 + C) \Big]_1^2 = (2^4 + C) - (1^4 + C)$$
$$= 16 + C - (1 + C) = 15$$

The value of $\int_1^2 4x^3 \, dx$ is 15, no matter which value of C is used. For this reason, when finding a definite integral, choose the antiderivative with $C = 0$. ◀ **2**

Example 1 illustrates the difference between the definite integral and the indefinite integral. A definite integral is a real number; an indefinite integral is a family of functions—all the functions that are antiderivatives of a function f.

The fundamental theorem of calculus receives its name from the fact that it is the key connection between differential calculus and integral calculus, which were originally developed separately without the knowledge of this connection. Although a proof of the fundamental theorem is beyond the scope of this book, we can give an informal argument that makes the theorem highly plausible in the case when $f(x) > 0$.

If f is a continuous function and $f(x) > 0$ on $[a, b]$, then the graph of f is an unbroken curve lying above the x-axis, as shown in Figure 13.12. We first use the graph of f to define a new function A by this rule: $A(x)$ is the area under the graph of $y = f(x)$ from a to x. Next, we show that this function A is an antiderivative of f, that is, $A' = f$.

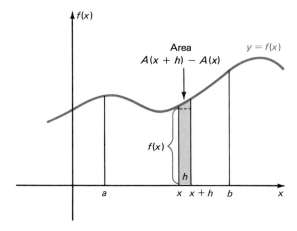

FIGURE 13.12

To do this, let h be a small positive number. Then $A(x + h) - A(x)$ (the difference between the area under the graph from a to $x + h$ and the area from a to x) is the shaded area of Figure 13.12. This area can be approximated by the area of the rectangle having width h and height $f(x)$. The area of the rectangle is $h \cdot f(x)$ and, hence,

$$A(x + h) - A(x) \approx h \cdot f(x).$$

Dividing both sides by h gives

$$\frac{A(x + h) - A(x)}{h} \approx f(x). \tag{*}$$

This approximation improves as h gets smaller and smaller (because the area of the rectangle then gets closer and closer to the actual shaded area). Taking the limit on both sides of (*) as h approaches 0, we have

$$\lim_{h \to 0} \frac{A(x + h) - A(x)}{h} = \lim_{h \to 0} f(x).$$

Now $f(x)$ is constant as h approaches 0, so the right-hand limit is just $f(x)$. By the definition of derivative, the left-hand limit is $A'(x)$. Therefore,

$$A'(x) = f(x).$$

Thus, A is an antiderivative of f, as we set out to show.

The next step is to show that the fundamental theorem is true when the antiderivative $A(x)$ is used. Since $A(x)$ is the area under the graph of f from a to x, we see that $A(a) = 0$ and $A(b)$ is the area under the graph from a to b. But this last area is the definite integral $\int_a^b f(x)\, dx$. Putting these facts together we have

$$\int_a^b f(x)\, dx = A(b) = A(b) - 0 = A(b) - A(a).$$

Therefore, the fundamental theorem is true when $A(x)$ is used as the antiderivative.

Finally, let $F(x)$ be *any* antiderivative of $f(x)$ on $[a, b]$. Then, as noted in Section 13.1, there is a constant C such that $F(x) - A(x) = C$, or equivalently, $F(x) = A(x) + C$. Therefore,

$$\int_a^b f(x)\, dx = A(b) - A(a)$$
$$= A(b) + C - A(a) - C = [A(b) + C] - [A(a) + C]$$
$$= F(b) - F(a).$$

Hence, the fundamental theorem is true for any antiderivative $F(x)$ of $f(x)$.

The variable used in the integrand does not matter; that is

$$\int_a^b f(x)\, dx = \int_a^b f(t)\, dt = \int_a^b f(u)\, du.$$

Each of these definite integrals represents the number $F(b) - F(a)$, where F is the antiderivative of f.

Key properties of definite integrals are listed below. Some of them are just restatements of properties from Section 13.1.

Properties of Definite Integrals

For any real numbers a and b for which the definite integrals exist,

1. $\displaystyle\int_a^a f(x)\ dx = 0;$

2. $\displaystyle\int_a^b k \cdot f(x)\ dx = k \cdot \int_a^b f(x)\ dx,$ for any real constant k (constant multiple of a function);

3. $\displaystyle\int_a^b [f(x) \pm g(x)]\ dx = \int_a^b f(x)\ dx \pm \int_a^b g(x)\ dx$ (sum or difference of functions);

4. $\displaystyle\int_a^b f(x)\ dx = \int_a^c f(x)\ dx + \int_c^b f(x)\ dx,$ for any real number c.

Geometrically, if $f(x) \geq 0$, since the distance from a to a is 0, the first property says that the "area" under the graph of f bounded by $x = a$ and $x = a$ is 0. Also, since $\int_a^c f(x)\ dx$ represents the darker region in Figure 13.13 and $\int_c^b f(x)\ dx$ represents the lighter region,

$$\int_a^b f(x)\ dx = \int_a^c f(x)\ dx + \int_c^b f(x)\ dx,$$

as stated in the fourth property. While the figure shows $a < c < b$, the property is true for any value of c where both $f(x)$ and $F(x)$ are defined. Of course, the properties apply even if $f(x) < 0$.

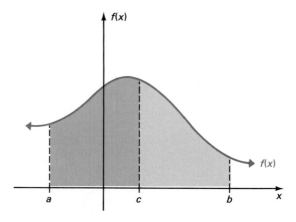

FIGURE 13.13

3 Evaluate each definite integral.

(a) $\displaystyle\int_1^3 (x + 3x^2)\, dx$

(b) $\displaystyle\int_2^4 (6k^2 - 2k + 1)\, dk$

(c) $\displaystyle\int_3^5 (3t^2 - 4t + 2)\, dt$

Answer:

(a) 30

(b) 102

(c) 70

4 Evaluate the following.

(a) $\displaystyle\int_0^4 e^x\, dx$

(b) $\displaystyle\int_3^5 \frac{dx}{x}$

(c) $\displaystyle\int_1^6 e^x\, dx$

(d) $\displaystyle\int_2^8 \frac{4}{x}\, dx$

Answer:

(a) 53.59815

(b) .51083

(c) 400.71051

(d) 5.54518

An algebraic proof is given here for the third property; proofs of the other properties are omitted. If $F(x)$ and $G(x)$ are antiderivatives of $f(x)$ and $g(x)$, respectively,

$$\int_a^b [f(x) + g(x)]\, dx = [F(x) + G(x)]\Big|_a^b$$
$$= [F(b) + G(b)] - [F(a) + G(a)]$$
$$= [F(b) - F(a)] + [G(b) - G(a)]$$
$$= \int_a^b f(x)\, dx + \int_a^b g(x)\, dx.$$

▶ **EXAMPLE 2** Evaluate $\displaystyle\int_2^5 (6x^2 - 3x + 5)\, dx$.

Use the properties above, and the fundamental theorem, along with the power rule from Section 13.1.

$$\int_2^5 (6x^2 - 3x + 5)\, dx = 6\int_2^5 x^2\, dx - 3\int_2^5 x\, dx + 5\int_2^5 dx \qquad \text{Constant multiple}$$

$$= 2x^3\Big]_2^5 - \frac{3}{2}x^2\Big]_2^5 + 5x\Big]_2^5 \qquad \text{Integrate}$$

$$= 2(5^3 - 2^3) - \frac{3}{2}(5^2 - 2^2) + 5(5 - 2) \qquad \text{Evaluate the limits}$$

$$= 2(125 - 8) - \frac{3}{2}(25 - 4) + 5(3)$$

$$= 234 - \frac{63}{2} + 15 = \frac{435}{2} \quad ◀ \boxed{3}$$

▶ **EXAMPLE 3** Find $\displaystyle\int_1^2 \frac{dy}{y}$.

Using a result from Section 13.1,

$$\int_1^2 \frac{dy}{y} = \ln |y|\Big]_1^2 = \ln |2| - \ln |1|$$

$$= \ln 2 - \ln 1 \approx .6931 - 0 = .6931 \quad ◀ \boxed{4}$$

▶ **EXAMPLE 4** Evaluate $\displaystyle\int_0^5 x\sqrt{25 - x^2}\, dx$.

Use substitution. Let $u = 25 - x^2$, so that $du = -2x\, dx$. With a definite integral, the limits should be changed, too. The new limits on u are found as follows.

If $x = 5$, then $u = 25 - 5^2 = 0$;

if $x = 0$, then $u = 25 - 0^2 = 25$.

5 Find $\displaystyle\int_0^2 \frac{x}{x^2+1}\, dx$.

Answer:

$\dfrac{1}{2} \ln 5$

Then

$$\int_0^5 x\sqrt{25-x^2}\, dx = -\frac{1}{2}\int_0^5 \sqrt{25-x^2}(-2x\, dx)$$

$$= -\frac{1}{2}\int_{25}^0 \sqrt{u}\, du \qquad\qquad \text{Substitute}$$

$$= -\frac{1}{2}\int_{25}^0 u^{1/2}\, du \qquad\qquad \text{Fractional exponent}$$

$$= -\frac{1}{2}\cdot\frac{u^{3/2}}{3/2}\bigg]_{25}^0 \qquad\qquad \text{Integrate}$$

$$= -\frac{1}{2}\cdot\frac{2}{3}[0^{3/2}-25^{3/2}] \qquad\qquad \text{Evaluate limits}$$

$$= -\frac{1}{3}(-125)$$

$$= \frac{125}{3}. \quad \blacktriangleleft \quad \boxed{5}$$

Caution Whenever you use the substitution method, be sure to replace x by its equivalent in terms of u *everywhere*. In particular, remember that the limits of integration refer to x and must be changed, too.

In the previous section we saw that if $f(x) > 0$ in $[a, b]$, the definite integral $\int_a^b f(x)\, dx$ gives the area below the graph of the function $y = f(x)$, above the x-axis, and between the lines $x = a$ and $x = b$. We shall now see how to find the area between the graph of a function f and the x-axis when the graph lies below the x-axis.

Look at the graph of $f(x) = x^2 - 4$ in Figure 13.14. The area bounded by the graph of f, the x-axis, and the vertical lines $x = 0$ and $x = 2$ lies below the x-axis. Using the fundamental theorem to find this area gives

$$\int_0^2 (x^2-4)\, dx = \left(\frac{x^3}{3}-4x\right)\bigg]_0^2 = \left(\frac{8}{3}-8\right)-(0-0) = \frac{-16}{3}.$$

The result is a negative number because $f(x)$ is negative for values of x in the interval $[0, 2]$. Since Δx is always positive, $f(x) < 0$ makes the product $f(x)\cdot\Delta x$ negative, so $\int_0^2 f(x)\, dx$ is negative. Since area is nonnegative, the required area is given by $|-16/3|$ or $16/3$. Using a definite integral, the area could be written as

$$\left|\int_0^2 (x^2-4)\, dx\right| = \left|-\frac{16}{3}\right| = \frac{16}{3}.$$

6 Find each area.

(a)

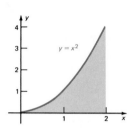

(b)

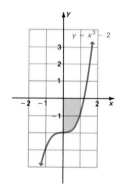

Answer:

(a) 8/3

(b) 7/4

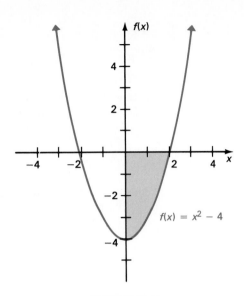

FIGURE 13.14

▶**EXAMPLE 5** Find the area of the region between the x-axis and the graph of $f(x) = x^2 - 3x$ from $x = 1$ to $x = 3$.

The region is shown in Figure 13.15. Since the region lies below the x-axis, the area is given by

$$\left| \int_1^3 (x^2 - 3x) \, dx \right|.$$

By the fundamental theorem,

$$\int_1^3 (x^2 - 3x) \, dx = \left(\frac{x^3}{3} - \frac{3x^2}{2} \right) \Big]_1^3 = \left(\frac{27}{3} - \frac{27}{2} \right) - \left(\frac{1}{3} - \frac{3}{2} \right) = -\frac{10}{3}.$$

The required area is $|-10/3| = 10/3$ square units. ◀ **6**

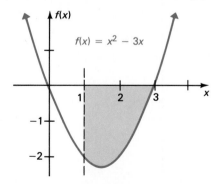

FIGURE 13.15

▶**EXAMPLE 6**   Find the area between the x-axis and the graph of $f(x) = x^2 - 4$ from $x = 0$ to $x = 4$.

Figure 13.16 shows the required region. Part of the region is below the x-axis. The definite integral over that interval will have a negative value. To find the area, integrate the negative and positive portions separately and take the absolute value of the first result before combining the two results to get the total area. Start by finding the point where the graph crosses the x-axis. This is done by solving the equation

$$x^2 - 4 = 0.$$

The solutions of this equation are 2 and -2. The only solution in the interval $[0, 4]$ is 2. The total area of the region in Figure 13.16 is given by

$$\left| \int_0^2 (x^2 - 4) \, dx \right| + \int_2^4 (x^2 - 4) \, dx = \left| \left(\frac{1}{3}x^3 - 4x \right) \right]_0^2 \right| + \left(\frac{1}{3}x^3 - 4x \right) \right]_2^4$$

$$= \left| \frac{8}{3} - 8 \right| + \left(\frac{64}{3} - 16 \right) - \left(\frac{8}{3} - 8 \right)$$

$$= 16 \text{ square units.} \quad ◀$$

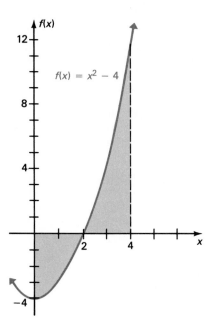

FIGURE 13.16

7 Find the area between the x-axis and the graph of $f(x) = x^2 - 4$ from

(a) $x = 1$ to $x = 5$ and

(b) $x = -2$ to $x = 3$.

Answer:

(a) 86/3

(b) 13

8 In Example 7, suppose a conservation campaign and higher prices cause the rate of consumption to be given by $C'(t) = \dfrac{1}{2}t + e^{.005t}$. Find the total amount of gas used in the nineties.

Answer:
About 35.25 trillion cubic feet

In Example 6, incorrectly using one integral over the entire interval to find the area would give

$$\int_0^4 (x^2 - 4)\, dx = \left(\frac{x^3}{3} - 4x\right)\Big]_0^4 = \left(\frac{64}{3} - 16\right) - 0 = \frac{16}{3},$$

which is not the correct area. This definite integral represents no area, but is just a real number. **7**

Finding Area

In summary, to find the area bounded by $y = f(x)$, the vertical lines $x = a$ and $x = b$, and the x-axis, use the following steps.

Step 1 Sketch a graph.

Step 2 Find any x-intercepts in $[a, b]$. These divide the total region into subregions.

Step 3 The definite integral will be *positive* for subregions above the x-axis and *negative* for subregions below the x-axis. Use separate integrals to find the areas of the subregions.

Step 4 The total area is the sum of the areas of all of the subregions.

In the last section, we saw that the area under a rate of change function $f'(x)$ from $x = a$ to $x = b$ gives the total value of $f(x)$ on $[a, b]$. Now we can use the definite integral to solve these problems.

▶ **EXAMPLE 7** The yearly rate of consumption of natural gas in trillions of cubic feet for a certain city is

$$C'(t) = t + e^{.01t},$$

when $t = 0$ corresponds to 1990. At this consumption rate, what is the total amount the city will use in the 10-year period of the nineties?

To find the consumption over the 10-year period starting in 1990, use the definite integral.

$$\int_0^{10} (t + e^{.01t})\, dt = \left(\frac{t^2}{2} + \frac{e^{.01t}}{.01}\right)\Big]_0^{10}$$
$$= (50 + 100e^{.1}) - (0 + 100)$$
$$\approx -50 + 100(1.10517) \approx 60.5$$

Therefore, a total of about 60.5 trillion cubic feet of natural gas will be used during the nineties if the consumption rate remains the same. ◀ **8**

13.4 EXERCISES

Evaluate each of the following definite integrals. (See Examples 1–4.)

1. $\displaystyle\int_{-2}^{4} (-dp)$

2. $\displaystyle\int_{-4}^{1} 6x\, dx$

3. $\displaystyle\int_{-1}^{2} (3t - 1)\, dt$

4. $\displaystyle\int_{-2}^{2} (4z + 3)\, dz$

5. $\displaystyle\int_{0}^{2} (5x^2 - 4x + 2)\, dx$

6. $\displaystyle\int_{-2}^{3} (-x^2 - 3x + 5)\, dx$

7. $\displaystyle\int_{4}^{9} 3\sqrt{u}\, du$

8. $\displaystyle\int_{2}^{8} \sqrt{2r}\, dr$

9. $\displaystyle\int_{0}^{1} 2(t^{1/2} - t)\, dt$

10. $\displaystyle\int_{0}^{4} -(3x^{3/2} + x^{1/2})\, dx$

11. $\displaystyle\int_{1}^{4} \left(5y\sqrt{y} + 3\sqrt{y}\right) dy$

12. $\displaystyle\int_{4}^{9} \left(4\sqrt{r} - 3r\sqrt{r}\right) dr$

13. $\displaystyle\int_{1}^{3} \frac{2}{x^2}\, dx$

14. $\displaystyle\int_{1}^{4} \frac{-3}{u^2}\, du$

15. $\displaystyle\int_{1}^{5} (5n^{-1} + n^{-3})\, dn$

16. $\displaystyle\int_{2}^{3} (3x^{-1} - x^{-4})\, dx$

17. $\displaystyle\int_{2}^{3} \left(2e^{-.1A} + \frac{3}{A}\right) dA$

18. $\displaystyle\int_{1}^{2} \left(\frac{-1}{B} + 3e^{.2B}\right) dB$

19. $\displaystyle\int_{1}^{2} \left(e^{5u} - \frac{1}{u^2}\right) du$

20. $\displaystyle\int_{.5}^{1} (p^3 - e^{4p})\, dp$

21. $\displaystyle\int_{-1}^{0} (2y - 3)^2\, dy$

22. $\displaystyle\int_{0}^{3} (4m + 2)^2\, dm$

23. $\displaystyle\int_{1}^{64} \frac{\sqrt{z} - 2}{\sqrt[3]{z}}\, dz$

24. $\displaystyle\int_{1}^{8} \frac{3 - y^{1/3}}{y^{2/3}}\, dy$

Use the definite integral to find the area between the x-axis and f(x) over the indicated interval.
Check first to see if the graph crosses the x-axis in the given interval. (See Examples 5 and 6.)

25. $f(x) = 2x + 3$; [8, 10]

26. $f(x) = 4x - 7$; [5, 10]

27. $f(x) = 2 - 2x^2$; [0, 5]

28. $f(x) = 9 - x^2$; [0, 6]

29. $f(x) = x^2 + 4x - 5$; [-1, 3]

30. $f(x) = x^2 - 6x + 5$; [-1, 4]

31. $f(x) = x^3$; [-1, 3]

32. $f(x) = x^3 - 2x$; [-2, 4]

33. $f(x) = e^x - 1$; [-1, 2]

34. $f(x) = 1 - e^{-x}$; [-1, 2]

35. $f(x) = \dfrac{1}{x}$; [1, e]

36. $f(x) = \dfrac{1}{x}$; [e, e²]

Use substitution to find the following definite integrals. (See Example 4.)

37. $\displaystyle\int_{0}^{1} 2x(x^2 + 1)^3\, dx$

38. $\displaystyle\int_{0}^{1} y^2(y^3 - 4)^3\, dy$

39. $\displaystyle\int_{0}^{4} \left(\sqrt{x^2 + 12x}\right)(x + 6)\, dx$

40. $\displaystyle\int_{6}^{8} \left(\sqrt{x^2 - 6x}\right)(x - 3)\, dx$

41. $\displaystyle\int_{-1}^{e-2} \frac{t}{t^2 + 2}\, dt$

42. $\displaystyle\int_{-2}^{0} \frac{-4x}{x^2 + 3}\, dx$

43. $\displaystyle\int_{0}^{1} ze^{2z^2}\, dz$

44. $\displaystyle\int_{1}^{2} x^2 e^{-x^3}\, dx$

Find the area of each shaded region.

45.

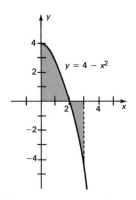

46.

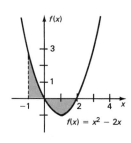

47.

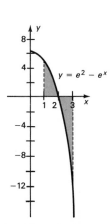

48.

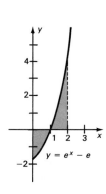

Work the following exercises. (See Example 7.)

49. Natural Science   A chemical plant is polluting the air at a rate of

$$p(t) = 15t^{3/2} + 10$$

(in appropriate units), where t is time in months. How much pollution is produced by the plant in 25 months?

50. Management   An oil tanker is leaking oil at the rate of $20t + 50$ barrels per hour, where t is time in hours after the tanker hits a hidden rock (when $t = 0$).
(a) Find the total number of barrels that the ship will leak on the first day.
(b) Find the number of barrels that the ship will leak on the second day.

51. Management   De Win Enterprises has found that its expenditure rate per day (in hundreds of dollars) on a certain type of job is given by

$$E(x) = 4x + 2,$$

where x is the number of days since the start of the job.
(a) Find the total expenditure if the job takes 10 days.
(b) How much will be spent on the job from the 10th to the 20th day?
(c) If the company wants to spend no more than $5000 on the job, in how many days must they complete it?

52. Management   De Win Enterprises (see Exercise 51) also knows that the rate of income per day (in hundreds of dollars) for the same job is

$$I(x) = 100 - x,$$

where x is the number of days since the job was started.
(a) Find the total income for the first 10 days.
(b) Find the income from the 10th to the 20th day.
(c) How many days must the job last for the total income to be at least $5000?

53. Natural Science   The rate at which a substance grows is given by

$$R(x) = 200e^x,$$

where x is the time in days. What is the total accumulated growth after 2.5 days?

54. Natural Science   A drug has a reaction rate in appropriate units of

$$R(t) = \frac{4}{t^2} + 1,$$

where t is measured in hours after the drug is administered. Find the total reaction to the drug
(a) from $t = 1/4$ to $t = 1$;
(b) from $t = 12$ to $t = 24$.

55. Natural Science   For a certain drug, the rate of reaction in appropriate units is given by

$$R'(t) = \frac{5}{t} + \frac{2}{t^2},$$

where t is measured in hours after the drug is administered. Find the total reaction to the drug over the following time periods.
(a) From $t = 1$ to $t = 12$
(b) From $t = 12$ to $t = 24$

56. Natural Science   The speed s of the blood in a blood vessel is given by

$$s = k(R^2 - r^2),$$

where R is the (constant) radius of the blood vessel, r is the distance of the flowing blood from the center of the blood vessel, and k is a constant. Total blood flow in millimeters per minute is given by

$$Q(R) = \int_0^R 2\pi sr \, dr.$$

(a) Find the general formula for Q in terms of R by evaluating the definite integral given above.
(b) Evaluate $Q(.4)$.

57. Management   Suppose that the rate of consumption of a natural resource is $c(t)$, where

$$c'(t) = ke^{rt}.$$

Here t is time in years, r is a constant, and k is the consumption in the year when $t = 0$. In 1990, an oil company sold 1.2 billion barrels of oil. Assume that $r = .04$.
(a) Set up a definite integral for the amount of oil that the company will sell in the next 10 years.
(b) Evaluate the definite integral of part (a).
(c) The company has about 20 billion barrels of oil in reserve. To find the number of years that this amount will last, solve the equation

$$\int_0^T 1.2e^{.04t} \, dt = 20.$$

(d) Rework part (c), assuming that $r = .02$.

58. Management A mine begins producing at time $t = 0$. After t years, the mine is producing at the rate of

$$P'(t) = 10t - \frac{15}{\sqrt{t}}$$

tons per year. Write an expression for the total output of the mine from year 1 to year T.

59. Management The rate of consumption of one natural resource in one country is

$$C'(t) = 72e^{.014t},$$

where $t = 0$ corresponds to 1990. How much of the resource will be used, altogether, from 1990 to year T?

60. Management In Exercise 57, the rate of consumption of oil (in billions of barrels) was given as

$$1.2e^{.04t},$$

where $t = 0$ corresponds to 1990. Find the total amount of oil used from 1990 to year T. At this rate, how much will be used in 5 years?

13.5 APPLICATIONS OF INTEGRALS

Given a function that represents the total maintenance charge for a machine from the time it is installed, the rate of maintenance at any time t is given by the derivative of the total maintenance charge. Now, the total maintenance function can be found by finding the antiderivative of the rate of maintenance function. As we have seen in this chapter, if we graph the function giving the rate of maintenance, then the total maintenance is given by the area under the curve. Example 1 shows how this works.

▶**EXAMPLE 1** Suppose a leasing company wants to decide on the yearly lease fee for a certain new typewriter. The company expects to lease the typewriter for 5 years, and it expects the *rate* of maintenance, $M(t)$ (in dollars), which it must supply, to be approximated by

$$M(t) = 10 + 2t + t^2,$$

where t is the number of years the typewriter has been used. How much should the company charge for maintenance on the typewriter for the 5-year period?

The definite integral can be used to find the total maintenance charge the company can expect over the life of the typewriter. Figure 13.17 shows the graph of $M(t)$. The total maintenance charge for the 5-year period will be given by the shaded area of the figure, which can be found as follows.

$$\int_0^5 (10 + 2t + t^2) \, dt = \left(10t + t^2 + \frac{t^3}{3} \right) \Bigg]_0^5$$

$$= 50 + 25 + \frac{125}{3} - 0$$

$$\approx 116.67$$

1 Find the total maintenance charge for a lease of

(a) 1 year;

(b) 2 years.

Answer:

(a) $11.33

(b) $26.67

2 Suppose the rate of change of cost in dollars to produce x items is given by

$$C(x) = x^2 - 8x + 15.$$

Find the total cost to produce the first 10 items.

Answer:
$83.33

3 Find the amount of savings over the first 6 years for the machine in Example 2.

Answer:
$54

The company can expect the total maintenance charge for 5 years to be about $117. Hence, the company should add about

$$\frac{\$117}{5} = \$23.40$$

to its annual lease price. ◀ **1** **2**

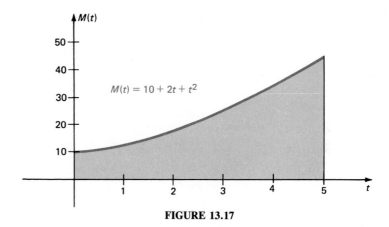

FIGURE 13.17

▶**EXAMPLE 2**   Elizabeth Wilson, who runs a factory that makes signs, has been shown a new machine that staples the signs to the handles. She estimates that the rate of savings, $S(x)$, from the machine will be approximated by

$$S(x) = 3 + 2x,$$

where x represents the number of years the stapler has been in use. If the machine costs $70, would it pay for itself in 5 years?

We need to find the area under the rate of savings curve shown in Figure 13.18 between the lines $x = 0$, $x = 5$, and the x-axis. Using a definite integral gives

$$\int_0^5 (3 + 2x)\, dx = 3x + x^2 \Big]_0^5 = 40.$$

The total savings in 5 years is $40, so the machine will not pay for itself in this time period. ◀ **3**

4 Find the number of years in which the machine in Example 2 will pay for itself if the rate of savings is $S(x) = 5 + 4x$.

Answer:
4.8 years

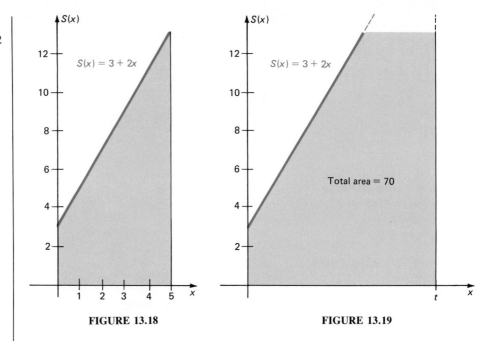

FIGURE 13.18   **FIGURE 13.19**

Since the machine in Example 2 costs a total of $70, it will pay for itself when the area under the savings curve of Figure 13.19 equals 70, or at a time t such that

$$\int_0^t (3 + 2x)\, dx = 70.$$

Evaluating the definite integral gives

$$3x + x^2 \Big]_0^t = (3t + t^2) - (3 \cdot 0 + 0^2) = 3t + t^2.$$

Since the total savings must equal 70,

$$3t + t^2 = 70.$$

Solve this quadratic equation to verify that the solutions are 7 and -10. Since -10 cannot be used here, the machine will pay for itself in 7 years.   **4**

▶ **EXAMPLE 3**   Wilson, of the previous example, believes that the rate of sales of her signs, $T(x)$, is given by

$$T(x) = 15 + 10x,$$

where x is the number of years she has been in business. When can she expect to sell her 1000th sign?

5 In Example 3 when will Wilson sell

(a) the 500th sign?

(b) the 1200th sign?

Answer:

(a) 8.61 years

(b) 14.06 years

6 Suppose the rate of sales in Example 3 is

$$T(x) = 40x - 350.$$

When will Wilson sell her 1000th sign?

Answer:
In 20 years

Let t be the time at which Wilson will sell her 1000th sign. The area under the sales curve between $x = 0$ and $x = t$ gives the total sales during that period. The total sales must be 1000, so

$$\int_0^t (15 + 10x)\, dx = 1000.$$

Since

$$\int_0^t (15 + 10x)\, dx = 15x + 5x^2 \Big]_0^t = 15t + 5t^2,$$

solve the equation

$$15t + 5t^2 = 1000$$
$$5t^2 + 15t - 1000 = 0.$$

Divide both sides of this equation by 5 to get

$$t^2 + 3t - 200 = 0,$$

from which (using the quadratic formula) the only meaningful solution is

$$t = 12.72 \text{ years.}$$

Wilson can expect to sell her 1000th sign about 12 years and 9 months after she goes into business. ◄ **5** **6**

In Section 13.3 we saw that a definite integral can be used to find the area *under* a curve. This idea can be extended to find the area *between* two curves, as shown in the next example.

►**EXAMPLE 4**   A company is considering a new manufacturing process in one of its plants. The new process provides substantial initial savings, with the savings declining with time x according to the rate-of-savings function

$$S(x) = 100 - x^2,$$

where $S(x)$ is in thousands of dollars. At the same time, the cost of operating the new process increases with time x, according to the rate-of-cost function (in thousands of dollars)

$$C(x) = x^2 + \frac{14}{3}x.$$

(a) For how many years will the company realize savings?

Figure 13.20 shows the graphs of the rate-of-savings and the rate-of-cost functions. The rate-of-cost (marginal cost) is increasing, while the rate-of-savings (marginal savings) is decreasing. The company should use this new process until the difference between these quantities is zero—that is, until the time at which these graphs intersect. The graphs intersect when

$$C(x) = S(x),$$

or

$$100 - x^2 = x^2 + \frac{14}{3}x.$$

Solve this equation as follows.

$$0 = 2x^2 + \frac{14}{3}x - 100$$

$$= 3x^2 + 7x - 150 \qquad \textbf{Multiply by 3/2}$$

$$= (x - 6)(3x + 25) \qquad \textbf{Factor}$$

Set each factor equal to 0 and solve both equations to get

$$x = 6 \quad \text{or} \quad x = -\frac{25}{3}.$$

Only 6 is a meaningful solution here. The company should use the new process for 6 years.

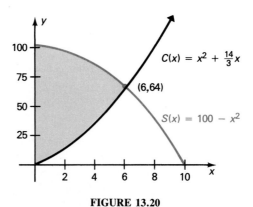

FIGURE 13.20

(b) What will be the net total savings during this period?

Since the total savings over the 6-year period is given by the area under the rate-of-savings curve and the total additional cost by the area under the rate-of-cost curve, the net total savings over the 6-year period is given by the area between the

7 In Example 4, find the total savings if pollution control regulations permit the new process for only 4 years.

Answer:
$320,000

8 If a company's marginal savings and marginal cost functions are

$$S(x) = 150 - 2x^2$$

and

$$C(x) = x^2 + 15x,$$

where $S(x)$ and $C(x)$ give amounts in thousands of dollars, find the following.

(a) Number of years until savings equal costs

(b) Total savings

Answer:

(a) 5

(b) $437,500

rate-of-cost and the rate-of-savings curves and the lines $x = 0$ and $x = 6$. This area can be evaluated with a definite integral as follows.

$$\text{Total savings} = \int_0^6 \left[(100 - x^2) - \left(x^2 + \frac{14}{3}x\right)\right] dx$$

$$= \int_0^6 \left(100 - \frac{14}{3}x - 2x^2\right) dx \qquad \text{Combine terms}$$

$$= 100x - \frac{7}{3}x^2 - \frac{2}{3}x^3 \Big]_0^6 \qquad \text{Integrate}$$

$$= 100(6) - \frac{7}{3}(36) - \frac{2}{3}(216) = 372.$$

The company will save a total of $372,000 over the 6-year period. ◀ **7** **8**

▶**EXAMPLE 5**   A farmer has been using a new fertilizer that gives him a better yield, but because it exhausts the soil of other nutrients he must use other fertilizers in greater and greater amounts, so that his costs increase each year. The new fertilizer produces a rate of increase in revenue (in hundreds of dollars) given by

$$R(t) = -.4t^2 + 8t + 10,$$

where t is measured in years. The rate of increase in yearly costs (also in hundreds of dollars) due to use of the fertilizer is given by

$$C(t) = 2t + 5.$$

How long can the farmer profitably use the fertilizer? What will be his net increase in revenue over this period?

The farmer should use the new fertilizer until the marginal costs equal the marginal revenue. Find this point by solving the equation $R(t) = C(t)$ as follows.

$$-.4t^2 + 8t + 10 = 2t + 5$$
$$-4t^2 + 80t + 100 = 20t + 50 \qquad \text{Multiply by 10}$$
$$-4t^2 + 60t + 50 = 0$$

By the quadratic formula the only positive solution is

$$t = 15.8.$$

The new fertilizer will be profitable for about 15.8 years.

To find the total amount of additional revenue over the 15.8-year period, find the area between the graphs of the rate of revenue and the rate of cost functions, as shown in Figure 13.21.

$$\text{Total savings} = \int_0^{15.8} [R(t) - C(t)] \, dt$$

$$= \int_0^{15.8} [(-.4t^2 + 8t + 10) - (2t + 5)] \, dt$$

$$= \int_0^{15.8} (-.4t^2 + 6t + 5) \, dt \qquad \qquad \textbf{Combine terms}$$

$$= \left(\frac{-.4t^3}{3} + \frac{6t^2}{2} + 5t \right) \Bigg]_0^{15.8} \qquad \qquad \textbf{Integrate}$$

$$= 302.01$$

The total savings will amount to about $30,000 over the 15.8-year period.

It is not realistic to say that the farmer will need to use the new process for 15.8 years—he will probably have to use it for 15 years or for 16 years. In this case, when the mathematical result is not in the domain of the function, it will be necessary to find the total savings after 15 years and after 16 years and then select the best result. ◄

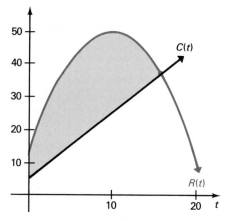

FIGURE 13.21

▶**EXAMPLE 6**   Suppose the price, in cents, for a certain product is

$$p(x) = 900 - 20x - x^2,$$

when the demand for the product is x units. Also, suppose the function

$$p(x) = x^2 + 10x$$

gives the price, in cents, when the supply is x units. The graphs of both functions are shown in Figure 13.22, along with the equilibrium point at which supply and demand are equal. To find the equilibrium supply or demand x, solve the equation

$$900 - 20x - x^2 = x^2 + 10x$$
$$0 = 2x^2 + 30x - 900$$
$$0 = x^2 + 15x - 450.$$

The only positive solution of the equation is $x = 15$.

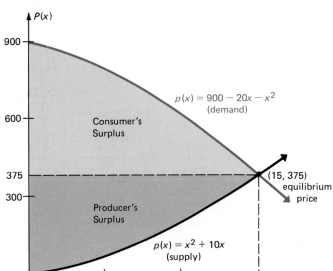

FIGURE 13.22

At the equilibrium point where supply and demand are both 15 units, the price is

$$p(15) = 900 - 20(15) - 15^2 = 375,$$

or \$3.75. As the demand graph shows, there are consumers who are willing to pay more than the equilibrium price, so they benefit from the equilibrium price. Some benefit a lot, some less, and so on. Their total benefit, called the **consumer's surplus,** is represented by the area shown in Figure 13.22. The consumer's surplus is

$$\int_0^{15} (900 - 20x - x^2)\, dx - \int_0^{15} 375\, dx.$$

Evaluating these definite integrals gives

$$\left(900x - 10x^2 - \frac{1}{3}x^3\right)\Bigg]_0^{15} - 375x\Bigg]_0^{15}$$

$$= \left[900(15) - 10(15)^2 - \frac{1}{3}(15)^3 - 0\right] - [(375)(15) - 0]$$

$$= 4500.$$

Here the consumer's surplus is 4500¢, or \$45.

9 Given the demand function $p(x) = 12 - .07x$ and the supply function $p(x) = .05x$, where $p(x)$ is the price in dollars, find

(a) the equilibrium point;

(b) the consumer's surplus;

(c) the producer's surplus.

Answer:

(a) $x = 100$ **(b)** $350   **(c)** $250

On the other hand, some suppliers would have offered the product at a price below the equilibrium price, so they too gain from the equilibrium price. The total of the supplier's gains, called the **producer's surplus,** is also represented by an area shown in Figure 13.22. The producer's surplus is given by

$$\int_0^{15} 375 \, dx - \int_0^{15} (x^2 + 10x) \, dx = 375x \Big]_0^{15} - \left(\frac{1}{3}x^3 + 5x^2\right)\Big]_0^{15}$$

$$= 5625 - \left[\frac{1}{3}(15)^3 + 5(15)^2\right]$$

$$= 3375.$$

The producer's surplus is $33.75. ◀ **9**

13.5 EXERCISES

Management   *Work the following exercises. (See Examples 1–5.)*

1. A car-leasing firm must decide how much to charge for maintenance on the cars it leases. After careful study, the firm decides that the rate of maintenance, $M(x)$, on a new car will approximate

$$M(x) = 60(1 + x^2),$$

where x is the number of years the car has been in use. What total maintenance charge can the company expect for a 2-year lease? What amount should be added to the monthly lease payments to pay for maintenance?

2. Using the function of Exercise 1, find the maintenance charge the company can expect during the third year. Find the total charge during the first 3 years. What monthly charge should be added to cover a 3-year lease?

3. A company is considering a new manufacturing process. It knows that the rate of savings from the process, $S(t)$, will be about

$$S(t) = 1000(t + 2),$$

where t is the number of years the process has been in use. Find the total savings during the first year. Find the total savings during the first 6 years.

4. Assume that the new process in Exercise 3 costs $16,000. About when will it pay for itself?

5. A company is introducing a new product. Production is expected to grow slowly because of difficulties in the start-up process. It is expected that the rate of production, $P(x)$, will be approximated by

$$P(x) = 1000e^{.2x},$$

where x is the number of years since the introduction of the product. Will the company be able to supply 20,000 units during the first 4 years?

6. About when will the company of Exercise 5 be able to supply its 15,000th unit?

7. Suppose a company wants to introduce a new machine that will yield a rate of savings given by

$$S(x) = 294 - x^2,$$

where x is the number of years of operation of the machine, but will also entail costs of

$$C(x) = x^2 + \frac{1}{2}x.$$

(a) For how many years will it be profitable to use this new machine?

(b) What are the total net savings during the first year of use of the machine?

(c) What are the total net savings over the entire period of profitable life of the machine?

8. A farmer has been using a new fertilizer that gives him a better yield. Because it exhausts the soil of other nutrients, he must use other fertilizers in greater and greater amounts, so that his costs also increase each year. The new fertilizer produces an increase in rate of revenue (in hundreds of dollars) of

$$K(t) = -.4t^2 + 8t + 10,$$

where t is measured in years. The yearly rate of costs due to use of the fertilizer increase according to the function

$$C(t) = 2t + 5.$$

(a) How long can the farmer profitably use the fertilizer?
(b) What will his net savings be over this period?

9. De Win Enterprises had an expenditure rate (in hundreds of dollars) of $E(x) = 4x + 2$, and an income rate (in hundreds of dollars) of $I(x) = 100 - x$ on a particular job, where x was the number of days from the start of the job. Their profit on that job will equal total income less total expenditure. Profit will be maximized if the job ends at the optimum time, which is the point where the two curves meet. Find the following.
(a) The optimum number of days for the job to last.
(b) The total income for the optimum number of days.
(c) The total expenditure for the optimum number of days.
(d) The maximum profit for the job.

10. A factory has installed a new process that will produce an increased rate of revenue (in thousands of dollars) of

$$R(t) = -t^2 + 15.5t + 16,$$

where t is measured in years. The new process produces additional costs (in thousands of dollars) at the rate of

$$C(t) = .25t + 4.$$

(a) When will it no longer be profitable to use this new process?
(b) Find the total net revenue.

Management *Work the following supply and demand exercises, where the price is given in dollars. (See Example 6.)*

11. Suppose the supply function for a certain commodity is given by

$$p(x) = 100 + 3x + x^2.$$

Suppose that supply and demand are in equilibrium when $x = 3$. Find the producer's surplus.

12. Find the consumer's surplus if the demand function for an item is given by

$$p(x) = 50 - x^2,$$

assuming supply and demand are in equilibrium at $x = 5$.

13. Find the consumer's surplus if the demand price for an item is given by

$$p(x) = -x^2 - 8x + 50,$$

if supply and demand are in equilibrium at $x = 3$.

14. Find the producer's surplus if the supply function for an item is

$$p(x) = \frac{1}{2}x^2 + 10x.$$

Assume supply and demand are in equilibrium when $x = 5$.

15. Find the producer's surplus if the supply function of some item is given by

$$p(x) = x^2 + 2x + 50.$$

Assume supply and demand are in equilibrium at $x = 20$.

16. Find the consumer's surplus if the demand for an item is given by

$$p(x) = -(x + 4)^2 + 66,$$

if supply and demand are in equilibrium at $x = 3$.

17. Find the consumer's surplus and the producer's surplus for an item having supply function

$$p(x) = x^2 + \frac{11}{4}x$$

and demand function

$$p(x) = 150 - x^2.$$

18. Suppose the supply function of a certain item is given by

$$p(x) = \frac{7}{5}x,$$

and the demand function is given by

$$p(x) = -\frac{3}{5}x + 10.$$

(a) Graph the supply and demand curves.
(b) Find the point at which supply and demand are in equilibrium.
(c) Find the consumer's surplus.
(d) Find the producer's surplus.

Work the following exercises.

19. Management In a recent inflationary period, costs of a certain industrial process, in millions of dollars, were increasing according to the function

$$i(t) = .45t^{3/2},$$

where $i(t)$ is the rate of increase in costs at time t measured in years. Find the total increase in costs during the first 4 years.

20. Management A new smog control device will reduce the output of oxides of sulfur from automobile exhausts. It is estimated that the rate of savings to the community from the use of this device will be approximated by

$$S(x) = -x^2 + 4x + 8,$$

where $S(x)$ is the rate of savings in millions of dollars after x years of use of the device. This new device cuts down on the production of oxides of sulfur, but causes an increase in the production of oxides of nitrogen. The additional costs to the community caused by this increase are approximated by

$$C(x) = \frac{3}{25}x^2,$$

where $C(x)$ is the additional cost in millions after x years.
(a) For how many years will it pay to use the new device?
(b) What will be the total savings over this period of time?

21. Management If a large truck has been driven x thousand miles, the rate of repair costs in dollars per mile is given by $R(x)$, where

$$R(x) = .05x^{3/2}.$$

Find the total repair costs if the truck is driven
(a) 100,000 miles;
(b) 400,000 miles.

22. Natural Science Pollution from a factory is entering a lake. The rate of concentration, $P(t)$, of the pollutant at time t, is given by

$$P(t) = 140t^{5/2},$$

where t is the number of years since the factory started spilling pollutant into the lake. Ecologists estimate that the lake can accept a total level of pollution of 4850 units before all the fish in the lake die. Can the factory operate for 4 years without killing all the fish in the lake?

23. Natural Science A program is started to clean up the lake in Exercise 22. The cleanup efforts result in changing the rate of concentration $P(t)$ of the pollutant at time t to

$$P(t) = 4000e^{-t/10},$$

where t is the number of years from the time the program is started. Suppose there are 2000 units of pollution in the lake when the cleanup program begins. Can the factory operate an additional 4 years without killing all the fish in the lake?

24. Natural Science From 1905 to 1920, most of the predators of the Kaibab Plateau of Arizona were killed by hunters. This allowed the deer population there to grow rapidly until they had depleted their food sources, which caused a rapid decline in population. The rate of change of this deer population during that time span is approximated by the function

$$D(t) = \frac{25}{2}t^3 - \frac{5}{8}t^4,$$

where t is the time in years ($0 \le t \le 25$).
(a) Find the function for the deer population if there were 4000 deer in 1905 ($t = 0$).
(b) What was the population in 1920?
(c) When was the population at a maximum?
(d) What was the maximum population?

25. Management After a new firm starts in business, it finds that its rate of profits (in hundreds of dollars) after t years of operation, is given by

$$P(t) = 6t^2 + 4t + 5.$$

Find the total profits in the first 3 years.

26. Management Find the profit in Exercise 25 in the fourth year of operation.

27. An oil tanker is leaking oil at the rate of $20t + 50$ barrels per hour, where t is time in hours after the tanker hits a hidden rock. Find the total number of barrels that the ship will leak on the first day.

28. Find the number of barrels that the ship of Exercise 27 will leak on the second day.

29. Management A worker new to a job will improve his efficiency with time so that it takes him fewer hours to produce an item with each day on the job up to a certain point. Suppose the rate of change of the number of hours it takes a worker in a certain factory to produce the xth item is given by

$$H(x) = 20 - 2x.$$

(a) What is the total number of hours required to produce the first 5 items?
(b) What is the total number of hours required to produce the first 10 items?

30. Natural Science After long study, tree scientists conclude that a eucalyptus tree will grow at the rate of $.2 + 4t^{-4}$ feet per year, where t is time in years. Find the number of feet that the tree will grow in the second year.

31. **Natural Science** Find the number of feet the tree in Exercise 30 will grow in the third year.

32. **Natural Science** For a certain drug, the rate of reaction in appropriate units is given by

$$R(t) = \frac{5}{t} + \frac{2}{t^2},$$

where t is measured in hours after the drug is administered. Find the total reaction to the drug
(a) from $t = 1$ to $t = 12$;
(b) from $t = 12$ to $t = 24$.

33. **Social Science** Suppose that all the people in a country are ranked according to their incomes, starting at the bottom. Let x represent the fraction of the community making the lowest income ($0 \le x \le 1$); $x = .4$, therefore, represents the lower 40% of all income producers. Let $I(x)$ rep-

resent the proportion of the total income earned by the lowest x of all people. Thus, $I(.4)$ represents the fraction of total income earned by the lowest 40% of the population. Suppose

$$I(x) = .9x^2 + .1x.$$

Find and interpret the following.
(a) $I(.1)$ (b) $I(.4)$ (c) $I(.6)$ (d) $I(.9)$

34. **Social Science** In Exercise 33, if income were distributed uniformly, we would have $I(x) = x$. The area between the curves $I(x) = x$ and the particular function $I(x)$ for a given country is called the *coefficient of inequality* for that country.
(a) Graph $I(x) = x$ and $I(x) = .9x^2 + .1x$ for $0 \le x \le 1$ on the same axes.
(b) Find the area between the curves.

13.6 TABLES OF INTEGRALS

As we have noted, there are many useful functions whose antiderivatives cannot be found by the methods we have discussed. In fact, some definite integrals cannot be evaluated exactly, but can be approximated by methods similar to the method of summing rectangles discussed in Section 13.3. Many useful integrals can be found from a table of integrals such as Table 9 at the back of the book. The next examples show how to use this table.

▶**EXAMPLE 1**   Find $\displaystyle\int \frac{1}{\sqrt{x^2 + 16}} \, dx$.

By inspecting the table, we see that if $a = 4$, this antiderivative is the same as entry 5 of the table. Entry 5 of the table is

$$\int \frac{1}{\sqrt{x^2 + a^2}} \, dx = \ln \left| \frac{x + \sqrt{x^2 + a^2}}{a} \right| + C.$$

Substituting 4 for a in this entry, we get

$$\int \frac{1}{\sqrt{x^2 + 16}} \, dx = \ln \left| \frac{x + \sqrt{x^2 + 16}}{4} \right| + C.$$

This last result could be verified by taking the derivative of the right-hand side of this last equation. ◀ **1**

1 Find the following.

(a) $\displaystyle\int \frac{4}{\sqrt{x^2 + 100}} \, dx$

(b) $\displaystyle\int \frac{-9}{\sqrt{x^2 - 4}} \, dx$

(c) $\displaystyle\int -6 \ln |x| \, dx$

Answer:

(a) $4 \ln \left| \dfrac{x + \sqrt{x^2 + 100}}{10} \right| + C$

(b) $-9 \ln \left| \dfrac{x + \sqrt{x^2 - 4}}{2} \right| + C$

(c) $-6x(\ln |x| - 1) + C$

2 Find the following.

(a) $\displaystyle\int \frac{1}{x^2 - 4}\, dx$

(b) $\displaystyle\int \frac{-6}{x\sqrt{25 - x^2}}\, dx$

(c) $\displaystyle\int \frac{5}{x\sqrt{36 + x^2}}\, dx$

Answer:

(a) $\dfrac{1}{4} \ln \left(\dfrac{x - 2}{x + 2}\right) + C;\ (x^2 > 4)$

(b) $\dfrac{6}{5} \ln \left(\dfrac{5 + \sqrt{25 - x^2}}{x}\right) + C;$

$(0 < x < 5)$

(c) $-\dfrac{5}{6} \ln \left|\dfrac{6 + \sqrt{36 - x^2}}{x}\right| + C$

3 Find the following.

(a) $\displaystyle\int \frac{3}{16x^2 - 1}\, dx$

(b) $\displaystyle\int \frac{-1}{100x^2 - 1}\, dx$

Answer:

(a) $\dfrac{3}{8} \ln \left|\dfrac{x - \frac{1}{4}}{x + \frac{1}{4}}\right| + C;$

$\left(x^2 > \dfrac{1}{16}\right)$

(b) $-\dfrac{1}{20} \ln \left|\dfrac{x - \frac{1}{10}}{x + \frac{1}{10}}\right| + C;$

$\left(x^2 > \dfrac{1}{100}\right)$

▶**EXAMPLE 2** Find $\displaystyle\int \frac{8}{16 - x^2}\, dx$.

Convert this antiderivative into the one given in entry 7 of the table by writing the 8 in front of the integral sign (permissible only with constants) and by letting $a = 4$. Doing this gives

$$8 \int \frac{1}{16 - x^2}\, dx = 8\left[\frac{1}{2 \cdot 4} \ln \left(\frac{4 + x}{4 - x}\right)\right] + C$$

$$= \ln \left(\frac{4 + x}{4 - x}\right) + C.$$

In entry 7 of the table, the condition $x^2 < a^2$ is given. Here $a = 4$, so the result given above is valid only for $x^2 < 16$, with the final answer written as

$$\int \frac{8}{16 - x^2}\, dx = \ln \left(\frac{4 + x}{4 - x}\right) + C, \quad \text{for } x^2 < 16.$$

Because of the condition $x^2 < 16$, the expression in parentheses is always positive, so that absolute value bars are not needed. ◀ **2**

▶**EXAMPLE 3** Find $\displaystyle\int \sqrt{9x^2 + 1}\, dx$.

This antiderivative seems most similar to entry 15 of the table. However, entry 15 requires that the coefficient of the x^2 term be 1. We can satisfy that requirement here by factoring out the 9.

$$\int \sqrt{9x^2 + 1}\, dx = \int \sqrt{9\left(x^2 + \frac{1}{9}\right)}\, dx$$

$$= \int 3\sqrt{x^2 + \frac{1}{9}}\, dx$$

$$= 3 \int \sqrt{x^2 + \frac{1}{9}}\, dx$$

Now, using entry 15 with $a = 1/3$,

$$\int \sqrt{9x^2 + 1}\, dx = 3\left[\frac{x}{2}\sqrt{x^2 + \frac{1}{9}} + \frac{\left(\frac{1}{3}\right)^2}{2} \cdot \ln \left|x + \sqrt{x^2 + \frac{1}{9}}\right|\right] + C$$

$$= \frac{3x}{2}\sqrt{x^2 + \frac{1}{9}} + \frac{1}{6} \ln \left|x + \sqrt{x^2 + \frac{1}{9}}\right| + C. \quad ◀ \ \mathbf{3}$$

13.6 EXERCISES

Use the table of integrals to find each antiderivative. (See Examples 1–3.)

1. $\int \dfrac{-4}{\sqrt{x^2 + 36}}\, dx$

2. $\int \dfrac{9}{\sqrt{x^2 + 9}}\, dx$

3. $\int \dfrac{6}{x^2 - 9}\, dx$

4. $\int \dfrac{-12}{x^2 - 16}\, dx$

5. $\int \dfrac{-4}{x\sqrt{9 - x^2}}\, dx$

6. $\int \dfrac{3}{x\sqrt{121 - x^2}}\, dx$

7. $\int \dfrac{-2x}{3x + 1}\, dx$

8. $\int \dfrac{6x}{4x - 5}\, dx$

9. $\int \dfrac{2}{3x(3x - 5)}\, dx$

10. $\int \dfrac{-4}{3x(2x + 7)}\, dx$

11. $\int \dfrac{4}{4x^2 - 1}\, dx$

12. $\int \dfrac{-6}{9x^2 - 1}\, dx$

13. $\int \dfrac{3}{x\sqrt{1 - 9x^2}}\, dx$

14. $\int \dfrac{-2}{x\sqrt{1 - 16x^2}}\, dx$

15. $\int \dfrac{4x}{2x + 3}\, dx$

16. $\int \dfrac{4x}{6 - x}\, dx$

17. $\int \dfrac{-x}{(5x - 1)^2}\, dx$

18. $\int \dfrac{-3}{x(4x + 3)^2}\, dx$

19. $\int x^4 \ln |x|\, dx$

20. $\int 4x^2 \ln |x|\, dx$

21. $\int \dfrac{\ln |x|}{x^2}\, dx$

22. $\int \dfrac{-2 \ln |x|}{x^3}\, dx$

23. $\int xe^{-2x}\, dx$

24. $\int xe^{3x}\, dx$

Use Table 9 to solve the following problems.

25. Management The rate of change of revenue in dollars from the sale of x units of small desk calculators is
$$R'(x) = 10\sqrt{x^2 + 4}.$$
Find the total revenue from the sale of the first 20 calculators.

26. Natural Science The rate of reaction to a drug is given by
$$r'(x) = 2x^2 e^{-x},$$
where x is the number of hours since the drug was administered. Find the total reaction to the drug from $x = 1$ to $x = 6$.

27. Natural Science The rate of growth of a microbe population is given by
$$m'(x) = 30xe^{2x},$$
where x is time in days. What is the total accumulated growth after 3 days?

28. Social Science The rate (in hours per item) at which a worker in a certain job produces the xth item is
$$h'(x) = \dfrac{1}{\sqrt{x^2 + 1}}.$$
What is the total number of hours it will take this worker to produce the first 7 items?

13.7 DIFFERENTIAL EQUATIONS AND APPLICATIONS

The rate of change with respect to time of many natural growth or decay processes can be expressed as a function of time by writing

$$\dfrac{dy}{dt} = kt \quad \text{or} \quad \dfrac{dy}{dt} = ky,$$

where t represents time, $y = f(t)$, and k is a constant. Any equation containing one or more of the derivatives of a function is called a **differential equation.**

1 Find the general solution of

(a) $dy/dx = 4x$;

(b) $dy/dx = -x^3$;

(c) $dy/dx = 2x^2 - 5x$.

Answer:

(a) $y = 2x^2 + C$

(b) $y = -\dfrac{1}{4}x^4 + C$

(c) $y = \dfrac{2}{3}x^3 - \dfrac{5}{2}x^2 + C$

In most differential equations, it is convenient to represent the derivative as dy/dx, rather than y' or $f'(x)$. For a second derivative, the notation d^2y/dx^2 is used. Some other examples of differential equations are

$$\frac{dy}{dx} = x^{1/2}, \quad \frac{dy}{dx} = 2y^2, \quad \text{and} \quad \frac{d^2y}{dx^2} = 3x^2 + 4.$$

To solve a differential equation such as

$$\frac{dy}{dt} = kt, \tag{1}$$

we take antiderivatives on each side of the equation. Since the left side of the equation is the derivative of y with respect to t, the antiderivative on the left is just $y + C_1$. On the right side,

$$\int kt \, dt = \frac{1}{2}kt^2 + C_2,$$

since t is the variable and k is a constant. The solution of equation (1) is thus

$$y + C_1 = \frac{1}{2}kt^2 + C_2$$

or

$$y = \frac{1}{2}kt^2 + C_2 - C_1.$$

Replacing the constant $C_2 - C_1$ with the single constant C gives

$$y = \frac{1}{2}kt^2 + C. \tag{2}$$

Equation (2) is called the **general solution** of differential equation (1). (From now on, we will add just one constant with the understanding that it represents the difference between the two constants obtained in the integrations.) **1**

▶ **EXAMPLE 1** The time dating of dairy products depends on the solution of a differential equation. The rate of growth of bacteria in such products increases with time. If y is the number of bacteria (in thousands) present at a time t (in days), then the rate of growth of bacteria can be expressed as dy/dt and we have

$$\frac{dy}{dt} = kt,$$

2 Suppose the maximum allowable value for y in Example 1 is 550. How should the product be dated?

Answer:
10 days from the date when $t = 0$

where k is an appropriate constant. Suppose, for some product, $k = 10$, so that

$$\frac{dy}{dt} = 10t. \tag{3}$$

Further, suppose at $t = 0$, $y = 50$ (in thousands). Find an equation for y in terms of t.

To solve the differential equation $dy/dt = 10t$, find an antiderivative for each side and add just one constant. This gives

$$y = \int 10t\ dt,$$

$$y = \frac{10t^2}{2} + C, \quad \text{or} \quad y = 5t^2 + C,$$

the general solution of equation (3). To find a **particular solution,** where the value of C is known, use the fact that $y = 50$ when $t = 0$. This is called an **initial condition** or **boundary condition.** By substitution, we have

$$50 = 5(0)^2 + C$$
$$C = 50.$$

Thus, the particular solution is

$$y = 5t^2 + 50. \quad \blacktriangleleft \quad \boxed{2}$$

The simplest kind of differential equation has the form

$$\frac{dy}{dx} = f(x).$$

A more general differential equation is one that can be written in the form

$$\frac{dy}{dx} = \frac{f(x)}{g(y)}.$$

Using differentials and multiplying on both sides by $g(y)\ dx$ gives

$$g(y)\ dy = f(x)\ dx.$$

In this form all terms involving y (including dy) are on one side of the equation, and all terms involving x (and dx) are on the other side. A differential equation in this form is said to be **separable,** since the variables x and y can be separated. A separable differential equation may be solved by integrating each side.

3 Find the general solution of

$$y^2 \frac{dy}{dx} = x + 2.$$

Answer:

$$y^3 = \frac{3}{2}x^2 + 6x + C$$

▶ **EXAMPLE 2**   Find the general solution of $y \dfrac{dy}{dx} = x^2$.

Begin by separating the variables to get

$$y \, dy = x^2 \, dx.$$

The general solution is found by taking antiderivatives on each side.

$$\int y \, dy = \int x^2 \, dx$$

$$\frac{y^2}{2} = \frac{x^3}{3} + C$$

$$y^2 = \frac{2}{3}x^3 + 2C = \frac{2}{3}x^3 + K$$

The constant K was substituted for $2C$ in the last step. The solution is left in implicit form, not solved explicitly for y.   ◀   **3**

▶ **EXAMPLE 3**   Find the general solution of $\dfrac{dy}{dx} = ky$, where k is a constant.

Separating variables leads to

$$\frac{1}{y} \, dy = k \, dx.$$

To solve this equation, take antiderivatives on each side.

$$\int \frac{1}{y} \, dy = \int k \, dx$$

$$\ln |y| = kx + C$$

Use the definition of logarithm to write the equation in exponential form as

$$|y| = e^{kx+C}.$$

By properties of exponents,

$$|y| = e^{kx} e^C.$$

Finally, use the definition of absolute value to get

$$y = e^{kx} e^C \quad \text{or} \quad y = -e^{kx} e^C.$$

Since e^C and $-e^C$ are constants, replace them with the constant M, which may have any nonzero real-number value, to get the single equation

$$y = Me^{kx}.$$

This equation, $y = Me^{kx}$, defines the exponential growth or decay function that was discussed previously in Chapter 4.   ◀

4 Find the particular solution of $\dfrac{dy}{dx} = .05y$ if y is 2000 when x is 0.

Answer:
$y = 2000e^{.05x}$

Recall that equations of the form $y = Me^{kx}$ arise in situations where the rate of change of a quantity is proportional to the amount present at time x; that is, where

$$\frac{dy}{dx} = ky.$$

The constant k is called the **growth rate constant,** while M represents the amount present at time $x = 0$. (A positive value of k indicates growth, while a negative value of k indicates decay.) **4**

As a model of population growth, the equation $y = Me^{kx}$ is not realistic over the long run for most populations. As shown by graphs of functions of the form $y = Me^{kx}$, with both M and k positive, growth would be unbounded. Additional factors, such as space restrictions or a limited amount of food, tend to inhibit growth of populations as time goes on. In an alternative model that assumes a maximum population of size N, the rate of growth of a population is proportional to how close the population is to that maximum. These assumptions lead to the differential equation

$$\frac{dy}{dx} = k(N - y),$$

the limited growth function mentioned in Chapter 4. Graphs of limited growth functions look like the graph in Figure 13.23, where y_0 is the initial population.

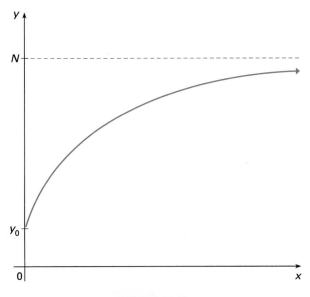

FIGURE 13.23

5 An animal population is growing at a constant rate of 4%. The habitat will support no more than 10,000 animals. There are 3000 animals present now. Write an equation giving the population y in x years.

Answer:
$y = 10{,}000 - 7000e^{-.04x}$

▶**EXAMPLE 4**   A certain area can support no more than 4000 mountain goats. There are 1000 goats in the area at present, with a growth constant of .20.

(a) Write a differential equation for the rate of growth of this population.

Let $N = 4000$ and $k = .20$. The rate of growth of the population is given by

$$\frac{dy}{dx} = .20(4000 - y).$$

To solve for y, first separate the variables.

$$\frac{dy}{4000 - y} = .2 \, dx$$

$$\int \frac{dy}{4000 - y} = \int .2 \, dx$$

$$-\ln (4000 - y) = .2x + C$$

$$\ln (4000 - y) = -.2x - C$$

$$4000 - y = e^{-.2x - C} = (e^{-.2x})(e^{-C})$$

The absolute value bars are not needed for $\ln (4000 - y)$ since y must be less than 4000 for this population so that $4000 - y$ is always nonnegative. Let $e^{-C} = B$. Then

$$4000 - y = Be^{-.2x}$$

$$y = 4000 - Be^{-.2x}.$$

Find B by using the fact that $y = 1000$ when $x = 0$.

$$1000 = 4000 - B$$

$$B = 3000$$

Notice that the value of B is the difference between the maximum population and the initial population. Substituting 3000 for B in the equation for y gives

$$y = 4000 - 3000e^{-.2x}.$$

(b) What will the goat population be in 5 years?

In 5 years, the population will be

$$y = 4000 - 3000e^{(-.2)(5)} = 4000 - 3000e^{-1}$$

$$= 4000 - 1103.6 = 2896.4,$$

or about 2900 goats. ◀ **5**

Marginal productivity is the rate at which production changes (increases or decreases) for a unit change in investment. Thus, marginal productivity can be expressed as the first derivative of the function which gives production in terms of investment.

6 In Example 5, if marginal productivity is changed to

$$P'(x) = 3x^2 + 2x,$$

with the same initial conditions,

(a) find an equation for production;

(b) find the increase in production if investment increases to $500,000.

Answer:

(a) $P(x) = x^3 + x^2 + 64$

(b) Production goes from 100 units to 214 units, an increase of 114 units.

▶ **EXAMPLE 5** Suppose the marginal productivity of a manufacturing process is given by

$$P'(x) = 3x^2 - 10,\qquad\qquad\text{(4)}$$

where x is the amount of the investment in hundreds of thousands of dollars. If the process produces 100 units per month with its present investment of $300,000 (that is, $x = 3$), by how much would production increase if the investment is increased to $500,000?

To obtain an equation for production, we can take antiderivatives on both sides of equation (4) to get

$$P(x) = x^3 - 10x + C.$$

To find C, use the given initial values: $P(x) = 100$ when $x = 3$.

$$100 = 3^3 - 10(3) + C$$
$$C = 103$$

Production is thus given by

$$P(x) = x^3 - 10x + 103,$$

and if investment is increased to $500,000, production becomes

$$P(5) = 5^3 - 10(5) + 103 = 178.$$

An increase to $500,000 in investment will increase production from 100 units to 178 units. ◀ **6**

13.7 EXERCISES

Find general solutions for the following differential equations. (See Example 2.)

1. $\dfrac{dy}{dx} = 8x - 7$

2. $\dfrac{dy}{dx} = -x + 2$

3. $\dfrac{dy}{dx} = -2x + 3x^2$

4. $\dfrac{dy}{dx} = 6x^2 - 4x$

5. $\dfrac{dy}{dx} = e^{4x}$

6. $\dfrac{dy}{dx} = e^{3x}$

7. $\dfrac{dy}{dx} = x\sqrt{1 - x^2}$

8. $\dfrac{dy}{dx} = x^2(1 + 2x^3)^{1/2}$

Find general solutions for the following differential equations. (See Examples 2 and 3.)

9. $y\dfrac{dy}{dx} = x$

10. $y\dfrac{dy}{dx} = x^2 - 1$

11. $\dfrac{dy}{dx} = 2xy$

12. $\dfrac{dy}{dx} = x^2y$

13. $\dfrac{dy}{dx} = 3x^2y - 2xy$

14. $(y^2 - y)\dfrac{dy}{dx} = x$

15. $\dfrac{dy}{dx} = \dfrac{y}{x}; x > 0$

16. $\dfrac{dy}{dx} = \dfrac{y}{x^2}$

17. $\dfrac{dy}{dx} = y - 5$

18. $\dfrac{dy}{dx} = 3 - y$

19. $\dfrac{dy}{dx} = y^2e^x$

20. $\dfrac{dy}{dx} = \dfrac{e^x}{e^y}$

Find particular solutions for the following equations. (See Example 1.)

21. $\dfrac{dy}{dx} = \dfrac{x^2}{y}$; $y = 3$ when $x = 0$

22. $x^2 \dfrac{dy}{dx} = y$; $y = -1$ when $x = 1$

23. $(2x + 3)y = \dfrac{dy}{dx}$; $y = 1$ when $x = 0$

24. $x \dfrac{dy}{dx} - y\sqrt{x} = 0$; $y = 1$ when $x = 0$

25. $\dfrac{dy}{dx} = \dfrac{y^2}{x}$; $y = 5$ when $x = e$

26. $\dfrac{dy}{dx} = x^{1/2}y^2$; $y = 12$ when $x = 4$

27. $\dfrac{dy}{dx} = \dfrac{2x + 1}{y - 3}$; $y = 4$ when $x = 0$

28. $\dfrac{dy}{dx} = \dfrac{x^2 + 5}{2y - 1}$; $y = 11$ when $x = 0$

Work the following problems. (See Examples 4 and 5.)

29. Management Suppose the marginal cost of producing x copies of a book is given by

$$\frac{dy}{dx} = \frac{50}{y\sqrt{x}}.$$

If $y = 150$ when $x = 4$, find the costs of producing the following numbers of books.
(a) 100 books   **(b)** 400 books   **(c)** 625 books.

30. Management Sales (in thousands) of a certain product are declining exponentially, with a decay constant of 25% a year.
(a) Write a differential equation to express the rate-of-sales decline.
(b) Find a general solution to the equation in part (a).

31. Management If inflation grows continuously at a rate of 6% per year, how long will it take for $1 to lose half its value?

32. Management Suppose that the gross national product (GNP) of a particular country increases exponentially, with a growth constant of 2% per year. Ten years ago, the GNP was 10^5 dollars. What will the GNP be in 5 years?

33. Management In a certain area, 1500 small business firms are threatened by bankruptcy. Assume the rate of change in the number of bankruptcies is proportional to the number of small firms that are not yet bankrupt. If the growth constant is 6% and if 100 firms are bankrupt initially, how many will be bankrupt in 2 years?

34. Management The sales of a new model of car are expected to follow the logistic curve

$$y = \frac{100,000}{1 + 100e^{-x}},$$

where x is the number of months since the introduction of the model. Find the following.

(a) The initial sales level
(b) The maximum sales level
(c) The month in which the maximum sales level will be achieved
(d) The rate at which sales are growing 6 months after the introduction of the model

35. Natural Science The amount of a tracer dye injected into the bloodstream decreases exponentially, with a decay constant of 3% per minute. If 6 cc are present initially, how many cc are present after 10 minutes? (Here k will be negative.)

36. Natural Science The rate at which the number of bacteria (in thousands) in a culture is changing after the introduction of a bactericide is given by

$$\frac{dy}{dx} = 50 - y,$$

where y is the number of bacteria (in thousands) present at time x. Find the number of bacteria present at each of the following times if there were 1000 thousand bacteria present at time $x = 0$.
(a) $x = 2$ **(b)** $x = 5$ **(c)** $x = 10$ **(d)** $x = 15$

37. Natural Science An isolated fish population is limited to 5000 by the amount of food available. If there are now 150 fish and the population is growing with a growth constant of 1% a year, find the expected population at the end of 5 years.

38. Natural Science A population of mites (in hundreds) increases exponentially, with a growth constant of 5% per week. At the beginning of an observation period there were 3000 mites. How many are present 4 weeks later?

39. Social Science Suppose the rate at which a rumor spreads—that is, the number of people who have heard the rumor over a period of time—increases with the number of people who have heard it. If y is the number of people who have heard the rumor, then

$$\frac{dy}{dt} = ky,$$

where t is the time in days.

(a) If y is 1 when $t = 0$, and y is 5 when $t = 2$, find k. Using the value of k from part (a), find y for each of the following times.

(b) $t = 3$ **(c)** $t = 5$ **(d)** $t = 10$

40. Social Science   A company has found that the rate at which a person new to the assembly line produces items is

$$\frac{dy}{dx} = 7.5e^{-.3y},$$

where x is the number of days the person has worked on the line. How many items can a new worker be expected to produce on the eighth day if he produces none when $x = 0$?

41. Physical Science   The amount of a radioactive substance decreases exponentially, with a decay constant of 5% a month.

(a) Write a differential equation to express the rate of change.

(b) Find a general solution to the differential equation from part (a).

(c) If there are 90 grams at the start of the decay process, find a particular solution for the differential equation for part (a).

(d) Find the amount left at 10 months.

42. Social Science   Assume the rate of change of the population of a certain city is given by

$$\frac{dy}{dt} = 6000e^{.06t},$$

where y is the population at time t, measured in years. If the population was 100,000 in 1970 (assume $t = 0$ in 1970), predict the population in 1990.

CHAPTER 13   SUMMARY

KEY TERMS AND SYMBOLS

13.1 $\int f(x)\ dx$ indefinite integral of f
antiderivative
integral sign
integrand
integration
power rule
constant multiple rule
sum or difference rule
integrals of exponential functions
integral of x^{-1}

13.2   differential
integration by substitution
general power rule
integral of e^u
integral of u^{-1}

13.3 $\int_a^b f(x)\ dx$ definite integral of f

Σ the sum of
limits of integration
total change in $f(x)$

13.4 $F(x)\Big]_a^b = F(b) - F(a)$
fundamental theorem of calculus

13.5   consumer's surplus
producer's surplus

13.6   tables of integrals

13.7   differential equation
general solution
particular solution
initial condition (boundary condition)
separable differential equation
marginal productivity

KEY CONCEPTS

$F(x)$ is an **antiderivative** of $f(x)$ if $F'(x) = f(x)$.

Indefinite Integral

If $F'(x) = f(x)$, then $\int f(x)\ dx = F(x) + C$, for any real number C.

Properties of Integrals

$\int k \cdot f(x)\ dx = k \cdot \int f(x)\ dx$, for any real number k.

$\int [f(x) \pm g(x)]\ dx = \int f(x)\ dx \pm \int g(x)\ dx$.

Rules For Integrals

General Power Rule For $u = f(x)$ and $du = f'(x)\ dx$,

$$\int u^n\ du = \frac{u^{n+1}}{n+1} + C; \qquad \int e^u\ du = e^u + C; \qquad \int u^{-1}\ du = \int \frac{du}{u} = \ln |u| + C.$$

The Definite Integral

If f is continuous on $[a, b]$, the definite integral of f from a to b is

$$\int_a^b f(x)\ dx = \lim_{n \to \infty} \sum_{i=1}^n f(x_i)\ \Delta x,$$

provided the limit exists, where $\Delta x = \dfrac{b - a}{n}$ and x_i is any value of x in the ith interval.

Total Change in $F(x)$

Let f be continuous on $[a, b]$ and $f(x) \ge 0$ for all x in $[a, b]$. If $f(x)$ is the rate of change of $F(x)$ for x in $[a, b]$, then the **total change** in $F(x)$ as x goes from a to b is given by $\int_a^b f(x)\ dx$.

Fundamental Theorem of Calculus

Let f be continuous on $[a, b]$, and let F be any antiderivative of f. Then

$$\int_a^b f(x)\ dx = F(b) - F(a) = F(x) \Big]_a^b.$$

A **differential equation** contains one or more of the derivatives of a function.

Find each indefinite integral.

1. $\int (2x^3 - x)\, dx$

2. $\int (x^2 + 5x)\, dx$

3. $\int (6x^2 - 2x + 5)\, dx$

4. $\int (-x^3 + 2x^2 - 7x)\, dx$

5. $\int (x^{3/2} - x + x^{2/5})\, dx$

6. $\int (5x^{1/4} - 2x^{1/2} + x^2)\, dx$

7. $\int (\sqrt{x})^3\, dx$

8. $\int (\sqrt{x})^{-3}\, dx$

9. $\int (1/x^3)\, dx$

10. $\int (-4/x^5)\, dx$

11. $\int \left(\dfrac{-2}{\sqrt{x}} - \dfrac{3}{x} \right) dx$

12. $\int \left(\dfrac{5}{\sqrt{x}} + \dfrac{6}{x} \right) dx$

13. $\int e^{-5x}\, dx$

14. $\int 2e^{4x}\, dx$

15. $\int (-5/x)\, dx$

16. $\int 3/(2x)\, dx$

Use substitution to find the following.

17. $\int 2x\sqrt{x^2 - 3}\, dx$

18. $\int x\sqrt{5x^2 + 6}\, dx$

19. $\int \dfrac{x^2\, dx}{(x^3 + 5)^4}$

20. $\int (x^2 - 5x)^4(2x - 5)\, dx$

21. $\int \dfrac{4x - 5}{2x^2 - 5x}\, dx$

22. $\int \dfrac{12(2x + 9)}{x^2 + 9x + 1}\, dx$

23. $\int \dfrac{x^3}{e^{3x^4}}\, dx$

24. $\int (3x^2 - 6x)^2(6x - 6)\, dx$

25. $\int -2e^{-5x}\, dx$

26. $\int e^{-4x}\, dx$

Find the cost functions for the following marginal cost functions.

27. $C'(x) = 6x + 2$; fixed cost is \$75

28. $C'(x) = .6x^2$; fixed cost is \$90

29. $C'(x) = x^{3/4} + 2x$; 16 units cost \$375

30. $C'(x) = 3x + 2/x^2$; 100 units cost \$9500

In the study of bioavailability, *a drug is given to a patient. The level of concentration of the drug is then measured periodically, producing* blood level curves, *such as the ones in the figure. The areas under the curves give the total amount of the drug available to the patient.**
Find the total area under the curve for

31. Formulation A;

32. Formulation B.

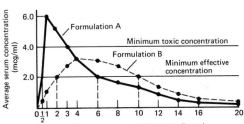

Evaluate the following definite integrals.

33. $\int_{1}^{2} (3x^2 + 5)\, dx$

34. $\int_{1}^{6} (2x^2 + x)\, dx$

35. $\int_{1}^{5} (3x^{-2} + x^{-3})\, dx$

36. $\int_{2}^{3} (5x^{-2} + x^{-4})\, dx$

37. $\int_{1}^{3} 2x^{-1}\, dx$

38. $\int_{1}^{6} 8x^{-1}\, dx$

39. $\int_{0}^{4} 2e^x\, dx$

40. $\int_{1}^{6} \dfrac{5}{2} e^{4x}\, dx$

---

*These graphs are taken from *Basics of Bioavailability*, by D. J. Chodos and A. R. DeSantos. The UpJohn Company. Copyright © 1978.

Work the following exercises.

41. Management The rate of change of sales of a new brand of tomato soup, in thousands, is given by

$$S(x) = \sqrt{x} + 2,$$

where x is the time in months that the new product has been on the market. Find the total sales after 9 months.

42. Natural Science The rate of change of a population of prairie dogs, in terms of the number of coyotes, x, which prey on them, is given by

$$P(x) = 25 - .1x.$$

Find the total number of prairie dogs as the coyote population grows from 100 to 200.

43. Find the area between the graphs of $f(x) = 5 - x^2$ and $g(x) = x^2 - 3$.

44. Management A company has installed new machinery which will produce a savings rate (in thousands of dollars) of

$$S(x) = 225 - x^2,$$

where x is the number of years the machinery is to be used. The rate of additional costs to the company due to the new machinery is expected to be

$$C(x) = x^2 + 25x + 150.$$

For how many years should the company use the new machinery? What will the net savings in thousands of dollars be over this period?

45. Natural Science The rate of change of the population of a rare species of Australian spiders is given by

$$\frac{dy}{dx} = -.2x,$$

where y is the number of spiders present at time x, measured in months. Find y when $x = 10$, if there were 100 spiders at $x = 0$.

46. Management A manufacturer of electronic equipment requires a certain rare metal. He has a reserve supply of 4,000,000 units which he will not be able to replace. If the rate at which the metal is used is given by

$$f(t) = 100,000e^{.03t},$$

where t is the time in years, how long will it be before he uses up the supply? (Hint: find an expression for the total amount used in t years and set it equal to the known reserve supply.)

Use the table of integrals to find the following.

47. $\displaystyle\int \frac{1}{\sqrt{x^2 - 64}}\, dx$

48. $\displaystyle\int \frac{5}{x\sqrt{25 + x^2}}\, dx$

49. $\displaystyle\int \frac{-2}{x(3x + 4)}\, dx$

50. $\displaystyle\int \sqrt{x^2 + 49}\, dx$

51. $\displaystyle\int 3x^2 \ln |x|\, dx$

52. $\displaystyle\int xe^x\, dx$

53. $\displaystyle\int \frac{12}{x^2 - 9}\, dx$

54. $\displaystyle\int \frac{15x}{2x - 5}\, dx$

Find general solutions for the following differential equations.

55. $\dfrac{dy}{dx} = 2x^3 + 6x$

56. $\dfrac{dy}{dx} = x^2 + 5x^4$

57. $\dfrac{dy}{dx} = 4e^x$

58. $\dfrac{dy}{dx} = \dfrac{1}{2x + 3}$

Find particular solutions for the following differential equations.

59. $\dfrac{dy}{dx} = x^2 - 5x;\; y = 1$ when $x = 0$

60. $\dfrac{dy}{dx} = 4x^3 + 2;\; y = 3$ when $x = 1$

61. $\dfrac{dy}{dx} = 5(e^{-x} - 1);\; y = 17$ when $x = 0$

62. $\dfrac{dy}{dx} = \dfrac{x}{x^2 - 3};\; y = 52$ when $x = 2$

Find general solutions for the following differential equations.

63. $\dfrac{dy}{dx} = \dfrac{3x + 1}{y}$

64. $\dfrac{dy}{dx} = \dfrac{e^x + x}{y - 1}$

65. $\dfrac{dy}{dx} = \dfrac{2y + 1}{x}$

66. $\dfrac{dy}{dx} = \dfrac{3 - y}{e^x}$

Find particular solutions for the following differential equations. (Some solutions may give y implicitly.)

67. $(5 - 2x)y = \dfrac{dy}{dx}$; $y = 2$ when $x = 0$

68. $\sqrt{x}\dfrac{dy}{dx} = xy$; $y = 4$ when $x = 1$

69. $\dfrac{dy}{dx} = \dfrac{1 - 2x}{y + 3}$; $y = 16$ when $x = 0$

70. $\dfrac{dy}{dx} = (3x + 2)^2 e^y$; $y = 0$ when $x = 0$

71. Management The marginal sales (in hundreds of dollars) of a computer software company are given by

$$\frac{dy}{dx} = 5e^{.2x},$$

where x is the number of months the company has been in business. Assume that sales were 0 initially.
(a) Find the sales after 6 months.
(b) Find the sales after 12 months.

72. Management A deposit of $10,000 is made to a savings account at 5% interest compounded continuously. Assume that continuous *withdrawals* of $1000 per year are made.
(a) Write a differential equation to describe the situation.
(b) How much will be left in the account after 1 year?

73. Management In Exercise 72, approximately how long will it take to use up the account?

74. Natural Science After use of an experimental insecticide, the rate of decline of an insect population is

$$\frac{dy}{dt} = \frac{-10}{1 + 5t},$$

where t is the number of hours after the insecticide is applied. Assume that there were 50 insects initially.
(a) How many are left after 24 hours?
(b) How long will it take for the entire population to die?

75. Natural Science A population of mites grows at a rate proportional to the number present, y. If the growth constant is 10% and 120 mites are present at time $t = 0$, in weeks, find the number present after 6 weeks.

76. Natural Science Find an equation relating x to y given the following equations, which describe the interaction of two competing species and their growth rates.

$$\frac{dx}{dt} = .2x - .5xy$$

$$\frac{dy}{dt} = -.3y + .4xy$$

Find the values of x and y for which both growth rates are 0. (Hint: solve the system of equations found by setting each growth rate equal to 0.)

► **Estimating Depletion Dates for Minerals**

It is becoming more and more obvious that the earth contains only a finite quantity of minerals. The "easy and cheap" sources of minerals are being used up, forcing an ever more expensive search for new sources. For example, oil from the North Slope of Alaska would never have been used in the United States during the 1930s since there was so much Texas and California oil readily available.

Mineral usage tends to follow an exponential growth curve. Thus, if q represents the rate of consumption of a certain mineral at time t, while q_0 represents consumption when $t = 0$, then

$$q = q_0 e^{kt},$$

where k is the growth constant. For example, the world consumption of petroleum in a recent year was about 19,600 million barrels, with the value of k about 6%. Letting $t = 0$ correspond to this base year, then $q_0 = 19,600$, $k = .06$, and

$$q = 19,600 e^{.06t}$$

is the rate of consumption at time t, assuming that all present trends continue.

Based on estimates of the National Academy of Science, 2,000,000 million barrels of oil are now in provable reserves or are likely to be discovered in the future. At the present rate of consumption, how many years would be necessary to deplete these estimated reserves? Use the integral calculus of this chapter to find out.

To begin, find the total quantity of petroleum that would be used between time $t = 0$ and some future time $t = t_1$. The figure shows a typical graph of the function $q = q_0 e^{kt}$.

Following the work in Section 13.3, divide the time interval from $t = 0$ to $t = t_1$ into n subintervals. Let the ith subinterval have width Δt_i. Let the rate of consumption for the ith subinterval be approximated by q_i^*. Thus, the approximate total consumption for the subinterval is given by

$$q_i^* \cdot \Delta t_i,$$

and the total consumption over the interval from time $t = 0$ to $t = t_1$ approximated by

$$\sum_{i=1}^{n} q_i^* \cdot \Delta t_i.$$

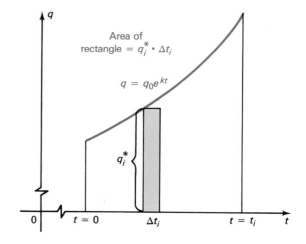

The limit of this sum as each of the Δt_i's approaches 0 gives the total consumption from time $t = 0$ to $t = t_1$. That is,

$$\text{Total consumption} = \lim_{\Delta t_i \to 0} \Sigma\ q_i^* \cdot \Delta t_i.$$

This limit is the definite integral of the function $q = q_0 e^{kt}$ from $t = 0$ to $t = t_1$, or

$$\text{Total consumption} = \int_0^{t_1} q_0 e^{kt}\ dt.$$

Evaluating this definite integral gives

$$\int_0^{t_1} q_0 e^{kt} \, dt = q_0 \int_0^{t_1} e^{kt} \, dt = q_0 \left(\frac{1}{k} e^{kt} \right) \Big]_0^{t_1}$$

$$= \frac{q_0}{k} e^{kt} \Big]_0^{t_1} = \frac{q_0}{k} e^{kt_1} - \frac{q_0}{k} e^0$$

$$= \frac{q_0}{k} e^{kt_1} - \frac{q_0}{k} \quad (1)$$

$$= \frac{q_0}{k} (e^{kt_1} - 1). \qquad \textbf{(1)}$$

Now return to the numbers given for petroleum: $q_0 = 19{,}600$ million barrels, where q_0 represents consumption in the base year; $k = .06$; and total petroleum reserves are estimated as $2{,}000{,}000$ million barrels. Thus, using equation (1),

$$2{,}000{,}000 = \frac{19{,}600}{0.06} (e^{.06t_1} - 1).$$

Multiply both sides of the equation by .06.

$$120{,}000 = 19{,}600 (e^{.06t_1} - 1)$$

Divide both sides of the equation by 19,600.

$$6.1 = e^{.06t_1} - 1$$

Add 1 to both sides.

$$7.1 = e^{.06t_1}$$

Take natural logarithms of both sides.

$$\ln 7.1 = \ln e^{.06t_1} = .06t_1 \ln e$$
$$= .06t_1 \quad \text{(since } \ln e = 1\text{)}$$

Finally,

$$t_1 = \frac{\ln 7.1}{.06}.$$

A calculator gives ln 7.1 as about 1.96. Thus,

$$t_1 = \frac{1.96}{.06} = 33.$$

By this result, petroleum reserves will last the world for 33 years.

The results of mathematical analyses such as this must be used with great caution. By the analysis above, the world would use all the petroleum that it wants in the thirty-second year after the base year, but there would be none at all in 34 years. This is not at all realistic. As petroleum reserves decline, the price will increase, causing demand to decline and supplies to increase.

EXERCISES

1. Find the number of years that the estimated petroleum reserves would last if used at the same rate as in the base year.

2. How long would the estimated petroleum reserves last if the growth constant was only 2% instead of 6%?

Estimate the length of time until depletion for each of the following minerals.

3. Bauxite (the ore from which aluminum is obtained), estimated reserves in base year 15,000,000 thousand tons, rate of consumption 63,000 thousand tons, growth constant 6%.

4. Bituminous coal, estimated world reserves 2,000,000 million tons, rate of consumption 2200 million tons, growth constant 4%.

This case uses some of the ideas of probability. The probability of an event is a number p, where $0 \le p \le 1$, such that p is the ratio of the number of ways that the event can happen divided by the total number of possible outcomes. For example, the probability of drawing a red card from a deck of 52 cards (of which 26 are red) is given by

$$P(\text{red card}) = \frac{26}{52} = \frac{1}{2}.$$

In the same way, the probability of drawing a black queen from a deck of 52 cards is

$$P(\text{black queen}) = \frac{2}{52} = \frac{1}{26}.$$

In this case the cost of a warranty program to a manufacturer is found. This costs depends on the quality of the products made. These variables are used.

c = constant product price, per unit, including cost of warranty (We assume the price charged per unit is constant, since this price is likely to be fixed by competition.)

m = expected lifetime of the product

w = length of the warranty period

N = size of a production lot, such as a year's production

r = warranty cost per unit

$C(t)$ = pro rata customer rebate at time t

$P(t)$ = probability of product failure at any time t

$F(t)$ = number of failures occurring at time t.

Assume that the warranty is of the pro rata customer rebate type, in which the customer is paid for the proportion of the warranty left. Hence, if the product has a warranty period of w and fails at time t, then the product worked for the fraction t/w of the warranty period. Hence, the customer is reimbursed for the unused portion of the warranty, or the fraction

$$1 - \frac{t}{w}.$$

Assuming the product cost c originally, and using $C(t)$ to represent the customer rebate at time t,

$$C(t) = c\left(1 - \frac{t}{w}\right).$$

For many different types of products, it has been shown by experience that

$$P(t) = 1 - e^{-t/m}$$

provides a good estimate of the probability of product failure at time t. The total number of failures at time t is given by the product of $P(t)$ and N, the total number of items per batch. If we use $F(t)$ to represent this total, we have

$$F(t) = N \cdot P(t) = N(1 - e^{-t/m}).$$

The total number of failures in some "tiny time interval" of width dt can be shown to be the derivative of $F(t)$,

$$F'(t) = \left(\frac{N}{m}\right)e^{-t/m},$$

*From "Determination of Warranty Reserves," by Warren W. Menke, from *Management Science,* Vol. 15, No. 10, June 1969. Copyright © 1969 The Institute of Management Sciences. Reprinted by permission.

while the cost for the failure in this "tiny time interval" is

$$C(t) \cdot F'(t) = c\left(1 - \frac{t}{w}\right)\left(\frac{N}{m}\right)e^{-t/m}.$$

The total cost for all failures during the warranty period is thus given by the definite integral

$$\int_0^w c\left(1 - \frac{t}{w}\right)\left(\frac{N}{m}\right)e^{-t/m}\,dt.$$

Using integration by parts (a method of integration not discussed), this definite integral can be shown to equal

$$Nc\left[-e^{-t/m} + \frac{t}{w} \cdot e^{-t/m} + \frac{m}{w}(e^{-t/m})\right]_0^w$$

or

$$Nc\left(1 - \frac{m}{w} + \frac{m}{w}e^{-w/m}\right).$$

This last quantity is the total warranty cost for all the units manufactured. Since there are N units per batch, the warranty cost per item is

$$r = \frac{1}{N}\left[Nc\left(1 - \frac{m}{w} + \frac{m}{w}e^{-w/m}\right)\right]$$

$$= c\left(1 - \frac{m}{w} + \frac{m}{w}e^{-w/m}\right).$$

For example, suppose a product which costs \$100 has an expected life of 24 months, with a 12-month warranty. Then we have $c = \$100$, $m = 24$, $w = 12$, with r, the warranty cost per unit, given by

$$r = 100(1 - 2 + 2e^{-.5})$$

$$= 100[-1 + 2(.6065)]$$

$$= 100(.2130) = 21.30.$$

EXERCISES

Find r for each of the following.

1. $c = \$50$, $m = 48$ months, $w = 24$ months

2. $c = \$1000$, $m = 60$ months, $w = 24$ months

3. $c = \$1200$, $m = 30$ months, $w = 30$ months

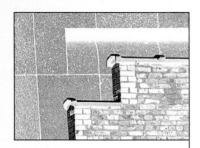

Multivariate Calculus

Many of the ideas developed for functions of one variable also apply to functions of more than one variable. In particular, the fundamental idea of derivative generalizes in a very natural way to functions of more than one variable.

14.1 FUNCTIONS OF SEVERAL VARIABLES

If a company produces x items at a cost of \$10 per item, then the total cost $C(x)$ of producing the items is given by

$$C(x) = 10x.$$

The cost is a function of one independent variable, the number of items produced. If the company produces two products, with x of one product at a cost of \$10 each, and y of another product at a cost of \$15 each, then the total cost to the firm is a function of *two* independent variables, x and y. By generalizing $f(x)$ notation, the total cost can be written as $C(x, y)$, where

$$C(x, y) = 10x + 15y.$$

When $x = 5$ and $y = 12$, the total cost is written $C(5, 12)$, with

$$C(5, 12) = 10 \cdot 5 + 15 \cdot 12 = 230.$$

Here is a general definition.

$z = f(x, y)$ is a **function of two independent variables** if a unique value of z is obtained from each ordered pair of real numbers (x, y). The variables x and y are **independent variables;** z is the **dependent variable.** The set of all ordered pairs of real numbers (x, y) such that $f(x, y)$ exists is the **domain** of f; the set of all values of $f(x, y)$ is the **range.**

1 Let $f(x, y) = x^3 - 4x^2 + xy$. Find

(a) $f(2, 4)$;

(b) $f(-2, 3)$.

Answer:

(a) 0

(b) -30

2 Let $f(x, y, z) = 3xy - 4xz + yz$. Find

(a) $f(2, 5, -1)$;

(b) $f(1, 4, 3)$

Answer:

(a) 33

(b) 12

▶**EXAMPLE 1** Let $f(x, y) = 4x^2 + 2xy + 3/y$ and find each of the following.

(a) $f(-1, 3)$.

Replace x with -1 and y with 3.

$$f(-1, 3) = 4(-1)^2 + 2(-1)(3) + \frac{3}{3} = 4 - 6 + 1 = -1$$

(b) $f(2, 0)$

Because of the quotient $3/y$, it is not possible to replace y with 0, so $f(2, 0)$ is undefined. Inspection shows that the domain of the function f consists of all ordered pairs (x, y) such that $y \neq 0$. ◀ **1**

▶**EXAMPLE 2** Let x represent the number of milliliters (ml) of carbon dioxide released by the lungs in 1 minute. Let y be the change in the carbon dioxide content of the blood as it leaves the lungs (y is measured in ml of carbon dioxide per 100 ml of blood). The total output of blood from the heart in one minute (measured in ml) is given by C, where C is a function of x and y such that

$$C = C(x, y) = \frac{100x}{y}.$$

Find $C(320, 6)$.

Replace x with 320 and y with 6 to get

$$C(320, 6) = \frac{100(320)}{6}$$

$$\approx 5333 \text{ ml of blood per minute.} \quad ◀$$

The definition given before Example 1 was for a function of two independent variables, but similar definitions could be given for functions of three, four, or more independent variables. Functions of more than one independent variable are called **multivariate functions.**

▶**EXAMPLE 3** Let $f(x, y, z) = 4xz - 3x^2y + 2z^2$. Find each of the following.

(a) $f(2, -3, 1)$

Replace x with 2, y with -3, and z with 1.

$$f(2, -3, 1) = 4(2)(1) - 3(2)^2(-3) + 2(1)^2 = 8 + 36 + 2 = 46$$

(b) $f(-4, 3, -2) = 4(-4)(-2) - 3(-4)^2(3) + 2(-2)^2$

$$= 32 - 144 + 8 = -104. \quad ◀ \quad \boxed{2}$$

3 Locate $P(3, 2, 4)$ on the coordinate system below.

Answer:

GRAPHING FUNCTIONS OF TWO INDEPENDENT VARIABLES Functions of one independent variable are graphed by using an x-axis and a y-axis to locate points in a plane. The plane determined by the x- and y-axes is called the xy-*plane*. A third axis is needed to graph functions of two independent variables—the z-axis, which goes through the origin in the xy-plane and is perpendicular to both the x-axis and the y-axis.

Figure 14.1 shows one possible way to draw the three axes. In Figure 14.1, the yz-plane is in the plane of the page, with the x-axis perpendicular to the plane of the page.

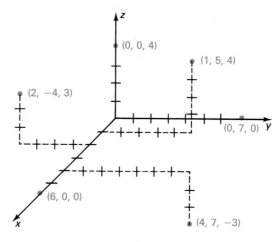

FIGURE 14.1

Just as we graphed ordered pairs earlier, we can now graph **ordered triples** of the form (x, y, z). For example, to locate the point corresponding to the ordered triple $(2, -4, 3)$, start at the origin and go 2 units along the positive x-axis. Then go 4 units in a negative direction (to the left), parallel to the y-axis. Finally, go up 3 units, parallel to the z-axis. The point representing $(2, -4, 3)$ and several other sample points are shown in Figure 14.1. The region of three-dimensional space where all coordinates are positive is called the **first octant.** **3**

In Chapter 2 we saw that the graph of $ax + by = c$ (where a and b are not both 0) is a straight line. This result generalizes to three dimensions.

Plane

The graph of

$$ax + by + cz = d$$

is a plane when a, b, c, and d are real numbers, with a, b, and c not all zero.

▶**EXAMPLE 4**   Graph $2x + y + z = 6$.

By the result in the box, the graph of this equation is a plane. Earlier, we graphed straight lines by finding x- and y-intercepts. A similar idea helps graph a plane. To find the x-intercept, the point where the graph crosses the x-axis, let $y = 0$ and $z = 0$.

$$2x + 0 + 0 = 6$$
$$x = 3$$

The point $(3, 0, 0)$ is on the graph. Letting $x = 0$ and $z = 0$ gives the point $(0, 6, 0)$, while $x = 0$ and $y = 0$ lead to $(0, 0, 6)$. The plane through these three points includes the triangular surface shown in Figure 14.2. This region is the first-octant part of the plane that is the graph of $2x + y + z = 6$.   ◀

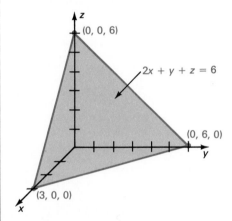

FIGURE 14.2   FIGURE 14.3

▶**EXAMPLE 5**   Graph $x + z = 6$.

To find the x-intercept, let $z = 0$, giving $(6, 0, 0)$. If $x = 0$, we get the point $(0, 0, 6)$. Because there is no y in the equation $x + z = 6$, there can be no y-intercept. A plane that has no y-intercept is parallel to the y-axis. The first-octant portion of the graph of $x + z = 6$ is shown in Figure 14.3.   ◀

▶**EXAMPLE 6**   (a) Graph $x = 3$.

This graph, which goes through $(3, 0, 0)$, can have no y-intercept and no z-intercept. It is, therefore, a plane parallel to the y-axis and the z-axis and, therefore, to the yz plane. The first-octant portion of the graph is shown in Figure 14.4.

4 Describe each graph and give any intercepts.

(a) $2x + 3y - z = 4$

(b) $x + y = 3$

(c) $z = 2$

Answer:

(a) A plane; $(2, 0, 0)$, $(0, 4/3, 0)$, $(0, 0, -4)$

(b) A plane parallel to the z-axis; $(3, 0, 0)$, $(0, 3, 0)$

(c) A plane parallel to the x-axis and the y-axis; $(0, 0, 2)$

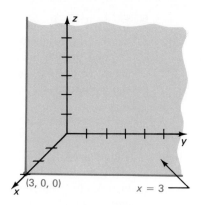

FIGURE 14.4

(b) Graph $y = 4$.

This graph goes through $(0, 4, 0)$ and is parallel to the xz-plane. The first-octant portion of the graph is shown in Figure 14.5.

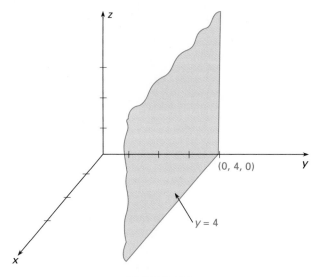

FIGURE 14.5

(c) Graph $z = 1$.

The graph is a plane parallel to the xy-plane, passing through $(0, 0, 1)$. Its first-octant portion is shown in Figure 14.6. ◀ **4**

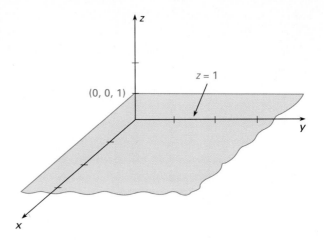

FIGURE 14.6

The graph of a function of one variable, $y = f(x)$, is a curve in the plane. If x_0 is in the domain of f, the point $(x_0, f(x_0))$ on the graph lies directly above or below the number x_0 on the x-axis, as shown in Figure 14.7.

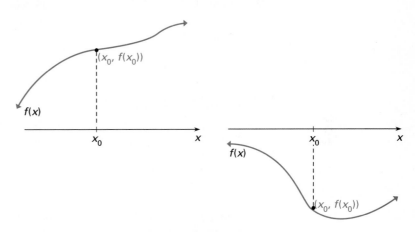

FIGURE 14.7

The graph of a function of two variables, $z = f(x, y)$ is a **surface** in three-dimensional space. If (x_0, y_0) is in the domain of f, the point $(x_0, y_0, f(x_0, y_0))$ lies directly above or below the point (x_0, y_0) in the xy-plane, as shown in Figure 14.8.

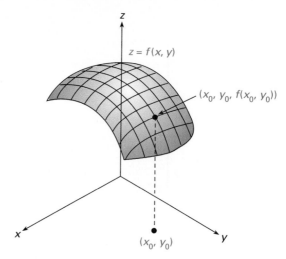

FIGURE 14.8

Although computer software is available for drawing the graphs of functions of two variables, you can often get a good picture of the graph without it by finding various **traces**—the curves that result when a surface is cut by a plane. The ***xy*-trace** is the intersection of the surface with the *xy*-plane. The ***yz*-trace** and ***xz*-trace** are defined similarly. You can also determine the intersection of the surface with planes parallel to the *xy*-plane. Such planes are of the form $z = k$, where k is a constant and the curves that result when they cut the surface are called **level curves.**

▶**EXAMPLE 7** Graph $z = x^2 + y^2$.

The *yz*-plane is the plane in which every point has first coordinate 0, so its equation is $x = 0$. When $x = 0$, the equation becomes $z = y^2$, which is the equation of a parabola in the *yz*-plane, as shown in Figure 14.9(a). Similarly, to find the intersection of the surface with the *xz*-plane (whose equation is $y = 0$), let $y = 0$ in the equation. It then becomes $z = x^2$, which is the equation of a parabola in the *xz*-plane (shown in Figure 14.9(a)). The *xy*-trace (the intersection of the surface with the plane $z = 0$) is the single point $(0, 0, 0)$ because $x^2 + y^2$ is never negative, and equal to 0 only when $x = 0$ and $y = 0$.

Next, we find the level curves by intersecting the surface with the planes $z = 1$, $z = 2$, $z = 3$, etc. (all of which are parallel to the *xy*-plane). In each case, the result is a circle

$$x^2 + y^2 = 1, \quad x^2 + y^2 = 2, \quad x^2 + y^2 = 3,$$

and so on, as shown in Figure 14.9(b). Drawing the traces and level curves on the same set of axes suggests that the graph of $z = x^2 + y^2$ is the bowl-shaped figure, called a **paraboloid,** that is shown in Figure 14.9(c). ◀

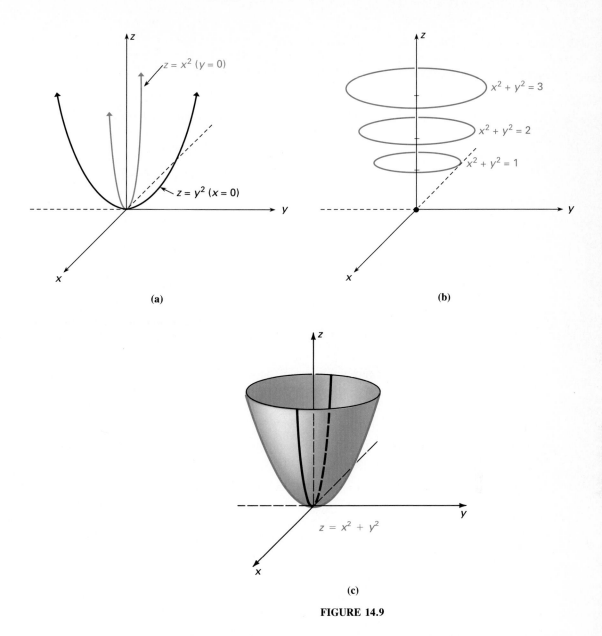

(a)

(b)

(c)

FIGURE 14.9

One application of level curves in economics occurs with production functions. A **production function** $z = f(x, y)$ is a function that gives the quantity z of an item produced as a function of x and y, where x is the amount of labor and y is the amount

5 Find the equation of the level curve for production of 100 items if the production function is $z = 5x^{1/4}y^{3/4}$.

Answer:
$$y = \frac{20^{4/3}}{x^{1/3}}$$

of capital (in appropriate units) needed to produce z units. If the production function has the special form $z = P(x, y) = Ax^a y^{1-a}$, where A is a constant and $0 < a < 1$, the function is called a **Cobb-Douglas production function.**

▶**EXAMPLE 8** Find the level curve at a production of 100 items for the Cobb-Douglas production function $z = x^{2/3}y^{1/3}$.

Let $z = 100$ and solve for y to get

$$100 = x^{2/3}y^{1/3}$$

$$\frac{100}{x^{2/3}} = y^{1/3}.$$

Now cube both sides to express y as a function of x.

$$y = \frac{100^3}{x^2}$$

$$= \frac{1,000,000}{x^2}.$$

The level curve of height 100 is shown graphed in three dimensions in Figure 14.10(a) and on the familiar xy-plane in Figure 14.10(b). The points of the graph correspond to those values of x and y that lead to production of 100 items. ◀ **5**

The curve in Figure 14.10 is called an *isoquant*, for *iso* (equal) and *quant* (amount). In Example 8, the "amounts" all "equal" 100.

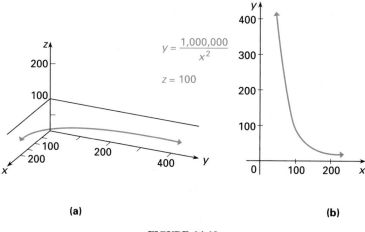

(a)   (b)

FIGURE 14.10

14.1 EXERCISES

1. Let $f(x, y) = 4x + 5y + 3$. Find the following. (See Example 1.)
   (a) $f(2, -1)$ (b) $f(-4, 1)$
   (c) $f(-2, -3)$ (d) $f(0, 8)$

2. Let $g(x, y) = -x^2 - 4xy + y^3$. Find the following.
   (a) $g(-2, 4)$ (b) $g(-1, -2)$
   (c) $g(-2, 3)$ (d) $g(5, 1)$

3. Let $h(x, y) = \sqrt{x^2 + 2y^2}$. Find the following.
   (a) $h(5, 3)$ (b) $h(2, 4)$
   (c) $h(-1, -3)$ (d) $h(-3, -1)$

4. Let $f(x, y) = \dfrac{\sqrt{9x + 5y}}{\log x}$. Find the following.

   (a) $f(10, 2)$ (b) $f(100, 1)$

   (c) $f(1000, 0)$ (d) $f\left(\dfrac{1}{10}, 5\right)$

Graph the first-octant portion of each of the following planes. (See Examples 4–6.)

5. $x + y + z = 6$ 6. $x + y + z = 12$

7. $2x + 3y + 4z = 12$ 8. $4x + 2y + 3z = 24$

9. $3x + 2y + z = 18$ 10. $x + 3y + 2z = 9$

11. $x + y = 4$ 12. $y + z = 5$

13. $x = 2$ 14. $z = 3$

Use a calculator with a y^x key to work the following problems.

15. **Natural Science** The oxygen consumption of a well-insulated mammal which is not sweating is approximated by

$$m = \frac{2.5(T - F)}{w^{.67}},$$

where T is the internal body temperature of the animal (in °C), F is the temperature of the outside of the animal's fur (in °C), and w is the animal's weight in kilograms.* Find m for the following data.
   (a) $T = 38°$, $F = 6°$, $w = 32$ kg
   (b) $T = 40°$, $F = 20°$, $w = 43$ kg

16. **Natural Science** The surface area of a human (in square meters) is approximated by

$$A = .202 \, W^{.425}H^{.725},$$

where W is the weight of the person in kilograms and H is the height in meters.* Find A for the following data.
   (a) $W = 72$, $H = 1.78$ (b) $W = 65$, $H = 1.40$
   (c) $W = 70$, $H = 1.60$
   (d) Find your own surface area.

17. **Management** Production of a precision camera is given by

$$P(x, y) = 100\left[\frac{3}{5}x^{-2/5} + \frac{2}{5}y^{-2/5}\right]^{-5},$$

where x is the amount of labor in work hours and y is the amount of capital. Find the following.
   (a) $P(32, 1)$ (b) $P(1, 32)$
   (c) If 32 work hours and 243 units of capital are used, what is the production output?

Management Find the level curve at a production of 500 for each of the production functions in Exercises 18–19. Graph each function on the xy-plane.

18. The production function z for the United States was once estimated as $z = x^7 y^3$, where x stands for the amount of labor and y stands for the amount of capital.

19. If x represents the amount of labor and y the amount of capital, a production function for Canada is approximately $z = x^4 y^{.6}$.

20. **Management** The number of cows that can graze on a certain ranch without causing overgrazing is approximated by

$$C(x, y) = 9x + 5y - 4,$$

where x is the number of acres of grass and y the number of acres of alfalfa. Find the following.
   (a) $C(50, 0)$ (b) $C(30, 4)$
   (c) How many cows may graze if there are 80 acres of grass and 20 acres of alfalfa?
   (d) If the ranch has 10 acres of alfalfa and 5 acres of grass, how many cows may graze?

By considering traces, match each equation in Exercises 21–26 with its graph in (a)–(f) below.

21. $z = x^2 + y^2$

22. $z^2 - y^2 - x^2 = 1$

23. $x^2 - y^2 = z$

24. $z = y^2 - x^2$

25. $\dfrac{x^2}{16} + \dfrac{y^2}{25} + \dfrac{z^2}{4} = 1$

26. $z = 5(x^2 + y^2)^{-1/2}$

(a)

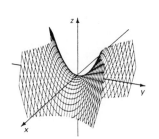

(b)

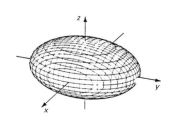

(c)

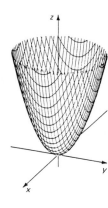

(d)

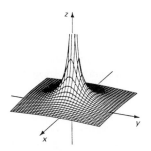

(e)

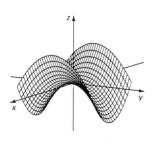

(f)

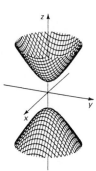

In Exercises 27–28, sketch the xy-, yz-, and xz-traces of the function $z = f(x, y)$. Then sketch the level curves when $z = 1$, $z = 2$, $z = 3$. Use this information to graph the first-octant portion of the graph of the function. (See Example 7.)

27. $z = 3 - x^2 - y^2$

28. $z = \sqrt{9 - x^2 - y^2}$

29. Let $f(x, y) = 9x^2 - 3y^2$, and find each of the following.

(a) $\dfrac{f(x + h, y) - f(x, y)}{h}$ 　　(b) $\dfrac{f(x, y + h) - f(x, y)}{h}$

30. Let $f(x, y) = 7x^3 + 8y^2$, and find each of the following.

(a) $\dfrac{f(x + h, y) - f(x, y)}{h}$ 　　(b) $\dfrac{f(x, y + h) - f(x, y)}{h}$

14.2 PARTIAL DERIVATIVES

A small firm makes only two products, radios and audiocassette recorders, with the profits of the firm given by

$$P(x, y) = 40x^2 - 10xy + 5y^2 - 80,$$

where x is the number of units of radios sold and y is the number of units of recorders sold. How will a change in x or y affect P?

Suppose that sales of radios have been steady at 10 units; only the sales of recorders vary. The management would like to find the marginal profit with respect to y, the number of recorders sold. Recall that marginal profit is given by the derivative of the profit function. Here, x is fixed at 10. Using this information, we begin by finding a new function, $f(y) = P(10, y)$. Let $x = 10$ to get

$$f(y) = P(10, y) = 40(10)^2 - 10(10)y + 5y^2 - 80$$
$$= 3920 - 100y + 5y^2.$$

The function $f(y)$ shows the profit from the sale of y recorders, assuming that x is fixed at 10 units. Find the derivative df/dy to get the marginal profit with respect to y.

$$\frac{df}{dy} = -100 + 10y$$

In this example, the derivative of the function $f(y)$ was taken with respect to y only; we assumed that x was fixed. To generalize, let $z = f(x, y)$. An intuitive definition of the *partial derivatives* of f with respect to x and y follows.

Partial Derivatives (Informal Definition)

The **partial derivative of f with respect to x** is the derivative of f obtained by treating x as a variable and y as a constant.

The **partial derivative of f with respect to y** is the derivative of f obtained by treating y as a variable and x as a constant.

The symbols $f_x(x, y)$ (no prime used), $\partial z/\partial x$, and $\partial f/\partial x$ are used to represent the partial derivative of $z = f(x, y)$ with respect to x, with similar symbols used for the partial derivative with respect to y. The symbol $f_x(x, y)$ is often abbreviated as just f_x, with $f_y(x, y)$ abbreviated f_y.

1 Find f_x and f_y.

(a) $f(x, y) = -x^2y + 3xy + 2xy^2$

(b) $f(x, y) = x^3 + 2x^2y + xy$

Answer:

(a) $f_x = -2xy + 3y + 2y^2$;
$f_y = -x^2 + 3x + 4xy$

(b) $f_x = 3x^2 + 4xy + y$;
$f_y = 2x^2 + x$

2 Find f_x and f_y.

(a) $f(x, y) = \ln(2x + 3y)$

(b) $f(x, y) = e^{xy}$

Answer:

(a) $f_x = \dfrac{2}{2x + 3y}$; $f_y = \dfrac{3}{2x + 3y}$

(b) $f_x = ye^{xy}$; $f_y = xe^{xy}$

Generalizing from the definition of derivative given earlier, partial derivatives of a function $z = f(x, y)$ are formally defined as follows.

Partial Derivatives (Formal Definition)

Let $z = f(x, y)$ be a function of two variables. Let all indicated limits exist. Then the **partial derivative of f with respect to x** is

$$f_x(x, y) = \lim_{h \to 0} \frac{f(x + h, y) - f(x, y)}{h};$$

the **partial derivative of f with respect to y** is

$$f_y(x, y) = \lim_{h \to 0} \frac{f(x, y + h) - f(x, y)}{h}.$$

If the indicated limits do not exist, then the partial derivatives do not exist.

Similar definitions could be given for functions of more than two independent variables.

▶ **EXAMPLE 1** Let $f(x, y) = 4x^2 - 9xy + 6y^3$. Find f_x and f_y.

To find f_x, treat y as a constant and x as a variable. The derivative of the first term, $4x^2$, is $8x$. In the second term, $-9xy$, the constant coefficient of x is $-9y$, so the derivative with x as the variable is $-9y$. The derivative of $6y^3$ is zero, since we are treating y as a constant. Thus,

$$f_x = 8x - 9y.$$

Now, to find f_y, treat y as a variable and x as a constant. Since x is a constant, the derivative of $4x^2$ is zero. In the second term, the coefficient of y is $-9x$ and the derivative of $-9xy$ is $-9x$. The derivative of the third term is $18y^2$. Thus,

$$f_y = -9x + 18y^2. \quad ◀ \boxed{1}$$

▶ **EXAMPLE 2** Let $f(x, y) = \ln(x^2 + y)$. Find f_x and f_y.

Recall the formula for the derivative of a natural logarithm function. If $g(x) = \ln x$, then $g'(x) = 1/x$. Using this formula and the chain rule,

$$f_x = \frac{1}{x^2 + y} \cdot D_x(x^2 + y) = \frac{1}{x^2 + y} \cdot 2x = \frac{2x}{x^2 + y},$$

and $$f_y = \frac{1}{x^2 + y} \cdot D_y(x^2 + y) = \frac{1}{x^2 + y} \cdot 1 = \frac{1}{x^2 + y}. \quad ◀ \boxed{2}$$

3 Let $f(x, y) = x^2 + xy^2 + 5y - 10$. Find the following.

(a) $f_x(2, 1)$

(b) $\dfrac{\partial f}{\partial y}(-1, 0)$

Answer:

(a) 5

(b) 5

The notation

$$f_x(a, b) \quad \text{or} \quad \frac{\partial f}{\partial x}(a, b)$$

represents the value of a partial derivative when $x = a$ and $y = b$, as shown in the next example.

▶**EXAMPLE 3** Let $f(x, y) = 2x^2 + 3xy^3 + 2y + 5$. Find the following.
(a) $f_x(-1, 2)$
First, find f_x by holding y constant.

$$f_x = 4x + 3y^3$$

Now let $x = -1$ and $y = 2$.

$$f_x(-1, 2) = 4(-1) + 3(2)^3 = -4 + 24 = 20$$

(b) $\dfrac{\partial f}{\partial y}(-4, -3)$
Since $\partial f/\partial y = 9xy^2 + 2$,

$$\frac{\partial f}{\partial y}(-4, -3) = 9(-4)(-3)^2 + 2 = 9(-36) + 2 = -322. \quad ◀ \quad \boxed{3}$$

The derivative of a function of one variable can be interpreted as the slope of the tangent line to the graph at that point. With some modification, the same is true of partial derivatives of functions of two variables. At a point on the graph of a function of two variables, $z = f(x, y)$, there may be many tangent lines, all of which lie in the same tangent plane, as shown in Figure 14.11.

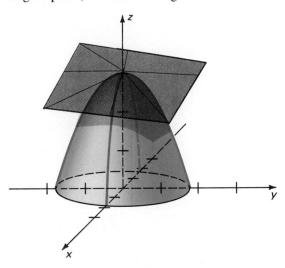

FIGURE 14.11

If we specify a particular direction, however, there will be only one tangent line in that direction. Partial derivatives can be used to find the slope of the tangent lines in the x- and y-directions as follows.

Figure 14.12 shows a surface $z = f(x, y)$ and a plane that is parallel to the xz-plane. The equation of the plane is $y = a$. (This corresponds to holding y fixed.) Since $y = a$ for points on the plane, any point on the curve that represents the intersection of the plane and the surface must have the form $(x, a, f(x, a))$. Thus, this curve can be described as $z = f(x, a)$. Since a is constant, $z = f(x, a)$ is a function of one variable. When the derivative of $z = f(x, a)$ is evaluated at $x = b$, it gives the slope of the line tangent to this curve at the point $(b, a, f(b, a))$, as shown in Figure 14.12. Thus, the partial derivative of f with respect to x, $f_x(b, a)$, gives the rate of change of the surface $z = f(x, y)$ in the x-direction at the point $(b, a, f(b, a))$. In the same way, the partial derivative with respect to y will give the slope of the line tangent to the surface in the y-direction at the point $(b, a, f(b, a))$.

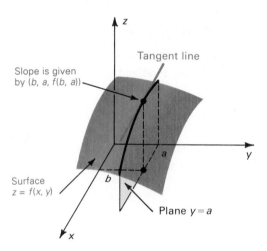

FIGURE 14.12

The derivative of $y = f(x)$ gives the rate of change of y with respect to x. In the same way, if $z = f(x, y)$, then f_x gives the rate of change of z with respect to x, provided that y is held constant.

▶**EXAMPLE 4** Suppose that the temperature of the water at the point on a river where a nuclear power plant discharges its hot waste water is approximated by

$$T(x, y) = 2x + 5y + xy - 40.$$

Here, x represents the temperature of the river water in degrees Celsius before it reaches the power plant and y is the number of megawatts (in hundreds) of electricity being produced by the plant.

4 Find the temperature of the water at the outlet if the water has an initial temperature of 10°C and if 400 megawatts of electricity are being produced.

Answer:
40°C

5 Use the function of Example 4 to find and interpret the following.

(a) $T_x(5, 4)$

(b) $T_y(8, 3)$

Answer:

(a) $T_x(5, 4) = 6$; the approximate increase in temperature if the input temperature increases from 5 to 6 degrees

(b) $T_y(8, 3) = 13$; the approximate increase in temperature if the production of electricity increases from 300 to 400 megawatts

(a) Find the temperature of the discharged water if the water reaching the plant has a temperature of 8°C and if 300 megawatts of electricity are being produced.

Here $x = 8$ and $y = 3$ (since 300 megawatts of electricity are being produced), with

$$T(8, 3) = 2(8) + 5(3) + 8(3) - 40 = 15.$$

The water at the outlet of the plant is at a temperature of 15°C.   **4**

(b) Find and interpret $T_x(9, 5)$.

First, find the partial derivative T_x.

$$T_x = 2 + y$$

This partial derivative gives the rate of change of T with respect to x. Replacing x with 9 and y with 5 gives

$$T_x(9, 5) = 2 + 5 = 7.$$

Just as marginal cost is the approximate cost of one more item, this result, 7, is the approximate change in temperature of the output water if input water temperature changes by 1 degree, from $x = 9$ to $x = 9 + 1 = 10$, while y remains constant.

(c) Find and interpret $T_y(9, 5)$.

The partial derivative T_y is

$$T_y = 5 + x.$$

This partial derivative gives the rate of change of T with respect to y, with

$$T_y(9, 5) = 5 + 9 = 14.$$

This result, 14, is the approximate change in temperature resulting from a 1-unit increase in production of electricity from $y = 5$ to $y = 5 + 1 = 6$ (from 500 to 600 megawatts), while the input water temperature x remains constant at 9°C.   ◀ **5**

As mentioned in the previous section, if $P(x, y)$ gives the output P produced by x units of labor and y units of capital, $P(x, y)$ is a production function. The partial derivatives of this production function have practical implications. For example, $\partial P/\partial x$ gives the marginal productivity of labor. This represents the rate at which the output is changing with respect to changes in labor for a fixed capital investment. That is, if the capital investment is held constant and labor is increased by 1 work hour, $\partial P/\partial x$ will yield the approximate change in the production level. Likewise, $\partial P/\partial y$ gives the marginal productivity of capital, which represents the rate at which the output is changing with respect to changes in capital for a fixed labor value. So if the labor force is held constant and the capital investment is increased by 1 unit, $\partial P/\partial y$ will approximate the corresponding change in the production level.

6 Suppose a production function is given by $P(x, y) = 10x^2y + 100x + 400y - 5xy^2$, where x and y are defined as in Example 5. Find the marginal productivity of labor and capital when $x = 30$ and $y = 50$.

Answer:
Marginal productivity of labor is 17,600. Marginal productivity of capital is -5600.

▶**EXAMPLE 5**   A company that manufactures computers has determined that its production function is given by

$$P(x, y) = 500x + 800y + 3x^2y - x^3 - \frac{y^4}{4},$$

where x is the size of the labor force (measured in work hours per week) and y is the amount of capital (measured in units of $1000) invested. Find the marginal productivity of labor and capital when $x = 50$ and $y = 20$, and interpret the results.

The marginal productivity of labor is found by taking the derivative of P with respect to x.

$$\frac{\partial P}{\partial x} = 500 + 6xy - 3x^2$$

$$\frac{\partial P}{\partial x}(50, 20) = 500 + 6(50)(20) - 3(50)^2$$

$$= -1000$$

Thus, if the capital investment is held constant at $20,000 and labor is increased from 50 to 51 work hours per week, production will decrease by about 1000 units. In the same way, the marginal productivity of capital is $\partial P/\partial y$.

$$\frac{\partial P}{\partial y} = 800 + 3x^2 - y^3$$

$$\frac{\partial P}{\partial y}(50, 20) = 800 + 3(50)^2 - (20)^3$$

$$= 300$$

If work hours are held constant at 50 hours per week and the capital investment is increased from $20,000 to $21,000, production will increase by about 300 units.   ◀   **6**

The second derivative of a function of one variable (the derivative of the derivative) is very useful in determining relative maxima and minima. **Second-order partial derivatives** (partial derivatives of a partial derivative) will play a similar role with functions of two variables. But the situation is slightly more complicated here. For instance, the function $f(x, y) = x^2 + xy^3$ has two first-order partial derivatives:

$$f_x = 2x + y^3 \quad \text{and} \quad f_y = 3xy^2.$$

We can take a partial derivative of $f_x = 2x + y^3$ in two ways—with respect to x (producing 2) or with respect to y (producing $3y^2$). And similarly, there are two partial derivatives of f_y. Thus, there will be *four* second-order partial derivatives of the original function f. The notation for second-order partial derivatives is as follows.

7 Let $f(x, y) = 4x^2y^2 - 9xy + 8x^2 - 3y^4$. Find all second partial derivatives.

Answer:
$f_{xx} = 8y^2 + 16$
$f_{yy} = 8x^2 - 36y^2$
$f_{xy} = 16xy - 9$
$f_{yx} = 16xy - 9$

8 Let $f(x, y) = 4e^{x+y} + 2x^3y$. Find all second partial derivatives.

Answer:
$f_{xx} = 4e^{x+y} + 12xy$
$f_{yy} = 4e^{x+y}$
$f_{xy} = 4e^{x+y} + 6x^2$
$f_{yx} = 4e^{x+y} + 6x^2$

Second-Order Partial Derivatives

For a function $z = f(x, y)$, if all indicated partial derivatives exists, then

$$\frac{\partial}{\partial x}\left(\frac{\partial z}{\partial x}\right) = \frac{\partial^2 z}{\partial x^2} = f_{xx} \qquad \frac{\partial}{\partial y}\left(\frac{\partial z}{\partial y}\right) = \frac{\partial^2 z}{\partial y^2} = f_{yy}$$

$$\frac{\partial}{\partial y}\left(\frac{\partial z}{\partial x}\right) = \frac{\partial^2 z}{\partial y \partial x} = f_{xy} \qquad \frac{\partial}{\partial x}\left(\frac{\partial z}{\partial y}\right) = \frac{\partial^2 z}{\partial x \partial y} = f_{yx}.$$

As seen above, f_{xx} is used as an abbreviation for $f_{xx}(x, y)$, with $f_{yy}, f_{xy},$ and f_{yx} used in a similar way. The symbol f_{xx} is read "the partial derivative of f_x with respect to x." Also, the symbol $\partial^2 z/\partial y^2$ is read "the partial derivative of $\partial z/\partial y$ with respect to y."

For many functions, including most functions found in applications and all the functions in this book, the second-order partial derivatives f_{xy} and f_{yx} are equal. Therefore, it is not necessary for us to be careful about finding these second-order partial derivatives in the correct order.

▶**EXAMPLE 6** Find all second-order partial derivatives for

$$f(x, y) = -4x^3 - 3x^2y^3 + 2y^2.$$

First find f_x and f_y.

$$f_x = -12x^2 - 6xy^3 \quad \text{and} \quad f_y = -9x^2y^2 + 4y$$

To find f_{xx}, take the partial derivative of f_x with respect to x.

$$f_{xx} = -24x - 6y^3$$

Take the partial derivative of f_y with respect to y; this gives f_{yy}.

$$f_{yy} = -18x^2y + 4$$

Find f_{xy} by starting with f_x, then taking the partial derivative of f_x with respect to y.

$$f_{xy} = -18xy^2$$

Finally, find f_{yx} by starting with f_y; take its partial derivative with respect to x.

$$f_{yx} = -18xy^2 \quad \blacktriangleleft \quad \boxed{7}$$

▶**EXAMPLE 7** Let $f(x, y) = 2e^x - 8x^3y^2$. Find all second-order partial derivatives.

Here $f_x = 2e^x - 24x^2y^2$ and $f_y = -16x^3y$. (Recall: if $g(x) = e^x$, then $g'(x) = e^x$.) Now find the second partial derivatives.

$$f_{xx} = 2e^x - 48xy^2 \qquad f_{xy} = -48x^2y$$
$$f_{yy} = -16x^3 \qquad f_{yx} = -48x^2y \quad \blacktriangleleft \quad \boxed{8}$$

9 Let $f(x, y, z) = xyz + x^2yz + xy^2z^3$. Find f_x, f_y, f_z, and f_{xz}.

Answer:
$f_x = yz + 2xyz + y^2z^3$
$f_y = xz + x^2z + 2xyz^3$
$f_z = xy + x^2y + 3xy^2z^2$
$f_{xz} = y + 2xy + 3y^2z^2$

Partial derivatives of multivariate functions with more than two independent variables are found in a way similar to that for functions with two independent variables. For example, to find f_x for $w = f(x, y, z)$, treat y and z as constants and differentiate with respect to x.

▶ **EXAMPLE 8** Let $f(x, y, z) = xy^2z + 2x^2y - 4xz^2$. Find f_x, f_y, f_z, f_{xy}, and f_{yz}.

$$f_x = y^2z + 4xy - 4z^2$$
$$f_y = 2xyz + 2x^2$$
$$f_z = xy^2 - 8xz$$

To find f_{xy}, differentiate f_x with respect to y.

$$f_{xy} = 2yz + 4x$$

In the same way, differentiate f_y with respect to z to get

$$f_{yz} = 2xy. \quad ◀ \quad 9$$

14.2 EXERCISES

Find each of the following partial derivatives. (See Examples 1–3.)

1. Let $z = f(x, y) = 12x^2 - 8xy + 3y^2$. Find each of the following.

(a) $\dfrac{\partial z}{\partial x}$ (b) $\dfrac{\partial z}{\partial y}$

(c) $\left(\dfrac{\partial f}{\partial x}\right)(2, 3)$ (d) $f_y(1, -2)$

2. Let $z = g(x, y) = 5x + 9x^2y + y^2$. Find each of the following.

(a) $\dfrac{\partial g}{\partial x}$ (b) $\dfrac{\partial g}{\partial y}$

(c) $\left(\dfrac{\partial z}{\partial y}\right)(-3, 0)$ (d) $g_x(2, 1)$

Find f_x and f_y for the functions defined as follows. Then find $f_x(2, -1)$ and $f_y(-4, 3)$. Leave the answers in terms of e in Exercises 7–10 and 15–16. (See Examples 1–3.)

3. $f(x, y) = -2xy + 6y^3 + 2$

4. $f(x, y) = 4x^2y - 9y^2$

5. $f(x, y) = 3x^3y^2$

6. $f(x, y) = -2x^2y^4$

7. $f(x, y) = e^{x+y}$

8. $f(x, y) = 3e^{2x+y}$

9. $f(x, y) = -5e^{3x-4y}$

10. $f(x, y) = 8e^{7x-y}$

11. $f(x, y) = \dfrac{x^2 + y^3}{x^3 - y^2}$

12. $f(x, y) = \dfrac{3x^2y^3}{x^2 + y^2}$

13. $f(x, y) = \ln(1 + 3x^2y^3)$

14. $f(x, y) = \ln(2x^5 - xy^4)$

15. $f(x, y) = xe^{x^2y}$

16. $f(x, y) = y^2e^{(x+3y)}$

Find all second-order partial derivatives. (See Examples 6 and 7.)

17. $f(x, y) = 6x^3y - 9y^2 + 2x$

18. $g(x, y) = 5xy^4 + 8x^3 - 3y$

19. $R(x, y) = 4x^2 - 5xy^3 + 12y^2x^2$

20. $h(x, y) = 30y + 5x^2y + 12xy^2$

21. $r(x, y) = \dfrac{4x}{x + y}$

22. $k(x, y) = \dfrac{-5y}{x + 2y}$

23. $z = 4xe^y$ **24.** $z = -3ye^x$ **25.** $r = \ln(x + y)$

26. $k = \ln(5x - 7y)$ **27.** $z = x \ln(xy)$ **28.** $z = (y + 1) \ln(x^3 y)$

Find f_x, f_y, f_z, and f_{yz} for the functions defined as follows. (See Example 8.)

29. $f(x, y, z) = x^2 + yz + z^4$ **30.** $f(x, y, z) = 3x^5 - x^2 + y^5$ **31.** $f(x, y, z) = \dfrac{6x - 5y}{4z + 5}$

32. $f(x, y, z) = \dfrac{2x^2 + xy}{yz - 2}$ **33.** $f(x, y, z) = \ln(x^2 - 5xz^2 + y^4)$ **34.** $f(x, y, z) = \ln(8xy + 5yz - x^3)$

Work the following exercises. (See Examples 4 and 5.)

35. Management Suppose that the manufacturing cost of a precision electronic calculator is approximated by

$$M(x, y) = 40x^2 + 30y^2 - 10xy + 30,$$

where x is the cost of electronic chips and y is the cost of labor. Find the following.
(a) $M_y(4, 2)$ (b) $M_x(3, 6)$
(c) $\left(\dfrac{\partial M}{\partial x}\right)(2, 5)$ (d) $\left(\dfrac{\partial M}{\partial y}\right)(6, 7)$

36. Management The revenue from the sale of x units of a tranquilizer and y units of an antibiotic is given by

$$R(x, y) = 5x^2 + 9y^2 - 4xy.$$

(a) Suppose $x = 9$ and $y = 5$. What is the approximate effect on revenue if x is increased to 10 while y is fixed?
(b) Suppose $x = 9$ and $y = 5$. What is the approximate effect on revenue if y is increased to 6 while x is fixed?

37. A car dealership estimates that the total weekly sales of its most popular model is a function of the car's list price, p, and the interest rate in percent, i, offered by the manufacturer. The approximate weekly sales are given by

$$f(p, i) = 132p - 2pi - .01p^2.$$

(a) Find the weekly sales if the average list price is \$9400 and the manufacturer is offering an 8% interest rate.
(b) Find and interpret f_p and f_i.
(c) What would be the effect on weekly sales if the price is \$9400 and interest rates rise from 8% to 9%?

38. Management Suppose the production function of a company is given by

$$P(x, y) = 100\sqrt{x^2 + y^2},$$

where x represents units of labor and y represents units of capital. (See Example 5.) Find the following when $x = 4$ and $y = 3$.
(a) The marginal productivity of labor
(b) The marginal productivity of capital

 39. Management A manufacturer estimates that production (in hundreds of units) is a function of the amounts x and y of labor and capital used, as follows.

$$f(x, y) = \left[\frac{1}{3}x^{-1/3} + \frac{2}{3}y^{-1/3}\right]^{-3}$$

(a) Find the number of units produced when 27 units of labor and 64 units of capital are utilized.
(b) Find and interpret $f_x(27, 64)$ and $f_y(27, 64)$.
(c) What would be the approximate effect on production of increasing labor by 1 unit?

40. Management A manufacturer of automobile batteries estimates that his total production in thousands of units is given by

$$f(x, y) = 3x^{1/3}y^{2/3},$$

where x is the number of units of labor and y is the number of units of capital utilized.
(a) Find and interpret $f_x(64, 125)$ and $f_y(64, 125)$ if the current level of production uses 64 units of labor and 125 units of capital.
(b) What would be the approximate effect on production of increasing labor to 65 units while holding capital at the current level?
(c) Suppose that sales have been good and management wants to increase either capital or labor by 1 unit. Which option would result in a larger increase in production?

41. Management The production function z for the United States was once estimated as

$$z = x^{.7}y^{.3},$$

where x stands for the amount of labor and y stands for the amount of capital. Find the marginal productivity of labor (find $\partial z/\partial x$) and of capital.

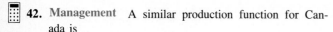

42. Management A similar production function for Canada is

$$z = x^{.4}y^{.6},$$

with x, y, and z as in Exercise 41. Find the marginal productivity of labor and of capital.

43. Natural Science The total number of matings per day between individuals of a certain species of grasshoppers is approximated by

$$M(x, y) = 2xy + 10xy^2 + 30y^2 + 20,$$

where x represents the temperature in °C and y represents the number of days since the last rain. Find the following.

(a) $\left(\dfrac{\partial M}{\partial x}\right)(20, 4)$ **(b)** $\left(\dfrac{\partial M}{\partial y}\right)(24, 10)$

(c) $M_x(17, 3)$ **(d)** $M_y(21, 8)$

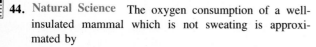

44. Natural Science The oxygen consumption of a well-insulated mammal which is not sweating is approximated by

$$m = m(T, F, w) = \frac{2.5(T - F)}{w^{.67}} = 2.5(T - F)w^{-.67},$$

where T is the internal body temperature of the animal (in °C), F is the temperature of the outside of the animal's fur (in °C), and w is the animal's weight in kilograms.

(a) Find m_T.

(b) Suppose $T = 38°$, $F = 12°$, and $w = 30$ kg. Find $m_T(38, 12, 30)$.

(c) Find m_w.

(d) Suppose $T = 40°$, $F = 20°$, and $w = 40$ kg. Find $m_w(40, 20, 40)$.

(e) Find m_F.

(f) Suppose $T = 36°$, $F = 14°$, and $w = 25$ kg. Find $m_F(36, 14, 25)$.

45. Natural Science The surface area of a human, in square meters, is approximated by

$$A(W, H) = .202W^{.425}H^{.725},$$

where W is the weight of the person in kilograms and H is the height in meters.

(a) Find $\partial A/\partial W$.

(b) Suppose $W = 72$ and $H = 1.8$. Find

$$\left(\frac{\partial A}{\partial W}\right)(72, 1.8).$$

(c) Find $\partial A/\partial H$.

(d) Suppose $W = 70$ and $H = 1.6$. Find

$$\left(\frac{\partial A}{\partial H}\right)(70, 1.6).$$

46. Natural Science In one method of computing the quantity of blood pumped through the lungs in 1 minute, a researcher first finds each of the following (in milliliters).

$b =$ quantity of oxygen used by body in 1 minute

$a =$ quantity of oxygen per liter of blood that has just gone through the lungs

$v =$ quantity of oxygen per liter of blood that is about to enter the lungs

In 1 minute,

     Amount of oxygen used

         = amount of oxygen per liter

            × number of liters of blood pumped.

If C is the number of liters pumped through the blood in 1 minute, then

$$b = (a - v) \cdot C \quad \text{or} \quad C = \frac{b}{a - v}.$$

(a) Find C if $a = 160$, $b = 200$, and $v = 125$.

(b) Find C if $a = 180$, $b = 260$, and $v = 142$. Find the following partial derivatives.

(c) $\partial C/\partial b$ **(d)** $\partial C/\partial v$

14.3 EXTREMA OF FUNCTIONS OF SEVERAL VARIABLES

One of the most important applications of calculus is in finding maxima and minima for functions. Earlier, we studied this idea extensively for functions of a single independent variable; now we will see that extrema can be found for functions of two variables. In particular, an extension of the second derivative test can be defined and used to identify maxima or minima. We begin with the definitions of relative maxima and minima.

Relative Maxima and Minima

Let (a, b) be the center of a circular region contained in the xy-plane. Then, for a function $z = f(x, y)$ defined for every (x, y) in the region,

1. $f(a, b)$ is a **relative maximum** if

$$f(a, b) \geq f(x, y)$$

for all points (x, y) in the circular region;
2. $f(a, b)$ is a **relative minimum** if

$$f(a, b) \leq f(x, y)$$

for all points (x, y) in the circular region.

As before, the word *extremum* is used for either a relative maximum or a relative minimum. Examples of a relative maximum and a relative minimum are given in Figures 14.13 and 14.14.

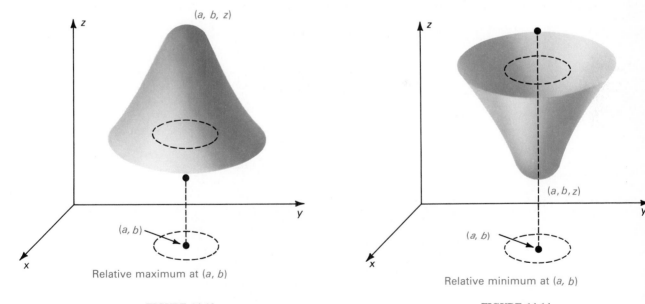

Relative maximum at (a, b)

FIGURE 14.13

Relative minimum at (a, b)

FIGURE 14.14

With functions of a single variable, we made a distinction between relative extrema and absolute extrema. The methods for finding absolute extrema are quite involved for functions of two independent variables, so we will discuss only relative extrema.

As suggested by Figure 14.15, at a relative maximum, the tangent line parallel to the x-axis has a slope of 0, as does the tangent line parallel to the y-axis. (Notice the similarity to functions of one variable.) That is, if the function $z = f(x, y)$ has a relative extremum at (a, b), then $f_x(a, b) = 0$ and $f_y(a, b) = 0$, as stated in the following theorem.

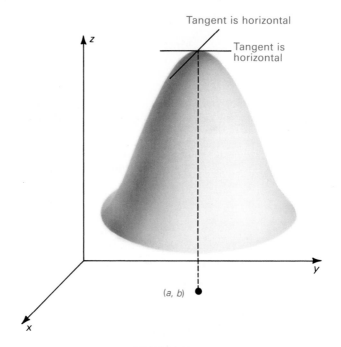

FIGURE 14.15

Location of Extrema

If a function $z = f(x, y)$ has a relative maximum or relative minimum at the point (a, b), and $f_x(a, b)$ and $f_y(a, b)$ both exist, then

$$f_x(a, b) = 0 \quad \text{and} \quad f_y(a, b) = 0.$$

Just as with functions of one variable, the fact that the slopes of the tangent lines are 0 is no guarantee that a relative extremum has been located. For example, Figure 14.16 shows the graph of $z = f(x, y) = x^2 - y^2$. Both $f_x(0, 0) = 0$ and $f_y(0, 0) = 0$, and yet $(0, 0)$ leads to neither a relative maximum nor a relative minimum for the function. The point $(0, 0, 0)$ on the graph of this function is called a **saddle point**; it is a minimum when approached from one direction but a maximum when approached from another direction. A saddle point is neither a maximum nor a minimum.

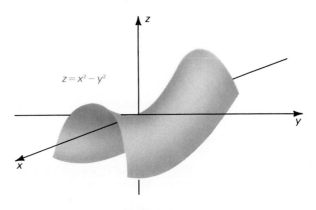

$z = x^2 - y^2$

FIGURE 14.16

The theorem on location of extrema suggests a useful strategy for finding extrema. First, locate all points (a, b) where $f_x(a, b) = 0$ and $f_y(a, b) = 0$. Then test each of these points separately, using the test given after the next example. For a function $f(x, y)$, the points (a, b) such that $f_x(a, b) = 0$ and $f_y(a, b) = 0$ are called **critical points.**

▶**EXAMPLE 1** Find all critical points for

$$f(x, y) = 6x^2 + 6y^2 + 6xy + 36x - 5.$$

We must find all points (a, b) such that $f_x(a, b) = 0$ and $f_y(a, b) = 0$. Here

$$f_x = 12x + 6y + 36 \quad \text{and} \quad f_y = 12y + 6x.$$

Set each of these two partial derivatives equal to 0.

$$12x + 6y + 36 = 0 \quad \text{and} \quad 12y + 6x = 0$$

These two equations make up a system of linear equations. We can use the substitution method to solve this system. First, rewrite $12y + 6x = 0$ as follows.

$$12y + 6x = 0$$
$$6x = -12y$$
$$x = -2y$$

1 Find all critical points for the following.

(a) $f(x, y) = 4x^2 + 3xy + 2y^2 + 7x - 6y - 6$

(b) $f(x, y) = 4x^2 + 6xy + y^2 + 34x + 8y + 5$

Answer:

(a) $(-2, 3)$

(b) $(1, -7)$

Now substitute $-2y$ for x in the other equation.

$$12x + 6y + 36 = 0$$
$$12(-2y) + 6y + 36 = 0$$
$$-24y + 6y + 36 = 0$$
$$-18y + 36 = 0$$
$$-18y = -36$$
$$y = 2$$

The equation $x = -2y$ leads to $x = -2(2) = -4$. The solution of the system of equations is $(-4, 2)$. Since this is the only solution of the system, $(-4, 2)$ is the only critical point for the given function. By the theorem above, if the function has a relative extremum, it will occur at $(-4, 2)$. ◄ **1**

The results of the next theorem can be used to decide whether $(-4, 2)$ in Example 1 leads to a relative maximum, a relative minimum, or neither. The proof of this theorem is beyond the scope of this course.

Test for Relative Extrema

For a function $z = f(x, y)$, assume f_{xx}, f_{yy}, and f_{xy} all exist. Let (a, b) be a point for which

$$f_x(a, b) = 0 \quad \text{and} \quad f_y(a, b) = 0.$$

Define the number M by

$$M = f_{xx}(a, b) \cdot f_{yy}(a, b) - [f_{xy}(a, b)]^2.$$

Then

1. $f(a, b)$ is a **relative maximum** if $M > 0$ and $f_{xx}(a, b) < 0$.

2. $f(a, b)$ is a **relative minimum** if $M > 0$ and $f_{xx}(a, b) > 0$.

3. $f(a, b)$ is a **saddle point** (neither a maximum nor a minimum) if $M < 0$.

4. If $M = 0$, the test gives **no information.**

The chart below summarizes the conclusions of the theorem.

	$f_{xx}(a, b) < 0$	$f_{xx}(a, b) > 0$
$M > 0$	Relative maximum	Relative minimum
$M = 0$	No information	
$M < 0$	Saddle point	

2 Find any relative maxima or minima for the functions defined in Problem 1 at the side.

Answer:

(a) Relative minimum at $(-2, 3)$

(b) Neither a relative minimum nor a relative maximum at $(1, -7)$

▶**EXAMPLE 2**   The previous example showed that the only critical point for the function

$$f(x, y) = 6x^2 + 6y^2 + 6xy + 36x - 5$$

is $(-4, 2)$. Does $(-4, 2)$ lead to a relative maximum, a relative minimum, or neither?

We can find out by using the test above. From Example 1,

$$f_x(-4, 2) = 0 \quad \text{and} \quad f_y(-4, 2) = 0.$$

Now find the various second partial derivatives used in finding M. From $f_x = 12x + 6y + 36$ and $f_y = 12y + 6x$,

$$f_{xx} = 12, \quad f_{yy} = 12, \quad \text{and} \quad f_{xy} = 6.$$

(If these second-order partial derivatives had not all been constants, we would have had to evaluate them at the point $(-4, 2)$.) Now

$$M = f_{xx}(-4, 2) \cdot f_{yy}(-4, 2) - [f_{xy}(-4, 2)]^2 = 12 \cdot 12 - 6^2 = 108.$$

Since $M > 0$ and $f_{xx}(-4, 2) = 12 > 0$, part 2 of the theorem applies, showing that $f(x, y) = 6x^2 + 6y^2 + 6xy + 36x - 5$ has a relative minimum at $(-4, 2)$. This relative minimum is $f(-4, 2) = -77$. ◀ **2**

▶**EXAMPLE 3**   Find all points where the function

$$f(x, y) = 9xy - x^3 - y^3 - 6$$

has any relative maxima or relative minima.

First find any critical points. Here

$$f_x = 9y - 3x^2 \quad \text{and} \quad f_y = 9x - 3y^2.$$

Place each of these partial derivatives equal to 0.

$$
\begin{array}{cc}
f_x = 0 & f_y = 0 \\
9y - 3x^2 = 0 & 9x - 3y^2 = 0 \\
9y = 3x^2 & 9x = 3y^2 \\
3y = x^2 & 3x = y^2
\end{array}
$$

In the first equation $(3y = x^2)$, notice that since $x^2 \geq 0$, $y \geq 0$. Also, in the second equation $(3x = y^2)$, $y^2 \geq 0$, so $x \geq 0$.

The substitution method can be used again to solve the system of equations

$$3y = x^2$$
$$3x = y^2.$$

3 Find any relative extrema for
$$f(x, y) = \frac{2\sqrt{2}}{3}x^3 - xy + \frac{1}{3}y^3$$
$$- 10.$$

Answer:

Relative minimum at $\left(\dfrac{1}{2}, \dfrac{\sqrt{2}}{2}\right)$

The first equation, $3y = x^2$, can be rewritten as $y = x^2/3$. Substitute this into the second equation to get

$$3x = y^2 = \left(\frac{x^2}{3}\right)^2$$

$$3x = \frac{x^4}{9}$$

Solve this equation as follows.

$$27x = x^4$$
$$x^4 - 27x = 0$$
$$x(x^3 - 27) = 0 \qquad \text{Factor}$$
$$x = 0 \quad \text{or} \quad x^3 - 27 = 0 \qquad \text{Set each factor equal to 0}$$
$$x^3 = 27$$
$$x = 3 \qquad \text{Take the cube root on each side}$$

Use these values of x, along with the equation $3x = y^2$, to find y.

If $x = 0$,	If $x = 3$,
$3x = y^2$	$3x = y^2$
$3(0) = y^2$	$3(3) = y^2$
$0 = y^2$	$9 = y^2$
$0 = y$	$3 = y \quad \text{or} \quad -3 = y.$

The points $(0, 0)$, $(3, 3)$ and $(3, -3)$ appear to be critical points; however, $(3, -3)$ does not have $y \geq 0$. The only possible relative extrema for $f(x, y) = 9xy - x^3 - y^3 - 6$ occur at the critical points $(0, 0)$ or $(3, 3)$. To identify any extrema, use the test. Here

$$f_{xx} = -6x, \quad f_{yy} = -6y, \quad \text{and} \quad f_{xy} = 9.$$

Test each of the possible critical points.

For $(0, 0)$:

$f_{xx}(0, 0) = -6(0) = 0$

$f_{yy}(0, 0) = -6(0) = 0$

$f_{xy}(0, 0) = 9$

$M = 0 \cdot 0 - 9^2 = -81.$

Since $M < 0$, there is a saddle point at $(0, 0)$.

For $(3, 3)$:

$f_{xx}(3, 3) = -6(3) = -18$

$f_{yy}(3, 3) = -6(3) = -18$

$f_{xy}(3, 3) = 9$

$M = -18(-18) - 9^2 = 243.$

Here $M > 0$ and $f_{xx}(3, 3) = -18 < 0$; there is a relative maximum at $(3, 3)$. ◀ **3**

▶**EXAMPLE 4**   A company is developing a new soft drink. The cost in dollars to produce a batch of the drink is approximated by

$$C(x, y) = 2200 + 27x^3 - 72xy + 8y^2,$$

where x is the number of kilograms of sugar per batch and y is the number of grams of flavoring per batch.

(a) Find the amounts of sugar and flavoring that result in minimum cost for a batch. Start with the following partial derivatives.

$$C_x = 81x^2 - 72y \quad \text{and} \quad C_y = -72x + 16y$$

Set each of these equal to 0 and solve for y.

$$81x^2 - 72y = 0 \qquad\qquad -72x + 16y = 0$$
$$-72y = -81x^2 \qquad\qquad 16y = 72x$$
$$y = \frac{9}{8}x^2 \qquad\qquad\qquad y = \frac{9}{2}x$$

From the equation on the left, $y \geq 0$. Since $(9/8)x^2$ and $(9/2)x$ both are equal to y, they are equal to each other. Set $(9/8)x^2$ and $(9/2)x$ equal and solve the resulting equation for x.

$$\frac{9}{8}x^2 = \frac{9}{2}x$$
$$9x^2 = 36x$$
$$9x^2 - 36x = 0$$
$$9x(x - 4) = 0$$
$$9x = 0 \quad \text{or} \quad x - 4 = 0$$

The equation $9x = 0$ leads to $x = 0$, which is not a useful answer for our problem. Substitute $x = 4$ into $y = (9/2)x$ to find y.

$$y = \frac{9}{2}x = \frac{9}{2}(4) = 18$$

Now check to see if the critical point $(4, 18)$ leads to a relative minimum. For $(4, 18)$,

$$C_{xx} = 162x = 162(4) = 648, \quad C_{yy} = 16, \quad \text{and} \quad C_{xy} = -72.$$

Also, $$M = (648)(16) - (-72)^2 = 5184.$$

Since $M > 0$ and $C_{xx}(4, 18) > 0$, the cost at $(4, 18)$ is a minimum.

(b) What is the minimum cost?

To find the minimum cost, go back to the cost function and evaluate $C(4, 18)$.

$$C(x, y) = 2200 + 27x^3 - 72xy + 8y^2$$
$$C(4, 18) = 2200 + 27(4)^3 - 72(4)(18) + 8(18)^2 = 1336$$

The minimum cost for a batch is $1336.00. ◀

14.3 EXERCISES

Find all points where the functions defined as follows have any relative extrema. Identify any saddle points. (See Examples 1–3.)

1. $f(x, y) = xy + x - y$

2. $f(x, y) = 4xy + 8x - 9y$

3. $f(x, y) = x^2 - 2xy + 2y^2 + x - 5$

4. $f(x, y) = x^2 + xy + y^2 - 6x - 3$

5. $f(x, y) = x^2 - xy + y^2 + 2x + 2y + 6$

6. $f(x, y) = x^2 + xy + y^2 + 3x - 3y$

7. $f(x, y) = x^2 + 3xy + 3y^2 - 6x + 3y$

8. $f(x, y) = 5xy - 7x^2 - y^2 + 3x - 6y - 4$

9. $f(x, y) = 4xy - 10x^2 - 4y^2 + 8x + 8y + 9$

10. $f(x, y) = x^2 + xy + 3x + 2y - 6$

11. $f(x, y) = x^2 + xy - 2x - 2y + 2$

12. $f(x, y) = x^2 + xy + y^2 - 3x - 5$

13. $f(x, y) = x^2 - y^2 - 2x + 4y - 7$

14. $f(x, y) = 4x + 2y - x^2 + xy - y^2 + 3$

15. $f(x, y) = 2x^3 + 3y^2 - 12xy + 4$

16. $f(x, y) = 5x^3 + 2y^2 - 60xy - 3$

17. $f(x, y) = x^2 + 4y^3 - 6xy - 1$

18. $f(x, y) = 3x^2 + 7y^3 - 42xy + 5$

19. $f(x, y) = e^{xy}$

20. $f(x, y) = x^2 + e^y$

21. $f(x, y) = 3xy - x^3 - y^3 + \dfrac{1}{8}$

22. $f(x, y) = \dfrac{3}{2}x - \dfrac{1}{2}x^3 - xy^2 + \dfrac{1}{16}$

23. $f(x, y) = x^4 - 2x^2 + y^2 + \dfrac{17}{16}$

24. $f(x, y) = 2y^3 - 3x^4 - 6x^2y + \dfrac{1}{16}$

25. $f(x, y) = x^4 - y^4 - 2x^2 + 2y^2 + \dfrac{1}{16}$

26. $f(x, y) = -x^4 + 4xy - 2y^2 + \dfrac{1}{16}$

Work the following exercises. (See Example 4.)

27. Management Suppose that the profit of a certain firm is approximated by

$$P(x, y) = 1000 + 24x - x^2 + 80y - y^2,$$

where x is the cost of a unit of labor and y is the cost of a unit of goods. Find values of x and y that maximize profit. Find the maximum profit.

28. Management The labor cost in dollars for manufacturing a precision camera can be approximated by

$$L(x, y) = \frac{3}{2}x^2 + y^2 - 2x - 2y - 2xy + 68,$$

where x is the number of hours required by a skilled craftsperson and y is the number of hours required by a semiskilled person. Find values of x and y that minimize the labor charge. Find the minimum labor charge.

29. Management The number of roosters that can be fed from x pounds of Super-Hen chicken feed and y pounds of Super-Rooster feed is given by

$$R(x, y) = 800 - 2x^3 + 12xy - y^2.$$

Find the number of pounds of each kind of feed that support the maximum number of roosters.

30. Management The total profit from one acre of a certain crop depends on the amount spent on fertilizer, x, and hybrid seed, y, according to the model

$$P(x, y) = -x^2 + 3xy + 160x - 5y^2 + 200y + 2,600,000.$$

Find values of x and y that lead to maximum profit. Find the maximum profit.

31. Management The total cost to produce x units of electrical tape and y units of packing tape is given by

$$C(x, y) = 2x^2 + 3y^2 - 2xy + 2x - 126y + 3800.$$

Find the number of units of each kind of tape that should be produced so that the total cost is a minimum. Find the minimum total cost.

32. Management The total revenue in thousands of dollars from the sale of x spas and y solar heaters is approximated by

$$R(x, y) = 12 + 74x + 85y - 3x^2 - 5y^2 - 5xy.$$

Find the number of each that should be sold to produce maximum revenue. Find the maximum revenue.

33. Management A rectangular closed box is to be built at minimum cost to hold 27 cubic meters. Since the cost will depend on the surface area, find the dimensions that will minimize the surface area of the box.

34. Management Find the dimensions that will minimize the surface area (and, hence, the cost) of a rectangular fish aquarium, open on top, with a volume of 32 cubic feet.

CHAPTER 14 SUMMARY

KEY TERMS AND SYMBOLS

14.1 $z = f(x, y)$ function of two independent variables

multivariate function
ordered triples
first octant
plane
surface
trace
level curves
paraboloid
production function
Cobb-Douglas production function

14.2 f_x or $\dfrac{\partial f}{\partial x}$ partial derivative of f with respect to x

$\dfrac{\partial}{\partial x}\left(\dfrac{\partial z}{\partial x}\right)$ or $\dfrac{\partial^2 z}{\partial x^2}$ or f_{xx} Second-order partial derivative of $\partial z/\partial x$ (or f_x) with respect to x

$\dfrac{\partial}{\partial y}\left(\dfrac{\partial z}{\partial x}\right)$ or $\dfrac{\partial^2 x}{\partial y \partial x}$ or f_{xy} Second-order partial derivative of $\partial z/\partial x$ (or f_x) with respect to y

14.3 saddle point

KEY CONCEPTS

The graph of $ax + by + cz = d$ is a **plane.**

The graph of $z = f(x, y)$ is a **surface** in three-dimensional space.

The graph of $z = ax^2 + by^2$ is a **paraboloid.**

The **partial derivative of f with respect to x** is the derivative of f found by treating x as a variable and y as a constant.

The **partial derivative of f with respect to y** is the derivative of f found by treating y as a variable and x as a constant.

Second-Order Partial Derivatives	For a function $z = f(x, y)$, if all indicated partial derivatives exist, then

$$\frac{\partial}{\partial x}\left(\frac{\partial z}{\partial x}\right) = \frac{\partial^2 z}{\partial x^2} = f_{xx} \qquad \frac{\partial}{\partial y}\left(\frac{\partial z}{\partial y}\right) = \frac{\partial^2 z}{\partial y^2} = f_{yy}$$

$$\frac{\partial}{\partial y}\left(\frac{\partial z}{\partial x}\right) = \frac{\partial^2 z}{\partial y \partial x} = f_{xy} \qquad \frac{\partial}{\partial x}\left(\frac{\partial z}{\partial y}\right) = \frac{\partial^2 z}{\partial x \partial y} = f_{yx}.$$

Let (a, b) be the center of a circular region in the xy plane. For a function $z = f(x, y)$, defined for every (x, y) in the region, $f(a, b)$ is a **relative maximum** if $f(a, b) \geq f(x, y)$ for all (x, y) in the circular region. $f(a, b)$ is a **relative minimum** if $f(a, b) \leq f(x, y)$ for all (x, y) in the circular region.

Location of Extrema

If $f(a, b)$ is a relative extremum, then $f_x(a, b) = 0$ and $f_y(a, b) = 0$.

Test for Relative Extrema

Let f_{xx}, f_{yy}, and f_{xy} all exist. Let (a, b) be a point for which $f_x(a, b) = 0$ and $f_y(a, b) = 0$. If $M = f_{xx}(a, b) \cdot f_{yy}(a, b) - [f_{xy}(a, b)]^2$, then

$f(a, b)$ is a **relative maximum** if $M > 0$ and $f_{xx}(a, b) < 0$.
$f(a, b)$ is a **relative minimum** if $M > 0$ and $f_{xx}(a, b) > 0$.
$f(a, b)$ is a **saddle point** if $M < 0$.

If $M = 0$, the test gives **no information.**

CHAPTER 14 REVIEW EXERCISES

Find $f(-1, 2)$ and $f(6, -3)$ for each of the following.

1. $f(x, y) = -4x^2 + 6xy - 3$ **2.** $f(x, y) = 3x^2y^2 - 5x + 2y$ **3.** $f(x, y) = \dfrac{x - 3y}{x + 4y}$ **4.** $f(x, y) = \dfrac{\sqrt{x^2 + y^2}}{x - y}$

Graph the first-octant portion of each plane.

5. $x + y + z = 4$ **6.** $x + y + 4z = 8$ **7.** $5x + 2y = 10$

8. $3x + 5z = 15$ **9.** $x = 3$ **10.** $y = 2$

Work the following exercise.

11. The charge in dollars for painting a sports car is given by

$$C(x, y) = 2x^2 + 4y^2 - 3xy + \sqrt{x},$$

where x is the number of hours of labor needed and y is the number of gallons of paint and sealer used. Find each of the following.
(a) $C(10, 5)$ **(b)** $C(15, 10)$ **(c)** $C(20, 20)$

12. Let $z = f(x, y) = -5x^2 + 7xy - y^2$. Find each of the following.
(a) $\dfrac{\partial z}{\partial x}$ **(b)** $\left(\dfrac{\partial z}{\partial y}\right)(-1, 4)$ **(c)** $f_{xy}(2, -1)$

13. Let $z = f(x, y) = \dfrac{x + y^2}{x - y^2}$. Find each of the following.
(a) $\dfrac{\partial z}{\partial y}$ **(b)** $\left(\dfrac{\partial z}{\partial x}\right)(0, 2)$ **(c)** $f_{xx}(-1, 0)$

Find f_x and f_y.

14. $f(x, y) = 9x^3y^2 - 5x$

15. $f(x, y) = 6x^5y - 8xy^9$

16. $f(x, y) = \sqrt{4x^2 + y^2}$

17. $f(x, y) = \dfrac{2x + 5y^2}{3x^2 + y^2}$

18. $f(x, y) = x^2 \cdot e^{2y}$

19. $f(x, y) = (y - 2)^2 \cdot e^{(x+2y)}$

20. $f(x, y) = \ln(2x^2 + y^2)$

21. $f(x, y) = \ln(2 - x^2y^3)$

Find f_{xx} and f_{xy}.

22. $f(x, y) = 4x^3y^2 - 8xy$

23. $f(x, y) = -6xy^4 + x^2y$

24. $f(x, y) = \dfrac{2x}{x - 2y}$

25. $f(x, y) = \dfrac{3x + y}{x - 1}$

26. $f(x, y) = x^2e^y$

27. $f(x, y) = ye^{x^2}$

28. $f(x, y) = \ln(2 - x^2y)$

29. $f(x, y) = \ln(1 + 3xy^2)$

Find all points where the functions defined below have any relative extrema. Find any saddle points.

30. $z = x^2 + 2y^2 - 4y$

31. $z = x^2 + y^2 + 9x - 8y + 1$

32. $f(x, y) = x^2 + 5xy - 10x + 3y^2 - 12y$

33. $z = x^3 - 8y^2 + 6xy + 4$

34. $z = \dfrac{1}{2}x^2 + \dfrac{1}{2}y^2 + 2xy - 5x - 7y + 10$

35. $f(x, y) = 3x^2 + 2xy + 2y^2 - 3x + 2y - 9$

36. $z = x^3 + y^2 + 2xy - 4x - 3y - 2$

37. $f(x, y) = 7x^2 + y^2 - 3x + 6y - 5xy$

---

38. Management The manufacturing cost in dollars for a medium-sized business computer is given by

$$c(x, y) = 2x + y^2 + 4xy + 25,$$

where x is the memory capacity of the computer in kilobytes and y is the number of hours of labor required. Find each of the following.

(a) $\dfrac{\partial c}{\partial x}(64, 6)$ **(b)** $\dfrac{\partial c}{\partial y}(128, 12)$

39. Management The production function z for one country is

$$z = x^{.6}y^{.4},$$

where x represents the amount of labor and y the amount of capital. Find the marginal productivity of each of the following.

(a) Labor   **(b)** Capital

40. Management The total cost in dollars to manufacture x solar cells and y solar collectors is

$$c(x, y) = x^2 + 5y^2 + 4xy - 70x - 164y + 1800.$$

(a) Find values of x and y that produce minimum total cost.

(b) Find the minimum total cost.

Appendix A   Graphing Calculators

Appendix B   Tables

Answers to Selected Exercises

Index

Index of Applications

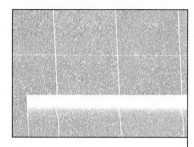

APPENDIX A

Graphing Calculators*

Graphing calculators are a recent result of the amazingly rapid evolution in computer technology aimed at packaging more power into smaller boxes. These new machines have powerful graphing capabilities in addition to the full range of features found on the latest programmable scientific calculators. Instead of a one-line display, graphing calculators typically can show up to eight lines of text. This makes it much easier to keep track of the steps of your work whether you are doing routine computations, entering a long mathematical function, or writing a program. Like programmable scientific calculators, graphing calculators are capable of doing many things we had previously come to expect only from computers. Programs can be written relatively easily and, after they are stored in memory, the programs are always available. Like computers, graphing calculators can be programmed to include graphic displays as part of the program.

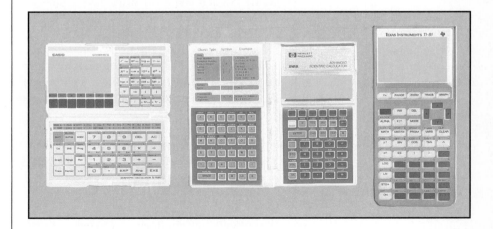

---

*Prepared by Jim Eckerman of American River College.

Graphing calculators have 49 or more keys. Most modern desk-top computers have 101 keys on their keyboards. With fewer keys, each key must be used for more actions, so you will find special mode-changing keys such as 2ND, SHIFT, ALPHA, and MODE. It takes some study to learn how to use graphing calculators, but they are much easier to master than most computers, and they can go wherever you do. Some models have exposed keys all in one box, while others, such as the Hewlett-Packard HP28S and the Casio FX7500G, come in clam-shell type cases that are quite durable. The Casio FX7500G has a sealed, touch-sensitive keyboard that is much less likely to be damaged by spilled liquids, dust, and so on. However, its display screen is slightly smaller than the other Casio models.

New models with added features, more memory, greater ease of use, and other improvements are frequently being introduced. In addition to the Casio models from the FX7000G to the FX8500G, there is the Sharp EL5000, the Texas Instruments TI-81, the Hewlett-Packard HP28S and HP48SX. At this time, the most recent entries in the graphing calculator market are the Texas Instruments TI-81 and the HP48SX. The TI-81, the Sharp EL5000, and the Casio FX7500G are very similar in both price and features.

At 2.4K, the TI-81 has less memory than the Sharp or the Casio, which have 8K and 4K, respectively. However, the TI-81 has pull-down display menus, can retain up to 37 different programs, and has larger keys that are more spread out to make the machine easier to use. In addition, the TI-81 has some other built-in features that must be programmed by the user into the Casio and Sharp machines. Although the Sharp has matrix operations, both it and the Casio must be programmed to draw graphs of parametrically defined equations, for example. Of more importance to students using this text is the box zoom feature. This is a menu choice on the TI-81, but must be programmed into the Casio and the Sharp machines.

The more expensive machines, such as the Casio FX8000G and FX8500G and the HP28S and HP48SX, offer more memory and the ability to attach peripherals such as printers and cassette storage devices. The HP machines also have what is called Symbolic Algebra Systems capability (SAS). Machines with SAS capability can solve equations algebraically, factor and multiply polynomials, find derivatives and integrals in calculus, and can perform many other symbolic operations. The HP machines are difficult to use due to their increased complexity and to the fact that they do not use standard algebraic entry of operations.

TASKS FOR WHICH THE FUNCTION GRAPHING CAPABILITY IS USEFUL

Some mathematical procedures that are quite difficult or even impossible to do algebraically can be done easily and to a very high degree of accuracy using a graphing calculator. Listed below are some of those procedures that are included in this text.

GRAPHING FUNCTIONS These machines can graph any mathematical function that can be put into rectangular, polar, or parametric form. The latter two are used primarily in technical fields, although occasional uses for the parametric form may occur in business and economics. (See Section 2.2 for graphs of functions in rectangular form.)

SOLVING EQUATIONS OF THE FORM $f(x) = k$ This can be done by graphing both $y = f(x)$ and the horizontal line $y = k$ in the same screen and zooming in on the point or points of intersection. As an example, consider trying to find the annual interest rate, x, that is required in order to have an initial deposit of $1000 grow to $5000 in value in 10 years if the interest is compounded monthly. The equation

$$5000 = 1000\left(1 + \frac{x}{12}\right)^{120}$$

cannot be solved algebraically (see Sections 4.4, 5.2, and 5.5). However, if we graph the two functions

$$y = 1000\left(1 + \frac{x}{12}\right)^{120} \quad \text{and} \quad y = 5000$$

in the same screen and then zoom in on the point of intersection of the two graphs, we can find the interest rate to any desired degree of accuracy. Using values of x from 0 to 0.2 and values of y from 0 to 6000, the picture you will see should look like Figure 1. After zooming in on the point of intersection repeatedly, you will see a picture similar to Figure 2. Notice that no units are shown along the axes. You must look at the RANGE screen to determine what the tic marks represent numerically. In Figure 1, the x-values shown are from 0 to 0.2 and the y-values are from 0 to 6000. In Figure 2, they are 0.162018 to 0.162038 and 4999.7 to 5000.3, respectively. By using TRACE, the x-coordinate of the point of intersection is 0.162028. Thus, the required annual interest rate is 16.2028%. This is accurate to six significant digits, and no algebraic manipulations were needed.

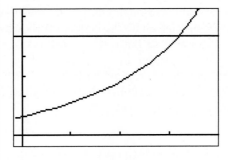

FIGURE 1

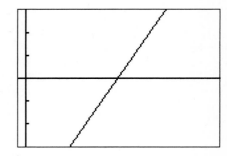

FIGURE 2

SOLVING SYSTEMS OF TWO EQUATIONS IN TWO UNKNOWNS   If both equations can be solved explicitly for one variable in terms of the other, then the two equations can be put into the forms $y = f(x)$ and $y = g(x)$. Then we can graph both in the same screen and zoom in on the point or points where the two curves intersect. (See Sections 6.1 and 7.1.)

FINDING MAXIMUM AND MINIMUM FUNCTION VALUES   Arrange the function in the form $y = f(x)$, graph it over its meaningful domain, and zoom in on the highest or lowest point. As an example, suppose y represents the profit in millions of dollars to be made on the production of x widgets. Suppose also that y is related to x according to the formula

$$y = 5 - 0.000000167(x - 6000)^2 + 0.000001\left(\frac{x - 6000}{100}\right)^3.$$

The graph of this function for $0 \leq x \leq 13{,}000$ and for $-1 \leq y \leq 6$ is shown in Figure 3.

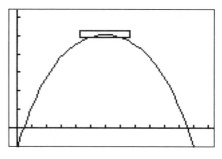

FIGURE 3

The rectangle is what you will see when using a box zoom. This method is built into the TI-80 and can be programmed into the Casio machines. This is the best way to find maximum and minimum points. Make sure that the branches of the curve exit the box through either the bottom (as shown) or the top. If this is not done, the graph will become so flat that you will not be able to tell where the highest (or lowest) point is located.

After zooming in repeatedly with the box placed so that the branches of the curve always exit the bottom of the box, a graph that appears very similar to the one in Figure 3 can be obtained, but with RANGE values close to

$$5999.8 \leq x \leq 6000.2 \quad \text{and} \quad 4.99999999 \leq y \leq 5.00000001.$$

Thus, the maximum profit (y) is 5 million dollars when 6000 widgets are produced and sold. (See Sections 3.2 and 12.1.)

SOME IMPORTANT CAPABILITIES TO CONSIDER

To get the most return on your investment, learn to use as many features of your machine as possible. (Of course, some of the features may not be of use to you. For example, the trigonometric function keys (SIN, COS, and TAN) would be of use to engineers and scientists, but not normally to accountants or management students.) A first session of two or three hours with your machine and your user's manual is essential. Several sessions of this length will be needed to become familiar enough with the HP machines to get started. Be sure to keep your manual handy, referring to it when needed. Listed below are topics that are important to focus on early.

1. Become familiar with the capabilities of the machine, the layout of the keyboard, how to adjust the contrast, and so on.
2. Learn how to obtain quick graphs of the basic functions such as $y = x^2$, $y = e^x$, and so forth. On the Casio machines, for instance, all you have to do to get the graph of $y = \log x$ is press three keys: $<$GRAPH–log–EXE$>$. In just a few seconds, the calculator will plot this graph in its most appropriate window. (Note: the word *window* is used to refer to a particular rectangular portion of the xy plane.)
3. Learn to set RANGE values to delineate a window that is appropriate for the function you are graphing before using the GRAPH key. This involves keying in the minimum and maximum values of x and y that will be displayed on the screen, along with the distance to be used between tic marks along the axes. If you do not do this, you will often find your graph screen blank! Usually one can quickly find a point on the graph by substituting either 0 or 1 for x. Use the RANGE command to set the x-values to the left and right of the x-coordinate of this point and set the y-values above and below the y-coordinate of this point. Then the zoom out feature can be used to see more of the graph.
4. Learn how to redraw graphs without reentering the function. Often you will need to change the RANGE settings several times before you get the window that is most appropriate for your function. All graphing calculators allow you to do this without reentering the function. If you plan to graph a particular function often, then you should write a program and store it in memory. For example, the following two-line program clears the screen of a Casio and then directs it to draw the graph of the function $y = \sqrt{x + 1}$.

CLS
GRAPH $Y = \sqrt{\ } (x + 1)$

Once entered into the machine's memory, this program can be run at any time. If you wish, you can include settings for the range within the program by inserting the line <RANGE −10,100,10,−1,10,1 > ahead of the <GRAPH Y = > line. The resulting graph will then appear similar to Figure 4.

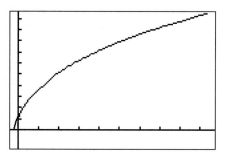

FIGURE 4

5. Learn how to use the ZOOM, TRACE, and PLOT features. All graphing calculators (except the earliest versions of the CASIO FX7000G model) have these abilities. The zoom feature allows a quick redrawing of your graph in either smaller windows (zoom in) or larger windows (zoom out). Thus, one can easily examine the behavior of a function within the close vicinity of a particular point or the general behavior as seen from farther away. With the TRACE feature, the cursor will follow the last curve plotted and the values of x and y will be displayed for each point plotted on the screen. With PLOT you can move the cursor to any point on the screen and either have the machine plot a single point or have the coordinates of the point displayed.

6. Learn how to **edit** expressions, text, and formulas. With the TI-81, pressing the Y = key accesses the Y = edit screen where functions to be graphed are entered. The Casio machines will display the formula for the function just graphed when you press one of the keys used to move the cursor. You may use these keys to move the cursor to wherever you would like to make a change, enter the change, and press EXE. The new graph will be plotted right over the old one. You can edit programs in a similar way and the changes will be recorded in memory. It will be helpful if you switch between INSERT editing and OVERSTRIKE editing as needed when editing.

7. Learn how to enter and read data in **scientific notation** form. This form is used when the numbers become too large or too small (too many zeros between the decimal point and the first significant digit) for the machine's display.

8. Learn how to **program** your machine. Many formulas are used often enough to justify automating the process of evaluating them. The quadratic formula and the formulas encountered in the mathematics of finance are good examples of such formulas. (See Sections 1.10, 3.1, and all of Chapter 5.) Programs that will ask for values of the variables, evaluate the formula, and then start over again can be written very easily. Other situations where programs can be very helpful include graphing a particular function repeatedly, repeated simulations of business situations, and processes that involve successive iterations such as the step-by-step amortization of a loan. There may be sample programs included in the user's manual that will be useful to you. In general, any set of procedures done on one of these machines can be programmed into the machine as long as there is enough memory available.

SOME SUGGESTIONS ON REDUCING FRUSTRATION

We all find ways to make even the simplest machines do the wrong things without our even trying. One of the more common problems with graphing calculators is getting a blank screen when a graph was expected. This usually results from not setting the RANGE values appropriately before graphing the function, although it could also easily result from incorrectly entering the function.

Another problem that is particularly annoying is interpreting cryptic error messages such as SYN ERROR and MA ERROR. (Keep that manual handy!) The most common mistakes are made entering formulas and using special functions. For example, on the Casio machines SYN ERROR means that a mistake was made when entering the function or operation, such as using the $(-)$ key instead of the $-$ key or having two right parentheses and only one left parenthesis. To confuse us further, these same machines think it is perfectly acceptable to have more left parentheses than right parentheses. For instance, the expression $5(3 - 4(2 + 7$ has two left parentheses and no right parentheses, but it will be evaluated as $5(3 - 4(2 + 7))$. The message MA ERROR appears when a number is too large or when a number is not allowed. Try to find the 101st power of 10 or the square root of -4 on any machine other than one of the Hewlett-Packards,* for example, and you will see this kind of error message. By pressing one of the CURSOR keys on the Casio you will see the cursor blinking at the location of the error in your expression. When the TI-81 detects an error, it displays a special menu that lists a code number and type of error and, for certain types of errors, offers the choice GO TO ERROR.

---

*The HP48SX displays 1.E101 in response to being asked to find the 101st power of 10, and it will display (0, 2) when asked to find the square root of -4. The "(0, 2)" represents the complex number $0 + 2i$. The HP can handle any real number smaller than the 500th power of 10. It displays 9.99999999999E499 for all numbers larger than the 500th power of 10.

When graphing **rational functions** you sometimes will see strange little blips in an otherwise smooth graph. This usually means there is a **vertical asymptote** at that location but you can't see the actual behavior there because your window is too large. Zoom in on the graph in the region of the irregularity to obtain a better view. (See Section 3.4.)

While studying mathematics it is important to learn the mathematical concepts well enough to make intelligent decisions about when to use and when not to use high-tech aids such as computers and graphing calculators. These machines make it easy to experiment with graphs of mathematical relations. One can learn much about the behavior of different types of functions by playing what-if games with the formulas. However, in a timed test situation, you may find yourself spending too much time working with the graphing calculator when a quick (and adequate) sketch with pencil and paper is appropriate.

These machines are fun to use, but they can be addictive; so set time limits for yourself, or you may find the machine more of a detriment than a help!

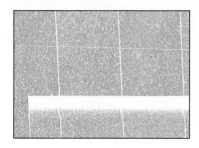

APPENDIX B

Tables

TABLE 1 POWERS OF e

x	e^x	e^{-x}	x	e^x	e^{-x}
0.00	1.00000	1.00000	1.60	4.95302	0.20189
0.01	1.01005	0.99004	1.70	5.47394	0.18268
0.02	1.02020	0.98019	1.80	6.04964	0.16529
0.03	1.03045	0.97044	1.90	6.68589	0.14956
0.04	1.04081	0.96078	2.00	7.38905	0.13533
0.05	1.05127	0.95122			
0.06	1.06183	0.94176	2.10	8.16616	0.12245
0.07	1.07250	0.93239	2.20	9.02500	0.11080
0.08	1.08328	0.92311	2.30	9.97417	0.10025
0.09	1.09417	0.91393	2.40	11.02316	0.09071
0.10	1.10517	0.90483	2.50	12.18248	0.08208
			2.60	13.46372	0.07427
0.11	1.11628	0.89583	2.70	14.87971	0.06720
0.12	1.12750	0.88692	2.80	16.44463	0.06081
0.13	1.13883	0.87810	2.90	18.17412	0.05502
0.14	1.15027	0.86936	3.00	20.08551	0.04978
0.15	1.16183	0.86071			
0.16	1.17351	0.85214	3.50	33.11545	0.03020
0.17	1.18530	0.84366	4.00	54.59815	0.01832
0.18	1.19722	0.83527	4.50	90.01713	0.01111
0.19	1.20925	0.82696			
			5.00	148.41316	0.00674
0.20	1.22140	0.81873	5.50	224.69193	0.00409
0.30	1.34985	0.74081			
0.40	1.49182	0.67032	6.00	403.42879	0.00248
0.50	1.64872	0.60653	6.50	665.14163	0.00150
0.60	1.82211	0.54881			
0.70	2.01375	0.49658	7.00	1096.63316	0.00091
0.80	2.22554	0.44932	7.50	1808.04241	0.00055
0.90	2.45960	0.40656	8.00	2980.95799	0.00034
1.00	2.71828	0.36787	8.50	4914.76884	0.00020
1.10	3.00416	0.33287			
1.20	3.32011	0.30119	9.00	8103.08392	0.00012
1.30	3.66929	0.27253	9.50	13359.72683	0.00007
1.40	4.05519	0.24659			
1.50	4.48168	0.22313	10.00	22026.46579	0.00005

TABLE 2   NATURAL LOGARITHMS

x	ln x	x	ln x	x	ln x
0.0		4.5	1.5041	9.0	2.1972
0.1	−2.3026	4.6	1.5261	9.1	2.2083
0.2	−1.6094	4.7	1.5476	9.2	2.2192
0.3	−1.2040	4.8	1.5686	9.3	2.2300
0.4	−0.9163	4.9	1.5892	9.4	2.2407
0.5	−0.6931	5.0	1.6094	9.5	2.2513
0.6	−0.5108	5.1	1.6292	9.6	2.2618
0.7	−0.3567	5.2	1.6487	9.7	2.2721
0.8	−0.2231	5.3	1.6677	9.8	2.2824
0.9	−0.1054	5.4	1.6864	9.9	2.2925
1.0	0.0000	5.5	1.7047	10	2.3026
1.1	0.0953	5.6	1.7228	11	2.3979
1.2	0.1823	5.7	1.7405	12	2.4849
1.3	0.2624	5.8	1.7579	13	2.5649
1.4	0.3365	5.9	1.7750	14	2.6391
1.5	0.4055	6.0	1.7918	15	2.7081
1.6	0.4700	6.1	1.8083	16	2.7726
1.7	0.5306	6.2	1.8245	17	2.8332
1.8	0.5878	6.3	1.8405	18	2.8904
1.9	0.6419	6.4	1.8563	19	2.9444
2.0	0.6931	6.5	1.8718	20	2.9957
2.1	0.7419	6.6	1.8871	25	3.2189
2.2	0.7885	6.7	1.9021	30	3.4012
2.3	0.8329	6.8	1.9169	35	3.5553
2.4	0.8755	6.9	1.9315	40	3.6889
2.5	0.9163	7.0	1.9459	45	3.8067
2.6	0.9555	7.1	1.9601	50	3.9120
2.7	0.9933	7.2	1.9741	55	4.0073
2.8	1.0296	7.3	1.9879	60	4.0943
2.9	1.0647	7.4	2.0015	65	4.1744
3.0	1.0986	7.5	2.0149	70	4.2485
3.1	1.1314	7.6	2.0281	75	4.3175
3.2	1.1632	7.7	2.0412	80	4.3820
3.3	1.1939	7.8	2.0541	85	4.4427
3.4	1.2238	7.9	2.0669	90	4.4998
3.5	1.2528	8.0	2.0794	95	4.5539
3.6	1.2809	8.1	2.0919	100	4.6052
3.7	1.3083	8.2	2.1041		
3.8	1.3350	8.3	2.1163		
3.9	1.3610	8.4	2.1281		
4.0	1.3863	8.5	2.1401		
4.1	1.4110	8.6	2.1518		
4.2	1.4351	8.7	2.1633		
4.3	1.4586	8.8	2.1748		
4.4	1.4816	8.9	2.1861		

TABLE 3 COMMON LOGARITHMS

n	0	1	2	3	4	5	6	7	8	9
1.0	.0000	.0043	.0086	.0128	.0170	.0212	.0253	.0294	.0334	.0374
1.1	.0414	.0453	.0492	.0531	.0569	.0607	.0645	.0682	.0719	.0755
1.2	.0792	.0828	.0864	.0899	.0934	.0969	.1004	.1038	.1072	.1106
1.3	.1139	.1173	.1206	.1239	.1271	.1303	.1335	.1367	.1399	.1430
1.4	.1461	.1492	.1523	.1553	.1584	.1614	.1644	.1673	.1703	.1732
1.5	.1761	.1790	.1818	.1847	.1875	.1903	.1931	.1959	.1987	.2014
1.6	.2041	.2068	.2095	.2122	.2148	.2175	.2201	.2227	.2253	.2279
1.7	.2304	.2330	.2355	.2380	.2405	.2430	.2455	.2480	.2504	.2529
1.8	.2553	.2577	.2601	.2625	.2648	.2672	.2695	.2718	.2742	.2765
1.9	.2788	.2810	.2833	.2856	.2878	.2900	.2923	.2945	.2967	.2989
2.0	.3010	.3032	.3054	.3075	.3096	.3118	.3139	.3160	.3181	.3201
2.1	.3222	.3243	.3263	.3284	.3304	.3324	.3345	.3365	.3385	.3404
2.2	.3424	.3444	.3464	.3483	.3502	.3522	.3541	.3560	.3579	.3598
2.3	.3617	.3636	.3655	.3674	.3692	.3711	.3729	.3747	.3766	.3784
2.4	.3802	.3820	.3838	.3856	.3874	.3892	.3909	.3927	.3945	.3962
2.5	.3979	.3997	.4014	.4031	.4048	.4065	.4082	.4099	.4116	.4133
2.6	.4150	.4166	.4183	.4200	.4216	.4232	.4249	.4265	.4281	.4298
2.7	.4314	.4330	.4346	.4362	.4378	.4393	.4409	.4425	.4440	.4456
2.8	.4472	.4487	.4502	.4518	.4533	.4548	.4564	.4579	.4594	.4609
2.9	.4624	.4639	.4654	.4669	.4683	.4698	.4713	.4728	.4742	.4757
3.0	.4771	.4786	.4800	.4814	.4829	.4843	.4857	.4871	.4886	.4900
3.1	.4914	.4928	.4942	.4955	.4969	.4983	.4997	.5011	.5024	.5038
3.2	.5051	.5065	.5079	.5092	.5105	.5119	.5132	.5145	.5159	.5172
3.3	.5185	.5198	.5211	.5224	.5237	.5250	.5263	.5276	.5289	.5302
3.4	.5315	.5328	.5340	.5353	.5366	.5378	.5391	.5403	.5416	.5428
3.5	.5441	.5453	.5465	.5478	.5490	.5502	.5514	.5527	.5539	.5551
3.6	.5563	.5575	.5587	.5599	.5611	.5623	.5635	.5647	.5658	.5670
3.7	.5682	.5694	.5705	.5717	.5729	.5740	.5752	.5763	.5775	.5786
3.8	.5798	.5809	.5821	.5832	.5843	.5855	.5866	.5877	.5888	.5899
3.9	.5911	.5922	.5933	.5944	.5955	.5966	.5977	.5988	.5999	.6010
4.0	.6021	.6031	.6042	.6053	.6064	.6075	.6085	.6096	.6107	.6117
4.1	.6128	.6138	.6149	.6160	.6170	.6180	.6191	.6201	.6212	.6222
4.2	.6232	.6243	.6253	.6263	.6274	.6284	.6294	.6304	.6314	.6325
4.3	.6335	.6345	.6355	.6365	.6375	.6385	.6395	.6405	.6415	.6425
4.4	.6435	.6444	.6454	.6464	.6474	.6484	.6493	.6503	.6513	.6522
4.5	.6532	.6542	.6551	.6561	.6571	.6580	.6590	.6599	.6609	.6618
4.6	.6628	.6637	.6646	.6656	.6665	.6675	.6684	.6693	.6702	.6712
4.7	.6721	.6730	.6739	.6749	.6758	.6767	.6776	.6785	.6794	.6803
4.8	.6812	.6821	.6830	.6839	.6848	.6857	.6866	.6875	.6884	.6893
4.9	.6902	.6911	.6920	.6928	.6937	.6946	.6955	.6964	.6972	.6981
5.0	.6990	.6998	.7007	.7016	.7024	.7033	.7042	.7050	.7059	.7067
5.1	.7076	.7084	.7093	.7101	.7110	.7118	.7126	.7135	.7143	.7152
5.2	.7160	.7168	.7177	.7185	.7193	.7202	.7210	.7218	.7226	.7235
5.3	.7243	.7251	.7259	.7267	.7275	.7284	.7292	.7300	.7308	.7316
5.4	.7324	.7332	.7340	.7348	.7356	.7364	.7372	.7380	.7388	.7396
n	0	1	2	3	4	5	6	7	8	9

TABLE 3   (CONTINUED)

n	0	1	2	3	4	5	6	7	8	9
5.5	.7404	.7412	.7419	.7427	.7435	.7443	.7451	.7459	.7466	.7474
5.6	.7482	.7490	.7497	.7505	.7513	.7520	.7528	.7536	.7543	.7551
5.7	.7559	.7566	.7574	.7582	.7589	.7597	.7604	.7612	.7619	.7627
5.8	.7634	.7642	.7649	.7657	.7664	.7672	.7679	.7686	.7694	.7701
5.9	.7709	.7716	.7723	.7731	.7738	.7745	.7752	.7760	.7767	.7774
6.0	.7782	.7789	.7796	.7803	.7810	.7818	.7825	.7832	.7839	.7846
6.1	.7853	.7860	.7868	.7875	.7882	.7889	.7896	.7903	.7910	.7917
6.2	.7924	.7931	.7938	.7945	.7952	.7959	.7966	.7973	.7980	.7987
6.3	.7993	.8000	.8007	.8014	.8021	.8028	.8035	.8041	.8048	.8055
6.4	.8062	.8069	.8075	.8082	.8089	.8096	.8102	.8109	.8116	.8122
6.5	.8129	.8136	.8142	.8149	.8156	.8162	.8169	.8176	.8182	.8189
6.6	.8195	.8202	.8209	.8215	.8222	.8228	.8235	.8241	.8248	.8254
6.7	.8261	.8267	.8274	.8280	.8287	.8293	.8299	.8306	.8312	.8319
6.8	.8325	.8331	.8338	.8344	.8351	.8357	.8363	.8370	.8376	.8382
6.9	.8388	.8395	.8401	.8407	.8414	.8420	.8426	.8432	.8439	.8445
7.0	.8451	.8457	.8463	.8470	.8476	.8482	.8488	.8494	.8500	.8506
7.1	.8513	.8519	.8525	.8531	.8537	.8543	.8549	.8555	.8561	.8567
7.2	.8573	.8579	.8585	.8591	.8597	.8603	.8609	.8615	.8621	.8627
7.3	.8633	.8639	.8645	.8651	.8657	.8663	.8669	.8675	.8681	.8686
7.4	.8692	.8698	.8704	.8710	.8716	.8722	.8727	.8733	.8739	.8745
7.5	.8751	.8756	.8762	.8768	.8774	.8779	.8785	.8791	.8797	.8802
7.6	.8808	.8814	.8820	.8825	.8831	.8837	.8842	.8848	.8854	.8859
7.7	.8865	.8871	.8876	.8882	.8887	.8893	.8899	.8904	.8910	.8915
7.8	.8921	.8927	.8932	.8938	.8943	.8949	.8954	.8960	.8965	.8971
7.9	.8976	.8982	.8987	.8993	.8998	.9004	.9009	.9015	.9020	.9025
8.0	.9031	.9036	.9042	.9047	.9053	.9058	.9063	.9069	.9074	.9079
8.1	.9085	.9090	.9096	.9101	.9106	.9112	.9117	.9122	.9128	.9133
8.2	.9138	.9143	.9149	.9154	.9159	.9165	.9170	.9175	.9180	.9186
8.3	.9191	.9196	.9201	.9206	.9212	.9217	.9222	.9227	.9232	.9238
8.4	.9243	.9248	.9253	.9258	.9263	.9269	.9274	.9279	.9284	.9289
8.5	.9294	.9299	.9304	.9309	.9315	.9320	.9325	.9330	.9335	.9340
8.6	.9345	.9350	.9355	.9360	.9365	.9370	.9375	.9380	.9385	.9390
8.7	.9395	.9400	.9405	.9410	.9415	.9420	.9425	.9430	.9435	.9440
8.8	.9445	.9450	.9455	.9460	.9465	.9469	.9474	.9479	.9484	.9489
8.9	.9494	.9499	.9504	.9509	.9513	.9518	.9523	.9528	.9533	.9538
9.0	.9542	.9547	.9552	.9557	.9562	.9566	.9571	.9576	.9581	.9586
9.1	.9590	.9595	.9600	.9605	.9609	.9614	.9619	.9624	.9628	.9633
9.2	.9638	.9643	.9647	.9652	.9657	.9661	.9666	.9671	.9675	.9680
9.3	.9685	.9689	.9694	.9699	.9703	.9708	.9713	.9717	.9722	.9727
9.4	.9731	.9736	.9741	.9745	.9750	.9754	.9759	.9763	.9768	.9773
9.5	.9777	.9782	.9786	.9791	.9795	.9800	.9805	.9809	.9814	.9818
9.6	.9823	.9827	.9832	.9836	.9841	.9845	.9850	.9854	.9859	.9863
9.7	.9868	.9872	.9877	.9881	.9886	.9890	.9894	.9899	.9903	.9908
9.8	.9912	.9917	.9921	.9926	.9930	.9934	.9939	.9943	.9948	.9952
9.9	.9956	.9961	.9965	.9969	.9974	.9978	.9983	.9987	.9991	.9996
n	0	1	2	3	4	5	6	7	8	9

TABLE 4 COMPOUND INTEREST

$(1 + i)^n$

n \ i	2.00%	2.50%	3.00%	3.50%	4.00%	4.50%	5.00%	5.50%
1	1.02000	1.02500	1.03000	1.03500	1.04000	1.04500	1.05000	1.05500
2	1.04040	1.05063	1.06090	1.07123	1.08160	1.09203	1.10250	1.11303
3	1.06121	1.07689	1.09273	1.10872	1.12486	1.14117	1.15763	1.17424
4	1.08243	1.10381	1.12551	1.14752	1.16986	1.19252	1.21551	1.23882
5	1.10408	1.13141	1.15927	1.18769	1.21665	1.24618	1.27628	1.30696
6	1.12616	1.15969	1.19405	1.22926	1.26532	1.30226	1.34010	1.37884
7	1.14869	1.18869	1.22987	1.27228	1.31593	1.36086	1.40710	1.45468
8	1.17166	1.21840	1.26677	1.31681	1.36857	1.42210	1.47746	1.53469
9	1.19509	1.24886	1.30477	1.36290	1.42331	1.48610	1.55133	1.61909
10	1.21899	1.28008	1.34392	1.41060	1.48024	1.55297	1.62889	1.70814
11	1.24337	1.31209	1.38423	1.45997	1.53945	1.62285	1.71034	1.80209
12	1.26824	1.34489	1.42576	1.51107	1.60103	1.69588	1.79586	1.90121
13	1.29361	1.37851	1.46853	1.56396	1.66507	1.77220	1.88565	2.00577
14	1.31948	1.41297	1.51259	1.61869	1.73168	1.85194	1.97993	2.11609
15	1.34587	1.44830	1.55797	1.67535	1.80094	1.93528	2.07893	2.23248
16	1.37279	1.48451	1.60471	1.73399	1.87298	2.02237	2.18287	2.35526
17	1.40024	1.52162	1.65285	1.79468	1.94790	2.11338	2.29202	2.48480
18	1.42825	1.55966	1.70243	1.85749	2.02582	2.20848	2.40662	2.62147
19	1.45681	1.59865	1.75351	1.92250	2.10685	2.30786	2.52695	2.76565
20	1.48595	1.63862	1.80611	1.98979	2.19112	2.41171	2.65330	2.91776
21	1.51567	1.67958	1.86029	2.05943	2.27877	2.52024	2.78596	3.07823
22	1.54598	1.72157	1.91610	2.13151	2.36992	2.63365	2.92526	3.24754
23	1.57690	1.76461	1.97359	2.20611	2.46472	2.75217	3.07152	3.42615
24	1.60844	1.80873	2.03279	2.28333	2.56330	2.87601	3.22510	3.61459
25	1.64061	1.85394	2.09378	2.36324	2.66584	3.00543	3.38635	3.81339
26	1.67342	1.90029	2.15659	2.44596	2.77247	3.14068	3.55567	4.02313
27	1.70689	1.94780	2.22129	2.53157	2.88337	3.28201	3.73346	4.24440
28	1.74102	1.99650	2.28793	2.62017	2.99870	3.42970	3.92013	4.47784
29	1.77584	2.04641	2.35657	2.71188	3.11865	3.58404	4.11614	4.72412
30	1.81136	2.09757	2.42726	2.80679	3.24340	3.74532	4.32194	4.98395
31	1.84759	2.15001	2.50008	2.90503	3.37313	3.91386	4.53804	5.25807
32	1.88454	2.20376	2.57508	3.00671	3.50806	4.08998	4.76494	5.54726
33	1.92223	2.25885	2.65234	3.11194	3.64838	4.27403	5.00319	5.85236
34	1.96068	2.31532	2.73191	3.22086	3.79432	4.46636	5.25335	6.17424
35	1.99989	2.37321	2.81386	3.33359	3.94609	4.66735	5.51602	6.51383
36	2.03989	2.43254	2.89828	3.45027	4.10393	4.87738	5.79182	6.87209
37	2.08069	2.49335	2.98523	3.57103	4.26809	5.09686	6.08141	7.25005
38	2.12230	2.55568	3.07478	3.69601	4.43881	5.32622	6.38548	7.64880
39	2.16474	2.61957	3.16703	3.82537	4.61637	5.56590	6.70475	8.06949
40	2.20804	2.68506	3.26204	3.95926	4.80102	5.81636	7.03999	8.51331
41	2.25220	2.75219	3.35990	4.09783	4.99306	6.07810	7.39199	8.98154
42	2.29724	2.82100	3.46070	4.24126	5.19278	6.35162	7.76159	9.47553
43	2.34319	2.89152	3.56452	4.38970	5.40050	6.63744	8.14967	9.99668
44	2.39005	2.96381	3.67145	4.54334	5.61652	6.93612	8.55715	10.54650
45	2.43785	3.03790	3.78160	4.70236	5.84118	7.24825	8.98501	11.12655
46	2.48661	3.11385	3.89504	4.86694	6.07482	7.57442	9.43426	11.73851
47	2.53634	3.19170	4.01190	5.03728	6.31782	7.91527	9.90597	12.38413
48	2.58707	3.27149	4.13225	5.21359	6.57053	8.27146	10.40127	13.06526
49	2.63881	3.35328	4.25622	5.39606	6.83335	8.64367	10.92133	13.78385
50	2.69159	3.43711	4.38391	5.58493	7.10668	9.03264	11.46740	14.54196

TABLE 4 (CONTINUED)

i \ n	6.00%	8.00%	10.00%	12.00%	14.00%	16.00%	18.00%
1	1.06000	1.08000	1.10000	1.12000	1.14000	1.16000	1.18000
2	1.12360	1.16640	1.21000	1.25440	1.29960	1.34560	1.39240
3	1.19102	1.25971	1.33100	1.40493	1.48154	1.56090	1.64303
4	1.26248	1.36049	1.46410	1.57352	1.68896	1.81064	1.93878
5	1.33823	1.46933	1.61051	1.76234	1.92541	2.10034	2.28776
6	1.41852	1.58687	1.77156	1.97382	2.19497	2.43640	2.69955
7	1.50363	1.71382	1.94872	2.21068	2.50227	2.82622	3.18547
8	1.59385	1.85093	2.14359	2.47596	2.85259	3.27841	3.75886
9	1.68948	1.99900	2.35795	2.77308	3.25195	3.80296	4.43545
10	1.79085	2.15892	2.59374	3.10585	3.70722	4.41144	5.23384
11	1.89830	2.33164	2.85312	3.47855	4.22623	5.11726	6.17593
12	2.01220	2.51817	3.13843	3.89598	4.81790	5.93603	7.28759
13	2.13293	2.71962	3.45227	4.36349	5.49241	6.88579	8.59936
14	2.26090	2.93719	3.79750	4.88711	6.26135	7.98752	10.14724
15	2.39656	3.17217	4.17725	5.47357	7.13794	9.26552	11.97375
16	2.54035	3.42594	4.59497	6.13039	8.13725	10.74800	14.12902
17	2.69277	3.70002	5.05447	6.86604	9.27646	12.46768	16.67225
18	2.85434	3.99602	5.55992	7.68997	10.57517	14.46251	19.67325
19	3.02560	4.31570	6.11591	8.61276	12.05569	16.77652	23.21444
20	3.20714	4.66096	6.72750	9.64629	13.74349	19.46076	27.39303
21	3.39956	5.03383	7.40025	10.80385	15.66758	22.57448	32.32378
22	3.60354	5.43654	8.14027	12.10031	17.86104	26.18640	38.14206
23	3.81975	5.87146	8.95430	13.55235	20.36158	30.37622	45.00763
24	4.04893	6.34118	9.84973	15.17863	23.21221	35.23642	53.10901
25	4.29187	6.84848	10.83471	17.00006	26.46192	40.87424	62.66863
26	4.54938	7.39635	11.91818	19.04007	30.16658	47.41412	73.94898
27	4.82235	7.98806	13.10999	21.32488	34.38991	55.00038	87.25980
28	5.11169	8.62711	14.42099	23.88387	39.20449	63.80044	102.96656
29	5.41839	9.31727	15.86309	26.74993	44.69312	74.00851	121.50054
30	5.74349	10.06266	17.44940	29.95992	50.95016	85.84988	143.37064
31	6.08810	10.86767	19.19434	33.55511	58.08318	99.58586	169.17735
32	6.45339	11.73708	21.11378	37.58173	66.21483	115.51959	199.62928
33	6.84059	12.67605	23.22515	42.09153	75.48490	134.00273	235.56255
34	7.25103	13.69013	25.54767	47.14252	86.05279	155.44317	277.96381
35	7.68609	14.78534	28.10244	52.79962	98.10018	180.31407	327.99729
36	8.14725	15.96817	30.91268	59.13557	111.83420	209.16432	387.03680
37	8.63609	17.24563	34.00395	66.23184	127.49099	242.63062	456.70343
38	9.15425	18.62528	37.40434	74.17966	145.33973	281.45151	538.91004
39	9.70351	20.11530	41.14478	83.08122	165.68729	326.48376	635.91385
40	10.28572	21.72452	45.25926	93.05097	188.88351	378.72116	750.37834
41	10.90286	23.46248	49.78518	104.21709	215.32721	439.31654	885.44645
42	11.55703	25.33948	54.76370	116.72314	245.47301	509.60719	1044.82681
43	12.25045	27.36664	60.24007	130.72991	279.83924	591.14434	1232.89563
44	12.98548	29.55597	66.26408	146.41750	319.01673	685.72744	1454.81685
45	13.76461	31.92045	72.89048	163.98760	363.67907	795.44383	1716.68388
46	14.59049	34.47409	80.17953	183.66612	414.59414	922.71484	2025.68698
47	15.46592	37.23201	88.19749	205.70605	472.63732	1070.34921	2390.31063
48	16.39387	40.21057	97.01723	230.39078	538.80655	1241.60509	2820.56655
49	17.37750	43.42742	106.71896	258.03767	614.23946	1440.26190	3328.26853
50	18.42015	46.90161	117.39085	289.00219	700.23299	1670.70380	3927.35686

TABLE 5 AMOUNT OF AN ANNUITY

$$s_{\overline{n}|i} = \frac{(1 + i)^{n-1}}{i}$$

i / n	1.00%	1.50%	2.00%	2.50%	3.00%	4.00%	5.00%
1	1.00000	1.00000	1.00000	1.00000	1.00000	1.00000	1.00000
2	2.01000	2.01500	2.02000	2.02500	2.03000	2.04000	2.05000
3	3.03010	3.04522	3.06040	3.07562	3.09090	3.12160	3.15250
4	4.06040	4.09090	4.12161	4.15252	4.18363	4.24646	4.31013
5	5.10101	5.15227	5.20404	5.25633	5.30914	5.41632	5.52563
6	6.15202	6.22955	6.30812	6.38774	6.46841	6.63298	6.80191
7	7.21354	7.32299	7.43428	7.54743	7.66246	7.89829	8.14201
8	8.28567	8.43284	8.58297	8.73612	8.89234	9.21423	9.54911
9	9.36853	9.55933	9.75463	9.95452	10.15911	10.58280	11.02656
10	10.46221	10.70272	10.94972	11.20338	11.46388	12.00611	12.57789
11	11.56683	11.86326	12.16872	12.48347	12.80780	13.48635	14.20679
12	12.68250	13.04121	13.41209	13.79555	14.19203	15.02581	15.91713
13	13.80933	14.23683	14.68033	15.14044	15.61779	16.62684	17.71298
14	14.94742	15.45038	15.97394	16.51895	17.08632	18.29191	19.59863
15	16.09690	16.68214	17.29342	17.93193	18.59891	20.02359	21.57856
16	17.25786	17.93237	18.63929	19.38022	20.15688	21.82453	23.65749
17	18.43044	19.20136	20.01207	20.86473	21.76159	23.69751	25.84037
18	19.61475	20.48938	21.41231	22.38635	23.41444	25.64541	28.13238
19	20.81090	21.79672	22.84056	23.94601	25.11687	27.67123	30.53900
20	22.01900	23.12367	24.29737	25.54466	26.87037	29.77808	33.06595
21	23.23919	24.47052	25.78332	27.18327	28.67649	31.96920	35.71925
22	24.47159	25.83758	27.29898	28.86286	30.53678	34.24797	38.50521
23	25.71630	27.22514	28.84496	30.58443	32.45288	36.61789	41.43048
24	26.97346	28.63352	30.42186	32.34904	34.42647	39.08260	44.50200
25	28.24320	30.06302	32.03030	34.15776	36.45926	41.64591	47.72710
26	29.52563	31.51397	33.67091	36.01171	38.55304	44.31174	51.11345
27	30.82089	32.98668	35.34432	37.91200	40.70963	47.08421	54.66913
28	32.12910	34.48148	37.05121	39.85980	42.93092	49.96758	58.40258
29	33.45039	35.99870	38.79223	41.85630	45.21885	52.96629	62.32271
30	34.78489	37.53868	40.56808	43.90270	47.57542	56.08494	66.43885
31	36.13274	39.10176	42.37944	46.00027	50.00268	59.32834	70.76079
32	37.49407	40.68829	44.22703	48.15028	52.50276	62.70147	75.29883
33	38.86901	42.29861	46.11157	50.35403	55.07784	66.20953	80.06377
34	40.25770	43.93309	48.03380	52.61289	57.73018	69.85791	85.06696
35	41.66028	45.59209	49.99448	54.92821	60.46208	73.65222	90.32031
36	43.07688	47.27597	51.99437	57.30141	63.27594	77.59831	95.83632
37	44.50765	48.98511	54.03425	59.73395	66.17422	81.70225	101.62814
38	45.95272	50.71989	56.11494	62.22730	69.15945	85.97034	107.70955
39	47.41225	52.48068	58.23724	64.78298	72.23423	90.40915	114.09502
40	48.88637	54.26789	60.40198	67.40255	75.40126	95.02552	120.79977
41	50.37524	56.08191	62.61002	70.08762	78.66330	99.82654	127.83976
42	51.87899	57.92314	64.86222	72.83981	82.02320	104.81960	135.23175
43	53.39778	59.79199	67.15947	75.66080	85.48389	110.01238	142.99334
44	54.93176	61.68887	69.50266	78.55232	89.04841	115.41288	151.14301
45	56.48107	63.61420	71.89271	81.51613	92.71986	121.02939	159.70016
46	58.04589	65.56841	74.33056	84.55403	96.50146	126.87057	168.68516
47	59.62634	67.55194	76.81718	87.66789	100.39650	132.94539	178.11942
48	61.22261	69.56522	79.35352	90.85958	104.40840	139.26321	188.02539
49	62.83483	71.60870	81.94059	94.13107	108.54065	145.83373	198.42666
50	64.46318	73.68283	84.57940	97.48435	112.79687	152.66708	209.34800

TABLE 5   (CONTINUED)

n \ i	6.00%	8.00%	10.00%	12.00%	14.00%	16.00%	18.00%
1	1.00000	1.00000	1.00000	1.00000	1.00000	1.00000	1.00000
2	2.06000	2.08000	2.10000	2.12000	2.14000	2.16000	2.18000
3	3.18360	3.24640	3.31000	3.37440	3.43960	3.50560	3.57240
4	4.37462	4.50611	4.64100	4.77933	4.92114	5.06650	5.21543
5	5.63709	5.86660	6.10510	6.35285	6.61010	6.87714	7.15421
6	6.97532	7.33593	7.71561	8.11519	8.53552	8.97748	9.44197
7	8.39384	8.92280	9.48717	10.08901	10.73049	11.41387	12.14152
8	9.89747	10.63663	11.43589	12.29969	13.23276	14.24009	15.32700
9	11.49132	12.48756	13.57948	14.77566	16.08535	17.51851	19.08585
10	13.18079	14.48656	15.93742	17.54874	19.33730	21.32147	23.52131
11	14.97164	16.64549	18.53117	20.65458	23.04452	25.73290	28.75514
12	16.86994	18.97713	21.38428	24.13313	27.27075	30.85017	34.93107
13	18.88214	21.49530	24.52271	28.02911	32.08865	36.78620	42.21866
14	21.01507	24.21492	27.97498	32.39260	37.58107	43.67199	50.81802
15	23.27597	27.15211	31.77248	37.27971	43.84241	51.65951	60.96527
16	25.67253	30.32428	35.94973	42.75328	50.98035	60.92503	72.93901
17	28.21288	33.75023	40.54470	48.88367	59.11760	71.67303	87.06804
18	30.90565	37.45024	45.59917	55.74971	68.39407	84.14072	103.74028
19	33.75999	41.44626	51.15909	63.43968	78.96923	98.60323	123.41353
20	36.78559	45.76196	57.27500	72.05244	91.02493	115.37975	146.62797
21	39.99273	50.42292	64.00250	81.69874	104.76842	134.84051	174.02100
22	43.39229	55.45676	71.40275	92.50258	120.43600	157.41499	206.34479
23	46.99583	60.89330	79.54302	104.60289	138.29704	183.60138	244.48685
24	50.81558	66.76476	88.49733	118.15524	158.65862	213.97761	289.49448
25	54.86451	73.10594	98.34706	133.33387	181.87083	249.21402	342.60349
26	59.15638	79.95442	109.18177	150.33393	208.33274	290.08827	405.27211
27	63.70577	87.35077	121.09994	169.37401	238.49933	337.50239	479.22109
28	68.52811	95.33883	134.20994	190.69889	272.88923	392.50277	566.48089
29	73.63980	103.96594	148.63093	214.58275	312.09373	456.30322	669.44745
30	79.05819	113.28321	164.49402	241.33268	356.78685	530.31173	790.94799
31	84.80168	123.34587	181.94342	271.29261	407.73701	616.16161	934.31863
32	90.88978	134.21354	201.13777	304.84772	465.82019	715.74746	1103.49598
33	97.34316	145.95062	222.25154	342.42945	532.03501	831.26706	1303.12526
34	104.18375	158.62667	245.47670	384.52098	607.51991	965.26979	1538.68781
35	111.43478	172.31680	271.02437	431.66350	693.57270	1120.71295	1816.65161
36	119.12087	187.10215	299.12681	484.46312	791.67288	1301.02703	2144.64890
37	127.26812	203.07032	330.03949	543.59869	903.50708	1510.19135	2531.68570
38	135.90421	220.31595	364.04343	609.83053	1030.99808	1752.82197	2988.38913
39	145.05846	238.94122	401.44778	684.01020	1176.33781	2034.27348	3527.29918
40	154.76197	259.05652	442.59256	767.09142	1342.02510	2360.75724	4163.21303
41	165.04768	280.78104	487.85181	860.14239	1530.90861	2739.47840	4913.59137
42	175.95054	304.24352	537.63699	964.35948	1746.23582	3178.79494	5799.03782
43	187.50758	329.58301	592.40069	1081.08262	1991.70883	3688.40213	6843.86463
44	199.75803	356.94965	652.64076	1211.81253	2271.54807	4279.54648	8076.76026
45	212.74351	386.50562	718.90484	1358.23003	2590.56480	4965.27391	9531.57711
46	226.50812	418.42607	791.79532	1522.21764	2954.24387	5760.71774	11248.26098
47	241.09861	452.90015	871.97485	1705.88375	3368.83801	6683.43257	13273.94796
48	256.56453	490.13216	960.17234	1911.58980	3841.47534	7753.78179	15664.25859
49	272.95840	530.34274	1057.18957	2141.98058	4380.28188	8995.38687	18484.82514
50	290.33590	573.77016	1163.90853	2400.01825	4994.52135	10435.64877	21813.09367

TABLE 6 PRESENT VALUE OF AN ANNUITY

$$a_{\overline{n}|i} = \frac{1 - (1 + i)^{-n}}{i}$$

i \ n	1.00%	1.50%	2.00%	2.50%	3.00%	4.00%	5.00%
1	0.99010	0.98522	0.98039	0.97561	0.97087	0.96154	0.95238
2	1.97040	1.95588	1.94156	1.92742	1.91347	1.88609	1.85941
3	2.94099	2.91220	2.88388	2.85602	2.82861	2.77509	2.72325
4	3.90197	3.85438	3.80773	3.76197	3.71710	3.62990	3.54595
5	4.85343	4.78264	4.71346	4.64583	4.57971	4.45182	4.32948
6	5.79548	5.69719	5.60143	5.50813	5.41719	5.24214	5.07569
7	6.72819	6.59821	6.47199	6.34939	6.23028	6.00205	5.78637
8	7.65168	7.48593	7.32548	7.17014	7.01969	6.73274	6.46321
9	8.56602	8.36052	8.16224	7.97087	7.78611	7.43533	7.10782
10	9.47130	9.22218	8.98259	8.75206	8.53020	8.11090	7.72173
11	10.36763	10.07112	9.78685	9.51421	9.25262	8.76048	8.30641
12	11.25508	10.90751	10.57534	10.25776	9.95400	9.38507	8.86325
13	12.13374	11.73153	11.34837	10.98318	10.63496	9.98565	9.39357
14	13.00370	12.54338	12.10625	11.69091	11.29607	10.56312	9.89864
15	13.86505	13.34323	12.84926	12.38138	11.93794	11.11839	10.37966
16	14.71787	14.13126	13.57771	13.05500	12.56110	11.65230	10.83777
17	15.56225	14.90765	14.29187	13.71220	13.16612	12.16567	11.27407
18	16.39827	15.67256	14.99203	14.35336	13.75351	12.65930	11.68959
19	17.22601	16.42617	15.67846	14.97889	14.32380	13.13394	12.08532
20	18.04555	17.16864	16.35143	15.58916	14.87747	13.59033	12.46221
21	18.85698	17.90014	17.01121	16.18455	15.41502	14.02916	12.82115
22	19.66038	18.62082	17.65805	16.76541	15.93692	14.45112	13.16300
23	20.45582	19.33086	18.29220	17.33211	16.44361	14.85684	13.48857
24	21.24339	20.03041	18.91393	17.88499	16.93554	15.24696	13.79864
25	22.02316	20.71961	19.52346	18.42438	17.41315	15.62208	14.09394
26	22.79520	21.39863	20.12104	18.95061	17.87684	15.98277	14.37519
27	23.55961	22.06762	20.70690	19.46401	18.32703	16.32959	14.64303
28	24.31644	22.72672	21.28127	19.96489	18.76411	16.66306	14.89813
29	25.06579	23.37608	21.84438	20.45355	19.18845	16.98371	15.14107
30	25.80771	24.01584	22.39646	20.93029	19.60044	17.29203	15.37245
31	26.54229	24.64615	22.93770	21.39541	20.00043	17.58849	15.59281
32	27.26959	25.26714	23.46833	21.84918	20.38877	17.87355	15.80268
33	27.98969	25.87895	23.98856	22.29188	20.76579	18.14765	16.00255
34	28.70267	26.48173	24.49859	22.72379	21.13184	18.41120	16.19290
35	29.40858	27.07559	24.99862	23.14516	21.48722	18.66461	16.37419
36	30.10751	27.66068	25.48884	23.55625	21.83225	18.90828	16.54685
37	30.79951	28.23713	25.96945	23.95732	22.16724	19.14258	16.71129
38	31.48466	28.80505	26.44064	24.34860	22.49246	19.36786	16.86789
39	32.16303	29.36458	26.90259	24.73034	22.80822	19.58448	17.01704
40	32.83469	29.91585	27.35548	25.10278	23.11477	19.79277	17.15909
41	33.49969	30.45896	27.79949	25.46612	23.41240	19.99305	17.29437
42	34.15811	30.99405	28.23479	25.82061	23.70136	20.18563	17.42321
43	34.81001	31.52123	28.66156	26.16645	23.98190	20.37079	17.54591
44	35.45545	32.04062	29.07996	26.50385	24.25427	20.54884	17.66277
45	36.09451	32.55234	29.49016	26.83302	24.51871	20.72004	17.77407
46	36.72724	33.05649	29.89231	27.15417	24.77545	20.88465	17.88007
47	37.35370	33.55319	30.28658	27.46748	25.02471	21.04294	17.98102
48	37.97396	34.04255	30.67312	27.77315	25.26671	21.19513	18.07716
49	38.58808	34.52468	31.05208	28.07137	25.50166	21.34147	18.16872
50	39.19612	34.99969	31.42361	28.36231	25.72976	21.48218	18.25593

TABLE 6   (CONTINUED)

n \ i	6.00%	8.00%	10.00%	12.00%	14.00%	16.00%	18.00%
1	0.94340	0.92593	0.90909	0.89286	0.87719	0.86207	0.84746
2	1.83339	1.78326	1.73554	1.69005	1.64666	1.60523	1.56564
3	2.67301	2.57710	2.48685	2.40183	2.32163	2.24589	2.17427
4	3.46511	3.31213	3.16987	3.03735	2.91371	2.79818	2.69006
5	4.21236	3.99271	3.79079	3.60478	3.43308	3.27429	3.12717
6	4.91732	4.62288	4.35526	4.11141	3.88867	3.68474	3.49760
7	5.58238	5.20637	4.86842	4.56376	4.28830	4.03857	3.81153
8	6.20979	5.74664	5.33493	4.96764	4.63886	4.34359	4.07757
9	6.80169	6.24689	5.75902	5.32825	4.94637	4.60654	4.30302
10	7.36009	6.71008	6.14457	5.65022	5.21612	4.83323	4.49409
11	7.88687	7.13896	6.49506	5.93770	5.45273	5.02864	4.65601
12	8.38384	7.53608	6.81369	6.19437	5.66029	5.19711	4.79322
13	8.85268	7.90378	7.10336	6.42355	5.84236	5.34233	4.90951
14	9.29498	8.24424	7.36669	6.62817	6.00207	5.46753	5.00806
15	9.71225	8.55948	7.60608	6.81086	6.14217	5.57546	5.09158
16	10.10590	8.85137	7.82371	6.97399	6.26506	5.66850	5.16235
17	10.47726	9.12164	8.02155	7.11963	6.37286	5.74870	5.22233
18	10.82760	9.37189	8.20141	7.24967	6.46742	5.81785	5.27316
19	11.15812	9.60360	8.36492	7.36578	6.55037	5.87746	5.31624
20	11.46992	9.81815	8.51356	7.46944	6.62313	5.92884	5.35275
21	11.76408	10.01680	8.64869	7.56200	6.68696	5.97314	5.38368
22	12.04158	10.20074	8.77154	7.64465	6.74294	6.01133	5.40990
23	12.30338	10.37106	8.88322	7.71843	6.79206	6.04425	5.43212
24	12.55036	10.52876	8.98474	7.78432	6.83514	6.07263	5.45095
25	12.78336	10.67478	9.07704	7.84314	6.87293	6.09709	5.46691
26	13.00317	10.80998	9.16095	7.89566	6.90608	6.11818	5.48043
27	13.21053	10.93516	9.23722	7.94255	6.93515	6.13636	5.49189
28	13.40616	11.05108	9.30657	7.98442	6.96066	6.15204	5.50160
29	13.59072	11.15841	9.36961	8.02181	6.98304	6.16555	5.50983
30	13.76483	11.25778	9.42691	8.05518	7.00266	6.17720	5.51681
31	13.92909	11.34980	9.47901	8.08499	7.01988	6.18724	5.52272
32	14.08404	11.43500	9.52638	8.11159	7.03498	6.19590	5.52773
33	14.23023	11.51389	9.56943	8.13535	7.04823	6.20336	5.53197
34	14.36814	11.58693	9.60857	8.15656	7.05985	6.20979	5.53557
35	14.49825	11.65457	9.64416	8.17550	7.07005	6.21534	5.53862
36	14.62099	11.71719	9.67651	8.19241	7.07899	6.22012	5.54120
37	14.73678	11.77518	9.70592	8.20751	7.08683	6.22424	5.54339
38	14.84602	11.82887	9.73265	8.22099	7.09371	6.22779	5.54525
39	14.94907	11.87858	9.75696	8.23303	7.09975	6.23086	5.54682
40	15.04630	11.92461	9.77905	8.24378	7.10504	6.23350	5.54815
41	15.13802	11.96723	9.79914	8.25337	7.10969	6.23577	5.54928
42	15.22454	12.00670	9.81740	8.26194	7.11376	6.23774	5.55024
43	15.30617	12.04324	9.83400	8.26959	7.11733	6.23943	5.55105
44	15.38318	12.07707	9.84909	8.27642	7.12047	6.24089	5.55174
45	15.45583	12.10840	9.86281	8.28252	7.12322	6.24214	5.55232
46	15.52437	12.13741	9.87528	8.28796	7.12563	6.24323	5.55281
47	15.58903	12.16427	9.88662	8.29282	7.12774	6.24416	5.55323
48	15.65003	12.18914	9.89693	8.29716	7.12960	6.24497	5.55359
49	15.70757	12.21216	9.90630	8.30104	7.13123	6.24566	5.55389
50	15.76186	12.23348	9.91481	8.30450	7.13266	6.24626	5.55414

TABLE 7   COMBINATIONS

n	$\binom{n}{0}$	$\binom{n}{1}$	$\binom{n}{2}$	$\binom{n}{3}$	$\binom{n}{4}$	$\binom{n}{5}$	$\binom{n}{6}$	$\binom{n}{7}$	$\binom{n}{8}$	$\binom{n}{9}$	$\binom{n}{10}$
0	1										
1	1	1									
2	1	2	1								
3	1	3	3	1							
4	1	4	6	4	1						
5	1	5	10	10	5	1					
6	1	6	15	20	15	6	1				
7	1	7	21	35	35	21	7	1			
8	1	8	28	56	70	56	28	8	1		
9	1	9	36	84	126	126	84	36	9	1	
10	1	10	45	120	210	252	210	120	45	10	1
11	1	11	55	165	330	462	462	330	165	55	11
12	1	12	66	220	495	792	924	792	495	220	66
13	1	13	78	286	715	1287	1716	1716	1287	715	286
14	1	14	91	364	1001	2002	3003	3432	3003	2002	1001
15	1	15	105	455	1365	3003	5005	6435	6435	5005	3003
16	1	16	120	560	1820	4368	8008	11440	12870	11440	8008
17	1	17	136	680	2380	6188	12376	19448	24310	24310	19448
18	1	18	153	816	3060	8568	18564	31824	43758	48620	43758
19	1	19	171	969	3876	11628	27132	50388	75582	92378	92378
20	1	20	190	1140	4845	15504	38760	77520	125970	167960	184756

For $r > 10$, it may be necessary to use the identity

$$\binom{n}{r} = \binom{n}{n-r}$$

TABLE 8   AREAS UNDER THE NORMAL CURVE

The column under A gives the proportion of the area under the entire curve which is between $z = 0$ and a positive value of z.

z	A	z	A	z	A	z	A
.00	.0000	.49	.1879	.98	.3365	1.47	.4292
.01	.0040	.50	.1915	.99	.3389	1.48	.4306
.02	.0080	.51	.1950	1.00	.3413	1.49	.4319
.03	.0120	.52	.1985	1.01	.3438	1.50	.4332
.04	.0160	.53	.2019	1.02	.3461	1.51	.4345
.05	.0199	.54	.2054	1.03	.3485	1.52	.4357
.06	.0239	.55	.2088	1.04	.3508	1.53	.4370
.07	.0279	.56	.2123	1.05	.3531	1.54	.4382
.08	.0319	.57	.2157	1.06	.3554	1.55	.4394
.09	.0359	.58	.2190	1.07	.3577	1.56	.4406
.10	.0398	.59	.2224	1.08	.3599	1.57	.4418
.11	.0438	.60	.2258	1.09	.3621	1.58	.4430
.12	.0478	.61	.2291	1.10	.3643	1.59	.4441
.13	.0517	.62	.2324	1.11	.3665	1.60	.4452
.14	.0557	.63	.2357	1.12	.3686	1.61	.4463
.15	.0596	.64	.2389	1.13	.3708	1.62	.4474
.16	.0636	.65	.2422	1.14	.3729	1.63	.4485
.17	.0675	.66	.2454	1.15	.3749	1.64	.4495
.18	.0714	.67	.2486	1.16	.3770	1.65	.4505
.19	.0754	.68	.2518	1.17	.3790	1.66	.4515
.20	.0793	.69	.2549	1.18	.3810	1.67	.4525
.21	.0832	.70	.2580	1.19	.3830	1.68	.4535
.22	.0871	.71	.2612	1.20	.3849	1.69	.4545
.23	.0910	.72	.2642	1.21	.3869	1.70	.4554
.24	.0948	.73	.2673	1.22	.3888	1.71	.4564
.25	.0987	.74	.2704	1.23	.3907	1.72	.4573
.26	.1026	.75	.2734	1.24	.3925	1.73	.4582
.27	.1064	.76	.2764	1.25	.3944	1.74	.4591
.28	.1103	.77	.2794	1.26	.3962	1.75	.4599
.29	.1141	.78	.2823	1.27	.3980	1.76	.4608
.30	.1179	.79	.2852	1.28	.3997	1.77	.4616
.31	.1217	.80	.2881	1.29	.4015	1.78	.4625
.32	.1255	.81	.2910	1.30	.4032	1.79	.4633
.33	.1293	.82	.2939	1.31	.4049	1.80	.4641
.34	.1331	.83	.2967	1.32	.4066	1.81	.4649
.35	.1368	.84	.2996	1.33	.4082	1.82	.4656
.36	.1406	.85	.3023	1.34	.4099	1.83	.4664
.37	.1443	.86	.3051	1.35	.4115	1.84	.4671
.38	.1480	.87	.3079	1.36	.4131	1.85	.4678
.39	.1517	.88	.3106	1.37	.4147	1.86	.4686
.40	.1554	.89	.3133	1.38	.4162	1.87	.4693
.41	.1591	.90	.3159	1.39	.4177	1.88	.4700
.42	.1628	.91	.3186	1.40	.4192	1.89	.4706
.43	.1664	.92	.3212	1.41	.4207	1.90	.4713
.44	.1700	.93	.3238	1.42	.4222	1.91	.4719
.45	.1736	.94	.3264	1.43	.4236	1.92	.4726
.46	.1772	.95	.3289	1.44	.4251	1.93	.4732
.47	.1808	.96	.3315	1.45	.4265	1.94	.4738
.48	.1844	.97	.3340	1.46	.4279	1.95	.4744

TABLE 8 (CONTINUED)

z	A	z	A	z	A	z	A
1.96	.4750	2.45	.4929	2.94	.4984	3.43	.4997
1.97	.4756	2.46	.4931	2.95	.4984	3.44	.4997
1.98	.4762	2.47	.4932	2.96	.4985	3.45	.4997
1.99	.4767	2.48	.4934	2.97	.4985	3.46	.4997
2.00	.4773	2.49	.4936	2.98	.4986	3.47	.4997
2.01	.4778	2.50	.4938	2.99	.4986	3.48	.4998
2.02	.4783	2.51	.4940	3.00	.4987	3.49	.4998
2.03	.4788	2.52	.4941	3.01	.4987	3.50	.4998
2.04	.4793	2.53	.4943	3.02	.4987	3.51	.4998
2.05	.4798	2.54	.4945	3.03	.4988	3.52	.4998
2.06	.4803	2.55	.4946	3.04	.4988	3.53	.4998
2.07	.4808	2.56	.4948	3.05	.4989	3.54	.4998
2.08	.4812	2.57	.4949	3.06	.4989	3.55	.4998
2.09	.4817	2.58	.4951	3.07	.4989	3.56	.4998
2.10	.4821	2.59	.4952	3.08	.4990	3.57	.4998
2.11	.4826	2.60	.4953	3.09	.4990	3.58	.4998
2.12	.4830	2.61	.4955	3.10	.4990	3.59	.4998
2.13	.4834	2.62	.4956	3.11	.4991	3.60	.4998
2.14	.4838	2.63	.4957	3.12	.4991	3.61	.4999
2.15	.4842	2.64	.4959	3.13	.4991	3.62	.4999
2.16	.4846	2.65	.4960	3.14	.4992	3.63	.4999
2.17	.4850	2.66	.4961	3.15	.4992	3.64	.4999
2.18	.4854	2.67	.4962	3.16	.4992	3.65	.4999
2.19	.4857	2.68	.4963	3.17	.4992	3.66	.4999
2.20	.4861	2.69	.4964	3.18	.4993	3.67	.4999
2.21	.4865	2.70	.4965	3.19	.4993	3.68	.4999
2.22	.4868	2.71	.4966	3.20	.4993	3.69	.4999
2.23	.4871	2.72	.4967	3.21	.4993	3.70	.4999
2.24	.4875	2.73	.4968	3.22	.4994	3.71	.4999
2.25	.4878	2.74	.4969	3.23	.4994	3.72	.4999
2.26	.4881	2.75	.4970	3.24	.4994	3.73	.4999
2.27	.4884	2.76	.4971	3.25	.4994	3.74	.4999
2.28	.4887	2.77	.4972	3.26	.4994	3.75	.4999
2.29	.4890	2.78	.4973	3.27	.4995	3.76	.4999
2.30	.4893	2.79	.4974	3.28	.4995	3.77	.4999
2.31	.4896	2.80	.4974	3.29	.4995	3.78	.4999
2.32	.4898	2.81	.4975	3.30	.4995	3.79	.4999
2.33	.4901	2.82	.4976	3.31	.4995	3.80	.4999
2.34	.4904	2.83	.4977	3.32	.4996	3.81	.4999
2.35	.4906	2.84	.4977	3.33	.4996	3.82	.4999
2.36	.4909	2.85	.4978	3.34	.4996	3.83	.4999
2.37	.4911	2.86	.4979	3.35	.4996	3.84	.4999
2.38	.4913	2.87	.4980	3.36	.4996	3.85	.4999
2.39	.4916	2.88	.4980	3.37	.4996	3.86	.4999
2.40	.4918	2.89	.4981	3.38	.4996	3.87	.5000
2.41	.4920	2.90	.4981	3.39	.4997	3.88	.5000
2.42	.4922	2.91	.4982	3.40	.4997	3.89	.5000
2.43	.4925	2.92	.4983	3.41	.4997		
2.44	.4927	2.93	.4983	3.42	.4997		

TABLE 9 INTEGRALS

(C is an arbitrary constant.)

1. $\int x^n\, dx = \dfrac{1}{n+1} x^{n+1} + C \quad$ (if $n \neq -1$)

2. $\int e^{kx}\, dx = \dfrac{1}{k} e^{kx} + C$

3. $\int \dfrac{a}{x}\, dx = a \ln |x| + C \quad (a \neq 0)$

4. $\int \ln |ax|\, dx = x\,(\ln |ax| - 1) + C$

5. $\int \dfrac{1}{\sqrt{x^2 + a^2}}\, dx = \ln \left| \dfrac{x + \sqrt{x^2 + a^2}}{a} \right| + C$

6. $\int \dfrac{1}{\sqrt{x^2 - a^2}}\, dx = \ln \left| \dfrac{x + \sqrt{x^2 - a^2}}{a} \right| + C$

7. $\int \dfrac{1}{a^2 - x^2}\, dx = \dfrac{1}{2a} \cdot \ln \left| \dfrac{a + x}{a - x} \right| + C \quad (x^2 < a^2)$

8. $\int \dfrac{1}{x^2 - a^2}\, dx = \dfrac{1}{2a} \cdot \ln \left| \dfrac{x - a}{x + a} \right| + C \quad (x^2 > a^2)$

9. $\int \dfrac{1}{x\sqrt{a^2 - x^2}}\, dx = -\dfrac{1}{a} \cdot \ln \left| \dfrac{a + \sqrt{a^2 - x^2}}{x} \right| + C \quad (0 < x < a)$

10. $\int \dfrac{1}{x\sqrt{a^2 + x^2}}\, dx = -\dfrac{1}{a} \cdot \ln \left| \dfrac{a + \sqrt{a^2 + x^2}}{x} \right| + C$

11. $\int \dfrac{x}{ax + b}\, dx = \dfrac{x}{a} - \dfrac{b}{a^2} \cdot \ln |ax + b| + C \quad (a \neq 0)$

12. $\int \dfrac{x}{(ax + b)^2}\, dx = \dfrac{b}{a^2(ax + b)} + \dfrac{1}{a^2} \cdot \ln |ax + b| + C \quad (a \neq 0)$

13. $\int \dfrac{1}{x(ax + b)}\, dx = \dfrac{1}{b} \cdot \ln \left| \dfrac{x}{ax + b} \right| + C \quad (b \neq 0)$

14. $\int \dfrac{1}{x(ax + b)^2}\, dx = \dfrac{1}{b(ax + b)} + \dfrac{1}{b^2} \cdot \ln \left| \dfrac{x}{ax + b} \right| + C \quad (b \neq 0)$

15. $\int \sqrt{x^2 + a^2}\, dx = \dfrac{x}{2} \sqrt{x^2 + a^2} + \dfrac{a^2}{2} \cdot \ln |x + \sqrt{x^2 + a^2}| + C$

16. $\int x^n \cdot \ln x\, dx = x^{n+1} \left[\dfrac{\ln |x|}{n + 1} - \dfrac{1}{(n + 1)^2} \right] + C \quad (n \neq -1)$

17. $\int x^n e^{ax}\, dx = \dfrac{x^n e^{ax}}{a} - \dfrac{n}{a} \cdot \int x^{n-1} e^{ax}\, dx + C \quad (a \neq 0)$

Answers to Selected Exercises

CHAPTER 1

SECTION 1.1 (page 10)

1. Natural number, whole number, integer, rational number, real number **3.** Integer, rational number, real number **5.** Rational number, real number **7.** Irrational number, real number **9.** Irrational number, real number **11.** True **13.** True **15.** False **17.** True **19.** Commutative **21.** Commutative **23.** Identity **25.** Associative **27.** Distributive and commutative **29.** Commutative properties of addition and multiplication

31. 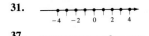 **33.** **35.**

37. **39.** **41.**

43. **45.** **47.**

49. 5 **51.** -58 **53.** -12 **55.** 2.22 **57.** -54 **59.** $-1/2$ **61.** 2 **63.** -4 **65.** 2 **67.** -4 **69.** 17 **71.** 4 **73.** -19 **75.** $=$ **77.** $<$ **79.** $=$ **81.** $=$ **83.** $=$ **85.** $=$ **87.** $=$ **89.** Yes **91.** Yes **93.** Let x be the percentage added to Gillette's market share. Then $x \geq 4$. **95.** Let x be the percent of Gillette's revenues from razors and blades. Then $x \geq 32$. **97.** Let x be the amount, in millions, spent on advertising. Then $x \leq 110$. **99.** 100; 0 **101.** 67; 11

SECTION 1.2 (page 20)

1. 4 **3.** 12 **5.** $-2/7$ **7.** $-7/8$ **9.** 40/7 **11.** 3 **13.** 12 **15.** 3/4 **17.** $-12/5$ **19.** No solution **21.** $-59/6$ **23.** $-9/4$ **25.** $x = -3a + b$ **27.** $x = (3a + b)/(3 - a)$ **29.** $x = (3 - 3a)/(a^2 - a - 1)$ **31.** $x = (2a^2)/(a^2 + 3)$ **33.** $V = k/P$ **35.** $g = (V - V_0)/t$ **37.** $B = (2A/h) - b$ or $(2A - bh)/h$ **39.** $R = (r_1 r_2)/(r_1 + r_2)$ **41.** .72 **43.** 6.53 **45.** -13.26 **47.** 68°F **49.** 15°C **51.** 37.8°C **53.** 104°F **55.** 13% **57.** $247 **59.** $4000 **61.** $205.41 **63.** $66.50 **65.** 5 cm **67.** $8000 **69.** $20,000 **71.** $70,000 for land that made a profit, $50,000 for land that produced a loss **73.** 38/3 pounds **75.** 400/3 liters

SECTION 1.3 (page 27)

1. $(-\infty, -3]$

3. $(-6, \infty)$

5. $(0, \infty)$

7. $(-\infty, 4]$

9. $(-\infty, -1]$

11. $(-\infty, 1)$

13. $(-1, \infty)$

15. $(-\infty, 1]$

17. $(1/5, \infty)$

19. $(-5, 6)$

21. $[7/3, 4]$

23. $[-11/2, 7/2]$

25. $(-\infty, 1]$ or $[6, \infty)$

27. $(-\infty, -2)$ or $[0, \infty)$

29. $(-\infty, 4)$ or $(8, \infty)$

31. $[-17/7, \infty)$

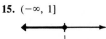

33. $[.3, \infty)$ **35.** $(-\infty, 1.5)$ **37.** $(-\infty, .5]$ **39.** The number is between $-8/5$ and $6/5$.   **41.** The number is greater than or equal to 2.   **43.** The number is at least 18.   **45.** 83 points   **47.** 4 summers   **49.** $[500, \infty)$ **51.** $[45, \infty)$ **53.** For positive values of x, C is always greater than R, so it is impossible to make a profit.

SECTION 1.4   (page 31)

1. 1, 3   **3.** $-1/3$, 1   **5.** 2/3, 8/3   **7.** -6, 14   **9.** 5/2, 7/2   **11.** $-4/3$, 2/9   **13.** $-7/3$, $-1/7$ **15.** $-3/5$, 11
17. $[-3, 3]$ **19.** $(-\infty, -1)$ or $(1, \infty)$ **21.** No solution   **23.** $[-10, 10]$

[number line: −3, 3]   *[number line: −1, 1]*   *[number line: −10, 10]*

25. $(-4, -1)$ **27.** $(-\infty, -2/3)$ or $(2, \infty)$ **29.** $(-\infty, -8/3]$ or $[2, \infty)$ **31.** $(-3/2, 13/10)$

[number line: −4, −1]   *[number line: $-\frac{2}{3}$, 2]*   *[number line: $-\frac{8}{3}$, 2]*   *[number line: $-\frac{3}{2}$, $\frac{13}{10}$]*

33. The number is between -6 and 6, inclusive.   **35.** The number is between $-7/4$ and $-5/4$, inclusive.   **37.** The number is less than or equal to -14 or greater than or equal to 10.   **39.** $|x - 2| \le 4$ **41.** $|z - 12| \ge 2$ **43.** $|k - 9| = 6$ **45.** If $|x - 2| \le .0004$, then $|y - 7| \le .00001$.

SECTION 1.5   (page 36)

1. 625   **3.** 64/125   **5.** -25 **7.** 64   **9.** 40   **11.** -162 **13.** 2^7 **15.** $(-5)^7$ or -5^7 **17.** -3^9 **19.** $(2z)^{11}$ **21.** $14m + 6$ **23.** $-x^2 + x + 9$ **25.** $-6y^2 + 3y + 10$ **27.** $-10x^2 + 4x - 2$ **29.** $-.327x^2 - 2.805x - 1.458$ **31.** $6p^2 - 15p$ **33.** $-18m^3 - 27m^2 + 9m$ **35.** $12z^3 + 14z^2 - 7z + 5$ **37.** $12k^2 - 20k + 3$ **39.** $6y^2 + 7y - 5$ **41.** $25r^2 + 5rs - 12s^2$ **43.** $.0036x^2 - .04452x - .0918$ **45.** $-p + 3$ **47.** $-x^2 - 15x - 27$ **49.** $5x^3 - 4x^2 - 8x + 1$ **51.** $x^2 - 7x + 40$

SECTION 1.6   (page 41)

1. $4(z + 1)$ **3.** $2(4x + 3y + 2z)$ **5.** $m(m^2 - 9m + 6)$ **7.** $8a(a^2 - 2a + 3)$ **9.** $5p^2(5p^2 - 4pq + 20q^2)$ **11.** $2(5x - 1)^2(20x - 3)$
13. $(x - 4)^3(9x^2 - 72x + 143)$ **15.** $(x + 5)(x - 1)$ **17.** $6(a - 10)(a + 2)$ **19.** $(x + 8)(x - 8)$ **21.** $3m(m + 3)(m + 1)$ **23.** $(b - 7)(b - 1)$
25. $(m - 3n)^2$ **27.** Cannot be factored   **29.** $(3m + 5)(3m - 5)$ **31.** Cannot be factored   **33.** $(2x + 1)(x - 3)$ **35.** $(3k - 4)(k + 2)$
37. $(7m + 2n)(3m + n)$ **39.** $(11a + 10)(11a - 10)$ **41.** $(5a + 3b)(a - 2b)$ **43.** $(y - 7z)(y + 3z)$ **45.** Cannot be factored   **47.** $(z + 7y)^2$
49. $2a^2(4a - b)(3a + 2b)$ **51.** $(3x - 1)(2x + 1)$ **53.** $(y + 5 + z)(y + 5 - z)$ **55.** $(m + n - 1)(m - n + 1)$ **57.** $x^2(x^2 + 9)^3(5x^2 + 18)$
59. $(a - 6)(a^2 + 6a + 36)$ **61.** $(2r - 3s)(4r^2 + 6rs + 9s^2)$ **63.** $(4m + 5)(16m^2 - 20m + 25)$ **65.** $(10y - z)(100y^2 + 10yz + z^2)$

SECTION 1.7   (page 47)

1. $m/4$ **3.** $z/2$ **5.** $5p/2$ **7.** 8/9   **9.** $3/(t - 3)$ **11.** $2(x + 2)/x$ **13.** $(m - 2)/(m + 3)$ **15.** $(x + 4)/(x + 1)$ **17.** $(2m + 3)/(4m + 3)$
19. $3k/5$ **21.** $25p^2/9$ **23.** $6/(5p)$ **25.** 2/9   **27.** 3/10   **29.** $2(a + 4)/(a - 3)$ **31.** $(k + 2)/(k + 3)$ **33.** $(m + 6)/(m + 3)$
35. $(m - 3)/(2m - 3)$ **37.** $5/(12y)$ **39.** 1   **41.** $(6 + p)/(2p)$ **43.** $(8 - y)/(4y)$ **45.** $137/(30m)$ **47.** $(3m - 2)/[m(m - 1)]$
49. $14/[3(a - 1)]$ **51.** $23/[20(k - 2)]$ **53.** $(7x + 9)/[(x - 3)(x + 1)(x + 2)]$ **55.** $y^2/[(y + 4)(y + 3)(y + 2)]$
57. $k(k - 13)/[(2k - 1)(k + 2)(k - 3)]$ **59.** $(x + 1)/(x - 1)$ **61.** $-1/[x(x + h)]$ **63.** $(2 + b - b^2)/(b - b^2)$

SECTION 1.8   (page 52)

1. 343   **3.** 1/8   **5.** 1/8   **7.** 1/5   **9.** 1/512   **11.** 8   **13.** 49/4   **15.** $1/3^6$ **17.** $1/2^3$ **19.** $1/6^2$ **21.** 4^3 **23.** $1/7^7$ **25.** 8^5 **27.** $1/10^8$ **29.** 5
31. x^2 **33.** $8k^3$ **35.** b^{10}/a^{20} **37.** $2^3/(5^2p^7)$ **39.** $r/7^2$ or $r/49$ **41.** $1/(12k^4)$ **43.** $x^3/(2^2y^3)$ **45.** m^7/p^2 **47.** $9a^2/(5b)$ **49.** 5   **51.** 6
53. 0   **55.** 73/8   **57.** 1/6   **59.** 9   **61.** 1/6   **63.** 3/2   **65.** $-5/9$ **67.** $b + a$ **69.** $1/(ab)$

SECTION 1.9   (page 59)

1. 9   **3.** 3   **5.** 4   **7.** 100   **9.** -25 **11.** 2/3   **13.** 4/3   **15.** 1/32   **17.** 4/3   **19.** 3/4   **21.** 5   **23.** $2^2 = 4$ **25.** $27^{1/3} = 3$ **27.** $4^2 = 16$
29. $1/6^{11/5}$ **31.** $1/4^{32/15}$ or $1/2^{64/15}$ **33.** $2^{5/6}p^{3/2}$ **35.** $1/(m^{14/15}n^{6/5})$ **37.** $x^{11/4} - 3x^{15/4}$ **39.** $6 + 2z$ **41.** $-3m^2 - 12m - 8$ **43.** 5   **45.** 6
47. -5 **49.** $5\sqrt{2}$ **51.** $-2\sqrt[4]{2}$ **53.** $-3\sqrt{5}/5$ **55.** $-\sqrt[3]{12}/2$ **57.** $\sqrt[4]{24}/2$ **59.** $24\sqrt{6}$ **61.** $2x^4z^2\sqrt{2z}$ **63.** $mn^3p^4\sqrt{n}$ **65.** $\sqrt{6x}/(3x)$
67. $(x^2y\sqrt{xy})/3$ **69.** $\sqrt[3]{2}$ **71.** $\sqrt[18]{x}$ **73.** $9\sqrt{3}$ **75.** $\sqrt{2}$ **77.** $-2\sqrt{7p}$ **79.** $2\sqrt{2}$ **81.** $7\sqrt[3]{3}$ **83.** -7 **85.** 10   **87.** $5\sqrt{6}$ **89.** $-3 -$
$3\sqrt{2}$ **91.** $3 + \sqrt{2}$ **93.** $[p(\sqrt{p} - 2)]/(p - 4)$ **95.** $\dfrac{(4a^2 - 12)\sqrt{a^2 - 4}}{a(a^2 - 4)}$ **97.** \$64   **99.** \$64,000,000   **101.** About 86 miles   **103.** About
211 miles   **105.** 29   **107.** 177

SECTION 1.10   (page 70)

1. 5, -4 **3.** -2, -3 **5.** 3, -1 **7.** 4   **9.** 5/2, -2 **11.** 4/3, $-1/2$ **13.** 5, 2   **15.** 4/3, $-4/3$ **17.** 0, 1   **19.** $\sqrt{29}$, $-\sqrt{29}$ **21.** $3 +$
$\sqrt{5}$, $3 - \sqrt{5}$ **23.** $(1 + \sqrt{19})/3$, $(1 - \sqrt{19})/3$ **25.** 2, -4 **27.** $(1 + \sqrt{2})/2$, $(1 - \sqrt{2})/2$ **29.** $(5 + \sqrt{13})/6 \approx 1.434$; $(5 - \sqrt{13})/6 \approx .232$
31. $(1 + \sqrt{33})/4 \approx 1.686$; $(1 - \sqrt{33})/4 \approx -1.186$ **33.** $5 + \sqrt{5} \approx 7.236$; $5 - \sqrt{5} \approx 2.764$ **35.** $(-6 + \sqrt{26})/2 \approx -.450$; $(-6 - \sqrt{26})/2$
≈ -5.550 **37.** 5/2, 1   **39.** 4/3, 1/2   **41.** -5, 2   **43.** No real number solutions   **45.** 0, -1 **47.** $-\sqrt{5}$, $\sqrt{5}$ **49.** $\sqrt{15}/3$, $-\sqrt{15}/3$
51. $\pm\sqrt{6 + 2\sqrt{29}}/2$ **53.** 11/3, 3/2   **55.** -1, $-15/2$ **57.** -5, 3/2   **59.** $-4 + 3\sqrt{2}$, $-4 - 3\sqrt{2}$ **61.** (a) $\dfrac{-r^2 + 3r - 12}{r(r - 2)}$ (b) 4, -3
63. (a) $\dfrac{-10k^2 + 17k - 1}{2k(k - 1)}$ (b) $-\dfrac{1}{5}$, $\dfrac{1}{2}$ **65.** 100 yd by 400 yd   **67.** 10 in by 13 in   **69.** 30 bicycles   **71.** (a) 2 sec (b) $\dfrac{1}{2}$ sec or $\dfrac{7}{2}$ sec (c) It
reaches the given height twice—once on the way up and once on the way down.   **73.** $\sqrt{2Sg}/g$ **75.** $(d^2\sqrt{kL})/L$ **77.** $\sqrt{(S - S_0 - k)g}/g$
79. $(-\pi h \pm \sqrt{\pi^2h^2 + 2\pi S})/2\pi$

SECTION 1.11   (page 77)

1. $(-2, 4)$

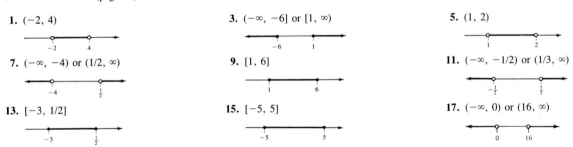

3. $(-\infty, -6]$ or $[1, \infty)$

5. $(1, 2)$

7. $(-\infty, -4)$ or $(1/2, \infty)$

9. $[1, 6]$

11. $(-\infty, -1/2)$ or $(1/3, \infty)$

13. $[-3, 1/2]$

15. $[-5, 5]$

17. $(-\infty, 0)$ or $(16, \infty)$

19. $(-\infty, -4]$ or $[0, 4]$ **21.** $(-7, 0)$ or $(7, \infty)$ **23.** $(-2, 0)$ or $(1/4, \infty)$ **25.** $(-5, 3]$ **27.** $(-\infty, -2)$ **29.** $(-\infty, 6)$ or $[15/2, \infty)$
31. $[-8, 5)$ **33.** $(-2, \infty)$ **35.** \$54   **37.** 5/4 months and 6 months   **39.** $0 < x < 5/3$ and $x > 10$ hundreds of dollars   **41.** $4 \le t \le 9.75$ sec

CHAPTER 1   REVIEW EXERCISES   (page 79)

1. 6   **3.** -12, -6, $-\sqrt{4}$ (or -2), 0, 6   **5.** $-\sqrt{7}$, $\pi/4$, $\sqrt{11}$ **7.** Natural, whole, integer, rational, real   **9.** ($\sqrt{36} = 6$) natural, whole,
integer, rational, real   **11.** Integer, rational, real   **13.** Irrational, real   **15.** Irrational, real   **17.** -7, -3, -2, 0, π, 8   **19.** $-|3 - (-2)|$,
$-|-2|$, $|6 - 4|$, $|8 + 1|$ **21.** -3 **23.** -1

25.

27.

29. -8 **31.** $-3/4$ **33.** 2   **35.** 6   **37.** $-11/3$ **39.** $-96/7$ **41.** -2 **43.** No solution   **45.** $x = 2/(9p - 1)$ **47.** $x = -6b - a - 6$
49. $x = (6 - 3m)/(1 + 2k - km)$ **51.** \$500   **53.** \$60,000 at 8%, \$40,000 at 5%   **55.** $(-7/13, \infty)$ **57.** $(-\infty, 1]$ **59.** $(1, \infty)$ **61.** $[4, 5]$
63. 3, -11 **65.** 23, -13 **67.** 15/8, $-13/8$ **69.** 4/3, $-2/7$ **71.** $[-7, 7]$ **73.** $(-\infty, -3)$ or $(3, \infty)$ **75.** No solution   **77.** $(-2/7, 8/7)$
79. $(-\infty, -4)$ or $(-2/3, \infty)$ **81.** $-7m^2 - m - 3$ **83.** $8r^4 - 8r^2 + 11r$ **85.** $-r^5 + r^4 + 19r^2$ **87.** $16y^2 + 42y - 49$ **89.** $21z^2 - 5zy -$
$50y^2$ **91.** $9k^2 - 30km + 25m^2$ **93.** $125x^3 - 150x^2 + 60x - 8$ **95.** $15w^3 - 22w^2 + 11w - 2$ **97.** $z(7z - 9z^2 + 1)$ **99.** $4p^3(3p^2 - 2p + 5)$
101. $(r + 7p)(r - 6p)$ **103.** $(3m + 1)(2m - 5)$ **105.** $(3m + 7)(m - 5)$ **107.** $(12p + 13q)(12p - 13q)$ **109.** $(2y - 1)(4y^2 + 2y + 1)$
111. $8p^2/5$ **113.** 4   **115.** $1/[2k^2(k - 1)]$ **117.** $17/(2r)$ **119.** $(5r + 3)/[(r - 1)r]$ **121.** $(q + p)/(pq - 1)$ **123.** 1/16   **125.** 1/125

127. 64/27 **129.** 6 **131.** 1/64 **133.** 729 **135.** 3/4 **137.** 25 **139.** $1/3^5$ **141.** $1/5^{2/3}$ **143.** $3^{7/2}a^{5/2}$ **145.** 3 **147.** −2 **149.** $2\sqrt{6}$ **151.** $3pq\sqrt[3]{2q^2}$ **153.** $n\sqrt{30m}/6m$ **155.** $\sqrt[6]{5}$ **157.** $-8\sqrt{3}$ **159.** $7s\sqrt[3]{2r^2}$ **161.** 4 **163.** $4 + 3\sqrt{15}$ **165.** $(-\sqrt{2} + \sqrt{6})/2$ **167.** $\sqrt{7}$, $-\sqrt{7}$ **169.** $-7 + \sqrt{5}$, $-7 - \sqrt{5}$ **171.** 1, 3 **173.** 1/2, −2 **175.** $1 + \sqrt{3}$, $1 - \sqrt{3}$ **177.** $(6 + \sqrt{58})/2$, $(6 - \sqrt{58})/2$ **179.** 5/2, −3 **181.** 7, −3/2 **183.** $\sqrt{3}/3$, $-\sqrt{3}/3$ **185.** $\sqrt{5}$, $-\sqrt{5}$ **187.** 5, −2/3 **189.** −5, −1/2 **191.** $r = (-Rp \pm E\sqrt{Rp})/p$ **193.** $s = (a \pm \sqrt{a^2 + 4K})/2$ **195.** (−3, 2) **197.** $(-\infty, -5]$ or $[3/2, \infty)$ **199.** $(-\infty, -3/2)$ or $(1/4, \infty)$ **201.** [0, 5/3] **203.** [−2, 0) **205.** (−1, 3/2) **207.** [−19, −5) or (2, ∞) **209.** 50 m by 225m; or 112.5 m by 100 m

CASE 1 (page 84)

1. 8000 **2.** 8000 **3.** 800 **4.** 800(10,000) = $8,000,000

CHAPTER 2
SECTION 2.1 (page 92)

1. Function **3.** Function **5.** Not a function **7.** Function **9.** Function **11.** Function **13.** Domain: $(-\infty, \infty)$; range: $(-\infty, \infty)$ **15.** Domain: $(-\infty, \infty)$; range: $[0, \infty)$ **17.** Domain: $(-\infty, \infty)$; range: $[0, \infty)$ **19.** Domain: all real numbers except 1; range: all nonzero real numbers **21.** (a) 14 (b) −7 (c) 2 (d) $3a + 2$ **23.** (a) −12 (b) 2 (c) −4 (d) $-2a - 4$ **25.** (a) 6 (b) 6 (c) 6 (d) 6 **27.** (a) 48 (b) 6 (c) 0 (d) $2a^2 + 4a$ **29.** (a) 5 (b) −23 (c) 1 (d) $-a^2 + 5a + 1$ **31.** (a) $\sqrt{7}$ (b) 0 (c) $\sqrt{3}$ (d) $\sqrt{a + 3}$ $(a \geq -3)$ **33.** (a) $2a - 3$ (b) $-2r - 3$ (c) $2m + 3$ (d) $2a + 2b - 3$ **35.** (a) $2a^2 - a$ (b) $2r^2 + r$ (c) $2m^2 + 11m + 15$ (d) $2a^2 + 4ab + 2b^2 - a - b$ **37.** (a) $a^3 + 1$ (b) $-r^3 + 1$ (c) $m^3 + 9m^2 + 27m + 28$ (d) $a^3 + 3a^2b + 3ab^2 + b^3 + 1$ **39.** Assume all denominators are nonzero. (a) $3/(a - 1)$ (b) $3/(-r - 1)$ (c) $3/(m + 2)$ (d) $3/(a + b - 1)$ **41.** Assume all radicands are nonnegative. (a) $\sqrt{2a}$ (b) $\sqrt{-2r}$ (c) $\sqrt{2m + 6}$ (d) $\sqrt{2a + 2b}$ **43.** (a) $80 (b) $80 (c) $80 (d) $120 (e) $160 **45.** (a) 1050 thousand dollars (b) 1100 thousand dollars (c) 1150 thousand dollars (d) 1200 thousand dollars **47.** $f(-3) = 9$; $f(0) = 2$; $f(2) = 4$; $f(4) = 6$; $f(10) = 10$

SECTION 2.2 (page 102)

1. Function **3.** Not a function **5.** Function

7.

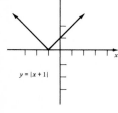

$y = |x + 1|$

9.

$y = |2 - x|$

11.

$y = |5x + 4|$

13.

$y = -|x|$

15.

$y = |x| + 4$

17.

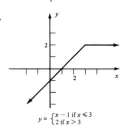

$y = \begin{cases} x - 1 & \text{if } x \leq 3 \\ 2 & \text{if } x > 3 \end{cases}$

19.

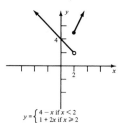

$y = \begin{cases} 4 - x & \text{if } x < 2 \\ 1 + 2x & \text{if } x \geq 2 \end{cases}$

21.

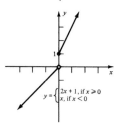

$y = \begin{cases} 2x + 1, & \text{if } x \geq 0 \\ x, & \text{if } x < 0 \end{cases}$

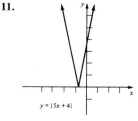

23. $y = \begin{cases} 2 + x & \text{if } x < -4 \\ -x & \text{if } -4 \leq x \leq 5 \\ 3x & \text{if } x > 5 \end{cases}$

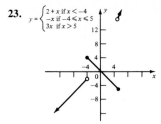

25.

$y = \begin{cases} |x| & \text{if } x > -2 \\ x & \text{if } x \leq -2 \end{cases}$

27.

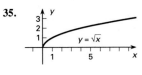

$y = [-x]$

29.

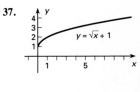

$y = [2x - 1]$

31.

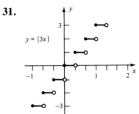

$y = [3x]$

33.

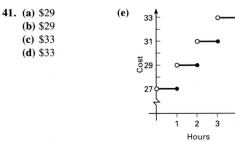

$y = [3x] - 1$

35.

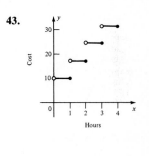

$y = \sqrt{x}$

37.

$y = \sqrt{x} + 1$

39.

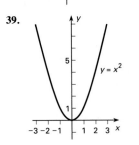

$y = x^2$

41. (a) \$29
(b) \$29
(c) \$33
(d) \$33

(e)

43.

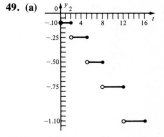

45.

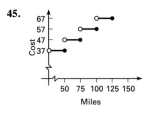

47. (a) 140
(b) 220
(c) 220
(d) 220
(e) 220
(f) 60
(g) 60

(h)

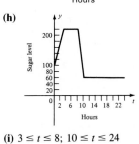

(i) $3 \leq t \leq 8$; $10 \leq t \leq 24$

49. (a)

(b) Domain: $0 \leq t \leq 16$;
range $\{-.10, -.25, -.50, -.75, -1.10\}$

51. (a) Yes (b) Years from 1984 to 1989 (c) Approx. [55, 1200]

SECTION 2.3   (page 113)

1.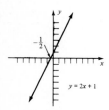

3.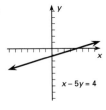

5.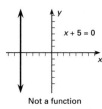

7.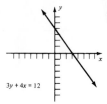

9.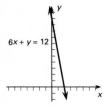

11.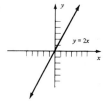

13.

15.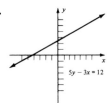

17.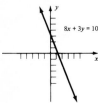

19.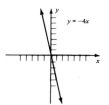

21. (image)

23.

25. $140
27. 10 items, 10 items

29. **(a)** $16 **(b)** $11 **(c)** $6 **(d)** 8
(e) 4 **(f)** 0
(g)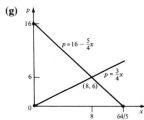
(h) 0 **(i)** 40/3 **(j)** 80/3
(k) See part (g) **(l)** 8 **(m)** 6

31. **(a)**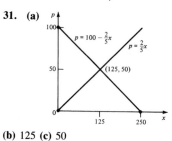
(b) 125 **(c)** 50

33. **(a)** $135 **(b)** $205 **(c)** $275 **(d)** $345
(e)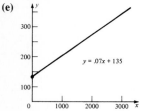

35. **(a)** $x = 200$ (200,000 policies)
(b)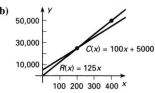
(c) $R(100) = \$12{,}500;\ C(100) = \$15{,}000$

SECTION 2.4 (page 123)

1. $-1/5$ **3.** $2/3$ **5.** $-3/2$ **7.** Undefined slope **9.** 0 **11.** Perpendicular **13.** Parallel **15.** Neither **17.** Neither **19.** 3; 4 **21.** -4; 8
23. $-3/4$; $5/4$ **25.** -3; 0 **27.** $-2/5$; 0 **29.** 0; 8 **31.** 0; -2 **33.** Undefined slope; no y-intercept

35.

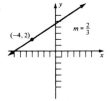

37.

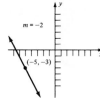

39.

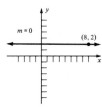

41.

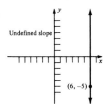

43.

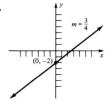

45.

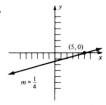

47. $4y = -3x + 16$ **49.** $2y = -x - 4$ **51.** $4y = 6x + 5$ **53.** $y = 2x + 9$ **55.** $y = -3x + 3$ **57.** $4y = x + 5$ **59.** $3y = 4x + 7$
61. $3y = -2x$ **63.** $x = -8$ **65.** $y = 3$ **67.** $y = 640x + 1100$; $m = 640$ **69.** $y = 2.5x - 70$; $m = 2.5$ **71.** $y = -1000x + 40{,}000$; $m = -1000$ **73.** **(a)** $h = 3.5r + 83$ **(b)** About 163.5 cm; about 177.5 cm **(c)** About 25 cm

SECTION 2.5 (page 131)

1. **(a)** 2600 **(b)** 2900 **(c)** 3200 **(d)** 2000 **(e)** 300 **3.** **(a)** 100 thousand **(b)** 70 thousand **(c)** 0 **(d)** -5 thousand; the number is decreasing
5. **(a)** $7y = 800{,}000x + 1{,}400{,}000$ **(b)** About \$429,000 **(c)** About \$1,230,000 **7.** **(a)** $f(x) = .3x + 10.3$ **(b)** .3% per year; they are the same
9. **(a)** 480 **(b)** 360 **(c)** 120 **(d)** June 29 **(e)** -20 **11.** If $C(x)$ is the cost of renting a saw for x hours, then $C(x) = 12 + x$. **13.** If $P(x)$ is
the cost (in cents) of parking for x half-hours, then $P(x) = 30x + 35$. **15.** $C(x) = 30x + 100$ **17.** $C(x) = 25x + 1000$ **19.** $C(x) = 50x + 500$ **21.** $C(x) = 90x + 2500$ **23.** **(a)** \$145 **(b)** \$145.125 **25.** **(a)** $C(x) = 10x + 500$ **(b)** $R(x) = 35x$ **(c)** (20, 700) **27.** **(a)** $C(x) = 18x + 250$ **(b)** $R(x) = 28x$ **(c)** (25, 700) **29.** **(a)** $C(x) = 100x + 2700$ **(b)** $R(x) = 125x$ **(c)** (108, 13,500) **31.** **(a)** \$100 **(b)** \$36 **(c)** \$24
33. **(a)** 1 **(b)** 2/3 **(c)** 6/10 or 3/5 **(d)** 4/10 or 2/5 **35.** 500 units; \$30,000 **37.** Break-even point is about 41 units; produce the item
39. Break-even point is about 467 units; do not produce the item **41.** Break-even is -200 units; product will never make a profit

43. **(a)–(b)**

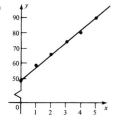

(c) $y = 8x + 50$

(d)

Year	Sales (actual)	Sales (predicted from equation of (c))	Difference, Actual Minus Predicted
0	48	50	-2
1	59	58	1
2	66	66	0
3	75	74	1
4	80	82	-2
5	90	90	0

(e) 106 **(f)** 122

CHAPTER 2   REVIEW EXERCISES   (page 136)

1. Domain: $(-\infty, \infty)$; range: $(-\infty, \infty)$

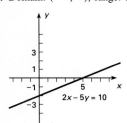
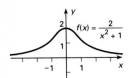

3. Domain: $(-\infty, \infty)$; range: $\left[-\dfrac{9}{8}, \infty\right)$

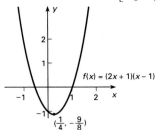

5. Domain: $(-\infty, \infty)$; range: $[-2, \infty)$

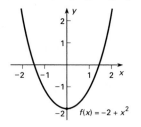

7. Domain: $(-\infty, \infty)$; range: $(0, 2]$

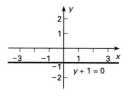

9. Domain: $(-\infty, \infty)$; range: $\{-1\}$

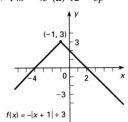

11. **(a)** 23 **(b)** -9 **(c)** $4p - 1$ **(d)** $4r + 3$ **13.** **(a)** -28 **(b)** -12 **(c)** $-p^2 + 2p - 4$ **(d)** $-r^2 - 3$ **15.** **(a)** -13 **(b)** 3 **(c)** -32 **(d)** 22
(e) $-k^2 - 4k$ **(f)** $-9m^2 + 12m$ **(g)** $-k^2 + 14k - 45$ **(h)** $12 - 5p$

17.

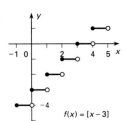

19.

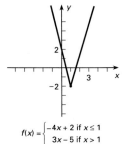

21.

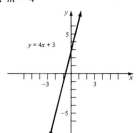

23.

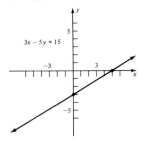

25.

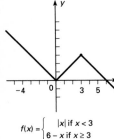

27. (a)

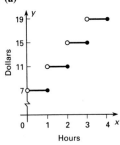

(b) Domain: $(0, \infty)$;
range: $\{7, 11, 15, 19, \ldots\}$

29. $m = 4$

31. $m = 3/5$

33. Undefined slope

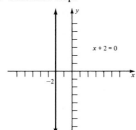

x + 2 = 0

35. *m* = 2

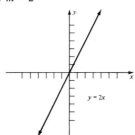

y = 2x

37. 3x − 4y = 22

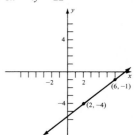

(6, −1)
(2, −4)

39. y = 3x + 13

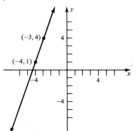

(−3, 4)
(−4, 1)

41. (a) 7/6; 9/2 (b) 2; 2 (c) 5/2, 1/2

(d)

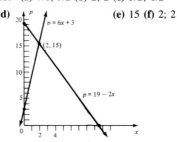

p = 6x + 3
(2, 15)
p = 19 − 2x

(e) 15 (f) 2; 2

43. 1/3 **45.** −2/11 **47.** −2/3 **49.** Undefined slope **51.** 1 **53.** −1/5 **55.** 3y = 2x − 13 **57.** 5x + 4y = 17 **59.** x = −1 **61.** 3y = 5x + 15 **63.** C(x) = 30x + 60 **65.** C(x) = 30x + 85 **67.** C(x) = 32x + 175 **69.** (a) 5 units (b) $200

CASE 2 (page 138)

1. 4.8 million units **2.** Portion of a straight line going through (3.1, 10.50) and (5.7, 10.67) **3.** In the interval under discussion (3.1 to 5.7 million units), the marginal cost always exceeds the selling price. **4.** (a) 9.87; 10.22 (b) portion of a straight line through (3.1, 9.87) and (5.7, 10.22) (c) .83 million units, which is not in the interval under discussion

CHAPTER 3
SECTION 3.1 (page 147)

1.

(e) A larger coefficient leads to a ''narrower'' parabola; a smaller coefficient leads to a ''broader'' one.

3.

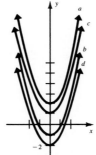

(e) They are shifted upward or downward, but otherwise have the same shape as the graph of f(x) = x².

5. (0, 0), x = 0

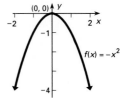

(0, 0)
−2
−1
f(x) = −x²
−4
2 x

7. $(0, 1)$, $x = 0$

9. $(0, -2)$, $x = 0$

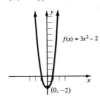

11. $(-2, 0)$, $x = -2$

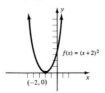

13. $(4, 0)$, $x = 4$

15. $(3, 0)$, $x = 3$

17. $(1, -3)$, $x = 1$

19. $(-4, 2)$, $x = -4$

21. $(2, 2)$, $x = 2$

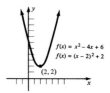

23. $(-6, -35)$, $x = -6$

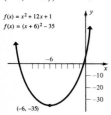

25. $(-1, -1)$, $x = -1$

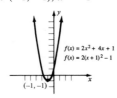

27. $(1, 3)$, $x = 1$

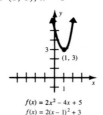

29. $(3, 3)$, $x = 3$

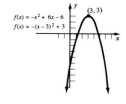

31. (a) $-3, 5$
(b) -15
(c) 1
(d) -16

33.

35.

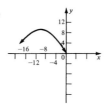

SECTION 3.2   (page 151)

1. (a) $C(x) = (x - 5)^2 + 15$
(c) 5 units
(d) Minimum cost is \$15

(b)

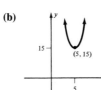

3. (a) $x = 3$ dimes or 30¢ **(b)** $p = 8$, or \$800   **5.** 16 ft; 2 sec   **7.** 80 ft by 160 ft

9. (a) 640 **(b)** 515 **(c)** 140
(d)–(e)
(f) 8
(g) 320

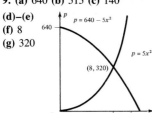

11. 8
13. (a) $R(x) = 150x - (x^2/4)$
(b) 300 **(c)** \$22,500

15. (a) $R(x) = x(500 - x)$
(b)

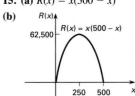

(c) 250 **(d)** \$62,500

17. (a) $100 - x$ **(b)** $50 + x$ **(c)** $R(x) = (100 - x)(50 + x)$ **(d)** 25 **(e)** \$5625   **19. (a)** 60 **(b)** 70 **(c)** 90 **(d)** 100 **(e)** 80 **(f)** 20   **21.** $6\sqrt{3}$ cm
23. (a) 2.3 or 17.7 **(b)** 15 **(c)** 10 **(d)** 120 **(e)** $0 < x < 2.3$ and $x > 17.7$ **(f)** $2.3 < x < 17.7$

SECTION 3.3   (page 158)

1.

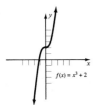

3.

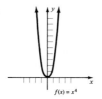

5.

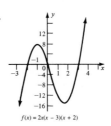

7.

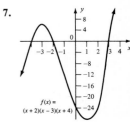

9.

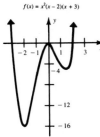

11.

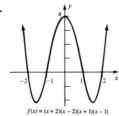

13.

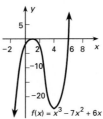

15.

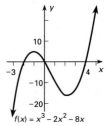

17.

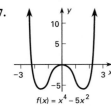

19.

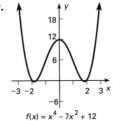

21.

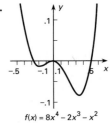

23. (a) 1.0 tenths of a percent, or .1%
(b) 2.0 tenths of a percent, or .2%
(c) 3.3 tenths of a percent, or .33%
(d) 3.1 tenths of a percent, or .31%
(e) .8 tenths of a percent, or .08%

(f)

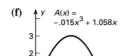

(g) Between 4 and 5 hours, closer to 5 hours
(h) From about $1\frac{1}{2}$ hours to about $7\frac{1}{2}$ hours

25. (a) 0, 108, 28, 10   **(b)**

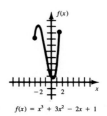

(c) Increasing for years 0 to 3 and from the 9th year on; decreasing for years 3 to 9.

27. Maximum of -10.013 when $x = .3$; minimum of -11.328 when $x = .8$ **29.** Maximum is 84 when $x = -2$; minimum is -13 when $x = -1$

31.

x	-3	-2.5	-2	-1.5	-1	$-.5$	0	.5	1	1.5
$f(x)$	7	9.125	9	7.375	5	2.625	1	.875	3	8.125

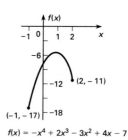

$f(x) = x^3 + 3x^2 - 2x + 1$

33.

x	-1	$-.5$	0	.5	1	1.5	2
$f(x)$	-17	-10.0625	-7	-5.5625	-5	-6.0625	-11

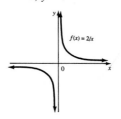

$f(x) = -x^4 + 2x^3 - 3x^2 + 4x - 7$

SECTION 3.4   (page 164)

1. $x = -2$, $y = 0$

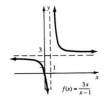

$f(x) = \frac{1}{x+2}$

3. $x = 3$, $y = 0$

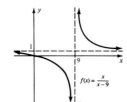

$f(x) = \frac{-4}{x-3}$

5. $x = 0$, $y = 0$

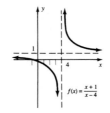

$f(x) = 2/x$

7. $x = -3/2$, $y = 0$

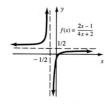

$f(x) = \frac{2}{3+2x}$

9. $x = 1$, $y = 3$

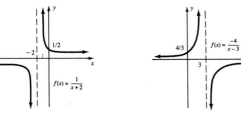

$f(x) = \frac{3x}{x-1}$

11. $x = 9$, $y = 1$

$f(x) = \frac{x}{x-9}$

13. $x = 4$, $y = 1$

$f(x) = \frac{x+1}{x-4}$

15. $x = -1/2$, $y = 1/2$

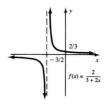

$f(x) = \frac{2x-1}{4x+2}$

17. $x = -4$, $y = -2/5$

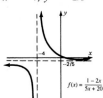

$$f(x) = \frac{1 - 2x}{5x + 20}$$

19. $x = -2$, $y = -1/3$

$$f(x) = \frac{-x - 4}{3x + 6}$$

21. **(a)** $12.50 **(b)** $10 **(c)** $6.25 **(d)** $5
(e) $3.85
(f)

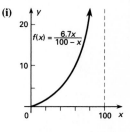

$$C(x) = \frac{500}{x + 30}$$

23. **(a)** $6700 **(b)** About $15,600 **(c)** $26,800 **(d)** $60,300 **(e)** $127,300 **(f)** $328,300 **(g)** $663,300 **(h)** No **(i)**

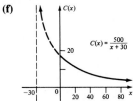

$$f(x) = \frac{6.7x}{100 - x}$$

25. **(a)** 6 min **(b)** 1.5 min **(c)** .6 min **(d)** $A = 0$
(f) W becomes negative. The waiting time approaches 0 as A approaches 3. The formula does not apply for $A > 3$ since there will be no waiting if people arrive more than 3 min apart.

(e)

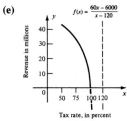

$$W = \frac{3(3 - A)}{A}$$

27. **(a)** About $10,000 **(b)** About $20,000

29. 500,000 gal of gasoline, 100,000 gal of oil

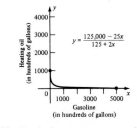

$$y = \frac{125,000 - 25x}{125 + 2x}$$

31. **(a)** $42.9 million **(b)** $40 million **(c)** $30 million **(d)** $0 million

(e)

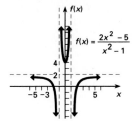

$$f(x) = \frac{60x - 6000}{x - 120}$$

33. Vertical asymptote $x = -3/2$; no horizontal asymptote

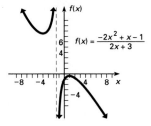

$$f(x) = \frac{-2x^2 + x - 1}{2x + 3}$$

35. Vertical asymptotes $x = 1$, $x = -1$; horizontal asymptote $y = 2$

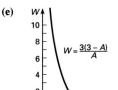

$$f(x) = \frac{2x^2 - 5}{x^2 - 1}$$

CHAPTER 3 REVIEW EXERCISES (page 167)

1. $(0, -4)$, $x = 0$

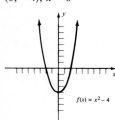

$f(x) = x^2 - 4$

3. $(1, 0)$, $x = 1$

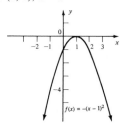

$f(x) = -(x - 1)^2$

5. $(-1, -5)$, $x = -1$

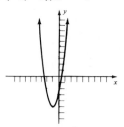

$f(x) = 3(x + 1)^2 - 5$

7. $(-3, -2)$, $x = -3$

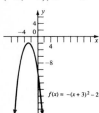

$f(x) = -(x + 3)^2 - 2$

9. $(2, -2)$, $x = 2$

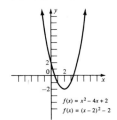

$f(x) = x^2 - 4x + 2$
$f(x) = (x - 2)^2 - 2$

11. $(3, 6)$, $x = 3$

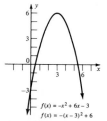

$f(x) = -x^2 + 6x - 3$
$f(x) = -(x - 3)^2 + 6$

13. $(1, -1)$, $x = 1$

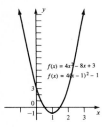

$f(x) = 4x^2 - 8x + 3$
$f(x) = 4(x - 1)^2 - 1$

15. $(-2, 4)$, $x = -2$

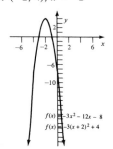

$f(x) = -3x^2 - 12x - 8$
$f(x) = -3(x + 2)^2 + 4$

17. $\left(\frac{5}{2}, \frac{17}{4}\right)$, $x = \frac{5}{2}$

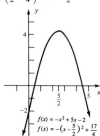

$f(x) = -x^2 + 5x - 2$
$f(x) = -\left(x - \frac{5}{2}\right)^2 + \frac{17}{4}$

19. Minimum of -3 when $x = 2$ **21.** Maximum of 11 when $x = -2$ **23.** Minimum of -1 when $x = 1$ **25.** $(0, \frac{5}{4})$, $(6, \infty)$ **27.** For any x in $(0, \frac{5}{3})$ or $(10, \infty)$ **29.** 50 m × 50 m; 2500 sq m

31.

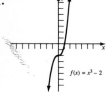

$f(x) = x^3 - 2$

33.

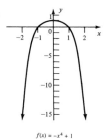

$f(x) = -x^4 + 1$

35.

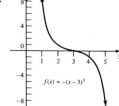

$f(x) = -(x - 3)^3$

37.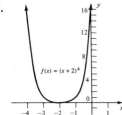

39.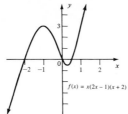

41.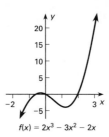

43.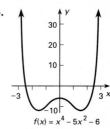

45.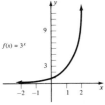

47.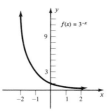

49.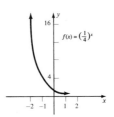

51. (a) About \$8200
(b) About \$113,000
(c) About 75%
53. (a) 2 **(b)** A loss **(c)** A profit

CHAPTER 4
SECTION 4.1   (page 174)

1. Exponential   **3.** Not exponential   **5.** 4/9   **7.** $\sqrt{6}/3$

9.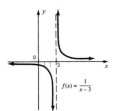

11.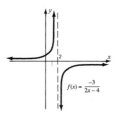

13.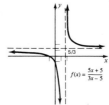

15.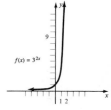

17.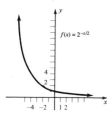

19.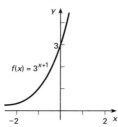

21.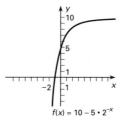

23.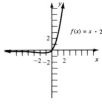

25. 2   **27.** −3   **29.** 2   **31.** 3/2   **33.** −2   **35.** 3   **37.** −5/4   **39.** 1/9   **41.** 4, −4   **43.** 2   **45.** 0, 5/3

47. (a)

t	0	1	2	3	4	5	6	7	8	9	10
y	1	1.05	1.10	1.16	1.22	1.28	1.34	1.41	1.48	1.55	1.63

(b)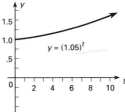

49. (a) About \$98,000 **(b)** Almost \$28.00
51. (a) 1,000,000 **(b)** About 1,410,000 **(c)** 2,000,000 **(d)** 4,000,000   **(e)**

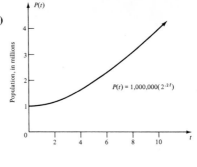

53. \$19,420.26
55. (a) \$100,000 **(b)** About \$225,000 **(c)** About \$1,818,000
(d) About \$2,488,000
57. (a) About 207 **(b)** About 235 **(c)** About 249
(d)

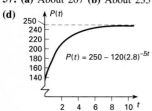

59. The curve formed is not a smooth curve normally generated by an exponential function.   **61. (a)** 7 **(b)** 9 **(c)** 15

SECTION 4.2   (page 181)

1. (a) 1,000,000 **(b)** About 1,040,000 **(c)** About 1,080,000 **(d)** About 1,220,000   **3. (a)** 500 g **(b)** 409 g **(c)** 335 g **(d)** 184 g   **5. (a)** 25,000
(b) About 30,500 **(c)** About 37,300 **(d)** About 68,000   **7.** These values were found using a calculator. **(a)** \$21,665.74 **(b)** \$29,836.49
(c) \$44,510.82   **9. (a)** 30 **(b)** About 42 **(c)** About 56 **(d)** About 59
11. (a) 0 **(b)** About 316 **(c)** About 432 **(d)** About 497 **(e)** Almost 500 **(f)** $y = 500$ **(g)**

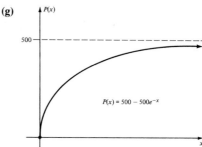

13. (a) About 72,400 **(b)** About 48,500   **15. (a)** 40 **(b)** 21 **(c)** 11 **(d)** 6 **(e)** 3 **(f)** 3   **17. (a)** \$100 million **(b)** About \$246 million **(c)** About
\$505 million   **19.** 125.36°C   **21.** 81.25°C   **23.** 117.5°C   **25.** 11.6 pounds per square inch

SECTION 4.3   (page 192)

1. $\log_2 8 = 3$ **3.** $\log_3 81 = 4$ **5.** $\log_{1/3} 9 = -2$ **7.** $2^7 = 128$ **9.** $5^{-2} = 1/25$ **11.** $10^4 = 10,000$ **13.** 3   **15.** -2 **17.** 2   **19.** 3
21. -2 **23.** 1/2
25. 1/9; 1/3; 1; 3; 27   **27.**   **29.**

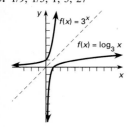   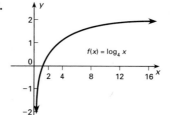   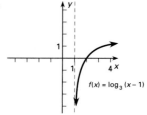

31. $\log_3 2 - \log_3 5$ **33.** $\log_9 7 + \log_9 m$ **35.** $1 + \log_3 x - \log_3 5 - \log_3 k$ **37.** $\log_k p + 2\log_k q - \log_k m$ **39.** Cannot be simplified with the properties of logarithms **41.** $\log_3 5 + (1/2)\log_3 2 - (1/4)\log_3 7$ **43.** 1.7810 **45.** -2.8069 **47.** 10.9768 **49.** .8451 **51.** log 5 **53.** ln 72 **55.** $\log\left(\dfrac{x^3}{y^2}\right)$ **57.** $\ln(3x^2 + 14x + 8)$ **59.** $\log\left(\dfrac{x^3\sqrt{x+2}}{(x+1)^4}\right)$ **61.** $3a$ **63.** $a + 3c$ **65.** $2c + 3a + 1$ **67.** 1/5 **69.** 3/2

71. 16 **73.** 8/5 **75.** 1/3 **77.** 10 **79.** No solution **81.** 3
83. (a) \$125 thousand **(b)** About \$211 thousand **(c)** About \$235 thousand **(d)** About \$307 thousand **(e)**
85. About \$60,000; about \$450,00

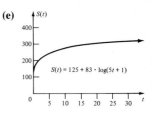

SECTION 4.4 (page 199)

1. 1.631 **3.** 1.069 **5.** $-.535$ **7.** $-.080$ **9.** 3.797 **11.** 2.386 **13.** $-.123$ **15.** No solution **17.** 1.104 **19.** 17.475 **21.** -2.141, 2.141 **23.** 1.386 **25.** 11 **27.** 5 **29.** 10 **31.** 5 **33.** 4 **35.** No solution **37.** 11 **39.** 3 **41.** No solution **43.** $-2, 2$ **45.** 1, 10 **47.** 2.423 **49.** $-.204$ **51. (a)** 5000 **(b)** About 4520 **(c)** About 3350 **(d)** $-(\ln .5)/.02 \approx 34.7$ seconds **53. (a)** About 23 days **(b)** About 8.7 days **(c)** About 3466 days **55.** About 1800 years **57.** About 18,600 years **59. (a)** About 67% **(b)** About 37% **(c)** About 23 days **(d)** About 46 days **61. (a)** 20 **(b)** 30 **(c)** 50 **(d)** 60 **63. (a)** 3 **(b)** 6 **(c)** 8 **65.** .629 **67.** 4.3 ml/min; 7.8 ml/min **69. (a)** 11% **(b)** 36% **(c)** 84%

CHAPTER 4 REVIEW EXERCISES (page 203)

1. -1 **3.** 2 **5.** **7.**

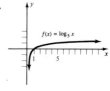

9. (a) 0 **(b)** About 55 **(c)** About 98 **(d)** About 100 **11.** $\log_2 64 = 6$ **13.** $\ln 1.09417 = .09$ **15.** $2^5 = 32$ **17.** $e^{4.41763} = 82.9$ **19.** 4 **21.** 4/5 **23.** 3/2 **25.** 1.2553 **27.** $-.0809$ **29.** 1.8245 **31.** 6.1800 **33.** $\log_5 21k^4$ **35.** $\log_2 (x^2/m^3)$ **37.** 1.416 **39.** -2.807 **41.** -3.305 **43.** 2.115 **45.** .747 **47.** 28.463 **49.** ±1.097 **51.** 8 **53.** 6 **55.** 2 **57.** No solution **59.** 3/2 **61.** 5/8 **63. (a)** .8 m **(b)** 1.2 m **(c)** 1.5 m **(d)** 1.8 m

CASE 3 (page 205)

1. .0315 **2.** λy_0 is about 98,000; painting is a forgery **3.** About 157,000; forgery **4.** About 14,000; cannot be modern forgery **5.** About 58,000; forgery **6.** About 127,000; forgery

CHAPTER 5

SECTION 5.1 (page 212)

1. \$120 **3.** \$2062.50 **5.** \$186.54 **7.** \$336 **9.** \$143.46 **11.** \$438.90 **13.** \$376.11 **15.** \$201.46 **17.** \$13,554.22 **19.** \$5180.17 **21.** \$14,920.20 **23.** \$6101.33 **25.** \$336.89 **27.** 12.77% **29.** 10.50% **31.** \$28,843.97 **33.** 9.25% **35.** \$1732.90 **37.** \$471.25, \$1028.75 **39.** \$5882.26 **41.** \$3738, 14.83% **43.** \$7318.98

SECTION 5.2 (page 221)

1. $20,974.31 **3.** $1903.00 **5.** $18,244.01 **7.** $5105.58 **9.** $60,484.63 **11.** $1167.56 **13.** $2551.12 **15.** $8265.24 **17.** $1067.82
19. $4872.35 **21.** $1000 now **23.** $21,665.74 **25.** $44,510.82 **27.** $25,424.98 **29.** 4.04% **31.** 8.16% **33.** 12.36%
35. $13,693.34 **37.** $7743.93 **39.** $142,886.40 **41.** $123,506.50 **43.** $23,182.70 **45.** $66.61 **47.** About $176,000 **49.** $136.87
51. (a) $14,288.98 **(b)** $3711.02 **(c)** $2883.46 **53.** About 6.6 years **55. (a)** 17.7 years **(b)** 11.9 years **(c)** 9.0 years **(d)** 6.1 years
57. (a) About 12 years **(b)** About 35 years

SECTION 5.3 (page 229)

1. 48 **3.** −648 **5.** 81 **7.** 64 **9.** 15 **11.** 156/25 **13.** −208 **15.** 15.91713 **17.** 21.82453 **19.** 22.01900 **21.** 20.02359
23. $437.46 **25.** $305,390.04 **27.** $671,994.58 **29.** $4,180,929.79 **31.** $180,307.41 **33.** $27,541.18 **35.** $516,397.05
37. $6294.79 **39.** $158,456.07 **41.** $13,486.56 **43.** $6517.42 **45.** $628.25 **47.** $952.33 **49. (a)** $149,850.69 **(b)** $137,895.79
(c) $11,954.90 **51.** $4168.30 **53.** $256,015.48 **55.** $67,940.98 **57. (a)** $226.11 **(b)** $245.77 **59.** $522.85 **61.** $112,796.87
63. $209,348.00 **65.** $775.08 **67.** $4462.94 **69.** $148.02 **71.** $23,023.98
73. (a) $1200 **(b)** $3511.58 except last payment, which is $3511.59. **(c)**

Payment Number	Amount of Deposit	Interest Earned	Total
1	$3511.58	$ 0	$ 3511.58
2	3511.58	105.35	7128.51
3	3511.58	213.86	10,853.95
4	3511.58	325.62	14,691.15
5	3511.58	440.73	18,643.46
6	3511.58	559.30	22,714.34
7	3511.58	681.43	26,907.35
8	3511.58	807.22	31,226.15
9	3511.58	936.78	35,674.51
10	3511.58	1070.24	40,256.33
11	3511.58	1207.69	44,975.60
12	3511.58	1349.27	49,836.45
13	3511.58	1495.09	54,843.12
14	3511.59	1645.29	60,000.00

SECTION 5.4 (page 238)

1. 9.71225 **3.** 12.65930 **5.** 14.71787 **7.** 5.69719 **9.** $6246.89 **11.** $7877.72 **13.** $153,724.51 **15.** $160,188.18 **17.** $111,183.87
19. $97,122.49 **21.** $68,108.64 **23.** $12,493.78 **25. (a)** $158.00 **(b)** $1584.00 **27.** $6699.00 **29.** $160.08 **31.** $5309.69
33. $11,727.32 **35.** $258.90 **37.** $476.81 **39.** $571.69 **41. (a)** $2349.51; $197,911.80 **(b)** $2097.30; $278,352.00
(c) $1965.82; $364,746.00

43.

Payment Number	Amount of Payment	Interest for Period	Portion to Principal	Principal at End of Period
0	—	—	—	$4000
1	$1207.68	$320.00	$ 887.68	3112.32
2	$1207.68	248.99	958.69	2153.63
3	$1207.68	172.29	1035.39	1118.24
4	$1207.70	89.46	1118.24	0

45.

Payment Number	Amount of Payment	Interest for Period	Portion to Principal	Principal at End of Period
0	—	—	—	$7184
1	$211.03	$107.76	$103.27	7080.73
2	211.03	106.21	104.82	6975.91
3	211.03	104.64	106.39	6869.52
4	211.03	103.04	107.99	6761.53
5	211.03	101.42	109.61	6651.92
6	211.03	99.78	111.25	6540.67

47. About $57,463 **49.** $7.61 **51.** $79.64 **53.** $35.24

55.

Payment Number	Amount of Payment	Interest for Period	Portion to Principal	Principal at End of Period
0	—	—	—	$37947.50
1	5783.49	$3225.54	$2557.95	35389.55
2	5783.49	3008.11	2775.38	32614.17
3	5783.49	2772.20	3011.29	29602.88
4	5783.49	2516.24	3267.25	26335.63
5	5783.49	2238.53	3544.96	22790.67
6	5783.49	1937.21	3846.28	18944.39
7	5783.49	1610.27	4173.22	14771.17
8	5783.49	1255.55	4527.94	10243.23
9	5783.49	870.67	4912.82	5330.41
10	5783.49	453.08	5330.41	0

SECTION 5.5 (page 240)

1. $57,513.15 **3.** $35,840.29 **5.** $12,125.91 **7.** $3903.93 **9.** $5054.47 **11.** $4502.36 **13.** $425.00 **15.** $1813.15 **17.** $19,356.63 **19.** $40,000.94; $7250.94 **21.** $69,331.75; $36,581.75 **23.** $44,313.15 **25.** $117,028.83 **27.** $62,197.46 **29.** $235,492.28 **31.** $21,384.28 **33.** $11,304.71 **35.** $633.59 **37.** $1054.05 **39.** $8702.22 **41.** $5842.10 **43.** $23,122.54

45. Payment Number	Amount of Payment	Interest for Period	Portion to Principal	Principal at End of Period
0	—	—	—	$8500.00
1	$1522.65	$510.00	$1012.65	7487.35
2	1522.65	449.24	1073.41	6413.94
3	1522.65	384.84	1137.81	5276.13
4	1522.65	316.57	1206.08	4070.05
5	1522.65	244.20	1278.45	2791.60
6	1522.65	167.50	1355.15	1436.45
7	1522.64	86.19	1436.45	0

47. 15% **49.** $8087.13 **51.** $2364.26 **53.** $107,892.82; $32,892.82 **55.** $5596.62 **57.** $4587.64 **59.** Monthly payment $1092.42; total interest $212,026

CHAPTER 5 REVIEW EXERCISES (page 244)

1. $848.16 **3.** $2686.84 **5.** $105.30 **7.** $23,622.05 **9.** $78,717.19 **11.** $665.54 **13.** $6056.35 **15.** 16.3% **17.** $1999.00 **19.** $43,988.40 **21.** $5506.73 **23.** $3576.92 **25.** $6002.84 **27.** $6289.05 **29.** $18,306.34 **31.** $2501.24 **33.** $12,025.46 **35.** $845.74 **37.** $3137.06 **39.** 2, 6, 18, 54, 162 **41.** −96 **43.** −120 **45.** 34.78489 **47.** $10,078.44 **49.** $148,204.84 **51.** $29,715.07 **53.** $27,320.71 **55.** $886.05 **57.** $5132.48 **59.** $2815.31 **61.** $47,988.11 **63.** $3322.43 **65.** $12,806.38 **67.** $2305.07 **69.** $717.21 **71.** **(a)** $822.89; $216,240.40 **(b)** $842.58; $172,774.00 **(c)** $960.13; $92,823.40

73. Payment Number	Amount of Payment	Interest for Period	Portion to Principal	Principal at End of Period
0	—	—	—	$5000
1	$985.09	$250.00	$735.09	4264.91
2	985.09	213.25	771.84	3493.07
3	985.09	174.65	810.44	2682.63
4	985.09	134.13	850.96	1831.67
5	985.09	91.58	893.51	938.16
6	985.07	46.91	938.16	0

CASE 4 (page 247)

1. $14,038 **2.** $9511 **3.** $8837 **4.** $3969

CHAPTER 6
SECTION 6.1 (page 261)

1. (3, 6) **3.** (−1, 4) **5.** (−2, 0) **7.** (1, 3) **9.** (4, −2) **11.** (2, −2) **13.** No solution **15.** All ordered pairs that are solutions of $4x − y = 9$ **17.** (12, 6) **19.** (7, −2) **21.** (1, 2, −1) **23.** (2, 0, 3) **25.** No solution **27.** (0, 2, 4) **29.** (1, 2, 3) **31.** (−1, 2, 1) **33.** (4, 1, 2) **35.** $((1 − z)/2, (11 − z)/2, z)$ for any real number z **37.** $(3 − z, 4 − z, z)$ for any real number z **39.** $(−3, −17 + z, z)$ for any real number z **41.** $a = -3$, $b = -1$ **43.** $a = 3/4$, $b = 1/4$, $c = -1/2$ **45.** 9000 pounds of grapefruit at \$1.60 per pound   **47.** 300 shares of USAir; 100 shares of BPAmerica   **49.** 260 skirts; 270 blouses   **51.** 5 model 201 and 8 model 301   **53.** \$3000 at 8%, \$6000 at 9%, \$1000 at 5%   **55.** 400/9 g of A; 400/3 g of B; 2000/9 g of C   **57. (a)** $y = .01x^2 - .3x + 4.24$ **(b)** 15 platters; \$1.99

SECTION 6.2   (page 271)

1. $\begin{bmatrix} 1 & 4 & 3 & | & 8 \\ 6 & 9 & 1 & | & 10 \\ 2 & 3 & 5 & | & 4 \end{bmatrix}$ **3.** $\begin{bmatrix} 1 & 2 & 4 & | & 6 \\ 0 & 5 & 2 & | & 9 \\ 4 & 7 & 8 & | & 15 \end{bmatrix}$ **5.** $\begin{bmatrix} 1 & 0 & -18 & | & -47 \\ 0 & 1 & 5 & | & 14 \\ 0 & 3 & 8 & | & 16 \end{bmatrix}$ **7.** $(-3, 4)$ **9.** $(1, 1)$ **11.** No solution   **13.** $(2, 4, 5)$

15. $(-1, 2, -2)$ **17.** $(-2, 1, 3)$ **19.** $(3, 2, -4)$ **21.** $(z - 3, -z + 5, z)$ for any real number z.   **23.** $(-2, -3, 4)$ **25.** Send 12 cars from I to A, 8 cars from II to A, 16 cars from I to B, and no cars from II to B.   **27.** 81 kg of the first chemical, 382.286 kg of the second, 286.714 kg of the third   **29.** 243 of A, 38 of B, 101 of C (truncated)   **31. (a)** Let x_1 be the units purchased from first supplier for Roseville; x_2 be the units from the first supplier for Akron; x_3 be the units from the second supplier for Roseville; and x_4 be the units from the second supplier for Akron. **(b)** $x_1 + x_2 = 75$; $x_3 + x_4 = 40$; $x_1 + x_3 = 40$; $x_2 + x_4 = 75$; $70x_1 + 90x_2 + 80x_3 + 120x_4 = 10750$ **(c)** $(40, 35, 0, 40)$

SECTION 6.3   (page 280)

1. False; corresponding elements not all equal   **3.** True   **5.** True   **7.** 2×2, square $\begin{bmatrix} 4 & -8 \\ -2 & -3 \end{bmatrix}$ **9.** 3×4 $\begin{bmatrix} 6 & -8 & 0 & 0 \\ -4 & -1 & -9 & -2 \\ -3 & 5 & -7 & -1 \end{bmatrix}$

11. 2×1, column $\begin{bmatrix} -2 \\ -4 \end{bmatrix}$ **13.** $x = 2$, $y = 4$, $z = 8$ **15.** $x = -15$, $y = 5$, $k = 3$ **17.** $z = 18$, $r = 3$, $s = 3$, $p = 3$, $a = 3/4$

19. $\begin{bmatrix} 9 & 12 & 0 & 2 \\ 1 & -1 & 2 & -4 \end{bmatrix}$ **21.** $\begin{bmatrix} 5 & 13 & 0 \\ 3 & 1 & 8 \end{bmatrix}$ **23.** Not possible   **25.** $\begin{bmatrix} 1 & 5 & 6 & -9 \\ 5 & 7 & 2 & 1 \\ -7 & 2 & 2 & -7 \end{bmatrix}$ **27.** $\begin{bmatrix} -12x + 8y & -x + y \\ x & 8x - y \end{bmatrix}$

29. $\begin{bmatrix} x & y \\ z & w \end{bmatrix} + \begin{bmatrix} r & s \\ t & u \end{bmatrix} = \begin{bmatrix} x + r & y + s \\ z + t & w + u \end{bmatrix}$ (a 2×2 matrix)   **31.** $\begin{bmatrix} x + (r + m) & y + (s + n) \\ z + (t + p) & w + (u + q) \end{bmatrix} = \begin{bmatrix} (x + r) + m & (y + s) + n \\ (z + t) + p & (w + u) + q \end{bmatrix}$

33. $\begin{bmatrix} m + 0 & n + 0 \\ p + 0 & q + 0 \end{bmatrix} = \begin{bmatrix} m & n \\ p & q \end{bmatrix}$ **35.** $\begin{bmatrix} 18 & 2.7 \\ 10 & 1.5 \\ 8 & 1.0 \\ 10 & 2.0 \\ 10 & 1.7 \end{bmatrix}$ $\begin{bmatrix} 18 & 10 & 8 & 10 & 10 \\ 2.7 & 1.5 & 1.0 & 2.0 & 1.7 \end{bmatrix}$ **37. (a)** $\begin{bmatrix} 2 & 1 & 2 & 1 \\ 3 & 2 & 2 & 1 \\ 4 & 3 & 2 & 1 \end{bmatrix}$ **(b)** $\begin{bmatrix} 5 & 0 & 7 \\ 0 & 10 & 1 \\ 0 & 15 & 2 \\ 10 & 12 & 8 \end{bmatrix}$ **(c)** $\begin{bmatrix} 8 \\ 4 \\ 5 \end{bmatrix}$

39. (a) $\begin{bmatrix} 88 & 48 & 16 & 112 \\ 105 & 72 & 21 & 147 \\ 60 & 40 & 0 & 50 \end{bmatrix}$ **(b)** $\begin{bmatrix} 110 & 60 & 20 & 140 \\ 140 & 96 & 28 & 196 \\ 66 & 44 & 0 & 55 \end{bmatrix}$ **(c)** $\begin{bmatrix} 198 & 108 & 36 & 252 \\ 245 & 168 & 49 & 343 \\ 126 & 84 & 0 & 105 \end{bmatrix}$

SECTION 6.4   (page 290)

1. 2×2; 2×2 **3.** 4×4; 2×2 **5.** 3×2; BA does not exist   **7.** AB does not exist; 3×2 **9.** $\begin{bmatrix} -4 & 8 \\ 0 & 6 \end{bmatrix}$ **11.** $\begin{bmatrix} 24 & -8 \\ -16 & 0 \end{bmatrix}$

13. $\begin{bmatrix} -22 & -6 \\ 20 & -12 \end{bmatrix}$ **15.** $\begin{bmatrix} 13 \\ 25 \end{bmatrix}$ **17.** $\begin{bmatrix} -2 & 10 \\ 0 & 8 \end{bmatrix}$ **19.** $\begin{bmatrix} -4 & 1 \\ 2 & -3 \end{bmatrix}$ **21.** $\begin{bmatrix} 3 & -5 & 7 \\ -2 & 1 & 6 \\ 0 & -3 & 4 \end{bmatrix}$ **23.** $\begin{bmatrix} 13 & 5 \\ 25 & 15 \end{bmatrix}$ **25.** $\begin{bmatrix} 13 \\ 29 \end{bmatrix}$ **27.** $\begin{bmatrix} 110 \\ 40 \\ -50 \end{bmatrix}$

29. $\begin{bmatrix} 22 & -8 \\ 11 & -4 \end{bmatrix}$ **31.** $AB = \begin{bmatrix} -3 & -9 \\ 2 & 6 \end{bmatrix}\begin{bmatrix} 4 & 6 \\ 2 & 3 \end{bmatrix} = \begin{bmatrix} -30 & -45 \\ 20 & 30 \end{bmatrix}$, but $BA = \begin{bmatrix} 4 & 6 \\ 2 & 3 \end{bmatrix}\begin{bmatrix} -3 & -9 \\ 2 & 6 \end{bmatrix} = \begin{bmatrix} 0 & 0 \\ 0 & 0 \end{bmatrix}$.

33. $(A + B)(A - B) = \begin{bmatrix} 1 & -3 \\ 4 & 9 \end{bmatrix}\begin{bmatrix} -7 & -15 \\ 0 & 3 \end{bmatrix} = \begin{bmatrix} -7 & -24 \\ -28 & -33 \end{bmatrix}$, but $A^2 - B^2 = \begin{bmatrix} -3 & -9 \\ 2 & 6 \end{bmatrix}\begin{bmatrix} -3 & -9 \\ 2 & 6 \end{bmatrix} - \begin{bmatrix} 4 & 6 \\ 2 & 3 \end{bmatrix}\begin{bmatrix} 4 & 6 \\ 2 & 3 \end{bmatrix} =$

$\begin{bmatrix} -9 & -27 \\ 6 & 18 \end{bmatrix} - \begin{bmatrix} 28 & 42 \\ 14 & 21 \end{bmatrix} = \begin{bmatrix} -37 & -69 \\ -8 & -3 \end{bmatrix}$.

35. $(PX)T = \begin{bmatrix} (mx + nz)r + (my + nw)t & (mx + nz)s + (my + nw)u \\ (px + qz)r + (py + qw)t & (px + qz)s + (py + qw)u \end{bmatrix}$; $P(XT)$ is the same, so that $P(XT) = (PX)T$

37. $PX = \begin{bmatrix} mx + nz & my + nw \\ px + qz & py + qw \end{bmatrix}$ (a 2 × 2 matrix)

39. $(k + h)P = \begin{bmatrix} (k + h)m & (k + h)n \\ (k + h)p & (k + h)q \end{bmatrix} = \begin{bmatrix} km + hm & kn + hn \\ kp + hp & kq + hq \end{bmatrix} = \begin{bmatrix} km & kn \\ kp & kq \end{bmatrix} + \begin{bmatrix} hm & hn \\ hp & hq \end{bmatrix} = kP + hP$

41. (a) $\begin{bmatrix} 47.5 & 57.75 \\ 27 & 33.75 \\ 81 & 95 \\ 12 & 15 \end{bmatrix}$ **(b)** [20 200 50 60]; [220 890 105 125 70] **(c)** [11,120 13,555]

43. (a) $C = \begin{bmatrix} 90,000 \\ 60,000 \\ 120,000 \end{bmatrix}$ **(b)** $PC = \begin{bmatrix} 72,000 \\ 48,000 \\ 96,000 \end{bmatrix}$ **(c)** $R = [1/2 \ \ 1/3 \ \ 1/6]$ **(d)** $\begin{bmatrix} 36,000 & 24,000 & 12,000 \\ 24,000 & 16,000 & 8,000 \\ 48,000 & 32,000 & 16,000 \end{bmatrix}$ **45.** $AX = B$ gives $2x_1 + 4x_2 = 2$

$x_1 + 3x_2 = -1$

$x_1 = 5, \quad x_2 = -2$

$AX = \begin{bmatrix} 2 & 4 \\ 1 & 3 \end{bmatrix}\begin{bmatrix} 5 \\ -2 \end{bmatrix} = \begin{bmatrix} 2 \\ -1 \end{bmatrix} = B$

47. $\begin{bmatrix} 44 & 75 & -60 & -33 & 11 \\ 20 & 169 & -164 & 18 & 105 \\ 113 & -82 & 239 & 218 & -55 \\ 119 & 83 & 7 & 82 & 106 \\ 162 & 20 & 175 & 143 & 74 \end{bmatrix}$ **49.** Cannot be found **51.** No

53. $\begin{bmatrix} -1 & 5 & 9 & 13 & -1 \\ 7 & 17 & 2 & -10 & 6 \\ 18 & 9 & -12 & 12 & 22 \\ 9 & 4 & 18 & 10 & -3 \\ 1 & 6 & 10 & 28 & 5 \end{bmatrix}\begin{bmatrix} -2 & -9 & 90 & 77 \\ -42 & -63 & 127 & 62 \\ 413 & 76 & 180 & -56 \\ -29 & -44 & 198 & 85 \\ 137 & 20 & 162 & 103 \end{bmatrix}\begin{bmatrix} -56 & -1 & 1 & 45 \\ -156 & -119 & 76 & 122 \\ 315 & 86 & 118 & -91 \\ -17 & -17 & 116 & 51 \\ 118 & 19 & 125 & 77 \end{bmatrix}$;

$\begin{bmatrix} 54 & -8 & 89 & 32 \\ 114 & 56 & 51 & -60 \\ 98 & -10 & 62 & 35 \\ -12 & -27 & 82 & 34 \\ 19 & 1 & 37 & 26 \end{bmatrix}\begin{bmatrix} -2 & -9 & 90 & 77 \\ -42 & -63 & 127 & 62 \\ 413 & 76 & 180 & -56 \\ -29 & -44 & 198 & 85 \\ 137 & 20 & 162 & 103 \end{bmatrix}$; yes

SECTION 6.5 (page 299)

1. $\begin{bmatrix} 2 & 4 \\ 0 & -1 \end{bmatrix}$ **3.** $\begin{bmatrix} 1 & 1/5 \\ 0 & 1 \end{bmatrix}$ **5.** $\begin{bmatrix} 1 & 5 & 6 \\ 0 & 13 & 11 \\ 4 & 7 & 0 \end{bmatrix}$ **7.** $\begin{bmatrix} -3 & 1 & -4 \\ 2 & 1 & 3 \\ -17 & 0 & -13 \end{bmatrix}$ **9.** Yes **11.** No **13.** No **15.** Yes **17.** $\begin{bmatrix} 0 & 1/2 \\ -1 & 1/2 \end{bmatrix}$

19. $\begin{bmatrix} 2 & 1 \\ 5 & 3 \end{bmatrix}$ **21.** None **23.** $\begin{bmatrix} 1 & 0 & 0 \\ 0 & -1 & 0 \\ -1 & 0 & 1 \end{bmatrix}$ **25.** $\begin{bmatrix} 15 & 4 & -5 \\ -12 & -3 & 4 \\ -4 & -1 & 1 \end{bmatrix}$ **27.** None **29.** $\begin{bmatrix} 7/4 & 5/2 & 3 \\ -1/4 & -1/2 & 0 \\ -1/4 & -1/2 & -1 \end{bmatrix}$

31. $\begin{bmatrix} 1/2 & 1/2 & -1/4 & 1/2 \\ -1 & 4 & -1/2 & -2 \\ -1/2 & 5/2 & -1/4 & -3/2 \\ 1/2 & -1/2 & 1/4 & 1/2 \end{bmatrix}$ **33.** $\begin{bmatrix} 1 & 0 \\ 0 & 1 \end{bmatrix} \cdot \begin{bmatrix} a & b \\ c & d \end{bmatrix} = \begin{bmatrix} a & b \\ c & d \end{bmatrix}$ **35.** $\begin{bmatrix} a & b \\ c & d \end{bmatrix} \cdot \begin{bmatrix} 0 & 0 \\ 0 & 0 \end{bmatrix} = \begin{bmatrix} 0 & 0 \\ 0 & 0 \end{bmatrix}$

37. $A^{-1} = \begin{bmatrix} d/(ad-bc) & -b/(ad-bc) \\ -c/(ad-bc) & a/(ad-bc) \end{bmatrix}$; $A^{-1}A = \begin{bmatrix} d/(ad-bc) & -b/(ad-bc) \\ -c/(ad-bc) & a/(ad-bc) \end{bmatrix} \cdot \begin{bmatrix} a & b \\ c & d \end{bmatrix}$

$$= \begin{bmatrix} \dfrac{ad-bc}{ad-bc} & \dfrac{db-bd}{ad-bc} \\ \dfrac{-ca+ac}{ad-bc} & \dfrac{-cb+ad}{ad-bc} \end{bmatrix} = \begin{bmatrix} 1 & 0 \\ 0 & 1 \end{bmatrix} = I$$

Answers to Exercises 39, 41, and 43 may vary slightly because of using different computers and different software.

39. $\begin{bmatrix} -.0447 & -.0230 & .0292 & .0895 & -.0402 \\ .0921 & .0150 & .0321 & .0209 & -.0276 \\ -.0678 & .0315 & -.0404 & .0326 & .0373 \\ .0171 & -.0248 & .0069 & -.0003 & .0246 \\ -.0208 & .0740 & .0096 & -.1018 & .0646 \end{bmatrix}$ **41.** $\begin{bmatrix} .0394 & .0880 & .0033 & .0530 & -.1499 \\ -.1492 & .0289 & .0187 & .1033 & .1668 \\ -.1330 & -.0543 & .0356 & .1768 & .1055 \\ .1407 & .0175 & -.0045 & -.1344 & .0655 \\ .0102 & -.0653 & .0993 & .0085 & -.0388 \end{bmatrix}$

43. $C^{-1}C = \begin{bmatrix} 1 & 1.596046 \times 10^{-7} & .89407 \times 10^{-7} & 0 & -.149012 \times 10^{-7} \\ .540167 \times 10^{-7} & 1 & .596046 \times 10^{-7} & 0 & 0 \\ -.484288 \times 10^{-7} & .372529 \times 10^{-7} & 1 & .596046 \times 10^{-7} & 0 \\ -.372529 \times 10^{-8} & 0 & -.149012 \times 10^{-7} & 1 & .745058 \times 10^{-8} \\ .149012 \times 10^{-7} & -.447035 \times 10^{-7} & 0 & 0 & 1 \end{bmatrix}$

SECTION 6.6   (page 309)

1. $\begin{bmatrix} 3 \\ 4 \end{bmatrix}$ **3.** $\begin{bmatrix} 36 & 6 \\ 28 & -2 \end{bmatrix}$ **5.** $\begin{bmatrix} -22 \\ -18 \\ 15 \end{bmatrix}$ **7.** $\begin{bmatrix} 4 & -11 \\ -10 & 11 \\ 4 & -5 \end{bmatrix}$ **9.** $\begin{bmatrix} -6 \\ -14 \end{bmatrix}$ **11.** $(-8, 6, 1)$ **13.** $(15, -5, -1)$ **15.** $(-31, 24, -4)$

17. $(-4, -3, 1)$ **19.** No inverse, no solution for system **21.** $(-7, -34, -19, 7)$

23. $\begin{bmatrix} 32/3 \\ 25/3 \end{bmatrix}$ Answers are rounded in Exercises 25 and 27. **25.** $\begin{bmatrix} 6.43 \\ 26.12 \end{bmatrix}$ **27.** $\begin{bmatrix} 6.67 \\ 20 \\ 10 \end{bmatrix}$

29. About 1079 metric tons of wheat; about 1428 metric tons of oil **31.** Gas $98 million; electricity $123 million **33.** About 1538.5 units of agriculture, 1282.1 units of manufacturing, and 1589.7 units of transportation

35. $\begin{bmatrix} 35 \\ 92 \end{bmatrix} \begin{bmatrix} -4 \\ 80 \end{bmatrix} \begin{bmatrix} 15 \\ 132 \end{bmatrix} \begin{bmatrix} -9 \\ 99 \end{bmatrix} \begin{bmatrix} 35 \\ 173 \end{bmatrix} \begin{bmatrix} 2 \\ 41 \end{bmatrix} \begin{bmatrix} -8 \\ 61 \end{bmatrix}$ **37.** $\begin{bmatrix} 76 \\ 77 \\ 96 \end{bmatrix} \begin{bmatrix} 62 \\ 67 \\ 75 \end{bmatrix} \begin{bmatrix} 88 \\ 108 \\ 97 \end{bmatrix} \begin{bmatrix} 141 \\ 160 \\ 168 \end{bmatrix} \begin{bmatrix} 147 \\ 166 \\ 174 \end{bmatrix} \begin{bmatrix} 105 \\ 120 \\ 123 \end{bmatrix} \begin{bmatrix} 111 \\ 131 \\ 119 \end{bmatrix} \begin{bmatrix} 92 \\ 119 \\ 94 \end{bmatrix} \begin{bmatrix} 75 \\ 93 \\ 79 \end{bmatrix} \begin{bmatrix} 181 \\ 208 \\ 208 \end{bmatrix}$

39. (a) 21 **(b)** 25 **41. (a)**

	S	J	NO	H
S	0	1	2	1
J	1	0	1	0
NO	2	1	0	1
H	1	0	1	0

(b) 2 **(c)** 3 **(d)** 2

CHAPTER 6 REVIEW EXERCISES (page 313)

1. $(-1, 3)$ **3.** $(6, -12)$ **5.** $(x, -x/2 + 15/8)$, dependent equations **7.** $(-1, 2, 3)$ **9.** 10 at 25¢, 12 at 50¢ **11.** $10,000 at 6%, $20,000 at 7%, $20,000 at 9% **13.** 50 from source I, 150 from source II, 100 from source III **15.** $(2z - 2, -z + 3, z)$ for any real number z **17.** $(-2, 0)$ **19.** $(0, 3, 3)$ **21.** No solution **23.** 2×2; $a = 2$, $b = 3$, $c = 5$, $q = 9$; square **25.** 1×4; $m = 12$, $k = 4$, $z = -8$, $r = -1$; row **27.** 2×2; $r = 2$, $m = 7/2$, $k = 11/2$, $x = -7$; square

29. $\begin{bmatrix} 8 & 8 & 8 \\ 10 & 5 & 9 \\ 7 & 10 & 7 \\ 8 & 9 & 7 \end{bmatrix}$ **31.** $\begin{bmatrix} -4 & -10 \\ 2 & 3 \\ -6 & -9 \end{bmatrix}$ **33.** $\begin{bmatrix} 13 & 20 \\ -5 & -3 \\ 16 & 25 \end{bmatrix}$ **35.** $\begin{bmatrix} -17 & 20 \\ 1 & -21 \\ -8 & -17 \end{bmatrix}$ **37.** Next day $\begin{bmatrix} 2310 & -1/4 \\ 1258 & -1/4 \\ 5061 & 1/2 \\ 1812 & 1/2 \end{bmatrix}$; two-day total $\begin{bmatrix} 4842 & -1/2 \\ 2722 & -1/8 \\ 10035 & -1 \\ 3566 & 1 \end{bmatrix}$

39. $\begin{bmatrix} 26 & 86 \\ -7 & -29 \\ 21 & 87 \end{bmatrix}$ **41.** $[9]$ **43.** $\begin{bmatrix} 138 & 646 \\ -43 & -209 \\ 129 & 627 \end{bmatrix}$ **45. (a)** $\begin{bmatrix} 1/4 & 1/3 \\ 1/2 & 1/3 \end{bmatrix}$ **(b)** 6 units of each size

47. Many correct answers, including $A = \begin{bmatrix} 0 & 0 \\ 1 & 1 \end{bmatrix}$, $B = \begin{bmatrix} 1 & 0 \\ 1 & 0 \end{bmatrix}$. **49.** $\begin{bmatrix} -1/4 & 1/6 \\ 0 & 1/3 \end{bmatrix}$ **51.** No inverse **53.** $\begin{bmatrix} 1/4 & 1/2 & 1/2 \\ 1/4 & -1/2 & 1/2 \\ 1/8 & -1/4 & -1/4 \end{bmatrix}$

55. No inverse **57.** $\begin{bmatrix} 6/7 & -5/7 \\ -1/7 & 2/7 \end{bmatrix}$ **59.** Not possible **61.** $\begin{bmatrix} -\dfrac{163}{133} & \dfrac{123}{133} \\ \dfrac{40}{133} & -\dfrac{31}{133} \end{bmatrix}$ **63.** $\begin{bmatrix} 18 \\ -7 \end{bmatrix}$ **65.** $\begin{bmatrix} -22 \\ -18 \\ 15 \end{bmatrix}$

67. $(2, 2)$ **69.** $(2, 1)$ **71.** $(-1, 0, 2)$ **73.** No inverse; no solution for the system **75.** 17.5 g of 12 carat, 7.5 g of 22 carat **77.** 4 lbs **79.** Boat, 15 kilometers per hour; current, 4 kilometers per hour **81.** $12,750 at 8%; $27,250 at 8½%; $10,000 at 11%

83. $\begin{bmatrix} 218.09 \\ 318.27 \end{bmatrix}$ **85. (a)** $\begin{bmatrix} 1 & -1/4 \\ -1/2 & 1 \end{bmatrix}$ **(b)** $\begin{bmatrix} 8/7 & 2/7 \\ 4/7 & 8/7 \end{bmatrix}$ **(c)** $\begin{bmatrix} 2800 \\ 2800 \end{bmatrix}$ **87.** Agriculture, $140,909; manufacturing, $95,455

89. (a) $\begin{bmatrix} 118 \\ -7 \end{bmatrix}\begin{bmatrix} 99 \\ -72 \end{bmatrix}\begin{bmatrix} 173 \\ -30 \end{bmatrix}\begin{bmatrix} 115 \\ -45 \end{bmatrix}\begin{bmatrix} 165 \\ -18 \end{bmatrix}\begin{bmatrix} 29 \\ -35 \end{bmatrix}\begin{bmatrix} 65 \\ -55 \end{bmatrix}$ **(b)** $\begin{bmatrix} \dfrac{1}{17} & -\dfrac{5}{17} \\ \dfrac{3}{17} & \dfrac{2}{17} \end{bmatrix}$

CASE 5 (page 317)

1. $PQ = \begin{bmatrix} 1 & 2 & 0 & 2 & 1 & 1 \\ 0 & 1 & 0 & 1 & 0 & 0 \\ 1 & 1 & 0 & 1 & 2 & 1 \end{bmatrix}$ **2.** None **3.** Yes, the third person

4. The second, fourth, and fifth persons in the third group each had six contacts in all.

CASE 6 (page 318)

1. (a) $A = \begin{bmatrix} .245 & .102 & .051 \\ .099 & .291 & .279 \\ .433 & .372 & .011 \end{bmatrix}$ $D = \begin{bmatrix} 2.88 \\ 31.45 \\ 30.91 \end{bmatrix}$ $X = \begin{bmatrix} x_1 \\ x_2 \\ x_3 \end{bmatrix}$ **(b)** $I - A = \begin{bmatrix} .755 & -.102 & -.051 \\ -.099 & .709 & -.279 \\ -.433 & -.372 & .989 \end{bmatrix}$ **(d)** $\begin{bmatrix} 18.2 \\ 73.2 \\ 66.8 \end{bmatrix}$

(e) $18.2 billion of agriculture, $73.2 billion of manufacturing, and $66.8 billion of household would be required (rounded to three significant digits).

2. (a)

$$A = \begin{bmatrix} .293 & 0 & 0 \\ .014 & .207 & .017 \\ .044 & .010 & .216 \end{bmatrix} \quad D = \begin{bmatrix} 138,213 \\ 17,597 \\ 1,786 \end{bmatrix} \quad \textbf{(b)} \quad I - A = \begin{bmatrix} .707 & 0 & 0 \\ -.014 & .793 & -.017 \\ -.044 & -.010 & .784 \end{bmatrix}$$

(d) Agriculture 195,000 thousand pounds; manufacture 26,000 thousand pounds; energy 13,600 thousand pounds

CHAPTER 7

SECTION 7.1   (page 329)

1.
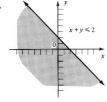

3.
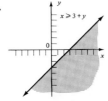

5.
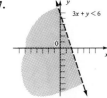

7.
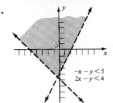

9.
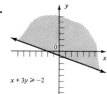

11.
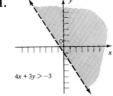

13.
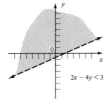

15.
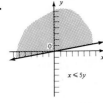

17.
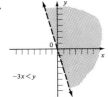

19.
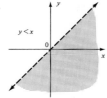

21.
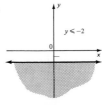

23.
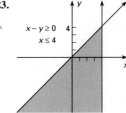

25.
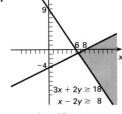

27.
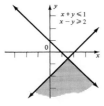

29.
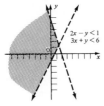

31.
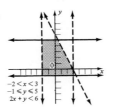

33.
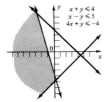

35.

37.
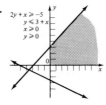

39.
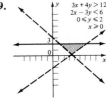

41.

	Number	Wheel	Kiln
Glazed	x	1/2	1
Unglazed	y	1	6
Maximum hours available		8	20

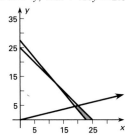

$(1/2)x - y \leq 8$; $x + 6y \leq 20$; $x \geq 0$; $y \geq 0$

43. $x \geq 3000$; $y \geq 5000$; $x + y \leq 10,000$

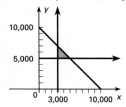

45. $x \geq 4y$; $.12x + .10y \geq 2.8$; $x + y \leq 25$; $x \geq 0$; $y \geq 0$

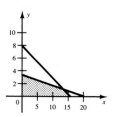

SECTION 7.2 (page 338)

1. Maximum of 65 at (5, 10); minimum of 8 at (1, 1)   **3.** Maximum of 9 at (0, 12); minimum of 0 at (0, 0)   **5.** No maximum; minimum of 18 at (3, 4)   **7.** Maximum of 25; minimum of 8   **9.** No maximum; minimum of 4   **11.** Maximum of 82; minimum of 0   **13.** Maximum of 42/5 at $x = 6/5$, $y = 6/5$ **15.** Maximum of 25 at $x = 10$, $y = 5$ **17.** Maximum of 235/4 at $x = 105/8$, $y = 25/8$ **19. (a)** $x = 18$, $y = 2$ **(b)** $x = 12/5$, $y = 39/5$ **(c)** $x = 0$, $y = 17/2$

SECTION 7.3 (page 345)

1. Let x be the number of product A made, and y the number of B. Then $2x + 3y \leq 45$.   **3.** Let x be the number of green pills, and y be the number of red. Then $4x + y \geq 25$.   **5.** Let x be the number of pounds of $6 coffee, and y the number of $5 coffee. Then $x + y \geq 50$.   **7.** 50 to plant I and 27 to plant II for a minimum cost of $1945. (It is not surprising that costs are minimized by sending the minimum number required.)   **9.** 0 units of policy A and 25/3 or $8\frac{1}{3}$ units of policy B for a minimum premium cost of $333.33.   **11.** 1600 type 1 bolts and 0 type 2 bolts for a maximum revenue of $160.   **13.** 150 kg half-and-half mix and 75 kg other mix for maximum revenue of $1260.   **15.** 50 gallons from dairy I and 50 gallons from dairy II for maximum butterfat of 3.45 percent.   **17.** $25 million in bonds and $15 million in mutual funds for maximum annual interest of $4.20 million.   **19.** 8/7 units of species I and 10/7 units of species II will meet requirements with a minimum of 6.57 units of energy; however, a predator probably can catch and digest only whole numbers of prey. This problem shows that it is important to consider whether a solution produces a realistic answer to a problem.   **21.** $3\frac{3}{4}$ servings of A and $1\frac{7}{8}$ servings of B for a minimum cost of $1.69.   **23. (b)**   **25. (c)**

SECTION 7.4 (page 354)

1. $x_1 + 2x_2 + x_3 = 6$ **3.** $2x_1 + 4x_2 + 3x_3 + x_4 = 100$ **5. (a)** 3 **(b)** x_3, x_4, x_5 **(c)** $4x_1 + 2x_2 + x_3 = 20$
$$5x_1 + x_2 + x_4 = 50$$
$$2x_1 + 3x_2 + x_5 = 25$$

7. (a) 2 **(b)** x_4, x_5 **(c)** $7x_1 + 6x_2 + 8x_3 + x_4 = 118$ **9.** $x_1 = 0$, $x_2 = 0$, $x_3 = 20$, $x_4 = 0$, $x_5 = 15$
$$4x_1 + 5x_2 + 10x_3 + x_5 = 220$$

11. $x_1 = 0$, $x_2 = 0$, $x_3 = 8$, $x_4 = 0$, $x_5 = 6$, $x_6 = 7$ **13.** $x_1 = 0$, $x_2 = 20$, $x_3 = 0$, $x_4 = 16$, $x_5 = 0$ **15.** $x_1 = 0$, $x_2 = 0$, $x_3 = 12$, $x_4 = 0$, $x_5 = 9$, $x_6 = 8$ **17.** $x_1 = 0$, $x_2 = 0$, $x_3 = 50$, $x_4 = 10$, $x_5 = 0$, $x_6 = 50$

19.

x_1	x_2	x_3	x_4	x_5	
2	3	1	0	0	6
4	1	0	1	0	6
5	2	0	0	1	15
-5	-1	0	0	0	0

21.

x_1	x_2	x_3	x_4	x_5	x_6	
2	4	1	1	0	0	18
1	6	2	0	1	0	45
5	7	3	0	0	1	60
-1	-5	-10	0	0	0	0

23.

x_1	x_2	x_3	x_4	x_5	x_6	
1	2	3	1	0	0	10
2	1	1	0	1	0	8
3	0	2	0	0	1	6
-6	-2	-3	0	0	0	0

25. If x_1 is the number of kg of half-and-half mix and x_2 is the number of kg of the other mix, find $x_1 \geq 0$, $x_2 \geq 0$, $x_3 \geq 0$, $x_4 \geq 0$ so that $(1/2)x_1 + (1/3)x_2 + x_3 = 100$, $(1/2)x_1 + (2/3)x_2 + x_4 = 125$, and $z = 6x_1 + 4.8x_2$ is maximized.

x_1	x_2	x_3	x_4	
$\frac{1}{2}$	$\frac{1}{3}$	1	0	100
$\frac{1}{2}$	$\frac{2}{3}$	0	1	125
-6	-4.8	0	0	0

27. If x_1 is the number of prams, x_2 is the number of runabouts, and x_3 is the number of trimarans, find $x_1 \geq 0$, $x_2 \geq 0$, $x_3 \geq 0$, $x_4 \geq 0$, $x_5 \geq 0$, $x_6 \geq 0$ so that $x_1 + 2x_2 + 3x_3 + x_4 = 6240$, $2x_1 + 5x_2 + 4x_3 + x_5 = 10{,}800$, $x_1 + x_2 + x_3 + x_6 = 3000$, and $z = 75x_1 + 90x_2 + 100x_3$ is maximized.

x_1	x_2	x_3	x_4	x_5	x_6	
1	2	3	1	0	0	6240
2	5	4	0	1	0	10,800
1	1	1	0	0	1	3000
-75	-90	-100	0	0	0	0

29. If x_1 is the number of Siamese cats and x_2 is the number of Persian cats, find $x_1 \geq 0$, $x_2 \geq 0$, $x_3 \geq 0$, $x_4 \geq 0$, $x_5 \geq 0$ so that $2x_1 + x_2 + x_3 = 90$, $x_1 + 2x_2 + x_4 = 80$, $x_1 + x_2 + x_5 = 50$, and $z = 12x_1 + 10x_2$ is maximized.

x_1	x_2	x_3	x_4	x_5	
2	1	1	0	0	90
1	2	0	1	0	80
1	1	0	0	1	50
-12	-10	0	0	0	0

SECTION 7.5 (page 363)

1. Maximum is 20 when $x_1 = 0$, $x_2 = 4$, $x_3 = 0$, $x_4 = 0$, $x_5 = 2$ **3.** Maximum is 8 when $x_1 = 4$, $x_2 = 0$, $x_3 = 8$, $x_4 = 2$, $x_5 = 0$
5. Maximum is 264 when $x_1 = 16$, $x_2 = 4$, $x_3 = 0$, $x_4 = 0$, $x_5 = 16$, $x_6 = 0$ **7.** Maximum is 20 when $x_1 = 0$, $x_2 = 4$, $x_3 = 0$, $x_4 = 12$
9. Maximum is 944 when $x_1 = 118$, $x_2 = 0$, $x_3 = 0$, $x_4 = 0$, $x_5 = 102$ **11.** Maximum is 60 when $x_1 = 0$, $x_2 = 30$, $x_3 = 0$, $x_4 = 10$, and
$x_5 = 0$ **13.** Maximum is 26,000 when $x_1 = 60$, $x_2 = 40$, $x_3 = 0$, $x_4 = 0$, $x_5 = 80$, $x_6 = 0$ **15.** Maximum is 250 when $x_1 = 0$, $x_2 = 0$, $x_3 = 0$, $x_4 = 50$, $x_5 = 0$, $x_6 = 50$ **17.** 6 churches and 2 labor unions for a maximum of $1000 per month   **19.** Make no jazz albums, 9 blues albums, and 4 reggae albums for a maximum weekly profit of $96.00.   **21.** Make no paper weights, 4 plaques, and 10 ornaments for a one-day profit of $46.00.   **23.** Make no 1-speed or 3-speed bicycles; make 2295 10-speed bicycles; maximum profit is $55,080.   **25. (a)** 3 **(b)** 4 **(c)** 3   **27.** 6700 trucks and 4467 fire engines for a maximum profit of $110,997

SECTION 7.6 (page 375)

1. $2x_1 + 3x_2 + x_3 = 8$
$x_1 + 4x_2 - x_4 = 7$
3. $x_1 + x_2 + x_3 + x_4 = 100$
$x_1 + x_2 + x_3 - x_5 = 75$
$x_1 + x_2 \qquad - x_6 = 27$
5. Change the objective function to maximize $z = -4y_1 - 3y_2 - 2y_3$.   **7.** Change the objective function to maximize $z = -y_1 - 2y_2 - y_3 - 5y_4$.   **9.** Maximum is 480 when $x_1 = 40$ and $x_2 = 0$ **11.** Maximum is 750 when $x_1 = 0$, $x_2 = 150$, and $x_3 = 0$ **13.** Maximum is 300 when $x_1 = 0$ and $x_2 = 100$ **15.** Minimum is 40 when $y_1 = 10$ and $y_2 = 0$ **17.** Minimum is 100 when $y_1 = 0$, $y_2 = 100$, and $y_3 = 0$ **19.** Ship 5000 barrels of oil from supplier 1 to distributor 2; ship 3000 barrels of oil from supplier 2 to distributor 1. Minimum cost is $175,000.

21. Make no commercial loans and $25,000,000 in home loans for a maximum return of $3,000,000.   **23.** 800,000 for whole tomatoes and 80,000 for sauce for a minimum cost of $3,460,000   **25.** Buy 1000 small and 500 large for a minimum cost of $210   **27.** $40,000 in government securities and $60,000 in mutual funds for maximum interest of $8800   **29.** 22 computers from W_1 to D_1, 0 computers from W_1 to D_2, 10 computers from W_2 to D_1, and 20 computers from W_2 to D_2 for a minimum cost of $628   **31.** Use $1\frac{2}{3}$ ounces of I, $6\frac{2}{3}$ ounces of II, $1\frac{2}{3}$ ounces of III for a minimum cost of $1.55 per gallon, so 10 ounces of additive should be used per gallon of gasoline.

SECTION 7.7 (page 386)

1. $\begin{bmatrix} 1 & 3 & 1 \\ 2 & 2 & 10 \\ 3 & 1 & 0 \end{bmatrix}$ **3.** $\begin{bmatrix} -1 & 13 & -2 \\ 4 & 25 & -1 \\ 6 & 0 & 11 \\ 12 & 4 & 3 \end{bmatrix}$ **5.** Maximize $z = 4x_1 + 6x_2$
subject to: $3x_1 + x_2 \leq 2$
$x_1 + 2x_2 \leq 4$
$x_1 \geq 0, x_2 \geq 0$

7. Maximize $z = 8x_1 + 12x_2$
subject to: $3x_1 + x_2 \leq 1$
$4x_1 + 5x_2 \leq 3$
$6x_1 + 2x_2 \leq 4$
$x_1 \geq 0, x_2 \geq 0$

9. Maximize $z = 18x_1 + 15x_2 + 20x_3$
subject to: $x_1 + 4x_2 + 5x_3 \leq 3$
$7x_1 + x_2 + 3x_3 \leq 4$
$x_1 \geq 0, x_2 \geq 0, x_3 \geq 0$

11. Maximize $z = 115x_1 + 200x_2 + 50x_3$
subject to: $x_1 + 2x_2 + x_3 \leq 1$
$2x_1 + x_2 \leq 1$
$3x_1 + 8x_2 + x_3 \leq 4$
$x_1 \geq 0, x_2 \geq 0, x_3 \geq 0$

13. $y_1 = 10, y_2 = 0$; minimum is 40

15. $y_1 = 0, y_2 = 100, y_3 = 0$; minimum is 100   **17.** $y_1 = 4/3, y_2 = 16/3$; minimum is 20/3   **19.** $y_1 = 3, y_2 = 0, y_3 = 1$; minimum is 43   **21.** $y_1 = 0, y_2 = 12, y_3 = 0$; minimum is 12   **23.** Make 15 tables and 45 chairs for a minimum cost of $4080.   **25.** 4 grams of meat byproducts, 4.8 grams of grain, and no soymeal at a minimum cost of 44¢.   **27. (a)** 1 bag of Feed 1, 2 bags of Feed 2 **(b)** $6.60 per day for 1.4 bags of Feed 1 and 1.2 bags of Feed 2

29. (a) Minimize $w = 100y_1 + 20{,}000y_2$ **(b)** $6300 when $x_1 = 52.5, x_2 = 0, x_3 = 0$ **(c)** $5700, $x_1 = 47.5, x_2 = 0, x_3 = 0$
subject to: $y_1 + 400y_2 \geq 120$
$y_1 + 160y_2 \geq 40$
$y_1 + 280y_2 \geq 60$
$y_1 \geq 0, y_2 \geq 0$

31. Use 5/3 oz of I, 20/3 oz of II, 5/3 oz of III for a minimum cost of $1.55 per gallon. 10 oz of the additive should be used per gallon of gasoline.

CHAPTER 7 REVIEW EXERCISES (page 390)

1.
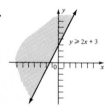
$y \geq 2x + 3$

3.
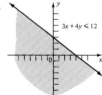
$3x + 4y \leq 12$

5.
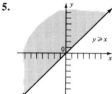
$y \geq x$

7.
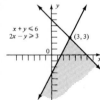
$x + y \leq 6$
$2x - y \geq 3$
$(3, 3)$

9.
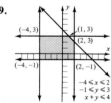
$(-4, 3)$ $(1, 3)$ $(2, 3)$ $(-4, -1)$ $(2, -1)$
$-4 \leq x \leq 2$
$-1 \leq y \leq 3$
$x + y \leq 4$

11.
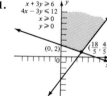
$x + 3y \geq 6$
$4x - 3y \leq 12$
$x \geq 0$
$y \geq 0$
$(0, 2)$ $\left(\frac{18}{5}, \frac{4}{5}\right)$

13. Let x = number of batches of cakes and y = number of batches of cookies. Then $x \geq 0$, $y \geq 0$, and $2x + (3/2)y \leq 15$
$$3x + (2/3)y \leq 13.$$

15. Minimum of 8 at (2, 1); maximum of 40 at (6, 7)   **17.** Maximum of 24 at (0, 6)   **19.** Minimum of 40 at any point on the line segment connecting (0, 20) and (10/3, 40/3)   **21.** Make 3 batches of cakes and 6 of cookies for a maximum profit of $210.   **23. (a)** Let x_1 = number of item A, x_2 = number of item B, and x_3 = number of item C she should buy. **(b)** $z = 4x_1 + 3x_2 + 3x_3$ **(c)** $5x_1 + 3x_2 + 6x_3 \leq 1200$
$$x_1 + 2x_2 + 2x_3 \leq 800$$
$$2x_1 + x_2 + 5x_3 \leq 500$$

25. (a) Let x_1 = number of gallons of fruity wine and x_2 = number of gallons of crystal wine to be made. **(b)** $z = 12x_1 + 15x_2$ **(c)** $2x_1 + x_2 \leq 1$
$$2x_1 + 3x_2 \leq 1$$
$$2x_1 + x_2 \leq$$

27. (a) $2x_1 + 5x_2 + x_3 = 50$ **(b)**

x_1	x_2	x_3	x_4	x_5	x_6	
2	5	1	0	0	0	50
1	3	0	1	0	0	25
4	1	0	0	1	0	18
1	1	0	0	0	1	12
−5	−3	0	0	0	0	0

$x_1 + 3x_2 + x_4 = 25$
$4x_1 + x_2 + x_5 = 18$
$x_1 + x_2 + x_6 = 12$

29. (a) $x_1 + x_2 + x_3 + x_4 = 90$ **(b)**

x_1	x_2	x_3	x_4	x_5	x_6	
1	1	1	1	0	0	90
2	5	1	0	1	0	120
1	3	0	0	0	1	80
−5	−8	−6	0	0	0	0

$2x_1 + 5x_2 + x_3 + x_5 = 120$
$x_1 + 3x_2 + x_6 = 80$

31. Maximum of 412/5 at (68/5, 0, 24/5, 0, 0) or maximum of 82.4 at (13.6, 0, 4.8, 0, 0)   **33.** Maximum of $70\frac{5}{8}$ at (5/6, 35/2, 85/12, 0, 0, 0)   **35.** Change the objective function to maximize $z = -10y_1 - 15y_2$.   **37.** Change the objective function to maximize $z = -7y_1 - 2y_2 - 3y_3$.   **39.** Minimum of 40 at any point on the segment connecting (0, 20) and (10/3, 40/3)   **41.** Minimum of 1957 at (47, 68, 0, 92, 35, 0, 0)   **43.** (9, 5, 8, 0, 0, 0); minimum is 62   **45.** None of A; 400 of B; none of C; maximum profit is $1200   **47.** Produce 36.25 gallons of fruity and 17.5 gallons of crystal for a maximum profit of $697.50   **49.** Build 3 atlantic and 3 pacific models for a minimum cost of $21,000.

CASE 7   (page 394)

1. 1: .95; 2: .83; 3: .75; 4: 1.00; 5: .87; 6: .94   **2.** $x_1 = 100$; $x_2 = 0$; $x_3 = 0$; $x_4 = 90$; $x_5 = 0$; $x_6 = 210$

CASE 8   (page 396)

1. (a) $x_2 = 58.6957$, $x_8 = 2.01031$, $x_9 = 22.4895$, $x_{10} = 1$, $x_{12} = .25$, $x_{13} = .25$ for a minimum cost of $12.55 **(b)** $x_2 = 71.7391$, $x_8 = 3.14433$, $x_9 = 23.1966$, $x_{10} = 1$, $x_{12} = .25$, $x_{13} = .25$ for a minimum cost of $15.05

CHAPTER 8

SECTION 8.1   (page 405)

1. $\notin$ **3.** $\in$ **5.** $\notin$ **7.** $=$ **9.** $\neq$ **11.** $\neq$ **13.** $\neq$ **15.** $=$ **17.** $\neq$ **19.** $=$ **21.** $\neq$ **23.** $=$ **25.** $\neq$ **27.** $=$ **29.** $=$ **31.** $\subseteq$ **33.** $\subseteq$
35. $\nsubseteq$ **37.** $\subseteq$ **39.** $\subseteq$ **41.** $\nsubseteq$ **43.** 64   **45.** 8   **47.** 8   **49.** 32   **51.** 1   **53.** 32   **55.** {3, 5}   **57.** U, or {2, 3, 4, 5, 7, 9}   **59.** {7, 9}
61. $\emptyset$ **63. (a)** {c, d, e, f, 3, 5} **(b)** {a, b, c, d, e, f, 1, 2, 3, 5} **(c)** {a, b, c, d, e, f, 1, 2, 3} **(d)** {a, b, c, d, e, f, 1, 2, 3, 5}   **65. (a)** {c, 3}
(b) {c, 3}   **67. (a)** {d, f, 5} **(b)** {b, d, e, f, 1, 3, 4, 5, 6} **(c)** {d, f, 5}   **69.** All students in this school not taking this course   **71.** All students
in this school taking accounting and zoology   **73.** All students in this school taking this course or zoology   **75. (a)** {s, d, c, g, i, m, h}
(b) {s, d, c} **(c)** {i, m, h} **(d)** {g} **(e)** {s, d, c, g, i, m, h} **(f)** {s, d, c}   **77.** A

SECTION 8.2   (page 413)

1.
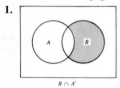
$B \cap A'$

3.
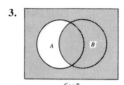
$A' \cup B$

5.
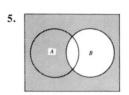
$B' \cup (A' \cap B')$

7. $\emptyset$

9.
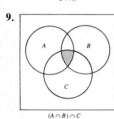
$(A \cap B) \cap C$

11.
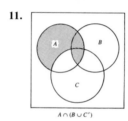
$A \cap (B \cup C')$

13.
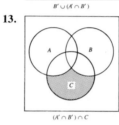
$(A' \cap B') \cap C$

15.
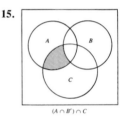
$(A \cap B') \cap C$

17. Yes; his data add up to 142 people.   **19. (a)** 89% **(b)** 66% **(c)** 4% **(d)** 1% **(e)** 4%   **21. (a)** 50 **(b)** 2 **(c)** 14   **23. (a)** 54 **(b)** 17 **(c)** 10
(d) 7 **(e)** 15 **(f)** 3 **(g)** 12 **(h)** 1   **25. (a)** 40 **(b)** 30 **(c)** 95 **(d)** 110 **(e)** 160 **(f)** 65   **27.** 9   **29.** 16

31.
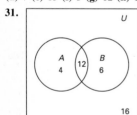

33.
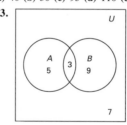

35.
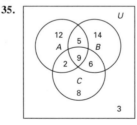

37.
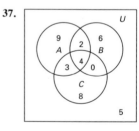

SECTION 8.3   (page 422)

1. {January, February, March, . . . , December}   **3.** {0, 1, 2, . . . , 80}   **5.** {go ahead, cancel}   **7.** {(h, 1), (h, 2), (h, 3), (h, 4), (h, 5),
(h, 6), (t, 1), (t, 2), (t, 3), (t, 4), (t, 5), (t, 6)}   **9.** {AB, AC, AD, AE, BC, BD, BE, CD, CE, DE} **(a)** {AC, BC, CD, CE} **(b)** {AB, AC,
AD, AE, BC, BD, BE, CD, CE} **(c)** {AC}   **11.** {(1, 2), (1, 3), (1, 4), (1, 5), (2, 3), (2, 4), (2, 5), (3, 4), (3, 5), (4, 5)} **(a)** {(2, 4)}
(b) {(1, 2), (1, 4), (2, 3), (2, 5), (3, 4), (4, 5)} **(c)** $\emptyset$ **13.** {(1, 1), (1, 2), (1, 3), (1, 4), (1, 5), (1, 6), (2, 1), (2, 2), (2, 3), (2, 4), (2, 5),
(2, 6), (3, 1), (3, 2), (3, 3), (3, 4), (3, 5), (3, 6), (4, 1), (4, 2), (4, 3), (4, 4), (4, 5), (4, 6), (5, 1), (5, 2), (5, 3), (5, 4), (5, 5), (5, 6),
(6, 1), (6, 2), (6, 3), (6, 4), (6, 5), (6, 6)} **(a)** {(3, 1), (3, 2), (3, 3), (3, 4), (3, 5), (3, 6)} **(b)** {(2, 6), (3, 5), (4, 4), (5, 3), (6, 2)} **(c)** $\emptyset$
15. (a) Worker is male. **(b)** Worker has worked 5 yrs or more. **(c)** Worker is female and has worked less than 5 yrs. **(d)** Worker has worked
less than 5 yrs or has contributed to a voluntary retirement plan. **(e)** Worker is female or does not contribute to a voluntary retirement plan.
(f) Worker has worked 5 yrs or more and does not contribute to a voluntary retirement plan.
17. 1/6   **19.** 2/3   **21.** 1/3   **23.** 1/13   **25.** 1/26   **27.** 1/52   **29.** 2/13   **31.** 1/2   **33.** 4/13   **35.** 1/5   **37.** 9/25   **39.** 21/25   **41.** 9/25
43. 5/11   **45.** 2/11

SECTION 8.4 (page 433)

1. No **3.** No **5.** Yes **7. (a)** 1/36 **(b)** 1/12 **(c)** 1/9 **(d)** 5/36 **9. (a)** 5/18 **(b)** 5/12 **(c)** 11/36 **11. (a)** 2/13 **(b)** 7/13 **(c)** 3/26 **(d)** 3/4
13. (a) 1/2 **(b)** 2/5 **(c)** 3/10 **15. (a)** 1/10 **(b)** 2/5 **(c)** 7/20 **17. (a)** .5 **(b)** .24 **(c)** .09 **(d)** .83 **19. (a)** 4/25 **(b)** 3/50 **(c)** 39/50 **21. (a)** .4
(b) .1 **(c)** .6 **(d)** .9 **23.** 84% **25.** .951 **27.** .473 **29.** .007 **31.** 1 to 5 **33.** 2 to 1 **35. (a)** 13 to 12 **(b)** 1 to 1 **37.** 11 to 4 **39.** 4/11
41. No; ''Odds in favor of a direct hit are very low.'' **43.** Possible **45.** Not possible because the sum of the probabilities is greater than
1. **47.** Not possible because one of the probabilities is negative. **49.** .3285 **51.** .7235 **53.** .2660 **55.** .1760 **57. (a)** .51 **(b)** .67
(c) .39 **(d)** .12 **59. (a)** .45 **(b)** .75 **(c)** .35 **(d)** .46 **61.** 1/4 **63.** 1/2 **65.** .90 **67.** The theoretical probability is 5/32. **69.** The
theoretical probability is 1/221 or .004525. **71.** The theoretical answer is 7/8. Using the Monte Carlo method, answers will vary.

SECTION 8.5 (page 448)

1. 0 **3.** 1 **5.** 1/6 **7.** 4/17 **9.** 11/51 **11.** .012 **13.** .245 **15.** .052 **17.** .06 **19.** .7456 **21.** 7/10 **23.** 2/3 **25.** 3/10 **27.** 1/10
29. 1/7 **31.** 1/2 **33.** 4/7 **35.** The probability of a customer not making a deposit given that the customer cashed a check is 3/8. **37.** The
probability of a customer not cashing a check given that the customer made a deposit is 2/7. **39.** 3/10 **41.** 1/3 **43.** 2/3 **45.** .02 **47.** 2/3
49. 1/4 **51.** 1/4 **53.** .527 **55.** .042 **57.** 6/7 or .857 **59.** Yes **61.** .05 **63.** .25 **65.** .287 **67.** .0183 **69.** .00013 **71.** .00367
73. .4955 **75.** .0361 **77.** Not very realistic **79.** .999985; fairly realistic

SECTION 8.6 (page 457)

1. 1/3 **3.** 2/41 **5.** 21/41 **7.** 8/17 **9.** 119/131 ≈ .908 **11.** 14/17 ≈ .824 **13.** 1/176 ≈ .006 **15.** 1/11 **17.** 5/9 **19.** 5/26 **21.** 82.4%
23. 178/343 ≈ .519 **25.** 72/73 ≈ .986 **27.** .146 **29.** .082

CHAPTER 8 REVIEW EXERCISES (page 460)

1. False **3.** False **5.** True **7.** False **9.** False **11.** ∅ **13.** {1, 2, 3, 4} **15.** $2^4 = 16$ **17.** {a, b, e} **19.** {c, d, g} **21.** {a, b, e, f}
23. U **25.** All employees in the accounting department who also have at least 10 years with the company **27.** All employees who are in
the accounting department or who have MBA degrees **29.** All employees who are not in the sales department and who have worked less than
10 years with the company

31.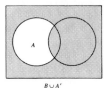

$B \cup A'$

33.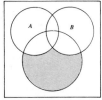

$A' \cap (B' \cap C)$

35. 52 **37.** 12 **39.** {1, 2, 3, 4, 5, 6} **41.** {0, .5, 1, 1.5, 2, . . . , 299.5, 300} **43.** {(3, R), (3, G), (5, R), (5, G), (7, R), (7, G),
(9, R), (9, G), (11, R), (11, G)} **45.** {(3, G), (5, G), (7, G), (9, G), (11, G)} **47.** $E' \cap F'$ **49.** 1/4 **51.** 3/13 **53.** 1/2 **55.** 1 **57.** 1
to 25 **59.** .86

61.

	N_2	T_2
N_1	N_1N_2	N_1T_2
T_1	T_1N_2	T_1T_2

63. 1/2 **65.** 5/36 **67.** 1/6 **69.** 1/6 **71.** 2/11 **73.** .66 **75.** .71 **77.** .3 **79.** 2/5 or .4 **81.** 3/4 or .75 **83. (a)** 3/11 **(b)** 7/39
85. 3/22 **87.** 2/3 **89.** 4/9

CASE 9 (page 463)

1. $1/200 = .0005$ **2.** $1/2000 = .0005$ **3.** $(1999/2000)^a$ **4.** $1 - (1999/2000)^a$ **5.** $(1999/2000)^{Nc}$ **6.** $1 - (1999/2000)^{Nc}$
7. $1 - .741 = .259$ **8.** $1 - .882 = .118$

CASE 10 (page 463)

1. .715; .569; .410; .321; .271 **2.** There are 94 other possible values for n, namely, 6, 7, 8, . . . , 99. Many of these are quite unlikely, but, for instance, 3rd and 10 is not uncommon. Hence, the outcomes listed do not exhaust the sample space.

CASE 11 (page 464)

1. .076 **2.** .542 **3.** .051

CHAPTER 9

SECTION 9.1 (page 475)

1. 12 **3.** 56 **5.** 8 **7.** 24 **9.** 792 **11.** 156 **13.** 6,375,600 **15.** 240,240 **17.** 352,716 **19.** 2,042,975 **21.** 30 **23.** (a) 27,600
(b) 35,152 **(c)** 1104 **25.** 15 **27.** 17,576,000 **29.** 22,464,000 **31.** 720 **33.** 120 **35.** 2730 **37.** 1326 **39.** 10 **41.** 50
43. Combinations; 6 **45.** Combinations; (a) 56 (b) 462 (c) 3080 (d) 8526 **47.** Combinations; (a) 220 (b) 220 **49.** Permutations;
479,001,600 **51.** Combinations; (a) 84 (b) 10 (c) 40 (d) 28 **53.** Combinations; (a) 21 (b) 6 (c) 11 **55.** Permutations; 5040
57. Combinations; (a) 10 (b) 0 (c) 1 (d) 10 (e) 30 (f) 15 (g) 0 **59.** (a) 840 (b) 180 (c) 420 **61.** (a) 362,880 (b) 6 (c) 1260 **63.** 2,598,960
65. 635,013,559,600 **67.** 27,839,970,224 **69.** Three initials give $26^3 = 17,576$; for 52,000 species at least 4 initials are needed.
71. (a) 17,576,000 (b) About 4 years

SECTION 9.2 (page 483)

1. 7/9 **3.** 5/12 **5.** .424 **7.** 1/6 **9.** 3/10 **11.** 1326 **13.** 33/221 **15.** 4/17 **17.** 10/17 **19.** $(1/26)^5$ **21.** $18,975/28,561 \approx .6644$
23. $1/649,740 \approx 1.539 \times 10^{-6}$ **25.** $1/4165 \approx .00024$ **27.** $1/\binom{52}{13}$ **29.** $\binom{4}{3}\binom{4}{3}\binom{44}{7}/\binom{52}{13}$ **31.** (a) .054 (b) .011 (c) .046 (d) .183 (e) .387
(f) .885 **33.** (a) .3297 (b) .4396 (c) .5455 **35.** (a) .0166 (b) .1079 (c) .3735 (d) .3597 (e) .4111 **37.** 3/8; 1/4 **39.** $120/343 \approx .3498$
41. 6 **43.** 362,880 **45.** 1/3

SECTION 9.3 (page 489)

1. 5/16 **3.** 1/32 **5.** 3/16 **7.** 13/16 **9.** .0000000005 **11.** .269 **13.** .875 **15.** 1/32 **17.** 13/16 **19.** .358 **21.** .246 **23.** .017
25. .243 **27.** .026 **29.** .879 **31.** .072 **33.** .035 **35.** .00079 **37.** .999922 **39.** .096 **41.** .104 **43.** .185 **45.** .256 **47.** .0025
49. (a) .1493 (b) .9116 **51.** (a) .000036 (b) .000052 (c) 0 **53.** (a) .148774 (b) .021344 (c) .978656

SECTION 9.4 (page 498)

1. Yes **3.** Yes **5.** No **7.** No **9.** Yes **11.** Yes **13.** No **15.** Regular **17.** Not regular **19.** Regular **21.** [2/5 3/5]
23. [4/11 7/11] **25.** [14/83 19/83 50/83] or [.169 .229 .602] **27.** [170/563 197/563 196/563] or [.302 .350 .348]
29.

	Works	Doesn't Work
Works	.95	.05
Doesn't Work	.70	.30

; long-range probability of line working correctly is 14/15

31. (a)
	Red	White
Red	.75	.25
White	.50	.50

(b) [.75 .25] (c) [.6670 .3330] (d) [2/3 1/3]

33. [51/209 88/209 70/209]   **35.** (a) 42,500; 5000; 2500 (b) 37,125; 8750; 4125 (c) 33,281; 11,513; 5206 (d) [28/59 22/59 9/59] or [.475 .373 .152]; in the long run, the probabilities of no accidents, one accident, and more than one accident are .475, .373, and .152, respectively.   **37.** 2/3

39. (a)
	Single	Multiple
Single	.90	.10
Multiple	.05	.95

(b) [.75 .25] (c) [68.8% 31.3%] (d) [1/3 2/3]   **41.** [1/3 1/3 1/3]

43. The first power is the given transition matrix;
$$\begin{bmatrix} .2 & .15 & .17 & .19 & .29 \\ .16 & .2 & .15 & .18 & .31 \\ .19 & .14 & .24 & .21 & .22 \\ .16 & .19 & .16 & .2 & .29 \\ .16 & .19 & .14 & .17 & .34 \end{bmatrix};\begin{bmatrix} .17 & .178 & .171 & .191 & .29 \\ .171 & .178 & .161 & .185 & .305 \\ .18 & .163 & .191 & .197 & .269 \\ .175 & .174 & .164 & .187 & .3 \\ .167 & .184 & .158 & .182 & .309 \end{bmatrix};$$
$$\begin{bmatrix} .1731 & .175 & .1683 & .188 & .2956 \\ .1709 & .1781 & .1654 & .1866 & .299 \\ .1748 & .1718 & .1753 & .1911 & .287 \\ .1712 & .1775 & .1667 & .1875 & .2971 \\ .1706 & .1785 & .1641 & .1858 & .301 \end{bmatrix};\begin{bmatrix} .17193 & .17643 & .1678 & .18775 & .29609 \\ .17167 & .17689 & .16671 & .18719 & .29754 \\ .17293 & .17488 & .17007 & .18878 & .29334 \\ .17192 & .17654 & .16713 & .18741 & .297 \\ .17142 & .17726 & .16629 & .18696 & .29807 \end{bmatrix}; .18719$$

45. (a) .847423 or about 85% (b) [.032 .0998125 .0895625 .778625]   **47.** [.219086 .191532 .252352 .178091 .158938]

SECTION 9.5 (page 512)

1. (a)
Number	0	1	2	3	4	5
Probability	0	0	.1	.3	.4	.2

(b)
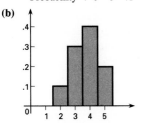

3. (a)
Number	0	1	2	3	4	5	6
Probability	0	.04	0	.16	.40	.32	.08

(b)
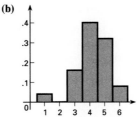

5. (a)
Number	0	1	2	3	4	5
Probability	.15	.25	.3	.15	.1	.05

(b)
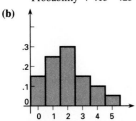

7.
Number of Heads	0	1	2	3	4
Probability	1/16	1/4	3/8	1/4	1/16

9.
Number of Aces	0	1	2	3
Probability	4324/5525 ≈ .783	1128/5525 ≈ .204	72/5525 ≈ .013	1/5525 ≈ .0002

11.
Number of Hits	0	1	2	3	4
Probability	.254	.415	.254	.069	.007

13.

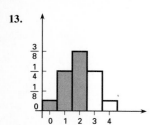

15.

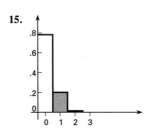

17.

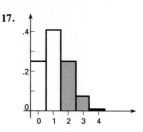

19. 3.6   **21.** 14.64   **23.** 2.7   **25.** 18   **27.** −$.64   **29.** 9/7 or 1.3   **31. (a)** 5/3 or 1.67 **(b)** 4/3 or 1.33   **33.** 1   **35.** No   **37.** −2.7¢
39. −20¢   **41. (a)** Yes; the probability of a match is still 1/2. **(b)** 40¢ **(c)** −40¢   **43.** $4500   **45.** 118   **47.** 3.51   **49. (a)** $50,000
(b) $65,000   **51.** −87.8¢

SECTION 9.6 (page 519)

1. (a) Buy speculative **(b)** Buy blue-chip **(c)** Buy speculative; $24,300 **(d)** Buy blue-chip   **3. (a)** Set up in the stadium **(b)** Set up in the gym
(c) Set up both; $1010
5. (a)

| | New Product | |
	Better	Not Better
Market new product	50,000	−25,000
Don't	−40,000	−10,000

(b) $5000 if they market the new product, −$22,000 if they don't; market the new product

7. (a)

	Strike	None
Bid $30,000	−5500	4500
Bid $40,000	4500	0

(b) Bid $40,000   **9.** Environment; 14.25

CHAPTER 9   REVIEW EXERCISES (page 522)

1. 720   **3.** $\binom{12}{3} = 220$ **5.** $1 \cdot 4 \cdot 3 \cdot 2 \cdot 1 = 24$ **7.** 140   **9.** $\binom{4}{3}/\binom{11}{3} = 4/165$ **11.** $\binom{4}{2}\binom{5}{1}/\binom{11}{3} = 2/11$ **13.** $\binom{5}{2}\binom{2}{1}/\binom{11}{3} = 4/33$ **15.** 25/102
17. 15/34   **19.** 5/16   **21.** 11/32   **23.** $\binom{20}{4}(.01)^4(.99)^{16} \approx .00004$ **25.** .81791 + .16523 + .01586 + .00096 + .00004 ≈ 1.0000   **27.** Yes
29. Yes   **31. (a)** [.54   .46] **(b)** [.6464   .3536] **(c)** 2/3 of the market   **33. (a)** [.195   .555   .250] **(b)** [.1945   .5555   .2500]
(c) [7/36   5/9   1/4] or [.1944   .5556   .2500]
35. (a)

Number	8	9	10	11	12	13	14
Probability	1/23	0	2/23	5/23	8/23	4/23	3/23

(b)

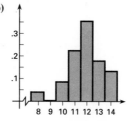

37. (a)

Number	2	3	4	5	6	7	8	9	10	11	12
Probability	1/36	1/18	1/12	1/9	5/36	1/6	5/36	1/9	1/12	1/18	1/36

(b)

39. .6   **41.** −83.3¢; no   **43.** 2.5   **45.** −6¢   **47.** $1.29
49. (a) Oppose **(b)** Oppose **(c)** Oppose; 2700 **(d)** Oppose; 1400

CASE 12 (page 527)

1. (a) $69.01 **(b)** $79.64 **(c)** $58.96 **(d)** $82.54 **(e)** $62.88 **(f)** $64.00   **2.** Stock only part 3 on the truck   **3.** p_1, p_2, p_3 are not the only events in the sample space.   **4.** $\binom{n}{0} + \binom{n}{1} + \binom{n}{2} + \ldots + \binom{n}{n}$

CHAPTER 10

SECTION 10.1 (page 534)

1. (a) About 17.5% **(b)** About 8% **(c)** 20–29

3. (a)

0–24	4
25–49	3
50–74	6
75–99	3
100–124	5
125–149	9

(b)

Interval	Frequency	Cumulative Frequency
0–24	4	4
25–49	3	7
50–74	6	13
75–99	3	16
100–124	5	21
125–149	9	30

(c)–(d)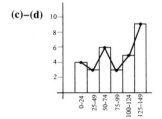

(e)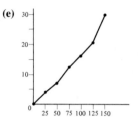

5. (a)–(b)

Interval	Frequency	Cumulative Frequency
70–74	2	2
75–79	1	3
80–84	3	6
85–89	2	8
90–94	6	14
95–99	5	19
100–104	6	25
105–109	4	29
110–114	2	31

(c)–(d)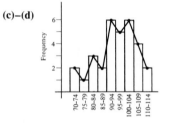

(e)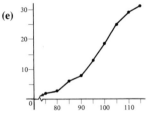

7.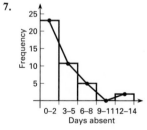

9. Answers vary

11.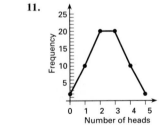

13. Answers vary

15.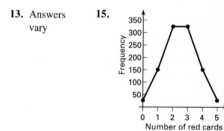

17. E: 16.9%   T: 9.1%   N: 7.0%   R: 7.0%   I: 6.6%   A: 6.2%   H: 5.8%
S: 5.8%   L: 5.0%   C: 4.5%   O: 4.1%   F: 3.7%   G: 3.7%   U: 3.7%
D: 2.1%   Y: 2.1%   M: 1.7%   P: 1.7%   W: 1.7%   Q: 1.2%   V: .4%
B: 0%   J: 0%   K: 0%   X: 0%   Z: 0%

SECTION 10.2 (page 542)

1. 16 **3.** 150.3 **5.** 27,955 **7.** 7.7 **9.** .1 **11.** 6.7 **13.** 17.2 **15.** 118.8 **17.** 51 **19.** 130 **21.** 49 **23.** 562 **25.** 29.1 **27.** 9
29. 64 **31.** 68 and 74 **33.** No mode **35.** 6.1 and 6.3 **37.** Mean = 86.2, modal class 125–149 **39.** Mean = 69.4, modal class 60–69
41. 55.5°F **43.** $3.29 **45.** Mean = 1.9, median = 1, mode = 1 **47.** Mean = 17.2, median = 17.5, no mode

SECTION 10.3 (page 548)

1. 6; $\sqrt{5} \approx 2.24$ **3.** 12; $\sqrt{22} \approx 4.7$ **5.** 53; $\sqrt{475.33} \approx 21.8$ **7.** 46; $\sqrt{258.4} \approx 16.1$ **9.** 24; $\sqrt{66} \approx 8.1$ **11.** 30; $\sqrt{139.7} \approx 11.8$
13. $\sqrt{2046.7} \approx 45.2$ **15.** $s = \sqrt{117.97} \approx 10.9$ **17.** 3/4 **19.** 24/25 **21.** At least 88.9% **23.** At least 96% **25.** No more than 11.1%
27. Forever Power, $\bar{x} = 26.2$, $s = 4.4$; Brand X, $\bar{x} = 25.5$, $s = 7.2$ **(a)** Forever Power **(b)** Forever Power **29. (a)** 12.5 **(b)** −3.0 **(c)** 4.9
(d) 4.2 **(e)** 15.5 **(f)** 7.55 to 23.45 **31.** Mean = 1.8158; standard deviation = .4451 **33.** Mean = 7.3571; standard deviation = .1326

SECTION 10.4 (page 558)

1. 49.38% **3.** 17.36% **5.** 45.64% **7.** 49.91% **9.** 7.7% **11.** 47.35% **13.** 92.42% **15.** 32.56% **17.** 1.64 or 1.65 **19.** −1.04
21. 5000 **23.** 4332 **25.** 642 **27.** 9918 **29.** 19 **31.** 15.87% **33.** .62% **35.** 84.13% **37.** 37.79% **39.** 2.27% **41.** 99.38%
43. 189 **45.** .0062 **47.** .4325 **49.** $62.64 and $41.86 **51.** .0823 **53.** 82 **55.** 70 **57.** .0256 **59.** Approx. 7:13 A.M.

SECTION 10.5 (page 565)

1. (a)

x	0	1	2	3	4	5	6
$P(x)$	.335	.402	.201	.054	.008	.001	.000

(b) 1.00 **(c)** .91 **3. (a)**

x	0	1	2	3
$P(x)$	.941	.058	.001	.000

(b) .06 **(c)** .24

5. (a)

x	0	1	2	3	4
$P(x)$	.0081	.0756	.2646	.4116	.2401

(b) 2.8 **(c)** .92
7. 12.5; 3.49 **9.** 51.2; 3.2 **11.** .1974 **13.** .1210 **15.** .0240 **17.** .9032 **19.** .8665 **21.** .0956 **23.** .0760 **25.** .6443 **27.** .0146
29. .1974 **31.** .0092 **33.** .0001 **35.** For Exercise 51: **(a)** .0001 **(b)** .0002 **(c)** 0; for Exercise 52: **(a)** .0469 **(b)** .9875 **(c)** .8810; for
Exercise 53: **(a)** .1684 **(b)** .0305 **(c)** .9695; for Exercise 54: **(a)** 0 **(b)** .0018

CHAPTER 10 REVIEW EXERCISES (page 568)

1. (a)

Interval	Frequency
450–474	5
475–499	6
500–524	5
525–549	2
550–574	2

(b)

(c)

3. 73 **5.** 34.9 **7.** Median = 44, mode = 46 **9.** 30–39
11. 18; $\sqrt{48.5} \approx 6.96$ **13.** $\sqrt{158.8} \approx 12.6$ **15. (a)** Stock 1: 8%, 7.9%; Stock 2: 8%, 2.6% **(b)** Stock 2 **17. (a)** At least 75% **(b)** No more
than 69.4% **19.** .3980 **21.** .3409 **23.** 2.27% **25.** 68.26%
27.

x	0	1	2	3	4
Probability	.9801	.0197	.0001485	.0000004975	6×10^{-10}

(b) mean = .02, σ = .141 **29. (a)** .0959 **(b)** .0028 **31.** 28.10%
33. 56.25% **35.** .0718 **37.** $\binom{100}{95}(.98)^{95}(.02)^5$ **39.** $\binom{100}{90}(.98)^{90}(.02)^{10} + \binom{100}{91}(.98)^{91}(.02)^9 + \cdots + \binom{100}{100}(.98)^{100}(.02)^0$

CASE 13   (page 571)

1. No; yes   **2.** No; yes

CHAPTER 11

SECTION 11.1   (page 586)

1. (a) 3 **(b)** 0   **3. (a)** Does not exist **(b)** $-1/2$ **5. (a)** 2 **(b)** Does not exist   **7. (a)** 1 **(b)** -1 **9.** **(a)** Does not exist **(b)** 1   **11.** .6805, .6919, .6930, .6931, .6932, .6933, .6944, .7055; $\lim_{x \to 3} f(x) \approx .693$ **13.** .2725, .3570, .3668, .3678, .3680, .3690, .3791, .4966; $\lim_{x \to 1} f(x) \approx$ .368   **15.** -9 **17.** 3/10   **19.** $-1/2$ **21.** -4 **23.** 7   **25.** 1/5   **27.** 1/4   **29.** 6/35   **31.** 2   **33.** Does not exist   **35.** 0   **37.** 1/4   **39.** 1/12   **41.** $1/4\sqrt{2}$ **43.** 0   **45.** 1/3   **47.** 0   **49.** 5/3   **51.** 4   **53.** 1/3   **55.** 0   **57.** 0   **59. (a)** 21 **(b)** 6.15 **(c)** 7.5 **(d)** 6   **61. (a)** 3 **(b)** Does not exist **(c)** 2 **(d)** 16 months   **63. (a)** .57 **(b)** .53 **(c)** .50 **(d)** .5   **65.** $-.24302$ **67.** 4.0140

SECTION 11.2   (page 597)

1. 5   **3.** 8   **5.** 1/3   **7.** $-1/3$ **9.** 20 ft/sec   **11.** 100 ft/sec   **13.** 60 ft/sec   **15.** 7 ft/sec   **17. (a)** 1 **(b)** 2/3 **(c)** 1/2 **(d)** 1/2   **19. (a)** About 1.8 per year **(b)** About 19.7 per year **(c)** About -40 per year **(d)** About 7.6 per year **(e)** About 0 per year **(f)** About -22 per year   **21. (a)** 7/3 ft/sec **(b)** 1 ft/sec **(c)** 2 ft/sec **(d)** 1 ft/sec **(e)** 11/6 ft/sec **(f)** 5/4 ft/sec   **23.** The average rate of change of market share is increasing in each period. **(a)** About .15 percentage points per year **(b)** About .13 percentage points per year **(c)** About .30 percentage points per year   **25. (a)** $500 per year **(b)** $375 per year   **27.** 11   **29. (a)** -25 boxes per dollar **(b)** -20 boxes per dollar **(c)** -30 boxes per dollar   **31. (a)** $-.3$ word per min **(b)** $-.9$ word per min

SECTION 11.3   (page 612)

1. 27   **3.** 1/8   **5.** 1/8   **7.** $y = 8x - 9$ **9.** $5x + 4y = 20$ **11.** $3y = 2x + 18$ **13.** 2   **15.** 1/5   **17.** 0   **19.** $f'(x) = -8x + 11; -5; 11; 35$ **21.** $f'(x) = 8; 8; 8; 8$ **23.** $f'(x) = 3x^2 + 3; 15; 3; 30$ **25.** $f'(x) = 2/x^2; 1/2;$ does not exist; 2/9   **27.** $f'(x) = -4/(x - 1)^2; -4; -4; -1/4$ **29.** $f'(x) = 1/(2\sqrt{x}); 1/(2\sqrt{2});$ does not exist; does not exist   **31.** 0   **33.** $-6, 6$ **35.** $-3, 0, 2, 3, 5$ **37. (a)** $P(x)$ is increasing **(b)** $P(x)$ is constant **(c)** $P(x)$ is decreasing   **39. (a)** $16 per table **(b)** $15.998 or $16 **(c)** The marginal revenue found in part (a) approximates the actual revenue from the sale of the 1001st item found in part (b).   **41. (a)** -8 **(b)** -20 **43. (a)** 30 **(b)** 20 **(c)** 10 **(d)** 0 **(e)** -10 **(f)** -30 **45.** From the left, our estimates are $-.005, .008, -.00125.$ **47.** $-.0435$ **49.** 4.2687   **51.** $-.10754$

SECTION 11.4   (page 625)

1. $f'(x) = 18x - 8$ **3.** $y' = 30x^2 - 18x + 6$ **5.** $y' = 4x^3 - 15x^2 + 18x$ **7.** $f'(x) = 9x^{.5} - 2x^{-.5}$ or $9x^{.5} - 2/x^{.5}$ **9.** $y' = -48x^{2.2} + 3.8x^{.9}$ **11.** $y' = 36t^{1/2} + 2t^{-1/2}$ or $36t^{1/2} + 2/t^{1/2}$ **13.** $y' = 4x^{-1/2} + (9/2)x^{-1/4}$ or $4/x^{1/2} + 9/(2x^{1/4})$ **15.** $g'(x) = -30x^{-6} + x^{-2}$ or $-30/x^6 + 1/x^2$ **17.** $y' = 8x^{-3} - 9x^{-4}$ or $8/x^3 - 9/x^4$ **19.** $y' = -20x^{-3} - 12x^{-5} - 6$ or $-20/x^3 - 12/x^5 - 6$ **21.** $f'(t) = -6t^{-2} + 16t^{-3}$ or $-6/t^2 + 16/t^3$ **23.** $y' = -36x^{-5} + 24x^{-4} - 2x^{-2}$ or $-36/x^5 + 24/x^4 - 2/x^2$ **25.** $f'(x) = -6x^{-3/2} - (3/2)x^{-1/2}$ or $-6/x^{3/2} - 3/(2x^{1/2})$ **27.** $p'(x) = 5x^{-3/2} - 12x^{-5/2}$ or $5/x^{3/2} - 12/x^{5/2}$ **29.** $y' = (-3/2)x^{-5/4}$ or $-3/(2x^{5/4})$ **31.** $y' = (-5/3)t^{-2/3}$ or $-5/(3t^{2/3})$ **33.** $dy/dx = -40x^{-6} + 36x^{-5}$ or $-40/x^6 + 36/x^5$ **35.** $(-9/2)x^{-3/2} - 3x^{-5/2}$ or $-9/(2x^{3/2}) - 3/x^{5/2}$ **37.** -28 **39.** 11/16   **41.** $-28; 28x + y = 34$ **43.** 79/6   **45.** $-11/27$ **47. (a)** 30 **(b)** 0 **(c)** -10 **49.** 10   **51. (a)** 100 **(b)** 1 **(c)** $-.01;$ the percent of acid is decreasing at the rate of .01 per day after 100 days   **53. (a)** The blood sugar level is decreasing at a rate of 4 points per unit of insulin. **(b)** The blood sugar level is decreasing at a rate of 10 points per unit of insulin.   **55. (a)** 264 **(b)** 510 **(c)** 97/4 or 24.25 **(d)** The number of matings per hour is increasing by about 24.25 matings at a temperature of 16°C.   **57. (a)** $8 **(b)** $192 **(c)** $428 **(d)** $760   **59. (a)** $v(t) = 6$ **(b)** 6; 6; 6   **61. (a)** $v(t) = 22t + 4$ **(b)** 4; 114; 224   **63. (a)** $v(t) = 12t^2 + 16t$ **(b)** 0; 380; 1360   **65. (a)** -32 ft/sec; -64 ft/sec **(b)** In 3 sec **(c)** -96 ft/sec

SECTION 11.5 (page 633)

1. $y' = 4x + 3$ **3.** $y' = 48x + 66$ **5.** $y' = 18x^2 - 6x + 4$ **7.** $y' = 9x^2 - 4x - 5$ **9.** $y' = 54x^5 - 36x^3 + 21x^2 - 7$ **11.** $y' = 4(2x - 5)$ or $8x - 20$ **13.** $y' = 4x(x^2 - 1)$ or $4x^3 - 4x$ **15.** $y' = 3x^{1/2}/2 + x^{-1/2}/2 + 2$ or $3\sqrt{x}/2 + 1/(2\sqrt{x}) + 2$ **17.** $y' = 10 + 3x^{-1/2}/2$ or $10 + 3/(2\sqrt{x})$ **19.** $y' = -3/(2x - 1)^2$ **21.** $f'(x) = 53/(3x + 8)^2$ **23.** $y' = -6/(3x - 5)^2$ **25.** $y' = -17/(4 + x)^2$ **27.** $f'(t) = (t^2 - 2t - 1)/(t - 1)^2$ **29.** $y' = (-x^2 + 4x + 1)/(x^2 + 1)^2$ **31.** $g'(x) = (-6x^4 - 4x^3 - 6x - 1)/(2x^3 - 1)^2$ **33.** $y' = (x^2 + 6x - 14)/(x + 3)^2$ **35.** $p'(t) = [-\sqrt{t}/2 - 1/(2\sqrt{t})]/(t - 1)^2$ or $(-t - 1)/[2\sqrt{t}(t - 1)^2]$ **37.** $y' = (5\sqrt{x}/2 - 3/\sqrt{x})/x$ or $(5x - 6)/(2x\sqrt{x})$ **39.** $f'(p) = (24p^2 + 32p + 29)/(3p + 2)^2$ **41.** $g'(x) = (10x^4 + 18x^3 + 6x^2 - 20x - 9)/[(2x + 1)^2(5x + 2)^2]$ **43. (a)** 108.80 **(b)** 198.40 **(c)** $9x + 18 + 8/x$ **(d)** $dC/dx = 9 - 8/x^2$ **45. (a)** $M'(d) = 2000/(3d + 10)^2$ **(b)** The new employee can assemble 7.8125 additional bicycles per day after 2 days of training and 3.2 additional bicycles per day after 5 days of training. **47. (a)** $s'(x) = m/(m + nx)^2$ **(b)** $s'(50) = 1/2560 \approx .000391$ **(c)** The amount of muscle contraction is increasing by .000391 millimeters when the concentration of the drug is 50 milliliters.

SECTION 11.6 (page 643)

1. 1122 **3.** 97 **5.** $256k^2 + 48k + 2$ **7.** $24x + 4$; $24x + 35$ **9.** $-64x^3 + 2$; $-4x^3 + 8$ **11.** $1/x^2$; $1/x^2$ **13.** $\sqrt{8x^2 - 4}$; $8x + 10$ **15.** If $f(x) = x^{1/3}$ and $g(x) = 3x - 7$, then $y = f[g(x)]$. **17.** If $f(x) = x^{1/2}$ and $g(x) = 9 - 4x$, then $y = f[g(x)]$. **19.** If $f(x) = (x + 3)/(x - 3)$ and $g(x) = \sqrt{x}$, then $y = f[g(x)]$. **21.** If $f(x) = x^2 + x + 5$ and $g(x) = x^{1/2} - 3$, then $y = f[g(x)]$. **23.** $y' = 4(2x + 9)$ **25.** $y' = 90(5x - 1)^2$ **27.** $y' = -144x(12x^2 + 4)^2$ **29.** $y' = 36(2x + 5)(x^2 + 5x)^3$ **31.** $y' = 36(2x + 5)^{1/2}$ **33.** $y' = (-21/2)(8x + 9)(4x^2 + 9x)^{1/2}$ **35.** $y' = 16(4x + 7)^{-1/2}$ or $16/\sqrt{4x + 7}$ **37.** $y' = -(2x + 4)(x^2 + 4x)^{-1/2}$ or $-(2x + 4)/\sqrt{x^2 + 4x}$ **39.** $y' = 12(2x + 3)(2x + 1)$ **41.** $y' = 3(x - 1)(x + 1)$ **43.** $y' = 70(x + 3)(2x - 1)^4(x + 2)$ **45.** $y' = [(3x + 1)^2(21x + 1)]/2\sqrt{x}$ **47.** $y' = -2(x - 4)^{-3}$ or $-2/(x - 4)^3$ **49.** $y' = [2(4x + 3)(4x - 7)]/(2x - 1)^2$ **51.** $y' = (-5x^2 - 36x + 8)/(5x + 2)^4$ **53.** $(3x^{1/2} - 1)/[4x^{1/2}(x^{1/2} - 1)^{1/2}]$ **55.** $D(c) = (-c^2 + 10c - 25)/25 + 500$ **57. (a)** \$101.22 **(b)** \$111.86 **(c)** \$117.59 **59. (a)** $-\$1050$ **(b)** $-\$457.06$ **61.** \$400 per additional worker **63.** $18a^2 + 24a + 9$ **65.** $A[r(t)] = A(t) = 4\pi t^2$; this function gives the area of the pollution in terms of the time since the pollutants were first emitted. **67. (a)** $-.5$ **(b)** $-1/54 \approx -.02$ **(c)** $-.011$ **(d)** $-1/128 \approx -.008$

SECTION 11.7 (page 650)

1. $y' = 4e^{4x}$ **3.** $y' = 12e^{-2x}$ **5.** $y' = -16e^{2x}$ **7.** $y' = -16e^{x+1}$ **9.** $y' = 2xe^{x^2}$ **11.** $y' = 12xe^{2x^2}$ **13.** $y' = 16xe^{2x^2-4}$ **15.** $y' = xe^x + e^x = e^x(x + 1)$ **17.** $y' = 2(x - 3)(x - 2)e^{2x}$ **19.** $y' = -1/(3 - x)$ or $1/(x - 3)$ **21.** $y' = (4x - 7)/(2x^2 - 7x)$ **23.** $y' = 1/[2(x + 5)]$ **25.** $y' = (12x + 17)/[(2x + 7)(3x - 2)]$ **27.** $y' = 22/[(2x + 4)(5x - 1)]$ **29.** $y' = 3(2x^2 + 5)/[x(x^2 + 5)]$ **31.** $y' = -3x(x + 2) - 3\ln(x + 2)$ **33.** $y' = x + 2x|\ln x|$ **35.** $y' = (2xe^x - x^2e^x)/e^{2x} = x(2 - x)/e^x$ **37.** $y' = (2x^3 - 1 + 6x^3 \ln|x|)/x$ **39.** $y' = (4x + 7 - 4x \ln|x|)/[x(4x + 7)^2]$ **41.** $y' = (1 - x \ln|x|)/(xe^x)$ **43.** $y' = (6x \ln|x| - 3x)/(\ln|x|)^2$ **45.** $y' = [4(\ln(x + 1))^3]/(x + 1)$ **47.** $y' = [x(\ln|x|)e^x - e^x]/[x(\ln|x|)^2]$ or $e^x[x \ln|x| - 1]/[x(\ln|x|)^2]$ **49.** $y' = [x(e^x - e^{-x}) - (e^x + e^{-x})]/x^2$ **51.** $y' = e^{x^2}/x + 2xe^{x^2} \ln|x|$ **53.** $y' = -20,000e^{-4x}/(1 + 10e^{-4x})^2$ **55.** $y' = 8000e^{-2x}/(9 + 4e^{-.2x})^2$ **57.** $y' = 1/(x \ln|x|)$ **59. (a)** $dR/dx = 100 + [50(\ln(x - 1))]/(\ln x)^2$ **(b)** \$112.48 **61. (a)** $C'(x) = 100$ **(b)** $P(x) = 50x/\ln x - 100$ **(c)** About \$12.48 **63. (a)** 20 **(b)** 1.34 **65. (a)** 100% **(b)** 94.1% **(c)** 88.5% **(d)** 83.3% **(e)** -3.05% **(f)** -2.87% **(g)** -2.70% **(h)** -2.54% **67.** $200e^{.4} \approx 298$; $200e^{1.6} \approx 991$ **69. (a)** 7.42 **(b)** 21.22 **(c)** .79 **71. (a)** .005 **(b)** .0007 **(c)** .000013 **(d)** $-.022$ **(e)** $-.0029$ **(f)** $-.000054$

SECTION 11.8 (page 659)

1. -1 **3.** -1 **5.** 0; 2 **7.** 1 **9.** Yes; no **11.** No; no; yes **13.** Yes; no; no **15.** Yes; no; yes **17.** Yes; no; yes **19.** Yes; yes; yes **21.** No; yes; yes **23. (a)** \$96 **(b)** \$150 **(c)** **(d)** At $x = 100$

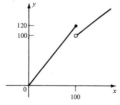

25. (a) \$30 **(b)** \$30 **(c)** \$25 **(d)** \$21.43 **(e)** \$22.50 **(f)** \$30 **(g)** \$25   **27.** $x = 6$, $x = 8$, $x = 12$, $x = 16$ **29.** $(-4, 6)$ **31.** $(-\infty, -2]$
33. $[-11, 9]$ **35.** $(-6, -2)$ **37.** $(-4, \infty)$ **39.** $(-\infty, 0)$ **41.** Yes; yes; no   **43.** No; yes; yes

CHAPTER 11   REVIEW EXERCISES   (page 665)

1. 4 **3.** -3 **5.** 4 **7.** 17/3 **9.** 8 **11.** -13 **13.** 1/6 **15.** 1/5 **17.** 3/4 **19.** 1/4 **21.** $-3/2$ **23.** 30 **25.** 9/77 **27.** $y' = 4$ **29.** $y' = -3x^2 + 7$ **31.** -2; $y + 2x = -4$ **33.** $-3/4$; $3x + 4y = -9$ **35.** 38; $y = 38x + 22$ **37.** 3/4; $4y = 3x + 7$

39. (a) \$150 **(b)** \$187.50 **(c)** \$189 **(d)**
(e) Discontinuous at $x = \$125$

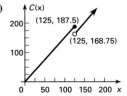

41. $y' = 10x - 7$ **43.** $y' = 14x^{4/3}$ **45.** $f'(x) = -3x^{-4} + (1/2)x^{-1/2}$ or $-3/x^4 + 1/(2x^{1/2})$ **47.** $y' = 15t^4 + 12t^2 - 7$ **49.** $y' = 12x^{1/2} - 6x^{-1/2}$ or $12x^{1/2} - 6/x^{1/2}$ **51.** $g'(t) = -20t^{1/3} - 7t^{-2/3}$ or $-20t^{1/3} - 7/t^{2/3}$ **53.** $y' = 9x^{-3/4} - 45x^{-7/4}$ or $9/x^{3/4} - 45/x^{7/4}$ **55.** $k'(x) = 15/(x + 5)^2$ **57.** $y' = (2 - x + 2\sqrt{x})/[2\sqrt{x}(x + 2)^2]$ **59.** $y' = (x^2 - 2x)/(x - 1)^2$ **61.** $f'(x) = 12(3x - 2)^3$ **63.** $y' = 1/\sqrt{2t - 5}$ **65.** $y' = 3(2x + 1)^2(8x + 1)$ **67.** $r'(t) = (-15t^2 + 52t - 7)/(3t + 4)^4$ **69.** $y' = (x^2 - 4x + 4)/(x - 2)^2 = 1$ **71.** $y' = -12e^{2x}$ **73.** $y' = -6x^2e^{-2x^3}$ **75.** $y' = 10xe^{2x} + 5e^{2x} = 5e^{2x}(2x + 1)$ **77.** $y' = 2x/(2 + x^2)$ **79.** $y' = [x - 3 - x \ln(3x)]/[x(x - 3)^2]$ **81.** $-1/[x^{1/2}(x^{1/2} - 1)^2]$ **83.** $dy/dt = (1 + 2t^{1/2})/[4t^{1/2}(t^{1/2} + t)^{1/2}]$ **85.** $-2/3$ **87.** None **89.** No; yes; yes **91.** Yes; no; yes; no **93.** Yes; no; yes **95.** Yes; yes; yes **97. (a)** 315 **(b)** 1015 **(c)** 1515 **99. (a)** $55/3 = 18\frac{1}{3}$ **(b)** $65/4 = 16\frac{1}{4}$ **(c)** 15 **101. (a)** -9.5 **(b)** -2.375

CHAPTER 12

SECTION 12.1   (page 679)

1. Relative minimum of -4 at $x = 1$ **3.** Relative maximum of 3 at $x = -2$ **5.** Relative maximum of 3 at $x = -4$; relative minimum of 1 at $x = -2$ **7.** Relative maximum of 3 at $x = -4$; relative minimum of -2 at $x = -7$ and $x = -2$ **9.** Relative minimum of -44 at $x = -6$ **11.** Relative maximum of -8 at $x = -3$; relative minimum of -12 at $x = -1$ **13.** Relative maximum of 827/96 at $x = -1/4$; relative minimum of $-377/6$ at $x = -5$ **15.** Relative maximum of 57 at $x = 2$; relative minimum of 30 at $x = 5$ **17.** Relative maximum of -4 at $x = 0$; relative minimum of -85 at $x = 3$ and $x = -3$ **19.** Relative maximum of 0 at $x = 8/5$ **21.** No relative extrema **23.** Relative minimum of 0 at $x = 0$ **25.** A relative maximum of $1/e$ at $x = -1$ **27.** A relative minimum of 2 at $x = 0$ **29.** Relative minimum of $-1/e$ at $x = 1/e$; by symmetry and absolute value there is a relative maximum of $1/e$ when $x = -1/e$ **31.** No absolute maximum; absolute minimum at x_1 **33.** Absolute maximum at x_1; absolute minimum at x_2 **35.** Absolute maximum at x_2; absolute minimum at x_3 **37.** Absolute maximum at $x = -1$; absolute minimum at $x = -5$ **39.** Absolute maximum at $x = -2$; absolute minimum at $x = 4$ **41.** Absolute maximum at $x = -2$; absolute minimum at $x = 3$ **43.** Absolute maximum at $x = 6$; absolute minimum at $x = -4$ and $x = 4$ **45.** Absolute maximum at $x = 6$; absolute minimum at $x = 4$ **47.** Absolute maximum at $x = \sqrt{2}$; absolute minimum at $x = 0$ **49.** Absolute maxima at $x = -3$ and $x = 3$; absolute minimum at $x = 0$ **51.** Absolute maximum at $x = 0$; absolute minima at $x = -1$ and $x = 1$ **53. (a)** \$6 **(b)** \$512 **55.** 1000 manuals; more than \$2.20 **57.** 6 months; 6% **59.** 25; 16.1 **61. (a)** 10 **(b)** 500 **63.** Absolute maximum is about 3.4, at about $x = 1.3$; absolute minimum is -7 at $x = 0$ **65.** Absolute maximum on $[-5, -4]$ is about 60.4 at $x = -4.5$ or $x = -4.4$; absolute minimum is 58 at $x = -5$. Absolute maximum on $[0, 1]$ is 4 at $x = 1$; absolute minimum is about 1.6 at $x = .4$ or $x = .5$.

SECTION 12.2   (page 687)

1. $f''(x) = 18x$; 0; 36; -54 **3.** $f''(x) = 36x^2 - 30x + 4$; 4; 88; 418 **5.** $f''(x) = 6(x + 4)$; 24; 36; 6 **7.** $f''(x) = 10/(x - 2)^3$; $-5/4$; $f''(2)$ does not exist; $-2/25$ **9.** $f''(x) = 2/(1 + x)^3$; 2; 2/27; $-1/4$ **11.** $f''(x) = -1/[4(x + 4)^{3/2}]$; $-1/32$; $-1/(4 \cdot 6^{3/2}) \approx -.0170$; $-1/4$ **13.** $f''(x) = (-6/5)x^{-7/5}$ or $-6/(5x^{7/5})$; $f''(0)$ does not exist; $-6/(5 \cdot 2^{7/5}) \approx -.4547$; $-6/[5 \cdot (-3)^{7/5}] \approx .2578$ **15.** $f''(x) = 2e^x$; 2; $2e^2$; $2e^{-3} = 2/e^3$ **17.** $f'''(x) = -24x$; $f^{(4)} = -24$ **19.** $f'''(x) = 240x^2 + 144x$; $f^{(4)} = 480x + 144$ **21.** $f'''(x) = 18(x + 2)^{-4}$ or $18/(x + 2)^4$; $f^{(4)}(x) = -72(x + 2)^{-5}$ or $-72/(x + 2)^5$ **23.** $f'''(x) = 40e^{2x}$; $f^{(4)}(x) = 80e^{2x}$ **25.** $f'''(x) = 2/x^3$; $f^{(4)}(x) = -6/x^4$ **27.** $f'''(x) = -1/x^2$; $f^{(4)}(x) = 2/x^3$ **29.** Relative minimum at -3 **31.** Relative minimum at 1 **33.** Relative minimum at 0; relative maximum at $-4/3$ **35.** Relative maximum at 1/2;

relative minimum at -3 **37.** Relative maximum at -1; relative minimum at $1/2$ **39.** Critical number at 0, but neither a relative maximum nor minimum there   **41.** Relative maximum at 0; relative minimum at -2 and 2   **43.** Relative maximum at $-\sqrt{3}$; relative minimum at $\sqrt{3}$ **45.** Relative maximum at -3; relative minimum at 3   **47.** No critical numbers; no maximum or minimum   **49.** $v(t) = 16t + 4$; $a(t) = 16$; $v(0) = 4$ cm/sec; $v(4) = 68$ cm/sec; $a(0) = 16$ cm/sec^2; $a(4) = 16$ cm/sec^2 **51.** $v(t) = -15t^2 - 16t + 6$; $a(t) = -30t - 16$; $v(0) = 6$ cm/sec; $v(4) = -298$ cm/sec; $a(0) = -16$ cm/sec^2; $a(4) = -136$ cm/sec^2 **53.** $v(t) = 6(3t + 4)^{-2}$ or $6/(3t + 4)^2$; $a(t) = -36(3t + 4)^{-3}$ or $-36/(3t + 4)^3$; $v(0) = 3/8$ cm/sec; $v(4) = 3/128$ cm/sec; $a(0) = -9/16$ cm/sec^2; $a(4) = -9/1024$ cm/sec^2 **55. (a)** -96 ft/sec **(b)** -160 ft/sec **(c)** -256 ft/sec **(d)** -32 ft/sec^2 **57.** 6 units; \$1072   **59.** 8 inches   **61. (a)** After 2 hours **(b)** $3/4\%$ **63. (a)** At 4 hours **(b)** 1160 million

SECTION 12.3   (page 697)

1. (a) $y = 100 - x$ **(b)** $P = x(100 - x)$ **(c)** $P' = 100 - 2x$; $x = 50$ **(d)** $\underline{50}$ and 50 **(e)** $50 \cdot 50 = 2500$ **3.** 100; 100   **5.** 100; 50   **7.** 10; 0 **9. (a)** $R(x) = 100{,}000x - 100x^2$ **(b)** 500 **(c)** 25,000,000¢   **11. (a)** $\sqrt{3200} \approx 56.6$ mph **(b)** \$45.25   **13. (a)** $1200 - 2x$ **(b)** $A(x) = 1200x - 2x^2$ **(c)** 300 m **(d)** 180,000 sq m   **15.** 405,000 sq m   **17.** 20,000 sq ft   **19. (a)** $40 - 2x$ **(b)** $100 + 5x$ **(c)** $R(x) = 4000 - 10x^2$ **(d)** Pick now **(e)** \$40 per tree   **21. (a)** 90 **(b)** \$405   **23.** 4 in by 4 in by 2 in   **25.** 3 ft by 6 ft by 2 ft   **27.** 40 mph gives the minimum cost of \$800. **29.** 1 mile from A   **31.** 10   **33.** 60   **35.** 5

SECTION 12.4   (page 716)

1. Increasing on $(1, \infty)$; decreasing on $(-\infty, 1)$ **3.** Increasing on $(-\infty, -2)$; decreasing on $(-2, \infty)$ **5.** Increasing on $(-\infty, -4)$ and $(-2, \infty)$; decreasing on $(-4, -2)$ **7.** Increasing on $(-7, -4)$ and $(-2, \infty)$; decreasing on $(-\infty, -7)$ and $(-4, -2)$

9. $(-6, \infty)$; $(-\infty, -6)$

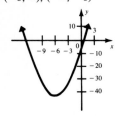

11. $(-\infty, 3/2)$; $(3/2, \infty)$

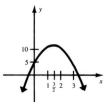

13. $(-\infty, -3)$ and $(4, \infty)$; $(-3, 4)$

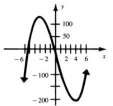

15. $(-\infty, -3/2)$ and $(4, \infty)$; $(-3/2, 4)$

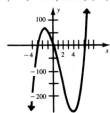

17. None; $(-\infty, \infty)$

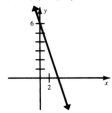

19. None; $(-\infty, -1)$, $(-1, \infty)$

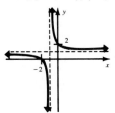

21. $(-4, \infty)$; $(-\infty, -4)$

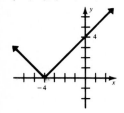

23. None; $(1, \infty)$

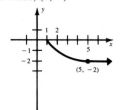

25. $(0, \infty)$; $(-\infty, 0)$

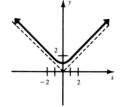

27. Concave upward on $(2, \infty)$; concave downward on $(-\infty, 2)$; point of inflection at $(2, 3)$ **29.** Concave upward on $(-\infty, -1)$ and $(8, \infty)$; concave downward on $(-1, 8)$; points of inflection at $(-1, 7)$ and $(8, 6)$ **31.** Concave upward on $(2, \infty)$; concave downward on $(-\infty, 2)$; no points of inflection **33.** Always concave upward; no points of inflection **35.** Always concave downward; no points of inflection **37.** Concave upward on $(-1, \infty)$; concave downward on $(-\infty, -1)$; point of inflection at $(-1, 44)$ **39.** Concave upward on $(-\infty, 3/2)$; concave downward on $(3/2, \infty)$; point of inflection at $(3/2, 525/2)$ **41.** Concave upward on $(5, \infty)$; concave downward on $(-\infty, 5)$; no points of inflection **43.** Concave upward on $(-10/3, \infty)$; concave downward on $(-\infty, -10/3)$; point of inflection at $(-10/3, -250/27)$

45. No points of inflection

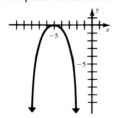

47. Point of inflection at $x = 1/3$

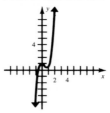

49. Point of inflection at $x = -3/2$

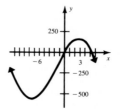

51. Point of inflection at $x = -7/12$

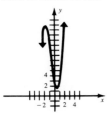

53. No points of inflection

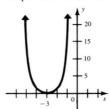

55. Points of inflection at $x = -\sqrt{3}$ and $x = \sqrt{3}$

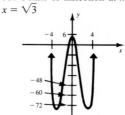

57. No points of inflection

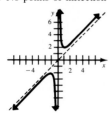

59. No points of inflection

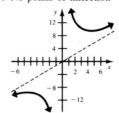

61. No points of inflection

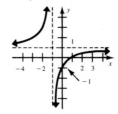

63. (a) Nowhere **(b)** Everywhere **65.** $[0, 1125)$ **67.** After 10 days **69. (a)** $(0, 3)$ **(b)** $(3, \infty)$ (Remember: x must be at least 0.) **71. (a)** Between car phones and Polaroid cameras **(b)** Near black-and-white TVs **(c)** Car phones and Polaroid cameras; the rate of growth of sales will now decline **(d)** Black-and-white TVs; the rate of decline of sales is starting to slow **73.** $(22, 6517.9)$

CHAPTER 12   REVIEW EXERCISES   (page 720)

1. Increasing on $(5/2, \infty)$; decreasing on $(-\infty, 5/2)$ **3.** Increasing on $(-4, 2/3)$; decreasing on $(-\infty, -4)$ and $(2/3, \infty)$ **5.** Never increasing; decreasing on $(-\infty, 4)$ and $(4, \infty)$ **7.** Relative maximum of -4 when $x = 2$ **9.** Relative minimum of -7 when $x = 2$ **11.** Relative maximum of 101 when $x = -3$; relative minimum of -24 when $x = 2$ **13.** Relative maximum of $-151/54$ at $1/3$; relative minimum of $-27/8$ at $-1/2$ **15.** Relative maximum of 1 when $x = 0$; relative minimum of $-2/3$ when $x = -1$ and $-29/3$ when $x = 2$ **17.** None **19.** Relative minimum of $-1/e$ when $x = -1$ **21.** Relative minimum of e^2 when $x = 2$ **23.** Absolute maximum of $29/4$ at $5/2$; absolute minimum of 5 at 1 and 4 **25.** Absolute maximum of 39 at -3; absolute minimum of $-319/27$ at $5/3$ **27.** $f''(x) = 36x^2 - 10$; 26; 314 **29.** $f''(x) = 14/(x + 1)^3$; $7/4$; $-7/4$ **31.** $f''(x) = 2e^{-x^2}(2x^2 - 1)$; $2e^{-1}$; $34e^{-9}$

33. Concave upward on $(-\infty, -1/12)$; concave downward on $(-1/12, \infty)$; point of inflection at $x = -1/12$

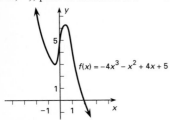

35. Concave upward on $(-\infty, \infty)$; no points of inflection

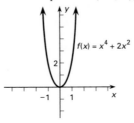

37. Concave upward on $(-\infty, 0)$; concave downward on $(0, \infty)$; no points of inflection

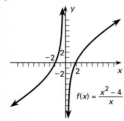

39. Concave upward on $(-\infty, 3)$; concave downward on $(3, \infty)$; no points of inflection

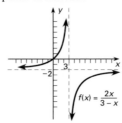

41. 12.5 and 12.5 **43.** (a) \$2000 (b) \$400 **45.** 2 m by 4 m by 4 m **47.** 225 m by 450 m **49.** 126 **51.** 225 m by 225 m

CASE 14 (page 722)

1. 12 knots (for a maximum profit of \$4400.00)

CASE 15 (page 723)

1. $-C_1/m^2 + DC_3/2$ **2.** $m = \sqrt{2C_1/(DC_3)}$ **3.** About 3.33 **4.** $m^+ = 4$ and $m^- = 3$ **5.** $Z(m^+) = Z(4) = \$11,400$; $Z(m^-) = Z(3) = \$11,300$ **6.** 3 months; 9 trainees per batch

CHAPTER 13

SECTION 13.1 (page 733)

1. $2x^2 + C$ **3.** $5t^3/3 + C$ **5.** $6k + C$ **7.** $z^2 + 3z + C$ **9.** $x^3/3 + 3x^2 + C$ **11.** $t^3/3 - 2t^2 + 5t + C$ **13.** $z^4 + z^3 + z^2 - 6z + C$ **15.** $10z^{3/2}/3 + C$ **17.** $2u^{3/2}/3 + 2u^{5/2}/5 + C$ **19.** $6x^{5/2} + 4x^{3/2}/3 + C$ **21.** $4u^{5/2} - 4u^{7/2} + C$ **23.** $-1/z + C$ **25.** $-1/(2y^2) - 2y^{1/2} + C$ **27.** $9/t - 2 \ln|t| + C$ **29.** $e^{2t}/2 + C$ **31.** $-15e^{-.2x} + C$ **33.** $3 \ln|x| - 8e^{-.5x} + C$ **35.** $\ln|t| + 2t^3/3 + C$ **37.** $e^{2u}/2 + u^2/8 + C$ **39.** $x^3/3 + x^2 + x + C$ **41.** $6x^{7/6}/7 + 3x^{2/3}/2 + C$ **43.** $C(x) = 2x^2 - 5x + 8$ **45.** $C(x) = .2x^3/3 + 5x^2/2 + 10$ **47.** $C(x) = 2x^{3/2}/3 - 8/3$ **49.** $C(x) = x^3/3 - x^2 + 3x + 6$ **51.** $C(x) = \ln|x| + x^2 + 7.45$ **53.** $P(x) = -x^2 + 20x - 50$ **55.** $5t^2 - 30t^{1/2}$ tons **57.** $f(x) = (3/5)x^{5/3}$

SECTION 13.2 (page 742)

1. $2(2x + 3)^5/5 + C$ **3.** $-2(y - 2)^{-2} + C$ **5.** $-(2m + 1)^{-2}/2 + C$ **7.** $-(x^2 + 2x - 4)^{-3}/3 + C$ **9.** $(z^2 - 5)^{3/2}/3 + C$ **11.** $-2e^{2p} + C$ **13.** $e^{2x^3}/2 + C$ **15.** $e^{2t-t^2}/2 + C$ **17.** $-e^{1/z} + C$ **19.** $-8(\ln|1 + 3x|)/3 + C$ **21.** $(\ln|2t + 1|)/2 + C$ **23.** $-(3v^2 + 2)^{-3}/18 + C$ **25.** $-(2x^2 - 4x)^{-1}/4 + C$ **27.** $[(1/r) + r]^2/2 + C$ **29.** $(x^3 + 3x)^{1/3} + C$ **31.** $(p + 1)^7/7 - (p + 1)^6/6 + C$ **33.** $2(5t - 1)^{5/2}/125 + 2(5t - 1)^{3/2}/75 + C$ **35.** $2(u - 1)^{3/2}/3 + 2(u - 1)^{1/2} + C$ **37.** $(x^2 + 1)^4/4 + C$ **39.** $(x^2 + 12x)^{3/2}/3 + C$ **41.** $[\ln(t^2 + 2)]/2 + C$ **43.** $e^{2z^2}/4 + C$ **45.** \$1,466,840 **47.** .5 million dollars

SECTION 13.3 (page 752)

1. 18 **3.** 65 **5.** 20 **7.** 8 **9.** 56 **11.** 32; 38 **13.** 15; 31/2 **15.** 20; 30 **17.** 16; 14 **19.** 12.8; 27.2 **21.** 2.67; 2.08 **23. (a)** 3
(b) 3.5 **(c)** 4 **25.** About 10,000,000 cars **27.** A concentration of about 19 units **29.** About 2000 feet **31.** About 4 **33.** About 12.14
35. About 1.19

SECTION 13.4 (page 764)

1. -6 **3.** 3/2 **5.** $28/3 \approx 9.33$ **7.** 38 **9.** 1/3 **11.** 76 **13.** 4/3 **15.** $5 \ln 5 + (12/25) \approx 8.527$ **17.** $20e^{-.2} - 20e^{-.3} + 3 \ln 3 - 3 \ln 2 \approx$
2.775 **19.** $e^{10}/5 - e^5/5 - 1/2 \approx 4375.1$ **21.** $49/3 \approx 16.33$ **23.** $447/7 \approx 63.857$ **25.** 42 **27.** 76 **29.** 24 **31.** 41/2 **33.** $e^2 + 3 + 1/e$
≈ 4.757 **35.** 1 **37.** 15/4 **39.** $512/3 \approx 170.67$ **41.** $[\ln (e^2 - 4e + 6) - \ln 3]/2$ or $\ln [(e^2 - 4e + 6)/3]^{1/2} \approx -.0880$ **43.** $(e^2 - 1)/4$
≈ 1.597 **45.** $23/3 \approx 7.67$ **47.** $e^3 - 2e^2 + e \approx 8.026$ **49.** 19,000 units **51. (a)** \$22,000 **(b)** \$62,000 **(c)** 4.5 days **53.** 2236.5
55. (a) 14.26 **(b)** 3.55 **57. (a)** $\int_0^{10} 1.2e^{.04t}\, dt$ **(b)** $30e^{.4} - 30 \approx 14.75$ billion **(c)** About 12.8 years **(d)** About 14.4 years
59. $(72/.014)(e^{.014T} - 1)$

SECTION 13.5 (page 774)

1. \$280; \$11.67 **3.** \$2500; \$30,000 **5.** No **7. (a)** 12 years **(b)** About \$293 **(c)** \$2340 **9. (a)** 19.6 days **(b)** \$176,792 **(c)** \$80,752
(d) \$96,040 **11.** \$31.50 **13.** \$54 **15.** \$5733.33 **17.** \$341.33; \$429.33 **19.** \$5.76 million dollars **21. (a)** \$2000 **(b)** \$64,000 **23.** No
25. \$8700 **27.** 6960 **29. (a)** 75 **(b)** 100 **31.** .32 ft **33. (a)** $I(.1) = .019$; the lower 10% of the income producers earn 1.9% of the total
income of the population **(b)** $I(.4) = .184$; the lower 40% of the income producers earn 18.4% of the total income of the population
(c) $I(.6) = .384$; the lower 60% of the income producers earn 38.4% of the total income of the population **(d)** $I(.9) = .819$; the lower 90% of
the income producers earn 81.9% of the total income of the population

SECTION 13.6 (page 779)

1. $-4 \ln |(x + \sqrt{x^2 + 36})/6| + C$ **3.** $\ln |(x - 3)/(x + 3)| + C$, $(x^2 > 9)$ **5.** $(4/3) \ln [(3 + \sqrt{9 - x^2})/x] + C$, $(0 < x < 3)$
7. $-2x/3 + 2 \ln |3x + 1|/9 + C$ **9.** $(-2/15) \ln |x/(3x - 5)| + C$ **11.** $\ln |(2x - 1)/(2x + 1)| + C$, $x^2 > 1/4$ **13.** $-3 \ln |(1 + \sqrt{1 - 9x^2})/(3x)|$
$+ C$ **15.** $2x - 3 \ln |2x + 3| + C$ **17.** $1/[25 (5x - 1)] - (\ln |5x - 1|)/25 + C$ **19.** $x^5[(\ln |x|)/5 - 1/25] + C$ **21.** $(1/x)(-\ln |x| - 1) + C$
23. $-xe^{-2x}/2 - e^{-2x}/4 + C$ **25.** About \$2070 **27.** About 15,100 microbes

SECTION 13.7 (page 785)

1. $y = 4x^2 - 7x + C$ **3.** $y = -x^2 + x^3 + C$ **5.** $e^{4x}/4 + C$ **7.** $-(1 - x^2)^{3/2}/3 + C$ **9.** $y^2 = x^2 + C$ **11.** $y = ke^{x^2}$ **13.** $y = ke^{(x^3 - x^2)}$
15. $y = Mx$ **17.** $y = Me^x + 5$ **19.** $y = -1/(e^x + C)$ **21.** $y^2 = 2x^3/3 + 9$ **23.** $y = e^{x^2 + 3x}$ **25.** $y = -5/(5 \ln |x| - 6)$
27. $y^2/2 - 3y = x^2 + x - 4$ **29. (a)** \$155.24 **(b)** \$161.55 **(c)** \$164.62 **31.** About 11.6 yr **33.** About 260 **35.** About 4.4 cc **37.** About
387 **39. (a)** $k \approx .8$ **(b)** 11 **(c)** 55 **(d)** 2981 **41. (a)** $dy/dt = -.05y$ **(b)** $y = Me^{-.05t}$ **(c)** $y = 90e^{-.05t}$ **(d)** 55 g

CHAPTER 13 REVIEW EXERCISES (page 789)

1. $x^4/2 - x^2/2 + C$ **3.** $2x^3 - x^2 + 5x + C$ **5.** $2x^{5/2}/5 - x^2/2 + 5x^{7/5}/7 + C$ **7.** $2x^{5/2}/5 + C$ **9.** $-x^{-2}/2 + C$ **11.** $-4x^{1/2} - 3 \ln |x| + C$
13. $-e^{-5x}/5 + C$ **15.** $-5 \ln |x| + C$ **17.** $2(x^2 - 3)^{3/2}/3 + C$ **19.** $-1/[9(x^3 + 5)^3] + C$ **21.** $\ln |2x^2 - 5x| + C$ **23.** $-e^{-3x^4}/12 + C$
25. $2e^{-5x}/5 + C$ **27.** $C(x) = 3x^2 + 2x + 75$ **29.** $C(x) = 4x^{7/4}/7 + x^2 + 321/7$ **31.** About 34 **33.** 12 **35.** 72/25 **37.** 2 ln 3, or ln 9 $\approx$
2.197 **39.** $2(e^4 - 1)$ **41.** 36,000 **43.** $64/3 \approx 21.33$ **45.** About 90 **47.** $\ln |(x + \sqrt{x^2 - 64})/8| + C$ **49.** $-(1/2) \ln |x|(3x + 4)| + C$
51. $3x^3[(\ln |x|)/3 - 1/9] + C$ or $x^3[3 \ln |x| - 1]/3 + C$ **53.** $2 \ln |(x - 3)/(x + 3)| + C$, $x^2 > 9$ **55.** $y = x^4/2 + 3x^2 + C$ **57.** $y = 4e^x + C$
59. $y = x^3/3 - 5x^2/2 + 1$ **61.** $y = -5e^{-x} - 5x + 22$ **63.** $y^2 = 3x^2 + 2x + C$ **65.** $y = (Cx^2 - 1)/2$ **67.** $y = 2e^{5x - x^2}$ **69.** $y^2 + 6y = 2x -$
$2x^2 + 352$ **71. (a)** \$5800 **(b)** About \$25,100 **73.** About 13.9 yr **75.** About 219

CASE 16 (page 793)

1. About 102 years **2.** About 55.6 years **3.** About 45.5 years **4.** About 90 years

CASE 17 (page 795)

1. 10.65 **2.** 175.8 **3.** 441.46

CHAPTER 14

SECTION 14.1 (page 805)

1. (a) 6 **(b)** -8 **(c)** -20 **(d)** 43 **3. (a)** $\sqrt{43}$ **(b)** 6 **(c)** $\sqrt{19}$ **(d)** $\sqrt{11}$

5.
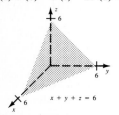
$x + y + z = 6$

7.
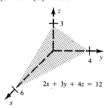
$2x + 3y + 4z = 12$

9.
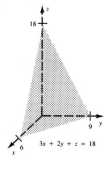
$3x + 2y + z = 18$

11.
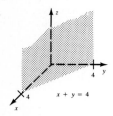
$x + y = 4$

13.
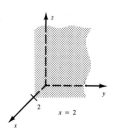
$x = 2$

15. (a) 7.85 **(b)** 4.02 **17. (a)** 1986.95 (rounded) **(b)** 595 (rounded) **(c)** 359,767.81 (rounded) **19.** $y = (500^{5/3})/x^{2/3} \approx 31{,}498/x^{2/3}$

21. (c) **23.** (e) **25.** (b)

27.
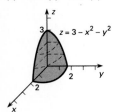
$z = 3 - x^2 - y^2$

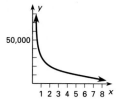

29. (a) $18x + 9h$ **(b)** $-6y - 3h$

SECTION 14.2 (page 814)

1. (a) $24x - 8y$ **(b)** $-8x + 6y$ **(c)** 24 **(d)** -20 **3.** $f_x = -2y$; $f_y = -2x + 18y^2$; 2; 170 **5.** $f_x = 9x^2y^2$; $f_y = 6x^3y$; 36; -1152 **7.** $f_x = e^{x+y}$; $f_y = e^{x+y}$; e^1 or e; e^{-1} or $1/e$ **9.** $f_x = -15e^{3x-4y}$; $f_y = 20e^{3x-4y}$; $-15e^{10}$; $20e^{-24}$ **11.** $f_x = (-x^4 - 2xy^2 - 3x^2y^3)/(x^3 - y^2)^2$; $f_y = (3x^3y^2 - y^4 + 2x^2y)/(x^3 - y^2)^2$; $-8/49$; $-1713/5329$ **13.** $f_x = 6xy^3/(1 + 3x^2y^3)$; $f_y = 9x^2y^2/(1 + 3x^2y^3)$; 12/11; 1296/1297 **15.** $f_x = e^{x^2y}(2x^2y + 1)$; $f_y = x^3e^{x^2y}$; $-7e^{-4}$; $-64e^{48}$ **17.** $f_{xx} = 36xy$; $f_{yy} = -18$; $f_{xy} = f_{yx} = 18x^2$ **19.** $R_{xx} = 8 + 24y^2$; $R_{yy} = -30xy + 24x^2$; $R_{xy} = R_{yx} = -15y^2 +$

$48yx$ **21.** $r_{xx} = -8y/(x + y)^3$; $r_{yy} = 8x/(x + y)^3$; $r_{xy} = r_{yx} = (4x - 4y)/(x + y)^3$ **23.** $z_{xx} = 0$; $z_{yy} = 4xe^y$; $z_{xy} = z_{yx} = 4e^y$
25. $r_{xx} = -1/(x + y)^2$; $r_{yy} = -1/(x + y)^2$; $r_{xy} = r_{yx} = -1/(x + y)^2$ **27.** $z_{xx} = 1/x$; $z_{yy} = -x/y^2$; $z_{xy} = z_{yx} = 1/y$ **29.** $f_x = 2x$; $f_y = z$; $f_z = y +$
$4z^3$; $f_{yz} = 1$ **31.** $f_x = 6/(4z + 5)$; $f_y = -5/(4z + 5)$; $f_z = -4(6x - 5y)/(4z + 5)^2$; $f_{yz} = 20/(4z + 5)^2$ **33.** $f_x = (2x - 5z^2)/(x^2 - 5xz^2 + y^4)$; $f_y =$
$4y^3/(x^2 - 5xz^2 + y^4)$; $f_z = -10xz/(x^2 - 5xz^2 + y^4)$; $f_{yz} = 40xy^3z/(x^2 - 5xz^2 + y^4)^2$ **35. (a)** 80 **(b)** 180 **(c)** 110 **(d)** 360   **37. (a)** \$206,800
(b) $f_p = 132 - 2i - .02p$; $f_i = -2p$; the rate at which sales revenue is changing per unit of change in price (f_p) or interest rate (f_i) **(c)** A sales
revenue drop of \$18,800   **39. (a)** 46.656 **(b)** $f_x(27, 64) = .6912$ and is the rate at which production is changing when labor changes by 1 unit
from 27 to 28 and capital remains constant; $f_y(27, 64) = .4374$ and is the rate at which production is changing when capital changes by 1 unit
from 64 to 65 and labor remains constant. **(c)** Production would be increased by about 69 units when labor is increased by 1 unit.
41. $.7x^{-.3}y^{.3}$ or $.7y^{.3}/x^{.3}$; $.3x^{.7}y^{-.7}$ or $.3x^{.7}/y^{.7}$ **43. (a)** 168 **(b)** 5448 **(c)** 96 **(d)** 3882   **45. (a)** $.08585\ W^{-.575}H^{.725}$ **(b)** .0112 **(c)** .14645
$W^{.425}H^{-.275}$ **(d)** .783

SECTION 14.3   (page 824)

1. Saddle point at $(1, -1)$ **3.** Relative minimum at $(-1, -1/2)$ **5.** Relative minimum at $(-2, -2)$ **7.** Relative minimum at $(15, -8)$
9. Relative maximum at $(2/3, 4/3)$ **11.** Saddle point at $(2, -2)$ **13.** Saddle point at $(1, 2)$ **15.** Saddle point at $(0, 0)$; relative minimum at
$(4, 8)$ **17.** Saddle point at $(0, 0)$; relative minimum at $(9/2, 3/2)$ **19.** Saddle point at $(0, 0)$ **21.** Saddle point at $(0, 0)$; relative maximum
at $(1, 1)$ **23.** Saddle point at $(0, 0)$; relative minima at $(1, 0)$ and $(-1, 0)$ **25.** Saddle points at $(0, 0)$, $(1, 1)$, $(1, -1)$, $(-1, 1)$, and
$(-1, -1)$; relative maxima at $(0, 1)$ and $(0, -1)$; relative minima at $(1, 0)$ and $(-1, 0)$ **27.** $P(12, 40) = 2744$ **29.** $x = 12$, $y = 72$
31. $C(12, 25) = 2237$ **33.** 3 m by 3 m by 3 m

CHAPTER 14   REVIEW EXERCISES   (page 826)

1. -19; -255 **3.** -1; $-5/2$
5.

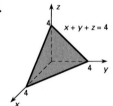

7.

9.

11. (a) $\$(150 + \sqrt{10})$ **(b)** $\$(400 + \sqrt{15})$ **(c)** $\$(1200 + 2\sqrt{5})$ **13. (a)** $4xy/(x - y^2)^2$ **(b)** $-1/2$ **(c)** 0   **15.** $f_x = 30x^4y - 8y^9$; $f_y = 6x^5 - 72xy^8$
17. $f_x = (-6x^2 + 2y^2 - 30xy^2)/(3x^2 + y^2)^2$; $f_y = (30x^2y - 4xy)/(3x^2 + y^2)^2$ **19.** $f_x = (y - 2)^2 \cdot e^{x+2y}$; $f_y = 2(y - 2)(y - 1) \cdot e^{x+2y}$ **21.** $f_x =$
$-2xy^3/(2 - x^2y^3)$; $f_y = -3x^2y^2/(2 - x^2y^3)$ **23.** $f_{xx} = 2y$; $f_{xy} = -24y^3 + 2x$ **25.** $f_{xx} = 2(3 + y)/(x - 1)^3$; $f_{xy} = -1/(x - 1)^2$ **27.** $f_{xx} =$
$2ye^x(2x^2 + 1)$; $f_{xy} = 2xe^x$ **29.** $f_{xx} = -9y^4/(1 + 3xy^2)^2$; $f_{xy} = 6y/(1 + 3xy^2)^2$ **31.** Relative minimum at $(-9/2, 4)$ **33.** Relative maximum at
$(-3/4, -9/32)$; saddle point at $(0, 0)$ **35.** Relative minimum at $(4/5, -9/10)$ **37.** Relative minimum at $(-8, -23)$ **39. (a)** $.6x^{-.4}y^{.4}$ or
$.6y^{.4}/x^{.4}$ **(b)** $.4x^{.6}y^{-.6}$ or $.4x^{.6}/y^{.6}$

Index

Absolute maximum, 676
Absolute minimum, 676
Absolute value, 9
  equation, 28
  inequality, 29
Absolute value function, 96
Acceleration, 685
Addition method for equation
  systems, 251
Addition of matrices, 277
Addition property
  of equality, 12
  of inequality, 23
Additive identity, 4
Additive inverse, 4
Additive inverse of a matrix,
  278
Amortization schedule, 235
Amortize, 234
Annuity, 223
  due, 227
  future value of, 224, 226
  ordinary, 224
  payment period, 224
  present value of, 232
  sinking fund, 228
  term of, 224
Antiderivative, 724
Applied problems, solving
  method, 17
Area
  between two curves, 763
  under a curve, 743
  under a normal curve, 556
Arithmetic mean, 536
Artificial variable, 375
Associative properties, 4
Asymptote, 160
Augmented matrix, 264
Average cost, 129, 632
Average rate of change, 127,
  591
Axis of a parabola, 141

Bank discount, 210
Base, 32
Basic feasible solutions, 351
Basic probability principle,
  420
Basic variables, 351
Bayes' formula, 452
Bernoulli processes, 486, 560
Binomial distribution, 560
Binomial problem, 486
Binomial trials, 486, 560
Boundary, 323
Boundary condition of a
  differential equation, 781
Break-even analysis, 130
Break-even point, 26, 130

Carbon-14 dating, 198
Cartesian coordinate system,
  93
Central tendency, measures
  of, 536
Certain event, 419
Chain rule, 636
Change in x, 115
Change in y, 115
Circular permutation, 485
Closed interval, 656
Cobb-Douglas production
  function, 804
Code theory, 307
Coefficient, 34
Coefficient matrix, 301
Column matrix, 275
Column vector, 275
Combinations, 469
Common denominator, 14
Common logarithm, 189
Common ratio of a sequence,
  223
Commutative properties, 4
Complement
  of an event, 426
  of a set, 402

Completing the square, 64,
  143
Complex fraction, 46
Composite function, 635
Compound amount, 214
Compounding periods, 215
Compound interest, 214
Concave downward, 706
Concave upward, 706
Conditional probability, 438
Conjugate, 59
Constant multiple rule for
  indefinite integrals, 728
Constant of integration, 725
Constant rule for derivatives,
  616
Constant times a function
  rule, 618
Constraints for functions, 331
  mixed, 367
Consumer's surplus, 773
Continuity
  at a point, 654
  on an open interval, 656
Continuous compounding, 217
Continuous distribution, 550
Continuous function, 652
  on an interval, 656
Coordinate system, 93
Corner point, 331
Corner point theorem, 334
Cost-analysis, 127
Cost-benefit function, 163
Cost function, 129
Counting numbers, 2
Critical number, 671
Critical point, 671, 819
Cumulative frequency
  distribution, 532
Cumulative frequency
  polygon, 532
Curve sketching, 712

Decision making, 517
Decreasing function, 700
Definite integral, 748
  properties of, 758
Degeneracies, 375
Degree
  of a polynomial, 34
  of a term, 34
Delta x, 115
Delta y, 115
Demand function, 109, 149,
  622
Demand matrix, 305
De Morgan's Laws, 416
Dependent events, 447, 478
Dependent system of
  equations, 251
Dependent variable, 86
Derivative, 604
  chain rule, 636
  constant rule, 616
  constant times a function
    rule, 618
  existence of, 610
  of exponential function,
    647
  general power rule, 638
  higher order, 683
  of natural logarithmic
    function, 649
  notation of, 615
  nth, 683
  partial, 807
  power rule, 617, 638
  product rule, 628
  quotient rule, 630
  rules for, 664
  second, 683
  sum or difference rule, 619
  third, 683
Deviations from the mean,
  545
Difference of two cubes, 41
Difference of two squares, 40
Differentiable function, 605

Differential equation, 779
   boundary condition of, 781
   general solution of, 780
   particular solution, 781
   separable, 781
Differential of a function, 734
Differentiation, 605
Dimension of a matrix, 275
Discontinuous function, 652
Discount interest, 210
Disjoint sets, 403
Distribution
   binomial, 560
   continuous, 550
   frequency, 501
   normal, 551
   probability, 433, 503, 550
   range of, 544
   skewed, 551
   standard normal, 553
Distributive property, 4
Domain of a function, 88,
   796
Duality, 378
   theorem of, 382
Duals, 380
   method of, 382

e, 645
Echelon method for equation
   systems, 254
Economic lot size, 694
Effective rate, 211, 217
Elements of a matrix, 264
Elements of a set, 398
Elimination method for
   equation systems, 251
Empirical probability, 431
Empty set, 400
Entry of a matrix, 264
Equality properties, 12
Equal matrices, 276
Equal sets, 399
Equation, 12
   absolute value, 28
   addition property of, 12
   dependent, 251
   differential, 779
   exponential, 171
   horizontal line, 107
   linear, 12, 105, 250
   matrix, 301
   multiplication property of,
      12
   quadratic, 61
   quadratic in form, 68
   solution of, 12
   vertical line, 108
Equations of a line, 123
Equilibrium demand, 113
Equilibrium price, 113
Equilibrium supply, 113
Equilibrium vector, 494
Equivalent system of
   equations, 254

Events
   certain, 419
   complement of, 426
   dependent, 447, 478
   impossible, 419
   independent, 447
   mutually exclusive, 424
   odds of, 430
   simple, 419
   union rule for, 426
Exact interest, 209
Expectation, mathematical,
   507
Expected value, 507
Experiment, 417
Exponent, 32, 34
   integers, 40
   negative, 49
   properties of, 51
   rational, 54
   zero, 49
Exponential equation, 171
Exponential function, 169,
   178, 647
   derivative of, 647
   indefinite integral of, 731
Exponential growth function,
   178
Extreme value theorem, 677
Extremum, 668, 676, 817

Factoring, 36
Factors of an integer, 36
Fair game, 511
Feasible region, 327
   unbounded, 333
Feasible solution, 327
First derivative test, 673
First octant, 798
Fixed cost, 127
Fixed probability vector, 494
FOIL, 34
Forgetting curve, 181
Fourth root, 53
Frequency distribution, 501
   cumulative, 532
   grouped, 529
Frequency polygon, 530
   cumulative, 532
Function, 87
   absolute value, 96
   composite, 635
   constraints of, 331
   continuous, 652
   cost, 129
   cost-benefit, 163
   decreasing, 700
   demand, 109, 149, 622
   differentiable, 605
   differential of, 734
   discontinuous, 652
   domain of, 88, 796
   exponential, 169, 178, 647
   $f(x)$ notation, 90

Function (continued)
   graph of, 93
   greatest integer, 99
   growth and decay, 178,
      782
   increasing, 700
   inverse, 185
   limits of, 573
   linear, 105, 125
   linear cost, 129
   logarithmic, 185, 194
   multivariate, 796
   objective, 331
   piecewise, 97
   polynomial, 110, 140
   probability, 504
   product-exchange, 164
   production, 803
   quadratic, 140
   range of, 88, 796
   rational, 160
   step, 99
   total change of, 751
   vertical line test for, 101
Functional notation, 90
Fundamental theorem of
   calculus, 755
Future value, 208, 226
Future value of annuity, 224,
   226
$f(x)$ notation, 90

Gauss-Jordan method, 265
General power rule
   for differentiation, 638
   for integration, 735
General solution of a
   differential equation, 780
Geometric sequence, 223
   sum of terms, 224
Graph
   of a function, 93
   of an inequality, 24
   of a multivariate function,
      798
Greater than, 5
Greatest common factor, 37
Greatest integer function, 99
Grouped frequency
   distribution, 529
Growth and decay function,
   178, 782
Growth rate constant, 783

Half-life, 198
Half-plane, 324
Histogram, 504
Horizontal asymptote, 161
Horizontal line equation, 107

Identity matrix, 293
Identity properties, 4
Identity properties of a
   matrix, 293

Impossible event, 419
Inconsistent system of
   equations, 251
Increasing function, 700
Indefinite integral, 726
   constant multiple rule, 728
   of exponential functions,
      731
   power rule for, 727
   properties of, 728
   sum or difference rule, 728
Independent events, 447
   product rule for, 447
Independent variable, 86, 796
Index of a radical, 56
Indicators, matrix, 350
Inequality, 23
   absolute value, 29
   graph of, 24
   linear, 23, 321
   properties of, 23
   quadratic, 72
   rational, 75
   system of, 327
Infinity, 581
Inflection point, 707
Initial condition of a
   differential equation, 781
Initial simplex tableau, 350
Input-output matrix, 303
Instantaneous rate of change,
   592
Instantaneous velocity, 593
Integers, 2
   exponents, 40
   factors of, 36
Integral
   definite, 748
   indefinite, 726
   sign of, 726
Integrand, 726
Integration, 726
   constant of, 725
   general power rule for, 735
   limits of, 748
   by substitution, 734
Interest, 207
   compound, 214
   discount, 210
   exact, 209
   ordinary, 209
   rate of, 207
   simple, 207
Intersection of sets, 403
Interval, closed and open, 656
Interval notation, 7
Inverse
   additive, 4
   of a function, 185
   of a matrix, 294
   multiplicative, 4
   properties, 4
Irrational numbers, 2

Learning curve, 180
Less than, 5
Level curves, 802
Like radicals, 58
Like terms, 34
Limit, 572
  of a function, 573
  at infinity, 581
  rules for, 577
  two-sided, 576
Limits of integration, 748
Linear cost function, 129
Linear equation, 12, 105, 250
  in n unknowns, 250
Linear function, 105, 125
Linear inequality, 23, 321
  boundary of, 323
  graphing, 322
Linear programming, 321
  basic feasible solutions, 351
  basic variables, 351
  constraints, 331
  corner point, 331, 334
  indicators, 350
  initial simplex tableau, 350
  objective function, 331
  pivoting, 352
  simplex method, 348, 360
  simplex tableau, 350
  slack variable, 349, 380
  solving by graphing, 331
  standard maximum form, 348
  standard minimum form, 370
Logarithm, 184
  common, 189
  natural, 189
  properties of, 187
Logarithmic function, 185, 194
  derivative of, 649
Lower limit of integration, 748

Marginal analysis, 620
Marginal average cost, 632
Marginal cost, 128, 597
Marginal productivity, 784
Marginal profit, 597
Marginal revenue, 597
Markov chain, 491
  regular, 494
  transition matrix, 492
Mathematical expectation, 507
Matrix, 264
  addition of, 277
  additive inverse of, 278
  augmented, 264
  coefficient, 301
  column, 275
  demand, 305
  dimension of, 275
  element of, 264

Matrix (continued)
  entry of, 264
  equal, 276
  equation, 301
  identity, 293
  identity property of, 293
  indicators, 350
  input-output, 303
  inverse, 294
  multiplication of, 283
  negative, 278
  operations on, 264
  order of, 275
  payoff, 517
  product, 283, 304
  production, 304
  properties of, 293
  row, 275
  row operations on, 264
  square, 275
  subtraction of, 278
  sum of, 277
  technological, 303
  transition, 492
  transpose of, 380
  zero, 278
Maturity value, 208
Maximization problems, 356
Maximum
  absolute, 676
  relative, 668, 817
Mean, 536
  deviations from, 545
Measures of central tendency, 536
Measures of variation, 543
Median, 539
Members of a set, 398
Minimization problems, 367
Minimum
  absolute, 676
  relative, 668, 817
Modal class, 541
Mode, 541
Multiplication of matrices, 283
Multiplication principle, 466
Multiplication property
  of equality, 12
  of inequality, 23
Multiplicative identity, 4
Multiplicative inverse, 4
Multivariate function, 796
  graph of, 798
  traces of, 802
Mutually exclusive events, 424

Natural logarithm, 189
Natural logarithmic function, derivative of, 649
Natural number, 2
Negative exponent, 49
Negative of a matrix, 278

Newton's Law of Cooling, 184
n-factorial, 467
Nominal rate, 217
Normal curve, 551
  area under, 556
Normal distribution, 551
nth root, 53
Numbers
  counting, 2
  integer, 2
  irrational, 2
  natural, 2
  rational, 2
  real, 2
  whole, 2
Number line, 1

Objective function, 331
Octant, 798
Odds of an event, 430
Ogive, 532
Open interval, 656
Operations on matrices, 264
Operations on sets, 404
Ordered pair, 93
Ordered triple, 798
Order of a matrix, 275
Order of operations, 7
Ordinary annuity, 224
Ordinary interest, 209
Origin, 93
Outcome of a trial, 417

Parabola, 141
Paraboloid, 802
Parameter, 256
Partial derivatives, 807
  second-order, 812
Particular solution of a differential equation, 781
Payment period of an annuity, 224
Payoff matrix, 517
Perfect square, 40
Period of compounding, 215
Permutation, 467
  circular, 485
Pi (π), 3, 745
Piecewise function, 97
Pivoting, 352
Plane, 798
Point of diminishing returns, 711
Point of inflection, 707
Point-slope form, 120
Polynomial, 33
  degree of, 34
Polynomial function, 110, 140
  of degree n, 153
Power, 32
Power property of exponents, 51

Power rule
  for derivatives, 617, 638
  general, 638
  for indefinite integrals, 727
Present value, 209, 219
  of an annuity, 232
Principal, 207
Probability, 416
  basic principle of, 420
  Bayes' formula, 452
  conditional, 438
  distribution, 433, 503, 550
  empirical, 431
  events. See Events
  expected value, 507
  function, 504
  odds, 430
  product rule of, 442
  properties of, 433
  sample space, 418
  trial, 417
  vector, 493
Producer's surplus, 774
Product-exchange function, 164
Product matrix, 283, 304
Product property of exponents, 33, 51
Product rule
  for derivatives, 628
  for independent events, 477
  of probability, 442
Production function, 803
  Cobb-Douglas, 804
Production matrix, 304
Properties
  associative, 4
  commutative, 4
  of definite integrals, 758
  distributive, 4
  of equality, 12
  of exponents, 51
  identity, 4
  of indefinite integrals, 728
  of inequality, 23
  inverse, 4
  of logarithms, 187
  of matrices, 293
  power, 51
  of probability, 433
  of radicals, 57
  of real numbers, 4
  of square roots, 63
  zero-factor, 62

Quadrant, 94
Quadratic equation, 61
  standard form, 62
Quadratic formula, 65
Quadratic function, 140
Quadratic inequality, 72
Quotient property of exponents, 51
Quotient rule for derivatives, 630

Radical, 56
  like, 58
  properties of, 57
  simplified form, 57
Radical sign, 56
Radicand, 56
Random experiment, 416
Random sample, 528
Random variable, 501
Range of a distribution, 544
Range of a function, 88, 796
Rate of change, 589
  average, 127, 591
  instantaneous, 592
Rate of interest, 207
Rational exponent, 54
Rational expression, 42
  operations with, 43
Rational function, 160
Rational inequality, 75
Rational numbers, 2
Rationalizing the denominator, 58
Real numbers, 2
  properties of, 4
Region of feasible solutions, 327
Regular Markov chain, 494
Regular transition matrix, 494
Relative extremum, 668, 817
  location of, 818
  test for, 820
Relative frequency, 502
Relative maximum, 668, 817
Relative minimum, 668, 817
Repeated trials, 486
Richter Scale, 189
Routing, 308
Row matrix, 275
Row operation, 264
  of a matrix, 264
Row vector, 275
Rules
  for derivatives, 664
  for limits, 577
  for rational expressions, 42

Saddle point, 819
Sample space, 418
Scalar, 283
Scrap value, 175
Secant line, 602
Second derivative, 683
Second derivative test, 686

Second-order partial derivative, 812
Separable differential equation, 781
Sequence, geometric, 223
Set, 398
  complement of, 402
  disjoint, 403
  elements of, 398
  empty, 400
  equal, 399
  intersection of, 403
  members of, 398
  operations, 404
  subset of, 400
  union of, 403
  universal, 399
Set-builder notation, 399
Shadow values, 385
Sigma notation, 536, 745
Simple discount note, 210
Simple event, 419
Simple interest, 207
Simplex method, 348, 360
Simplex tableau, 350
Simplified form of a radical, 57
Sinking fund annuity, 228
Skewed distribution, 551
Slack variable, 349, 380
Slope
  of a curve, 603
  of a line, 115
  of a tangent line, 603
Slope-intercept form, 119
Solution, 12
  for an equation, 12
  feasible, 327
  for a specified variable, 12
  of a system, 250, 331
Square matrix, 275
Square root, 53
Square root property, 63
Standard deviation, 545
Standard maximum form, 348
Standard minimum form, 370
Standard normal distribution, 553
Stated rate, 217
States of nature, 517
Step function, 99
Stochastic processes, 491
Subset, 400

Substitution method for integration, 734
Subtraction of matrices, 278
Sum of terms, of a geometric sequence, 224
Sum of two cubes, 41
Sum of two matrices, 277
Sum or difference rule for derivatives, 619
Sum or difference rule for indefinite integrals, 728
Summation notation, 536, 745
Supply function, 109, 149
Surplus variable, 367
System of equations, 249
  addition method for solving, 251
  dependent, 251
  echelon method for solving, 254
  elimination method for solving, 251
  equivalent, 254
  Gauss-Jordan method, 265
  inconsistent, 251
  solution of, 250, 331
  transformations, 254
System of inequalities, 327
  feasible region of, 327
  solution of, 331

Tangent line, 602
  slope of, 603
Technological matrix, 303
Term, 34
  of an annuity, 224
  degree of, 34
  like, 34
  unlike, 34
Third derivative, 683
Time of a loan, 207
Total change of a function, 751
Traces for a multivariate function, 802
Transformations, 254
Transition matrix, 492
  regular, 494
Transpose of a matrix, 380
Tree diagram, 401
Trial, 417
  binomial, 486, 560
  outcome, 417
  of probability, 417
Two-sided limit, 576

Union of sets, 403
Union rule for events, 426
Universal set, 399
Unlike terms, 34
Upper limit of integration, 748

Variable, 12, 34
  artificial, 375
  basic, 351
  dependent, 86
  independent, 86, 796
  random, 501
  slack, 349, 380
  surplus, 367
Variance, 545
Vector
  column, 275
  equilibrium, 494
  fixed, 494
  of gross output, 304
  probability, 493
  row, 275
Velocity, 593
  instantaneous, 593
Venn diagram, 402, 407
Vertex of parabola, 141
Vertical asymptote, 160
Vertical line equation, 108
Vertical line test, 101

Weighted average, 508
Whole numbers, 2
Word problems. See Applied problems

x-axis, 93
x-coordinate, 94
x-intercept, 95, 146

y-axis, 93
y-coordinate, 94
y-intercept, 95, 146

Zero exponent, 49
Zero-factor property, 62
Zero matrix, 278
z-scores, 553

Index of Applications

Management

Accelerated mortgages, 246
Advertising, 168, 282, 444
Airline dependability, 451
Amortization, 235, 242
Amortization schedule, 237, 239
Annuity, 233
Appliance reliability, 457
Approximate annual interest rate, 22
Average cost, 129, 165, 168, 588, 681
Average cost per unit of production, 632, 633, 634
Average profit, 634, 667
Average rate of change of revenue and income, 137
Average rate of change of sales, 133

Backward-bending supply curves, 104
Bakery income, 365
Bank discount, 214, 242, 244
Bankruptcy, 569, 786
Battery life, 548
Blending a soft drink, 378
Blending chemicals, 378, 388
Blending gasoline, 378, 388
Blending milk, 346
Blending seed, 377
Break-even analysis, 134, 137, 153, 168
Break-even point, 81, 115, 131

Car payments, 239
Car reliability, 651
Car rental, 661
Car restoration, 165
Catalog sales, 175, 597
Cattle ranching, 805
Chicken production, 824

Christmas card production, 519
Citrus farming, 518
Cobb-Douglas production function, 804
Commodities profit, 168
Commodity market, 77
Comparing investments, 212, 213, 222
Compound amount, 175
Concert preparation, 519
Construction, 347
Consumer demand, 588
Contests, 515, 516
Continuously compounded interest, 182
Continuous withdrawals, 791
Contractor bidding, 520
Contractor costs, 290, 393
Cost analysis, 130, 292, 390, 661
Cost function, 731, 823
Credit, 457
Credit cards, 523

Defective item production, 456
Delinquent taxes, 213
Demand function, 600, 613, 623, 644, 740
Depletion dates for minerals, 793
Depreciation, 644
Depreciation, sum-of-the-years'-digits method, 292
Dining out, 114
Divorce settlement, 242
Doubling time, 222
Dry cleaning, 492

Economic lot size, 697, 700, 721
Economy versus investment, 457
Elasticity of demand, 104

Electricity consumption, 753
Employee absence, 535
Employee productivity, 588, 634
Equilibrium supply, 113
Estimated revenue, 183
Estimated salaries, 183
Estimating occupancy, 515
Estimating profit, 515
Estimating sales, 134
Expenditures, 765, 775

Fast-food consumption, 490
Feed cost, 385, 387, 388
Finance, 331, 346, 376
Flying Delta airlines, 311
Franchise fees, 114
Fuel consumption of a kiln, 263
Fuel cost, 700
Fuel efficiency, 634

Gillette Sensor razors, 11
Gross National Product, 786

Hot dog production, 523
Housing starts, 717

Import cars, 599
Income of an apartment complex, 151
Individual Retirement Accounts, 231
Inflation, 775, 786
Input-output analysis, 306, 310, 311, 316
Insurance, 345, 499, 512, 515
Interest on investments, 262, 377
Inventory, 132, 274, 277, 279, 527, 571
Investment, 81
Investment habits, 416
Investment interest, 313
Investments, 20, 22, 519

Job qualifications, 457

Labor costs, 824
Labor-management relations, 525
Laffer curve, 166
Land development, 519
Law of diminishing returns, 712, 718
Leasing, 767, 774
Leontief's model of the American economy, 318
Light bulb life, 558
Loan interest, 213, 262
Loan payments, 246
Loan repayment, 212, 213

Machine maintenance, 742
Machine repairs, 520
Machinery overhaul, 520
Making ice cream, 397
Manufacturing cost, 815, 827
Manufacturing process, 774
Marginal cost, 133, 139, 622, 627, 786
Marginal productivity, 785, 815
Marginal profit, 624, 626, 627, 667
Marginal revenue, 626, 644, 651, 667
Marginal revenue product, 643, 644, 651
Marginal sales, 791
Market share, 499
Marketing, 520
Maximizing profit, 699, 721
Maximizing revenue, 692, 698, 699
Maximum profit, 148, 151, 341, 345, 346, 355, 356, 365, 366, 391, 627, 679, 681, 688
Maximum revenue, 152, 345, 346, 355
Medical expenses, 132
Merit pay, 495
Mine production, 766

Minimizing cost, 693, 698, 700
Minimum cost, 151
Mining output, 734
Mortgage defaults, 458
Mortgage payments, 239

Natural gas consumption, 763
Natural resource consumption, 766
New car sales, 786

Oil consumption, 765, 766
Oil distribution, 315
Oil pollution, 765, 776
Oil price and ship speed, 722

Package design, 721, 825
Packaging, 273
Packaging design, 699
Performance index, 194
Personnel screening, 490
Pork bellies futures, 549
Present value, 248
Price to earnings ratio, 405
Pricing, 700
Prime interest rate, 661
Product life cycle, 718
Product-exchange function, 164, 166
Production, 588
Production costs, 283, 377, 387
Production function, 805, 812
Production rates, 262
Production requirements, 314
Production scheduling, 315, 329, 330, 331
Productivity, 827
Profit, 598, 599
Profit function, 706, 717, 734, 742, 824
Profit sharing, 242
Profitability of fertilizer, 772, 775
Projected sales, 263

Quality control, 457, 462, 481, 490, 497, 498, 558, 559, 565, 569

Rate of change of account balance, 641, 644
Rate of change of cost, 596, 613
Rate of change of revenue, 779
Rating sales accounts, 515
Rental charge, 93, 103
Repair costs, 776
Revenue, 93, 292
Revenue function, 599, 613, 641, 733, 742, 815, 825
Rule of 72, 222
Rule of 78, 22

Sales, 131, 179, 182, 193, 282, 499, 598, 752, 790, 815
Sales analysis, 92, 127, 132
Sales decay, 182
Sales decline, 786
Sales function, 626, 651, 667
Sales promotion, 549
Savings, 213, 222, 242
Savings annuity, 227, 230, 231, 245
Savings interest, 222
Scrap value, 175
Seed production model, 84
Service charge, 99, 103
Shipping decisions, 330
Shopping center rentals, 282
Simple interest rate, 213
Sinking fund, 229, 231, 232, 245
Smog control, 776
Solar heaters, 61, 124
Steel costs, 393
Stock reports, 314
Stock returns, 568
Storage capacity, 343
Supply and demand, 113, 114, 137, 150, 151, 168, 262, 774, 775
Supply costs, 377
Supply of metal, 790

Telephone service, 311
Television advertising, 457
Time management, 330, 390
Timing income, 699
Total cost, 827
Toy production, 388
Training program, 500, 667, 723
Transportation costs, 272, 273, 344, 346, 374, 376, 377
Tread life, 557

Use of materials, 699
Utility company management, 412

Velocity and acceleration of revenue, 689

Warranty cost, 795
Wheat price and production, 543
Worker efficiency, 776
Worker error, 456
Worker production, 180, 182

Natural Science

Acid concentration, 626
Alcohol concentration, 159, 718, 753
Animal activity, 314
Animal feed, 250
Animal growth, 282
Ant population, 652
Ant population growth, 131
Antismoking campaign, 520
Atmospheric pressure, 184

Bacteria food requirements, 272
Bacteria population, 179, 182, 200, 537, 600, 614, 645, 689, 786
Bacterial growth, 131
Birth weight, 559
Blending nutrients, 365, 377
Blood acidity, 549
Blood antigens, 415
Blood cell velocity, 595
Blood cholesterol level, 436
Blood clotting, 559
Blood flow, 765, 816
Blood level curves, 789
Blood pressure medication, 451
Blood sugar level, 626
Blood vessel volume, 627
Body surface area, 805, 816
Breeding pheasants, 698
Buffalo analysis, 440

Calcium usage, 645
Carbon-14 dating, 199
Carbon dioxide in the blood, 797
Cardiac output, 159
Catfish population, 711
Cavities, 491
Chemical reaction, 688
Chemical spraying, 569
Chlorophyll production, 153
Color blindness, 415, 450, 491
Complexity of an organism, 176
Concentration of a solute, 734
Contagion, 317
Cost-benefit function, 163, 165, 168

Decibel rating, 201
Deer population, 159, 776
Dietetics, 262, 282, 292, 347, 376, 387
Dissolving chemicals, 181
Drug concentration, 589, 689, 718
Drug dosage, 165
Drug effectiveness, 280, 490, 566
Drug reaction, 765, 779
Drug screening, 463
Drug sensitivity, 742
Drug sequencing, 477
Drug side effects, 491

Effect of insecticide, 791
Effects of pollution on fish population, 124
Effects of radiation, 491
Escherichia coli population, 176

Fertilizer composition, 262
Fish food requirements, 273
Fish population, 182, 644, 786
Flu inoculations, 491
Food web, 311
Fox population, 193

Galápagos plant species, 61
Genetics, 415, 437, 448, 450, 499, 504
Grasshopper matings, 816
Growth of a mite population, 791
Growth of a substance, 765
Growth of a tumor, 625
Growth of goat population, 784

Hazardous waste, 407
Health care, 346, 347, 376
Heart attacks, 524
Height, 534
Height estimation, 204
Hepatitis blood test, 458
Home range, 535

Insect classification, 477
Insect mating patterns, 627, 652
Insulin and blood sugar levels, 104
Interaction of two competing species, 791

Level of pollutants, 104
Lice population, 181
Logistic function, 183

Maximum permitted level of pollutants, 201
Medical diagnosis, 464
Microbe concentration, 682
Microbe population, 779
Mite population, 786
Mixing plant foods, 272
Moisture, 202
Mortality rates, 565
Mosquito population, 151, 688
Muscle reaction, 634

Nerve impulse, 151
Newton's Law of Cooling, 184
Nuclear power plant discharge, 812

Octane rating, 23
Offspring, 516
Oil pollution, 644
Oil pressure in a reservoir, 159
Organic waste, 582
Oxygen consumption, 173, 201, 805, 816
Oxygen inhalation, 753

Peach harvest, 414
Pollution, 681, 765, 776
Pollution concentration, 652
Population growth, 132, 718
Prairie dog population, 790
Predator food requirements, 346

Radioactive contamination, 200
Radioactive decay, 175, 179, 181, 198, 200
Radioactivity decay, 652
Radius bone and human height, 125
Rat diets, 568
Richter Scale, 201

Salmon spawning, 681
Seeding storms, 516
Shellfish population, 614
Sickle cell anemia, 462
Snow depth, 754
Spider population, 790
Spread of infection, 718
Stocking a lake with fish, 366
Surgery survival rates, 490

Temperature, 21, 126, 534, 543, 598, 662
Thermal inversion, 644
Thyroid problems, 406
Tibia bone and human height, 125
Time dating of dairy products, 781
Toxemia test, 458
Tracer dye, 786
Tree growth, 777

Visible distance from plane to the horizon, 61
Vitamin A deficiency, 490
Vitamin requirements, 559

Weather, 500
Weight lifting, 202

Physical Science

Circuit gain, 166

Gasoline mileage, 682

Height of a projectile, 72, 77, 151
Horizontal distance, 72

Nova, 659

Projective height, 688

Radioactive decay, 787

Total distance, 813

Velocity, 168, 598
Velocity and acceleration, 685, 688
Velocity of an object, 77, 594, 600, 627

Social Science

Airline usage, 458
Alcohol consumption, 415
Art forgeries, 206

Billboard "Hot 100" survey, 741
Body types, 434

Charitable contributions, 364
Class size, 132
Code theory, 308, 311, 316
College majors, 500
Commuting times, 569
Cooking habits, 414
Country-western music, 414
Credit card charges, 436

Drinking and driving, 451
Drug reaction, 645

Education, 521, 530, 559
Electoral college, 61
Estimation of time, 134
Evolution of languages, 201

Forgetting curve, 181

Habit strength, 652
Housing patterns, 500

Income, 777
Information content, 682

Land use, 500
Language, 434
Learning, 176
Learning skills, 661
Legislative turnover, 204
Legislative voting, 589
Letter frequency, 536
Life insurance, 436
Lodge visitors, 681

Magazine readership, 411
Making a first down, 463
Medical school applications, 183
Memory, 635, 706
Morse code, 466
Music expenditures, 434

Park design, 721
Politics, 365
Population, 534
Population decline, 200
Population growth, 175, 178, 181, 787
Postage costs, 136
Production rate of a worker, 779

Reelection strategy, 521, 525
Routing, 309, 311, 316
Rumors, 182

Social service costs, 313
Social status, 11
Spread of a rumor, 787
Stimulation effect, 132

Television viewing, 461
Testing, 490
Traffic analysis, 432
Traffic control, 274
TV-watching habits, 436
Typing speed, 182

Unemployment analysis, 414

Voting, 458
Voting for House of Representatives, 124

Waiting time, 165
Worker productivity, 787

We would appreciate it if you would take a few minutes to answer these questions. Then cut the page out, fold it, seal it, and mail it. No postage is required. Thank you!

Which chapters were taught?

How much algebra did you have before this course?

How long ago?

Years in high school (circle)   0   1/2   1   1-1/2   2 or more   _____ last 2 years

Courses in college   0   1   2   3   _____ last 3-5 years

_____ 5 years or more

We would appreciate knowing of any errors you found in the book. (Please supply page numbers.)

How would you characterize the quality of the explanations? If there were any which were unclear, please give us the page numbers.

What did you think of the examples (were there enough, right amount of detail, etc.) and exercises If there is room for improvement, please let us know where.

How helpful were the marginal problems, chapter summaries, and cases in the text? Do you have any creative ideas on what else might help you study still more effectively?

What is your major or career goal? _____

What additional math courses, if any, do you plan on taking ?_____

What kind of calculator do you own (model and type): _____ Did you use it in class?_____

What did you like most about the book?

FOLD HERE

What did you like the least about the book?

Name _____

College _____ State _____

FOLD HERE

BUSINESS REPLY MAIL

FIRST CLASS   PERMIT NO. 1537   NEW YORK, NY

Postage will be paid by

HarperCollins Publishers
College Division   Attn: MATH GROUP
1900 East Lake Avenue
Glenview, Illinois   60025

**We would appreciate it if you would take a few minutes to answer these questions.
Then cut the page out, fold it, seal it, and mail it. No postage required. Thank you!**

For what course (give title and brief description) and student population did you consider this book?
Annual enrollment _____ .

What chapters did you look at in considering this book?

How would you assess the organization and coverage?

How would you characterize the quality of the exposition? Where, if any, can improvements be made?

How would you assess the quality and quantity of the examples/applications? Where, if any,
can improvements be made?

How would you assess the pedagogical quality of the chapters (marginal problems, summaries, etc.)?
Any suggestions for improvement?

Your current book(author/title) is _____.

How does this book compare?

Do you have any comments on the supplements? Ideas for additions?

If you adopted this book, what chapters would you cover? Are you willing to serve as a possible reviewer?
Do you have (or are you aware of someone who has) any plans to write a book?

What did you like most about the book?

- FOLD HERE -

What did you like the least about the book?

College _____   State _____

- FOLD HERE -

The **derivative** of the function f is the function denoted f' whose value at the number x is defined to be the number

$$f'(x) = \lim_{h \to 0} \frac{f(x+h) - f(x)}{h},$$

provided this limit exists.

Rules for Derivatives Assume all indicated derivatives exist.

Constant Function If $f(x) = k$, where k is any real number, then

$$f'(x) = 0.$$

Power Rule If $f(x) = x^n$, for any real number n, then

$$f'(x) = n \cdot x^{n-1}.$$

Constant Times a Function Let k be a real number. Then the derivative of $y = k \cdot f(x)$ is

$$y' = k \cdot f'(x).$$

Sum or Difference Rule If $y = f(x) \pm g(x)$, then

$$y' = f'(x) \pm g'(x).$$

Product Rule If $f(x) = g(x) \cdot k(x)$, then

$$f'(x) = g(x) \cdot k'(x) + k(x) \cdot g'(x).$$

Quotient Rule If $f(x) = \dfrac{g(x)}{k(x)}$, and $k(x) \neq 0$, then

$$f'(x) = \frac{k(x) \cdot g'(x) - g(x) \cdot k'(x)}{[k(x)]^2}.$$

Chain Rule If y is a function of u, say $y = f(u)$, and if u is a function of x, say $u = g(x)$, then $y = f(u) = f[g(x)]$, and

$$\frac{dy}{dx} = \frac{dy}{du} \cdot \frac{dy}{dx}.$$

Chain Rule (alternate form) Let $y = f[g(x)]$. Then

$$y' = f'[g(x)] \cdot g'(x).$$

Generalized Power Rule Let u be a function of x, and let $y = u^n$ for any real number n. Then

$$y' = n \cdot u^{n-1} \cdot u'.$$

Natural Logarithmic Function If $y = \ln |g(x)|$, then

$$y' = \frac{g'(x)}{g(x)}.$$

Exponential Function If $y = e^{g(x)}$, then

$$y' = g'(x) \cdot e^{g(x)}.$$